Lecture Notes in Artificial Intelligence 10235

Subseries of Lecture Notes in Computer Science

More information about this series at http://www.springer.com/series/1244

Jinho Kim · Kyuseok Shim
Longbing Cao · Jae-Gil Lee
Xuemin Lin · Yang-Sae Moon (Eds.)

Advances in Knowledge Discovery and Data Mining

21st Pacific-Asia Conference, PAKDD 2017
Jeju, South Korea, May 23–26, 2017
Proceedings, Part II

Editors
Jinho Kim
Kangwon National University
Chuncheon
Korea (Republic of)

Kyuseok Shim
Seoul National University
Seoul
Korea (Republic of)

Longbing Cao
University of Technology Sydney
Sydney, NSW
Australia

Jae-Gil Lee
KAIST
Daejeon
Korea (Republic of)

Xuemin Lin
University of New South Wales
Sydney, NSW
Australia

Yang-Sae Moon
Kangwon National University
Chuncheon
Korea (Republic of)

ISSN 0302-9743 ISSN 1611-3349 (electronic)
Lecture Notes in Artificial Intelligence
ISBN 978-3-319-57528-5 ISBN 978-3-319-57529-2 (eBook)
DOI 10.1007/978-3-319-57529-2

Library of Congress Control Number: 2017938164

LNCS Sublibrary: SL7 – Artificial Intelligence

Printed on acid-free paper

This Springer imprint is published by Springer Nature
The registered company is Springer International Publishing AG
The registered company address is: Gewerbestrasse 11, 6330 Cham, Switzerland

PC Chairs' Preface

It is our great pleasure to introduce the proceedings of the 21st Pacific-Asia Conference on Knowledge Discovery and Data Mining (PAKDD 2017).

We received a record-breaking number of 458 submissions from 36 countries all over the world. This highest number of submissions is very encouraging because it reflects the improving status of PAKDD. To rigorously review the submissions, we conducted a double-blind review following the tradition of PAKDD and constructed the largest ever committee consisting of 38 Senior Program Committee (SPC) members and 196 Program Committee (PC) members. Each valid submission was reviewed by three PC members and meta-reviewed by one SPC member who also led the discussion with the PC members. We, the PC co-chairs, considered the recommendations from the SPC members and looked into each submission as well as its reviews to make the final decisions. Borderline papers were thoroughly discussed by us before final decisions were made.

As a result, 129 out of 458 papers were accepted, yielding an acceptance rate of 28.2%. Among them, 45 papers were selected as long-presentation papers, and 84 papers were selected as regular-presentation papers. Mining social networks or graph data was the most popular topic in the accepted papers. The review process was supported by the Microsoft CMT system. During the three main conference days, these 129 papers were presented in 23 research sessions. A long-presentation paper was given 25 minutes for presentation, and a regular-presentation paper was given 15 minutes for presentation. These two types of papers, however, are not distinguished in the proceedings.

We would like to thank all SPC members, PC members, and external reviewers for their hard work to provide us with thoughtful and comprehensive reviews and recommendations. Also, we would like to express our sincere thanks to Yang-Sae Moon for compiling all accepted papers and for working with the Springer team to produce the proceedings.

We hope that the readers of the proceedings find the content interesting and rewarding.

April 2017

Longbing Cao
Jae-Gil Lee
Xuemin Lin

General Chairs' Preface

Welcome to the proceedings of the 21st Pacific-Asia Conference on Knowledge Discovery and Data Mining. PAKDD has successfully brought together researchers and developers since 1997, with the purpose of identifying challenging problems facing the development of advanced knowledge discovery. After 14 years since PAKDD 2003 in Seoul, PAKDD was held again in Korea, during May 23–26, 2017, in Jeju Island.

We are very grateful to the many authors who submitted their work to the PAKDD 2017 technical program. The technical program was enhanced by three keynote speeches, delivered by Sang Cha from Seoul National University, Rakesh Agrawal from Data Insights Laboratories, and Dacheng Tao from the University of Sydney. In addition to the main technical program, the offerings of this conference were further enriched by three tutorials as well as four international workshops on leading-edge topics.

We would like to acknowledge the key contributions by Program Committee co-chairs, Longbing Cao, Jae-Gil Lee, and Xuemin Lin. We would like to extend our gratitude to the workshop co-chairs, U. Kang, Ee-Peng Lim, and Jeffrey Xu Yu; the tutorial co-chairs, Dongwon Lee, Yasushi Sakurai, and Hwanjo Yu; the contest co-chairs, Nitesh Chawla, Younghoon Kim, and Young-Koo Lee; the publicity co-chairs, Sang-Won Lee, Guoliang Li, Steven Whang, and Xiaofang Zhou; the registration co-chairs, Min-Soo Kim and Wookey Lee; the local Arrangements co-chairs, Joonho Kwon, Jun-Ki Min, Chan Jung Park, and Young-Ho Park; the Web chair, Ha-Joo Song; the finance co-chairs, Jaewoo Kang and Jaesoo Yoo; the treasury chair, Chulyun Kim; and the proceedings chair, Yang-Sae Moon. We would like to express our special thanks to our honorary chair, Kyu-Young Whang, for providing valuable advice on all aspects of the conference's organization.

We are grateful to our sponsors that include: platinum sponsors — Asian Office of Aerospace Research & Development/Air Force Office of Scientific Research, Mirhenge, Naver, NCSOFT, Seoul National University Big Data Institute and SK Holdings C&C; gold sponsors — KISTI (Korea Institute of Science and Technology Information); silver sponsors — Daumsoft, Douzone, HiBrainNet, Korea Data Agency, and SK Telecom; and publication sponsors — Springer for their generous and valuable support. We are also thankful to the PAKDD Steering Committee for its guidance and Best Paper Award, Student Travel Award, and Early Career Research Award sponsorship. In addition, we would like to express our gratitude to the KIISE Database Society of Korea for hosting this conference. Finally, we thank the student volunteers and everyone who helped us in organizing PAKDD 2017.

April 2017

Jinho Kim
Kyuseok Shim

Organization

Organizing Committee

Honorary Chair

Kyu-Young Whang	DGIST/KAIST, Korea

General Co-chairs

Jinho Kim	Kangwon National University, Korea
Kyuseok Shim	Seoul National University, Korea

Program Committee Co-chairs

Longbing Cao	University of Technology Sydney, Australia
Jae-Gil Lee	KAIST, Korea
Xuemin Lin	University of New South Wales, Australia

Tutorial Co-chairs

Dongwon Lee	Pennsylvania State University, USA
Yasushi Sakurai	Kumamoto University, Japan
Hwanjo Yu	POSTECH, Korea

Workshop Co-chairs

U Kang	Seoul National University, Korea
Ee-Peng Lim	Singapore Management University, Singapore
Jeffrey Xu Yu	The Chinese University of Hong Kong, SAR China

Publicity Co-chairs

Sang-Won Lee	Sungkyunkwan University, Korea
Guoliang Li	Tsinghua University, China
Steven Euijong Whang	Google Research, USA
Xiaofang Zhou	University of Queensland, Australia

Finance Co-chairs

Jaewoo Kang	Korea University, Korea
Jaesoo Yoo	Chungbuk National University, Korea

Treasury Chair

Chulyun Kim	Sookmyung Women's University, Korea

Proceedings Chair

Yang-Sae Moon	Kangwon National University, Korea

Contest Co-chairs

Nitesh Chawla	University of Notre Dame, USA
Younghoon Kim	Hanyang University, Korea
Young-Koo Lee	Kyung Hee University, Korea

Local Arrangements Co-chairs

Joonho Kwon	Pusan National University, Korea
Jun-Ki Min	Korea University of Technology and Education, Korea
Chan Jung Park	Jeju National University, Korea
Young-Ho Park	Sookmyung Women's University, Korea

Registration Co-chairs

Min-Soo Kim	DGIST, Korea
Wookey Lee	Inha University, Korea

Web Chair

Ha-Joo Song	Pukyong National University, Korea

Steering Committee

Co-chairs

Tu Bao Ho	Japan Advanced Institute of Science and Technology, Japan
Ee-Peng Lim	Singapore Management University, Singapore

Treasurer

Graham Williams	Togaware, Australia (see also under Life Members)

Members

Tu Bao Ho	Japan Advanced Institute of Science and Technology, Japan (Member since 2005, Co-chair 2012–2014, Chair 2015–2017, Life Member since 2013)
Ee-Peng Lim	Singapore Management University, Singapore (Member since 2006, Co-chair 2015–2017)
Thanaruk Theeramunkong	Thammasat University, Thailand (Member since 2009)
P. Krishna Reddy	International Institute of Information Technology, Hyderabad (IIIT-H), India (Member since 2010)
Joshua Z. Huang	Shenzhen Institutes of Advanced Technology, Chinese Academy of Sciences, China (Member since 2011)

Longbing Cao	University of Technology Sydney, Australia (Member since 2013)
Jian Pei	Simon Fraser University, Canada (Member since 2013)
Myra Spiliopoulou	Otto von Guericke University of Magdeburg, Germany (Member since 2013)
Vincent S. Tseng	National Cheng Kung University, Taiwan (Member since 2014)
Tru Hoang Cao	Ho Chi Minh City University of Technology, Vietnam (Member since 2015)
Gill Dobbie	University of Auckland, New Zealand (Member since 2016)

Life Members

Hiroshi Motoda	AFOSR/AOARD and Osaka University, Japan (Member since 1997, Co-chair 2001–2003, Chair 2004–2006, Life Member since 2006)
Rao Kotagiri	University of Melbourne, Australia (Member since 1997, Co-chair 2006–2008, Chair 2009–2011, Life Member since 2007, Treasury Co-sign since 2006)
Huan Liu	Arizona State University, USA (Member since 1998, Treasurer 1998–2000, Life Member since 2012)
Ning Zhong	Maebashi Institute of Technology, Japan (Member since 1999, Life Member since 2008)
Masaru Kitsuregawa	Tokyo University, Japan (Member since 2000, Life Member since 2008)
David Cheung	University of Hong Kong, SAR China (Member since 2001, Treasurer 2005–2006, Chair 2006–2008, Life Member since 2009)
Graham Williams	Australian National University, Australia (Member since 2001, Treasurer since 2006, Co-chair 2009–2011, Chair 2012–2014, Life Member since 2009)
Ming-Syan Chen	National Taiwan University, Taiwan, ROC (Member since 2002, Life Member since 2010)
Kyu-Young Whang	Korea Advanced Institute of Science and Technology, Korea (Member since 2003, Life Member since 2011)
Chengqi Zhang	University of Technology Sydney, Australia (Member since 2004, Life Member since 2012)
Zhi-Hua Zhou	Nanjing University, China (Member since 2007, Life Member since 2015)
Jaideep Srivastava	University of Minnesota, USA (Member since 2006, Life Member since 2015)
Takashi Washio	Institute of Scientific and Industrial Research, Osaka University, Japan (Member since 2008, Life Member since 2016)

Past Members

Hongjun Lu	Hong Kong University of Science and Technology, SAR China (Member 1997–2005)
Arbee L.P. Chen	National Chengchi University, Taiwan, ROC (Member 2002–2009)
Takao Terano	Tokyo Institute of Technology, Japan (Member 2000–2009)

Senior Program Committee

James Bailey	The University of Melbourne, Australia
Peter Christen	The Australian National University, Australia
Guozhu Dong	Wright State University, USA
Patrick Gallinari	LIP6, Université Pierre et Marie Curie, France
Joshua Huang	Shenzhen University, China
Seung-won Hwang	Yonsei University, Korea
George Karypis	University of Minnesota, USA
Latifur Khan	The University of Texas at Dallas, USA
Sang-Wook Kim	Hanyang University, Korea
Byung Suk Lee	University of Vermont, USA
Jiuyong Li	University of South Australia, Australia
Nikos Mamoulis	University of Hong Kong, SAR China
Wee Keong Ng	Nanyang Technological University, Singapore
Wen-Chih Peng	National Chiao Tung University, Taiwan
Vincenzo Piuri	Università degli Studi di Milano, Italy
Rajeev Raman	University of Leicester, UK
P. Krishna Reddy	International Institute of Information Technology, Hyderabad (IIIT-H), India
Dou Shen	Baidu, China
Masashi Sugiyama	RIKEN/The University of Tokyo, Japan
Kai Ming Ting	Federation University, Australia
Hanghang Tong	Arizona State University, USA
Vincent S. Tseng	National Chiao Tung University, Taiwan
Jianyong Wang	Tsinghua University, China
Wei Wang	University of California, Los Angeles, USA
Takashi Washio	Osaka University, Japan
Xindong Wu	University of Louisiana at Lafayette, USA
Xing Xie	Microsoft Research Asia, China
Hui Xiong	Rutgers University, USA
Yue Xu	Queensland University of Technology, Australia
Hayato Yamana	Waseda University, Japan
Jin Soung Yoo	Indiana University-Purdue University Fort Wayne, USA
Jeffrey Yu	The Chinese University of Hong Kong, SAR China
Osmar Zaiane	University of Alberta, Canada
Zhao Zhang	Soochow University, China

Yanchun Zhang	Victoria University, Australia
Yu Zheng	Microsoft Research Asia, China
Ning Zhong	Maebashi Institute of Technology, Japan
Xiaofang Zhou	The University of Queensland, Australia

Program Committee

Aijun An	York University, Canada
Enrique Muñoz Ballester	Università degli Studi di Milano, Italy
Gustavo Batista	University of Sao Paulo, Brazil
Johannes Blömer	Paderborn University, Germany
Kevin Bouchard	Université du Québec à Chicoutimi, Canada
Krisztian Buza	Rheinische Friedrich-Wilhelms-Universität Bonn, Germany
K. Selcuk Candan	Arizona State University, USA
Tru Hoang Cao	Ho Chi Minh City University of Technology, Vietnam
Wei Cao	HeFei University of Technology, China
Tanmoy Chakraborty	University of Maryland, College Park, USA
Jeffrey Chan	RMIT University, Australia
Chia-Hui Chang	National Central University, Taiwan
Muhammad Aamir Cheema	Monash University, Australia
Chun-Hao Chen	Tamkang University, Taiwan
Enhong Chen	University of Science and Technology of China, China
Shu-Ching Chen	Florida International University, USA
Ling Chen	University of Technology Sydney, Australia
Meng Chang Chen	Academia Sinica, Taiwan
Yi-Ping Phoebe Chen	La Trobe University, Australia
Songcan Chen	Nanjing University of Aeronautics and Astronautics, China
Zhiyuan Chen	University of Maryland Baltimore County, USA
Zheng Chen	Microsoft Research Asia, China
Silvia Chiusano	Politecnico di Torino, Italy
Jaegul Choo	Korea University, Korea
Kun-Ta Chuang	National Cheng Kung University, Taiwan
Bruno Cremilleux	Université de Caen, France
Alfredo Cuzzocrea	ICAR-CNR and University of Calabria, Italy
Xuan-Hong Dang	UC Santa Barbara, USA
Zhaohong Deng	Jiangnan University, China
Anne Denton	North Dakota State University, USA
Lipika Dey	Tata Consultancy Services, India
Bolin Ding	Microsoft Research, USA
Gillian Dobbie	The University of Auckland, New Zealand
Xiangjun Dong	Qilu University of Technology, China
Dejing Dou	University of Oregon, USA
Vladimir Estivill-Castro	Griffith University, Australia
Xuhui Fan	University of Technology Sydney, Australia

Ki Yong Lee	Sookmyung Women's University, Korea
Kyumin Lee	Utah State University, USA
Yue-Shi Lee	Ming Chuan University, Taiwan
Sunhwan Lee	IBM Almaden Research Center, USA
Wang-Chien Lee	Pennsylvania State University, USA
SangKeun Lee	Korea University, Korea
Philippe Lenca	IMT Atlantique, France
Carson K. Leung	University of Manitoba, Canada
Zhenhui Li	Pennsylvania State University, USA
Dancheng Li	Northeastern University, China
Gang Li	Deakin University, Australia
Jianmin Li	Tsinghua University, China
Ming Li	Nanjing University, China
Sheng Li	Northeastern University, USA
Xiaoli Li	Institute for Infocomm Research, Singapore
Xuelong Li	Chinese Academy of Science, China
Yaliang Li	University at Buffalo, USA
Yidong Li	Beijing Jiaotong University, China
Zhixu Li	Soochow University, China
Hsuan-Tien Lin	National Taiwan University, Taiwan
Jerry Chun-Wei Lin	Harbin Institute of Technology Shenzhen, China
Jiajun Liu	Renmin University, China
Bin Liu	IBM T.J. Watson Research Center, USA
Wei Liu	University of Technology Sydney, Australia
Woong-Kee Loh	Gachon University, Korea
Shuai Ma	Beihang University, China
Giuseppe Manco	Università della Calabria, Italy
Florent Masseglia	Inria, France
Yasuko Matsubara	Kumamoto University, Japan
Xiangfu Meng	Liaoning Technical University, China
Jun-Ki Min	Korea University of Technology and Education, Korea
Nguyen Le Minh	JAIST, Japan
Yasuhiko Morimoto	Hiroshima University, Japan
Miyuki Nakano	Advanced Institute of Industrial Technology, Japan
Wilfred Ng	Hong Kong University of Science and Technology, SAR China
Ngoc-Thanh Nguyen	Wroclaw University of Technology, Poland
Xuan Vinh Nguyen	University of Melbourne, Australia
Kouzou Ohara	Aoyama Gakuin University, Japan
Salvatore Orlando	Ca' Foscari University of Venice, Italy
Satoshi Oyama	Hokkaido University, Japan
Jia-Yu Pan	Google, USA
Shirui Pan	University of Technology Sydney, Australia
Guansong Pang	University of Technology Sydney, Australia
Noseong Park	University of North Carolina, Charlotte, USA
Gabriella Pasi	University of Milano-Bicocca, Italy

Dhaval Patel	IBJ T.J. Watson Research Center, USA
Dinh Phung	Deakin University, Australia
Santu Rana	Deakin University, Australia
P. Krishna Reddy	International Institute of Information Technology Hyderabad, India
Chandan Reddy	Virginia Tech, USA
Patricia Riddle	University of Auckland, New Zealand
P.S. Sastry	Indian Institute of Science, Bangalore, India
Jiwon Seo	Ulsan National Institute of Science and Technology, Korea
Hong Shen	Adelaide University, Australia
Bin Shen	Zhejiang University, China
Chuan Shi	Beijing University of Posts and Telecommunications, China
Arnaud Soulet	François Rabelais University of Tours, France
Fabio Stella	University of Milano-Bicocca, Italy
Mahito Sugiyama	Osaka University, Japan
Yuqing Sun	Shangdong University, China
Yasuo Tabei	Tokyo Institute of Technology, Japan
Ichigaku Takigawa	Hokkaido University, Japan
Ming Tang	Chinese Academy of Sciences, China
David Taniar	Monash University, Australia
Xiaohui (Daniel) Tao	The University of Southern Queensland, Australia
Khoat Than	Hanoi University of Science and Technology, Vietnam
Hiroyuki Toda	NTT Cyber Solutions Laboratories, NTT Corporation, Japan
Ranga Raju Vatsavai	North Carolina University, USA
Zhangyang Wang	Texas A&M University, USA
Lizhen Wang	Yunnan University, China
Ruili Wang	Massey University (Albany Campus), New Zealand
Shoujin Wang	Advanced Analytics Institute, University of Technology Sydney, Australia
Jason Wang	New Jersey Institute of Technology, USA
Yang Wang	University of New South Wales, Australia
Wei Wang	University of New South Wales, Australia
Xin Wang	University of Calgary, Canada
Lijie Wen	Tsinghua University, China
Steven Euijong Whang	Google Research, USA
Joyce Jiyoung Whang	Sungkyunkwan University, Korea
Raymond Chi-Wing Wong	Hong Kong University of Science and Technology, SAR China
Brendon Woodford	University of Otago, New Zealand
Lin Wu	University of Queensland, Australia
Jia Wu	University of Technology Sydney, Australia
Xintao Wu	University of Arkansas, USA
Yuni Xia	Indiana University - Purdue University Indianapolis (IUPUI), USA

Guandong Xu	University of Technology Sydney, Australia
Congfu Xu	Zhejiang University, China
Bing Xue	Victoria University of Wellington, New Zealand
Takehiro Yamamoto	Kyoto University, Japan
Yusuke Yamamoto	Kyoto University, Japan
Jianye Yang	UNSW, Australia
Ming Yang	Nanjing Normal University, China
Shiyu Yang	UNSW, Australia
Min Yao	Zhejiang University, China
Ilyeop Yi	KAIST, Korea
Hongzhi Yin	University of Queensland, Australia
Ming Yin	Harvard University, USA
Yang Yu	Nanjing University, China
Long Yuan	UNSW, Australia
Xiaodong Yue	Shanghai University, China
Se-Young Yun	Los Alamos National Laboratory, USA
Yifeng Zeng	Teesside University, UK
Fan Zhang	University of Technology Sydney, Australia
Junping Zhang	Fudan University, China
Xiuzhen Zhang	RMIT University, Australia
Yating Zhang	Kyoto University, Japan
Ying Zhang	University of New South Wales, Australia
Du Zhang	Macau University of Science and Technology, SAR China
Min-Ling Zhang	Southeast University, China
Wenjie Zhang	University of New South Wales, Australia
Zhongfei Zhang	Binghamton University, USA
Peixiang Zhao	Florida State University, USA
Yong Zheng	Illinois Institute of Technology, USA
Shuigeng Zhou	Fudan University, China
Xiangmin Zhou	RMIT University, Australia
Chengzhang Zhu	National University of Defense Technology, China
Xingquan Zhu	Florida Atlantic University, USA
Arthur Zimek	University of Southern Denmark, Denmark

Additional Reviewers

Enzo Acerbi
Weiling Cai
Minsoo Choy
Thomas Devogele
Van Nguyen Do
Khan Chuong Duong
Laurent Etienne
Li Gao
Viet Huynh
Zhao Kang
Daehoon Kim
Jungeun Kim
Sundong Kim
Nicolas Labroche
Trung Le
Chengjun Li
Dominique Li
Wentao Li

Chun-Yi Liu	Linh Ngo Van
Jing Lv	Yanran Wang
Kiem-Hieu Nguyen	Yisen Wang
Oanh Nguyen	Kuoliang Wu
Thin Nguyen	Hongyu Xu
Thuong Nguyen	Wanqi Yang
Tu Nguyen	Xuesong Yang
Vu Nguyen	Haichao Yu
Linshan Shen	Hanchao Yu
Fengyi Song	Chen Zhang
Hwanjun Song	Chenwei Zhang
Gabriele Sottocornola	Yuhai Zhao

Sponsors

Platinum Sponsors

AFOSR/
AOARD

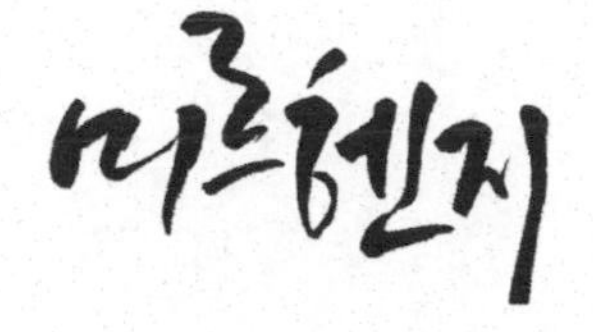

NAVER

Gold Sponsor

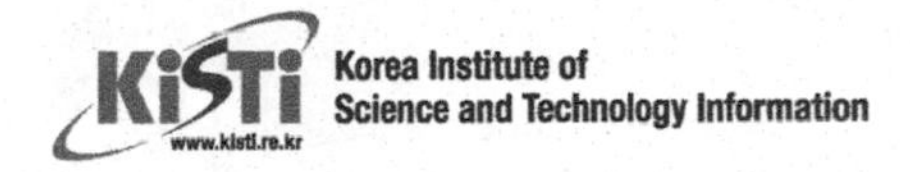

Silver Sponsors

Daumsoft
MINING MINDS

DOUZONE

hbn HiBrain.Net

K data 한국데이터진흥원
Korea Data Agency

Publication Sponsor

Contents – Part II

Clustering and Anomaly Detection

Recommender Systems

Feature Selection

Text and Opinion Mining

Clustering and Matrix Factorization

Dynamic, Stream Data Mining

Novel Models and Algorithms

Contents – Part I

Classification and Deep Learning

Social Network and Graph Mining

Clustering and Anomaly Detection

A Targeted Retraining Scheme of Unsupervised Word Embeddings for Specific Supervised Tasks

Pengda Qin, Weiran Xu(✉), and Jun Guo

School of Information and Communication Engineering,
Beijing University of Posts and Telecommunications, Beijing, China
{qinpengda,xuweiran,guojun}@bupt.edu.cn

Abstract. This paper proposes a simple retraining scheme to purposefully adjust unsupervised word embeddings for specific supervised tasks, such as sentence classification. Different from the current methods, which fine-tune word embeddings in training set through the supervised learning procedure, our method treats the labels of task as implicit context information to retrain word embeddings, so that every required word for the intended task obtains task-specific representation. Moreover, because our method is independent of the supervised learning process, it has less risk of over-fitting. We have validated the rationality of our method on various sentence classification tasks. The improvements of accuracy are remarkable, when only scarce training set is available.

Keywords: Word embedding · Unsupervised learning · Task-specific

1 Introduction

Recent studies have confirmed that word embedding is beneficial to improve the performance of standard NLP tasks, such as POS tagging and NER [5], sentiment analysis [6,7,10], relation classification [17,22] and machine translation [23]. Particularly, along with the development of deep learning in NLP, word embedding is naturally and admittedly treated as input initialization for neural network. In general, word embeddings are produced from massive unlabeled corpus. Such embeddings have a plausible nature that the mutually countered words, such as "good" and "bad", have similar vector representations. This nature, however, often brings some supervised tasks negative effect, such as in sentiment analysis. What's more, same word that occurs in different tasks may convey different sentiment polarities. For example, two sentences with the word "infectious" are given below:

- Although it bangs a very cliched drum at times, this crowd-pleaser's fresh dialogue, energetic music, and good-natured spunk are often **infectious**. (*MR dataset*)
- Which **infectious** disease kills the most people worldwide? (*TREC dataset*)

J. Kim et al. (Eds.): PAKDD 2017, Part II, LNAI 10235, pp. 3–14, 2017.
DOI: 10.1007/978-3-319-57529-2_1

For different tasks (dataset *MR* and *TREC* are described in Sect. 4.1), word "infectious" reflects *positive* tendency for the first sentence, but *negative* for the second one. Consequently, indiscriminate word embedding set for diverse tasks is imperfect. To overcome this issue, for a given supervised task, word embeddings are often trimmed with label information.

Prevailing solution is to modify pre-trained word embeddings through the training process of corresponding supervised task. It back-propagates the error of label prediction to word embedding layer [4,7,20]. Unfortunately, this solution is ineffective for the words **out-of-training-set (OOTS)**[1] and **out-of-vocabulary (OOV)**. Moreover, for the words that can be trimmed, if training dataset is on a small scale, over-fitting problem often exists.

In this paper, we present a novel strategy to obtain suitable task-specific word embeddings for all task-required words, solely relying on task dataset. We utilize pre-trained unsupervised word embeddings as initial word representations, and retrain them for the specific task. Label information is treated as the implicit context information to predict central word together with the explicit context words. Meanwhile, prediction dataset is trained but without label information. During this procedure, all words for task training or predicting, can gain rational task-specific updates in the same vector space. Because the method is separated from the process of learning pre-trained embeddings and training supervised task, it effectively alleviates the over-fitting problem. We have validated the proposed method on 4 sentence classification tasks. Compared with baselines, our method yields competitive performance, even when the train corpus is on a small scale.

The rest of this paper is organized as follows. In Sect. 2, we provide a brief review of related works, especially methods for specific domain or tasks. Section 3 introduces our word embedding retraining schemes, Task-specific CBOW and Task-specific SG. Datasets and the setup of experimental evaluation are presented in Sect. 4. In Sect. 5, we give the detailed analysis of our proposed methods. Finally, the concluding remarks is given in Sect. 6.

2 Related Works

In terms of application field, related researches of word embedding can be divided into three parts. **Generic Word Embedding** is learned from massive unlabeled corpus [3,13,15]. The most useful and practical models are continuous bag-of-words (CBOW) model, continuous skip-gram (SG) model and Global Vector (Glove) model. **Semi-generic Word Embedding** aims at specific domain [2,19,21,24,25]. A probabilistic model [12] is built to capture both semantic similarities (in unsupervised way) and word sentiment (supervised by labeled dataset). Topic-enriched multi-prototype word embeddings (TMWE) [16] incorporates topic information into capturing tweet context for learning twitter-domain word embedding. Unlike the proposed method, researches above need collect abundant labeled dataset as priori. Our work belongs to **Task-specific**

[1] Words exist in vocabulary but merely present on prediction corpus.

Word Embedding, which focus on fine-grained objective [1,8]. It concentrates on learning word embeddings for intended task from small labeled datasets. The simple solution is that, when training a convolutional neural network for sentence classification [7], word embeddings are fine-tuned by system back-propagation procedure on labeled datasets. Subsequently, based on this strategy, an improved approach [1] is proposed to insert sub-space projection operation to word embeddings layer, which fits the task complexity and adapts additional words existed in pre-trained vocabulary. Compared to our method, this obtaining process of task-specific embeddings depends on task training process, while our method is an independent procedure. The re-embedding method [8] exploits an independent procedure to gain task-specific embeddings. They re-embed words supervised by task labels, with regularization of a distortion matrix. Due to the requirement of labels and source embeddings, words that do not occur in labeled corpus or vocabulary are not tailored. The proposed work is different in that we retrain word embeddings according to the principle of co-occurrence of context words. Every required word, even OOTS and OOV words, can obtain feasible vector representations in the same task-specific space.

3 Retraining Word Embeddings for Specific Task

The proposed approaches are inspired by Distributed Memory Model of Paragraph Vectors (PV-DM) model [9] and implemented based on word2vec[2] [13]. Motivated by original CBOW and SG, two word embedding retraining schemes are correspondingly developed, Task-specific CBOW and Task-specific SG. What makes CBOW and SG so popular is their high efficiency and effectivity. Correlation intensity between words are directly measured by the inner product value between word embeddings. Hierarchical Softmax and Negative Sampling are two computational optimization algorithm. Let $\Psi \in R^{|V| \times K}$ be the word embedding matrix, where K is the dimension of embedding and $|V|$ is the vocabulary magnitude. Given a text sequence $s_j = \{w_1, w_2, ..., w_i, ...\}$, ψ_i is the corresponding distributed representation of word w_i.

Original CBOW. The core concept of CBOW is to build a log-linear classifier to correctly classify central (middle) word with future and past words in fixed-length window (context) as input. The objective function can be optimized by maximizing the average log probability,

$$L_{CBOW} = \beta \sum_{s_j \in S} \sum_{w_i \in s_j} \log p(w_i | C_i), \tag{1}$$

where C_i stands for combination of context word distributed representations, β represents the fixed coefficient of average operation.

[2] http://radimrehurek.com/gensim/models/word2vec.html.

Original SG. Compared with CBOW, SG seeks to maximize classification of each context word based on central word in the same sentence,

$$L_{SG} = \beta \sum_{s_j \in S} \sum_{w_i \in s_j} \sum_{-T \leq n \leq T, n \neq 0} \log p(w_{i+n}|w_i), \tag{2}$$

where T denotes the one-side context length.

PV-DM Model. PV-DM model follows the idea of original CBOW. Paragraph vectors are treated as another word and learned by participating the prediction task of next word given contexts sampled from the paragraph. They works as memory containers to remember information from history to future in specific paragraph. However, it is noteworthy that the ultimate goal of PV-DM is to generate paragraph embeddings, but the proposed methods devote to optimize word embeddings.

3.1 Task-Specific CBOW and SG

With regard to different tasks, the required word semantic meanings mainly depend on the label information of task. So, how to reasonably integrate label information into word embedding is the crucial point for improving task performance. With this goal, we propose two methods, namely Task-specific CBOW (TS-CBOW) and Task-specific SG (TS-SG).

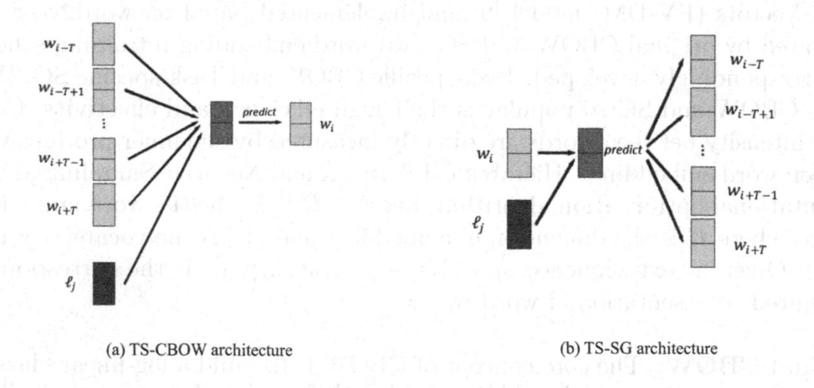

Fig. 1. The architecture of TS-CBOW and TS-SG.

Enlightened by PV-DM model, we leverage the mutual prediction between word and label to inject task-specific information into word embedding. For a specific task, every label is mapped to a unique vector with the same dimension of word embedding, and randomly initialized. Word embedding matrix $\Psi \in R^{|V| \times K}$ is initialized by pre-trained word embedding set. At this time, $|V|$ is the vocabulary magnitude of task-specific corpus. Embeddings of OOV words are randomly initialized. $\Phi \in R^{|L| \times K}$ represents the label matrix. $|L|$ denotes the number of labels. Given a text segment $s_j = \{w_1, w_2, ..., w_i, ...\}$, ℓ_j is the label of s_j. Following this notation, ψ_i, ϕ_j are the respective vector representation of word w_i and label ℓ_j.

Task-Specific CBOW. For training set S_l, along with fixed-length contexts sampled from the sliding window over sentences, specific label vector is shared across all contexts from the sentences assigned with the same label. Word embedding matrix Ψ is unique for the entire task corpus. Label information is leveraged as implicit context element, and we combine such information with context words jointly predict the central word (Fig. 1). Therefore, label vectors act as memory units to remember the characteristics of corresponding classes. Labels are merely available for training set S_l; therefore, there does not exist straightforward label information for prediction dataset S_p. Due to the existence of OOTS words, if we ignore this part, it will undoubtedly cause the inconsistency of word embedding space. This inconsistency problem may be the serious barrier to task performance. In order to resolve this dilemma, we adopt a simple strategy. Words that occur in prediction dataset are trained simultaneously; however, original objective of CBOW is retained. Due to mutual prediction between words, despite a lack of labels, OOTS words (that are not directly retrained with label) still can be embedded in task-specific feature from the words with label information. Therefore, the objective is changed into

$$\begin{aligned} L_{TS\text{-}CBOW} = \beta \sum_{s_j \in S_l} \sum_{w_i \in s_j} \log p(w_i | C_i + \ell_j) \\ + \beta \sum_{s_j \in S_p} \sum_{w_i \in s_j} \log p(w_i | C_i). \end{aligned} \tag{3}$$

The combination way is the weighted vector addition operation, described as

$$\hat{r}(C_i + \ell_j) = \eta_w \sum_{n=-T, n \neq 0}^{T} \psi_{i+n} + \eta_l \phi_j, \tag{4}$$

where T denotes the number of words in one-sided context, η_w and η_l respectively represent the weighting factors (scalar) for word part and label, $\hat{r}(\cdot)$ implies the vector representation.

Embeddings of words, labels and weighting factors (η_w, η_l) are trimmed simultaneously via back-propagation. The acknowledged success of word2vec and PV-DM model demonstrates that, in spite of learning from unlabeled corpus, CBOW model is capable of compelling central word to remember the information of context word. Therefore, it is convinced that TS-CBOW can impel word embedding to obtain the corresponding label information.

In essence, for most tasks, the majority of words have equal probability of occurrence in different label corpus. Hence, for this part, the modification of word embedding is not obvious; however, words that occur more frequently with one kind of label can be embedded in increasingly crucial label information. This part of words are so-called trigger words. For example, word "excellent" for positive movie review, "conflict" for attack event, "practical" for positive product review.

Task-Specific SG. Adopting the similar concept, the integration of label information and central word is employed to predict context words (Fig. 1):

$$\begin{aligned} L_{TS\text{-}SG} = \beta \sum_{s_j \in S_l} \sum_{w_i \in s_j} \sum_{-T \leq n \leq T, n \neq 0} \log p(w_{i+n} | w_i + \ell_j) \\ + \beta \sum_{s_j \in S_p} \sum_{w_i \in s_j} \sum_{-T \leq n \leq T, n \neq 0} \log p(w_{i+n} | w_i), \end{aligned} \tag{5}$$

$$\hat{r}(w_i + \ell_j) = \eta_w \psi_i + \eta_l \phi_j. \tag{6}$$

Despite deriving from similar concept, there still exists some distinctions between two proposed methods. Relative to TS-CBOW, TS-SG has opposite prediction order. What's more, different context words are exerted on different weights when predicting in TS-SG. These distinctions lead to the difference in performance. With respect to this comparison, detailed analyses are presented in Sect. 5.1.

Share Embeddings. We select Negative Sampling as optimization algorithm. In this case, the original CBOW and SG models [13] need initialize two sets of word embedding. One is the output version, the other represents the sampled negative words. In order to generate desirable word embedding, large unlabeled corpus is indispensable. By contrast, in proposed approaches, only scarce labeled dataset is available. So, retraining negative word embeddings needs more time and may bring uncertainties. Moreover, our retraining process is based on pre-trained embeddings, which means the initial state already involves priori semantic knowledge. Consequently, the pre-trained embeddings are shared with these two sets.

4 Datasets and Experimental Setup

The effectiveness of proposed methods is reflected by the improvement of the tasks with word embedding as input. We use a simple and practical convolutional neural network[3] [7] as classification system, and make verification on several commonly used sentence classification tasks. Some details of experiments are described below.

4.1 Datasets

We employ four typical benchmarks to verify our models.

- **MR**: Dataset that involves movie reviews assigned with positive/negative label [14].
- **SST-1**: Stanford Sentiment Treebank. It is an extension of MR [18] with 5 fine-grained labels (very positive, positive, neutral, negative, very negative).

[3] https://github.com/yoonkim/CNN_sentence.

Table 1. Summary statistics for the experimental datasets. ℓ: Number of labels. N_S: Dataset size. $|V|$: Vocabulary size. $|V_{OOTS}|$: Number of OOTS words. $|V_{OOV}|$: Number of OOV words. Obviously, the training dataset of these four tasks are in small scale.

Dataset	ℓ	N_S	$\lvert V\rvert$	$\lvert V_{OOTS}\rvert$	$\lvert V_{OOV}\rvert$
MR	2	10662	18765	**735**	**2317**
SST-2	2	9613	16198	**1360**	**1369**
SST-1	5	11855	17833	**1421**	**1571**
TREC	6	5952	8772	**265**	**1241**

- **SST-2**: Same as SST-1 but with binary labels (remove neutral reviews). Notably, because of emphasizing on scarce dataset, we use sentence-level dataset with 11,855 items instead of phase-level dataset with 239,231 items for both SST-1 and SST-2. So, the performance of baseline in this paper is slightly less than Kim's results [7].
- **TREC**: Question classification dataset with 6 types [11].

4.2 Pre-trained Embedding

In order to present a more comprehensive verification of the proposed methods, we select two word embedding sets as pre-trained embedding sets, which are trained by different methods and different corpus:

- `GoogleNews-vectors-negative300.bin`[4] is learned by Mikolov's word2vec from part of Google News dataset (about 100 billion words). It contains 300-dimensional vectors for 3 million words and phrases.
- `glove.840B.300d.zip`[5] is generate by Pennington's Glove method [15]. Training corpus has 840 billion words, and vocabulary size is 2.2 million. Similarly, the dimension size is 300.

Naturally, the dimension of label embeddings in our experiments is 300.

4.3 Training and Hyperparameter Setting

In common with original word2vec, there are four hyperparameters involved as shown in Table 2. For task-specific word embedding, due to retraining based on pre-trained embeddings, the original optimal word2vec hyperparameter setup is relatively large. So we decrease the value of ξ and α for the two proposed models.

As for classification system, experimental results are obtained by early stopping on dev sets. As for datasets without pre-defined dev set, we randomly extract 10% of training data as the dev set.

[4] https://code.google.com/archive/p/word2vec/.
[5] http://nlp.stanford.edu/projects/glove/.

Table 2. Hyperparameter setup of TS-CBOW and TS-SG

Hyperparameter	TS-CBOW	TS-SG
Negative sample ξ	2	2
Iteration number γ	4	2
Learning rate α	10^{-3}	10^{-3}
Context window size T	5	5

5 Results and Discussion

The most intuitive evaluation of our models is the performance on NLP tasks. Table 3 illustrates the accuracy obtained by different methods.

Row **static** means embeddings will not be fine-tuned via training classification system, which straightforwardly reflects the intrinsic quality of word embeddings for specific tasks. The obtained task-specific word embeddings from our methods are also applied in this case. With the comparison, both two proposed methods achieve obvious improvements against pre-trained word embeddings. It indicates that, for intended task, our task-specific word embeddings have more excellent property than unsupervised version. Both binary and multiple classification can obtain new state-of-the-art results.

Compared with **static** case, row **non-static** shows that fine-tuning pre-trained embeddings via training procedure of task indeed yields improvements. However, our methods still have distinct superiority. Such superiority demonstrates that task-specific information is more effectively embedded in word embeddings through our retraining schemes, in spite of training on scarce dataset.

The same outstanding performance in various embedding set (Table 3) also provides evidence that the proposed methods have good generalization ability.

An more intuitive evaluation of our proposed methods is presented in Table 4, which illustrates the change of relevant words after retraining process for MR task. From the first entry, our models are capable of generating task-specific embeddings for words that occur in training dataset. According to Table 1, the number of OOTS and OOV words is non-ignorable. For these two parts, essential content is also evidently captured to reflect movie review sentiment tendency.

5.1 Comparison Between TS-CBOW and TS-SG

The distinctions described in Sect. 3.1 determine that there exist some property distinctions between them. In terms of the achieved improvement of tasks, both models have their merits. With different released versions of pre-trained embedding set, winners are not always similar for the same sentence classification task. In time efficiency, even though TS-CBOW needs twice the iteration number of TS-SG, TS-SG has more than twice the computational complexity of TS-CBOW in every context window under our parameter setting. It means that, under the same scale of training corpus, TS-SG spends more time. In conclusion, both

Table 3. Accuracy of our models against original method for two word embedding sets. "-" means using pre-trained word embedding to initialize input. **static**: Word embeddings are kept static and only the classification system parameters are learned. **non-static**: Word embeddings are fine-tuned as parameters via back-propagation.

GoogleNews-vectors-negative300.bin					
Word embed	Model	MR	SST-2	SST-1	TREC
Static	-	80.60	84.91	45.20	92.60
Non-static	-	81.26	85.85	46.06	93.30
Static	TS-SG	82.02	85.74	**46.96**	**93.60**
Static	TS-CBOW	**82.80**	**86.07**	46.52	93.30
glove.840B.300d.zip					
Word embed	Model	MR	SST-2	SST-1	TREC
Static	-	79.47	84.08	43.34	92.20
Non-static	-	80.41	84.96	45.02	93.20
Static	TS-SG	81.63	**85.18**	45.20	**93.60**
Static	TS-CBOW	**82.11**	84.58	**46.56**	92.80

Table 4. A list of representative words from the 20 closest-ranked (cosine-distance) words extracted from pre-trained (pre) and task-specific embedding (TS) set. TS embedding set is learned from MR dataset.

IN TRAIN	
Good	pre: great, bad, terrific
	TS: excellent, pretty, unbelievable
OUT OF TRAIN	
Opulent	pre: luxurious, opulence, palatial
	TS: sumptuous, lavish, ostentatious
OUT OF VOCAB	
Unsuspenseful	TS: humourless
Inhospitability	TS: individuals

proposed methods can bring performance improvement. But when dealing with large-scale training corpus, if paying more attention to time efficiency, there is no lack of reason to select TS-CBOW.

5.2 Impact of Parameter Tuning

Experimental results presented in Table 3 are obtained under the optimal setting for selected corpus. Proved through the experiments with different hyper-parameter setups, the performance of TS-CBOW and TS-SG, up to a certain extent, is affected by hyper-parameter tuning.

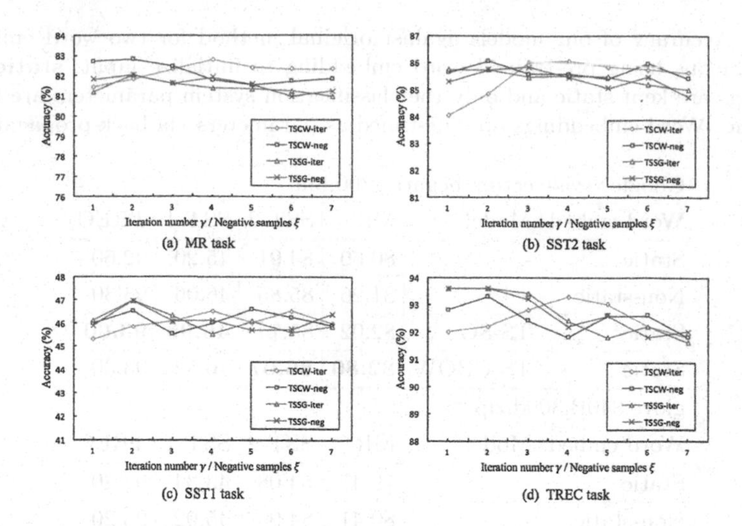

(a) MR task (b) SST2 task (c) SST1 task (d) TREC task

Fig. 2. Influence of hyper-parameter tuning for 4 targeted sentence classification tasks. **TSCW** and **TSSG** respectively denote TS-CBOW and TS-SG methods. **iter** and **neg** respectively represent the operation of adjusting γ and ξ. In order to save space, they share the same horizontal coordinate.

- Figure 2 shows how the accuracy of sentence classification governed by number of iteration γ and negative samples ξ. We analysis the influences for all four tasks. Overall, little value is enough to achieve remarkable performances. For iteration γ, larger value can not achieve sensible improvement and contributes to unnecessary waste of time. Through observing experiments on various datasets, corpus with larger scale need a slightly lower number of iteration. With regard to ξ, larger value is redundant for performance and even counterproductive.
- In terms of learning rate α, larger numerical value causes undesirable semantic deviation; however, less value may reduce the speed of convergence.
- For pre-trained embedding set, even generated from different corpus or different methods, better performances are still achieved by the proposed methods.

5.3 Analysis of Label Embeddings

Label embeddings are trained in the same space with word embeddings, so the relative position between them indicates the relative semantic relationship. In the case of MR task, the nearest embedding of positive label is negative label, and vice versa. It makes sense that they are parallel concepts in formula and most contexts of these opposite labels are same. However, similarity does not mean identity. Different trigger word sets of different sentiment polarities determine different information involved in label embeddings, which is the key point to generate desirable task-specific word embeddings.

6 Conclusion

We present an effective method to adapt unsupervised word embedding to the specific need of the intended supervised tasks. With input initialized by task-specific word embedding, corresponding sentence classification tasks achieve significant improvement. This method involves two advantages. On the one hand, every required word yields rational task-specific distributed representation, even for OOTS and OOV words. On the other hand, our method is independent of task training procedure, which effectively alleviates over-fitting problem.

Acknowledgments. This work was supported by 111 Project of China under Grant no. B08004, the National Natural Science Foundation of China (61273217, 61300080), the Ph.D. Programs Foundation of Ministry of Education of China (20130005110004).

References

1. Astudillo, R.F., Amir, S., Lin, W., Silva, M., Trancoso, I.: Learning word representations from scarce and noisy data with embedding sub-spaces. In: Proceedings of the Association for Computational Linguistics (ACL), Beijing, China (2015)
2. Bansal, M., Gimpel, K., Livescu, K.: Tailoring continuous word representations for dependency parsing. In: ACL, vol. 2, pp. 809–815 (2014)
3. Bengio, Y., Schwenk, H., Senécal, J.S., Morin, F., Gauvain, J.L.: Neural probabilistic language models. In: Holmes, D.E., Jain, L.C. (eds.) Innovations in Machine Learning, pp. 137–186. Springer, Heidelberg (2006)
4. Chen, Y., Xu, L., Liu, K., Zeng, D., Zhao, J.: Event extraction via dynamic multi-pooling convolutional neural networks. In: Proceedings of the 53rd Annual Meeting of the Association for Computational Linguistics and the 7th International Joint Conference on Natural Language Processing. vol. 1, pp. 167–176 (2015)
5. Collobert, R., Weston, J., Bottou, L., Karlen, M., Kavukcuoglu, K., Kuksa, P.: Natural language processing (almost) from scratch. J. Mach. Learn. Res. **12**, 2493–2537 (2011)
6. Kalchbrenner, N., Grefenstette, E., Blunsom, P.: A convolutional neural network for modelling sentences. arXiv preprint arXiv:1404.2188 (2014)
7. Kim, Y.: Convolutional neural networks for sentence classification. arXiv preprint arXiv:1408.5882 (2014)
8. Labutov, I., Lipson, H.: Re-embedding words. In: ACL, vol. 2, pp. 489–493 (2013)
9. Le, Q.V., Mikolov, T.: Distributed representations of sentences and documents. arXiv preprint arXiv:1405.4053 (2014)
10. Li, R., Shindo, H.: Distributed document representation for document classification. In: Cao, T., Lim, E.-P., Zhou, Z.-H., Ho, T.-B., Cheung, D., Motoda, H. (eds.) PAKDD 2015. LNCS (LNAI), vol. 9077, pp. 212–225. Springer, Cham (2015). doi:10.1007/978-3-319-18038-0_17
11. Li, X., Roth, D.: Learning question classifiers. In: Proceedings of the 19th International Conference on Computational Linguistics-Volume 1, pp. 1–7. Association for Computational Linguistics (2002)
12. Maas, A.L., Daly, R.E., Pham, P.T., Huang, D., Ng, A.Y., Potts, C.: Learning word vectors for sentiment analysis. In: Proceedings of the 49th Annual Meeting of the Association for Computational Linguistics: Human Language Technologies-Volume 1, pp. 142–150. Association for Computational Linguistics (2011)

13. Mikolov, T., Sutskever, I., Chen, K., Corrado, G.S., Dean, J.: Distributed representations of words and phrases and their compositionality. In: Advances in Neural Information Processing Systems, pp. 3111–3119 (2013)
14. Pang, B., Lee, L.: Seeing stars: exploiting class relationships for sentiment categorization with respect to rating scales. In: Proceedings of the 43rd Annual Meeting on Association for Computational Linguistics, pp. 115–124. Association for Computational Linguistics (2005)
15. Pennington, J., Socher, R., Manning, C.D.: GloVe: global vectors for word representation. In: EMNLP, vol. 14, pp. 1532–1543 (2014)
16. Ren, Y., Zhang, Y., Zhang, M., Ji, D.: Improving twitter sentiment classification using topic-enriched multi-prototype word embeddings. In: Thirtieth AAAI Conference on Artificial Intelligence (2016)
17. dos Santos, C.N., Xiang, B., Zhou, B.: Classifying relations by ranking with convolutional neural networks. In: Proceedings of the 53rd Annual Meeting of the Association for Computational Linguistics and the 7th International Joint Conference on Natural Language Processing, vol. 1, pp. 626–634 (2015)
18. Socher, R., Perelygin, A., Wu, J.Y., Chuang, J., Manning, C.D., Ng, A.Y., Potts, C.: Recursive deep models for semantic compositionality over a sentiment treebank. In: Proceedings of the conference on Empirical Methods in Natural Language Processing (EMNLP), vol. 1631, p. 1642. Citeseer (2013)
19. Tang, D., Wei, F., Yang, N., Zhou, M., Liu, T., Qin, B.: Learning sentiment-specific word embedding for twitter sentiment classification. In: ACL, vol. 1, pp. 1555–1565 (2014)
20. Xu, K., Feng, Y., Huang, S., Zhao, D.: Semantic relation classification via convolutional neural networks with simple negative sampling. arXiv preprint arXiv:1506.07650 (2015)
21. Yang, H., Hu, Q., He, L.: Learning topic-oriented word embedding for query classification. In: Cao, T., Lim, E.-P., Zhou, Z.-H., Ho, T.-B., Cheung, D., Motoda, H. (eds.) PAKDD 2015. LNCS (LNAI), vol. 9077, pp. 188–198. Springer, Cham (2015). doi:10.1007/978-3-319-18038-0_15
22. Zeng, D., Liu, K., Lai, S., Zhou, G., Zhao, J., et al.: Relation classification via convolutional deep neural network. In: COLING, pp. 2335–2344 (2014)
23. Zhang, M., Liu, Y., Luan, H., Sun, M., Izuha, T., Hao, J.: Building earth movers distance on bilingual word embeddings for machine translation. In: Thirtieth AAAI Conference on Artificial Intelligence (2016)
24. Taghipour, K., Ng, H.T.: Semi-supervised word sense disambiguation using word embeddings in general and specific domains. In: Conference of the North American Chapter of the Association for Computational Linguistics: Human Language Technologies, pp. 314–323 (2015)
25. Yin, Y., Wei, F., Dong, L., Xu, K., Zhang, M., Zhou, M.: Unsupervised word and dependency path embeddings for aspect term extraction (2016)

A Neural Joint Model for Extracting Bacteria and Their Locations

Fei Li[1], Meishan Zhang[2], Guohong Fu[2], and Donghong Ji[1(✉)]

[1] School of Computer, Wuhan University, Wuhan, China
{lifei_csnlp,dhji}@whu.edu.cn

[2] School of Computer Science and Technology, Heilongjiang University, Harbin, China

Abstract. Extracting *Lives_In* relations between bacteria and their locations involves two steps, namely bacteria/location entity recognition and *Lives_In* relation classification. Previous work solved this task by pipeline models, which may suffer error propagation and cannot utilize the interactions between these steps. We follow the line of work using joint models, which perform two subtasks simultaneously to obtain better performances. A state-of-the-art neural joint model for relation extraction in the Automatic Content Extraction (ACE) task is adapted to our task. Furthermore, we propose two strategies to improve this model. First, a novel relation is suggested in the second step to detect the errors in the first step, thus this relation can correct some errors in the first step. Second, we replace the original greedy-search decoding with beam-search, and train the model with early-update techniques. Experimental results on a standard dataset for this task show that our adapted model achieves better precisions than other systems. After adding the novel relation, we gain a nearly 2% improvement of F1 for *Lives_In* relation extraction. When beam-search is used, the F1 is further improved by 6%. These demonstrate that our proposed strategies are effective for this task. However, additional experiments show that the performance improvement in another dataset of bacteria and location extraction is not significant. Therefore, whether our methods are effective for other relation extraction tasks needs to be further investigated.

Keywords: Bacteria · Biotope · Relation extraction · Joint model

1 Introduction

The information of bacteria and their surviving environments is useful in many areas such as food safety and health sciences. Therefore, extracting bacteria and their locations has received much research attention in the biomedical natural language processing (BioNLP) community [2,6,14]. Taking a sentence "The vibrios are ubiquitous to oceans." in the guideline of the Bacteria Biotope (BB) task at BioNLP shared task (BioNLP-ST) 2016 [3] as an example, the task aims to extract bacteria entity mentions (e.g., vibrios), location entity mentions (e.g., oceans), and *Lives_In* relations (e.g., {vibrios, oceans}) from this sentence.

J. Kim et al. (Eds.): PAKDD 2017, Part II, LNAI 10235, pp. 15–26, 2017.
DOI: 10.1007/978-3-319-57529-2_2

This is a typical relation extraction task that involves two steps. First, entity mentions are recognized and second, each pair of entity mentions is examined, deciding whether a *Lives_In* relation exists. The first step can be treated as a named entity recognition (NER) task [7], and the second step can be casted as a relation classification task [19]. We focus on the line of work using neural networks, which have achieved state-of-the-art performances for both tasks.

Recently, Miwa and Bansal [13] proposed a neural joint model for relation extraction in the ACE task[1], which can be adapted to our task. Compared with pipeline models that handle NER and relation classification separately, joint models can alleviate the problem of error propagation [9]. For example, if the bacteria or location entity of a *Lives_In* relation is not correctly recognized, this relation will be definitely lost. Another advantage of joint models is that they can utilize the interactions between two steps. Miwa and Bansal [13] implicitly performed it by building the features of the second task based on the outputs of the first task, and jointly training these features. To enhance the interactions explicitly, we add a special relation called *Invalid_Entity*, which means that some entities related to such relation may be incorrectly recognized. If an entity is only associated with *Invalid_Entity* relations, it will be removed from final results of entity recognition. Thus, even if there are some wrongly-recognized entities, we can still correct them by the second step.

Moreover, Miwa and Bansal [13] exploited a greedy left-to-right manner to predict entity recognition labels incrementally, which may suffer error propagation among these labels, i.e., the error in the prior prediction can induce new errors in the subsequent predictions. In this paper, we use beam-search, which has been successfully applied in other tasks [9,21], to alleviate this problem.

We adapt the model of Miwa and Bansal [13] as our baseline, and verify our strategies gradually in the BB task at BioNLP-ST 2016 [3], which is a standard competition for *Lives_In* relation extraction between bacteria and location entities. Results show that our baseline can achieve state-of-the-art performances for this task. By adding the *Invalid_Entity* relation, we gain a nearly 2% improvement of F1. When beam-search is used, the F1 is further improved by 6%.

2 Related Work

Extracting *Lives_In* relations between bacteria and location entities belongs to the line of work on relation extraction. Prior work usually used two-step pipeline models to handle this task [2,4,14]. First, all possible bacteria/location entities are recognized using sequence labeling models. Then *Lives_In* relations are extracted between bacteria/location entity pairs using binary classification models. We do not exploit this framework because it can easily suffer the error propagation problem. Moreover, the useful interaction information between two steps is unable to be incorporated.

Our work falls into the line of work using joint models for relation extraction. Roth and Yih [16] proposed a joint inference framework based on integer

[1] https://www.ldc.upenn.edu/collaborations/past-projects/ace.

linear programming to extract entities and relations. Li and Ji [9] exploited a single transition-based model to accomplish entity recognition and relation classification simultaneously. Kordjamshidi et al. [6] proposed a structured learning model to extract biomedical entities and their relationships. Very recently, Miwa and Bansal [13] proposed a neural model based on long short-term memories (LSTMs) [5] to perform relation extraction jointly. This model captures both word sequence and dependency structure information by stacking tree-structured recurrent neural networks (RNNs) on sequential RNNs, which allows the model to share parameters between two submodules of entity recognition and relation classification. Such method utilizes the correlations between the relevant subtasks for mutual benefit, and outperforms state-of-the-art feature-based model [9,16]. We follow the work of Miwa and Bansal [13], with extensions of a novel interaction mechanism and beam-search [9,21].

Our work is also related to neural network models of NER [7], relation classification [8,12,17–19] and relation extraction [10]. For NER, Lample et al. [7] exploited RNNs to extract features, which are similar with our neural network structures for NER. For relation classification, Zeng et al. [19] leveraged convolutional neural networks (CNNs) to classify relations with lexical, sentence and word position features. Li et al. [8] used the similar framework and features, but focused on *Lives_In* relation classification between bacteria and their locations. In particular, our neural network structures of relation classification are similar with [12,18], which exploited RNNs over the shortest dependency path between two target entities to extract neural features. For relation extraction, prior work focused on distant supervised methods using Freebase [10], whose methods and tasks are essentially different from ours.

3 Baseline

We follow the work of Miwa and Bansal [13] to build our baseline for extracting bacteria and their locations. Figure 1 shows an example of the analysis process when a sentence "The vibrios are ubiquitous to oceans." is given.

3.1 Bacteria/Location Entity Recognition

The model casts bacteria/location entity recognition as a sequence labeling problem. The output sequence labels are defined to recognize three entity types in our task with a *BILOU* scheme [7], where *B-Bacteria*/*B-Habitat*/*B-Geographical*, *I-Bacteria*/*I-Habitat*/*I-Geographical* and *L-Bacteria*/*L-Habitat*/*L-Geographical* denote the beginning, following and last words of bacteria/habitat/geographical entities. *U-Bacteria*/*U-Habitat*/*U-Geographical* denote the only words of corresponding entities, and *O* denotes that the word does not belong to any type of entities. Following the task definition [3], we consider that both habitat and geographical entities are location entities.

Our model predicts the entity label of each word from left to right. Given an input sentence $w_1/t_1, w_2/t_2, \ldots, w_n/t_n$, where w denotes a word and t denotes

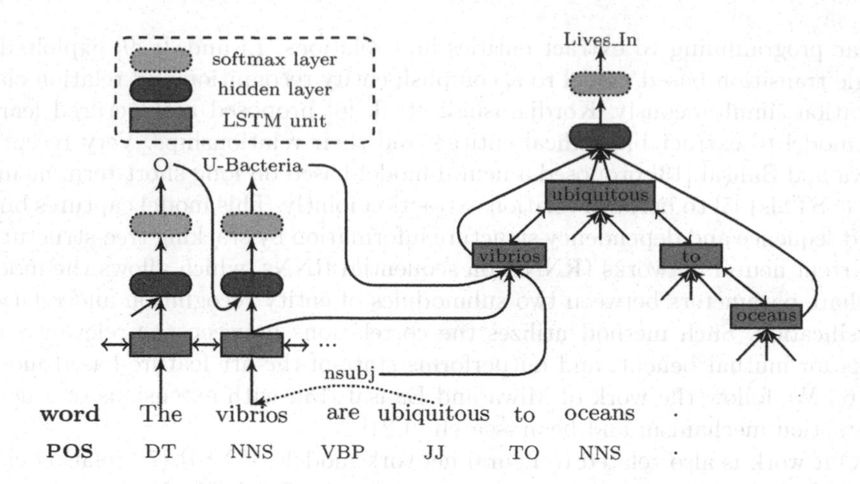

Fig. 1. "vibrios" and "oceans" are bacteria and location entities. "POS" denotes part-of-speech tags. The dotted arrow line denotes a "nsubj" dependency type between "vibrios" and its governor "ubiquitous". "O" and "U-Bacteria" denote entity labels, and "Lives_In" denotes a relation label. The left part recognizes bacteria/location entities by tagging each word with an entity label from left to right incrementally. The right part determines whether a *Lives_In* relation exists between a pair of bacteria/location entities by building a dependency tree and extracting features from it.

its POS tag. We represent each w_i/t_i by concatenating their embeddings, namely $x_i = [e(w_i); e(t_i)]$. A bi-directional LSTM-RNN is built based on $x_1, x_2, \ldots, x_n$, and outputs $h_1, h_2, \ldots, h_n$. h_i is selected as one source of features to predict the entity label l_i of w_i/t_i. The label l_{i-1} of last word is selected as another source. Finally, we concatenate h_i and $e(l_{i-1})$, and use a feed-forward neural network with a hidden layer s_i and a softmax layer to compute the scores of all entity labels. The label with the highest score is selected as l_i for w_i/t_i.

3.2 *Lives_In* Relation Extraction

Once entity recognition is finished, we start binary relation classification to determine whether a *Lives_In* relation exists between a pair of bacteria and location entities. The key idea of the classification is to build a dependency tree whose root is the lowest common ancestor of two target entities, and model the shortest dependency path between the ancestor and target entities.

As shown in the right part of Fig. 1, given two target entities a (e.g., vibrios), b (e.g., oceans) and their lowest common ancestor c (e.g., ubiquitous) in the dependency tree. The shortest dependency paths can be formally represented by $\{a, a_1, \ldots, a_m, c, b_n, \ldots, b_1, b\}$ (e.g., {vibrios, ubiquitous, to, oceans}), where $a_1, \ldots, a_m$ or $b_1, \ldots, b_n$ denotes the words occurred on the path between a and c, or b and c, respectively. The path can be divided into two parts, where $\uparrow seq_a = \{a, a_1, \ldots, a_m, c\}$ (e.g., {vibrios, ubiquitous}) and $\uparrow seq_b = \{b, b_1, \ldots, b_n, c\}$ (e.g., {oceans, to, ubiquitous}) are bottom-up sequences, and $\downarrow seq_a =$

$\{c, a_m, \ldots, a_1, a\}$ (e.g., {ubiquitous, vibrios}) and $\downarrow seq_b = \{c, b_n, \ldots, b_1, b\}$ (e.g., {ubiquitous, to, oceans}) are top-down sequences.

Features are extracted from these sequences by LSTMs. The input of each LSTM unit is a concatenation of three parts, $x_i = [h_i; e(l_i); e(d_i)]$, where h_i is the output of the LSTM unit for entity recognition in Sect. 3.1. $e(l_i)$ and $e(d_i)$ are the entity label and dependency type embeddings. The last outputs of LSTMs computing along $\uparrow seq_a, \uparrow seq_b, \downarrow seq_a$ and $\downarrow seq_b$ are $\uparrow h_a, \uparrow h_b, \downarrow h_a$ and $\downarrow h_b$. Finally, $\uparrow h_a, \uparrow h_b, \downarrow h_a$ and $\downarrow h_b$ are fed into a hidden layer s_{ab}, and a softmax layer is used to compute the scores of all relation labels. The label with the highest score is selected as the relation type of target entities.

3.3 Training

Both parts of the neural network in Fig. 1 employ the same training algorithm based on stochastic gradient decent, so we describe their training in one section for conciseness. The final training objective based on cross-entropy losses is

$$L(\theta) = -\sum_i \log p_y + \frac{\lambda}{2} \| \theta \|_2^2, \tag{1}$$

where θ denotes all the model parameters, y denotes the gold label of a training example, p_y denotes the probability predicted by our model, and λ denotes the regularization parameter of L_2 regularization term. We exploit back propagation to compute the gradients of model parameters.

4 Our Method

4.1 *Invalid_Entity* Relation

In our baseline, the two subtasks, entity recognition and *Lives_In* relation extraction, have their own neural network structures, respectively. The two sub-networks share several common inputs, thus the two subtasks are mutually affected. In addition, the training losses of relation classification network can be propagated back into the entity recognition network. All these interactions are performed implicitly through the sharing of model parameters, because parameter weights of both sub-networks are influenced by losses of both subtasks.

However, we aim to make the upper relation classification task help the entity recognition task explicitly. In the baseline model, the relation classification submodule handles two categories, namely *Lives_In* relation and not *Lives_In* relation. It is built upon the assumption that the given entity pair is a real bacteria/location pair, which cannot be corrected when the entity recognition submodule makes errors. In order to handle this case, we add a relation *Invalid_Entity* to the relation classification submodule. This relation indicates that at least one of two target entities recognized in the first step is incorrect. If an entity is only associated with *Invalid_Entity* relations, it will be removed from final results

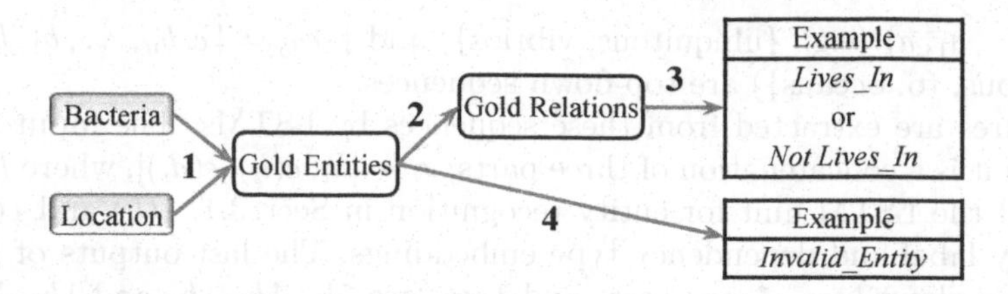

Fig. 2. Training procedure of relation classification. 1: Given a pair of recognized bacteria and location entities, we match them with gold entities. 2: If both of them can be matched, we search their gold relation type in gold relations by entities. 3: A training example is built with the gold relation type, namely *Lives_In* or not *Lives_In*. 4: If any of them cannot be matched with gold entities, this entity pair is impossible to be associated with any gold relation. Therefore, a training example is built with the *Invalid_Entity* relation type.

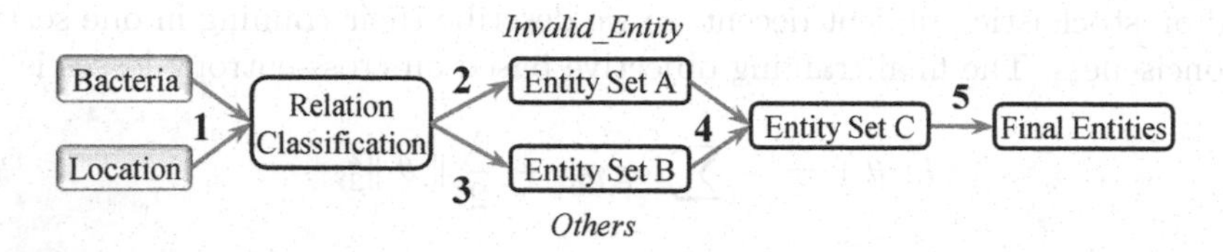

Fig. 3. Decoding procedure of relation classification. 1: Given a pair of recognized bacteria and location entities, our model predicts their relation. 2: If the predicted relation is *Invalid_Entity*, we add two target entities into the entity set A. 3: If the predicted relation is others (either *Lives_In* or not *Lives_In*), we add two target entities into the entity set B. 4: The entity set C denotes the set difference (A-B) of set A and B, so the entities in the set C are only associated with *Invalid_Entity* relations. 5: Entities in the set C will be removed from the final results of entity recognition, and the relations related to the entities in the set B can be used as the final results of relation classification.

of entity recognition. This relation can further help us to correct several errors made by the entity recognition submodule.

After adding the *Invalid_Entity* relation, the training procedure of entity recognition does not change but that of relation classification changes as shown in Fig. 2. Similarly, the decoding procedure of relation classification changes correspondingly as shown in Fig. 3.

4.2 Beam-Search

During entity recognition, our baseline model exploited a greedy left-to-right manner to assign an entity label to each word. The prediction of next step requires the entity label of current step. Thus, when the current step is incorrect, it could influence the result of the next step. This kind of error propagation is less severe than that of pipeline models, because the parameters of joint models are trained jointly and the errors could be considered implicitly to some extent.

```
1: agenda ← { }
2: for word in sentence
3:     beam ← { }
4:     for candidate in agenda
5:         for label in entity labels
6:             score ← COMPUTE(candidate, word, label)
7:             beam ← NEWCANDIDATE(candidate, word, label, score)
8:     agenda ← TOPK(beam)
9: best ← BEST(agenda)
10: entities ← CREATEENTITY(best)
```

Fig. 4. Beam-search decoding for entity recognition.

For each word (i.e., step) in a given sentence, we firstly fetch a history candidate prediction in *agenda* (line 4), and then give a score for each entity label based on the candidate and the current word (line 6). After that, a new candidate prediction is generated and added to *beam* (line 7). After all the candidates in *agenda* have been iterated, we rank the candidates in *beam* (line 8) by accumulating the entity label score of each word in each candidate, formally by

$$score(candidate) = \sum_{l_i \in L} score(l_i) = \sum_{l_i \in L} w \cdot f(l_i), \tag{2}$$

where $L = \{l_1, l_2, \ldots, l_n\}$ denotes the entity label sequence of the current candidate, w denotes the model parameters and f denotes the feature extraction function. K-best candidates are stored back into *agenda* for next step (line 8). After the last step, we use the best candidate prediction and create entities based on it (lines 9–10). The advantage of beam-search is that we have multiple choices at each step, in case that the optimal local prediction is incorrect. The candidate predictions are ranked by global scores, thus error propagation can be alleviated.

We exploit the early-update strategy [9,21] during training, which has been widely used with beam-search. The updating of model parameters is performed at the time when gold-standard results cannot be recovered by the predicted candidates in the beam. Thus only the losses of partial results are used for back propagation. In Fig. 4, the early-update strategy is applied immediately after fetching the k-best candidates in the beam at each step (line 8).

5 Experiments

5.1 Experimental Settings

We conduct experiments on a standard dataset from the BB task at BioNLP-ST 2016 [3], which includes an open competition named *BB-event+ner*. In this competition, gold entities are not given, so participants need to perform both bacteria/location entity recognition and *Lives_In* relation extraction. The dataset consists of 161 documents from PubMed, and we follow the official method to split

Table 1. Hyper-parameter settings, where D denotes vector dimensions.

Type	Hyper-parameter
Training	$\alpha = 0.01$, $\lambda = 10^{-8}$
Embedding	$D(e(w_i)) = 200$
	$D(e(t_i))$, $D(e(d_i))$ or $D(e(l_i)) = 25$
Entity Network Structure	$D(h_i) = 200$, $D(s_i) = 100$
Relation Network Structure	$D(\uparrow h_a, \uparrow h_b, \downarrow h_a \text{or} \downarrow h_b) = 100$
	$D(s_{ab}) = 100$

the dataset into training, development and test sets. In particular, we remove the discontinuous and nested entities, in order to fit our models.

For the development set, we use precision (P), recall (R) and F1-score (F1) to evaluate the performances of entity recognition and relation extraction. A recognized entity is counted as true positive (TP) if its boundary and type match those of a gold entity. An extracted relation is counted as TP if its relation type is correct, and the boundaries and types of its related entities match those of the entities in a gold relation. For the test set, we use the official evaluation service[2], which shows only the overall performance (P, R, F1) of relation extraction.

Hyper-parameters are tuned based on the development set. In Table 1, α and λ denotes the learning rate and L_2 regularization parameter. "Entity Network Structure" and "Relation Network Structure" denote the structures of neural networks for entity recognition and relation classification, respectively. "Embedding" denotes the basic features we used. We use pre-trained biomedical word embeddings [15] to initial our word embeddings and other kinds of embeddings are randomly initialized in the range $(-0.01, 0.01)$.

Given a document, we split it into sentences and then tokenize these sentences. All the tokens are transformed into lowercase forms and numbers are replaced by zeroes. Stanford CoreNLP toolkit [11] is used for POS tagging and dependency parsing. Neural networks are implemented based on LibN3L [20].

5.2 Development Results

As shown in Table 2, our model improves F1 in bacteria/location entity recognition by 0.6% after adding the *Invalid_Entity* relation. The performance improvement is mainly due to the growth of precision. A likely reason may be that some incorrectly-recognized entities are removed. This demonstrates that the relation classification submodule can help the entity recognition submodule to correct some errors through the *Invalid_Entity* relation. In addition, the F1-score of *Lives_In* relation extraction also increases, from 16.3% to 20.3%. It demonstrates that the *Invalid_Entity* relation can help to boost relation extraction as well. The

[2] http://bibliome.jouy.inra.fr/demo/BioNLP-ST-2016-Evaluation/index.html.

Table 2. Developmental results (%) of the baseline and our proposed methods.

Method	Entity recognition			Relation extraction		
	P	R	F1	P	R	F1
Baseline	63.6	47.9	54.7	24.2	12.3	16.3
+Invalid_Entity	68.8	46.2	55.3	25.0	17.2	20.3
+Beam (4)	**69.7**	**51.8**	**59.4**	**27.8**	**20.9**	**23.9**

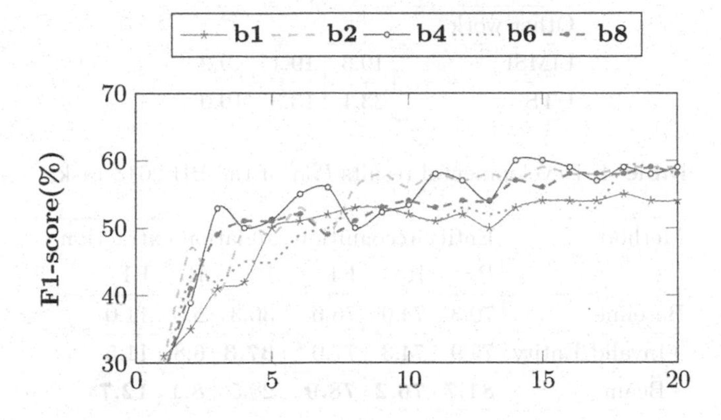

Fig. 5. F1 against the training epoch using different beam sizes for entity recognition.

possible reason is that this relation can divide the non-*Lives_In* relations more reasonably. Overall, the results demonstrate that *Invalid_Entity* can boost the performances of both entity recognition and relation extraction.

Figure 5 shows the development results of beam-search, namely the F1 scores of entity recognition with respect to the training epoches. We experiment with five beam settings, including beam 1, 2, 4, 6 and 8, where beam 1 denotes the baseline greedy search. With beam-search (the beam size is larger than 1), the performance of entity recognition outperforms the baseline method. According to Fig. 5, we set the final beam size by 4, which achieves the best performance. In Table 2, we also show the concrete developmental results of both bacteria/location entity recognition and *Lives_In* relation extraction. The recall values of relation extraction are greatly boosted by beam-search, which is similar with our *Invalid_Entity* strategy. Actually, we do not use beam-search in the relation classification phase, thus the main benefit comes from entity recognition, which brings better performances for the overall relation extraction as well. Overall, beam-search can give a further increase of 3.6% in F1 for relation extraction.

5.3 Final Results

Table 3 shows the final overall relation extraction results of our models. The baseline model can obtain 20.7% of F1, and after adding the *Invalid_Entity*

Table 3. Final results (%) on the test set.

Method	Relation extraction		
	P	R	F1
Our models			
Baseline	**46.9**	13.2	20.7
+Invalid_Entity	46.1	14.9	22.5
+Beam	46.1	**20.7**	**28.5**
Other work			
LIMSI	19.3	19.1	19.2
UTS	33.1	13.3	19.0

Table 4. Developmental results (%) of the BB 2013 task.

Method	Entity recognition			Relation extraction		
	P	R	F1	P	R	F1
Baseline	79.3	74.0	76.6	36.3	6.5	11.0
+Invalid_Entity	79.9	74.3	77.0	**37.3**	6.8	11.5
+Beam	**81.7**	**76.2**	**78.9**	28.5	**8.1**	**12.7**

relation, F1 is boosted by 1.8%. When beam-search is applied, we can have a further improvement of 6%, which demonstrates our proposed strategies are useful. In particular, we find our strategies can mainly contribute to the recall values, which is consistent with the finding on the development set. Considering the low proportion of *Lives_In* relations, the recall is highly important.

Moreover, we show the performances of the top-two systems for this task, namely LIMSI and UTS, which both leverage pipeline models. LIMSI [4] uses conditional random field (CRF) and post-processing rules to identify mentions of bacteria and locations, and support vector machine (SVM) to classify *Lives_In* relations between two entity mentions. UTS [3] relies on two independent SVMs to perform entity recognition and relation classification, respectively. From Table 3, we can see that they suffer either lower precision or recall.

5.4 Additional Experiments

We also additionally evaluated our method on the subtask 3 of the bacteria biotope (BB) task [1] in the BioNLP 2013 shared task. The BB 2013 task is similar with the BB 2016 task [3], and we focused on the extraction of *Localization* relations which represent the same meaning as *Lives_In* relations. The BB 2013 task includes 78, 27 and 26 documents as training, development and test sets. Since the official evaluation service is unavailable, we used the development set for evaluation. The experimental settings of the BB 2013 task is identical to those of the BB 2016 task, and the development results are shown in Table 4.

If the *Invalid_Entity* relation is added, the precision, recall and F1-score of entity recognition increase slightly (0.6%, 0.3% and 0.4%), and those of relation extraction also rise by 1.0%, 0.3% and 0.5% respectively. By utilizing beam-search, the precision, recall and F1 of entity recognition further increase by 1.8%, 1.9% and 1.9%, and the performance of relation extraction is generally improved except the precision, declining by 8.8%. Overall, the performance improvement in the BB 2013 task is not as apparent as that in the BB 2016 task.

6 Conclusion

To extract bacteria and their habitats, we employed a state-of-the-art system for joint entity and relation extraction. To enhance this system, two extensions were made. First, we added the *Invalid_Entity* relation to model the conditions with incorrectly recognized bacteria/location entities. Then we applied beam-search to replace the greedy decoding. Experimental results on a benchmark dataset showed that both of our extensions could improve the performance significantly. We demonstrate that implicit parameter sharing for joint models is not enough and greedy decoding also influences the performance of joint models. However, additional experiments on another dataset showed that the performance improvements were not obvious. Therefore, we need to evaluate our method on more relation extraction tasks to further demonstrate its effectiveness.

Acknowledgments. This work is supported by the National Natural Science Foundation of China (No. 61373108), the National Philosophy Social Science Major Bidding Project of China (No. 11&ZD189). This work is also supported by Humanities and Social Science Foundation of Ministry of Education of China (16YJCZH004), and the China Postdoctoral Science Foundation (2014T70722).

References

1. Bossy, R., Golik, W., Ratkovic, Z., Valsamou, D., Bessières, P., Nédellec, C.: Overview of the gene regulation network and the bacteria biotope tasks in BioNLP 2013 shared task. BMC Bioinform. **16**(10), S1 (2015)
2. Claveau, V.: IRISA participation to BioNLP-ST13: lazy-learning and information retrieval for information extraction tasks. In: Proceedings of the BioNLP Shared Task 2013 Workshop (2013)
3. Deléger, L., Bossy, R., Chaix, E., Ba, M., Ferré, A., Bessières, P., Nédellec, C.: Overview of the bacteria biotope task at BioNLP shared task 2016. In: Proceedings of the 4th BioNLP Shared Task Workshop (2016)
4. Grouin, C.: Identification of mentions and relations between bacteria and biotope from PubMed abstracts. In: Proceedings of the 4th BioNLP Shared Task Workshop (2016)
5. Hochreiter, S., Schmidhuber, J.: Long short-term memory. Neural Comput. **9**(8), 1735–1780 (1997)
6. Kordjamshidi, P., Roth, D., Moens, M.F.: Structured learning for spatial information extraction from biomedical text: bacteria biotopes. BMC Bioinform. **16**, 129 (2015)

7. Lample, G., Ballesteros, M., Subramanian, S., Kawakami, K., Dyer, C.: Neural architectures for named entity recognition. In: Proceedings of the 2016 Conference of the NAACL, pp. 260–270 (2016)
8. Li, H., Zhang, J., Wang, J., Lin, H., Yang, Z.: DUTIR in BioNLP-ST 2016: utilizing convolutional network and distributed representation to extract complicate relations. In: Proceedings of the 4th BioNLP Shared Task Workshop (2016)
9. Li, Q., Ji, H.: Incremental joint extraction of entity mentions and relations. In: Proceedings of the 52nd ACL, pp. 402–412 (2014)
10. Lin, Y., Shen, S., Liu, Z., Luan, H., Sun, M.: Neural relation extraction with selective attention over instances. In: Proceedings of the 54th Annual Meeting of the ACL, pp. 2124–2133 (2016)
11. Manning, C.D., Surdeanu, M., Bauer, J., Finkel, J., Bethard, S.J., McClosky, D.: The Stanford coreNLP natural language processing toolkit. In: Proceedings of 52nd ACL, pp. 55–60, September 2014
12. Mehryary, F., Björne, J., Pyysalo, S., Salakoski, T., Ginter, F.: Deep learning with minimal training data: TurkuNLP entry in the BioNLP shared task 2016. In: Proceedings of the 4th BioNLP Shared Task Workshop (2016)
13. Miwa, M., Bansal, M.: End-to-end relation extraction using LSTMs on sequences and tree structures. In: Proceedings of the 54th Annual Meeting of the ACL, pp. 1105–1116 (2016)
14. Nguyen, N., Tsuruoka, Y.: Extracting bacteria biotopes with semi-supervised named entity recognition and coreference resolution. In: Proceedings of the BioNLP Shared Task 2011 Workshop, pp. 94–101 (2011)
15. Pyysalo, S., Ginter, F., Moen, H., Salakoski, T., Ananiadou, S.: Distributional semantics resources for biomedical text processing. In: LBM (2013)
16. Roth, D., Yih, W.: Global inference for entity and relation identification via a linear programming formulation. In: Introduction to Statistical Relational Learning (2007)
17. Wang, L., Cao, Z., de Melo, G., Liu, Z.: Relation classification via multi-level attention CNNs. In: Proceedings of the 54th Annual Meeting of the ACL, pp. 1298–1307 (2016)
18. Xu, Y., Mou, L., Li, G., Chen, Y., Peng, H., Jin, Z.: Classifying relations via long short term memory networks along shortest dependency paths. In: Proceedings of the EMNLP, pp. 1785–1794 (2015)
19. Zeng, D., Liu, K., Lai, S., Zhou, G., Zhao, J.: Relation classification via convolutional deep neural network. In: Proceedings of the 25th COLING, pp. 2335–2344 (2014)
20. Zhang, M., Yang, J., Teng, Z., Zhang, Y.: LibN3L: a lightweight package for neural NLP. In: Proceedings of the Tenth International Conference on Language Resources and Evaluation, pp. 23–28 (2016)
21. Zhang, M., Zhang, Y., Fu, G.: Transition-based neural word segmentation. In: Proceedings of the 54th Annual Meeting of the ACL, pp. 421–431 (2016)

Advanced Computation of Sparse Precision Matrices for Big Data

Abdelkader Baggag(✉), Halima Bensmail(✉), and Jaideep Srivastava

Qatar Computing Research Institute, Hamad Bin Khalifa University, Doha, Qatar
{abaggag,hbensmail}@hbku.edu.qa

Abstract. The precision matrix is the inverse of the covariance matrix. Estimating large sparse precision matrices is an interesting and a challenging problem in many fields of sciences, engineering, humanities and machine learning problems in general. Recent applications often encounter high dimensionality with a limited number of data points leading to a number of covariance parameters that greatly exceeds the number of observations, and hence the singularity of the covariance matrix. Several methods have been proposed to deal with this challenging problem, but there is no guarantee that the obtained estimator is positive definite. Furthermore, in many cases, one needs to capture some additional information on the setting of the problem. In this paper, we introduce a criterion that ensures the positive definiteness of the precision matrix and we propose the inner-outer alternating direction method of multipliers as an efficient method for estimating it. We show that the convergence of the algorithm is ensured with a sufficiently relaxed stopping criterion in the inner iteration. We also show that the proposed method converges, is robust, accurate and scalable as it lends itself to an efficient implementation on parallel computers.

1 Introduction

Recent applications often encounter high dimensionality with a limited number of data points leading to a number of covariance parameters that greatly exceeds the number of observations. Examples include marketing, e-commerce, and warehouse data in business; microarray, and proteomics data in genomics and heath sciences; and biomedical imaging, functional magnetic resonance imaging, tomography, signal processing, high-resolution imaging, and functional and longitudinal data. In biological sciences, one may want to classify diseases and predict clinical outcomes using microarray gene expression or proteomics data, in which hundreds of thousands of expression levels are potential covariates, but there are typically only tens or hundreds of subjects. Hundreds of thousands of single-nucleotide polymorphisms are potential predictors in genome-wide association studies. The dimensionality of the variables' spaces grows rapidly when interactions of such predictors are considered.

Large-scale data analysis is also a common feature of many problems in machine learning, such as text and document classification and computer vision.

J. Kim et al. (Eds.): PAKDD 2017, Part II, LNAI 10235, pp. 27–38, 2017.
DOI: 10.1007/978-3-319-57529-2_3

For a $p \times p$ covariance matrix $\mathbf{\Sigma}$, there are $p(p+1)/2$ parameters to estimate, yet the sample size n is often small. In addition, the positive-definiteness of $\mathbf{\Sigma}$ makes the problem even more complicated. When $n > p$, the sample covariance matrix is positive-definite and unbiased, but as the dimension p increases, i.e., when $p \gg n$, which is the case in this study, the sample covariance matrix tends to become unstable and can fail to be consistent, since it becomes singular.

Due to its importance, there has been an active line of work on efficient optimization methods for solving the ℓ_1 regularized Gaussian maximum likelihood estimator problem (MLE). Among these methods, we cite the projected subgradients [7], the alternating linearization, the inexact interior point method, the greedy coordinate descent method and G-LASSO which is a block coordinate descent method. For typical high-dimensional statistical problems, optimization methods typically suffer sub-linear rates of convergence. This would be too expensive for the Gaussian MLE problem, since the number of matrix entries scales quadratically with the number of nodes.

Sparse modeling has been widely used to deal with high dimensionality. The main assumption is that the p dimensional parameter vector is sparse, with many components being exactly zero or negligibly small. Such an assumption is crucial in identifiability, especially for the relatively small sample size. Although the notion of sparsity gives rise to biased estimation in general, it has proved to be effective in many applications. In particular, variable selection can increase the estimation accuracy by effectively identifying important predictors and can improve the model interpretability.

To estimate the inverse of the covariance matrix, a thresholding approach is proposed as the following,

$$\hat{\mathbf{M}} = \arg\min_{\mathbf{M}} \frac{1}{2}\|\hat{\mathbf{\Sigma}}\mathbf{M} - \mathbf{I}\|_F^2 + \lambda\|\mathbf{M}\|_1, \tag{1}$$

where $\mathbf{M}$ is the precision matrix, i.e., the inverse of $\mathbf{\Sigma}$. This equation emphasizes the fact that the solution may not be unique. Such nonuniqueness always occurs since $\mathbf{\Sigma} = \mathbf{X}^T\mathbf{X}$ where $\mathbf{X} \in \mathbb{R}^{n \times p}$ is the sample data matrix and rank $(\mathbf{X}) \leq n \ll p$, which is our case. However, there is no guarantee that the thresholding estimator is always positive definite [5]. Although the positive definite property is guaranteed in the asymptotic setting with high probability, the actual estimator can be an indefinite matrix, especially in real data analysis.

To overcome these limitations, i.e. to achieve simultaneously sparsity, positive semi-definiteness, and add structure or constraints on the coefficient of the precision matrix in a graphical model setting for example, a natural solution is to add the positive semi-definite constraint and penalize the ℓ_1 norm of a matrix $\mathbf{D}$ multiplied by the precision matrix estimate $\mathbf{M}$, where $\mathbf{D} \in \mathbb{R}^{p \times p}$, i.e.,

$$\hat{\mathbf{M}} = \arg\min_{\mathbf{M}} \frac{1}{2}\|\hat{\mathbf{\Sigma}}\mathbf{M} - \mathbf{I}\|_F^2 + \lambda\|\mathbf{D}\mathbf{M}\|_1. \tag{2}$$

Before discussing our contribution in that sense, let us introduce G-LASSO, which is a popular penalized approach for estimating the precision matrix.

1.1 Tikhonov and G-LASSO

The log-likelihood of a sample $\mathbf{x}_i$ in a Gaussian model with mean $\boldsymbol{\mu}$ and precision matrix $\mathbf{M}$ is given by

$$\mathcal{L}(\mathbf{x}_i, \boldsymbol{\mu}, \hat{\boldsymbol{\Sigma}}) = \log\det\mathbf{M} - (\mathbf{x}_i - \boldsymbol{\mu})^T\mathbf{M}(\mathbf{x}_i - \boldsymbol{\mu}), \tag{3}$$

and since this average log-likelihood depends only on the sample covariance matrix $\hat{\boldsymbol{\Sigma}}$, we define

$$\mathcal{L}(\mathbf{M}, \hat{\boldsymbol{\Sigma}}) = \sum_i \mathcal{L}(\mathbf{x}_i, \boldsymbol{\mu}, \hat{\boldsymbol{\Sigma}}) = \log\det\mathbf{M} - \langle\hat{\boldsymbol{\Sigma}}, \mathbf{M}\rangle. \tag{4}$$

A very well known technique for the estimation of precision matrices is Tikhonov regularization. Given a sample covariance matrix $\hat{\boldsymbol{\Sigma}} \succeq \mathbf{0}$, the *Tikhonov-regularized problem* is defined as

$$\max_{\mathbf{M}\succ 0} \quad \mathcal{L}(\mathbf{M}, \hat{\boldsymbol{\Sigma}}) - \lambda\ \mathrm{Trace}(\mathbf{M}), \tag{5}$$

or equivalently:

$$\max_{\mathbf{M}\succ 0} \quad \log\det\mathbf{M} - \langle\hat{\boldsymbol{\Sigma}}, \mathbf{M}\rangle - \lambda\ \mathrm{Trace}(\mathbf{M}). \tag{6}$$

This can be seen as imposing a matrix variate Gaussian prior on $\mathbf{M}^{1/2}$ with both row and column covariance matrices equal to $\mathbf{I}$ to make the solution well defined, assuming positiveness of $\mathbf{M}$. The optimal solution of (6) is given by

$$\mathbf{M} = (\hat{\boldsymbol{\Sigma}} + \lambda\mathbf{I})^{-1}. \tag{7}$$

Since the ℓ_1-norm is the tightest convex upper bound of the cardinality of a matrix, several methods based on ℓ_1-regularized log-determinant program have been proposed. This is called G-LASSO, or Graphical LASSO, after the algorithm that efficiently computes the solution,

$$\max_{\mathbf{M}\succ 0} \quad \log\det\mathbf{M} - \langle\hat{\boldsymbol{\Sigma}}, \mathbf{M}\rangle - \lambda\|\mathbf{M}\|_1. \tag{8}$$

Similar to Eq. (8), Bien and Tibshirani in [3] suggested to add to the likelihood a LASSO penalty on $\|\mathbf{D} \odot \boldsymbol{\Sigma}\|$, where $\mathbf{D}$ is an arbitrary matrix with non-negative elements and $\odot$ denotes the element-wise multiplication, i.e., the Hadamard product. So the regularization problem becomes

$$\min_{\mathbf{M}\succ 0} \quad -\log\det\mathbf{M} + \langle\hat{\boldsymbol{\Sigma}}, \mathbf{M}\rangle + \lambda\|\mathbf{D}\odot\mathbf{M}\|_1, \tag{9}$$

and one may tighten the constraint $\mathbf{M} \succ 0$ to $\mathbf{M} \succ \delta\mathbf{I}$ for some $\delta > 0$, so that the above equation becomes

$$\min_{\substack{\mathbf{M}\succ\delta\mathbf{I}\\ \delta>0}} \quad -\log\det\mathbf{M} + \langle\hat{\boldsymbol{\Sigma}}, \mathbf{M}\rangle + \lambda\|\mathbf{D}\odot\mathbf{M}\|_1. \tag{10}$$

The asymptotic properties of the estimator has been studied in [13]. Cai and Zhou in [5] proposed a constrained ℓ_1-minimization procedure for estimating sparse precision matrices by solving the optimization problem

$$\min \quad \|\mathbf{M}\|_1 \quad \text{subject to} \quad \|\hat{\mathbf{\Sigma}}\mathbf{M} - \mathbf{I}\|_\infty \leq \lambda. \tag{11}$$

Equation (11) is obtained from a decomposition of series of Dantzig selector problems. In the above, the symmetry condition is not enforced on $\mathbf{M}$ and as a result the solution is not symmetric in general. Cai and Zhou in [5] established the rates of convergence of CLIME, which is the algorithm derived from the above regularization problem, under both the entry-wise ℓ_1- and the Frobenius-norm and proved that CLIME estimator is positive definite with high probability. However, in practice there is no guarantee that CLIME is always positive definite, especially when applied in real applications.

This warrants our study and our contributions.

1.2 Summary of the Paper and Contributions

In this paper, we propose a new way of estimating large sparse precision matrices, which is significant in the area of learning, vision and mining. This requires seeking a sparse solution and also ensuring positive definiteness. We, therefore, present an optimization model that encourages both sparsity and positive definiteness, and propose an implementation based on the inner-outer alternating direction method of multipliers (ADMM). The method is validated using numerical experiments.

The emphasis of the paper is on introducing a new criterion that insures the positive-definiteness of the precision matrix by adding a tuning parameter $\delta > 0$ in the constraints. This additional constraint will guard against positive semi-definiteness. We add structure on the coefficient of the precision matrix and we derive an efficient inner-outer ADMM form to obtain an optimal solution. We perform the ADMM steps, a variant of the standard Augmented Lagrangian method, that uses partial updates, but with three innovations that enable finessing the caveats detailed above. Also, we show that the proposed algorithm converges with a very relaxed stopping criterion in the inner iteration. It is scalable and with better accuracy properties than existing methods.

2 The Proposed Method

Given a data matrix $\mathbf{X} \in \mathbb{R}^{n \times p}$, a sample of n realizations from a p-dimensional Gaussian distribution with zero mean and covariance matrix $\mathbf{\Sigma} = (\boldsymbol{X}^T\boldsymbol{X}) \in \mathbb{R}^{p \times p}$ and $p \gg n$. The goal is to estimate the precision matrix $\mathbf{M}$, i.e., the inverse of the covariance matrix, $\mathbf{\Sigma}$. Let $\hat{\mathbf{\Sigma}}$ be the empirical covariance matrix. The Graphical LASSO problem minimizes the ℓ_1-regularized negative log-likelihood

$$f(\mathbf{M}) = -\log\det\mathbf{M} + \langle\mathbf{M}, \hat{\mathbf{\Sigma}}\rangle + \lambda\|\mathbf{M}\|_1, \tag{12}$$

subject to the constraint that $\mathbf{M} \succ \mathbf{0}$. This is a semi-definite programming problem in the variable $\mathbf{M}$. When we take the derivative of $f(\mathbf{M})$ with respect to $\mathbf{M}$ and we set it to zero, we obtain the optimality condition

$$-\mathbf{M}^{-1} + \hat{\boldsymbol{\Sigma}} + \lambda\,\boldsymbol{\Gamma} = \mathbf{0}, \tag{13}$$

where $\boldsymbol{\Gamma}$ is a matrix of componentwise signs of $\mathbf{M}$, i.e.,

$$\begin{cases} \gamma_{jk} = \text{sign}(m_{jk}) & \text{if } m_{jk} \neq 0 \\ \gamma_{jk} \in [-1, 1] & \text{if } m_{jk} = 0 \end{cases} \tag{14}$$

Taking the norm, we get

$$\|\mathbf{M}^{-1} - \hat{\boldsymbol{\Sigma}}\|_\infty \leq \lambda. \tag{15}$$

Therefore, our interest is to construct a sparse precision matrix estimator via convex optimization that performs well in high-dimensional settings and is positive definite in finite samples, see e.g. [14]. Positive definiteness is desirable when the covariance estimator is applied to methods for supervised learning. Many of these methods either require a positive definite covariance estimator, or use optimization that is convex only if the covariance estimator is nonnegative definite, e.g., quadratic discriminant analysis and covariance regularized regression, see e.g. [16].

We define our estimator as a solution to the following problem

$$\hat{\mathbf{M}} = \arg\min_{\substack{\mathbf{M} \succeq \delta\mathbf{I} \\ \delta \succeq 0}} \frac{1}{2}\|\hat{\boldsymbol{\Sigma}}\mathbf{M} - \mathbf{I}\|_F^2 + \lambda\|\mathbf{D} \odot \mathbf{M}\|_1, \tag{16}$$

where $\hat{\boldsymbol{\Sigma}}$ is the empirical covariance matrix, $\mathbf{D} = (d_{ij})_{1 \leq i,j \leq p}$ is an arbitrary matrix with non-negative elements where $\mathbf{D}$ can take different forms: it can be the matrix of all ones or it can be a matrix with zeros on the diagonal to avoid shrinking diagonal elements of $\mathbf{M}$. Furthermore, we can take $\mathbf{D}$ with elements $d_{ij} = 1_{i \neq j}|\hat{\mathbf{M}}_{ij}^{\text{init}}|$, where $\hat{\mathbf{M}}^{\text{init}}$ is an initial estimator of $\mathbf{M}$. The later choice of $\mathbf{D}$ corresponds to a precision matrix analogue of the Adaptive LASSO penalty.

To derive our inner-outer ADMM algorithm, as in [1], we will first introduce a new variable $\boldsymbol{\Theta} \in \mathbb{R}^{p \times p}$ and an equality constraint as follows

$$\hat{\mathbf{M}} = \arg\min_{\boldsymbol{\Theta}, \mathbf{M}} \left\{ \frac{1}{2}\|\hat{\boldsymbol{\Sigma}}\mathbf{M} - \mathbf{I}\|_F^2 + \lambda\|\mathbf{D} \odot \mathbf{M}\|_1 \;\; | \;\; \mathbf{M} = \boldsymbol{\Theta}, \; \boldsymbol{\Theta} \geq \delta\mathbf{I} \right\}. \tag{17}$$

The solution to (17) gives the solution to (16). To deal with the problem (17), we have to minimize its augmented Lagrangian function, $\mathcal{L}_\varrho(\boldsymbol{\Theta}, \mathbf{M}, \boldsymbol{\Lambda})$ for some given penalty parameter ϱ, i.e.,

$$\mathcal{L}_\varrho(\boldsymbol{\Theta}, \mathbf{M}, \boldsymbol{\Lambda}) = \frac{1}{2}\|\hat{\boldsymbol{\Sigma}}\mathbf{M} - \mathbf{I}\|_F^2 + \lambda\|\mathbf{D} \odot \mathbf{M}\|_1 - \langle \boldsymbol{\Lambda}, \boldsymbol{\Theta} - \mathbf{M} \rangle + \frac{1}{2}\varrho\|\boldsymbol{\Theta} - \mathbf{M}\|_F^2, \tag{18}$$

where $\boldsymbol{\Lambda} \in \mathbb{R}^{p \times p}$ is the ensemble of Lagrange multipliers associated with the constraint. Hence at iteration k, the ADMM algorithm consists of three steps, namely the $\boldsymbol{\Theta}$-Step, the $\mathbf{M}$-Step and the dual-update step.

First step: In the first step of the ADMM algorithm, we fix $\mathbf{M}$ and $\boldsymbol{\Lambda}$ and minimize the augmented Lagrangian function over $\boldsymbol{\Theta}$ to get

$$\boldsymbol{\Theta}^{k+1} = \arg\min_{\boldsymbol{\Theta}\succeq\delta\mathbf{I}} \mathcal{L}_\varrho(\boldsymbol{\Theta}, \mathbf{M}^k, \boldsymbol{\Lambda}^k) = \text{Proj}_{\boldsymbol{\mathcal{K}}}(\mathbf{M}^k + \frac{1}{\varrho}\boldsymbol{\Lambda}^k),$$

where $\text{Proj}_{\boldsymbol{\mathcal{K}}}(\bar{\mathbf{A}})$ denotes the projection of the matrix $\bar{\mathbf{A}} = \mathbf{M}^k + \frac{1}{\varrho}\boldsymbol{\Lambda}^k$ onto the convex cone $\boldsymbol{\mathcal{K}} = \{\boldsymbol{\Theta} \succeq \delta\mathbf{I}\}$.

Lemma 1. *Let $Proj_{\boldsymbol{\mathcal{K}}}(\bar{\mathbf{A}})$ be the projection of the matrix $\bar{\mathbf{A}}$ onto the convex cone $\boldsymbol{\mathcal{K}} = \{\boldsymbol{\Theta} \succeq \delta\mathbf{I}\}$. If $\bar{\mathbf{A}}$ has the eigen-decomposition $\bar{\mathbf{A}} = \mathbf{V}\ Diag(\lambda_1, \lambda_2, \cdots, \lambda_n)\ \mathbf{V}^T$, then*

$$Proj_{\boldsymbol{\mathcal{K}}}(\bar{\mathbf{A}}) = \mathbf{V}\ Diag(\max(\delta, \lambda_1), \max(\delta, \lambda_2), \cdots, \max(\delta, \lambda_n))\ \mathbf{U}^T. \quad (19)$$

Proof. This result can be traced back to early statisticians, e.g. see [15], who noticed that if a matrix $\mathbf{A}$ has an eigendecomposition

$$\mathbf{A} = \mathbf{U}\ \text{Diag}(\mu_1, \mu_2, \cdots, \mu_n)\ \mathbf{U}^T, \quad (20)$$

where $\mu_1 \geq \cdots \geq \mu_n$ are the eigenvalues of $\mathbf{A}$ and $\mathbf{U}$ is the corresponding orthonormal matrix of eigenvectors; then the projection of $\mathbf{A}$ onto the closed convex cone $\boldsymbol{\mathcal{S}}^+$, formed by the set of positive semidefinite matrices, is

$$\text{Proj}_{\boldsymbol{\mathcal{S}}^+}(\mathbf{A}) = \mathbf{U}\ \text{Diag}(\max(0, \mu_1), \max(0, \mu_2), \cdots, \max(0, \mu_n))\ \mathbf{U}^T. \quad (21)$$

Therefore, the extension to $\text{Proj}_{\boldsymbol{\mathcal{K}}}(\bar{\mathbf{A}})$ is straightforward. □

Since then this projection has been widely used, e.g. see [9]. Notice that the numerical cost of computing this projection is essentially that of computing the spectral decomposition of the matrix to project, i.e., $\bar{\mathbf{A}}$.

Second step: In the second step, we fix $\boldsymbol{\Theta}$ and $\boldsymbol{\Lambda}$ and minimize the augmented Lagrangian over $\mathbf{M}$, i.e.,

$$\mathbf{M}^{k+1} = \arg\min_{\mathbf{M}} \mathcal{L}_\varrho(\boldsymbol{\Theta}^{k+1}, \mathbf{M}, \boldsymbol{\Lambda}^k), \quad (22)$$

where this second step requires a special care since the concept of an inner iteration is introduced at this level of the procedure.

Third step: The final step of the ADMM algorithm is to update the dual variable $\boldsymbol{\Lambda}$, i.e.,

$$\boldsymbol{\Lambda}^{k+1} = \boldsymbol{\Lambda}^k - (\boldsymbol{\Theta}^{k+1} - \mathbf{M}^{k+1}). \quad (23)$$

2.1 The Inner Iteration

It can be shown that

$$\mathbf{M}^{k+1} = \arg\min_{\mathbf{M}} \frac{1}{2}\left\{\|\hat{\boldsymbol{\Sigma}}\mathbf{M}\|_F^2 + \varrho\|\mathbf{M}\|_F^2 - 2\langle\mathbf{M}, \hat{\boldsymbol{\Sigma}} + \varrho\boldsymbol{\Theta}^{k+1} - \boldsymbol{\Lambda}^k\rangle\right\} + \lambda\|\mathbf{D}\odot\mathbf{M}\|_1,$$

for which we are not able to derive a closed form for $\mathbf{M}$. To overcome this problem, we propose to derive a new ADMM to update $\mathbf{M}$ (the inner iteration). To do this, we reparametrize the $\mathbf{D} \odot \mathbf{M}$ with $\mathbf{\Gamma} \in \mathbb{R}^{p \times p}$ and we add an equality constraint $\mathbf{D} \odot \mathbf{M} = \mathbf{\Gamma}$, then we minimize the following

$$\begin{cases} \text{Minimize} & \frac{1}{2}\left\{\|\hat{\mathbf{\Sigma}}\mathbf{M}\|_F^2 + \varrho\|\mathbf{M}\|_F^2 - 2\langle \mathbf{M}, \hat{\mathbf{\Sigma}} + \varrho\mathbf{\Theta}^{k+1} - \mathbf{\Lambda}^k\rangle\right\} + \lambda\|\mathbf{\Gamma}\|_1, & (24) \\ \text{subject to} & \mathbf{D} \odot \mathbf{M} = \mathbf{\Gamma}. & (25) \end{cases}$$

The augmented Lagrangian $\mathcal{K}_\varrho(\mathbf{M}, \mathbf{\Gamma}, \mathbf{\Delta})$ associated with this problem is

$$\begin{aligned} \mathcal{K}_\varrho(\mathbf{M}, \mathbf{\Gamma}, \mathbf{\Delta}) = \frac{1}{2}\left\{\|\hat{\mathbf{\Sigma}}\mathbf{M}\|_F^2 + \varrho\|\mathbf{M}\|_F^2 - 2\langle \mathbf{M}, \hat{\mathbf{\Sigma}} + \varrho\mathbf{\Theta}^{k+1} - \mathbf{\Lambda}^k\rangle\right\} \\ + \lambda\|\mathbf{\Gamma}\|_1 - \langle \mathbf{\Delta}, \mathbf{\Gamma} - \mathbf{D} \odot \mathbf{M}\rangle + \frac{1}{2}\varrho\|\mathbf{\Gamma} - \mathbf{D} \odot \mathbf{M}\|_F^2 \end{aligned} \tag{26}$$

where $\mathbf{\Delta} \in \mathbb{R}^{p \times p}$ is the matrix containing the ensemble set of Lagrange multipliers associated with the constraint. As before, the ADMM here, i.e., the inner iteration, consists of the following three steps:

$$\begin{cases} \mathbf{M}_k^{j+1} & = \arg\min_{\mathbf{M}} \mathcal{K}_\varrho(\mathbf{M}, \mathbf{\Gamma}^j, \mathbf{\Delta}^j) & (27) \\ \mathbf{\Gamma}^{j+1} & = \arg\min_{\mathbf{\Gamma}} \mathcal{K}_\varrho(\mathbf{M}_k^{j+1}, \mathbf{\Gamma}, \mathbf{\Delta}^j) & (28) \\ \mathbf{\Delta}^{j+1} & = \mathbf{\Delta}^j - (\mathbf{\Gamma}^{j+1} - \mathbf{D} \odot \mathbf{M}_k^{j+1}) & (29) \end{cases}$$

<u>The intermediate $\mathbf{M}$-Step</u>: It can be shown that $\mathbf{M}_k^{j+1}$ is the minimizer of $\mathcal{K}_\varrho(\mathbf{M}, \mathbf{\Gamma}^j, \mathbf{\Delta}^j)$, i.e.,

$$\frac{\partial}{\partial \mathbf{M}}\left[\mathcal{K}_\varrho(\mathbf{M}_k^{j+1}, \mathbf{\Gamma}^j, \mathbf{\Delta}^j)\right] = \mathbf{0}, \tag{30}$$

which is equivalent to

$$\begin{aligned} \left(\hat{\mathbf{\Sigma}}^T\hat{\mathbf{\Sigma}} + \varrho\mathbf{I}\right)\mathbf{M}_k^{j+1} + \varrho\mathbf{D} \odot \mathbf{D} \odot \mathbf{M}_k^{j+1} + (\mathbf{\Delta}^j - \varrho\mathbf{\Gamma}^j) \odot \mathbf{D} \\ - \left(\hat{\mathbf{\Sigma}} + \varrho\mathbf{\Theta}^{k+1} - \mathbf{\Lambda}^k\right) = \mathbf{0}, \end{aligned} \tag{31}$$

and therefore, given by the previous expression, $\mathbf{M}_k^{j+1}$ has finally a closed form, despite the additional but straightforward computational effort at this level. This additional step, can be considered as a start-up for the original ADMM algorithm. This is very important when dealing with complex problems and large datasets. This step is solved via a fixed-point iteration, see [2].

<u>The intermediate $\mathbf{\Gamma}$-Step</u>: To deal with this $\mathbf{\Gamma}$-Step, we define an entry-wise soft-thresholding rule for all the off-diagonal elements of a matrix $\mathbf{A}$ as

$$\boldsymbol{\mathcal{S}}(\mathbf{A}, \kappa) = \{s(a_{jl}, \kappa)\}_{1 \le j,l \le p} \tag{32}$$

with $s(a_{jl}, \kappa) = \text{sign}(a_{jl}) \max(|a_{jl}| - \kappa, 0)\, 1_{\{j \neq l\}}$, see e.g. [6,11,12,17]. Then the $\mathbf{\Gamma}$-Step has a closed form given by

$$\mathbf{\Gamma}^{j+1} = \boldsymbol{\mathcal{S}}(\frac{1}{\varrho}\mathbf{\Delta}^{j} + \mathbf{D} \odot \mathbf{M}_{k}^{j+1}, \lambda). \tag{33}$$

2.2 The Inner-Outer ADMM Algorithm

Assembling the different expressions, we get the following inner-outer alternating directions method of multipliers algorithm. The advantage of this algorithm is that there is no need to solve the inner iteration with a higher accuracy. It is sufficient to stop the inner iteration with an accuracy of 10^{-3}, whereas in the outer iteration, we require a tighter accuracy of the order of 10^{-6}. This saves a lot computational time. Moreover, the motivation of using ADMM is the fact that it lends itself to an efficient implementation on parallel computers. Therefore, both the inner and outer iteration can easily be parallelized.

Algorithm 1. Inner-Outer ADMM

1: Input: Initialize the variables $\mathbf{\Theta}_0, \mathbf{M}_0, \mathbf{\Lambda}_0, \mathbf{\Gamma}^0, \mathbf{\Delta}^0$
2: Output: A precision matrix $\mathbf{M}$
3: Select a penalty scalar ϱ and a tuning parameter λ
4: $\mathbf{B} \leftarrow (\hat{\mathbf{\Sigma}}^T \hat{\mathbf{\Sigma}} + \varrho \boldsymbol{I})^{-1}$
5: **for** $k \leftarrow 0, 1, 2, \cdots$ **until convergence do**
6: $\quad \mathbf{\Theta}_{k+1} \leftarrow \text{Proj}_{\mathcal{K}}(\mathbf{M}_k + \frac{1}{\varrho}\mathbf{\Lambda}_k)$
7: $\quad \mathbf{A}_k \leftarrow (\hat{\mathbf{\Sigma}} - \mathbf{\Lambda}_k + \varrho \mathbf{\Theta}_{k+1})$
8: $\quad$ **for** $j \leftarrow 0, 1, 2, \cdots$ **until convergence do**
9: $\quad\quad \mathbf{A} \leftarrow \mathbf{A}_k - (\mathbf{\Delta}^j - \varrho \mathbf{\Gamma}^j) \odot \boldsymbol{D} - \varrho \mathbf{D} \odot \mathbf{D} \odot \mathbf{M}_k^j$
10: $\quad\quad \mathbf{M}_k^{j+1} \leftarrow \mathbf{B} \times \mathbf{A}$
11: $\quad\quad \mathbf{\Gamma}^{j+1} \leftarrow \boldsymbol{\mathcal{S}}(\frac{1}{\varrho}\mathbf{\Delta}^j + \mathbf{D} \odot \mathbf{M}_k^{j+1}, \lambda)$
12: $\quad\quad \mathbf{\Delta}^{j+1} \leftarrow \mathbf{\Delta}^j - (\mathbf{\Gamma}^{j+1} - \mathbf{D} \odot \mathbf{M}_k^{j+1})$
13: $\quad\quad$ Test of convergence of the inner iteration: $\|\mathbf{\Delta}^{j+1} - \mathbf{\Delta}^j\| \leq \varepsilon_{\text{inner}} = 10^{-3}$
14: $\quad$ **end for**
15: $\quad \mathbf{M}_{k+1} \leftarrow \lim_{j \to \infty} \mathbf{M}_k^j$
16: $\quad \mathbf{\Lambda}_{k+1} \leftarrow \mathbf{\Lambda}_k - (\mathbf{\Theta}_{k+1} - \mathbf{M}_{k+1})$
17: $\quad$ Test of convergence of the outer iteration: $\|\mathbf{\Lambda}_{k+1} - \mathbf{\Lambda}_k\| \leq \varepsilon_{\text{outer}} = 10^{-6}$
18: **end for**
19: **return M**

3 Experiments

The ADMM algorithm is known to be scalable. Hence both the inner and outer iterations can be easily parallelized, thus making the proposed algorithm efficient for big data. To show the convergence and efficiency of our approach, it is sufficient to use simulations on synthetic data and on a real example to compare the performance of our estimator with Graphical LASSO.

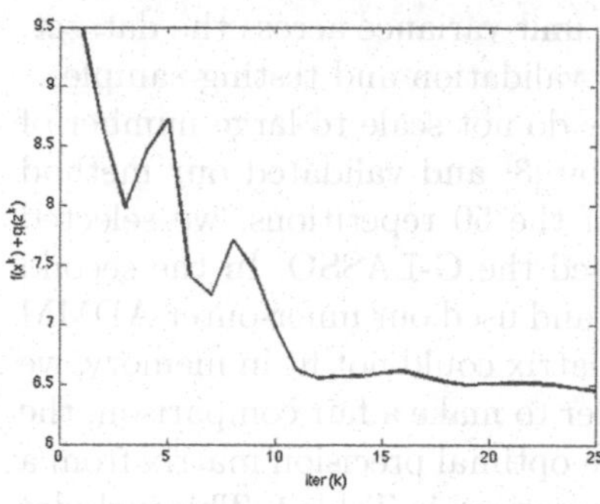

Fig. 1. Plot of the objective function for $\lambda = 0.01$

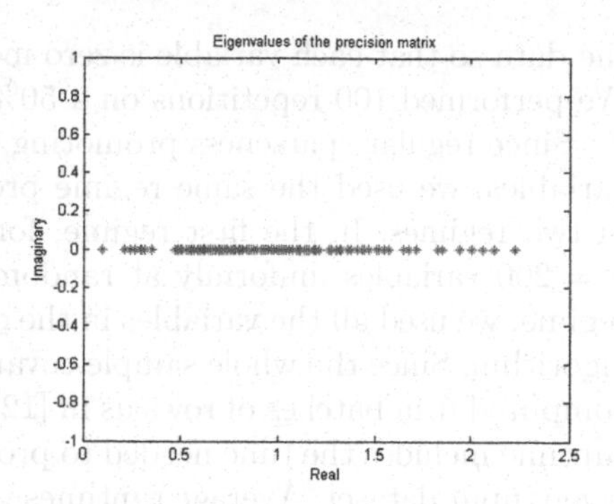

Fig. 2. Eigenvalue distribution of the estimated precision matrix.

3.1 Validation on Synthetic Data

In order to validate our approach, we used the same simulation structure as in [5]. We generated $n = 1000$ samples from a $p = 600$-dimensional normal distribution with correlation structure of the form $\sigma(x_i, x_j) = 0.6|i - j|$. This model has a banded structure, and the values of the entries decay exponentially as they move away from the diagonal. We generated an independent sample of size 1000 from the same distribution for validating the tuning parameter λ. Using the training data, we compute a series of estimators with 50 different values of λ and use the one with the smallest likelihood loss on the validation sample, where the likelihood loss is defined by, see e.g. [8],

$$\mathcal{L}(\hat{\mathbf{\Sigma}}, \mathbf{M}) = -\log\det(\mathbf{M}) + \langle \hat{\mathbf{\Sigma}}, \mathbf{M} \rangle. \quad (34)$$

We mention that all the experiments are conducted on a PC with 4 GB of RAM, 3 Ghz CPU using Matlab 2009a.

Results for $\lambda = 0.01$ are summarized in Fig. 1. The convergence is achieved in 25 steps and needs just 0.54 s. After a few steps of fluctuations (≈ 12 iterations), the objective function stabilizes and converges to its optimal value where the eigenvalues of the precision matrix estimated by our algorithm are real and positive. This proves the positive definiteness of the obtained precision matrix as shown in Fig. 2.

3.2 Validation on Real Data

For experimental validation, we used 4 cancer datasets publicly available at the Gene Expression Omnibus, see e.g., http://www.ncbi.nlm.nih.gov/geo/. For a fair comparison with the other method of estimating the inverse covariance matrix, we follow the same analysis scheme used by [8]. The datasets are: Liver cancer (GSE1898), Colon cancer (GSE29638), Breast cancer (GSE20194) and Prostate cancer (GSE17951) with sample size $n = 182; 50; 278$ and 154 respectively and number of genes $p = 21794; 22011; 22283$ and 54675. We preprocessed the data so that each variable is zero mean and unit variance across the dataset. We performed 100 repetitions on a 50% – 50% validation and testing samples.

Since regular sparseness promoting methods do not scale to large number of variables, we used the same regime proposed by [8] and validated our method in two regimes. In the first regime, for each of the 50 repetitions, we selected $n = 200$ variables uniformly at random and used the G-LASSO. In the second regime, we used all the variables in the dataset, and used our inner-outer ADMM algorithm. Since the whole sample covariance matrix could not fit in memory, we computed it in batches of rows as in [12]. In order to make a fair comparison, the runtime includes the time needed to produce the optimal precision matrix from a given input dataset. Average runtimes were summarized in Table 1. This includes the time to solve each optimization problem and also the time to compute the covariance matrix (if needed). Our method is considerably faster than the G-LASSO method as shown in Table 1.

Table 1. Runtimes for gene expression datasets. Our method is considerably faster than G-LASSO.

Dataset	Graphical LASSO	Our estimator
GSE1898	3.8 min	1.0 min
GSE29638	3.8 min	2.6 min
GSE20194	3.8 min	2.5 min
GSE17951	14.9 min	4.8 min

3.3 Validation on CGH Data

Alterations in the genome that lead to changes in DNA sequence copy number are a characteristic of solid tumors and are found in association with developmental abnormalities and/or mental retardation. Comparative genomic hybridization (CGH) can be used to detect and map these changes therefore knowledge of copy number aberrations can have immediate clinical use in diagnosis, and in some cases provide useful prognostic information.

In a typical CGH measurement, total genomic DNA is isolated from test and reference cell populations, differentially labeled, and hybridized to a representation of the genome that allows the binding of sequences at different genomic locations to be distinguished. Array CGH has been implemented using a wide variety of techniques such as BAC array, i.e., produced from bacterial artificial chromosomes; cDNA microarray which is made from cDNAs and oligo array, made from oligonucleotides (Affy, Agilent, Illumina) to name a few. The output from array CGH experiment is a log2 ratio of the copy number in the test versus the reference. The goal of this experiment is to identify genome regions with DNA copy number alterations.

The glioma data from Bredel et al. [4] contain samples representing primary GBMs, a particular malignant type of brain tumor. We investigate the performance of Fused LASSO and ADMM LASSO methods on the array CGH profiles of the GBM samples examined in Lai et al. [10]. To generate a more challenging situation where both local amplification and large region loss exist in the same chromosome, we paste together the following 2 array regions: (1) chromosome 7 in GBM29 from 40 to 65 Mb and (2) chromosome 13 in GBM31. The performance of the two methods on this pseudo chromosome is illustrated in Fig. 3. We can see that the proposed method using ADMM LASSO successfully identified both the local amplification and the big chunk of copy number loss.

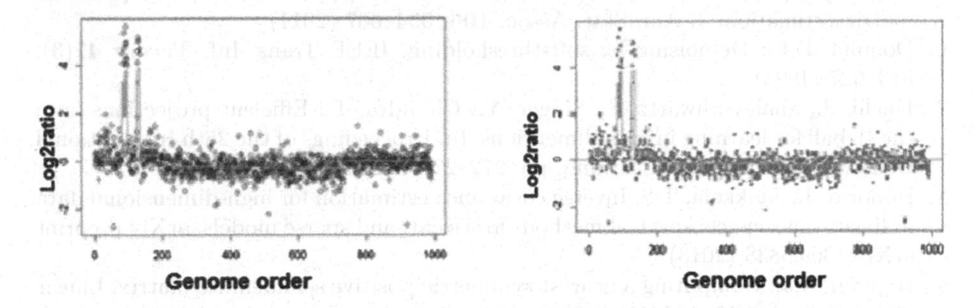

Fig. 3. Estimated copy number; Left: from Fused LASSO regression shows copy number alteration regions. Right: using ADMM LASSO algorithm

4 Conclusion and Future Work

The sparse precision matrix estimator has been shown to be useful in many applications. Penalizing the matrix is a tool with good asymptotic properties for estimating large sparse covariance and precision matrices. However, its positive definiteness property and unconstrained structure can be easily violated in practice, which prevents its use in many important applications such as graphical models, financial assets and comparative genomic hybridization.

In this paper, we have expressed the precision matrix estimation equation in a convex optimization framework and considered a natural modification by imposing the positive definiteness and problem-solving constraints. We have developed a fast alternating direction method of multipliers (ADMM) to solve the constrained optimization problem and the resulting estimator retains the sparsity and positive definiteness properties simultaneously. We are at the phase of demonstrating the general validity of the method and its advantages over correlation networks based on competitive precision matrix estimators with computer-simulated reaction systems, to be able to demonstrate strong signatures of intracellular pathways and provide a valuable tool for the unbiased reconstruction of metabolic reactions from large-scale metabolomics data sets.

References

1. Anbari, M.E., Alam, S., Bensmail, H.: COFADMM: a computational features selection with alternating direction method of multipliers. Procedia Comput. Sci. **29**, 821–830 (2014)

2. Baggag, A., Sameh, A.: A nested iterative scheme for indefinite linear systems in particulate flows. Comput. Methods Appl. Mech. Eng. **193**(21–22), 1923–1957 (2004)
3. Bien, J., Tibshirani, R.: Sparse estimation of a covariance matrix. Biometrika **98**, 807–820 (2011)
4. Bredel, M., Bredel, C., Juric, D., Harsh, G.R., Vogel, H., Recht, L.D., Sikic, B.I.: High-resolution genome-wide mapping of genetic alterations in human glial brain tumors. Cancer Res. **65**(10), 4088–4096 (2005)
5. Cai, T., Zhou, H.: A constrained l_1 minimization approach to sparse precision matrix estimation. J. Am. Stat. Assoc. **106**, 594–607 (2011)
6. Donoho, D.L.: De-noising by soft-thresholding. IEEE Trans. Inf. Theory **41**(3), 613–627 (1995)
7. Duchi, J., Shalev-Shwartz, S., Singer, Y., Chandra, T.: Efficient projections onto the l1-ball for learning in high dimensions. In: Proceedings of the 25th International Conference on Machine Learning, pp. 272–279. ACM (2008)
8. Honorio, J., Jaakkola, T.S: Inverse covariance estimation for high-dimensional data in linear time, space: spectral methods for riccati and sparse models. arXiv preprint arXiv:1309.6838 (2013)
9. Higham, N.J.: Computing a nearest symmetric positive semidefinite matrix. Linear Algebra Appl. **103**, 103–118 (1988)
10. Lai, W.R., Johnson, M.D., Kucherlapati, R., Park, P.J.: Comparative analysis of algorithms for identifying amplifications and deletions in array CGH data. Bioinform. **21**(19), 3763–3770 (2005)
11. Maurya, A.: A well-conditioned and sparse estimation of covariance and inverse covariance matrices using a joint penalty. J. Mach. Learn. Res. **17**, 1–28 (2016)
12. Mazumder, R., Hastie, T.: Exact covariance thresholding into connected components for large-scale graphical lasso. J. Mach. Learn. Res. **13**(Mar), 781–794 (2012)
13. Ravikumar, P., Wainwright, M., Raskutti, G., Yu, B.: High-dimensional covariance estimation by minimizing l_1-penalized log-determinant divergence. Electron. J. Statist. **5**, 935–980 (2011)
14. Rothman, A.J., Bickel, P.J., Levina, E., Zhu, J., et al.: Sparse permutation invariant covariance estimation. Electron. J. Stat. **2**, 494–515 (2008)
15. Schwertman, N.C., Allen, D.M.: Smoothing an indefinite variance-covariance matrix. J. Stat. Comput. Simul. **9**(3), 183–194 (1979)
16. Witten, D.M., Tibshirani, R.: Covariance-regularized regression and classification for high dimensional problems. J. R. Stat. Soc.: Ser. B (Stat. Methodol.) **71**(3), 615–636 (2009)
17. Xue, L., Ma, S., Zou, H.: Positive-definite 1-penalized estimation of large covariance matrices. J. Am. Stat. Assoc. **107**(500), 1480–1491 (2012)

Accurate Recognition of the Current Activity in the Presence of Multiple Activities

Weihao Cheng[(✉)], Sarah Erfani, Rui Zhang, and Ramamohanarao Kotagiri

School of CIS, The University of Melbourne, Parkville, Australia
weihaoc@student.unimelb.edu.au,
{sarah.erfani,rui.zhang,kotagiri}@unimelb.edu.au

Abstract. Sensor based activity recognition (AR) has gained extensive attention in recent years due to the ubiquitous presence of smart devices, such as smartphones and smartwatches. One of the major challenges posed by AR is to reliably recognize the current activity, when a given window of time series data contains several activities. Most of the traditional AR methods assume the entire window corresponds to a single activity, which may cause high error rate in activity recognition. To overcome this challenge, we propose a Weighted Min-max Activity Recognition Model (WMARM), which reliably predicts the current activity by finding an optimal partition of the time series matching the occurred activities. WMARM can handle the time series containing an arbitrary number of activities, without having any prior knowledge about the number of activities. We devise an efficient dynamic programming algorithm that solves WMARM in $\mathcal{O}(n^2)$ time complexity, where n is the length of the window. Extensive experiments conducted on 5 real datasets demonstrate about 10%–30% improvement on accuracy of WMARM compared to the state-of-the-art methods.

1 Introduction

Sensor based activity recognition (AR) has become an important research topic in recent years due to the ubiquitous presence of the smart devices, such as smartphones and smartwatches. The main goal of AR is to identify the current activity of a user, e.g., walking, running or being stationary, based on the sensor readings, e.g., acceleration. There are many applications of AR in our daily life [10], such as fitness tracking, safety monitoring, and context-aware behavior. Most of the existing AR methods [2, 8, 12] use segmented time series to train classifiers for recognition, where each sample represents a single activity. In practice, such AR systems utilize a window to capture the data stream of sensors in a fixed time duration, and supply the captured time series data to a trained classifier to predict the current activity. However, a window of the data stream may contain more than one activity causing transitions at arbitrary time positions, see Fig. 1 for an example. Simply using a time series containing multiple activities for classification that expects input containing single activity can lead to a poor recognition accuracy. A trivial approach is to use a small window, so that

J. Kim et al. (Eds.): PAKDD 2017, Part II, LNAI 10235, pp. 39–50, 2017.
DOI: 10.1007/978-3-319-57529-2_4

there is a high probability to capture the exact time series of the current activity with a minimal chance of transition taking place. But the trade-off is that the fewer data points will result in lower recognition performance. Therefore, accurate recognition of the current activity in the presence of multiple activities is a challenging task.

There are a few related studies, which aim to minimize the effects induced by activity transitions [13], or recognize the transitions [7,14]. Rednic et al. [13] reported that activity transitions can cause rapid fluctuations in classifier output. They utilized filters to stabilize the prediction, but the approach is unable to identify the current activity. Some AR systems [7,14] learn a classifier to recognize the activity transitions in time series. As the transition can provide information of the activities sequence, this approach can be utilized to infer the current activity. However, it is unsuitable to handle time series containing several transitions, such as stand-walk-run, since there will be a factorial number of classes that should be trained. For example, if there are N different activities, and the system demands to handle at most m transitions, then the total number of required classes is $\sum_{r=1}^{m+1} P_r^N$, where P_r^N stands for the number of permutations selecting r ordered objects from N objects. As a consequence, learning transitions is not an efficient approach for current activity recognition.

To address this difficult problem, an idea is to divide the observed window of time series into segments matching activities transitions. Thereby, the clean time series of the current activity, which is represented by the last segment, can be obtained for recognition. However, the existing time series segmentation methods [1,3,4,6,9,15,17] have at least one of the following drawbacks: (1) optimal solution is not guaranteed; (2) requiring the input of an exact or a maximum number of transitions; (3) only focusing on segmentation without considering activity recognition performance. A detailed discussion is later provided in the Related Work section. To address these drawbacks, we propose a *Weighted Minmax Activity Recognition Model (WMARM)*, which reliably predicts the current activity by finding an optimal partition of the time series matching the occurred activities. WMARM calculates a set of segments that the maximum value of the recognition errors on those segments is minimized, and the current activity is recognized based on the last segment. WMARM can handle time series containing an arbitrary number of transitions without having any prior knowledge about the number of transitions. WMARM can also be extended by imposing weights on the segments to improve recognition accuracy. Since the search space size of WMARM is $\mathcal{O}(2^n)$, we provide an efficient algorithm using dynamic programming to solve the model in $\mathcal{O}(n^2)$ time complexity, where n is the length of the window. Moreover, we propose a computationally efficient implementation of WMARM that the time series is divided into frames for coarse-grained processing. We conduct extensive experiments on 5 real datasets. The results demonstrate the superior performance of WMARM compared to the state-of-the-art methods when handling time series that contains one or more activity transitions. We also measure the execution time of WMARM algorithm on a smartphone, and the results indicate that the model can be effectively used on such resource constrained devices.

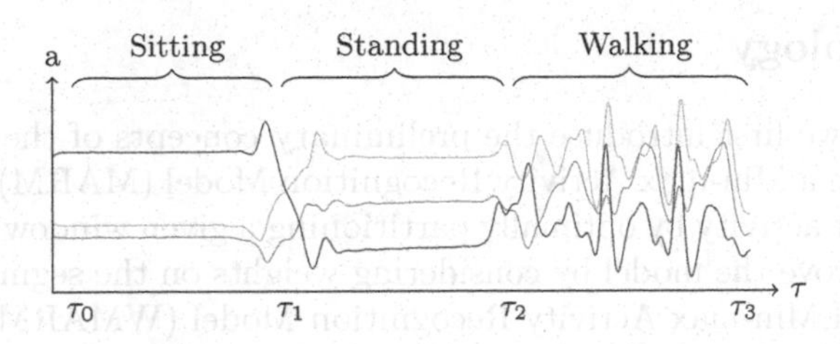

Fig. 1. The time series shown in the curves are 3-axis acceleration signals. There are 3 activities captured in the time series with transition points τ_0, τ_1, τ_2, τ_3, where τ_0 and τ_3 are two end points. Sitting and standing are the previous activities, walking is the current activity that we expect to recognize.

2 Related Work

In this section, we provide a brief survey of the relevant contributions to activity recognition, activity transition processing, and time series segmentation. Traditional sensor based AR systems [2,8,12] train classifiers with segmented samples assuming that each of them contains only one activity. However, they usually fail to recognize the current activity when the input time series contains two or more activities. Rednic et al. [13] focused on reducing the fluctuations caused by activity transitions. They used the Exponentially Weighted Voting filter to avoid spurious prediction, but the method is unable to detect the current activity during transitions. Some works focused on learning and recognizing the transitions [7,14]. However, such approaches normally require training a factorial number of classes regarding the number of transitions are considered. Therefore, they are not efficient for current activity recognition.

Time series segmentation aims to divide a 1-dimension sequence into several homogeneous segments, and existing methods can be summarized into following categories: (1) Heuristic based methods [6] use top-down, bottom-up, sliding window or hybrid ways for time series dividing. The results of heuristic methods are not stable since the optimal solution cannot be guaranteed. (2) LASSO based methods [9] solve the segmentation problem via a least-square regression with a ℓ_1-penalty. The methods require the number of maximum transitions as input. (3) Clustering based methods [17] divide the subsequences in a time series into K-clusters by using K-mean approach. The methods require the number of patterns as input. (4) Dynamic programming based methods [1,3,4,15] obtain an optimal partition of the time series by revealing an optimal structure of the problem. There are two kinds of dynamic programming approaches. One is for handling K-segmentation problem [1,3,15], where the number of transitions is required. The other one [4] can handle an arbitrary number of transitions, but it does not take into account the recognition performance.

In this paper, we propose a current activity recognition model based on time series segmentation via dynamic programming. The model reliably recognizes the current activity while possessing efficient execution time.

3 Methodology

In this section, we first introduce the preliminary concepts of the methodology. We then propose a Min-max Activity Recognition Model (MARM), which recognizes the current activity by optimally partitioning a given window of time series. We further improve the model by considering weights on the segments, and propose a Weighted Min-max Activity Recognition Model (WMARM). Both of the models can be solved using dynamic programing in $\mathcal{O}(n^2)$ time complexity, where n is the length of the window. Finally, we propose an efficient implementation of WMARM for obtaining high performance on resource constraint devices.

3.1 Problem Statement

Let $X = \{x_1, x_2, ..., x_n\}$ be a time series observed by a window. We define $X_{i:j} = \{x_i, x_{i+1}, ..., x_{j-1}, x_j\}$ $(1 \leq i \leq j \leq n)$ as a subsequence of X containing data points from x_i to x_j. Suppose there is a set of m transition points $\boldsymbol{\tau} = \{\tau_1, \tau_2, ..., \tau_m\}$ in the time series X. We define $\tau_0 = 0$, $\tau_{m+1} = n$ and $0 = \tau_0 < \tau_1 < \tau_2 < ... < \tau_m < \tau_{m+1} = n$. Therefore, the transitions points divide the time series X into $m+1$ segments $\{X_{1:\tau_1}, X_{\tau_1+1:\tau_2}, ..., X_{\tau_m+1:n}\}$, where each segment $X_{\tau_i+1:\tau_{i+1}}$ represents a single activity that is different from its neighbors. For the reliable prediction of the current activity, we expect to locate those transition points that the observed time series can be well-divided into clean segments. As a consequence, the current activity can be exhibited by the last segment and identified accurately.

3.2 Min-Max Activity Recognition Model (MARM)

Suppose we have a hypothesis $P(y \mid Z)$, which outputs the probability of the activity y represented by the time series Z. We define an error function of Z as:

$$\mathcal{E}(Z) = 1 - \max_{y} P(y \mid Z). \tag{1}$$

The function $\mathcal{E}(Z)$ returns the probability error of the predicted activity $\hat{y}$, where $\hat{y}$ holds the highest probability and is represented as:

$$\hat{y} = \underset{y}{\operatorname{argmax}} P(y \mid Z). \tag{2}$$

Thus, given a time series segment $X_{\tau_i+1:\tau_{i+1}}$, we can obtain the corresponding error $\mathcal{E}(X_{\tau_i+1:\tau_{i+1}})$ and the activity prediction $\hat{y} = \operatorname{argmax}_y P(y \mid X_{\tau_i+1:\tau_{i+1}})$. We propose a segmentation function $F(\boldsymbol{\tau})$ of the transition points $\boldsymbol{\tau}$ as follows:

$$F(\boldsymbol{\tau}) = \max_{\tau_i \in \boldsymbol{\tau} \bigcup \{\tau_0\}} \{\mathcal{E}(X_{\tau_i+1:\tau_{i+1}})\}. \tag{3}$$

The function $F(\boldsymbol{\tau})$ returns the maximum error of the segments corresponding to $\boldsymbol{\tau}$. Then, we propose MARM as:

$$\boldsymbol{\tau}^* = \underset{\boldsymbol{\tau}}{\operatorname{argmin}} \{F(\boldsymbol{\tau})\}, \tag{4}$$

where we aim to find an optimal solution $\boldsymbol{\tau}^* = \{\tau_1^*, \tau_2^*, ..., \tau_m^*\}$ such that the maximum error of those segments is minimized. After obtaining $\boldsymbol{\tau}^*$, the current activity is represented by the last segment $X_{\tau_{m+1}^*:n}$ and is predicted as:

$$\hat{y}^* = \underset{y}{\text{argmax}} P(y \mid X_{\tau_{m+1}^*:n}). \tag{5}$$

The intuitive explanation of solving MARM is to properly place the transition points by forcing down the upper bound of the recognition errors. Since the space size of valid $\boldsymbol{\tau}$ is $\mathcal{O}(2^n)$, exhaustive searching the solution is infeasible. However, we can employ dynamic programming to solve the problem in $\mathcal{O}(n^2)$ inspired from the work of [4]. We claim that the problem of optimizing our model exhibits optimal substructure, i.e., optimal solutions to a problem incorporate optimal solutions to related subproblems. Let X_k be the simplified notation of the time series $X_{1:k}$, and $X_0 = \emptyset$. Let $\boldsymbol{\tau}_k^*$ be an optimal solution on X_k, and $\boldsymbol{\tau}_0^* = \emptyset$. We propose the dynamic programming functional equation (DPFE) to solve MARM (Eq. 4) as follows:

$$F_{X_l}(\boldsymbol{\tau}_l^*) = \min_{0 \leq k < l} \{\max \{F_{X_k}(\boldsymbol{\tau}_k^*), \mathcal{E}(X_{k+1:l})\}\} \; (0 < l \leq n), \tag{6}$$

where $\boldsymbol{\tau}_1^*, \boldsymbol{\tau}_2^*, ..., \boldsymbol{\tau}_{l-1}^*$ are the previous optimal solutions that have already been obtained. Then, $\boldsymbol{\tau}_l^*$ is calculated as follows:

$$\boldsymbol{\tau}_l^* = \boldsymbol{\tau}_p^* \bigcup \{p\}, \tag{7}$$

where p is the last transition point in $\boldsymbol{\tau}_l^*$ and is obtained by:

$$p = \underset{0 \leq k < l}{\text{argmin}} \{\max \{F_{X_k}(\boldsymbol{\tau}_k^*), \mathcal{E}(X_{k+1:l})\}\}. \tag{8}$$

The DPFE in Eq. 6 indicates the optimal substructure that an optimal solution $\boldsymbol{\tau}_l^*$ to the problem regarding X_l is derived from the optimal solutions $\boldsymbol{\tau}_1^*, ..., \boldsymbol{\tau}_{l-1}^*$ to the subproblems regarding $X_1, ..., X_{l-1}$, which are the prefixes of the time series X. We show the correctness of the DPFE in Theorem 1.

Theorem 1. *$\boldsymbol{\tau}_l^*$ obtained by Eqs. 7 and 8 is an optimal solution to $F_{X_l}(\boldsymbol{\tau})$.*

Proof. We assume $\boldsymbol{\tau}_l^*$ is not an optimal solution of $F_{X_l}(\boldsymbol{\tau})$, and claim that $\boldsymbol{\tau}_l^+$ is an optimal solution. Suppose p is the last transition point of $\boldsymbol{\tau}_l^*$, then

$$F_{X_l}(\boldsymbol{\tau}_l^*) = \max \{F_{X_p}(\boldsymbol{\tau}_p^*), \mathcal{E}(X_{p+1:n})\}, \tag{9}$$

where $\boldsymbol{\tau}_p^* = \boldsymbol{\tau}_l^* - \{p\}$. Let q be the last transition point of $\boldsymbol{\tau}_l^+$, then

$$F_{X_l}(\boldsymbol{\tau}_l^+) = \max \{F_{X_q}(\boldsymbol{\tau}_q^+), \mathcal{E}(X_{q+1:n})\}, \tag{10}$$

where $\boldsymbol{\tau}_q^+ = \boldsymbol{\tau}_l^+ - \{q\}$. Since $\boldsymbol{\tau}_l^+$ is an optimal solution and $\boldsymbol{\tau}_l^*$ is not, then $F_{X_l}(\boldsymbol{\tau}_l^+) < F_{X_l}(\boldsymbol{\tau}_l^*)$. But we have:

$$\begin{aligned} F_{X_l}(\boldsymbol{\tau}_l^+) &= \max \{F_{X_q}(\boldsymbol{\tau}_q^+), \mathcal{E}(X_{q+1:l})\} \\ &\geq \max \{F_{X_q}(\boldsymbol{\tau}_q^*), \mathcal{E}(X_{q+1:l})\} \\ &\geq \max \{F_{X_p}(\boldsymbol{\tau}_p^*), \mathcal{E}(X_{p+1:l})\} = F_{X_l}(\boldsymbol{\tau}_l^*), \end{aligned} \tag{11}$$

Algorithm 1. MARM Algorithm

Input: (1) The time series X of length n.
Output: (1) The set of transition points $\boldsymbol{\tau}^*$; (2) The predicted current activity $\hat{y}^*$.
1: $\boldsymbol{\tau}_0^* = \emptyset$
2: $\hat{y}^* = Unkown$
3: $F_{X_0}(\boldsymbol{\tau}_0^*) = 0$
4: **while** $l = 1, 2, ..., n$ **do**
5: $p = \underset{0 \leq k < l}{\text{argmin}}\{\max\{F_{X_k}(\boldsymbol{\tau}_k^*), \mathcal{E}(X_{k+1:l})\}\}$ ▷ Using Eq. 8.
6: $\boldsymbol{\tau}_l^* = \boldsymbol{\tau}_p^* \bigcup \{p\}$ ▷ Using Eq. 7.
7: **if** $l == n$ **then**
8: $\hat{y}^* = \underset{y}{\text{argmax}} P(y \,|\, X_{p+1:l})$ ▷ Predicting the current activity.
9: **end if**
10: **end while**
11: $\boldsymbol{\tau}^* = \boldsymbol{\tau}_n^*$
12: **return** $\boldsymbol{\tau}^*, \hat{y}^*$

which is a contradiction. Therefore, $\boldsymbol{\tau}_l^*$ is an optimal solution of $F_{X_l}(\boldsymbol{\tau})$ on the time series X_l.

Based on the proposed DPFE, we can use dynamic programming to obtain an optimal solution $\boldsymbol{\tau}_n^* \equiv \boldsymbol{\tau}^*$ that minimizes $F_{X_n}(\boldsymbol{\tau}) \equiv F(\boldsymbol{\tau})$. We present the algorithm of solving MARM in Algorithm 1. We explain and analyze the algorithm in terms of time complexity: In lines 4–10, we iteratively calculate $\boldsymbol{\tau}_l^*$ from $l = 1$ to n, and each $\boldsymbol{\tau}_l^*$ is calculated in lines 5–6 with $\mathcal{O}(n)$ time complexity. In summary, the final solution $\boldsymbol{\tau}^*$ can be found in $\mathcal{O}(n^2)$ time complexity. When calculating $\boldsymbol{\tau}_n^*$, the last segment $X_{p+1:n}$ is exhibited, and the current activity is predicted as $\hat{y}^*$, which is shown in line 8.

3.3 Weighted Min-Max Activity Recognition Model (WMARM)

MARM finds a set of optimal segments on the observed time series X, and obtains the prediction of the current activity represented by the last segment. Normally, we would like to have a more reliable prediction for the current activity. Therefore, we place emphasis on reducing the error for the last segment. To deliver a more accurate prediction, we propose a new segmentation function $F_{LA}(\boldsymbol{\tau})$ which imposes weights on the last segment and previous segments. Let $p \equiv \tau_m$ be the last transition point in $\boldsymbol{\tau}$, $F_{LA}(\boldsymbol{\tau})$ is defined as follows:

$$F_{LA}(\boldsymbol{\tau}) = \max\{(1-\mu) \cdot F_{X_p}(\boldsymbol{\tau} - \{p\}),\, \mu \cdot \mathcal{E}(X_{p+1:n})\}, \tag{12}$$

in which a weight parameter $\mu \in [0, 1]$ is multiplied to the error of the last segment $X_{p+1:n}$, and $1 - \mu$ to the maximum error of the previous $m-1$ segments obtained by $F_{X_p}(\boldsymbol{\tau} - \{p\})$. We propose WMARM based on $F_{LA}(\boldsymbol{\tau})$ as follows:

$$\boldsymbol{\tau}^* = \underset{\boldsymbol{\tau}}{\text{argmin}}\{F_{LA}(\boldsymbol{\tau})\}. \tag{13}$$

By setting μ properly, the accuracy of prediction can be improved. If μ is set to 0.5, the model is equivalent to the original MARM without weight. WMARM (Eq. 13) can be solved with Theorem 2.

Theorem 2. *Given $\boldsymbol{\tau}_1^*, \boldsymbol{\tau}_2^*, ..., \boldsymbol{\tau}_{n-1}^*$, which are the optimal solutions of $F_{X_1}(\boldsymbol{\tau})$, $F_{X_2}(\boldsymbol{\tau})$, ..., $F_{X_{n-1}}(\boldsymbol{\tau})$, respectively. An optimal solution $\boldsymbol{\tau}^*$ of $F_{LA}(\boldsymbol{\tau})$ can be calculated as:*

$$\boldsymbol{\tau}^* = \boldsymbol{\tau}_p^* \bigcup \{p\}, \tag{14}$$

where p is the last transition point of $\boldsymbol{\tau}^*$ and is obtained by:

$$p = \underset{0 \leq k < n}{\operatorname{argmin}} \{\max\{(1-\mu) \cdot F_{X_k}(\boldsymbol{\tau}_k^*),\ \mu \cdot \mathcal{E}(X_{k+1:n})\}\}. \tag{15}$$

Proof. We assume $\boldsymbol{\tau}^*$ is not an optimal solution, and claim that $\boldsymbol{\tau}^+$ is an optimal solution, then

$$F_{LA}(\boldsymbol{\tau}^*) = \max\{(1-\mu) \cdot F_{X_p}(\boldsymbol{\tau}_p^*),\ \mu \cdot \mathcal{E}(X_{p+1:n})\}, \tag{16}$$

where $\boldsymbol{\tau}_p^* = \boldsymbol{\tau}^* - \{p\}$. Let q be the last transition point of $\boldsymbol{\tau}_l^+$, then

$$F_{LA}(\boldsymbol{\tau}^+) = \max\{(1-\mu) \cdot F_{X_q}(\boldsymbol{\tau}_q^+),\ \mu \cdot \mathcal{E}(X_{q+1:n})\}, \tag{17}$$

where $\boldsymbol{\tau}_q^+ = \boldsymbol{\tau}^+ - \{q\}$. Since $\boldsymbol{\tau}^+$ is an optimal solution and $\boldsymbol{\tau}^+$ is not, then $F_{LA}(\boldsymbol{\tau}^+) < F_{LA}(\boldsymbol{\tau}^*)$. But we have:

$$\begin{aligned} F_{LA}(\boldsymbol{\tau}^+) &= \max\{(1-\mu) \cdot F_{X_q}(\boldsymbol{\tau}_q^+),\ \mu \cdot \mathcal{E}(X_{q+1:n})\} \\ &\geq \max\{(1-\mu) \cdot F_{X_q}(\boldsymbol{\tau}_q^*),\ \mu \cdot \mathcal{E}(X_{q+1:n})\} \\ &\geq \max\{(1-\mu) \cdot F_{X_p}(\boldsymbol{\tau}_p^*),\ \mu \cdot \mathcal{E}(X_{p+1:n})\} \\ &= F_{LA}(\boldsymbol{\tau}^*), \end{aligned} \tag{18}$$

which is a contradiction. Therefore, $\boldsymbol{\tau}^*$ is an optimal solution of $F_{LA}(\boldsymbol{\tau})$ on the time series X.

To find the exact solution of WMARM, we present a dynamic programming algorithm in Algorithm 2. Similar to Algorithm 1, Algorithm 2 computes $\boldsymbol{\tau}_1^*, \boldsymbol{\tau}_2^*, ..., \boldsymbol{\tau}_{n-1}^*$ in $\mathcal{O}(n^2)$, as shown in lines 4–7. Calculating the last transition point p needs to iteratively examine the optimal value of $F_{X_k}(\boldsymbol{\tau}_k^*)$ from $k = 0$ to $n-1$, as shown in line 8, which needs $\mathcal{O}(n)$ time complexity. Then, the final solution $\boldsymbol{\tau}^*$ is obtained by combining p into $\boldsymbol{\tau}_p^*$, shown in line 9. In summary, the total time complexity of Algorithm 2 is $\mathcal{O}(n^2)$. Finally, the last segment $X_{p+1:n}$ is exhibited when calculating $\boldsymbol{\tau}^*$, and the current activity is predicted as $\hat{y}^*$, which is shown in line 10.

Algorithm 2. WMARM Algorithm

Input: (1) The time series X of length n.
Output: (1) The set of transition points $\boldsymbol{\tau}^*$; (2) The predicted current activity $\hat{y}^*$.
1: $\boldsymbol{\tau}_0^* = \emptyset$
2: $\hat{y}^* = Unkonwn$
3: $F_{X_0}(\boldsymbol{\tau}_0^*) = 0$
4: **while** $l = 1, 2, ..., n-1$ **do**
5: $p = \underset{0 \leq k < l}{\text{argmin}}\{\max\{F_{X_k}(\boldsymbol{\tau}_k^*), \mathcal{E}(X_{k+1:l})\}\}$ ▷ Using Eq. 8.
6: $\boldsymbol{\tau}_l^* = \boldsymbol{\tau}_p^* \bigcup \{p\}$ ▷ Using Eq. 7.
7: **end while**
8: $p = \underset{0 \leq k < n}{\text{argmin}}\{\max\{(1-\mu) \cdot F_{X_k}(\boldsymbol{\tau}_k^*), \mu \cdot \mathcal{E}(X_{k+1:n})\}\}$ ▷ Using Eq. 15.
9: $\boldsymbol{\tau}^* = \boldsymbol{\tau}_p^* \bigcup \{p\}$ ▷ Using Eq. 14.
10: $\hat{y}^* = \underset{y}{\text{argmax}} P(y \mid X_{p+1:n})$ ▷ Predicting the current activity.
11: **return** $\boldsymbol{\tau}^*$, $\hat{y}^*$

3.4 Efficient Implementation of WMARM

WMARM partitions the time series on data point level, which results in $\mathcal{O}(n^2)$ time complexity. However, the input time series for activity recognition may contain hundreds of data points, which produces a large amount of computation regarding the quadratic complexity. In practice, we can divide the time series into several frames of size h, and we treat the frame as the basic element in the time series. Therefore, running WMARM on the frame level reduces the amount of computation by a factor of h^2. Moreover, the computations of obtaining the hypothesis $P(y \mid Z)$ in the training phase is also reduced as a result of coarse-grained time series.

4 Empirical Evaluation

In this section, we evaluate the performance of WMARM in terms of accuracy on a desktop platform. The experiment scripts are written in Python 2.7 on 64-bit Ubuntu 14.04 LTS operating system. We also evaluate the execution time of the proposed WMARM algorithm on an iPhone 6 with iOS 9.0 system. The source code is written in Objective-C and C++.

Datasets: The experiments are conducted on 5 datasets: (1) Human Activity Sensing Consortium (HASC) 2011 [5]; (2) Human Activity Recognition on Smartphones Dataset (HARSD)[1]; (3) Actitracker dataset (ACTR) [8]; (4) Daily Sport Activities dataset (DSA)[1](see footnote 1); (5) Smartphone-Based Recognition of Human Activities and Postural Transitions Data Set (HAPT)[1](see footnote 1). We use the acceleration data of those datasets [11].

Experimental Settings: We use 4-fold cross validation for evaluating, and repeated 5 times (the datasets are randomly partitioned each time). We use

[1] https://archive.ics.uci.edu/ml/datasets.html.

3/4 data to generate clean samples (contain only one activity) for training the classifiers, and 1/4 data to generate several 5 s time series samples where each sample is randomly formed by $K+1$ segments of different activities with K transitions ($K = 0, 1, 2, 3$). The length of each activity segment is randomly selected with no less than 0.5 s. We extract the 1/3 lowest frequency Fourier coefficients of given time series as features. We set $h = n/10$ for the efficient implementation of WMARM. We train 10 classifiers on time series of sizes from 0.5 s to 5.0s (every 0.5 s), respectively, to form the hypothesis $P(y \mid Z)$. Each classifier is a random forest with 10 estimators.

Baselines: We use 5 baseline methods in comparison with WMARM: (1) Naive: We simply use the entire time series for predicting without segmentation. (2) GIR: We use Global Iterative Replacement (GIR) algorithm [3] to segment the time series, where the last segment is used for predicting. (3) OPM: We use Optimal Partitioning Method (OPM) [4] to segment the time series. (4) RNN: We use Recurrent Neural Network (RNN) [16] with LSTM + Softmax layers to predict the current activity. We extract features on every 0.5 s time series frames to form the sequential input of RNN. (5) MSG: We manually obtain the true segment of the current activity for predicting, and we denote this method as 'ManualSegment' (MSG).

4.1 Measuring the Accuracy of Current Activity Recognition

To evaluate the accuracy of WMARM for current activity recognition, we compare WMARM with the baselines using datasets: HASC, HARSD, ACTR, and DSA. In this experiment, we set the weight $\mu = 0.7$ for WMARM since we experimentally show later that this value of μ obtains the best accuracy among the cross-validation. According to the results shown in Table 1, WMARM outperforms Naive, GIR, OPM, and RNN in all cases, except for $K = 0$ on ACTR dataset, since WMARM always find the optimal segments for recognizing the current activity. Generally, it should be expected that no algorithm can obtain a better accuracy than MSG. However, WMARM outperforms MSG when $K = 0$ on HASC, HARSD, and DSA datasets. This is because that WMARM finds the best fitted segment for recognition instead of the true segment, thereby a part of noise can be excluded. To statistically compare the performance of WMARM with the baselines, we conduct the Wilcoxon signed-rank test on their results (80 pairs for each test). The returned p-values represent the lowest level of significance of a hypothesis that results in rejection. This value allows one to determine whether two methods have significantly different performance. We set the significance level $\alpha = 0.05$ for the comparison. For $K = 1, 2, 3$, the returned p-values ranging from 7.747e-15 to 1.199e-13, reject the null hypothesis for the comparisons: WMARM vs. all the baselines except for MSG, which indicate superior performance of WMARM against those methods. For $K = 0$, only the p-value 2.477e-06 of WMARM vs. GIR, rejects of the null hypothesis, which indicates the similar performances of the two methods.

Table 1. Results of accuracy on datasets: HASC, HARSD, ACTR, and DSA. K is the number of transitions in time series.

	HASC						HARSD					
K	Naive	GIR	OPM	RNN	WMARM	MSG	Naive	GIR	OPM	RNN	WMARM	MSG
0	88.00%	85.00%	88.55%	17.05%	**89.50%**	88.00%	81.30%	81.00%	82.85%	20.95%	**83.65%**	81.30%
1	35.90%	59.35%	46.25%	37.35%	**72.65%**	83.45%	32.35%	53.60%	45.25%	46.90%	**77.00%**	80.60%
2	22.45%	41.65%	28.45%	40.70%	**56.00%**	73.30%	20.40%	34.10%	29.20%	53.75%	**67.40%**	79.05%
3	16.05%	28.20%	19.85%	43.50%	**46.45%**	69.25%	15.90%	23.85%	22.85%	58.95%	**62.25%**	77.80%
	ACTR						DSA					
K	Naive	GIR	OPM	RNN	WMARM	MSG	Naive	GIR	OPM	RNN	WMARM	MSG
0	75.30%	73.25%	**75.50%**	18.65%	71.30%	75.30%	82.00%	76.00%	82.15%	7.00%	**84.60%**	82.00%
1	30.90%	51.60%	46.15%	43.55%	**63.70%**	72.85%	25.85%	63.75%	32.75%	19.25%	**65.25%**	77.30%
2	23.65%	37.55%	32.20%	50.15%	**61.15%**	70.45%	13.15%	47.80%	18.85%	22.30%	**52.30%**	73.90%
3	16.30%	25.65%	23.10%	50.00%	**54.95%**	67.90%	7.40%	31.85%	12.80%	28.50%	**45.10%**	70.05%

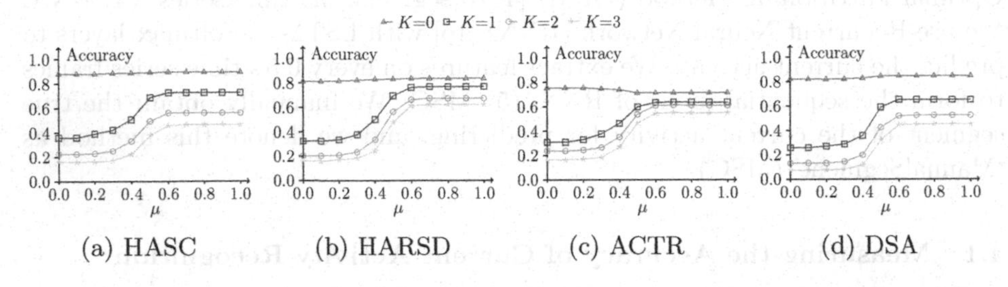

Fig. 2. Accuracy of WMARM with respect to μ on datasets: HASC, HARSD, ACTR, and DSA. For $K = 0$, the accuracy slightly improves on HASC, HARSD, and DSA, and slightly drops on ACTR. For $K = 1, 2, 3$, the accuracy reaches maximum around $\mu = 0.7(\pm 0.1)$, and then slightly decreases by less than 1% or becomes stable.

4.2 Evaluating the Impact of μ on Accuracy

We conduct experiments to evaluate the performance of WMARM with different settings of μ. According to the results shown in Fig. 2, when there is no transition in the data, i.e., $K = 0$, the accuracy is slightly affected by the weight μ since the optimal result should be only one segment. However, when there are transitions in the data, i.e., $K > 0$, the accuracy of WMARM significantly improves with respect to the increase of μ when $\mu < 0.7$, since the model emphasis more on the last segment, i.e., the current activity. The accuracy normally reaches the maximum around $\mu = 0.7(\pm 0.1)$, then slightly decreases by less than 1%, or becomes stable in a few cases. Since over emphasizing the weight on the last segment may impair the segmentation results on the previous segments, so that the prediction accuracy on the last segment is affected by the previous segments.

4.3 Measuring the Accuracy on Actual Transitions

In the previous experiments, we use splicing testing samples in order to study the performance of WMARM. In this experiment, we evaluate WMARM on actual transitions resulting from user's changing activities, for example changing naturally from running to walking. The samples are extracted from HAPT dataset [14] which provides several long time series containing a protocol of activities. We randomly select 70% of the data for training and the rest for testing, and repeat the experiment 10 times. The training samples are extracted during the activities, and the testing samples are extracted between transitions. WMARM obtains 75.41% accuracy, which outperforms the baselines: Naive (30.34%), GIR (49.44%), OPM (39.97%), and RNN (55.82%), except for MSG (76.69%). To explore the statistical significance of the performances of the methods on handling actual transitions, we conduct the Wilcoxon signed-rank test on their results (10 pairs). We set the significance level $\alpha = 0.05$ for the comparison. The p-values of WMARM vs. Naive/GIR/OPM/RNN (0.003346 for all), reject the null hypothesis for the accuracy measurements, implying a significant improvement of WMARM over those methods.

4.4 Evaluating the Execution Time on Smartphone

To evaluate the execution time of the WMARM algorithm, we develop an iOS app on iPhone 6 using Objective-C, and implement the WMARM algorithm as an internal function using C++. The app captures the 3-axis acceleration data with 100 samples per second, which is supplied to WMARM algorithm for processing. We observe a total execution time for 500 runs, and calculate the average time. WMARM algorithm only costs averagely 0.0153 s for one execution, which is not expensive for running AR systems on smartphones. Naive method costs averagely 0.0012 s for one execution, but its accuracy is much lower than WMARM.

5 Conclusions

In this paper, we highlight a problem normally presented in activity recognition (AR) that traditional methods usually fail to recognize the current activity in the presence of multiple activities. To solve this problem, we devise a Weighted Min-max Activity Recognition Model (WMARM), which predicts the current activity by optimally partitioning the observed window of time series matching the activities presented. WMARM considers weights on the partitioned segments to obtain reliable recognition accuracy. WMARM can also effectively process the time series containing an arbitrary number of transitions without any prior knowledge about the number of transitions. Instead of exhaustively searching the optimal solution of WMARM in exponential space, we propose an efficient dynamic programming algorithm that computes the model in $\mathcal{O}(n^2)$ time complexity, where n is the length of the window. Moreover, we present an efficient implementation of WMARM that the computation cost can be further reduced. Extensive experiments on 5 real datasets have demonstrated the superior performance of WMARM on handling time series with one or more activity transitions. The results show about 10%-30% improvement on the accuracy of current activity recognition compared to state-of-the-art methods. The experiment on iPhone 6 shows the prominent computational efficiency of WMARM.

References

1. Bellman, R.: On the approximation of curves by line segments using dynamic programming. Commun. ACM **4**(6), 284 (1961)
2. Hemminki, S., Nurmi, P., Tarkoma, S.: Accelerometer-based transportation mode detection on smartphones. In: Proceedings of Conference on Embedded Networked Sensor Systems, p. 13 (2013)
3. Himberg, J., Korpiaho, K., Mannila, H., Tikanmaki, J., Toivonen, H.T.: Time series segmentation for context recognition in mobile devices. In: Proceedings of International Conference on Data Mining, pp. 203–210 (2001)
4. Jackson, B., Scargle, J.D., Barnes, D., Arabhi, S., Alt, A., Gioumousis, P., Gwin, E., Sangtrakulcharoen, P., Tan, L., Tsai, T.T.: An algorithm for optimal partitioning of data on an interval. Signal Proces. Lett. **12**(2), 105–108 (2005)
5. Kawaguchi, N., Ogawa, N., Iwasaki, Y., Kaji, K., Terada, T., Murao, K., Inoue, S., Kawahara, Y., Sumi, Y., Nishio, N.: HASC challenge: gathering large scale human activity corpus for the real-world activity understandings. In: Proceedings of Augmented Human International Conference, p. 27 (2011)
6. Keogh, E., Chu, S., Hart, D., Pazzani, M.: Segmenting time series: a survey and novel approach. Data Min. Time Ser. Databases **57**, 1–22 (2004)
7. Khan, A.M., Lee, Y.K., Lee, S.Y., Kim, T.S.: A triaxial accelerometer-based physical-activity recognition via augmented-signal features and a hierarchical recognizer. Trans. Inf. Technol. Biomed. **14**(5), 1166–1172 (2010)
8. Kwapisz, J.R., Weiss, G.M., Moore, S.A.: Activity recognition using cell phone accelerometers. SigKDD Explor. Newsl. **12**(2), 74–82 (2011)
9. Levy-leduc, C., Harchaoui, Z.: Catching change-points with lasso. In: Proceedings of Advances in Neural Information Processing Systems, pp. 617–624 (2008)
10. Lockhart, J.W., Pulickal, T., Weiss, G.M.: Applications of mobile activity recognition. In: Proceedings of Conference on Ubiquitous Computing, pp. 1054–1058 (2012)
11. Nguyen, T., Gupta, S.K., Venkatesh, S., Phung, D.Q.: A bayesian nonparametric framework for activity recognition using accelerometer data. In: Proceedings of International Conference on Pattern Recognition, pp. 2017–2022 (2014)
12. Reddy, S., Mun, M., Burke, J., Estrin, D., Hansen, M., Srivastava, M.: Using mobile phones to determine transportation modes. Trans. Sens. Netw. (TOSN) **6**(2), 13 (2010)
13. Rednic, R., Gaura, E., Kemp, J., Brusey, J.: Fielded autonomous posture classification systems: design and realistic evaluation. In: Proceedings of ACIS International Conference on Software Engineering, Artificial Intelligence, Networking and Parallel/Distributed Computing, pp. 635–640 (2013)
14. Reyes-Ortiz, J.L., Oneto, L., Samà, A., Parra, X., Anguita, D.: Transition-aware human activity recognition using smartphones. Neurocomputing **171**, 754–767 (2016)
15. Rosman, G., Volkov, M., Feldman, D., Fisher III, J.W., Rus, D.: Coresets for k-segmentation of streaming data. In: Proceedings of Advances in Neural Information Processing Systems, pp. 559–567 (2014)
16. Sak, H., Senior, A.W., Beaufays, F.: Long short-term memory recurrent neural network architectures for large scale acoustic modeling. In: Proceedings of Interspeech, pp. 338–342 (2014)
17. Tseng, V.S., Chen, C.H., Huang, P.C., Hong, T.P.: Cluster-based genetic segmentation of time series with dwt. Pattern Recogn. Lett. **30**(13), 1190–1197 (2009)

Modeling Information Sharing Behavior on Q&A Forums

Biru Cui[1](✉), Shanchieh Jay Yang[2], and Christophan M. Homan[3]

[1] Computing and Information Sciences, Rochester Institute of Technology, Rochester, NY 14623, USA
{bxc2868,jay.yang}@rit.edu

[2] Computer Engineering, Rochester Institute of Technology, Rochester, NY 14623, USA

[3] Computer Science, Rochester Institute of Technology, Rochester, NY 14623, USA
cmh@cs.rit.edu

Abstract. Q&A forums pool massive amounts of crowd expertise from a broad spectrum of geographical, cultural, and disciplinary knowledge toward specific, user-posed questions. Existing studies on these forums focus on how to route questions to the best answerers based on content or predict whether a question will be answered, but few of them investigated the inherent knowledge sharing relationship among users. We study knowledge sharing among users of StackOverflow, a popular Q&A forum, where the knowledge sharing process is related to the time elapsed since a question was posted, the reputation of the questioner, and the content of the posted text. Taking these factors into consideration, the paper proposes *time-based information sharing model* (TISM), where the likelihood a user will share or provide knowledge to another is modeled as a continuous function of time, reputation, and post length. With the resulting knowledge sharing network learned by TISM, we are able to predict for a given question the number of responses over time, who will answer the question and who will provide the accepted answer. Our experiments show that predictions using TISM outperform NetRate, query likelihood language, random forest, and linear regression models.

Keywords: StackOverflow · Knowledge sharing · Question answering

1 Introduction

Q&A forums are arguably one of the earliest online venues for the crowdsourcing of information seeking tasks. Many existing studies on these sites have used content analysis to improve the quality of answers by matching questions to the best answer providers or predicting whether a question will be answered in a classification approach. In this paper, we model the inherent knowledge sharing relationships among users of the Q&A forum, and use the resulting *knowledge sharing network* to predict for a given question the number of responses

J. Kim et al. (Eds.): PAKDD 2017, Part II, LNAI 10235, pp. 51–63, 2017.
DOI: 10.1007/978-3-319-57529-2_5

over time, the (online) identity of who will respond to the question, and which response will contain the answer accepted by the original questioner.

The process of question answering (the process of peer-to-peer information sharing in Q&A forums) is itself collective, complex, and dynamic. Each question to be answered is different from all previous questions. So the knowledge gleaned from the answers to one question cannot directly answer the next one. Once a question is solved, the new response on the question thread will dramatically decrease, *i.e.*, a question thread's life time is related to when the question is solved. By taking into account the user account's reputation, post body text, and temporal factor, we proposed *time-based information sharing model* (TISM) and tested on two topic areas, "mechanics" and "security," found on StackOverflow, a popular Q&A forum. The test results are compared to NetRate [4], regression methods such as random forest, linear regression, and a text mining algorithm query likelihood language (QLL).

The rest of the paper is organized as follows: Sect. 2 discusses the related works on Q&A forum and information diffusion. Section 3 gives an introduction of the StackOverflow dataset and discusses its properties. Section 4 introduces the TISM. Section 5 evaluates the performance of TISM in comparison to NetRate, RandomForest, linear regression, and QLL. Section 6 concludes the paper.

2 Related Work

Yang *et al.* [9] studied the problem of predicting whether a question will receive an answer. Anderson *et al.* [1] focused on predicting whether a StackOverflow thread will have long-lasting value, *i.e.*, receive high or low pageviews in the future, and whether a question will receive satisfied answer. Asaduzzaman *et al.* [2] investigated how long it takes a question to be in answered. Hanrahan *et al.* [6] modeled question difficulty as a function of the time it takes for a question to receive an acceptable answer and users' expertise. To route questions to appropriate answerers, Chang *et al.* [3] and Li *et al.* [7] compare several matching algorithms; Zhou *et al.* [10] and Liu *et al.* [8] examine classification approaches to route question predict the questioner's satisfaction, respectively.

Among above studies, the classification is a common approach being utilized, and some studies investigated how to select features to improve performance. The classification is able to find the relation between the features and the result, but ignores the inherent process of how knowledge being shared between users. We study similar prediction problems as these works; the main difference is that we try to model the inherent information sharing network, and use that to solve these problems.

On the other hand, information sharing is similar to the information diffusion process that information flows from the information container to the information seeker. To model information diffusion, many studies focus on learning the links between users. Goyal *et al.* [5] estimated link weight based as the ratio of similar actions between two persons. Gomez *et al.* [4] proposed an algorithm where each edge is associated with a continuous likelihood function.

3 Q&A Forums

A Q&A forum is composed of a number of Q&A threads, where each begins with one question and is followed by answers, comments, and votes; the relevant actors are questioner, answerer, commenter and voter. Typical relationships among the Q&A forum are: questioner-answerer, questioner-commenter, answerer-commenter[1]. A post published by an actor is referred to as either a question, an answer or a comment. The study makes the assumption:

Assumption 1. *An answering or a commenting action in a Q&A thread indicates knowledge being shared from the answerer to the questioner, the commenter to the questioner, or the commenter to the answerer; an answer or a comment is only published if it provides new knowledge or information to the Q&A thread.*

This study utilizes the StackOverflow[2], and on two particular topics: "mechanics" and "security". "Mechanics" comprises questions related to automotive mechanism and maintenance. "Security" is about cyber security and secure coding. The mechanics dataset thus represents a relatively more traditional body of knowledge, where we hypothesize that knowledge evolved relatively slower than in computer security. Table 1 lists the relevant statistics.

Table 1. Datasets summary

Dataset	#Posts	#Comments	#Votes	#Users	Start-date	End-date
Mechanics	15463	19175	46975	8417	2010-08-08	2015-03-08
Security	62901	98841	377874	53728	2010-11-26	2015-08-16

We define a *knowledge sharing network* as the set of all actors who have posted (questions or responses) on StackOverflow, where two actors share a tie if one actor responds to post by the other. Similar to other social networks, this knowledge sharing network also follows a power-law degree distribution.

4 TISM: Time-Based Information Sharing Model

Given the *knowledge sharing network*, it is still unknown what is the link weight which requires to model the knowledge sharing process. In principle, modeling the knowledge sharing process is to learn the probability a sharing action (answering or commenting) happens. During the knowledge sharing process, time is an important factor that a later answer may take the advantage of existing answers and is more likely to be accepted. Also the time difference from posting is able to reflect the strength of knowledge sharing relationship. On the other hand, actor's reputation is based on the history of whether the actor's posts

[1] The study did not include voter since voter's information is not published by StackOverflow dataset.

[2] https://archive.org/details/stackexchange.

received positive responses; and the post body text also affect whether it will receive answers. By taking the reputation, post body text and the temporal factor into account, we propose "TISM (Time-based Information Sharing Model)" to reveal the information sharing process between actors in the Q&A forum.

Given a size N network (N actors) and a thread set C, each thread includes a question, multiple answers and comments, tagged with time-stamps. Thus, $C = \{t^1, t^2, \cdots, t^{|C|}\}$, where $t^c = \{t_1^c, t_2^c, \cdots, t_N^c\}$. t_i^c is the time-stamp of i posts an answer or comment in the thread c. The link from node i to node j means the likelihood node i shares knowledge to j when i issued a post (it can either be an answer or a comment) after j. Firstly, i can only share knowledge to j if i published a post after j; secondly, difference on expertise level (represented by the reputation difference) could be a factor affects whether the answering/commenting happened between nodes; thirdly, it is also affected by the knowledge level obtained by the question, where the knowledge level of a question is represented by the questioner reputation and body text. Since votes happened after the answers or the comments, votes information is not included in the model. Thus, the conditional likelihood function is:

$$f_{ji}(t_i|t_j, \delta_r(j,i), k_j^c) = \begin{cases} f_{ji}(t_{j,i}, \delta_r(j,i), k_j^c) & t_i > t_j \\ 0 & t_i <= t_j \end{cases} \quad (1)$$

$$\text{where } k_j^c = r_j * v_j^c$$

where $t_{j,i}$ is the time difference between j and i published their posts; $\delta_r(i,j)$ is the reputation difference between j and i; r_j is the absolute reputation of j; v_j^c is the number of unique words of j's post in Q&A thread c. k_j^c is the knowledge level contained in post j of thread c, which is the product of r_j and v_j^c.

In general, f can be in the form of exponential function of the product of the selected features, such that

$$f_{ji}(t_{j,i}, \delta_r(j,i), k_j^c) = \exp -(\alpha_{ji}^{(1)} * t_{j,i} + \alpha_{ji}^{(2)} * \delta_r(j,i) + \alpha_{ji}^{(3)} * k_j^c) \quad (2)$$

Since reputation difference is relatively constant,

$$\begin{aligned} 1 &= \int_{-\infty}^{+\infty} f_{ji}(t_{j,i}, \delta_r(j,i), k_j^c) \\ &= e^{-\alpha_{ji}^{(2)}\delta_r(j,i)} \int_0^{+\infty} e^{-\alpha_{ji}^{(1)} * t} dt \int_0^{+\infty} e^{-\alpha_{ji}^{(3)} * k} dk = e^{-\alpha_{ji}^{(2)}\delta_r(j,i)} \frac{1}{\alpha_{ji}^{(1)}} \frac{1}{\alpha_{ji}^{(3)}} \end{aligned} \quad (3)$$

so, $e^{-\alpha_{ji}^{(2)}\delta_r(j,i)} = \alpha_{ji}^{(1)}\alpha_{ji}^{(3)}$, and f_{ji} depends only on variable t and k, and Eq. (2) can be simplified as:

$$f_{ji}(t_{j,i}, k_j^c) = \alpha_{ji}\beta_{ji} \exp -(\alpha_{ji} t_{j,i} + \beta_{ji} k_j^c) \quad (4)$$

Accordingly, the likelihood that node i does not share knowledge to j by the end of the thread is defined as a survival function:

$$S_{ji}(t^c - t_j) = 1 - \int_0^{t^c - t_j} f_{ji}(t_{j,i}, k_j^c)$$
$$= 1 - \int_0^{t^c - t_j} \alpha_{ji} e^{-\alpha_{ji} t} dt \int_0^{+\infty} \beta_{ji} e^{-\beta_{ji} k} dk = e^{-\alpha_{ji}(t^c - t_j)} \tag{5}$$

t^c is the end time (the time elapsed since the question to the last post in the thread) of the Q&A thread c. Knowledge k is positive (product of reputation and number of unique words of the post), so it is integral from 0 to ∞. This survival function means, for any question or answer j published, i does not share knowledge to j for any time from t_j to the end time of this thread t^c.

According to the Assumption 1, once a post is published, it provides some knowledge to previous ones. It is also possible an actor published multiple comments/answers in a Q&A thread. In many cases, the subsequent comments published by the same actor are to explain her/his previous answer or comment, such that the knowledge provided could be overlapping. So, for each actor, only her/his first answer or comment is included. Therefore, the likelihood an actor i provides knowledge to a Q&A thread is the likelihood either the actor published an answer or a comment on existing answering/commenting links.

$$\ell_i^+(c) = f_{oi}(t_{o,i}^c, k_o^c)^{I(E_{oi})} \prod_{j:j \neq o, t_j^c < t_i^c} f_{ji}(t_{j,i}^c, k_j^c)^{I(E_{ji})}$$
$$= \prod_{j:t_j^c < t_i^c} f_{ji}(t_{j,i}^c, k_j^c)^{I(E_{ji}^c)} \tag{6}$$

where o is the questioner such that $t_o^c < t_i^c$ for $\forall i : i \neq o$; $I(E_{ji}^c)$ represents whether there is an answering or commenting action from i to j in thread c; $t_{j,i}^c$ is the time difference between j's and i's post. The likelihood a node i does not share any knowledge (does not publish any answer or comment) to the Q&A thread is:

$$\ell_i^-(c) = \prod_{j:t_j \leq t^c} S_{ji}(t^c - t_j) \tag{7}$$

It is the product of the survival function that for each node j, if j published a post but i did not share knowledge to j.

Therefore, the total likelihood for a thread is:

$$\ell(c) = \prod_{i:t_i^c \leq t^c} \ell_i^+(c) \prod_{i:t_i^c > t^c} \ell_i^-(c) \tag{8}$$

and the likelihood of all threads is

$$L(C; A, B) = \prod_{c \in C} \ell(c, A, B) \tag{9}$$

relevant log likelihood is

$$
\begin{aligned}
&-LL(C;A,B)\\
&=-\sum_{c\in C}\log\ell(c,A,B)=-\sum_{c\in C}\Big(\sum_{i:t_i^c\le t^c}\log\ell_i^+(c)+\sum_{i:t_i^c>t^c}\log\ell_i^-(c)\Big)\\
&=-\sum_{c\in C}\Big(\sum_{i:t_i^c\le t^c}\sum_{j:t_j^c\le t_i^c}(\log\alpha_{ji}+\log\beta_{ji}-\alpha_{ji}t_{j,i}-\beta_{ji}k_j^c)^{I(E_{ji}^c)}\\
&+\sum_{i:t_i^c>t^c}\sum_{j:t_j^c\le t^c}-\alpha_{ji}(t^c-t_j)\Big)
\end{aligned}
\tag{10}
$$

The problem is transferred to

$$
\text{minimize}_{A,B}-\sum_{c\in C}\log\ell(c,A,B) \tag{11}
$$

$$
\begin{aligned}
&\text{subject to } \alpha_{j,i}\ge 0,\beta_{j,i}\ge 0,i,j=1,\dots,N,i\ne j\\
&\quad\text{where } A:=A_j|j=1,\dots,n;A_j:=\alpha_{j,i}|i=1,\dots,n,i\ne j\\
&\quad\text{where } B:=B_j|j=1,\dots,n;B_j:=\beta_{j,i}|i=1,\dots,n,i\ne j
\end{aligned}
$$

The whole optimization is computation intensive, but it can be accelerated by running optimization on each node in parallel, *i.e.*, optimizing A_j, B_j on the set C for each j independently. Each dataset is tested in two cases: including answers only; including both answers and comments. The time stamp of the question is set as 0 (the beginning of the thread), while the time stamp of the following answers and comments are set as the time elapsed since the question is posted. All time stamps are normalized by the maximum time stamp across all threads. The reputation is normalized according to the maximum actor's reputation in the dataset; while the number of unique words is normalized according to the maximum number of post's body text's unique words across all posts.

The optimal α and β from both datasets also follow the power law. This means that in the network, the knowledge is shared faster on a small portion of links; or in other words, there are a small number of users who are more active than others to share their knowledge. This observation is the same as the assumption proposed in [1] that users are organized as a *reputation pyramid* where the high reputation users are on the top and answer questions quicker than low reputation users who are on the bottom of the pyramid.

5 Prediction Based on the Knowledge Sharing Network

Given the *knowledge sharing network* learned by TISM, after actor j published a question, if there is direct link from actor i to j, the probability i shares knowledge to j by posting an answer or a comment by time T is:

$$
P_{ji}^{(1)}=\int_0^T f_{ji}dt \tag{12}
$$

where f_{ji} is the likelihood function for the link from i to j. f_{ji} is defined as Eq. 4.

If i is 2-hop away from j, by time T, the probability of i joining j's thread by posting a comment on an answer of j's question is:

$$P_{ji}^{(2)} = 1 - \prod_{l}(1 - P_{jl}^{(1)} P_{li}^{(1)}) = 1 - \prod_{l}(1 - \int_{0}^{T} f_{jl}(t, k_j) f_{li}(T - t, k_l) dt) \quad (13)$$

where l is the bridge node that there is a link from l to j and a link from i to l. For each node i 2-hop away from the questioner j, if the bridge node l is indeed an answerer of j's question, applying the actual l's reputation and l's answer, otherwise, randomly sample a knowledge level for k_l.

To evaluate the performance of TISM, we apply it to solve several predictive analysis tasks: how many people will be involved in a Q&A thread; how many people will be involved in a Q&A thread over time; who will answer the question; and who will provide the accepted answer.

5.1 Task1: Thread Size Prediction

The first question is to know how many people will be involved in a Q&A thread. There are mainly two type of approaches to solve the problem.

- Use regression methods to determine the relation between the properties of the initial question/questioner and the thread size;
- Learn the link weight and then estimate the thread size.

Thread Size Estimation with Regression. When an actor raised a question, the only known information are from the question and the questioner. We select features: q_reput, #questions, title_total, title_nostop, title_unique, body_total, body_nostop, body_unique[3]. Linear Regression and Random Forest are utilized to train the regression models.

Thread Size Prediction with Link Weight. The second method is to learn the link weight and then estimate the thread size. We use TISM and NetRate [4] to learn the link weight, where both methods recover the link weight as a likelihood function.

Given j published a question, the thread size is the summation of the probability that each actor joins j's thread. By including answers only, the thread size is the summation of the probability of any actor i answers j's question:

$$TS_j = \sum_{i \neq j} P_{ji}^{(1)} \quad (14)$$

[3] q_reput: questioner's reputation; #questions: number of questions published by the questioner; title_total: #words of title; title_nostop: #non-stopwords in the title; title_unique: #unique non-stopwords in the title; body_total: #words of body text; body_nostop: #non-stopwords in the body text; body_unique: #unique non-stopwords in the body text.

where TS_j is the thread size of j's question; $P_{ji}^{(1)}$ is the probability i answers j's question, and there is direct link from i to j in the TISM or NetRate learned network.

By including both answers and comments, the thread size is:

$$TS_j = \sum_{i \neq j} (1 - (1 - P_{ji}^{(1)})(1 - P_{ji}^{(2)})) \tag{15}$$

In this case, $P_{ji}^{(1)}$ is the probability i answers or comments j's question, *i.e.*, i can directly reach j; and $P_{ji}^{(2)}$ is the probability i comments on answers of j's question, *i.e.*, i is 2-hop away from the j in the TISM or NetRate learned network.

Result. The experiment runs 10-fold cross validation. All Q&A threads (including questions both answered and not answered) are randomly split into 10 folds. Each round selects 1 fold of threads as the testing data, and the remaining as the training data. Each method trains the regression model or learns the link weights from the training data, and tests on the testing data. The mean square error (MSE) is the difference between the prediction and the ground truth on all testing threads. Table 2 shows the mean MSE on all 10 rounds on "mechanics" and "security", respectively. RF and LM represent the case of using RandomForest and LienarRegression. "Ans" means only including answers in a thread such that the thread size is the total number of answerers in the thread; while "AnsComt" includes both answers and comments in a thread that the thread size is the total number of actors (answerer and commenter) in the thread. Both regression models, NetRate and TISM are trained separately for "Ans" and "AnsComt". In "AnsComt" case, NetRate and TISM will include more edges which from commenting actions.

Table 2. MSE of estimation on Q&A thread size

	Dataset	RF	LM	NetRate	TISM
Ans	Mech	1.59	1.58	6.33	**1.29**
AnsComt	Mech	4.37	4.39	21.30	**3.86**
Ans	Secu	2.70	2.80	12.30	**2.58**
AnsComt	Secu	13.00	13.27	44.55	**12.90**

As shown in Table 2, TISM outperforms other methods in all cases. Two regression methods perform similarly. NetRate does not work well in this situation since it only takes the fact of whether an actor has posts in a thread into account, but not leveraging the existence of explicit answering/commenting actions. In addition, NetRate's link fucntion only depends on the time which ignores the variety of question/answer published. For example, for the same questioner, answerer's behavior could be different with different question published.

5.2 Task2: Thread Size Prediction over Time

The second prediction task is to estimate the thread size over time. Table 2 shows the MSE of thread size prediction until the end of all threads. Comparing to the regression models, TISM has the advantage that the probability is a time function, which can show how thread size changes over time. Given a time T, the probability of a node will join the thread by time T can be derived as discussed in Sect. 5. Different from the regression method in Sect. 5.1 which is based on question and questioner features, to estimate the thread size over time with regression method, the temporal factor are also included. Such that the regression model is trained with features: q_reput, #questions, title_total, title_nostop, title_unique, body_total, body_nostop, body_unique, time, and the response (thread size).

Also, different from the Sect. 5.1 where the threads are randomly split, in this test, all Q&A threads are sorted in the ascending order of when the question is raised. All three methods (RF,LM,TISM) use the first 90% threads to train the model, and the remaining 10% threads for the testing. Only answers are included in this experiment, *i.e.*, the thread size is the #answerers. The MSE at each time point is the difference between the estimated thread size and the actual thread size.

Figure 1 shows the MSE of three methods over time. The x-axis is the normalized time and "10" is the maximum time of all threads in the dataset.

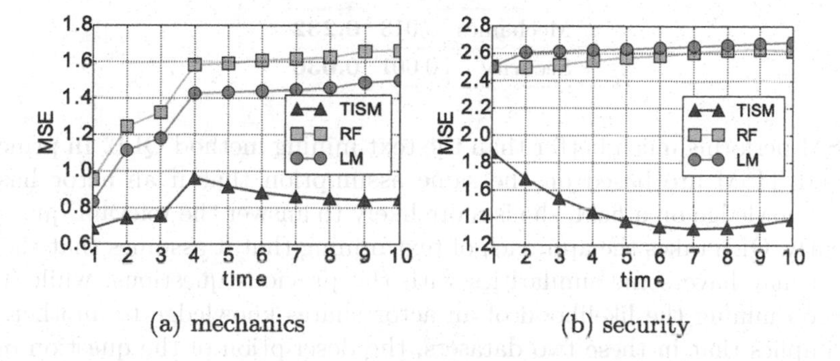

(a) mechanics (b) security

Fig. 1. MSE of prediction over time

In both datasets, TISM outperforms RF and LM. TISM has a better estimation from the beginning to the end of the threads. One interesting observation in "mechanics" is that there is a turn point at time "4". This is because most threads received their accepted answer before time "4" such that the actual thread size increases until time "4" and keeps constant after that. Though regression method is also able to reflect the thread size change by including the time feature, TISM's exponential function models the thread size change smoother. MSE at time "10" in Fig. 1 is also better than the overall MSE in Table 2. This is due to that all threads are sorted in time order such that the trained TISM network from training threads is able to provide more guidance in predicting future testing threads.

5.3 Task3: Who Will Provide Answers

The third task is to predict who will provide answers to a question. Since the thread size is the number of actors joining the thread, it is necessary to know how these methods work on the prediction of whether each actor will join a Q&A thread. This question is the same as asking how to routing question to answerers. QLL [7] is utilized as the baseline method. QLL calculates the probability an actor will answer a question according to the similarity of the question against all questions answered by the actor. For each questioner and its question, TISM is applied to estimate the probability any other actor except the questioner will join the thread. These actors are ranked based on the probability they will join a thread from the highest to the lowest. The actual actor in each thread is then identified with a rank among all actors. The higher the rank of the actual actor, the better the prediction, *i.e.*, which means the prediction is closer to the ground truth. The final result is evaluated with Mean Reciprocal Rank (MRR)[7] which is defined as the mean of the reverse of the rank.

Same as Sect. 5.2, we use first 90% threads as training data, and left 10% for the testing. Table 3 shows the prediction result of TISM and QLL on "mechanics" and "security", respectively.

Table 3. Estimation on answerer

Dataset	QLL	TISM
Mechanics	0.013	**0.232**
Security	0.001	**0.036**

TISM performs much better than the text mining method QLL. In principle, QLL and TISM are based on the same assumption that if an actor has the specific knowledge on a field, she is more likely to answer the relevant questions. Differently, QLL takes the approach of text mining that it assumes that the new question may have some similarities with the previous questions; while TISM focuses on mining the likelihood of an actor shares knowledge to another. The result implies that in these two datasets, the description of the question or the selection of words has more uncertainties; while the interests and knowledge buildup of a person evolves gradually so that an actor's question is more likely to be answered by the same actor who answered his question before. Comparing the performance for the two datasets, both QLL and TISM perform worse in "security". This is due to that the "security" dataset has more users, making prediction more difficult. On the other hand, it also implies that the "security" dataset is more volatile that both the words and the actors on the community change relatively rapidly.

5.4 Task4: Who Will Provide the Accepted Answer

The fourth task is to predict who will provide the accepted answer. The question-solving is a process of accumulating answers until received the final proper one.

TISM can estimate the probability an actor joining a thread by a time T; on the other hand, it can also be used to derive *when* will an actor join the thread with a probability P. From Eq. (12),

$$T = -\frac{1}{\alpha_{qi}}\log(1 - \frac{P_{qi}}{1 - e^{-\beta_{qi}k_q^c}}) \tag{16}$$

where k_q^c is the knowledge level of the question raised by q in thread c, and P_{qi} is the probability i will provide an answer to questioner q.

As [1], whether an answer will be accepted is dependent on *when* the answer is published, how many answers already exist and the average score of existing answers. Since the answer score is unknown during prediction, we only use the first two features to build the classifier, *i.e.*, when an answer is published (t), and #existing answers (e). Table 4 shows the algorithm to find the rank of the probability each actor will provide the accepted answer, where A is the actor set; Pr is the predefined probability used to derive the time t; g is the ground-truth actor who provided the accepted answer to question q. The high rank of p_g the better performance. In this case, Pr is set as 0.5 which is to estimate when an actor will provide an answer to the question over 50% chance.

Table 4. Alg: estimating who will provide accepted answer

```
Algorithm Who will provide an accepted answer
Init Train Classifier(T, E) for which answer
    will be accepted based on when the answer is posted t
    and number of existing answers e.
Func Rank(q, Pr, A)
    for each actor a_i ∈ A:
        t_i = Est(q, a_i, Pr) according to Eq.16
        p_i = 0
    e = 0
    for each t_i in ascending order:
        p_i = Classifer(t_i, e)
        e = e + 1
    return probability vector P of p_i
Main
    P = Rank(q, Pr, A)
    find rank of p_g in P
```

We test the two datasets using the algorithm shown in Table 4, which is built upon TISM. QLL is applied as the baseline that the rank of a user is based on the similarity between the new coming question and the questions the user ever provided the accepted answers. The result is shown in Table 5.

Comparing to the results shown in Sect. 5.3, TISM achieves decent performance in estimating who will provide an accepted answer, which is related to the question of predicting who will provide an answer since it provides a set of candidates, and the selected one is one of these answerers.

Table 5. Estimation on answerer in MRR

Dataset	QLL	TISM
Mechanics	0.0013	**0.165**
Security	0.0003	**0.035**

6 Conclusion

In this paper, we model the information sharing behavior in a Q&A forum: StackOverflow. By taking the knowledge level of the question and questioner, and the temporal factors, we propose TISM where the knowledge sharing likelihood or link weights are learned from the history data and described as a continuous function of time and knowledge level. The experiment results show, by using the *knowledge sharing network* learned from TISM, we outperform in both predicting the thread size over time and the individual action including who will answer the question and who will provide the accepted answer.

References

1. Anderson, A., Huttenlocher, D., Kleinberg, J., Leskovec, J.: Discovering value from community activity on focused question answering sites: a case study of stack overflow. In: Proceedings of the 18th ACM SIGKDD International Conference on Knowledge Discovery and Data Mining, KDD 2012, pp. 850–858. ACM, New York (2012)
2. Asaduzzaman, M., Mashiyat, A.S., Roy, C.K., Schneider, K.A.: Answering questions about unanswered questions of stack overflow. In: Proceedings of the 10th Working Conference on Mining Software Repositories, MSR 2013, pp. 97–100. IEEE Press, Piscataway (2013)
3. Chang, S., Pal, A.: Routing questions for collaborative answering in community question answering. In: Proceedings of the IEEE/ACM International Conference on Advances in Social Networks Analysis and Mining, ASONAM 2013, pp. 494–501. ACM, New York (2013)
4. Gomez-Rodriguez, M., Balduzzi, D., Schölkopf, B.: Uncovering the temporal dynamics of diffusion networks. In: Proceedings of the 28th International Conference on Machine Learning, ICML, Bellevue, Washington, USA, 28 June–2 July, pp. 561–568 (2011)
5. Goyal, A., Bonchi, F., Lakshmanan, L.V.: Learning influence probabilities in social networks. In: Proceedings of WSDM, pp. 241–250 (2010)
6. Hanrahan, B.V., Convertino, G., Nelson, L.: Modeling problem difficulty and expertise in stackoverflow. In: Proceedings of the ACM Conference on Computer Supported Cooperative Work Companion, CSCW 2012, pp. 91–94, ACM, New York (2012)
7. Li, B., King, I.: Routing questions to appropriate answerers in community question answering services. In: Proceedings of the 19th ACM International Conference on Information and Knowledge Management, CIKM 2010, pp. 1585–1588, ACM, New York (2010)

8. Liu, Y., Bian, J., Agichtein, E.: Predicting information seeker satisfaction in community question answering. In: Proceedings of the 31st Annual International ACM SIGIR Conference on Research and Development in Information Retrieval, SIGIR 2008, pp. 483–490. ACM, New York (2008)
9. Yang, L., Bao, S., Lin, Q., Wu, X., Han, D., Su, Z., Yu, Y.: Analyzing and predicting not-answered questions in community-based question answering services. In: Proceedings of the Twenty-Fifth AAAI Conference on Artificial Intelligence, AAAI 2011, pp. 1273–1278. AAAI Press (2011)
10. Zhou, T.C., Lyu, M.R., King, I.: A classification-based approach to question routing in community question answering. In: Proceedings of the 21st International Conference on World Wide Web, WWW 2012 Companion, pp. 783–790. ACM, New York (2012)

Effective Multiclass Transfer for Hypothesis Transfer Learning

Shuang Ao, Xiang Li, and Charles X. Ling(✉)

Department of Computer Science, Western University, London, Canada
{sao,lxiang2}@uwo.ca, cling@csd.uwo.ca

Abstract. In this paper, we investigate the visual domain adaptation problem under the setting of *Hypothesis Transfer Learning* (HTL) where we can only access the source model instead of the data. However, previous studies of HTL are limited to either leveraging the knowledge from certain type of source classifier or low transfer efficiency on a small training set. In this paper, we aim at two important issues: effectiveness of the transfer on small target training set and compatibility of the transfer model for real-world HTL problems. To solve these two issues, we proposed our method, Effective Multiclass Transfer Learning (EMTLe). We demonstrate that EMTLe, which uses the prediction of the source models as the transferable knowledge can exploit the knowledge of different types of source classifiers. We use the transfer parameter to weigh the importance the prediction of each source model as the auxiliary bias. Then we use the bi-level optimization to estimate the transfer parameter and demonstrate that we can effectively obtain the optimal transfer parameter with our novel objective function. Empirical results show that EMTLe can effectively exploit the knowledge and outperform other HTL baselines when the size of the target training set is small.

1 Introduction

Domain adaptation for image recognition tries to exploit the knowledge from a source domain with plentiful data to help learn a classifier for the target domain with a different distribution and little labeled training data. In domain adaptation, the source and target domains share the same label but their data are drawn from different distributions.

In domain adaptation, the knowledge of the source domain can be transferred by 3 different approaches: *instance transfer*, *model transfer* and *feature representation transfer* [13]. In this paper, we focus on the model transfer approach. Some recent works show that exploiting the knowledge from the source model can boost the performance of the target model effectively [11,16]. Moreover, in some real applications, we can only obtain the source models and it is difficult to access their training data for different reasons such as the data credential. Recently, a framework called Hypothesis Transfer Learning (HTL) [10] has been proposed to handle this situation. HTL assumes only source models trained on the source domain can be utilized and there is no access to source data, nor any knowledge about the relatedness of the source and target distributions.

J. Kim et al. (Eds.): PAKDD 2017, Part II, LNAI 10235, pp. 64–75, 2017.
DOI: 10.1007/978-3-319-57529-2_6

Previous research [2,3] shows that without carefully measuring the distribution similarity between the source and target data, the source knowledge could not be exploited effectively or even hurt the learning process (called *negative transfer*) [13]. However, as we are not able to access the source data in an HTL setting, how to effectively and safely exploit the knowledge from the source model could be an important issue in HTL, especially when target data is relatively small (Effectiveness issue). Moreover, the source models from different domains can be trained with different kinds of classifiers. For example most models trained from ImageNet are deep convolutional neural networks while some models of the VOC recognition task could be SVMs or ensemble models. Therefore, a practical HTL algorithm should be compatible with different types of source classifiers (Compatibility issue). Previous work is limited to either leveraging the knowledge from certain type of source classifiers [7,16] or low transfer efficiency in a small training set [8]. To the best of our knowledge, none of the previous work in HTL is able to solve these two issues at the same time.

In this paper, we propose our method, called Effective Multiclass Transfer Learning (EMTLe), that can solve these two issues simultaneously. In this paper, we introduce our strategy that uses the class prediction of the source model as the transferable knowledge to help the classification. Specifically, we use the weighted class probabilities produced by the source models to adjust the prediction from the target model. Here we call the weight of each source model *transfer parameter* which essentially controls the amount of knowledge transferred from the specific model. Moreover, compared to the previous work such as MKTL [8], EMTLe has fewer hyperparameters to estimate. Therefore, it is easier for EMTLe to learn a good target model especially on a small training set.

To estimate the transfer parameter, we introduce bi-level optimization [14], which has been widely used for many different hyperparameter optimization problems recently. Specifically, on the low-level optimization problem, we use a least-square SVMs to train a model on the target data and on the high level, we introduce our novel multi-class hinge loss with ℓ_2 penalty that can better estimate the transfer parameter when training set is small. Moreover, we show that our bi-level optimization transfer parameter estimation problem is a strongly convex optimization problem and demonstrate that our method EMTLe can find the $O(\log(t)/t)$ optimal solution with t iterations.

We perform comprehensive experiments on 4 real-world datasets from two benchmark datasets (3 from Office and 1 from Caltech256). We show that EMTLe can effectively transfer the knowledge with different types of source models and outperforms the baseline methods under the HTL setting.

2 Related Work

As we focus on the model transfer approach under the HTL setting, in this section, we review some important methods using this approach. A model transfer approach assumes that the parameters of the model for the source task can be transferred to the target task. Two types of learning methods are generally used for model knowledge transfer, generative probabilistic method and max margin method.

Generative probabilistic method can predict the target domain by combining the source distribution to generate a posterior distribution. Li et al. [7] used Bayesian transfer learning approach to learn the common prior for object recognition. Davis et al. [6] used an approach based on a form of second-order Markov logic to compensate for the domain shift. Wang et al. [17] proposed a method to change the marginal and conditional distributions smoothly to transfer the knowledge between tasks.

Alternatively, max margin methods try to use the hyperplane parameter to transfer the knowledge between source and target domains. Yang et al. [19] proposed Adaptive SVMs transferring parameters by incorporating the auxiliary classifier trained from the source domain. In addition to Yang's work, Ayatar et al. [1] proposed PMT-SVM that can determine the transfer regularizer automatically according to the target data. Tommasi et al. [16] proposed Multi-KT that can utilize the parameters from multiple source models for the target classes. Kuzborskij et al. [11] proposed a similar method to learn new categories by leveraging the known source models. Luo et al. [8] proposed MKTL and used feature augmentation method to leverage the source model.

Our work corresponds to the context above. In this paper, we propose EMTLe based on the model transfer approach. Specifically, we focus on how to exploit the knowledge from the predictions of the source models.

3 Using the Source Knowlege as the Auxiliary Bias

Some previous work such as MKTL [8] suggests that using the prediction from the source model as the source knowledge can greatly release the constraint of the type of the source model. However, with complex feature augmentation method, there are many hyperparameters to be estimated which makes it inefficient with small training set. In this paper, we adopt the idea of using the source model prediction as the transferable knowledge and propose our transfer strategy.

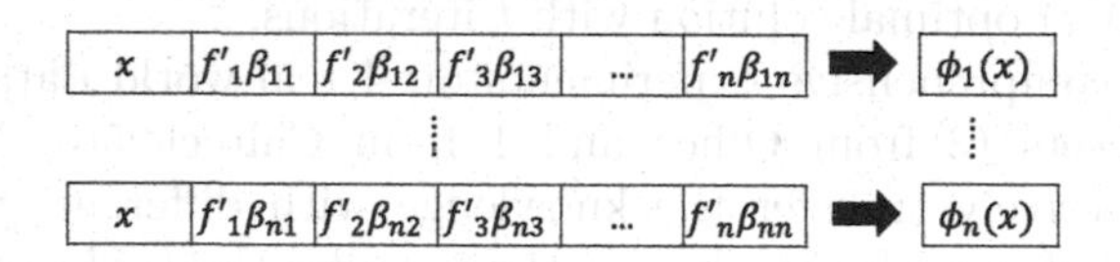

Fig. 1. Illustration of feature augmentation in MKTL. f_i' is the output of the i-th source model and β_{in} is the hyperparameter (need to be estimated) to weigh the augmented feature. $\phi_n(x)$ is augmented feature for the n-th binary model.

Suppose we have to recognize a image from one of the N visual classes and there are N experts each of who can only provide the probability of this image for one certain class (binary source model). After we make our decision for one example (prediction from target model), the experts provide their own decisions as well (probabilities from the source models). Their decisions can provide extra information regarding this example as the auxiliary bias and adjust our final prediction. As each of the experts is a specialist in one class, we should weigh their decisions as well due to the bias of their predictions (see Fig. 2).

Unlike previous work [1,16,18] which has to use the specific parameter of the source model as the source knowledge, our strategy is more compatible with different types of classifiers. Compared to MKTL [8], we only have to estimate N hyperparameters for the N-class problem while there are $N \times N$ hyperparameters in MKTL (see Fig. 1). Therefore, it is easier to estimate the transfer parameters with our strategy and EMTLe can perform better especially when the size of the training set is small. In addition, there are two advantages of our strategy: (1) It is an effective and easy way to align the knowledge from different types of source classifiers. (2) The auxiliary bias term is naturally normalized in the same dimension as the class probabilities are always in the interval $[0, 1]$. As EMTLe can select more types of source classifiers, this makes it more practical in a real HTL scenario.

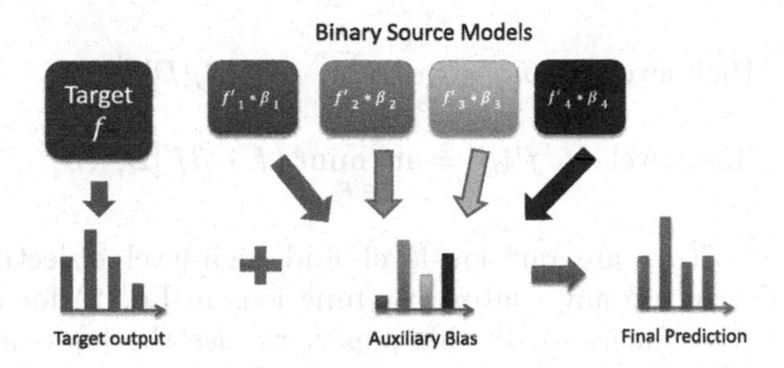

Fig. 2. Demonstration of using the source class probability as the auxiliary bias to adjust the output of the target model.

Here, the weight of each source model reflects the relatedness between the source model and our target domain. The more related they are, the better decision the source model can make and the larger weight we should apply to it. Specifically, in this paper, we call the weight *transfer parameter*. Therefore, for any target data $D = \{x, y\}$ and the given source models $f' = \{f'_1, ..., f'_N\}$, our goal is to find the target model f:

$$f = \arg\min_{f \in \mathcal{F}} \ell\,(f + \beta f' | D, \beta) \tag{1}$$

where $\beta = [\beta_1, ..., \beta_N]$ is the transfer parameter and $\ell(\cdot, \cdot)$ is the loss function to learn the target model. It is obvious that assigning the proper transfer parameter to the source model can significantly improve the performance of our final prediction. From Eq. (1) we can see that, once we have determined the value of the transfer parameter β, we are able to find the target model f and solve the learning problem. However, the transfer parameter in Eq. (1) is a hyperparameter and we cannot solve it directly. Therefore, we introduce our bi-level optimization method for transfer parameter estimation in the next section.

4 Bi-level Optimization for Transfer Parameter Estimation

As we discussed before, the transfer parameter in Eq. (1) is a hyperparameter that cannot be solved directly. Here we use bi-level optimization (**BO**) [14], a popular method that is used in hyperparameter optimization to estimate the transfer parameter. In BO, the low-level optimization problem is to learn the target model and the high-level problem is another cross-validation (CV) hyperparameter optimization problem corresponding to the model learned at the low-level. Suppose we use K-fold CV on the high-level problem. For the i-th fold CV, the target set D is split into training set D_i^{tr} and validation set D_i^{val}. The transfer parameter can be optimized with the following BO function:

$$\begin{aligned} &\text{High level} \quad && \beta = \arg\min_{\beta} \sum_{i}^{K} \mathcal{L}(f^i(\beta)|D_i^{val}) \\ &\text{Low level} \quad && f^i(\beta) = \arg\min_{f \in \mathcal{F}} \ell\left(f + \beta f' | D_i^{tr}, \beta\right) \end{aligned} \tag{2}$$

Here, $\ell(\cdot,\cdot)$ and $\mathcal{L}(\cdot,\cdot)$ are our low-level and high-level objective functions respectively. We can use any convex loss functions in Eq. (2) for optimization (e.g. SVM objective function). In this paper, we use the leave-one-out cross-validation (**LOOCV**) in the high-level problem. Previous research [10] suggests that LOOCV can increase the robustness of the estimated hyperparameter especially on the small dataset. In previous studies [12,14], BO is a non-convex problem and can only obtain the approximate solution. However, we will show that problem (2) is strongly convex and we are able to obtain its optimal solution.

4.1 Low-Level Optimization Problem

To better illustrate our learning scenario, we define our learning process as follows. Suppose we have N visual categories and can obtain N source binary classifiers $f' = \{f'_1, ..., f'_N\}$ from the source domain. We want to train a target function f consisting of N binary classifiers $f = \{f_1, ..., f_N\}$ using the target training set D and the source models f'. Specifically, in our BO problem Eq. (2), for the low-level optimization, we consider the scenario where we have to train N binary linear target models $f_i = w_i x + b_i$ so that for any $\{x_i, y_i\}_{i=1}^{l} \in D$, the adjusted result satisfies $f(x) + f'(x)\beta = y$. Let $D^{\backslash i} = D \backslash \{x_i, y_i\}$. Then, we use mean square loss in the low-level objective function to optimize each target model f_n with any given transfer parameter β:

$$\begin{aligned} \text{Low-level:} \quad & f^{\setminus i}(\beta) : \min_{w,b} \sum_{n}^{N} \frac{1}{2}||w_n||^2 + \frac{C}{2}\sum_{j} e_{jn}^2 \\ \text{s.t.} \quad & f_n(x) = w_n x + b_n; \quad x_j \in D^{\setminus i} \\ & e_{jn} = Y_{jn} - f_n(x_j) - \beta_n f'_n(x_j) \end{aligned} \tag{3}$$

Here, Y is an encoded matrix of y using the one-hot strategy where $Y_{in} = 1$ if $y_i = n$ and 0 otherwise.

The reason why we use the objective function (3) is that it can provide an unbiased closed form Leave-one-out error estimation for each binary model f_n [4]. As a result, the high-level problem becomes a convex problem and we are able to estimate our transfer parameter easier.

Let $K(X, X)$ be the kernel matrix and C be the penalty parameter in Eq. (3). We have:

$$\psi = \left[K(X, X) + \frac{1}{C}\mathrm{I}\right] \tag{4}$$

Let ψ^{-1} be the inverse of matrix ψ and ψ_{ii}^{-1} is the ith diagonal element of ψ^{-1}. $\hat{Y}_{in}$, the LOO estimation of binary model $f_n^{\setminus i}$ for sample x_i, can be written as [4]:

$$\hat{Y}_{in} = Y_{in} - \frac{\alpha_{in}}{\psi_{ii}^{-1}} \quad \text{for} \quad n = 1, ..., N \tag{5}$$

where the matrix $\boldsymbol{\alpha} = \{\alpha_{in} | i = 1, ...l; n = 1, ..., N\}$ can be calculated as:

$$\boldsymbol{\alpha} = \psi^{-1}Y - \psi^{-1}f'(X)diag(\boldsymbol{\beta}) \tag{6}$$

4.2 High-Level Optimization Problem

For the high level optimization problem, we use multi-class hinge loss [5] with ℓ_2 penalty in our objective function.

$$\begin{aligned} \text{High-level:} \quad & \beta : \min \frac{\lambda}{2}\sum_{n}^{N} \|\beta_n\|^2 + \sum_{i} \xi_i \\ \text{s.t.} \quad & 1 - \varepsilon_{ny_i} + \hat{Y}_{in} - \hat{Y}_{iy_i} \leq \xi_i \end{aligned} \tag{7}$$

Here, $\varepsilon_{ny_i} = 1$ if $n = y_i$ otherwise 0. λ is used to balance the ℓ_2 penalty and our multi-class hinge loss. Compared to the previous work [11,16] which uses the multi-class hinge loss without the ℓ_2 penalty, there are two main advantages for our high-level objective function:

1. When the training set is small, our LOOCV estimation could have a large variance. It is important to add the ℓ_2 penalty to reduce the variance and improve the generalization ability of the estimated transfer parameter.
2. It is clear that $\hat{Y}$ is a linear function w.r.t. β. With the ℓ_2 penalty, the high-level optimization problem (7) becomes a strongly convex optimization problem w.r.t. the transfer parameter β. Therefore, we can obtain an $O(\log(t)/t)$ optimal solution with t iterations using Algorithm 1 (see proof of Theorem 1 in Appendix).

Algorithm 1. EMTLe

Input: $\lambda, \psi, Y, f', T,$
Output: $\beta = \{\beta^1, ..., \beta^n\}$
1: $\beta^0 = 1, \alpha' = \psi^{-1}Y, \alpha'' = \psi^{-1}f'$
2: **for** $t = 1$ to T **do**
3: $\hat{Y} \leftarrow Y - (\psi^{-1} \circ I)^{-1} (\alpha' - \alpha'' diag(\beta))$
4: **for** $i = 1$ to l **do**
5: $\Delta_\beta = \lambda\beta$
6: $l_{ir} = \max(1 - \varepsilon_{y_i r} + \hat{Y}_{ir} - \hat{Y}_{iy_i})$
7: **if** $l_{ir} > 0$ **then**
8: $\Delta_\beta^{y_i} \leftarrow \Delta_\beta^{y_i} - \frac{\alpha''_{iy_i}}{\psi_{ii}^{-1}}$, $\Delta_\beta^r \leftarrow \Delta_\beta^r + \frac{\alpha''_{ir}}{\psi_{ii}^{-1}}$
9: **end if**
10: **end for**
11: $\beta^t \leftarrow \beta^{(t-1)} - \frac{\Delta_\beta}{\lambda \times t}$
12: **end for**

5 Experiments

In this section, we show empirical results of our algorithm for different transferring situations on two image benchmark datasets: Office and Caltech.

5.1 Dataset and Baseline Methods

Office contains 31 classes from 3 subsets (Amazon, Dslr and Webcam) and Caltech contains 256 classes. We select 13 shared classes from two datasets[1]. The input features of all examples are extracted using AlexNet [9]. We compare our algorithm EMTLe with two kinds of baselines. The first one is the methods without leveraging any source knowledge (no transfer baselines), including two methods. **No transfer:** SVMs trained only on target data. Any transfer algorithm that performs worse than it suffers from negative transfer. **Batch:** We combine the source and target data, assuming that we have full access to all data, to train the SVMs. The result of the Batch method is expected to outperform other methods under the HTL setting as it can access the source data. The second kind of baseline consists of two previous transfer methods in HTL, **MKTL** [8] and **Multi-KT** [16]. Similar to EMTLe, both of them use the LOOCV method to estimate the relatedness of the source model and target domain, but they use their own convex objective function without the ℓ_2 penalty terms. We use linear kernel for all methods in all our experiments.

5.2 Transfer from Single Source Domain

In this subsection, following the experiment protocol in [8,16] for fair comparison, we perform 12 groups of experiments under the setting of HTL. For each

[1] 13 classes include: backpack, bike, helmet, bottle, calculator, headphone, keyboard, laptop, monitor, mouse, mug, phone and projector.

Table 1. Statistics of the datasets and subsets

Dataset	Subsets	# classes	# examples	# features
Office	Amazon (A)	13	1173	4096
	Dslr (D)	13	224	4096
	Webcam (W)	13	369	4096
Caltech256	Caltech (C)	13	1582	4096

experiment, one of the 4 (sub)datasets is selected as the source, while another dataset is used as the target. We evaluate the performance of EMTLe when all source models are of the same type. As Multi-KT can only leverage knowledge when the source model is SVM, All source models are trained with linear SVMs. The size of each target dataset is varied from 1 to 5 to see how EMTLe and other baselines behave under the extremely small dataset. We use a heuristic way to set the value of λ in Eq. (7):

$$\lambda = 2e^{err_n - err_s} \tag{8}$$

where err_n and err_s denote the performance of "No transfer" and the source model on the training set. We perform each experiment 10 times and report the average result in Fig. 3.

Observation and Discussion: EMTLe can significantly outperform other baselines especially with a small training set. As we have discussed above, when the training set is small, with the transfer parameter estimated by our ℓ_2 penalty in our high-level objective functions, EMTLe has a strong generalization ability and performs better on the test data. As the training size increases, the variance of training data decreases and the affect of the ℓ_2 penalty term become less significant. Therefore, EMTLe and the other two HTL baselines show similar performance. It is interesting to see that MKTL even falls into negative transfer even with 5 training examples per class in some experiments. We found that, MKTL is more sensitive to the variance of the training data. Its performance is not as stable as Multi-KT and EMTLe over the 10 experiments. Because MKTL needs to learn more hyperparameters than Multi-KT and EMTLe, even though the training size increases, it may not be able to obtain a good model. In some experiments, we can see that EMTLe can even outperform the Batch method which can access more information and is expected to outperform the other methods under the setting of HTL.

5.3 Transfer from Multiple Source Domains

As we mentioned, EMTLe can exploit knowledge from different types of source classifiers which could greatly extend our choice of the source domain under the HTL setting. In this subsection we show that EMTLe can successfully transfer the knowledge from two different types of source classifiers. Meanwhile, MKTL and "No Transfer" are used as our baseline.

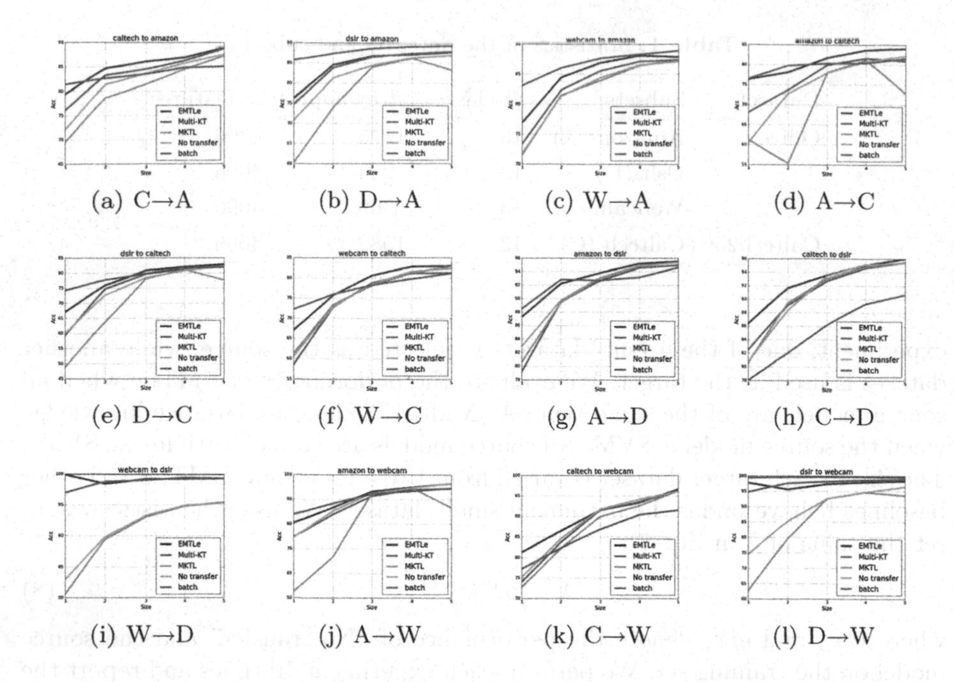

(a) C→A (b) D→A (c) W→A (d) A→C

(e) D→C (f) W→C (g) A→D (h) C→D

(i) W→D (j) A→W (k) C→W (l) D→W

Fig. 3. Recognition accuracy for HTL domain adaptation from a single source. 5 different sizes of target training sets are used in each group of experiments. A, D, W and C denote the 4 subsets in Table 1 respectively.

In this experiment, we assume that there is no single source domain that can cover all 13 classes in our target domain and we have to select source models from different source domains. Specifically, the 13 classes are selected from two different domains separately (6 from DSLR and 7 from Webcam) according to Table 2. Similar to our previous experiment configurations, we only use Caltech and Amazon as the target domains. We show the experiment results in Fig. 4.

Table 2. The selected classes of the two source domains and the classifier type of the source model.

	Class	Classifier
DSRL	monitor, bike, helmet, calcu, headphone, projector	Logistic
Webcam	keyboard, mouse, phone, backpack, mug, bottle, laptop	SVMs

Observation and Discussion: In our multi-source scenario, it is more difficult to leverage the knowledge from the source models as the models are trained from different domains separately. From the results we can see that, EMTLe can still exploit the knowledge from the source models despite the types of the source classifiers while MKTL can hardly leverage the source knowledge. EMTLe uses a

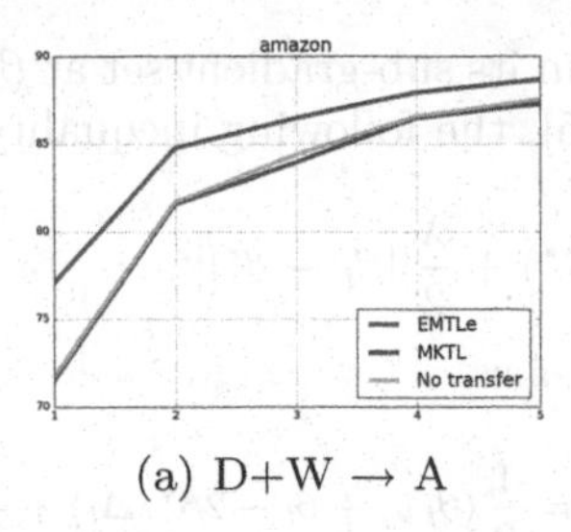

(a) D+W → A

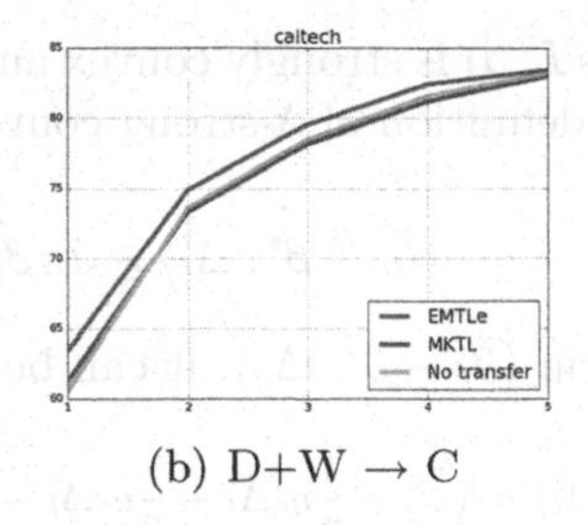

(b) D+W → C

Fig. 4. Recognition accuracy for multi-model & multi-source experiment on two target datasets.

simple way to leverage the source models and BO can help us better estimate the transfer parameter. However, MKTL uses a sophisticated feature augmentation and has more hyperparameters to estimate. Without sufficient training data, it is difficult for MKTL to measure the importance of each source model and exploit the knowledge from the models.

6 Conclusion

In this paper, we propose a method, EMTLe that can effectively transfer the knowledge under the HTL setting. We focus on the effectiveness and compatibility issues for HTL problems. We propose our auxiliary bias strategy to let our model exploit the knowledge from different types of source classifiers. The transfer parameter of EMTLe is estimated by bi-level optimization method using our novel high-level objective function which allows our model to better exploit the knowledge from source models. Experiment results demonstrate that EMTLe can effectively transfer the knowledge even though the size of training data is extremely small.

Acknowledgments. We thank the anonymous reviewers for their valuable comments to improve this paper. This work is supported by Natural Sciences and Engineering Research Council of Canada (NSERC).

Appendix

Theorem 1. *Let $L(\beta)$ be a λ-strongly convex function and β^* be its optimal solution. Let $\beta_1, ..., \beta_{T+1}$ be a sequence such that $\beta_1 \in B$ and for $t > 1$, we have $\beta_{t+1} = \beta_t - \eta_t \Delta_t$, where Δ_t is the sub-gradient of $L(\beta_t)$ and $\eta_t = 1/(\lambda t)$. Assume we have $||\Delta_t|| \leq G$ for all t. Then we have:*

$$L(\beta_{T+1}) \leq L(\beta^*) + \frac{G^2(1 + \ln(T))}{2\lambda T} \tag{9}$$

Proof. As $L(\beta)$ is strongly convex and Δ_t is in its sub-gradient set at β_t, according to the definition of λ-strong convexity [15], the following inequality holds:

$$\langle \beta_t - \beta^*, \Delta_t \rangle \geq L(\beta_t) - L(\beta^*) + \frac{\lambda}{2}||\beta_t - \beta^*||^2 \tag{10}$$

For the term $\langle \beta_t - \beta^*, \Delta_y \rangle$, it can be written as:

$$\langle \beta_t - \beta^*, \Delta_t \rangle = \left\langle \beta_t - \frac{1}{2}\eta_t \Delta_t + \frac{1}{2}\eta_t \Delta_t - \beta^*, \Delta_t \right\rangle = \frac{1}{2}\langle \beta_{t+1} + \beta_t - 2\beta^*, \Delta_t \rangle + \frac{1}{2}\eta_t \Delta_t^2 \tag{11}$$

Then we have:

$$||\beta_t - \beta^*||^2 - ||\beta_{t+1} - \beta^*||^2 = \langle \beta_{t+1} + \beta_t - 2\beta^*, \eta_t \Delta_t \rangle \tag{12}$$

Using the assumption $||\Delta_t|| \leq G$, we can rearrange (10) and plug (11) and (12) into it, we have:

$$Diff_t = L(\beta_t) - L(\beta^*) \leq \frac{\lambda(t-1)}{2}||\beta_t - \beta^*||^2 - \frac{\lambda t}{2}||\beta_{t+1} - \beta^*||^2 + \frac{1}{2}\eta_t G^2 \tag{13}$$

Due to the convexity, for each pair of $L(\beta_t)$ and $L(\beta_{t+1})$ for $t = 1, ..., T$, we have the following sequence $L(\beta^*) \leq L(\beta_T) \leq L(\beta_{T-1}) \leq ... \leq L(\beta_1)$. For the sequence $Diff_t$ for $t = 1, ..., T$, we have:

$$\sum_{t=1}^{T} Diff_t = \sum_{t=1}^{T} L(\beta_t) - TL(\beta^*) \geq T\left[L(\beta_T) - L(\beta^*)\right] \tag{14}$$

Next, we show that

$$\begin{aligned} \sum_{t=1}^{T} Diff_t &= \sum_{t=1}^{T} \left\{ \frac{\lambda(t-1)}{2}||\beta_t - \beta^*||^2 - \frac{\lambda t}{2}||\beta_{t+1} - \beta^*||^2 + \frac{1}{2}\eta_t G^2 \right\} \\ &= -\frac{\lambda T}{2}||\beta_{T+1} - \beta^*||^2 + \frac{G^2}{2\lambda}\sum_{t=1}^{T}\frac{1}{t} \leq \frac{G^2}{2\lambda}\sum_{t=1}^{T}\frac{1}{t} \leq \frac{G^2}{2\lambda}(1 + \ln(T)) \end{aligned} \tag{15}$$

Combining (14) and rearranging the result, we have:

$$L(\beta_{T+1}) \leq L(\beta^*) + \frac{G^2(1 + \ln(T))}{2\lambda T}$$

References

1. Aytar, Y., Zisserman, A.: Tabula rasa: model transfer for object category detection. In: 2011 IEEE International Conference on Computer Vision (ICCV), pp. 2252–2259. IEEE (2011)
2. Ben-David, S., Blitzer, J., Crammer, K., Kulesza, A., Pereira, F., Vaughan, J.W.: A theory of learning from different domains. Mach. Learn. **79**(1–2), 151–175 (2010)

3. Ben-David, S., Blitzer, J., Crammer, K., Pereira, F., et al.: Analysis of representations for domain adaptation. Adv. Neural Inf. Process. Syst. **19**, 137 (2007)
4. Cawley, G.C.: Leave-one-out cross-validation based model selection criteria for weighted LS-SVMs. In: International Joint Conference on Neural Networks, IJCNN 2006, pp. 1661–1668. IEEE (2006)
5. Crammer, K., Singer, Y.: On the algorithmic implementation of multiclass kernel-based vector machines. J. Mach. Learn. Res. **2**, 265–292 (2002)
6. Davis, J., Domingos, P.: Deep transfer via second-order Markov logic. In: Proceedings of the 26th Annual International Conference on Machine Learning, pp. 217–224. ACM (2009)
7. Fei-Fei, L., Fergus, R., Perona, P.: One-shot learning of object categories. IEEE Trans. Pattern Anal. Mach. Intell. **28**(4), 594–611 (2006)
8. Jie, L., Tommasi, T., Caputo, B.: Multiclass transfer learning from unconstrained priors. In: 2011 IEEE International Conference on Computer Vision (ICCV), pp. 1863–1870. IEEE (2011)
9. Krizhevsky, A., Sutskever, I., Hinton, G.E.: Imagenet classification with deep convolutional neural networks. In: Advances in Neural Information Processing Systems, pp. 1097–1105 (2012)
10. Kuzborskij, I., Orabona, F.: Stability and hypothesis transfer learning. In: Proceedings of the 30th International Conference on Machine Learning, pp. 942–950 (2013)
11. Kuzborskij, I., Orabona, F., Caputo, B.: From n to n+1: multiclass transfer incremental learning. In: 2013 IEEE Conference on Computer Vision and Pattern Recognition (CVPR), pp. 3358–3365. IEEE (2013)
12. Maclaurin, D., Duvenaud, D., Adams, R.P.: Gradient-based hyperparameter optimization through reversible learning. In: Proceedings of the 32nd International Conference on Machine Learning (2015)
13. Pan, S.J., Yang, Q.: A survey on transfer learning. IEEE Trans. Knowl. Data Eng. **22**(10), 1345–1359 (2010)
14. Pedregosa, F.: Hyperparameter optimization with approximate gradient. In: Proceedings of the 33nd International Conference on Machine Learning, ICML 2016, New York City, NY, USA, 19–24 June 2016, pp. 737–746 (2016)
15. Rockafellar, R.T.: Convex Analysis. Princeton University Press, Princeton (2015)
16. Tommasi, T., Orabona, F., Caputo, B.: Learning categories from few examples with multi model knowledge transfer. Pattern Anal. Mach. Intell. **36**(5), 928–941 (2014)
17. Wang, X., Huang, T.K., Schneider, J.: Active transfer learning under model shift. In: Proceedings of the 31st International Conference on Machine Learning (ICML 2014), pp. 1305–1313 (2014)
18. Yang, J., Yan, R., Hauptmann, A.G.: Adapting SVM classifiers to data with shifted distributions. In: 2007 Seventh IEEE International Conference on Data Mining Workshops, ICDM Workshop 2007, pp. 69–76. IEEE (2007)
19. Yang, J., Yan, R., Hauptmann, A.G.: Cross-domain video concept detection using adaptive SVMs. In: Proceedings of the 15th International Conference on Multimedia, pp. 188–197. ACM (2007)

Clustering Based on Dominant Set and Cluster Expansion

Jian Hou[1,2](✉) and Weixue Liu[3]

[1] College of Engineering, Bohai University, Jinzhou 121013, China
[2] ECLT, Università Ca' Foscari Venezia, Venezia 30124, Italy
dr.houjian@gmail.com
[3] College of Information Science, Bohai University, Jinzhou 121013, China

Abstract. While numerous clustering algorithms can be found in the literature, existing algorithms are usually afflicted by two major problems. First, the majority of clustering algorithms requires user-specified parameters as input, and their clustering results rely heavily on these parameters. Second, many algorithms generate clusters of only spherical shapes. In this paper we try to solve these two problems based on dominant set and cluster expansion. We firstly use a modified dominant sets clustering algorithm to generate initial clusters which are parameter independent and usually smaller than the real clusters. Then we expand the initial clusters based on two density based clustering algorithms to generate clusters of arbitrary shapes. In experiments on various datasets our algorithm outperforms the original dominant sets algorithm and several other algorithms. It is also shown to be effective in image segmentation experiments.

1 Introduction

As an important unsupervised learning approach, data clustering has wide application in various fields and has received much attention for decades. A lot of clustering algorithms have been proposed from different perspectives in the literature. In addition to the well-known k-means algorithm, the normalized cuts algorithm (NCuts) [19] and DBSCAN (Density-Based Spatial Clustering of Applications with Noise) [8] have also obtained successful applications. The affinity propagation (AP) algorithm [3] passes affinity message among the data and identify the cluster centers and members iteratively. By defining dominant set as a graph-theoretic concept of a cluster, the dominant sets algorithm (DSets) extracts the clusters sequentially and determines the number of clusters by itself. In recent developments, [25] presented a method to build the robust similarity matrix in spectral clustering, which is shown to generate very good results on some datasets. A density peak based clustering algorithm (DP) presented in [18] is shown to generate excellent clustering results on various datasets, on condition that the cluster centers are identified correctly.

Although a vast amount of clustering algorithms have been proposed from a wide variety of perspectives in the literature, there are stillsome common

J. Kim et al. (Eds.): PAKDD 2017, Part II, LNAI 10235, pp. 76–87, 2017.
DOI: 10.1007/978-3-319-57529-2_7

problems afflicting existing algorithms. Perhaps one of the most frequently encountered problem is the parameter dependence. Many algorithms require one or more parameters as input, and their clustering results depend heavily on the parameters. For example, the k-means, NCuts and spectral clustering algorithms need to be fed the number of clusters. DBSCAN is able to determine the number of clusters automatically, but it involves two density related parameters *Eps* and *MinPts* instead. Similarly, AP requires the preference values of each data as input, and the DSets algorithm relies on appropriate similarity matrices to generate satisfying results. Moreover, the DP algorithm uses correctly identified cluster centers to accomplish the clustering process, where multiple parameters and even human intervention may be involved. In general, the appropriate parameters must be carefully tuned based on *a priori* knowledge of the dataset, human experience, or some specialized algorithms, e.g., the algorithms to determine the number of clusters [16]. In this case, the inappropriately selected parameters may degrade the clustering performance significantly. While in some cases it may be reasonable to adjust the clustering results with different parameters, we also note that on many datasets the optimal cluster results are unique, e.g., the datasets with unique ground truth clustering results in the UCI machine learning repository. Therefore we believe that it is quite necessary to reduce the dependence on user-specified parameters. Another problem is that many algorithms generate only spherical clusters. The algorithms with this problem include k-means, NCuts, spectral clustering and dominant sets, etc. Although DBSCAN and DP are able to detect clusters of arbitrary shapes, their results are dependent on appropriate parameters.

In order to solve the above two problems, we make an in-depth study of existing algorithms and find that the DSets, DBSCAN and DP algorithms have some complementary properties. The DSets algorithm uses only the pairwise similarity matrix of data as input and extracts the clusters sequentially, where the number of clusters is obtained naturally in the clustering process. If the pairwise similarity matrix is independent of any parameters, then the DSets clustering results are parameter independent. Unfortunately, the DSets algorithm can only generate spherical clusters, and this limits its performance on datasets with clusters of arbitrary shapes. In contrast, the DBSCAN the DP algorithms are able to generate clusters of arbitrary shapes, but they involve user-specified parameters to guarantee the clustering quality. In addition, DBSCAN also extracts clusters sequentially, and both DBSCAN and DP generate clusters in a cluster-expansion manner. These properties make it possible to combine these algorithms to make use of their merits while avoiding the drawbacks.

Based on the discussion above, in this paper we present a new clustering algorithm based on the DSets, DBSCAN and DP algorithms. Following DSets and DBSCAN, our algorithm extracts the clusters one by one and determines the number of clusters automatically. In detecting each cluster, we use a modified DSets algorithm to generate the initial small clusters, which are then expanded with the ideas from DBSCAN and DP. Specifically, we use histogram equalization to transform the pairwise similarity matrix before DSets clustering, and make the initial clusters invariant to parameters. In the cluster expansion with DBSCAN and DP, the required parameters are determined adaptively based on the initial clusters. In this way the parameter dependence problem is relieved. Since both DBSCAN and DP are able to generate clusters of arbitrary shapes, the cluster expansion step based on these two algorithms enables our algorithm to possess such an ability. In experiments our algorithm is shown to outperform the original DSets, DBSCAN, DP and some other algorithms in comparison.

The remaining of this paper is organized as follows. In Sect. 2 we introduce the DSets, DBSCAN and DP algorithms briefly, based on which the new clustering algorithm is presented in details in Sect. 3. We use extensive experiments and comparisons to validate the proposed algorithm in Sect. 4, and summarize the conclusions in Sect. 5.

2 Related Works

2.1 Dominant Set

The DSets algorithm is based on the definition of dominant set as a graph-theoretic concept of a cluster. By regarding a dominant set as a cluster, the DSets algorithm extracts the clusters one by one and determines the number of clusters automatically, similar to the sequential clustering manner of DBSCAN. Different from the k-means-like algorithms where the cluster members are influenced by the number of clusters, the data in a dominant set are totally determined by the pairwise similarity matrix. Specifically, a non-parametric internal coherency criterion based on the internal similarity in a dominant set is used to screen the data to be included into the dominant set. The extraction of a dominant set can be regarded as a process of maximizing the dominant set size while preserving the internal coherency. Since the data in a dominant set are highly similar to each other and are less similar to the outside data, a dominant set satisfies the basic property required of a cluster and can therefore be regarded as a cluster. A dominant set can be extracted with game dynamics, e.g., the replicator dynamics [17] or the infection and immunization dynamics [4].

The DSets clustering results are determined only by the pairwise similarity matrix. Therefore if the data to be clustered are represented in the form of the pairwise similarity matrix, we can regard the DSets algorithm as parameter independent. By extracting the clusters sequentially, the number of clusters are determined automatically. Since a dominant set preserves internal coherency, only the data with high pairwise similarity will be grouped into a cluster. This means that the outliers have a small chance to be grouped into any cluster. The dominant set concept can also be used to obtain overlapping clusters [5,20]. Due to these nice properties, the DSets algorithm has received successful application in various tasks [11–13,15,21,22]. However, as a dominant set requires each member to be very similar to all the other members, the DSets algorithm can only detect spherical clusters. Evidently this limits its performance on some datasets with clusters of non-spherical shapes.

2.2 DBSCAN

DBSCAN is one of the most popular density based clustering algorithm. It uses a neighborhood radius *Eps* and the minimum number *MinPts* of points in the neighborhood to denote the density threshold, and include data with sufficient density into clusters. The DBSCAN algorithm uses a density constraint to determine if a set of data belong to the same cluster. In this case, one data only needs to be similar to the nearest neighbors. As a result, DBSCAN is able to detect clusters of arbitrary shapes. The outliers are distributed far from other data and the local density is quite small. Therefore they won't be admitted into any cluster and are regarded as noise. One problem with DBSCAN is that the density threshold defined by the fixed *Eps* and *MinPts* may not be able to deal with the large density difference of different clusters. Some other density based clustering algorithms following DBSCAN include [1,2].

2.3 DP

The DP algorithm [18] is proposed to identify cluster centers based on the density peaks, and then determine the cluster members based on the density relationship among the data. We firstly calculate the local density ρ_i of each data i, based on which we obtain δ_i which is the distance between i and its nearest neighbor with higher local density, i.e.,

$$\delta_i = \min_{j \in S, \rho_j > \rho_i} d_{ij}. \tag{1}$$

where S is the set of data to be clustered. We assume that the cluster centers correspond to density peaks with both high ρ's and high δ's, whereas the non-center data are with either small ρ's or small δ's. Obviously this assumption holds in many practical cases. Under this assumption, we build a decision graph with ρ and δ denoting the two axes, where we will find that the cluster centers are far from both axes and isolated from the non-center data. This observation is then used to identify the cluster centers. After that, the labels of the non-center data are assigned to be the same as their nearest neighbors with higher density, and the clustering process is accomplished.

The DP algorithm is very simple and able to generate clusters of arbitrary shapes. On condition that the cluster centers are identified correctly, this algorithm can be very effective. However, it is not easy to discriminate between the "high" and "low" of ρ's and δ's, and [18] failed to provide a satisfying solution to the cluster center identification problem. In addition, the density kernels and involved parameters also influence the clustering results evidently. In summary, the DP algorithm has been shown to be a promising clustering approach, but there are still some important problems to be solved.

3 Our Approach

From Sect. 2 we see that the DSets algorithm uses only the pairwise similarity matrix as input and does not involve any parameter, but it can only generate

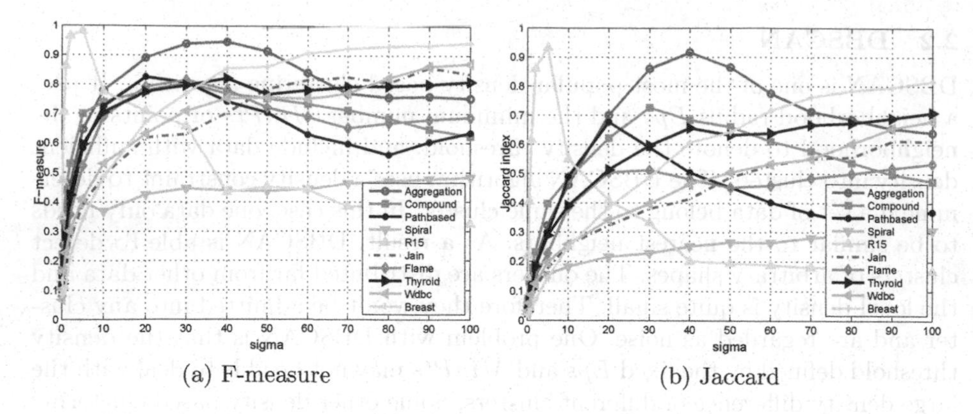

(a) F-measure (b) Jaccard

Fig. 1. The DSets clustering results on ten datasets with different σ's.

spherical clusters. In contrast, the DBSCAN and DP algorithms are able to generate clusters of arbitrary shapes, on condition that appropriate parameters are determined beforehand. This observation motivates us to combine these algorithms to merge their merits and avoid their drawbacks. Specifically, we use the non-parametric DSets algorithm to generate initial clusters, and then expand initial clusters based on DBSCAN and DP to generate clusters of arbitrary shapes, where the parameters required in expansion are determined based on the initial clusters. In this way the whole clustering process are parameter independent and clusters of arbitrary shapes can be obtained.

The DSets algorithm uses only the pairwise similarity matrix as input. Therefore if the data to be clustered are represented in the form of the pairwise similarity matrix, the DSets algorithm can be said to be parameter independent. However, in many tasks the data are represented as feature vectors in a feature space, and we need to build the pairwise similarity matrix from the data vectors. With two data x and y, their similarity is usually calculated in the form of $s(x, y) = exp(-d(x, y)/\sigma)$ where $d(x, y)$ is the Euclidean distance and σ is the decay parameter. In this case the parameter σ is involved in DSets clustering process and it is found to influence the DSets clustering results. For example, we apply the DSets algorithm to ten datasets, including Aggregation [10], Compound [24], Pathbased [6], Spiral [6], R15 [23], Jain [14], Flame [9] and three UCI datasets Thyroid, Wdbc and Breast. Since the data in some datasets are not 2D, we use F-measure and Jaccard index to evaluate the clustering results. The variance of the clustering results with respect to σ's is reported in Fig. 1, where it is quite evident that σ has a significant influence on DSets clustering results. For the ease of expression, in Fig. 1 the horizontal axes denote only the coefficients of σ's, and the real values of σ's are the product of the horizontal axes and $\overline{d}$, which denotes the average of the pairwise distances.

While Fig. 1 shows the influence of σ's on DSets clustering results, it also shows that the best-performing σ's vary widely for different datasets. Since we still have no algorithm to determine the best-performing σ for a given dataset, in this paper we resort to a different approach to solve the parameter dependence problem. Noticing that σ's influence the similarity matrices which then impacts on the clustering results, we choose to remove the influence of σ's on the pairwise similarity matrices. In this paper, this purpose is achieved by the histogram equalization transformation of the similarity matrices.

3.1 Histogram Equalization Transformation

Histogram equalization is originally used in image enhancement to increase the overall intensity contrast in an image. Given the pixels in an image, we quantize the intensity range into N bins and build the histogram $H = \{h_k\}, k = 1, \cdots, N$, where h_k denotes the number of pixels in the k-th bin. After histogram equalization transformation, the pixels are assigned new values based on the proportion of pixels in the bins with smaller intensity values. Specifically, the pixels in the k-th bin will be assigned a new intensity value as

$$g_k = L \sum_{j=1}^{k} \frac{h_j}{n}, \tag{2}$$

where L is the maximum intensity value and n is the number of all pixels. After the transformation, the pixel intensity values are distributed in the whole range more evenly and the intensity histogram becomes more flat, and this is the reason why the transformation is called histogram equalization. In our application, the similarity values in the pairwise similarity matrix are the data to be transformed and L equals to 1. If the number of bins is sufficiently large so that each bin contains only data of the same value, we see that the new similarity values are totally determined by the ordering of the original similarity values.

From $s(x, y) = exp(-d(x, y)/\sigma)$ we see that σ's influences only the absolute magnitude of each similarity value. Since the distance $d(x, y)$ are not influenced by σ's, we know that the relative magnitude of the similarity values, or the sorting of the similarity values, is invariant to σ's. Therefore if we apply histogram equalization to transform the similarity matrix, we find that the new similarity matrix is not influenced by σ's, only if the number of histogram bins is large enough. Since the influence of σ's on the similarity matrices is eliminated, it is natural that σ's no longer impact on the DSets clustering results. In practice, one bin usually contains more than one similarity value and different σ's will still cause slight variances in clustering results. Our experiments show that the variations can be ignored if the number N of bins is large enough. In implementation we use $N = 100$ and $\sigma = \overline{d}$. Corresponding to Fig. 1, we report the results obtained after histogram equalization in Fig. 2. Evidently the influence of σ's has been eliminated effectively. For ease of expression, we use DSets-histeq to denote the DSets algorithm where the similarity matrix is transformed by histogram equalization.

While Fig. 1 indicate that the influence of σ's has been eliminated by histogram equalization transformation, we also notice that the DSets-histeq clustering results are usually not satisfactory. Our explanation for this observation

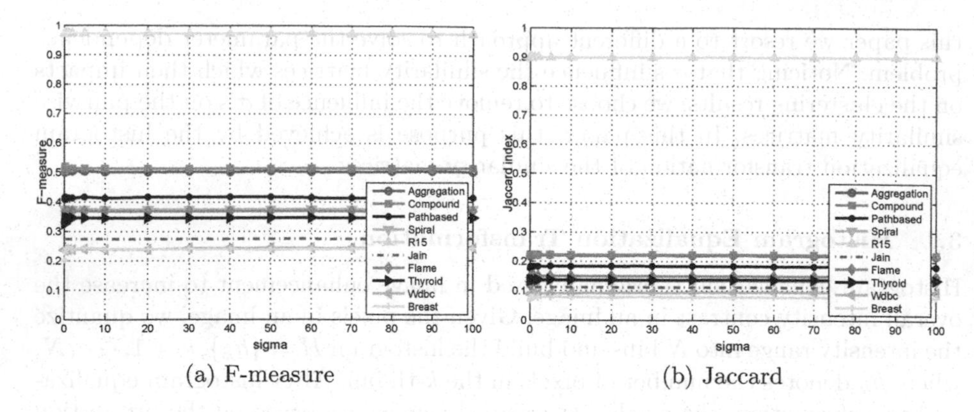

Fig. 2. The DSets clustering results on ten datasets with different σ's, where the similarity matrices are transformed by histogram equalization.

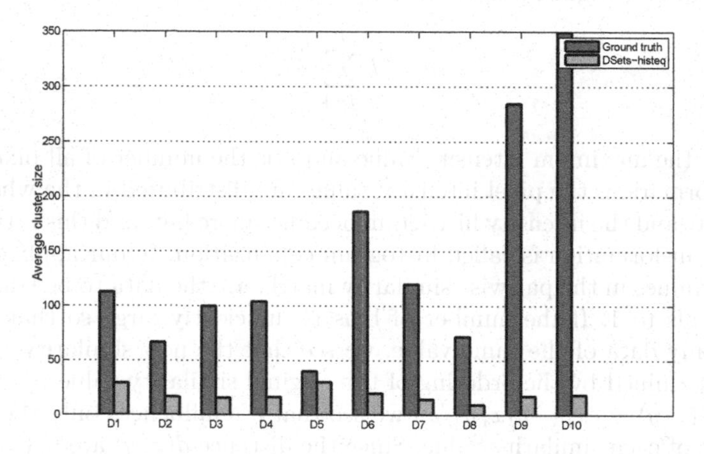

Fig. 3. Comparison of the average cluster sizes from DSets-histeq and the ground truth.

is that the histogram equalization increases the similarity value contrast and causes over-small clusters. As a result, the obtained clusters are usually smaller than the real ones, as shown in Fig. 3, where we use D1 to D10 to denote the Aggregation, Compound, Pathbased, Spiral, R15, Jain, Flame, Thyroid, Wdbc and Breast, respectively. While this effect is discouraging, it also provides us with an opportunity to improve clustering results by expanding clusters and obtaining clusters of arbitrary shapes.

3.2 Cluster Expansion by DBSCAN

The clusters from DSets-histeq are usually smaller than the real ones, and these clusters are of only spherical shapes. Since both factors imply unsatisfactory clustering results, we regard these clusters as initial ones and expand them to improve clustering results. On the other hand, the DBSCAN algorithm extracts clusters in a cluster expansion manner, and it is able to generate clusters of arbitrary shapes. These properties make DBSCAN a perfect choice for the cluster expansion step of our algorithm. However, we also notice that the DBSCAN algorithm requires two parameters *Eps* and $MinPts$ as input. In the following we introduce how to determine the parameters based on the initial clusters.

The DBSCAN algorithm uses *Eps* and $MinPts$ to measure the local density, and any data with at least $MinPts$ data in the *Eps*-neighborhood are regarded as satisfying the density requirement. If we fix $MinPts$ as recommended in [8], then the problem is to find the maximum distance to the $MinPts$-th nearest neighbor which corresponds to the smallest density. Since we already have the initial cluster as a subset of the final cluster, we can determine *Eps* with the data in the initial cluster. Specifically, with each data in the initial cluster C, we calculate the distance to its $MinPts$-th nearest neighbor and set *Eps* as the minimum of these distances, i.e.,

$$Eps = \max_{i \in C} d(i, i_{MinPts}) \tag{3}$$

with $d(i, i_{MinPts})$ denoting the distance between i and its $MinPts$-th nearest neighbor in the initial cluster. The $MinPts$ is fixed to be 3 in this paper, which is close to the value 4 recommended in [8] and shown to be the best-performing one in experiments.

3.3 Cluster Expansion by DP

In the cluster expansion based on DBSCAN, we use the initial cluster to determine the density threshold and the parameters required by DBSCAN. In this way we guarantee that after cluster expansion, the local density is preserved. Noticing that by definition the dominant set imposes a very high density constraint on the data included. As a result, the high density constraint may restrict the cluster sizes and cause small clusters. Since the cluster expansion based on DBSCAN fails to relax the density constraint, we need some other methods to solve the problem. In this paper we use another cluster expansion step based on the DP algorithm.

In the DP algorithm, after the cluster centers are identified, the non-center data are assigned the same labels as their nearest neighbors with higher density. This method of labeling non-center data is very important as it builds a connection between the labels of high-density data and of low-density ones. If the label of data i_1 is not known, we only need to check the label of i_2, which is the nearest one in the neighbors with higher density of i_1. Here the density is calculated based on the nearest neighbors. Similarly, if the label of i_2 is not available, we continue to check of the label of i_3. In this way we will finally reach one cluster center (density peak) and obtain the label. Motivated by this method, following the DBSCAN-based expansion, we continue to expand the cluster as follows. We firstly sort the outside data according to their distance to the cluster. Starting from the nearest outside data j_1, we check its nearest neighbor j_2 with higher density. If j_2 is in the cluster, then j_1 is included into

Table 1. Clustering results of different clustering algorithms, with F-measure.

	D1	D2	D3	D4	D5	D6	D7	D8	D9	D10	Average
k-means	0.83	0.68	0.70	0.35	0.82	0.79	0.84	0.83	0.84	0.96	0.77
NCuts	0.99	0.70	0.96	0.58	0.99	0.63	0.99	0.64	0.84	0.64	0.80
SPRG	0.73	0.64	0.56	0.37	0.93	0.86	0.60	0.97	0.91	0.97	0.75
DBSCAN	0.90	0.92	0.72	0.89	0.73	0.85	0.96	0.68	0.69	0.87	0.82
AP	0.82	0.77	0.66	0.38	0.29	0.64	0.74	0.50	0.79	0.96	0.65
DP-cutoff	0.99	0.82	0.69	0.64	0.99	0.90	1.00	0.55	0.61	0.67	0.78
DP-Gaussian	0.99	0.69	0.68	1.00	0.99	0.87	0.79	0.51	0.77	0.66	0.80
Ours	0.85	0.83	0.94	0.81	0.81	0.96	0.99	0.81	0.82	0.92	0.87

the cluster. Otherwise, we continue to check the nearest neighbor j_3 with higher density. In this searching process, if the distance between j_k and j_{k+1} is above a threshold th_δ, the searching is terminated and the data j_1 is not included into the cluster. The threshold th_δ reflects the discrimination between the δ distance of center and non-center data, and is obtained by a histogram of all the δ's.

4 Experiments

4.1 Data Clustering

We use the aforementioned ten datasets in experiments and compare our algorithm with seven other algorithms, including k-mean, NCuts, SPRG [25], DBSCAN, AP, DP-cutoff (DP with the cutoff kernel) and DP-Gaussian (DP with the Gaussian kernel). For k-means, NCuts and SPRG, the required number of clusters is set as the ground truth, and we report the average results of ten runs. With DBSCAN, the parameter $MinPts$ are selected from 2, 3, $\cdots$, 10 and Eps is calculated with the method proposed in [7]. With AP, we use the code from [18] to calculate the range of preference value and select the best-performing sample in the range. With DP-cutoff and DP-Gaussian we select the cluster centers based on $\gamma = \rho\delta$ and the ground truth number of clusters. In summary, all the five algorithms used in comparison benefit from carefully selected parameters. The comparison results with F-measure and Jaccard index are reported in Tables 1 and 2, respectively, where we use D1 to D10 to denote the 10 datasets in the order of Aggregation, Compound, Pathbased, Spiral, R15, Jain, Flame, Thyroid, Wdbc and Breast.

From the comparison we see that while some algorithms generate very good results on some datasets, their results on others are not satisfactory. In contrast, our algorithm shows great generality and generates quite good results on all datasets, and outperforms all the other algorithms in average results.

4.2 Image Segmentation

We then apply our algorithm to image segmentation tasks. The images are taken from the BSD500 dataset and we use the R, G and B channels to build the

Table 2. Clustering results of different clustering algorithms, with Jaccard index.

	D1	D2	D3	D4	D5	D6	D7	D8	D9	D10	Average
k-means	0.64	0.46	0.50	0.20	0.65	0.53	0.59	0.64	0.65	0.87	0.57
NCuts	0.98	0.46	0.85	0.30	0.99	0.42	0.97	0.40	0.65	0.39	0.64
SPRG	0.49	0.42	0.34	0.20	0.83	0.63	0.41	0.90	0.74	0.89	0.59
DBSCAN	0.81	0.87	0.53	0.63	0.40	0.91	0.90	0.57	0.53	0.78	0.69
AP	0.73	0.69	0.47	0.21	0.14	0.39	0.47	0.29	0.50	0.86	0.47
DP-cutoff	0.98	0.71	0.49	0.39	0.96	0.71	1.00	0.29	0.42	0.48	0.64
DP-Gaussian	0.99	0.47	0.49	1.00	0.99	0.65	0.52	0.29	0.58	0.40	0.64
Ours	0.75	0.73	0.80	0.54	0.60	0.95	0.97	0.74	0.61	0.79	0.75

Fig. 4. Image segmentations results with our algorithm.

feature vector and calculate the similarity matrix. The segmentation results of some example images are shown in Fig. 4. Although the number of segments is not specified, we observe that our algorithm generates reasonable segmentation results.

4.3 Discussion

In this paper we try to eliminate the parameter dependence problem and present a parameter independent clustering algorithm. In fact, the first step of the algorithm, i.e., DSets-histeq, is really parameter independent, where the only parameter N can be fixed reasonably. However, in order to obtain clusters of arbitrary shapes, we use DBSCAN and DP in two cluster expansion steps, where new parameters are introduced. While it is true that the data in initial clusters can be used in determining the new parameters, there still exists parameters which cannot be determined automatically in the current algorithm. In DBSCAN based expansion we have to fixed $MinPts$ to be 3, and in DP based expansion the density calculation also involves an user-specified parameter, which is the number of nearest neighbors. Although these parameters are not determined automatically, they have explicit meaning and can be estimated reasonably, in contrast to the σ in the DSets algorithm. In other words, our algorithm does relieve the parameter dependence problem evidently. On the other hand, we are working on better density calculation methods and cluster expansion methods, in order to reduce the parameter dependence further.

Another problem worth mentioning is the relative large computation load of the algorithm. Compared with the original DBSCAN algorithm, the DSets algorithm is a little computationally expensive. In addition, the histogram equalization, density calculation and point-by-point expansion adds to the computation load of the algorithm. Therefore it is necessary to explore more efficient approaches to improve the current algorithm.

5 Conclusions

In this paper we propose to combine the DSets algorithm with DBSCAN and DP algorithm to solve the problems afflicting existing clustering algorithms. We firstly use histogram equalization to transform the pairwise similarity matrix before DSets clustering, thereby solving the parameter dependence problem of the DSets algorithm. Then we use the DBSCAN and DP algorithms to expand the initial clusters from the DSets algorithm and obtain clusters of arbitrary shapes. In the cluster expansion process, the parameters required by DBSCAN and DP are determined based on the initial clusters. By making use of the merits and avoid the drawbacks of DSets, DBSCAN and DP algorithms, our algorithm has less dependence on parameters and is able to generate clusters of arbitrary. Data clustering and image segmentation experiments and comparison with other algorithms validate the effectiveness of our algorithm.

Acknowledgement. This work is supported in part by the National Natural Science Foundation of China under Grant No. 61473045 and by China Scholarship Council.

References

1. Achtert, E., Bohm, C., Kroger, P.: DeLi-CLu: boosting robustness, completeness, usability, and efficiency of hierarchical clustering by a closest pair ranking. In: International Conference on Knowledge Discovery and Data Mining, pp. 119–128 (2006)
2. Ankerst, M., Breunig, M.M., Kriegel, H.P., Sander, J.: Optics: ordering points to identify the clustering structure. In: ACM SIGMOD International Conference on Management of Data, pp. 49–60 (1999)
3. Brendan, J.F., Delbert, D.: Clustering by passing messages between data points. Science **315**, 972–976 (2007)
4. Bulo, S.R., Pelillo, M., Bomze, I.M.: Graph-based quadratic optimization: a fast evolutionary approach. Comput. Vis. Image Underst. **115**(7), 984–995 (2011)
5. Bulo, S.R., Torsello, A., Pelillo, M.: A game-theoretic approach to partial clique enumeration. Image Vis. Comput. **27**(7), 911–922 (2009)

6. Chang, H., Yeung, D.Y.: Robust path-based spectral clustering. Pattern Recogn. **41**(1), 191–203 (2008)
7. Daszykowski, M., Walczak, B., Massart, D.L.: Looking for natural patterns in data: part 1. density-based approach. Chemometr. Intell. Lab. Syst. **56**(2), 83–92 (2001)
8. Ester, M., Kriegel, H.P., Sander, J., Xu, X.W.: A density-based algorithm for discovering clusters in large spatial databases with noise. In: International Conference on Knowledge Discovery and Data Mining, pp. 226–231 (1996)
9. Fu, L., Medico, E.: Flame, a novel fuzzy clustering method for the analysis of DNA microarray data. BMC Bioinform. **8**(1), 1–17 (2007)
10. Gionis, A., Mannila, H., Tsaparas, P.: Clustering aggregation. ACM Trans. Knowl. Disc. Data **1**(1), 1–30 (2007)
11. Hou, J., Gao, H., Li, X.: DSets-DBSCAN: a parameter-free clustering algorithm. IEEE Trans. Image Process. **25**(7), 3182–3193 (2016)
12. Hou, J., Liu, W., Xu, E., Cui, H.: Towards parameter-independent data clustering and image segmentation. Pattern Recogn. **60**, 25–36 (2016)
13. Hou, J., Pelillo, M.: A simple feature combination method based on dominant sets. Pattern Recogn. **46**(11), 3129–3139 (2013)
14. Jain, A.K., Law, M.H.C.: Data clustering: a user's dilemma. In: Pal, S.K., Bandyopadhyay, S., Biswas, S. (eds.) PReMI 2005. LNCS, vol. 3776, pp. 1–10. Springer, Heidelberg (2005). doi:10.1007/11590316_1
15. Zemene, E., Pelillo, M.: Interactive image segmentation using constrained dominant sets. In: Leibe, B., Matas, J., Sebe, N., Welling, M. (eds.) ECCV 2016. LNCS, vol. 9912, pp. 278–294. Springer, Cham (2016). doi:10.1007/978-3-319-46484-8_17
16. Monti, S., Tamayo, P., Mesirov, J., Golub, T.: Consensus clustering: a resampling-based method for class discovery and visualization of gene expression microarray data. Mach. Learn. **52**(1–2), 91–118 (2003)
17. Pavan, M., Pelillo, M.: Dominant sets and pairwise clustering. IEEE Trans. Pattern Anal. Mach. Intell. **29**(1), 167–172 (2007)
18. Rodriguez, A., Laio, A.: Clustering by fast search and find of density peaks. Science **344**, 1492–1496 (2014)
19. Shi, J., Malik, J.: Normalized cuts and image segmentation. IEEE Trans. Pattern Anal. Mach. Intell. **22**(8), 167–172 (2000)
20. Torsello, A., Bulo, S.R., Pelillo, M.: Beyond partitions: allowing overlapping groups in pairwise clustering. In: International Conference on Pattern Recognition, pp. 1–4 (2008)
21. Tripodi, R., Pelillo, M.: Document clustering games. In: The 5th International Conference on Pattern Recognition Applications and Methods, pp. 109–118 (2016)
22. Vascon, S., Mequanint, E.Z., Cristani, M., Hung, H., Pelillo, M., Murino, V.: Detecting conversational groups in images and sequences: a robust game-theoretic approach. Comput. Vis. Image Underst. **143**, 11–24 (2016)
23. Veenman, C.J., Reinders, M., Backer, E.: A maximum variance cluster algorithm. IEEE Trans. Pattern Anal. Mach. Intell. **24**(9), 1273–1280 (2002)
24. Zahn, C.T.: Graph-theoretical methods for detecting and describing gestalt clusters. IEEE Trans. Comput. **20**(1), 68–86 (1971)
25. Zhu, X., Loy, C.C., Gong, S.: Constructing robust affinity graphs for spectral clustering. In: IEEE International Conference on Computer Vision and Pattern Recognition, pp. 1450–1457 (2014)

Recommender Systems

Friend Recommendation Considering Preference Coverage in Location-Based Social Networks

Fei Yu[1](✉), Nan Che[2], Zhijun Li[1], Kai Li[1], and Shouxu Jiang[1](✉)

[1] School of Computer Science and Technology,
Harbin Institute of Technology, Harbin, China
{yf,lizhijun_os,jsx}@hit.edu.cn, likai.1991.cs@gmail.com

[2] School of Software, Harbin University of Science and Technology,
Harbin, China
chenan_1980@163.com

Abstract. Friend recommendation (FR) becomes a valuable service in location-based social networks. Its essential purpose is to meet social demand and demand on obtaining information. The most of current existing friend recommendation methods mainly focus on the preference similarity and common friends between users for improving the recommendation quality. The similar users are likely to have similar preferences of point-of-interests (POIs), the kinds of information they provided are limited and redundant, can not cover all of the target user's preferences of POIs. This paper aims to improve amount of information on users' preferences through FR. We give a definition of friend recommendation considering preference coverage problem (FRPCP), and it is also one NP-hard problem. This paper proposes the greedy algorithm to solve the problem. Compared to the existing typical recommendation approaches, the large-scale LBSN datasets validate recommendation quality and significant increase in the degree to preferences coverage.

Keywords: LBSN · Friend recommendation · Power-law distribution · Preference coverage

1 Introduction

As the mobile internet is gradually into people's life, study and job. Recently, location-based social networks (LBSNs) have been widely applied, such as Foursquare, Gowalla, weibo, QQ, etc. Compared with the traditional social networks (SNs), the significant advantage of LBSNs is that LBSNs can link with the behaviour habits in the physical world and social networks. Meanwhile, people have more and more demands on location-based services. Wherein friend recommendation (FR) has become the main service application in LBSNs. The essential goal is to meet users' social demand and demand on obtaining information.

J. Kim et al. (Eds.): PAKDD 2017, Part II, LNAI 10235, pp. 91–105, 2017.
DOI: 10.1007/978-3-319-57529-2_8

Currently, the most existing work focused on the social demand and improving recommendation quality. Excessive pursuit of recommendation quality, however, can lead to ignoring the user's information demand. At the same time, the information types are single and too redundant.

So this paper considers sacrificing a small amount of recommendation quality, in order to meet user demand on obtaining information covering user's preferences, and to avoid the excessive type single and redundant information. The essential difference between researches on the recommendation problem in LBSNs and in the traditional social networks is that recommendations algorithms in LBSNs take full advantage of the user's physical behaviour information. Because people's behaviour in the virtual social network can not truly reflect their real character preferences and behavioural habit in the physical space. Meanwhile, LBSNs also provide users with a social platform. So the friend recommendation problem in LBSNs becomes one of current problems to be solved.

Friends recommendation algorithm on traditional social networks from the recommended object can be divided into two categories: (1) friends offline; (2) strangers. Friends recommendation algorithm in traditional social networks from the type of mined information can be divided into two categories: (1) topology information (e.g. common friends) [1]; (2) non-topology information (e.g. personal profiles, etc.). At present, friend recommendation algorithms mainly use collaborative filtering [2,3], the random walk [4], genetic algorithm [4], weighted Voronoi Diagrams algorithms [5]. Social Survey by Marketing Letter said: The access to obtain information for people is more likely to get it from their own friends. Thus friend recommendation can help people to obtain information. At the same time, users in a social network have social demand (to broaden the circle of friends) and obtaining information demand [6]. The existing friend recommendation algorithms mainly focused on broadening your social circle. Meanwhile users influence and offer information each others in LBSNs. The conventional friend recommendation methods mainly consider the similar between

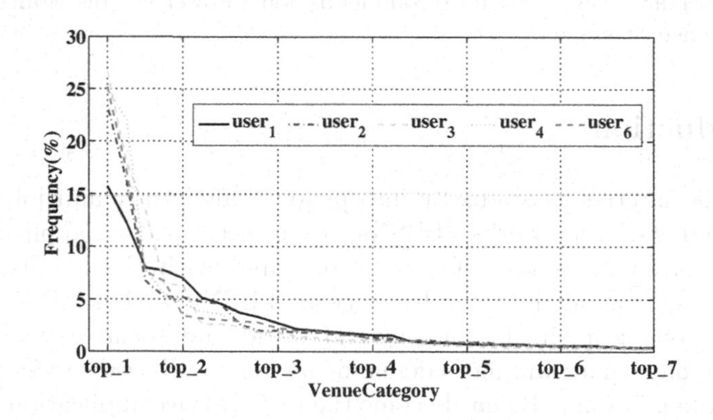

Fig. 1. The distribution of users' POI preferences in LBSNs, and the $top_1, top_2, \cdots$ in the x-axis represent the top-N favorite POI categories each user (e.g. $user_1, user_2, user_3, user_4, user_6$)

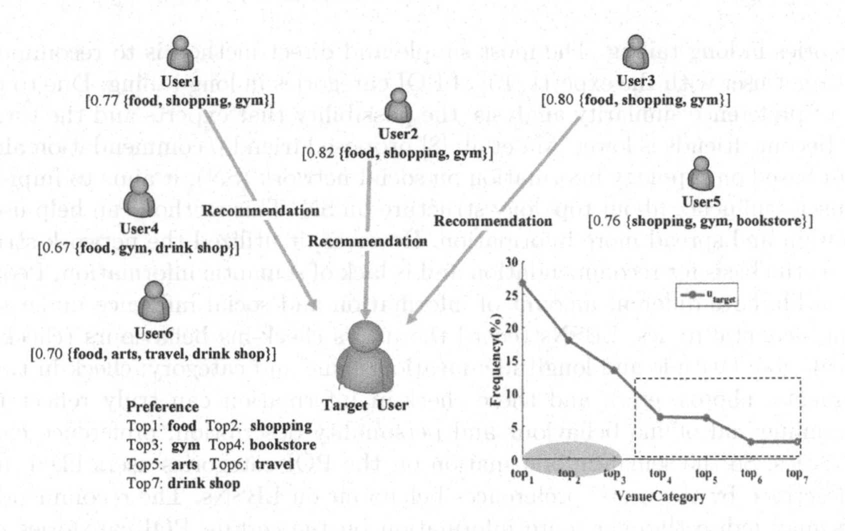

Fig. 2. The traditional friend recommendation. There are six users and the target users, and each user is denoted by one tuple in which the number scores its similarity to target user and the term describes these POI categories of information that these users can provide the target user with.

users' preferences, check-in behaviours. But the more similar the user are as the target users' preference on POI category, the less users provide the target users with the information on POI categories in long tailing. Therefore, the traditional recommendation algorithms are not able to address the issues raised above. Our goal is to improve the amount of information on the POI categories the user like better (not best). Generally, the distribution of person's preference obeys one long-tailing distribution [10]. Through the experiment we validate the user's preferences to obey power-law distribution, as shown in Fig. 1. In another words, users have little knowledge on POI category preference in the long tailing. We give one example of friend recommendation in Fig. 2. Based on the preference similarity only, the system recommends the potential friend set $\{user_1, user_2, user_3\}$ for target user u_T. However, the POI preference category of this recommendation result is rather monotonic, and they only covered a part $\{top_1, top_2, top_3\}$ of target user's preference. The black dotted box parts are not covered. In contrast, the use list $\{user_2, user_5, user_6\}$, would be a better recommendation list compared with the above list, and it achieves better similarity scores and simultaneously provides the variety $\{top_1, top_2, \cdots, top_7\}$ of information for target user. Thus, the recommendation for the target user based on preference similarity hardly provides much information on these POI categories in the long tailing. As the recommendation results are increasing more and more, the information from these results becomes homogeneous and all users already known. So, these recommendations are bad for diversities of information acquisition.

So this paper considers the above users' demands during friend recommendation simultaneously, and is in order to overcome the limitation of the POI categories in long tailing. The most simple and direct method is to recommend the target user with the experts [15] of POI categories in long tailing. Due to the lack of preference similarity analysis, the possibility that experts and the target user become friends is lower. Ma et al. [8] proposed friend recommendation algorithm based on topology information on social network (SN), it aims to improve the user's influence about topology structure on SN. This method can help users to obtain and spread more information. However, it utilized the network structure as the basis for recommendation and is lack of semantic information. People in real life have different amount of information and social influence under different semantic topics. LBSNs record the user's check-ins behaviours (check-in information: latitude and longitude; location name and category; check-in time, comments, photos, etc.), and these check-in information can truly reflect the user online and offline behaviour and personality description, preference characteristics. So the semantic information on the POI categories users liked may be described by the users' preferences behaviour on LBSNs. The recommended users may indirectly offer more information on the certain POI categories the target user like better or best. Wan et al. [9] provided a recommendation method to recommend friends based on the informational utility in traditional social network. But this method does not consider the distribution feature of the users' preferences, and is lack of the check-in behaviour analysis in physical world. The information distribution of every user's preferences on POI categories obey power-law distributions, and it is with a long trailing, as shown in the Fig. 1.

Most friend recommendation methods recommend the target user with the users owning the high similarity on POI preferences. Their main goal is to improve the recommendation quality, but the effectiveness of recommendation quality in the recall ratio and precision ratio is not significant, the increase amplitude is only about one percent. Thus, under users' tolerate recommendation quality, enriching user information demand has greater application value compared with simply improving the recommendation quality. We consider simultaneously the recommendation quality and users' information demand. Because these recommended users' POI preference distributions all obey the power-law distribution with a long tailing, they and the target user relatively infrequently check in the some POI categories in the long tailing. We see this situation as users have less information on these POI categories in the long tailing. This paper is in order to overcome the trailing through the friend recommendation. These recommended potential friends are the similar as the target user' preferences on POI categories, and they can provide more information on POI categories in long tail for the target user. The key is that these information can fully coverage the target users' POI category preference.

Our contribution could be summarized as follows:

- **The Problem:** We provide a novel friend recommendation problem considering preference coverage in LBSNs. This problem is one optimization problem. Meanwhile, and it is also NP-hard.
- **Design a preference coverage measure:** We utilize the Shannon entropy to evaluate the degree to the preferences coverage through FR.
- **Performance:** This paper provides a greedy algorithm (called FRPC) to the optimization problem.

- **Discovery:** we have conducted comprehensive experiments on two massive real datasets (Foursquare, Gowalla), and the experimental results shows that our algorithm on recommendation quality is relatively the consistency as the two typical recommendation methods. In terms of the degree to preference coverage, this method has significant advantages.

The remainder of this paper is organized as follows: Sect. 2 gives a formal definition of the friend recommendation considering preference coverage, the descriptions of related definitions and symbols. Then we presents how to solve the above problem. The experimental results are shown in Sect. 3. Section 4 concludes this paper.

2 Friend Recommendation

In this section, a few core definitions are firstly described in detail and then the FRPC algorithm is presented.

2.1 Notations and Definitions

We firstly introduce relevant definitions and notations used throughout this paper. The notations used in this paper are summarized in Table 1.

Table 1. Symbol description

Symbol	Description
$G(U,E)$	Social networks G, nodes set U, edges set E
$< G,C >$	LBSNs, where C represents users' check-in recorders
C	$C = \{(u,l,t)\}$, user u checks in location l at the time t
L	Locations set in LBSNs $L = \{l_1, l_2, \cdots\}$
l	$l = (lon, lat, a)$, l is a location, a is POI category
M, N	Number of users, POI categories
N_u	Number of POI categories checked in by user u
A	The POI categories set, $a_i \in A$
$P(u, a_i)$	Probability that user u likes POI categories a_i
$sim(u_i, u_j \mid a_i)$	Similarity about POI category between u_i, u_j
u_T	The target user in recommendation system
U_{re}	The recommended users set for target user u_T

Definition 1 ***(LBSNs).*** *An LBSNs $< G,C >$ consist of a social network $G =< U,E >$, where U is the set of users, $E = \{(u_i, u_j)|$ one social connection from u_i to u_j, $u_i, u_j \in U, u_i \neq u_j\}$ and the set of check-in records $C = \{(u,l,t)\}$. (u,l,t) represents a check-in record where a user u checks in location l at time t, and $l \in L$. A location l presents: $l = (lon, lat, a)$, wherein lon is longitude, lat is latitude, a is one POI category.*

In this section, we describe the users' preference on POI category in LBSNs.

Definition 2 ***(User Preference).*** *Given a user u, her/his intrinsic check-in behaviors' preferences in LBSNs are which POI categories users always check in, the frequency degree denotes as $P(u, a)$.*

This paper examine how the distribution fit the power-law curve by Kolmogorov-Smirnov test [12]. This method estimates the three parameters as the following:

(1) x_{min}: we convert the POI categories to the corresponding numerical value based on the order of frequency of POI categories checked-in. The best fitted cutoff value, thus the values more than x_{min} can fit a power-law distribution;
(2) $\hat{\alpha}, \hat{\beta}$: the parameters decides the slop of the best fitted distribution, thus the values more than x_{min} obey the power-law distribution $P(u, x) = \hat{\alpha}x^{-\hat{\beta}}$;
(3) p-value: the statistical significance of the goodness of the power-law fitting [13], if the p-value is more than 0.05, the fitted distribution is a significant good fitting.

2.2 The Similarity Between Users

In this section, we describe how to compute the preferences similarity between users.

Let $A_T = \{a_{(1)}, a_{(2)}, \cdots, a_{(k)}\}$ be a set of the target user's favorite POI categories. $sim(u_T, u_l), u_l \in U$ is the traditional similarity metric (as Eq. 1) between users u_T and u_l. $P(u, a_i)$ is the degree of the user u's preference on POI category a. $E(u_T)$ and $E(u_l)$ respectively represent the mathematical expectation of user u_T and u_l check-in behaviours, $u_l \in U$. In the following equation, A_T means the set of POI category that user u_T and $u_l, u_l \in U$ common preferences.

$$sim(u_T, u_l) = \frac{\sum_{a_j \in A_T}(P(u_T, a_j) - E(u_T))(P(u_l, a_j) - E(u_j))}{\sqrt{\sum_{a_j \in A_T}(P(u_T, a_j) - E(u_T))^2}\sqrt{\sum_{a_j \in A_T}(P(u_l, a_j) - E(u_j))^2}} \quad (1)$$

Note that the similarity function in Eq. 1 is one of the similarity measure methods between users. Our proposed method can be applicable if we use other related measure approaches to compute the similarity between users.

2.3 Preference Coverage

Persons always prefer to obtain information from their friends, however, most friend recommendation methods focus on expanding the social circle and ignore the users' demand on obtaining information from friends. This paper mainly provides one novel recommendation method for obtaining more information. Our goal is that the informations gained from the recommended users. Since they cover the target user u_T's preferences on POI categories, so these users can improve amount of information on u_T's potential POI categories in the long tailing. Preference coverage considers how to select a list of users U_{re} owning the similar POI preference categories as u_T.

This paper computes the preference coverage of a list of users in the follows as:

$$I(U) = \sum_{a_i \in A_T} w_{u_T}^{a_i} cov^{a_i}(U) \quad (2)$$

wherein $w_{u_T}^{a_i}(> 0), a_i \in A_T$, is the degree of the u_T's demand on obtaining information of category a_i, generally denoted as $w_{u_T}^{a_i} = -\log(P(u_T, a_i))$. Based on the point-information entropy theory, the demand degree to obtaining information about each POI category user liked is represented by the point-information entropy under each POI category. $cov^{a_i}(U)$ is used for calculating the degree to which POI category a_i is covered by the information the least one user $u_l \in U$ provided. So, this section computes $cov^{a_i}(U)$ as the following equation:

$$cov^{a_i}(U) = 1 - \prod_{u_l \in U} [1 - cov^{a_i}(u_l)] \quad (3)$$

where $cov^{a_i}(u_l)$ is the degree to which POI category a_i is covered by the information the user u_l provided. The amount of information on a certain POI category user owned is depended on the degree of user's preferences on POI categories, and its specific calculation formula is as described above. Thus we compute the $cov^{a_i}(u_l)$, in Eq. 4.

$$cov^{a_i}(u_l) = P(u_l, a_i) \quad (4)$$

2.4 FRPC Algorithm

In this section, we firstly give one definition of friend recommendation considering preference coverage (FRPC). Then this paper call this problem as **FRPCP**, and describes this problem with a formula and regards it as an optimization problem. Lastly, we provide a greedy algorithm to solve this problem.

Definition 3 (FRPCP). Given a LBSN $< G, C >$, the target user u_T, the candidate recommended users with the similar preferences on POI categories $A_u = \{a_1, a_2, \cdots\}$ as u_T, denoted as U_{Cre}, and the target user u_T's preference list denoted $A_T = \{a_{(1)}, a_{(2)}, \cdots, a_{(k)}\}$, $\#A_T = k$.

$$U_{re} = \arg \max_{U_j \subseteqq U_{Cre}, \#U_j = k} \sigma(U_j)$$

$$s.t. \quad U_{Cre} = \arg_{u_i \in U}\{\|A_{u_i} \cap A_T)\| = k\}, \#U_{re} = k.$$

Wherein $\sigma(U_j) = \gamma \times I(U_j) + (1 - \gamma) sim(U_j, u_T)$, and the parameter γ is determined by the user's requires on either meeting the preference similarity or meeting information demand. γ is set to the small value when the user wants to find some users with the high preferences similar as the u_T. However, the value of γ is depended on the user's personal demand. And the candidate recommended users with the similar preferences on POI categories $A_u = \{a_1, a_2, \cdots\}$ as u_T, as the following:

$$U_{Cre} = \arg_{u_i \in U}\{\|A_{u_i} \cap A_T)\| = k\} \quad (5)$$

The value k is determined by the number of the target user' preference about POI categories. As is shown in Fig. 2, we find the distribution of POI categories checked-in by u_T obey the power-law distribution. The value k is calculated by Eq. 7:

$$k = \arg\min_{k \in \mathbb{R}}\{\sum_{i=1}^{k} P(u_T, a_{(i)}) \geq \varepsilon\} \quad (6)$$

where the value of ε is set in advance, and it is generally set to 0.78. By means of the statistical analysis, we know that, if the value of ε is more than 0.78, the number of users with the similar POI preferences as the target user u_T is relatively less and fail to help the friend recommendation for meeting well demand on obtaining information. This paper gives a suggestion of k value. The output of this algorithm is the user set U_{re}. The goal is to generate potential friend list with certain better accuracy and help u_T to obtain more information compared with the state-of-the-art recommendation methods.

We observe that our friend recommendation is one complex combinatorial optimization problem, and it owns two objectives. Due to the monotone submodularity of the objective function, we give one greedy algorithm (called FRPC-A) to solve this issue. Firstly, this paper gives the proof of $sim(U, u_T)$ and $I(U)$'s monotone submodularity respectively.

Lemma 1. *The similarity function $sim(U, u_T)$ is monotone and submodular.*

Proof. Let $U_1 \subseteq U_2 \subseteq U$ and $u_j \in U$. For $\forall u_j \notin U_1$, we can compute

$$sim(U_1 \cup u_j, u_T) - sim(U_1, u_T) = [sim(U_1, u_T) - sim(u_j, u_T)] - sim(U_1, u_T) = sim(u_j, u_T)$$

As the similarity function is nonnegative. $sim(\cdot, u_T)$ is monotone. In the same way, there is that $sim(U_1 \cup u_j, u_T) - sim(U_1, u_T) \geq [sim(U_2, u_T) - sim(u_j, u_T)] - sim(U_2, u_T)$. Thus, the similarity function is submodular. ■

Lemma 2. *The preference coverage function $I(U)$ is monotone and submodular.*

Proof. Let $U_1 \subseteq U_2 \subseteq U$ and $u_j \in U$. For $\forall u_j \notin U_1$, we can compute

$$cov^{a_i}(U_1 \cup u_j) - cov^{a_i}(U_1) = \prod_{u_l \in U_1} [1 - cov^{a_i}(u_l)] -$$

$$\prod_{u_l \in U_1 \cup u_j} [1 - cov^{a_i}(u_l)] = cov^{a_i}(u_j) \prod_{u_l \in U_1} [1 - cov^{a_i}(u_l)]$$

Since the function $cov^{a_i}(\cdot) \in [0, 1]$, we get $cov^{a_i}(U_1 \cup u_j) \geq cov^{a_i}(U_1)$. So, $cov^{a_i}(\cdot)$ is monotone. Next, we prove the submodular property.

$$[cov^{a_i}(U_1 \cup u_j) - cov^{a_i}(U_1)] - [cov^{a_i}(U_2 \cup u_j) - cov^{a_i}(U_2)]$$
$$= cov^{a_i}(u_j) \prod_{u_l \in U_1} [1 - cov^{a_i}(u_l)](1 - \prod_{u_l \in U_2 - U_1} [1 - cov^{a_i}(u_l)])$$

Then we get $[cov^{a_i}(U_1 \cup u_j) - cov^{a_i}(U_1)] \geq [cov^{a_i}(U_2 \cup u_j) - cov^{a_i}(U_2)]$, thus $cov^{a_i}(\cdot)$ is submodular. Since $I(\cdot)$ is a linear combination consisted of $cov^{a_i}(\cdot)$, its weight coefficient is $w_{u_T}^{a_i} > 0$ decided by the target user u_T. The linear combination with nonnegative weights is monotone and submodular. Thus $I(\cdot)$ is monotone and submodular. ∎

Algorithm 1. FRPC-A algorithm

Input: LBSN $< G, C >$, the target user u_T, U_{Cre}, k, γ
Output: A list of users, U_{re}, with $\#U_{re} = k$
1: Initialize $U_{re} \Leftarrow \phi$
2: Compute $\sigma(u_i)$ for each $u_i \in U_{Cre}$;
3: Rank U_{Cre} in decreasing order of $sim(u_i, u_T)$;
4: **for** $j = 1$ **to** k **do**
5: $u_j \leftarrow \arg\max_{u_j}[\sigma(U_{re} \cup u_j) - \sigma(U_{re})]$;
6: $U_{re} \leftarrow U_{re} \cup u_j$
7: **return** U_{re}

Based on the two lemmas, we can prove the monotone submodularity of $\sigma(U)$ in Lemma 3.

Lemma 3. *The objective function $\sigma(U)$ is monotone and submodular.*

Proof. Since the function $\sigma(U)$ if a linear combination of $sim(U, u_t)$ and $I(U)$, and the two function are monotone and submodular. Meanwhile the weight coefficient $\gamma \in [0, 1]$, so $1 - \gamma \geq 0$. Based on the analysis of the above, the function $\sigma(U)$ is also monotone and submodular. ∎

Based on the above three lemmas, we get the following Theorem 1.

Theorem 1. *The friend recommendation considering preference coverage problem (FRPCP) in Eq. 5 is NP-hard.*

This paper give one greedy algorithm (called the friend recommendation algorithm considering preferences coverage, FRPC-A), and it is described by the Algorithm 1 in detail.

In Algorithm 1, the FRPC-A algorithm starts with one empty set (line 1), then puts the user with largest marginal score increase into the user set U_{re} until $\#U_{re} = k$ (line 5–6). Finally, our algorithm gets a near-optimal solution of the FRPC with a $(1 - \frac{1}{e})$ approximation [19] on the optimal score.

3 Experimental Evaluation

In this section, we firstly describe the setting of experiments including the real LBSN datasets (Foursquares, Gowalla datasets), comparative methods. Then this paper's experimental results reflect the recommendation quality of our approach.

3.1 Experimental Settings

Data Sets. LBSNs (such as Foursquare, Gowalla) are new social sites and social platforms where users can publish and participate in social events, and upload photos and share check-in locations which users may be interested in. This paper utilizes the real Foursquare and Gowalla datasets. The datasets respectively consist of 4,163 users, 36,907 users; 124,436 locations, 221,142 locations; 636 POI categories, 596 POI categories; 483,813 check-in times, 1048,575 check-in times; the time span: 12/07/2009–7/21/2013, 1/18/2010–8/11/2011; the friendship pairs contain indirect social connections among these users: 32,512 pairs, 231,148 pairs.

The Methods for Comparison. FRPC algorithm is compared with two typical friend recommendation methods: (a) Common Friend Recommendation (CFR) [17]; (b) Preference Similarity Recommendation (PSR) [18]. Wherein CFR recommends the target user with the users owning the same common friends as the target user, the basic idea of PSR is only based on the preference similarity between users.

3.2 Performance of Methods

Performance Metrics. The friend recommendation approaches are in order to recommend the target user with the list of users based on the σ value. This paper utilizes the Shannon entropy [14] as a measure method to the degree of the preference coverage through the friend recommendation.

$$div@k = -\frac{\sum_{a_i \in A_T} \frac{m(U_{re},a_i)}{k} \ln(\frac{m(U_{re},a_i)}{k})}{\ln k}, u_j \in U_{re} \tag{7}$$

where $m(U_{re}, a_i) = I(P(u_j, a_i) > P(u_T, a_i), \exists u_j \in U_{re})$ represents the number of the recommended users who mainly provide the information on the POI category a_i for the target user.

Effectiveness of Methods. In this section, we verify the effectiveness of our friend recommendation method considering preference coverage, and compare with the typical friend recommendation approaches in location-based social networks (LBSNs). We utilize the recall ratio: Recall@k and the precision ratio: Pre@k [15,16] to evaluate the friend recommendation quality, and it is important to find out how many users actually are the real friends of the target user through the recommendation approaches.

$$Pre@k = \frac{\#\{U_{re} \cap U_{u_T@friend}\}}{k} \tag{8}$$

$$Recall@k = \frac{\#\{U_{re} \cap U_{u_T@friend}\}}{\#U_{u_T@friend}} \tag{9}$$

where $U_{u_T@friend}$ are the friends of the target user u_T in LBSNs. The two metrics for the entire friend recommendation system are computed by averaging the above two metrics value for all the users respectively.

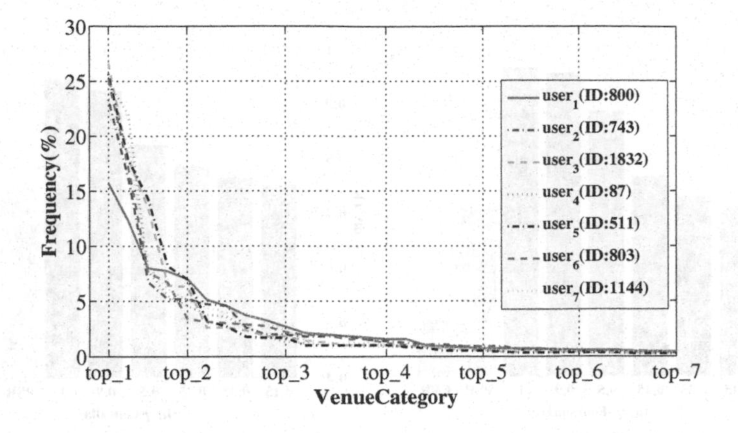

Fig. 3. The distribution of POI category preferences for 7 target users

Experimental Results. In this section, we utilize the *div@k*, *Pre@k* and *Recall@k* to compare the performance and effectiveness between FRPC-A algorithm and thc several existing friend recommendation approaches in LBSNs. We randomly select the number $N = 200$ of users in LBSNs as the target users U_{target}. As shown in Fig. 3, seven users are regarded as the target users in the friend recommend system, and their distribution of POI category preferences obey the power-law distribution. This paper utilizes the fitted distribution functions (the power-law distribution), and estimate parameters in the functions as the following Table 2. Next, we compare the recommendation quality and the degree of preferences coverage of our algorithm FRPC-A and the two typical friend recommendation methods PSR and CFR with the settings of $\gamma = 0.15, 0.35, 0.45, 0.5, 0.9, 1$ respectively in Foursquare and Gowalla. In Fig. 4, our method has the significant advantage in the degree of preference coverage compared with PSR and CFR. Since the CFR is based on topology information in social networks, it have no information on POI categories. Next, we importantly compare the methods (FRPC, PSR) about POI category preference in Pre@k and Recall@k. As shown in Figs. 5 and 6, although there was so little difference between FRPC-A algorithm and PSR algorithm in the precision ratio and recall ratio, the PSR and CFR hardly provide the more information to coverage the target's preferences. Users' demand on obtaining information is the essential purpose of meeting new friends [6]. Currently, most researches focus on modeling users' preferences in LBSN, but they ignore the angle of users obtaining information through the friend recommendation and the degree to the information from the recommended users covering the target users' preferences. Thus, this paper provides a novel recommendation method to solve the preference coverage problem.

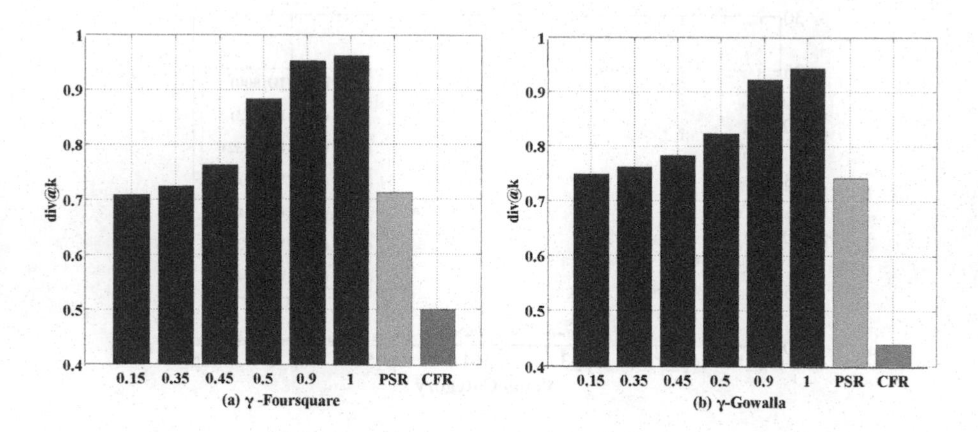

Fig. 4. Comparison of the degree to preference coverage of three methods

Table 2. The distribution of users' preferences $P(u, x) = \alpha x^{-\beta}$

User-ID	Coefficients (with 95% confidence bounds)
800	$\alpha = 18(16.69, 19.31), \beta = 0.855(0.7986, 0.9113)$
743	$\alpha = 24.77(23.6, 25.95), \beta = 1.04(0.9888, 1.092)$
1832	$\alpha = 29.04(27.27, 30.81), \beta = 1.083(1.011, 1.154)$
87	$\alpha = 27.84(25.59, 30.08), \beta = 1.077(0.9828, 1.17)$
511	$\alpha = 28.42(26.36, 30.49), \beta = 1.054(0.9723, 1.135)$
803	$\alpha = 26.68(25.42, 27.94), \beta = 1.078(1.023, 1.133)$
1144	$\alpha = 24.68(22.76, 26.61), \beta = 0.9992(0.92, 1.078)$

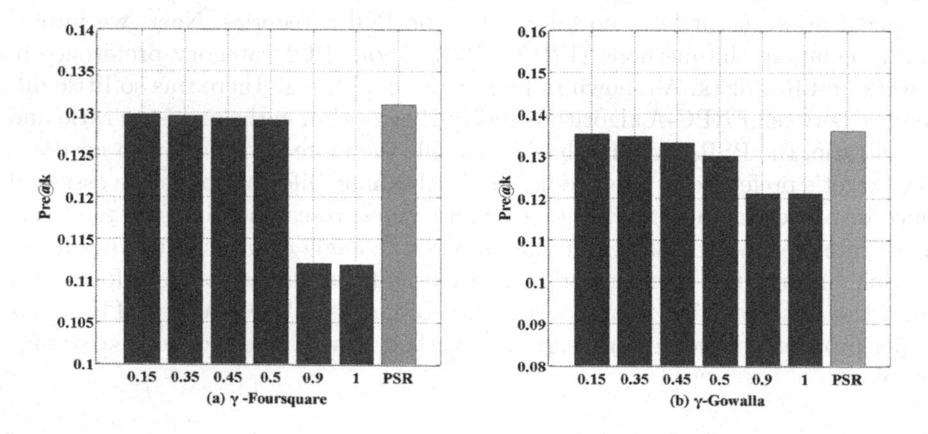

Fig. 5. Comparison of $Pre@k$ of three methods

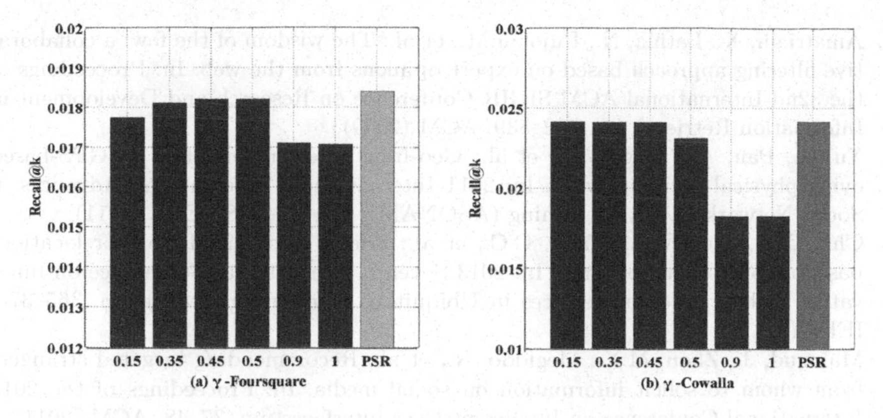

Fig. 6. Comparison of *Recall@k* of three methods

4 Conclusion

This paper pays attention to the friend recommendation considering preference coverage in location-based social networks. Firstly, we discuss the similarity between users' preference on POI categories, and describe the model of the preference coverage in detail. Based on the above models, this paper provide a novel friend recommendation problem as one optimization problem. Then we give a greedy algorithm **FRPC-A** to solve the recommendation issue. Our method recommends the target user with the users who can provide more information on u_T's POI category preferences, and avoid the situation that users own less information about POI categories in long tailing. Compared with the existing typical friend recommendation methods in LBSNs, the large scale of LBSNs dataset verifies the degree to the preference coverage through the friend recommendation and the effectiveness of our method. Especially, our approach shows the significant advantage in the degree of the preference coverage. As for future work, we plan to utilize multi-source information (such as the comment information, check-in time, and model the periodic of check-in behaviors for improving the effectiveness of the recommendation.

Acknowledgments. This work was supported in part by the National Science Foundation grants IIS-61370214, IIS-61300210.

References

1. Kim, S.: Friend recommendation with a target user in social networking services. In: 2015 31st IEEE International Conference on Data Engineering Workshops (ICDEW), pp. 235–239. IEEE (2015)
2. Bian, L., Holtzman, H.: Online friend recommendation through personality matching, collaborative filtering. In: Proceedings of UBICOMM, pp. 230–235 (2011)

3. Amatriain, X., Lathia, N., Pujol, J.M., et al.: The wisdom of the few: a collaborative filtering approach based on expert opinions from the web. In: Proceedings of the 32nd International ACM SIGIR Conference on Research and Development in Information Retrieval, pp. 532–539. ACM (2009)
4. Yu, X., Pan, A., Tang, L.A., et al.: Geo-friends recommendation in GPS-based cyber-physical social network. In: 2011 International Conference on Advances in Social Networks Analysis, Mining (ASONAM), pp. 361–368. IEEE (2011)
5. Chu, C.H., Wu, W.C., Wang, C.C., et al.: Friend recommendation for location-based mobile social networks. In: 2013 Seventh International Conference on Innovative Mobile, Internet Services in Ubiquitous Computing (IMIS), pp. 365–370. IEEE (2011)
6. Mahmud, J., Zhou, M.X., Megiddo, N., et al.: Recommending targeted strangers from whom to solicit information on social media. In: Proceedings of the 2013 International Conference on Intelligent User Interfaces, pp. 37–48. ACM (2011)
7. Cheng, Z., Caverlee, J., Barthwal, H.: Who is the barbecue king of Texas? a geo-spatial approach to finding local experts on Twitter. In: Proceedings of the 37th International ACM SIGIR Conference on Research and Development in Information Retrieval, pp. 335–344. ACM (2014)
8. Ma, G., Liu, Q., Chen, E., Xiang, B.: Individual influence maximization via link recommendation. In: Dong, X.L., Yu, X., Li, J., Sun, Y. (eds.) WAIM 2015. LNCS, vol. 9098, pp. 42–56. Springer, Cham (2015). doi:10.1007/978-3-319-21042-1_4
9. Wan, S., Lan, Y., Guo, J., et al.: Informational friend recommendation in social media. In: Proceedings of the 36th International ACM SIGIR Conference on Research and Development in Information Retrieval, pp. 1045–1048. ACM (2013)
10. Noulas, A., Scellato, S., Mascolo, C., et al.: An empirical study of geographic user activity patterns in foursquare. ICwSM **11**, 70–573 (2011)
11. Chen, X., Zeng, Y., Cong, G., et al.: On information coverage for location category based point-of-interest recommendation. In: AAAI, pp. 37–43 (2015)
12. Clauset, A., Shalizi, C.R., Newman, M.E.J.: Power-law distributions in empirical data. SIAM Rev. **51**(4), 661–703 (2009)
13. Liu, X., He, Q., Tian, Y.: Event-based social networks: linking the online and offline social worlds. In: Proceedings of the 18th ACM SIGKDD International Conference on Knowledge Discovery and Data Mining, pp. 1032–1040. ACM (2012)
14. Cover, T.M., Thomas, J.A.: Elements of Information Theory. Wiley, Hoboken (2012)
15. Cheng, C., Yang, H., King, I., et al.: Fused matrix factorization with geographical and social influence in location-based social networks. In: Aaai, vol. 12, pp. 17–23 (2012)
16. Ye, M., Yin, P., Lee, W.C., et al.: Exploiting geographical influence for collaborative point-of-interest recommendation. In: Proceedings of the 34th International ACM SIGIR Conference on Research, Development in Information Retrieval, pp. 325–334. ACM (2011)
17. Chen, J., Geyer, W., Dugan, C.: Make new friends, but keep the old: recommending people on social networking sites. In: Proceedings of the SIGCHI Conference on Human Factors in Computing Systems, pp. 201–210. ACM (2009)

18. Bao, J., Zheng, Y., Mokbel, M.F.: Location-based and preference-aware recommendation using sparse geo-social networking data. In: Proceedings of the 20th International Conference on Advances in Geographic Information Systems, pp. 199–208. ACM (2012)
19. Gharan, S.O., Vondrk, J.: Submodular maximization by simulated annealing. In: Proceedings of the Twenty-Second Annual ACM-SIAM Symposium on Discrete Algorithms, SIAM 2011, pp. 1098–1116 (2011)

Contrast Pattern Based Collaborative Behavior Recommendation for Life Improvement

Yan Chen[1](✉), Margot Lisa-Jing Yann[1], Heidar Davoudi[1], Joy Choi[1], Aijun An[1], and Zhen Mei[2]

[1] Department of Electrical Engineering and Computer Science, York University, Toronto, Canada
{ychen,lisayan,davoudi,aan}@cse.yorku.ca, ysjoychoi@gmail.com
[2] Manifold Data Mining Inc., Toronto, Canada
zhen@manifolddatamining.com

Abstract. Positive attitudes and happiness have major impacts on human health and in particular recovery from illness. While contributing factors leading human beings to positive emotional states are studied in psychology, the effects of these factors vary and change from one person to another. We propose a behaviour recommendation system that recommends the most effective behaviours leading users with a negative mental state (i.e. unhappiness) to a positive emotional state (i.e., happiness). By leveraging the contrast pattern mining framework, we extract the common contrasting behaviours between happy and unhappy users. These contrast patterns are aligned with user behaviours and habits. We find the *personalized* behaviour recommendation for those with negative emotional states by placing the problem into the nearest neighborhood collaborative filtering framework. A real dataset of people with heart disease or diabetes is used in our recommendation system. The experiments conducted show that the proposed method can be effective in the health-care domain.

1 Introduction

The pursuit of happiness can be characterized as a psychological factor and a life goal for human beings. The emerging field of sentiment analysis and opinion mining provides a means of computational analysis of emotion, affect, subjective experience and perception. These factors have a direct effect on human behaviours and attitudes. However, how the factors affect people psychologically is not apparent.

Personalized health-care can improve the patients' health experience and prognosis; early intervention can significantly reduce the health-care cost caused by related and predictable emergent conditions. Our goal is to provide recommendations of the most effective behaviours leading to positive psychological attitudes for high-risk patients with chronic diseases, in order to reduce health-care cost by reducing, e.g., the incidence of acute treatment related to mismanagement of disease conditions.

J. Kim et al. (Eds.): PAKDD 2017, Part II, LNAI 10235, pp. 106–118, 2017.
DOI: 10.1007/978-3-319-57529-2_9

The targeted behaviour recommender system faces several challenges. First, traditional recommender techniques work based on the notion of implicit/explicit rating for a set of items. These ratings are not available for a behaviour recommendation system, but rather each user is characterized as a set of attribute-values. Second, latent factor models can capture underlying reasons behind the user behaviour/preference even though it could be quite difficult for them to recommend behaviours that cannot be characterized by the latent factors. Third, recommendation systems are usually evaluated based on the standard measures such as precision@N and Mean Square Error (MSE). However, due to different problem settings and lack of the ratings we need an intuitive evaluation protocol for this type of recommendation systems.

Due to these challenges for personalized health-care, we need a novel methodology to provide effective recommendation, and to give a personalized evaluation system for improving health. Thus, our focus is on analyzing the high-risk patients with chronic disease related lifestyle and social conditions, as well as identifying the difference that exists between positive and negative attitudes. We use contrast pattern mining on a rich dataset that includes a population of patients with heart disease or diabetes, to identify group behavioural factors that reflect an individual's emotional state of unhappiness or happiness. With the contrast patterns, we generate recommendations based on the existing differences among the population. This information of contrast patterns describing the difference is used to build a behavioural recommendation system that provides recommendations for individuals with attitudes to make certain behaviour changes. In order to find the most relevant recommendations, the k-nearest neighbours (k-NN) algorithm is applied to identify the most effective behaviours for the user from the contrast patterns found.

The major goal of our proposed recommendation system is to discover and recommend the behaviours to improve the quality of life of users. As such, the problem of behaviour recommendation can be defined so as to provide users with recommendations based on the differences extracted between groups of people, in order to improve their lifestyle and life satisfaction.

The contributions of this paper are as follows:

- We define and formulate the problem of behaviour recommendation and design an effective solution for it.
- We apply contrast pattern mining to identify the transitional patterns as effective recommendations (i.e., behaviours), to suggest users to become members of a class of interest (i.e., happy people).
- A simple intuitive protocol based on standard evaluation methods is designed to assess the effectiveness of these types of recommenders.
- We conduct the experimental evaluation and show the effectiveness of proposed model in a health-care domain.

2 Related Work

Recommendation systems have been widely utilized in different domains to meet user interests and boost user satisfaction. For example, recently Abel proposed a recommendation system to help people find a job [1], or Backstrom and Leskovec [3] suggested a recommender system to find friends from social networks (i.e., Facebook). However, to date, most of recommendation systems have been applied in e-commerce and news domains [3,11,15]. Approaches for recommendation systems are usually divided in three broad categories: *collaborative filtering* [15], *content-based* [12] and *hybrid* [4] approaches. In collaborative filtering, we recommend items in which people with similar tastes and preferences are interested. Furthermore, the collaborative filtering techniques can be categorized into two general classes of *neighborhood* and *model-based* methods. In neighborhood-based (i.e., memory-based/heuristic-based) methods, user ratings for items stored in the system are directly used to generate the list of recommendations or predict the ratings for new items. Two major approaches in this framework are *user-based* collaborative filtering [15] whereas interest of a user for an item is estimated based on the rating for this item by other users (i.e., neighbors), and *item-based* approaches [11] which predict the rating of a user for an item based on the rating of the user for similar items. In contrast to neighborhood-based methods, model-based approaches exploit the users ratings to learn a predictive model. Bayesian Clustering [4], Latent Semantic Analysis [8], Latent Dirichlet Allocation [17], and Maximum Entropy are instances of this category. On the other hand, content-based recommender systems [9] recommend the items that are similar to the ones that she/he was interested in the past, and hybrid approaches refer to the class of algorithms that combine collaborative and content-based schemes to achieve better performance.

Another related area is contrast pattern mining. Contrast patterns are those that are significantly different among different classes, times, locations or/and other dimensions of interest. They have been utilized in different tasks and applications such as building the accurate and robust classifiers [14], detecting malware [16], or diagnosing disease [10]. The contrast patterns reflecting different frequencies in two datasets sometimes are refereed as *diverging patterns* [2], or *emerging patterns* [13]. For example, An et al. [2] consider a pattern as the diverging if its respective supports in two datasets and its diverging ratio (defined based on the distance between the four-dimensional vectors representing pattens) is more than certain thresholds. Ramamohanarao and Bailey [13] suggested different types of emerging patterns such as jumping emerging patterns (which exist in one dataset and are absent in another one), constrained emerging patterns (whose supports are more and less than specific thresholds in the first and second dataset accordingly). They argued while jumping emerging patterns can represent the sharp contrast between two datasets, they are susceptible to noise, so in many cases constrained emerging patterns would be the better choices. Webb et al. [18] proposed that contrast pattern mining can be seen as a special case of the general rule learning task where contrast patterns and groups for which they are characteristic are the antecedents and consequents of the rules respectively. This formulation allows any standard rule discovery algorithm to be adapted for the contrast pattern mining problem.

All aforementioned recommendation methods work based on the notion of implicit/explicit ratings of users for items. However, for the behavior recommendation problem such ratings are not available. Moreover, to best of our knowledge, there is no work on using contrast pattern mining for recommendation purpose. In contrast, in our problem setting, users are characterized based on set of attribute-values and belong to one of two disjoint classes (i.e., happy and sad people). The goal is to recommend a set of most effective transitional patterns (i.e., behaviors) which make a user likely to become a member of the class of interest (i.e., happy people).

3 Methodology

In this section, the dataset used in our system is first described. Then we present the overall framework of our recommendation system. The detailed steps for generating recommendations will be discussed in different subsections.

3.1 Dataset Description

The dataset comes from 2011/2012 Canadian Community Health Care Survey Data, which includes 16836 patients with diabetes and heart disease. The attributes (and their respective values) of patients in the dataset are captured with more than 100 survey questions. These questions are classified into seven categories, namely, geo-demographics, lifestyles, adherence, health-care experience, mental health, social connections and supports, and quality of life.

In the original dataset, the data is first discretized and transformed into transaction dataset with itemsets, where each item is an attribute-value pair. Furthermore, based on the characteristics of the attributes, we categorize the attributes into three different types:

- **static** attributes: cannot be changed, e.g. gender, age, or suffering from heart disease.
- **mutable** attributes: can be changed, e.g. alcohol use, volunteering activities, characteristic and habitual behaviors that signify mood or attitude.
- **swing** attributes: can or cannot be changed depending on willingness, ability to undertake cognitive behavioral change or other factors.

Not all of the attributes in this dataset have significant affect on the "happiness" of people. The attributes are filtered to remove the insignificant ones. Only 30 of the attributes are left in the dataset. Weka [7] is used here for this purpose, with the built-in "AttributeSelection" filter.

Table 1 shows some examples of attributes and values in this dataset.

3.2 Overall Framework

Figure 1 shows the overall framework of our recommendation system. Generally, our system includes the following steps for the process of generating recommendations:

Table 1. Attributes and values examples

Category	Attribute	Value
Life style	Daily consumption (fruits/vegetables)	5-10 Times/day
Life style	Smoke	Daily
Mental health	Satisfaction with life in General	Satisfied
Mental health	Perceived life stress	Not at all
Quality of life	Pain	No pain
Geo-demographics	Health Region	City of Toronto
Healthcare experience	No. of consultations with medical doctor	Not at all

a. Generate contrast patterns;
b. For each individual in dataset, find all the matching contrast patterns;
c. Find k-nearest neighbors for the current user;
d. Provide recommendation to the current user from its neighbors' matching contrast patterns;

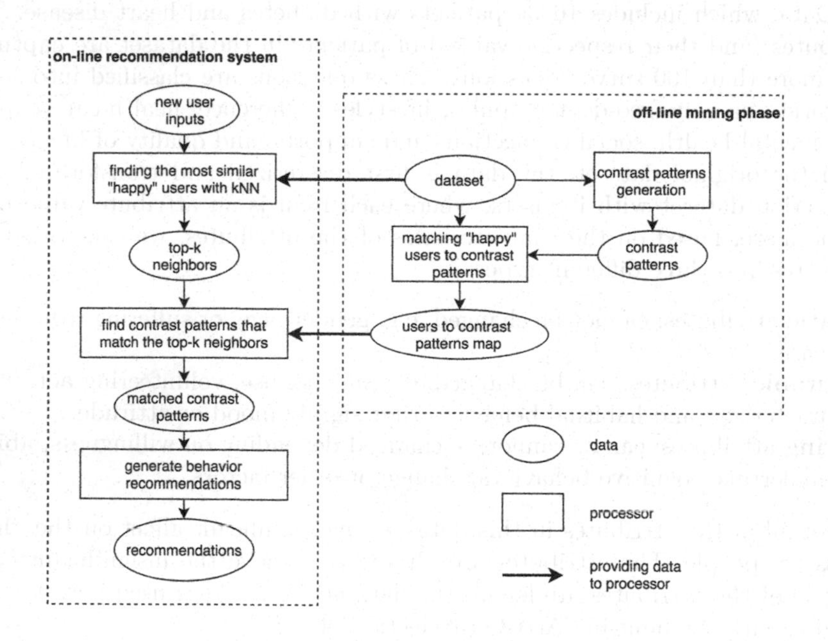

Fig. 1. Recommendation system flow chart

We have two main stages in our recommendation. The off-line stage includes contrast pattern mining and the process of matching users to contrast patterns. In order to distinguish the underlying differences between happy and unhappy, we use contrast pattern mining on the dataset to identify groups of behavior factors that may change people's feeling from unhappy to happy or vice versa. The contrast patterns extracted from the dataset can be applied to all the users in the dataset and need to be *personalized* for each individual user. We refer to these patterns as *global* contrast patterns in this paper. With such information, we utilize neighborhood-based collaborative filtering framework to customize recommendations for each user to adopt contrasting groups of behaviors. Last, in our on-line recommendation system, upon completion of a questionnaire by a user, user similarity assessment is performed using the k-nearest neighbors algorithm to identify people that are similar to the user. Subsequently, *global* contrast patterns (i.e., the contrast patterns found over the whole dataset) that match the identified similar people are used to generate personalized behavior recommendations that will have positive impacts on the user.

Pseudo-code of our behavior recommendation system is provided in Fig. 2.

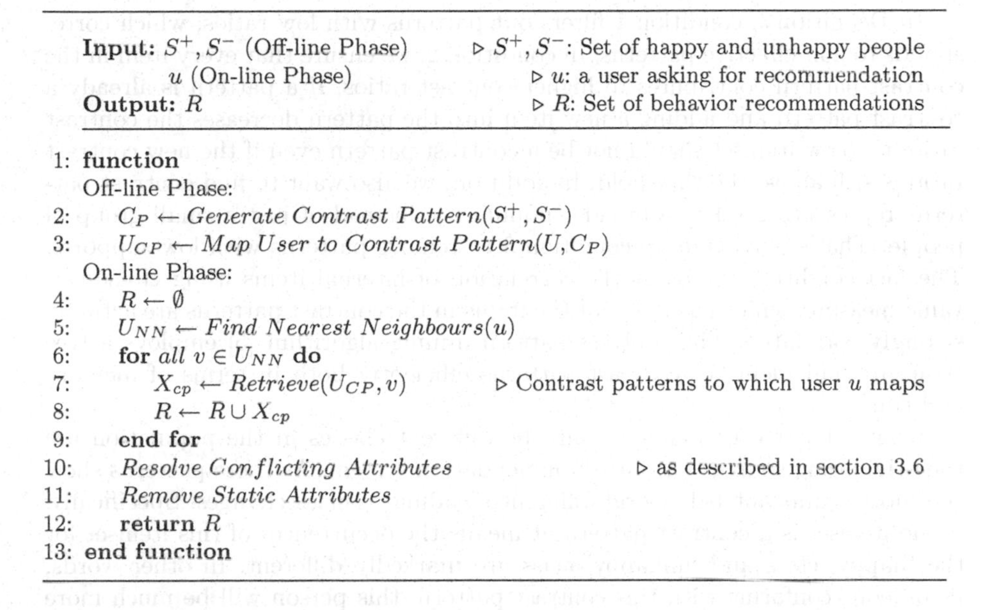

Input: S^+, S^- (Off-line Phase) ▷ S^+, S^-: Set of happy and unhappy people
u (On-line Phase) ▷ u: a user asking for recommendation
Output: R ▷ R: Set of behavior recommendations

1: **function**
Off-line Phase:
2: $C_P \leftarrow$ *Generate Contrast Pattern*(S^+, S^-)
3: $U_{CP} \leftarrow$ *Map User to Contrast Pattern*(U, C_P)
On-line Phase:
4: $R \leftarrow \emptyset$
5: $U_{NN} \leftarrow$ *Find Nearest Neighbours*(u)
6: **for** *all* $v \in U_{NN}$ **do**
7: $X_{cp} \leftarrow$ *Retrieve*(U_{CP}, v) ▷ Contrast patterns to which user u maps
8: $R \leftarrow R \cup X_{cp}$
9: **end for**
10: *Resolve Conflicting Attributes* ▷ as described in section 3.6
11: *Remove Static Attributes*
12: **return** R
13: **end function**

Fig. 2. Pseudo-code of our behavior recommendation system.

3.3 Contrast Patterns Generation

Before we describe the process of generating contrast patterns, some definitions will be given first.

Let $I^* = \{I_1, I_2, ..., I_m\}$ be a set of items. An itemset X is a set of items $\{I_{e_1}, I_{e_2}, ..., I_{e_Z}\}$, where Z is the length of X, denoted by $|X|$. A dataset D is a list of transactions $\{T_1, T_2, ..., T_n\}$, where each transaction $T_d \in D$ is an itemset.

Definition 1 (Support of itemset X in dataset D). The support of itemset X in D is defined as the fraction of itemsets in D, which contain itemset X.

Definition 2 (Contrast ratio of itemset X in datasets D_1 and D_2). The contrast ratio of itemset X in D_1 and D_2 is defined as $\frac{support\ of\ X\ in\ D_2}{support\ of\ X\ in\ D_1}$.

Please note that the order of the datasets in the Definition 2 affects the contrast ratio. Inverting the order will also invert the ratio value.

In our recommendation system, we use the definition of contrast pattern as well as the mining algorithm for contrast pattern mining from Fan and Ramamohanarao [5].

Definition 3 (Contrast pattern in datasets $D_1$and D_2). An itemset X is a contrast pattern in datasets D_1 and D_2, if and only if

1. *Contrast Ratio of* $X \geq threshold_1$ *(denoted as* θ *in Sect.* 4),
2. *Contrast Ratio of* $X \geq$ *Contrast Ratio of* $Y \quad \forall\, Y \subseteq X$,
3. *Support of* X *in* $D_1 \geq threshold_2$ *&* *Support of* X *in* $D_2 \geq threshold_2$,
4. $\chi^2 \geq threshold_3$.

In Definition 3, condition 1 filters out patterns with low ratios, which corresponds to non-effective patterns. In condition 2, we ensure that every item in the contrast pattern contributes to higher contrast ratios. If a pattern is already a contrast pattern and adding a new item into the pattern decreases the contrast ratio, the new itemset should not be a contrast pattern even if the new contrast ratio is still above the threshold. In addition, we also want to find contrast patterns representing relative broader popularity, instead of just a small group of people. That's why we have condition 3 to remove patterns with low supports. The last condition evaluates the correlation of internal items using chi-square value measure, which ensures that the items in the contrast patterns are actually strongly correlated. The contrast pattern mining algorithm [5] employs a tree structure and identifies contrast patterns efficiently both in terms of memory and time.

Contrast patterns found among the different classes in the population are the originating sources for our recommendation system. As these patterns show the most significant behavioral difference leading to class changes. Specifically, if one itemset is a contrast pattern, it means the occurrences of this itemset for the 'happy' class and 'unhappy' class are markedly different. In other words, if someone conforms with this contrast pattern, this person will be much more likely to belong to one class than the other. For example, if percentages of people who 'smoke' and 'consume little fruit in their diet' appears in the positive class and negative class are 1% and 12% respectively, we can conclude that, people who 'smoke' and 'consume little fruit' are 12 times more possible to express a negative attitude. This "smoke and consume little fruit" is a contrast pattern and its contrast ratio is 12. Given these contrast patterns, each contrast pattern is converted into a set of recommendations. For example, given the the contrast pattern of "smoke and consume little fruit", one individual with negative 'unhappy' class label is given advice to avoid "smoke and consume little fruit" as an alternate lifestyle choice. In general, numerous recommendations are generated based on the contrast patterns found among the population. Specifically, over 4,000 contrast itemsets are generated from our dataset, most of which have more than two items in each pattern.

3.4 Matching "Happy" Users to Contrast Patterns

Given the contrast patterns for the positive and negative class, we need to map users to the contrast patterns. As such, each user in the training dataset is compared to all contrast patterns. The matching process is fairly simple. A user is matched to a contrast pattern if she/he has all attribute-values of the contrast pattern. For example, an individual maps to the contrast pattern of 'smoke and consume little fruit' if he/she 'smokes' and 'consumes little fruit'.

3.5 Finding the Most Similar "Happy" Users with kNN

The *global* contrast patterns show the significant differences between two population of users even though they are not personalized for individual users and, consequently cannot be used directly for the recommendation purpose. As such, we adopt the neighborhood-based collaborative filtering approach to find the set of personalized recommendation candidates. In particular, with a new user's inputs, k 'happy' users who are most similar to her/him are identified. We use Pearson's correlation coefficient to calculate the similarities between two individuals as it can handle missing values and grade-inflation well [6] (there are a lot of missing values in the dataset we use since users usually do not answer all questions).

3.6 Behavioral Recommendations Generation

Given the k nearest neighbors of a new user, the set of contrast patterns to which each neighbor user maps is obtained using the results from Sect. 3.4. Next, the union of all contrast patterns of the k neighbors is considered as the initial set of personalized recommendation candidates. It is possible that we have some conflicting attribute-values (e.g., job type attributes is set to both part time and full time), in that case, we keep the the attribute-value with the higher contrast ratio rate. In case that the ratio rates are the same, the conflicting attribute-values are chosen randomly. Furthermore, we remove the static attributes (as they are not appropriate for behavior recommendation) and the existing attributes (those that the user already has) from the set of recommendation candidates and generate the final set of recommendations. The rationale is that while there are some behaviors which are prevalent among happy users, certain behaviors are common among specific group of happy users. As such, The more similar user U_1 is to happy user U_2, the more likely that user U_2 behaviors can be applied to user U_1 to achieve happiness. For example, if user U_1 is similar to user U_2 and U_3, and user U_2 and U_3 are mapped to set of contrast patterns S_2 and S_3 respectively, the initial set of recommendation candidates is $S_1 \cup S_2$.

It is worth to mention that if an attribute is a swing attribute, we provide an extra question in our recommendation system asking whether the user is able/willing to make changes on this attribute. The attribute is then categorized into static or mutable according to users' answers.

4 Experimental Results and Analysis

In this section, our recommendation system will be evaluated. The evaluation focuses mainly on the effectiveness of our off-line training approach. The implementation details of the on-line recommendation system are also mentioned at the end of this section.

4.1 Evaluation Design and Protocol

Recommendation systems traditionally are evaluated based on the classification performance measurement (e.g., Precision @ 10), rank-based performance measurement, or rate-based performance measurement (e.g., MSE), depending on the targeted task (e.g., predicting the top recommendation item or rates). However, for the behavior recommendation task such ranks and the ground truth are not available. In fact, we do not know whether the recommendations of particular behaviors will make users happy or not in the real life. As such, one major challenge in the proposed behavior recommendation system is how to evaluate it.

In order to address the evaluation problem, and measure the effectiveness of our recommendation system, we develop a classification system to calculate the possibility of being 'happy' for a user before and after applying the recommendations. If the possibility of being happy increases after applying the recommendations, it means the recommendations are effective. We compare different classification techniques and choose an ensemble method as user class (i.e., happy or unhappy) predictor. The classification algorithm is ensemble of AdaBoost, Random Forest, J48, Bayesian Network, and Logistic Regression. The final results are based on the voting method (the same weight for all the algorithms are used). For this part we use the standard implementation of these algorithms Weka [7] with their respective default parameters. Using 10-fold cross validation, the overall accuracy of the ensemble method reaches 71.8%.

4.2 Performance Evaluation

To effectively evaluate our recommendation system, we designed two other approaches as baseline, called *RANDKN* and *TOPCP*, for comparison purposes. The *RANDKN* approach does not use k-NN to find the most similar users to the new user. Instead, it tries to find the same number of contrast patterns as in k-NN randomly. For example, for a new user, in our recommendation system, if 20 contrast patterns are generated and applied to the user, the system also choose 20 contrast patterns in *RANDKN*, but randomly. *TOPCP* uses the same strategy as *RANDKN*, but instead of choosing contrast patterns randomly, *TOPCP* ranks the contrast patterns by their ratios and chooses the same number of contrast patterns as in our recommendation system with highest ratios.

In the evaluation, we first use leave-one-out to generate recommendations for each user in the dataset. Then the recommendations are applied to the user.

Table 2. Effectiveness of recommendation systems (A: percentage of people having probability of happy improved; B: percentage of unhappy instances having probability of happy improved).

Algorithm	A	B
Our method	66.03%	95.82%
RANDKN	60.82%	91.94%
TOPCP	58.95%	89.69%

Thereafter, a new dataset containing all the itemsets after applying the recommendations is generated. The ensemble method described previously is used to evaluate the classification of each user as type '1' (happy) or type '2' (unhappy) on the new dataset. The possibility of being classified as type '1' (happy) for each date sample is computed. The results are shown in Table 2.

As shown in Table 2, we can see that 95.82% of the unhappy users may become happier after applying the recommended changes using our method. Comparing the *RANDKN* (random selected k neighbors) and *TOPCP* (selected 20 highest-ratio neighbors), our method performs better than both of them. The fact that *TOPCP* performs worse than *RANDKN* is because the contrast patterns with the highest ratios can have significant amount of overlaps, which decreases the number of item choices in the recommendations. This further justifies that our approach of using k-NN to identify most similar neighbors in order to further obtain related contrast pattern items.

In the experiments, we choose $k = 7$ and $\theta = 4.0$. Note that we do not set the value of θ too low, since lower θ values lead to less effective patterns. θ can also not be set too high; otherwise, not enough patterns are given as recommendation to users. Thus, $\theta = 4.0$ is chosen so that the patterns are enough for recommendations and effective in the same time. The k value for k-NN method is also carefully selected. If a higher k is chosen, the user is recommended with more patterns. Too many patterns are not practical for users to take on all the recommendations. But too few also does not provide enough information for the user to obtain possible recommendations to change to be happier. We run experiments on using different k values, and compared the number of distinct contrast patterns, the number of item changes, percentage of people having probability of happy improved and percentage of unhappy instances having probability of happy improved. The results are in Fig. 3.

4.3 Implementation

The initial version of implemented system uses Django as web framework and MySQL as data storage layer. However, it may take up to several minutes to generate recommendations for a single user on a server with a Intel(R) Xeon(R) CPU E5-2620 v3 CPU and 64 GB of RAM. The reason is that the system needs to scan the complete dataset for every user to find the k nearest neighbors. Also, the

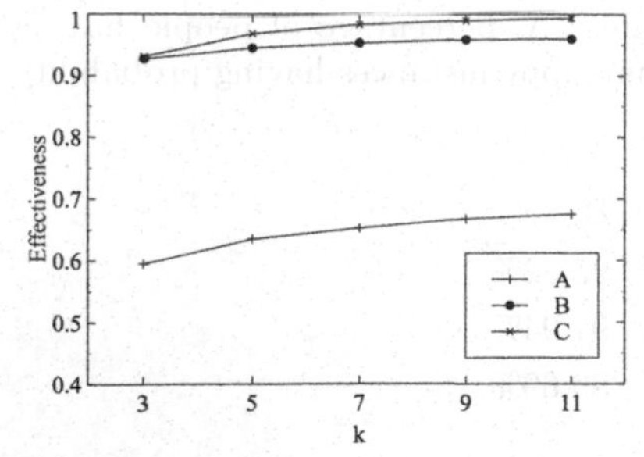

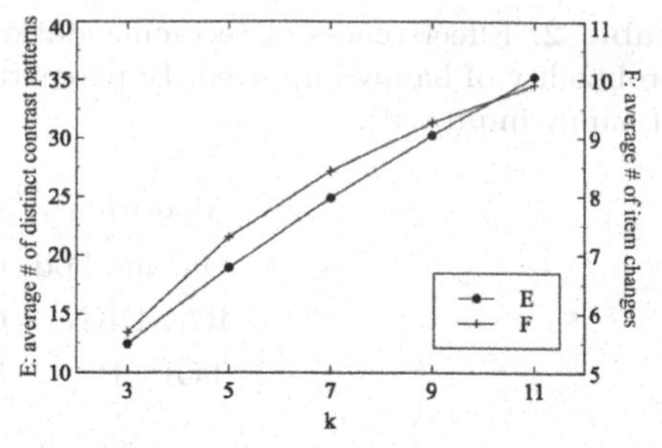

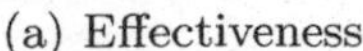

(a) Effectiveness
A: percentage of people having probability of happy improved;
B: percentage of unhappy instances having probability of happy improved;
C: percentage of unhappy instances having probability of happy over 50%

(b) Recommendation size
E: average # of distinct contrast patterns;
F: average # of item changes

Fig. 3. Performance of recommendation system with different k values

program design only uses a single thread, which means that only a small fraction of the computation power of the CPU is used. To solve the performance issue mentioned above, we refactor the implementation to use all cores of the CPU on the server. Furthermore, we design to enable our recommendation system to be able to scale across different machines in order to allow large numbers of users to use this website in the same time. In the refactored website, we use Play framework and Akka to distribute the computation across a cluster of servers. Redis has been used for the data storage to allow us to retrieve all the data from memory instead of disk to reduce the processing time. After applying all changes, the system takes approximately 1–5 s for each user to obtain its recommendations online.

5 Conclusions

The impact of positive attitudes is an acknowledged factor in people's health. However, the contributing factors to happiness depends on characteristics which are unique and subjective. In this paper, we proposed a personalized behaviour recommendation that recommends the most effective behaviours for changing users emotional state from negative (i.e., unhappy) to positive (i.e., happy). We showed that contrast pattern mining served effectively as the transitional patterns from the negative to positive class. The contrast pattern mining framework is adopted and combined with the collaborative filtering to produce the personalized behaviour recommendation. The experiments on an actual dataset showed that our proposed method performed well and was effective in the health-care domain.

Acknowledgement. This work is funded by the Big Data Research, Analytics, and Information Network (BRAIN) Alliance (established by the Ontario Research Fund - Research Excellence Program), Manifold Data Mining Inc., and Natural Sciences and Engineering Research Council of Canada (NSERC). We would like to thank Manifold for providing the dataset used in this research. In particular, we thank Ted Hains of Manifold and Jianhong Wu of York University for their insights and collaboration in our joint project.

References

1. Abel, F.: We know where you should work next summer: job recommendations. In: Proceedings of the 9th ACM Conference on Recommender Systems, pp. 230–230. ACM (2015)
2. An, A., Wan, Q., Zhao, J., Huang, X.: Diverging patterns: discovering significant frequency change dissimilarities in large databases. In: Proceedings of the 18th ACM Conference on Information and Knowledge Management, pp. 1473–1476. ACM (2009)
3. Backstrom, L., Leskovec, J.: Supervised random walks: predicting and recommending links in social networks. In: Proceedings of the Fourth ACM International Conference on Web Search and Data Mining, pp. 635–644. ACM (2011)
4. Burke, R.: Hybrid recommender systems: survey and experiments. User Model. User-Adap. Inter. **12**(4), 331–370 (2002)
5. Fan, H., Ramamohanarao, K.: Efficiently mining interesting emerging patterns. In: Dong, G., Tang, C., Wang, W. (eds.) WAIM 2003. LNCS, vol. 2762, pp. 189–201. Springer, Heidelberg (2003). doi:10.1007/978-3-540-45160-0_19
6. Garren, S.T.: Maximum likelihood estimation of the correlation coefficient in a bivariate normal model with missing data. Stat. Probab. Lett. **38**(3), 281–288 (1998)
7. Hall, M., Frank, E., Holmes, G., Pfahringer, B., Reutemann, P., Witten, I.H.: The WEKA data mining software: an update. SIGKDD Explor. Newsl. **11**(1), 10–18 (2009)
8. Karacapilidis, N., Hatzieleftheriou, L.: A hybrid framework for similarity-based recommendations. Int. J. Bus. Intell. Data Min. **1**(1), 107–121 (2005)
9. Kim, W., Kerschberg, L., Scime, A.: Learning for automatic personalization in a semantic taxonomy-based meta-search agent. Electron. Commer. Res. Appl. **1**(2), 150–173 (2002)
10. Li, J., Yang, Q.: Strong compound-risk factors: efficient discovery through emerging patterns and contrast sets. IEEE Trans. Inf. Technol. Biomed. **11**(5), 544–552 (2007)
11. Linden, G., Smith, B., York, J.: Amazon.com recommendations: item-to-item collaborative filtering. IEEE Internet Comput. **7**(1), 76–80 (2003)
12. Pazzani, M.J., Billsus, D.: Content-based recommendation systems. In: Brusilovsky, P., Kobsa, A., Nejdl, W. (eds.) The Adaptive Web. LNCS, vol. 4321, pp. 325–341. Springer, Heidelberg (2007). doi:10.1007/978-3-540-72079-9_10
13. Ramamohanarao, K., Bailey, J.: Emerging patterns: mining and applications. In: Proceedings of International Conference on Intelligent Sensing and Information Processing, 2004, pp. 409–414. IEEE (2004)
14. Ramamohanarao, K., Bailey, J., Fan, H.: Efficient mining of contrast patterns and their applications to classification. In: 3rd International Conference on Intelligent Sensing and Information Processing, pp. 39–47. IEEE (2005)

15. Ricci, F., Rokach, L., Shapira, B.: Introduction to Recommender Systems Handbook. Springer, Heidelberg (2011)
16. Sun, X., Huang, Q., Zhu, Y., Guo, N.: Mining distinguishing patterns based on malware traces. In: 3rd IEEE International Conference on Computer Science and Information Technology (ICCSIT), vol. 2, pp. 677–681. IEEE (2010)
17. Trewin, S.: Knowledge-based recommender systems. Encycl. Libr. Inf. Sci. **69**(Suppl. 32), 180 (2000)
18. Webb, G.I., Butler, S., Newlands, D.: On detecting differences between groups. In: Proceedings of the Ninth ACM SIGKDD International Conference on Knowledge Discovery and Data Mining, pp. 256–265. ACM (2003)

Exploiting Location Significance and User Authority for Point-of-Interest Recommendation

Yonghong Yu[1,2](✉), Hao Wang[1], Shuanzhu Sun[3], and Yang Gao[1]

[1] State Key Lab for Novel Software Technology, Nanjing University, Nanjing, People's Republic of China
yuyh.nju@gmail.com, {wanghao,gaoy}@nju.edu.cn
[2] College of Tongda, Nanjing University of Posts and Telecommunications, Nanjing, People's Republic of China
[3] Jiangsu Frontier Electric Technology Co. Ltd., Nanjing, People's Republic of China

Abstract. With the rapid growth of location-based social networks (LBSNs), point-of-interest (POI) recommendation has become indispensable. Several approaches have been proposed to support personalized POI recommendation in LBSNs. However, most of the existing matrix factorization based methods treat users' check-in frequencies as ratings in traditional recommender systems and model users' check-in behaviors using the Gaussian distribution, which is unsuitable for modeling the heavily skewed frequency data. In addition, little methods systematically consider the effects of location significance and user authority on users' final check-in decision processes. In this paper, we integrate probabilistic factor model and location significance to model users' check-in behaviors, and propose a location significance and user authority enhanced probabilistic factor model. Specifically, a hybrid model of HITS and PageRank is adapted to compute user authority and location significance. Moreover, user authorities are used to weight users' implicit feedback. Experimental results on two real world data sets show that our proposed approach outperforms the state-of-the-art POI recommendation algorithms.

Keywords: Point-of-interest recommendation · Probabilistic factor model · Location significance · User authority

1 Introduction

Recently, location-based social networks (LBSNs) have become very popular and attract lots of attention from industry and academia. Typical location-based social networks include Foursquare, Gowalla, Facebook Place, and GeoLife, etc. In LBSNs, users can build connections with their friends, upload photos, and share their locations via check-ins at points of interest (e.g., restaurants, tourists spots, and stores, etc.). Besides providing users with social interaction platforms, it is more desirable for LBSNs to make use of the check-in history of users and other side information to mine users' preferences and recommend interesting places which users would prefer. The task of recommending interesting places

J. Kim et al. (Eds.): PAKDD 2017, Part II, LNAI 10235, pp. 119–130, 2017.
DOI: 10.1007/978-3-319-57529-2_10

is referred as point-of-interest (POI) recommendation. POI recommender systems [1] have played an important role in LBSNs since they not only meet users' personalized demands, but also help LBSNs providers to increase revenues by providing users with intelligent location services, such as location-aware advertisements.

Intuitively, by treating POIs as "items" (e.g., movie, music, book and so on) in traditional recommender systems, a direct idea of generating POI recommendations is to employ classic collaborative filtering (CF) methods, which are widely used for building recommender systems. Based on this intuition, several POI recommendation algorithms are proposed by extending traditional CF methods [2–5]. However, the task of POI recommendation is not completely equivalent to traditional recommendation tasks. Several unique characteristics distinguish POI recommendation from traditional item recommendation tasks. For example, the preferences of users in LBSNs are reflected by the frequencies of check-in at locations and users' check-in frequencies are heavily skewed; users always prefer to visit nearby locations rather than distant ones because users' check-in activities require physical interactions between users and POIs [3,6]; social influence has limited effects on users' check-in behaviors [3,7], etc.

Several POI recommendation algorithms have been proposed by extending traditional recommendation approaches with some of the above unique characteristics [2,3,8–10]. For instance, Ye et al. [2] proposed a variant of user-based CF [11] for POI recommendation, which exploits social influence, geographical influence as well as user preferences. Cheng et al. [3] fused geographical influence, social influence and matrix factorization for POI recommendation. Differing from [2,3], which model geographical influence from a user's perspective, Liu et al. [8] proposed IRenMF, which exploits geographical influence from a location's perspective. Lian et al. [9] integrated the spatial clustering property into weighted matrix factorization [12] to make POI recommendations. Li et al. [10] proposed a two-step framework for POI recommendation problem, which considers three types of friends, i.e., social friends, location friends, and neighboring friends.

However, most of existing methods simplify users' check-in frequencies at a location, i.e., regardless how many times a user checks-in a location, they use binary values to indicate whether a user has visited a POI [2,4,8–10]. In other words, the user-POI check-in frequency matrix is substituted with a 0/1 rating matrix. Intuitively, a user's check-in frequency reflects the degree of the user's preferences for POIs. The larger the number of check-ins, the more preferred. Hence, the simplified scheme can not accurately capture users' preferences for POIs. Moreover, most of existing matrix factorization based POI recommendation algorithms [3,8–10,13] treat users' check-in frequencies as ratings in traditional recommender systems and utilize the Gaussian distribution to model users' check-in behaviors. In fact, though Gaussian distribution might be suitable for modeling users' rating behaviors, it is not a good choice for modeling the check-in frequency data, which are heavily skewed. We randomly select users from

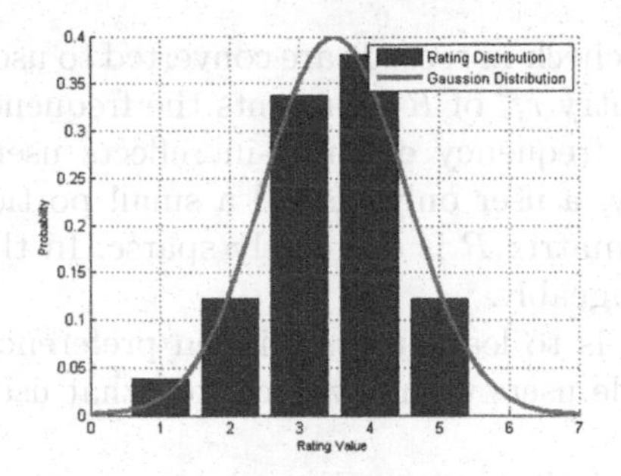

(a) The rating distribution of a randomly selected user in MovieLens100K

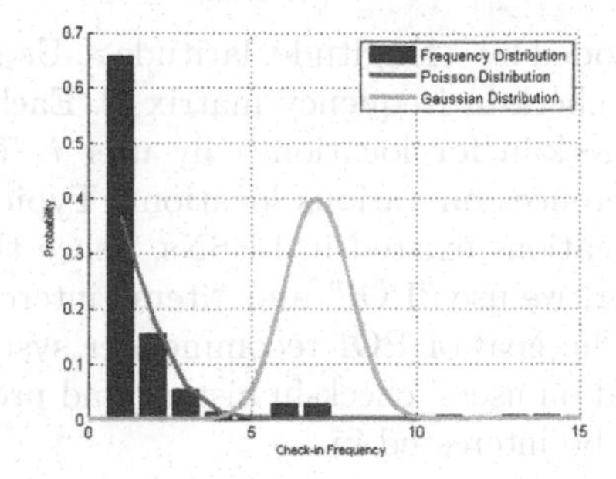

(b) The check-in frequency distribution of a randomly selected user in Foursquare

Fig. 1. The rating and check-in frequency distribution in MovieLens100K and Foursquare

MovieLens100K[1] and Foursquare[2], then plot their rating and check-in frequency distributions in Fig. 1. As shown in Fig. 1, the Gaussian distribution fits ratings well in MovieLens100K, but fails to model the check-in frequency. Moreover, few studies consider the effects of location significance on the final check-in decision process, which is usually complex and can be affected by many factors. Obviously, besides user individual preferences, the significance of a location also affects whether a user will visit the location. The check-in frequency for a location should vary with the significance of the location. Further, as reported in [14], users are more likely to adopt recommendations for other users with high authorities, the reviews and feedback derived from high authority users are more convincing. However, a few works take into account user authority for POI recommendation.

In this paper, we propose a location significance and user authority enhanced probabilistic factor model to overcome the above problems. Specifically, a hybrid model of HITS [15] and PageRank [16] is adapted to compute user authority and location significance by exploiting the mutual reinforcement between users' travel experience and locations' significance as well as the mutual influence between locations. Then we integrate probabilistic factor model and location significance to model users' check-in behaviors and deal with the skewed check-in frequency data. Furthermore, user authorities are used to weight users' feedback in the objective function. Experimental results on two real world data sets show that our proposed approach can model users' check-in behaviors better, and outperforms the state-of-the-art POI recommendation algorithms.

2 Preliminary Knowledge

2.1 Problem Definition

In a typical LBSNs, the POI recommender system consists a set of M users $\mathcal{U} = \{u_1, u_2, \ldots, u_M\}$, and a set of N locations $\mathcal{L} = \{l_l, l_2, \ldots, l_N\}$. Each location is

[1] http://grouplens.org/datasets/movielens/.
[2] http://www.ntu.edu.sg/home/gaocong/datacode.htm.

geocoded by <longitude, latitude>. Users' check-in records are converted to user-POI check-in frequency matrix R. Each entry r_{ij} of R represents the frequency of check-in for location j by user i. The frequency of check-in reflects users' preferences on various locations. Typically, a user only visited a small portion of locations existed in LBSNs, hence the matrix R is extremely sparse. In this paper, we use "POI" and "item" interchangeably.

The goal of POI recommender system is to learn users' hidden preferences based on users' check-in history and provide users with new locations that users may be interested in.

2.2 Poisson Factor Model

Poisson factor model (PFM) [17] is a generative probabilistic model, which assumes that each observed element r_{ij} follows the Poisson distribution with expectation f_{ij}: $r_{ij} \sim Poisson(f_{ij})$. The expected value matrix $F \in \mathbb{R}^{M \times N}$ is factorized into user latent feature matrix $U \in \mathbb{R}^{K \times M}$ and item latent feature matrix $V \in \mathbb{R}^{K \times N}$: $F \sim U^T V$. Besides assuming the Poisson distribution generates observed elements, PFM places Gamma priors over u_{ik} and v_{jk},

$$p(u_{ik}|\alpha_k, \beta_k) = \frac{u_{ik}^{\alpha_k - 1} exp(-u_{ik}/\beta_k)}{\beta_k^{\alpha_k} \Gamma(\alpha_k)}, \qquad p(v_{jk}|\alpha_k, \beta_k) = \frac{v_{jk}^{\alpha_k - 1} exp(-v_{jk}/\beta_k)}{\beta_k^{\alpha_k} \Gamma(\alpha_k)}, \tag{1}$$

where $\Gamma(.)$ is the Gamma function. α_k and β_k are the shape and rate parameters of Gamma distribution, respectively.

The objective function of PFM is formulated as:

$$\begin{aligned} \ell' = \min_{U,V} & \sum_{i=1}^{M} \sum_{k=1}^{K} (u_{ik}/\beta_k - (\alpha_k - 1) \ln(u_{ik}/\beta_k)) \\ & + \sum_{j=1}^{N} \sum_{k=1}^{K} (v_{jk}/\beta_k - (\alpha_k - 1) \ln(v_{jk}/\beta_k)) + \sum_{i=1}^{M} \sum_{j=1}^{N} (f_{ij} - r_{ij} \ln f_{ij}) + const. \end{aligned} \tag{2}$$

PFM applies the stochastic gradient descent algorithm (SGD) technique to learn user latent feature matrix U and item latent feature matrix V.

3 Our Proposed Method

3.1 Location Significance and User Authority Enhanced PFM

To model users' overall check-in behaviors, we assume that the following two factors affect the final check-in decision processes of user i: (1) the individual preferences of user i over location j; (2) the significance of location j. The individual preferences are usually captured by the inner product of user latent feature vector and item latent feature vector, which are inferred from latent factor models. Typical latent factor models include probabilistic matrix factorization model [18] and Poisson factor model [17]. Probabilistic matrix factorization is a classical Gaussian matrix factorization model. The classical Gaussian matrix factorization assumes that users' check-ins frequency data follows the Gaussian distribution, which does not fit the heavily skewed frequency data well. Alternatively, the Poisson distribution is suitable for fitting the skewed frequency data, as shown in Fig. 1(b), and guarantees that the number of check-ins is positive. In order to more accurately fit user check-in frequency data, we choose Poisson factor model to model users' check-in behaviors.

Given user latent feature vector U_i and item latent feature vector V_j, which are both learned by factorizing the expected value matrix F, the individual preference of user i for location j is defined as $U_i^T V_j$. Besides user individual preferences, the significance of a location also affects whether a user will visit this location. Intuitively, users tend to visit some popular locations and ignore those locations with less attractiveness. The more significant the location, the larger the number of check-ins at this location. Let $p(j)$ be the significance of location j, the probability of visiting location j by user i will monotonically increase as $p(j)$ increases.

By using a linear combination model to integrate user individual preferences with location significance, the expected value of check-in frequency f_{ij} at location j by user i is computed as follows:

$$f_{ij} = U_i^T V_j \times p(j). \tag{3}$$

With the expected value f_{ij}, the user-POI check-in frequency r_{ij} is modeled as:

$$r_{ij} \sim Poisson(r_{ij} | U_i^T V_j . p(j)). \tag{4}$$

Since we assume each observed element of R is independent, the conditional distribution of the user-POI check-in frequency matrix R given expected value matrix F can be defined as follows:

$$p(R|F) = \prod_{i=1}^{M} \prod_{j=1}^{N} [Poisson(r_{ij} | U_i^T V_j . p(j))]^{I_{ij}} \tag{5}$$

where I_{ij} is the indicator function that is equal to 1 if user i has visited location j and equal to 0 otherwise.

The log-posterior distribution $p(U, V | R, \alpha, \beta)$ is defined as:

$$\begin{aligned} &log(p(U, V|R, \alpha, \beta)) \propto log(p(R|F)p(U|\alpha, \beta)p(V|\alpha, \beta)) \\ &= \sum_{i=1}^{M}\sum_{k=1}^{K}((\alpha_k - 1)\ln(u_{ik}/\beta_k) - u_{ik}/\beta_k) + \sum_{j=1}^{N}\sum_{k=1}^{K}((\alpha_k - 1)\ln(v_{jk}/\beta_k) - v_{jk}/\beta_k) \\ &+ \sum_{i=1}^{M}\sum_{j=1}^{N} I_{ij}\left(r_{ij}\ln(p(j).U_i^T V_j) - p(j).U_i^T V_j\right) \end{aligned} \tag{6}$$

So far, we have describe how to extend basic Poisson factor model with location significance. In our proposed method, we also assume that user authority plays an important role in POI recommendation. As reported in [14], users are more likely to adopt recommendations for other users with high authorities, the reviews and feedback derived from high authority users are more convincing. Hence, users with different authority should have different weights in the third term of Eq. 6. Specifically, a high user authority will require the error between $r_{ij} \ln(p(j).U_i^T V_j)$ and $U_i^T V_j.p(j)$ as far as possible small, while a low user authority will relax this requirement. Hence, we further extend our proposed model by using user's authority to weight the corresponding response r_{ij}. The third term of Eq. 6 is changed to:

$$\sum_{i=1}^{M}\sum_{j=1}^{N} I_{ij} a(i) \left(r_{ij} \ln(p(j) \sum_{k=1}^{K} u_{ik} v_{jk}) - p(j) \sum_{k=1}^{K} u_{ik} v_{jk} \right). \tag{7}$$

where $a(i)$ is the authority value of user i.

Unifying user authority and location significance, the objective function of our propose POI recommendation algorithm is formulated as:

$$\mathbf{L} = \min_{U,V} \sum_{i=1}^{M}\sum_{k=1}^{K} (u_{ik}/\beta_k - (\alpha_k - 1)\ln(u_{ik}/\beta_k)) + \sum_{j=1}^{N}\sum_{k=1}^{K} (v_{jk}/\beta_k - (\alpha_k - 1)\ln(v_{jk}/\beta_k))$$
$$+ \sum_{i=1}^{M}\sum_{j=1}^{N} I_{ij} a(i) \left(p(j) U_i^T V_j - r_{ij} \ln(p(j) U_i^T V_j) \right). \tag{8}$$

We apply the SGD method to seek a local minimum of the objective function $\mathbf{L}$. The derivatives of $\mathbf{L}$ with respect to u_{ik} and v_{jk} are computed as:

$$\begin{aligned} \frac{\partial \mathbf{L}}{\partial u_{ik}} &= \frac{1}{\beta_k} - \frac{\alpha_k - 1}{u_{ik}} + a(i) \sum_{j=1}^{N} I_{ij} (1 - \frac{r_{ij}}{p(j) U_i^T V_j}) v_{jk} p(j) \\ \frac{\partial \mathbf{L}}{\partial v_{jk}} &= \frac{1}{\beta_k} - \frac{\alpha_k - 1}{v_{jk}} + p(j) \sum_{i=1}^{M} I_{ij} (1 - \frac{r_{ij}}{p(j) U_i^T V_j}) u_{ik} a(i). \end{aligned} \tag{9}$$

3.2 Computing Location Significance and User Authority

Location significance and user authority are two key components of our proposed approach since they to some extend determine the number of visits at a location and the confidence level of a user's implicit feedback, i.e., a user's check-in frequency. This section describes how to compute location significance and user authority based on users' check-ins in LBSN.

Inspired by locations rank method [19], which extracts significant semantic locations from GPS data, we adapt the hybrid model of HITS and PageRank to compute more reasonable user authority and location significance by exploiting both the links between users and locations and the links between POIs. The hybrid model can be viewed as a random walk on a two-layered graph, which consists of the user-POI graph $\mathbb{G}_{UL}$ and the POI-POI graph $\mathbb{G}_{LL}$. The hybrid model builds a Markov chain by using $\mathbb{G}_{UL}$ and $\mathbb{G}_{LL}$ as well as three transition probabilities, i.e., $p(u_k|l_i)$, $p(l_i|u_k)$ and $p(l_i|l_j, u_k)$. The $p(u_k|l_i)$ is the transition probability from a location node l_i to a user node u_k, $p(l_i|u_k)$ is the transition probability from a user node u_k to a location node l_i, and $p(l_i|l_j, u_k)$ is the transition probability to location node l_i for user u_k given u_k at location l_j. These transition probabilities are computed as follows:

$$p(u_k|l_i) = \epsilon\frac{Num(u_k, l_i)}{Num(l_i)} + (1-\epsilon)\frac{1}{M} \qquad p(l_i|u_k) = \sum_{j=1}^{N} p(l_j|u_k)p(l_i|l_j, u_k)$$
$$p(l_i|l_j, u_k) = \alpha\frac{Num(l_i, l_j, u_k)}{Num(l_j, u_k)} + (1-\alpha)\frac{1}{N} \tag{10}$$

where $Num(l_i)$ indicates the total number of check-ins at l_i, $Num(u_k, l_i)$ represents the number of check-ins of user u_k at l_i and $Num(l_i, l_j, u_k)$ counts the times of user u_k co-visiting l_i and l_j. ϵ and α are the "teleport probabilities".

Given transition probability matrix $T_{UL} \in \mathbb{R}^{M\times N}$ with elements $p(u_k|l_i)$ and probability matrix $T_{LU} \in \mathbb{R}^{N\times M}$ with elements $p(l_i|u_k)$, the hybrid model can be describes as follows:

$$w_{loc}^{k+1} = T_{LU}.w_{user}^{k}, \qquad w_{user}^{k+1} = T_{UL}.w_{loc}^{k+1} \tag{11}$$

where $w_{loc}^{k+1} = [p(l_1)^{k+1} \quad p(l_2)^{k+1} \quad \dots \quad p(l_N)^{k+1}]^T$ and $w_{user}^{k+1} = [p(u_1)^{k+1} \quad p(u_2)^{k+1} \quad \dots \quad p(u_M)^{k+1}]^T$. In this paper, $p(u_i)$ is equal to $a(i)$, which denotes the authority value of user i. $p(l_j)$ is equal to $p(j)$, which denotes the location significance of location j.

The power iteration algorithm is applied to compute the user authority vector w_{user} and location significance vector w_{loc} in the hybrid model.

4 Experiments and Evaluation

4.1 Datasets and Evaluation Metrics

We choose two publicly available datasets[3]: Foursquare and Gowalla to evaluate the performance of our proposed method. General statistics of Foursquare and Gowalla are summarized in Table 1.

Since POI recommendation algorithms provide each target user with top-N highest ranked POIs, we employ two widely used rank metrics to evaluate the performance of different POI recommendation algorithms, i.e., Precision@N and Recall@N, where N is the length of ranked recommendation list of POIs.

[3] http://www.ntu.edu.sg/home/gaocong/datacode.htm.

4.2 Baseline Methods

We compare our proposed method with the following state-of-the-art POI recommendation approaches:

- UserKNN: This method is the user-based collaborative filtering [11]. In UserKNN, the similarity between users is computed by cosine similarity.
- ItemKNN: This method is the item-based collaborative filtering [20]. In ItemKNN, the similarity between items is computed by cosine similarity.
- PMF: PMF [18] can be viewed as a probabilistic extension of the SVD model. PMF has been exploited for POI recommendation in [3].
- WRMF: This method is the weighted matrix factorization [12]. WRMF has been evaluated for POI recommendation in [9].
- BPR-MF: BPR-MF adopts a Bayesian Personalized Ranking criterion [21] for item ranking.
- Geo-MF: Geo-MF [9] incorporates the spatial clustering property into weighted matrix factorization to make POI recommendations.
- PFM: This method is proposed by Ma et al. [17]. PFM focuses on Web site recommendation and models frequency data using the Poisson distribution.

The main parameter settings of all comparison methods are listed in Table 2. Note that we set parameters of each method according to respective references or based on our experiments. Under these parameter settings, each method achieves its best performance. The number of dimensions K of latent feature vectors is set to 10 in all our experiments. The "teleport probability" ϵ and α are set to 0.85, following the PageRank and HITS algorithms.

We conduct a five-fold cross validation by randomly extracting different training and test sets at each time, which accounts for 80% of visited POIs and 20% of visited POIs for each user, respectively. Finally, we report the average results.

4.3 Comparison with Baselines

Fig. 2 reports the results of POI recommendation quality for all compared algorithms. From Fig. 2, we can observe that: except for PMF, UserKNN and ItemKNN perform worse than other factorization based models. PMF achieves the worst performance among all compared methods. This observation is consistent with the results reported in [22]. The reason is that PMF assumes that a

Table 1. Statistics of Foursquare and Gowalla

Statistics	Foursquare	Gowalla
Num. of check-ins	194,108	456,967
Num. of users, N	2,321	10,162
Num. of POIs, M	5,596	24,237
Sparsity	99.18%	99.88%
Avg. POIs peruUser	45.57	30.27

Table 2. Parameter settings of comparison methods

Methods	Parameter settings
UserKNN	The size of similar neighborhood: 30
ItemKNN	The size of similar neighborhood: 30
PMF	$\lambda_U = \lambda_V = 0.001$
WRMF	$\lambda = 0.001, \alpha = 1$
BPR-MF	$\lambda_\Theta = 0.001$
PFM	$\alpha_k = 20, \beta_k = 0.2, k = 1, \ldots, K$
GeoMF	$\gamma = 0.01, \alpha = 10$
Our method	$\alpha_k = 20, \beta_k = 0.2, k = 1, \ldots, K$

user's implicit feedback follows the Gaussian distribution, which is not suitable for modeling the check-in frequency. In addition, BPRMF performs worse than WRMF and GeoMF. This is because the underlying assumption of BPRMF is the Gaussian distribution over a user's check-in frequency, although BPRMF, WRMF and GeoMF all consider POI recommendation problem as One-Class Collaborative Filtering (OCCF) problem. Moreover, we can see that PFM is generally superior to PMF, BPRMF and WRMF in terms of precision and recall on both data sets, which indicates that the Poisson distribution is more suitable for modeling uses' check-in frequency than the Gaussian distribution.

Furthermore, our proposed method consistently outperforms other methods, which either utilize the Gaussian distribution to model users' check-in behaviors or ignore the effects of location significance and user authority on users' check-in decision processes. Our proposed method improves the Precision@5 of GeoMF by 14% and 3% on Foursquare and Gowalla, respectively. In terms of Recall@5, the improvements of our proposed method are 13.5% and 5% on Foursquare and Gowalla data sets, respectively. This observation confirms our assumption utilizing the Poisson distribution to model users' check-in behaviors as well as taking into account the effects of location significance and user authority on users' check-in decision processes can improve the POI recommendation quality.

4.4 Impact of Location Significance and User Authority

We conduct another group of experiments to investigate the contribution of location significance and user authority to our proposed method by eliminating the corresponding components. We evaluate the following three reduced methods:

- \Sig: This reduced method eliminates the impact of location significance by setting $p(j) = 1$ in Eq. 8.
- \Auth: This reduced method eliminates the impact of user authority by set $a(i) = 1$ in Eq. 8.
- \Sig-Auth: This reduced method eliminates the impacts of both location significance and user authority by set $p(j) = 1, a(i) = 1$ in Eq. 8, which is equivalent to PFM.

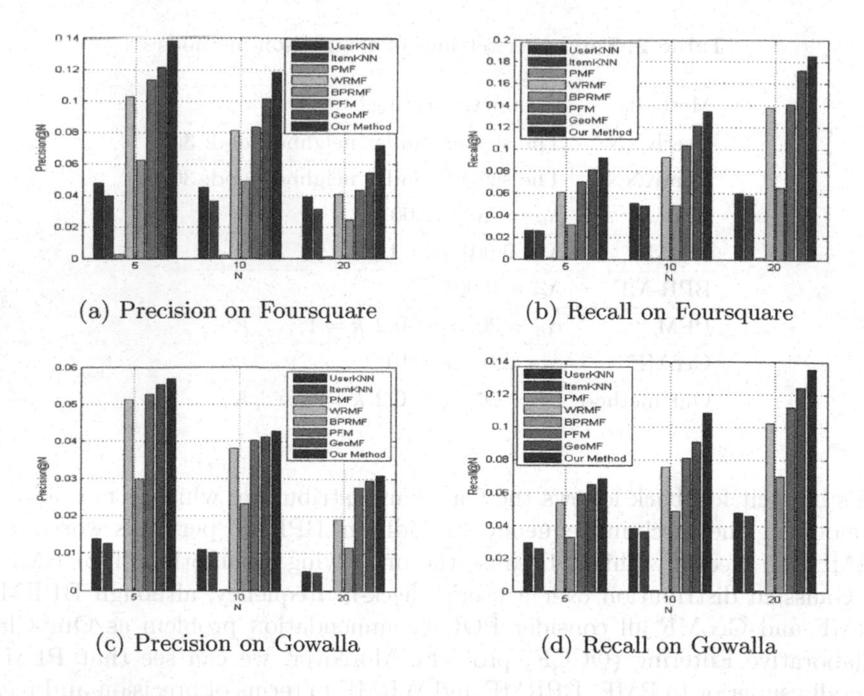

(a) Precision on Foursquare (b) Recall on Foursquare

(c) Precision on Gowalla (d) Recall on Gowalla

Fig. 2. Performance comparison on POI recommendation (K = 10)

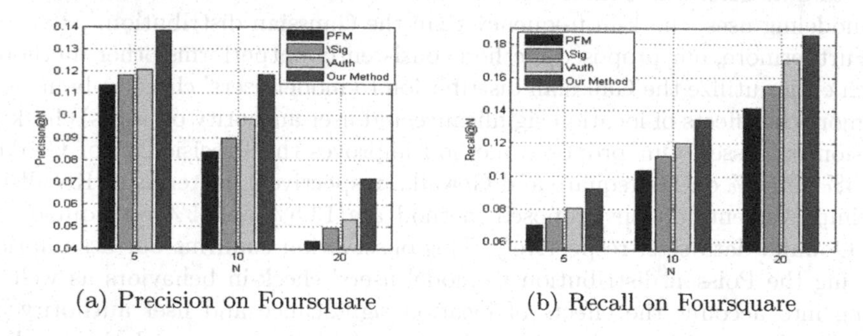

(a) Precision on Foursquare (b) Recall on Foursquare

Fig. 3. Performance comparison with three reduced methods

Since the experimental results on both data sets show similar trends, we only plot the experimental results on Foursquare in Fig. 3. From Fig. 3, we can see that two reduced methods, i.e., \Auth and \Sig, perform better than PFM, indicating that both location significance and user authority are beneficial for PFM. Moreover, the reduced method \Auth outperforms \Sig, which suggests that location significance contributes more than user authority on the performance improvement of our proposed method.

4.5 Impact of Parameters α_k and β_k

In this section, we perform a group of experiments to investigate the impacts of α_k and β_k on the performance of our proposed method by changing the values

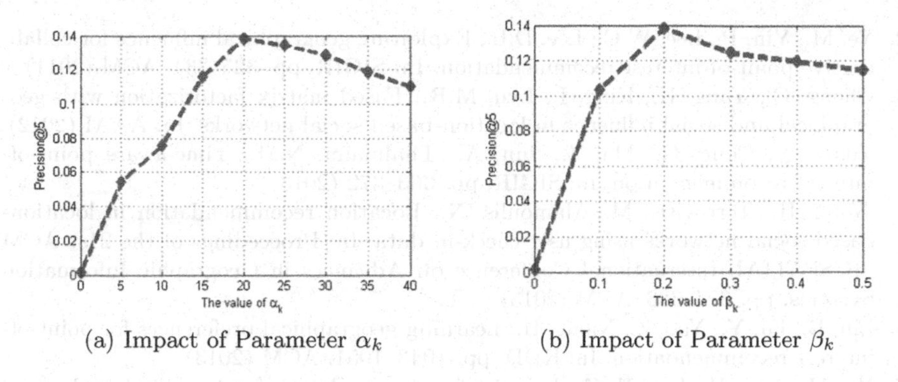

(a) Impact of Parameter α_k (b) Impact of Parameter β_k

Fig. 4. Impact of parameters α_k and β_k

of α_k from 0 to 40 given $\beta_k = 0.2$ or varying the values of β_k from 0 to 0.5 given $\alpha_k = 20$. We only plot the impacts of α_k and β_k on Precisson@5 for Foursquare in Fig. 4 since Recall@5 shows similar trends. We can see that parameters α_k and β_k significantly affect the POI recommendation quality. As the α_k/β_k increases, the values of Precision@5 first move upwards, the recommendation quality improves. After α_k/β_k reaches a certain threshold, Precision@5 begin to drop down as α_k/β_k increases.

5 Conclusion

In this paper, we propose a location significance and user authority enhanced probabilistic factor model for supporting POI recommendation in LBSNs. We first adapt a hybrid model of HITS and PageRank to compute user authority and location significance, which considers the mutual reinforcement between users' travel experience and locations' significance as well as the mutual influence between locations. Then we integrate probabilistic factor model with location significance to model users' check-in behaviors. Finally, we use user authorities to weight the confidence levels of users' feedback. Experimental results on two data sets show that our proposed approach outperforms other state-of-the-art POI recommendation algorithms.

Acknowledgments. The authors would like to acknowledge the support for this work from the National Natural Science Foundation of China (Grant Nos. 61432008, 61503178, 61403208), the Natural Science Foundation of Jiangsu Province of China (BK20150587) and NUPTSF (Grant No. NY217114).

References

1. Yu, Y., Chen, X.: A survey of point-of-interest recommendation in location-based social networks. In: Workshops at the Twenty-Ninth AAAI Conference on Artificial Intelligence (2015)

2. Ye, M., Yin, P., Lee, W.C., Lee, D.L.: Exploiting geographical influence for collaborative point-of-interest recommendation. In: SIGIR, pp. 325–334. ACM (2011)
3. Cheng, C., Yang, H., King, I., Lyu, M.R.: Fused matrix factorization with geographical and social influence in location-based social networks. In: AAAI (2012)
4. Yuan, Q., Cong, G., Ma, Z., Sun, A., Thalmann, N.M.: Time-aware point-of-interest recommendation. In: SIGIR, pp. 363–372 (2013)
5. Wang, H., Terrovitis, M., Mamoulis, N.: Location recommendation in location-based social networks using user check-in data. In: Proceedings of the 21st ACM SIGSPATIAL International Conference on Advances in Geographic Information Systems, pp. 374–383. ACM (2013)
6. Liu, B., Fu, Y., Yao, Z., Xiong, H.: Learning geographical preferences for point-of-interest recommendation. In: KDD, pp. 1043–1051. ACM (2013)
7. Ye, M., Yin, P., Lee, W.C.: Location recommendation for location-based social networks. In: SIGSPATIAL, pp. 458–461. ACM (2010)
8. Liu, Y., Wei, W., Sun, A., Miao, C.: Exploiting geographical neighborhood characteristics for location recommendation. In: CIKM, pp. 739–748. ACM (2014)
9. Lian, D., Zhao, C., Xie, X., Sun, G., Chen, E., Rui, Y.: GeoMF: joint geographical modeling and matrix factorization for point-of-interest recommendation. In: SIGKDD, pp. 831–840. ACM (2014)
10. Li, H., Ge, Y., Zhu, H.: Point-of-interest recommendations: learning potential check-ins from friends. In: KDD. ACM (2016)
11. Breese, J.S., Heckerman, D., Kadie, C.: Empirical analysis of predictive algorithms for collaborative filtering. In: UAI, pp. 43–52. Morgan Kaufmann Publishers Inc. (1998)
12. Pan, R., Zhou, Y., Cao, B., Liu, N.N., Lukose, R., Scholz, M., Yang, Q.: One-class collaborative filtering. In: ICDM, pp. 502–511. IEEE (2008)
13. Gao, H., Tang, J., Hu, X., Liu, H.: Exploring temporal effects for location recommendation on location-based social networks. In: RecSys, pp. 93–100. ACM (2013)
14. Massa, P.: A survey of trust use and modeling in real online systems. In: Trust E-Services: Technologies, Practices and Challenges, pp. 51–83. Idea Group Inc. (2007)
15. Kleinberg, J.M.: Authoritative sources in a hyperlinked environment. J. ACM (JACM) **46**(5), 604–632 (1999)
16. Page, L., Brin, S., Motwani, R., Winograd, T.: The pagerank citation ranking: bringing order to the web (1999)
17. Ma, H., Liu, C., King, I., Lyu, M.R.: Probabilistic factor models for web site recommendation. In: SIGIR, pp. 265–274. ACM (2011)
18. Mnih, A., Salakhutdinov, R.: Probabilistic matrix factorization. In: NIPS, pp. 1257–1264 (2007)
19. Cao, X., Cong, G., Jensen, C.S.: Mining significant semantic locations from GPS data. PVLDB **3**(1–2), 1009–1020 (2010)
20. Sarwar, B., Karypis, G., Konstan, J., Riedl, J.: Item-based collaborative filtering recommendation algorithms. In: WWW, pp. 285–295. ACM (2001)
21. Rendle, S., Freudenthaler, C., Gantner, Z., Schmidt-Thieme, L.: BPR: Bayesian personalized ranking from implicit feedback. In: UAI, pp. 452–461. AUAI Press (2009)
22. Li, X., Cong, G., Li, X.L., Pham, T.A.N., Krishnaswamy, S.: Rank-GeoFM: a ranking based geographical factorization method for point of interest recommendation. In: SIGIR, pp. 433–442. ACM (2015)

Personalized Ranking Recommendation via Integrating Multiple Feedbacks

Jian Liu[1], Chuan Shi[1,4](✉), Binbin Hu[1], Shenghua Liu[2], and Philip S. Yu[3]

[1] Beijing Key Lab of Intelligent Telecommunications Software and Multimedia, Beijing University of Posts and Telecommunications, Beijing, China
fullback@yeah.net, {shichuan,hubinbin}@bupt.edu.cn

[2] Chinese Academy of Sciences, Institute of Computing Technology, Beijing, China
liushenghua@ict.ac.cn

[3] University of Illinois at Chicago, Chicago, USA
psyu@uic.edu

[4] Beijing Advanced Innovation Center for Imaging Technology, Capital Normal University, Beijing, China

Abstract. Recently, recommender system has attracted a lot of attentions, which helps users to find items of interest through utilizing the user-item interaction information and/or content information associated with users and items. The interaction information (i.e., feedback) between users and items are widely exploited to build recommendation models. The feedback data in recommender systems usually comes in the form of both explicit feedback (e.g., rating) and implicit feedback (e.g., browsing histories, click logs). Although existing works have begun to utilize either explicit or implicit feedback for better recommendation, they did not make best use of these feedback information together. In this paper, we first study the personalized ranking recommendation problem by integrating multiple feedbacks, i.e., one type of explicit feedback and multiple types of implicit feedbacks. Then we propose a unified and flexible personalized ranking framework MFPR to integrate multiple feedbacks. Moreover, as there are no readily available training data, an explicit feedback based training data generation algorithm is designed to generate item pairs with more accurate partial order consistent with the multiple feedbacks for the proposed ranking model. Extensive experiments on two real-world datasets validate the effectiveness of the MFPR model, and the integration of multiple feedbacks making up better complementary information significantly improves recommendation performance.

Keywords: Recommender system · Multiple feedbacks · Explicit feedback · Implicit feedback · Bayesian Personalized Ranking

1 Introduction

In recent years, recommender systems have attracted much attention from multiple disciplines. The interaction information (i.e., feedback) between users and

J. Kim et al. (Eds.): PAKDD 2017, Part II, LNAI 10235, pp. 131–143, 2017.
DOI: 10.1007/978-3-319-57529-2_11

items are widely exploited to build recommendation models. The feedback data in recommender systems usually comes in the form of explicit or implicit feedback [4]. Explicit feedback is the interaction information that directly expresses user preferences to items, such as the rating information of users to items. While implicit feedback indirectly reflects user opinions and can imply user probable preferences [9], such as the "collect" and "share" of users to items. Figure 1 shows a toy example of multiple feedbacks in Douban Book. The rating (1–5 scales) is the explicit feedback and there are two types of implicit feedbacks. Thereinto, the "wish" means the user wishes to read the book but has not begun yet; the "reading" means the user is currently in reading process. It is obvious that the explicit feedback (i.e., rating) is critical for recommendation, while the implicit feedbacks also provide important supplementary information.

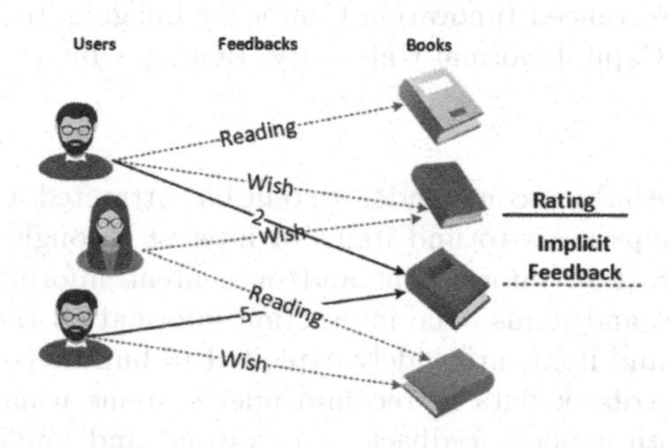

Fig. 1. A toy example of multiple feedback between users and books in Douban Book

Many methods exploit the feedbacks to build recommender systems. Figure 2 shows how those methods utilize these information. Traditional collaborative filtering usually utilizes explicit feedback information (i.e., ratings) [5,7,14] (see Fig. 2(a)). Since implicit feedback information is widely and cheaply available, researchers began to exploit the implicit feedback. Some works considered to use one single type of implicit feedback [6,10,13] (see Fig. 2(b)), and Costa Fortes et al. [2] combined several types of implicit feedbacks using a simple ensemble approach not long ago (see Fig. 2(c)). In addition, SVD++ [7] was designed to combine rating information with a single type of implicit feedback for more accurate rating prediction, as shown in Fig. 2(d). Unfortunately, all these works have

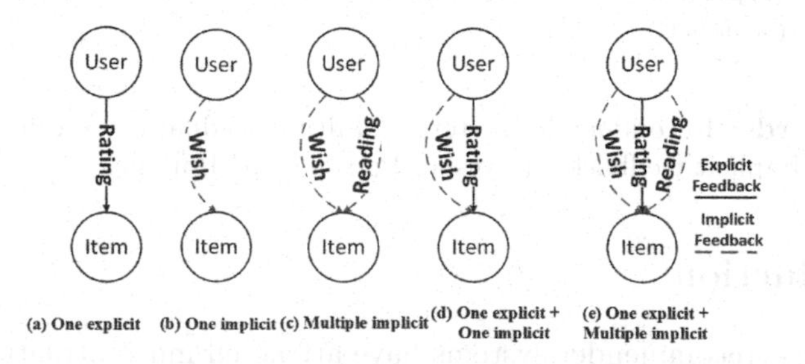

Fig. 2. The schemas of utilizing feedback information

not utilized comprehensive feedback information in recommender systems. In this paper, we propose to solve the personalized ranking problem by integrating multiple feedbacks, as shown in Fig. 2(e). For convenience, multiple feedbacks mean one type of explicit feedback and multiple types of implicit feedbacks in the following sections. In many review web sites, such as Yelp and Dianping, users are required to give a rating score (i.e., explicit feedback) to a business, and they can also have other interactions with businesses, such as "checking in" and "viewing". Obviously, our problem setting is a general framework to utilize feedback information, and existing problems are special cases of our problem setting. In addition, from recommendation perspective, the predicted ranking over an item is much more meaningful than the predicted rating. Thus in this work, we focus on developing a personalized ranking model that integrates multiple feedbacks. Although many methods have been proposed to utilize the feedbacks, these models are usually designed for special problem settings, and they cannot be directly applied in multiple-feedback setting.

However, integration of multiple feedbacks faces two challenges. (1) Design a unified ranking model integrating multiple feedbacks. In order to make the best use of these feedback information, we need to design an effective mechanism to handle relations between explicit and implicit feedbacks as well as relations among implicit feedbacks. (2) Generate training samples. As a ranking method, we need to generate preference pairs or lists for training. However, there are multiple types of feedbacks. What kind of feedbacks could we utilize for better preference pair or sequence?

The major contributions of our paper are summarized as follows: (1) We first try to solve the personalized ranking recommendation problem by integrating multiple feedbacks. The problem widely exists in real recommender system, and it is a general problem setting to encompass existing works. (2) We propose a Bayesian Personalized Ranking (BPR) based model MFPR to integrate multiple feedbacks. Moreover, as there are no readily available training data for this problem, an effective algorithm is designed to generate the training data that is more consistent with multiple feedbacks for the MFPR model. (3) We crawl comprehensive Douban Book and Dianping datasets[1] including ratings and multiple types of implicit feedbacks.

2 Preliminary

2.1 Explicit and Implicit Feedback and Problem Formulation

Formally, when the data is in the form of explicit feedback with single implicit feedback, each user u is associated with two types of item sets: implicit item set $N(u)$ and explicit feedback set $E(u)$. *Explicit feedback* is intentionally provided by users to directly express user preferences (e.g., likes or dislikes) to items. For an item $i \in E(u)$, the rating given by user u to item i is denoted as R_{ui}. *Implicit feedback* reflects user opinions indirectly and can imply user probable preferences

[1] The datasets are available at https://github.com/7thsword/MFPR-Datasets.

[9]. For an item $i \in N(u)$, the implicit feedback does not necessarily mean that user u likes the item i.

When data consists of explicit feedback with multiple types of implicit feedbacks, each user is associated with single explicit feedback and τ types of implicit feedbacks ($\tau \geq 2$). For user u, the explicit item set is still denoted as $E(u)$ which contains items user u has rated (i.e., rating) on, and the implicit item sets are denoted as $N^1(u), N^2(u), \cdots, N^\tau(u)$ where $N^t(u)$ contains items user u has expressed the t-type implicit feedback on($t = 1, \cdots, \tau$).

Let $\mathcal{U}$ and $\mathcal{I}$ denote the set of users and items respectively. We define a ranking recommendation problem on multiple feedback data $R_d = \{\mathcal{U}, \mathcal{I}, E_f, I_f\}$. E_f, defined as $E_f = \{E(u)|u \in \mathcal{U}\}$, denotes the explicit feedback data consisting of all users' explicit item sets. I_f, defined as $I_f = \{N^t(u)|u \in \mathcal{U}, t = 1, \cdots, \tau\}$, denotes the implicit feedback data consisting of all users' implicit item sets. Hence, as shown in Fig. 2(e), our task is to design a model for better personalized ranking recommendation through making full use of the explicit feedback data E_f and the implicit feedback data I_f.

2.2 Base Learner Integrating Explicit and Implicit Feedback

The explicit feedback (i.e. rating) is very important for recommendation but rare, and the implicit feedback is popular in real systems. Some researchers began to consider the integration of explicit and implicit feedback for more accurate rating prediction. Assume that there are m users and n items (i.e., $|\mathcal{U}| = m$, $|\mathcal{I}| = n$). Given a rating matrix $R = (R_{ui})^{m \times n}$, where R_{ui} denotes the score user u has rated on item i. The predicted rating $\hat{R}_{ui}$ user u may give to item i in SVD++ [7] can be modeled as:

$$\hat{R}_{ui} = (p_u + |N(u)|^{-\frac{1}{2}} \sum_{k \in N(u)} \gamma_k) q_i^{\mathrm{T}}, \tag{1}$$

where $p_u \in \mathbb{R}^d$ is the explicit latent vector of user u, $q_i \in \mathbb{R}^d$ is the explicit latent vector of item i and $d \ll min(m, n)$. $\gamma_k \in \mathbb{R}^d$ is the implicit latent vector of item k and $N(u)$ is the implicit item set as mentioned above. Here a user u is modeled as $p_u + |N(u)|^{-\frac{1}{2}} \sum_{k \in N(u)} \gamma_k$, and the complemented sum term $|N(u)|^{-\frac{1}{2}} \sum_{k \in N(u)} \gamma_k$ represents the perspective of implicit feedback. SVD++ treats the explicit and implicit feedback differently. It makes best use of explicit feedback and adds implicit feedback as supplements.

Unfortunately, these existing models cannot be directly applied to our problem setting. Although SVD++ also considers explicit and implicit feedbacks, it just integrates one type of implicit feedback. In addition, SVD++ is originally designed for the rating prediction problem. Since predicting exact ratings is not necessary for recommendation, we propose to use ranking framework.

3 Personalized Ranking with Multiple Feedbacks

The explicit and implicit feedbacks have different characteristics, we need to treat them differently. Through adapting the Bayesian Personalized Ranking framework [13], we first design a *P*ersonalized *R*anking model which integrates explicit and one *S*ingle implicit *F*eedbacks (called SFPR). Then we extend the SFPR model to integrate more implicit feedbacks and propose a unified *M*ultiple *F*eedbacks based *P*ersonalized *R*anking model (called MFPR).

3.1 The SFPR Model

Firstly, we design a ranking model to combine explicit feedback and one type of implicit feedback. Assume that a training set $\mathcal{T}_r$ consists of triples of the form (u, i, j) with $i \succ j$ denoting that user u prefers item i to item j. Note that the generation of training set $\mathcal{T}_r$ is an important issue and it will be discussed in Sect. 4. The Bayesian formulation of finding the correct personalized ranking is to maximize the following posterior probability:

$$p(\theta|\mathcal{T}_r) \propto p(\mathcal{T}_r|\theta)p(\theta), \tag{2}$$

where θ is the parameter of a certain base learner and $p(\theta)$ is the prior probability of base leaner parameter. We use $p(i \succ j; u|\theta)$ to denote the probability that user u prefers item i over item j under the model expressed by θ. With the assumption that each triple $(u, i, j) \in \mathcal{T}_r$ is independent, the likelihood function can be expanded as follows:

$$p(\mathcal{T}_r|\theta) = \prod_{(u,i,j)\in\mathcal{T}_r} p(i \succ j; u|\theta). \tag{3}$$

Since the SVD++ can effectively differentiates explicit and implicit feedback and fully utilize the explicit feedback, we utilize the SVD++ as our base learner. Then the individual probability $p(i \succ j; u|\theta)$ can be modeled as:

$$p(i \succ j; u|\theta) = \sigma(\hat{R}_{ui} - \hat{R}_{uj}), \tag{4}$$

where σ is the logistic sigmoid function $\sigma(x) = \frac{1}{1+e^{-x}}$.

For convenience, we simplify $\hat{R}_{ui} - \hat{R}_{uj}$ in Eq. 4 as $\hat{x}_{uij}$. Note that $\hat{x}_{uij}$ is a real-valued function of θ which captures ranking relation between item i and item j with the given user u. Assume that $p(\theta)$ is a Gaussian distribution with zero mean and variance-covariance matrix $\sum_\theta = \lambda_\theta I$. Now we can estimate parameter θ of the base learner through maximizing the posterior probability in Eq. 2 as follows:

$$\begin{aligned} \max_\theta \mathcal{L} &= ln\ p(\theta|\mathcal{T}_r) \\ &= ln\ p(\mathcal{T}_r|\theta)p(\theta) \\ &= \sum_{(u,i,j)\in\mathcal{T}_r} ln\ p(i \succ j; u|\theta) - \lambda_\theta\|\theta\|^2 \\ &= \sum_{(u,i,j)\in\mathcal{T}_r} ln\ \sigma(\hat{x}_{uij}) - \lambda_\theta\|\theta\|^2, \end{aligned} \tag{5}$$

where $\lambda_\theta\|\theta\|^2$ is a L2 regularization term which can be derived from the Gaussian distribution $p(\theta)$ mentioned above.

3.2 Learning SFPR Model

The objective function Eq. 5 is differentiable, gradient ascent based algorithms can be employed as optimizer. The gradient of Eq. 5 with respect to the parameter θ is:

$$\begin{aligned} \frac{\partial \mathcal{L}}{\partial \theta} &= \sum_{(u,i,j)\in \mathcal{T}_r} \frac{\partial}{\partial \theta} ln\sigma(\hat{x}_{uij}) - \lambda_\theta \frac{\partial}{\partial \theta}\|\theta\|^2 \\ &\propto \sum_{(u,i,j)\in \mathcal{T}_r} \frac{1}{1+e^{\hat{x}_{uij}}} \frac{\partial}{\partial \theta}\hat{x}_{uij} - \lambda_\theta \theta. \end{aligned} \tag{6}$$

We adopt stochastic gradient ascent (SGA) to optimize the model SFPR. Then with a training sample (u, i, j), the model parameter θ can be updated as:

$$\theta \leftarrow \theta + \eta(\frac{1}{1+e^{\hat{x}_{uij}}} \frac{\partial}{\partial \theta}\hat{x}_{uij} - \lambda_\theta \theta), \tag{7}$$

where η is the given learning rate and generally tuned via cross validation. The gradient of $\hat{x}_{uij}$ with respect to each model parameter has to be known before gradient ascent process. $\hat{x}_{uij} = \hat{R}_{ui} - \hat{R}_{uj}$ is defined above and we can get the derivatives:

$$\frac{\partial \hat{x}_{uij}}{\partial \theta} = \begin{cases} q_i - q_j & \text{if } \theta = p_u, \\ p_u + |N(u)|^{-\frac{1}{2}} \sum_{k\in N(u)} \gamma_k & \text{if } \theta = q_i, \\ -(p_u + |N(u)|^{-\frac{1}{2}} \sum_{k\in N(u)} \gamma_k) & \text{if } \theta = q_j, \\ |N(u)|^{-\frac{1}{2}}(q_i - q_j) & \text{if } \theta = \gamma_k. \end{cases}$$

The predicted $\hat{R}_{ui}$ in SFPR model cannot be regarded as the usual predicted rating (i.e. 1 to 5 scales). Here, we call $\hat{R}_{ui}$ the predicted ranking score, which implies that degree of user u prefers item i. The larger the ranking score is, the higher preference it implies.

3.3 The MFPR Model

The proposed SFPR is designed to integrate single explicit feedback and single implicit feedback. Here we extend the SFPR model to integrate more implicit feedbacks. When considering multiple feedbacks, as mentioned in Sect. 2.1, each user u is associated with an explicit item set $E(u)$ and τ types of implicit item sets $N^1(u), N^2(u), \cdots, N^\tau(u)$. For integrating multiple implicit feedbacks, our extended preference predictor can be designed as

$$\hat{R}_{ui} = (p_u + \frac{1}{\tau}\sum_{t=1}^{\tau} |N^t(u)|^{-\frac{1}{2}} \sum_{k\in N^t(u)} \gamma_k^t) q_i^{\mathrm{T}}, \tag{8}$$

where $\gamma_k^t \in \mathbb{R}^d$ represents the implicit latent vector of item k under the t-th implicit feedback. The model in Eq. 8 can be seen as a more general version of the SFPR model. Now we have the $\hat{x}_{uij} = \hat{R}_{ui} - \hat{R}_{uj}$ as:

$$\hat{x}_{uij} = (p_u + \frac{1}{\mathcal{T}} \sum_{t=1}^{\mathcal{T}} |N^t(u)|^{-\frac{1}{2}} \sum_{k \in N^t(u)} \gamma_k^t)(q_i - q_j)^{\mathrm{T}}. \quad (9)$$

Similarly, we apply SGA to solve the optimization problem.

4 Training Set Generation Algorithm

The MFPR model is fed with training data in the form of (u, i, j) with $i \succ j$ denoting that user u prefers item i over item j. Since the preference partial pairs significantly affect performances [1], it is an important issue that how we can effectively generate (u, i, j) from multiple feedbacks. For those traditional personalized ranking models utilizing only one or more types of implicit feedbacks, such as BPR-MF in [13] and the approach in [2], their training set generation algorithms just take implicit feedbacks into account. Specifically, they draw partially ordered item pairs from the cartesian product of user's interacted items (items belong to user's implicit item set) and user's non-interacted items (items do not belong to user's implicit item set). However, in terms of multiple feedbacks, such training set generation algorithm is inapplicable for MFPR. Besides implicit feedbacks, there are quality rating information in our problem setting, which can better reflect user preference. Hence, we need to design a new training data generation algorithm.

Burgess and Shaked et al. [1] have proved that if the ranking probabilities of every adjacent document pair in a permutation of all documents to be ranked are known, then the ranking probabilities of any document pair can be derived. Inspired by this conclusion, we design the training set generation algorithm which utilizes the most significant preference information in the multiple feedbacks: rating information. For each user u, we randomly split his or her explicit item set $E(u)$ into two subsets $E_{tr}(u)$ and $E_{te}(u)$ with the given split ratio, where $E_{tr}(u)$ is designed for constructing training set $\mathcal{T}_r$ and $E_{te}(u)$ is for test set $\mathcal{T}_e$. When constructing $\mathcal{T}_r$, we first get a random permutation of $E_{tr}(u)$. Then, for every adjacent item pair (i, j) in the permutation: (1) if $R_{ui} > R_{uj}$, put the triple (u, i, j) into $\mathcal{T}_r$; (2) if $R_{ui} < R_{uj}$, put the triple (u, j, i) into $\mathcal{T}_r$; (3) if $R_{ui} = R_{uj}$, skip and continue to check next adjacent pair. Through the process for every user, we can get the training set $\mathcal{T}_r$ eventually. And the similar process is done for the test set $\mathcal{T}_e$.

Figure 3 gives a toy example for user u. We have explicit item set $E_{tr}(u) = \{6, 8, 9, 11, 17\}$ and the corresponding ratings are $R_{u,6} = 4$, $R_{u,8} = 3$, $R_{u,9} = 2$, $R_{u,11} = 5$ and $R_{u,17} = 4$. Assume that a random permutation of E_{tr} is $P_{tr} = \{11, 8, 17, 6, 9\}$, then we in turn check every adjacent item pairs $(11, 8)$, $(8, 17)$, $(17, 6)$, $(6, 9)$ of the permutation. Finally, the triples $(u, 11, 8)$, $(u, 17, 8)$ and $(u, 6, 9)$ are selected and put into the training set $\mathcal{T}_r$.

We name this algorithm as *IPPE* which means that *I*tem *P*airs with partial order are obtained from checking adjacent items in a *P*ermutation of *E*xplicit item set. The IPPE method considers every adjacent item pair, rather than any item pair. This strategy significantly reduces the size of training samples without much sacrifice in recommendation performance.

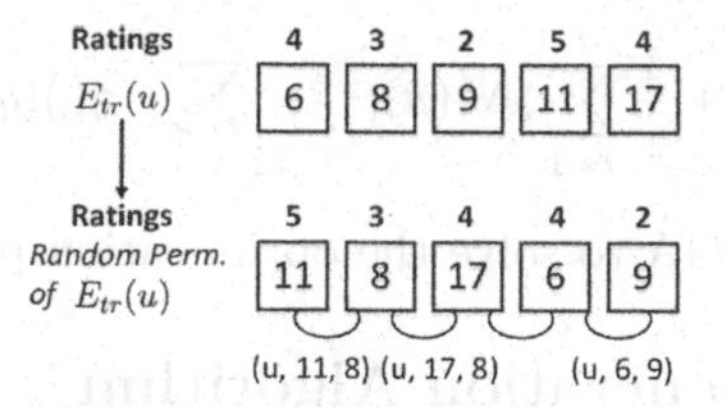

Fig. 3. The toy example of generating training data for user u

5 Experiment

5.1 Datasets

In this paper, we focus on exploiting multiple feedbacks. As far as we know, it is difficult to obtain such public datasets. Hence, we crawled two real-world datasets for the experiments.

The Douban Book dataset[2] contains 190,590 ratings (1–5 scales) involving 12,850 users and 22,040 books. The ratings to books are considered as explicit feedback. There are 6 types of implicit feedbacks: "wish", "reading", "read", "tag", "comment" and "rated". All these implicit feedbacks are recorded using a binary matrix ("1" for done and "0" for not). Note that the "rated" implicit feedback comes from rating information through degrading the rating matrix into a binary matrix ("1" means "rated" and "0" for "not rated").

The Dianping dataset[3] contains 188,813 ratings (1–5 scales) involving 10,549 users and 17,707 restaurants. There are four types of ratings in Dianping, including overall rating (1–5 scales) and ratings (1–5 scales) on taste, environment and service. We use the overall ratings as explicit feedback and degrade overall, taste, environment and service ratings into "1" if rating $\geq$3 otherwise "0". Then four types of implicit feedbacks are obtained: "good taste", "good environment", "good service" and "good overall". The details can be seen in Table 1.

5.2 Comparison Methods and Evaluation Metrics

We compare the performance of the proposed SFPR and MFPR with five representative methods:

- Most Popular (MP). This baseline ranks items according to their popularity and is non-personalized.
- SVD [7]. This method is a typical matrix factorization based model. It is a rating prediction model and the input data needs only the rating information. We rank items using the predicted ratings in our experiments.
- BPR-MF [13]. This pairwise ranking method was introduced by Rendle et al. and is a state-of-the-art personalized ranking model using only one type of implicit feedback.

[2] http://book.douban.com.
[3] http://www.dianping.com.

Table 1. Statistics of datasets

Dataset	Type	Feedbacks (A–B)	No. of A	No. of B	No. of (A–B)
Douban book	Explicit	User-rating	12850	22040	190590
	Implicit	User-wish	11107	16406	162565
		User-reading	9776	12787	71662
		User-read	12029	20014	174726
		User-tag	8487	19942	162070
		User-comment	8776	18888	151758
		User-rated	12850	22040	190590
Dianping	Explicit	User-rating	10549	17707	188813
	Implicit	User-good taste	10473	14043	122060
		User-good env.	10293	12135	90350
		User-good service	10354	13271	105846
		User-good overall	10425	14283	125173

- Ensemble of BPRMF (EN-BPRMF) [2]. This method is an ensemble approach to unify different types of implicit feedbacks based on BPR-MF. In experiments, we ensemble all types of implicit feedbacks using this approach.
- SVD++ [7]. This method is also a matrix factorization based rating prediction model and the first to integrate rating information with one type of implicit feedback. We rank items using the predicted ratings.
- Factorization Machine (FM) [11]. This method is a general predictor which works with any real valued feature vector and combines the advantages of support vector machines with factorization models. We integrate rating information and all types of implicit feedbacks into the feature vector. It is a rating prediction model and we rank items using the predicted rating.

Since BPR-MF, SVD++ and SFPR need one type of implicit feedback, we choose the "read" feedback in Douban Book and the "good overall" feedback in Dianping for them. The reason is that the best performance is achieved in these conditions. In addition, some baselines are obtained from open resources. FM is from libFM [12], while MP and BPR-MF are from MyMediaLite [3].

We use two evaluation metrics, which are widely used to evaluate ranking performance. *Zero-One Error* [8] is the average ratio of correctly ordered item pairs of triples (u, i, j) in test set $\mathcal{T}_e$:

$$\varepsilon_{0/1} = \frac{1}{|\mathcal{T}_e|} \sum_{(u,i,j) \in \mathcal{T}_e} [\hat{x}_{uij}(R_{ui} - R_{uj}) > 0], \tag{10}$$

where $\hat{x}_{uij}$ is the difference between predicted ranking score $\hat{R}_{ui}$ and $\hat{R}_{uj}$ as defined above. And $[c]$ denotes a condition indicator that return 1 iff c is true otherwise 0.

NDCG@k [8] is designed to take into count the order of items in the recommendation list. To define $NDCG_u@k$ for a user u, $DCG_u@k = \sum_{i=1}^{k} \frac{2^{R_{ui}}-1}{log_2(i+1)}$

should be given formally first, thereinto i ranges over positions in the recommended list of user u, and we use the observed rating R_{ui} to weigh the degree user u prefers item i. $NDCG_u@k$ is the ratio of $DCG_u@k$ to ideal DCG for that user:

$$NDCG_u@k = \frac{DCG_u@k}{IDCG_u@k}, \tag{11}$$

where $IDCG_u@k$ is the maximum possible DCG when the recommended items are just in descending order by user u preference. $NDCG@k$ is the mean value of $NDCG_u@k$ over all users, reflecting model performance of recommended list at the top k ranking.

5.3 Effectiveness

This section will validate the effectiveness of the proposed SFPR and MFPR compared to those baselines. For Douban Book and Dianping datasets, we generate training set $\mathcal{T}_r$ and test set $\mathcal{T}_e$ using different split ratios 30%, 50%, 70%, respectively. The random split was carried out 5 times independently in all experiments and we report the mean values of $\varepsilon_{0/1}$ and $NDCG$.

Parameters of all methods are tuned to the optimal values through cross validation on the datasets. For fair comparison, we set the same number of latent dimension $d = 10$ for all matrix factorization based methods. We select $\varepsilon_{0/1}$, $NDCG@5$ and $NDCG@10$ as evaluation metrics. We also record the improvement ratio on these evaluation metrics of all methods compared to the SVD. Moreover, we also conduct the t-test experiments with 95% confidence, which shows that the $\varepsilon_{0/1}$ and the $NDCG$ improvements is statistically stable and non-contingent.

The experimental results are shown in Table 2, the main findings from the experimental comparisons are summarized as follows: (1) MFPR achieves the best performance in all conditions, which validates the significant benefits of integrating both explicit feedback and multiple implicit feedbacks. The experiments also confirm that better performance will be achieved through integrating more feedback information. For example, SFPR outperforms BPR-MF due to integration of ratings, SVD++ outperforms SVD because of implicit feedback, and the superiority of MFPR to SFPR comes from more implicit feedbacks. Note that MFPR and FM both utilize all feedback information, while MFPR always has better performance. The reason lies in that MFPR designs an effective mechanism treating explicit and implicit feedbacks differently, while FM handles all feedbacks equally. In all, exploiting and integrating multiple feedbacks is really helpful to improve the performance in the personalized ranking recommendation task. (2) When considering different training data ratios, we can find that the improvements of those models integrating explicit feedback with implicit feedbacks (i.e., SVD++, FM, SFPR and MFPR) are more significant for less training data. This indicates that integrating implicit feedbacks into models can effectively alleviate data sparsity of rating information. Specifically, FM outperforms SVD++ and MFPR outperforms SFPR because of integrating

Table 2. Performance comparisons on Douban Book and Dianping (d=10, the baseline of improvement ratio is SVD)

Datasets	Training	Metric	MP	SVD	BPR-MF	EN-BPRMF	SVD++	FM	SFPR	MFPR
Douban book	30%	$\epsilon_{0/1}$	0.5210	0.5251	0.5314	0.5372	0.6089	0.6145	0.6270	**0.6307**
		Improve	−0.66%		1.20%	2.30%	15.96%	17.03%	19.41%	20.11%
		NDCG@5	0.7831	0.7879	0.7845	0.7861	0.8291	0.8288	0.8371	**0.8399**
		Improve	−0.78%		−0.43%	−0.23%	5.23%	5.19%	6.24%	6.60%
		NDCG@10	0.8301	0.8332	0.8318	0.8323	0.8656	0.8691	0.8706	**0.8726**
		Improve	−0.37%		−0.17%	−0.11%	3.89%	4.31%	4.49%	4.73%
	50%	$\epsilon_{0/1}$	0.5225	0.5909	0.5299	0.5374	0.6396	0.6399	0.6605	**0.6636**
		Improve	−11.58%		−10.32%	−9.05%	8.24%	8.29%	11.78%	12.30%
		NDCG@5	0.7969	0.8347	0.7989	0.7994	0.8516	0.8500	0.8564	**0.8611**
		Improve	−4.53%		−4.29%	−4.23%	2.02%	1.83%	2.60%	3.16%
		NDCG@10	0.8478	0.8747	0.8493	0.8494	0.8887	0.8864	0.8927	**0.8959**
		Improve	−3.08%		−2.90%	−2.89%	1.60%	1.34%	2.06%	2.42%
	70%	$\epsilon_{0/1}$	0.5239	0.6242	0.5312	0.5397	0.6558	0.6582	0.6676	**0.6756**
		Improve	−16.07%		−14.90%	−13.54%	5.06%	5.45%	6.95%	8.23%
		NDCG@5	0.8338	0.8791	0.8403	0.8409	0.8874	0.8875	0.8895	**0.8932**
		Improve	−5.15%		−4.41%	−4.35%	0.94%	0.96%	1.18%	1.60%
		NDCG@10	0.8814	0.9110	0.8821	0.8824	0.9172	0.9164	0.9196	**0.9220**
		Improve	−3.25%		−3.17%	−3.14%	0.68%	0.59%	0.94%	1.21%
Dianping	30%	$\epsilon_{0/1}$	0.5967	0.5922	0.5999	0.6072	0.6118	0.6220	0.6248	**0.6253**
		Improve	0.59%		1.30%	2.53%	3.31%	5.03%	5.50%	5.59%
		NDCG@5	0.8214	0.8178	0.8225	0.8261	0.8293	0.8365	0.8377	**0.8387**
		Improve	0.44%		0.57%	1.01%	1.41%	2.29%	2.43%	2.56%
		NDCG@10	0.8619	0.8594	0.8630	0.8658	0.8692	0.8689	0.8721	**0.8752**
		Improve	0.29%		0.42%	0.74%	1.14%	1.11%	1.48%	1.84%
	50%	$\epsilon_{0/1}$	0.5965	0.6191	0.6009	0.6062	0.6304	0.6307	0.6345	**0.6367**
		Improve	−3.65%		−2.94%	−2.08%	1.83%	1.87%	2.49%	2.84%
		NDCG@5	0.8628	0.8727	0.8643	0.8674	0.8774	0.8778	0.8801	**0.8815**
		Improve	−1.13%		−0.96%	−0.61%	0.54%	0.58%	0.85%	1.01%
		NDCG@10	0.8924	0.8999	0.8940	0.8961	0.9044	0.9040	0.9056	**0.9076**
		Improve	−0.83%		−0.66%	−0.42%	0.50%	0.46%	0.63%	0.86%
	70%	$\epsilon_{0/1}$	0.5987	0.6348	0.6006	0.6103	0.6411	0.6437	0.6468	**0.6498**
		Improve	−5.69%		−5.39%	−3.86%	0.99%	1.40%	1.89%	2.36%
		NDCG@5	0.8858	0.8982	0.8875	0.8891	0.9012	0.8996	0.9015	**0.9029**
		Improve	−1.38%		−1.19%	−1.01%	0.33%	0.16%	0.37%	0.50%
		NDCG@10	0.9099	0.9196	0.9110	0.9126	0.9217	0.9209	0.9219	**0.9234**
		Improve	−1.05%		−0.94%	−0.76%	0.23%	0.14%	0.25%	0.41%

more implicit feedbacks. More combined implicit feedbacks mean more supplementary information for ratings. Thus, it is desirable to achieve much better recommendation performance through integrating comprehensive multiple feedbacks, particularly when rating information is insufficient. (3) From the results, we can also note that pairwise methods are more suitable for personalized ranking recommendation. Specifically, SVD, SVD++ and FM are rating prediction models, also known as pointwise methods, while SFPR and MFPR are pairwise ranking models. It is obvious that SFPR and MFPR outperform those three pointwise models. Specially, SFPR uses the same base learner as SVD++. Note that the other two pairwise ranking models (i.e. BPR-MF and EN-BPRMF) fail to defeat those pointwise models. We think the reason lies in that BPR-MF and EN-BPRMF utilize only implicit feedback, so they fail to generate accurate partial order item pairs as training set. In contrast, the proposed SFPR and

MFPR generate item pairs with more accurate ranking order as training set from explicit feedback.

5.4 Impact of Different Training Set Generation Algorithms

In this section, we verify the effectiveness of the designed training set generation algorithm IPPE. In order to validate the superiority of the IPPE, we compare it with the following two baseline methods. Following the idea of BPR-MF in [13], for user u, we make cartesian product of $E_{tr}(u)$ with user's unknown items to construct training set. We name this approach as *IPUC* which means *I*tem *P*airs of partial order are obtained from *U*nknown item related *C*artesian product. We also consider a variation of the IPPE method. From $E_{tr}(u)$ of each user u, we randomly sample two items each time and generate the item pair with partial order according to their observed ratings. In order to produce the similar training data size as the IPPE, the random process for each user u was conducted $|E_{tr}(u)|$ times. We name this approach as *IPRE* which means *I*tem *P*airs of partial order are obtained from checking *R*andom pairs in *E*xplicit item set. And we retain the same generation strategy for test set as the IPPE for these two approaches.

We apply these three different training set generation algorithms in SFPR and MFPR. As shown in Fig. 4, SFPR based on the methods IPUC, IPRE and IPPE are named as SFPR_{UC}, SFPR_{RE}, SFPR_{PE} respecitvely. It is similar for MFPR. We conduct experiments on both Douban Book and Dianping datasets, where the "read" feedback and the "good overall" feedback are still chosen for the SFPR. We can observe that models with IPPE have much better performance than those with IPUC. Specifically, SFPR_{UC} and MFPR_{UC} have very bad performance, which degrades as BPR-MF in Table 2. Since the method IPPE makes full use of the rating information and thus the corresponding training set $\mathcal{T}_r$ consists of item pairs with more accurate partial order. On the contrary, the approach IPUC just discards the item orders implied by rating information and deals with the rating as ordinary implicit feedback. Moreover, we observe that SFPR_{PE} and MFPR_{PE} outperform SFPR_{RE} and MFPR_{RE} slightly but stably. This shows that sampling adjacent items pairs from random permutations

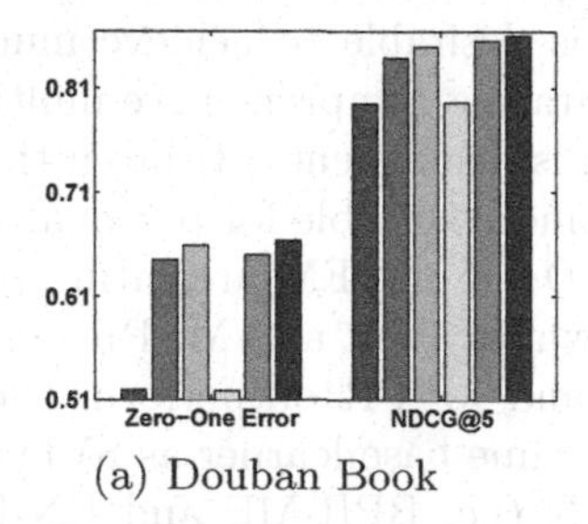

(a) Douban Book

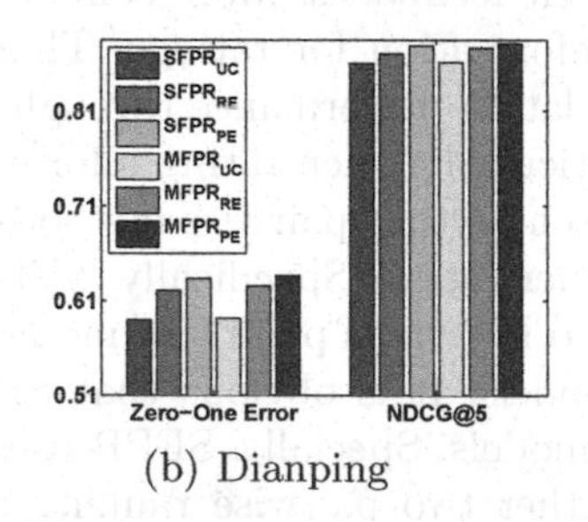

(b) Dianping

Fig. 4. Performance under different training set generation algorithms on Douban Book and Dianping

outperforms that sampling item pairs randomly. In summary, for such multiple-feedback data, the proposed IPPE method is more effective to generate training set for the personalized ranking models.

6 Conclusion

In this paper, we study the personalized ranking recommendation by integrating multiple feedbacks, and propose a unified multiple feedbacks personalized ranking framework MFPR. Extensive experiments on two real-world datasets conform the superiority of MFPR. Moreover, we also have designed a delicate algorithm IPPE to generate training data with more accurate partial order for the proposed ranking model. The empirical evaluation results suggest that IPPE through checking adjacent items in a permutation is superior to IPUC and IPRE.

Acknowledgments. This work is supported in part by the National Natural Science Foundation of China (Nos. 61375058, 61572467), National Key Basic Research and Department (973) Program of China (No. 2013CB329606), and the Co-construction Project of Beijing Municipal Commission of Education.

References

1. Burges, C., Shaked, T., Renshaw, E., Lazier, A., Deeds, M., Hamilton, N., Hullender, G.: Learning to rank using gradient descent. In: ICML, pp. 89–96. ACM (2005)
2. da Costa, A.F., Manzato, M.G.: Ensemble learning in recommender systems: combining multiple user interactions for ranking personalization. In: PBSMW, pp. 47–54. ACM (2014)
3. Gantner, Z., Rendle, S., et al.: Mymedialite: a free recommender system library. In: RecSys, pp. 305–308. ACM (2011)
4. He, R., McAuley, J.: VBPR: visual Bayesian personalized ranking from implicit feedback. arXiv preprint arXiv:1510.01784 (2015)
5. Hoyer, P.O.: Non-negative matrix factorization with sparseness constraints. JMLR **5**, 1457–1469 (2004)
6. Hu, Y., Koren, Y., Volinsky, C.: Collaborative filtering for implicit feedback datasets. In: ICDM, pp. 263–272. IEEE (2008)
7. Koren, Y.: Factorization meets the neighborhood: a multifaceted collaborative filtering model. In: SIGKDD, pp. 426–434. ACM (2008)
8. Lee, J., Bengio, S., et al.: Local collaborative ranking. In: WWW, pp. 85–96. ACM (2014)
9. Oard, D.W., Kim, J., et al.: Implicit feedback for recommender systems. In: AAAI, pp. 81–83 (1998)
10. Pan, R., Zhou, Y., et al.: One-class collaborative filtering. In: ICDM, pp. 502–511. IEEE (2008)
11. Rendle, S.: Factorization machines. In: ICDM, pp. 995–1000. IEEE (2010)
12. Rendle, S.: Factorization machines with libFM. ACM Trans. Intell. Syst. Technol. **3**(3), 57:1–57:22 (2012)
13. Rendle, S., Freudenthaler, C., et al.: BPR: Bayesian personalized ranking from implicit feedback. In: UAI, pp. 452–461. AUAI Press (2009)
14. Salakhutdinov, R., Mnih, A.: Probabilistic matrix factorization. Citeseer (2011)

Fairness Aware Recommendations on Behance

Natwar Modani[1(✉)], Deepali Jain[1], Ujjawal Soni[2], Gaurav Kumar Gupta[3], and Palak Agarwal[4]

[1] BigData Experience Lab, Adobe Research, Basel, Switzerland
{nmodani,deepjain}@adobe.com
[2] IIT Madras, Chennai, India
ujjawals@cse.iitm.ac.in
[3] IIT Roorkee, Roorkee, India
gup22uec@iitr.ac.in
[4] IIT Kanpur, Kanpur, India
palakag@iitk.ac.in

Abstract. Traditionally, recommender systems strive to maximize the user acceptance of the recommendations, while more recently, diversity and serendipity have also been addressed. In two-sided platforms, the users can have two personas, consumers who would like relevant and diverse recommendations, and creators who would like to receive exposure for their creations. If the new creators do not get adequate exposure, they tend to leave the platform, and consequently, less content is generated, resulting in lower consumer satisfaction. We propose a re-ranking strategy that can be applied to the scored recommendation lists to improve exposure distribution across the creators (thereby improving the *fairness*), without unduly affecting the relevance of recommendations provided to the consumers. We also propose a different notion of diversity, which we call representative diversity, as opposed to dissimilarity based diversity, that captures level of interest of the consumer in different categories. We show that our method results in recommendations that have much higher level of fairness and representative diversity compared to the state-of-art recommendation strategies, without compromising the relevance score too much. Interestingly, higher diversity and fairness leads to increased user acceptance rate of the recommendations.

1 Introduction

The typical objective of the recommender systems is to maximize the user acceptance of the recommendations, treating the acceptance of recommendation as a proxy to maximizing the utility from the consumers' point-of-view. Hence, the focus in all recommender systems has been to improve the prediction accuracy.

In a two-sided creative content discovery platform, e.g., Behance [1], the users can have two personas; consumers that consume the items, and creators who produce/supply the items. Such a platform needs to satisfy both the personas in order to be successful. Consumers satisfaction with the recommendations is based on, and can be measured by the traditional metrics (relevance of the

J. Kim et al. (Eds.): PAKDD 2017, Part II, LNAI 10235, pp. 144–155, 2017.
DOI: 10.1007/978-3-319-57529-2_12

recommendations, the level of diversity and chance of serendipitous discovery). On the other hand, the creators look for opportunities to reach out to a wide set of audience in order to be noticed and appreciated for their creations. If the creators (especially the new ones) do not get adequate exposure, they tend to leave the platform (or become inactive), and consequently, less content is generated on the platform, resulting in lower consumer satisfaction. Hence, for the two-sided platforms, while the relevance of the recommendations to the consumers remain a high priority, providing adequate exposures to the creators also plays an important role in creating a thriving community. The current state-of-art collaborative filtering techniques have been shown to favour popular items [5], thereby increasing the chances of new creators not receiving adequate exposure.

Diversity in the recommendations is recognized as an important consideration. The current notion of diversity is based on (dis)similarity of items, and hence, a uniform strategy is adopted for all consumers to introduce diversity in the recommendation results. Such strategies do not recognize the fact that different consumers have different level of interest in different categories of items. We propose a new notion of diversity, which we call 'representative diversity' that captures level of interest of the consumer in different categories.

In this paper, we focus on Behance [1], a creative content discovery platform. We propose a re-ranking strategy that can be applied to the scored recommendation lists to improve exposure distribution across the creators, without unduly affecting the relevance of recommendations provided to the consumers, and provides representatively diverse results. We define 'Creative Capital' as a notion of value of the creators, based on their contributions to the platform, measured in terms of number of projects created, number of views and appreciations received on their projects along with the recency of such events. 'Desired Exposure' is the ideal amount of exposure to be given to the creator based on the creative capital, and is defined as a sublinear function of contribution of the creators to address the fairness requirement. Fairness is defined as inverse of Jensen-Shannon Divergence (JS-Divergence) between the desired distribution and the actual obtained distribution of the exposures for the creators. Similarly, representative diversity is defined as inverse of JS-Divergence between the desired distribution and the actual obtained distribution of the exposures for the categories. We show that our method results in recommendation lists that have much higher level of fairness and representative diversity compliance compared to the state-of-art recommendation strategies, while the relevance score is not compromised too much. In fact, our experimental results on real data show that improvement in fairness and diversity tends to increase the user acceptance rate of the recommendations (which is the most relevant metric), even though the cumulative relevance score as assigned by the recommender systems is marginally lower.

2 Related Work

Over the years, many different recommendation techniques have been developed, mainly categorized into three types:

1. **Content-Based Filtering:** In these type of recommender systems, items (projects in case of Behance) with similar features to the ones already liked

by the consumer are recommended [9]. For creative images of Behance, Fang et al. [6] have proposed a feature learning paradigm to learn image similarities. Content-based filtering techniques are fair for all creators, i.e., projects of established popular creators as well as less popular (or upcoming creators) have equal chance of being recommended. But these techniques are limited to recommend items similar to those already liked by the consumer, hence less diverse and serendipitous recommendations.
2. **Collaborative Filtering:** These recommender systems predict relevant items to be recommended to a consumer using the history of items liked by other consumers. There is vast literature in Collaborative filtering (CF), including Item-based CF [10], user-based (nearest neighbors) CF [13], Matrix Factorization [7,8], and other techniques. CF techniques solves the problem of diverse and serendipitous recommendations to some extent. Though CF techniques perform better than content-based filtering, they tend to favor popular projects [5]. Since recommendations provide exposure to projects, this in-turn increases the likelihood of those projects being appreciated. This creates a clear rich-getting-richer scenario.
3. **Hybrid Recommender Systems:** To improve performance of recommender systems, content-based and collaborative filtering techniques are sometimes combined in the form of Hybrid Recommender Systems [3]. These methods deal with the cold start problem better than collaborative filtering by recommending new items through content filtering. A major limitation of these systems is the requirement of rich content and meta-data of the items. Moreover, these systems tend to be computationally more complex than either of the two approaches and hence, less scalable.

Diversity [2,4,12] has also been considered in some research, but they focus on diversifying the recommendations and do not consider consumer's diversity preferences. To the best of our knowledge, fairness for the creators on a two-sided platform is not studied as yet. Our method of ensuring fairness resembles the idea of the lottery scheduling method in CPU time allocation [11].

3 Definitions

As we will work in the context of 'Behance' as the application domain, we will start by discussing it briefly. Behance [1] is a creative content discovery platform. Users of Behance can have two personas; creators, who create 'projects' and publish them on Behance, and consumers, who view projects created by the creators. The projects can have one or more of the 137 creative fields associated with them, which can be thought of as categories on Behance. Every click of a consumer on a project to open it is counted towards number of views on the project. The consumers can also appreciate projects, which is another metric associated with the project. Consumers can also follow their favorite creators. We will now define the various notions we will use in the rest of paper.

Positional Value: Since the recommendations are ranked lists, and items at the lower ranking are less likely to receive attention of the consumers, we associate a *positional value* with each rank in the recommendation list. We take the positional value for rank 1 as 1 unit, and determine the positional value in relative terms by observing the relative click-through rates. Due to lack of space, we will not present detailed results, but we observed a near exponential decrease in the click-through rate for the items in various positions. Accordingly, we take the positional value of rank k based on the best fit to the data as:

$$pv(k) = e^{-\frac{k-1}{45}} \tag{1}$$

Creative Capital: Now we define 'Creative Capital' for the Behance creators, which is a measure of their contribution to the platform factoring in the recency of contributions. One can imagine that the creators who create more projects contribute more to the platform. However, higher quality projects should carry more weight. The quality of the projects can be estimated by popularity of the projects, which can be captured in term of the number of views and appreciations. Further, since our focus here is on recommendations, projects that are recent, or have received views and/or appreciation recently should carry more weight than projects that are old, and have not received much user attention recently. Accordingly, we define 'Creative Capital' $C_u(t)$ as follows:

$$C_u(t) = \gamma \times C_u(t-1) + \omega_p \times \Delta n_p(t) + \omega_a \times \Delta n_a(t) + \omega_v \times \Delta n_v(t) \tag{2}$$

A creator u *earns* creative capital by creating projects or receiving views and appreciations for projects created by her. The creative capital at the previous time step $C_u(t-1)$ is decayed with by a factor γ and carried over. Here, ω_p, ω_a and ω_v are the weights of each project creation, appreciation and view respectively. Also, $\Delta n_p(t)$ is the number of projects created by this creator between $(t-1)$ and t. Similarly, $\Delta n_a(t)$ and $\Delta n_v(t)$ are the number of appreciations and views received on his projects from $(t-1)$ to t, respectively.

Desired Exposure Distribution: We had noted that due to favoring popular items, collaborative filtering techniques tend to create rich-gets-richer scenario. To avoid this situation (which is key to ensure fairness), we allocate the exposures to the creators based on a sub-linear function of their creative capital. Please note that we want the creators who contribute more to receive more exposures to maintain incentive compatibility (i.e., there should always be incentive to produce more of high quality work, assuming that having more exposure is the incentive), and hence, the exposures should be a monotonic function of the creative capital. Hence, we define the deserved exposure for a creator u as:

$$E_u = \theta \times {C_u}^{\alpha}, 0 < \alpha < 1 \tag{3}$$

where θ is a normalization factor such that $\sum E_u = 1$ for all users u.

Fairness: Let the amount of exposures provided to the creations of a creator u be denoted by A_u, and the desired exposure distribution for the creator be E_u. Then, we can think of fractional exposure provided to creators (by normalizing across all creators) and exposure distributions as probability distributions over the creators. We define a fairness of a recommender system as inverse of JS-Divergence between these two distributions. Low value of JS-Divergence means that the actual exposure distribution is close to the desired exposure distribution, and hence the system is fair (so the fairness score is high), and a high JS-Divergence implies that the actual exposure distribution is significantly different than the desired exposure distribution, and hence the system is not fair.

$$F = \frac{1}{JSD(E||A)} \tag{4}$$

where, $JSD(E||A)$ is Jensen-Shannon divergence between two probability distributions E and A.

Representative Diversity: Different consumers on a platform have different appetite for different categories of items. We allocate the exposure to be given to the items from a category g for a consumer based on their (normalized) interest in that category. One challenge in such a strategy is that the user may not have explored the items of the platform enough for us to learn her preferences completely. Hence, we keep the exposure allocation for the category as a weighted average of the consumer's preference for the category and global preference of the category. The weight is based on the number of observations available for the consumer. As we gather more and more data about the consumer's preference, the global preference's weight keeps decreasing.

$$E_g(u) = \beta \times (\lambda_u p_g^u + (1 - \lambda_u)G_g) \tag{5}$$

where $E_g(u)$ is the exposure fraction allocated to category g for consumer u, $0 \leq \lambda_u \leq 1$ is the degree of certainty about estimate of consumer u's preferences, p_g^u is the estimated preference of consumer u for category g, and G_g is the global preference for category g. Also, β is a normalizing factor to ensure that $\sum_g E_g(u) = 1$. Clearly, λ_u is a function of amount of data available about consumer u's preferences.

We define the diversity compliance of the recommender system for a consumer as inverse of JS-Divergence of the desired exposure distribution for the categories E^c and the actual exposure distribution A^c for that consumer.

$$DC(u) = \frac{1}{JSD(E^c(u)||A^c(u))} \tag{6}$$

The global diversity compliance is defined as

$$GDC = \sum_u \{W(u) \times DC(u)\} / \sum_u W(u) \tag{7}$$

where $W(u)$ is the importance of consumer u, which we take as the sum of positional value of all exposures provided to the user u.

A Note About Simplification: In Behance, a project can be created by collaboration amongst multiple creators. Also, the project can have multiple categories associated with it. In the above description, we have given all formula considering only the case where each project is created by one creator and is associated with one category. This is done for ease of reading. While implementing our system, we have assigned partial credit to the creators and categories for such projects. Our experimental results are given for partial credit assignments.

3.1 The Final Objective Function

Recall that our aim is to provide "*relevant* and *representatively diverse* recommendations to the consumers, that provide *fair exposure* to the creators". Hence, we define our overall objective function as a combination of the user relevance, fairness to creators and representative diversity across categories. Suppose the relevance of an item i for a consumer u is given by r_{ui}, which may be based on the underlying recommendation algorithm (e.g., Collaborative Filtering). We define the overall relevance R_u for the user u as $R_u = \sum_k pv(k) \times r_{ui}$, where $pv(k)$ is the positional value of rank k, and r_{ui} is the relevance of the item i, which is recommended in position k in the recommendation list. The final relevance score for the recommender system across all users R_{all} is given as

$$R_{all} = \sum_u W(u) \times R_u \tag{8}$$

where $W(u)$ is the importance of consumer u as in Eq. (7), which we again take as the sum of positional value of all exposures provided to the user u.

Finally, we are ready to define our overall objective function:

$$O = (w_1 + R_{all})^{w_r} \times (w_2 + F)^{w_f} \times (w_3 + GDC)^{w_d} \tag{9}$$

This form of objective function ensures that none of the factors can be ignored completely. The various weights $(w_1, w_r, w_2, w_f, w_3, w_d)$ control the importance of the different factors. We would like to give higher importance to relevance and fairness compared to the diversity, and hence we select $w_1 = 0, w_r = 1, w_2 = 0$, $w_f = 1$, and $w_3 = 1, w_d = 1$. This results in simplified objective function

$$O = R_{all} \times F \times (1 + GDC) \tag{10}$$

Given that, we would not know a-priori which consumers are likely to visit the platform on a given day, we would like to make the recommendations in such a way, that the solution has a high value of objective function on an ongoing basis, and not only at the end of one round of execution. In the next section, we give a heuristic approach for ongoing optimization of this objective function, as due to JS-Divergence in the objective function for our problem formulation, it is not possible to devise an efficient exact or approximation algorithm.

4 Algorithm for Generating Recommendations

We will first outline an optimization approach for a general resource allocation problem and then illustrate how to translate it to the present context of re-ranking recommendations.

Consider a set of resource requesters, along with a prespecified share of resource eligibility for each requester. The resource become available in chunks in an online fashion. When a resource chunk becomes available, it needs to be allocated to one requester (without dividing it). The goal is to allocate resource chunks in such a fashion, that at every time, the resource distribution over all requesters is as close to the prespecified resource eligibility share as possible.

We propose the following greedy algorithm to solve the given problem. For every resource chunk $r(t)$, calculate the value of allocating the resource to each requester u as

$$V_u = E_u \times \frac{(\sum_v A_v(t-1) + r(t))}{(A_u(t-1) + r(t))} \tag{11}$$

where E_u is the pre-specified share of resource eligibility for requester u, $A_u(t-1)$ is the already allocated resource units to requester u until time $(t-1)$. Now, there are two strategies possible. First is a deterministic strategy, where we allocate $r(t)$ to the requester such that the value is the highest. Second is a probabilistic strategy, where we allocate the resource to the requesters with probability equal to the normalized value.

One can see both the fairness and representative diversity as resource allocation problem described above. Our overall objective function is a combination of three components. Hence, we use this method to generate two of the factors which we use for the re-ranking strategy, while the third component is based on the relevance as assigned by the underlying recommendation algorithm.

First, we generate a rating or relevance scores $r_{u,i}$ using state-of-art collaborative filtering techniques for all project-consumer pairs. We also compute the global popularity ratings g_i for all projects as the average of all observed rating for the project. We then follow the following steps for recommending projects to each consumer u, for whom, we need to generate k_u recommendations:

1. Create a candidate pool of projects to recommend by taking all the projects for which the rating is positive (i.e., $r_{u,i} > 0$).
2. If the pool is smaller than the number of projects to be recommended, add all the other projects (ones with $r_{u,i} = 0$) to the pool.
3. Then calculate goodness of all the projects in pool as follows:

$$G_{u,i} = r_{u,i} \times V_F(c(i)) \times V_D(g(i)) \tag{12}$$

   where, $V_F(c(i))$ is value of allocating the exposure to the creator of project i (refer Eq. 11), $V_D(g(i))$ is the value of allocating the exposure to the category that project i belongs to, and $r_{u,i}$ is the relevance rating of project i to the user u as mentioned earlier.
4. Now select project with maximum goodness (we will call this as deterministic approach) or select a project probabilistically from the list with probability of selection equal to its normalized goodness (we will call this as probabilistic approach).
5. Remove the selected project from the list and continue recommending from remaining projects until all recommendations are done.

While computing the relevance, the rating of projects with $r_{ui} = 0$ is taken as $1/5$ of the lowest value of r_{ui} from the top-k projects.

5 Experimental Results

First, we will validate our motivation by analyzing the churn rate of the creators to show that creators who do not get adequate views in the beginning tend to churn with higher probability. We will show that the Creative Capital is a good metric to capture the contribution of the creators for the platform. Both these analysis are on the full Behance dataset. We will then describe our data set for recommendation re-ranking. We study the performance of various state-of-art collaborative filtering techniques to choose the baseline relevance assignment approach. We will compare the performance of the proposed approaches and baseline approaches on the three axis, fairness, relevance, and diversity. Finally, we will also evaluate the various approaches for the precision and recall based performance. Given the space limitation, we will not present detailed results and plot in all cases, and only quote the results in the running text.

5.1 Churn Rate Analysis

To illustrate the need to address the fairness, we have done an analysis of the churn rate of creators on Behance. A creator is said to have churned if he stops publishing any new projects. We calculated the number of creators who got only a small number of views/appreciations in their initial 12 months, and computed the churn rate as the fraction of creators who stopped publishing projects after this initial period. We found that the churn rate for creators who get up to 5 views during the initial 12 months is approximately 2.5 times than the creators who got at least 100 views in the first 12 months. However, the churn rate does not change significantly for the creators that received at least 100 views. If we assume that the relation between views received and churn rate remains the same, then the re-ranking strategy proposed in this paper that marginally reduces number of exposures for highly popular creators and distributes those among less popular creators for fairer exposure, results in 12% reduction in churn rate. This experiment clearly highlights the importance of giving fair opportunities to creators for their projects to be viewed to reduce churn-rate.

5.2 Creative Capital Analysis

As explained in Sect. 3 Eq. (2), we computed the 'Creative Capital' as a function of number of projects created and number of views and appreciations received, along with recency of such signal. We used the following parameter values: $\gamma = 0.98$, $\omega_p = 50$, $\omega_a = 5$ and $\omega_v = 1$. These weights are inversely proportional to the relative frequency of occurrence of respective events in order to give equal importance to each of these. The intent of this metric was to capture the perceived contribution of the creators to the platform. Typically, people tend to follow the creators based on their contribution. As we did not use the follower information for defining this metric, we can use it for cross validating the metric. If the metric is indeed a good indicator of creator's contribution, the increase in number of followers should coincide with the increase in creative capital.

Accordingly, we calculated the Pearson Correlation Coefficient between increase in creative capital, $C_u(t) - C_u(t-1)$ and corresponding $\Delta n_f(t)$ (increase in the number of followers of u from $t-1$ to t). The average correlation was observed to be 0.7457 which establishes the validity of Creative Capital as a measure of worthiness of a creator.

5.3 Data Set

Behance has an active user base of multiple million users, with about one quarter of the users being creators. The number of projects created by these creators is also in millions. To evaluate the recommendation performance, we work with a sample of data that has 638 creators, having 2,000 projects, and 1,400 consumers. The total number of project views and appreciations were 28,000 and 9,800, respectively. We split the data such that approximately 80% views and appreciations go into train and 20% in test sets.

5.4 Baseline

As collaborative filtering techniques have been shown to outperform other recommendation approaches, we take collaborative filtering techniques as the baseline for comparison. Since there are many collaborative filtering techniques proposed in literature, we first conducted experiments to determine which of these techniques perform the best for our dataset. We implemented nearest neighbour, item-item and matrix factorization based collaborative techniques, and checked for accuracy of the recommendations provided. We found that item-item jaccard nearest neighbor based CF algorithm performed the best with approximately 5% better accuracy in top-k recommendations for a broad range of k. Hence, we take item-item CF as our main baseline and call it 'Traditional' baseline. We also take randomized strategy as baseline called 'Baseline Random', as randomness would likely result in high degree of fairness and diversity. To ensure that the recommendations are not completely irrelevant, we also created hybrid baselines called 'Baseline Hybrid', where first 50% of the recommendations are the ones with the highest predicted ratings and the rest are chosen randomly.

5.5 Fairness, Diversity and Relevance

We now evaluate the performance of our two approaches (probabilistic and deterministic), and the results are compared against traditional CF approach and other baselines.

First, we look into fairness. There are two aspects of fairness; first, the strategy should allocate the exposures to the creators in a manner consistent with the objective of giving fair exposure to all creators. Second, the recommendation algorithm should follow the exposure allocation while performing the recommendations. The left hand side of the plot in Fig. 1 shows the number of people (on y-axis) who will be given a certain amount of exposure (on x-axis). Here, we have

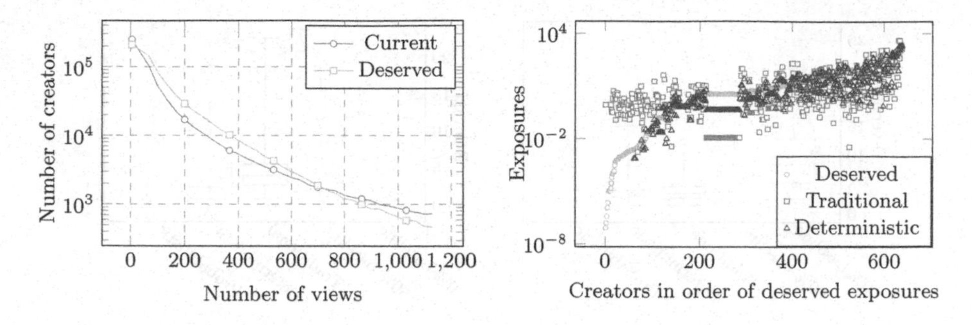

Fig. 1. Fairness results: Left - view allocation; Right - actual exposure distribution

taken $\alpha = 0.75$ for determining the deserved/allocated exposures (ref. Eq. (3)). As one can see, the number of people who receive large number of exposures is reduced, and the number of people who get moderate exposures is increased. This shows that our method allocates the exposure in a fairer manner compared to the current state-of-art collaborative filtering techniques. The right hand side plot in the same Fig. 1 shows the deserved exposure allocation, the exposure provided by our method, and the exposure provided by the collaborative filtering technique. One can clearly see that our method adheres much more closely to the allocated exposures as compared to the collaborative filtering. The correlation between the deserved and actual exposure provided by our deterministic method is 0.8682, whereas the correlation for the item-item CF with deserved exposure is 0.6573. This clearly shows that our method has good intent (left plot) and good execution (right plot) for fairness to creators. Table 1 reports the fairness numbers achieved by various methods, which clearly shows that our proposed approaches achieve nearly twice as good fairness compared to traditional and randomized baselines.

Figure 2 compares the diversity in the categories of the projects recommended and the relevance for the consumers. The figure on the left shows that our models (especially probabilistic without beyond k) perform better than the traditional approach. The randomized baseline approaches are expected to perform well because picking random projects would lead to increase in diversity. The figure on the right shows that while our models perform very well on fairness and diversity fronts, as expected it lags behind in terms of relevance, as the improvement in fairness has been achieved at the cost of drop in relevance. However, we find that the average loss in relevance was about 9% only, whereas the average improvement in fairness was 97.1%, over and above the considerable increase in

Table 1. Fairness value achieved by various methods

Method	Baseline random	Baseline hybrid	Baseline IICF	Deterministic	Probabilistic
Fairness	4.11	3.85	2.97	6.56	6.03

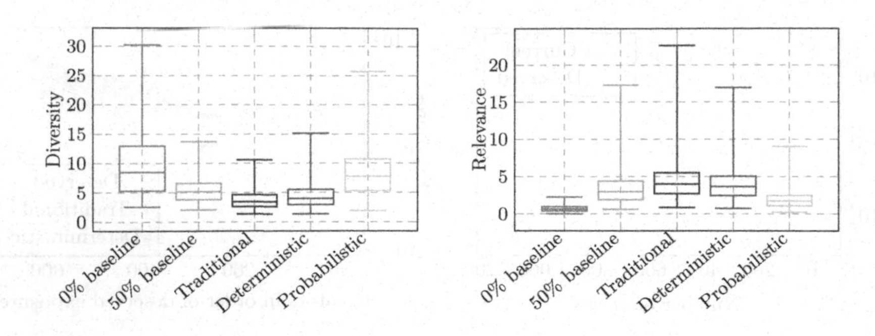

Fig. 2. Diversity and relevance distribution

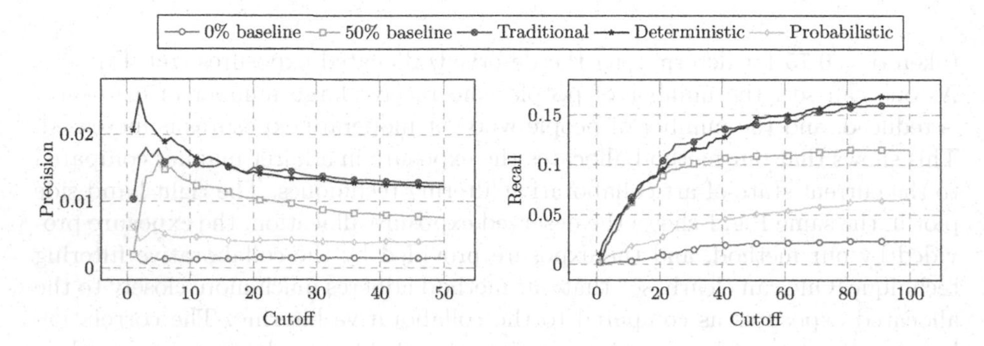

Fig. 3. Precision and recall results for different methods

diversity. We also see that the random (and hybrid) baseline performs poorly on the relevance front even though it performed fairly well in terms of fairness. This means that randomized approach are not viable alternatives.

Finally, Fig. 3 compares the precision and recall of the results for all the approaches. where precision and recall at cutoff k are defined as:

$$P(k) = |a \cap p_k|/k \qquad R(k) = |a \cap p_k|/|a|$$

where a is the set of projects that the consumer has appreciated and p_k is the set of top k projects recommended to the consumer. As we can see our models have higher precision and recall than the baseline models, including even the best performing CF technique. The deterministic approaches perform the best in general. The high precision and recall for the traditional method is expected.

6 Conclusions

In this paper, we addressed an important issue of fairness to the creators while providing relevant and diverse recommendations to the consumers on a two-sided platform. We showed that by sacrificing a small amount of relevance, one can achieve a much higher degree of fairness and diversity in the recommendations.

Further, we also showed that in terms of the precision and recall, which are the most relevant metrics, our proposed approach outperforms the state-of-art collaborative filtering techniques. There are some interesting research directions as a follow up of this work, including more robust definition of Creative Capital and approximation guarantee algorithms.

References

1. http://www.behance.net
2. Adomavicius, G., Kwon, Y.: Toward more diverse recommendations: item re-ranking methods for recommender systems. In: Workshop on Information Technologies and Systems (2009)
3. Burke, R.: Hybrid recommender systems: survey and experiments. User Model. User-Adapt. Interact. **12**(4), 331–370 (2002)
4. Carbonell, J., Goldstein, J.: The use of MMR, diversity-based reranking for reordering documents and producing summaries. In: Proceedings of the 21st Annual International ACM SIGIR Conference on Research and Development in Information Retrieval, pp. 335–336. ACM (1998)
5. Celma, Ò, Cano, P.: From hits to niches? Or how popular artists can bias music recommendation and discovery. In: Proceedings of the 2nd KDD Workshop on Large-Scale Recommender Systems and the Netflix Prize Competition, p. 5. ACM (2008)
6. Fang, C., Jin, H., Yang, J., Lin, Z.: Collaborative feature learning from social media. In: Proceedings of the IEEE Conference on Computer Vision and Pattern Recognition, pp. 577–585 (2015)
7. Koren, Y.: Factorization meets the neighborhood: a multifaceted collaborative filtering model. In: Proceedings of the 14th ACM SIGKDD International Conference on Knowledge Discovery and Data Mining, pp. 426–434. ACM (2008)
8. Koren, Y., Bell, R., Volinsky, C., et al.: Matrix factorization techniques for recommender systems. Computer **42**(8), 30–37 (2009)
9. Pazzani, M.J., Billsus, D.: Content-based recommendation systems. In: Brusilovsky, P., Kobsa, A., Nejdl, W. (eds.) The Adaptive Web. LNCS, vol. 4321, pp. 325–341. Springer, Heidelberg (2007). doi:10.1007/978-3-540-72079-9_10
10. Sarwar, B., Karypis, G., Konstan, J., Riedl, J.: Item-based collaborative filtering recommendation algorithms. In: Proceedings of the 10th International Conference on World Wide Web, pp. 285–295. ACM (2001)
11. Waldspurger, C.A., Weihl, W.E.: Lottery scheduling: flexible proportional-share resource management. In: Proceedings of the 1st USENIX Conference on Operating Systems Design and Implementation, p. 1. USENIX Association (1994)
12. Zhang, M., Hurley, N.: Avoiding monotony: improving the diversity of recommendation lists. In: Proceedings of the 2008 ACM Conference on Recommender Systems, pp. 123–130. ACM (2008)
13. Zhao, Z.-D., Shang, M.-S.: User-based collaborative-filtering recommendation algorithms on hadoop. In: Third International Conference on Knowledge Discovery and Data Mining, WKDD 2010, pp. 478–481. IEEE (2010)

A Performance Evaluation Model for Taxi Cruising Path Recommendation System

Huimin Lv, Fang Fang, Yishi Zhao, Yuanyuan Liu, and Zhongwen Luo(✉)

College of Information Engineering,
China University of Geosciences (Wuhan), Beijing, China
lvhuimin320@gmail.com, ffang1014@163.com,
{zhaoyishi,luozw}@cug.edu.cn,
yuanyuanliu@mails.ccnu.edu.cn

Abstract. Recommending an appropriate route to reduce taxi drivers' mileage spent without a fare is a long-standing challenge. The current solution has been to get the best route which has optimal performance, and the performance usually combined the conditional probability for getting a passenger and the cruising distance. However, the main reference has some limitation. To eliminate the limitation, a novel model is proposed to evaluate the candidate route performance. And based on this new model, a recommendation system is tested. Firstly, by mining the knowledge of the historical taxi trajectory, we extract the temporal probabilistic recommending points. Then based on it, the evaluation model is presented to estimate the performance of each candidate route. Finally, a route recommendation algorithm is used to get the optimal route for taxi drivers. And as the result, the experiment is performed on real-world taxi trajectories data set, and shows the effectiveness of the proposed model for evaluating the performance.

Keywords: Evaluation model · Mobile recommendation systems · Taxi drivers

1 Introduction

Nowadays, taxi service plays an important role in public transportation service in large cities. However, there are often a huge number of taxis cruising around the city with no passengers. The vacant taxis not only waste energy but also result in a traffic jam. So, a recommendation system to improve the performance of taxis is needed. And the advances of various technologies provide the possibility.

Indeed, most of the existent mobile recommendation systems are using the integration of the conditional probability and the cruising distance or others such as income to measure the performance of the route and then recommend the best one to the taxi [1–4]. However, we find that the living performance evaluation method is wrong in some cases. Using the existing methods will result in sending the taxi to a lower performance route. Hence, a new performance evaluation method is proposed in this paper. In addition, since taxi trajectories are big spatial-temporal data and how to extract the useful information like the mobility pattern of the passengers with consideration of the time factor is also challenging.

J. Kim et al. (Eds.): PAKDD 2017, Part II, LNAI 10235, pp. 156–167, 2017.
DOI: 10.1007/978-3-319-57529-2_13

To that end, in this paper, we propose a recommendation system based on historical trajectory taxi data. The key idea is that it utilizes a new route model to evaluate a candidate route and then provide an algorithm to find a potential passenger with the minimum cruising miles. Specifically, the contributions of this paper are as follows:

1. A novel model for evaluating the candidate route is proposed. The model computes the potential cruising distance along the route for picking up per passenger. The main difference of the new model from the traditional one is considering the passenger number of the route. A recommendation system for taxi drivers to minimize their cruising driving distance for taking per passenger is presented.
2. To verify the effectiveness of the model, we conduct extensive experiments on a real-world data set. And the result shows that the new model is more effective.

The remainder of this paper is organized as follows. Section 2 shows some related works. In Sect. 3, we formulate the problem of route recommendations for taxi drivers and introduce some preliminaries in the paper. Section 4 presents the generation of the temporal probabilistic recommending pick-up points. In Sect. 5, the recommending model is discussed in details. Section 6 shows some experimental results and the paper is concluded in Sect. 7.

2 Related Works

In the literature, a mass of research has been devoted to the recommendation system [5–9]. Based on the massive data of the taxis' trajectory, the route recommending system's main target is to provide the more efficient driving route for taxi drivers by finding the behavior pattern of the experienced taxi drivers, the potential flowing direction of the crowds, etc. Li et al. [10] pay attention to the prediction of the movement of human beings. They present an adaptive hot extraction algorithm to cluster the pick-up/drop-off events of the passengers. Awasthi et al. [11] propose a rule-based method to evaluate the fastest path in the city. In order to get the fastest route, they build a statistical model using the traffic log. Gonzalez et al. [12] develop an adaptive fastest path algorithm by considering the speed patterns mined from historical GPS trajectory data. Ge et al. [1] develop a mobile recommendation system to recommend a taxi driver with the shortest potential travel distance route for finding a passenger. Then some concern the carpool service [13, 14] to save energy and seek the balance of demand between the taxi drivers and passengers. In the T-Share system, users submit request of taking a taxi with the location of getting on and off, the number of passengers and the expected time to the destination through the phone. System maintains all states of the taxi in real-time in the back, and after receiving a request, search out the best cab which satisfies the conditions of the new user and the passengers already in the cab. In addition, other works care about the optimization of calculation. Statistics show that the time complexity of existing recommendation methods are usually exponentially [15]. Trestian et al. [16] use the orthogonal kd-tree. Yang et al. [4] propose a new kds-tree structure which is a binary tree and extended from kd-tree and ball-tree. In this article, we focus on the recommendation of the shortest potential cruising distance for taxi drivers. Different from the earlier studies, we propose a novel model to evaluate the performance of each candidate route and then based on it, get the optimal route for taxi drivers.

3 Problem Definition

Definition 1. Picking-up rate: picking-up rate is the probability of finding a passenger at one pick-up point.

Definition 2. Given a set of N potential pick-up points $C = \{C_1, C_2, \ldots, C_N\}$, a route R is a sequence of connected pick-up points, i.e., $R = (C_1 \rightarrow C_2 \rightarrow \ldots \rightarrow C_K)$ $(1 \leq K \leq N)$, The length of route R denotes as $|\vec{R}| = K$, the number of pick-up points. R_{set} is the set of R, which is generated from C. Independence probability set $P_R = (P_1, P_2, \ldots, P_K)$ denotes the set of picking-up rate of pick-up point and distance subset D_R represents the set of distance between each pair of pick-up points en route R.

Note that C_0 always denotes the current position of a taxi PoTaxi in this paper.

Definition 3. The taxi mobile routing recommendation problem is to recommend a profitable route to a taxi driver so that the potential cruising distance to a possible passenger is minimized.

As the calculating the potential cruising driving distance depends on the current position of the taxi, time period, route R, and the corresponding picking-up rate set P_R and the distance set D_R, the potential cruising distance function can be denoted as:

$$F(\text{PoTaxi}, T, R, P_R, D_R)$$

Note that, in this paper, we limit the length of route R to be K. This is because the calculating constraints and considering the practical applications.

So, the taxi mobile routing recommendation problem can be formulized as:

$$\min_{R \in R_{set}} F(\text{PoTaxi}, T, R, P_R, D_R)$$

Almost all the current existing researches of evaluating the performance of the route are using the integration of the conditional probability for getting a passenger and the cruising distance. However, this is wrong. Because it only takes the potential driving distance into account without considering the probability of picking up passengers. In other words, they do not consider the number of the passengers along the route. The potential cruising distance of finding a passenger will be the correct evaluation standard. So, in this paper, we put forth a novel model, the potential cruising distance function, which is not only considering the driving distance also the probability of picking up passengers, to evaluate the route from a taxi to a potential passenger, which will be discussed in detail later.

First of all, let's focus on a demonstration of recommending. Figure 1 shows an illustration example of two candidate paths.

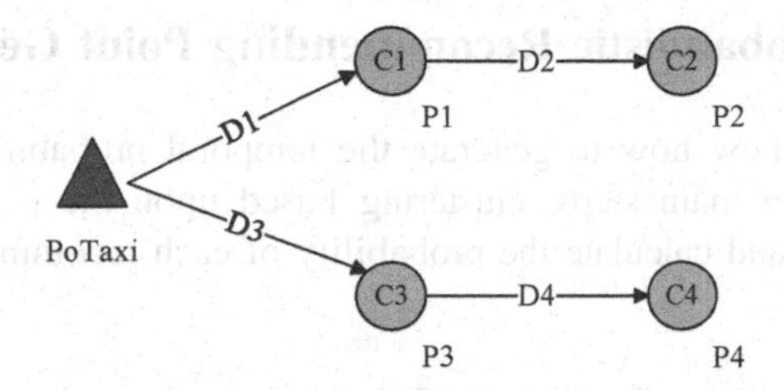

Fig. 1. An example of two candidate path

In this graph, node PoTaxi (C_0) represents the current position of an empty taxi at time period t, node C_i (i = 1, 2, 3, 4) denote the recommending pick-up point with the estimated picking-up rate P_i (i = 1, 2, 3, 4) respectively. D_i (i = 1, 2, 3, 4) indicate the distance between node C_{i-1} and node C_i. In addition, there are two candidate driving routes $R_1 = \{PoTaxi, C_1, C_2\}$ and $R_2 = \{PoTaxi, C_3, C_4\}$. Note that the nodes in the path are sequential and assumed to be different from each other. This is because we do not allow taxi drivers to drive back and forth.

Nowadays, almost all the methods of calculating the potential cruising distance from the taxi to a potential passenger are integrating the conditional probability with the cruising distance. For example, in Fig. 1, the potential cruising distance of route R_1 could be $P_1D_1 + (1 - P_1)P_2(D_1 + D_2)$, for route R_2, it will be $P_3D_3 + (1 - P_3)P_4(D_3 + D_4)$. In some cases, it makes sense, just like the probability is almost similar but the distance is very different. In other cases, however, this is not really applicable. As Fig. 2 shows, under this circumstance, the method is wrong.

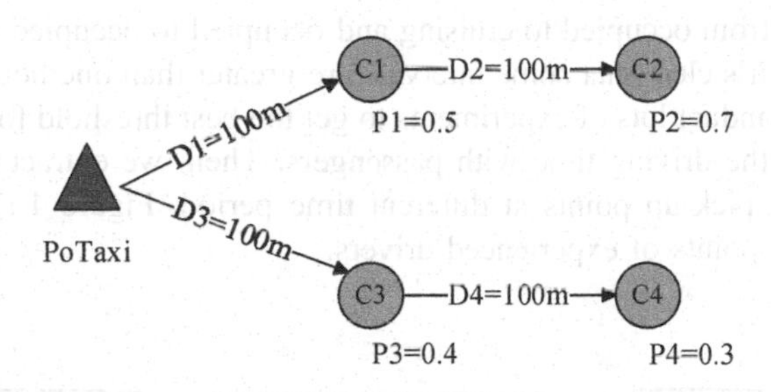

Fig. 2. A concrete example of routes

According to the previous method, the calculated potential cruising distance respectively is 120 m and 76 m, so the better route will be $R_2 = \{PoTaxi, C_3, C_4\}$. However, it is inconsistent with the facts, and clearly that we should choose route R_1 rather than R_2 in any cases unless you do not want to make a profit. The potential cruising distance from the taxi to a potential passenger cannot be simply integrating the conditional probability with the cruising distance. Furthermore, we should also take the probability of picking up passengers into account.

4 Temporal Probabilistic Recommending Point Generation

In this section, we show how to generate the temporal probabilistic recommending points. There are two main steps: clustering based upon the pick-up points of the experienced drivers, and calculate the probability of each recommended point.

4.1 Clustering Based on the Pick-up Points of the Experienced Drivers

To generate the recommended points, firstly, the experienced drivers are extracted from a large number of taxi track data. Then we can get their pick-up and drop-off points at different time period. Secondly, calculating the pair wise driving distance of these pick-up points of different time period using the Google Maps Distance Matrix API. Finally, clustering based on the calculated driving distance.

The driving time and the driving occupancy rates are the main factors to extract the experienced drivers, while the state of the driver is important to calculate the driving time and the driving occupancy rates. We consider drivers with plenty of driving time and high driving occupancy rates to be experienced. In general, there is three status of the driver's driving state: occupied, cruising and out-of-service state. Driver's driving time is the time when the state is not out-of-service, and the occupancy rate is the ratio of driving time of occupied to total driving time. Assume that there are two continuous GPS points of a driver, the state of the two points are occupied, but the time interval is greater than an hour, can we expect this time interval as occupied driving time? Figure 3(a) shows the distributions of the time interval of two continuous GPS points of more than 500 drivers in San Francisco over a period of about 30 days. Figure 3(b) and (c) show the distributions of the time interval of two continuous GPS points when the state changes from occupied to cruising and occupied to occupied of these drivers. From the figures, it's clear that some intervals are greater than one hour. Based on this observation, we conduct lots of experiments to get the best threshold for calculating the driving time and the driving time with passengers. Then, we extract the experienced drivers with their pick-up points at different time period. Figure 4 shows the distribution of pick-up points of experienced drivers.

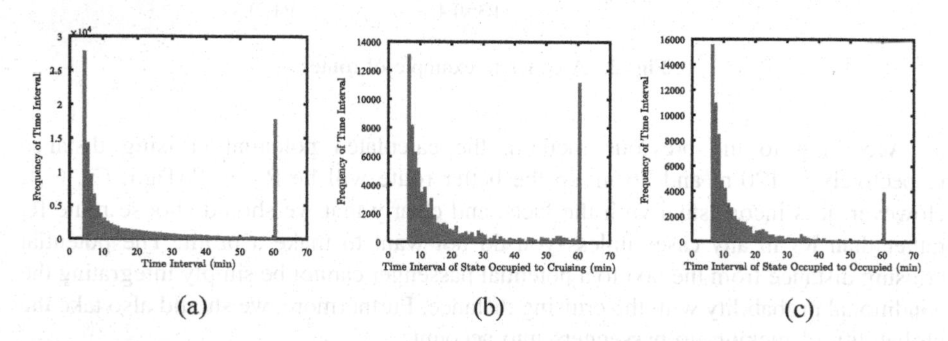

Fig. 3. Distributions of time interval of two continuous GPS points: (a) all status; (b) occupied to cruising; (c) occupied to occupied

As Fig. 4 shows, different time periods have various numbers of pick-up events. In other words, there is different pick-up probability. And we can find that the trend is in accordance with our common sense, where the picking-up events happen less in the midnight (02:00–04:00) and higher during the night (18:00–22:00). After obtaining the historical pick-up points at different time periods, we use driving distance to cluster these points into N clusters for different time periods. Using driving distance rather than simply the spherical distance or Euclidean distance can make more accurate recommended results. Furthermore, we use the Cluto [17] for clustering by using vcluster clustering programs with parameters "–clmethod = direct". Eventually, the center of each cluster is the recommending points we needed.

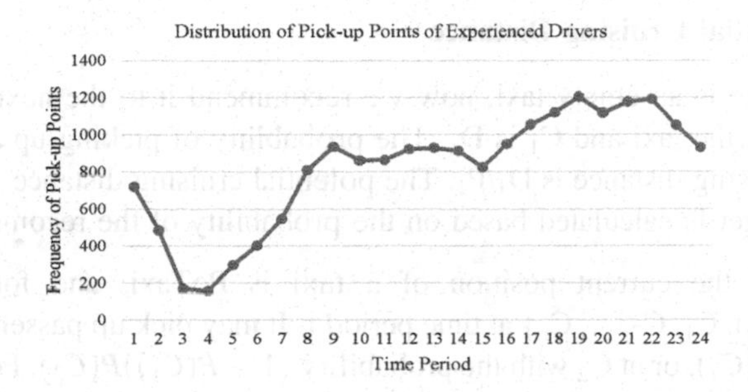

Fig. 4. Distribution of pick-up points of experienced drivers. The size of the timeslot is one hour, where 1 stands for 00:00–00:59, 2 stands for 01:00–02:59, etc.

4.2 Calculation of Probability of Recommended Points

To generate the probability of each recommended point, we measure the number of taxis, which pick up passengers when passing by the cluster while unoccupied. After getting the clusters, we should obtain the temporal-spatial coverage of each cluster. For each point in each cluster, we get the distance to the center of the cluster, then obtain the average distance for each cluster. The temporal-spatial coverage defines as a circle with radius of the average distance.

Definition 4. The probability of finding a passenger for each cluster c at time period t can be estimated as:

$$\mathrm{P(c,t)} = \frac{|states(cruising \rightarrow occupied)|}{|states(cruising)|}$$

where |status(cruising)| denotes the number of cruising taxis which passed by cluster c at time period t, and |status(cruising→occupied)| is the number of these cruising taxis which passed by cluster c at time period t and changed their state from cruising to occupied.

Since the probability of picking-up is very sensitive to time, for time period t, we divide it into several small ones. Then we calculate |status(cruising→occupied)| and

|status(cruising)| for each cluster respectively, and finally get the probability of finding a passenger for each cluster c at time period t.

5 Optimal Route Recommendation

In this section, we introduce the technical details for searching a route with the shortest potential cruising distance, which is we consider the optimal route. Firstly, we will show the model for measuring the potential cruising distance of each candidate route. Next, a recommending algorithm to get the optimal route will be presented.

5.1 The Potential Cruising Distance

Assume that there is an empty taxi, now we recommend it to the next place C_1. The distance between the taxi and C_1 is D_1. The probability of picking-up at C_1 is P_1. So, the potential cruising distance is D_1/P_1. The potential cruising distance from a taxi to a potential passenger is calculated based on the probability of the recommended points.

Definition 5. If the current position of a taxi is PoTaxi, and follow the route $R = \{PoTaxi\ (C_0), C_1, C_2, \ldots, C_n\}$ at time period t. It may pick up passengers at C_1 with the probability $P(C_1)$, or at C_2 with the probability $(1 - P(C_1))P(C_2)$. For each pick-up point C_i, the picking-up rate is following as:

$$P(C_i|R, t) = \begin{cases} P(C_i, t); & i = 1 \\ P(C_i, t) \prod_{j=1}^{i-1} (1 - P(C_j, t)). & i > 1 \end{cases}$$

In addition, we use $D(C_j, C_{j+1})$ to represent the driving distance between pick-up point C_j and pick-up point C_{j+1}. Thus,

Definition 6. The potential cruising distance function F can be defined as:

$$F = \frac{\sum_{i=1}^{n-1} \left(P(C_i|R,t) \sum_{j=0}^{i-1} D(C_j, C_{j+1}) \right) + P(C_n|R,t)/p(C_n,t) \sum_{j=0}^{n-1} D(C_j, C_{j+1})}{\sum_{i=1}^{n} P(C_i|R,t)}$$

By observing the form of the potential cruising distance function, we can simplify this formula and re-write as:

$$F = \frac{\sum_{i=1}^{n} \frac{P(C_i|R,t) D(C_{i-1}, C_i)}{P(C_i,t)}}{\sum_{i=1}^{n} P(C_i|R,t)}$$

To clearly explain our potential cruising distance function, we illustrate it via an example. Figure 5 shows an example of a recommended cruising route PoTaxi → C_1 → C_2 with the corresponding probability $\{P_1, P_2\}$, driving distance $\{D_1, D_2\}$

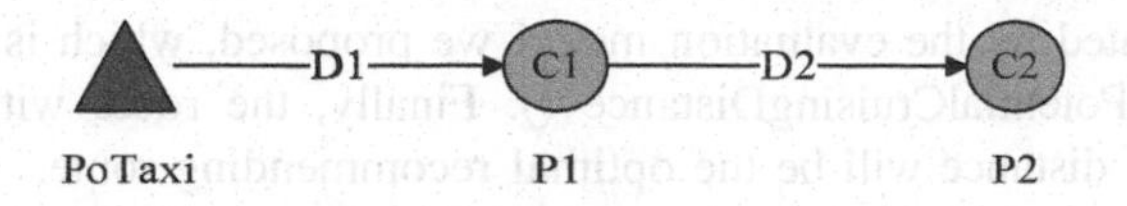

Fig. 5. An example of a recommended cruising route

respectively. The potential driving distance could be $(P_1D_1 + (1 - P_1)(D_1 + D_2))$ while it may have passengers of $(P_1 + (1 - P_1)P_2)$. Therefore, the potential cruising distance to a potential passenger will be:

$$\frac{P_1D_1 + (1 - P_1)(D_1 + D_2)}{P_1 + (1 - P_1)P_2}$$

5.2 Optimal Route Recommendation

In this subsection, we introduce the method for recommending the route with the shortest potential cruising distance to target taxi. Once the capacities of all the road are obtained based on the evaluation model proposed above, we can recommend a trajectory to a taxi given its current location and time.

Figure 6 shows the pseudo-code of the recommending algorithm. Given the current location (PoTaxi) and time (T) of the taxi; first of all, we can obtain the set of the recommended cluster nodes of current time (C_{set}), the set of the probability of the cluster nodes of the current time (P_{set}) which is corresponding to the Cset, and the driving distance matrix of the cluster nodes (D_{set}). Then, based on the above mentioned and the length of the route (k), all the candidate route can be gotten by the function GetCandidateRouteSet (). Next, for all candidate routes, the potential cruising distance

```
Algorithm GetRecommendingRoute (PoTaxi, T, k, Cset, Pset, Dset) {
    ShortestCruisingDistance = + ∞;
    CandidateRouteSet = GetCandidateRouteSet (k);
    for each route in CandidateRouteSet {
        PotentialCruisingDistance = GetPotentialCruisingDistance (PoTaxi,
T, Cset, Pset, Dset);
        if ShortestCruisingDistance > PotentialCruisingDistance
            ShortestCruisingDistance = PotentialCruisingDistance;
        end if
    The route with the minimum ShortestCruisingDistance is the optimal
recommending route;
    end for
}
```

Fig. 6. The recommending algorithm

should be calculated by the evaluation model we proposed, which is encapsulated in the function GetPotentialCruisingDistance (). Finally, the route with the minimum potential cruising distance will be the optimal recommending route.

6 Experiments

In this section, to demonstrate the effectiveness of the proposed evaluation model and evaluate the performance of the proposed recommendation system, we have done extensive experiments on real-world data sets.

6.1 Experiment Data

In this paper, we train our system using the real-world data sets collected in the San Francisco Bay Area in 30 days, which provided by the Exploratorium-the museum of science, art and human perception through the cabspotting project. The mobility traces are the records of more than 500 taxis' driving states in consecutive time. Each record can be expressed as a tuple: (unique taxi ID, latitude, longitude, status, time stamp).

In the experiments, we obtain the experienced drivers by exploring the important properties of the drivers: driving time and driving occupancy rate. Figure 7 shows the distributions of the driving time and the driving occupancy rate. From Fig. 4 we can see that the picking-up events occur most frequently during the time period 18:00–19:00, and during 14:00–15:00, the gradient has a sharp change. So, we will focus on this two time period in the experiment. In total, 1203 pick-up points of experience drivers and 561573 points of all taxis are obtained during 18:00–19:00, and 822 pick-up points of experience drivers and 509362 points of all taxis are obtained during 14:00–15:00. All potential pick-up points are clustered into 10 clusters. Table 1 shows the information of the 10 temporal probabilistic pick-up points during 18:00–19:00. And Table 2 shows the information of the 10 temporal probabilistic pick-up points during 14:00–15:00. Note that the latitude and longitude represent the corresponding centroid of the cluster, the $P(C_i)$ represents the picking-up rate in the cluster C_i.

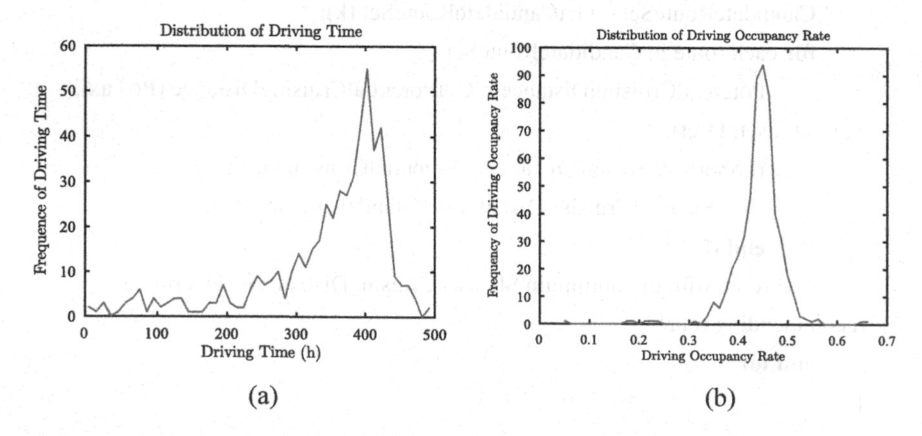

Fig. 7. Distribution of: (a) Driving time; (b) Driving occupancy rate

Table 1. Description of the 10 clusters during 18:00–19:00

No.	C_1	C_2	C_3	C_4	C_5	C_6	C_7	C_8	C_9	C_{10}
Latitude	37.78647	.80349	.79091	.79240	.76676	.79456	.77573	.77438	.75038	.43327
Longitude	−122.40942	.41193	.40027	.42260	.42574	.43721	.39663	.45873	.43327	.38711
P(Ci)	0.8795	0.7039	0.8888	0.8713	0.7856	0.7383	0.5831	0.6935	0.8377	0.4419

Table 2. Description of the 10 clusters during 14:00–15:00

No.	C_1	C_2	C_3	C_4	C_5	C_6	C_7	C_8	C_9	C_{10}
Latitude	37.61475	.70423	.75358	.79087	.78653	.76744	.80369	.79360	.77130	.77758
Longitude	−122.38618	.41852	.43296	.40164	.41253	.44328	.41476	.43607	.42185	.39720
P(Ci)	0.4955	0.2967	0.8182	0.7764	0.7974	0.6861	0.6739	0.6852	0.5736	0.5866

6.2 Effectiveness

In this section, we compare the proposed model in the paper with the PTD function [1]. Here, we show the optimal driving routes recommended by the PTD function and our new model. Figure 8 shows the potential recommending points (the red points) within the time period 18:00–19:00 (a) and 14:00–15:00 (b) and the assumed position of the empty taxi to be recommended (the green point). Tables 3 and 4 shows the results of the recommendation during the time period 18:00–19:00 and 14:00–15:00 respectively.

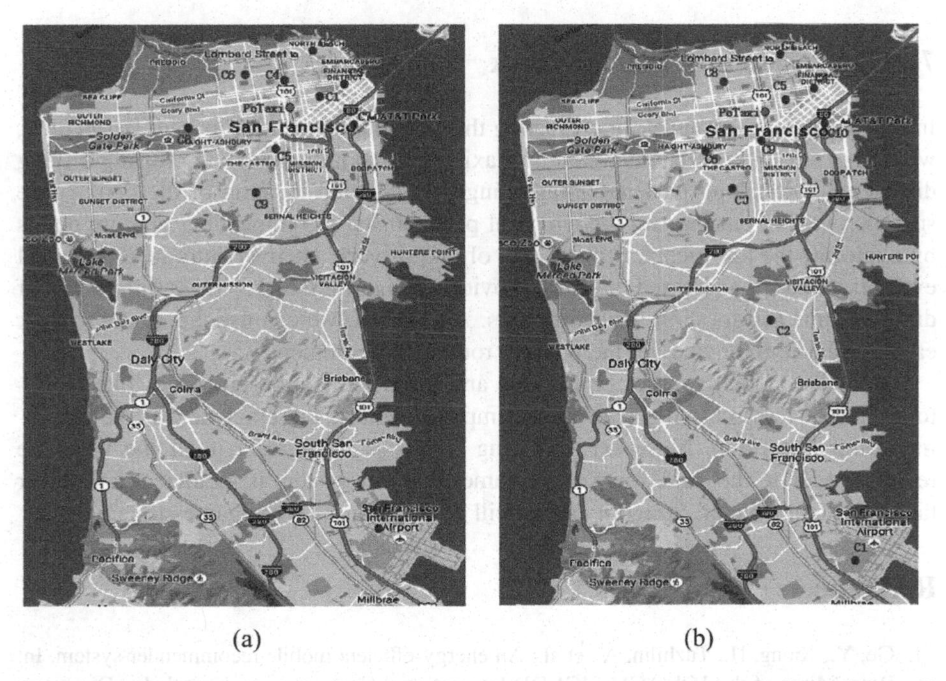

Fig. 8. Route recommendation. The red points denote the potential recommending points, and the green point denotes the target taxi (Color figure online)

Table 3. The results of the recommendation during the time period 18:00–19:00

Routes	Our Method	PTD
k = 3	PoTaxi→C_1→C_3→C_4	PoTaxi→C_1→C_3→C_7
k = 4	PoTaxi→C_1→C_3→C_4→C_6	PoTaxi→C_1→C_3→C_4→C_6
k = 5	PoTaxi→C_1→C_3→C_4→C_6→C_2	PoTaxi→C_1→C_3→C_4→C_6→C_2

Table 4. The results of the recommendation during the time period 14:00–15:00

Routes	Our Method	PTD
k = 3	PoTaxi→C_5→C_4→C_7	PoTaxi→C_5→C_4→C_{10}
k = 4	PoTaxi→C_5→C_4→C_7→C_8	PoTaxi→C_5→C_4→C_{10}→C_2
k = 5	PoTaxi→C_5→C_4→C_7→C_8→C_6	PoTaxi→C_5→C_4→C_7→C_8→C_2

In the experiment, during the time period 18:00–19:00, the driving distance between C3 and C4 is 2355 m, and the driving distance between C3 and C7 is 2643 m. As it can be seen from Table 3, when k = 3, the optimal route is PoTaxi → C1 → C3→ C7 generated by the PTD method while PoTaxi → C1 → C3 → C4 is generated by our proposed method during the time period 18:00–19:00. Obviously, our method works much better than the PTD. Because of the picking-up rate of C4 is higher than C7 and the distance between C3 and C4 is smaller than the distance between C3 and C7. And the potential driving distance of our method is about 1336 m while the potential driving distance of PTD is about 1350 m.

7 Conclusion and Future Work

In this paper, a novel model for evaluating the candidate route is proposed. Based on it, we design a recommendation system for taxi drivers to minimize their cruising driving distance before taking passengers regarding the time and location of the taxi. To be specific, we first put forward the temporal probabilistic recommending pick-up points by exploring the historical trajectory data of taxi drivers. Then we introduce the novel evaluation model, and based on it, we provide an algorithm to get the optimal route of different time and location for taxi drivers. As a result, we can use the model to rank each candidate route and get the optimal route for recommending.

Since the model is more complicated and loses some good properties like monotonic, future research will focus on the improvement of the efficiency of the recommendation algorithm. Moreover, choosing routes is like game, if all the taxis are recommended to the same route at the same time, the system is inefficient and fewer taxis will be a winner. So more efforts will be studying the taxi game strategy.

References

1. Ge, Y., Xiong, H., Tuzhilin, A., et al.: An energy-efficient mobile recommender system. In: Proceedings of the 16th ACM SIGKDD International Conference on Knowledge Discovery and Data Mining, pp. 899–908. ACM (2010)

2. Hu, H., Wu, Z., Mao, B., Zhuang, Y., Cao, J., Pan, J.: Pick-Up tree based route recommendation from taxi trajectories. In: Gao, H., Lim, L., Wang, W., Li, C., Chen, L. (eds.) WAIM 2012. LNCS, vol. 7418, pp. 471–483. Springer, Heidelberg (2012). doi:10.1007/978-3-642-32281-5_45
3. Hwang, R.H., Hsueh, Y.L., Chen, Y.T.: An effective taxi recommender system based on a spatio-temporal factor analysis model. Inf. Sci. **314**, 28–40 (2015)
4. Yang, W., Wang, X., Rahimi, S.M., Luo, J.: Recommending profitable taxi travel routes based on big taxi trajectories data. In: Cao, T., Lim, E.-P., Zhou, Z.-H., Ho, T.-B., Cheung, D., Motoda, H. (eds.) PAKDD 2015. LNCS (LNAI), vol. 9078, pp. 370–382. Springer, Cham (2015). doi:10.1007/978-3-319-18032-8_29
5. Letchner, J., Krumm, J., Horvitz, E.: Trip router with individualized preferences (trip): incorporating personalization into route planning. In: Proceedings of the National Conference on Artificial Intelligence, Menlo Park, CA, vol. 21, no. 2, p. 1795. MIT Press, Cambridge, AAAI Press, London, 1999, 2006
6. Malviya, N., Madden, S., Bhattacharya, A.: A continuous query system for dynamic route planning. In: 2011 IEEE 27th International Conference on Data Engineering, pp. 792–803. IEEE (2011)
7. Li, B., Zhang, D., Sun, L., et al.: Hunting or waiting? Discovering passenger-finding strategies from a large-scale real-world taxi dataset. In: 2011 IEEE International Conference on Pervasive Computing and Communications Workshops (PERCOM Workshops), pp. 63–68. IEEE (2011)
8. Yuan, J., Zheng, Y., Zhang, L., et al.: Where to find my next passenger. In: Proceedings of the 13th International Conference on Ubiquitous Computing, pp. 109–118. ACM (2011)
9. Qu, M., Zhu, H., Liu, J., et al.: A cost-effective recommender system for taxi drivers. In: Proceedings of the 20th ACM SIGKDD International Conference on Knowledge Discovery and Data Mining, pp. 45–54. ACM (2014)
10. Li, X., Pan, G., Wu, Z., et al.: Prediction of urban human mobility using large-scale taxi traces and its applications. Front. Comput. Sci. **6**(1), 111–121 (2012)
11. Awasthi, A., Lechevallier, Y., Parent, M., et al.: Rule based prediction of fastest paths on urban networks. In: Proceedings of 2005 IEEE Intelligent Transportation Systems 2005, pp. 978–983. IEEE (2005)
12. Gonzalez, H., Han, J., Li, X., et al.: Adaptive fastest path computation on a road network: a traffic mining approach. In: Proceedings of the 33rd International Conference on Very Large Data Bases, pp. 794–805. VLDB Endowment (2007)
13. Ma, S., Zheng, Y., Wolfson, O.: T-share: a large-scale dynamic taxi ridesharing service. In: 2013 IEEE 29th International Conference on Data Engineering (ICDE), pp. 410–421. IEEE (2013)
14. Ma, S., Zheng, Y., Wolfson, O.: Real-time city-scale taxi ridesharing. IEEE Trans. Knowl. Data Eng. **27**(7), 1782–1795 (2015)
15. Linden, G., Smith, B., York, J.: Amazon.com recommendations: item-to-item collaborative filtering. IEEE Internet Comput. **7**(1), 76–80 (2003)
16. Trestian, I., Huguenin, K., Su, L., et al.: Understanding human movement semantics: a point of interest based approach. In: Proceedings of the 21st International Conference on World Wide Web, pp. 619–620. ACM (2012)
17. Karypis, G.: CLUTO. http://glaros.dtc.umn.edu/gkhome/views/cluto

MaP2R: A Personalized Maximum Probability Route Recommendation Method Using GPS Trajectories

Ge Cui and Xin Wang(✉)

Department of Geomatics Engineering, University of Calgary,
2500 University Drive, N.W., Calgary, AB T2N 1N4, Canada
{cuig, xcwang}@ucalgary.ca

Abstract. Personalized travel route recommendation refers to the planning of an optimal travel route between two geographical locations based on the road networks and users' travel preferences. In this paper, we extract users' travel behaviours from their historical GPS trajectories and propose a personalized maximum probability route recommendation method called MaP2R. MaP2R utilizes the concepts of appearance behaviour and transition behaviour to describe users' travel behaviours and applies matrix factorization and Laplace smoothing method to estimate users' travel behaviour probabilities. When making recommendation, a route with the maximum probability of a user's travel behaviours is generated based on Markov property and searched through a generated behaviour graph. The experimental results on a real GPS trajectory dataset show that the proposed MaP2R achieves better results for travel route recommendations compared with the existing methods.

Keywords: GPS trajectories · Personalized travel route recommendation · Collaborative filtering

1 Introduction

Travelling is a critical component in daily life, and route recommendation is the most popular service for travelling. Current route recommendation services generally consider a certain metric such as the distance or traveling time, and provide the shortest or quickest path between locations to users; however, the recommended path would not often be chosen in real travel [2, 3, 5].

Personalized travel route recommendation is an active and important research topic. It refers to the planning of an optimal travel route between two geographical locations based on the road networks and users' travel preferences. However, users' specific travel preferences are influenced by many factors, such as distance, traffic volume, travelling time, weather, and many other hidden factors. Thus, it is difficult to determine a comprehensive user travel preference metric to develop personalized travel route recommendation. With advances in Global Positioning System (GPS) technology and the popularity of the mobile devices, massive amounts of human movement data in GPS trajectories have been collected, which could assist understanding users' preferences.

J. Kim et al. (Eds.): PAKDD 2017, Part II, LNAI 10235, pp. 168–180, 2017.
DOI: 10.1007/978-3-319-57529-2_14

There are, however, a few challenges when using GPS trajectories for personalized travel route recommendation. GPS readings are recorded over time. How to extract and describe users' travel behaviours from GPS trajectories is an issue that must be addressed first. Moreover, as a user generally travels on only a few routine routes daily, each user's GPS trajectories cover limited road segments of the road network. Hence, a user's travel behaviour probability estimation on the road segments that he/she has never travelled is significant for route recommendation, especially for the cases when users intend to travel to unfamiliar locations. More importantly, despite we can estimate users' travel behaviour probabilities, how to search the optimal route out of the multiple routes can be taken between two locations is also critical.

In this paper, we propose a personalized Maximum Probability Route Recommendation method, called MaP2R, to extract travel behaviours from historical trajectories and provide users the personalized maximum probability route by considering the Markov property of users' adjacent travel behaviours. Markov property plays an important role in movement models [1, 2, 5], and ignorance of the dependency relationship between travel behaviours may hamper the model and produce detours in the route planning.

Specifically, the contributions of the paper are summarized as follows:

First, we propose the concepts of appearance behaviour and transition behaviour based on the users' historical GPS trajectories and the road network, and then extract the frequencies of the two behaviours to present users' travel behaviours. MaP2R treats each trip of the GPS trajectories as a statement of preference and learns users' preferences through travel behaviours implicitly.

Second, we address the sparseness of user trajectory coverage on the road network by applying matrix factorization to estimate the frequencies of users' missing travel behaviours, and utilizing Laplace smoothing method to estimate the probability of users' missing travel behaviours.

Third, to search the maximum probability route between two locations, Markov property of adjacent appearance behaviours is considered, and route search is implemented in a proposed behaviour graph.

Last, experimental studies are conducted based on a real trajectory dataset. The experimental results show that the proposed MaP2R method outperforms the shortest distance route method and the frequent route method.

This paper is organized as follows: Sect. 2 introduces the related works on route recommendation and applications using GPS trajectories. Sections 3 and 4 provide preliminary and a detailed discussion on the proposed MaP2R method for personalized route recommendation. Section 5 presents experimental results based on the GPS trajectory dataset in Beijing, China. Section 6 gives the conclusion and discusses future work.

2 Related Works

Different studies have been conducted utilizing GPS trajectories to guide human mobility. Studies have investigated location and travel recommendations based on human movement GPS trajectories. Zheng et al. [8] proposed a user-centred CF method to make mobile recommendations for locations and activities based on historical trajectories. Zheng and Xie [9] inferred the interest level of a location and a user's travel experience to give both generic and personalized recommendations for interesting locations and travel sequences by mining multiple users' GPS traces. Chen et al. [1] proposed a method to discover popular routes from trajectories. They distinguished intersections by clustering based on direction and density as nodes in a transfer network. Then, a popularity value was assigned to each node versus destination, and the route with the largest popularity was taken as the most popular route.

Very few research works, however, have examined personalized travel route recommendation using GPS traces in the past. TRIP proposed in [5] calculates an inefficient ratio of routes for each driver based on GPS traces and takes the ratio as a metric in an A* algorithm for a personalized route recommendation. It is a very simple and effective method but it does not work well when the users have little or no trajectories. In addition, it does not consider the similarity of travel behaviours among users. Dai et al. [2] construct a preference vector to describe users' preference and find the reference trajectories from users having similar preference for recommendation. The proposed personalized route recommendation (PRR) can make analysis of user's specific preferences (such as travel time, fuel consumption or other costs) and provide the personalized route recommendation. But the personal preferences information required by the algorithm can hardly be acquired from users.

3 Preliminary

Definition 1 Road network. The road network is a graph, $G = (V, E)$, where V is a set of vertices representing the terminal points of road segments, and E is a set of edges representing the road segments. Vertex $v \in V$ is a terminal point of road segment. An edge $e \in E$ is a road segment with a starting vertex $e.start$ and an end vertex $e.end$, where $e.start \in V$ and $e.end \in V$.

Definition 2 GPS-reading. A GPS-reading p is a 3-tuple denoted as: $p = (t, lat, lng)$, where t is the timestamp of the GPS-reading, and lat and lng are the latitude and longitude of the location of the GPS-reading at time t.

Definition 3 GPS trajectory. A GPS trajectory is a sequence of GPS-readings $trj = (p_1, p_2, p_3, \ldots, p_m)$ where $p_i.t - p_{i-1}.t > 0, 1 < i \leq m$.

Definition 4 Route. Given road network $G = (V, E)$, a route from vertex v_i to vertex v_j is a sequence of connected road segments $R = (v_i, e_1, e_2, e_3, \ldots, e_n, v_j)$ which starts at vertex v_i, and ends at vertex v_j where $v_i, v_j \in V, e_i \in E$ and e_i is the i-th road segment in R, $e_i \neq e_j$ if $i \neq j$, and $e_1.start = v_i, e_n.end = v_j$.

In this paper, we propose and extract two types of travel behaviours from users' GPS trajectories, called appearance behaviour and transition behaviour, defined as follows.

Definition 5 Appearance behaviour. Given a road network $G = (V, E)$ and the set of time intervals $T = \{t_1, t_2, \ldots, t_r\}$ of a day, the appearance behaviour b of a user is a tuple denoted as: $b = (e, t)$, where $e \in G.E$ and $t \in T$.

The concept of appearance behaviours describes the location and time of a user's movements. Given user u and his/her appearance behaviour b, $frq(u, b)$ is the frequency that user u has the behaviour b.

An appearance behaviour could be followed by another appearance behaviour on the given road network for a time interval. To describe the sequential relationship between the appearance behaviours, we propose a concept called transition behaviour.

Definition 6 Transition behaviour. A transition behaviour $tb = (b_i \rightarrow b_j)$ is an ordered tuple to describe the appearance behaviour $b_i = (e_i, t_i)$ is followed by the appearance behaviour $b_j = (e_j, t_j)$ if $e_i.end = e_j.start$ and $t_i = t_j$. It can also be denoted as $tb_{i \rightarrow j}$ in short.

Given a set of appearance behaviours B, a set of transition behaviours TB can be generated by considering all possible transition behaviours between adjacent appearance behaviours in B. In the other words, for $\forall tb_k \in TB, tb_k = (b_i \rightarrow b_j)$ where $b_i \in B$ and $b_j \in B$. $frq(u, tb_k)$ is the frequency of the transition behaviour tb_k by user u. In the following discussion, we will use the travel behaviour to refer to the above two types of behaviours for conciseness.

4 Personalized Maximum Probability Route Recommendation

In this section, we propose a personalized maximum probability route recommendation (MaP2R) algorithm and discuss it in detail. As mentioned, MaP2R assumes that the route a user actually takes is preferred by the user over any other route he/she could have taken between the same endpoints. Therefore, MaP2R extracted and estimated the probabilities of travel behaviours on the road network from the users' historical GPS trips and deals with the personalized route recommendation problem by searching route with the maximum travel behaviour probability. Specifically, given the origin o and the destination d, $P(R|u, t)$ is the probability that the route R is preferred given the user u at the time t in the road network, can be represented as follows:

$$P(R|u,t) = P(e_1, e_2, e_3, \ldots, e_n|u,t) = P(e_1, e_2, e_3, \ldots, e_n, t|u)/P(t|u) \tag{1}$$

Where $e_1.start = o$ and $e_n.end = d$. Since the probability $P(t|u)$ is constant when u and t are given, to find the personalized maximum probability route is equal to maximize $P(e_1, e_2, e_3, \ldots, e_n, t|u)$.

$$P(e_1, e_2, e_3, \ldots, e_n, t|u) = P(b_1, b_2, b_3, \ldots, b_n|u), \text{where } b_i = (e_i, t) \tag{2}$$

According to the rule of conditional probability,

$$P(b_1, b_2, b_3, \ldots, b_n|u) = P(b_1|u)P(b_2|b_1, u)\ldots P(b_n|b_1, \ldots, b_{n-1}, u) \quad (3)$$

As Markov property is often used to describe the behaviours in movement models [1, 2, 5], this study assumes the probability of user's current appearance behaviour b_i depends on the last appearance behaviour b_{i-1}. Therefore, Eq. (3) becomes:

$$P(b_1, b_2, b_3, \ldots, b_n|u) = P(b_1|u)\ldots P(b_i|b_{i-1}, u)\ldots P(b_n|b_{n-1}, u) \quad (4)$$

where $P(b_1|u)$ is the appearance behaviour probability starting from the origin o by u, and $P(b_i|b_{i-1}, u)$ is the transition probability from the appearance behaviour b_{i-1} to the next travel behaviour b_i and is represented as $P(tb_{i-1\to i}|u)$ for conciseness in the below. Therefore, we can define the personalized travel route recommendation problem as follows:

Definition 7 Personalized maximum probability route recommendation problem. Given time interval t, origin vertex o and destination vertex d, the personalized maximum probability route recommendation problem is to find the maximum probability route $R = (e_1, e_2, e_3, \ldots, e_n)$ in the road network $G = (V, E)$, $e_1.start = o$ and $e_n.end = d$ so that

$$R = Argmax_{e_1,\ldots,e_n \in G.E}(P(b_1|u)P(tb_{1\to 2}|u)\ldots P(tb_{n-1\to n}|u)) \quad (5)$$

MaP2R includes four steps: data preparation, frequency calculation for travel behaviours, probability estimation for travel behaviour and maximum probability travel behaviour route search. In data preparation, trajectories are firstly split into sub-trajectories that are trips with origin and destination points by using the method in [7]. The second task of the data preparation step is to match the GPS trips to the road network by applying the map matching method in [6]. After matching trips to the road network, users' travel behaviours are extracted from the routes and their frequencies are counted. In the following, we will focus on the next three steps of MaP2R.

4.1 Matrix Factorization for Estimation of Missing Travel Frequency

As mentioned, users generally travel on very limited routes daily, covering only a small number of road segments in a city. To estimate the frequency of missing travel behaviours of each user, matrix factorization [4] is used in this study. The first step is the generation of user-appearance behaviour matrix and the user-transition behaviour matrix.

Definition 8 User-Appearance Behaviour Matrix. Given a set of appearance behaviours $B = \{b_1, b_2, \ldots, b_l\}$ and a set of m users $U = \{u_1, u_2, \ldots, u_m\}$, the pairs of (u, b) are used to construct a user-appearance behaviour matrix $UB_{m\times l}$. The element in the user-appearance behaviour matrix $UB_{m\times l}$ is the frequency of the pair (u_i, b_j), i.e. $frq(u_i, b_j)$, denoted as $UB_{i,j}$, i.e. $UB_{i,j} = frq(u_i, b_j)$.

To estimate the frequency of missing appearance behaviour of each user, users and their appearance behaviours are characterized by two vectors of latent factors. To be more specific, each user u_i is associated with a vector of latent factors, $p_{u_i} = (f^1_{u_i}, f^2_{u_i}, \ldots, f^k_{u_i})$, and each appearance behaviour b_j is associated with vector $q_{b_j} = (f^1_{b_j}, f^2_{b_j}, \ldots, f^k_{b_j})$, where k is the length of the vectors. The predicted frequency $\widehat{UB}_{i,j}$ of appearance behaviour b_j by user u_i is approximated by the dot product of p_{u_i} and q_{b_j}:

$$\widehat{UB}_{i,j} = q^T_{b_j} p_{u_i} \tag{6}$$

The vectors of latent factors are learned by minimizing the regularized squared error of the set of known appearance behaviour frequencies:

$$L = \sum\nolimits_{(u_i,b_j)\in S} (UB_{i,j} - q^T_{b_j} p_{u_i})^2 + \lambda(\|q_{b_j}\|^2 + \|p_{u_i}\|^2) \tag{7}$$

where S is the set of the (u_i, b_j) pairs for which appearance behaviour b_j by user u_i is known, $UB_{i,j}$ is the frequency of appearance behaviour b_j by user u_i, and the constant λ controls the extent of regularization. After learning the vectors of latent factors p_{u_i} and q_{b_j} with alternating least squares method, the frequencies of user's appearance behaviours can be predicted in Eq. (6). Similarly, the matrix factorization for estimation of users' transition behaviour frequencies can be implemented after constructing a user-transition behaviour matrix.

4.2 Estimation for Travel Behaviour Probability

The initial appearance behaviour probability describes the probability $P(b|u)$ of the first appearance behaviour that a user would take given a starting vertex o in the road network. Given the set S_o of the appearance behaviours starting from o, the probability $P(b|u)$ of the appearance behaviour b starting from o by user u can be calculated as the frequency of the appearance behaviour b over the total frequency of the possible initial appearance behaviours $b_i \in S_o$.

There may, however, exist very few road segments starting from o where users have never travelled, according to their historical GPS trajectories. In this case, the Laplace smoothing method is used to estimate the probability of the missing initial appearance behaviours in Eq. (8), in order to avoid assigning a zero probability to any initial appearance behaviour, so that it is possible for users to travel to any road segment in the road network.

$$P(b|u) = \begin{cases} \frac{\widehat{frq}(u,b)+\alpha}{\sum_{b_i\in S_o} \widehat{frq}(u,b_i)+\alpha * d}, & \widehat{frq}(u,b) > 0 \\ \frac{\alpha}{\sum_{b_i\in S_o} \widehat{frq}(u,b_i)+\alpha * d}, & otherwise \end{cases}, b_i = (e_i, t) \text{ and } e_i.start = o \tag{8}$$

where $P(b|u)$ is the probability of u's initial appearance behaviour b on road segment e at time t. For any appearance behaviour $b_i = (e_i, t)$ in the set S_o, $e_i.start = o$. $\widehat{frq}(u, b_i)$

is the estimated appearance behaviour frequency using matrix factorization and α is the smoothing parameter.

Given the current appearance behaviour b_i and the set S' of the next appearance travel behaviours, the transition behaviour probability $P(tb_{i\to j}|u)$ measures the likeliness of transferring from the current appearance behaviour b_i to the next adjacent appearance behaviour b_j by the user u.

If all transition behaviours in S' have never been conducted by any user, a uniform value β will be assigned to the probability of each transition behaviour in S' based on priori knowledge.

$$P(tb_{i\to j}|u) = \beta \tag{9}$$

Otherwise, the transition probability from appearance behaviour b_i to b_j is estimated as the frequency of the transition behaviour $tb_{i\to j}$ over the total frequency of the possible transition behaviours $tb_{i\to k}$ and $b_k \in S'$.

Similarly, to avoid assigning a zero probability to any transition behaviour, Laplace smoothing method is utilized as follows:

$$P(tb_{i\to j}|u) = \begin{cases} \frac{\widehat{frq}(u,tb_{i\to j}) + \alpha}{\sum_{k=1}^{d} \widehat{frq}(u,tb_{i\to k}) + \alpha * d}, & \widehat{frq}(u,tb_{i\to j}) > 0 \\ \frac{\alpha}{\sum_{k=1}^{d} \widehat{frq}(u,tb_{i\to k}) + \alpha * d}, & otherwise \end{cases} \tag{10}$$

where d is the number of all appearance behaviours in S', α is the smoothing parameter based on a priori knowledge. Equations (9) and (10) can assign a non-zero probability to any transition behaviours that never happened in users' GPS trajectories.

4.3 Maximum Probability Route Search

Appearance behaviours and transition behaviours are utilized to describe the maximum probability route in Eq. (4). In the following, a behaviour graph is defined and can be generated from the road network and probabilities of travel behaviours, which will be used for searching the maximum travel behaviour probability route.

Definition 9 Behaviour Graph. Given time t, user u, road network G, a start vertex v_{start} and an end vertex v_{end}, a behaviour graph is denoted by $G' = (V', E')$ where the set of vertices V' includes three parts, a set of vertices V^* where each element presents an appearance behaviour of the user, the start vertex $\{v_{start}\}$, and the end vertex $\{v_{end}\}$, i.e. $V' = V^* \cup \{v_{start}\} \cup \{v_{end}\}$; the set of edges E' is also constituted of three parts $E' = E^* \cup E_{start} \cup E_{end}$, where E^* is a set of edges in which each element presents a transition behaviour from one appearance behaviour vertex to another, E_{start} includes all edges starting from v_{start} to its possible adjacent appearance behaviour vertices and E_{end} includes edges connecting from appearance behaviour vertices to v_{end}. There is a weight ϑ on each edge in a behaviour graph associated with travel behaviour probabilities.

Figure 1 gives an example of a behaviour graph from a road network. The behaviour graph describes all possible appearance behaviours and transition behaviours from a start vertex v_{start} to an end vertex v_{end} for the given time and user.

As defined, the personalized route recommendation problem requires the maximization of $P(b_1|u)P(tb_{1\to2}|u)P(tb_{2\to3}|u)\ldots P(tb_{n-1\to n}|u)$ in Eq. (4). In order to find the maximum travel behaviour probability route, the multiplication of probabilities is transformed to the summarization format required by the typical route planning algorithms, let $L = \frac{1}{P(b_1|u)P(tb_{1\to2}|u)P(tb_{2\to3}|u)\ldots P(tb_{n-1\to n}|u)}$; thus, the problem in this paper is equivalent to find the minimization of L.

Taking the logarithm for both sides:

$$\ln L = \ln\left(\frac{1}{P(b_1|u)}\prod_{i=1}^{n-1}\frac{1}{P(tb_{i\to i+1}|u)}\right) = \ln\frac{1}{P(b_1|u)} + \sum_{i=1}^{n-1}\ln\frac{1}{P(tb_{i\to i+1}|u)} \tag{11}$$

Let $\vartheta(v_{start}, b_k) = \ln\frac{1}{P(b_k|u)}$, $\vartheta(b_i, b_j) = \ln\frac{1}{P(tb_{i\to j}|u)}$ and $\vartheta(b_k, v_{end}) = 0$, the personalized maximum probability route recommendation problem is to find a minimum weight path in the generated behaviour graph, which can be solved with Dijkstra's algorithm.

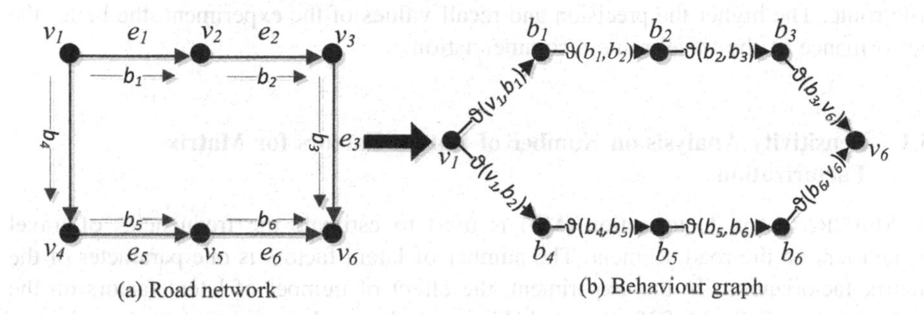

Fig. 1. An example showing the relationship between the road work and the behaviour graph

5 Experiments

In this section, MaP2R is evaluated on a real world GPS trajectory dataset, Geolife [8]. The dataset contains 17,621 trajectory files of 182 users. Out of them, 22 drivers' GPS trajectories are extracted and divided into 728 trips. The study area is in the central district of Beijing, China, ranging from 39.69° N to 40.11° N and from 116.09° E to 116.62° E. Totally 43,381 road segments and 38,485 nodes are in the road network of the area. The travel behaviours are generated using one hour interval.

For each of the following experiments, the dataset is separated into training datasets and testing datasets. The training dataset contains 80% randomly chosen trips of each user; and, the testing dataset includes all the remaining trips. All algorithms are implemented in C#. The experiments are conducted on a 2.5 GHz Core i7 PC with 16 GB of RAM.

Both precision and recall are used to evaluate the performance of the recommended travel routes. Since the lengths of road segments are different, the precision and recall are defined in terms of the number of road segments and the distance, respectively, as shown below.

$$Precision_{\#\ of\ road\ segement} = \frac{\#\ of\ correct\ recommended\ road\ segments}{\#\ of\ road\ segments\ on\ recommended\ route}$$

$$Precision_{distance} = \frac{distance\ of\ correct\ recommended\ road\ segments}{distance\ of\ the\ recommended\ route}$$

$$Recall_{\#\ of\ road\ segment} = \frac{\#\ of\ correct\ recommended\ road\ segments}{\#\ of\ road\ segments\ on\ true\ route}$$

$$Recall_{distance} = \frac{distance\ of\ correct\ recommended\ road\ segments}{distance\ of\ true\ route}$$

The true route is the route that the user actually travelled based on trajectories. The two precision values of the recommended travel route measure the percentage of the correct road segments in the recommended route. The recalls of the recommended travel route measure the percentage of the correctly recommended road segments in the true route. The higher the precision and recall values of the experiment, the better the performance of the travel route recommendation.

5.1 Sensitivity Analysis on Number of Latent Factors for Matrix Factorization

In MaP2R, matrix factorization (MF) is used to estimate the frequencies of travel behaviours on the road segment. The number of latent factors is one parameter of the matrix factorization. In this experiment, the effect of number of latent factors on the performance of the MaP2R is tested. Figure 2 shows that root mean squared error (RMSE) and running time with respect to the number of latent factors in matrix factorization ranging from 5 to 15. It could be observed that with the increase of the number of the latent factors, the RMSE values of both the user-appearance behaviour matrix and the user-transition behaviour matrix slightly change, but then keep at a

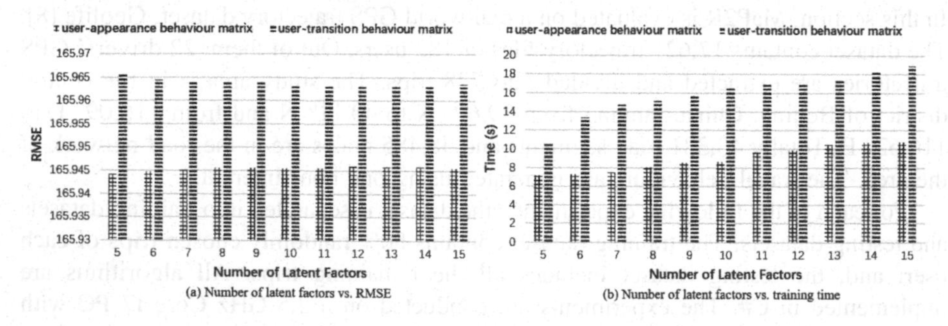

Fig. 2. The effect of number of latent factors in MF vs. (a) RMSE and (b) training time

steady level while the training time for MaP2R increases with the number of latent factors. Hence, we choose the number of latent factors as 5 in the following experiments.

5.2 Overall Performance vs. Route Distance

In this experiment, we evaluate the performances of MaP2R and compare with the shortest distance route method (SDR) and the frequent route method (FR) in terms of the trip lengths. SDR recommends the route with the shortest distance. FR first estimates the appearance behaviour probabilities of all road segments and then recommends a route with the highest consecutive appearance probabilities from the origin to the destination. The route of FR usually contains frequently traveled road segments.

We first separate trips in the testing dataset into four different groups based on the trip length. Each distance range group contains 35 trips. The distance ranges for the four groups are: Group 1: (2.05–14.58 km), Group 2: (14.58–19.21 km), Group 3: (19.21–28.74 km) and Group 4: (28.74–56.72 km). The number of latent factors is set as 5, the regularization parameter λ as 0.1, and the smoothing parameter α and β as 0.01, respectively. Figure 3 illustrates the precision and recall values of MaP2R, SDR and FR on four groups.

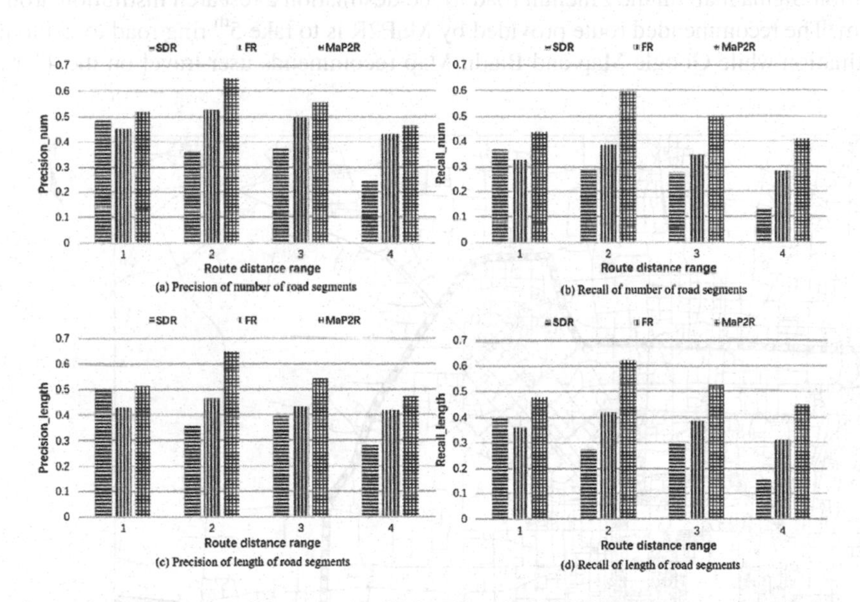

Fig. 3. Performance comparison among SDR, FR and MaP2R for four groups of trips with different lengths (a) precision of number of road segment; (b) recall of number of road segments; (c) precision of length of road segments; (d) recall of length of road segments.

The results show that MaP2R outperforms the other two methods in both precision and recall in all four groups. When the trip is short, i.e. for trips in Group 1, the performances of the three methods do not have much discrepancy. SDR method is slightly better than FR method because when the distance between the origin and the destination is small, users prefer to the shortest distance path rather than passing through the high frequency road segments to reach the destination, which is reasonable in reality. With the increase of the distance of trips, MaP2R and FR outperform SDR because MaP2R and FR both consider users' preference from historical trajectories while SDR does not consider the factor. Moreover, MaP2R outperforms FR by 3 ~ 18% in precision and 13 ~ 21% in recall for the Groups 2 ~ 4. The reason is that MaP2R considers user's preference for each travel behaviour and the dependencies between the travel behaviours, which could better reflect users' travel preference, but FR only considers high frequency of road segments separately. Therefore, these results demonstrate that the route from MaP2R method has larger correspondence with users' preference compared to SDR method and FR method.

5.3 Case Study for Performance

In this experiment, we compare the route recommendations from the proposed MaP2R with the two most popular online route recommendation applications, i.e., Google Map and Baidu Map.

Figure 4 shows a case of Geolife dataset that a user intends to travel from the origin location Sigma mall on the Zhichun road to the destination a research institution around 10am. The recommended route provided by MaP2R is to take 5th ring road to get to the destination while Google Map and Baidu Map recommends user travel on the 4th ring

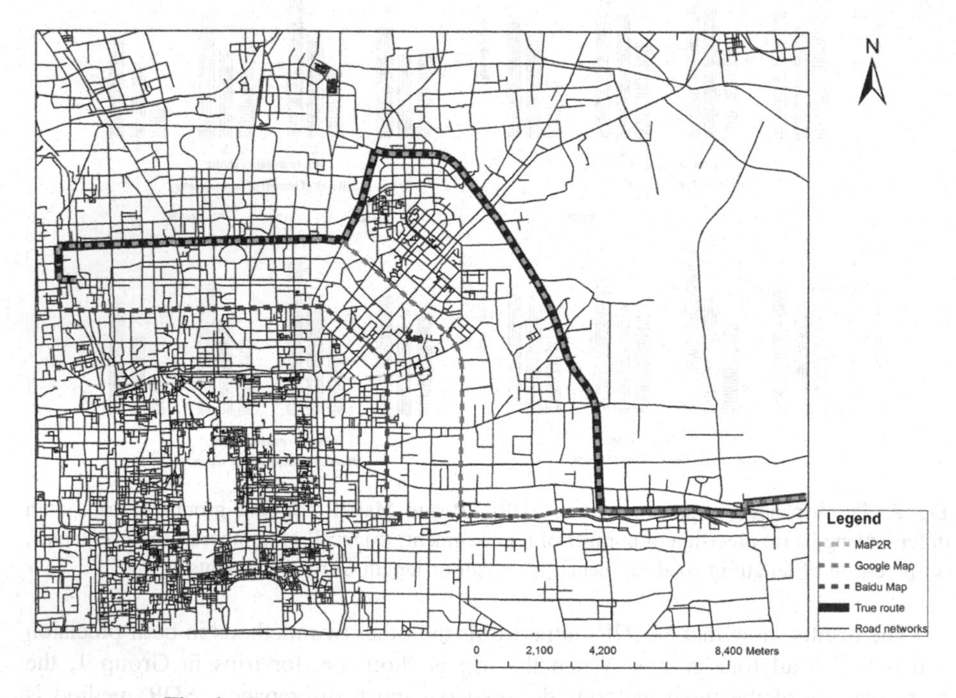

Fig. 4. Comparison among MaP2R, Google map and Baidu map

road and the 3rd ring road, respectively. The route recommended by MaP2R is same as the user's real travel route but the routes recommended by Google Map and Baidu Map have a large discrepancy with the real route. The reason is that Google Map and Baidu Map would recommend all users the same route in the figure from Sigma mall to the research institution based on certain criterion, but MaP2R would learn users' preference from historical trajectories and recommend the route high corresponding with users' preferences. In fact, the target user is found indeed prefer to travel on 5th ring road by scrutinizing this users' historical GPS trajectories manually.

6 Conclusions and Future Works

In this paper, we propose a personalized maximum probability route recommendation method (MaP2R) based on historical GPS trajectories. In this method, distance, traveling time, road safety and any other factors in route planning are all considered as hidden factors. Two concepts, appearance behaviour and transition behaviour, are proposed to describe users' travel behaviours. Moreover, m Matrix factorization and Laplace smoothing method is used to estimate the frequencies and probabilities of users' travel behaviours. Finally, the route with the maximum travel behaviour probability is recommended to the user. The experiment results show that MaP2R outperforms the shortest distance path method and the frequent path method. In future, we will continue to explore the spatiotemporal correlations between travel behaviours and integrate with the current estimation model. More experiments will be conducted to compare MaP2R with other personalized route recommendation methods. The cold start problem also will be addressed to improve MaP2R.

Acknowledgements. The research is supported by the Natural Sciences and Engineering Research Council of Canada Discovery Grant to Xin Wang and National Natural Science Foundation of China (No. 41271387).

References

1. Chen, Z., Shen, H.T., Zhou, X.: Discovering popular routes from trajectories. In: 27th IEEE International Conference on Data Engineering, pp. 900–911. IEEE Press (2011)
2. Dai, J., Yang, B., Guo, C., Ding, Z.: Personalized route recommendation using big trajectory data. In: IEEE 31st International Conference on Data Engineering, pp. 543–554. IEEE Press (2015)
3. Delling, D., Wagner, D.: Pareto paths with SHARC. In: Vahrenhold, J. (ed.) SEA 2009. LNCS, vol. 5526, pp. 125–136. Springer, Heidelberg (2009). doi:10.1007/978-3-642-02011-7_13
4. Koren, Y., Bell, R., Volinsky, C.: Matrix factorization techniques for recommender systems. Computer **42**, 30–37 (2009)
5. Letchner, J., Krumm, J., Horvitz, E.: Trip router with individualized preferences (TRIP): incorporating personalization into route planning. In: Proceedings of the 18th AAAI Conference on Innovative Applications of Artificial Intelligence, pp. 1795–1800. AAAI Press, Palo Alto (2006)

6. Newson, P., Krumm, J.: Hidden Markov map matching through noise and sparseness. In: Proceedings of the 17th ACM SIGSPATIAL International Conference on Advances in Geographic Information Systems. pp. 336–343. ACM Press, New York (2009)
7. Yang, W., Wang, X., Rahimi, S.M., Luo, J.: Recommending profitable taxi travel routes based on big taxi trajectories data. In: Cao, T., Lim, E.-P., Zhou, Z.-H., Ho, T.-B., Cheung, D., Motoda, H. (eds.) PAKDD 2015. LNCS (LNAI), vol. 9078, pp. 370–382. Springer, Cham (2015). doi:10.1007/978-3-319-18032-8_29
8. Zheng, V.W., Cao, B., Zheng, Y., Xie, X., Yang, Q.: Collaborative filtering meets mobile recommendation: a user-centered approach. In: Proceedings of the 24th AAAI Conference on Artificial Intelligence, pp. 236–241. AAAI Press, Palo Alto (2010)
9. Zheng, Y., Xie, X.: Learning travel recommendations from user-generated GPS traces. ACM Trans. Intell. Syst. Technol. (TIST) **2**, 1–29 (2011)

Feature Selection

SNE: Signed Network Embedding

Shuhan Yuan[1], Xintao Wu[2], and Yang Xiang[1(✉)]

[1] Tongji University, Shanghai, China
{4e66,shxiangyang}@tongji.edu.cn
[2] University of Arkansas, Fayetteville, AR, USA
xintaowu@uark.edu

Abstract. Several network embedding models have been developed for unsigned networks. However, these models based on skip-gram cannot be applied to signed networks because they can only deal with one type of link. In this paper, we present our signed network embedding model called SNE. Our SNE adopts the log-bilinear model, uses node representations of all nodes along a given path, and further incorporates two signed-type vectors to capture the positive or negative relationship of each edge along the path. We conduct two experiments, node classification and link prediction, on both directed and undirected signed networks and compare with four baselines including a matrix factorization method and three state-of-the-art unsigned network embedding models. The experimental results demonstrate the effectiveness of our signed network embedding.

1 Introduction

Representation learning [1], which aims to learn the features automatically based on various deep learning models [15], has been extensively studied in recent years. Traditionally, supervised learning tasks require hand-designed features as inputs. Deep learning models have shown great success in automatically learning the semantic representations for different types of data, like image, text and speech [6,8,12]. In this paper, we focus on representation learning of networks, in particular, signed networks. Several representation learning methods of unsigned networks have been developed recently [9,26,30,31]. They represent each node as a low-dimensional vector which captures the structure information of the network.

Signed networks are ubiquitous in real-world social systems, which have both positive and negative relationships. For example, Epinions[1] allows users to mark their trust or distrust to other users on product reviews and Slashdot[2] allows users to specify other users as friends or foes. Most unsigned network embedding models [9,26,30] are based on skip-gram [20], a classic approach for training word embeddings. The objective functions used in unsigned network embedding approaches do not incorporate the sign information of edges. Thus, they cannot

[1] http://www.epinions.com/.
[2] https://slashdot.org/.

J. Kim et al. (Eds.): PAKDD 2017, Part II, LNAI 10235, pp. 183–195, 2017.
DOI: 10.1007/978-3-319-57529-2_15

simply migrate to signed networks because the negative links change the theories or assumptions on which unsigned network embedding models rely [28].

In this paper, we develop a signed network embedding model called SNE. To the best of our knowledge, this is the first research on signed network embedding. Our SNE model adopts the log-bilinear model [21,22], uses node representations of all nodes along a given path, and further incorporates two signed-type vectors to capture the positive or negative relationship of each edge along the path. Our SNE significantly outperforms existing unsigned network embedding models which assume all edges are from the same type of relationship and only use the representations of nodes in the target's neighborhood. We conduct two experiments to evaluate our model, node classification and link prediction, on both an undirected signed network and a directed signed network built from real-world data. We compare with four baselines including a matrix factorization method and three state-of-the-art network embedding models designed for unsigned networks. The experimental results demonstrate the effectiveness of our signed network embedding.

2 Preliminary

In this section, we first introduce the skip-gram model, one of commonly used neural language models to train word embeddings [2]. We then give a brief overview of several state-of-the-art unsigned network embedding models based on the skip-gram model.

Skip-Gram Model. The skip-gram is to model the co-occurrence probability $p(w_j|w_i;\theta)$ that word w_j co-occurs with word w_i in a context window. The co-occurrence probability is calculated based on softmax function:

$$p(w_j|w_i;\theta) = \frac{\exp(\mathbf{v}_{w_j}^T \mathbf{v}_{w_i})}{\sum_{j' \in \mathcal{V}} \exp(\mathbf{v}_{w_{j'}}^T \mathbf{v}_{w_i})}, \tag{1}$$

where $\mathcal{V}$ is the set of all words and $\mathbf{v}_{w_j}$ and $\mathbf{v}_{w_i} \in \mathbb{R}^d$ are word embeddings for w_j and w_i, respectively. The parameters θ, i.e., $\mathbf{v}_{w_i}$, $\mathbf{v}_{w_j}$, are trained by maximizing the log likelihood of predicting context words in a corpus:

$$J = \sum_{i}^{|\mathcal{V}|} \sum_{j \in context(i)} \log p(w_j|w_i), \tag{2}$$

where $context(i)$ is the set of context words of w_i.

Network Embedding. Network embedding aims to map the network $G = (V, E)$ into a low dimensional space where each vertex is represented as a low dimensional real vector. The network embedding treats the graph's vertex set V as the vocabulary $\mathcal{V}$ and treats each vertex v_i as a word w_i in the skip-gram approach. The corpus used for training is composed by the edge set E, e.g., in [30], or a set of truncated random walks from the graph, e.g., in [9,26].

To train the node vectors, the objective of previous network embedding models is to predict the neighbor nodes $N(v_i)$ of a given source node v_i. However, predicting a number of neighbor nodes requires modeling the joint probability of nodes, which is hard to compute. The conditional independence assumption, i.e., the likelihood of observing a neighbor node is independent of observing other neighbor nodes given the source node, is often assumed [9]. Thus, the objective function is defined as:

$$J = \sum_{v_i \in V} \log p(N(v_i)|v_i) = \sum_{v_i \in V} \sum_{v_i' \in N(v_i)} \log p(v_i'|v_i), \tag{3}$$

where $p(v_i'|v_i)$ is softmax function similar to Eq. 1 except that the word vectors are replaced with node vectors.

3 SNE: Signed Network Embedding

We present our network embedding model for signed networks. For each node's embedding, we introduce the use of both source embedding and target embedding to capture the two potential roles of each node.

3.1 Problem Definition

Formally, a signed network is defined as $G = (V, E_+, E_-)$, where V is the set of vertices and E_+ (E_-) is the set of positive (negative) edges. Each edge $e \in E_+ \cup E_-$ is represented as $e_{uv} = (u, v, \varepsilon_{uv})$, where $u, v \in V$ and ε_{uv} indicates the sign value of edge e, i.e., $\varepsilon_{uv} = 1$ if $e \in E_+$ and $\varepsilon_{uv} = -1$ if $e \in E_-$. In the scenario of signed directed graphs, e_{uv} is a directed edge where node node u denotes the source and v denotes the target. Our goal is to learn node embedding for each vertex in a signed network while capturing as much topological information as possible. For each vertex v_i, its node representation is defined as $\bar{\mathbf{v}}_{v_i} = [\mathbf{v}_{v_i} : \mathbf{v'}_{v_i}]$ where $\mathbf{v}_{v_i} \in \mathbb{R}^d$ denotes its source embedding and $\mathbf{v'}_{v_i} \in \mathbb{R}^d$ denotes its target embedding.

3.2 Log-Bilinear Model for Signed Network Embedding

We develop our signed network embedding by adapting the log-bilinear model such that the trained node embedding can capture node's path and sign information. Recall that existing unsigned network embedding models are based on the skip-gram which only captures node's neighbour information and cannot deal with the edge sign.

Log-Bilinear Model. Given a sequence of context words $g = w_1, \ldots, w_l$, the log-bilinear model firstly computes the predicted representation for the target word by linearly combining the feature vectors of words in the context with the position weight vectors:

$$\hat{\mathbf{v}}_g = \sum_{j=1}^{l} \mathbf{c}_j \odot \mathbf{v}_{w_i}, \tag{4}$$

where $\odot$ indicates the element-wise multiplication and $\mathbf{c}_j$ denotes the position weight vector of the context word w_i. A score function is defined to measure the similarity between the predicted target word vector and its actual target word vector:

$$s(w_i, g) = \hat{\mathbf{v}}_g^T \mathbf{v}_{w_i} + b_{w_i}, \tag{5}$$

where b_{w_i} is the bias term. The log-bilinear model then trains word embeddings $\mathbf{v}$ and position weight vectors $\mathbf{c}$ by optimizing the objective function similar to the skip-gram.

SNE Algorithm. In our signed network embedding, we adopt the log-bilinear model to predict the target node based on its paths. The objective of the log-bilinear model is to predict a target node given its predecessors along a path. Thus, the signed network embedding is defined as a maximum likelihood optimization problem. One key idea of our signed network embedding is to use signed-type vector $\mathbf{c}_+ \in \mathbb{R}^d$ ($\mathbf{c}_- \in \mathbb{R}^d$) to represent the positive (negative) edges. Formally, for a target node v and a path $h = [u_1, u_2, \ldots, u_l, v]$, the model computes the predicted target embedding of node v by linearly combining source embeddings ($\mathbf{v}_{u_i}$) of all source nodes along the path h with the corresponding signed-type vectors ($\mathbf{c}_i$):

$$\hat{\mathbf{v}}_h = \sum_{i=1}^{l} \mathbf{c}_i \odot \mathbf{v}_{u_i}, \tag{6}$$

where $\mathbf{c}_i \equiv \mathbf{c}_+$ if $\varepsilon_{u_i u_{i+1}} = 1$, or $\mathbf{c}_i \equiv \mathbf{c}_-$ if $\varepsilon_{u_i u_{i+1}} = -1$ and $\odot$ denotes element-wise multiplication. The score function is to evaluate the similarity between the predicted representation $\hat{\mathbf{v}}_h$ and the actual representation $\mathbf{v}'_v$ of target node v:

$$s(v, h) = \hat{\mathbf{v}}_h^T \mathbf{v}'_v + b_v, \tag{7}$$

where b_v is a bias term.

To train the node representations, we define the conditional likelihood of target node v generated by a path of nodes h and their edge types q based on softmax function:

$$p(v|h, q; \theta) = \frac{exp(s(v, h))}{\sum_{v' \in V} exp(s(v', h))}, \tag{8}$$

where V is the set of vertices, and $\theta = [\mathbf{v}_{u_i}, \mathbf{v}'_v, \mathbf{c}, b_v]$. The objective function is to maximize the log likelihood of Eq. 8:

$$J = \sum_{v \in V} \log p(v|h, q; ; \theta). \tag{9}$$

Algorithm 1 shows the pseudo-code of our signed network embedding. We first randomly initialize node embeddings (Line 1) and then use random walk to generate the corpus (Line 2). Lines 4–11 show how we specify $\mathbf{c}_i$ based on the sign of edge $e_{u_i u_{i+1}}$. We calculate the predicted representation of the target node by combining source embeddings of nodes along the path with the edge type vectors (Line 12). We calculate the score function to measure the similarity between the predicted representation and the actual representation of the target

Algorithm 1. The SNE algorithm

Input : Signed graph $G = (V, E_+, E_-)$, embedding dimension d, length of path l, length of random walks L, walks per nodes t
Output: Representation of each node $\hat{\mathbf{v}}_i = [\mathbf{v}_i : \mathbf{v'}_i]$

```
1  Initialization: Randomly initialize the source and target node embeddings v_i
   and v'_i of each node V
2  Generate the corpus based on uniform random walk
3  for each path [u_1, u_2, ..., u_l, v] in the corpus do
4      for j = 1 to l do
5          if ε_{u_i u_{i+1}} == 1 then
6              c_i ≡ c_+
7          end
8          else
9              c_i ≡ c_-
10         end
11     end
12     compute v̂_h by Eq. 6
13     compute s(v, h) by Eq. 7
14     compute p(v|h, q; θ) by Eq. 8
15     update θ with Adagrad
16 end
```

node (Line 13) and compute the conditional likelihood of target node given the path (Line 14). Finally, we apply the Adagrad method [7] to optimize the objective function (Line 15). The procedures in Lines 4–15 repeat over each path in the corpus.

For a large network, the softmax function is expensive to compute because of the normalization term in Eq. 8. We adopt the sampled softmax approach [11] to reduce the computing complexity. During training, the source and target embeddings of each node are updated simultaneously. Once the model is well-trained, we get node embeddings of a signed network. We also adopt the approach in [26] to generate paths efficiently. Given each starting node u, we uniformly sample the next node from the neighbors of the last node in the path until it reaches the maximum length L. We then use a sliding window with size $l + 1$ to slide over the sequence of nodes generated by random walk. The first l nodes in each sliding window are treated as the sequence of path and the last node as the target node. For each node u, we repeat this process t times.

4 Experiments

To compare the performance of different network embedding approaches, we focus on the quality of their output, i.e., node embeddings. We use the generated node embeddings as input of two data mining tasks, node classification and link prediction. For node classification, we assume each node in the network is associated with a known class label and use node embeddings to build classifiers. In link prediction, we use node embeddings to predict whether there is a positive, negative, or no edge between two nodes. In our signed network embedding, we use the whole node representation $\bar{\mathbf{v}}_{v_i} = [\mathbf{v}_{v_i} : \mathbf{v}'_{v_i}]$ that contains both source embedding $\mathbf{v}_{v_i}$ and target embedding $\mathbf{v}'_{v_i}$. This approach is denoted as SNE_{st}. We also use only the source node vector $\mathbf{v}_{v_i}$ as the node representation. This approach is denoted as SNE_s. Comparing the performance of SNE_{st} and SNE_s on both directed and undirected networks expects to help better understand the performance and applicability of our signed network embedding.

Baseline Algorithms. We compare our SNE with the following baseline algorithms.

- SignedLaplacian [14]. It calculates eigenvectors of the k smallest eigenvalues of the signed Laplacian matrix and treats each row vector as node embedding.
- DeepWalk [26]. It uses uniform random walk (i.e., depth-first strategy) to sample the inputs and trains the network embedding based on skip-gram.
- LINE [30]. It uses the breadth-first strategy to sample the inputs based on node neighbors and preserves both the first order and second order proximities in node embeddings.
- Node2vec [9]. It is based on skip-gram and adopts the biased random walk strategy to generate inputs. With the biased random walk, it can explore diverse neighborhoods by balancing the depth-first sampling and breath-first sampling.

Datasets. We conduct our evaluation on two signed networks. (1) The first signed network, *WikiEditor*, is extracted from the UMD Wikipedia dataset [13]. The dataset is composed by 17015 vandals and 17015 benign users who edited the Wikipedia pages from Jan 2013 to July 2014. Different from benign users, vandals edit articles in a deliberate attempt to damage Wikipedia. One edit may be reverted by bots or editors. Hence, each edit can belong to either *revert* or *no-revert* category. The WikiEditor is built based on the co-edit relations. In particular, a positive (negative) edge between users i and j is added if the majority of their co-edits are from the same category (different categories). We remove from our signed network those users who do not have any co-edit relations with others. Note that in WikiEditor, each user is clearly labeled as either benign or vandal. Hence, we can run node classification task on WikiEditor in addition to link prediction. (2) The second signed network is based on the Slashdot Zoo dataset[3]. The Slashdot network is signed and directed. Unfortunately, it does not contain node label information. Thus we only conduct link prediction. Table 1 shows the statistics of these two signed networks.

Table 1. Statistics of WikiEditor and Slashdot

	WidiEditor	Slashdot
Type	Undirected	Directed
# of users (+, −)	21535 (7852, 13683)	82144 (N/A, N/A)
# of links (+, −)	348255 (269251, 79004)	549202 (425072, 124130)

[3] https://snap.stanford.edu/data/.

Parameter Settings. In our SNE methods, the number of randomly sampled nodes used in the sampled softmax approach is 512. The dimension of node vectors d is set to 100 for all embedding models except SignedLaplacian. SignedLaplacian is a matrix factorization approach. We only run SignedLaplacian on WikiEditor. This is because Slashdot is a directed graph and its Laplacian matrix is non-symmetric. As a result, the spectral decomposition involves complex values. For WikiEditor, SignedLaplacian uses 40 leading vectors because there is a large eigengap between the 40th and 41st eigenvalues. For other parameters used in DeepWalk, LINE and Node2vec, we use their default values based on their published source codes.

4.1 Node Classification

We conduct node classification using the WikiEditor signed network. This task is to predict whether a user is benign or vandal in the WikiEditor signed network. In our SNE training, the path length l is 3, the maximum length of random walk path L is 40, and the number of random walks starting at each node t is 20. We also run all baselines to get their node embeddings of WikiEditor. We then use the node embeddings generated by each method to train a logistic regression classifier with 10-fold cross validation.

Classification Accuracy. Table 2 shows the comparison results of each method on node classification task. Our SNE_{st} achieves the best accuracy and outperforms all baselines significantly in terms of accuracy. This indicates that our SNE can capture the different relations among nodes by using the signed-type vectors **c**. All the other embedding methods based on skip-gram have a low accuracy, indicating they are not feasible for signed network embedding because they do not distinguish the positive edges from negative edges. Another interesting observation is that the accuracy of SNE_s is only slightly worse than SNE_{st}. This is because WikiEditor is undirected. Thus using only source embeddings in the SNE training is feasible for undirected networks.

Visualization. To further compare the node representations trained by each approach, we randomly choose representations of 7000 users from WikiEditor and map them to a 2-D space based on t-SNE approach [19]. Figure 1 shows the projections of node representations from DeepWalk, Node2vec, and SNE_{st}. We observe that SNE_{st} achieves the best and DeepWalk is the worst. Node2vec performs slightly better than DeepWalk but the two types of users still mix together in many regions of the projection space.

Table 2. Accuracy for node classification on WikiEditor

	SignedLaplacian	DeepWalk	Line	Node2vec	SNE_s	SNE_{st}
Accuracy	63.52%	73.78%	72.36%	73.85%	79.63%	**82.07%**

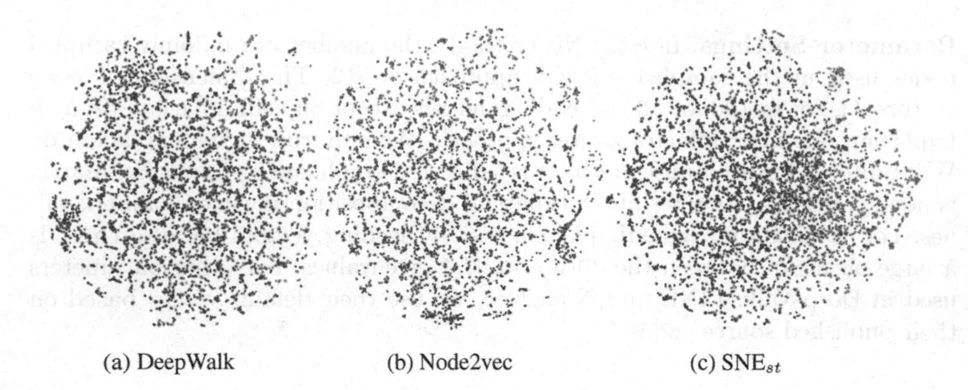

Fig. 1. Visualization of 7000 users in WikiEditor. Color of a node indicates the type of the user. Blue: "Vandals", red: "Benign Users". (Color figure online)

4.2 Link Prediction

In this section, we conduct link prediction on both WikiEditor and Slashdot signed graphs. We follow the same procedure as [9] to make link prediction as a classification task. We first use node representations to compose edge representations and then use them to build a classifier for predicting whether there is a positive, negative or no edge between two nodes. Given a pair of nodes (u, v) connected by an edge, we use an element-wise operator to combine the node vectors $\mathbf{v}_u$ and $\mathbf{v}_v$ to compose the edge vector $\mathbf{e}_{uv}$. We use the same operators as [9] and show them in Table 3. We train and test the one-vs-rest logistic regression model with 10-fold cross validation by using the edge vectors as inputs.

Table 3. Element-wise operators for combining node vectors to edge vectors

Operator	Definition
Average	$\mathbf{e}_{uv} = \frac{1}{2}(\mathbf{v}_u + \mathbf{v}_v)$
Hadamard	$\mathbf{e}_{uv} = \mathbf{v}_u * \mathbf{v}_v$
L1_Weight	$\mathbf{e}_{uv} = \|\mathbf{v}_u\text{-}\mathbf{v}_v\|$
L2_Weight	$\mathbf{e}_{uv} = \|\mathbf{v}_u\text{-}\mathbf{v}_v\|^2$

For Slashdot, we set the path length $l = 1$ in our SNE training, which corresponds to the use of the edge list of the Slashdot graph. This is because there are few paths with length larger than 1 in Slashdot. For WikiEditor, we use the same node representations adopted in the previous node classification task. We also compose balanced datasets for link prediction as suggested in [9]. We keep all the negative edges, randomly sample the same number of positive edges, and then randomly generate an equal number of fake edges connecting two nodes. At last, we have 79004 edges for each edge type (positive, negative, and fake) in WikiEditor and we have 124130 edges for each type in Slashdot.

Table 4. Comparing the accuracy for link prediction

Dataset	Approach	Hadamard	Average	L1_Weight	L2_Weight
WikiEditor (undirected)	SignedLaplacian	0.3308	0.5779	0.5465	0.3792
	DeepWalk	0.7744	0.6821	0.4515	0.4553
	Line	0.7296	0.6750	0.5205	0.4986
	Node2vec	0.7112	0.6491	0.6787	0.6809
	SNE_s	**0.9391**	**0.6852**	**0.8699**	0.8775
	SNE_{st}	**0.9399**	0.6043	0.8495	**0.8871**
Slashdot (directed)	DeepWalk	0.6907	**0.6986**	0.5877	0.5827
	Line	0.5823	0.6822	0.6158	0.6087
	Node2vec	0.6560	0.6475	0.4595	0.4544
	SNE_s	0.4789	0.5474	0.6078	0.6080
	SNE_{st}	**0.9328**	0.5810	**0.8358**	**0.8627**

Experimental Results. Table 4 shows the link prediction accuracy for each approach with four different operators. We observe that our SNE with Hadamard operator achieves the highest accuracy on both WikiEditor and Slashdot. SNE also achieves good accuracy with the L1_Weight and L2_Weight. For the Average operator, we argue that it is not suitable for composing edge vectors from node vectors in signed networks although it is suitable in unsigned networks. This is because a negative edge pushes away the two connected nodes in the vector space whereas a positive edge pulls them together [14]. When examining the performance of all baselines, their accuracy values are significantly lower than our SNE, demonstrating their infeasibility for signed networks.

We also observe that there is no big difference between SNE_{st} and SNE_s on WikiEditor whereas SNE_{st} outperforms SNE_s significantly on Slashdot. This is because WikiEditor is an undirected network and Slashdot is directed. This suggests it is imperative to combine both source embedding and target embedding as node representation in signed directed graphs.

4.3 Parameter Sensitivity

Vector Dimension. We evaluate how the dimension size of node vectors affects the accuracy of two tasks on both WikiEditor and Slashdot. For link prediction, we use SNE_{st} with Hadamard operation as it can achieve the best performance as shown in the last section. Figure 2a shows how the accuracy of link prediction varies with different dimension values of node vectors used in SNE_{st} for both datasets. We can observe that the accuracy increases correspondingly for both datasets when the dimension of node vectors increases. Meanwhile, once the accuracy reaches the top, increasing the dimensions further does not have much impact on accuracy any more.

Sample Size. Fig. 2b shows how the accuracy of link prediction varies with the sample size used in SNE_{st} for both datasets. For WikiEditor, we tune the sample size by changing the number of random walks starting at each node (t). In our experiment, we set $t = 5, 10, 15, 20, 25$ respectively and calculate the corresponding sample sizes. For Slashdot, we directly use the number of sampled edges in our training as the path length is one. For both datasets, the overall trend is similar. The accuracy increases with more samples. However, the accuracy becomes stable when the sample size reaches some value. Adding more samples further does not improve the accuracy significantly.

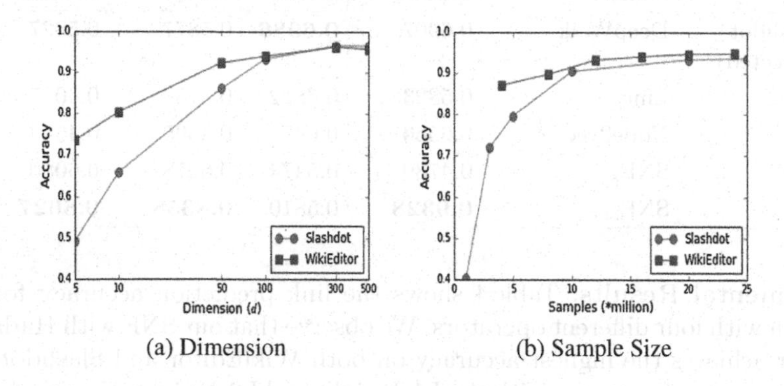

(a) Dimension (b) Sample Size

Fig. 2. The sensitivity of SNE on the WikiEditor and Slashdot

Path Length. We use WikiEditor to evaluate how the path length l affects the accuracy of both node classification and link prediction. From Fig. 3a, we observe that slightly increasing the path length in our SNE can improve the accuracy of node classification. This indicates that the use of long paths in our SNE training can generally capture more network structure information, which is useful for node classification. However, the performance of the SNE_s and SNE_{st} decreases when the path length becomes too large. One potential reason is that

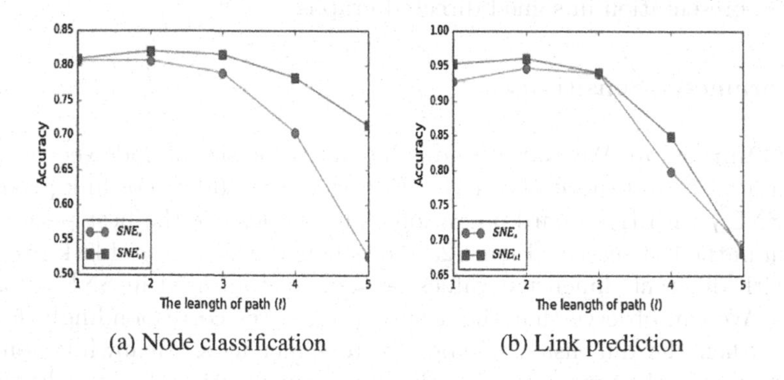

(a) Node classification (b) Link prediction

Fig. 3. The sensitivity of SNE on the WikiEditor by changing the path length (l)

SNE uses only two signed-type vectors for all nodes along paths and nodes in the beginning of a long path may not convey much information about the target node. In Fig. 3b, we also observe that the accuracy of link prediction decreases when the path length increases. For link prediction, the performance depends more on local information of nodes. Hence the inclusion of one source node in the path can make our SNE learn the sufficient local information.

5 Related Work

Signed Network Analysis. Mining signed network attracts increasing attention [5,14,16,27,28]. The balance theory [10] and the status theory [16] have been proposed and many algorithms have been developed for tasks such as community detection, link prediction, and spectral graph analysis of signed networks [5,14,18,28,34,35,38]. Spectral graph analysis is mainly based on matrix decomposition which is often expensive and hard to scale to large networks. It is difficult to capture the non-linear structure information as well as local neighborhood information because it simply projects a global matrix to a low dimension space formed by leading eigenvectors.

Network Embedding. Several network embedding methods including DeepWalk [26], LINE [30], Node2vec [9], Deep Graph Kernels [36] and DDRW [17] have been proposed. These models are based on the neural language model. Several network embedding models are based on other neural network model. For example, DNR [33] uses the deep auto-encoder, DNGR [3] is based on a stacked denoising auto-encoder, and the work [23] adopts the convolutional neural network to learn the network feature representations. Meanwhile, some works learn the network embedding by considering the node attribute information. In [32,39] the authors consider the node label information and present semi-supervised models to learn the network embedding. The heterogeneous network embedding models are studied in [4,25,29,37]. HOPE [24] focuses on preserving the asymmetric transitivity of a directed network by approximating high-order proximity of a network. Unlike all the works described above, in this paper, we explore the signed network embedding.

6 Conclusion

In this paper, we have presented SNE for signed network embedding. Our SNE adopts the log-bilinear model to combine the edge sign information and node representations of all nodes along a given path. Thus, the learned node embeddings capture the information of positive and negative links in signed networks. Experimental results on node classification and link prediction showed the effectiveness of SNE. Our SNE expects to keep the same scalability as DeepWalk or Node2vec because SNE adopts vectors to represent the sign information and uses linear operation to combine node representation and signed vectors. In our future work, we plan to examine how other structural information (e.g., triangles or motifs) can be preserved in signed network embedding.

Acknowledgments. The authors acknowledge the support from the National Natural Science Foundation of China (71571136), the 973 Program of China (2014CB340404), and the Research Project of Science and Technology Commission of Shanghai Municipality (16JC1403000, 14511108002) to Shuhan Yuan and Yang Xiang, and from National Science Foundation (1564250) to Xintao Wu. This research was conducted while Shuhan Yuan visited University of Arkansas. Yang Xiang is the corresponding author of the paper.

References

1. Bengio, Y., Courville, A., Vincent, P.: Representation learning: a review and new perspectives. TPAMI **35**(8), 1798–1828 (2013)
2. Bengio, Y., Ducharme, R., Vincent, P., Jauvin, C.: A neural probabilistic language model. J. Mach. Learn. Res. **3**, 1137–1155 (2003)
3. Cao, S., Lu, W., Xu, Q.: Deep neural networks for learning graph representations. In: AAAI (2016)
4. Chang, S., Han, W., Tang, J., Qi, G.-J., Aggarwal, C.C., Huang, T.S.: Heterogeneous network embedding via deep architectures. In: KDD (2015)
5. Chiang, K.-Y., Hsieh, C.-J., Natarajan, N., Tewari, A., Dhillon, I.S.: Prediction, clustering in signed networks: a local to global perspective. arXiv:1302.5145 [cs] (2013)
6. Collobert, R., Weston, J., Bottou, L., Karlen, M., Kavukcuoglu, K., Kuksa, P.: Natural language processing (almost) from scratch. J. Mach. Learn. Res. **12**, 2493–2537 (2011)
7. Duchi, J., Hazan, E., Singer, Y.: Adaptive subgradient methods for online learning and stochastic optimization. J. Mach. Learn. Res. **12**, 2121–2159 (2011)
8. Graves, A., Mohamed, A.-R., Hinton, G.: Speech recognition with deep recurrent neural networks. In: arXiv:1303.5778 [cs] (2013)
9. Grover, A., Leskovec, J.: node2vec: scalable feature learning for networks. In: KDD (2016)
10. Heider, F.: Attitudes and cognitive organization. J. Psychol. **21**(1), 107–112 (1946)
11. Jean, S., Cho, K., Memisevic, R., Bengio, Y.: On using very large target vocabulary for neural machine translation. arXiv:1412.2007 [cs] (2014)
12. Krizhevsky, A., Sutskever, I., Hinton, G.E.: Imagenet classification with deep convolutional neural networks. In: NIPS (2012)
13. Kumar, S., Spezzano, F., Subrahmanian, V.: VEWS: a Wikipedia vandal early warning system. In: KDD (2015)
14. Kunegis, J., Schmidt, S., Lommatzsch, A., Lerner, J., De Luca, E.W., Albayrak, S.: Spectral analysis of signed graphs for clustering, prediction and visualization. In: SDM (2010)
15. LeCun, Y., Bengio, Y., Hinton, G.: Deep learning. Nature **521**(7553), 436–444 (2015)
16. Leskovec, J., Huttenlocher, D., Kleinberg, J.: Signed networks in social media. In: Proceedings of the SIGCHI Conference on Human Factors in Computing Systems (2010)
17. Li, J., Zhu, J., Zhang, B.: Discriminative deep random walk for network classification. In: ACL (2016)
18. Li, Y., Wu, X., Lu, A.: On spectral analysis of directed signed graphs. CoRR, abs/1612.08102 (2016)

19. Maaten, L.V.D., Hinton, G.: Visualizing data using t-SNE. J. Mach. Learn. Res. **9**, 2579–2605 (2008)
20. Mikolov, T., Corrado, G., Chen, K., Dean, J.: Efficient estimation of word representations in vector space. In: ICLR (2013)
21. Mnih, A., Hinton, G.: A scalable hierarchical distributed language model. In: NIPS (2008)
22. Mnih, A., Kavukcuoglu, K.: Learning word embeddings efficiently with noise-contrastive estimation. In: NIPS (2013)
23. Niepert, M., Ahmed, M., Kutzkov, K.: Learning convolutional neural networks for graphs. In: ICML (2016)
24. Ou, M., Cui, P., Pei, J., Zhu, W.: Asymmetric transitivity preserving graph embedding. In: KDD (2016)
25. Pan, S., Wu, J., Zhu, X., Zhang, C., Wang, Y.: Tri-party deep network representation. In: IJCAI (2016)
26. Perozzi, B., Al-Rfou, R., Skiena, S.: DeepWalk: online learning of social representations. In: KDD (2014)
27. Tang, J., Aggarwal, C., Liu, H.: Node classification in signed social networks. In: SDM (2016)
28. Tang, J., Chang, Y., Aggarwal, C., Liu, H.: A survey of signed network mining in social media. arXiv:1511.07569 [physics] (2015)
29. Tang, J., Qu, M., Mei, Q.: PTE: Predictive text embedding through large-scale heterogeneous text networks. In: KDD (2015)
30. Tang, J., Qu, M., Wang, M., Zhang, M., Yan, J., Mei, Q.: Line: large-scale information network embedding. In: WWW (2015)
31. Tian, F., Gao, B., Cui, Q., Chen, E., Liu, T.-Y.: Learning deep representations for graph clustering. In: AAAI (2014)
32. Tu, C., Zhang, W., Liu, Z., Sun, M.: Max-margin deepwalk: discriminative learning of network representation. In: IJCAI (2016)
33. Wang, D., Cui, P., Zhu, W.: Structural deep network embedding. In: KDD (2016)
34. Wu, L., Wu, X., Lu, A., Li, Y.: On spectral analysis of signed and dispute graphs. In: ICDM, pp. 1049–1054 (2014)
35. Wu, L., Ying, X., Wu, X., Lu, A., Zhou, Z.-H.: Spectral analysis of k-balanced signed graphs. In: Huang, J.Z., Cao, L., Srivastava, J. (eds.) PAKDD 2011. LNCS (LNAI), vol. 6635, pp. 1–12. Springer, Heidelberg (2011). doi:10.1007/978-3-642-20847-8_1
36. Yanardag, P., Vishwanathan, S.: Deep graph kernels. In: KDD (2015)
37. Yang, C., Liu, Z., Zhao, D., Sun, M., Chang, E.Y.: Network representation learning with rich text information. In: IJCAI (2015)
38. Yang, Y., Lichtenwalter, R.N., Chawla, N.V.: Evaluating link prediction methods. Knowl. Inf. Syst. **45**(3), 751–782 (2015)
39. Yang, Z., Cohen, W.W., Salakhutdinov, R.: Revisiting semi-supervised learning with graph embeddings. In: ICML (2016)

mHUIMiner: A Fast High Utility Itemset Mining Algorithm for Sparse Datasets

Alex Yuxuan Peng, Yun Sing Koh(✉), and Patricia Riddle

The University of Auckland, Auckland, New Zealand

ypen260@aucklanduni.ac.nz, {ykoh,pat}@cs.auckland.ac.nz

Abstract. High utility itemset mining is the problem of finding sets of items whose utilities are higher than or equal to a specific threshold. We propose a novel technique called mHUIMiner, which utilises a tree structure to guide the itemset expansion process to avoid considering itemsets that are nonexistent in the database. Unlike current techniques, it does not have a complex pruning strategy that requires expensive computation overhead. Extensive experiments have been done to compare mHUIMiner to other state-of-the-art algorithms. The experimental results show that our technique outperforms the state-of-the-art algorithms in terms of running time for sparse datasets.

Keywords: High-utility itemset mining · Transaction utility

1 Introduction

The problem of frequent itemset mining (FIM) [1,4] is to find a set of itemsets that appear frequently in a transaction database. Classic FIM assumes that an item can only appear once in a transaction and every distinct item in the database has the same weight or importance. It is obvious that this assumption is not always true in real world scenarios. For example, consider a transaction database of a supermarket. Multiple identical items can appear in the same transaction. The profit and price of each item can also be different. The item with a higher profit or price should be assigned a higher weight. FIM algorithms generate a set of frequent itemsets, but these itemsets may contribute low profit or revenue to a company. To address these issues, the problem of *high-utility itemset mining* (HUIM) was proposed [9]. In HUIM, an item can appear more than once in a transaction, and each item has a utility value (weight). The goal of HUIM is to find a set of itemsets whose utility values are higher than a specific threshold. The techniques found in FIM algorithms cannot usually be directly used in HUIM problems. This is due to the fact that the downward closure property does not hold in a HUIM problem.

HUIM algorithms such as Two-Phase [7], UPGrowth [8] and IHUP [2] all involve two phases in the mining process. These algorithms usually generate a large number of candidates and the process of computing the exact utility value

J. Kim et al. (Eds.): PAKDD 2017, Part II, LNAI 10235, pp. 196–207, 2017.
DOI: 10.1007/978-3-319-57529-2_16

for each of the candidates can be very expensive. HUI-Miner [5] is a one-phase algorithm that generates high-utility itemsets without a candidate-generation process. HUI-Miner proposed a structure called a *utility-list*. This process of creating the utility-list can be expensive, and HUI-Miner does not have a mechanism to prune out the unnecessary constructions of utility-lists. FHM [3] introduced a pruning strategy to avoid the unnecessary utility-list constructions. However, this approach is not efficient for databases that are sparse. Another algorithm called EFIM [10] introduced an array-based utility counting technique to compute the overestimation of utility for pruning purposes. It also uses database projection and transaction merging to reduce the cost of the database scan. EFIM has been shown to be more efficient than previous algorithms both in terms of running time and memory consumption. However, experiments show that the performance advantage of EFIM shrinks on sparse datasets. An example of a sparse dataset is the transaction database of a supermarket. A supermarket usually has a vast variety of products available. However, each transaction contains only a tiny portion of the products available. Considering market basket analysis is a very important application of HUIM, a new algorithm that achieves better performance on sparse datasets is necessary.

We propose a novel high-utility itemset mining algorithm, mHUIMiner (modified HUI-Miner), that provides the best running time on sparse datasets, while maintaining a comparable performance to other state-of-the-art algorithms on dense datasets. Unlike what the name suggests this goes beyond minor modifications to the existing HUI-Miner. This technique avoids unnecessary utility-list constructions in HUI-Miner by incorporating a tree structure. It also does not have a complex pruning strategy that requires expensive computational overhead, which usually does not achieve economies of scale in sparse datasets. We also present performance comparisons of the proposed mHUIMiner against other state-of-the-art high-utility itemset mining algorithms. Based on experiments on a set of real-world transaction datasets, it shows that mHUIMiner is the fastest on sparse datasets. We also conducted experiments on a set of synthetic datasets with various densities. The results show that as the density decreases, mHUIMiner still performs efficiently.

The rest of the paper is organised as follows. We provide the problem definition and related works in Sect. 2. We describe the proposed algorithm in Sect. 3. In Sect. 4, we present the experimental results and evaluation of our proposed algorithm. Finally, we conclude our work in Sect. 5.

2 Preliminaries and Related Work

Let $I = \{i_1, i_2, i_3, \ldots, i_n\}$ be a set of single items. A transaction database DB usually consists of a transaction table and a utility table. The transaction table contains a set of transactions $\{T_1, T_2, T_3, \ldots, T_k\}$, where T_{id} is the unique transaction identifier for each transaction. Each transaction is a subset of I and a count value is associated with each item in the transaction. The utility table stores all the utility values for each item i in I.

Table 1. Transaction table (top) and utility table (bottom)

Tid	Transactions									
T1	c	2	b	1	e	1				
T2	a	3	e	2	g	1	b	4		
T3	a	1	b	2	c	3	d	4	e	5
T4	f	3	g	1						
T5	b	1	a	1	d	1				

Item	a	b	c	d	e	f	g
Profit	5	1	3	4	2	1	2

The *internal utility* of item i in transaction T of database DB is the count value of i. It represents the quantity of item i in transaction T. We denote *internal utility* of item i in transaction T as $iu(i,T)$. In Table 1, the *internal utility* of item c in transaction T_1 is 2. It means that item c appears twice in the transaction T_1. The *external utility* of item i in database DB is the utility value (e.g. unit profit) of i in the utility table. It indicates the importance or weight of an item. The *external utility* is assumed to be non-negative in the scope of our research. We denote *external utility* of item i as $eu(i)$. For example, in Table 1, the *external utility* of item c is 3 according to the utility table. The utility of item i in transaction T is defined as $u(i,T) = iu(i,T) \times eu(i)$. It measures the total utility of item i in a single transaction T. For example, the utility of item c in transaction T_1 is $u(c,T_1) = iu(c,T_1) \times eu(c) = 2 \times 3 = 6$. An *itemset* X is a subset of a transaction T. The utility of itemset X in transaction T is defined as $u(X,T) = \sum_{i \in X \wedge X \subseteq T} u(i,T)$. It is the sum of the total utility in the transaction T of every item that is in the itemset X, where X is a subset of transaction T. For example, the utility of itemset $\{cb\}$ in transaction T_1 is $u(cb,T_1) = u(c,T_1) + u(b,T_1) = 2 \times 3 + 1 \times 1 = 7$. And the utility of itemset $\{cbe\}$ in transaction T_1 is $u(cbe,T_1) = u(c,T_1) + u(b,T_1) + u(e,T_1) = 2 \times 3 + 1 \times 1 + 1 \times 2 = 9$. The *utility of itemset X in transaction database DB* is defined as $u(X,DB) = \sum_{T \in DB} u(X,T)$. This measures the total utility of an itemset X over the transaction database DB.

The goal of *high utility itemset mining* is to find all the *high utility itemsets* in a transaction database. Let *minutil* be a user-specified minimum utility threshold. We say an itemset X is a *high utility itemset* if $u(X,DB)$ is greater than or equal to *minutil*. The *transaction utility*, denoted as $tu(T)$, is the sum of the utilities of all the items in transaction T, where $tu(T) = \sum_{i \in T} u(i,T)$. Table 2 is the transaction utility table. The *transaction weighted utility* of an itemset X is defined as $TWU(X) = \sum_{T \in DB \wedge X \subseteq T} tu(T)$. Intuitively, $TWU(X)$ is the sum of the transaction utilities for all the transactions that contain itemset X. For example, the transaction weighted utility of item b, $TWU(b) = tu(T_1) + tu(T_2) + tu(T_3) + tu(T_5) = 9 + 25 + 42 + 10 = 86$. Table 3 is the transaction weighted utility table for the transaction database. The transaction weighted utility (TWU) of an itemset is an overestimation of the exact utility of this itemset, $TWU(X) \geq u(X)$. TWU is anti-monotonic, i.e. $TWU(X) \geq TWU(Y)$ if

Table 2. Transaction utility

Tid	T1	T2	T3	T4	T5
TU	9	25	42	5	10

Table 3. Transaction-weighted utility

Item	b	a	e	d	c	g	f
TWU	86	77	76	52	51	30	5

$X \subset Y$. It means that if the TWU of itemset X is smaller than the user entered threshold, there is no need to consider all the supersets of X, because the TWUs of the supersets of X are guaranteed to be smaller as well.

Two phase algorithms such as IHUP [2], Two-Phase [7] and UPGrowth [8] use *transaction weighted utility* (TWU) [6] as a measure to prune the search space. Recently, more efficient one-phase algorithms, such as HUI-Miner [5] and FHM [3], have been introduced. These algorithms utilise a utility-list structure to maintain utility related information for each itemset, and they mine high-utility itemsets in a single phase without candidate generation. Let the *utility-tuple* of itemset X in transaction T be denoted as *(tid, iutil, rutil)*. *tid* is the transaction identifier of transaction T. *iutil* is the utility of itemset X in transaction T, namely $u(X,T)$. *rutil* is the remaining utility of itemset X in transaction T. Suppose items in transaction T are sorted based on a certain order, *rutil* of X is the sum of utilities for all the items in T after X. The *Utility-list* of an itemset X in database DB is a set of tuples such that there is a utility-tuple for each transaction that contains X. The tuples in a utility-list are ordered based on transaction identifiers. $sum(iutils)$ is the sum of *iutils* of all utility-tuples in a utility-list. It is equivalent to *the total utility of itemset X in the transaction database.* More formally, $sum(iutils) = u(X, DB) = \sum_{T \in DB} u(X,T)$. $sum(iutils) + sum(rutils)$ is the sum of *iutils* and *rutils* of all utility-tuples in a utility-list. Similar to TWU, $sum(iutils) + sum(rutils)$ is an overestimation of an itemset's utility in the transaction database. However, it can be easily proven that $sum(iutils) + sum(rutils) \leq$ TWU. This means that $sum(iutils) + sum(rutils)$ can prune out more of the search space than TWU.

HUI-Miner is more efficient than those two-phase algorithms, but it is not without drawbacks. Suppose the utility-lists of two itemsets have m and n utility-tuples respectively. Then, the total number of comparisons needed in the new utility-list construction procedure is at most $(m + n)$. This procedure can be expensive if the utility-lists are very large. The HUI-Miner algorithm does not have a good mechanism to prune the constructions of utility-lists. This can cause the algorithm to run a lot of unnecessary operations to try creating utility-lists for itemsets that are not even in the database. This problem becomes more serious for sparse transaction databases. In this context, a sparse transaction database is a database that has a large number of distinct items while the average number of items in a transaction is relatively small.

3 mHUIMiner

In this section, we present our *mHUIMiner* (modified HUI-Miner) algorithm which solves the problem in the HUI-Miner algorithm. This algorithm is a modified HUI-Miner that integrates the IHUP-tree structure into the original HUI-

Miner algorithm. A nice property of the IHUP-tree structure is that a path in the tree corresponds to a transaction in the database. It means that the tree contains all the information about the composition of all the transactions in the database. This information tells us which itemsets or patterns actually exist in the database. Hence, if we mine itemsets along the paths of the tree incrementally, we can avoid expanding the current itemset into one that does not exist in the database. Next, we will describe the details of mHUIMiner algorithm and demonstrate how the algorithm works using simple examples based on the transaction database in Table 1.

Algorithm 1. mHUIMiner

Input: *DB*: a transaction database, *minutil*: a user-specified threshold
Output: a set of high-utility itemsets

1 scan *DB* to calculate the *TWU* for each single item;
2 scan *DB* again to create a global $IHUP_{TWU} - Tree\ T$ along with its header table *T.headerTable* and a global *hashmap* to store utility-list *UL* for every single item;
3 **for** *item i from the bottom of T.headerTable* **do**
4   get utility-list *ULi* from the global *hashmap* for item *i*;
5   **if** *sum(ULi.iutils)+sum(ULi.rutils)* $\geq$ *minutil* **then**
6   create local prefix tree *Ti* and its header table for item *i*;
7   call Mining(*i*, *Ti*, *ULi*, *minutil*);
8   **end**
9 **end**

The main procedure (Algorithm 1) of mHUIMiner algorithm takes as input a transaction database and a user-specified threshold *minutil*. The algorithm scans the database for the first time to calculate the TWU value for each distinct item in the database. We create a global tree to maintain transaction information and initial utility-lists for all the distinct items. During the tree building process, the header table of the tree is also created. The header table contains TWU values for all the items that are in the global tree and it is sorted in descending order of TWU values. During the second scan of the original database, any item whose TWU is smaller than the threshold *minutil* is discarded. This means that the tree and single-item utility-list will only contain items whose TWU values are equal to or larger than the threshold. Unlike most of the other algorithms using similar tree structures, the tree in our algorithm does not store any utility related information. The purpose of the tree is solely to guide the itemset mining and expansion process. All utility related information we need for high-utility itemset mining is captured in the utility-lists. We have tried to store utility-tuples inside the nodes of the tree. However, this means that we have to construct utility-lists for single-item itemsets on-the-fly during the mining process. Because we construct a new itemset by adding an item to the current itemset, the utility-list of the same item would be constructed every time when it is needed. When we tested this approach it was very inefficient, so we have chosen to use two data

Table 4. Example of revised transactions

Tid	Item	Util.	Item	Util.	Item	Util.	Item	Util.	Item	Util.
T1	c	(6)	e	(2)	b	(1)				
T2	e	(4)	a	(15)	b	(4)				
T3	c	(9)	d	(16)	e	(10)	a	(5)	b	(2)
T5	d	(4)	a	(5)	b	(1)				

Table 5. Global header table

Item	TWU
b	86
a	77
e	76
d	52
c	51

structures to store different types of information. Starting from the bottom of the header table, we test if the sum of all the *iutils* and *rutils* of the item's utility-list is smaller than *minutil*. We ignore an item if this sum is smaller than the threshold. For the remaining items, local prefix trees and local header tables for these items are created, and the *Mining* procedure of the algorithm is called. Note that the local header table is similar to the global header table, except that the local header table only contains items that are in the local prefix tree.

Suppose that the user specified *minutil* is 40. Table 3 contains all the TWU values computed during the first database scan. Because the TWU values of items g and f are smaller than 40, they can be discarded from the transactions. Table 4 shows all the revised transactions. Revised transactions only contain items whose TWU values are greater than 40. And all revised transactions are ascendingly sorted based on TWU values. Figure 1 shows the global tree built using the revised transactions and Table 5 is the header table of the global tree. We start from the bottom of the header table. The last item in the header table is c. The utility-list of c is shown in Fig. 4. According to the utility-list, the sum of *iutils* and *rutils* of c is 51. Since the *minutil* is smaller than 51, we will create the local prefix tree and its local header table for c. The prefix tree of c can be seen in Fig. 2. Then the mining procedure is called. However, if *minutil* were larger than 51, we would ignore item c and move on to the next item, d, in this case.

The *Mining* procedure (Algorithm 2) takes as input a prefix tree, a current itemset, the utility-list of the current itemset and the threshold. It checks if the sum of all *iutils* of the utility-list is larger than or equal to the threshold. If the sum of *iutils* of the current itemset is larger than the threshold, this itemset is a high-utility itemset. Then it decides whether to expand current itemset by comparing the sum of all *iutils* and *rutils* against the threshold. If it is larger than or equal to *minutil*, we expand the current itemset by adding one new item from the header table. The prefix tree and the local header table of this new itemset are created. Then the *Construct* procedure (Algorithm 3) is called to create the utility-list for this new itemset. Finally, we call the *Mining* process recursively.

Suppose the current itemset is $\{c\}$. According to the utility-list of $\{c\}$, the sum of *iutils* is 15. $\{c\}$ is not a high-utility itemset because *minutil* is 40. $\{c\}$ would be a high-utility itemset if the sum of *iutils* were larger than or equal to *minutil*. Recall that the sum of all the *iutils* and *rutils* of c is 51. Since *minutil*

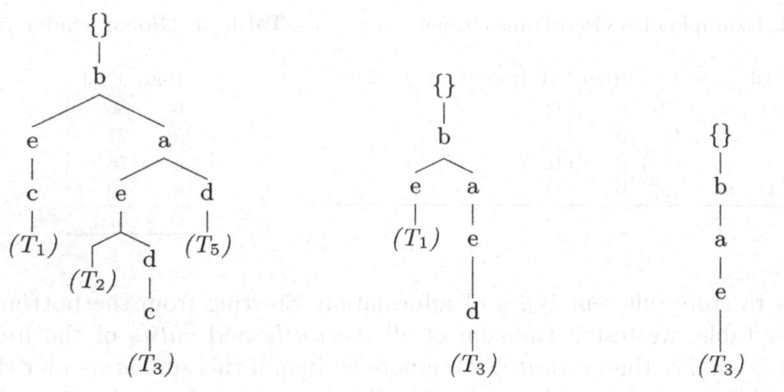

Fig. 1. Global tree **Fig. 2.** Local tree of c **Fig. 3.** Local tree of cd

Tid	iutil	rutil
T1	6	3
T3	9	33

{c}

Fig. 4. Utility-list of c

Tid	iutil	rutil
T3	25	17

{cd}

Fig. 5. Utility-list of cd

is less than 51, we create the new itemsets by adding single items from its local prefix tree (Fig. 2) to $\{c\}$. In this case, the new itemsets would be $\{cd\}, \{ce\}$, $\{ca\}$ and $\{cb\}$. Utility-lists and prefix trees of these new itemsets are generated. For example, the local prefix tree of $\{cd\}$ is shown in Fig. 3 and the utility-list of $\{cd\}$ is shown in Fig. 5.

The *Mining* process is called again to test if $\{cd\}, \{ce\}, \{ca\}$ and $\{cb\}$ are high-utility itemsets and whether we should expand them further. We repeat this process recursively. The algorithm will terminate when we cannot generate a non-empty prefix tree. As we can see, an itemset expands by adding one single item from its local prefix tree. This is crucial in making sure that we do not spend time on itemsets that do not exist in the database.

The *Construct* procedure (Algorithm 3) takes as input the utility-list for the current itemset and the utility-list for a single item. First, an empty utility-list for the new itemset is created. Common transactions are identified by comparing transaction identifiers. For every common transaction, a new utility-tuple is created and appended to the utility-list. Our mHUIMiner incorporates a tree structure to guide the itemset expansion process, this ensures that it never needs to consider an itemset that does not exist in the transaction database. The source code is published on GitHub: github.com/superRookie007/mHUIMiner.

Algorithm 2. Mining

Input: p: current itemset, Tp: current itemset's prefix tree, ULp: utility list of the current itemset, $minutil$: user-specified threshold
Output: high-utility itemsets

1 **if** *sum(ULp.iutils)* $\geq$ *minutil* **then**
2 output p as a high-utility itemset;
3 **end**
4 **if** *sum(ULp.iutils)+sum(ULp.rutils)* $\geq$ *minutil* **then**
5 **for** *item x from the bottom of Tp.headerTable* **do**
6 create prefix-tree Tpx and local header table for new itemset px;
7 get utility-list ULx from global *hashmap* for item x;
8 $ULpx$ = Construct(ULp, ULx);
9 call Mining(px, Tpx, $ULpx$, $minutil$);
10 **end**
11 **end**

Algorithm 3. Construct

Input: ULp: the utility-list for itemset P, ULx: the utility-list for item x
Output: $ULpx$, the utility-list for new itemset px

1 $ULpx = NULL$;
2 **for** *element Ep in ULp* **do**
3 **for** *element Ex in ULx* **do**
4 **if** *Ex.tid==Ep.tid* **then**
5 $Epx = < Ep.tid, Ep.iutil + Ex.iutil, Ex.rutil >$;
6 **end**
7 append Epx to $ULpx$;
8 **end**
9 **end**
10 **return** $ULpx$;

4 Evaluations

Extensive experiments have been done to evaluate the performance of the proposed mHUIMiner algorithm against other state-of-the-art algorithms. The competing algorithms include IHUP [2], FHM [3], HUIMiner [5] and EFIM [10]. All algorithms were implemented in Java. Experiments were performed on a machine with 2.2 GHz Intel Core i5 CPU and 8 GB of RAM running Windows 10. Note that all these bench algorithms and mHUIMiner are deterministic and all these algorithms generate the same high-utility itemsets and rules on the same dataset with the same threshold setting. Thus we will not be investigating rule quality as it does not differ between these datasets.

At first, we compare the proposed mHUIMiner against competing algorithms over standard real life datasets. The datasets and their characteristics are shown in Table 6. These datasets are obtained from the SPMF website. Suppose a transaction database has N distinct items and M different items per transaction on average, the database *density* is defined as M/N.

Table 6. Characteristics of standard datasets

Dataset	# transactions	# distinct items	Avg. # of items per trans	Density
Accidents	340183	468	33.8079	7.2239%
BMS	59602	497	2.5106	0.5052%
Chainstore	1112949	46086	7.2266	0.0157%
Chess	3196	75	37.0000	49.3333%
Foodmart	4141	1559	4.4238	0.2838%
Kosarak	990002	41270	8.1000	0.0196%
Mushroom	8124	119	23	19.3277%
Retail	88162	16470	10.3057	0.0626%

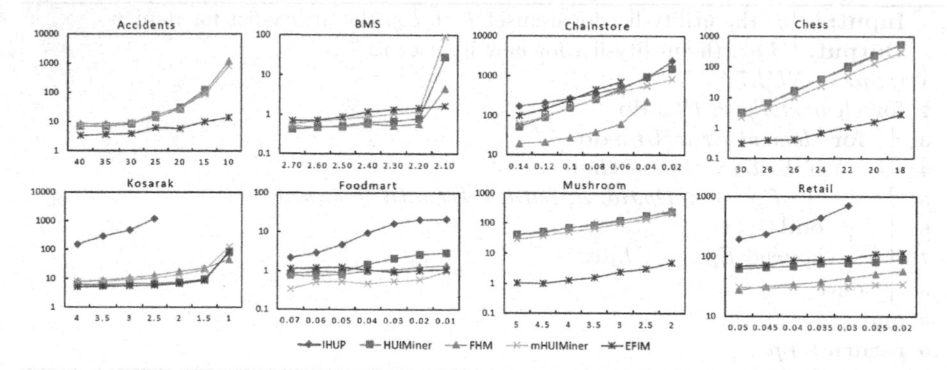

Fig. 6. Running time. Vertical-axis: execution time (sec.), horizontal-axis: minimum utility (%)

We ran all the algorithms on each of the datasets while gradually decreasing the *minutil* threshold, until an out-of-memory error was raised or a timeout happened. We terminated experiments whose running times were over 20 min. The experimental results can be found in Fig. 6 (running times) and Fig. 7 (memory consumption). It can be seen that in general, the performance of mHUIMiner is similar to that of HUIMiner and FHM. It is not surprising considering all of these algorithms use the same utility-list method in their mining process. Within the group of these three algorithms, mHUIMiner is faster than the others on *chess, foodmart, mushroom and retail.* And for *accidents and chainstore*, as the *minutil* threshold gets smaller, mHUIMiner starts to run faster than FHM and HUIMiner. The IHUP algorithm only managed to finish the experiments on *chainstore* and *foodmart* datasets without overtime or memory error. The EFIM algorithm performed very well in terms of running time for dense datasets such as *accidents, chess and mushroom.* But it was outperformed by mHUIMiner over sparser datasets such as *chainstore, foodmart and retail.* EFIM is much more memory efficient than the other algorithms over datasets *chess, mushroom and retail.* This is because while other algorithms rely on complex tree or utility-list structures to maintain information, EFIM generates projected databases that are often very small in size due to transaction merging.

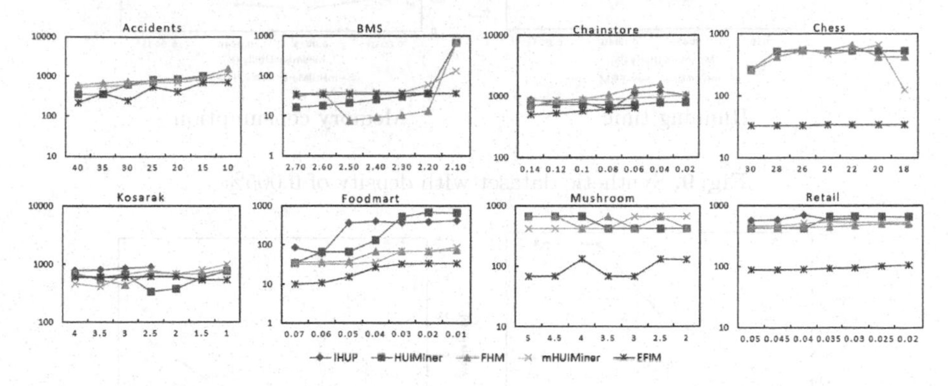

Fig. 7. Memory consumption. Vertical-axis: memory (MB), horizontal-axis: minimum utility (%)

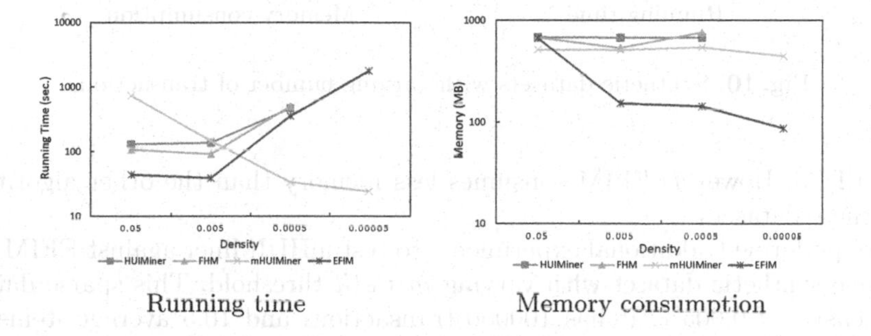

Fig. 8. Synthetic datasets with varying density

It seems that the mHUIMiner algorithm does provide a slight advantage over FHM and HUIMiner in terms of running time on most of these real datasets. And while EFIM is very fast on dense datasets, it tends to be outperformed by mHUIMiner over sparser datasets. We generated additional synthetic datasets with different density to test this hypothesis. All these synthetic datasets were generated using the dataset generator in SPMF. All these datasets have the same 100000 transactions and 10.5 average items per transaction. But the density varies from 5% to 0.005%. We also kept *minutil* threshold constant at 0.004% for all the tests. We terminated experiments whose running times were over 30 min. The results are depicted in Fig. 8. It can be seen that mHUIMiner is the slowest when density is high, but as the density decreases mHUIMiner becomes substantially more efficient in terms of running time. When density is at 0.005%, both FHM and HUIMiner ran out of time and mHUIMiner is over 70 times faster

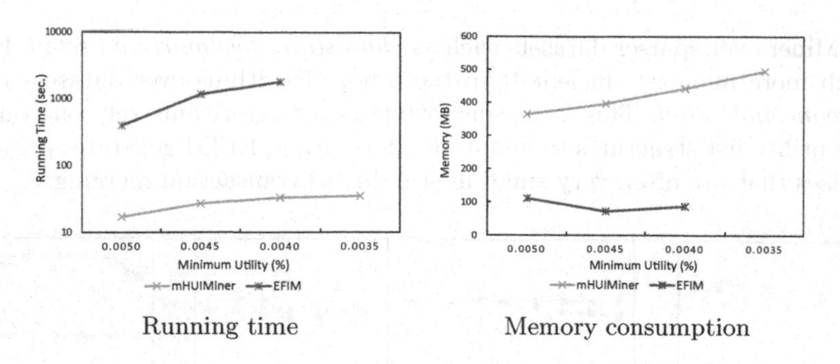

Running time Memory consumption

Fig. 9. Synthetic dataset with density of 0.005%

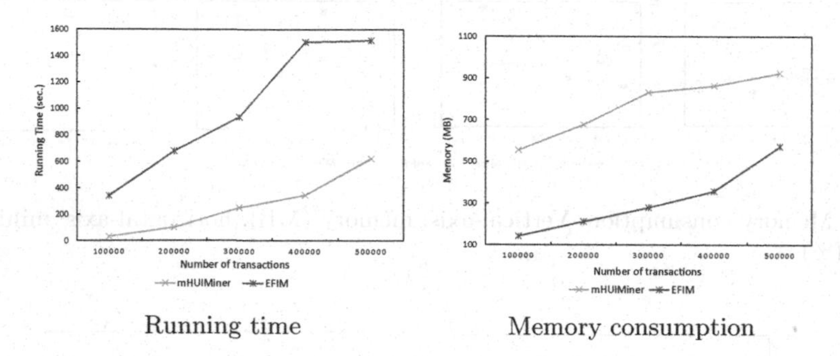

Running time Memory consumption

Fig. 10. Synthetic datasets with varying number of transactions

than EFIM. However, EFIM consumes less memory than the other algorithms over these datasets.

We performed additional experiments to test mHUIMiner against EFIM over a sparse synthetic dataset with varying *minutil* threshold. This sparse dataset has density of 0.005%. It has 100000 transactions and 10.5 average items per transaction. Again, we set the time limit as 30 min. The results are depicted in Fig. 9. It can be seen that mHUIMiner is much faster than EFIM on this sparse dataset, however the memory consumption of mHUIMiner is higher. Moreover, when *minutil* was set to 0.0035%, EFIM did not manage to finish in time.

To confirm that mHUIMiner's running time and memory consumption do not grow exponentially according to the number of transactions in the dataset, we performed tests on datasets with different number of transactions. All of these datasets have the same number of distinct items (20000) and the same average number of items per transaction (10.5). We also used the same *minutil* threshold of 0.005% for all the tests. The number of transactions varies from 100000 to 500000. The results are depicted in Fig. 10. It is obvious that even though both running time and memory consumption of mHUIMiner increase as the input size increases, they do not grow exponentially.

5 Conclusions and Future Work

We proposed a novel mHUIMiner algorithm for high-utility itemset mining, which outperforms all of the compared state-of-the-art algorithms in terms of running time for sparse datasets. Currently our mHUIMiner algorithm is designed to work on static databases, but it can be adapted to mine high-utility itemsets over incremental databases. Ideally, the incremental algorithm should be able to reuse previous data structures and mining results to avoid repetitive computation after a database is updated. Finally, the current algorithm assumes that utility values are non-negative. Obviously, this is not always true in real world applications.

References

1. Agrawal, R., Imieliński, T., Swami, A.: Mining association rules between sets of items in large databases. In: ACM SIGMOD Record, vol. 22, pp. 207–216. ACM (1993)
2. Ahmed, C.F., Tanbeer, S.K., Jeong, B.S., Lee, Y.K.: Efficient tree structures for high utility pattern mining in incremental databases. IEEE Trans. Knowl. Data Eng. **21**(12), 1708–1721 (2009)
3. Fournier-Viger, P., Wu, C.-W., Zida, S., Tseng, V.S.: FHM: faster high-utility itemset mining using estimated utility co-occurrence pruning. In: Andreasen, T., Christiansen, H., Cubero, J.-C., Raś, Z.W. (eds.) ISMIS 2014. LNCS (LNAI), vol. 8502, pp. 83–92. Springer, Cham (2014). doi:10.1007/978-3-319-08326-1_9
4. Han, J., Cheng, H., Xin, D., Yan, X.: Frequent pattern mining: current status and future directions. Data Min. Knowl. Disc. **15**(1), 55–86 (2007)
5. Liu, M., Qu, J.: Mining high utility itemsets without candidate generation. In: Proceedings of the 21st ACM International Conference on Information and Knowledge Management, pp. 55–64. ACM (2012)
6. Liu, Y., Liao, W.k., Choudhary, A.: A fast high utility itemsets mining algorithm. In: Proceedings of the 1st International Workshop on Utility-Based Data Mining, pp. 90–99. ACM (2005)
7. Liu, Y., Liao, W.k., Choudhary, A.: A two-phase algorithm for fast discovery of high utility itemsets. In: Ho, T.B., Cheung, D., Liu, H. (eds.) PAKDD 2005. LNCS (LNAI), vol. 3518, pp. 689–695. Springer, Heidelberg (2005). doi:10.1007/11430919_79
8. Tseng, V.S., Shie, B.E., Wu, C.W., Philip, S.Y.: Efficient algorithms for mining high utility itemsets from transactional databases. IEEE Trans. Knowl. Data Eng. **25**(8), 1772–1786 (2013)
9. Yao, H., Hamilton, H.J., Butz, C.J.: A foundational approach to mining itemset utilities from databases. In: SDM, vol. 4, pp. 215–221. SIAM (2004)
10. Zida, S., Fournier-Viger, P., Lin, J.C.-W., Wu, C.-W., Tseng, V.S.: EFIM: a highly efficient algorithm for high-utility itemset mining. In: Sidorov, G., Galicia-Haro, S.N. (eds.) MICAI 2015. LNCS (LNAI), vol. 9413, pp. 530–546. Springer, Cham (2015). doi:10.1007/978-3-319-27060-9_44

Partial Tree-Edit Distance: A Solution to the Default Class Problem in Pattern-Based Tree Classification

Maciej Piernik(✉) and Tadeusz Morzy

Institute of Computing Science, Poznan University of Technology,
ul. Piotrowo 2, 60-965 Poznan, Poland
maciej.piernik@cs.put.poznan.pl

Abstract. Pattern-based tree classifiers are capable of producing high quality results, however, they are prone to the problem of the default class overuse. In this paper, we propose a measure designed to address this issue, called partial tree-edit distance (PTED), which allows for assessing the degree of containment of one tree in another. Furthermore, we propose an algorithm which calculates the measure and perform an experiment involving pattern-based classification to illustrate its usefulness. The results show that incorporating PTED into the classification scheme allowed us to significantly improve the accuracy on the tested datasets.

Keywords: Tree-subtree similarity · Tree classification · Tree-edit distance

1 Introduction

Rooted, ordered, labeled tree is a popular data structure which finds various applications in many different domains. One of the most important issues concerning this structure is similarity computation. This problem has several practical applications such as XML document similarity [1], comparison of RNA secondary structures [2], natural language processing [3], or data integration [4,5]. In this paper, we focus on the problem of tree similarity in the context of pattern-based tree classification.

The process of pattern-based tree classification is as follows. First, a training dataset of trees is mined for patterns (frequent subtrees), separately for each class. Next, based on the discovered patterns, a classifier is constructed as a set of rules: *pattern* → *class*, where *pattern* denotes a frequent subtree mined from the training dataset with a given *class*. Once a classifier is created, each new unclassified tree is tested against every rule for pattern containment (e.g., with subtree matching) and assigned to the class from which it contains the highest percentage of patterns.

Pattern-based tree classification is a straightforward method capable of producing high quality results [6]. However, there are two problematic cases for this

J. Kim et al. (Eds.): PAKDD 2017, Part II, LNAI 10235, pp. 208–219, 2017.
DOI: 10.1007/978-3-319-57529-2_17

approach: (1) if a classified tree contains the same percentage of patterns from two or more classes, (2) if a classified tree does not contain any pattern. The first case can be resolved by creating a ranking of patterns and weighing them according to their importance. In the second case, however, a common approach is to use a so called *default class*, which assigns all unmatched trees to a single class, e.g., a majority class in the training dataset. Clearly, such a situation should be avoided as there is a high chance of deteriorating the classification quality. For traditional, transactional data, other solutions than the default class exist [7], most of which are based on partial similarity of a rule with a classified object. In the tree processing domain, however, to the best of our knowledge, an accurate tree-subtree similarity measure is not available.

In this paper, we analyze the above described problem of tree-subtree similarity defined as follows. Given two trees P (a pattern tree) and D (a document tree), find how much does P need to be modified to become a subtree of D. In order to answer this question, we propose a new distance measure, called *partial tree-edit distance* (PTED), along with an algorithm to calculate it. We will show that incorporating PTED into the classification scheme significantly improves upon the existing methods of dealing with the default class problem.

The remainder of this paper is organized as follows. Section 2 gives the background of related work for the proposed measure. Section 3 introduces necessary notation and definitions. In Sect. 4 we describe and define the partial tree-edit distance measure. Section 5 presents an efficient algorithm which calculates the proposed measure. In Sect. 6 we empirically evaluate the algorithm in terms of time complexity and illustrate the usefulness of the proposed measure with an experiment involving pattern-based tree classification. Finally, in Sect. 7 we conclude the article and draw lines of further research.

2 Related Work

One of the first attempts at solving the tree-subtree similarity problem was proposed by Zhang and Shasha [8]. The authors present a generalization of tree-edit distance, which can be stated as follows. Given trees T_1 and T_2, what is the minimum distance between T_1 and T_2 when zero or more subtrees can be removed from T_2 at no cost. This problem is similar to the question stated in this paper, however, it works closely only when the root node of T_1 is mapped to the root node of T_2. Furthermore, it allows for all edit operations to appear in both trees while we allow only for the pattern tree to be modified. Given our motivation, we need our measure to identify subtrees anywhere in the hierarchy of a tree and only by modifying a pattern tree in the least invasive way.

Another problem, similar to the one stated in this paper, was explored by Augsten et al. [5]. The problem concerned finding the best k matches of a small query tree in a large document tree. This approach, however, is also unsuited for our problem because it focuses on subtrees spanning to the bottom (leaf nodes) of a document tree while we need our measure to identify subtrees of any shape and depth.

Cohen and Or [9] recently proposed a framework for solving the subtree similarity-search problem, along with an indexing structure to enhance the efficiency of the searching [10]. Their solution is generic and allows for a wide variety of similarity measures to be used. However, the aim and scope of the framework is different to the problem addressed in this paper. The authors focus on finding several similar subtrees using some subtree similarity measure while we focus on the sole problem of how to measure the subtree similarity. Therefore, the scope of our work is different to that of Cohen and Or, nevertheless, complementary.

Much effort has also been put into tree pattern matching, which is a more general problem than the one stated in this paper [1]. An interesting approach to tree pattern matching, called tree pattern relaxation, was proposed by Amer-Yahia et al. [11]. The authors propose four relaxations of pattern constraints which allow for approximate pattern matching. This approach, however, requires specifically constructed weighted patterns, what makes it unsuited for our problem since our patterns are simple trees.

Another pattern matching related problem is approximate tree matching with variable length don't cares [12]. In 1993 Zhang et al. adopted the idea of VLDC's from string matching to tree matching. However, this approach, similarly to tree pattern relaxation, requires patterns of a specific structure. This requirement makes this method unusable in our case, since we are focusing on simple subtrees.

All of these solutions tackle similar issues to the one stated in this article. However, their detailed characteristics showcase that they are unsuited for our particular problem.

3 Preliminaries

A *tree* T is a connected graph with $|T|$ nodes and $|T|-1$ edges. We call a tree T *rooted* if all edges in T are directed away from one designated node, called a *root* node. We denote a tree T rooted at a node x by T_x and a root node of a tree T by r^T. If two nodes x and y are connected with an edge and x is closer to the root node than y, then x is a *parent* of y and y is a *child* of x. Nodes without any children are called *leaf* nodes. Children of the same node x are called *siblings* and the number of all children of x is denoted by $|x|$. We also designate a special node λ, called *empty node*.

A rooted tree T is *ordered* if there exists a total order among all nodes in T. In our approach, we order the nodes according to the pre-order traversal. The fact that a node x appears in a tree before a node y is expressed by $x < y$.

A tree T is *labeled* if every node in this tree $x \in T$ has a label assigned to it, symbolized by $l(x)$. For convenience, hereinafter, a rooted, ordered, labeled tree will be referred to as *tree*.

An ordered set of trees is called a *forest*. A forest F containing trees rooted at all children nodes of a node x is denoted by F_x. The rightmost tree of a forest F is denoted by $\bar{F}$. A forest F without a tree T is symbolized by $F - T$ and the number of nodes in all trees in forest F is symbolized by $|F|$.

A tree S whose nodes and edges form subsets of nodes and edges of another tree T is called a *subtree* of T. We denote that S is a subtree of T by $S \subseteq T$.

Let us now define the edit operations which can be performed on tree nodes. In general, there are three basic edit operations: *insertion*, *deletion*, and *relabeling*. By inserting a node x into a tree T at a node y, x becomes a child of the parent of y, taking y's place in the sibling order, while y becomes a child of x. When deleting a node x from a tree T, all children of x become the children of the parent of x. Consequently, when x is a root node, the result is a forest F_x.

4 Partial Tree-Edit Distance

4.1 Conceptual Description

To illustrate how partial tree-edit distance works, first let us consider an example presented in Fig. 1. In this example, by T^i we will denote the i-th node (according to the pre-order traversal) in tree T. As the question stated in this paper implies, the task is to determine how many operations need to be performed on P for it to become a subtree of D. Looking at the example, clearly, P is not a subtree of D. However, as illustrated with the grey areas, there is a part of P which can be directly mapped into D. Namely, nodes P^2, P^4, and P^5 can be mapped into D^1, D^5, and D^{11}, respectively, as they have the same labels. As a result of this mapping, we also have to map P^3 into D^4. This time, however, we need to use the relabeling operation as the labels are different. Finally, as nodes P^1, P^6, and P^7 have no corresponding nodes in D, they have to be removed using the deletion operation. Therefore, the total number of edit operations required to transform P into a subtree of D is 4 (1 relabeling, 3 deletions).

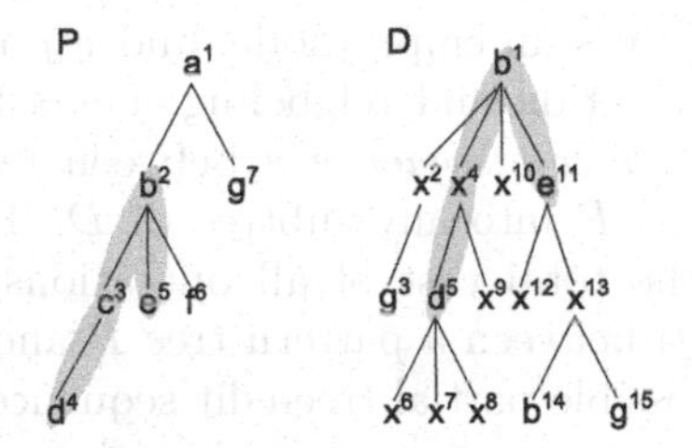

Fig. 1. Example of fitting a pattern tree P into a document tree D. The nodes in P covered by the grey area are relabeled to the corresponding nodes covered by the grey area in D, while the nodes in P uncovered by the grey area are deleted. Numbers represent the order of pre-order traversal.

So far, we have only used relabeling and deletion. Furthermore, we only deleted the root node (P^1) and the leaf nodes (P^6, P^7). Let us now discuss the possible consequences of using other edit operations, namely, deletion of inner nodes and insertion of inner and non-inner nodes. Inserting a non-inner node into P does not make sense, since it could only increase the number of operations needed to fit P into D. That is why, in partial tree-edit distance insertion of non-inner nodes is forbidden. Deleting or inserting an inner node results in a children nodes' transfer, so the internal structure of a tree is altered. In our case however, given the pattern-based classification motivation, allowing for such operations to appear would alter the inner structure of the patterns. Since patterns are frequent subtrees, by deleting non-inner nodes we are guaranteed to obtain structures which are at least as frequent as the base structure thanks to the anti-monotonicity property of the support measure (any subtree of a frequent subtree will have equal or higher support). However, allowing for an inner node to be inserted or deleted from a pattern results in a subtree of which frequency we know absolutely nothing about, therefore, it cannot be called a pattern anymore. It may even happen that such a pattern does not appear in the dataset at all. Therefore, insertion and deletion of inner nodes into a pattern may lead to wrong class assignments and deteriorate the overall classification quality.

It is worth noting that other applications may benefit from allowing these additional edit operations to appear and exploring such operations constitutes an interesting topic for a future research. However, given our main motivation, they are out of the scope of this paper.

Given the above, partial tree-edit distance is defined around two edit operations: deletion of non-inner nodes and relabeling. Both of these operations have an associated cost, which can be universally expressed with the following formula:

$$c(x,y) = \begin{cases} 0 & x = \lambda \\ w_d & y = \lambda \\ w_r & otherwise \end{cases} \tag{1}$$

where x and y are nodes, λ is an empty node, and w_d and w_r are user-defined weights associated with deletion and relabeling. Let s be a sequence of these two operations. *Partial tree-edit sequence* s between two trees P and D is a sequence which transforms P into any subtree of D. The cost $c(s)$ of partial tree-edit sequence s is the total cost of all operations in s. *Partial tree-edit distance (PTED)* $\Delta(P, D)$ between a pattern tree P and a document tree D is the minimal cost of all possible partial tree-edit sequences between P and D.

As we can see, the measure works as a combination of subtree matching and tree-edit distance, producing a distance equal 0 when a pattern appears in a tree, and a value between 0 and the size of a pattern, otherwise.

4.2 Formal Definition

Definition 1. *A partial mapping m between a pattern tree P and a document tree D is a subset of $P \times (D \cup \{\lambda\})$, such that: (1) each node from P appears in m exactly once, (2) each node from D appears in m at most once, (3) for any $(x, x'), (y, y') \in m$ where $x' \neq \lambda$ and $y' \neq \lambda$: x is a parent of $y \Leftrightarrow x'$ is a parent of y', (4) for any $(x, x'), (y, y') \in m$ where x is a sibling of y and x' is a sibling of y': $x < x' \Leftrightarrow y < y'$.*

Each element in the mapping $(x, x') \in m$ represents an edit operation and has an associated cost $c(x, x')$, as defined in Eq. 1. An element where $x' = \lambda$ represents a deletion while an element where $x' \neq \lambda$ represents a relabeling. The cost $c(m)$ of a partial mapping m is a sum of costs of all elements in m.

Definition 2. *Partial tree-edit distance $\Delta(P, D)$ between a pattern tree P and a document tree D is the minimal cost of all possible partial mappings between P and D.*

Now, we will introduce a recursive formula which calculates partial tree-edit distance. The formula works in two stages. The purpose of the first stage, performed by a main function Δ and defined in Eq. 2, is to place P at each possible position in D.

$$\Delta(P, D) = \min_{x \in P, y \in D} \Big\{ \delta(\{T_x\}, \{T_y\}) + \sum_{\{z \in P : z \notin T_x\}} c(z, \lambda) \Big\} \tag{2}$$

Next, for each placement of P in D, the second stage takes place. The goal of the second stage, performed by an auxiliary function δ defined in Eq. 3, is to check how well does P fit in D, at a given placement. The function accepts two forests G and H as parameters and recursively considers 3 cases: ignoring the rightmost tree of H, deleting the rightmost tree of G, and fitting the rightmost tree of G into the rightmost tree of H.

$$\delta(G, H) = \min \begin{cases} \delta\left(G, H - \bar{H}\right) \\ \delta\left(G - \bar{G}, H\right) + \delta\left(\{\bar{G}\}, \emptyset\right) \\ \delta\left(G - \bar{G}, H - \bar{H}\right) + \delta\left(F_{r^{\bar{G}}}, F_{r^{\bar{H}}}\right) + c(r^{\bar{G}}, r^{\bar{H}}) \end{cases} \tag{3}$$

Equation 4 defines the boundary conditions of the auxiliary function δ. The first two cases reflect the fact that the cost of fitting an empty pattern into any tree equals 0, while the third case signifies that the cost of fitting any non-empty pattern into an empty tree equals the cost of removing the whole pattern.

$$\begin{aligned} \delta(\emptyset, \emptyset) &= 0 \\ \delta(\emptyset, H) &= 0 \\ \delta(G, \emptyset) &= \delta\left(G - \bar{G}, \emptyset\right) + \delta\left(F_{r^{\bar{G}}}, \emptyset\right) + c(r^{\bar{G}}, \lambda) \end{aligned} \tag{4}$$

5 Algorithm

In this section we propose an algorithm which calculates partial tree-edit distance. Similarly to the formal definition, the algorithm consists of two main components: (1) the main loop Δ which places P at every possible position in D and (2) the auxiliary function δ which checks the quality of each placement. The algorithm for the main loop is a trivial implementation of Eq. 2, so we will skip the pseudocode for this step. The auxiliary function is implemented with a dynamic programming algorithm, given in Algorithm 1.

Algorithm 1. Partial tree edit distance algorithm: $\delta(T_v, T_w)$

```
Require: trees T_v and T_w,
Ensure: a cost of a partial mapping m between T_v and T_w with restriction (v, w) ∈ m
 1: tab ← [|v| + 1, |w| + 1]
 2: for j = 0..|w| do
 3:     tab[0, j] ← 0;
 4: end for
 5: for i = 1..|v| do
 6:     tab[i, 0] ← tab[i − 1, 0] + (|T_{v_i}|) · w_d;
 7: end for
 8: for i = 1..|v| do
 9:     for j = 1..|w| do
10:         tab[i, j] ← min{
               tab[i − 1, j] + (|T_{v_i}|) · w_d,
               tab[i, j − 1],
               tab[i − 1, j − 1] + δ(T_{v_i}, T_{w_j})
            };
11:     end for
12: end for
13: return tab[|v|, |w|] + (l(v) = l(w) ? 0 : w_r);
```

The algorithm accepts two trees T_v and T_w as parameters and outputs the minimal cost of a partial mapping between T_v and T_w, given that v is mapped into w. Variable *tab* stores the intermediate results of mapping the children nodes of v into the children nodes of w, so it is an $\mathcal{R}^{|v|+1\times|w|+1}$ matrix (Line 1). In Lines 2–4 the top row in the matrix is initialized to 0. This reflects the fact that the subtrees in the right tree can be removed without any cost (ignored). In practice, it fulfills the second boundary condition from Eq. 4. In Lines 5–7, the left column is initialized with the cumulative cost of deleting consecutive subtrees of v ($tab[i, 0]$ = cost of removing $T_{v_1}..T_{v_i}$). These values fulfill the third boundary condition from Eq. 4. Lines 8–12 contain the main loop of the auxiliary function. It scans through all children nodes of v and w and for each pair v_i, w_j stores a temporary result $tab[i, j]$ which holds the minimal cost of mapping $v_1 \dots v_i$ into $w_1 \dots w_j$. This cost is computed in Line 10 as the minimum of 3 expressions, reflecting the 3 options in Eq. 3:

- $tab[i-1, j] + (|v_i| + 1) \cdot w_d$ accounts for removing the rightmost subtree from the left tree;
- $tab[i, j-1]$ accounts for ignoring the rightmost subtree from the right tree;
- $tab[i-1, j-1] + \delta(T_{v_i}, T_{w_j})$ accounts for mapping the rightmost subtree of the left tree into the rightmost subtree of the right tree.

In the end, $tab[|v|, |w|]$ holds the minimal cost of mapping the children of v into the children of w (with descendants). Finally, in Line 13, by adding the cost of mapping v into w we obtain the total cost of the minimal partial mapping between T_v and T_w with v mapped into w. This concludes the algorithm.

Let us now analyze the complexity of the presented algorithm. It is easy to notice that the auxiliary function algorithm is an adoption of the algorithm for the Levenstein distance between two sequences [13], which has a quadratic complexity. Here however, the auxiliary function is called within the main loop which is also quadratic in time, so the overall complexity is $O(n^4)$. However, the auxiliary function runs only as deep as the height of the smaller tree, so since pattern trees are usually much smaller than document trees, the algorithm will usually be more efficient than the complexity suggests.

6 Experimental Evaluation

6.1 Datasets and Experimental Setup

During the experiments, we used both real and synthetic datasets containing XML documents represent as rooted, ordered, labeled trees. For the time complexity evaluation, we generated a dataset of 20 documents ranging between 100 and 2000 in the number of elements. To generate this dataset, we used the software developed by Zaki [14].

To test the applicability of PTED in pattern-based classification, we used the datasets created by Zaki and Aggarwal [6]. The synthetic datasets DS1-4, were generated by the aforementioned authors and are composed of a training and a testing set each, containing between 60000 and 100000 documents. The real datasets CS1-3, each consisting of around 8000 documents, contain web logs categorized into two classes (for a detailed description see [6]). Since they were not divided into training and testing sets, we used each for both purposes and cross-validated them with one another. By CSXY we will denote the CSX set used for training and CSY for testing. This gives us a total of 10 tests: 4 on synthetic and 6 on real data. The minimal frequency of a subtree required to consider it a pattern was 0.1% for DS datasets and 1% for CS datasets.

All classifiers were evaluated using a weighted accuracy measure [6], defined as follows:

$$Accuracy = \sum_{c \in C} \left(w_c \cdot \frac{|\mathcal{D}_c^{test}|}{|\mathcal{D}_c|} \right) \tag{5}$$

where C is the set of all classes, $\mathcal{D}_c^{test}$ is the set of documents correctly assigned to class c, $\mathcal{D}_c$ is the set of documents which should be assigned to class c, and w_c is a weight associated with each of the classes. Similarly as Zaki and Aggarwal [6], we analyzed three weighting models:

- *proportional*: $w_c = |\mathcal{D}_c|/|\mathcal{D}|$ — classes weighted proportionally to their distribution in the training dataset,
- *equal*: $w_c = 1/|\mathcal{C}|$ — all classes weighted equally,
- *inverse*: $w_c = \frac{1/\mathcal{D}_c}{\sum_{c' \in C} 1/|\mathcal{D}_{c'}|}$ — classes weighted inversely to their distribution in the training dataset.

Additionally, we used the Friedman test [15] to determine whether the compared approaches performed significantly differently and a post-hoc Nemenyi test [15] to check if the proposed solution significantly improved the quality of classification.

6.2 Time Complexity Evaluation

To assess the time complexity of the proposed algorithm we used the generated dataset containing 20 documents of increasing sizes. For each pair of documents, we calculated partial tree-edit distance 100 times and measured the average computing time. The results of this test are presented in Fig. 2.

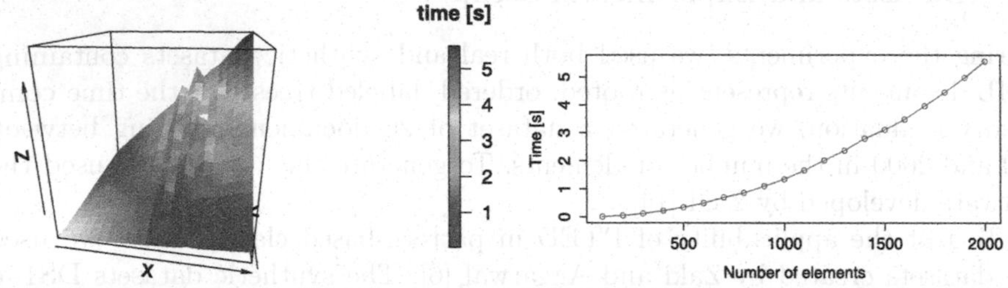

(a) PTED processing time for varying pattern and document sizes. Axes X and Y represent the sizes of pattern and document trees while the Z axis represents processing time.

(b) PTED processing time for increasing pattern and document size.

Fig. 2. Time complexity of the algorithm calculating partial tree-edit distance.

Figure 2(a) illustrates how much time it takes to calculate PTED for trees of various sizes. First, let us observe how the algorithm behaves when both pattern and document trees are expanded. As we can see, with increasing sizes of both trees, processing time presents a polynomial growth. This is reflected in the spine on the 3D chart (the diagonal line w.r.t. X and Y axes) which is extracted and visualized in Fig. 2(b) to facilitate the observation. When increasing the size of only one of the trees, the increase in processing time is much slower. Considering the purpose of our measure, this is a very important observation. Since we are assessing the degree of containment of one tree in another, the left (pattern) tree should be usually much smaller than the right (document) tree. This is certainly true in the practical example involving pattern-based classification presented in Sect. 6.3, as the largest patterns discovered in all experiments contained only 11 nodes. Additionally, it is worth noticing that the chart is symmetrical w.r.t. the X/Y diagonal. This means that the algorithm behaves the same regardless of which tree is bigger.

6.3 Practical Application

To empirically evaluate the usefulness of PTED, we performed an experiment involving pattern-based tree classification. Our goal of is to illustrate the importance of the default class problem and show how partial tree-edit distance can be used to address it. We compare four ways of dealing with this problem. In the first three approaches, we use different methods for determining the default class, as proposed by Zaki and Aggarwal in the state-of-the-art XRules classifier [6]. All three approaches determine the dafault class based on the class distribution in the documents from the training dataset which are not covered by any rule (do not contain any of the discovered frequent subtrees). Moreover, each method maximizes the accuracy measure from Eq. 5 w.r.t. one of the three weights: proportional, equal, and inverse. Analogously to the accuracy measure, given that $\mathcal{D}_c$ represents the training documents with class c and $\bar{\mathcal{D}}_c$ represents a portion of these documents uncovered by any of the rules, the method for determining the default class is defined as follows:

$$Class(D) = \arg\max_{c \in \mathcal{C}} \left(w_c \cdot \frac{|\bar{\mathcal{D}}_c|}{|\mathcal{D}_c|} \right) \tag{6}$$

where w_c is one of the three previously defined weights: proportional, equal, or inverse.

In the last approach, we use partial tree-edit distance to assign each ambiguous document D to one of the classes according to the following formula:

$$Class(D) = \arg\max_{c \in \mathcal{C}} \left(\sum_{P \in \mathcal{P}_c} \left(1 - \frac{\Delta(P, D)}{|P|} \right) \right) \tag{7}$$

where $\mathcal{C}$ is a set of classes and $\mathcal{P}_c$ is a set of patterns with class c. Intuitively, this formula measures the similarity of D with all patterns in each class and assigns it to the class with the highest cumulative similarity.

Table 1 presents the results of this experiment. The first column represents the datasets used in each test (the values in square brackets [DC%] will be explained later) while the following columns present the accuracies achieved by each of the described approaches. The results of the proportional and equal methods are presented in a single column as they produced the same outcome on every dataset. Each method was evaluated with three variants of the accuracy measure and the differences in the results were tested for statistical significance. In order to determine whether by using PTED we were able to significantly improve the quality of classification, for every dataset we ranked each algorithm's performance from 1 to 3, where 1 is the highest and 3 is the lowest score. In cases when one or more of the algorithms were tied, average ranks were assigned (e.g., if two algorithms were tied at the 2nd place, each was granted a rank of 2.5). Once created, the ranking (presented in the "Avg. rank" row) was used to perform the Friedman test [15]. The null-hypothesis for this test is that there is no difference in the performance between the tested methods. Moreover, in case of rejecting this null-hypothesis we used the Nemenyi post-hoc test [15] to verify whether the performance of the best approach is statistically different from the remaining approaches.

The results clearly illustrate that by using partial tree-edit distance we were able to improve the classification quality in almost every test, regardless of the applied accuracy measure. This outcome is partially confirmed by the Friedman

Table 1. Comparison of methods for handling unclassified examples in a pattern-based classifier. Bold indicates the best result.

Approach	Prop./Eq.	Inv.	PTED	Prop./Eq.	Inv.	PTED	Prop./Eq.	Inv.	PTED
Dataset [DC%]	Proportional accuracy [%]			Equal accuracy [%]			Inverse accuracy [%]		
DS1 [56]	**53.37**	47.74	47.74	**52.35**	50.47	**52.35**	**56.95**	47.57	**56.95**
DS2 [70]	**62.38**	34.23	34.23	48.74	**52.47**	48.74	**63.25**	42.56	**63.25**
DS3 [74]	54.03	54.03	**59.93**	54.03	54.03	**59.93**	54.03	54.03	**59.93**
DS4 [63]	**54.02**	**54.02**	**54.02**	**53.43**	**53.43**	**53.43**	**52.85**	**52.85**	**52.85**
CS12 [47]	72.32	72.32	**72.44**	64.04	64.04	**64.43**	55.76	55.76	**56.43**
CS21 [49]	**72.78**	**72.78**	**72.78**	62.62	62.62	**62.64**	52.47	52.47	**52.50**
CS13 [48]	72.26	72.26	**72.69**	63.33	63.33	**64.33**	54.40	54.40	**55.96**
CS31 [50]	72.63	72.63	**73.61**	62.68	62.68	**67.27**	52.73	52.73	**60.93**
CS23 [47]	73.61	73.61	**73.64**	63.60	63.60	**63.66**	53.59	53.59	**53.67**
CS32 [50]	73.17	73.17	**73.71**	62.79	62.79	**67.07**	52.42	52.42	**60.44**
Avg. rank	2.10	2.40	**1.50**	2.35	2.35	**1.30**	2.25	2.55	**1.20**

statistical test. The critical value of the Friedman statistic for the analyzed setting at $\alpha = 0.05$ is 3.560 and the F scores for the proportional, equal, and inverse accuracy tests are 2.392, 5.229, and 9.090, respectively. Therefore, the analyzed approaches perform significantly differently according to the two latter measures, but not according to the first one. The additional post-hoc Nemenyi test reveals that the critical distance (difference in average ranks) required to deem an approach significantly superior to others equals 1.048 at $\alpha = 0.05$, so PTED is indeed significantly better than any of the three default class strategies according to equal and inverse accuracy.

In order to emphasize the gravity of the default class problem, we additionally measured how many times the default class had to be used in the analyzed datasets. The numbers in the square brackets in the first column of Table 1 ([DC%]) present the percentage of documents from the test set which were uncovered by any pattern from the classifier. In every test, this problem concerned around half or more documents (e.g., for test DS3 which contains 100000 test documents there were 73906 documents without any matching pattern). By using partial tree-edit distance we are able to treat each of these cases individually instead of assigning them arbitrarily to the same class.

7 Conclusions

In this paper, we introduced a new measure for assessing the tree-subtree similarity, called partial tree-edit distance (PTED), which describes to what extent one tree is included in another. We also proposed an algorithm which calculates the proposed measure in polynomial time. Furthermore, we performed an experiment involving pattern-based tree classification using partial tree-edit distance to illustrate the usefulness of the measure. The results show that by using PTED we were able to significantly improve the classification quality over the classical pattern-based approach.

The measure proposed in this paper opens several possibilities of future research. It could be used to improve the quality of approximate subtree matching, XML querying, ranking, clustering, or classification. Encouraged by the results achieved in our experiments, we plan on developing a new pattern-based XML classification algorithm designed around partial tree-edit distance.

Acknowledgments. This research is partly funded by the Polish National Science Center under Grant No. 2015/19/B/ST6/02637.

References

1. Hachicha, M., Darmont, J.: A survey of XML tree patterns. IEEE Trans. Knowl. Data Eng. **25**(1), 29–46 (2013)
2. Dulucq, S., Tichit, L.: RNA secondary structure comparison: exact analysis of the Zhang-Shasha tree edit algorithm. Theor. Comput. Sci. **306**(1–3), 471–484 (2003)
3. Kouylekov, M., Magnini, B.: Combining lexical resources with tree edit distance for recognizing textual entailment. In: Quiñonero-Candela, J., Dagan, I., Magnini, B., d'Alché-Buc, F. (eds.) MLCW 2005. LNCS (LNAI), vol. 3944, pp. 217–230. Springer, Heidelberg (2006). doi:10.1007/11736790_12
4. Augsten, N., Bohlen, M., Dyreson, C., Gamper, J.: Approximate joins for data-centric XML. In: IEEE 24th International Conference on Data Engineering, ICDE 2008, pp. 814–823 (2008)
5. Augsten, N., Barbosa, D., Bohlen, M., Palpanas, T.: Efficient top-k approximate subtree matching in small memory. IEEE Trans. Knowl. Data Eng. **23**(8), 1123–1137 (2011)
6. Zaki, M.J., Aggarwal, C.C.: XRules: an effective algorithm for structural classification of XML data. Mach. Learn. **62**(1–2), 137–170 (2006)
7. Stefanowski, J.: Algorithms of rule induction for knowledge discovery. Habilitation thesis (2001)
8. Zhang, K., Shasha, D.: Simple fast algorithms for the editing distance between trees and related problems. SIAM J. Comput. **18**(6), 1245–1262 (1989)
9. Cohen, S., Or, N.: A general algorithm for subtree similarity-search. In: Proceedings of the 30th International Conference on Data Engineering, ICDE 2014, pp. 928–939 (2014)
10. Cohen, S.: Indexing for subtree similarity-search using edit distance. In: Proceedings of the 2013 ACM SIGMOD International Conference on Management of Data, SIGMOD 2013, pp. 49–60 (2013)
11. Amer-Yahia, S., Cho, S.R., Srivastava, D.: Tree pattern relaxation. In: Jensen, C.S., Šaltenis, S., Jeffery, K.G., Pokorny, J., Bertino, E., Böhn, K., Jarke, M. (eds.) EDBT 2002. LNCS, vol. 2287, pp. 496–513. Springer, Heidelberg (2002). doi:10.1007/3-540-45876-X_32
12. Zhang, K., Shasha, D., Wang, J.T.L.: Approximate tree matching in the presence of variable length don't cares. J. Algorithms **16**, 33–66 (1993)
13. Levenshtein, V.: Binary codes capable of correcting deletions, insertions and reversals. Sov. Phys. Dokl. **10**, 707 (1966)
14. Zaki, M.J.: Efficiently mining frequent trees in a forest: algorithms and applications. IEEE Trans. Knowl. Data Eng. **17**(8), 1021–1035 (2005)
15. Demsar, J.: Statistical comparisons of classifiers over multiple data sets. J. Mach. Learn. Res. **7**, 1–30 (2006)

A Domain-Agnostic Approach to Spam-URL Detection via Redirects

Heeyoung Kwon[1], Mirza Basim Baig[1], and Leman Akoglu[2(✉)]

[1] Computer Science, Stony Brook University, Stony Brook, USA
{heekwon,mbaig}@cs.stonybrook.edu
[2] H. John Heinz III College, Carnegie Mellon University, Pittsburgh, USA
lakoglu@cs.cmu.edu

Abstract. Web services like social networks, video streaming sites, etc. draw numerous viewers daily. This popularity makes them attractive targets for spammers to distribute hyperlinks to malicious content. In this work we propose a new approach for detecting spam URLs on the Web. Our key idea is to leverage the properties of URL redirections widely deployed by spammers. We combine the redirect chains into a *redirection graph* that reveals the underlying infrastructure in which the spammers operate, and design our method to build on key characteristics closely associated with the modus operandi of the spammers. Different from previous work, our approach exhibits three key characteristics; (1) *domain-independence*, which enables it to generalize across different Web services, (2) *adversarial robustness*, which incurs difficulty, risk, or cost on spammers to evade as it is tightly coupled with their operational behavior, and (3) *semi-supervised detection*, which uses only a few labeled examples to produce competitive results thanks to its effective usage of the redundancy in spammers' operations. Evaluation on large Twitter datasets shows that we achieve above 0.96 recall and 0.70 precision with false positive rate below 0.07 with only 1% of labeled data.

1 Introduction

Web services are ubiquitous: social networks (e.g. Facebook, Twitter), review sites (e.g., Yelp, Amazon), video streaming sites (e.g. YouTube, Hulu), blogs, forums, etc. draw billions of viewers daily. The widespread adoption of these services makes them attractive for spammers to distribute harmful content (scam, phishing, malware, etc.) through links they post on these sites to such content. As a result, detecting and filtering malicious content effectively becomes crucial for the quality and trustiness of the Web.

IP blacklisting—a popular solution for social network operators and URL shortening services—has been found to provide false positive rates ranging between 0.5 to 26.9%, and false negative rates between 40.2 to 98.1% [16,17], which is quite inaccurate. Blacklisting is also quite slow to keep up with the speed and scale that Web services are being consumed today. Alternative solutions focus on identifying suspicious *accounts* operated by spammers that behave

J. Kim et al. (Eds.): PAKDD 2017, Part II, LNAI 10235, pp. 220–232, 2017.
DOI: 10.1007/978-3-319-57529-2_18

in automated or fraudulent ways [2,8,18]. These, however, have limited ability to detect spam disributed through *compromised* accounts. In fact, 97% of accounts participating in spam campaigns on Facebook [4] and 86% on Twitter [5] have been found to involve compromised accounts. Moreover, they incur detection delays as they require a history of mis-activity committed by an account. Thus, it is essential to build solutions that can make fine-grained, i.e., *URL-level decisions*, which could enable services to filter individual posts rather than shutting down user accounts. It is also desirable to have solutions that are *generalizable* to different kinds of Web services, i.e. that spot spam URLs regardless of the context, platform, or domain in which they appear.

We propose a general and robust solution for detecting malicious URLs. Our key realization is the widespread usage of *redirect chains* by spammers to distribute spam on the Web [6,21]. Our main contributions are as follows.

- We develop a new graph-based approach for spotting malicious URLs that appear on the Web. Our method leverages the underlying redirection network used by the spammers. In particular, we build a graph, called the Redirect Chain Graph (RCG), based on the redirect paths of the URLs and use its structural properties to design and extract indicative features of spam.
- Our features fall under three main groups (resource sharing, heterogeneity, and flexibility) and capture the very nature of spammers' operational behaviors. These are hard to alter by the spammers without incurring monetary or management cost. As such, our features have higher *adversarially robustness.*
- Our approach relies solely on the redirection infrastructure and does not use any domain-specific information, which makes it *context/content-agnostic.* As such, it can detect spam URLs in various domains, including URLs shared on any online site, URLs returned as online search results, and so on.
- In a fully supervised setting, our approach performs extremely well. When compared to context-aware supervised detection that uses user account and post content features, our context-free features perform equally well, despite ignoring all domain-specific information.
- Finally, we propose a *semi-supervised* method, designed for more realistic scenarios where labeled data is scarce. By carefully exploiting the redundancy present in spammers' infrastructures, our proposed method requires only a few labeled examples to achieve desirably high performance to be applicable in the real-world.

In contrast, numerous existing methods, such as [9,11,12,19,20], either (*i*) utilize easy-to-evade information (low robustness), (*ii*) rely on context-dependent information (low generality), and/or (*iii*) require large collections of labeled data for training (low applicability in practice).

2 Redirection Infrastructure

Many studies have shown the pervasive use of redirects by spammers [1,6,9, 11,21]. In this section, we introduce the interconnected architecture of redirect chains, which provides the main motivation for our graph-based approach.

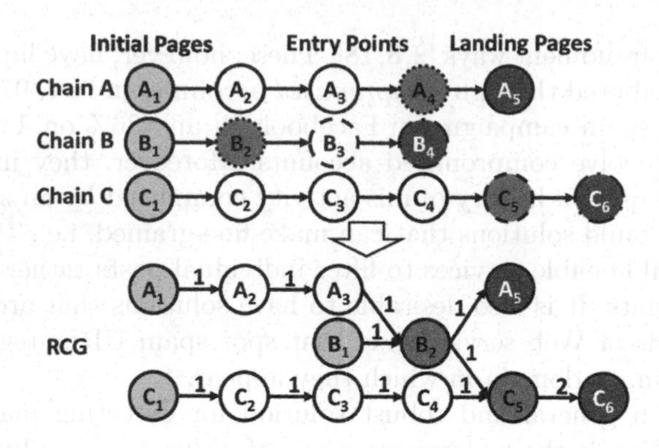

Fig. 1. 3 example redirect chains and their RCG. Chains may contain the same URL(s) (e.g., A_4 & B_2, B_3 & C_5, B_4 & C_6), yielding the interconnected network RCG.

Definition 1 (Redirect Chain). *A redirect chain* C *consists of an ordered set of URLs,* $C = \{U_1, U_2, \ldots, U_l\}$*, starting with an initial URL* U_1*, followed by URLs automatically and conditionally redirected in a sequence, and landing on a final URL* U_l. $l = |C|$ *denotes the length of the chain.*

Initial URL (often shortened by e.g. `bit.ly`) is the one displayed to users on a site, whereas landing page is where the user ends up after clicking the initial URL (cf. Fig. 1).

As data preprocessing, we group the domain names of the URLs that appear on the same IP. For example, if http://123.com/hi.html, http://xyz.com/hi.html, and http://xyz.com/hello.html are all co-located at the same-IP address, then we replace the first two URLs with `http://[123.com,xyz.com]/hi.html` and the third with `http://[123.com,xyz.com]/hello.html`. This grouping helps us unify malicious URLs that use several domain names so as to bypass blacklisting. Moreover, a (grouped) URL may be located at multiple different IPs, a list of which we also store. As such, each URL is associated with a list of domain names as well as a list of IPs.

Our key motivation for a graph-based solution is due to the following observation: (malicious) redirect chains deployed by spam campaigns contain several URLs in common, i.e. *shared* across chains, creating a network structure as shown in Fig. 1. In other words, the (malicious) redirection infrastructure of spammers is highly *inter-connected.*

Provided that the redirect chains are likely to share several URLs, it is beneficial to study them collectively, rather than individually. As such, we merge the redirect chains of URLs[1] to create the Redirect Chain Graph (RCG).

[1] Note that what is posted on a Web service are the initial URLs. We run a crawler to go through the redirects to extract the chains.

Definition 2 (Redirect Chain Graph (RCG)). *Given a set of redirect chains* $\mathcal{C} = \{C_1, \ldots, C_m\}$*, we decompose each chain* $C_i = \{U_{i,1}, \ldots, U_{i,l_i}\}$ *into a set of directed edges between consecutive URLs,* $E_i = \{e(U_{i,j}, U_{i,j+1}) \mid j \in \{1, \ldots, l_i - 1\}\}$*. The* $RCG = (V, E)$ *then consists of all the edges across chains, where* $E = \biguplus E_i$ *and* $V = \bigcup C_i$ *for* $1 \leq i \leq m$*.*

Note that the RCG nodes are the unique set of URLs across all chains, whereas the edges are allowed to reappear (hence the multiset addition $\biguplus$). As such, RCG is a directed and weighted graph, where edge weights depict the number of chains that a redirection step appears in. Notice for example the edge weight 2 in Fig. 1.

Finally we introduce the *entry-point URLs.* Those are the nodes with large in-weight in the RCG, considered as "directors"—they are central pages that aggregate user traffic and direct them to one of several malicious pages (sort of routers). As such, entry-points are critical in functionality but hard to identify without aggregate graph analysis—the entry point does not serve the actual spam, as such it is more difficult to spot and shut down. We characterize each chain by its entry-point URL, as defined below.

Definition 3 (Entry-Point URL). *Given a redirect chain* $C = \{U_1, U_2, \ldots, U_l\}$*, let* w_j *denote URL* j*'s in-weight in the RCG. The entry-point of* C *is the URL* U_k *with the largest in-weight, where* $w_k = \max\{w_1, \ldots, w_l\}$*.*

3 Feature Design Using Redirects

There exists a vast body of work that use information derived from the user account that a URL originated from, the URL itself or its page content [8,9,13, 20]. We choose not to use such information for the reasons we discuss below.

Rationale to Exclude Account and URL and Page-Content Information: First reason is to ensure a general solution. Such features are derived from meta data on the specific site the URL appears in (e.g., number of followers). Relying on contextual information would make it hard to cross Web service boundaries due to potentially disparate contextual information across sites. Second, context-aware solutions require personally identifiable information from user accounts, which may not be desirable due to privacy concerns.

As for content, spammers can use feedback from classifiers to fine-tune their URLs and page content in an attempt to evade detection by the classifiers, e.g., by spoofing sufficient benign features with high weights as studied in adversarial classification [3,10]. For example, they can avoid using spam terms, adjust URL length and character distribution, and modify links and plugins to imitate non-spam pages and URLs, while remaining sufficiently effective in eliciting response from the target users.

Our features (Table 1) fall into three main categories, characterizing spammer operations that reflect (1) shared resources, (2) heterogeneity, and (3) flexibility.

(1) Shared resources-driven features. Spammers would ideally deliver each copy of malicious content through a dedicated independent channel, such

Table 1. Features introduced (3 categories). RC: redirect chain, CC: connected component of RCG a RC resides in. TREE: BFS-tree of CC, rooted at entry-URL. TLD: top-level domain. Node degrees & edge weights are based on RCG.

Feature name	Description
Shared resources-driven (17 features)	
EntryURLiw	In-weight (freq.) of entry-point URL
EntryURLid	In-degree of entry-point URL
AvgURLiw	Mean in-weight of URLs in RC
AvgURLid	Mean in-degree of URLs in RC
ChainWeight	Total weight of edges in RC
CCsize	Number of nodes in CC
CCdensity	Edge density of CC
MaxRCLen	Max. length of RCs in CC
MinRCLen	Min. length of RCs in CC
TreeHeight	Height of TREE (root: entry-URL)
MaxLevelWidth	Max. node count at TREE levels
ImbalanceH	Horizontal imbalance of TREE
ImbalanceV	Vertical imbalance of TREE
MaxLdURLDom	Max. domain count of CC landing URLs
AvgLdURLDom	Mean domain count of CC landing URLs
MaxURLDom	Max domain name count per URL in RC
AvgURLDom	Mean domain name count per URL in RC
Heterogeneity-driven (12 features)	
GeoDist	Total geo-distance (km's) of hops in RC
MaxGeoDist	Max. geo-distance (km's) across hops in RC
XContinentHops	Number of cross-continent hops in RC
CntContinent	Number of unique continents in RC
XCountryHops	Number of cross-country hops in RC
CntCountry	Number of unique countries in RC
XIPHops	Number of cross-IP hops in RC
CntIP	Number of unique IPs in RC
XDomainHops	Number of cross-domain hops in RC
CntDomain	Number of unique domains in RC
XTLD	Number of cross-TLD hops in RC
CntTLD	Number of unique TLDs in RC
Flexibility-driven (10 features)	
ChainLen	Length (#URLs) of RC
EntryURLDist	Distance from initial to entry URL in RC
CntInitURL	Number of initial URLs in RCG
CntInitURLDom	Total domain name count in initial URLs
CntLdURL	Number of final landing URLs in RCG
MaxIPperURL	Max. IP count each URL in RC appears in
AvgIPperURL	Mean IP count each URL in RC appears in
MaxIPperLdURL	Max. IP count landing URLs in CC appear in
AvgIPperLdURL	Mean IP count landing URLs in CC appear in
RatioCheapTLD	Fraction of non-.com/.mil/etc. URLs in RC

that if a server fails or is shut down, it has minimal effect. Avoiding to reuse components in their infrastructure (domain names, servers, etc.), however, would increase their costs and limit profits. As such, spammers often reuse their underlying hosting infrastructure for significant periods [1,13].

The first type of sharing occurs due to the same URLs being reused across different redirect chains. As discussed in Sect. 2, nodes with large in-weight in the RCG (e.g., the entry points) are those URLs that are reused to route traffic for many redirect paths. As such, for each given URL[2], we identify the connected component of the RCG it resides in and extract features based on structural graph properties, e.g., in-weight of its entry point URL, average in-degree and in-weight of nodes in its chain, density of its RCG component, etc. We also treat the RCG component as a tree, rooted or "hung" at the entry point URL. This tree is obtained by a breadth-first search traversal of the RCG component starting at the entry-URL. Intuitively, a small number of unique entry points in a large RCG is suspicious; implying a few URLs shared among many chains. Tree-based features such as level width and horizontal imbalance capture this, as few entry points cause large fan-out.

A second type of sharing occurs due to the same servers hosting many different domain names. To evade and stay ahead of domain blacklisting, spammers run through many domain names. To reduce operating costs, they host them on the same server (IP address), all serving the same malicious content. We leverage this domain co-location property based on domain counts both in landing URLs of the RCG component as well as in all the URLs in the redirect chain of a URL.

(2) Heterogeneity-driven features: The operational infrastructure of spammers consists of a variety of heterogeneous agents, including various compromised servers and bot machines from various geo-locations, besides their own hosting servers. This kind of heterogeneity arises naturally and is crucial for their operations. First, it would require high maintenance to ensure all compromised machines are of a single type or all reside in close geo-locations. Moreover it would be risky if everything resided on one machine or all machines were at the same geo-location, as the infrastructure would have a few failure points. Based on this insight, we design features that quantify infrastructure heterogeneity.

These features mainly leverage geo-spatial and domain name heterogeneity. For example, given the sequence of URLs in a redirect chain, we quantify the total distance in km's traversed. Similarly, we count the number of transfers between different continents, countries, and IPs. We also count the cross-domain hops—contrary to a hodgepodge of IPs and domain names in spam redirect chains, benign sites have the opposite incentive to keep visitors within their own domain. Compromised IPs or sites would also come from various kinds of top-level domains (TLDs), such as `.edu`, `.org`, `.com`, etc., therefore we also keep a count of transfers between different TLDs.

[2] Note that the given URLs are the observed ones posted on the Web, also referred to as the initial URLs in this work.

(3) Flexibility-driven features: Finally, we derive features from operational properties that allow spammers flexibility, through which their maintenance overhead or expenses are reduced.

To have the advantage of luring as many users as possible, spammers use multiple different initial URLs (even though they redirect to the same malicious content) to make their posts look different. We capture this by keeping count of the initial URLs in the RCG component of a given URL, as well as the total number of domain names that they host.

Using multiple landing URLs (serving the same content), on the other hand, provides redundancy; if a landing page goes down, others can still distribute malicious content. Another way that spammers achieve redundancy is by having copies of the same URLs across multiple different IPs, which we capture through features associated with the number of IP addresses that URLs appear in.

Using long redirect chains helps with dynamicity and selectivity, which is hard to evade for spammers, if they want to be flexible in how they replace machines and how they choose who to spam. Specifically, a series of redirects provides them with the flexibility to modify intermediate steps (plug-in & plug-out), as well as the flexibility to hide malicious, "bullet-proof" landing URLs behind layers of redirection. The location of entry point URLs also plays a key role—since these pages have to conditionally redirect visitors to different landing URLs, suspicious entry point URLs are often located early in the chains.

Another case of flexibility is related to top-level domain (TLD) names. Spammers tend not to invest on trustworthy but costly TLDs such as `.com` and `.net`, or try to compromise often bullet-proof TLDs such as `.mil` and `.gov`—especially given that most URLs are only for redirection purposes and not for delivering content. As a result, they resort to acquiring or attacking cheap TLDs.

4 Spam Detection

After crawling the redirect chains, constructing the RCG, and extracting for each URL in the dataset the 39 features as listed in Table 1, our next step is spam detection. In this work, we study both supervised and semi-supervised detection, with a note that the latter presents a more realistic scenario.

Supervised Detection. When a large body of labeled URLs is available, one can build classifiers. In this work, we analyze the performance of our feature categories and characterize the most discriminative ones. In addition, we extract context-based features from user accounts and keywords appearing alongside the URLs to build context-aware classifiers, which we compare to our models.

Note that acquiring labels for each URL is quite time consuming, as annotators need to set up virtual sandboxes and analyze the URL, landing page content, behaviors a click triggers in their system, etc. Moreover, since spam is rare as compared to normal URL traffic, a reasonably large number of URLs needs to be labeled to ensure representative amount of spam labels in the training data.

Semi-supervised Detection. Due to the challenges with supervised detection, we design an approach that utilizes only a small set of labeled examples. Our method achieves comparable performance to fully supervised methods. As such, it is both applicable under the most realistic scenarios where labeled data is scarce as well as desirably effective in detecting spam.

In particular, we leverage the user–URL graph to formulate the problem as a network-based classification task, which we solve using label propagation based inference. More formally, we consider the bipartite graph $G = (N, E)$ in which n user nodes $U = \{u_1, \ldots, u_n\}$ are connected to m URL nodes $V = \{v_1, \ldots, v_m\}$, $N = U \cup V$, through 'post' relations in E. To define a classification task on this network, we utilize pairwise Markov Random Fields (MRFs) [7]. An MRF model consists of an undirected graph where each node i is associated with a random variable Y_i that can be in one of a finite number of states (i.e., class labels). In our case, the domain of labels for URLs is $\mathcal{L}_V = \{spam, benign\}$ and it is $\mathcal{L}_U = \{spammer, non\text{-}spammer\}$ for users. In pairwise MRFs, the label of a node is assumed to be dependent only on its neighbors and independent of other nodes in the graph. As such, the joint probability of labels is written as a product of individual and pairwise factors, respectively parameterized over the nodes and the edges;

$$P(\mathbf{y}) = \frac{1}{Z} \prod_{Y_i \in N} \phi_i(y_i) \prod_{(Y_i, Y_j) \in E} \psi_{ij}(y_i, y_j) \quad (1)$$

where $\mathbf{y}$ denotes an assignment of labels to all nodes, and y_i refers to node i's assigned label. Individual factors $\phi : \mathcal{L} \to \mathbb{R}^+$ are called *prior* potentials, and represent class probabilities for each node initialized based on prior knowledge. Pairwise factors $\psi : \mathcal{L}_U \times \mathcal{L}_V \to \mathbb{R}^+$ are called *compatibility* potentials, and capture the likelihood of a node labeled y_i to be connected to a node with y_j.

As we consider a semi-supervised setting, only a small set of the URL labels is available. For the known spam URLs we set the priors as $\phi_i(spam) = 1 - \epsilon$ and $\phi_i(benign) = \epsilon$, and vice versa for the known benign URLs. To set the priors for the unknown URLs, we learn a classifier using the available labeled data and employ it to assign class probabilities, i.e. priors, to the unknown URLs in the graph. For the users we set unbiased priors, i.e., $\phi_i(spammer) = 0.5$ and $\phi_i(non\text{-}spammer) = 0.5$, as we do not want to rely on any context-specific information (e.g., profile data such as ratio of followers to followees) to estimate such priors.

On the other hand, we instantiate the compatibility potentials so as to enforce homophily among connected nodes. Homophily captures the insight that URLs posted by spammers are spam and those shared by regular users are benign, with high probability, where ψ_{ij}'s are set as follows.

ψ_{ij}	URLs	
Users	*spam*	*benign*
spammer	$1 - \epsilon$	ϵ
non-spammer	ϵ	$1 - \epsilon$

We note that ϵ's in ϕ_i's for URLs with known labels account for the uncertainty in the labels associated with annotator agreement. ϵ's in ψ_{ij}'s capture the slight probability that non-spammers unknowingly can post spam URLs (e.g., retweet) and that spammers can post benign URLs for camouflage.

Provided the model parameters, the classification task is to infer the best assignment $\mathbf{y}$ to the nodes such that the joint probability in Eq. (1) is maximized. This is a combinatorially hard problem that is intractable for large graphs. Therefore, we use an approximate inference algorithm called Loopy Belief Propagation (LBP) [22]. LBP is an iterative algorithm where connected nodes exchange messages. A message m_{ij} captures the *belief* of i about j, specifically the probability distribution over the labels of j. Intuitively, it is what i 'believes' j's label probabilities are, given the current label distribution and the priors of i. The key idea is that after certain number of iterations of message passes between the nodes, the 'conversations' likely come to a consensus, which determines the marginal class probabilities of all the unknown variables. Although convergence is not theoretically guaranteed, LBP converges quickly in practice [15].

5 Experiments

5.1 Data Description

In this work we detect spam links posted on Twitter. Using the Twitter Streaming API, we collected 15,828,532 Twitter posts by 1,080,466 unique users during a period from May 2–September 10, 2014. This interval captures major world events such as the World Cup and the ongoing search for the Malaysia Airlines Flight 370. Those serve as attractive means to spread spam, where e.g., users are lured to click a malicious link that supposedly points to a video that shows the four goals that Germany scored against Brazil in six minutes during the semi-finals, but instead triggers a drive-by-download exploit. We identified 3,871,911 (initial) URLs from 3,764,395 (≈24%) of the posts that contained links, i.e., a small fraction of posts contained multiple URLs.

We built a crawler and extracted the redirection chains for all the URLs. The chain lengths vary from 1 to 46, with more than 99% being less than 6. We combined the redirect chains into RCG, a unified (weighted, directed) graph, which contains 4,874,256 nodes and 3,839,633 edges.

To construct a labeled URL set, we used a crawler to first identify a set of suspended Twitter users. In particular, if a user profile page input to our crawler automatically redirected to http://www.twitter.com/suspended, we label the account as a malicious one. This provided us with 88,147 suspended users. After removing these, we sampled another 1,000 users for human labeling. Five annotators were provided with the links to the profile pages of these users. Each annotator labeled each user by manually analyzing their tweets, number of followers/followees, temporal behavior, etc. At the end, a user is labeled by majority voting, which provided us with 216 spam users and 784 non-spammers.

We labeled URLs using the labeled users, where URLs inherit the majority label of the users who posted them (labeling URLs through users posting them is extremely pure: 99.06% have majority fraction 100%, i.e., all spammer or none). As a result, we obtained 459,822 labeled URLs, out of which 191,726 are spam.

For supervised detection, we built a balanced dataset (50% spam) by considering the 191,726 chains from the unique spam URLs, and randomly sampled the same number of URLs from those labeled as benign for a total of 383,452 chains. For semi-supervised detection, where we leverage the user–URL bipartite graph, we constructed a graph with 784 users from each class and all the URLs shared by those users for a total of 315,120 ground-truth URLs, with 16% consisting of spam. We experimented with 1% or 5% of this labeled set as input to our semi-supervised approach. Note that our method not only leverages a much smaller labeled set, but also works in an imbalanced setting as in practice.

5.2 Detection Results

We use two metrics to compare the detection methods; (1) average precision, which is the area under the precision–recall (PR) plot, denoted as AP, and (2) area under the ROC curve (false-positive vs. true-positive rate), denoted as AUC.

Supervised Detection. Our experiments with linear SVM, Logistic Regression, and Decision Tree (DT) show that non-linear DT significantly outperforms both. This suggests that the decision boundary of our task is complex. We use the DT model in the remaining supervised detection experiments.

Feature Contribution Analysis. To investigate the importance of individual features, we quantify their discriminative power. In particular, we use the sum of information gains weighted by the number of samples split by each feature at the internal tree nodes [14] based on the DT model trained on the entire dataset.

Figure 2 shows the ranking of features by the aforementioned importance score. We note that (*i*) the GeoDist feature is considerably the most informative one, and that the scores drop quickly. This suggests that in practice only a small subset of all the features could be enough to build accurate models. We also notice that (*ii*) the majority (5/10, 11/20) of the informative ones are from the shared-resources-driven features (green bars), which are mainly derived based on the RCG structure. Moreover, (*iii*) the top features come from a mix of all three feature categories, suggesting that they carry non-redundant information.

In Fig. 3 we demonstrate the performances achieved by individual feature categories. In agreement with observation (*ii*) above, (S)hared-resources-driven features perform slightly better than other categories. In addition, potentially due to (*iii*) that different feature groups carry non-redundant signals, using all the features holistically (from all S+H+F categories) yields the best result.

Context-Free vs. Context-Aware Detection. Next we ask: how do context-driven features perform? To investigate this question, we build models (a) based solely on context-based features and (b) integrating them with our original set.

The context-based features are mainly derived from the account that posted the URL and the post content, including the account's age, number of hashtags,

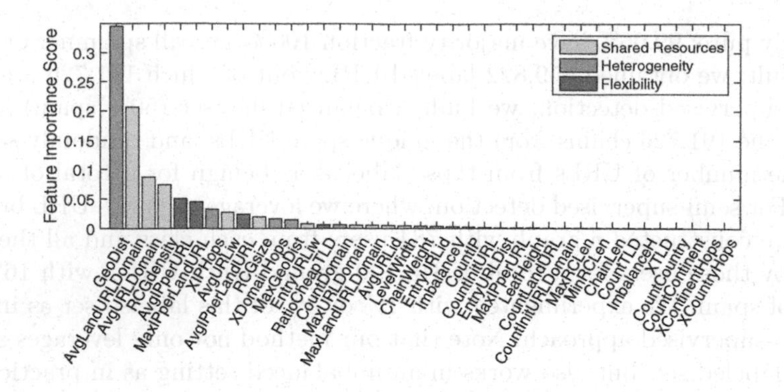

Fig. 2. Feature ranking by discriminative role (best in color).

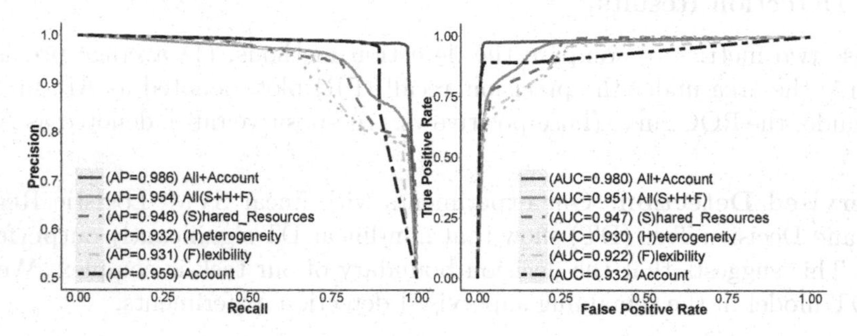

Fig. 3. Supervised detection. Context-free features achieve competitive performance.

number of mentioned other users, the follower–followee ratio, total number of posts, fraction of posts containing a URL, and the keywords used in the posts (including hashtags and @mentions).

In Fig. 3, we observe that the context-based model (labeled Account) is slightly better than our model with All features in terms of AP and slightly worse with respect to AUC. We conclude that our features are equally discriminative, even when no domain-specific or potentially missing, hidden, or private information is used.

Semi-supervised Detection. Under this setting, we randomly sample 1% or 5% of the URLs and reveal their labels (results are averaged over 10 runs). We then under-sample the majority class to obtain a balanced dataset, on which we train a classifier to assign class priors to the unlabeled URLs.

Semi-supervised results are given in Fig. 4. First, we investigate the performance of semi-supervised classifiers alone (without the LBP on the user–URL graph), namely linear SVM, polynomial SVM, and the decision tree (DT) classifiers. As before, both non-linear classifiers outperform linear SVM, where DT is superior to polynomial SVM.

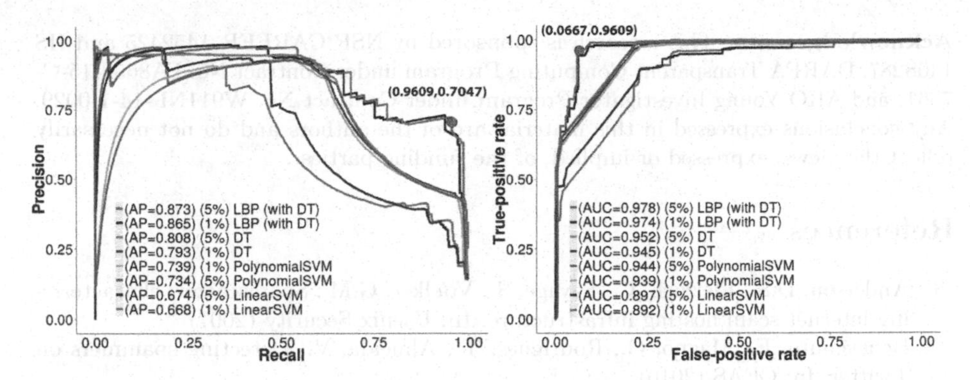

Fig. 4. Semi-supervised detection. LBP (with DT) achieves competitive performance. Red circles depicts values at classification threshold 0.5. (color figure online)

Next, we employ our semi-supervised method. In particular, we leverage the user–URL graph in which we initialize the class priors of (1% or 5%) labeled URLs as $(1-\epsilon, \epsilon)$ for spam URLs, and vice versa for the benign. The class priors of unlabeled URLs are set to the class probabilities from DT, and of the users as $(0.5, 0.5)$, i.e. unbiased. As we see in Fig. 4, incorporating relational information through LBP significantly improves the detection performance. Perhaps more importantly, the performance is desirably high; at a small false positive rate of 0.0667, recall and precision are above 0.95 and 0.70, respectively.

6 Conclusion

We considered the problem of detecting spam URLs that appear on the Web in various contexts. Our main goal has been to build an effective solution that is at the same time (*i*) context-free, (*ii*) adversarially robust, and (*iii*) semi-supervised; such that it is generalizable across Web service boundaries, costly to evade by spammers, and applicable in the face of label scarcity, respectively.

To achieve these goals, we utilize the URL redirect chains and the underlying network that they form to design three categories of domain-agnostic features. Our features are closely tied to the operational characteristics of the spammers, particularly related to their (1) reusing and sharing of resources, (2) heterogeneous hosting infrastructure, and (3) flexibility. Intuitively, evading detection by changing their behavior would incur considerable monetary or management cost upon the spammers. Evaluations on a large Twitter collection with millions of URL posts show that our context-free features yield quite similar performance against context-aware features. Moreover, our semi-supervised detection algorithm produces competitive results, at above 0.96 recall and 0.70 precision with false positive rate below 0.07, even with very limited supervision.

We publicly share our Twitter URL data collection (including ground truth labels, redirect chains, and the RCG) as well as our redirect chain crawler at http://bit.ly/2jvdiFI.

Acknowledgments. This research is sponsored by NSF CAREER 1452425 and IIS 1408287, DARPA Transparent Computing Program under Contract No. FA8650-15-C-7561, and ARO Young Investigator Program under Contract No. W911NF-14-1-0029. Any conclusions expressed in this material are of the authors and do not necessarily reflect the views, expressed or implied, of the funding parties.

References

1. Anderson, D.S., Fleizach, C., Savage, S., Voelker, G.M.: Spamscatter: characterizing internet scam hosting infrastructure. In: Usenix Security (2007)
2. Benevenuto, F., Magno, G., Rodrigues, T., Almeida, V.: Detecting spammers on Twitter. In: CEAS (2010)
3. Dalvi, N., Domingos, P., Mausam, Sanghai, S., Verma, D.: Adversarial classification. In: KDD, pp. 99–108 (2004)
4. Gao, H., Hu, J., Wilson, C., Li, Z., Chen, Y., Zhao, B.Y.: Detecting and characterizing social spam campaigns. In: IMC (2010)
5. Grier, C., Thomas, K., Paxson, V., Zhang, C.M.: @spam: the underground on 140 characters or less. In: CCS, pp. 27–37 (2010)
6. Gyöngyi, Z., Garcia-Molina, H.: Web spam taxonomy. In: AIRWeb (2005)
7. Kindermann, R., Snell, J.L.: MRFs and their applications (1980)
8. Lee, K., Caverlee, J., Webb, S.: Uncovering social spammers: social honeypots + machine learning. In: SIGIR (2010)
9. Lee, S., Kim, J.: WarningBird: detecting suspicious URLs in Twitter stream. In: NDSS (2012)
10. Lowd, D., Meek, C.: Adversarial learning. In: KDD, pp. 641–647 (2005)
11. Lu, L., Perdisci, R., Lee, W.: SURF: detecting and measuring search poisoning. In: CCS, pp. 467–476 (2011)
12. Ma, J., Saul, L.K., Savage, S., Voelker, G.M.: Beyond blacklists: learning to detect malicious Web sites from suspicious URLs. In: KDD, pp. 1245–1254 (2009)
13. Ma, J., Saul, L.K., Savage, S., Voelker, G.M.: Identifying suspicious URLs: an application of large-scale online learning. In: ICML (2009)
14. Neville, P.G.: Decision Trees for Predictive Modeling. SAS Institute Inc., Cary (1999)
15. Pandit, S., Chau, D.H., Wang, S., Faloutsos, C.: Netprobe: a fast and scalable system for fraud detection in online auction networks. In: WWW (2007)
16. Ramachandran, A., Feamster, N., Vempala, S.: Filtering spam with behavioral blacklisting. In: CCS (2007)
17. Sinha, S., Bailey, M., Jahanian, F.: Shades of grey: on the effectiveness of reputation-based blacklists. In: Malicious & Unwanted Softw, IEEE (2008)
18. Stringhini, G., Kruegel, C., Vigna, G.: Detecting spammers on social networks. In: ACSAC, pp. 1–9 (2010)
19. Stringhini, G., Kruegel, C., Vigna, G.: Shady paths: leveraging surfing crowds to detect malicious web pages. In: CCS, pp. 133–144 (2013)
20. Thomas, K., Grier, C., Ma, J., Paxson, V., Song, D.: Design and evaluation of a real-time URL spam filtering service. In: IEEE Symposium on Security and Privacy (2011)
21. Wu, B., Davison, B.D.: Cloaking, redirection: a preliminary study. In: AIRWeb, pp. 7–16 (2005)
22. Yedidia, J.S., Freeman, W.T., Weiss, Y.: Understanding belief propagation and its generalizations. In: Exploring AI in the New Millennium (2003)

Automatic and Effective Mining of Coevolving Online Activities

Thinh Minh Do(✉), Yasuko Matsubara, and Yasushi Sakurai

Kumamoto University, Kumamoto, Japan
do@dm.cs.kumamoto-u.ac.jp

Abstract. Given a large collection of time-evolving online user activities, such as Google Search queries for multiple keywords of various categories (celebrities, events, diseases, etc.), which consist of d keywords/activities, for l countries/locations of duration n, how can we find patterns and rules? How do we go about capturing non-linear evolutions of local activities and forecasting future patterns? We also aim to achieve good monitoring of the data sequences statistically, and detection of the patterns immediately. In this paper, we present Δ-SPOT, a unifying analytical non-linear model for analysing large scale web search data, which is sense-making, automatic, scalable and free of parameters. Δ-SPOT can also forecast long-range future dynamics of the keywords/queries. Besides, we also provide an efficient and effective fitting algorithm, which leads to novel discoveries and sense-making features, and contribute to the need of monitoring multiple co-evolving data sequences.

1 Introduction

Online news, blogs, SNS and many other web search services have been speedily developing and playing a very important part in information searching. Our goal is to detect patterns, rules and outliers in a huge set of web search data, consisting of tuples of the form: *(query, location, time)*. Intuitively, given a large collection of online activities, which consists of d keywords in l locations of duration n with complex dynamics, we want to find global and local patterns, detect external shocks (important events in reality), and forecast future activities. Especially, we want to capture all these features automatically and effectively. Besides, we also introduce an incremental online algorithm, Δ-STREAM, which enhances the robustness of Δ-SPOT for data streams monitoring.

Contributions. In this paper, we propose Δ-SPOT, a unifying analytical non-linear model for large-scale online user activities. Briefly, our proposed model, Δ-SPOT, has the following desirable properties:

1. **Sense-making:** Our method can detect external shocks which are related to real-time events, such as the annual sporting events, or the release of new products.
2. **Automatic:** Thanks to our modeling framework, Δ-SPOT is fully automatic, requiring no manual tuning, where the goal is to minimize the modeling cost.
3. **Scalable:** Our method scales linearly to the input size.
4. **Parameter-free:** Δ-SPOT requires no parameters or specialized tuning.

J. Kim et al. (Eds.): PAKDD 2017, Part II, LNAI 10235, pp. 233–246, 2017.
DOI: 10.1007/978-3-319-57529-2_19

Table 1. Capabilities of approaches. Only our approach meets all specifications.

	SI/++	AR/++	FUNNEL	Δ-SPOT/Δ-STREAM
Non-linear	√		√	√
Outliers detection			√	√
Online activities				√
Cyclic events/shocks				√
Local analysis			√	√
Parameter-free			√	√
Forecasting		√	√	√
Online processing				√

2 Related Work

We provide a survey of the related literature, which falls into following categories:

Pattern Discovery in Time Series. Traditional approaches typically use linear methods, such as auto-regression (AR), linear dynamical systems (LDS), TBATS [8] and their variants [3,5,6,17]. TriMine [12] is a scalable method for forecasting complex time-stamped events, while, [9] developed AutoPlait, which is a fully-automatic mining algorithm for co-evolving sequences.

Social Activity Analysis. The work described in [13] studied the rise and fall patterns in the information diffusion process through online social media. FUNNEL [14] is a non-linear model for spatially co-evolving epidemic tensors, while, EcoWeb [10] is the first attempt to bridge the theoretical modeling of a biological ecosystem and user activities on the Web. [11] developed CompCube, which operates on large collections of co-evolving activities and summarizes them succinctly with respect to multiple aspects (i.e., keyword, location, time). Gruhl et al. [2] explored online "chatter" (e.g., blogging) activity, and measured the actual sales ranks on Amazon.com, while the work of Ginsberg et al. [1] examined a large number of search engine queries tracking influenza epidemics.

Contrast to the Competitors. Table 1 illustrates the relative advantages of our method. Only our Δ-SPOT matches all requirements, while other methods, such as the SI model (and SIR, SIRS, *SKIPS* [16], etc.), the traditional AR, ARIMA and related forecasting methods including AWSOM [15], PLiF [7], TriMine [12], and FUNNEL [14], fail to acquire this.

3 Proposed Model

3.1 Intuition Behind Our Method

We have a collection of sequences with d unique keywords, l locations/countries with duration n. We can treat this set of $d \times l$ sequences as a 3rd-order tensor,

Table 2. Symbols and definitions

Symbol	Definition
d	Number of keywords/queries
l	Number of locations/countries
n	Duration of sequences
$\mathcal{X}$	3rd-order tensor ($\mathcal{X} \in \mathbb{N}^{d \times l \times n}$)
$\boldsymbol{x}_{ij}$	Local-level sequence of keyword i in location j i.e., $\boldsymbol{x}_{ij} = \{x_{ij}(t)\}_{t=1}^{n}$
$\bar{\boldsymbol{x}}_i$	Global-level sequence of keyword i i.e., $\bar{\boldsymbol{x}}_i = \sum_{j=1}^{l} \boldsymbol{x}_{ij}$
$S_{ij}(t)$	Count of (S)usceptibles i in location j at time t
$I_{ij}(t)$	Count of (I)nfectives i in location j at time t
$V_{ij}(t)$	Count of (V)igilants i in location j at time t
$\mathbf{B_G}$	Base global matrix ($d \times 4$) i.e., $d \times \{\mathbb{N}, \beta, \delta, \gamma\}$
$\mathbf{B_L}$	Base local matrix ($d \times l$) i.e., $\mathbf{B_L} = \{b^{(L)}{}_{ij}\}_{i,j=1}^{d,l}$
$\mathbf{R_G}$	Growth effect global matrix ($d \times 2$) i.e., $d \times \{t_\eta, \eta_0\}$
$\mathbf{R_L}$	Growth effect local matrix ($d \times l$) i.e., $\mathbf{R_L} = \{r^{(L)}{}_{ij}\}_{i,j=1}^{d,l}$
$\mathcal{S}$	External shock tensor i.e., $\mathcal{S} = \{\mathbf{s}_1, \mathbf{s}_2, \ldots, \mathbf{s}_k\}$
$\mathcal{C}$	Cyclic external shock candidate set i.e., $\mathcal{C} = \{\mathbf{c}_1, \mathbf{c}_2, \ldots, \mathbf{c}_k\}$
$\mathcal{F}$	Complete set of parameters i.e., $\mathcal{F} = \{\mathbf{B_G}, \mathbf{B_L}, \mathbf{R_G}, \mathbf{R_L}, \mathcal{S}\}$

i.e., $\mathcal{X} \in \mathbb{N}^{d \times l \times n}$, where the element $x_{ij}(t)$ of $\mathcal{X}$ shows the total number of entries of the i-th keyword in the j-th country at time-tick t. We refer to each sequence of the i-th keyword in the j-th location: $\boldsymbol{x}_{ij} = \{x_{ij}(t)\}_{t=1}^{n}$, as a "local/country"-level web search sequence. Similarly, we can turn these local sequences into "global/world"-level web search sequences: $\bar{\boldsymbol{x}}_i = \{\bar{x}_i(t)\}_{t=1}^{n}$, where $\bar{x}_i(t)$ shows the total count of the i-th keyword at time-tick t, i.e., $\bar{x}_i(t) = \sum_{j=1}^{l} x_{ij}(t)$.

3.2 Δ-SPOT - with a Single Sequence

The model we propose has nodes (=users) of three classes (Table 2):

(a) **Susceptible.** Nodes who can get influenced by the external event or their neighboring nodes who have searched for it. In other words, citizens of this class are ready to search for the keywords.
(b) **Infective.** Nodes who have already got "infected" by the event's influence and searched for the keywords, also capable of influencing other available nodes (share the story about the event), namely, transmitting the interest in the topic to the citizens in the Susceptible class.
(c) **Vigilant.** Nodes in this class do not get condition to search for the information (no network connection, no free time to care about the topic), or do not pay attention to the news, so they are immune to the influence of the trend.

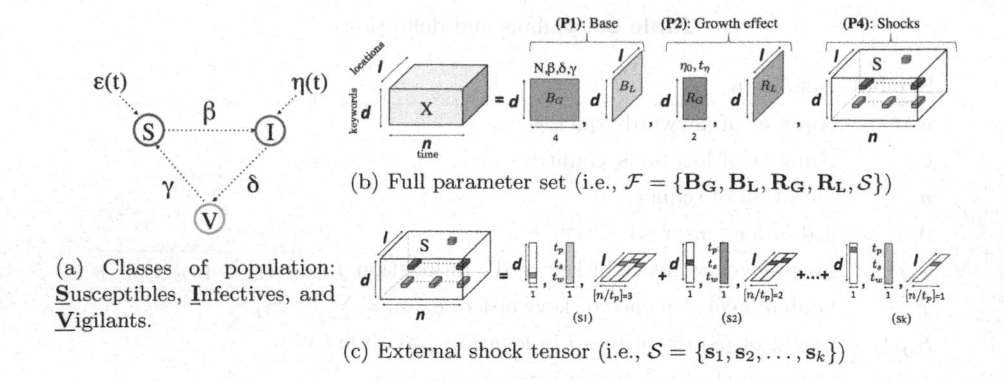

(a) Classes of population: Susceptibles, Infectives, and Vigilants.

(b) Full parameter set (i.e., $\mathcal{F} = \{\mathbf{B_G}, \mathbf{B_L}, \mathbf{R_G}, \mathbf{R_L}, \mathcal{S}\}$)

(c) External shock tensor (i.e., $\mathcal{S} = \{\mathbf{s}_1, \mathbf{s}_2, \ldots, \mathbf{s}_k\}$)

Fig. 1. (a) Δ-SPOT diagram, (b) Δ-SPOT structure: important properties extracted from tensor $\mathcal{X}$, also (c) external shock tensor $\mathcal{S}$ consists of a set of k components.

Figure 1(a) shows a diagram of the relationship between online users of three above classes in the social network, under the influence of external shock effect $\epsilon(t)$ and growth effect $\eta(t)$. Here, β represents the rate of effective contacts between citizens in Infective and Susceptible classes, δ is the rate at which infected citizens lost interest in the topic and stop searching for it, and γ is the immunization loss probability for a change in status: being ready to search for the topic. Consequently, to handle all properties of online user activities, we introduce our model, namely, Δ-SPOT:

Model 1 (Δ-SPOT-single). *Δ-SPOT can be described as these equations:*

$$\begin{aligned} S(t+1) &= S(t) - \beta S(t)\epsilon(t)I(t)(1+\eta(t)) + \gamma V(t) \\ I(t+1) &= I(t) + \beta S(t)\epsilon(t)I(t)(1+\eta(t)) - \delta I(t) \\ V(t+1) &= V(t) + \delta I(t) - \gamma V(t) \end{aligned} \quad (1)$$

The growth effect $\eta(t)$, starting at time t_η, is defined as: $\eta(t) = \begin{cases} 0 & (t < t_\eta) \\ \eta_0 & (t \geq t_\eta) \end{cases}$

We also introduce the temporal susceptible rate $\epsilon(t)$:

$$\epsilon(t) = 1 + \sum_{i=1}^{k} f(t; \mathbf{s}_i), where f(t; \mathbf{s}) = \begin{cases} \epsilon_0 & (t_s + t_p\lceil t/t_p \rceil < t < t_s + t_p\lceil t/t_p \rceil + t_w) \\ 0 & (else) \end{cases}$$

and k is the number of shocks. If $k = 0$, then $\epsilon(t) = 1$.

Here, t_p is the periodicity of the event (if $t_p = \infty$, the event is non-cyclic), t_s is the starting point of the event, t_w is the duration of the event, ϵ_0 is the strength of the external shock.

3.3 Δ-SPOT - with Multi-evolving Sequences

Next, we want to extract the main trends and external patterns of co-evolving sequences $\mathcal{X} \in \mathbb{N}^{d \times l \times n}$, and make a good representation of $\mathcal{X}$. Figure 1(b)–(c) shows our modeling framework. Given a tensor $\mathcal{X}$, it extracts important patterns with respect to the following aspects, base properties of global and local trends $\mathbf{B_G}$, $\mathbf{B_L}$, population growth effect $\mathbf{R_G}$, $\mathbf{R_L}$, and external shock events $\mathcal{S}$.

(P1) Base Trends and Global Influence. We assume that the parameters are the same for all l countries. For example, the potential infection rate of each keyword (e.g., "Harry Potter") is the same for US and JP.

(P2) Area Specificity. What is the difference of users reaction for keyword "Ebola" between the U.S. (US) and Nepal (NP)? Our answer is: their behavior is similar, except for the "local sensitivity" of the sequence. Specifically, we share the parameters of the global-level matrices for all l countries. with one exception, N_{ij}, which describes the total popularity size of keyword i in the j-th country. Specifically, we set the invariant, $N_{ij} = S_{ij}(t) + I_{ij}(t) + V_{ij}(t)$ This parameter corresponds to the fraction of individuals who are likely to be infected by the trend. For example, US has more users than NP, because they have more capacities for network connection.

(P3) Population Growth Effect. The growth effect appears due to the launch of new products and services that raise the interest of users, which should have the same starting time all over the world. Many sequences consists of the population growth effect, and Δ-SPOT should not detect and filter them as normal external shocks.

(P4) External Shocks. To describe each external shock, we create a new parameter set, namely external shock tensor $\mathcal{S}$, which consists of a set of k external shocks, as described in Fig. 1(c). i.e., $\mathcal{S} = \{\mathbf{s}_1, \mathbf{s}_2, \ldots, \mathbf{s}_k\}$ A single external shock $\mathbf{s}$ can be described as three components: $\mathbf{s} = \{\mathbf{s^{(D)}}, \mathbf{s^{(N)}}, \mathbf{s^{(L)}}\}$. Here, the $(d \times 1)$ size component $\mathbf{s^{(D)}}$ represents the external view for d keywords/queries; the (3×1) size component $\mathbf{s^{(N)}}$ describes the periodicity (t_p), the starting time (t_s), and the duration (t_w) of the external event; finally, the $(\lceil n/t_p \rceil \times l)$ size component $\mathbf{s^{(L)}}$ expresses the strength ϵ_0 of the external shock in l countries, where $\lceil n/t_p \rceil$ is the number of shocks belonging to one specified event.

4 Algorithm

4.1 Model Quality and Data Compression

We propose an intuitive coding scheme, which is based on the minimum description length (MDL) principle. Here, it follows the assumption that the more we can compress data, the more we can detect its hidden patterns.

Model Description Cost. The description complexity of model parameter set consists of two terms. Firstly, the number of keywords d, locations l, and time-ticks n require $\log^*(d) + \log^*(l) + \log^*(n)$ bits.[1] Secondly, the model parameter set of the global base $(\mathbf{B_G})$, global growth effect $(\mathbf{R_G})$, and local base, growth

[1] Here, $\log^*$ is the universal code length for integers.

Algorithm 1. Δ-SPOT($\mathcal{X}$)

1: **Input:** Tensor $\mathcal{X}$ ($d \times l \times n$)
2: **Output:** Full parameters, i.e., $\mathcal{F} = \{\mathbf{B_G}, \mathbf{B_L}, \mathbf{R_G}, \mathbf{R_L}, \mathcal{S}\}$
3: $\{\mathcal{F}_\mathcal{G}\}$ =GlobalFit ($\mathcal{X}$); /* Parameter fitting for global-level sequences */
4: $\{\mathcal{F}_\mathcal{L}\}$ =LocalFit ($\mathcal{X}, \mathcal{F}_\mathcal{G}$); /* Parameter fitting for local-level sequences */
5: **return** $\mathcal{F} = \{\mathcal{F}_\mathcal{G}, \mathcal{F}_\mathcal{L}\}$;

effect ($\mathbf{B_L}, \mathbf{R_L}$), matrices require $d \times 4$, $d \times 2$, $d \times l$ parameters, respectively, i.e., $Cost_M(\mathbf{B_G}) + Cost_M(\mathbf{R_G}) + Cost_M(\mathbf{B_L}) + Cost_M(\mathbf{R_L}) = c_F \cdot d(4+2+l)$, where c_F is the floating point cost[2]. Similarly, the model description cost of the external shock tensor $\mathcal{S} = \{\mathbf{s}_1, \mathbf{s}_2, \ldots, \mathbf{s}_k\}$ consists of: the number of external shocks k ($\log^*(k)$ bits required), the shock-keyword vector $\mathbf{s}^{(\mathbf{D})}$ ($\log(d)$ bits required), the shock-time vector $\mathbf{s}^{(\mathbf{N})} = \{t_p, t_s, t_w\}$ ($3 \cdot \log(n)$ required), and the shock-location matrix $\mathbf{s}^{(\mathbf{L})}$ ($|\mathbf{s}^{(\mathbf{L})}| \cdot (\log(d) + \log(l) + \log(n) + c_F)$ required). Note that, for each shock $\mathbf{s}$, it requires $Cost_M(\mathbf{s}) = Cost_M(\mathbf{s}^{(\mathbf{D})}) + Cost_M(\mathbf{s}^{(\mathbf{N})}) + Cost_M(\mathbf{s}^{(\mathbf{L})})$. Consequently, the model cost of the external shock tensor $\mathcal{S} = \{\mathbf{s}_1, \cdots, \mathbf{s}_k\}$ is $Cost_M(\mathcal{S}) = \log^*(k) + \sum_{i=1}^{k} Cost_M(\mathbf{s}_i)$.

Data Coding Cost. Given the full parameter set $\mathcal{F}$, we can encode the data $\mathcal{X}$ $Cost_C(\mathcal{X}|\mathcal{F}) = \sum_{i,j,t=1}^{d,l,n} \log_2 p^{-1}_{Gauss(\mu,\sigma^2)}(x_{ij}(t) - I_{ij}(t))$, where, $x_{ij}(t)$ is the elements in $\mathcal{X}$, and $I_{ij}(t)$ is the estimated count of infections (i.e., Model 1).[3]

Data Compression Equation. Consequently, the total code length for $\mathcal{X}$ with respect to a given parameter set $\mathcal{F}$ can be described in the following equation, which we want to minimize as our next goal:

$$\begin{aligned} Cost_T(\mathcal{X}; \mathcal{F}) = \log^*(d) + \log^*(l) + \log^*(n) + Cost_M(\mathbf{B_G}) + Cost_M(\mathbf{B_L}) \\ + Cost_M(\mathbf{R_G}) + Cost_M(\mathbf{R_L}) + Cost_M(\mathcal{S}) + Cost_C(\mathcal{X}|\mathcal{F}) \end{aligned} \quad (2)$$

4.2 Multi-layer Optimization

The idea is that we split parameter set $\mathcal{F}$ into two subsets, i.e., $\mathcal{F}_\mathcal{G}$ and $\mathcal{F}_\mathcal{L}$, each of which corresponds to a global/local-level parameter set, and try to fit the parameter sets separately. Algorithm 1 shows an overview of Δ-SPOT to find the full parameter set given a tensor $\mathcal{X}$.

Global-Level Parameter Fitting. Given a tensor $\mathcal{X}$, our sub-goal is to find the optimal global parameter set: $\mathcal{F}_\mathcal{G}$, to minimize the cost function (i.e., Eq. 2). As shown in Algorithm 2, we provide a detailed algorithm of the global-level fitting. Given a tensor $\mathcal{X}$, it creates a set of d global sequences: $\{\bar{x}_i\}_{i=1}^{d}$. The goal is to fit the global-level parameter set, as well as find the appropriate number of external-shocks. We apply the *Levenberg-Marquardt (LM)* [4] algorithm to

[2] We used 4×8 bits in our setting.

[3] Here, μ and σ^2 are the mean and variance of the distance between the original and estimated values.

Algorithm 2. GLOBALFIT($\mathcal{X}$)

1: **Input:** Tensor $\mathcal{X}$
2: **Output:** Set of global-level parameters $\mathcal{F_G}$ i.e., $\mathcal{F_G} = \{\mathbf{B_G}, \mathbf{R_G}, \mathcal{S}\}$
3: **for** $i = 1 : d$ **do**
4: Create $\bar{\boldsymbol{x}}_i$ from $\mathcal{X}$; /* Global sequence $\bar{\boldsymbol{x}}_i$ of i-th keyword */
5: $\mathbf{s}_i = \emptyset$; /* Initialize external shocks for keyword i */
6: **while** improving the cost **do**
7: $\mathbf{b^{(G)}}_i = \arg\min_{\mathbf{b^{(G)}}'_i} Cost_C(\bar{\boldsymbol{x}}_i|\mathbf{b^{(G)}}'_i, \mathbf{r^{(G)}}_i, \mathbf{s}_i)$; /* Base */
8: $\mathbf{r^{(G)}}_i = \arg\min_{\mathbf{r^{(G)}}'_i} Cost_C(\bar{\boldsymbol{x}}_i|\mathbf{b^{(G)}}_i, \mathbf{r^{(G)}}'_i, \mathbf{s}_i)$; /* Growth */
9: $\mathbf{s}_i = \emptyset$; /* Initialize values */
10: /* Find external shocks for keyword i */
11: **while** improving the cost **do**
12: $\mathbf{s} = \arg\min_{\mathbf{s}'} Cost_C(\bar{\boldsymbol{x}}_i|\mathbf{b^{(G)}}_i, \mathbf{r^{(G)}}_i, \{\mathbf{s}_i \cup \mathbf{s}'\})$; $\mathbf{s}_i = \mathbf{s}_i \cup \mathbf{s}$;
13: **end while**
14: **end while**
15: $\mathbf{B_G} = \mathbf{B_G} \cup \mathbf{b^{(G)}}_i$; $\mathbf{R_G} = \mathbf{R_G} \cup \mathbf{r^{(G)}}_i$; $\mathcal{S} = \mathcal{S} \cup \mathbf{s}_i$; /* Update parameter set of i-th keyword */
16: **end for**
17: **return** $\mathcal{F_G} = \{\mathbf{B_G}, \mathbf{R_G}, \mathcal{S}\}$;

Algorithm 3. LOCALFIT($\mathcal{X}, \mathbf{B_G}, \mathbf{R_G}, \mathcal{S}$)

1: **Input:** (a) Tensor $\mathcal{X}$, (b) global-level parameter set $\mathcal{F_G}$
2: **Output:** Set of local-level parameters, i.e., $\mathcal{F_L}$
3: **while** improving the cost **do**
4: /* For each local sequence $\boldsymbol{x}_{ij}$ of i-th keyword in j-th country */
5: **for** $i = 1 : n$ **do**
6: **for** $j = 1 : l$ **do**
7: $b^{(L)}{}_{ij} = \arg\min_{b^{(L)}{}'_{ij}} Cost_C(\boldsymbol{x}_{ij}|\mathbf{B_G}, \mathbf{R_G}, b^{(L)}{}'_{ij}, \mathcal{S})$;
8: $r^{(L)}{}_{ij} = \arg\min_{r^{(L)}{}'_{ij}} Cost_C(\boldsymbol{x}_{ij}|\mathbf{B_G}, \mathbf{R_G}, r^{(L)}{}'_{ij}, \mathcal{S})$;
9: **end for**
10: **end for**
11: **for each** external shock $\mathbf{s}$ in $\mathcal{S}$ **do**
12: Update $\mathbf{s}$ to minimize the cost /* Local participation rate */
13: **end for**
14: **end while**
15: **return** $\mathcal{F_L} = \{\mathbf{B_L}, \mathbf{R_L}, \mathcal{S}\}$;

minimize the cost function. Note that the extra tensor $\mathcal{S}$ consists of k entries $\{\mathbf{s}_1, \mathbf{s}_2, \ldots, \mathbf{s}_k\}$, Algorithm 2 can find only the global-level entry, which consists of (*keyword*, *time*). We will introduce the local-level parameter fitting algorithm in Algorithm 3, to describe how the local-level entries can be computed. Also, the cost function (2) includes the cost of local-level parameters such as $\mathbf{B_L}$, $\mathbf{R_L}$ but these terms are independent of the global model fitting. Hence, we can simply consider them to be constant.

Local-Level Parameter Fitting. Given a set of $d \times l$ local-level sequences, $\{\boldsymbol{x}_{ij}\}_{i,j=1}^{d,l} \in \mathcal{X}$, and a set of global-level parameters, $\mathcal{F}_{\mathcal{G}}$, our next goal is to fit the individual parameters of each keyword in each country, that is, $\mathcal{F}_{\mathcal{L}} = \{\mathbf{B_L}, \mathcal{S}\}$. We propose an iterative optimization algorithm (see Algorithm 3). Our algorithm searches for the optimal solution with respect to (a) the base local matrix $\mathbf{B_L}$ and (b) the local-level external shocks $\mathcal{S}$, so that the total coding cost is minimized.

5 Online Processing

In this section, we describe our online algorithm, namely, Δ-STREAM, which is an effective method of monitoring data sequences. Algorithm 4 shows an overview of Δ-STREAM. Given a new tensor $\mathcal{X}'$, our first task is to find the appropriate parameter set in both global level ($\mathcal{F}'_{\mathcal{G}}$), and local level ($\mathcal{F}'_{\mathcal{L}}$). The initial values of all parameters are the same to the parameters of $\mathcal{F} = \{\mathbf{B_G}, \mathbf{B_L}, \mathbf{R_G}, \mathbf{R_L}, \mathcal{S}\}$ Then we use them to update the original sequence's global parameter set $\mathcal{F}_{\mathcal{G}}$ and local parameter set $\mathcal{F}_{\mathcal{L}}$. We introduce a new parameter, the cyclic external shock candidate set $\mathcal{C}$ to further reduce the processing time. The candidate set $\mathcal{C}$ includes multiple cyclic shocks with different period, time-shift and duration. A single external shock candidate $\mathbf{c}$ consists of three components which are similar to a normal external shock: $\mathbf{c} = \{\mathbf{c^{(D)}}, \mathbf{c^{(N)}}, \mathbf{c^{(L)}}\}$.

Here, the idea is to keep track of the most suitable judgment for the new shock. We compare the shock-time vector $\mathbf{s^{(N)}} = \{t_p, t_s, t_w\}$ of all external shocks

---

Algorithm 4. Δ-STREAM($\mathcal{X}'$,$\mathcal{C}$)

1: **Input:** A subsequence $\mathcal{X}'$ ($d \times l \times n'$) of duration $n_{\mathcal{X}'}$, and a set of candidates $\mathcal{C}$
2: **Output:** An optimal parameter set for $\mathcal{X}'$ i.e., $\mathcal{F}' = \{\mathbf{B'_G}, \mathbf{B'_L}, \mathbf{R_G}', \mathbf{R_L}', \mathcal{S}'\}$;
   and an update of parameter set, i.e., $\mathcal{F} = \{\mathbf{B_G}, \mathbf{B_L}, \mathbf{R_G}, \mathbf{R_L}, \mathcal{S}\}$
3: $\mathcal{S}' = \emptyset$; /* Initialize external shocks for new sequence */
4: $\{\mathbf{B'_G}, \mathbf{R_G}'\}$ =GLOBALFIT ($\mathcal{X}'$); /* Parameter fitting for global-level sequences */
5: **for** each $\mathbf{c}_i$ in $\mathcal{C}$ **do**
6:    Generate the next shock consisting of $\mathbf{c^{(D)}}$, $\mathbf{c^{(N)}}$ and calculate $\mathbf{c^{(L)}}$
7:    $\mathcal{S}' = \mathcal{S}' \cup \{\mathbf{c^{(D)}}, \mathbf{c^{(N)}}, \mathbf{c^{(L)}}\}$;
8: **end for**
9: $\{\mathcal{S}\} = \{\mathcal{S}\} \cup \{\mathcal{S}'\}$; /* Merge the external shocks tensor */
10: **for** every shock in $\mathcal{S}$ **do**
11:    **if** there exists a new cyclic event $\mathbf{s}'$ **then**
12:       /* Add the new candidate to $\mathcal{C}$ */
13:       $\mathcal{C} = \mathcal{C} \cup \{\mathbf{s}'\}$;
14:    **end if**
15: **end for**
16: $\mathcal{F}'_{\mathcal{L}}$ =LOCALFIT ($\mathcal{X}'$, $\mathcal{F}'_{\mathcal{G}}$); /* Parameter fitting for local-level sequences */
17: $\{\mathcal{F}_{\mathcal{G}}\} = \{\mathcal{F}_{\mathcal{G}}\} \cup \{\mathcal{F}'_{\mathcal{G}}\}$; /* Update the global parameter set */
18: $\{\mathcal{F}_{\mathcal{L}}\} = \{\mathcal{F}_{\mathcal{L}}\} \cup \{\mathcal{F}'_{\mathcal{L}}\}$; /* Update the local parameter set */
19: **return** $\mathcal{F} = \{\mathcal{F}_{\mathcal{G}}, \mathcal{F}_{\mathcal{L}}\}$; $\mathcal{F}' = \{\mathcal{F}'_{\mathcal{G}}, \mathcal{F}'_{\mathcal{L}}\}$;

---

to detect the cyclic ones. A cyclic event is defined to consist of multiple external shocks with the same periodicity t_p and duration t_w. If a new captured shock forms with the old ones a potential cyclic event (with specified period and duration), a new candidate is added to $\mathcal{C}$. For example, if we have detected the shocks that form a periodical events, with the same duration (i.e., for keyword "Grammy", we have found the annual shocks ($t_p = 52$) in February, of one-week duration ($t_w = 1$)), we form a cyclic shock candidate and insert it into the cyclic external shock candidate set $\mathcal{C}$. When dealing with a new sequence, the next shock of the cyclic event is automatically generated, and fit with its strength (height). If there are no suitable shocks in the new sequence, the strength will be zero. If the new shock satisfies the condition of cyclic pattern, it will be added to the corresponding cyclic event. Or else, if it, with another old shock, form a promising cyclic event, a new candidate will be added to $\mathcal{C}$.

6 Experiments

In this section we show the effectiveness of Δ-SPOT with real dataset, by demonstrating three following properties: *Sense-making*, *Accuracy*, and *Scalability.*

Dataset Description. We performed experiments on three real datasets:

1. *GoogleTrends*: This dataset consists of the volume of searches for queries (i.e., keywords) in various topics on Google[4] from January 2004 to January 2015 (in weekly basis), collected in 232 countries.
2. *Twitter*: We used more than 7 million Twitter[5] posts covering an 8-month period from June 2011 to January 2012.
3. *MemeTracker*: This dataset covers three months of blog activity from August 1 to October 31 2008[6], It contains more than 70,000 short quoted phrases ("memes"), each of which consists of the number of mentions over time.

6.1 Sense-Making

In this experiment, we demonstrate how effective Δ-SPOT can be in terms of data fitting, external events detection and other important properties. Our objective is to fit the popularity size of the Infective class. We demonstrate the global fitting results of several sequences collected from three above datasets: Fig. 2(a)–(b) show the results of model fitting on two trending keywords of *GoogleTrends* dataset; Fig. 2(c) shows the results of a popular hashtag "#apple"; and Fig. 2(d) shows the results of a "meme".[7] In all above figures, we show the original sequences (i.e., black dots) and estimated sequences: $I(t)$ (i.e., red line) in linear-linear scales. Also, we made several important observations:

[4] http://www.google.com/insights/search/.
[5] http://twitter.com/.
[6] http://memetracker.org/.
[7] Meme#3: "yes we can yes we can".

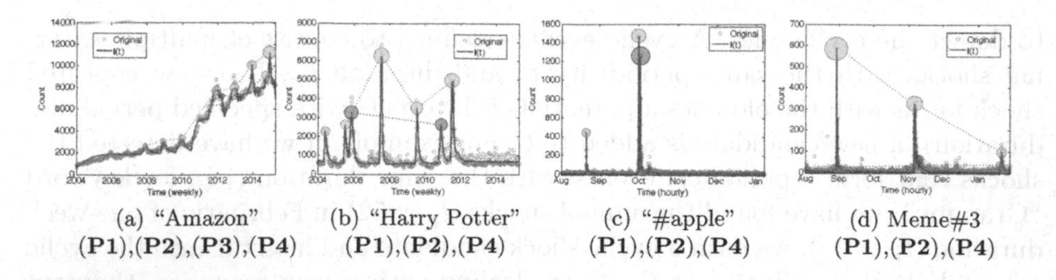

(a) "Amazon" (P1),(P2),(P3),(P4) (b) "Harry Potter" (P1),(P2),(P4) (c) "#apple" (P1),(P2),(P4) (d) Meme#3 (P1),(P2),(P4)

Fig. 2. Global fitting results for 4 sequences in (a)–(b) *GoogleTrends*, (c) *Twitter*, and (d) *MemeTracker* dataset, of different topics. (Color figure online)

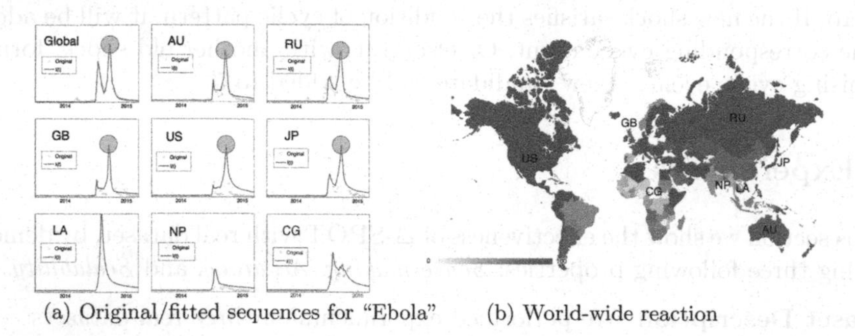

(a) Original/fitted sequences for "Ebola" (b) World-wide reaction

Fig. 3. Local fitting power of Δ-SPOT for the keyword "Ebola" which refers to the Ebola Virus bursting in 2014 (shown in green circles). (a) It can capture the local similar behaviors in Australia (AU), Russia (RU), the U.K. (GB), the U.S. (US) and Japan (JP). It can also capture local outliers in Laos (LA), Nepal (NP) and DR Congo (CG), in comparison to the global trend. And we have a clearer observation in (b) the world map of user reaction to the disease burst in 2014. (Color figure online)

(P1) Base Trends and Global Influence. As shown in Fig. 2, our proposed model successfully captures long-range non-linear dynamics of user activities, as well as fit the data sequences in high accuracy.

(P2) Area Specificity. Figure 3(a) shows the local fitting results for keyword "Ebola" of *GoogleTrends* dataset; in which, we detected some countries (AU,RU,GB,US,JP) that behave similar to the global trend (i.e., the world reaction to the burst of Ebola Virus in 2014, shown in green circles). Besides, we can also detect several outliers from the countries which have less capacities of network connection (LA,NP,CG).

(P3) Population Growth Effect. In Fig. 2(a), we detected the growth effect in keyword "Amazon", which starts from time-tick 343. In fact, this relates to the development of online service since 2011, which leads to the quick rise in the number of users who search for the Amazon services, including online shopping, media downloads and other cloud infrastructure services.

(P4) External Shock Events. Δ-SPOT can capture important external events relating to the keywords, including the cyclic events. Furthermore,

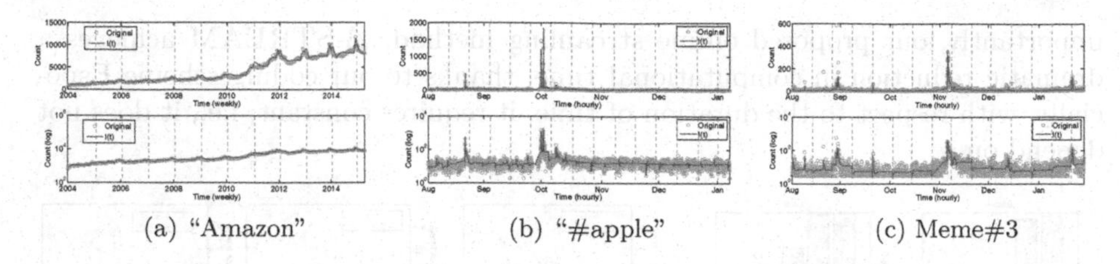

(a) "Amazon" (b) "#apple" (c) Meme#3

Fig. 4. Online processing results for 3 queries: for each new coming subsequence, Δ-STREAM captures all important features, including the stream dynamics and patterns, as well as updates the external events.

Δ-SPOT can capture the cyclic external events of different periodicity and duration. For better visualization, we mark the cyclic external shocks by the same color circles. Δ-SPOT guarantees not to miss any important external events that happen during the sequences.

Moreover, we execute the online process experiments to evaluate the fitting capacity of Δ-STREAM. Figure 4(a) shows the result of keyword "Amazon", where we set the window size of one-year-length (i.e., wd = 52 time-ticks). Δ-STREAM can capture the correct increasing pattern of the web search data stream, as well as detect the annual external events relating to the keyword. Whereas, in Fig. 4(b)–(c), we set the window size of one-week-length (i.e., wd = 168 time-ticks). Δ-STREAM capture the basic trend of the web search data stream, also detect some external shocks during the scan.

6.2 Accuracy

We used the fitting result for keyword "Amazon" of *GoogleTrends* dataset, and compared Δ-SPOT with the standard *SIRS* model, *SKIPS* [16], and multi time-evolving sequences mining model, *FUNNEL* [14]. Figure 5(a) shows the root-mean-square error (RMSE) between the original and estimated counts of the global sequences $\{\bar{\boldsymbol{x}}_i(t)\}_{i,t}^{d,n}$. Similarly, Fig. 5(b) shows the results of the local counts $\{\boldsymbol{x}_{ij}(t)\}_{i,j,t}^{d,l,n}$, (i.e., each keyword in each country, at each time-tick). A lower value indicates a better fitting accuracy. As shown in the figures, the *SIRS* model and *SKIPS* failed to capture the complicated patterns of data sequences, *FUNNEL* cannot detect cyclic external events, while our method achieved those properties with high fitting accuracy. We also evaluated the accuracy of Δ-STREAM in terms of global/local fitting. Δ-STREAM still provides better fitting accuracy compared to other methods.

6.3 Scalability

We made the evaluation of the scalability of Δ-SPOT, and verified the complexity of our method, which we discussed in Sect. 4. As shown in Fig. 6, Δ-SPOT is linear with respect to (a) keywords d, (b) countries l, and (c) duration n. More importantly, our proposed online streaming method, Δ-STREAM achieves a dramatic reduction in computational time, thanks to our coding scheme Especially, with respect to the duration of time, it requires constant; i.e., it does not depend on n.

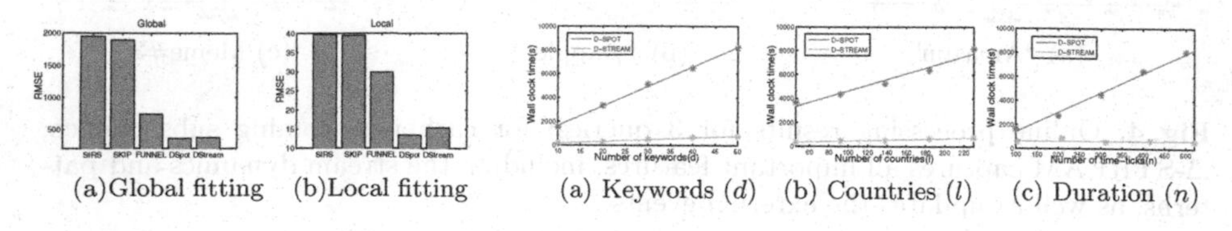

(a)Global fitting (b)Local fitting

Fig. 5. Fitting accuracy

(a) Keywords (d) (b) Countries (l) (c) Duration (n)

Fig. 6. Scalability of Δ-SPOT

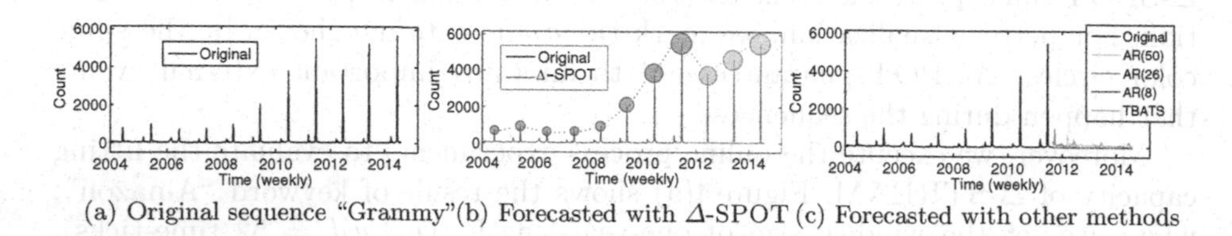

(a) Original sequence "Grammy" (b) Forecasted with Δ-SPOT (c) Forecasted with other methods

Fig. 7. Forecasting result: we train the model parameters using first 400 time-ticks of the sequences and do forecasting the remaining part. (Color figure online)

7 Δ-SPOT at Work

In this section, we discuss the most important and challenging task of Δ-SPOT, namely, forecasting the future dynamics of co-evolving activities. The goal here is to predict the future dynamics of the online activities, including the basic pattern and the external events of the sequences. As described in Sect. 4, Δ-SPOT can detect the exact periodicity of the cyclic events. Given the external shock tensor $\mathcal{S}$, Δ-SPOT automatically generates the next shocks of the cyclic events in terms of the time and duration, respectively. Here, we use the regression function to estimate the strength of those shocks, given the strength of the previous shocks. Figure 7 shows results of our forecasting in relation to keyword "Grammy". We train the model parameters by using the 400 time-ticks of the sequence (solid black lines in the figures), and then do forecasting on the following years (solid red lines). The result shows that Δ-SPOT can predict the time-tick, the duration and the relative strength of incoming external events, which refer to the annual Grammy Awards, held every February. We also make a comparison between Δ-SPOT and other competing methods, including the Auto Regressive (AR) model, and TBATS model. For AR, We applied several regression coefficients: $r = 8, 26, 50$. In Fig. 7 (a,b,c), we show the original sequences, the forecast results of Δ-SPOT and AR with TBATS, respectively. It is clearly shown that our method can predict the next three external shocks, which refer to the next three Grammy Awards. On the other side, AR and TBATS fails to capture the accurate future dynamics, also cannot detect the future external events.

8 Conclusion

In this paper, we presented Δ-SPOT, an intuitive model for mining large scale time-evolving online activities, and its extension Δ-STREAM for data stream monitoring. Through the extensive experiments on real datasets, Δ-SPOT demonstrates all the following desirable properties:

1. It is **effective**: it can detect important hidden events that match the reality. The online algorithm Δ-STREAM can dramatically speed up the processing time as well as achieves high accuracy.
2. It is **automatic**: it requires no training set and no domain expertise.
3. It is **scalable**: Δ-SPOT is linear to the data size (i.e., $O(dln)$).
4. It is **practical**: Δ-SPOT can undertake long-range forecasting and outperforms existing methods.

Acknowledgement. This work was supported by JSPS KAKENHI Grant-in-Aid for Scientific Research Number JP15H02705, JP16K12430, JP26280112, PRESTO JST, the MIC/SCOPE #162110003 and JP26730060.

References

1. Ginsberg, J., Mohebbi, M., Patel, R., Brammer, L., Smolinski, M., Brilliant, L.: Detecting influenza epidemics using search engine query data. Nature **457**, 1012–1014 (2009)
2. Gruhl, D., Guha, R., Kumar, R., Novak, J., Tomkins, A.: The predictive power of online chatter. In: KDD, pp. 78–87 (2005)
3. Jain, A., Chang, E.Y., Wang, Y.-F.: Adaptive stream resource management using kalman filters. In: SIGMOD, pp. 11–22 (2004)
4. Levenberg, K.: A method for the solution of certain non-linear problems in least squares. Q. J. Appl. Math. **II**(2), 164–168 (1944)
5. Li, L., Liang, C.-JM., Liu, J., Nath, S., Terzis, A., Faloutsos, C.: Thermocast: a cyber-physical forecasting model for data centers. In: KDD (2011)
6. Li, L., McCann, J., Pollard, N., Faloutsos, C.: Dynammo: mining and summarization of coevolving sequences with missing values. In: KDD (2009)
7. Li, L., Prakash, B.A., Faloutsos, C.: Parsimonious linear fingerprinting for time series. PVLDB **3**(1), 385–396 (2010)
8. Livera, A.M.D., Hyndman, R.J., Snyder, R.D.: Forecasting time series with complex seasonal patterns using exponential smoothing. J. Am. Stat. Assoc. **106**(496), 1513–1527 (2011)
9. Matsubara, Y., Sakurai, Y., Faloutsos, C.: AutoPlait: automatic mining of co-evolving time sequences. In: SIGMOD, pp. 193–204 (2014)
10. Matsubara, Y., Sakurai, Y., Faloutsos, C.: The web as a jungle: non-linear dynamical systems for co-evolving online activities. In: WWW (2015)
11. Matsubara, Y., Sakurai, Y., Faloutsos, C.: Non-linear mining of competing local activities. In: WWW (2016)
12. Matsubara, Y., Sakurai, Y., Faloutsos, C., Iwata, T., Yoshikawa, M.: Fast mining and forecasting of complex time-stamped events. In: KDD, pp. 271–279 (2012)

13. Matsubara, Y., Sakurai, Y., Prakash, B.A., Li, L., Faloutsos, C.: Rise, fall patterns of information diffusion: model and implications. In: KDD, pp. 6–14 (2012)
14. Matsubara, Y., Sakurai, Y., van Panhuis, W.G., Faloutsos, C.: FUNNEL: automatic mining of spatially coevolving epidemics. In: KDD, pp. 105–114 (2014)
15. Papadimitriou, S., Brockwell, A., Faloutsos, C.: Adaptive, hands-off stream mining. In: VLDB, pp. 560–571 (2003)
16. Stone, L., Olinky, R., Huppert, A.: Seasonal dynamics of recurrent epidemics. Nature **446**, 533–536 (2007)
17. Tao, Y., Faloutsos, C., Papadias, D., Liu, B.: Prediction and indexing of moving objects with unknown motion patterns. In: SIGMOD, pp. 611–622 (2004)

Keeping Priors in Streaming Bayesian Learning

Anh Nguyen Duc, Ngo Van Linh, Anh Nguyen Kim, and Khoat Than(✉)

Hanoi University of Science and Technology, No. 1, Dai Co Viet Road, Hanoi, Vietnam
nguyenanh.nda@gmail.com, {linhnv,anhnk,khoattq}@soict.hust.edu.vn

Abstract. Exploiting prior knowledge in the Bayesian learning process is one way to improve the quality of Bayesian model. To the best of our knowledge, however, there is no formal research about the influence of prior in streaming environment. In this paper, we address the problem of using prior knowledge in streaming Bayesian learning, and develop a framework for keeping priors in streaming learning (KPS) that maintains knowledge from the prior through each minibatch of streaming data. We demonstrate the performance of our framework in two scenarios: streaming learning for latent Dirichlet allocation and streaming text classification in comparison with methods that do not keep prior.

Keywords: Streaming learning · Prior knowledge · Bayesian model

1 Introduction

Incorporating prior knowledge into Bayesian models is one of the essential problems that has attracted a lot of interests from researchers. Many works have showed that the priors, such as language or semantic knowledge, are valuable to make an improvement in the quality of Bayesian models [4,7,12]. This prior information guides the model to meet user's specific need. For instance, the Zipf's law in the natural language domain states that the frequency of any word is approximately proportional to the inverse of its rank in the frequency table [10,11]. According to [12], the authors have succeeded in capturing the Zipf's law with topic model that outperforms latent Dirichlet allocation model (LDA) [2] in terms of perplexity. In text classification, [7] used an asymmetrical prior which gave high weighted value for seed words of each class to gain better performance. In sentiment analysis, Aspect and Sentiment Unification Model (ASUM) [4] also exploited a word list that consists of a set of positive and negative words (e.g. good and bad) to determine the sentiment of each document.

In streaming data, the prior knowledge is not adequately noticed, although there are several probabilistic models and effective inference methods [3,8,14]. In batch learning, the influence of prior knowledge gradually reduce when the amount of data becomes bigger. On the other hand, the streaming data is often processed in minibatches, which are small collections of data, therefore, we believe that keeping the appropriate prior's impact through every minibatch will

J. Kim et al. (Eds.): PAKDD 2017, Part II, LNAI 10235, pp. 247–258, 2017.
DOI: 10.1007/978-3-319-57529-2_20

rapidly improve the quality of the learning model. Moreover, it is not straightforward to know how many data that a method needs to overcome lack of the prior knowledge.

Existing methods have difficulties in incorporating prior knowledge in streaming data. In inference algorithms [3,14], the prior is only appeared in the initialization stage, hence, its impact will be lost quickly after a few minibatches, and not effectively prove it's value. In particular, a result illustrated this view can be found in [6], in which the authors used seed words of sentiment aspect to form a prior, and then maintained it in streaming ASUM learning as a heuristics. However, it had no formal explanation for applying to general probabilistic model.

In this paper, we propose a general framework for keeping prior knowledge in streaming Bayesian learning to investigate the influence of prior knowledge. This framework emphasizes the role of the prior in the streaming learning by using it through every minibatch; therefore the prior's effect can be maintained in the entire learning process, not just the initial stage. We conduct the experiments in two scenarios: streaming learning for LDA and streaming text classification. Comparing with the framework that does not keep prior, KPS gives a better quality in predictive capacity and coherence of topics within the first scenario and in accuracy for the other.

In the rest of paper, Sect. 2 reviews an existed streaming learning framework with some discussions then explicitly describes our KPS framework. Sections 3 and 4 are two case studies. Finally, Sect. 5 concludes our work.

2 Streaming Learning with Prior Knowledge

In this section, at first, we review the streaming variational Bayes (SVB) framework by Broderick et al. [3] for learning a Bayesian model from a data stream with discussion about some properties of SVB. We then present our novel framework for encoding prior knowledge into SVB named *Keeping Priors in streaming Bayesian learning.*

2.1 Streaming Variational Bayes

Streaming data is considered as a sequence of minibatches data $\{C_i\}$. The problem of learning the Bayesian model's parameters $\{\Phi_i\}$, which continually generates the data stream, often leads to optimize the posterior probability of these parameters given observed data and previous information. According to [3], the authors have introduced a general framework for streaming computation of Bayesian posterior (Fig. 1).

Given prior η, presuming that $b-1$ minibatches have been processed, the posterior after b minibatches can be calculated by:

$$p(\Phi_b|\Phi_{b-1}, C_b) \propto p(C_b|\Phi_b)p(\Phi_b|\Phi_{b-1}) \quad (1)$$

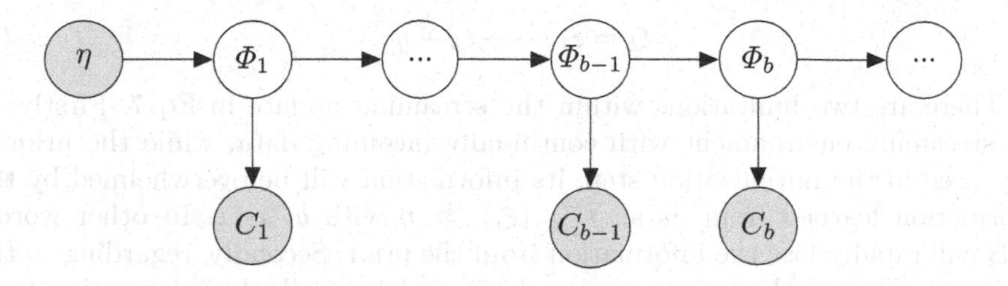

Fig. 1. Graphical representation for streaming learning in SVB

The posterior in Eq. 1 is often intractable to precisely compute. Suppose that, given $p(\Phi)$ and data minibatch C, we have an approximate algorithm A that returns a distribution in form of exponential family: $q(\Phi) = A(C, p(\Phi)) = \exp\{\xi \cdot T(\Phi)\})$, where ξ is the natural parameter and $T(\Phi)$ is the sufficient statistics. The posterior now can be continuously updated by:

$$p(\Phi_b|\Phi_{b-1}, C_b) \approx q_b(\Phi) = A(C_b|q_{b-1}(\Phi)) \tag{2}$$

With the approximate posterior after $b-1$ minibatches: $q(\Phi_{b-1}) = \exp\{\xi_{b-1} \cdot T(\Phi)\})$, we have:

$$p(C_b|\Phi_b) \propto \frac{p(\Phi_b|C_b)}{p(\Phi_b)} \approx \exp\{\tilde{\xi}_b \cdot T(\Phi)\} \tag{3}$$

and:

$$p(\Phi_b|\Phi_{b-1}) \approx q(\Phi_{b-1}) = \exp\{\xi_{b-1} \cdot T(\Phi)\}) \tag{4}$$

Equation 3 can be viewed as the information from minibatch C_b and Eq. 4 is the probability of the transition to the next state. Together, the final form of streaming update is:

$$q(\Phi_b) = \exp\{(\tilde{\xi}_b + \xi_{b-1}) \cdot T(\Phi)\} \tag{5}$$

or:

$$\xi_b = \tilde{\xi}_b + \xi_{b-1} \tag{6}$$

The initialization value is $q_0(\phi) = p(\Phi) = p(\eta)$ or $\xi_0 = \eta$. The Eqs. 5 and 6 restore the form of the SVB update presented in [3].

While SVB has an appealing feature that it provides a streaming update method without revisiting the old data, the approach only uses little information from the prior. In general, data can not be exactly described by any mathematical model, which mean most models are mis-specified. Within these models, combining the prior knowledge about data will guide model to be learned more accurately. Even in unstable environment, the prior knowledge also keeps some valuable meaning, therefore, maintaining the impact of the prior knowledge needs to be adequately considered. However, SVB does not fully cover the ability to use the information from the prior. To reveal this shortcoming, we can write Eq. 6 by:

$$\xi_b = \tilde{\xi}_1 \cdots + \tilde{\xi}_b + \eta \tag{7}$$

There are two limitations within the streaming update in Eq. 7. Firstly, in the streaming environment with continually incoming data, while the prior is only used in the initialization step, its information will be overwhelmed by the information learned from data: $\sum_{i=1}^{b}(\tilde{\xi}_i) \gg \eta$ with $b \gg 1$. In other words, SVB will rapidly lose the information from the prior. Secondly, regarding to the beginning stage of the learning process, where b is small, the information from little data is not enough for the model. To quickly improve the quality of the model, emphasizing the impact of prior is a core problem, but SVB again does not address this problem.

The KPS framework attacks to these weaknesses of SVB by maintaining the impact from the prior information in the streaming learning process.

2.2 Proposal Framework

The main idea of our framework is that we simply let the prior directly impact to the model within each minibatch (Fig. 2). So that, each minibatch will use one more information from the prior. By this way, the meaning of the prior will be emphasized through the learning process.

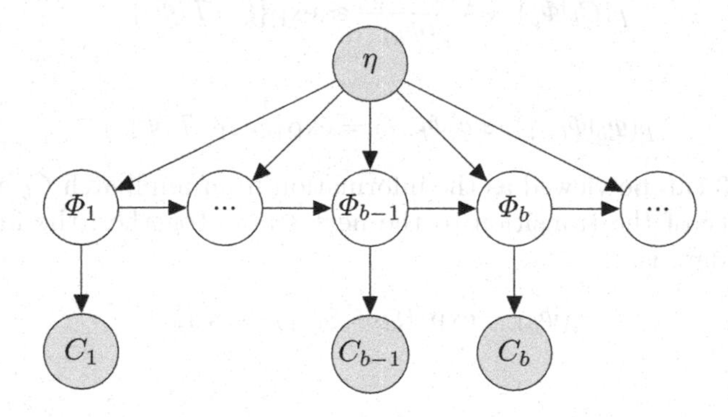

Fig. 2. Graphical representation for keeping prior in streaming Bayesian learning (KPS)

The posterior can be rewritten as:

$$p(\Phi_b|C_b, \Phi_{b-1}, \eta)$$

By simple Bayes transformation:

$$\begin{aligned} p(\Phi_b|C_b, \Phi_{b-1}, \eta) &\propto p(C_b|\Phi_b, \Phi_{b-1}, \eta)p(\Phi_b|\Phi_{b-1}, \eta) \\ &= p(C_b|\Phi_b)p(\Phi_b|\Phi_{b-1}, \eta) \end{aligned}$$

The element $p(\Phi_b|\Phi_{b-1},\eta)$ differs from Eq. 4 in that the changing of model is not only from the previous information after $b-1$ minibatches, but also from the prior information η. We hypothesize that:

$$p(\Phi_b|\Phi_{b-1},\eta) \approx q(\Phi_b|\Phi_{b-1})q(\Phi_b|\eta) \tag{8}$$

with $q(\Phi_b|\eta)$ is the impact of prior information into the current minibatch: $q(\Phi_b|\eta) = \exp\{f_b(\eta) \cdot T(\Phi)\}$, where $f_b(\eta)$ is a function that describes how the prior impacts on each minibatch. In this paper, we propose a simple form:

$$f_b(\eta) = \rho_b.\eta \tag{9}$$

where:

$$\rho_b = (1+b)^{-\kappa} \tag{10}$$

To constantly maintain the impact of the prior in streaming learning, we propose to set $\kappa = 0$ that leads to $f(\eta)_b = \eta$, this setting is named *strongly keeping prior in streaming* (S-KPS). Otherwise, with $\kappa > 0$, the information from the prior will be decreased through minibatches, so-called *dimly keeping prior in streaming* (D-KPS).

Finally, the posterior can be updated continually as the following:

$$q_b = \exp\{(\tilde{\xi}_b + \xi_{b-1} + f_b(\eta)) \cdot T(\Phi))\} \tag{11}$$

or:

$$\xi_b = \tilde{\xi}_b + \xi_{b-1} + f_b(\eta) \tag{12}$$

The only difference between Eqs. 6 and 12 is the $f_b(\eta)$ element that describes the impact of the prior. We can interpret the updating equation in Eq. 12 to mean that the parameter of the next minibatch contains 3 parts: the first one is the information from current minibatch, the second one is the information from previous step ξ_{b-1}, and the last one is the impact of the prior.

For the next two sections, we evaluate the performance of KPS framework against SVB which does not keep prior in two case studies: streaming learning for LDA and streaming text classification, respectively. Each case study is organized as follows: at the beginning, the model is briefly summarized. Then we adopt SVB and KPS framework to get streaming learning algorithms for the model. In the next step, we describe how to extract prior knowledge. The final part is experiments with settings and results.

3 Case Study 1: Streaming Learning for LDA

LDA (Fig. 3) is a generative model for modeling text data [2]. It assumes that a corpus is composed from K topics $\beta = (\beta_1, \beta_2, \ldots, \beta_K)$, each of which is drawn from a Dirichlet distribution: $\beta \sim Dirichlet(\eta)$. A document d is a mixture θ of those topics and is presumed to arises from the following generative process:

1. Draw topic mixture $\theta|\alpha \sim Dirichlet(\alpha)$
2. For the ith word of d:
   - Draw topic index $z_i|\theta \sim Multinomial(\theta)$
   - Draw word $w_i|z_i, \beta \sim Multinomial(\beta_{z_i})$

β is the distribution of words over vocabulary within each topic and the prior parameter η contains information about this distribution. The prior η is usually set as a symmetric value because we have no prior knowledge about which words are outstandings others in each topic. However, in linguistic data, we can have more information about the words such as Zipf's law, so the prior can bring more information with form of an asymmetric vector distribution over words. This case study exploits this idea to improve the quality of LDA model.

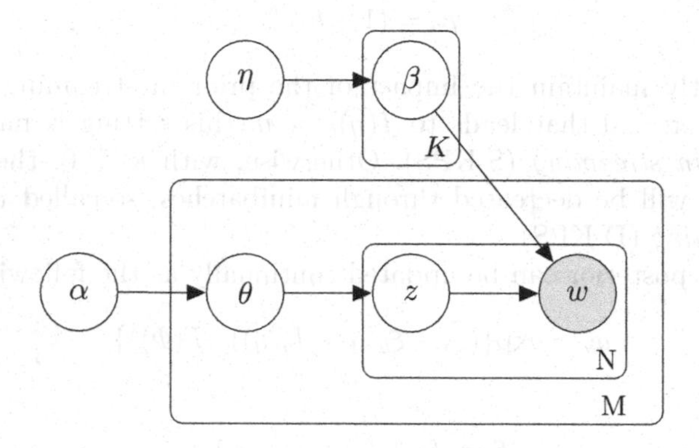

Fig. 3. Graphical representation for LDA model

Algorithm 1. SVB for LDA

Input: Prior η, hyper-parameter α, sequence of minibatches $C_1, C_2, ...$
Output: λ
Initialize : $\lambda_0 \leftarrow \eta$
for each minibatch C in $C_1, C_2, ...$ **do**
  for each document d in C **do**
    $\Phi_d \leftarrow VBInference(d, \lambda)$
  end for
  $\tilde{\lambda_b} \leftarrow \sum_{dinC} \Phi_{dvk} n_{dv}$
  $\lambda_b \leftarrow \lambda_{b-1} + \tilde{\lambda_b}$
end for

Algorithm 2. KPS for LDA

Input: Prior η, hyper-parameter α, sequence of minibatches $C_1, C_2, ...$
Output: λ
Initialize : $\lambda_0 \leftarrow \eta$
for each minibatch C in $C_1, C_2, ...$ **do**
  for each document d in C **do**
    $\Phi_d \leftarrow VBInference(d, \lambda)$
  end for
  $\tilde{\lambda_b} \leftarrow \sum_{dinC} \Phi_{dvk} n_{dv}$
  $\lambda_b \leftarrow \lambda_{b-1} + \tilde{\lambda_b} + f_b(\eta)$
end for

Streaming Learning Algorithm: With corpus data $C = \{C_i\}$, the posterior is:

$$p(z, \theta, \beta | C, \alpha, \eta)$$

Using variational Bayes as the approximate algorithm with (Φ, γ, λ) are variational variable of (z, θ, β), respectively, we apply SVB and KPS framework to get two streaming learning algorithms for LDA as in Algorithm 1 and Algorithm 2. The $VBInference$ here is a variational Bayes procedure that receives the global model parameter λ to infer the topic distribution Φ of document d [2]. The Algorithm 1 is the same as SSU algorithm in [3]. The only different between the two algorithms is the impacted element of the prior $f_b(\eta)$.

Prior in Use: Relating to the distribution of word in natural language, Zipf's law give us an interesting property that the frequencies of words in a specific language followed a power-law distribution given by: $p(w) \propto r_w^{-l}$, in which $p(w)$ is the proportion of word w in the language. r_w is the rank of the word in the descending sorted frequencies, which means the most frequency word has ranking $r = 1$. The parameter l depends on the specific language. We use Zipf's law as the prior knowledge in KPS algorithm as follows:

$$\eta_w \propto p(w). \tag{13}$$

Evaluation Metric: We use log predictive probability [3] to evaluate the predictive capacity and NPMI [9] to check the coherence of learned topics.

Table 1. Dataset for streaming LDA

Dataset	Vocab size	Training size	Testing size
Grolier	15,726	23,044	1000
Pubmed	141,044	100,000	10,000
Nytimes	102,660	200,000	10,000

Table 2. Dataset for text classification

Dataset	Num of class	Vocab size	Training size	Testing size
News20	20	62,061	16,000	3,900
Cade12	12	193,997	27,322	4000

Data and Settings: We use 3 datasets: Grolier, Nytimes and Pubmed with information in Table 1. We simulate the streaming data by dividing the dataset into sequential minibatches with batchsize respectively: 500, 5000 and 10000 for Grolier, Pubmed and Nytimes. The number of topics is set equally with $K = 100$, and the hyperparameter $\alpha = 0.01$. The prior η is taken from Eq. 13 with a heuristic parameter $l = 1.07$ [5] and the ranking of word's frequencies is downloaded from top 100,000 most frequently-used English words text[1], only the words appeared in vocabulary of dataset are used. We deploy 2 versions of KPS are: S-KPS which sets $\kappa = 0$ and D-KPS with $\kappa = 0.7$.

Result: The results are shown in Fig. 4a and b. These results demonstrate that 2 versions of KPS outperform the SVB method in both predictive capacity and coherence. Within log predictive probability metric, all results have the same pattern: there only a slightly different between D-KPS and S-KPS. And from the beginning minibatch data, the result of KPS is better than SVB, it is because

[1] https://gist.github.com/h3xx/1976236.

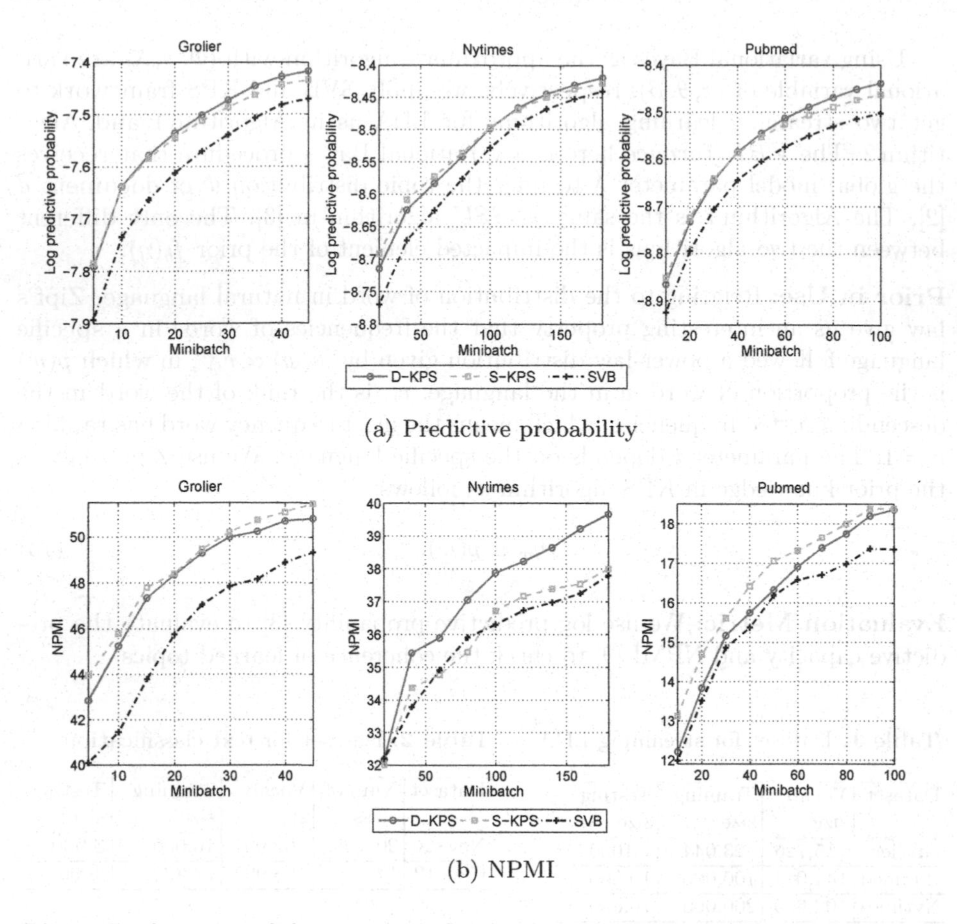

(a) Predictive probability

(b) NPMI

Fig. 4. Performance of three methods for streaming learning LDA, higher is better.

KPS uses more information from prior knowledge, so it can quickly improve the quality of model even the amount of data is small. At the later minibatch, the result of D-KPS tends to higher than S-KPS, we can explain as: when data is bigger, the model can learn information itself from data that contains prior information, so the keeping prior may lead to overfiting, slowly decrease the impact of prior will help to avoid this problem.

4 Case Study 2: Streaming Text Classification

In [1] the authors introduced Mview-LDA for classification, and within this case study, we use a version of Mview-LDA represented in [13]. The idea of Mview-LDA is that: each document belongs to one of J classes with probability contributed by a multinominal distribution $\chi \sim Multinomial(\pi)$. Each class contains K local topics $\{\beta_{jk}^{(l)}\}_{k=1}^{K}$ with the distribution over topics $\theta^{(l)}$

defined by a Dirichlet distribution $Dirichlet(\alpha_j^{(l)})$. Besides, they assume that there exists of R global topics $\{\beta_r^{(g)}\}_{r=1}^{R}$ that shared by all classes with distribution: $Dirichlet(\alpha^{(g)})$. The binary variable δ decides a word is belonged to global or shared topics (Fig. 5).

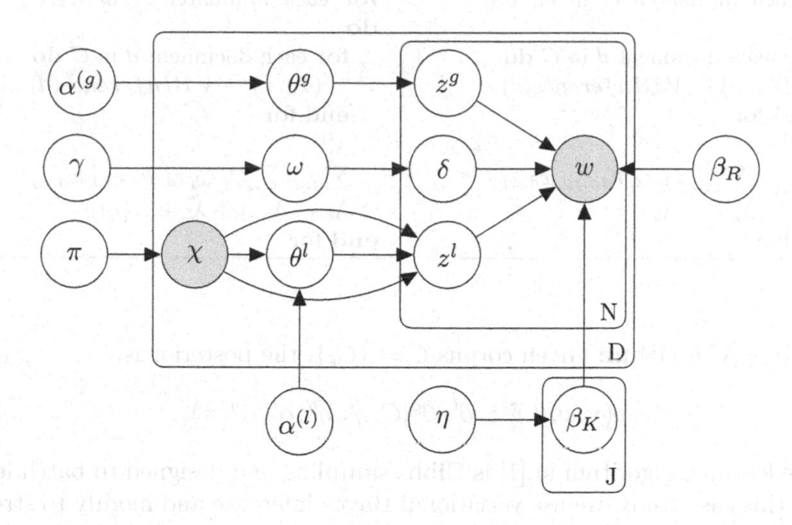

Fig. 5. Graphical representation for Mview-LDA

The generative process of a document d in Mview-LDA as follows:

1. Draw a class: $\chi \sim Multinomial(\pi)$
2. Draw local topic proportion: $\theta_\chi^{(l)} \sim Dirichlet(\alpha_\chi^{(l)})$
3. Draw global topic proportion: $\theta^{(g)} \sim Dirichlet(\alpha^{(g)}$,
4. Draw Bernoulli parameter $\omega \sim Beta(\gamma)$
5. For each word w of document d:
   - Draw a binary indicator $\delta \sim Bernoulli(\omega)$
   - If $\delta = 1$, word w belongs to local topic:
     - Draw a local topic $z_\eta^{(l)} \sim Multinominal(\theta_\eta^{(l)})$
     - Draw word $w \sim Mutinomial(\beta_{z_\chi^{(l)}})$
   - If $\delta = 0$, word w belongs to global topic:
     - Draw a global topic: $z^{(g)} \sim Multinomial(\theta^{(g)})$
     - Draw word $w \sim Multinomial(\beta_{z^{(g)}})$

Let each local topic contributed by a Dirichlet prior parameter $\eta_j^{(l)}$. Note that local topics within each class contain features words of its class, so if we set the weighted value of these words larger than the others and use it as a prior, we will provide more information into the model to specify class, therefore it can increase the quality of learning process.

Algorithm 3. SVB for Mview-LDA

Input: Prior η, sequence of minibatch $C_1, C_2, ...$
Output: λ
Initialize : $\lambda_0 \leftarrow \eta$
for each minibatch C in $C_1, C_2, ...$ **do**
  for each document d in C **do**
    $(\Phi_d, \tau) \leftarrow VBInference(d)$
  end for
  $\tilde{\lambda}_b^{(l)} \leftarrow \sum_{d \in C} \sum_{i=1}^{N_d} \zeta_{dj} \tau_{di} \Phi_{d,i,j,k} w_{d,i,j}$
  $\lambda_b \leftarrow \lambda_{b-1} + \tilde{\lambda}_b$
end for

Algorithm 4. KPS for Mview-LDA

Input: Prior η, sequence of minibatch $C_1, C_2, ...$
Output: λ
Initialize : $\lambda_0 \leftarrow \eta$
for each minibatch C in $C_1, C_2, ...$ **do**
  for each document d in C **do**
    $(\Phi_d, \tau) \leftarrow VBInference(d)$
  end for
  $\tilde{\lambda}_b^{(l)} \leftarrow \sum_{d \in C} \sum_{i=1}^{N_d} \zeta_{dj} \tau_{di} \Phi_{d,i,j,k} w_{d,i,j}$
  $\lambda_b \leftarrow \lambda_{b-1} + \tilde{\lambda}_b + f_b(\eta))$
end for

Learning Algorithm: Given corpus $C = \{C_i\}$, the posterior is:

$$p(\chi, \omega, \theta, \delta, z, \beta^l, \beta^g | C, \eta^l, \eta^g, \alpha^l, \alpha^g, \pi) \tag{14}$$

The learning algorithm in [1] is Gibbs sampling and designed to batch learning. In this case study, we use variational Bayes inference and modify to streaming classification learning. Let $\lambda^{(l)}, \tau, \Phi$ be the approximate variantional variables of $\beta^{(l)}, \delta, z$ respectively. Because we aim to keep the information from prior $\eta^{(l)}$ of the local topics, we only apply KPS to $\beta^{(l)}$. Simplify $\eta^{(l)}$ to η, adopting SVB and KPS leads to 2 versions for streaming updating $\lambda^{(l)}$ in Algorithm 3, and Algorithm 4. The indicator $\zeta_{dj} = 1$ when document d belongs to class j and $\zeta_{dj} = 0$ for others, the $VBInference$ procedure here receives model parameters to infer the topics ϕ and its contribution τ to local or global topics for each document d.

Prior in Use: With the idea from feaLDA [7], we extract the feature words of each class and use them as the prior knowledge for its class. At first, we calculate TF.IDF for words in each class then select top T words with highest values as seed words. The seed words of class j then used to initialize prior Dirichlet distribution $\eta_j^{(l)}$ by assigning a value $0 < s < 1$. The other values are set to a small value $0 < \epsilon < s$.

Evaluation Metric: The classification accuracy is used in this evaluation.

Data and Settings: We use 2 labeled datasets: Cade 12 and News20[2] with information in Table 2.

With News20 dataset, we set minibatch size equal to 1000, the number of topics in each local group $K = 10$ and the number of global group: $R = 8$. For Cade 12 dataset, minibatch size is 2000, $K = 15$ and $R = 4$. The Dirichlet distribution prior $\eta^{(l)}$ for each group is taken with $s = 0.5$ and $\epsilon = 0.01$, the number of top words for each group: $T = 5000$. The other Dirichlet prior parameters are

[2] http://ana.cachopo.org/datasets-for-single-label-text-categorization.

set equal to 0.01. For S-KPS, the parameter $\kappa = 0$ and D-KPS has $\kappa = 0.7$. We also give the SVM results with bag of word representation as another baseline to evaluate the quality of classification model.

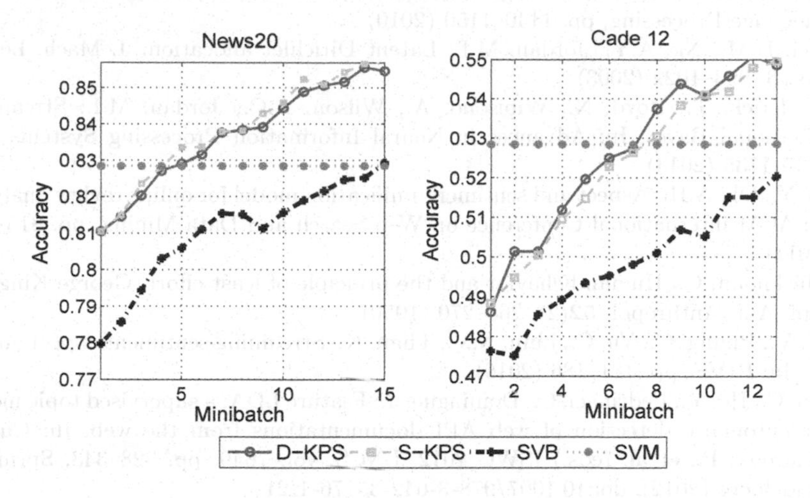

Fig. 6. Accuracy of three streaming classification algorithms with SVM baseline

Results: The results are presented in Fig. 6. Again, the result of D-KPS and S-KPS are better than that of the SVB without keeping prior. When more minibatches are observed, the accuracy of streaming classification is increased. However, at the latter minibatches, while KPS gives higher accuracy, SVB method shows a lower accuracy against SVM baseline, it suggests that providing prior knowledge into streaming learning can markedly improve the performance of the learned model.

5 Conclusion

In this paper, we have introduced the keeping priors in streaming learning framework which has tried to make the best use of prior knowledge in the entire streaming learning process, in a view that incoming data is processed in infinite number of collections. If the prior information is good and impactful enough, this method can shine its wise value and entail its effect. We have illustrated this idea in two scenarios: streaming LDA and streaming text classification. In both cases, KPS has showed better results than the old method without keeping prior.

Acknowledgement. This research is funded by Vietnam National Foundation for Science and Technology Development (NAFOSTED) under grant number 102.05-2014.28, and by the Air Force Office of Scientific Research (AFOSR), Asian Office of Aerospace Research & Development (AOARD), and US Army International Technology Center, Pacific (ITC-PAC) under award number FA2386-15-1-4011.

References

1. Ahmed, A., Xing, E.P.: Staying informed: supervised and semi-supervised multi-view topical analysis of ideological perspective. In: Empirical Methods in Natural Language Processing, pp. 1140–1150 (2010)
2. Blei, D.M., Ng, A.Y., Jordan, M.I.: Latent Dirichlet allocation. J. Mach. Learn. Res. **3**, 993–1022 (2003)
3. Broderick, T., Boyd, N., Wibisono, A., Wilson, A.C., Jordan, M.I.: Streaming variational Bayes. In: Advances in Neural Information Processing Systems, pp. 1727–1735 (2013)
4. Jo, Y., Oh, A.H.: Aspect and sentiment unification model for online review analysis. In: ACM international Conference on Web Search and Data Mining, pp. 815–824 (2011)
5. Kluckhohn, C.: Human behavior and the principle of least effort. George Kingsley Zipf. Am. Anthropol. **52**(2), 268–270 (1950)
6. Le, V., Phung, C., Vu, C., Linh, N.V., Than, K.: Streaming sentiment-aspect analysis. In: RIVF, pp. 181–186 (2016)
7. Lin, C., He, Y., Pedrinaci, C., Domingue, J.: Feature LDA: a supervised topic model for automatic detection of web API documentations from the web. In: Cudré-Mauroux, P., et al. (eds.) ISWC 2012. LNCS, vol. 7649, pp. 328–343. Springer, Heidelberg (2012). doi:10.1007/978-3-642-35176-1_21
8. McInerney, J., Ranganath, R., Blei, D.M.: The population posterior and Bayesian inference on streams. In: International Conference on Neural Information Processing Systems (2015)
9. Mimno, D., Wallach, H.M., Talley, E., Leenders, M., McCallum, A.: Optimizing semantic coherence in topic models. In: Empirical Methods in Natural Language Processing, pp. 262–272 (2011)
10. Newman, M.E.J.: Power laws, Pareto distributions and Zipf's law. Contemp. Phys. **46**(5), 323–351 (2005)
11. Piantadosi, S.T.: Zipfs word frequency law in natural language: a critical review and future directions. Psychon. Bull. Rev. **21**(5), 1112–1130 (2014)
12. Sato, I., Nakagawa, H.: Topic models with power-law using Pitman-Yor process. In: ACM SIGKDD International Conference on Knowledge Discovery and Data Mining, pp. 673–682 (2010)
13. Van Linh, N., Anh, N.K., Than, K., Dang, C.N.: An effective and interpretable method for document classification. Knowl. Inf. Syst. **50**(3), 763–793 (2016)
14. Viet, H., Phung, D., Venkatesh, S.: Streaming variational inference for dirichlet process mixtures. In: Asian Conference on Machine Learning (2015)

Text and Opinion Mining

Efficient Training of Adaptive Regularization of Weight Vectors for Semi-structured Text

Tomoya Iwakura(✉)

Fujitsu Laboratories Ltd., 1-1, Kamikodanaka 4-chome,
Nakahara-ku, Kawasaki 211-8588, Japan
iwakura.tomoya@jp.fujitsu.com

Abstract. We propose an efficient training method of Confidence Weighted Learning (CWL) algorithms for semi-structured text and its application to Adaptive Regularization of Weight Vectors (AROW), which is a CWL algorithm. CWL algorithms are online learning algorithms that combines large margin training and confidence weighting of features. CWL algorithms learn confidence weights of features, therefore, it is difficult to apply kernel methods that implicitly expand features. If we expand features in advance, it leads to increased memory usage. To solve the problem, we propose a training method that dynamically extracted features from semi-structured text. In addition, we propose a pruning method for improved training speed. The pruning skips training samples classified correctly more than or equal to certain times. We compared our method using word-strings as semi-structured texts with AROW that expands all the features in advance. Experimental results of text classification tasks on an Amazon data set show that our training method contributes to improved memory usage and two to three times faster training speed while maintaining accuracy for learning longer n-grams.

1 Introduction

Online learning algorithms represent a family of fast and simple machine learning techniques. These include Perceptron [14] and Passive-Aggressive (PA) algorithms [4]. Recent years have seen the second-order online learning algorithms, which learn confidences of features to improve online learning performance. For example, a family of Confidence Weighted Learning (CWL) algorithms such as Confidence-Weighted (CW) [5], soft-CW [7] and Adaptive Regularization of Weights (AROW) [6] have been proposed and these algorithms have shown better accuracy on a wide range of applications.

One of the characteristics of these algorithms are learning of confidence weights of features. Less confident features are updated more aggressively than more confident ones. Therefore, in order to learn confidence weights of features such as word n-grams on document classification tasks, we have to give such features in advance by expanding them from a given input, compared with Perceptron or PA with kernel methods like string kernels [12,13] does not. As a result, CWL algorithms require more memory usage by using of longer word n-grams.

J. Kim et al. (Eds.): PAKDD 2017, Part II, LNAI 10235, pp. 261–272, 2017.
DOI: 10.1007/978-3-319-57529-2_21

To solve the problem, we propose a training method that dynamically extracts features from semi-structured text. Our method dynamically extracts features from a given current training sample when these are necessary, therefore, we can save memory usage. In addition, we propose a pruning method for improved training speed. The pruning skips training samples classified correctly more than or equal to certain times.

We evaluate the proposed methods using word-strings as semi-structured texts with AROW. The experimental results of text classification tasks on an Amazon data set show that our training method contributes to improved memory usage and two to three times faster training speed than those of AROW that expands all the features in advance while maintaining accuracy for learning longer n-grams.

2 AROW

This section describes a confidence weighted learning method called AROW and the difficulty of application of kernel methods.

2.1 An Overview of AROW

We describe the problem treated by our paper as follows. The goal is to induce a mapping:

$$f(\mathbf{x}) = \mathbf{x} \cdot \boldsymbol{\mu}, \tag{1}$$

where an input $\mathbf{x}$ and a weight vector $\boldsymbol{\mu}$ is d dimensional vector. The μ is induced from $D = \{(\mathbf{x}_1, y_1), ..., (\mathbf{x}_m, y_m)\}$ and a given learning algorithm, where $y_i \in \{1, -1\}$ and $\mathbf{x}_i \in \mathcal{R}^d$.

Input parameters r
Training samples: $D = \{(\mathbf{x}_1, y_1), ..., (\mathbf{x}_m, y_m)\}$
Initialize $\boldsymbol{\mu}_0 = \mathbf{0}$, $\boldsymbol{\Sigma}_0 = \boldsymbol{I}$
For $t = 1, ..., \mathrm{T}$
- Receive a training example from D as $\mathbf{x}_t$
- Compute margin m_t and confidence v_t
  $m_t = \boldsymbol{\mu}_{t-1} \cdot \boldsymbol{x}_t$
  $v_t = \boldsymbol{x}_t^{\mathrm{T}} \boldsymbol{\Sigma}_{t-1} \boldsymbol{x}_t$
- Receive true label y_t
- If $m_t y_t < 1$, update parameters:
  $\boldsymbol{\mu}_t = \boldsymbol{\mu}_{t-1} + \alpha_t \boldsymbol{\Sigma}_{t-1} y_t \boldsymbol{x}_t \qquad \boldsymbol{\Sigma}_t = \boldsymbol{\Sigma}_{t-1} - \beta_t \boldsymbol{\Sigma}_{t-1} \boldsymbol{x}_t \boldsymbol{x}_t^{\mathrm{T}} \boldsymbol{\Sigma}_{t-1}$
  $\beta_t = \frac{1}{v_t + r} \qquad \alpha_t = \max(0, 1 - y_t\, \boldsymbol{x}_t^{\mathrm{T}} \boldsymbol{\mu}_{t-1}) \beta_t$
- Else:
  $\boldsymbol{\mu}_t = \boldsymbol{\mu}_{t-1} \quad \boldsymbol{\Sigma}_t = \boldsymbol{\Sigma}_{t-1}$

Fig. 1. AROW with squared hinge

Figure 1 shows a pseudo code of AROW. The $\boldsymbol{\mu}_t$ is a d dimensional weight vector and $\boldsymbol{\Sigma}_t$ is a $d \times d$ matrix that maintains confidences of features at t-th round. When updating a model, $\boldsymbol{\mu}_t$ and $\boldsymbol{\Sigma}_t$ are updated. In our implementation, a diagonalized version, in which $\boldsymbol{\Sigma}_t$ is diagonalized, was used.

2.2 Difficulty of Application of Kernel Method to AROW

In order to handle semi-structured data with kernel methods [4], a following function is induced from given training data D:

$$f(\mathbf{x}) = \sum_{i=1}^{m} \alpha_i K(\mathbf{x}_i, \mathbf{x}), \tag{2}$$

where α_i is an importance of i-th training sample.

Kernel methods implicitly expand features in a given kernel function K. By doing so, semi-structured data and structured data can be handled as in d dimensional vectors. However, we see from Eq. (2) that use of kernel methods in CWL algorithms is difficult because CWL algorithms learn confidences of features.

3 Proposed Method

We describe our AROW training method with dynamic feature expansion and a pruning method for improving training speed with efficient implementation.

3.1 Training with a Dynamic Feature Expansion and a Pruning

Our method expands n-gram features dynamically. In order to improve training speed, we focus on that only features extracted from current training sample are necessary for updating a model. Therefore, we propose expanding features from current training sample dynamically. In addition, in order to obtain further improved training speed, we propose a pruning of correctly classified training samples more than certain times.

Figure 2 shows a pseudo algorithm. Let $\mathbf{X}$ be a word-sequence in this paper. The $\phi(\mathbf{X}, L)$ converts $\mathbf{X}$ into d dimensional vectors by expanding features dynamically up to maximum size L.

For example, "I eat corn soup" are given as an input $\mathbf{X}$ with $L = 2$ for n-gram feature extraction, "I","eat","corn","soup","I-eat","eat-corn" and"corn-soup" are expanded. When updating models, a weight vector $\boldsymbol{\mu}_t$ and confidence of features $\boldsymbol{\Sigma}_t$ are updated as in the original AROW.

In our implementation, memory space is allocated when new features are added to a model. In contrast, the naive implementation of the original AROW requires memory space of all the features from the beginning of training.

Our method also incorporates a pruning method of training samples. In our preliminary experiments, about correctly classified training samples in a training,

Input parameters r
Training samples: $D = \{(\mathbf{X}_1, y_1), ..., (\mathbf{X}_m, y_m)\}$
Initialize $\boldsymbol{\mu}_0 = \mathbf{0}$, $\boldsymbol{\Sigma}_0 = \boldsymbol{I}$
L: Maximum n-gram size
$ok[\mathbf{X}_t]$: the number of times $\mathbf{X}_t$ is correctly classified.
p : a threshold for pruning
For t = 1, ..., T
- Receive a training example from D as $\mathbf{X}_t$
- If $p \leq ok[\mathbf{X}_t]$ continue ; # skip this sample
  - Compute margin m_t and confidence v_t
    $m_t = \boldsymbol{\mu}_{t-1} \cdot \phi(\boldsymbol{X}_t, L)$
    $v_t = \phi(\boldsymbol{X}_t, L)^{\mathrm{T}} \boldsymbol{\Sigma}_{t-1} \phi(\boldsymbol{X}_t, L)$
  - Receive true label y_t
  - If $m_t y_t < 1$, update parameters:
    $\boldsymbol{\mu}_t = \boldsymbol{\mu}_{t-1} + \alpha_t \boldsymbol{\Sigma}_{t-1} y_t \phi(\boldsymbol{X}_t, L)$ $\quad \boldsymbol{\Sigma}_t = \boldsymbol{\Sigma}_{t-1} - \beta_t \boldsymbol{\Sigma}_{t-1}\, \phi(\boldsymbol{X}_t, L)\phi(\boldsymbol{X}_t, L)^{\mathrm{T}} \boldsymbol{\Sigma}_{t-1}$
    $\beta_t = \frac{1}{v_t + r}$ $\quad \alpha_t = \max(0, 1 - y_t\, \phi(\boldsymbol{X}_t, L)^{\mathrm{T}} \boldsymbol{\mu}_{t-1})\beta_t$
- Else :
    $ok[\mathbf{X}_t]{+}{+}$ $\quad \boldsymbol{\mu}_t = \boldsymbol{\mu}_{t-1}$ $\quad \boldsymbol{\Sigma}_t = \boldsymbol{\Sigma}_{t-1}$

Fig. 2. AROW based on dynamic feature expansion with a pruning of training samples

only 5.8% on average were used for updating models after their correct classification. Therefore, we introduce a pruning method that skips training samples classified correctly more than or equal to p times. The $ok[\mathbf{X}_t]$ maintains the number of times that $\mathbf{X}_t$ was correctly classified in a training in terms of the hinge loss used in AROW training. If $p \leq ok[\mathbf{X}_t]$, we skip the training with $\mathbf{X}_t$.

3.2 Efficient Implementation

Our algorithm dynamically expands features. Therefore, a naive implementation requires to expand all features from a current sample at each training phase. However, in order to calculate $\boldsymbol{\mu}_t \cdot \phi(\boldsymbol{X}_t, L)$, only features of current training sample included in a current model $\boldsymbol{\mu}_t$ are necessary. In order to avoid time-consuming feature expansion, we propose to use only features included in a model.

For efficient calculation of $\boldsymbol{\mu}_t \cdot \phi(\boldsymbol{X}_t, L)$, we maintain a model as a trie. By using a trie-based model, when partial features are included in $\boldsymbol{\mu}_t$, we continue to expand features for generating new n-gram features. We use double-array [1], which is an implementation method of trie.

Figure 3 shows a matching with a given input "corn soup" and a model represented by a trie includes "coral", "coring", "corn" and "corn soup". The blank indicates an existence of the following words and the "#" indicates the end of an entry in the trie.

Our method does not enumerate all combinations for classification. First, we check whether "corn" are matched in the trie from the root. Here, "corn" and "#" after "corn" were found, therefore, "corn" was used as a feature. Then, we found "blank" after "corn" and match "soup#" with the trie from the next

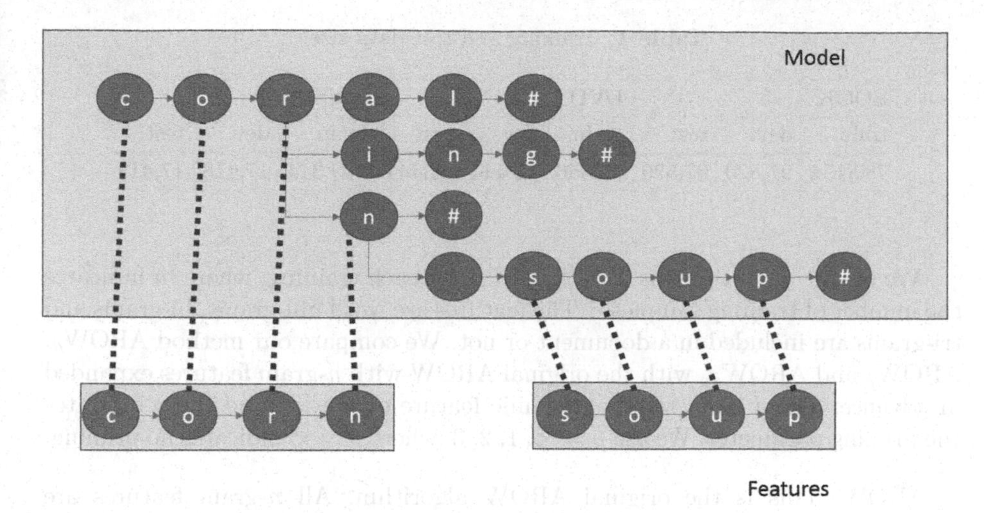

Fig. 3. Matching with a trie-based model.

of "blank", therefore, we use "corn soup" as a feature. By using this method, without expanding all features in advance, we can expand only features included in a current model.

In addition, we cache $\phi(\boldsymbol{X}_t, L)$ after $\boldsymbol{X}_t$ was used for updating models. This is because all features extracted from $\boldsymbol{X}_t$ are included in a model after a model update with $\boldsymbol{X}_t$. By caching $\phi(\boldsymbol{X}_t, L)$, we can avoid time-consuming feature generation.

4 Experiments

4.1 Experimental Setting

We use three most frequent categories of an Amazon data [2] described in Table 1. The goal is to classify a product review as either positive or negative. We used the file, all.review, for each domain in the data set for this evaluation. By following the paper [2], review texts that have ratings more than three are used as positive reviews, and review texts that have ratings less than three are used as negative reviews. We used the text data represented by word sequences. Each training data is 80% of samples in all.review of each category, and the test and the development data are 10%. Parameters are decided in terms of F-measure on negative reviews of the development data, and we evaluate F-measure obtained with the decided parameters. The number of positive reviews in the data set is much larger than negative reviews. Therefore, we evaluated the F-measure of the negative reviews.

Table 1. Training and test data size.

BOOK			DVD			MUSIC		
train	dev.	test	train	dev.	test	train	dev.	test
780,154	97,520	97,520	99,550	12,444	12,444	139,344	17,418	17,418

We use 10 $\times$ m times as T with $r = 1$ for each training, where m indicates the number of training samples. [1] The features are word uni-grams, bi-grams and tri-grams are included in a document or not. We compare our method AROW$_d$, AROW$_p$ and AROW$_{dp}$ with the original AROW with n-gram features expanded in advance. The d indicates the dynamic feature expansion and the p indicates the pruning parameter. We use $p = \infty, 1, 2, 3$, where $p = \infty$ indicates no pruning.

- AROW: This is the original AROW algorithm. All n-gram features are expanded in advance.
- AROW$_d$: This is the AROW algorithm with only the dynamic expansion described in Fig. 2. The n-gram features are dynamically expanded. This is the same as the following AROW$_{dp = \infty}$.
- AROW$_p$: This is the AROW algorithm with only the pruning method of Fig. 2. This algorithm does not use the dynamic expansion method. Therefore, all n-gram features are expanded in advance. AROW$_{p = \infty}$ is the same as the AROW.
- AROW$_{dp}$: This is the AROW algorithm described in Fig. 2. The n-gram features are dynamically expanded and the pruning method is used.

4.2 Accuracy

Table 2 shows that the accuracy of each pruning parameter. Even if we apply the pruning method, accuracy was maintained. The accuracy was measured on each test data with the number of the iteration over given training samples in training that showed the best accuracy on the corresponding development data. We see that even if we use the pruning method, our method maintained accuracy. AROW$_{p = 1}$ indicates slightly lower accuracy, however, AROW$_{p = 2,3}$ showed competitive or better accuracy in most cases. AROW$_{p = 2}$ showed three times better, five times the same and one time worse accuracy compared with those of AROW$_{p = \infty}$. AROW$_{p = 3}$ showed two times better, six times the same and one time worse accuracy compared with those of AROW$_{p = \infty}$.

We guess that one of the reasons of the improved accuracy is alleviation of overfitting. Use of the pruning skips correctly classified samples more than

[1] In our implementation, we randomly shuffled the training samples at the beginning and then use each of them in the shuffled order. After processing all the shuffled training samples, we shuffled the training samples again and use each of them in the shuffled order. Therefore, each training sample was used 10 times.

Table 2. Accuracy on test data obtained with different p. The each item indicates the following: F-measure (Recall, Precision, # of iteration for attaining the best accuracy). The bold font indicates the best and the underlined one indicates the worst F-measure for each L. Each $^+$ or $^-$ indicates better or worse accuracy than the baseline ($p = \infty$), respectively.

BOOK				
$L \setminus p$	∞	1	2	3
1	**90.11** (88.69,91.57, 10)	<u>89.60</u>$^-$ (88.76,90.47, 9)	90.05$^-$ (88.77,91.37, 10)	90.08$^-$ (88.73,91.47, 10)
2	**94.12** (92.77,95.50, 2)	<u>94.09</u>$^-$ (92.82,95.39, 2)	**94.12** (92.77,95.50, 2)	**94.12** (92.77,95.50, 2)
3	**94.14** (92.62,95.71, 2)	<u>94.13</u>$^-$ (92.73,95.58, 2)	**94.14** (92.62,95.71, 2)	**94.14** (92.62,95.71, 2)
DVD				
$L \setminus p$	∞	1	2	3
1	85.15 (82.72,87.73, 5)	<u>84.79</u>$^-$ (82.61,87.09, 3)	**85.18**$^+$ (83.06,87.41, 4)	85.15 (82.72,87.73, 5)
2	<u>88.13</u> (84.61,91.97, 6)	88.29$^+$ (84.78,92.09, 2)	**88.30**$^+$ (84.72,92.20, 2)	88.16$^+$ (84.61,92.02, 6)
3	**88.38** (84.04,93.20, 2)	<u>88.33</u>$^-$ (84.04,93.08, 2)	**88.38** (84.04,93.20, 2)	**88.38** (84.04,93.20, 2)
MUSIC				
$L \setminus p$	∞	1	2	3
1	80.42 (76.95,84.22, 6)	<u>79.66</u>$^-$ (76.60,82.97, 4)	80.42 (76.81,84.39, 7)	**80.48**$^+$ (77.09,84.18, 6)
2	**83.99** (78.54,90.27, 2)	<u>83.72</u>$^-$ (78.54,89.63, 2)	**83.99** (78.54,90.27, 2)	**83.99** (78.54,90.27, 2)
3	<u>83.25</u> (76.26,91.65, 7)	**83.28**$^+$ (76.81,90.95, 5)	83.26$^+$ (76.32,91.58, 7)	<u>83.25</u> (76.26,91.65, 7)

certain times. Therefore, less confident training samples in terms of the hinge loss are also skipped, which may be used as update in further training iteration. As a result, we guess that we can obtain better accuracy by avoiding the updates for fitting to such training samples.

4.3 Memory Usage

Table 3 lists the maximum memory usage of each AROW training. The results of AROW and AROW$_p$ are the column "A" ones because these algorithms allocate required memory from the beginning of each training. In terms of memory usage, our method showed generally lower memory usage than AROW.

There are no big difference of memory consumption between AROW$_{dp}$ obtained with $L = 2$ and AROW with $L = 1$. Some of the reasons would be the following two. The first reason is following nature of the double array trie. The double array trie might generate different trie for the same keys if we insert the keys in different order. As a result, longer size of memory is allocated to

Table 3. Maximum memory usage (GB). Each A indicates original AROW and p indicates the pruning parameter used in AROW$_{dp}$. L indicates the maximum length of n-gram. Each bold font indicates the best memory usage for each L on each data set.

	BOOK					DVD					MUSIC				
$L \setminus p$	A	∞	1	2	3	A	∞	1	2	3	A	∞	1	2	3
1	15.75	15.46	**14.56**	15.41	15.46	**2.10**	2.16	2.16	2.16	2.16	**3.50**	**3.50**	**3.50**	**3.50**	**3.50**
2	19.23	15.98	16.03	**15.95**	15.97	3.11	2.45	**2.44**	2.49	2.53	3.89	**3.50**	**3.50**	**3.50**	**3.50**
3	28.23	19.49	**16.44**	18.63	18.42	5.45	3.22	**3.18**	3.52	3.22	5.73	3.91	**3.90**	4.18	3.91

a trie even if the same keys are inserted. Therefore, even if we use small number of features, the original AROW might use lower memory usage than those of AROW_{dp}. The other is our implementation. Our implementation allocates memory by a fixed size if a trie requires new memory space. Therefore, if the allocated memory is enough, even if the number of keys is different, the memory size of the trie might be the same.

4.4 Training Speed

Table 4 shows training time with each parameter and data set. AROW_{dp}, which uses the dynamic feature expansion and the pruning, shows larger improvement of training speed compared with AROW_p without the dynamic expansion for training longer n-grams. For example, we see that we obtain approximately two to three faster training speed with AROW_{dp} for training with $L = 3$. In addtion, $\text{AROW}_{dp\ =\ 2,3}$ maintain competitive accuracy as shown in Table 2. These results indicate that our method contribute to improved training speed while maintaining accuracy.

Table 4. Training time (hours) for 10 times iteration. The bold font ones indicate faster training time than those of AROW with the same L.

	BOOK							
	AROW_{dp}				AROW_p			
$L \setminus p$	∞	1	2	3	∞	1	2	3
1	0.49	**0.08**	0.13	0.14	0.09	**0.08**	**0.08**	**0.06**
2	1.36	**0.26**	**0.43**	0.50	0.49	**0.38**	**0.43**	**0.40**
3	2.24	**0.68**	**0.92**	**1.15**	2.02	**1.75**	**1.83**	**1.92**
	DVD							
	AROW_{dp}				AROW_p			
$L \setminus p$	∞	1	2	3	∞	1	2	3
1	0.06	0.01	0.02	0.02	0.01	0.01	0.01	0.01
2	0.22	**0.04**	**0.04**	0.08	0.05	0.04	0.05	0.04
3	**0.29**	**0.11**	**0.15**	**0.18**	0.31	**0.30**	**0.30**	**0.30**
	MUSIC							
	AROW_{dp}				AROW_p			
$L \setminus p$	∞	1	2	3	∞	1	2	3
1	0.08	0.01	0.02	0.03	0.01	0.01	0.01	0.01
2	0.27	**0.03**	**0.08**	**0.07**	0.10	0.10	0.10	0.07
3	0.48	**0.12**	**0.15**	**0.17**	0.44	**0.39**	**0.36**	**0.35**

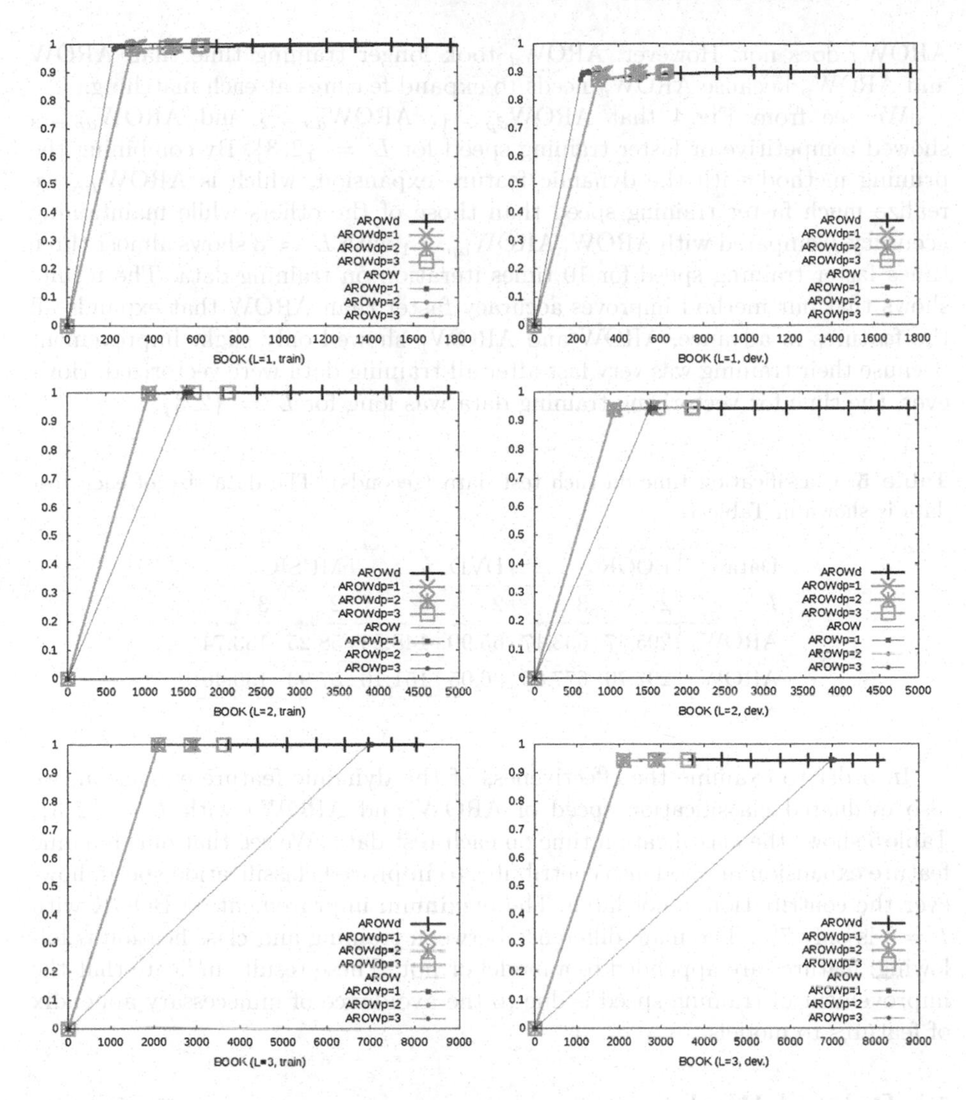

Fig. 4. Transition of accuracy. The x-axis indicates training time and the y-axis indicates accuracy. The left figure is for training and the right one is for development data.

Figure 4 shows accuracy of each training time on BOOK. The processing time includes feature generation and training time. The wide lines indicate transition of accuracy with AROW_d and AROW_{dp}. The others are those of AROW and AROW_p. As for the results with $L = 1$, AROW and AROW_p, which use expanded features in advance, showed faster training speed. However, training of AROW_d with bi-grams and tri-grams was faster than AROW and AROW_p in early stage because AROW and AROW_p expand features in advance in contrast

AROW_d does not. However, AROW_d took longer training time than AROW and AROW_p because AROW_d needs to expand features at each matching.

We see from Fig. 4 that $\text{AROW}_{dp=1}$, $\text{AROW}_{dp=2}$ and $\text{AROW}_{dp=3}$ showed competitive or faster training speed for $L = \{2,3\}$. By combining the pruning method with the dynamic feature expansion, which is AROW_{dp}, we realize much faster training speed than those of the others while maintaining accuracy. Compared with AROW, $\text{AROW}_{dp=1}$ with $L = 3$ shows almost three times faster training speed for 10 times iteration on training data. The results shows that our method improves accuracy faster than AROW that expands all the features in advance. AROW and AROW_p showed only slight improvement because their training was very fast after all training data were vectorized. However, the time for vectorizing training data was long for $L = \{2,3\}$.

Table 5. Classification time on each test data (seconds). The data size of each test data is shown in Table 1.

Data	BOOK		DVD		MUSIC	
L	2	3	2	3	2	3
AROW_d	295.47	633.17	65.90	148.24	58.25	133.74
AROW	297.70	677.06	66.05	151.16	57.93	136.46

In order to examine the effectiveness of the dynamic feature expansion, we also evaluated classification speed of AROW and AROW_d with $L = \{2,3\}$. Table 5 shows the classification time on each test data. We see that our dynamic feature expansion method also contributes to improved classification speed, however the contribution is not large. The maximum improvement on BOOK with $L = 3$ is only 7%. The main difference between training and classification is following; features are appended to a model or not. These results indicate that the improvement of training speed is due to the avoidance of unnecessary appendix of features to models.

5 Related Work

In order to improve training speed of kernel methods, a combination of partially expansion of features and the calculation of the reminders in a kernel method was proposed [18]. For improving classification speed of kernel-based methods, a conversion of a kernel to a linear classifier, which selects features from training samples consisting of kernel-based models [9], was proposed. In order to train models from semi-structured data, a feature selection method [15] was also proposed. In stead of expanding features from semi-structured or structured data in advance other than kernel methods, boosting-based approaches, which select features in batch training, have been proposed [8,10]. In addition, to improve the training speed of the boosting approaches, pruning methods while maintaining the convergence property have also been proposed [17].

Compared with these methods, we expand features dynamically in online learning while maintaining fast training speed. In addition, we can save memory and improve training speed by avoiding generation of unnecessary features.

6 Extension of Our Method

Our method can be applied to tree, graph data, learning combination of features and sequence labeling such as CRF by defining $\phi(\mathbf{X}, L)$ of Fig. 2.

- Tree: For example, by using the emulation method of [19], we can expand sub-trees up to a given size and update model. In addition, as described in [19], each tree can be represented by a string, we can use a trie structure for efficient matching.
- Graph: For example, by using the enumeration method of [16], we can obtain sub-graph features represented by DFS code. When updating a model, enumerating sub-graph up to a given maximum size and update models.
- Sequence Labeling: When training sequence labeling algorithms such as CRF [11] with stochastic gradient decent and structured perceptron [3], we can apply our dynamic feature expansion method.

7 Conclusion

We have proposed a training method with dynamic feature expansion and a pruning. The experimental results on an Amazon data showed that our proposed method improved training speed, memory usage and classification speed when using n-gram features while maintaining competitive accuracy, faster training speed and lower memory usage.

References

1. Aoe, J.: An efficient digital search algorithm by using a double-array structure. IEEE Trans. Softw. Eng. **15**(9), 1066–1077 (1989)
2. Blitzer, J., Dredze, M., Pereira, F.: Biographies, bollywood, boom-boxes and blenders: domain adaptation for sentiment classification. In: ACL, pp. 440–447 (2007)
3. Collins, M.: Discriminative training methods for hidden Markov models: theory and experiments with perceptron algorithms. In: Proceedings of EMNLP 2002, pp. 1–8 (2002)
4. Crammer, K., Dekel, O., Keshet, J., Shalev-Shwartz, S., Singer, Y.: Online passive-aggressive algorithms. J. Mach. Learn. Res. **7**, 551–585 (2006)
5. Crammer, K., Dredze, M., Pereira, F.: Confidence-weighted linear classification for text categorization. J. Mach. Learn. Res. **13**, 1891–1926 (2012)
6. Crammer, K., Kulesza, A., Dredze, M.: Adaptive regularization of weight vectors. Mach. Learn. **91**(2), 155–187 (2013)
7. Hoi, S.C.H., Wang, J., Zhao, P.: Exact soft confidence-weighted learning. In: Proceedings of ICML 2012 (2012)

8. Kudo, T., Maeda, E., Matsumoto, Y.: An application of boosting to graph classification. In: NIPS 2004, pp. 729–736 (2004)
9. Kudo, T., Matsumoto, Y.: Fast methods for kernel-based text analysis. In: Proceedings of the ACL 2003, pp. 24–31 (2003)
10. Kudo, T., Matsumoto, Y.: A boosting algorithm for classification of semi-structured text. In: EMNLP 2004, pp. 301–308 (2004)
11. Lafferty, J.D., McCallum, A., Pereira, F.C.N.: Conditional random fields: probabilistic models for segmenting and labeling sequence data. In: Proceedings of ICML 2001, pp. 282–289 (2001)
12. Leslie, C., Eskin, E., Noble, W.S.: The spectrum kernel: a string kernel for SVM protein classification. In: Proceedings of the 7th Pacific Symposium on Biocomputing, pp. 564–575 (2002)
13. Lodhi, H., Saunders, C., Shawe-Tayor, J., Cristianini, N., Watkins, C.: Text classification using string kernels. J. Mach. Learn. Res. **2**, 419–444 (2002)
14. Rosenblatt, F.: The perceptron: a probabilistic model for information storage and organization in the brain. **65**(6), 386–408 (1958)
15. Suzuki, J., Isozaki, H., Maeda, E.: Convolution kernels with feature selection for natural language processing tasks. In: Proceedings of ACL 2004, pp. 119–126 (2004)
16. Yan, X., Han, J.: gSpan: Graph-based substructure pattern mining (2002)
17. Yoshikawa, H., Iwakura, T.: Fast training of a graph boosting for large-scale text classification. In: Booth, R., Zhang, M.-L. (eds.) PRICAI 2016. LNCS (LNAI), vol. 9810, pp. 638–650. Springer, Cham (2016). doi:10.1007/978-3-319-42911-3_53
18. Yoshinaga, N., Kitsuregawa, M.: Kernel slicing: scalable online training with conjunctive features. In: Proceedings of COLING 2010, pp. 1245–1253 (2010)
19. Zaki, M.: Efficiently mining frequent trees in a forest. In: Proceedings of SIGKDD 2002, pp. 71–80 (2002)

Behavior-Based Location Recommendation on Location-Based Social Networks

Seyyed Mohammadreza Rahimi, Xin Wang(✉), and Behrouz Far

University of Calgary, Calgary, AB, Canada
{smrahimi,xcwang,far}@ucalgary.ca

Abstract. Location recommendation makes suggestions of nearby locations based on user's locational preferences and spatial movement patterns. In this paper, we propose two novel location recommendation methods called Behavior Factorization (BF) and Latent Behavior Analysis (LBA). Both methods utilize behavioral and spatio-temporal patterns in user movements to make location recommendation. Experiments on a real-world dataset shows that the proposed methods outperform existing location recommendation methods in terms of both precision and recall. Comparing LBA and BF, it is observed that LBA achieves better results since it utilizes the number of times each pattern has happened in the dataset.

Keywords: Location-based social networks · Recommendation systems · Spatio-temporal data analysis

1 Introduction

Location recommendation makes suggestions of nearby locations based on user's locational preferences and spatial movement patterns. It has been an active research area recently. Current location recommendation methods mainly focus on the geographical reachability of locations [1–3] and utilize additional information—such as time [2,3], social relationships [1,2] and location category [3]—to further improve the recommendation. However, these methods do not effectively utilize the similar patterns in users' movements in the recommendation process.

Additionally, location recommendation methods are also subject to the sparsity and cold-start problems [4]. The sparsity of data is common in the recommendation systems, it happens since users are bound to their local vicinity and cannot visit locations that are beyond their reachability. Cold start problem happens when a new user or a new location is added to the system. In such situations, due to lack of information, the recommendation system cannot give proper recommendations to the new user or recommend a newly added location to the users.

In this paper, we intend to enhance the current location recommendation methods by better integrating temporal information and utilizing history of

J. Kim et al. (Eds.): PAKDD 2017, Part II, LNAI 10235, pp. 273–285, 2017.
DOI: 10.1007/978-3-319-57529-2_22

similarly behaving users. Temporal features play an important role in location recommendation [5] and people are more likely to conduct the same activity around the similar time of the day [3,6]. We also assume that the category of a location reflects the activities happening in those locations [3]. Thus we can combine temporal and location category information to introduce the concept of behavior. By finding and utilizing the similar behaviors among users enables us to predict unseen behaviors.

In this study, we propose two behavior-based location recommendation methods called Behavior Factorization (BF) and Latent Behavior Analysis (LBA) that predict the category of location the user is likely to visit based on his/her past behavior and the behaviors of similar users, and then recommend matching nearby locations to the user. The problem of sparseness is resolved by aggregating check-ins to a higher level (i.e. the category of locations). It also treats all locations of the same category as similar locations, and finds common behaviors among users to solve the cold start users problem. Specifically, our contributions in this study are summarized as follows:

- Formally defining the term behavior in the context of location recommendation and constructing a user-behavior graph based on the user behaviors. The proposed concept of behavior and user-behavior graph enables the proposed location recommendation methods to effectively reduce the sparseness of the check-in dataset, while preserving the essential information about user preferences that help find similar users and behaviors.
- Proposing two novel location recommendation methods—namely Behavior Factorization (BF) and Latent Behavior Analysis (LBA)—that utilize the user-behavior graph. Not only these methods provide more effective location recommendations, but also they effectively address the problem of cold-start locations and cold-start users. BF and LBA methods address the cold-start location problem by finding the category matching the preference of the user, and recommending the closest location of that category to the user. They also address the problem of the cold-start users by finding behavior patterns and other users that show similar behaviors. The common behaviors of the similar behaving users are recommended to the user. Compared to BF, LBA utilizes the number of times a user has shown a behavior so it achieves better recommendations.
- Conducting experiments on a real-world LBSN check-in dataset. To evaluate the performance of the BF and LBA methods, we use a real world check-in dataset collected from Gowalla[1]. We discover that both BF and LBA methods outperform two existing location recommenders. It is also observed that LBA achieves better performance compared to BF.

The rest of this paper is organized as follows: in Sect. 2, a literature review on location recommendation methods is reported. In Sect. 3, the step-by-step instructions to build BF and LBA recommendation systems are discussed. The experimental configuration and results are discussed in Sect. 4. Finally, the study is concluded in Sect. 5.

[1] A location-based social networking website that in operation until 2012.

2 Related Works

Location recommendation aims to predict locations that users would like to visit. Information such as the local time, location history and preferences of the users can be used. Location recommendation has bee approached from different perspectives, we will discuss three common approaches in the following.

The first approach is to treat locations as items and then apply the existing item recommendation methods to location recommendation, such as [7–9]. However, these methods do not consider the reachability of locations when making recommendations.

The second approach is to model the physical interaction between users and locations. As stated by the Tobler's First Law of Geography —"Everything is related to everything else, but near things are more related than distant things"—the physical interactions between users and location is a differentiating factor of location recommendation from traditional item recommendation systems [10]. Geographical constraints determine the reachability of a location. Recommending a location that is spatially unreachable to a user is a failed recommendation. [1,11,12] include spatial features in their recommendation model and improve over the first group of location recommendation methods by adding the spatial features. However, they do not take into account the temporal factors in the location recomemndation process.

More recent studies utilize temporal information in addition to spatial information. Utilizing temporal information in location recommendation can result in better representation of user behavior since temporal features play an important role in location recommendation [5]. [2,3,13] utilize temporal features in their recommendation models. With further investigation, it can be argued that these models under-utilize the temporal information as they do not consider the shared behaviors of users.

3 Behavior-Based Location Recommendation

Location based social networks collect users' locational information in the form of a check-in. A check-in shows the location of the user at a certain time.

Definition 1. *A* ***check-in*** *is a tuple consisting of a user u, a location l and a time stamp t, it is formally represented as $c = (u, l, t)$, which shows that a user u has visited a location l at time t. Location can in turn be interpreted into latitude and longitude.*

Check-in history of the users is the main source of information for location recommendation. However, to recommend suitable locations to a user at a given time, a recommendation method needs to answer two questions. (1) What activity does the user prefer to do at the moment? The recommendation should be based on the activity the user wants. (2) Where is the user's location? User's location information helps identify locations in the user's vicinity.

As mentioned, knowing the user activity is a key to location recommendation. In this study, we propose to use the check-in history of similarly behaving users to predict the behavior of the target user. To bring this intuition into location recommendation, we formalize a concept called behavior and it is defined in Definition 2.

Definition 2. *A **behavior** is a tuple containing a category and a time interval, denoted as $b = (cat, ti)$, where b is a behavior, cat is a location category and ti is a time interval.*

The definition enables us to find similar and dissimilar user as well as common behaviors among users. If two users visit locations of the same category at similar times, they are considered as users with similar behaviors.

In this section, two novel behavior-based location recommendation methods are proposed that utilizes similar behaviors to make location recommendations to the target user. They predict the likelihood of showing behaviors by the given user. Then they use the predicted behaviors to filter and rank nearby locations and make an effective recommendation. The following subsection will discuss them in detail.

3.1 Behavior Factorization

Behavior Factorization (BF) consists of two main components, (1) a BF model builder and (2) a location recommender. The BF model builder component is responsible for processing check-ins, and making spatial and behavioral models. The spatial model is used to model the reachability of locations to the user and the behavioral model predicts the activity of the users. The location recommender component, on the other hand, uses the behavioral and spatial models along with the user information including the current location, time, and home location to make personalized location recommendations.

BF Model Builder. BF makes location recommendations using the shared behaviors of the users. As shown in Algorithm 1, three steps are taken to build a BF model: First, behaviors are extracted from the check-in history of the users and a user-behavior graph is built. A user-behavior graph is a bipartite graph that represents the relationships between users and behaviors in a more concise way compared to the check-ins. Second, a behavior model is built by feeding the user-behavior graph into a collaborative filtering method, i.e. matrix factorization. Matrix factorization uses the user-behavior graph to make predictions about user behaviors based on his/her past behaviors and similar users' behavior. Finally, a spatial component is built using the check-in data to model the spatial reachability of locations for different users. In the following, we will illustrate each step in detail.

As the first step of the BF model builder, a behavior graph is built using the check-in data. The behavior graph contains the user-behavior information that are presented implicitly in the check-in data.

Algorithm 1. BFModelBuilder(checkins)

1: $behaviorGraph \leftarrow buildBehaviorGraph(checkins)$;
2: $behaviorModel \leftarrow buildBehaviorFactorizationModel(behaviorGraph)$;
3: $spatialModel \leftarrow findSpatialProbabilityDistribution(checkins)$;

Definition 3. *A* ***behavior graph*** *is an undirected bipartite graph, denoted as* $G = (B, U, E)$*, where* B *is the set of behavior nodes and* U *is the set of user nodes.* E *is the set of edges that connect users and his/her corresponding behaviors.*

In BF, the behavior graph is an unweighted graph. To build the graph, check-ins are processed one by one and the corresponding users are created. Using the time of the check-in and the type of the location checked-in, the behavior node and the edge connecting the user and behavior are created.

The second step of the BF model builder is to build the behavior model. The behavioral model is built based on the behavior graph generated in the first step. It is a probabilistic model that quantifies the probability of an edge existing between a user and a behavior in the user-behavior graph. This model is responsible for collaborative features of BF. Behavior model uses the information such as similar behaviors and similar behaving users to predict the behavior of the target user.

In BF, Matrix Factorization (MF) approach is used to build the behavior model [14]. In this study, each user and behavior is modeled as a vector of latent features. User u is represented by $f_u = (f_u^1, f_u^2, \ldots, f_u^k)$ and behavior b is modeled as $q_b = (q_b^1, q_b^2, \ldots, q_b^k)$ where k is the number of latent features. The probability of user u showing behavior b is estimated as the dot product of corresponding latent feature vectors shown in Eq. (1).

$$\hat{p_b}(u, b) = f_u \cdot q_b \tag{1}$$

where $\hat{p_b}(u, b)$ represents the estimated probability of user u showing behavior b.

The objective is to find feature vectors that minimize the regularized estimation error—Eq. (2)—on the observed check-ins.

$$argmin_{p,q} \sum_{e_{u,b}^w \neq nil} (e_{u,b}^w - f_u \cdot q_b)^2 + \lambda(||q_b||_F^2 + ||f_u||_F^2) \tag{2}$$

where $e_{u,b}^w$ represents the weight of the edge connecting user u to behavior b on the behavior graph and λ is a regularization term.

Using Stochastic Gradient Descent to find f and q that satisfy the Eq. (2), we get the following update functions.

$$f_u = f_u + \gamma(e_{u,b} f_u + \lambda q_b) \tag{3}$$

$$q_b = q_b + \gamma(e_{u,b} q_b + \lambda f_u) \tag{4}$$

where $e_{u,b} = e_{u,b}^w - f_u \cdot q_b$ and γ is a parameter used to adjust the magnitude of parameter modification. To learn the latent user and behavior features, we need

---

Algorithm 2. buildBehaviorFactorizationModel(behaviorGraph)

1: For each user u, initialize f_u as a vector of size K with values of 1
2: For each behavior b, initialize q_b as a vector of size K with values of $\frac{1}{K}$
3: **repeat**
4:     **for all** edges of *behaviorGraph* **do**
5:         $e_{u,b} \leftarrow e^w_{u,b} - f_u \cdot q_b$
6:         $f_u \leftarrow f_u + \gamma(e_{u,b} f_u + \lambda q_b)$ ▷ Eq. (3)
7:         $q_b \leftarrow q_b + \gamma(e_{u,b} q_b + \lambda f_u)$ ▷ Eq. (4)
8:     **end for**
9: **until** f and q converge.
10: **return** BFModel(f,q)

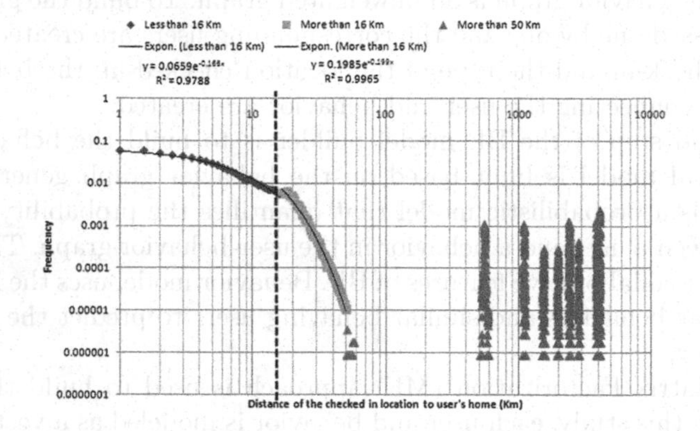

Fig. 1. Spatial probability distribution for Gowalla dataset

to iterate on the user-behavior graph until the latent feature vectors converge. Algorithm 2 shows the training procedure of BF. Having the latent feature vectors f and q, we can estimate the probability of showing any behavior by any user using Eq. (1).

The third step of building the BF model is to model the reachability of recommend locations to the user. The reachability of locations to the user is quantified using a spatial component. To find the spatial probability distribution, the distances of check-ins to the user's home location are calculated and plotted. The user's home location can be derived using the method proposed by Cho *et al.* [2]. As an example, Fig. 1 shows the probability distribution of the distances of checked in locations to each user's home for the Gowalla dataset that was used for the experiments in this paper.

Figure 1 shows that the probability of check-in decreases with increasing the distance of location to user's home and it follows a power law distribution. It also observed that, (1) for distances greater than 50 km, the frequency of check-in varies randomly, this means that for distances greater than 50 Km there is no relationship between the distance and check-in probability. (2) Based on the slope of the linear relationship in for distances less than 50 km, we can further divide this part into two smaller parts, less than 16 km and greater than 16 km.

For this specific example, the probability distribution is shown in Eq. (5) and it can be used as the spatial component for the BF location recommender for this dataset.

$$p_s(u, l|home_u) = \begin{cases} 0.0886e^{-0.166 \cdot dist(l, home_u)} & dist(l, home_u) \leq 16km \\ 0.3122e^{-0.204 \cdot dist(l, home_u)} & 16km < dist(l, home_u) \leq 50km \\ 0 & 50km < dist(l, h) \end{cases} \quad (5)$$

Please note that the spatial component can be different from one dataset to another dataset. Using the similar approach, a spatial probability distribution function can be found for any check-in dataset.

Location Recommender. The location recommender uses the behavior model and the spatial models built by the BF model builder. As mentioned, the query of a location recommender contains a user and a time, and the output is a list of locations suitable for the user for the given time. The behavior-based check-in probability is calculated based on the input. This probability value is used to rank locations for recommendation.

Definition 4. *The behavior-based probability of user u checking in to location l at the given time t is:*

$$p(u, l|t) = \hat{p}(u, b) * p_s(u, l|home_u) \quad (6)$$

where $p(u, l|t)$ is the probability of the user u checking into location l at time t. $\hat{p}(u, b)$ is the behavior probability of the user u showing behavior b, which is made using the category of location l and time interval of t. The value of behavior probability is calculated using Eq. (1). $p_s(u, l|home_u)$ is the spatial probability of user u checking in to location l given the home location of the user and is estimated using Eq. (5).

To make a recommendation to the user, it is needed to calculate the value of $p(u, l|t)$ for the candidate locations and current time. Locations with higher estimated probabilities are then recommended to the user.

3.2 Latent Behavior Analysis (LBA)

Behavior factorization helps find the hidden interactions between users and behaviors and enables us to predict the behavior of the user for the given time. However, it uses the binary—i.e. existing or not existing—user-behavior interactions.

The Latent Behavior Analysis (LBA) method for location recommendation is based on this intuition. In LBA, the spatial model and behavior graph are built in the same was as in BF, but the behavior recommender model (line 2, Algorithm 1) is built in a different way. In the following we will discuss how the behavior recommender model is built in LBA.

In LBA, a set of latent intermediate nodes—denoted by z_i—are introduced to the behavior graph. Collaborative feature of LBA roots in these intermediate

Algorithm 3. buildLatentBehaviorAnalysisModel(behaviorGraph)

1: **while** $p(z_i), p(u|z_i)$ and $p(b|z_i)$ have not converged **do**
2: **for all** edges of *behaviorGraph* **do** ▷ Expectation Step
3: **for** $i = 1$ to K **do**
4: $p(z_i|u,b) \leftarrow \frac{p(z_i)p(u|z_i)p(b|z_i)}{\sum_{j=1}^{K} p(z_j)p(u|z_j)p(b|z_j)}$ ▷ Eq. (8)
5: **end for**
6: **end for**
7: **for all** edges of *behaviorGraph* **do** ▷ Maximization Step
8: **for** $i = 1$ to K **do**
9: $p(u|z_i) \leftarrow \sum_{b \in Behaviors} n(u,b)p(z_i|u,b)$ ▷ Eq. (9)
10: $p(b|z_i) \leftarrow \sum_{u \in Users} n(u,b)p(z_i|u,b)$ ▷ Eq. (10)
11: $p(z_i) \leftarrow \sum_{b \in Behaviors} \sum_{u \in Users} n(u,b)p(z_i|u,b)$ ▷ Eq. (11)
12: **end for**
13: **end for**
14: **end while**
15: **return** LBAModel(f,q)

nodes. Adding the intermediate nodes, the user-behavior graph becomes a tripartite graph where users and behaviors are connected to intermediate nodes and there is not a direct connection between the users and behaviors. The probability of user showing a behavior can then be estimated using Eq. (7) [15].

$$\hat{p_b}(u,b) = \sum_{i=1}^{K} p(z_i)p(u|z_i)p(b|z_i) \tag{7}$$

where u and b are a user and a behavior, respectively. z_i is the i-th intermediate node and K is the number of latent intermediate nodes. To utilize Eq. (7), we first need to learn the values of $p(z_i), p(u|z_i)$ and $p(b|z_i)$.

To learn these values, the Expectation Maximization approach [15] is used. In the expectation step, we find the value of $p(z_i|u,b)$ using Eq. (8).

$$p(z_i|u,b) = \frac{p(z_i)p(u|z_i)p(b|z_i)}{\sum_{j=1}^{K} p(z_j)p(u|z_j)p(b|z_j)} \tag{8}$$

Using $p(z_i|u,b)$ in the maximization step, we can update equations the probability values of $p(z_i), p(u|z_i)$ and $p(b|z_i)$ using the update functions in Eqs. (9), (10) and (11) [15].

$$p(u|z) = \sum_{b \in Behaviors} n(u,b)p(z|u,b) \tag{9}$$

$$p(b|z) = \sum_{u \in Users} n(u,b)p(z|u,b) \tag{10}$$

$$p(z) = \sum_{b \in Behaviors} \sum_{u \in Users} n(u,b)p(z|u,b) \tag{11}$$

where $n(u, b)$ is the number of times user u has shown behavior b. We use expectation and maximization steps until the values of $p(z_i), p(u|z_i)$ and $p(b|z_i)$ converge. Algorithm 3 shows the pseudocode of building the LBA model. Using the learnt probability values and Eq. (7), we can estimate the probability of showing any behavior by any user.

4 Experiments

In this section, the proposed methods are evaluated on a real-world location-based social network dataset. All recommendation methods are implemented in Java and the experiments were conducted on a 2.3 GHz Core i5 Mac computer with 8 GBs of RAM.

The dataset chosen for this study is collected from Gowalla. The details of data crawler can be found in [7]. This dataset contains 5,462 users, 5,999 locations and 104,851 check-ins. In the experiments, whole dataset is first separated into a training dataset and a testing dataset. The testing dataset contains one randomly chosen check-in of each user, the training dataset, on the other hand, contains all the remaining check-ins. To remove the effect of random selection, five different testing and training dataset pairs are generated. All experiments are performed on all five pairs of datasets and the average value is reported. In all cases, the response time for recommendation with LBA, BF, PMM and USG is less than 0.5 s. Since, this run time is within the acceptable range for all the methods, we focus on the recommendation quality for the rest of this section.

4.1 Performance Measures

Precision and recall are chosen as performance measures. Precision measures the quality of recommendation. It measures the ratio of the given recommendations that are correct. Recall, on the other hand, measures the performance of the location recommendation methods in retrieving the correct recommendations by measuring the ratio of the correct answers that are covered in the set of recommendations. Precision and Recall are calculated using Definition 5.

Definition 5. *Given a set of recommended locations, the precision and recall of the recommendation is respectively defined as:*

$$Precision = \frac{|RecommendedLocations \cap CorrectLocations|}{|RecommendedLocations|} \tag{12}$$

$$Recall = \frac{|RecommendedLocations \cap CorrectLocations|}{|CorrectLocations|} \tag{13}$$

4.2 Comparison of the Location Recommenders

In this experiment, we compare the precision and recall of two proposed location recommendation methods together with two of the most well-known location

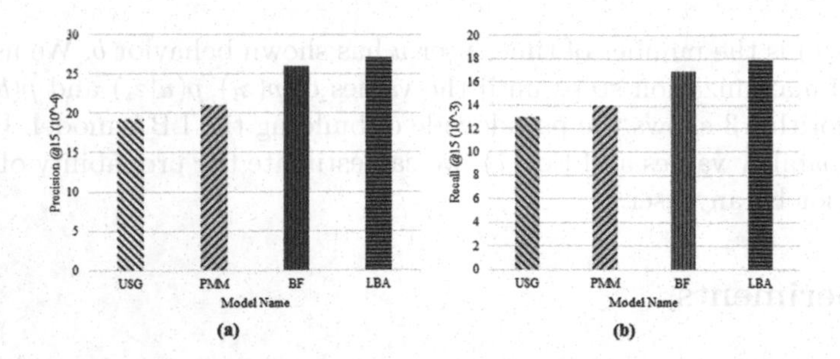

Fig. 2. (a) Precision values of locations recommenders recommending 15 locations. (b) Recall values of locations recommenders recommending 15 locations.

recommendation methods, namely Periodic Mobility Model (PMM) proposed in [2] and User-based CF, Social influence, Geographical influence (USG) model proposed in [1].

Figure 2 depict the precision and recall values for each method for the case of recommending 15 locations. As shown in Fig. 2, the proposed methods, BF and LBA, outperform the two baseline models, i.e. USG and PMM, in terms of precision and recall. Specifically, compared to USG, LBA improves the precision and recall by 41% and 37%, respectively. This proves that integrating temporal influence and utilizing similarly behaving users results in better location recommendations. It is also observed that the LBA outperforms BF by 4% and 5% in terms of precision and recall.

4.3 Performance of the Recommenders on Cold-Start Users

One of the objectives of this study is to provide better recommendations for the cold-start users. In this section, the performance of different location recommendation methods for the cold-start users is compared. To do so, a special testing dataset is generated from Gowalla dataset. The dataset only contains 587 users who have less than five check-ins in the training dataset. The precision of the recommendation is only measured for those users. The precision and recall of location recommendation methods recommending 15 locations is given in Fig. 3.

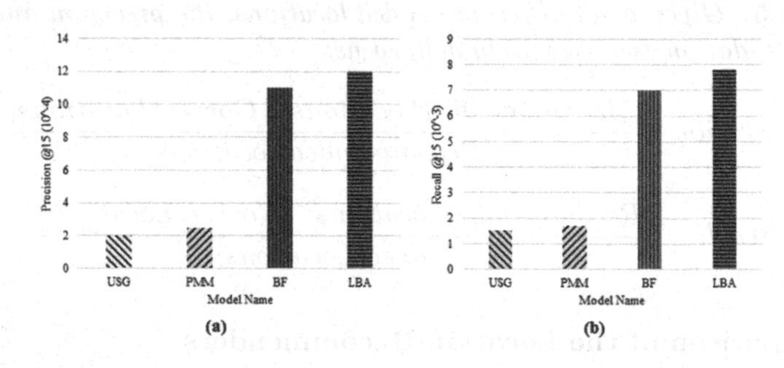

Fig. 3. (a) Precision of location recommenders on cold-start users recommending 15 locations. (b) Recall of location recommenders on cold-start users recommending 15 locations.

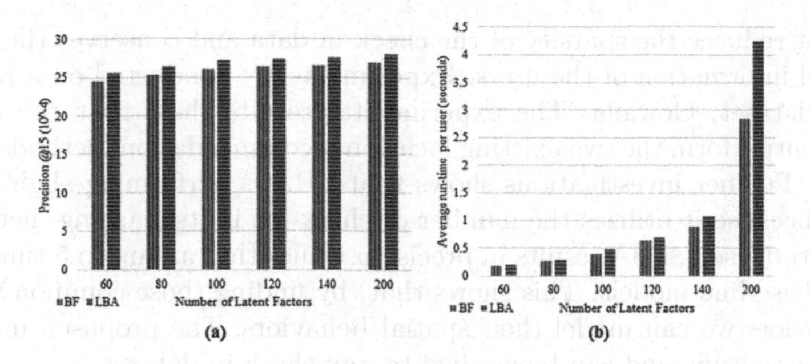

Fig. 4. (a) Precision of BF and LBA location recommendation vs. number of latent factors (b) Average run-time of BF and LBA per user vs. number of latent factors.

It is also observed that BF and LBA models are much better performing compared to two existing models when considering the cold-start users, scoring up to five times the precision. This is because BF and LBA utilize information from similar users that more active. This enables us to predict unseen behaviors of the cold-start users from the similarities they have with active users.

The figure also shows that when cold-start users are considered, the precision of each model is less than the precision of the model when it makes recommendations for all users. The reason is that the recommendation methods do not have enough information about the cold-start users to draw correct conclusions about their preferences.

4.4 Sensitivity of BF and LBA to the Number of Latent Factors

BF and LBA are based on latent factors, and the number of latent factors is passed to those models as a parameter. In this experiment, the effect of number of latent factors on the performance of the two methods is tested. The precision of the two models using 60, 80, 100, 120, 140 and 200 latent factors is shown in Fig. 4(a). As shown in the figure, the precision increases with the increase of the number of latent factors, but the improvement is not significant after 100 number of factors. With the increase of number of latent factors, the training time of the model increases almost exponentially. As shown in Fig. 4(b), for latent factors more than 100, the average run-time per user for both BF and LBA is greater than half a second. Based on Fig. 4(a) and (b), we can argue that the increased performance of increasing the number of factors beyond 100 is not enough to justify the higher running time of training the model, thus we chose 100 as the number of factors for training BF and LBA.

5 Conclusions and Future Works

In this paper, after a review of existing location recommendation methods, two new location recommendation methods are proposed, namely BF and LBA. These two methods utilize the concept of behavior proposed to build a behavior graph that reduces the sparsity of the check-in data and conserves the essential behavioral information of the users. Experiments are conducted on a real-world check-in dataset, Gowalla. The experimental results show that the proposed methods outperform the two existing location recommendation methods of PMM and USG. Further investigations shows that LBA is performing slightly better than BF because it utilizes the number of check-ins in its learning method. For the cold-start users, LBA results in precision values that are up to 5 times higher than the base-line models. This shows that, by finding those common temporal user behaviors we can model their spatial behaviors. The proposed models are not dataset specific and can be applied to any check-in dataset.

Continuing the research, testing the models on different check-in datasets especially those with higher number of check-ins will help in testing the superiority of the proposed models. Additionally, further investigation of temporal features of user check-ins might result in improved models that will lead to generation of better location recommendations. Developing more complicated spatial models that enables the finding of user-based and temporal spatial behaviors is also a promising path to continue this research.

Aknowledgement. The research is supported by the Natural Sciences and Engineering Research Council of Canada Discovery Grant to Xin Wang and National Natural Science Foundation of China (No. 41271387).

References

1. Ye, M., Ying, P., Lee, W.-C., Lee, D.L.: Exploiting geographical influence for collaborative point-of-interest recommendation. In: Proceedings of the ACM International Conference on Research & Development on Information Retrieval, Beijing, China, pp. 325–344 (2011)
2. Cho, E., Myers, S.A., Leskovec, J.: Friendship and mobility: user movement in location-based social networks. In: Proceedings of the 17th ACM SIGKDD Conference on Knowledge Discovery and Data Mining, San Diego, California, USA, pp. 1082–1090 (2011)
3. Rahimi, S.M., Wang, X.: Location recommendation based on periodicity of human activities and location categories. In: Pei, J., Tseng, V.S., Cao, L., Motoda, H., Xu, G. (eds.) PAKDD 2013. LNCS (LNAI), vol. 7819, pp. 377–389. Springer, Heidelberg (2013). doi:10.1007/978-3-642-37456-2_32
4. Bobadilla, J., Ortega, F., Hernando, A., Bernal, J.: A collaborative filtering approach to mitigate the new user cold start problem. Knowl. Based Syst. **26**, 225–238 (2012)
5. Zhao, S., King, I., Lyu, M.: A survey of point-of-interest recommendation in location-based social networks. CoRR. Vol abs/1607.00647 (2016). http://arxiv.org/abs/1607.00647
6. Eagle, N., Alex, S.: Eigenbehaviors: identifying structure in routine. Behav. Ecol. Sociobiol. **63**(7), 1057–1066 (2009)
7. Zhou, D., Wang, B., Rahimi, S.M., Wang, X.: A study of recommending locations on location-based social network by collaborative filtering. In: Kosseim, L., Inkpen, D. (eds.) AI 2012. LNCS (LNAI), vol. 7310, pp. 255–266. Springer, Heidelberg (2012). doi:10.1007/978-3-642-30353-1_22

8. Zhang, J., Chow, C., Li, Y.: iGeoRec: a personalized and efficient geographical location recommendation framework. IEEE Trans. Serv. Comput. **8**(5), 701–714 (2015)
9. Yu, Y., Wang, C., Gao, Y.: Attributes coupling based item enhanced matrix factorization technique for recommender systems. arXiv preprint arXiv:1405.0770 (2014)
10. Yu, Y., Chen, X.: A survey of point-of-interest recommendation in location-based social networks. In: Workshops at AAAI (2015)
11. Lian, D., Zhao, C., Xie, X., Sun, G., Chen, E., Rui, Y.: GeoMF: joint geographical modeling and matrix factorization for point-of-interest recommendation. In: Proceedings of the 20th ACM SIGKDD International Conference on Knowledge Discovery and Data Mining, New York, NY, USA (2014)
12. Yuan, F., Guo, G., Jose, J., Chen, L., Yu, H., Chen, L.: Joint geo-spatial preference and pairwise ranking for point-of-interest recommendation. In: Proceeding of: 28th International Conference on Tools with Artificial Intelligence, San Jose, California, USA (2016)
13. Gao, H., Tang, J., Hu, X., Liu, H.: Exploring temporal effects for location recommendation on location-based social networks. In: Proceedings of the 7th ACM Conference on Recommender Systems - RecSys 2013 (2013)
14. Koren, Y., Bell, R., Volinsky, C.: Matrix factorization techniques for recommender systems. IEEE Comput. **42**(8), 30–37 (2009)
15. Hofmann, T.: Probabilistic latent semantic analysis. In: Proceedings of Uncertainty in Artificial Intelligence, UAI 1999, Stockholm, Sweden (1999)

Integer Linear Programming for Pattern Set Mining; with an Application to Tiling

Abdelkader Ouali[1,2](✉), Albrecht Zimmermann[2], Samir Loudni[2], Yahia Lebbah[1], Bruno Cremilleux[2], Patrice Boizumault[2], and Lakhdar Loukil[1]

[1] Laboratory of LITIO, University of Oran 1, 31000 Oran, Algeria
abdelkader.ouali@unicaen.fr
[2] Normandie Univ, UNICAEN, ENSICAEN, CNRS, GREYC, 14000 Caen, France

Abstract. Pattern set mining is an important part of a number of data mining tasks such as classification, clustering, database tiling, or pattern summarization. Efficiently mining pattern sets is a highly challenging task and most approaches use heuristic strategies. In this paper, we formulate the pattern set mining problem as an optimization task, ensuring that the produced solution is the best one from the entire search space. We propose a method based on integer linear programming (ILP) that is exhaustive, declarative and optimal. ILP solvers can exploit different constraint types to restrict the search space, and can use any pattern set measure (or combination thereof) as an objective function, allowing the user to focus on the optimal result. We illustrate and show the efficiency of our method by applying it to the tiling problem.

1 Introduction

Pattern mining is one of the fundamental tasks in the process of knowledge discovery, and a range of techniques have been developed for producing extensive collections of patterns. However, resulting pattern collections are generally too large, difficult to exploit, and unstructured – without interpretable relations between patterns. This explains the interest of the community to *pattern sets* [18]. Instead of evaluating and selecting patterns individually, pattern sets (i.e., sets of patterns) assemble local patterns to provide knowledge from a high-level viewpoint, using quality measures that evaluate, and constraints that constrain, the entire set. Examples of problems related to pattern sets include concept-learning, database tiling, data compression, or clustering, to cite a few. However, as the number of possible pattern sets is exponential in the size of the set of local patterns, which is itself huge, the computational efficiency of pattern set mining is a very challenging task. For specific quality measures, such as joint entropy, relatively tight upper bounds can be derived to prune candidate sets [13]. Unfortunately, such pruning strategies are limited to very few cases. In practice, most approaches use a step-wise strategy in which first all local patterns are computed, then heuristically post-processed according to an objective function to be optimized. Therefore only a *single* pattern set is returned; examples are [2,4,14,20]. Obviously this process does not ensure the optimality of

J. Kim et al. (Eds.): PAKDD 2017, Part II, LNAI 10235, pp. 286–299, 2017.
DOI: 10.1007/978-3-319-57529-2_23

the returned pattern set according to the objective function. To the best of our knowledge, only the algorithm proposed in [18] proceeds by exhaustive search while pruning parts of the search space by using pattern set constraints. However, these pruning effects can be weak, and the number of patterns being very large, this method only works for small pattern collections.

In this paper, we formulate the pattern set mining problem as an optimization task, ensuring that the produced solution is the best from the entire search space. In a sense, we return to the spirit of the original idea of pattern set mining [18] based on a complete method. However, we produce only the best solution, avoiding being drowned by patterns. What's more, we use constraint programming (CP) techniques since CP solvers can exploit a wider range of constraints than data mining approaches that are typically locked into a rather rigid search strategy. Modeling constraints independently from the search strategy also allows them to accomodate a variety of constraints, and therefore adapt the resulting pattern sets to the need of the user.

The key contribution of this paper is a method based on *integer linear programming* (ILP) that is (1) *exhaustive*, avoiding the loss of interesting solutions, (2) *declarative*, allowing us to make the most of provided constraints instead of being tied to a particular search strategy, (3) *optimal*, always returning the best solution according to an optimization criterion that satisfies the given constraints. Any measure that can be used as a constraint in an ILP model can also be chosen as an objective function to be optimized. This allows the user to prioritize particular aspects of the solution. Measures to be optimized can be combined, as long as they can be expressed as a linear term. This is once again an advantage over traditional mining, where a change would typically require the explicit redefinition of the search strategy. Our approach allows us to provide the first *practically useful* algorithm for addressing this problem setting. As an illustration, we experimentally address the tiling problem but our approach is broad enough to cover and leverage many pattern mining problems such as clustering [1,5], classification [14], or pattern summarization [21].

The rest of the paper is structured as follows. Section 2 recalls preliminaries. Section 3 describes our approach. Section 4 introduces several complex queries, and Sect. 6 shows the results of solving queries for the k-tiling problem on different data sets. We discuss related work in Sect. 5.

2 Preliminaries

2.1 Local Patterns

Let $\mathcal{I}$ be a set of n distinct items, an itemset (or *pattern*) is a non-null subset of $\mathcal{I}$. The language of itemsets $\mathcal{L}$ corresponds to $2^{\mathcal{I}} \setminus \emptyset$. A *transactional dataset* $\mathcal{D}$ is a multi-set of m itemsets of $\mathcal{L}$, with each $t_i, 1 \leq i \leq m$ called a *transaction*. We assume the database $\mathcal{D}$ is represented as a binary matrix of size $n \times m$ with $\mathcal{D}_{ti} = 1 \leftrightarrow (t, i) \in \mathcal{D}$.

Let $\mathcal{D}$ a database, $\phi \in \mathcal{L}$ be a pattern, and $match : \mathcal{L} \times \mathcal{L} \mapsto \{true, false\}$ a matching operator. The cover of ϕ w.r.t $\mathcal{D}$, denoted by $cov(\phi, \mathcal{D})$, is the set of transactions in $\mathcal{D}$ that ϕ matches: $cov(\phi, \mathcal{D}) = \{t \in \mathcal{D} \mid match(\phi, t) = true\}$. The support of ϕ is the size of its cover: $sup(\phi, \mathcal{D}) = |cov(\phi, \mathcal{D})|$. The tile of a pattern ϕ contains all tuples that are covered by the pattern: $tile(\phi, \mathcal{D}) = \{(t, i) \mid t \in cover(\phi, D), i \in \phi\}$. These tuples form a tile or rectangle of 1's in the database D. The area of $tile(\phi, \mathcal{D})$ is equal to its cardinality: $area(\phi, \mathcal{D}) = |tile(\phi, \mathcal{D})| = |\phi| \cdot sup(\phi, \mathcal{D})$.

A pattern ϕ is said to be *more general than* a pattern ψ ($\phi \preceq \psi$) (resp., ψ is *more specific* than ϕ) iff $\forall t \in \mathcal{L} : match(\psi, t) \Rightarrow match(\phi, t)$, i.e. if ψ matches any transaction t then ϕ matches it as well. A pattern ϕ is strictly more general than a pattern ψ ($\phi \prec \psi$), if $\phi \preceq \psi$ and $\neg(\psi \preceq \phi)$.

The local pattern mining problem consists of finding a theory $Th(\mathcal{L}, \mathcal{D}, q) = \{\phi \in \mathcal{L} \mid q(\phi, \mathcal{D}) \textit{ is true}\}$, where $q(\phi, \mathcal{D})$ a selection predicate that states the constraints under which the pattern ϕ is a solution w.r.t. the database $\mathcal{D}$. A common example is the minimum support constraint $sup(\phi, \mathcal{D}) \geq \theta$, which is satisfied by all patterns ϕ whose support in the database D exceeds a given minimal threshold θ. Combined with $\mathcal{L}$, this gives rise to the *frequent itemset mining* problem. An exact condensed representation of the frequent itemsets consists of the closed patterns [17]. A closed pattern is one whose specializations have a smaller cover than the pattern itself: $closed(\phi) \Leftrightarrow \forall \psi, \phi \preceq \psi : cov(\psi) \subset cov(\phi)$.

2.2 Pattern Set Mining

Pattern sets are simply sets of patterns. The task of pattern set mining entails discovering a set of patterns that satisfies a set of constraints involving not only individual patterns, as in the local pattern mining setting, but the whole set of patterns. Hereafter, we will denote by $\mathbf{L}$ the set of all the possible pattern sets that can be enumerated given a language $\mathcal{L}$, i.e., $\mathbf{L} = 2^{\mathcal{L}}$. The individual/local patterns occurring in a pattern set will be denoted using lower case characters such as $\phi, \ldots, \psi$ and for patterns sets, we will employ upper case characters such as $\Phi, \ldots, \Psi$.

More formally, the problem of pattern set mining can be formulated as the problem of computing the theory $\mathbf{Th}(\mathbf{L}, \mathcal{D}, p) = \{\Phi \in \mathbf{L} \mid p(\Phi, \mathcal{D}) \textit{ is true}\}$, where $p(\Phi, \mathcal{D})$ a selection predicate that states the constraints under which the pattern set Φ is a solution w.r.t. the database $\mathcal{D}$. In addition, as the number of pattern sets can become very large, we will study how to find the best pattern set with respect to an optimisation criterion $f(\Phi)$, i.e. $\text{argmax}_{\Phi \in \mathbf{Th}(\mathbf{L}, \mathcal{D}, p)} f(\Phi)$. In classical pattern set mining, this is achieved by dynamically increasing the threshold of the constraint involving f. When using the ILP framework (see Sect. 3), this can be achieved by using f as an objective function to guide the search.

2.3 Categories of Constraints

This section discusses several categories of constraints that can be specified at the level of the pattern set as a whole.

Coverage Constraints deal with defining and measuring how well a pattern set covers the data. Let $\Phi \in \mathbf{L}$ be a pattern set and $\mathcal{D}$ a database,

- *Pattern set cover.* The cover of Φ, denoted as $cov(\Phi, \mathcal{D})$, is the set of transactions in $\mathcal{D}$ that Φ covers: $cov(\Phi, \mathcal{D}) = \bigcup_{\phi \in \Phi} cov(\phi, \mathcal{D})$. With this definition, Φ is interpreted as the disjunction of the individual patterns ϕ it contains.
- *Support of pattern set.* The support of Φ, denoted as $sup(\Phi, \mathcal{D})$, is calculated in the same way as for individual patterns: $sup(\Phi, \mathcal{D}) = |cov(\Phi, \mathcal{D})|$.
- *Size of pattern set.* The size of Φ, denoted as $size(\Phi)$, is the number of patterns that Φ contains: $size(\Phi) = |\Phi|$.
- *Area of pattern set.* The area of a pattern set was studied in the context of large tile mining [6]. The area of Φ, denoted as $area(\Phi, \mathcal{D})$, is defined as the area of all the tiles of the individual patterns ϕ it contains: $area(\Phi, \mathcal{D}) = |\bigcup_{\phi \in \Phi} tile(\phi, \mathcal{D})|$.
- *Generality of pattern set.* A pattern set Φ is more general than a pattern set Ψ, denoted as $\Phi \preceq \Psi$, iff for all pattern $\psi \in \Psi$, there exists a pattern $\phi \in \Phi$ s.t. $\phi \preceq \psi$.

Discriminative Constraints. Given a database $\mathcal{D}$ organized into possibly overlapping subsets $\mathcal{D}_1, \ldots, \mathcal{D}_n \subseteq \mathcal{D}$, the discriminative constraints can be used to measure and optimize how well a pattern set discriminates between examples of subsets $\mathcal{D}_i$. Discriminative measures are typically defined by comparing the number of examples covered by the pattern set for a subset $\mathcal{D}_i$, to the total number of examples covered in $\mathcal{D}$.

- *Representativeness of a pattern set.* Representativeness indicates how characteristic the examples covered by the pattern set are for a subset $\mathcal{D}_i$. $rep(\Phi, \mathcal{D}_i, \mathcal{D}) = sup(\Phi, \mathcal{D}_i)/sup(\Phi, \mathcal{D})$.
- *Accuracy of a pattern set.* Let a dataset D partitioned into subsets $\mathcal{D}_1, \ldots, \mathcal{D}_n$, where each subset $\mathcal{D}_i$ contains transactions from class i. The accuracy of a pattern ϕ is defined as $acc(\phi) = \frac{\max_{\mathcal{D}_i \in \mathcal{D}} sup(\phi, \mathcal{D}_i)}{sup(\phi, \mathcal{D})}$. The accuracy of an entire pattern set is harder to quantify. We can however *approximate* it as $acc(\Phi) = \frac{\sum_{\phi \in \Phi} acc(\phi) \times sup(\phi, \mathcal{D})}{\sum_{\phi \in \Phi} sup(\phi, \mathcal{D})}$

Redundancy Constraints can be used to constrain or minimize the redundancy between different patterns. One way to measure this redundancy is to count the number of transactions covered by multiple patterns in the pattern set Φ:

$red(\Phi, \mathcal{D}) = |\{t \in \mathcal{D} \mid \exists(\phi, \psi) \in \Phi, t \in ovlp(\phi, \psi, \mathcal{D})\}|$, where $ovlp(\phi, \psi, \mathcal{D}) = cov(\phi, \mathcal{D}) \cap cov(\psi, \mathcal{D})$ denotes the overlap between the two patterns ϕ, ψ [18].

3 Mining Pattern Sets Using ILP

Throughout the remainder of this paper we employ the Integer Linear Programming (ILP) framework for representing and solving pattern set mining problems. ILP [15] is one of the most widely used methods for handling optimization problems, due to its rigorousness, flexibility and extensive modeling capability.

This framework has been shown (1) to allow for the use of a wide range of constraints, (2) to offer a higher level of problem formalization and modeling, and (3) to work for conceptual clustering [16]. Moreover, modern ILP solvers are very efficient with improved search heuristics.

3.1 Resolution Approach

Finding a good pattern set is often a hard task; many pattern set mining tasks and their optimization versions, such as the k-tiling [6] or the concept learning [11], are NP-hard. Hence, there are no straightforward algorithms for solving such tasks in general, giving rise to a wide ranges of approaches (Two-step vs one-step) and search strategies (Exact vs heuristic). In this paper, we adopt a **two-step** approach:

(i) A *local mining step* mines the set of local patterns $Th(\mathcal{L}, \mathcal{D}, q)$ that satisfy a set of constraints.
(ii) An *ILP mining step* post-processes these patterns with the **ILP exact** solving technique to obtain the best pattern set in $\mathbf{Th}(\mathbf{L}, \mathcal{D}, p)$ under the given constraints.

Our motivation for adopting a two-step approach is two-fold: First, there exist efficient miners [22] to find local patterns. Second, in the second step, the formulated ILP model (see Sect. 3.2) is very close to the well known partitioning (and covering) problems which are extensively studied within the integer programming community [10]. Modern ILP solvers, such as Cplex [8], are efficient on such problems. Our ILP model is detailed in the next section.

3.2 ILP Models

This section presents the ILP model of a pattern set, and the ILP formulations for constraints presented in Sect. 2.3.

Modeling a Pattern Set. Let $\mathcal{P}$ be the set of ℓ patterns. Hence, the pattern set mining problem can be modeled as an ILP using ℓ boolean variables $x_p (p \in \mathcal{P})$, where $(x_p = 1)$ iff the pattern p belongs to the unknown pattern set Φ that we are looking for.

Coverage Constraints. Let $\mathcal{D}$ be a database with m transactions. We introduce m boolean variables y_t, $(t \in \mathcal{D})$ such that $(y_t = 1)$ iff there exists at least one pattern $\phi \in \Phi$ such that pattern ϕ matches t. So, we have $sup(\Phi, \mathcal{D}) = \sum_{t \in \mathcal{D}} y_t$. With some given threshold θ, the coverage constraint $(C_{cov}^{\theta, \lessgtr})$ is defined on such boolean variables. Table 1 shows the formulation of the coverage constraints. Constraints $C_{x,y}$ establish the relationship between variables x and y, and state that each transaction t must belong to at most $|\mathcal{P}|$ patterns. Note that $(y_t = 0)$ iff there exists no pattern $\phi \in \Phi$ such that the pattern ϕ matches t.

Let $(a_{t,p})$ be an $m \times \ell$ binary matrix where $(a_{t,p} = 1)$ iff $match(p, t) = true$, i.e., the pattern p matches the transaction t. For the area constraint, we need to compute the number of ones in the binary matrix that are covered by the set of patterns. We can model this by introducing a temporary variable $q_{t,i}$ for every tuple (t, i), such that $(q_{t,i} = 1)$ iff there exists at least one tile (pattern) $\phi \in \Phi$ such that $t \in cover(\phi, D)$ and $i \in \phi$. Let $cq^p_{t,i}$ be a binary matrix associated to each pattern p where $(cq^p_{t,i} = 1)$ iff p covers both transaction t and item i. Constraints $(C_{x,q})$ establish the relationship between variables q and x.

Let Φ_0 a given pattern set. We model the generality constraint $(C^{\Phi_0}_{gen})$ as follows: $(\Phi \preceq \Phi_0)$ iff for any pattern $\phi \in \Phi_0$, there exists (at least) one pattern $p \in \Phi$ s.t. $p \preceq \phi$, i.e. $x_p = 1$. For the specialisation constraint $(C^{\Phi_0}_{spe})$, $(\Phi_0 \preceq \Phi)$ iff for any pattern $p \in \Phi$, there exists (at least) one pattern $\phi \in \Phi_0$ s.t. $\phi \preceq p$. "For any pattern $p \in \Phi$" is modeled by stating that the number of patterns $p \in \Phi$ verifying the property must be greater or equal to $size(\Phi) = \sum_{p \in \mathcal{P}} x_p$.

Handling Aggregates. Table 1 shows how constraints involving aggregates (e.g. *sum*, *avg*, *min*, *max*) can be modeled using ILP. For example, the constraint

Table 1. ILP formulations of constraints discussed in Sect. 2.3.

Constraint name	Notation	ILP formulation
$(C^{\theta,\lesseqgtr}_{size})$	$size(\Phi) \lesseqgtr \theta$	$\sum_{p \in \mathcal{P}} x_p \lesseqgtr \theta$
$(C^{\theta,\lesseqgtr}_{cov})$ $(C_{x,y})$	$sup(\Phi, \mathcal{D}) \lesseqgtr \theta$	$\sum_{t \in \mathcal{D}} y_t \lesseqgtr \theta$ $y_t \leq \sum_{p \in \mathcal{P}} a_{t,p} \cdot x_p \leq \lvert\mathcal{P}\rvert \cdot y_t, \forall t \in \mathcal{D}$
$(C^{\theta,\lesseqgtr}_{area})$ $(C_{x,q})$	$area(\Phi, \mathcal{D}) \lesseqgtr \theta$	$\sum_{i \in \mathcal{I}, t \in \mathcal{D}} q_{t,i} \lesseqgtr \theta$ $q_{t,i} \leq \sum_{p \in \mathcal{P}} cq^p_{t,i} x_p \leq \lvert\mathcal{P}\rvert . q_{t,i}, \forall t \in \mathcal{D}, \forall i \in \mathcal{I}$
$(C^{\Phi_0}_{gen})$	$\Phi \preceq \Phi_0$	$\sum_{\{p:\ \forall p \in \mathcal{P}\ \mid\ p \preceq \phi\}} x_p \geq 1, \quad \forall \phi \in \Phi_0$
$(C^{\Phi_0}_{spe})$	$\Phi_0 \preceq \Phi$	$\sum_{\{p:\ \forall p \in \mathcal{P}\ \mid\ \exists \phi \in \Phi_0,\ \phi \preceq p\}} x_p \geq \sum_{p \in \mathcal{P}} x_p$
$(C^{\theta,\lesseqgtr}_{redd})$ $(C_{x,u})$	$red(\Phi, \mathcal{D}) \lesseqgtr \theta$	$\sum_{t \in \mathcal{D}} u_t \lesseqgtr \theta$ $2u_t \leq \sum_{p \in \mathcal{P}} a_{t,p} x_p \leq y_t + \lvert\mathcal{P}\rvert . u_t, \forall t \in \mathcal{D}$
$(C^{\theta,\lesseqgtr,\mathcal{D}_i}_{rep})$ $(C_{x,y'})$	$rep(\Phi, \mathcal{D}_i, \mathcal{D}) \lesseqgtr \theta$	$\sum_{t \in \mathcal{D}_i} y'_t \lesseqgtr \theta \times \sum_{t \in \mathcal{D}} y_t$ $y'_t \leq \sum_{p \in \mathcal{P}} a_{t,p} \cdot x_p \leq \lvert\mathcal{P}\rvert \cdot y'_t, \quad \forall t \in \mathcal{D}_i$
$(C^{\theta,\lesseqgtr}_{avg})$	$avg(sup(\Phi, \mathcal{D})) \lesseqgtr \theta$	$\sum_{p \in \mathcal{P}} f_p \cdot x_p - \theta \sum_{p \in \mathcal{P}} x_p \lesseqgtr 0$
$(C^{\theta,\lesseqgtr}_{sum})$	$sum(sup(\Phi, \mathcal{D})) \lesseqgtr \theta$	$\sum_{p \in \mathcal{P}} f_p \cdot x_p \lesseqgtr 0$

($C_{sum}^{\theta,\lesseqgtr}$) expresses that the sum of the supports over all patterns in Φ should be $\lesseqgtr$ than θ. It can be modeled using a linear constraint, where f_p is the support value of a local pattern p. Similarly, we can constraint the average taken over *all* patterns in Φ ($C_{avg}^{\theta,\lesseqgtr}$).

Redundancy Constraints. To deal with redundancy, we need to know transactions that are multiply covered. Thus, we introduce boolean variables (u_t), $(t \in \mathcal{D})$ s.t. $(u_t = 1)$ iff transaction t is matched by at least 2 patterns. The total number of such transactions is $\sum_{t \in \mathcal{D}} u_t$. Table 1 gives the modelisation of the redundancy constraint ($C_{redd}^{\theta,\lesseqgtr}$), while constraints ($C_{x,u}$) establish the relationship between intermediate variables (u and y) and decision variables x.

Our definition of redundancy is similar to that proposed in [19] yet different from the (pairwise) redundancy proposed in [18]. The latter was adopted mainly due to its effectiveness for pruning in a level-wise mining algorithm. The differences between the two formalizations of redundancy are briefly sketched here.

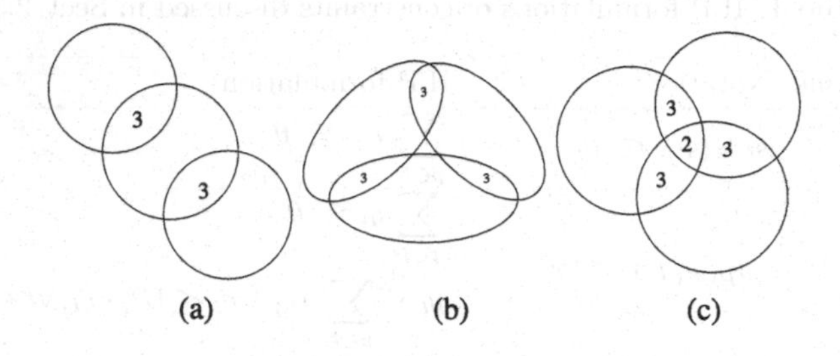

Fig. 1. Three patterns overlapping

Consider the three cases of overlapping patterns shown in Fig. 1. The numbers in overlapping areas denote the number of transactions in the overlap. All three cases would be considered the same by a constraint measuring the maximal pairwise overlap between patterns, such as used in [18]. The global redundancy measure we employ evaluates to 6 for case (a), 9 for case (b), and 11 for case (c), capturing the actual situation much better. Notably, summing pairwise redundancies, another option proposed in that paper, will lead to a result of 15 for (c), overstating the redundancy. For this reason, we claim that our modelisation is more appropriate than the one chosen in [18].

Representativeness Constraints. Let $\mathcal{D}_i \subseteq \mathcal{D}$ be a partial data set. We introduce $|\mathcal{D}_i|$ extra boolean variables $y'_t, (t \in \mathcal{D}_i)$ such that $(y'_t = 1)$ iff there exists at least one pattern $\phi \in \Phi$ such that pattern ϕ matches t. The modelisation of the representativeness constraint ($C_{rep}^{\theta,\lesseqgtr,\mathcal{D}_i}$) is shown in Table 1. Constraints ($C_{x,y'}$) establish the relationship between variables (y'_t) and (x_p).

4 Queries and How to Model Them

This section provides three examples of complex queries and shows how to model them as a combination of constraints presented in Sect. 2.3.

As a first query (Q_1), we show the quintessential pattern set mining task: given the result of a local pattern mining operation, we aim to find a (relatively) small subset of patterns that is representative of the entire result. To this end, we want to select patterns that have very little overlap and together cover as much of the data as possible. There are two ways of modeling this query as an ILP problem:

$$\begin{cases} \texttt{Maximize } \sum_{t\in\mathcal{D}} y_t \\ (C^{\theta,\leq}_{redd}) \\ (C_{x,y,u}) \\ x_p \in \{0,1\}, \forall p \in \mathcal{P} \\ y_t \in \{0,1\}, u_t \in \{0,1\}, \forall t \in \mathcal{D} \end{cases} \quad (1) \qquad \begin{cases} \texttt{Minimize } \sum_{t\in\mathcal{D}} u_t \\ (C^{\theta,\geq}_{cov}) \\ (C_{x,y,u}) \\ x_p \in \{0,1\}, \forall p \in \mathcal{P} \\ y_t \in \{0,1\}, u_t \in \{0,1\}, \forall t \in \mathcal{D} \end{cases} \quad (2)$$

(1) *maximize support* ($z = \sum_{t\in\mathcal{D}} y_t$) subject to maximum redundancy constraints;
(2) *minimize redundancy* ($z = \sum_{t\in\mathcal{D}} u_t$) subject to minimum support constraints.

Constraints ($C_{x,y}$) and ($C_{x,u}$) governing the new variables u and y can be merged and will be denoted as the following linear constraints ($C_{x,y,u}$): $y_t + u_t \leq \sum_{p\in\mathcal{P}} a_{t,p}x_p \leq y_t + |\mathcal{P}|\,.\,u_t, y_t \geq u_t, \forall t \in \mathcal{D}$

Our second query (Q_2) is a refinement of the first one, by imposing a generality constraint. For instance, we aim to summarize a set of subgraphs mined from molecular data, so that a non-data miner, e.g. a chemist, has only a small set of fragments to evaluate. In that case, the practitioner might already have an idea what fragments she would like to see, and wants to see the rest fleshed out. This can be achieved by requiring that a pattern set include a particular pattern (or syntactically related patterns), i.e. that is *more general* than another one. In the ILP case as well, this just requires a generality constraint to be added.

Our last query (Q_3) concerns the *k-tiling*. The task consists of finding k tiles maximizing the area ($z = \sum_{t\in\mathcal{D},i\in\mathcal{I}} q_{t,i}$). Equation (3) depicts our first ILP model `M1`. The number of tiles k can also be defined as a variable whose value will be determined by the ILP solver. This can be done by specifying a lower bound k_{min} and/or an upper bound k_{max} on the value of k. Note that tiles can be overlapping, but every tuple (t, i) covered is only counted once. This encoding requires $(n \times m + 2)$ constraints and $((n \times m) + \ell + 1)$ variables. This constitutes a major limitation when it comes to handling very large databases. Thus, we propose a second ILP model `M2` that approximates the k-tiling by summing the areas v_p ($p \in \mathcal{P}$) of the individual tiles it contains. In this case, each tuple (t, i) covered may be counted more than once.

$$\begin{cases} \texttt{Maximize } \sum_{t \in \mathcal{D}, i \in \mathcal{I}} q_{t,i} \\ (C_{x,q}) \\ k = \sum_{p \in \mathcal{P}} x_p \\ k_{min} \leq k \leq k_{max} \\ k \in \mathbb{N}, \\ x_p \in \{0,1\}, \\ \quad p \in \mathcal{P} \\ q_{i,t} \in \{0,1\}, \\ \quad i \in \mathcal{I} \wedge t \in \mathcal{D} \end{cases} \tag{3}$$

$$\begin{cases} \texttt{Maximize } \sum_{p \in \mathcal{P}} v_p \cdot x_p \\ (1)\ y_t \leq \sum_{p \in \mathcal{P}} a_{t,p} \cdot x_p \leq \delta_o \cdot y_t, & \forall t \in \mathcal{D} \\ (2)\ \sum_{t \in \mathcal{D}} y_t \geq \theta_t \\ (3)\ z_i \leq \sum_{p \in \mathcal{P}} w_{i,p} \cdot x_p \leq \gamma_o \cdot z_i, & \forall i \in \mathcal{I} \\ (4)\ \sum_{i \in \mathcal{I}} z_i \geq \theta_i \\ k = \sum_{p \in \mathcal{P}} x_p \\ k_{min} \leq k \leq k_{max} \\ k \in \mathbb{N}, \\ x_c \in \{0,1\}, & c \in \mathcal{P} \\ y_t \in \{0,1\}, & t \in \mathcal{D} \\ z_i \in \{0,1\}, & i \in \mathcal{I} \end{cases} \tag{4}$$

Equation (4) depicts our second ILP model M2. It consists of finding a set of tiles covering both the set of transactions and the set of items, with small overlaps on transactions and on items. In this way, we allow to control the redundancy on tuples (t, i) that are multiply counted in the tiling. Constraint (2) states that at least θ_t transactions must be covered, while Constraint (1) states that each transaction t cannot occur in more than δ_o closed patterns. Let $w_{i,p}$ be an $n \times \ell$ binary matrix where $(w_{i,p} = 1)$ iff the item i belongs to the tile or pattern p. Constraint (4) states that at least θ_i items must be covered, while Constraint (3) states that each item i cannot occur in more than γ_o tiles. In the experimental section, due to the space limitation, we focus on questions related to the k-tiling problem.

5 Related Work

Pattern set mining is an important part of a number of data mining tasks such as classification [14], clustering [1,5], pattern summarization [21], or database tiling [6]. Due to the highly combinatorial nature of the problem, most methods proceed in two phases. First, an exhaustive algorithm generates the whole collection of local patterns satisfying given constraints and the second phase produces pattern sets by selecting smaller subsets of relevant patterns from the whole collection of local patterns, often by using heuristics to manage computational complexity. Unlike these works, our method does not use heuristics to provide a pattern set but relies on a solid formalization in the ILP paradigm, while the search is guided by the optimization of an objective function.

There are very few attempts on searching pattern sets according to complete approaches. The original idea of pattern set mining was proposed in [18]. The authors formally introduce a variety of constraints at the level of the pattern sets. Unfortunately, the pruning techniques are weak and the algorithm remains limited to small collections of patterns. Specific settings have been proposed by [13] (fixing the size of pattern sets and relying on the anti-monotonicity of a particular quality function) or [3] (using a dedicated global constraint on the attributes). The approach in [6] is heuristic but does *not* use post-processing, instead iteratively mining tiles, taking already found ones into account.

The constraint programming framework was investigated to accomplish the pattern set mining task by modeling pattern sets with constraints [7,12]. However, these methods require to fix the number of local patterns included in a pattern set, a strong limitation in practice, and tend to have scaling problems. Recent contributions also employ more specialized systems such as satisfiability solvers [9] and integer linear programming techniques [1,16]. These methods address particular problem settings whereas we propose a declarative and exhaustive method based on ILP returning the best solution according to an optimization criterion, and which is able to handle a wide variety of constraints and therefore various different pattern set mining tasks.

6 Experiments

For a better understanding of the suitability of the ILP approach, we focus our experiments on one prototypical task for pattern set mining: the k-tiling problem (NP-Hard). The experimental evaluation is designed to address the following questions:

1. How do the running times of our approach (`ILP`) compare to the only existing CP approach, proposed by Guns et al. (`KPatternSet`), for the k-tiling problem?
2. Given the space requirements of the ILP model `M1`, how do the obtained k-tilings compare qualitatively with those resulting from the (approximating) ILP model `M2`?
3. In light of the exact nature of our approach, how do the resulting k-tilings compare qualitatively with those resulting from (`k-LTM`) [6].[1]
4. How do the k-tilings with overlapping tiles compare qualitatively with those having non-overlapping tiles?

Experimental Protocol. Experiments were carried out on the same datasets which were used in [7] and available from the UCI repository. Table 2 shows the characteristics of these datasets. All experiments were conducted on AMD Opteron 6174 with 2.60 GHz of CPU and 256 GB of RAM.[2] We used closed patterns to represent tiles since they cover a larger area than their non-closed counterparts. We used `LCM` to extract the set of all closed patterns and `CPLEX` v.12.4 to solve the different ILP models. For all methods, a time limit of 24 h has been used. As `M2` requires setting the parameters for the coverage and non-overlap relations, we propose the following three settings :

- `M2a` with settings allowing similar amounts of coverage and overlap as `k-LTM`;
- `M2b` with settings allowing the coverage of all transactions ($\theta_i = m$) with the maximum overlap ($\delta_0 = |\mathcal{P}|$) and the coverage of at least one item ($\theta_i = 1$) without any overlap for items ($\gamma_0 = 1$);
- `M2c` with settings allowing the coverage of all transactions ($\theta_i = m$) without any overlap ($\delta_0 = 1$) and the coverage of at least one item ($\theta_i = 1$) with the maximum overlap for items ($\gamma_0 = |\mathcal{P}|$).

Table 2. Comparing the different approaches. (TO: TimeOut; - : no solution ; (1): ILP-M1; (2): ILP-M2a; (3): ILP-M2b; (4): KPatternSet; (5): k-LTM).

$\mathcal{D}$	Items	Trans.	k	Recall					Time (s)				
				(1)	(2)	(3)	(4)	(5)	(1)	(2)	(3)	(4)	(5)
Zoo-1	34	101	3	0.48	0.38	0.38	0.48	0.46	8.99	5.87	1.57	2,418	
			4	0.57	0.47	0.48	TO	0.56	14.38	4.53	3.67	TO	
			5	0.63	0.54	0.52	TO	0.62	25.11	4.49	2.99	TO	
			6	0.68	0.55	0.55	TO	0.67	34.96	10.82	2.98	TO	
			7	0.73	0.57	0.58	TO	0.71	22.85	5.81	3.49	TO	
			8	0.78	0.68	0.61	TO	0.75	25.56	6.96	3.16	TO	
			9	0.82	0.69	0.64	TO	0.79	35.07	8.04	3.59	TO	
			10	0.85	0.69	0.66	TO	0.82	26.21	6.91	3.23	TO	0.99
Lymph	66	148	3	0.43	0.42	0.35	TO	0.42	299.91	50.92	152.54	TO	
			4	0.48	0.38	0.39	TO	0.47	954.74	143.84	152.13	TO	
			5	0.52	0.44	0.42	TO	0.51	1,629	97.88	152.72	TO	
			6	0.55	0.37	0.45	TO	0.54	4,261	175.01	152.52	TO	
			7	0.58	0.40	0.46	TO	0.57	6,027	155.19	153.15	TO	
			8	0.61	0.40	0.48	TO	0.59	5,197	47.68	146.13	TO	
			9	0.63	0.40	0.49	TO	0.61	7,313	45.51	148.06	TO	
			10	0.65	0.44	0.50	TO	0.63	28,779	43.56	148.51	TO	1321
Primary-tumor	29	336	3	0.46	0.40	0.36	0.46	0.46	405.91	33.25	82.26	41,496	
			4	0.54	0.48	0.47	TO	0.53	680.64	74.99	88.60	TO	
			5	0.59	0.47	0.55	TO	0.57	3,807	43.93	68.65	TO	
			6	0.64	0.44	0.60	TO	0.61	3,863	92.28	58.76	TO	
			7	0.68	0.46	0.64	TO	0.65	6,407	154.82	61.58	TO	
			8	0.71	0.50	0.67	TO	0.68	86,429	41.75	53.93	TO	
			9	0.74	0.54	0.70	TO	0.71	86,425	65.50	65.13	TO	
			10	0.77	0.52	0.73	TO	0.73	86,423	43.70	58.23	TO	4.9
Soybean	47	630	3	0.44	0.42	0.33	0.44	0.43	127.94	15.31	8.35	32,485	
			4	0.50	0.47	0.37	TO	0.49	147.11	24.38	6.96	TO	
			5	0.55	0.49	0.41	TO	0.53	141.70	22.10	6.43	TO	
			6	0.59	0.47	0.43	TO	0.58	144.27	19.25	6.08	TO	
			7	0.62	0.47	0.45	TO	0.60	157.41	14.11	5.90	TO	
			8	0.65	0.45	0.47	TO	0.62	358.05	16.30	9.65	TO	
			9	0.67	0.46	0.48	TO	0.64	342.79	14.65	8.83	TO	
			10	0.69	0.48	0.49	TO	0.66	734.20	9.42	6.28	TO	7
Tic-Tac-Toe	24	958	3	0.15	0.15	0.11	TO	0.15	24.49	8.93	5.98	TO	
			4	0.20	0.20	0.16	TO	0.20	23.67	6.11	5.05	TO	
			5	0.25	0.25	0.21	TO	0.25	25.81	5.43	5.82	TO	
			6	0.29	0.29	0.27	TO	0.29	24.45	5.92	4.91	TO	
			7	0.33	0.33	0.33	TO	0.33	43.17	5.53	2.70	TO	
			8	0.38	0.38	0.38	TO	0.38	44.76	5.44	3.82	TO	
			9	0.42	0.42	0.42	TO	0.42	42.76	4.87	3.50	TO	
			10	0.46	0.46	0.46	TO	0.46	41.98	5.16	3.35	TO	0.1
Vote	45	435	3	0.32	0.30	0.06	TO	0.31	3,020	81.75	34.36	TO	
			4	0.37	0.35	0.19	TO	0.36	1,947	84.99	46.09	TO	
			5	0.42	0.38	0.33	TO	0.40	4,269	732.10	49.42	TO	
			6	0.46	0.39	0.40	TO	0.44	3,651	83.48	43.23	TO	
			7	0.50	0.38	0.45	TO	0.47	13,831	282.56	39.13	TO	
			8	0.53	0.37	0.49	TO	0.51	8,671	819.33	33.03	TO	
			9	0.56	0.42	0.53	TO	0.54	15,152	881.76	36.10	TO	
			10	0.60	0.43	0.57	TO	0.57	13,974	701.37	30.53	TO	3.3
Mushroom	112	8124	3	TO	0.34	-	TO	0.34	TO	431.82	-	TO	
			4	TO	0.40	-	TO	0.40	TO	667.61	-	TO	
			5	TO	0.46	-	TO	0.46	TO	780.05	-	TO	
			6	TO	0.48	-	TO	0.50	TO	572.54	-	TO	
			7	TO	0.49	-	TO	0.53	TO	502.91	-	TO	
			8	TO	0.44	-	TO	0.56	TO	1,302	-	TO	
			9	TO	0.48	-	TO	0.58	TO	763.20	-	TO	
			10	TO	0.51	-	TO	0.60	TO	479.45	-	TO	666.4
hepatitis	66	137	3	TO	0.27	0.25	TO	0.28	TO	1,261	604.16	TO	
			4	TO	0.33	0.33	TO	0.34	TO	910.38	1,007.34	TO	
			5	TO	0.37	0.38	TO	0.39	TO	1,310	725.38	TO	
			6	TO	0.42	0.42	TO	0.44	TO	887.81	781.88	TO	
			7	TO	0.43	0.46	TO	0.48	TO	1,327	616.90	TO	
			8	TO	0.45	0.49	TO	0.52	TO	2,372	736.19	TO	
			9	TO	0.45	0.52	TO	0.55	TO	3,046	625.16	TO	
			10	TO	0.45	0.54	TO	0.57	TO	16,295	656.40	TO	6103

To assess the quality of a k-tiling Φ, we define the *recall* of Φ, measured by the fraction of all ones in the binary matrix $\mathcal{D}$ belonging to $area(\Phi, \mathcal{D})$, which should be as large as possible: $recall(\Phi, \mathcal{D}) = \frac{\sum_{(t,i) \in area(\Phi,\mathcal{D})} \mathcal{D}_{ti}}{\sum_t \sum_i \mathcal{D}_{ti}}$.

(a) **Comparing ILP-M1 with KPatternSet.** Table 2 compares the performance compares of ILP-M1 and KPatternSet(in terms of CPU-times) for various values of k on different datasets. We also report the corresponding value of recall. The CPU-times of ILP-M1 include those for the preprocessing step. ILP-M1 clearly outperforms KPatternSet on all datasets: KPatternSetgoes over the timeout for $k \geq 4$. For the value of k for which an optimal k-tiling can be found, ILP-M1 is up to several orders of magnitude faster than KPatternSet.

(b) **Comparing M1 with M2a and M2b.** ILP-M1 finds the optimal solution on most of the datasets, but ILP-M2a remains relatively close in terms of recall, particularly for ($k \leq 5$). In addition, ILP-M2a is much faster, particularly on Lymph and Vote (speed-up of up to 660). The main limitation of M1 remains its space requirement. For the three most difficult datasets – Mushroom, Hepatitis and Anneal – ILP-M1 fails to find a solution. Comparing ILP-M2a with ILP-M2b, the latter shows clearly higher recall for Hepatitis, Primary

[1] We use the implementation available at https://people.mmci.uni-saarland.de/~jilles/prj/tiling/.

[2] The k-LTM implementation is Windows-only, and run times therefore only roughly indicate its behavior.

Table 3. Qualitative comparison. ((3): ILP-M2b; (3'): ILP-M2c; (5): k-LTM).

$\mathcal{D}$	k	Recall		Redundant Trans.			Redundant Items			Coverage Items		
		(3)	(3')	(3)	(3')	(5)	(3)	(3')	(5)	(3)	(3')	(5)
Mushroom	3	-	0.23	-	0.00	0.55	-	0.50	0.31	-	0.08	0.16
	4	-	0.31	-	0.00	0.64	-	0.32	0.28	-	0.16	0.18
	5	-	0.36	-	0.00	0.81	-	0.42	0.29	-	0.22	0.24
	6	-	0.42	-	0.00	0.91	-	0.44	0.41	-	0.21	0.25
	7	-	0.46	-	0.00	0.92	-	0.59	0.55	-	0.24	0.25
	8	-	0.51	-	0.00	0.93	-	0.66	0.53	-	0.24	0.26
	9	-	0.54	-	0.00	0.93	-	0.69	0.51	-	0.24	0.27
	10	-	0.56	-	0.00	0.93	-	0.53	0.50	-	0.32	0.28
Soybean	3	0.33	0.28	0.85	0.00	0.59	0.00	0.30	0.41	0.16	0.20	0.25
	4	0.37	0.34	0.90	0.00	0.64	0.00	0.54	0.37	0.20	0.26	0.34
	5	0.41	0.40	0.93	0.00	0.74	0.00	0.56	0.47	0.22	0.36	0.36
	6	0.43	0.44	0.96	0.00	0.85	0.00	0.57	0.44	0.24	0.46	0.38
	7	0.45	0.48	0.98	0.00	0.85	0.00	0.52	0.52	0.26	0.42	0.40
	8	0.47	0.52	1.00	0.00	0.90	0.00	0.50	0.57	0.26	0.52	0.40
	9	0.48	0.54	1.00	0.00	0.93	0.00	0.62	0.55	0.28	0.52	0.42
	10	0.49	0.66	1.00	0.00	0.95	0.00	0.69	0.52	0.28	0.52	0.44
Primary-tumor	3	0.36	0.15	0.87	0.00	0.60	0.00	0.33	0.25	0.23	0.19	0.41
	4	0.47	0.22	0.90	0.00	0.71	0.00	0.38	0.38	0.35	0.26	0.44
	5	0.55	0.28	0.94	0.00	0.82	0.00	0.33	0.42	0.42	0.39	0.48
	6	0.60	0.33	0.96	0.00	0.84	0.00	0.50	0.53	0.48	0.39	0.51
	7	0.64	0.38	0.98	0.00	0.89	0.00	0.44	0.62	0.52	0.52	0.55
	8	0.67	0.42	1.00	0.00	0.91	0.00	0.42	0.75	0.52	0.61	0.55
	9	0.70	0.45	1.00	0.00	0.92	0.00	0.45	0.87	0.52	0.65	0.55
	10	0.73	0.48	1.00	0.00	0.95	0.00	0.45	0.77	0.58	0.71	0.62
Anneal	3	-	0.36	-	0.00	TO	-	0.47	TO	-	0.34	TO
	4	-	0.46	-	0.00	TO	-	0.64	TO	-	0.35	TO
	5	-	0.52	-	0.00	TO	-	0.74	TO	-	0.37	TO
	6	-	0.57	-	0.00	TO	-	0.67	TO	-	0.46	TO
	7	-	0.59	-	0.00	TO	-	0.84	TO	-	0.47	TO
	8	-	0.62	-	0.00	TO	-	0.71	TO	-	0.56	TO
	9	-	0.64	-	0.00	TO	-	0.78	TO	-	0.59	TO
	10	-	0.65	-	0.00	TO	-	0.82	TO	-	0.59	TO

Tumor and Vote. This is because ILP-M2b allows overlapping tiles with high redundancy (see Table 3). In addition ILP-M2b is faster than ILP-M2a on 51 instances (out of 64) with a speed-up between 1 and 4 for 27, and between 7 to 25 for 9 instances. Note that on Mushroom dataset, no k-tiling exists with M2b. These results show that ILP-M2a and ILP-M2b achieve good recall compared to ILP-M1 with less space requirement, hence the interestingness of using ILP-M2b as a faster alternative for approximating the optimal k-tiling.

(c) **Comparing ILP with k-LTM.** k-LTM differs from our approach in three points: (1) using a heuristic is faster but may lead to suboptimal solutions, (2) mining iteratively, k-LTM can take information about already found tiles into account, and (3) a k-LTM $k+1$-tiling will always be a superset of a k-tiling – ILP can find different solutions. As Table 2 shows, ILP-M1 always achieves better recall than k-LTM, yet requires more time to find the optimum, (k-LTM running times are shown in the last column). For the most difficult dataset Anneal, neither method find a solution. Comparing ILP-M2b with k-LTM, k-LTM has a slight advantage (three data sets). While complete search beats iterative mining, it *does* help gaining an advantage over heuristic post-processing.

(d) **Comparing M2c with M2b and k-LTM.** In our last experiment, we mine k tiles without any overlap (i.e. M2c) and compare them to those resulting from M2b and k-LTM. Table 3 shows four distinct cases of recall and redundancy for the three approaches. Col. 4 reports the redundancy of the k-tiling measured by $red(\Phi, \mathcal{D})/sup(\Phi, \mathcal{D})$. Col. 5 (resp 6) denotes the percentage of redundant (resp. covered) items. Generally, ILP-M2b and k-LTM achieve higher recall than ILP-M2c. This is not surprising as the tilings found by ILP-M2b consist of large, transaction-overlapping tiles, contrary to those of ILP-M2c (see Col. 4). ILP-M2c makes up for this by covering more *items* than the other approaches, and mining item-overlapping tiles. This tuning of output characteristics is a strength of the declarative approach.

7 Conclusion

Pattern set mining has become an indispensable data mining task to control the overly large result sets of local pattern mining operations. In this work, we have for the first time presented a practically useful approach that retains the richness of the constraint language of the original pattern set mining approach [18]. Our method is declarative, based on the techniques developed in ILP, allowing to choose particular (combinations of) pattern set measures as objective functions to be optimized. This permits the user to prioritize particular aspects of a pattern set, while constraining others. Existing ILP solver guarantee to return the *best* pattern set, according to the given optimization criterion, that satisfies a user-defined set of constraints. Experiments have illustrated and shown the efficiency of our approach through the example of the tiling problem but our approach is broad enough to cover and leverage many pattern mining problems such as clustering, classification, or pattern summarization. The flexibility of our approach is clearly a major step towards developing the interactive data mining systems that are requested in data science. Having defined this framework, further work will consist of properly specifying the models corresponding to different data mining tasks. It will be necessary to formulate new constraints and models, and fine-tune them to achieve best results. We also plan to exploit column generation techniques to enhance the scalability of our solving step.

References

1. Babaki, B., Guns, T., Nijssen, S.: Constrained clustering using column generation. In: Simonis, H. (ed.) CPAIOR 2014. LNCS, vol. 8451, pp. 438–454. Springer, Cham (2014). doi:10.1007/978-3-319-07046-9_31
2. Bringmann, B., Zimmermann, A.: One in a million: picking the right patterns. Knowl. Inf. Syst. **18**(1), 61–81 (2009)
3. Cagliero, L., Chiusano, S., Garza, P., Bruno, G.: Pattern set mining with schema-based constraint. Knowl.-Based Syst. **84**, 224–238 (2015)
4. Cheng, H., Yan, X., Han, J., Hsu, C.-W.: Discriminative frequent pattern analysis for effective classification. In: ICDE 2007, Istanbul, Turkey, April 15, pp. 716–725 (2007)
5. Ester, M., Kriegel, H.-P., Sander, J., Xu, X.: A density-based algorithm for discovering clusters in large spatial databases with noise. In: KDD 1996, Portland, pp. 226–231 (1996)
6. Geerts, F., Goethals, B., Mielikäinen, T.: Tiling databases. In: Suzuki, E., Arikawa, S. (eds.) DS 2004. LNCS (LNAI), vol. 3245, pp. 278–289. Springer, Heidelberg (2004). doi:10.1007/978-3-540-30214-8_22
7. Guns, T., Nijssen, S., De Raedt, L.: k-pattern set mining under constraints. IEEE Trans. Knowl. Data Eng. **25**(2), 402–418 (2013)
8. IBM/ILOG, Inc. ILOG CPLEX: High-performance software for mathematical programming and optimization (2016)
9. Jabbour, S., Sais, L., Salhi, Y.: The top-k frequent closed itemset mining using top-k SAT problem. In: Blockeel, H., Kersting, K., Nijssen, S., Železný, F. (eds.) ECML PKDD 2013. LNCS (LNAI), vol. 8190, pp. 403–418. Springer, Heidelberg (2013). doi:10.1007/978-3-642-40994-3_26

10. Jünger, M., Liebling, T.M., Naddef, D., Nemhauser, G.L., Pulleyblank, W.R., Reinelt, G., Rinaldi, G., Wolsey, L.A. (eds.): 50 Years of Integer Programming 1958–2008 - From the Early Years to the State-of-the-Art. Springer, Heidelberg (2010)
11. Kearns, M.J., Vazirani, U.V.: An Introduction to Computational Learning Theory. MIT Press, Cambridge (1994)
12. Khiari, M., Boizumault, P., Crémilleux, B.: Constraint programming for mining n-ary patterns. In: Cohen, D. (ed.) CP 2010. LNCS, vol. 6308, pp. 552–567. Springer, Heidelberg (2010). doi:10.1007/978-3-642-15396-9_44
13. Knobbe, A.J., Ho, E.K.Y.: Maximally informative k-itemsets and their efficient discovery. In: ACM SIGKDD 2006, Philadelphia, PA, USA, pp. 237–244 (2006)
14. Liu, B., Hsu, W., Ma, Y.: Integrating classification and association rules mining. In: Proceedings of Fourth International Conference on Knowledge Discovery & Data Mining (KDD 1998), pp. 80–86, New York. AAAI Press (1998)
15. Andrzej, J.O.: Integer and combinatorial optimization. Int. J. Adapt. Control Signal Process. **4**(4), 333–334 (1990)
16. Ouali, A., Loudni, S., Lebbah, Y., Boizumault, P., Zimmermann, A., Loukil, L.: Efficiently finding conceptual clustering models with integer linear programming. In: IJCAI 2016, New York, NY, USA, 9–15 July 2016, pp. 647–654 (2016)
17. Pasquier, N., Bastide, Y., Taouil, R., Lakhal, L.: Discovering frequent closed itemsets for association rules. In: Beeri, C., Buneman, P. (eds.) ICDT 1999. LNCS, vol. 1540, pp. 398–416. Springer, Heidelberg (1999). doi:10.1007/3-540-49257-7_25
18. De Raedt, L., Zimmermann, A.: Constraint-based pattern set mining. In: SIAM 2007, 26–28 April 2007, Minneapolis, Minnesota, USA, pp. 237–248 (2007)
19. Shima, Y., Hirata, K., Harao, M.: Extraction of frequent few-overlapped monotone DNF formulas with depth-first pruning. In: Ho, T.B., Cheung, D., Liu, H. (eds.) PAKDD 2005. LNCS (LNAI), vol. 3518, pp. 50–60. Springer, Heidelberg (2005). doi:10.1007/11430919_8
20. Vreeken, J., van Leeuwen, M., Siebes, A.: KRIMP: mining itemsets that compress. Data Min. Knowl. Discov. **23**(1), 169–214 (2011)
21. Xin, D., Cheng, H., Yan, X., Han, J.: Extracting redundancy-aware top-k patterns. In: ACM SIGKDD 2006, Philadelphia, PA, USA, 20–23 August 2006, pp. 444–453 (2006)
22. Xindong, W., Vipin, K.: The Top Ten Algorithms in Data Mining, vol. 1. Chapman & Hall/CRC, Boca Raton (2009)

Secured Privacy Preserving Data Aggregation with Semi-honest Servers

Zhigang Lu[1] and Hong Shen[1,2](✉)

[1] School of Computer Science, The University of Adelaide, Adelaide, Australia
{zhigang.lu,hong.shen}@adelaide.edu.au
[2] School of Data and Computer Science, Sun Yat-Sen University, Guangzhou, China

Abstract. With the large deployment of smart devices, the collections and analysis of user data significantly benefit both industry and people's daily life. However, it has showed a serious risk to people's privacy in the process of the above applications. Recently, combining multiparty computation and differential privacy was a popular strategy to guarantee both computational security and output privacy in distributed data aggregation. To decrease the communication cost in traditional multiparty computation paradigm, the existing work introduces several trusted servers to undertake the main computing tasks. But we will lose the guarantee on both security and privacy when the trusted servers are vulnerable to adversaries. To address the privacy disclosure problem caused by the vulnerable servers, we provide a two-layer randomisation privacy preserved data aggregation framework with semi-honest servers (we only take their computation ability but do not trust them). Differing from the existing approach introduces differential privacy noises globally, our framework randomly adds random noises but maintains the same differential privacy guarantee. Theoretical and experimental analysis show that to achieve same security and privacy insurance, our framework provides better data utility than the existing approach.

Keywords: Differential privacy · Secured Multiparty Computation · Data aggregation

1 Introduction

The application and development of Internet of Thing (IoT), such as Radio Frequency Identification (RFID) and wireless sensor networks, have a greatly positive impact on the way we live in the recent decades [6]. For example, the modern recommender systems provide us the accurate recommendations by collecting and analysing our profiles, shopping histories, visited locations, and other users who have similar preference with us.

In many real-life applications of IoT, the source data are distributed in different entities. To learn some important insights (such as statistics, patterns/relationships, global/local optimised solutions) from the distributed data, we hope to create a joint computation over the distributed source data from

J. Kim et al. (Eds.): PAKDD 2017, Part II, LNAI 10235, pp. 300–312, 2017.
DOI: 10.1007/978-3-319-57529-2_24

each entity. Such joint computation brings two main privacy challenges: computational security, output privacy [12]. Specifically, the computational security is to guarantee that every computational entity only learns the knowledge which can be learned from the computation outputs [8]; while, the output privacy is to decrease the probability of inferring user's privacy from the computation outputs.

To ensure computational security, the techniques to achieve Secure Multi-party Computation (SMC or MPC) [12] paradigm are the most famous ones. Generally, SMC allows multiple parties jointly compute a function over each party's private data without a trusted third party, while each party is oblivious to other parties' private inputs. After thirty years development, there are a number of famous techniques to ensure the requirements of SMC, such as Secret sharing, Oblivious Transfer (OT), Zero-knowledge proof, and Homomorphic encryption [2]. However, SMC is vulnerable to inference on explicit output. Because the final output of SMC has no difference with the corresponding non privacy preserving computation, adversaries can still infer people's privacy based on the explicit output and their auxiliary information.

To achieve the output privacy, some popular techniques (data perturbation) in the research field include data randomisation [3] and data anonymisation [16], which perturb the original data to satisfy a required privacy preservation assurance. Among all the techniques of data randomisation and anonymisation, *Differential privacy* [3] is the most powerful one because of its strongest assumption on adversary's background knowledge and its ingenious settings of the privacy budget ϵ for the trade-off between data utility and privacy. However, differential privacy cannot provide any computational security assurance in a distributed environment.

Since both SMC and differential privacy cannot address the above two privacy challenges independently, a straitforward way will be combining SMC and differential privacy. Currently, a lot of researchers have applied the combination of SMC and differential privacy scheme into privacy preserving data aggregation [1,4–6,12,13,15]. Particularly, in [5,12], the schemes worked on a full connected network, which will result in a very high communication cost. While, in [1,5,6,12,13,15], the schemes assumed the computation entities are semi-honest, which will disclose a specific user(entity)'s privacy when there is a collusion of the other $n-1$ entities (malicious adversaries). [4] addressed the problems of malicious adversaries and high communication cost by introducing several trusted computation servers which undertake the main computation task. However, once the trusted servers are vulnerable to adversaries, user's privacy will be in a clear risk of privacy leaking.

Contributions. In this paper, basically, we enhance the classic techniques of SMC paradigm with differential privacy, which provides the privacy preserving data analysis on private data from distributed data curators against malicious adversary with semi-honest (rather than honest in the existing work) servers in lower communication complexity. The main contributions are listed below:

- We provide a new privacy preserving data aggregation scheme which keeps the advantages of the existing work (secure, privacy preservation, and low communication cost) but addresses the privacy disclosure problem caused by the trusted computation server, and improves the data utility under the same privacy and security guarantee.
- We provide a theoretical evaluation to analyse the performance of privacy, and utility on our privacy preserving data aggregation.
- We provide an experimental evaluation to analyse the performance of data utility across various experimental settings.

Organisation. The rest of this paper is organised as follows: In Sect. 2, we discuss the existing hybrid privacy preserving methods (combination of SMC and differential privacy) on data aggregation for both advantages and disadvantages. Next, in Sect. 3, we will give a brief introduction on the preliminaries of this paper. Then we propose our novel privacy preserving data aggregation scheme which addresses the privacy disclosure problem in existing work in Sect. 4. Afterwards, in Sect. 5, we provide the experimental evaluation on the existing work and our framework. Finally, in Sect. 6, we conclude this paper.

2 Literature Review

In this section, we briefly summarise both the advantages and disadvantages of the current work on privacy preserving data aggregation by combining SMC and differential privacy. Basically, the SMC paradigm is mainly applied in the joint computation when sharing private data with other data curators, which guarantees the security of computation. While, differential privacy is usually used for preserve the privacy leaking from the computation output.

Rastogi et al. [13] proposed the first schemes to ensure a private data aggregation, with the combination of SMC and differential privacy, named PASTE. Particularly, PASTE follows a simple idea: locally, each party adds differential privacy noise, then encrypts the privacy preserved data; the trusted aggregator sums up all parties' encrypted data then transfers it back to each party; the parties decrypt the message and add differential noise again, then send the updated value to aggregator; finally, the aggregator calculates the mean of the values from the parties as the final output. This method guaranteed the computational security and output privacy together against semi-honest adversary. Unfortunately, a collusion of $n-1$ parties will easily infer the victim's privacy.

Ács et al. [1] proposed an addition-based encryption scheme. By applying Diffie-Hellman key exchange protocol, two entities will jointly generate the encryption keys together. The tricky in this scheme is that the summation of the keys in each pair is 0, which guarantees the low noise added into the original data. Every entity keeps r encryption keys which are paired with the keys in other r entities. As a result for summation function, all encryption keys are summed up, they cancel out and no decryption is necessary [7]. This method successfully decreased the total noise added into the overall summation; however, since the noise added to each party's private input is not enough, if $n-1$

parties work together (minus their sum from the overall output), they can figure out the victim's private information with high probability.

Shi et al. [15] proposed a scheme which requests a set of computation entities to send encrypted and noise added private data to a semi-honest data aggregator periodically. The data aggregator in this scheme can only learn the answer to a query but nothing else because of the security guarantee by homomorphic encryption in each period. This method is similar to the method in [13] but without the last two steps (so more efficient than [13]). However, [15] achieved a $(\sum_i \epsilon_i, \delta)$-Differential Privacy with a global sensitivity, which means a poor privacy guarantee because of the large value of the overall privacy budget and the use of global sensitivity.

Goryczka et al. [6] provided an enhanced scheme to [15], where a fault tolerance algorithm was added. In [6], if all neighbours of a party i faulted, then sending the recovery key to the aggregator would reveal the contributed value x_i. Therefore, the party subtracts x_i from the recovery key, which will remove it from the aggregated result as well. That is, in an extreme case, this method will not guarantee any utility of the final aggregation output.

Eigner et al. [4] proposed a method which works with trusted computation servers, called PrivaDA, which inspired our work. Different from [15], to enhance the data utility while preventing any inference from the final output, PrivaDA introduced more servers which generated the noise for each participants. Between the computation servers and the participants, Shamir secret sharing scheme wass applied. However, since the participants send encrypted original data to the computation servers, once these trusted servers are hacked, we will lose all the privacy guarantee.

Elahi et al. [5] applied secret sharing scheme to guarantee the privacy of each party's original data among several servers. Then the privacy preserved data will be aggregated at a trusted server by simple summation. This method works in a full graph network, then there will be a $\mathcal{O}(n^2)$ communication complexity which is worse than all of the above schemes.

Pettai et al. [12] provided a new method where random sampling was introduced. This method firstly applied secret sharing scheme to share partial inputs with all involved parties. Then each party returns its local output with a fixed threshold to avoid leaking the real output. Finally, all parties aggregate the output for the final result. This method is stronger in privacy preserving; however, the network model is a full connected network, that is, the communication complexity will be $\mathcal{O}(n^2)$ when sharing any messages in this network.

3 Preliminaries

The notion of privacy is important to a privacy preserving algorithm, as it offers the standard to evaluate whether a(n) algorithm/scheme/framework is privacy preserved. In this section, we will introduce two notions of privacy mentioned in Sect. 1 briefly: Secure Multiparty Computation and Differential Privacy.

Secure Multiparty Computation. Secure Multiparty Computation (SMC or MPC) is also known as Secure Function Evaluation (SFE), which is firstly and formally introduced by Andrew Yao [2]. Originally, SMC concerns the question: Can multiple parties jointly compute a function over their own private data while ensuring no party learns others' private data after the computation? More generally, SMC requires that a secure computation only allows to learn knowledge from the computation output. There are several classic techniques to achieve SMC: secret sharing, homomorphic encryption, oblivious transfer, and zero-knowledge proof [2]. In this paper, we mainly use secret sharing as the SMC technique to ensure the computational security.

Shamir [14] firstly invented a secret sharing scheme, that is, a secret curator splits his/her secret value x into n parts: $x_1, x_2, \ldots, x_n$, then shares x_i with participants i. The secret can be recovered only when k out of n x_is are collected together, where k is a sufficiently large number to n. Therefore, the secret sharing guarantees that if the n computation entities do not cooperate to cheat the system, the secret value x will be secure in the computation process.

Differential Privacy. Differential privacy is one of the most popular notion of privacy in current research field of data randomisation [8,9], which was firstly introduced and defined by Dwork et al. [3]. Informally, differential privacy is a scheme that minimises the sensitivity of output for a given statistical operation on two neighbouring (differentiated in one record to protect) datasets. Specifically, differential privacy guarantees the presence or absence of any record in a database will be concealed to the adversary.

In differential privacy, the basic setting is a pair of neighbouring dataset X and X', where X' contains the information of all the entries except one record in a database X. A formal definition of Differential Privacy is shown as follow:

Definition 1 (ϵ-Differential Privacy [3]**).** *A randomised mechanism $\mathcal{T}$ is ϵ-differential privacy if for all neighbouring datasets X and X', and for all outcome sets $S \subseteq Range(\mathcal{T})$, $\mathcal{T}$ satisfies:*

$$\Pr[\mathcal{T}(X) \in S] \leq exp(\epsilon) \cdot \Pr[\mathcal{T}(X') \in S]$$

where ϵ is the privacy budget.

The privacy budget ϵ is set by the database curator. Theoretically, a smaller ϵ denotes a higher privacy guarantee because the privacy budget ϵ reflects the magnitude of difference between two neighbouring datasets.

There are two main applications of the randomised mechanism $\mathcal{T}$: the Laplace mechanism [3] and the Exponential mechanism [11]. The Laplace mechanism adds random noise with Laplace distribution in the result/process of numeric computation, while the exponential mechanism introduces a score function $q(X, x)$ which reflects how appealing the pair (X, x) is, where X denotes a dataset, x is the respond. In this paper, we mainly use the Laplace mechanism to achieve differential privacy.

4 Privacy Preserving Data Aggregation with Semi-honest Servers

In this section, we will introduce our privacy preserving data aggregation framework and show the theoretical evaluations on the performance of security, privacy, and utility of our framework.

4.1 Adversary Model

Generally, we categorise the adversary into two models (semi-honest adversary and malicious adversary) according to the willingness to follow a privacy preserving protocol.

Semi-honest Adversary. A semi-honest adversary is also known as a *honest-but-curious* (HBC) or *passive* adversary. That is, a semi-honest adversary always follows the protocol faithfully, but tries to infer extra information (especially the private information) from both the process and output of a protocol.

Malicious Adversary. A malicious adversary is also known as a *active* adversary who aims to cheat the protocol arbitrarily to disclose the targeted victim's privacy. The malicious adversary can work alone or together to enhance their abilities.

4.2 Our Framework

This work is inspired by [4]. Because the network model in [4] is not a full connect graph, that is, the low communication cost will be sufficiently low: $\mathcal{O}(n)$, where n is the number of users. What is more, the computation security is guaranteed by the application of secret sharing between the users and servers from the architecture in [4]. Based on the above two advantages of [4], our framework is also formed by two layers, the basic settings are showed in Fig. 1.

Specifically, in Fig. 1, there are n data curators with private data as the inputs to a linear data aggregation function. We assume that there are at most $n-1$ malicious parties who work together for obtaining a victim's private value. There are c semi-honest servers who assist the joint computation. Differing from [4], we assume the computation servers are semi-honest rather honest. It is quite straightforward to take this assumption because user's privacy/secret will be easily disclosed by a cooperation of k out of c servers, where k is the security bound in secret sharing scheme (in Sect. 3). So we cannot send the original data to any servers directly. Instead, we have to upload the differentially private data to these servers to ensure the privacy of user data. However, a new problem will be popped out if we only add noise at the user layer. As we care about the quality of data, we have to decrease the extent that we perturb the original data, then a cooperation of adversaries can also discover the victim's privacy from the final output with high probability. Thus, we also need to add differentially private noise at the server layer. But since we add noises at both user and server

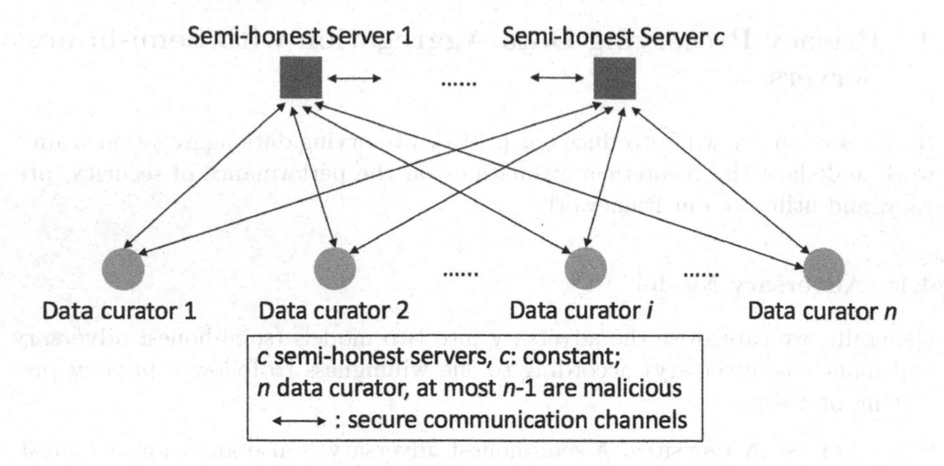

Fig. 1. System settings

layer, the overall data utility will be harmed a lot because of the extra noises. Therefore, in this paper, we introduce two probabilities at the two layers for adding the noise. That it, the data utility will be controlled by tuning the two probabilities. Algorithm 1 shows how our framework works.

Algorithm 1. Privacy Preserving Data Aggregation.

Input:

Original user data set: $\mathcal{U} = \{u_1, u_2, \ldots, u_n\}$;
Data aggregation request function (linear function): $f(\mathcal{U})$;
Number of users: n; Number of servers: c;
Differential privacy budget: ϵ_u^i, for each user i; ϵ_s^j, for each server j;
Noise adding probability: $\Pr_u^i$, for each user i; $\Pr_s^j$, for each server j.

Output:

Privacy preserved data aggregation result: $\mathcal{R}$.

1: User i splits u_i into c parts: $u_i = \sum_{j=1}^{c} \omega_i^j u_i$, where $\sum_{j=1}^{c} \omega_i^j = 1$;
2: User i generates a differential privacy noise Nu_i with ϵ_u^i and local sensitivity $\Delta sum(\omega_i^j u_i)$, splits $Nu_i = \sum_{j=1}^{c} \omega_i^j Nu_i$;
$u_i^{j\prime} = u_i^j + \omega_i^j Nu_i$ with probability $\Pr_u^i$, $u_i^{j\prime} = u_i^j$ with probability $1 - \Pr_u^i$;
3: User i sends $u_i^{j\prime}$ to Server j;
4: Server j receives a set $s_j = \{u_1^{j\prime}, u_2^{j\prime}, \ldots, u_n^{j\prime}\}$ from each User i;
5: Server j generates a differential privacy noise Ns_j with ϵ_s^j and local sensitivity $\Delta f(s_j)$, then calculates $f(s_j)' = f(s_j) + Ns_j$ with probability $\Pr_s^j$, $f(s_j)' = f(s_j)$ with probability $1 - \Pr_s^j$;
6: All Server j construct a Secured Multiparty Computation for $\mathcal{R} = f(f(s_j)')$;
7: **return** $\mathcal{R}$;

From Line 1 to Line 3, the computation entities at the user layer apply differential privacy and secret sharing scheme to guarantee both the security and privacy of their private data. A noise adding probability, $\Pr_u^i$, is used for a User i to decide whether a noise will be added. From Line 4 to Line 6, the semi-honest servers apply differential privacy and the basic SMC paradigm, so that none of the data at each server will be released to other servers. Similar to what happened at the user layer, each Server j will also take a noise adding probability, $\Pr_s^j$, to add a differentially private noise randomly.

4.3 Theoretical Evaluation

Privacy and Security Guarantee

Lemma 1. *The proposed privacy preserving data aggregation framework is ϵ-differentially private.*

Proof. Because our framework works with two layers, we start from user layer:

Since each User i adds differential privacy noise with probability $\Pr_u^i$, we will have an expectation for the noise adding:

$$u_i' = u_i + \Pr_u^i \times Lap((\max_i\{\omega_i^j u_i\}), \epsilon_u^i) + (1 - \Pr_u^i) \times 0,$$

then we will have:

$$\frac{\Pr(u_i = x_i)}{\Pr(u_i' = x_i)} = \frac{noise(x_i - u_i)}{noise(x_i - u_i')} = \frac{\Pr_u^i \times Lap((\max_i\{\omega_i^j u_i\}), \epsilon_u^i, (x_i - u_i))}{\Pr_u^i \times Lap((\max_i\{\omega_i^j u_i\}), \epsilon_u^i, (x_i - u_i'))} < exp(\epsilon_u^i)$$

(as at the user layer, the computational function is actually summation, for each User i, the local sensitivity is the maximum value among $\omega_i^j u_i$), that is,

$$\Pr(u_i = x_i) < \Pr(u_i' = x_i) \times exp(\epsilon_u^i).$$

Since, at user layer, we have a parallel composition, according to Theorem 4 in [10], we will have a $(\max_i\{\epsilon_u^i\})$-differential privacy at this layer. Similarly, we will have a $(\max_j\{\epsilon_s^j\})$-differential privacy at the server layer as the function f in this paper is linear function, it is easy to maintain the same property with the user layer.

Overall, because our framework works with two layers sequentially, based on Theorem 3 in [10], our framework will guarantee a $(\max_i\{\epsilon_u^i\} + \max_j\{\epsilon_s^j\})$-differential privacy. □

Actually, Lemma 1 provides us an interesting property that the noise adding probabilities at both user and server layer do not impact the privacy performance of our framework. Therefore, we can safely tune the two probabilities to achieve better data utility but keeping the same privacy guarantee as the existing privacy preserving data aggregation schemes.

Lemma 2. *The proposed privacy preserving data aggregation framework is secure.*

The proof of Lemma 2 is the same with [4], please refer to Sect. 5 in [4].

Utility Ensurence. In this section, we will use the variance of Laplace distribution (noise generation distribution for Differential Privacy in this paper) to measure the performance of computation utility between our framework with PrivaDA (Laplacian mechanism) in [4]. It would be straightforward that the worst case for noise addition can be denoted by the variance of Laplace distribution. Then we will have the following relationship to achieve better data utility than PrivaDA while guaranteeing the same differential privacy.

$$\sum_{i=1}^{n}(\Pr_u^i \frac{\Delta f}{\epsilon_u^i}) + \sum_{j=1}^{c}(\Pr_s^j \frac{\Delta f}{\epsilon_s^j}) \leq \sum_{j=1}^{c} \frac{\Delta f}{\epsilon_s^j + \max_i\{\epsilon_u^i\}},$$

that is,

$$\frac{n}{c}\min_i\{\Pr_u^i\}(1 + \frac{\max_j\{\epsilon_s^j\}}{\max_i\{\epsilon_u^i\}}) + \min_j\{\Pr_s^j\}(1 + \frac{\max_i\{\epsilon_u^i\}}{\max_j\{\epsilon_s^j\}}) \leq 1,$$

let $ratio = \frac{\max_i\{\epsilon_u^i\}}{\max_i\{\epsilon_u^i\}+\max_j\{\epsilon_s^j\}} \in (0,1)$, we will have:

$$\frac{n}{c} \times \frac{\min_i\{\Pr_u^i\}}{ratio} + \frac{\min_j\{\Pr_s^j\}}{1 - ratio} \leq 1. \tag{1}$$

In Sect. 5, we will use Eq. 1 to find the value of privacy budget *ratio* to achieve better data utility in the experiments.

5 Experimental Evaluation

5.1 Evaluation Metric

In this paper, we use Mean Absolute Error (MAE) and Mean Relative Error (MRE), to measure the data utility in the experiments:

$$MAE = \frac{1}{N}\sum_{i=1}^{N}|R_i - R|, \quad MRE = \frac{1}{N}\sum_{i=1}^{N}\frac{|R_i - R|}{R} \tag{2}$$

where N is the overall number of experiments, R_i is the output of the ith experiment on a privacy preserving data aggregation scheme, R is the ground truth, which is the output of the original data aggregation function. Specifically, in our experiments, we set $N = 10000$. Clearly, a lower MAE (MRE) means a better performance on data utility.

5.2 Experimental Results

Experimental Settings. According to our analysis in Sect. 2, in our experiments, we compare the performance of data utility from the following schemes: PrivaDA [4], PASTE [13], our original idea (Orig, add differentially private noise at both user and server layer), and the proposed idea in this paper (ThisWork). To simplify the experiments, we assign same privacy budget (in all of the four above schemes) and noise addition probability for all participants in same layer (in ThisWork). Then, to achieve the same differential privacy guarantee, in Orig and ThisWork, we set the privacy budget as $\epsilon_u + \epsilon_s = \epsilon$, where ϵ is the privacy budget for PrivaDA and PASTE. Especially, we have $\epsilon_u = ratio \times \epsilon$. The number of users $n = 10^4$, the number of servers $c = 5$.

In the experiments, the values of *ratio* we used are very approximate to the theoretically optimal one by Eq. 1, but not exactly matched because of the randomised mechanism involved in the experiments. Furthermore, we do not use the real-world dataset because our objective is to evaluation the performance of the above schemes with the numeric properties of the input data. Therefore, we generate the input data (positive value) by a standard normal distribution (mean = 500, variance = 1).

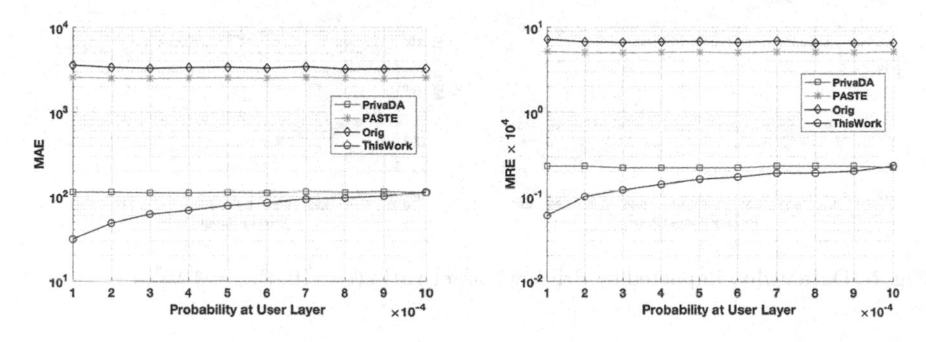

Fig. 2. Data utility impacted by Probability at User Layer ($\epsilon = 10$, $\Pr_s = 10^{-2}$)

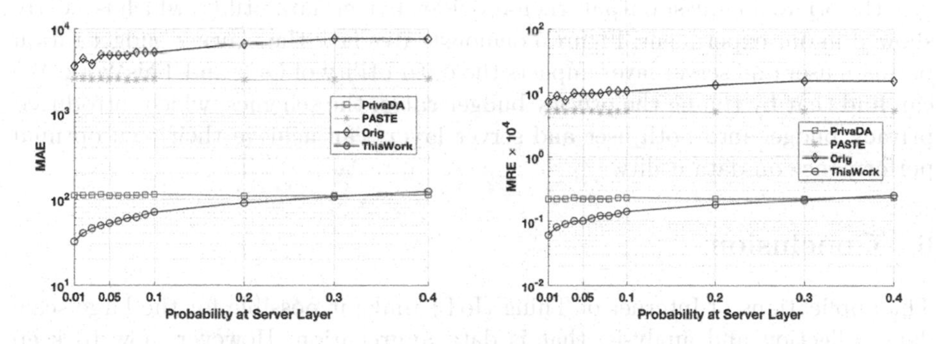

Fig. 3. Data utility impacted by Probability at Server Layer ($\epsilon = 10$, $\Pr_u = 10^{-4}$)

Experimental Results Analysis. Figures 2 and 3 show how the noise adding probability at user and server layer impacts the data utility. As we can see, the noise adding probabilities have positive impact on data utility. It is clear that by tuning these two probabilities, we can control the overall output quality to achieve better data utility than the existing schemes. While, Fig. 4 illustrates

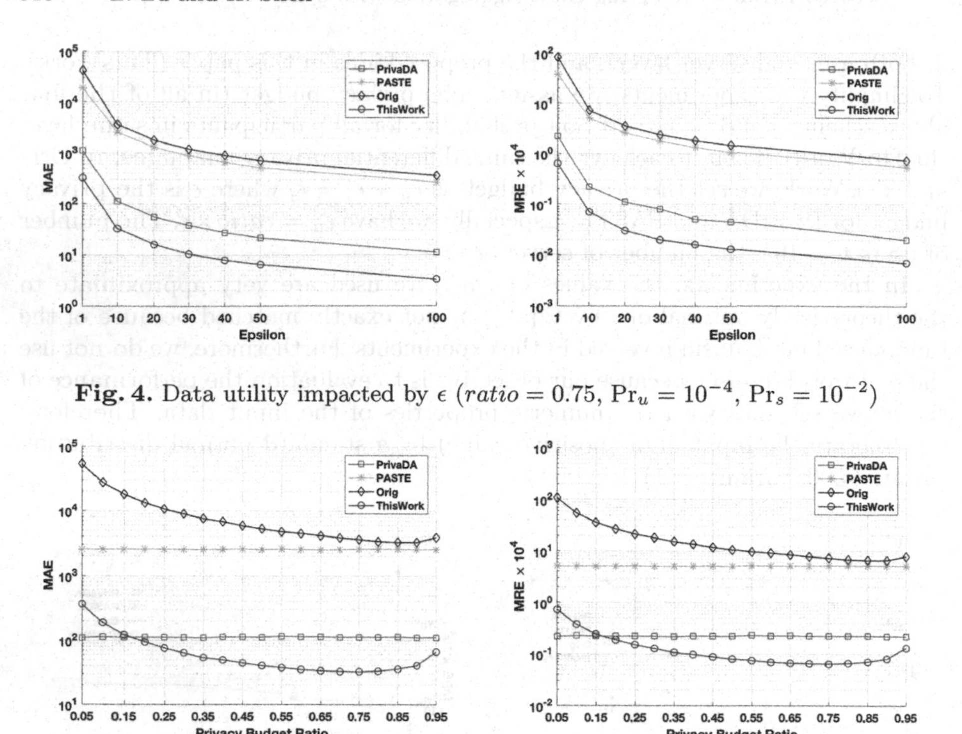

Fig. 4. Data utility impacted by ϵ ($ratio = 0.75$, $\mathrm{Pr}_u = 10^{-4}$, $\mathrm{Pr}_s = 10^{-2}$)

Fig. 5. Data utility impacted by Privacy budget ratio ($\epsilon = 10$, $\mathrm{Pr}_u = 10^{-4}$, $\mathrm{Pr}_s = 10^{-2}$)

how the privacy budget ϵ in differential privacy impacts the data utility. As an important property of differential privacy, a larger ϵ denotes to less noise added into the original request output, then provides better data utility, which is clearly showed in our experiment. Figure 5 demonstrates how the privacy budget ration between user and server layer impacts the data utility of Orig and ThisWork. We can find that by tuning the privacy budget ratio, the schemes, which introduced privacy budget into both user and server layer, can achieve their own optimal performance on data utility.

6 Conclusion

The applications of Internet of Thing (IoT) make it possible for the large scale data collection and analysis, that is data aggregation. However, how to keep people's privacy while ensuring acceptable data utility is a big challenge in our era. In this paper, to overcome the privacy disclosure problem in the existing work caused by the vulnerable trusted computation server, we provide a novel privacy preserving data aggregation scheme works on semi-honest computation servers. Both theoretical and experimental evaluation show that to achieve the same computational security and output privacy as the previous works, our scheme provides better output data utility.

Acknowledgements. The authors would like to thank the anonymous reviewers for their valuable comments. This work is supported by Australian Government Research Training Program Scholarship, Australian Research Council Discovery Project DP150104871, and Research Initiative Grant of Sun Yat-Sen University under Project 985. The corresponding author is Hong Shen.

References

1. Ács, G., Castelluccia, C.: I have a DREAM! (DiffeRentially privatE smArt Metering). In: Filler, T., Pevný, T., Craver, S., Ker, A. (eds.) IH 2011. LNCS, vol. 6958, pp. 118–132. Springer, Heidelberg (2011). doi:10.1007/978-3-642-24178-9_9
2. Cramer, R., Damgård, I., Nielsen, J.B.: Multiparty computation, an introduction. In: Catalano, D., Cramer, R., Di Crescenzo, G., Damgård, I., Pointcheval, D., Takagi, T. (eds.) Contemporary cryptology. Advanced Courses in Mathematics - CRM Barcelona, pp. 41–87. Birkhäuser Basel, Basel (2009)
3. Dwork, C.: Differential privacy. In: Bugliesi, M., Preneel, B., Sassone, V., Wegener, I. (eds.) ICALP 2006. LNCS, vol. 4052, pp. 1–12. Springer, Heidelberg (2006). doi:10.1007/11787006_1
4. Eigner, F., Kate, A., Maffei, M., Pampaloni, F., Pryvalov, I.: Differentially private data aggregation with optimal utility. In: Proceedings of the 30th Annual Computer Security Applications Conference, pp. 316–325. ACM (2014)
5. Elahi, T., Danezis, G., Goldberg, I.: PrivEx: private collection of traffic statistics for anonymous communication networks. In: Proceedings of the 2014 ACM SIGSAC Conference on Computer and Communications Security, pp. 1068–1079. ACM (2014)
6. Goryczka, S., Xiong, L.: A comprehensive comparison of multiparty secure additions with differential privacy. Trans. Dependable Secure Comput. (2015). Preliminary version - Goryczka, S., Xiong, L., Sunderam, V.: Secure multiparty aggregation with differential privacy: a comparative study. In: Proceedings of the Joint EDBT/ICDT 2013 Workshops, EDBT 2013, Genoa, Italy, pp. 155–163. ACM, New York (2013). doi:10.1145/2457317.2457343
7. Goryczka, S., Xiong, L., Fung, B.C.: Privacy for collaborative data publishing. IEEE Trans. Knowl. Data Eng. **26**(10), 2520–2533 (2014)
8. Gupta, A., Ligett, K., McSherry, F., Roth, A., Talwar, K.: Differentially private combinatorial optimization. In: Proceedings of the 21st Annual ACM-SIAM Symposium on Discrete Algorithms, SODA 2010, pp. 1106–1125 (2010)
9. Lu, Z., Shen, H.: A security-assured accuracy-maximised privacy preserving collaborative filtering recommendation algorithm. In: Proceedings of the 19th International Database Engineering & Applications Symposium, pp. 72–80. ACM (2015)
10. McSherry, F.: Privacy integrated queries. In: Proceedings of the 2009 ACM SIGMOD International Conference on Management of Data (SIGMOD). ACM (2009)
11. McSherry, F., Talwar, K.: Mechanism design via differential privacy. In: 48th Annual IEEE Symposium on Foundations of Computer Science, pp. 94–103. IEEE (2007)
12. Pettai, M., Laud, P.: Combining differential privacy and secure multiparty computation. In: Proceedings of the 31st Annual Computer Security Applications Conference, pp. 421–430. ACM (2015)
13. Rastogi, V., Nath, S.: Differentially private aggregation of distributed time-series with transformation and encryption. In: Proceedings of the 2010 ACM SIGMOD International Conference on Management of Data, pp. 735–746. ACM (2010)

14. Shamir, A.: How to share a secret. Commun. ACM **22**(11), 612–613 (1979)
15. Shi, E., Chan, H., Rieffel, E., Chow, R., Song, D.: Privacy-preserving aggregation of time-series data. In: Annual Network & Distributed System Security Symposium (NDSS). Internet Society (2011)
16. Sweeney, L.: *k*-anonymity: a model for protecting privacy. Int. J. Uncertain. Fuzziness Knowl.-Based Syst. **10**(05), 557–570 (2002)

Efficient Pedestrian Detection in the Low Resolution via Sparse Representation with Sparse Support Regression

Wenhua Fang[1(✉)], Jun Chen[1,2,3,4,5], and Ruimin Hu[1,2,3,4,5]

[1] National Engineering Research Center for Multimedia Software, School of Computer, Wuhan University, Wuhan 430072, China
fangwh@whu.edu.cn
[2] State Key Laboratory of Software Engineering, Wuhan University, Wuhan, China
[3] Hubei Provincial Key Laboratory of Multimedia and Network Communication Engineering, Wuhan University, Wuhan, China
[4] Research Institute of Wuhan University in Shenzhen, Shenzhen, China
[5] Collaborative Innovation Center of Geospatial Technology, Wuhan, China

Abstract. We propose a novel pedestrian detection approach in the extreme Low-Resolution (LR) images via sparse representation. Pedestrian detection in the extreme LR images is very important for some specific applications such as abnormal event detection and video forensics from surveillance videos. Although the pedestrian detection in High-Resolution (HR) images has achieved remarkable progress, it is still a challenging task in the LR images, because the discriminative information in the HR images usually disappear in the LR ones. It makes the precision of the detectors in the LR images decrease by a large margin. Most of the traditional methods enlarge the LR image by the linear interpolation methods. However, it can not preserve the high frequency information very well, which is very important for the detectors. For solving this problem, we reconstruct the LR image in the high resolution by sparse representation. In our model, the LR and HR dictionaries are established respectively in the training stage, and the representative coefficients mapping relations are determined. Moreover, for improving the speed of feature extraction, the feature reconstruction in the LR images is converted to the sparse linear combination between the coefficients and the response of the atoms in HR dictionary by the LR-HR mapping, no matter how complex the feature extraction is. Experiments on the four challenging datasets: Caltech, INRIA, ETH and TUD-Brussels, demonstrate that our proposed method outperforms the state-of-the-art approaches and is much efficient with more than 10 times speedup.

1 Introduction

Pedestrian detection is one of the most challenging problems in computer vision. It is difficult due to the significant amount of variation between images belonging to the same object category. Other factors, such as changes in viewpoint and

J. Kim et al. (Eds.): PAKDD 2017, Part II, LNAI 10235, pp. 313–323, 2017.
DOI: 10.1007/978-3-319-57529-2_25

Fig. 1. Red and green bounding boxes represent missing objects and detecting objects respectively. (Color figure online)

scale, illumination, partial occlusions and multiple instances further complicate the problem of object detection. It is very difficult to find the positions of all concerned objects in a low resolution image. More specifically, the goal is to find the bounding box for each object. One common approach is to use a sliding window to scan the image exhaustively in scale-space, and classify each window individually [1,3,14,17,19]. According to Dalal [4], detection performance in INRIA dataset is 90% when the sliding window is (64×128) and resolution of image is 640×480 and the size of object is similar to (64×128). However, the performance dropped significantly to 40% when the sliding window is (16×32). Because the proper size of the pedestrian in the image is unknown previously, it is prone to missing when the resolution of the object is low. Objects in LR are always missed in the traditional detection method [4] as shown in Fig. 1. Moreover, pedestrian detection in the LR image is also important as well as in the HR image, such as in criminal investigation and public security field. Additionally, object detection is much more difficult in LR image than in HR one because traditional discriminative features in high resolution cannot be extracted in low resolution. The performance of the pedestrian detectors, including the state-of-the-art method [9,15], in LR will decrease by a large margin. Traditional multi-scale pedestrian detection methods in LR images are resizing the detection windows or/and images [7]. And magnifying the image from LR image by linear interpolation will lose the high frequency details such as edges, which are discriminative feature for classification. For retaining the critical detailed information during the magnification, the sparse representation in Super Resolution, is used to reconstruct the image and then the object detection methods are performed in the image. In this paper, we improve the perceptual image quality from LR to HR, among which the resolution enhancement technology is called super-resolution. We can get high-resolution and high quality image with more details economically, not by imposing higher requirements on hardware devices and sensors. We use a manifold regularized regression framework for super-resolution as shown in Fig. 2. The sparse representation is relaxed for LR to HR sparse support domain regression, which is flexible in using the information

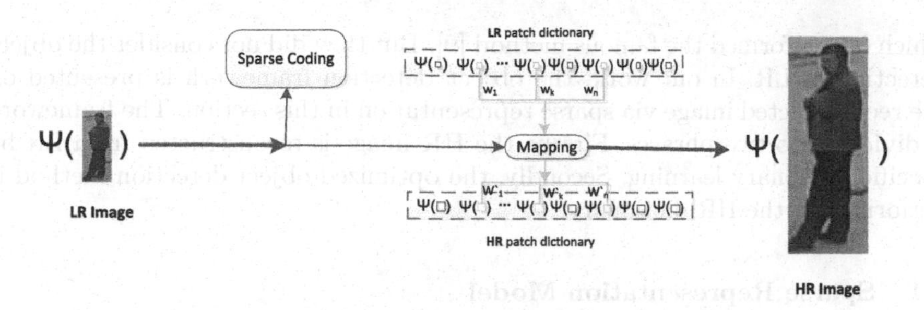

Fig. 2. Flowchart of the proposed method. Note that the red patches denote the sparse support domain of the input LR-HR patches on the LR-HR dictionaries, and we use Ψ to denote the operation of feature extraction. (Color figure online)

of local training samples. Note that image patches have regular structures where accurate estimation of pixel values via regression is possible. Accordingly, the proposed method has more power and flexibility to describe different image patterns. In addition, the proposed method simultaneously considers the manifold regularization, thus capturing the intrinsic geometrical structure of the dictionary. We use a mapping function from low resolution (LR) patches to high-resolution (HR) patches will be learned by a local regression algorithm called sparse support regression, which can be constructed from the support bases of the LR-HR dictionary. Then, we can also use the two important components (feature and classifier) in the pedestrian detection from HR we can get from the above. First, features capture the most discriminative information of pedestrians. Second, a classifier decides whether a candidate window shall be detected as enclosing a pedestrian and SVM (Support vector machine) is often used. The connection between features and classifier components is usually achieved using manual parameter configuration. The HOG feature is individually designed with its parameters manually tuned given the linear SVM classifier [4]. There are two main contributions of this paper: (1) a object detection framework in extreme low resolution is proposed by detecting objects in reconstructed high resolution image; (2) a mapping function from low resolution patches to high resolution patches will be learned by a local regression algorithm called sparse support regression, which can be constructed from the support based of the LR-HR dictionary. Our low resolution detector produced better performance for pedestrian detection than state-of-the-art methods. Experiments on four challenging datasets: Caltech, INRIA, ETH and TUD-Brussels, demonstrate that our proposed method outperforms the state-of-the-art approaches and is much efficient (more than 10 times acceleration).

2 Sparse Representation for Object Detection

Sparse representation is usually used to reconstruct the image in super resolution and is rarely used for object detection. Ren et al. [15] firstly adopted sparse representation to extract the histogram of sparse codes feature for object detection, which outperformed the famous method [9]. But they did not consider the object detection in LR. In our work, the object detection framework is presented on the reconstructed image via sparse representation in this section. The framework is divided into two phrases. Firstly, the HR image is reconstructed from LR by specific dictionary learning. Secondly, the optimized object detection method is performed in the HR dictionary.

2.1 Sparse Representation Model

Given a set of image patches $Y = [y_1, \cdots, y_n]$, finding a dictionary $D = [d_1, \cdots, d_m]$ and an associated sparse code matrix $X = [x_1, \cdots, x_n]$ by minimizing the reconstruction error

$$\min_{D,X} ||\mathrm{Y} - \mathrm{DX}||_F^2 s.t. \forall_i, ||x_i||_0 \leq K \tag{1}$$

where x_i are the columns of X, the zero-norm $||\cdot||_0$ counts the non-zero entries in the sparse code x_i, and K is a predefined sparsity level.

2.2 LR-HR Pairwise Dictionary

Given a set of LR and HR training image patch pairs, $\{(x_1, y_1), \cdots, (x_N, y_N)\} \subset R^d \times R^D$, d and D are the dimensions of one LR and one HR patch respectively. Define $X = [x_1, \cdots, x_N]$ and $Y = [y_1, \cdots, y_N]$, each column of which is a patch sample. Thus the matrixes X and Y can be viewed as the LR and HR patch dictionaries respectively.

Considering that the manifold assumption (two manifolds spanned by the feature spaces of the LR and HR patches are locally similar) may not be tenable, we learn a much more stable LR-HR mapping in the support domain for super-resolution. Thus it can be transformed to a regression problem.

Our another important goal is to encode the geometry of the HR patch manifold, which is much more credible and discriminated compared with that of the LR one [11], and preserve the geometry for the reconstructed HR patch space. This will ensure that the local geometric structure of the reconstructed HR patch manifold is consistent with that of the original HR one. Based on the above discussions, our MSSR (Manifold regularized Sparse Support Regression) algorithm for image super-resolution should be equipped with two properties: (1) The shared support of each LR patch and HR patch has an explicit regression relationship; (2) The local geometrical information on the original HR patch dictionary is preserved. In the following part, we will describe how we formulate MSSR with these two desired properties.

2.3 Sparse Support Regression

Instead of assuming that each pair of HR and LR patches has the same sparse representation, in our proposed MSSR method, this strong regularization of sparse representation is relaxed for sparse support regression, and the sparse coefficient vectors of one LR and HR patch pair share the same support, i.e., the same indices of nonzero elements.

Given a set of LR and HR training patches (dictionary pairs), $\{(x_1, y_1), \cdots, (x_N, y_N)\} \subset R^d \times R^D$ for an unseen LR patch x_t, we try to learn a mapping function $f(x, P) = Px$, from the LR patch to the HR one to minimize the following regularized cost function for the regression,

$$\varepsilon(P) = \sum_{i \in S} (Px_i - y_i)^2 + \alpha \|P\|_H^2 \tag{2}$$

where α is a regularization parameter, P is a $D \times d$ matrix to be learned, $\|P\|_H^2$ is the induced norm of f in the reproducing kernel Hilbert space (RKHS) space H, and S is the support of the coding coefficients θ^* of the unseen patch x_t on LR training patches X:

$$\theta^* = \arg\min_{\theta} \|x_t - X\theta\|_2 + \lambda_1 \|\theta\|_1 \tag{3}$$

Thus, $S = \text{support}(\theta^*)$. In Eq. (2), $\|\theta\|_1$ denotes the l_1 norm of θ and the parameter λ_1 balances the coding error of x_t and the sparsity of θ The solution of Eq. (2) can be achieved by convex optimization methods referring to [13]. The support of one vector is referring to the indices of nonzero elements in the vector. Defining X_s and Y_s as $X_s = \{x_i | i \in S\}$ and $Y_s = \{y_i | i \in S\}$ respectively and using Fibonacci norm to represent the smoothness of H, we can rewrite Eq. (1) as the following matrix form:

$$\varepsilon(P) = \|PX_S - Y_S\|_F^2 + \alpha \|P\|_F^2 \tag{4}$$

2.4 Mining the Geometry on HR Patch Dictionary

This section targets on the second property, which is to preserve the local geometrical information on the HR patch dictionary. Note that the neighborhood relation, which guides the formulation of sparseness, is defined on the manifold rather than the Euclidean space.

Researchers have proposed various methods to measure the similarity between data points [12,16], e.g., pair-wise distance based similarity and reconstruction coefficient based similarity. Since the former is suitable for discriminant analysis problems, such as recognition and clustering. Alternatively, reconstruction coefficient based similarity is datum-adaptive, and thus more suitable for image super-resolution. LLE is one of the representative works for reconstruction coefficient similarity estimation. It calculates the coefficient for each data through k-NN searching, thus k sparsity. The performance of LLE graph will decrease rapidly when the datas are non-uniformly sampled from underlying manifold, and this situation is very common in practice.

Recently, some researchers have demonstrated that the sparse structure of one manifold can be explored by the l_i graph [16], resulting in many benefits for machine learning and image processing problem. Let y_i be the i-th HR patch, which is under consideration now. We want to identify its neighbors on the smooth manifold rather than the entire Euclidean space. On the smooth patch manifold space, the patch can be well sparsely approximated by a linear combination of a few nearby patches. Thus, it has a sparse representation over the support domain Y_s. For any HR patch y_i, it can be sparsely approximate by the data matrix Y_s except y_i:

$$W_i^* = \arg\min_{\mathrm{W}_i} \|y_i - Y_S \mathrm{W}_i\|_2 + \lambda_2 \|\mathrm{W}_i\|_1 \tag{5}$$

where W_i denotes the i-th column of the matrix W whose diagonal elements are zeros, and λ_2 is the parameter balancing the coding error of y_i and the sparsity of W_i.

2.5 Feature Extraction via Sparse Representation

Feature Extraction can be regarded as the linear combination of the representative coefficients and the response of the items of the learned dictionary. Denoting the feature pyramid of an image I as Φ, and $I = [P_1, \cdots, P_N]$, and D_j in $D = [D_1, \cdots, D_K]$ is the atom of D (Dictionary), we have $\Psi * P_i \approx \Psi * (\sum_j \alpha_{ij} D_j) = \sum_j \alpha_{ij} (\Psi * D_j)$, where $*$ denotes the convolution operator. Concretely, we can recover individual part filter responses via sparse matrix multiplication (or lookups) with the activation vector replacing the heavy convolution operation as shown in Eq. 6.

$$\begin{bmatrix} \Psi * P_1 \\ \Psi * P_2 \\ \vdots \\ \vdots \\ \vdots \\ \vdots \\ \Psi * P_N \end{bmatrix} \approx \begin{bmatrix} \alpha_1 \\ \alpha_2 \\ \vdots \\ \vdots \\ \vdots \\ \vdots \\ \alpha_N \end{bmatrix} \begin{bmatrix} \Psi * D_1 \\ \Psi * D_2 \\ \vdots \\ \Psi * D_K \end{bmatrix} = AM \tag{6}$$

For efficient pedestrian detection, the extraction of some features should be made appropriate adjustments. Take HOG feature for example. It is composed of concatenated blocks. Each block includes 2×2 cells, and each cell is the 8×8 pixels of the image. So the block is 16×16 pixels. The concatenation of histograms of the blocks has two strategies: overlap and non-overlap. In the overlap manner, the sliding step width is usually the width of the cell. In the non-overlap manner, the sliding step width is the width of the block. So the dimension of the feature of the non-overlap is smaller than that of the overlap. But the performance of the feature will be lost by nearly 1% [4]. So the standard HOG feature chooses the overlap manner for better performance. For high acceleration, in this paper, we choose the non-overlap manner.

3 Experiments

For evaluating our method, we conduct the experiments on four challenging pedestrian datasets: Caltech [6], INRIA [4], ETH [8] and TUD-Brussels [21]. And we resize the images to 1/4 of the original. The state-of-the-art and classic pedestrian detectors are chosen to test our framework: HOG [4], ChnFtrs [18], ACF [5], HOGLBP [20], LatSvmV2 [10] and VeryFast [2]. In the experiments, the training and testing data setting is as same as in [5]. We first discuss the relation of the performance versus the sparsity degree, the size of the dictionary, the size of atom. And then we evaluate our method.

3.1 Dictionary Learning vs Performance

Because our method is based on sparse coding, how to select the parameters of dictionary learning directly affects the performance of feature reconstruction. For choosing the best parameters, we conduct some experiments on INRIA Person Dataset and the type of synthesized feature is HOG. INRIA Person Dataset consists of 1208 positive training images (and their reflections) of standing people, cropped and normalized to 64×128, as well as 1218 negative images and 741 test images. This dataset is an ideal setting, as it is what HOG was designed and optimized for, and training is straightforward.

Sparsity Level and Dictionary Size. Figure 4 shows the average precision on INRIA when we change the sparsity level along with the dictionary size using 5×5 patches. We observe that when the dictionary size is small, a patch cannot be well represented with a single codeword. However, when the dictionary size grows and includes more structures in its codes, the $K = 1$ curve catches up, and performs very well. Therefore we use $K = 1$ in all the following experiments.

Patch Size and Dictionary Size. Next we investigate whether our synthesized features can capture richer structures using larger patches. Figure 3 shows the

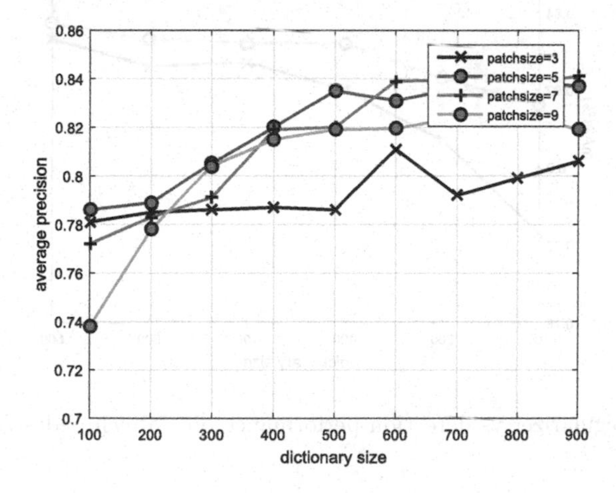

Fig. 3. The patch size vs detection performance on Caltech pedestrian dataset.

average precision as we change both the patch size and the dictionary size. While 3×3 codes barely show an edge, 7×7 codes work much better. However, 9×9 patches, may be too large for our setting and do not perform well.

Regularizer. With $K = 1$, one can also use different regularizers to learn a dictionary. Figure 5 compares the detection accuracy with Lasso penalty vs Elastic net penalty on 7×7 patches. The Elastic net penalty is better because it include more constraints to learn discriminative representation.

In the following experiments, we set the size of the dictionary, the sparsity degree to be 600, 1 respectively. We set the size of the atom of the dictionary to be 7×7 for better performance.

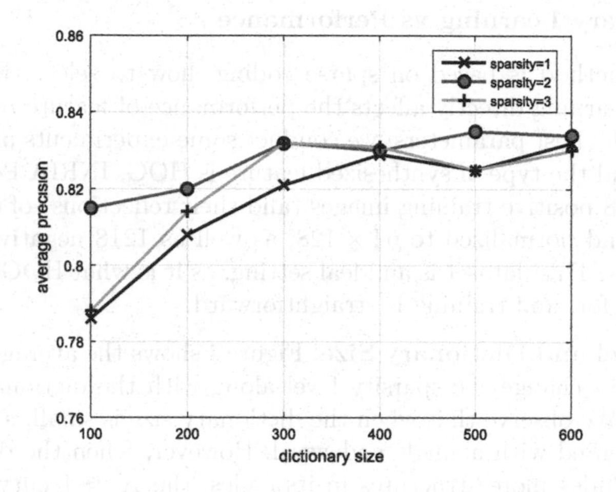

Fig. 4. The sparsity vs detection performance on Caltech pedestrian dataset.

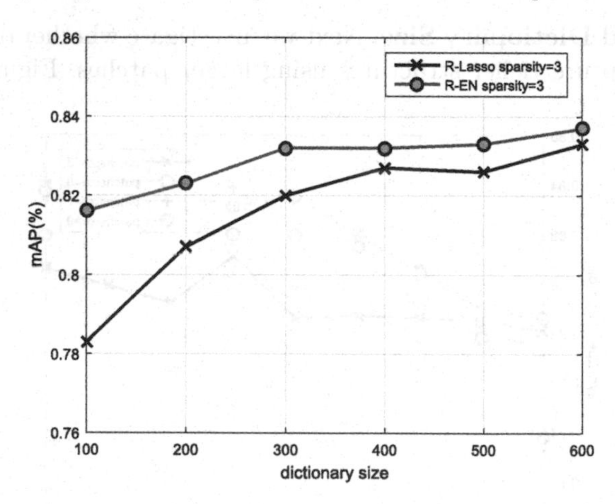

Fig. 5. The regularizer vs detection performance on Caltech pedestrian dataset.

Table 1. Missing rates of pedestrian detectors (Sparse Coding vs Bicubic Linear Interpolation) on four challenging datasets. "SC" denotes sparse coding, and "BLC" stands for bicubic linear interpolation.

Detectors	INRIA [4]		Caltech [6]		TUD-Brussels [21]		ETH [8]	
	SC	BLI	SC	BLI	SC	BLI	SC	BLI
HOG [4]	0.36	0.49	0.58	0.60	0.78	79	54	65
ChnFtrs [18]	0.22	0.28	0.46	0.57	0.50	0.61	0.47	0.58
ACF [5]	0.17	0.18	0.33	0.45	40	0.51	0.40	0.52
HOGLBP [20]	0.29	0.40	0.58	0.68	0.72	0.81	0.45	0.56
LatSvmV2 [10]	0.10	0.23	0.53	0.64	0.60	0.71	0.41	0.52
VeryFast [2]	0.11	0.17	0.43	0.54	0.51	0.62	0.45	0.56

3.2 Performance Comparison

We just pay attention to whether the performance is lost and the degree of performance loss. As shown in Table 1, we can see the performance comparison of the detectors between the sparse coding reconstruction and the bicubic linear interpolation reconstruction. We resize the width and height of the images by half. As can be seen from the table, the performance degradation is very large, about ten percent. We think the reason is that our method is based on reconstruction error minimum and sparsity constraints, and can preserve the discriminative information for pedestrian detection.

3.3 Speed Comparison

The speed of the detector is more important than performance in the real-world applications. In this section, we will show the speed comparison of the above origin detectors and the corresponding synthesized detectors. Because the speed of the detector depends on the resolution.

We just do the statistics and analysis on the INRIA dataset because the results on the other datasets are the same as that on this dataset. The resolution of the image is 640×480 in INRIA testing set. As shown in Table 2, acceleration

Table 2. Speed comparison of pedestrian detectors (origin vs Our feature extraction) on INRIA person dataset. The unit of speed is the frame of per second.

Detectors	INRIA [4]	
	Origin	Synthesizer
HOG [4]	0.23	**96.5**
ChnFtrs [18]	16.4	**121.2**
ACF [5]	31.9	**125.4**
HOGLBP [20]	0.06	**120.4**
LatSvmV2 [10]	0.6	**108.5**
VeryFast [2]	50	**110.2**

of the synthesized detector is very obvious. Take the detector HOGLBP [20] for example. The speedup ratio is up to 2000. The speed of original veryfast detector [2] is 50 fps because it is accelerated by GPU. But the speed of our synthesized detector is 110 fps. From the table, experiment results confirm our conjecture that the runtime of our synthesized detector depends on the decomposition of the image based on the dictionary.

4 Conclusions

In this work we proposed a novel pedestrian detection approach in extremely low resolution image using sparse representation and divided the object detection into online and off-line stages. In the off-line stages, the patch based feature extraction on dictionary atoms was conducted. And in the online stage, the final visual feature is combined linearly by above patch based feature. Our work is first clearly demonstrate the advantages of sparse LR-HR dictionaries for pedestrian detection. Our studies show that large structures in large patches, when are captured in a large dictionary, generally improve pedestrian detection, calling for future work on designing and learning even richer representation. We design a novel sparse regression algorithm for pedestrian detection which can enhance the learning performance. It is experimentally shown that our methods can produce more faithful details and higher objective quality for all state-of-the-art pedestrian detection approaches.

Acknowledgement. The research was supported by National High Technology Research and Development Program of China (2015AA016306), National Nature Science Foundation of China (61231015, 61671336, 61671332, 61562048), Natural Science Fundation of JiangSu Province (BK20160386), the EU FP7 QUICK project under Grant Agreement (PIRSES-GA-2013-612652), the Technology Research Program of Ministry of Public Security (2016JSYJA12), the Fundamental Research Funds for the Central Universities (2042014kf0250, 2014211020203).

References

1. Aytar, Y., Zisserman, A.: Tabula rasa: model transfer for object category detection. In: IEEE International Conference on Computer Vision, pp. 2252–2259 (2011)
2. Benenson, R., Mathias, M., Timofte, R., Gool, L.V.: Pedestrian detection at 100 frames per second. In: IEEE Conference on Computer Vision and Pattern Recognition, vol. 157, pp. 2903–2910 (2012)
3. Cevikalp, H., Triggs, B.: Efficient object detection using cascades of nearest convex model classifiers. In: IEEE Conference on Computer Vision and Pattern Recognition, pp. 3138–3145 (2012)
4. Dalal, N., Triggs, B.: Histograms of oriented gradients for human detection. In: IEEE Conference on Computer Vision and Pattern Recognition, pp. 886–893 (2005)
5. Dollar, P., Appel, R., Belongie, S., Perona, P.: Fast feature pyramids for object detection. IEEE Trans. Pattern Anal. Mach. Intell. **36**(8), 1532–1545 (2014)

6. Dollar, P., Wojek, C., Schiele, B., Perona, P.: Pedestrian detection: an evaluation of the state of the art. IEEE Trans. Pattern Anal. Mach. Intell. **34**(4), 743–761 (2012)
7. Dollar, P., Belongie, S., Perona, P.: The fastest pedestrian detector in the west. In: Proceedings of the British Machine Vision Conference, BMVC 2010, Aberystwyth, UK, 31 August - 3 September, pp. 1–11 (2010)
8. Ess, A., Leibe, B., Gool, L.V.: Depth and appearance for mobile scene analysis. In: IEEE International Conference on Computer Vision, pp. 1–8 (2007)
9. Felzenszwalb, P.F., Girshick, R.B., Mcallester, D.: Cascade object detection with deformable part models. In: IEEE Conference on Computer Vision and Pattern Recognition, pp. 2241–2248 (2010)
10. Felzenszwalb, P.F., Girshick, R.B., Mcallester, D., Ramanan, D.: Object detection with discriminatively trained part-based models. IEEE Trans. Pattern Anal. Mach. Intell. **32**(9), 1627–1645 (2014)
11. Jiang, J., Hu, R., Han, Z., Huang, K.: Efficient single image super-resolution via graph embedding. In: IEEE International Conference on Multimedia and Expo, pp. 610–615 (2012)
12. Jiang, J., Hu, R., Wang, Z., Han, Z.: Manifold regularized sparse support regression for single image super-resolution. In: IEEE International Conference on Acoustics, Speech and Signal Processing, vol. 32, pp. 1429–1433 (2013)
13. Lee, H., Battle, A., Raina, R., Ng, A.Y.: Efficient sparse coding algorithms. In: Advances in Neural Information Processing Systems, pp. 801–809 (2007)
14. Rahtu, E., Kannala, J., Blaschko, M.: Learning a category independent object detection cascade. In: IEEE International Conference on Computer Vision, vol. 23, pp. 1052–1059 (2011)
15. Ren, X., Ramanan, D.: Histograms of sparse codes for object detection. In: IEEE Conference on Computer Vision and Pattern Recognition, vol. 9, pp. 3246–3253 (2013)
16. Shen, B., Si, L.: Non-negative matrix factorization clustering on multiple manifolds (2010)
17. Sun, M., Savarese, S.: Articulated part-based model for joint object detection and pose estimation. In: IEEE Conference on Computer Vision and Pattern Recognition, vol. 23, pp. 723–730 (2011)
18. Tu, Z., Perona, P., Belongie, S.: Pasadena: integral channel features. **2**(3), 5–11 (2009)
19. Vijayanarasimhan, S., Grauman, K.: Efficient region search for object detection. **42**(7), 1401–1408 (2011)
20. Wang, X., Han, T.X., Yan, S.: An HOG-LBP human detector with partial occlusion handling. In: IEEE International Conference on Computer Vision, pp. 32–39 (2009)
21. Wojek, C., Walk, S., Schiele, B.: Multi-cue onboard pedestrian detection. In: IEEE Conference on Computer Vision and Pattern Recognition, pp. 794–801 (2009)

Multi-task Representation Learning for Enhanced Emotion Categorization in Short Text

Anirban Sen[1](✉), Manjira Sinha[2], Sandya Mannarswamy[2], and Shourya Roy[3]

[1] Computer Science and Engineering Department, IIT Delhi, New Delhi, India
anirban@cse.iitd.ac.in

[2] Conduent Labs India, Bangalore, India
{Manjira.Sinha,Sandya.Mannarswamy}@conduent.com

[3] Big Data Labs, American Express, New York, USA
Shourya.Roy@gmail.com

Abstract. Embedding based dense contextual representations of data have proven to be efficient in various NLP tasks as they alleviate the burden of heavy feature engineering. However, generalized representation learning approaches do not capture the task specific subtleties. In addition, often the computational model for each task is developed in isolation, overlooking the interrelation among certain NLP tasks. Given that representation learning typically requires a good amount of labeled annotated data which is scarce, it is essential to explore learning embedding under supervision of multiple related tasks jointly and at the same time, incorporating the task specific attributes too. Inspired by the basic premise of multi-task learning, which supposes that correlation between related tasks can be used to improve classification, we propose a novel technique for building jointly learnt task specific embeddings for emotion and sentiment prediction tasks. Here, a sentiment prediction task acts as an auxiliary input to enhance the primary emotion prediction task. Our experimental results demonstrate that embeddings learnt under supervised signals of two related tasks, outperform embeddings learnt in a uni-tasked setup for the downstream task of emotion prediction.

Keywords: Multi-tasking · Emotion prediction · Representation learning · Joint learning

1 Introduction

In order to develop efficient machine-learning models for complex NLP tasks, it is desirable to explore approaches that can discover features automatically from the data and make the learning algorithm less dependent on fragile hand-crafted feature-engineering. In recent times, dense, low-dimensional vector space

This work was done when the first author was an intern at Conduent Labs India.
S. Roy—This work was done when the author was at Conduent Labs India.

J. Kim et al. (Eds.): PAKDD 2017, Part II, LNAI 10235, pp. 324–336, 2017.
DOI: 10.1007/978-3-319-57529-2_26

representations of the text have been explored in various NLP learning tasks [3] including that of emotion classification. However, typically word embeddings have been built using contextual information, which has the disadvantage that two contexts could be very closely related but could convey extremely different emotions leading to poor performance by an emotion classifier using generalized word embeddings as input. This problem has been addressed by means of learning task sensitive embeddings. For instance, sentiment specific word embeddings have been learnt and have been shown to benefit performance in sentiment classification task [20].

While task specific word embeddings have been shown to be of benefit in downstream tasks [12,20], not many of the prior works have proposed using auxiliary related tasks to improve the learnt word embeddings' quality. *Most of the prior works on task specific embeddings have been in the context of a single task learning.* Given that learning embeddings require labeled annotated data for the main task, which is hard to come by, it makes sense to explore building embeddings which can be learnt under supervision of multiple related tasks in parallel. Given the basic premise of multi-task learning [4], which supposes that correlation between related tasks can be utilized to improve classification by learning tasks in parallel, vector representations can be jointly learnt using multiple related task objectives to improve classification performance. We build on this premise and propose a novel technique for building jointly learnt embeddings sensitive to both emotion and sentiment, which are related tasks.

Emotion annotated data with explicit emotional cues are scarce; instead intrinsic emotion cues can be present through the underlying sentiment of the text. Hence, emotion classification can benefit by using input text embeddings which are built using the signals carried not just by emotions, but also by the underlying sentiments of the text. Our work proposes the novel idea of learning text embedding representations jointly using the emotion and sentiment signals, and use these jointly learnt embeddings for improving the downstream task of fine grained emotion prediction. Furthermore, while this specific embodiment of our invention is on jointly learnt embeddings sensitive to emotion and sentiment, our invention proposes the broader idea of utilizing related joint tasks training for representation learning which can be applied to other domains.

There has been prior work on joint learning of related tasks [5,6], where the supervised signals from both the main and the auxiliary tasks have been used in model parameter learning. However, there has not been much prior work on joint learning of representations or embeddings on two related tasks, which results in improved text embedding representations, to be used as input for multiple downstream tasks. Our paper addresses this issue in the setting of emotion prediction, and shows that jointly learnt embeddings outperform individual task embeddings in this prediction task. The rest of the paper is organized as follows: In Sect. 2, we provide a brief summary of related work. In Sect. 3, we describe our approach for joint learning of embeddings supervised under emotion and sentiment signals. In Sect. 4, we report our experimental results and conclude in Sect. 5.

2 Related Work

There has been extensive work on emotion and sentiment classification of text over the last decade, both being studied as independent tasks on their own. Given the extensive prior research in the area of emotion and sentiment analysis, we provide only a brief outline of some of the relevant work in this space, without being exhaustive.

Approaches to emotion detection in text can be broadly classified into keyword based approaches [9,13], linguistic rule based approaches [2,15,16] and machine learning based approaches [17,19]. ML techniques can be broadly classified into supervised learning based approaches [17,23], and unsupervised learning based approaches [1,7,19]. In supervised learning based approaches, various hand-crafted features for identification of emotion content of text are extracted and used as input to a classifier. Similarly, sentiment analysis has been a much crowded area of research in NLP community with different sentiment classification schemes being proposed such as unsupervised learning methods [22], supervised approaches [17], semi-supervised learning methods [11], just to cite a few. Both sentiment and emotion classification approaches have typically depended on extensive hand-crafted features for their performance gains. While such hand-crafted features have provided good performance in emotion classification of text, it comes at the cost of extensive human effort and is typically not scalable due to lack of portability across domains and datasets. As we discussed in the introduction, alternative input representations [3,14,18] such as word embedding representations have gained popularity in ML based emotion and sentiment classification systems. In addition to general word embeddings, task specific word embeddings have been proposed and used in sentiment analysis [20]. A natural extension of this is to learn emotion specific word embeddings and use them as input for emotion prediction.

Most of the prior work on emotion and sentiment classification have typically studied each of these two tasks in isolation. There have been very few works studying the joint learning of emotion and sentiment. Gao [6] studied joint learning of sentiment and emotion using an extra data set that is annotated with both sentiment and emotion labels. They trained two separate classifiers for sentiment and emotion and during the testing phase, each sample is classified by both classifiers and the probabilities belonging to each sentiment or emotion label is obtained. The jointly labeled data set is used to estimate the transformation probability between the two kinds of labels and the transformation probability is leveraged to transfer the classification labels to benefit the two tasks from each other. Wang et al. [5] propose a joint sentiment emotion classification framework using Integer Linear Programming. They map different emotions to the sentiment categories of positive, negative, and neutral, and apply integer linear programming to maximize the similarity of the two classifiers' output labels to the posterior probabilities of the output labels subject to certain constraints.

Our current work is inspired by these earlier works on joint learning of sentiment and emotion. While there has been prior work on using sentiment specific word embeddings as input for sentiment classification task, we are not aware of any prior work which learns word embeddings jointly sensitive to both sentiment and emotion and uses the learnt embeddings as input in the downstream sentiment and emotion prediction task. *In this paper, we propose a novel representation of jointly learnt emotion-sentiment specific word embeddings, and show that the jointly learnt embeddings lead to performance improvements in the emotion prediction task as opposed to using the individually learnt emotion or sentiment specific word embedding (and also over pre-trained general word embeddings).* Given the close relation between sentiment and emotion classification tasks, using a jointly learnt word embedding as input to the model serves as an appropriate input representation for improving the prediction performance.

3 Approach

The three major steps of our approach are (a) Preprocessing of tweets (b) Embedding learning and (c) Tweet emotion prediction using the learnt embeddings, which are described in details in the following subsections.

3.1 Preprocessing of Tweets

In this step, we perform preprocessing of the raw tweet data. It includes converting tweets to lower case, stopword removal, @ mention removal, and removal of hyperlinks from tweets. Additionally, we also join negative modifiers with the modified words such as *dont_go, not_available, shouldnot_show*, etc. The step of joining negative modifiers with the modified words is taken in order to retain the context in which the word has been used. The quality of the learnt embeddings might suffer if the context of a word is not captured, especially in case of negative modifiers. The ideal way to identify a modifier-modified relation is to use a dependency parser. However, in our data, we found that almost all of the negative modifiers are adjacent to the modified words. Hence, we join all negative modifiers with their adjacent words in order to save the overhead associated with running a dependency parser.

3.2 Embedding Learning

In this step, we generate unigram embeddings as explained subsequently.

Emotion+sentiment specific word embeddings (ES-SWE): This is the multitasking or joint learning setup where we simultaneously train a neural network on both emotion and sentiment classes of the tweets present in the training dataset. The resulting embeddings are called Emotion and Sentiment Specific Word Embeddings (ES-SWE). We show the setup for this multi-tasking framework in Fig. 1. In this step, our two-task neural network model uses each 5-gram of a tweet to map its words into fixed dimensional vector representations. These vectors or embeddings can be subsequently used to perform emotion prediction.

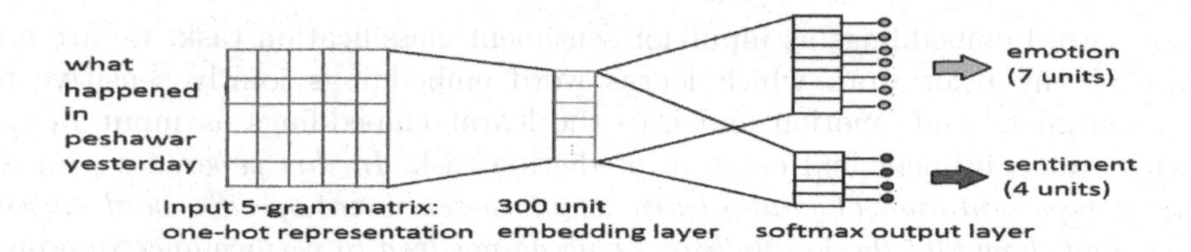

Fig. 1. Architecture of the multi-tasking neural network used for embedding learning

Given each 5-gram of a tweet, this neural network model predicts two vectors: one for emotion and another for sentiment. We used 5-grams as inputs for the embedding learning network following the methodology of previous works like [21], and also to ensure that the inputs are of fixed length, so as to generate accurate unigram embeddings. We have used a 3-layer neural network. The input layer of this network is of shape 5x$|V|$ (where $|V|$ is the size of the vocabulary, and each unigram is represented as a 1-hot encoding over $|V|$. For our training sets, $|V|$ is approximately 9000). The first hidden layer is the embedding-learning layer of size 300 units[1]. Each unigram is fully connected to the 300 units and after the completion of network training, the 300 updated connection-weights represent the 300-dimensional embedding for the unigram. The next layer is a fully connected output layer. We have seven and four output classes for emotion and sentiment categorization, respectively. We use the *Adam* update algorithm [8] to train the network, with a learning rate of 0.001. To prevent over-fitting, we also use L2-regularization in this network with the parameter value of 0.5. We consider a linear combination of the cross-entropy losses[2] for the prediction tasks. The cross-entropy losses for emotion and sentiment prediction tasks for the softmax layer are shown in Eq. 1 below:

$$\begin{aligned} loss_{emo}(t) &= -\sum_{k=0}^{6} f_k^g(t).log(f_k^p(t)) \\ loss_{sen}(t) &= -\sum_{k=0}^{3} f_k^g(t).log(f_k^p(t)) \end{aligned} \quad (1)$$

where t is the 5-gram, $f_k^g(t)$ is the gold k-dimensional multinomial class distribution of the 5-gram. The number of dimensions is 4 for sentiment and 7 for emotion, corresponding to our 4 sentiment classes and 7 emotion categories respectively[3]. The sentiment or emotion label of a 5-gram is the same as that

[1] *Lasagne* package has been used for the implementation [10].

[2] For a brief introduction to cross-entropy loss, please refer to http://eprints.eemcs.utwente.nl/7716/01/fulltext.pdf.

[3] sentiment classes are [*positive, negative, neutral, none*] and the emotion classes are [*happiness, sadness, anger, love, hope, amusement, excitement*].

of the containing tweet. $f_k^p(t)$ is the predicted class distribution. The combined loss of the embedding learning classifier is provided in Eq. 2 below:

$$loss_{esswe}(t) = \lambda.loss_{emo}(t) + (1-\lambda).loss_{sen}(t) \tag{2}$$

For our experiments, λ is set at 0.7 empirically. The intuitive explanation for the relatively more weight given to the emotion prediction loss is that emotion categorization being a 7-class problem is more fine-grained and complex than the 4-class sentiment prediction problem.

We also generate the ***Emotion specific word embeddings (E-SWE)*** in a uni-task environment for comparative evaluation. To obtain the E-SWE embeddings, we train a neural network similar to the one described above, the only difference being that the model predicts only emotions, and hence, considers only the cross-entropy emotion loss.

3.3 Emotion Prediction of Tweets

In this step, the word-embeddings obtained from the previous step are used as inputs to appropriate prediction algorithms for obtaining the emotion-sentiment label of the data. We have empirically compared the performance of 12 different combinations of prediction algorithms and input representations to establish the credibility of our approach (refer to Fig. 2). Section 4 provides detailed result descriptions for the different setups. For SVM, GMM, and K-Means based prediction, the tweet level representation is obtained from the learnt unigram embeddings by taking the mean of the word-wise embedding of the constituent words of the tweet. For the CNN based experiments, we use a multi-tasking convolutional neural network to handle inputs (tweets) of varying lengths. The input to this network is the term vector matrix of a tweet, which is of shape nx300. Here, n is the maximum tweet length[4]. We use zero padding of input whenever the length of the tweet is smaller than n. We then apply a 2D-convolutional layer with filters of dimensions cx300 where c is the width of the filter, and 300 is the height. Max-pooling is then applied on the top of the convolutional layer output to extract crucial local features that form a vector of fixed length. Nonlinear rectified linear unit (ReLU) is then applied to this output, which is then fed to a fully connected hidden layer of 200 units. Finally, the terminal output layers of this network are of size seven units and four units, for emotion and

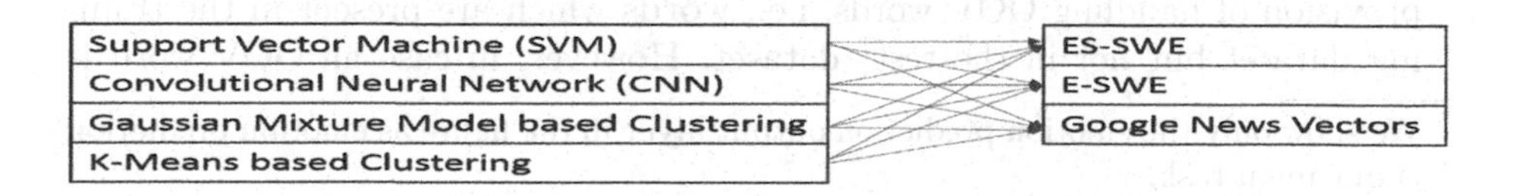

Fig. 2. Different experimental settings

[4] We omitted a few tweets that were either too long or too short, depending on the range to which majority (90%) of the tweet-lengths belonged.

sentiment prediction, respectively, with softmax activation. We use the *Adam* update algorithm with a learning rate of 0.001 for this network too. We feed the tweet matrix (constituted by the unigram embeddings), and the emotion (and sentiment) labels to this CNN to train it. The CNN classifier architecture[5] for emotion prediction is shown in Fig. 3. For the Gaussian Mixture Model and K-Means, we performed the clustering separately for sentiment and emotion, using the open source *scikit-learn* Python library. For the GMM, we specified four Gaussian components for sentiment, and seven components for emotion. The model was iterated for 2000 iterations, with the covariance type as *diag*. We used exactly the same number of clusters to perform the clustering using the K-Means algorithm. The models were made to fit on the training data, and tested on the test data. For clustering based prediction, it is essential to see the goodness of the clusters formed in a clustering algorithm after the data fit, as the test accuracy will fully depend on how good or homogeneous the clusters are. Hence, we measured the cluster purity[6] of the clusters (components) formed by GMM and K-Means on the fit data. These cluster purity values are provided in Table 2. As can be seen, the purity values for the clusters are quite high in almost all of the cases (above 90%) for the 90:10 data split. We obtained similar purity values even for the 70:30 data split, which we do not show exclusively owing to space constraint. Hence, we assumed that the majority emotion (or sentiment) belonging to a cluster as the cluster label, i.e., if a cluster had majority of the points (tweet vectors) in it annotated as emotion e (or sentiment s), we assumed that the cluster belonged to emotion e (or sentiment s).

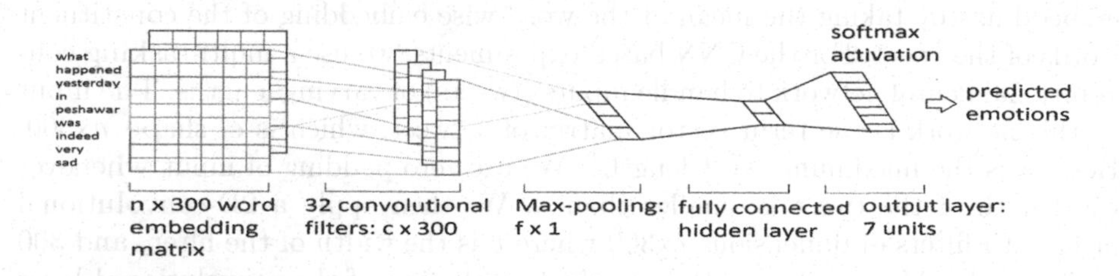

Fig. 3. CNN architecture for tweet classification

- **Considering out of vocabulary (OOV) words:** One of the major advantages of our CNN and clustering based models is that these models have a provision of handling OOV words, i.e., words which are present in the training dataset but not in the test[7] dataset. However, in case an OOV word is

[5] We show only the emotion prediction output layer in the figure as emotion prediction is our main task.

[6] Cluster purity was measured using the standard measure of calculating the fraction of majority labels of the data points among the total number of data points in the cluster.

[7] Embeddings can only be learnt from annotated training data but not from the test data.

Algorithm 1. Replacing OOV words

```
    Input: OOV word, Set of ES-SWE obtained from training data
    Output: Nearest neighbor of OOV word with an ES-SWE
 1: procedure REPLACE_OOV(word, set_of_embeddings)
 2:     if word is present in Google News corpus then
 3:         nbr_list ← FIND_TOP_TEN_NEAREST_NEIGHBORS(word)
 4:         for neighbor in nbr_list do
 5:             if neighbor in set_of_embeddings then
 6:                 return neighbor
 7:             end if
 8:         end for
 9:     end if
10: end procedure
```

encountered in the test set, the system would not be able to find its embedding in the set of learnt embeddings. In such situations, we replaced that word's embedding with the embedding of the nearest neighbor of the word, which is present in the set of learnt embeddings. The nearest neighbor query was performed using a KD-Tree structure, using the Google News vectors. Algorithm 1 describes the process of OOV word handling. Inputs to this algorithm are the set of learnt ES-SWE embeddings from the training set and the OOV word. The output is the nearest neighboring word of the OOV word, which has an ES-SWE representation learnt from the training set.

An important point to note here is that joining negative modifiers does not change the number of non-standard (negative modifier joined) OOV words much in the test set, in our case. However, we plan to address this issue of handling non-standard OOV unigrams in our future work, where we can obtain the word embeddings from a customized large corpus containing such non-standard words, and use them to handle such negative-modifier-joined OOV words (instead of using the Google News vectors).

4 Experimental Evaluation

In this section we discuss our dataset, experimental setup and results obtained by applying our technique on our data set.

4.1 Data

For our experiments, we use the LREC emotweet dataset [24]. In this dataset, a total of 17015 tweets are annotated for both emotion and sentiment. Out of the 29 classes of emotions present in the data, we selected the seven emotion classes:

happiness, sadness, anger, love, hope, amusement, and excitement[8]. We did not work with all 29 classes of emotions as it would have made the prediction task too complicated as many emotion classes were extremely close. An important point to be noted is that for our experiments, we exclude 1317 tweets that are annotated with multiple emotion classes[9]. The four sentiment classes are: *positive*, *negative*, *neutral*, and *none* (*none* indicating that for a certain tweet, the sentiment alignment could not be understood). This brings our used dataset size to around 4.3K (4366 tweets).

4.2 Experimental Results

We present our experimental results in Tables 1 and 2 for the different configurations (for 90:10 training-test split), and summarize our results next.

Jointly learnt task specific embeddings outperform general domain word embeddings for the main task of emotion prediction: As we can see from Tables 1 and 2, ES-SWE performs much better than general word vector embeddings (pre-trained Google News vectors) on all of the models for the task of emotion prediction. For sentiment prediction too, the clustering models perform the best for ES-SWE vectors. For sentiment prediction, the performance of ES-SWE slightly deteriorates for the SVM and CNN classifiers, compared to Google News vectors. The deterioration in sentiment prediction for the CNN can be attributed to the fact that since emotion prediction was our primary goal in this work, we gave less weightage to the sentiment loss (while training the CNN, a high value of λ was chosen). There is an increase of 4% in the emotion classification accuracy

Table 1. Emotion and sentiment detection test accuracies in percentage for different SVM and CNN based classifiers for 90:10 training-test split (CV refers to the cross-validation accuracies)

	Emotion		Sentiment	
Learning algorithm (feature)	CV accuracy	Test accuracy	CV accuracy	Test accuracy
SVM (Google News)	61.78	53.58	81.52	**82.00**
SVM (ES-SWE)	98.98	**54.81**	99.43	80.77
SVM (E-SWE)	99.14	49.08	99.04	78.94
CNN (Google News)	89.10	50.00	94.87	**81.84**
CNN (ES-SWE)	99.15	**54.00**	99.37	79.79
CNN (E-SWE)	98.00	48.54	99.27	79.59

[8] The emotion classes were selected based on two criteria: to maintain the class balance, and to ensure that each sentiment class has more than one emotion classes associated with it. The latter criteria ensures that the emotion class cannot be implied just from the sentiment class. This serves as a motivation of joint learning.

[9] We removed multi-emotion tweets in order to ensure unambiguity in the embedding learning phase. Considering multi-emotion tweets can be taken up as a future work to see if it enhances or deteriorates the quality of the learnt embeddings.

Table 2. Emotion and sentiment detection test accuracies in percentage for GMM and K-Means for 90:10 training-test split

	Emotion		Sentiment	
Learning algorithm (feature)	Cluster purity	Test accuracy	Cluster purity	Test accuracy
GMM (Google News)	0.3456	18.81	0.6809	71.17
GMM (ES-SWE)	0.9906	**56.44**	0.9917	**81.60**
GMM (E-SWE)	0.9905	49.69	0.9192	75.87
KMeans (Google News)	0.3685	20.25	0.6917	71.17
KMeans (ES-SWE)	0.9904	**57.46**	0.9937	**80.37**
KMeans (E-SWE)	0.9937	50.10	0.9208	74.44

with ES-SWE using CNN, and an increase of 1.23% using SVM, compared to the general embeddings. Also, we find that for GMM and K-Means, pre-trained Google News vectors perform poorly (around 19% for GMM, and 20% for K-Means). An early indicator of this poor performance can be seen in the cluster purity values reported in Table 2. Since Google News vectors are not obtained by specific training on sentiment or emotion prediction tasks, the clusters formed by GMM or K-Means are not homogeneous w.r.t emotion or sentiment classes of the tweet vectors (purity values lie in the 34-37% range) using Google News vectors. Task specific ES-SWE or E-SWE capture the emotion or sentiment alignment much better, on the other hand. Hence, clusters obtained using the tweet vectors formed from ES-SWE or E-SWE are much more homogeneous, resulting in better emotion prediction.

Jointly learnt embeddings (multi-tasking) outperform embeddings learnt in a uni-tasked setup in the main task of emotion prediction: We show the bar chart of the emotion prediction test accuracies in Fig. 4, wherein the green colored bars indicate the test accuracy with multi-task trained embeddings, showing their superiority over uni-task trained embeddings as input to the classifier. The highest test accuracy using jointly learnt embeddings is shown by K-Means (57.46%), followed by the GMM (56.44%). These observations support our second hypothesis. Also, from Tables 1, 2, and Fig. 4, we obtain an interesting observation: embeddings learnt jointly (ES-SWE) perform the best in a

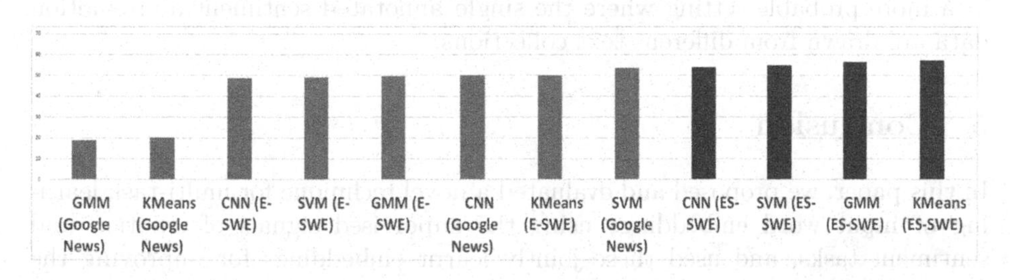

Fig. 4. Emotion classification test accuracy with different embeddings

semi-supervised clustering approach (using GMM or K-Means) in our setup. In fact, both of the models show a drop of around 7%, when uni-task embedding (E-SWE) are used for the clustering. We also ran the models on a 70:30 split of data. Owing to space constraint, we report here the test accuracies only for the clustering models for the 70:30 split[10]. For GMM, the test accuracies for emotion and sentiment prediction were 53.17% and 79.86%, respectively. For K-Means, they were 52.02% and 78.19%. ntuitively, we expect uni-tasked embeddings like E-SWE to perform the best in emotion prediction, since these embeddings are exclusively formed using the task at hand. The reason behind this performance gain of our multi-tasked embeddings is primarily that they capture both emotion and sentiment alignment of the tweets, and these tasks are interrelated. E-SWE word embeddings are based only on the emotion alignment of the tweet. Thus, these embeddings are a bit too biased towards the emotion class to which a tweet belongs, and it ignores the role of sentiment completely. Due to this reason, while predicting the emotion for an unseen (new) tweet, it heavily depends just on the emotional alignment of its constituent words, which at times misleads the classifier. On the other hand, considering both the emotion and sentiment alignment of the words aids us in getting an improved learning for the embeddings, and during emotion prediction, the role of both of these alignments of the constituent words come into play.

To the best of our knowledge, this is the first work on task specific fine tuned representation learning in a emotion-sentiment multi-task setting. One of the related work existing in literature is of Gao et al. [6]. As stated in Sect. 2, they have developed a weighted transition probability based approach for enhancing emotion and sentiment prediction in Chinese texts. According to them, their proposed model can learn from a combination of single-annotated sentiment and emotion data, along with a joint annotated one. However, a direct comparison with them is not appropriate due to the following reasons: firstly, for extracting sentiment-emotion conditional probabilities, they have relied on very less shared data, which may not be exhaustive enough to represent the emotion-sentiment relatedness of the entire data, and secondly, their approach requires much more annotated data than ours. In a practical setting, unless domain adaptation measures are considered, a model needs sufficient in-domain annotated data, which in-turn increases annotation dependency. The work in [6] mentions no such transfer learning measures in their technique to account for distributional difference in a more probable setting where the single annotated sentiment and emotion data are drawn from different text collections.

5 Conclusion

In this paper, we proposed and evaluated a novel technique for multi-task learning of input word embeddings using the supervised signals of emotion and sentiment tasks, and used these jointly learnt embeddings for improving the

[10] The clustering models again showed the best results for the 70:30 split, among all of the other models.

performance of emotion prediction. The main contribution of this paper is that we showed that our jointly learnt embeddings outperform both general pre-trained embeddings and uni-tasked embeddings for the main task of emotion prediction. While our current work was in the context of emotion prediction, our technique for jointly learnt embeddings from multiple related tasks has broad applicability. Hence, we plan to experiment next with a setting where embeddings are jointly learnt sensitive to emotion-topic, and utilized for prediction of emotion or topic downstream. Our jointly learnt embeddings can also be used to incorporate more classification tasks parallely in future, apart from emotion and sentiment prediction. While our current work uses unigram word embeddings, it is straight forward to extend this to n-gram embeddings, and we plan to experiment with this also as part of future work.

References

1. Agrawal, A., An, A.: Unsupervised emotion detection from text using semantic and syntactic relations. In: 2012 IEEE/WIC/ACM International Conferences on Web Intelligence and Intelligent Agent Technology (WI-IAT), vol. 1, pp. 346–353. IEEE (2012)
2. Chaumartin, F.R.: UPAR7: a knowledge-based system for headline sentiment tagging. In: Proceedings of the 4th International Workshop on Semantic Evaluations, SemEval 2007, pp. 422–425 (2007)
3. Collobert, R., Weston, J., Bottou, L., Karlen, M., Kavukcuoglu, K., Kuksa, P.: Natural language processing (almost) from scratch. J. Mach. Learn. Res. **12**, 2493–2537 (2011)
4. Evgeniou, T., Pontil, M.: Regularized multi-task learning. In: Proceedings of the tenth ACM SIGKDD International Conference on Knowledge Discovery and Data Mining, pp. 109–117 (2004)
5. Feng, S., Wang, D., Yu, G., Gao, W., Wong, K.F.: Extracting common emotions from blogs based on fine-grained sentiment clustering. Knowl. Inf. Syst. **27**(2), 281–302 (2011)
6. Gao, W., Li, S., Lee, S.Y.M., Zhou, G., Huang, C.R.: Joint learning on sentiment and emotion classification. In: Proceedings of the 22nd ACM International Conference on Conference on Information & Knowledge Management, pp. 1505–1508 (2013)
7. Kim, S.M., Valitutti, A., Calvo, R.A.: Evaluation of unsupervised emotion models to textual affect recognition. In: Proceedings of the NAACL HLT 2010 Workshop on Computational Approaches to Analysis and Generation of Emotion in Text, CAAGET 2010, pp. 62–70 (2010)
8. Kingma, D., Ba, J.: Adam: a method for stochastic optimization. arXiv preprint arXiv:1412.6980 (2014)
9. Krcadinac, U., Pasquier, P., Jovanovic, J., Devedzic, V.: Synesketch: an open source library for sentence-based emotion recognition. IEEE Trans. Affect. Comput. **4**(3), 312–325 (2013)
10. Lasagne Contributors: Embedding Layer (2014–2015). http://lasagne.readthedocs.io/en/latest/modules/layers/embedding.html. Accessed 11 Nov 2016
11. Li, S., Huang, C.R., Zhou, G., Lee, S.Y.M.: Employing personal/impersonal views in supervised and semi-supervised sentiment classification. In: Proceedings of the 48th Annual Meeting of the Association for Computational Linguistics, ACL 2010, pp. 414–423 (2010)

12. Liu, Y., Liu, Z., Chua, T.S., Sun, M.: Topical word embeddings. In: Proceedings of the Twenty-Ninth AAAI Conference on Artificial Intelligence, AAAI 2015, pp. 2418–2424 (2015)
13. Ma, C., Prendinger, H., Ishizuka, M.: Emotion estimation and reasoning based on affective textual interaction. In: Tao, J., Tan, T., Picard, R.W. (eds.) ACII 2005. LNCS, vol. 3784, pp. 622–628. Springer, Heidelberg (2005). doi:10.1007/11573548_80
14. Mikolov, T., Chen, K., Corrado, G., Dean, J.: Efficient estimation of word representations in vector space. arXiv preprint arXiv:1301.3781 (2013)
15. Mostafa Al Masum, S., Prendinger, H., Ishizuka, M.: Emotion sensitive news agent: an approach towards user centric emotion sensing from the news. In: Proceedings of the IEEE/WIC/ACM International Conference on Web Intelligence, WI 2007, pp. 614–620 (2007)
16. Neviarouskaya, A., Prendinger, H., Ishizuka, M.: Recognition of affect, judgment, and appreciation in text. In: Proceedings of the 23rd International Conference on Computational Linguistics, COLING 2010, pp. 806–814 (2010)
17. Pang, B., Lee, L., Vaithyanathan, S.: Thumbs up?: sentiment classification using machine learning techniques. In: Proceedings of the ACL-02 Conference on Empirical Methods in Natural Language Processing, vol. 10, pp. 79–86. Association for Computational Linguistics (2002)
18. Pennington, J., Socher, R., Manning, C.D.: Glove: global vectors for word representation. In: Proceedings of the 2014 Conference on Empirical Methods in Natural Language Processing, EMNLP 2014, Doha, Qatar, 25–29 October 2014, pp. 1532–1543 (2014)
19. Strapparava, C., Mihalcea, R.: Learning to identify emotions in text. In: Proceedings of the 2008 ACM Symposium on Applied Computing, pp. 1556–1560. ACM (2008)
20. Tang, D., Wei, F., Qin, B., Liu, T., Zhou, M.: Coooolll: a deep learning system for twitter sentiment classification. In: Proceedings of the 8th International Workshop on Semantic Evaluation (SemEval 2014), pp. 208–212, August 2014
21. Tang, D., Wei, F., Yang, N., Zhou, M., Liu, T., Qin, B.: Learning sentiment-specific word embedding for twitter sentiment classification. In: ACL, vol. 1, pp. 1555–1565 (2014)
22. Turney, P.D.: Thumbs up or thumbs down?: semantic orientation applied to unsupervised classification of reviews. In: Proceedings of the 40th Annual Meeting on Association for Computational Linguistics, pp. 417–424. Association for Computational Linguistics (2002)
23. Wu, Y., Kita, K., Matsumoto, K., Kang, X.: A joint prediction model for multiple emotions analysis in sentences. In: Gelbukh, A. (ed.) CICLing 2013. LNCS, vol. 7817, pp. 149–160. Springer, Heidelberg (2013). doi:10.1007/978-3-642-37256-8_13
24. Yan, J.L.S., Turtle, H.R., Liddy, E.D.: EmoTweet-28: a fine-grained emotion corpus for sentiment analysis, pp. 1149–1156 (2016)

Fine-Grained Emotion Detection in Contact Center Chat Utterances

Shreshtha Mundra[1](✉), Anirban Sen[2], Manjira Sinha[1], Sandya Mannarswamy[1], Sandipan Dandapat[3], and Shourya Roy[4]

[1] Conduent Labs, Bangalore, India
{Shreshtha.Mundra,Manjira.Sinha,Sandya.Mannarswamy}@conduent.com
[2] CSE, IIT Delhi, New Delhi, India
anirban@cse.iitd.ac.in
[3] Microsoft IDC, Bangalore, India
sadandap@microsoft.com
[4] Big Data Labs, American Express, New York City, USA
shourya.roy@gmail.com

Abstract. Contact center chats are textual conversations involving customers and agents on queries, issues, grievances etc. about products and services. Contact centers conduct periodic analysis of these chats to measure customer satisfaction, of which the chat emotion forms one crucial component. Typically, these measures are performed at chat level. However, retrospective chat-level analysis is not sufficiently actionable for agents as it does not capture the variation in the emotion distribution across the chat. Towards that, we propose two novel weakly supervised approaches for detecting fine-grained emotions in contact center chat utterances in real time. In our first approach, we identify novel contextual and meta features and treat the task of emotion prediction as a sequence labeling problem. In second approach, we propose a neural net based method for emotion prediction in call center chats that does not require extensive feature engineering. We establish the effectiveness of the proposed methods by empirically evaluating them on a real-life contact center chat dataset. We achieve average accuracy of the order 72.6% with our first approach and 74.38% with our second approach respectively.

Keywords: Emotion detection · Contact center chat utterances

1 Introduction

Contact center is a general term for help desks, information lines and customer service centers. They provide support over multiple channels such as online chat, email-based and voice call to solve product and services-related issues, queries, and requests. To ensure good customer service, contact centers conduct various quality monitoring over these conversations. Emotion detection is one such

(During the research, all authors were part of Conduent Labs India).

J. Kim et al. (Eds.): PAKDD 2017, Part II, LNAI 10235, pp. 337–349, 2017.
DOI: 10.1007/978-3-319-57529-2_27

quality feature, which is used to measure customer satisfaction. However, such retrospective analysis are often performed at the whole chat level and therefore, leave out many details. Table 1 presents an example online chat conversation from a telecommunication contact center with emotion tagged corresponding to each of the agent and customer turns. It can be observed that leaving out fine changes in emotion pattern and only performing one-dimensional analysis will not provide sufficient actionable insights. One point is to be noted here that emotion prediction is quite different from sentiment prediction and they do not always correspond one to one [14]. While emotion detection has been studied for text and spoken conversations, but rarely for chat conversations over only text channel.

In this paper, we propose two novel methods for fine-grain emotion prediction from call center chat utterances[1]. Towards this, we noted that, unlike sentiment, emotion categories (tagset) are not standard and depend on applications. Popular emotion classes (Ekman and Keltner [5] and Plutchik's [18] model) are not suitable for call center chats and propose a new tag-set customized for call center conversations (as annotated in Table 1).

Table 1. An example conversation from our dataset

09:50:14	AGENT:	Hi there, thanks for contacting Mobile Product Support. My name is AGENT, how can I assist you today?	**courteous**
09:51:28	CUST	How you doing AGENT I need help manually actavating this SERVICE and have been trying to contact sombody from customer service for 2 hrs think you can help me?	**unhappy**
09:52:05	AGENT	Certainly, what's going on?	**assurance**
...	...	...	...
09:58:01	AGENT	Then the phone is communicating with the carrier. They may need to manually activate the number on their side	**no emotion**
09:59:27	CUST	yea i no thats why i contacted you guys because they are putting me constantly on hold. Looks like you guys are suggesting me to go back to them!	**unhappy**

In our first approach, We model emotion detection in conversation utterances as sequence labeling problem to exploit sequential nature of conversations using conditional random fields (CRF) [9]. While CRF is a known and well studied technique, application of CRF models with standard bag-of-words (BoW) does

[1] Utterances here referred to textual communications corresponding to each turn from either of the parties.

not lead to satisfactory performance (reported as baseline result in our case Sect. 4). We identified a number of content and meta (non-content) features relevant for emotion detection task which we describe in detail in Sect. 3.1.

Our second technique proposes a neural network based approach to predict the emotion category of each turn in the chat, using the novel idea of feeding the emotion vectors which represent the emotional content of each turn as input to the neural network. This approach bypass the need for extensive feature engineering. Earlier work on emotion vectors [1] have used the noisy emotion vectors directly for emotion prediction task. We learn the vectors for each word from a large un-annotated corpus to incorporate more context specific characteristics and reduce noise, and then feed the emotion vectors to a neural network which performs the emotion classification of each conversational turn.

Lexicon-based approaches have shown advantages of using tagset specific lexicons in emotion categorization task [8], however, developing such lexicons, which are customized to the concerned domain require extensive human expertise. To mitigate this problem and make the process efficient, for both of our approaches, we develop a minimally supervised technique to build emotion lexicon for new domains in contact centers, refer to Sect. 3.

The rest of the paper is organized as follows: In Sect. 2, we provide a brief summary of relevant related work. In Sect. 3, we describe our two approaches for emotion detection in call centre chats. In Sect. 4, we empirically show how our techniques perform on the call centre chat data set and share our observations. We then conclude in Sect. 5 with a short summary of our proposed future work.

2 Related Work

Emotion detection is part of the broader area of Affective Computing, whose focus is to recognize and express emotions [17]. Affect Detection systems can be employed on different modalities and channels such as voice, text, video etc. Given the context of our paper, we confine our summary to relevant work on detecting emotion in text. There exists considerable research on emotion detection in text. Due to space constraints, we provide here only a very brief summary and refer the interested reader to the detailed survey found in [2].

Approaches to emotion detection in text can be broadly classified into keyword based approaches [8,10], linguistic rule based approaches [3,13,15] and machine learning based approaches [16,19]. These categories can be further differentiated in terms of their employing an emotion lexicon or not. While keyword based approaches are simplistic and yield reasonable results, they have certain disadvantages. They do not generalize across multiple domains; Keyword association with emotion is subjective and is error-prone. Since they operate at simple word level, they cannot capture emotions which are expressed through syntactic/semantic relations and context. Linguistic rule based approaches use complex rules for associating syntax, semantic and contextual relations with affective words [3,13,15]. However they suffer from the fact that designing and modifying such emotion recognition rules is a complex task and the rules may not generalize across domains.

To overcome the disadvantages of keyword and rule based approaches, machine learning based techniques for emotion detection in text have been proposed. ML techniques can be broadly classified into supervised learning based approaches [16,20], and unsupervised learning based approaches [1,6,19]. Our work is focused on proposing and evaluating two interesting design points on this spectrum of approaches for emotion detection. Our first approach is a supervised sequence labeling approach for emotion detection with a rich set of content and meta (non-content) features, employing a emotion lexicon which can be built with minimal supervision. Our second approach is a neural network based approach, wherein emotion vectors representing the emotional content in the utterances are fed as input to neural net. Unlike the complex and rich set of features used in our first approach, our second technique does not require heavy feature engineering and the emotion vectors fed as input to neural net are built using a simple technique outlined in an earlier research work [1]. We find that our second technique while being lightweight in feature engineering can still provide reasonably good results compared to a feature-heavy approach.

3 Our Approach

Given a sequence of utterances from call center conversations, our challenge is to find the corresponding sequence of emotions from our emotion tag set. We propose two novel approaches to addressing this problem, building upon earlier relevant work in this space. While these are independent approaches, exploring two totally different design choice points among the wide spectrum of techniques available for emotion detection in text, they follow the general architecture shown in Fig. 1, which is typical of the machine learning based approaches for emotion detection. Our first approach which we term as **Emotion Detection using Minimal Supervised Emotion Lexicon Generation (ED-MSEL)** is characterized by novel contextual and meta-features and has the unique advantage of being supported by an emotion lexicon generated with minimal supervision. Our second approach which we term as **Emotion Detection using Neural**

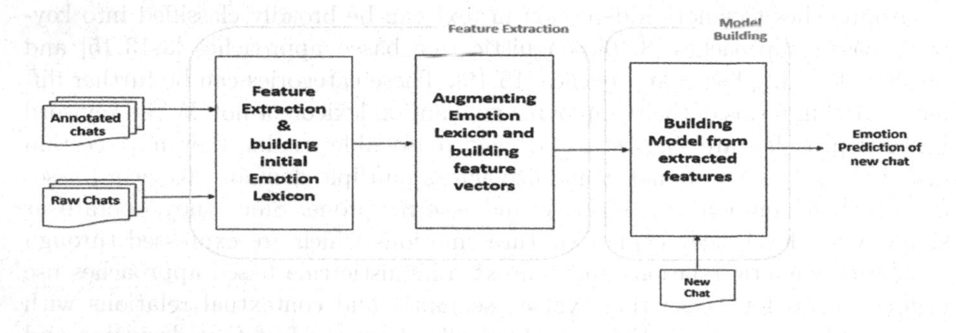

Fig. 1. General architecture

Network Driven by Emotion Vectors (ED-NNEV) builds on the idea of emotion vectors first proposed in [1], and uses these emotion vectors as input to a neural network architecture for emotion detection. In Sect. 3.1, we first describe our **ED-MSEL** technique and subsequently in Sect. 3.2, we describe our **ED-NNEV** approach.

Tagset: As mentioned before, we noted that generic emotion tagset [5,18] containing classes Happiness, Anger, Fear, Surprise, Disgust and Sadness are not suitable for our problem context of call centre conversations. Emotions like fear, surprise or disgust are not very common in customer chats. Already existing state of the art emotional taggers do not perform well on such kind of text due to the fact that the expressed emotion does not map well to the existing tagset. For example, the customer utterance: *"You suggest that I delete my email account, but how does that help?"* will be predicted as showing the emotion class anger by a state of the art emotion tagger[2]. However, clearly this is not the best possible mapping since the customer is not angry at the agent, but only disagreeing with the suggestion made by him. Hence it would be appropriate to tag the text with disapproval as the customer is expressing disapproval to the expressed suggestion. Exploring the call centre chat data, and after considerable deliberation with various call center personnel, we have come up with a novel contact center specific 8-class emotion tagset viz. Happiness (Ha), Assurance (As), Approval (Ag), Courteousness (Co), Apology(Ap), Unhappiness (Uh), Disapproval (Di) and No-Emotion (Ne). While some of our tags such as Apology/Assurance/Approval are related in spirit to dialogue acts in conversations, we found that the emotional overtones carried by these tags mapped more accurately to our call centre data set conversation as we saw in example above, rather than the standard emotions such as fear/anger/surprise which do not arise typically in professional call centre conversations.

3.1 Emotion Detection with Minimal Supervised Emotion Lexicon Generation (ED-MSEL)

Figure 2 shows the block diagram for our first approach. We modeled emotion prediction as a sequential labeling problem using Conditional Random Field (CRF) [9].[3]

Features: Table 2 represents the different content and meta features used in our CRF-based models. While both sets of features have been designed to capture conversational aspect of the problem, particularly meta-features exploit sequence and interrelationship between utterances.

Emotion Lexicon: Inspired from the previous work [8], we develop a technique for building word-emotion lexicon using a small amount of domain-specific labeled conversation (D) and a larger amount of unlabeled chats (R) from the

[2] https://tone-analyzer-demo.mybluemix.net/.

[3] We have used open source implementation CRF++ http://taku910.github.io/crfpp.

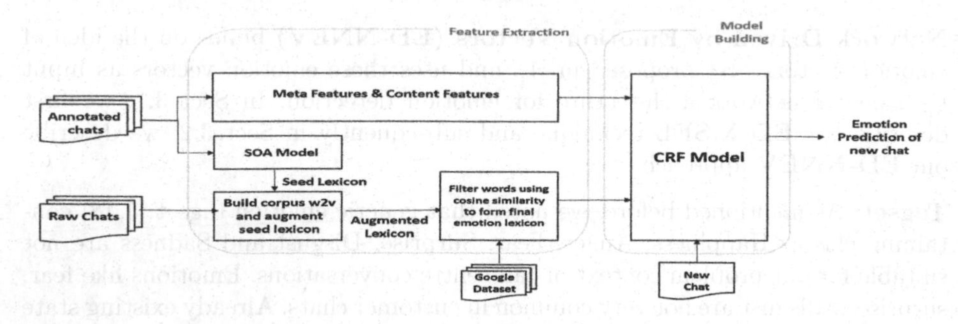

Fig. 2. Block diagram of ED-MSEL approach

Table 2. Features used for emotion categorization

Content-based features	
BOW	Bag of content words whose term frequency ≥ 5
Polar word	Number of positive and negative words in a utterance (using Synesketch [8] lexicon)
Emoticon	Number of positive and negative emoticons
Subsentential	Number of items separated by punctuation
Lex	Presence of lexical items: *but, too* or any *negations*
Prev tag	Emotion category of the previous utterance
Meta features	
A/C label	Boolean indicating Agent(A) or Customer(C) utterance
Delay	Time difference between two consecutive utterances
Position	Utterance sequence number in a conversation
Length	Number of words in a utterance
Segment type	Utterance segments indicating *general statement, query, problem statement, response to a query*

same domain. First, we find word-emotion association from D to identify the set of words which have **strength of association** (SOA) above a threshold with certain emotion categories [12]. The threshold is determined empirically. We use point-wise mutual information (PMI) to estimate the SOA between a word-emotion (w, e) pair based on the Eq. 1. Typically, affect bearing words in a sentence are few and are mainly either of Noun, Adjective, Verb, and Adverb (NAVA) [1]. This is the SOA Model referred to in the block diagram.

$$SOA(w, e) = PMI(w, e) - PMI(w, \neg e) \tag{1}$$

We add the (w, e) pair into our initial lexicon (**IL**) if its SOA is above a threshold. Once we have the initial lexicon IL constructed using the above step, we then use the unannotated corpus data R to augment the Initial Lexicon with new words using vector representation of words. To perform the augmentation, we build the word embeddings of the words present in the unannotated corpus data R using the word2vec method proposed in [11]. Using similarity measures of currently existing words in the Initial Lexicon to the word embeddings in R, we augment the Initial Lexicon as follows: If a novel word $w_j \in$ (R-D) has high cosine similarity with a word $w_i \in IL$ above an empirically determined threshold, we assign same emotional strength of w_i to w_j and add it to the augmented lexicon (**AL**).

During the augmentation step performed above, we essentially measure the contextual domain similarity of words. However sometimes domain similarity measure can result in a non-emotion carrying word getting mapped to a emotion category due to it's high co-occurrence with the emotion carrying word in the corpus. A trivial example is the word pair "really sorry" which due to its high co-occurrence statistics in the specific corpus, results in the word "really" getting mapped to the "apology" category due to its co-occurrence with the word "sorry" which carries the apologetic emotion. While such spurious co-occurrences are infrequent, they need to be filtered out. We perform this filtering by checking whether such a word pair also has some degree of similarity in the general domain and not just in the domain specific corpus. For obtaining the general domain word similarity, we used the pre-trained word vector from Google[4]. For the word pairs which were augmented in the augmentation step, we check their similarity score in the general domain word embedding and using an experimentally arrived similarity threshold, we filter out the words whose similarity in the general domain word embedding fall below the specified threshold. This enables us to remove the spurious word-emotion pairs from AL.

We use the emotion lexicon constructed as mentioned above to create a lexicon driven vector for each conversational turn in the chat. For each turn, we count the number of words in that turn belonging to each of the eight emotion categories and construct a eight dimensional vector for that turn. These Emotion Lexicon based emotion vectors together with the Content-based features and meta features are together used as a feature vector for each turn. This feature vector is fed to the CRF classifier for classifying the emotion category of each turn. We report the experimental results of our first approach in Sect. 4.

Next we describe our second approach for emotion detection which is based on neural networks.

3.2 Emotion Detection Using Neural Network Driven by Emotion Vectors (ED-NNEV)

We propose a neural network based approach to predict the emotion category of each turn in the conversation, using the novel idea of feeding the emotion vectors which represent the emotional content of each turn as input to the neural network. The idea of emotion vectors was first proposed in the paper [1], where they proposed an unsupervised emotion detection scheme using the emotion

[4] https://code.google.com/p/word2vec/.

vectors. While our approach leverages their basic idea of emotion vectors, instead of using the emotion vectors directly as the predictor of emotion class of a turn, we use the emotion vectors as input to a neural network architecture, which is trained using a small amount of annotated data and is used to predict the emotion class of each turn. Also, while the earlier work [1] required that emotion representative initial seed words need to be provided manually for each emotion category, we automatically build the emotion lexicon automatically using the in-domain annotated chat data (note that our dataset consists of a small amount of annotated data (D) and a large corpus of unannotated data (R)).

Our approach consists of the following major steps. First it uses in-domain annotated chat data to build the emotion lexicon. It then extracts the affect bearing words in the corpus, and creates the word level emotion vectors for these affect bearing words by computing their semantic relatedness to the emotion representative words in the emotion lexicon. It then computes the emotion vector corresponding to each turn by aggregating the word-wise emotion vectors obtained, which is then fed to the neural network as input. The neural network which has been trained using the small annotated data set, is then used to predict the emotion category associated with each turn. The overall block diagram is shown in Figure 3. We describe these steps briefly next.

Emotion Lexicon: Following are the steps to build the emotion lexicon, similar to that in approach 1:

(1) We first build word vectors using the standard word2vec method for all the words in our whole corpus (both annotated and unannoated) and this file is our w2v file.
(2) We first collect all NAVA words in the annotated corpus. For each NAVA word, we count the number of times it has occurred in a turn with a particular emotion category, and take the maximum of this count. For example, if the word glad has occurred in turns annotated in the happiness (HA) category the maximum number of times, we take HA as the representative emotion for glad. These NAVA words are then updated in their corresponding emotion bags, and act as the seed words for the next step.
(3) After updating the emotion lexicon with the seed words (from the annotated chat data), we now calculate the inner product similarity of each of these seed words with the words of the word vector file (w2v file created in step1), trained using the complete (annotated + un-annotated) dataset. We empirically set a similarity threshold in order to select only the relevant and sufficiently similar words. These words are also inserted into their corresponding emotion classes in the lexicon. In case the same word is similar to more than one emotion class (meaning the same word is similar to more than one seed words belonging to different emotion classes), we use the simplistic approach of keeping it only in the emotion class where it was first assigned based on similarity. This step completes the process of building the emotion lexicon.

Word Level and Turn Level Emotion Vectors: Similar to the approach followed in [1], we calculate the word level emotion vectors using PMI of each

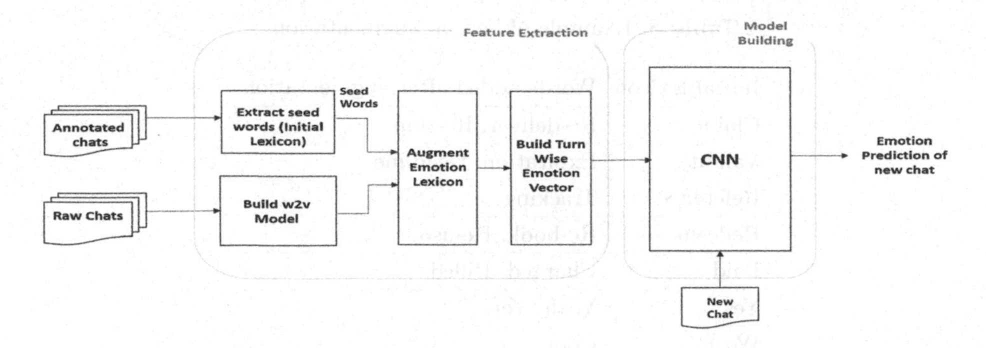

Fig. 3. Block diagram of ED-NNEV approach

NAVA word with all of the emotion representative words. To obtain the emotion score of each NAVA word (w.r.t. a particular emotion class), we take the geometric mean of the PMIs of that NAVA word with all representative words in that emotion class. Since our approach for generating emotion vectors is similar to that followed in [1], we do not describe it here in detail, due to space constraints. Finally, for a turn, we concatenate these word-level emotion vectors (in our case, these vectors will be 8-dimensional owing to the eight emotion classes) to form the turn level emotion vector matrix.

Classification Using a Convolution Neural Network: We use a convolution neural network consisting of one 2D convolution operation and one pooling operation. The size of the convolution filter is 5×8 and the width of the max-pool layer is 6. The size of the input layer of the network is 70 units (calculated from the maximum length of a turn in our data), and the size of the output layer is 8 units. The size of the fully connected hidden layer is 250 units. We use zero-padding in case an input turn has a length smaller than 70 words. We use the *Adam* update algorithm [7] with a cross-entropy loss function to train the neural network with an initial learning rate of 0.001. The *Softmax* activation function is used in the output layer for the prediction. The input to the network is the 70×8 emotion vector matrix corresponding to each turn of the annotated chat dataset. The outputs are the 8 emotion classes. These classes are Happiness (Ha), Unhappiness (Uh), Approval (Ag), Disapproval (Di), Assurance (As), Apology (Ap), Courteousness (Co) and No-emotion (Ne).

4 Experiments and Results

We selected a real-world dataset of contact center chats about mobile phones and manually annotated about 300 chats, i.e. total 2537 (150 Ha, 241 Uh, 177 Ag, 37 Di, 113 As, 95 Ap, l621 Co, 1117 Ne) utterance segments from agents and customers combined according to the emotion tagset defined in Sect. 3. About a fifth of the annotation exercise were repeated by 3 annotators independently to measure inter-annotator agreement (IAA) which came out to be 0.78 using

Table 3. Example of lexicon augmentation

Initial lexicon	Words added after augmentation
Claim	Re-deliver, Re-ship
Validity	Expiration, Deadline
Reference	Tracking
Redeem	Re-book, Re-use
Paid	Charged, Billed
Yes	Yeah, Yep
Wow	Cool

Cohen's kappa [4]. Additionally, we used another 7000 chats (5,17,352 utterances) to build the augmented emotion lexicon.

Table 3 shows examples of words which have been added to the lexicon after augmentation. Normal thesaurus search for these words does not return the augmented words as synonyms. Thus we are able to capture the task specific context.

Table 4 shows relative performances of CRF based emotion classification system under different settings as described in the previous sections. Accuracy numbers refer to the ratio of correctly labeled segments to total number of segments. Baseline performance refers to CRF with Bag-of-Word (BoW) features. Using Content and Meta features improve performance significantly over baseline demonstrating usefulness of the features identified. Use of AL gives improvement of another 2.9% over the *content+meta* feature-based model. Talking specifically of the Customer's turn, use of AL over *content+meta* feature-based model gives improvement of 5.6% (from 56.8 to 62.4%) in accuracy. This is because AL essentially reduce the data sparsity problem by extracting relevant word-emotion pairs from large unlabeled data. The best overall accuracy achieved for emotion detection is 72.6% against the human agreement of 78% (IAA) which is reasonable considering this was a 8-class classification task. For our call center domain chat data set, we find that the size of Initial Lexicon was 268 words and the size of the final Augmented Lexicon was 762 words. Too demonstrate advantage of using CRF over other supervised techniques, we have also evaluated the performance of a Support Vector Machine (SVM) with same set of features. We have found that SVM's accuracy was about 8% less than CRF which we believe is primarily owing to inability to consider sequence information (emotion of an utterance indeed depends on emotion of prior utterance(s)). Intrigued reader might have made an interesting observation that usage of sophisticated features and AL improve accuracy on customer segments significantly by 22% but only by 5% for agent segments. We opine that this is because agents' utterances are homogeneous in nature and lack variation with respect to number of emotion categories (in fact, majority of agents' utterances are *Ne* and *Co* statements).

The ED-NNEV approach shows an improvement in the cross validation accuracy (74.38%). We believe that this is primarily because of the property of the neural network classifier to inherently learn features for the classification task. The novelty of this approach is the fact that it does not require explicit feature engineering to perform the classification (unlike ED-MSEL), and the classifier performs the classification task commendably, given only the emotion vector representations. We have extensively experimented with different parameter combinations and found that the network performs best for dropout = 0.2 and hidden layer size 250.

• **Dataset 2**

We have mentioned in earlier section that our main assumption behind developing two models ED-MSEL and ED-NNEV, is that if presented with a smaller dataset, explicit feature engineering based approach that is ED-MSEL will be more appropriate. To test this hypothesis, we have considered a second dataset of call center chat utterances collected from an e-commerce domain. This dataset contains 559 annotated data instances across our 8 emotion classes (32 Ha, 95 Uh, 26 Ag, 2 Di, 24 As, 45 Ap, 126 Co, 206 Ne). We have applied both our models on this dataset and obtained accuracy of **55%** for ED-MSEL and **51%** for ED-NNEV respectively.

By this experiment, we emphasize on the fact that both of our approaches are equally important, depending on the scenario under consideration. While for small datasets, we can leverage the advantage of explicit feature engineering, for larger ones, we can perform satisfactorily using a neural network based classifier.

Table 5 shows a detail class-specific view of the proposed system along with distribution of utterance segments belonging to each class. Identification of two large classes, viz. *Ne* and *Co*, is contributing significantly to the overall performance. Smaller and more specific classes need additional attention on the basis of other confusing classes (as shown in the last column). It is intuitive that negative emotion classes (*Ap*, *Uh*, *Di*) are often confused with other negative emotion classes whereas positive emotion (*Ha*, *As*, *Ag*, *Co*) are confused with other positive emotion classes. This hints that a two-level classification approach might be worth exploring to segregate positive instances from negative followed by finer grained classification.

Table 4. Leave one out cross validation emotion categorization accuracy

Features	Accuracy (%)
Baseline (BOW)	59.9
Content	63.9
Content+Meta	69.7
ED-MSEL (with CRF)	**72.6**
ED-NNEV	**74.38**
ED-MSEL (with SVM)	64.4

Table 5. Accuracy for different emotion types on the two approaches.

Emotion(#)	ED-MSEL			ED-NNEV		
	Precision	Recall	Top confusions	Precision	Recall	Top confusions
Assurance(As)	71.67	69.35	Ne(9), Co(6)	97.34	88.50	Ne(14)
Approval(Ag)	50.00	47.73	Ha(14), Ne(14)	85.39	82.16	Ne(24), Ha(7)
Courteous(Co)	79.90	81.60	Ne(30), Ha(25)	93.75	100	-
Happiness(Ha)	39.25	41.58	Co(23), Ag(13)	78.84	53.24	Ne(21), Ag(15)
Apology(Ap)	77.78	46.67	Co(6), Ne(3)	100	100	–
Disapproval(Di)	61.11	47.83	Ne(8), Ag(2)	92.30	44.44	Ne(14),Uh(1)
Unhappiness(Uh)	60.40	49.59	Ne(51), Ha(6)	24/28	46.15	Ne(25)
No-emotion(Ne)	78.12	82.84	Co(34), Ha(30)	85.60	97	Ag(10), Ha(3)

5 Conclusion and Future Work

In this paper, we have proposed and evaluated two novel approaches for emotion detection in short task-oriented chats using a new tagset specific for contact center conversation. With the emergence of automatic conversational agents in enterprises and consumer devices we believe that this work is well timed as such agents are expected to be *emotionally correct* in their responses. Towards that the first step is to identify speakers' emotion which has been the motivation for our work. Future work include improving the technique by introducing a two-level classification (sentiment plus emotion classification) as well as validation of accuracy numbers on other similar datasets. We also intend to perform our experiments on larger datasets (when available), and eventually develop a unified emotion classification model which leverages the benefits of both of the approaches. Finally, we want to gradually develop our notion of turn-level emotion classification into a holistic approach for chat-level customer satisfaction analysis. This would require us to develop suitable aggregation mechanisms (from turn-level to chat-level), which we will be working on in future.

References

1. Agrawal, A., An, A.: Unsupervised emotion detection from text using semantic and syntactic relations. In: 2012 IEEE/WIC/ACM International Conferences on Web Intelligence and Intelligent Agent Technology (WI-IAT), vol. 1, pp. 346–353. IEEE (2012)
2. Canales, L.: Emotion detection from text: a survey (2014)
3. Chaumartin, F.R.: Upar7: a knowledge-based system for headline sentiment tagging. In: Proceedings of the 4th International Workshop on Semantic Evaluations, pp. 422–425. SemEval 2007, Association for Computational Linguistics, Stroudsburg (2007). http://dl.acm.org/citation.cfm?id=1621474.1621568
4. Cohen, J.: Weighted kappa: nominal scale agreement provision for scaled disagreement or partial credit. Psychol. Bull. **70**(4), 213 (1968)

5. Ekman, P., Keltner, D.: Universal facial expressions of emotion. Calif. Ment. Health Res. Dig. 8(4), 151–158 (1970)
6. Kim, S.M., Valitutti, A., Calvo, R.A.: Evaluation of unsupervised emotion models to textual affect recognition. In: Proceedings of the NAACL HLT 2010 Workshop on Computational Approaches to Analysis and Generation of Emotion in Text, pp. 62–70. CAAGET 2010, Association for Computational Linguistics, Stroudsburg (2010). http://dl.acm.org/citation.cfm?id=1860631.1860639
7. Kingma, D., Ba, J.: Adam: a method for stochastic optimization. arXiv preprint (2014). arXiv:1412.6980
8. Krcadinac, U., Pasquier, P., Jovanovic, J., Devedzic, V.: Synesketch: an open source library for sentence-based emotion recognition. IEEE Trans. Affect. Comput. 4(3), 312–325 (2013)
9. Lafferty, J., McCallum, A., Pereira, F.C.: Conditional random fields: probabilistic models for segmenting and labeling sequence data (2001)
10. Ma, C., Prendinger, H., Ishizuka, M.: Emotion estimation and reasoning based on affective textual interaction. In: Tao, J., Tan, T., Picard, R.W. (eds.) ACII 2005. LNCS, vol. 3784, pp. 622–628. Springer, Heidelberg (2005). doi:10.1007/11573548_80
11. Mikolov, T., Chen, K., Corrado, G., Dean, J.: Efficient estimation of word representations in vector space. arXiv preprint (2013). arXiv:1301.3781
12. Mohammad, S.M.: Emotional tweets. In: Proceedings of the First Joint Conference on Lexical and Computational Semantics, vol. 1: Proceedings of the Main Conference and the Shared Task, vol. 2: Proceedings of the Sixth International Workshop on Semantic Evaluation, pp. 246–255. Association for Computational Linguistics (2012)
13. Mostafa Al Masum, S., Prendinger, H., Ishizuka, M.: Emotion sensitive news agent: an approach towards user centric emotion sensing from the news. In: Proceedings of the IEEE/WIC/ACM International Conference on Web Intelligence, pp. 614–620. WI 2007. IEEE Computer Society (2007). http://dx.doi.org/10.1109/WI.2007.129
14. Munezero, M.D., Montero, C.S., Sutinen, E., Pajunen, J.: Are they different? Affect, feeling, emotion, sentiment, and opinion detection in text. IEEE Trans. Affect. Comput. 5(2), 101–111 (2014)
15. Neviarouskaya, A., Prendinger, H., Ishizuka, M.: Recognition of affect, judgment, and appreciation in text. In: Proceedings of the 23rd International Conference on Computational Linguistics, pp. 806–814. COLING 2010. Association for Computational Linguistics, Stroudsburg (2010). http://dl.acm.org/citation.cfm?id=1873781.1873872
16. Pang, B., Lee, L., Vaithyanathan, S.: Thumbs up?: sentiment classification using machine learning techniques. In: Proceedings of the ACL-2002 Conference on Empirical Methods in Natural Language Processing, vol. 10, pp. 79–86. Association for Computational Linguistics (2002)
17. Picard, R.W.: Affective computing (1995)
18. Plutchik, R.: Emotions and Life: Perspectives from Psychology, Biology, and Evolution. American Psychological Association, Ann Arbor (2003)
19. Strapparava, C., Mihalcea, R.: Learning to identify emotions in text. In: Proceedings of the 2008 ACM Symposium on Applied Computing, pp. 1556–1560. ACM (2008)
20. Wu, Y., Kita, K., Matsumoto, K., Kang, X.: A joint prediction model for multiple emotions analysis in sentences. In: Gelbukh, A. (ed.) CICLing 2013. LNCS, vol. 7817, pp. 149–160. Springer, Heidelberg (2013). doi:10.1007/978-3-642-37256-8_13

Dependency-Tree Based Convolutional Neural Networks for Aspect Term Extraction

Hai Ye[1], Zichao Yan[1], Zhunchen Luo[2(✉)], and Wenhan Chao[1]

[1] School of Computer Science and Engineering, Beihang University, Beijing, China
{yehai,yanzichao,chaowenhan}@buaa.edu.cn

[2] China Defense Science and Technology Information Center, Beijing, China
zhunchenluo@gmail.com

Abstract. Aspect term extraction is one of the fundamental subtasks in aspect-based sentiment analysis. Previous work has shown that sentences' dependency information is critical and has been widely used for opinion mining. With recent success of deep learning in natural language processing (NLP), recurrent neural network (RNN) has been proposed for aspect term extraction and shows the superiority over feature-rich CRFs based models. However, because RNN is a sequential model, it can not effectively capture tree-based dependency information of sentences thus limiting its practicability. In order to effectively exploit sentences' dependency information and leverage the effectiveness of deep learning, we propose a novel dependency-tree based convolutional stacked neural network (DTBCSNN) for aspect term extraction, in which tree-based convolution is introduced over sentences' dependency parse trees to capture syntactic features. Our model is an end-to-end deep learning based model and it does not need any human-crafted features. Furthermore, our model is flexible to incorporate extra linguistic features to further boost the model performance. To substantiate, results from experiments on SemEval2014 Task4 datasets (reviews on restaurant and laptop domain) show that our model achieves outstanding performance and outperforms the RNN and CRF baselines.

Keywords: Aspect term extraction · Dependency information · Tree-based convolution · Deep learning

1 Introduction

Aspect-based sentiment analysis (or opinion mining) aims to identify the opinions in a given document. To achieve this goal, six subtasks should be considered and aspect term extraction is one of the important subtasks [1]. Aspect terms are attributes (or properties) of the entity that opinion expresses on. For example, given the product review "I love the way the entire suite of software works together", the aspect term is "suite of software".

The task of aspect term extraction is usually regarded as a sequence labeling problem, in which each word in sentence is labeled by conventionally used

J. Kim et al. (Eds.): PAKDD 2017, Part II, LNAI 10235, pp. 350–362, 2017.
DOI: 10.1007/978-3-319-57529-2_28

BIO tagging scheme. In this paper, we also regard aspect term extraction as a sequence labeling problem. Conditional random fields (CRFs) and its variants like semi-CRFs have been successfully applied to this problem. However, these CRFs based models are feature-rich models which need much human-crafted feature engineering effort to work well.

In recent years, deep learning has become the popular and effective method to deal with the tasks in computer vision (CV) and natural language processing (NLP). [2] first applied deep learning to NLP tasks including part-of-speech tagging, chunking, etc. Meanwhile, deep recurrent neural network (RNN[1]) based models have been proposed for aspect term extraction [3]. Unlike the CRFs based models, these RNN based models do not need any manually features. The experimental results on SemEval2014 Task4 datasets show the superiority of RNN based models over traditional CRFs based models.

Previous work has shown that leveraging syntactic features is helpful for opinion mining. Dependency parse tree is one of the important syntactic features and has been widely applied to aspect-based sentiment analysis [4–8]. [4] applied dependency path features to opinion target extraction. Recently, [8] employed an unsupervised method to incorporate dependency context features into embeddings for aspect term extraction. These works manifest that leveraging dependency information of sentences may be helpful and necessary for aspect term extraction.

Although these RNN based models mentioned above can solve the shortcomings that CRFs based models have, they can not make full use of dependency information of sentences that is critical for opinion mining. Because RNN belongs to sequential models, it codes words one by one along the sentence and can only capture the linear context features, thus ignoring tree-based syntactic features over a long path. This drawback may limit its practicability for aspect term extraction.

So, in order to exploit sentences' dependency information and leverage the effectiveness of deep learning for aspect term extraction, we propose a novel dependency-tree based convolutional stacked neural network (DTBCSNN) to extract aspect terms without any human-crafted feature engineering effort. DTBCSNN consists of three main parts: a dependency-tree based convolutional layer (DTBCL), a stacked neural network (SNN[2]) and an inference layer. DTBCL is applied to effectively capture the sentences' dependency information and its core notion is tree-based convolution. **Tree-based convolution** has been explored in a lot of works [9–11]. It can effectively exploit sentences' syntactic features over the parse trees, capturing the relations between words in a long distance. So we adopt tree-based convolution to exploit dependency information of sentences. Specifically, DTBCL first does convolution operation over the fixed-depth subtrees of a parsed dependency tree. Then the output hidden features from DTBCL are propagated to the SNN to learn tag score distributions

[1] In this paper, RNN refers to recurrent neural network.

[2] In order to simplify the description of our model, we define the several hidden neural networks being stacked together as SNN.

for tags. The inference layer is to find tag path with highest scores based on the learned tag score distributions. Though our model is an end-to-end model, it is flexible and can incorporate extra linguistic features to further boost the model performance. We conduct experiments on SemEval2014 Task4 datasets and the experimental results show the superiority of our model over the RNN and CRF baselines.

To sum up, our contributions in this paper can be encapsulated as follows:

- Novel tree-based convolution combined with a neural network is introduced to effectively leverage sentences' dependency information critical but not fully exploited by previous deep learning based models for aspect term extraction.
- Our model is an end-to-end deep learning based model that does not need any manually features. Furthermore, our model is flexible enough to incorporate linguistic features to boost model performance.
- We conduct extensive experiments to evaluate the model sensitivities to architectures, adding linguistic features and word embeddings.

2 Related Work

Among previous work on aspect terms or opinion targets extraction, there are typical methods worth to mention. [12] applied part-of-speech tagging parser to label words and phrases to extract hot (frequent) features for mining customer reviews. Then [1] used association mining method to extract product features. Following up, [13] proposed to use human-defined opinion word seeds and rules from dependency parsing to extract opinion targets iteratively. This kind of problem could also be regarded as a sequence labeling problem and then a classifier is applied. Hidden Markov Models (HMMs) [14] and conditional random fields (CRFs) [4,15] are usually the chosen ideal models and the wining systems [16,17] in SemEval2014 Task4 datasets are CRFs based models. Topic model techniques can also be applied to this kind of problem using Latent Dirichlet Allocation (LDA) [18,19].

The topic of sentiment analysis has been explored by deep learning in recent years and has witnessed state-of-the-art performance in this domain [21,22]. There are also some work on aspect term or opinion expression extraction using deep learning models. [20] originally combined the deep recurrent neural networks (RNN) and pre-trained word embeddings for opinion expression extraction. Afterwards, motivated by this work, [3] proposed a similar recurrent neural network model and push it further, a set of different types of RNN models were explored. These proposed RNN models outperform traditional feature-rich CRF model. However, recurrent neural network can not effectively capture tree-based syntactic information. As a result, RNN model may not so well fit the aspect term extraction problem.

Tree-based convolution has been studied by [9], which aims to capture sentences' syntactic features to solve certain problems where syntactic information is needed. Tree-based convolution achieved extraordinary performance on sentiment analysis and question classification [9], which manifests the effectiveness of this kind of approach. Furthermore, tree-based convolution is successfully applied to programming language processing [10].

Though inspired by [9], our model is totally different from theirs as follows: (1) We deal with different problems. Ours is aspect term extraction but theirs are sentiment polarity analysis and question classification. (2) We use different model architectures. To fit the problem, we connect the output features of DTBCL over a fixed-size window instead of applying a pooling layer. And we also combine DTBCL with SNN. Similar to [2], we use an inference layer instead of softmax layer [3] for the task of choosing labels, so we apply a stacked neural network after DTBCL to learn tag score distributions for inference layer. We will detail our model in the following sessions.

3 Dependency-Tree Based Convolutional Stacked Neural Networks

Overall, our model can be divided into three major components: (1) dependency-tree based convolutional layer; (2) stacked neural network; (3) inference layer. Figure 1 right shows the overall architecture of DTBCSNN. Following, we will give out detail discussions about our model.

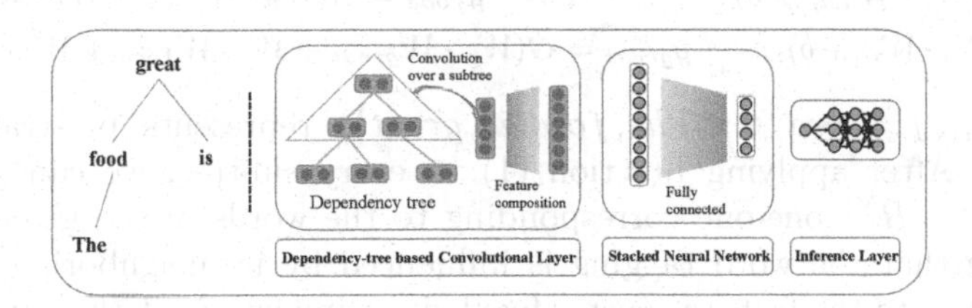

Fig. 1. Left: Example of a dependency parse tree for sentence "The food is great". Right: Illustration of the overall architecture of the Dependency-tree based Convolutional Stacked Neural Network (DTBCSNN). The main components of DTBCSNN are dependency-tree based convolutional layer, stacked neural network and inference layer.

3.1 Dependency-Tree Based Convolutional Layer for Incorporating Dependency Information

Dependency-tree based convolutional layer (DTBCL) is the core component of our model, which aims to capture sentences' dependency information.

To apply dependency-tree based convolution, a sentence should firstly be converted into a dependency parse tree. Each of its nodes represents the original word of the sentence and is initialized by pre-trained word embeddings. Non-leaf nodes can be seen as parent nodes governing a set of child nodes, which have the relationship with their corresponding child nodes called parent-child relation (PCR). In this paper, we regard the different relationships between parent nodes and child nodes like *nsubj*, *nmod*, *conj*, etc. as one shared relation PCR.

Every word in the sentence can be seen as a parent node connecting with its child nodes. Figure 1 left shows an example of dependency parse tree of sentence "The food is great". We can see from the parse tree that word "great" has child nodes "food" and "is" and, word "food" has word "The" as its child node. Then we use a two-layer fixed-depth feature detector to slide over the sentences' dependency parse trees to capture features which are each corresponding to the words in the sentence.

Suppose that we are given a subtree: a parent node p with its child nodes c_1,c_2,...,c_n, the output feature y of the subtree can be calculated by the function:

$$y = G(W_p \cdot p + \sum_{i=1}^{n} W_c \cdot c_i + b) \tag{1}$$

where W_p, $W_c \in R^{N_f \times N_{embed}}$, W_p is the weight matrix for parent nodes and W_c is the weight matrix for child nodes based on the relation PCR; b is the bias; p, $c_i \in R^{N_{embed}}$, $b \in R^{N_f}$. (N_f is the dimension of output feature y; N_{embed} is the word embedding dimension; n represents the number of child nodes of parent p.) $G(\cdot)$ is an activation function and we use the *ReLU* function in this paper.

We take the parse tree from Fig. 1 left for example. The features for each word can be calculated by function (1) as follows:

$$y_{The} = G(W_p \cdot W_{The} + b) \qquad y_{food} = G(W_p \cdot W_{food} + W_c \cdot W_{The} + b)$$
$$y_{is} = G(W_p \cdot W_{is} + b) \qquad y_{great} = G(W_p \cdot W_{great} + W_c \cdot W_{food} + W_c \cdot W_{is} + b)$$

where W_{word} ($word \in \{The, food, is, great\}$) represents pre-trained word embedding. After applying function (1) to every subtree, we can get output features $\{f\} \in R^{N_f}$ one-one corresponding to the words in the given sentence. Considering that the word tagging is influenced by its neighboring words, we further aggregate the features over a fixed-size window to get the compositional features $\{f_1\}$. Specifically,

$$f_{1,i} = \begin{pmatrix} f_{\llcorner i - N_{win}/2 \lrcorner} \\ \vdots \\ f_{\llcorner i + N_{win}/2 \lrcorner} \end{pmatrix} \tag{2}$$

where $f_{1,i} \in \{f_1\}$; $\{f_1\} \in R^{N_{f_1}}$, $N_{f_1} = N_{win} \cdot N_f$. ($N_{win}$ is the size of window.) The features with indexes exceeding the boundary of the sentence are padded with zero vectors.

3.2 Stacked Neural Networks for Tag Score Distributions Learning

After getting the features that leverage the sentences' dependency information, we propagate these features to the stacked neural network (SNN). SNN is applied to learn tag score distributions for inference layer.

A SNN with L layers can be seen as a composition function $H_\theta(\cdot)$ with parameters θ:

$$H_\theta(\cdot) = H_\theta^L(H_\theta^{L-1}(\ldots H_\theta^1 \ldots)) \tag{3}$$

where H_θ^l is defined for layer l ($1 \leq l \leq L$). In this paper, we apply two hidden layers as our stacked neural networks. The architecture of two hidden layers has been applied in [2] for tag score distributions learning. Motivated by this, we adopt two hidden layers for this paper.

For each hidden feature $f_{1,i} \in \{f_1\}$, we apply SNN to learn the tag score distributions. Concretely, the score distribution $f_{s,i}$ for $f_{1,i}$ can be calculated by function (4):

$$f_{s,i} = W^{(2)} \cdot f_{2,i} + b^{(2)} = W^{(2)} \cdot g(W^{(1)} \cdot f_{1,i} + b^{(1)}) + b^{(2)} \tag{4}$$

where ($W^{(1)} \in R^{N_{f_2} \times N_{f_1}}$, $b^{(1)} \in R^{N_{f_2}}$) and ($W^{(2)} \in R^{N_{f_s} \times N_{f_2}}$,$b^{(2)} \in R^{N_{f_s}}$) are the parameter matrixs and bias for the first hidden layer and second hidden layer respectively. (N_{f_1} and N_{f_2} are the output vector dimensions of the first hidden layer and second hidden layer respectively; N_{f_s} is the size of tag score distribution vector and also is the number of tags.) $g(\cdot)$ is an activation function and we use *Sigmoid* function in this paper. After applying SNN to each hidden feature $f_{1,i} \in \{f_1\}$, we get the tag score distributions $\{f_s\} \in R^{N_{f_s}}$.

3.3 Inference Layer

After SNN, we can get score distributions over tags for every word in a sentence. The inference layer is used to find a specific tag path with a highest score representing the most possible BIO labels for the sentence. More explanation about the physical meanings of inference layer can be found in [2].

Suppose we are given the tag score distributions $\{f_s\} \in R^{N_{f_s}}$ for the sentence with size n and learned transition matrix $A \in R^{N_{f_s} \times (N_{f_s}+1)}$, we can get the tag path $[t_1 : t_n]^*$ by function (5):

$$[t_1 : t_n]^* = argmax_{[t_1:t_n]} \sum_{i=1}^{n} A_{t_{i-1},t_i} + f_{s,t_i} \tag{5}$$

where t_i is the tag for the word in the i_{th} position of the sentence; A_{t_{i-1},t_i} represents the transition possibility from tag t_{i-1} to tag t_i; f_{s,t_i} is the score for tag t_i in i_{th} position. We use the *Viterbi* algorithm to solve function (5).

3.4 Training Method

As discussed above, we get the parameters of our model: $\Theta = \{W_p, W_c, b, W^{(1)}, b^{(1)}, W^{(2)}, b^{(2)}, A\}$. The object function is:

$$\sum_{\forall d \in S, d \rightarrow t} log\ p(t|d, \theta) \tag{6}$$

where S is training set; t is the golden tag path for $d \in S$; $log\ p(t|d,\theta) = log\ exp\{s(d,t,\theta)\} \ / \sum_{\forall t'} exp\{s(d,t',\theta)\} = s(d,t,\theta) - log \sum_{\forall t'} exp\{s(d,t',\theta)\}$, which is a sentence-level likelihood ($s(d,t',\theta)$ is the score for tag path t')

(Because of space limitation, we can not discuss sentence-level likelihood in detail but more explanation about it can be found in [2]). We use a variation of stochastic gradient descent called AdaGrad and backpropagation algorithm to update parameters Θ.

4 Boosting the Model by Adding Linguistic Features

Though our model DTBCSNN is an end-to-end deep learning based model without any human-crafted features, it can easily incorporate linguistic features to further boost model performance. Our model does not need to change the dimensions of word embeddings or stacked neural network settings, only requiring to append the linguistic feature vectors to the output feature vectors of the first hidden layer and learn extra parameters for linguistic features. The number of linguistic features that may be useful for aspect term extraction such as POS tags, sentiment lexicon is large. In this paper, we choose POS tags and chunk information [3] as the linguistic features.

5 Experiments

5.1 Experimental Settings

Pre-processing. We use the common BIO coding method to label our dataset, in which "B" represents "beginning of aspect term", "I" represents "inside of aspect term" and "O" is for "outside of aspect term". The predicted segmentation with "B" at the beginning followed by "O" is regarded as an aspect term.

Datasets. In this paper, we adopt the datasets from SemEval2014 Task4[3] (reviews on restaurant and laptop domain) whose specific description is shown in Table 1. For each domain, we only use the training set to train our model and then apply the test set to evaluate the trained model. In all of the experiments, we train and test our model in a unified manner.

Table 1. Statistics of SemEval2014 datasets. #S means "sentence" and #T means "aspect term".

	Restaurant		Laptop		Total
	Training	Test	Training	Test	
#S	3,041	800	3,045	800	7,686
#T	3,693	1,134	2,358	654	7,839

Evaluation Metrics. Exact evaluation metric is applied in our paper. This means that only the predicted aspect term whose boundary matches the golden

[3] http://alt.qcri.org/semeval2014/task4/index.php?id=data-and-tools.

boundary can be seen a right one. We use F1 scores to evaluate the model performance.

Model Settings. We firstly use pre-trained word embeddings to initialize the word vectors. Considering that the performance of model can benefit from the word embeddings trained by the same domain corpora, we train word embeddings with gensim[4] on Yelp Challenge dataset[5] for restaurant domain and on Amazon dataset[6] for laptop domain. We select all restaurant reviews from Yelp datasets and only the electronic reviews from Amazon datasets. We tuned word embedding dimension in $\{100, 150, 200, 250, 300, 350, 400\}$ and determined the size of 300 for the two domains. The window size is 3 tuned from $\{1, 3, 5\}$. For laptop domain, N_f is 300 tuned from $\{200, 250, 300, 350, 400\}$ and N_{f_2} is 250 tuned from $\{200, 250, 300, 350\}$; for restaurant domain, N_f is 250 and N_{f_2} is 200 tuned the same way as laptop domain. During the process, Stanford Parser [23] is used to get dependency parse trees for sentences.

We combine mini-batch AdaGrad and early stopping to train our model. The mini-batch size is 25 tuned from $\{10, 15, 20, 25, 30\}$. In order to use the method of early stopping, 10% data from the training set are randomly selected as validation set and the rest of them as training set. When each epoch training is finished, we use validation set to evaluate the model, to keep whichever parameters that entail the best performance on the validation set. By the way the early stopping steps are set to 10 tuned from $\{5, 10, 15, 20\}$, meaning that if the performance fails to exceed the best result over 10 times, the training will be stopped. After going through the steps mentioned above, it is time to use test set to evaluate the trained model on the best parameters.

Adding Linguistic Features. We use POS tags and chunk information as the extra linguistic features. POS tags contains four types including *noun*, *adjective*, *verb* and *adverb*, while chunk contains five classes: *NP*, *VP*, *PP*, *ADJP* and *ADVP*. These are all coded as binary features.

5.2 Baseline Methods

We compare our model with the following baselines:

- **CRF.** A linear-chain CRF with commonly used linguistic features including current word, context information, POS tag, positions, stylistics and prefixes and suffixes between one to four characters [3].
- **Elman-RNN.** An Elman type recurrent neural network on the top of word embeddings proposed by [3]. Elman-RNN contains a lookup-table layer, a hidden layer and an output layer.
- **Elman-RNN+F.** The above Elman-RNN adding the same linguistic features as ours.

[4] https://pypi.python.org/pypi/gensim.
[5] http://www.yelp.com/datasetchallenge.
[6] http://jmcauley.ucsd.edu/data/amazon/links.html.

- **LSTM-RNN.** An LSTM network which is the another type of recurrent neural network proposed by [3]. It shares a same architecture with Elman-RNN.
- **LSTM-RNN+F.** The above LSTM network adding the same linguistic features as ours.
- **HIS_RD.** The top system [17] on the laptop domain for SemEval2014 Task4. It is a CRF based model leveraging lexical features, syntactic information, etc.
- **DLIREC.** The winning system [16] on the restaurant domain for SemEval2014 Task4. It is another CRF based model and it considers various lexical, syntactic, semantic features of the sentences.

We mainly compare our model with the RNN based models proposed by [3]. In their paper, they have proposed numerous types of RNN based models including Jordan-RNN, Elman-RNN, LSTM, Bi-Elman-RNN, Bi-LSTM, etc. Their experimental results show that Elman-RNN and LSTM achieve higher performance over other models, so we adopt Elman-RNN and LSTM as the baselines ignoring the other inferior models.

5.3 Final Results and Analysis

Table 2 shows the results of the mentioned baselines and our model. We use the same linguistic features which are used in [3]. In [3], they used different dimensional word embeddings for a specific model to evaluate performance, so we report the best result for a specific model from their paper. The followings are the analysis and conclusions:

- Comparing to the traditional linear CRF model, the deep learning based models (DTBCSNN and RNNs) apparently achieve much better performance especially in the laptop domain. This demonstrates the effectiveness of deep learning and its superiority over linear CRF.

Table 2. Model comparison results in terms of F1 scores (%).

Models	Restaurant	Laptop
CRF	77.28	68.66
Elman-RNN	80.37	74.43
Elman-RNN+F	81.66	74.25
LSTM-RNN	79.79	73.52
LSTM-RNN+F	81.37	75.00
HIS_RD	79.62	74.55
DLIREC	**84.01**	73.78
DTBCSNN	82.26	74.70
DTBCSNN+F	83.97	**75.66**

- Our model is more outstanding than RNN based models. Quantitatively, from results on the Table 2, DTBCSNN can achieve a result of 82.26% in restaurant domain and outperforms all of the RNNs based models even including those adding extra linguistic features. In laptop domain, DTBCSNN can achieve 74.70% which is better than HIS_RD and almost every RNNs based models except LSTM-RNN+F. These results indicate the limits of the RNNs based models and the effectiveness of our model by leveraging syntactic features for aspect term extraction.
- Adding linguistic features is ascertained to boost our model, as DTBCSNN+F achieves 83.97% exceeding DTBCSNN by 1.71% in the restaurant domain and exceeds 0.96% than DTBCSNN in the laptop domain. Comparing with the wining systems HIS_RD and DLIREC, DTBCSNN+F can achieve a performance close to DLIREC which is slightly lower by 0.04% in the restaurant domain but DTBCSNN and DTBCSNN+F all outperform HIS_RD in the laptop domain.

5.4 Development Experiments for Model Analysis

In this part, we do extensive experiments to evaluate the model architecture, with added linguistic features and word embeddings.

Effect of DTBCL. DTBCL refers to the first part of our model that is dependency-tree based convolutional layer. We do experiment excluding DTBCL to see how much influence it can engender to the model. We use DTBCSNN as the baseline. After dropping DTBCL, the rest of the model is a stacked neural network (SNN). The result is in the Table 3, from which, we can see that the performance is damaged after dropping DTBCL on both dataset domain. The score on restaurant domain is reduced by 1.84% and 4.11% on laptop domain, which demonstrates the importance and effectiveness of tree-based convolution to capture syntactic features for aspect term extraction in our model.

Table 3. Effect of DTBCL in terms of F1 scores (%).

Models	Restaurant	Laptop
SNN	80.42	70.59
DTBCSNN	82.26	74.70

Effect of Adding Linguistic Features. We divide the two linguistic features mentioned above into isolated ones and add each feature to DTBCSNN to see the performances. The results are in Table 4, from which we can find out that the performance of DTBCSNN is improved after adding one certain linguistic feature on both domains and adding all the linguistic features is better than adding only one single feature. Based on the results, we can safely conclude that (1) our model can easily incorporate linguistic features to improve the model

Table 4. Effect of adding linguistic features in terms of F1 scores (%).

Models	Restaurant	Laptop
DTBCSNN	82.26	74.70
DTBCSNN+POS	82.60	75.43
DTBCSNN+chunk	82.53	75.08
DTBCSNN+F	83.97	75.66

performance; (2) we can explore more linguistic features and then combine them to further boost our model.

Effect of Word Embedding Dimensions. In an attempt to evaluate the model sensitivity to the dimension of word embeddings, we vary the dimension from 50 to 400 with interval as 50. Restaurant domain uses word embeddings trained on Yelp dataset and laptop domain applies word embeddings trained on Amazon dataset. After conducting the experiment on DTBCSNN model, results are shown in Fig. 2, which tells us that the highest score is achieved at around dimension 300 and that after dimension 150, the performance does not vary so much. The results verify that DTBCSNN is not so sensitive to the word embedding dimensions on the condition that the dimension is in the appropriate range e.g. from 150 to 400.

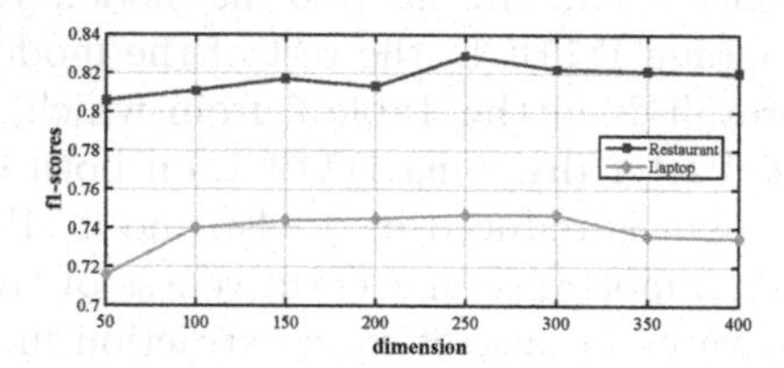

Fig. 2. Effect of word embedding dimensions.

Effect of Word Embedding Types. In order to see the effect of word embedding types to the model, we use word embeddings trained on Yelp datasets (YWE), word embeddings trained on Amazon datasets (AWE) and Google word embeddings[7] (GWE) to do the experiment with DTBCSNN. Specifically, we test the model with YWE and GWE on restaurant domain and with AWE and GWE on laptop domain. The results are shown in Table 5. From the results, we can see that the performance of the model declines when using GWE as the word embeddings. On restaurant domain, the performance reduces by 0.86% compared to YWE and on laptop domain it drops by 2.07% compared to AWE, which tells us that using word embeddings trained by the same domain corpora is necessary for the problem.

[7] https://code.google.com/archive/p/word2vec/.

Table 5. Effect of word embedding types in terms of F1 scores (%).

Type	Restaurant	Type	Laptop
Yelp	82.26	Amazon	74.70
Google	81.40	Google	72.63

6 Conclusion

In this paper, we propose a novel dependency-tree based convolutional stacked neural network, aiming to leverage dependency information of sentences and the effectiveness of deep learning for aspect term extraction. We apply tree-based convolution to capture dependency information for this problem. Our model does not need any manually features and is flexible to incorporate extra linguistic features to further boost model performance. The results of experiments on datasets show the superiority of our model to baselines.

Acknowledgements. This work was supported by National Natural Science Foundation of China (No. 61602490) and the National High-tech Research and Development Program (863 Program) (No. 2014AA015105). Thanks for the anonymous reviewers for their valuable comments.

References

1. Liu, B., Hu, M., Cheng, J.: Opinion observer: analyzing and comparing opinions on the web. In: WWW, pp. 342–351 (2005)
2. Collobert, R., Weston, J., Bottou, L., Karlen, M., Kavukcuoglu, K., Kuksa, P.P.: Natural language processing (almost) from scratch. J. Mach. Learn. Res. **12**, 2493–2537 (2011)
3. Liu, P., Joty, S.R., Meng, H.M.: Fine-grained opinion mining with recurrent neural networks and word embeddings. In: EMNLP, pp. 1433–1443 (2015)
4. Jakob, N., Gurevych, I.: Extracting opinion targets in a single and cross-domain setting with conditional random fields. In: EMNLP, pp. 1035–1045 (2010)
5. Johansson, R., Moschitti, A.: Syntactic and semantic structure for opinion expression detection. In: CoNLL, pp. 67–76 (2010)
6. Johansson, R., Moschitti, A.: Extracting opinion expressions and their polarities-exploration of pipelines and joint models. In: ACL, pp. 101–106 (2011)
7. Li, F., Han, C., Huang, M., Zhu, X., Xia, Y., Zhang, S., Yu, H.: Structure-aware review mining and summarization. In: COLING, pp. 653–661 (2010)
8. Yin, Y., Wei, F., Dong, L., Xu, K., Zhang, M., Zhou, M.: Unsupervised word and dependency path embeddings for aspect term extraction. In: IJCAI, pp. 2979–2985 (2016)
9. Mou, L., Peng, H., Li, G., Xu, Y., Zhang, L., Jin, Z.: Discriminative neural sentence modeling by treebased convolution. In: EMNLP, pp. 2315–2325 (2015)
10. Mou, L., Li, G., Zhang, L., Wang, T., Jin, Z.: Convolutional neural networks over tree structures for programming language processing. In: AAAI, pp. 1287–1293 (2016)

11. Ma, M., Huang, L., Zhou, B., Xiang, B.: Dependency-based convolutional neural networks for sentence embedding. In: ACL, pp. 174–179 (2015)
12. Hu, M., Liu, B.: Mining and summarizing customer reviews. In: KDD, pp. 168–177 (2004)
13. Qiu, G., Liu, B., Bu, J., Chen, C.: Opinion word expansion and target extraction through double propagation. Comput. Linguist. **37**(1), 9–27 (2011)
14. Jin, W., Ho, H.H.: A novel lexicalized hmm-based learning framework for web opinion mining. In: ICML, pp. 465–472. ACM (2009)
15. Yang, B., Cardie, C.: Extracting opinion expressions with semi-Markov conditional random fields. In: EMNLP, pp. 1335–1345 (2012)
16. Toh, Z., Wang, W.: DLIREC: aspect term extraction and term polarity classification system. In: SemEval, pp. 235–240 (2014)
17. Chernyshevich, M.: IHS R&D Belarus: cross-domain extraction of product features using CRF. In: SemEval, pp. 309–313 (2014)
18. Titov, I., McDonald, R.T.: Modeling online reviews with multi-grain topic models. In: WWW, pp. 111–120 (2008)
19. Moghaddam, S., Ester, M.: On the design of LDA models for aspect-based opinion mining. In: CIKM, pp. 803–812 (2012)
20. Irsoy, O., Cardie, C.: Opinion mining with deep recurrent neural networks. In: EMNLP, pp. 720–728 (2014)
21. Socher, R., Perelygin, A., Wu, J.Y., Chuang, J., Manning, C.D., Ng, A.Y., Potts, C.: Recursive deep models for semantic compositionality over a sentiment treebank. In: EMNLP, pp. 1631–1642 (2013)
22. Kim, Y.: Convolutional neural networks for sentence classification. In: EMNLP, pp. 1746–1751 (2014)
23. Klein, D., Manning, C.D.: Accurate unlexicalized parsing. In: ACL, pp. 423–430 (2003)

Topic Modeling over Short Texts by Incorporating Word Embeddings

Jipeng Qiang[1,2,3(✉)], Ping Chen[3], Tong Wang[3], and Xindong Wu[2,4]

[1] Yangzhou University, Yangzhou 225009, China
qjp2100@163.com
[2] Hefei University of Technology, Hefei 230009, China
[3] University of Massachusetts Boston, Boston, MA 02155, USA
[4] University of Louisiana at Lafayette, Lafayette, LA 70504, USA

Abstract. Inferring topics from the overwhelming amount of short texts becomes a critical but challenging task for many content analysis tasks. Existing methods such as probabilistic latent semantic analysis (PLSA) and latent Dirichlet allocation (LDA) cannot solve this problem very well since only very limited word co-occurrence information is available in short texts. This paper studies how to incorporate the external word correlation knowledge into short texts to improve the coherence of topic modeling. Based on recent results in word embeddings that learn semantically representations for words from a large corpus, we introduce a novel method, Embedding-based Topic Model (ETM), to learn latent topics from short texts. ETM not only solves the problem of very limited word co-occurrence information by aggregating short texts into long pseudo-texts, but also utilizes a Markov Random Field regularized model that gives correlated words a better chance to be put into the same topic. The experiments on real-world datasets validate the effectiveness of our model comparing with the state-of-the-art models.

Keywords: Topic modeling · Short text · Word embeddings

1 Introduction

Topic modeling has been proven to be useful for automatic topic discovery from a huge volume of texts. Topic model views texts as a mixture of probabilistic topics, where a topic is represented by a probability distribution over words. Many topic models such as Latent Dirichlet Allocation (LDA) have demonstrated great success on long texts (news article and academic paper) [2,5]. In recent years, knowledge-based topic models have been proposed, which ask human users to provide some prior domain knowledge to guide the model to produce better topics instead of purely relying on how often words co-occur in different contexts. For example, two recently proposed models, i.e., a quadratic regularized topic model based on semi-collapsed Gibbs sampler [10] and a Markov Random Field regularized Latent Dirichlet Allocation model based on Variational Inference [18], share the idea of incorporating the correlation between words.

J. Kim et al. (Eds.): PAKDD 2017, Part II, LNAI 10235, pp. 363–374, 2017.
DOI: 10.1007/978-3-319-57529-2_29

With the rapid development of the World Wide Web, short text has been an important information source not only in traditional web site, e.g., web page title and image caption, but in emerging social media, e.g., tweet, status message, and question in Q&A websites. Compared with long texts, topic discovery from short texts has the following three challenges: only very limited word co-occurrence information is available, the frequency of words plays a less discriminative role, and the limited contexts make it more difficult to identify the senses of ambiguous words [15]. Therefore, long text topic models cannot work very well on short texts [4,20]. Finally, how to extract topics from short texts remains a challenging research problem [16]. Three major heuristic strategies have been adopted to deal with how to discover the latent topics from short texts. One follows the simple assumption that each text is sampled from only one latent topic which is totally unsuited to long texts, but it can be suitable for short texts compared to the complex assumption that each text is modeled over a set of topics [19,21]. Therefore, many models for short texts were proposed based on this simple assumption [4,20]. Zhao et al. [21] proposed a Twitter-LDA model by assuming that one tweet is generated from one topic. But, the problem of very limited word co-occurrence information in short texts has not been solved yet. The second strategy takes advantage of various heuristic ties among short texts to aggregate them into long pseudo-texts before topic inference that can help improve word co-occurrence information [8,17]. For example, some models aggregated all the tweets of a user as a pseudo-text [17]. As these tweets with the same hashtag may come from a topic, Mehrotra et al. [8] aggregated all tweets into a pseudo-text based on hashtags. However, these schemes are heuristic and highly dependent on the data, which is not fit for short texts such as news titles, advertisements or image captions. The last scheme directly aggregates short texts into long pseudo-texts through clustering methods [15], in which the clustering method will face this same problem of very limited word co-occurrence information.

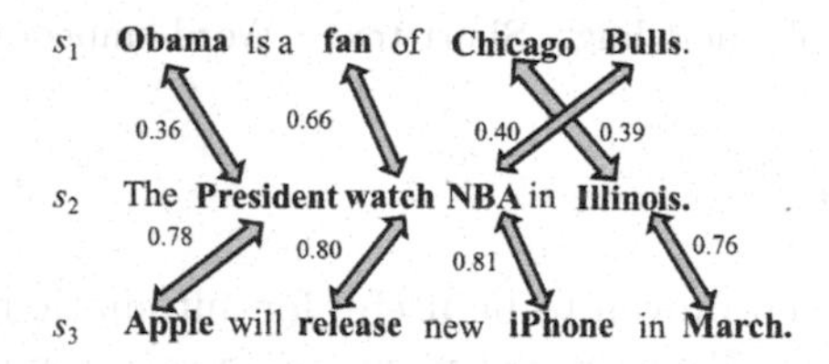

Fig. 1. An illustration of the relationship among short texts.

Figure 1 shows an example to explain the shortcomings of existing short text topic models. There are three short texts, and non-stop words are marked in bold. The shortest distances between two words from different short texts are labeled using the arrows, in which the distance is computed by word embeddings [13]. We can see s_1 and s_2 probably include two topics. 'Obama' and 'President' are likely to come from the same topic, and 'NBA' and 'Bulls' are from another topic. The simple assumption that each text is sampled from only one latent topic is unsuited to these texts. And if we directly aggregate the three short texts into two long pseudo-texts, it is very hard to decide how to aggregate these texts since they do not share the same words. But, it is very clear that s_1 is more similar to s_2 than s_3.

To overcome these inherent weaknesses and keep the advantages of three strategies, we propose a novel method, Embedding-based Topic Model (ETM), to discover latent topics from short texts. Our method leverages recent results by word embeddings that obtain vector representations for words [9]. ETM has the following three steps. ETM firstly builds distributed word embeddings from a large corpus, and then aggregates short texts into long pseudo-texts by incorporating the semantic knowledge from word embeddings, thus alleviates the problem of very limited word co-occurrence information in short texts. Finally, ETM discovers latent topics from pseudo-texts based on the complex assumption that each text of a collection is modeled over a set of topics. ETM adopts a Markov Random Field regularized model based on collapsed Gibbs sampling which utilizes word embeddings to improve the coherence of topic modeling. Within a long pseudo-text, if two words are labeled as similar according to word embedding, a binary potential function is defined to encourage them to share the same latent topic. Experiments demonstrate that ETM can discover more prominent and coherent topics than the baselines.

2 Algorithm

Our model includes three steps. First, we build distributed word embeddings for the vocabulary of the collection. Different from Word2Vec [9] that only utilizes local context windows, Pennington et al. later introduced a new global log-bilinear regression model, Glob2Vec [13], which combines global word-word co-occurrence counts and local context windows. Therefore, we adopt Glob2Vec to learn word vector representation. Second, we aggregate short texts into long pseudo-texts by incorporating the semantic knowledge from word embeddings. A new metric, Word Mover's Distance (WMD) [7], to compute the distance between two short texts. Third, we adopt a Markov Random Field regularized model based on collapsed Gibbs Sampling to improve the coherence of topic modeling. The framework of ETM is shown in Fig. 2.

2.1 Aggregate Short Texts into Long Pseudo-texts

After obtaining word embeddings of each word, we use the typical cosine distance measure for the distance between words, i.e., for word vector v_x and word vector v_y, we define the distance $d(v_x, v_y) = 1 - \frac{v_x}{\|v_x\|_2} \times \frac{v_y}{\|v_y\|_2}$. Consider a collection of short texts, $S = \{s_1, s_2, \ldots, s_i, \ldots, s_n\}$, for a vocabulary of V words, where s_i represents the ith text. We assume each text is represented as a normalized bag-of-words (nBOW) vector, $\mathbf{r}_i \in \mathbb{R}^V$ is the vector of s_i, a V-dimension vector, $r_{i,j} = \frac{c_{i,j}}{\sum_{v=1}^{V} c_{i,v}}$ where $c_{i,j}$ denotes the occurrence times of the jth word of the vocabulary in text s_i. We can see that a nBOW vector is very sparse as only a

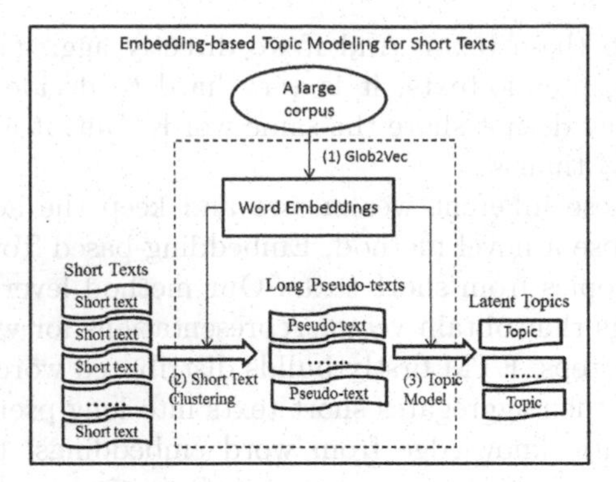

Fig. 2. Embedding-based topic model for short texts

few words appear in each text. For example, given three short texts in Fig. 1, if we adopt these metrics (e.g., Euclidean distance, Cosine Similarity) to measure distance between two texts, it is hard to find their difference. Therefore, we introduce WMD to compute the distance between texts. WMD computes the minimum cumulative cost that words from one text need to travel to match exactly the words of the other text as the distance of texts, in which the distance bewteen words is computed by word embeddings.

Let $\mathbf{r}_i$ and $\mathbf{r}_j$ be the nBOW representation of s_i and s_j. Each word of $\mathbf{r}_i$ can be allowed to travel to the word of $\mathbf{r}_j$. Let $T \in \mathbb{R}^{m \times m}$ be a flow matrix, where $T_{u,v}$ represents how much of the weight of word u of $\mathbf{r}_i$ travels to word v of $\mathbf{r}_j$. To transform all weights of $\mathbf{r}_i$ into $\mathbf{r}_j$, we guarantee that the entire outgoing flow from vertex u equals to $r_{i,u}$, namely $\sum_v T_{u,v} = r_{i,u}$. Correspondingly, the amount of incoming flow to vertex v must equal to $r_{j,v}$, namely, $\sum_u T_{u,v} = r_{j,v}$. At last, we can define the distance of two texts as the minimum cumulative cost required to flow from all words of one text to the other text, namely, $\sum_{u,v} T_{u,v} d(u,v)$. The best average time complexity of solving the WMD problem is O(m^3log$m$), where m is the number of unique words in the text. To speed up the optimization problem, we relax the WMD optimization problem and remove one of the two constraints. Consequently, the optimization becomes,

$$\min_{T \geq 0} \sum_{u,v}^{m} T_{u,v} d(u,v) \quad s.t. \sum_{v}^{m} T_{u,v} = r_{i,u} \forall u \in \{1, 2, ..., m\} \tag{1}$$

The optimal solution is the probability of each word in one text is moved to the most similar word in the other text. The time complexity of WMD can be reduced to O($m$log$m$). Once the distance between texts have been computed, we aggregate short texts into long pseudo-texts based on K-Means algorithm [1].

2.2 Topic Inference by Incorporating Word Embeddings

Model Description: We adopt the MRF model to learn the latent topics which can incorporate word distances into topic modeling for encouraging words labeled similarly to share the same topic assignment [18]. Here, we continue to use word embeddings to compute the distance between words. We can see from Fig. 3, MRF model extends the standard LDA model [2] by imposing a Markov Random Field on the latent topic layer.

Suppose the corpus contains K topics and long pseudo-texts with L texts over V unique words in the vocabulary. Following the standard LDA, Φ is represented by a $K \times V$ matrix where the kth row ϕ_k represents the distribution of words in topic k, Θ is represented by a $L \times K$ where the lth row θ_l represents the topic distribution for the lth long pseudo-texts, α and β are hyperparameters, z_{li} denotes the topic identities assigned to the i_{th} word in the lth long pseudo-text.

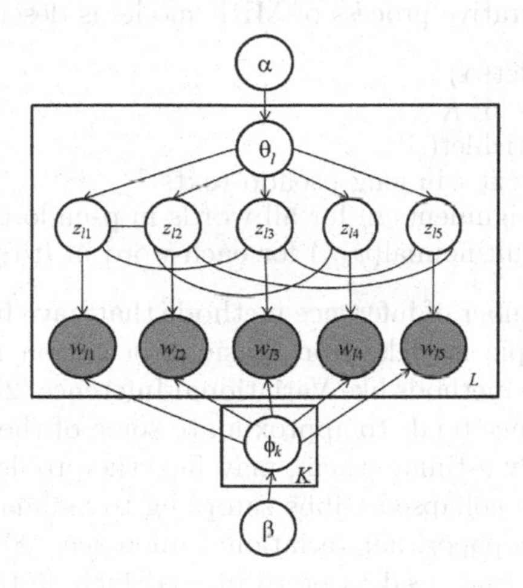

Fig. 3. Markov random field regularized model

The key idea is that if the distance between two words in one pseudo-text is smaller than a threshold, they are more likely to belong to the same topic. For example, in Fig. 1, 'President' and 'Obama' ('Bulls' and 'NBA') are likely to belong to the same topic. Based on this idea, MRF model defines a Markov Random Field over the latent topic. Given a long pseudo-text l consisting of n_l words $\{w_{li}\}_{i=1}^{n_l}$. If the distance between any word pair (w_{li},w_{lj}) in l is smaller than a threshold, MRF model creates an undirected edge between their topic assignments (z_{li},z_{lj}). Finally, MRF creates an undirected graph G_l for the lth pseudo-text, where nodes are latent topic assignments $\{z_{li}\}_{i=1}^{n_l}$ and edges connect the topic assignments of correlated words. For example, in Fig. 3, G_l is consisted of five nodes $(z_{l1},z_{l2},z_{l3},z_{l4},z_{l5})$ and five edges $\{(z_{l1},z_{l2},), (z_{l1},z_{l3},), (z_{l2},z_{l4},), (z_{l2},z_{l5},), (z_{l3},z_{l5})\}$.

The same to LDA, MRF model uses the unary potential for z_{li} as $p(z_{li} \mid \theta_l)$. The difference is MRF model defines binary potential over each edge (z_{li}, z_{lj}) of G_l as $exp\{\mathcal{I}(z_{li} = z_{lj})\}$, which produces a large value if the two topic assignments are the same and generates a small value if the two topic assignments are different, where $\mathcal{I}(\cdot)$ is the indicator function. Hence, similar words in one pseudo-text have a high probability to be put into the same topic. The joint probability of all topic assignments $\mathbf{z}_l = \{z_{li}\}_{i=1}^{n_l}$ in MRF model can be calculated as

$$p(\mathbf{z}_l \mid \theta_l, \lambda) = \prod_{i=1}^{n_l} p(z_{li} \mid \theta_l) exp\{\lambda \frac{\sum_{(li,lj)\in\mathcal{P}_l} \mathcal{I}(z_{li} = z_{lj})}{\mid \mathcal{P}_l \mid}\} \tag{2}$$

where $\mathcal{P}_l$ represents all edges of G_l and $\mid \mathcal{P}_l \mid$ is the number of all edges. Here, λ is a user-specified parameter that controls the tradeoff between unary potential and binary potential. If $\lambda = 0$, MRF model is reduced to LDA. Different from LDA that topic label z_{li} is determined by topic distribution θ_l, z_{li} in MRF depends on both θ_l and the topic assignments of similar words in the lth pseudo-text.

Formally, the generative process of MRF model is described as follows.

(1) Draw $\Theta \sim$ Dirichlet(α)
(2) For each topic $k \in [1, K]$
  (a) draw $\phi_k \sim$ Dirichlet(β)
(3) For each pseudo-text l in long pseudo-texts
  (a) draw topic assignments $\mathbf{z}_l$ for all words in pseudo-text l using Eq. (2)
  (b) draw $w_{li} \sim$ Multinomial($\phi_{z_{li}}$) for each word in lth pseudo-text

There have been a number of inference methods that have been used to estimate the parameters of topic models, from basic expectation maximization [6], to approximate inference methods like Variational Inference [2] and Gibbs sampling [5]. Variational Inference tends to approximate some of the parameters, such as Φ and Θ, not explicitly estimate them, may face the problem of local optimum. Therefore, we will use collapsed Gibbs sampling to estimate parameters under Dirichlet priors in this paper, not variational inference [18].

These parameters that need to be estimated include the topic assignments of $\mathbf{z}$, the multinomial distribution parameters Φ and Θ. Using the technique of collapsed Gibbs sampling, we only need to sample the topic assignments of $\mathbf{z}$ by integrating out ϕ and θ according to the following condition distribution:

$$p(z_{li} = k \mid \mathbf{z}_{l,-li}, \mathbf{w}_{l,-li}) = (n_{l,-li}^{k} + \alpha) \frac{n_{k,-li}^{w_{li}} + \beta}{n_{k,-li} + V\beta} exp(\lambda \frac{\sum_{j\in\mathcal{N}_{li}} (z_{lj} = k)}{\mid \mathcal{N}_{li} \mid}) \tag{3}$$

where z_{li} denotes the topic assignment for word w_{li} in the lth pseudo-text, $\mathbf{z}_{l,-li}$ denotes the topic assignments for all words except w_{li} in the lth pseudo-text, $n_{l,-li}^{k}$ is the number of times assigned to topic k excluding w_{li} in the lth pseudo-text, $n_{k,-li}^{w_{li}}$ is the number of times word w_{li} assigned to topic k excluding w_{li}, $n_{k,-li}$ is the number of occurrences of all words V that belongs to topic k excluding w_{li}, $\mathcal{N}_{li}$ denotes the words that are labeled to be similar to w_i in the lth pseudo-text, and $\mid \mathcal{N}_{li} \mid$ is the number of words in $\mathcal{N}_{li}$.

Parameter Estimation: There are three types of variables ($\mathbf{z}$, Φ and Θ) to be estimated for our model ETM. For the lth pseudo-text, the joint distribution of all known and hidden variables is given by the hyperparameters:

$$p(\mathbf{z}_l, \theta_l, \mathbf{w}_l, \Phi \mid \alpha, \beta, \lambda) = p(\Phi|\beta) \cdot \prod_{li=1}^{n_l} p(w_{li} \mid \phi_{z_{li}}) \cdot p(\mathbf{z}_l \mid \theta_l, \lambda) \cdot p(\theta_l \mid \alpha) \tag{4}$$

We can obtain the likelihood of the lth pseudo-text $\mathbf{w}_l$ of the joint event of all words by integrating out ϕ and θ and summing over z_{li}.

$$p(\mathbf{w}_l \mid \alpha, \beta, \lambda) = \int\int p(\theta_l \mid \alpha) \cdot p(\Phi|\beta) \cdot \prod_{li=1}^{n_l} p(w_{li} \mid \phi_{z_{li}}, \Phi, \lambda) \tag{5}$$

Finally, the likelihood of all pseudo-texts $\mathbf{W} = \{\mathbf{w}_l\}_{l=1}^{L}$ is determined by the product of the likelihood of the independent pseudo-texts:

$$p(\mathbf{W} \mid \alpha, \beta, \lambda) = \prod_{l=1}^{L} p(\mathbf{w}_l \mid \alpha, \beta, \lambda) \tag{6}$$

We try to formally derive the conditional distribution $p(z_{li} = k \mid \mathbf{z}_{l,-li}, \mathbf{w}_{l,-li})$ used in our ETM algorithm as follows.

$$p(z_{li} = k \mid \mathbf{z}_{l,-li}, \mathbf{w}_{l,-li}) = \frac{p(\mathbf{w}, \mathbf{z} \mid \alpha, \beta, \lambda)}{p(\mathbf{w}, \mathbf{z}_{l,-li} \mid \alpha, \beta, \lambda)} \propto \frac{p(\mathbf{w}, \mathbf{z} \mid \alpha, \beta, \lambda)}{p(\mathbf{w}_{l,-li}, \mathbf{z}_{l,-li} \mid \alpha, \beta, \lambda)} \tag{7}$$

From the graphical model of ETM, we can see

$$p(\mathbf{w}, \mathbf{z} \mid \alpha, \beta, \lambda) = p(\mathbf{w} \mid \mathbf{z}, \beta) p(\mathbf{z} \mid \alpha, \lambda) \tag{8}$$

The same to LDA, the target distribution $p(\mathbf{w} \mid \mathbf{z}, \beta)$ is obtained by integrating over ϕ,

$$p(\mathbf{w} \mid \mathbf{z}, \beta) = \prod_{z_{li}=1}^{K} \frac{\Delta(\mathbf{n}_{z_{li}} + \beta)}{\Delta(\beta)}, \mathbf{n}_{z_{li}} = \{n_{z_{li}}^{(w)}\}_{w=1}^{V} \tag{9}$$

where $n_{z_{li}}^{(w)}$ is the number of word w occurring in topic z_{li}. Here, we adopt the Δ function in Heinrich (2009), and we can have $\Delta(\beta) = \frac{\prod_{w=1}^{V} \Gamma(\beta)}{\Gamma(V\beta)}$ and $\Delta(\mathbf{n}_{z_{li}} + \beta) = \frac{\prod_{w \in \mathbf{w}} \Gamma(n_k^w + \beta)}{\Gamma(n_k + V\beta)}$, where Γ denotes the gamma function.

According to Eq. (3), we can get

$$p(\mathbf{z}_l \mid \theta_l, \lambda) = exp\{\lambda \frac{\sum_{(li,lj)\in\mathcal{P}_l} \sum_{k=1}^{K} (z_{li} z_{lj})}{\mid \mathcal{P}_l \mid}\} \prod_{k=1}^{K} \theta_k^{n_l^k} \tag{10}$$

Similarly, $p(\mathbf{z}_l \mid \alpha, \lambda)$ can be obtained by integrating out Θ as

$$\begin{aligned} p(\mathbf{z} \mid \alpha, \lambda) &= \int p(\mathbf{z} \mid \Theta, \lambda) p(\Theta \mid \alpha) \\ &= \prod_{l=1}^{L} exp\{\lambda \frac{\sum_{(li,lj)\in\mathcal{P}_l} \sum_{k=1}^{K} (z_{li} z_{lj})}{\mid \mathcal{P}_l \mid}\} \frac{\Delta(\mathbf{n}_l + \alpha)}{\Delta(\alpha)} \end{aligned} \tag{11}$$

where $p(\Theta \mid \alpha)$ is a Dirichlet distribution, and $\mathbf{n}_l = \{n_l^{(k)}\}_{k=1}^{K}$.

Finally, we put the joint distribution $p(\mathbf{w}, \mathbf{z} \mid \alpha, \beta, \lambda)$ into Eq. (11), the conditional distribution in Eq. (3) can be derived

$$
\begin{aligned}
p(z_{li} = k \mid \mathbf{z}_{l,-li}, \mathbf{w}_{l,-li}) &\propto \frac{p(\mathbf{w}, \mathbf{z} \mid \alpha, \beta, \lambda)}{p(\mathbf{w}_{l,-li}, \mathbf{z}_{l,-li} \mid \alpha, \beta, \lambda)} \\
&\propto \frac{\Delta(\mathbf{n}_l + \alpha)}{\Delta(\mathbf{n}_{l,-li} + \alpha)} \frac{\Delta(\mathbf{n}_{z_{li}} + \beta)}{\Delta(\mathbf{n}_{z_{l,-li}} + \beta)} exp(\lambda \frac{\sum_{j \in \mathcal{N}_{li}} (z_{lj} = k)}{\mid \mathcal{N}_{li} \mid}) \\
&\propto (n_{l,-li}^{k} + \alpha) \frac{n_{k,-li}^{w_{li}} + \beta}{n_{k,-li} + V\beta} exp(\lambda \frac{\sum_{j \in \mathcal{N}_{li}} (z_{lj} = k)}{\mid \mathcal{N}_{li} \mid})
\end{aligned}
\tag{12}
$$

3 Experiments

Datasets and Setup: We study the empirical performance of ETM on two short text datasets, Tweet2011 and GoogleNews[1]. Similar to existing papers [20], we utilize Google news as a dataset to evaluate the performance of topic models. We took a snapshot of the Google news on April 27, 2015, and crawled the titles of 6,974 news articles belonging to 134 categories. For each dataset, we conduct the same preprocessing with this paper [14]. We compare our model ETM with the following baselines. Three short text topic models: Unigrams [12], DMM [20], and BTM [4]. Two Long text topic models: LDA [5] and MRF-LDA [18]. For the baselines, we chooses the parameters according to their original papers. For LDA, Unigrams and BTM, both hyperparameters α and β are set to $50/K$ and 0.01. For DMM and ETM, both hyperparameters α and β are set to 0.1. For MRF-LDA, $\alpha = 0.5$ and $\lambda = 1$. For ETM, λ is set to 1. For our model and MRF-LDA, words pairs with distance lower than 0.4 are labeled as correlated.

A lot of metrics have been proposed for measuring the coherence of topics in texts [11]. Most conventional metrics try to estimate the likelihood of held-out testing data based on parameters inferred from training data. However, this likelihood is not necessarily a good indicator of the quality of extracted topics [3]. Similar to [18], we also evaluate our model in a qualitative and quantitative manner. And we validate topic models on short text clustering and short text classification. Due to the space limit, we omit some experiments. First, we discuss some exemplar topics learned by the six methods on the two datasets. Each topic is visualized by the top ten words. Then, we evaluate our model based on the coherence measure (CM) to assess how coherent the learned topics are. For each topic, we choose the top 10 candidate words and ask human annotators to judge whether they are relevant to the corresponding topic. To do this, annotators need to judge whether a topic is interpretable or not. If not, the 10 words of the topic are labeled as irrelevant; otherwise these words are identified by annotators as relevant words for this topic. Coherence measure (CM) is defined as the ratio between the number of relevant words and the total number of candidate words.

[1] http://news.google.com.

Qualitative Evaluation: On Tweet2011 dataset, there is no category information for each tweet. Manual labeling might be difficult due to the incomplete and informal content of tweets. Fortunately, some tweets are labeled by their authors with hashtags in the form of '#keyword' or '@keyword'. We manually choose 10 frequent hashtags as labels and collect documents with their hashtags. These hashtags are 'NBA', 'NASA', 'Art', 'Apple', 'Barackobama', 'Worldprayr', 'Starbucks', 'Job', 'Travel', 'Oscars', respectively. On GoogleNews dataset, the four topics are events on April 27, 2015, which are "Nepal earthquake", "Iran nuclear", "Indonesia Bali", and "Yemen airstrikes".

Table 1 shows some topics learned by the six models. Each topic is visualized by the top ten words. Words that are noisy and lack of representativeness are highlighted in bold. From Tabel 1, our model ETM can learn more coherent topics with fewer noisy and meaningless words than all baseline models. Long text topic modelings (LDA and MRF-LDA) that model each text as a mixture of topics does not fit for short texts, as short text suffers from the sparsity of word co-occurrence patterns. MRF-LDA incorporating word correlation knowledge cannot improve the coherence of topic modeling since binary potential of MRF cannot work when short text only consists of a few words. In addition, MRF-LDA based on variational inference may face the problem of local optimum. Consequently, the top 10 words of yemeb of LDA and Apple of MRF-LDA are not relevant to the corresponding topic.

The existing short text topic models suffer from two problems. On one hand, the frequency of words in short text plays a less discriminative role than long text, making it hard to infer which words are more correlated in each text. On the other hand, these models bring in little additional word co-occurrence information and cannot alleviate the sparsity problem. As a consequence, the topics extracted from these three short text topic models are not satisfying. For example, Unigrams cannot identify topic "iran", BTM cannot identify topic "yemen", and the learned topics of DMM consists of meaning-less words such as *going*, *today*, etc.

Our method ETM incorporates the word correlation knowledge provided by words embedding over the latent topic to cluster short texts to generate long pseudo-text. In this condition, the frequency of words in pseudo-text plays an important role to discover the topics based on this assumption each text is modeled as a mixture of topics. After aggregating short texts into long pseudo-texts, more similar words are in one text than the original text. Therefore, the Markove Random Field regularized model can paly an important in learning latent topics from pseudo-texts, which uses the word correlation knowledge over the latent topic to encourage correlated words to share the same topic label. Hence, although similar words may not have high co-occurrence in the corpus, they remain have a high probability to be put into the same topic. Consequently, from Table 1 we can see that the topics learned by our model are far better than those learned by the baselines. The learned topics have high coherence and contain fewer noisy and irrelevant words. Our model also can recognize the topic words that only have a few occurrences in the collection. For instance, the word *flight* from topic "NASA", *writer* from topic "Art", and *tablet* of topic "Apple" can only be recognized by ETM.

Table 1. Topics learned from Tweet2011 and GoogleNews dataset

Data	Class	Method	Top 10 words
Tweet2011	NBA	LDA	Game lebron kobe player lakers team coach **going** james points
		MRF-LDA	Game lebron kobe player **museum** lakers play **tonight** james **better**
		Unigrams	Game lebron kobe player lakers team **going** james play allen
		DMM	Game lebron kobe player lakers team james points **going lead**
		BTM	Game kobe lebron lakers team player scored points **going** james
		ETM	Game lebron kobe player lakers team points james play allen
	NASA	LDA	Space shuttle launch nasa atlantis **live** video weather **watch check**
		MRF-LDA	Space shuttle **great** launch **good** nasa **store watch today** atlantis
		Unigrams	Space shuttle launch nasa atlantis **check live watch** weather crew
		DMM	Space shuttle launch nasa atlantis **live check** video weather **today**
		BTM	Space shuttle launch nasa atlantis **live** crew weather **watch** image
		ETM	Space shuttle launch nasa flight weather atlantis crew image ares
	Art	LDA	Artist museum **great check** photo **blog** artists gallery painting modern
		MRF-LDA	**Time** Artist **video twitter blog year record coming** work artists
		Unigrams	Artist museum **good** artists painting photo **blog check** gallery exhibition
		DMM	Artist museum **check** photo painting artists exhibition modern gallery **blog**
		BTM	Artist **great** museum **check miami** painting artists gallery **blog free**
		ETM	Artist museum writer painting gallery artists exhibition modern photo arts
	Apple	LDA	Apple iphone store **time** steve jobs snow **best good google**
		MRF-LDA	**Apple iphone check team live love follow star coach going**
		Unigrams	Apple iphone store **time good** ipod jobs video snow steve
		DMM	Apple iphone store **time** steve snow jobs **google great good**
		BTM	Apple iphone **good** steve video store **time** jobs ipod **going**
		ETM	Apple iphone store video ipod **twitter** tablet steve **blog google**
GoogleNews	Nepal	LDA	Nepal death israel quake rescue aid israeli israelis relief help
		MRF-LDA	Nepal death israel quake rescue aid everest israeli israelis relif
		Unigrams	Nepal quakes toll death quake everest tops aid rises israelis
		DMM	Nepal israelis quake toll israel rescue death aid everest israeli
		BTM	Nepal aids quake rescue israel toll death aid everest israeli
		ETM	Nepal israelis quake toll israel rescue death aid everest israeli
	Iran	LDA	Iran nuclear meet kerry zarif talks **good** israel **foreign deal**
		MRF-LDA	Iran meet kerry nuclear zarif **deal victim** talks powers arms
		Unigrams	**Nepal quakes israel rescue quake aid israeli relief help good**
		DMM	Iran nuclear meet kerry zarif israel weapon npt **deal foreign**
		BTM	**Yeman** Iran nuclear meet **saudi kerry** arms talks zarif **good**
		ETM	Iran nuclear meet kerry zarif israel weapon npt talks powers
	Bali	LDA	Bali Indonesia execution excecutions chan marries andrew duo death **deal**
		MRF-LDA	Bali Indonesia death **toll** execution executions chan andrew marries **nuclear**
		Unigrams	Bali Indonesia execution executions chan marries andrew death duo drug
		DMM	Bali Indonesia execution executions chan marries andrew death duo drug
		BTM	Bali chan executions andrew marries sukumaran **final** duo myuran **ahead**
		ETM	Bali Indonesia execution executions chan marries andrew death duo drug
	Yemen	LDA	**Yemen Nepal toll death saudi quake quakes Iran strikes tops**
		MRF-LDA	Yemen Saudi talks **drug iran** strikes yemeni war saudis **babies**
		Unigrams	Yemen **Iran nuclear meet** kerry saudi **zarif** talks arms **israel**
		DMM	Yemen Saudi **Iran** strikes yemeni saudis talks strike **tops** houthis
		BTM	**Nepal Bali chan drug aids arms chaims death duo pair**
		ETM	Yemen Saudi strikes yemeni saudis strikes war talks houthis arms

Table 2. CM (%) on Tweet2011 and GooleNews (*Ai* represents the *i*th annotator)

Method	Tweet2011					GoogleNews				
	A1	A2	A3	A4	Mean	A1	A2	A3	A4	Mean
LDA	54	42	45	67	52 ± 11.2	95	95	79	95	91 ± 8
MRF-LDA	44	46	46	57	48.2 ± 5.9	91	85	80	74	82.5 ± 7.2
Unigrams	66	45	56	59	56.5 ± 8.7	88	73	79	94	83.5 ± 9.3
DMM	70	49	50	60	57.2 ± 9.8	94	93	90	93	92.5 ± 1.7
BTM	62	45	50	77	58.5 ± 14.2	80	85	75	78	79.5 ± 4.2
ETM	**72**	**62**	**73**	**83**	**72.5 ± 8.5**	**96**	**96**	**94**	**96**	**95.5 ± 1.0**

Quantitative Evaluation: Table 2 shows the coherence measure of topics inferred on Tweet2011 and GoogleNews datasets, respectively. We can see our model ETM significantly outperforms the baseline models. On Tweet2011 dataset, ETM achieves an average coherence measure of 72.5%, which is larger than long text topic models (LDA and MRF-LDA) with a large margin. Compared to short text topic models, ETM still has a big improvement. In GoogleNews dataset, our model is also much better than the baselines.

4 Conclusion

We propose a novel model, Embedding-based Topic Modeling (ETM), to discover the topics from short texts. ETM first aggregates short texts into long pseudo-texts by incorporating the semantic knowledge from word embeddings, then infers topics from long pseudo-texts using Markov Random Field regularized model, which encourages words labeled as similar to share the same topic assignment. Therefore, by incorporating the semantic knowledge ETM can alleviate the problem of very limited word co-occurrence information in short texts.

Acknowledgement. This research is partially supported by the National Key Research and Development Program of China (2016YFB1000900), and the National Natural Science Foundation of China (No. 61503116).

References

1. Aggarwal, C.C., Zhai, C.: A survey of text clustering algorithms. In: Mining Text Data, pp. 77–128 (2012)
2. Blei, D.M., Ng, A.Y., Jordan, M.I.: Latent dirichlet allocation. JMLR **3**, 993–1022 (2003)
3. Chang, J., Gerrish, S., Wang, C., Boyd-Graber, J.L., Blei, D.M.: Reading tea leaves: how humans interpret topic models. In: NIPS, pp. 288–296 (2009)
4. Cheng, X., Yan, X., Lan, Y., Guo, J.: BTM: topic modeling over short texts. TKDE **26**(12), 2928–2941 (2014)

5. Griffiths, T., Steyvers, M.: Finding scientific topics. Proc. Nat. Acad. Sci. **101**, 5228–5235 (2004)
6. Hofmann, T.: Probabilistic latent semantic indexing. In: SIGIR, pp. 50–57 (1999)
7. Kusner, M.J., Sun, Y., Kolkin, N.I., Weinberger, K.Q.: From word embeddings to document distances. In: ICML, pp. 957–966 (2015)
8. Mehrotra, R., Sanner, S., Buntine, W., Xie, L.: Improving LDA topic models for microblogs via tweet pooling and automatic labeling. In: SIGIR, pp. 889–892 (2013)
9. Mikolov, T., Sutskever, I., Chen, K., Corrado, G.S., Dean, J.: Distributed representations of words and phrases and their compositionality. In: NIPS, pp. 3111–3119 (2013)
10. Newman, D., Bonilla, E.V., Buntine, W.: Improving topic coherence with regularized topic models. In: NIPS, pp. 496–504 (2011)
11. Newman, D., Lau, J.H., Grieser, K., Baldwin, T.: Automatic evaluation of topic coherence. In: NAACL, pp. 100–108 (2010)
12. Nigam, K., McCallum, A.K., Thrun, S., Mitchell, T.: Text classification from labeled and unlabeled documents using EM. Mach. Learn. **39**(2–3), 103–134 (2000)
13. Pennington, J., Socher, R., Manning, C.D.: Glove: Global vectors for word representation. In: EMNLP, pp. 1532–1543 (2014)
14. Qiang, J., Chen, P., Ding, W., Wang, T., Fei, X., Wu, X.: Topic discovery from heterogeneous texts. In: ICTAI (2016)
15. Quan, X., Kit, C., Ge, Y., Pan, S.J.: Short and sparse text topic modeling via self-aggregation. In: ICAI, pp. 2270–2276 (2015)
16. Wang, X., Wang, Y., Zuo, W., Cai, G.: Exploring social context for topic identification in short and noisy texts. In: AAAI (2015)
17. Weng, J., Lim, E.-P., Jiang, J., He, Q.: Twitterrank: finding topic-sensitive influential twitterers. In: WSDM, pp. 261–270 (2010)
18. Xie, P., Yang, D., Xing, E.P.: Incorporating word correlation knowledge into topic modeling. In: NACACL (2015)
19. Yan, X., Guo, J., Lan, Y., Xu, J., Cheng, X.: A probabilistic model for bursty topic discovery in microblogs. In: AAAI, pp. 353–359 (2015)
20. Yin, J., Wang, J.: A dirichlet multinomial mixture model-based approach for short text clustering. In: SIGKDD, pp. 233–242 (2014)
21. Zhao, W.X., Jiang, J., Weng, J., He, J., Lim, E.-P., Yan, H., Li, X.: Comparing Twitter and traditional media using topic models. In: Clough, P., Foley, C., Gurrin, C., Jones, G.J.F., Kraaij, W., Lee, H., Mudoch, V. (eds.) ECIR 2011. LNCS, vol. 6611, pp. 338–349. Springer, Heidelberg (2011). doi:10.1007/978-3-642-20161-5_34

Mining Drug Properties for Decision Support in Dental Clinics

Wee Pheng Goh[1(✉)], Xiaohui Tao[1], Ji Zhang[1], and Jianming Yong[2]

[1] Faculty of Health, Engineering and Sciences,
University of Southern Queensland, Toowoomba, Australia
{weepheng.goh,xtao,ji.zhang}@usq.edu.au
[2] Faculty of Business, Education, Law and Arts,
University of Southern Queensland, Toowoomba, Australia
jianming.yong@usq.edu.au

Abstract. The rise of polypharmacy requires from health providers an awareness of a patient's drug profile before prescribing. Existing methods to extract information on drug interactions do not integrate with the patient's medical history. This paper describes state-of-the-art approaches in extracting the term frequencies of drug properties and combining this knowledge with consideration of the patient's drug allergies and current medications to decide if a drug is suitable for prescription. Experimental evaluation of our models association of the similarity ratio between two drugs (based on each drug's term frequencies) with the similarity between them yields a superior accuracy of 79%. Similarity to a drug the patient is allergic to or is currently taking are important considerations as to the suitability of a drug for prescription. Hence, such an approach, when integrated within the clinical workflow, will reduce prescription errors thereby increasing the health outcome of the patient.

Keywords: Adverse relationship · Drug allergy · Drug properties · Knowledge-base · Personalised prescription · Similarity ratio · Term frequency

1 Introduction

With the increase in volume and complexity of data encountered by dentists, the use of decision support systems to aid decision-making is becoming more necessity than luxury. In order for decision support systems to be readily accepted by dentists, they should have, among other features, the ability to provide assistance on drug prescription based on individual patient profile [6,7].

Predicting drug-drug interactions (DDI) at point-of-care to reduce prescription error is important as an adverse event can lead to serious health consequences for the patient and result in expensive legal suits for the practitioner. A common cause for hospital admission worldwide is adverse drug reactions, with incidence being as high as 24% [11]. Naturally, many such admissions could have

J. Kim et al. (Eds.): PAKDD 2017, Part II, LNAI 10235, pp. 375–387, 2017.
DOI: 10.1007/978-3-319-57529-2_30

been avoided if more care was taken in drug prescription, such as by considering the patient's drug allergies.

A recent work by [2] derives similarity within a drug-pair by comparing textual description in DrugBank with those in MeSH. Although the experiment reported favourable results with metformin, a drug for treating diabetes, the focus was on drug repositioning to treat other conditions. So far, many methods have been developed to extract information on DDI [3,4], but these methods do not integrate with the patient's medical history within the clinical workflow. Having identified this gap in existing research, this paper describes state-of-the-art approaches in determining if a drug-pair is similar as well as using such information to support the dentist's prescription decision. In this study, we propose a three-tier conceptual framework, consisting of the knowledge layer, the data layer and the user layer. Data mining is performed within the data layer, with knowledge extracted from the knowledge layer and presented to the user layer for decision-making. The way that a drug-pair's similarity is associated with its properties forms the unique approach of this model.

As with many decision support systems that are developed based on knowledge discovered from data mining [13], this paper describes a model which computes the similarity within a drug-pair for predicting its suitability before the dentist prescribes it to the patient. By utilising neighborhood similarities and textual data from currently available open source datasets to predict the relationship of a drug-pair, knowledge obtained from data mining is used for prescription support of health care professionals. Based on a novel data-driven text mining technique, clusters of drugs which have adverse interactions, together with their properties, are collected by identifying their field markers from the web content. This information allows a similarity ratio to be computed which indicates if a drug-pair can be safely prescribed to the patient. Our work performs well compared to other methods of prediction, with a F-measure of 69% with drug properties gathered from textual data obtained through bio-medical sources. This model is easily utilised in predicting a drug's suitability for prescription, by considering the patient's drug allergies to avoid allergic reactions, and the drugs the patient is currently taking to avoid adverse DDI.

This study will help provide strategies in research agenda and priorities such as methodologies for knowledge reasoning and inference in the context of a dental clinic. Research outcomes of this project, especially in this climate of increasing polypharmacy, will help reduce the risk of prescribing drugs that may cause the patient to suffer an adverse reaction and thus improve the quality of treatment in clinics. This system which delivers information on interacting drug-pairs based on the patient's drug profile will also benefit those who are involved in clinical education relating to drug dispensing, such as in medicine, nursing and pharmacy. The practical use of data mining techniques in supporting the dentists prescription of drugs has great potential to extend to the wider medical domain, since there is also a need for doctors to ensure the safe prescription of drugs to patients. By considering the patient's individual conditions in decision-making support, our work will make significant contributions to the transformation of the current health care industry to one that is evidence-based and personalised.

The rest of paper is organised as follows: Sect. 2 discusses the related work in data mining and how our model differs in the way the drug-drug relationship is detected and deployed for use. Section 3 introduces the framework of our model and Sect. 4 outlines the parameters used for evaluating our work. This is followed by Sect. 5 which discusses and compares the results with other approaches, and Sect. 6 presents the conclusions obtained.

2 Related Work

Many systems have been developed using data mining techniques to explore DDI. In fact, such techniques are evolving quickly to improve the accuracy of the experiments, though in most situations results may not be sufficient to derive DDI [15]. A recent work by [1] attempts to determine DDI by identifying neutral candidates, negation cues and scopes from bio-medical articles. Features extracted from these articles include linguistic definition of negation, the position of the drugs discussed in the sentence and the linguistic-based confidence level of an interaction. By using datasets from DrugBank, it is reported that the results achieved an F-measure of 68.4%. Text mining techniques were also recently used to predict protein interactions from bio-medical literature [10].

Another common way of examining DDI is to extract relevant information from text. For example, Tari *et al.* [14] has developed a method that combines text mining and automated reasoning to predict enzyme-specific DDI. [16] also uses text mining techniques to create features based on relevant information such as genes and disease names extracted from drug databases to augment limited domain knowledge. These features are then used to build a logistic regression model to predict DDI.

Another study to extract information on DDI from bio-medical text was proposed by Bui et al. [3]. DDI pairs are mapped according to their syntactic structure followed by the generation of feature vectors for these DDI pairs. These feature vectors are then used for the generation of a predictive model which classify the drug-pair as interacting or not interacting [3].

Though these studies use data mining methods to extract relevant information for the prediction of DDI, unfortunately, these works are only confined to two tiers, the knowledge layer and data mining layer, as compared to the three-tier framework in this paper.

The crucial need to use the knowledge obtained from data mining motivates us to develop the three-tier conceptual framework proposed in this paper. Although our system is similar to that proposed by [4] in terms of using information from the patient, the unique approach adopted in this paper goes one step further in using such information to support the decision-making process for the dentist at point-of-care within the clinical workflow. In this model, an additional user layer is introduced. This layer provides an important interface between the user and the knowledge mined from bio-medical data sources. Moreover, state-of-the-art approaches adopted in the data layer allow the efficient extraction of features. These features relate the similarity of a drug-pair in terms of the shared difference in their term frequencies. Experimental results show that this approach performs favourably compared to other existing models.

3 Proposed Method

The aim of this study is to propose a unique approach in supporting the dentist in drug prescription, with consideration of the drugs that the patient is currently taking and the drugs that the patient is allergic to. In order to advise the dentist if the drug to be prescribed is suitable, a three-tier conceptual model is used: the knowledge layer, data mining layer and user layer (Fig. 1). Essentially, each layer is defined by the task that they are responsible for. At each layer, data is transformed and processed which culminates as advice to the dentist as to whether the drug is safe for prescription, and to suggest alternative drugs.

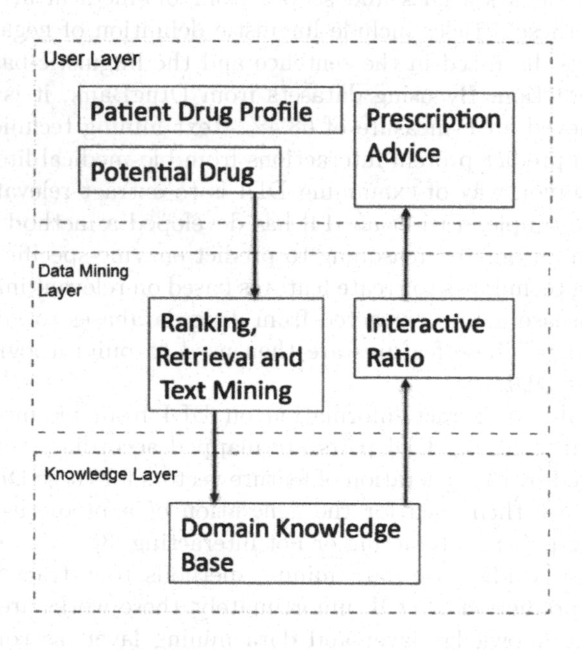

Fig. 1. Three-tier framework

3.1 Knowledge Layer

Many decision support systems depend on information retrieved from the web for further processing and inference-making to arrive at a decision [5]. In this layer, a taxonomy of drugs will be generated using information gathered from the web with expert domain knowledge. The publicly available website http://www.drugs.com contains independent and accurate information on more than 24,000 prescription drugs pertaining to interactions, dosages and other important information for both patients and professionals. It is maintained in collaboration with the US Food and Drug Administration (FDA), which acknowledges that such partnerships "are part of FDAs effort to ensure the public has easy access to reliable, useful information that can help people protect and improve their health"[1]. Each drug is being described from a different perspective to suit both patients (under the heading "Overview") and health professionals (under the heading "Professionals") while information on side-effects are found under the heading "Side Effects". In this study, term frequencies from each of these descriptions are extracted so that the similarity ratio can be computed. Essentially, this layer consists of the term frequencies of each drug within the drug taxonomy $\mathcal{T}$ for the respective drug properties described under the headings "Overview", "Professional" and "Side Effects".

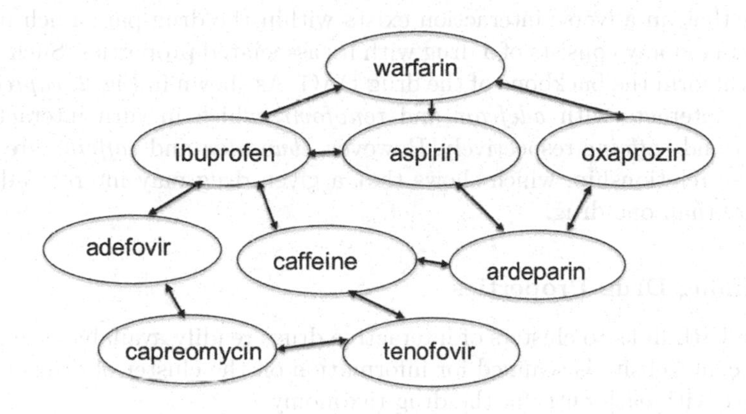

Fig. 2. Subset of drug taxonomy

Definition 1 (Drugs). *Let* $\mathbb{D} = \{d_1, d_2$ *be the domain set of unique drugs. An element* $d \in \mathbb{D}$ *is formalised as a 3-tuple* $\langle label, properties, adverse(d)\rangle$, *where*

- *label is the identity of the drug d;*
- *properties are the attribute set of d;*
- *adverse(d) is a function returning a set of drugs that are adverse to d in terms of treatment effect.* □

Definition 2 (Drug Taxonomy). *Let* $\mathcal{T}$ *be the conceptual taxonomy of all drugs.* $\mathcal{T}$ *consists of the domain of drugs linked by their semantic relations, and is defined as a 3-tuple* $\mathcal{T} := \langle \mathbb{D}, \mathbb{R}, \mathcal{H}_{\mathbb{D}}^{\mathbb{R}} \rangle$, *where*

- $\mathbb{D}$ *is the domain set of drugs* $\mathbb{D} = \{d_1, d_2, \ldots, d_{|\mathbb{D}|}\}$;
- $\mathbb{R}$ *is a set of semantic relations with a single element* $\mathbb{R} = \{r^-\}$, *where* $r^-(d_i, d_j)$ *defines the adverse relationship between* d_i *and* d_j;
- $\mathcal{H}_{\mathbb{D}}^{\mathbb{R}}$ *is the taxonomical structure constructed by all* $d \in \mathbb{D}$ *linked by* $r \in \mathbb{R}$. □

[1] http://www.fda.gov/NewsEvents/Newsroom/PressAnnouncements/ucm212844.htm.

The ability to store interactive drug pairs within a network of nodes and edges allows the knowledge layer to be represented by a directed acyclic graph (DAG). Each drug is represented as a vertex on the graph. The edges that connect a pair of vertices show the interactions between the drug pair. The cluster of drugs that has adverse interactions with a given drug can be known from the drug taxonomy. Since this project aims to find the similarity between a drug pair based on the term frequencies of each drug, the DAG allows such information to be obtained efficiently by having an algorithm to traverse through the network for the existence of the drug pair. Moreover, features of the drugs stored within each node of the DAG contributes to the Data Layer for mining the properties in terms of their term frequencies (see next section).

Figure 2 shows a subset of the major DDI in the drug taxonomy. Note that nodes in the taxonomy are connected to one another through arrows which indicates that an adverse interaction exists within the drug-pair. Each node on the drug taxonomy consists of a drug with its associated properties. Such a chain of DDI will form the backbone of the drug DAG. As shown in Fig. 2, *capreomycin* adversely interacts with *adefvoir* and *tenofovir*, which in turn interacts with *ibuprofen* and *caffeine* respectively. However, *ibuprofen* and *caffeine* are also in an adverse relationship, which shows that a given drug may interact adversely with more than one drug.

3.2 Mining Drug Properties

With the URL links to clusters of interactive drugs readily available, the content of drugs.com website is scanned for information on the cluster of drugs that are interactive with each entry in the drug taxonomy.

Each web page contains field markers to delimit the relevant content on interactive drugs, and these markers are used to extract the cluster of drugs that adversely interact with each drug in the dataset. Information on the cluster of adversely interactive drugs is crucial as it provides the ground truth in deciding whether a drug-pair has an adverse relationship. Such information are contained in the drug taxonomy $\mathcal{T}$ within the knowledge layer of the conceptual framework. Besides mining the information related to clusters of DDI for each drug in the drug taxonomy, the underlying properties for each drug is also obtained to provide information on the similarity of a drug-pair. These properties are term frequencies tf^v, tf^p and tf^s mined respectively from the "Overview", "Professional" and "Side Effects" tab of each drug in the drug taxonomy. These knowledge will then be used to compute the cosine similarity of the drug pair - refer to Fig. 3 for the flow of the experimental design.

In this paper, we are only interested in the content in terms of the properties of each drug within the drug taxonomy $\mathcal{T}$. This will enable us to compute the similarity based on their term frequencies. Such properties of the drugs are obtained from the content provided under "Professional", "Side Effects" and "Overview". To determine the similarity within a drug pair, information on the term frequencies of each drug is collected. Given the numerous techniques and algorithms available in data mining [8,12], the approach adopted in this layer is focused on speed and accuracy with easy interfacing between the knowledge layer and the user layer. Thus, at this data mining layer, the properties associated with the drugs residing on the drug taxonomy backbone are mined, namely the term frequencies tf^v, tf^p and tf^s respectively for "Overview", "Professional" and "Side Effects". Before term frequencies and document frequencies are computed from the properties of the drug, data is pre-processed, including stop word removal and stemming. The idea behind our novel approach is to use the similarity ratio of the feature vectors of term frequencies to decide if the drug-pair is similar, and subsequently using this knowledge to advise the dentist if it is suitable for prescription, taking into consideration the patient's individual medical status.

Algorithm 1. Computing Drug Similarity

input : $\mathbb{D}$: the set of drugs in Drugbank
output: similarity ratio s of drug pairs
1 **foreach** $d_i \in \mathbb{D}$ **do**
2 | compute feature vector $\overrightarrow{f_i}$ for d_i
3 | $\overrightarrow{f_i} = (t_1^v, n_1), (t_2^v, n_2)..(t_k^v, n_k)$ where n is $tf * idf$ of t^v, k = number of unique terms in d_i
4 **end**
5 **foreach** $\{d_1, d_2\} \subset \mathbb{D}$ **do**
6 | compute similarity ratio $s(d_1, d_2)$ as Eq. 1
7 **end**

To arrive at a decision as to whether a drug is suitable for prescription, a similarity ratio is required as a measure of how similar two drugs are with regards to their term frequencies. These term frequencies are obtained after performing data mining on the properties of each drug, associated with "Professional", "Overview" and "Side Effects". Feature vectors are then constructed based on *tf*idf*, and determining the difference between the feature vectors of a drug-pair enables the similarity ratio to be calculated. If p and q are the feature vectors, then the cosine similarity based on properties from, say, "Overview" is given by:

$$s^v(p,q) = \frac{\sum_{i=1}^{k} p_i.q_i}{\sqrt{a \cdot b}} \quad \text{where} \quad a = \sum_{i=1}^{k} {p_i}^2, \quad b = \sum_{i=1}^{k} {q_i}^2 \tag{1}$$

As seen in Algorithm 1, this cosine similarity s is computed after gathering the features within the properties of each drug, from which average values are computed. With these values as a guide, a threshold value θ is set to maximise the F-measure for each of the drug properties. Section 5 explains the approach used in deciding the threshold value.

For a given drug-pair, the similarity ratio taken from each of the three properties can also be combined and used to check if the drug to be prescribed is similar to the drugs that the patient is allergic to. In our work, the individual ratio of each property is weighted according to the maximum F-measure as indicated in

Table 3. For example, if F^v_{max}, F^p_{max} and F^s_{max} is the maximum F-measure for drug property "Overview", "Professional" and "Side Effect" respectively, then the weight w_1 against the similarity ratio for "Overview" is given by:

$$w_1 = \frac{F^v_{max}}{(F^v_{max} + F^p_{max} + F^s_{max})} \quad (2)$$

w_2 and w_3 can also be calculated in a similar manner. Thus, the overall similarity ratio $s(p, q)$ for drug-pair with feature vector p and q is given by:

$$s(p, q) = w_1 * s^v(p, q) + w_2 * s^p(p, q) + w_3 * s^s(p, q) \quad (3)$$

where s^v is the similarity ratio for the drug property from "Overview", s^p is that from "Professional", and s^s is that from "Side Effects".

If the similarity ratio exceeds the average threshold value θ, the model will return a false, indicating that the drug to be prescribed is similar to the drug that patient is allergic to.

3.3 User Layer

The importance of a user-friendly user layer in transforming patient profiles and data in a knowledge base into usable and useful knowledge for the dentist cannot be over-emphasised. A good interface is also crucial in the technology diffusion process to enable high user acceptance and absorption rates. In fact, a poorly designed user interface can reduce the performance and benefits to clinicians [9], forming a barrier against system adoption.

Table 1. Features of conceptual framework

User layer	Data layer	Knowledge layer
• Efficient mapping of user requirements • User friendly interface	• Efficient choice of programming approach • Implementation of data mining • Algorithm design	• Bio medical data sources, drug taxonomy • Drug properties

In this system, the data layer essentially performs the mining and ranking of drug pairs while the user layer consists of the drug profile of the patients and the drug which the dentist is going to prescribe. The user layer also presents the results after computation of similarity ratio is completed at the data layer, acting as a supporting tool to the dentist in deciding whether the drug in question is safe for prescription. If the drug to be prescribed is found to adversely interact with drugs that the patient is currently taking or is similar to the drugs that the patient is allergic to, then it is in this layer that an alternative drug for prescription is presented to the dentist.

As highlighted in Table 1 for the three layers in the framework, user requirements in the user layer need to be efficiently mapped onto the data layer to enable useful and relevant information to be extracted for further computing of the similarity ratio.

4 Experimental Evaluation

We choose an experimental approach to assess the accuracy and efficiency of the proposed method. It tested the hypothesis that similar drug-pairs have a higher similarity ratio compared to those dissimilar pairs.

Precision, recall and F-measure were used to evaluate the performance of our model. Precision indicates how accurately the model predicted drug-pairs as similar, while recall indicates how accurately similar drug-pairs were predicted. Accuracy was also used to measure the percentage of correct predictions combining both the similar and dissimilar predictions.

Table 2. Baseline models

	Tari [14]	Yan [16]	Proposed model
Aim	Discover drug interaction	Predict drug interaction	Personalised drug prediction
Source	Drug bank and MeSH	Drug bank and MeSH	Drug bank
Method	Combine text mining and reasoning approach based on biological entities	Compose feature vectors based on names of disease and genes	Create feature vectors from textual drug description
Accuracy	77.7%	69%	79%

Our work was evaluated against other works to highlight how adoption of this novel approach results in superior performance. The work of [14] predicted DDI by parsing bio-medical text for syntactic and semantic information on biological entities like induction and inhibition of enzymes by drugs. These relations were then mapped with the general knowledge about drug metabolism and interactions to derive the DDI.

Just like our work, DrugBank was also used by [16]. However, one of the methods in their preparation of data was to represent each drug by a vector of drug targets. The values in each vector are either 1 or 0, depending on whether the drug target is associated with the given drug. In our work, we chose to construct feature vectors of *tf*idf* from textual information related to the properties of each drug. Table 2 shows a summary of the experiment methods of the baseline models.

5 Results and Discussions

With the unique three-tier conceptual framework where knowledge is extracted from the knowledge base and delivered to the data layer, the ensuing results demonstrate the efficiency and robustness of our model. Not only is the algorithm able to compute the similarity of the drug-pair based on the hypothesis that a drug-pair is similar if the cosine similarity ratio between their frequency terms is high but such information can also be adopted as a decision support tool for the health professional in drug prescription.

By computing the similarity ratio between drug-pairs, their average values are obtained as a guide to set the threshold θ in order to maximise the F-measure. As shown in Table 3, a range of values for θ are applied for each of the drug properties "Overview", "Professional" and "Side Effects". For example, θ of 0.45 is used as a threshold to compute the recall, precision and F-measure for features gathered from the drug property "Professional" as the maximum value of F score occurs at this value. Figure 4 shows the recall, precision and F scores achieved with drug properties gathered from "Overview", "Professional" and "Side Effects". As indicated in Fig. 4, the recall rate of 96% is achieved from drug properties obtained from "Side Effects", showing that our model performed much better than other methods of prediction. In contrast, the work by [14] achieved 48.5% with predictions based on the inhibition properties of drugs in

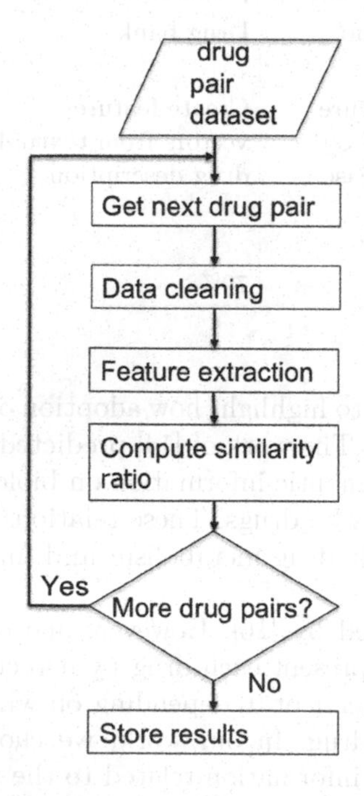

Fig. 3. Experimental design

Table 3. F-measure at different θ

θ levels	Overview	Professional	SideEffect
0.45	0.61	**0.60**	0.67
0.48	**0.60**	0.56	0.66
0.50	0.60	0.57	0.67
0.53	0.56	0.51	0.68
0.55	0.53	0.52	0.69
0.58	0.51	0.54	0.69
0.60	0.54	0.52	0.68
0.63	0.56	0.46	**0.73**
0.65	0.53	0.44	0.70

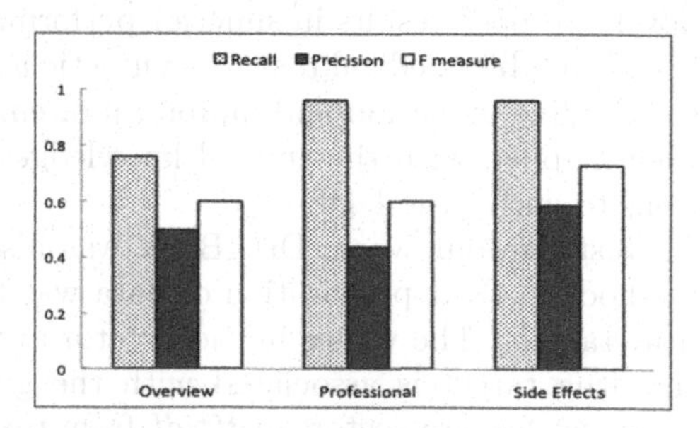

Fig. 4. Performance comparison against different drug properties

the knowledge base. In terms of accuracy, the percentage of correct predictions combining both the similar and dissimilar predictions, our system comes out at 79% compared to 69% where drug predictions were based on the relationship between drug targets [16].

Table 4. Sample result of recommender

Drug currently taking	Drug allergy	Drug prescribed	Drug recommended
Warfarin	Penicillin	Clindamycin	Clindamycin
		Amoxicillin	Troleandomycin
		Acetaminophen	Acetaminophen
		Ibuprofen	Dextropropoxyphene

To illustrate the conceptual framework of this study, the same model can be used to decide if the drug is suitable for prescription. Based on the overall similarity (as explained in Sect. 3.2) from the three properties of the drug-pair, the system can detect if the drug is similar to the drugs that the patient is allergic to. This approach highlights the usefulness of our framework where knowledge generated from the data layer can be applied to the user layer and becomes useful to the user, or with regards to this case, as a decision support tool for the health professional. This novel strategy supports the aim of the study in allowing us to support the dentist to make the right prescription by ensuring the drug is not in adverse relationship with the drugs the patient is taking, and is also dissimilar to the drugs the patient is allergic to.

Consider for example a patient with a cardiac condition taking *warfarin* and with an allergy to penicillin. In a scenario where this patient requires an antibiotic, the dentist may consider the commonly prescribed drug *amoxicillin*. Use of our model will note the similarity of *amoxicillin* to the drug of allergy, *penicillin*, and will thus recommend an alternative drug. Not only should this drug be dissimilar to *penicillin*, it should also not be in adverse relationship with *warfarin*. As shown in Table 4, one such drug is *troleandomycin*. On the other hand, if the dentist is initially considering the prescription of *clindamycin*, our model will evaluate this drug as safe (not similar to *penicillin* and not in adverse relationship with *warfarin*) and thus nothing new will be suggested. This is the same scenario when an analgesic is required in the same patient for the prescription of *acetaminophen*. Conversely, inputting *ibuprofen* into our model will lead to a suggestion of *dextropropoxyphene* instead, as the former is in adverse relationship with *warfarin* and thus not safe to be prescribed. On receiving the suggestion of the alternative drug, it is then for the dentist to decide whether this is an appropriate drug to prescribe after further consideration of the duration and dosage of the patients current drugs.

6 Conclusions

This paper has presented a novel approach in advising the suitability of a drug for prescription by predicting the similarity of a drug-pair and in practical terms, integrating this prediction with the patient's personal medical status by considering the patient's drug allergies to avoid allergic reactions, and the drugs the patient is currently taking to avoid adverse DDI. The included experimental proof also demonstrates that the three-tier approach adopted in our research design performs well and thus can readily be implemented within the clinical workflow of a dental clinic. In future work, we will investigate the drug taxonomy for more complex semantic relations existing between drugs, for example, *neutral* and *advantageous*, and have the clinic decision support system to be supported by a more comprehensive, valuable knowledge base.

Acknowledgments. This research is partially supported by Glory Dental Surgery Pte Ltd, Singapore (http://glory.sg) and undertaken collaboratively with their panel of dentists. We would like to thank Ms Elizabeth Goh for enriching the authors' understanding of dentists' requirements.

References

1. Bokharaeian, B., Diaz, A., Chitsaz, H.: Enhancing extraction of drug-drug interaction from literature using neutral candidates, negation, and clause dependency. PLoS ONE **11**(10), 1–20 (2016)
2. Brown, A.S., Patel, C.J.: MeSHDD: literature-based drug-drug similarity for drug repositioning. J. Am. Med. Inform. Assoc. **27**, 1–5 (2016)
3. Bui, Q., Sloot, P., van Mulligen, E., Kors, J.: A novel feature-based approach to extract drug-drug interactions from biomedical text. Bioinformatics **30**(23), 3365–3371 (2014)
4. Casillas, A., Prez, A., Oronoz, M., Gojenola, K., Santiso, S.: Learning to extract adverse drug reaction events from electronic health records in Spanish. Expert Syst. Appl. **61**, 235–245 (2016)
5. Domingues, M., Jorge, A., Soares, C., Rezende, S.: Web mining for the integration of data mining with business intelligence in web-based decision support systems. In: Azevedo, A., Santos, M. (eds.) Integration of Data Mining in Business Intelligence Systems, pp. 120–142 (2015)
6. Goh, W.P., Tao, X., Zhang, J., Yong, J.: A study of drug interaction for personalised decision support in dental clinics. In: 2015 IEEE/WIC/ACM Workshop Proceedings on International Conferences on Web Intelligence (WI), vol. 1, pp. 88–91 (2015)
7. Goh, W.P., Tao, X., Zhang, J., Yong, J.: Decision support systems for adoption in dental clinics: a survey. Knowl.-Based Syst. **104**, 195–206 (2016)
8. Han, L., Kamber, M., Pei, J.: Data Mining Concepts and Techniques. Morgan Kaufmann Publishers, Burlington (2011)
9. Horsky, J., Phansalkar, S., Desai, A., Bell, D., Middleton, B.: Design of decision support interventions for medication prescribing. Int. J. Med. Inform. **82**(6), 492–503 (2013)

10. Li, A., Zang, Q., Sun, D., Wang, M.: A text feature-based approach for literature mining of lncRNA-protein interactions. Neurocomputing **206**, 73–80 (2016)
11. Lieber, N.S.R., RibeiroII, E.: Adverse drug reactions leading children to the emergency department. Revista Brasileira de Epidemiologia **15**, 265–274 (2012)
12. Lu, J., Hales, A., Rew, D., Keech, M., Frohlingsdorf, C., Mills-Mullett, A., Wette, C.: A case study from breast cancer research. In: Proceedings of the 6th International Conference on Information Technology in Bio- and Medical Informatics, pp. 56–70 (2015)
13. Mukherjee, S., Varde, A.S., Javidi, G., Sheyban, E.: Predictive analysis of engine health for decision support. SIGKDD Explor. Newsl. **15**(2), 39–49 (2014)
14. Tari, L., Anwar, S., Liang, S., Cai, J., Baral, C.: Discovering drug-drug interactions: a text-mining and reasoning approach based on properties of drug metabolism. Bioinformatics **26**(18), 547 (2010)
15. Wu, H.Y., Chiang, C.W., Li, L.: Text mining for drug-drug interaction. Methods Mol. Biol. **1159**, 47–75 (2014)
16. Yan, S., Jiang, X., Chen, Y.: Text mining driven drug-drug interaction detection. In: Proceedings of 2013 IEEE International Conference on Bioinformatics and Biomedicine, pp. 349–354 (2013)

PURE: A Novel Tripartite Model for Review Sentiment Analysis and Recommendation

Yue Xue[1(✉)], Liutong Xu[1], Hai Huang[1], and Yao Cheng[2]

[1] Telecommunications Software Engineering Group, School of Computer Science, Beijing University of Posts and Telecommunications, Beijing, China
xueyue1015@sina.com

[2] Beijing Huatong Xinlian Technology Co. Ltd, Beijing, China

Abstract. Nowadays, more and more users like to leave online reviews. These reviews, which are based on their experiences on a set of service or products, often express different opinions and sentiments. Correlated topic model (CTM), an effective text mining model, can reduce the dimension without losing important information. However, traditional analyses based on CTM still have some problems. In this paper, we propose the Product-User-Review tripartite sEntiment model (PURE), which is based on content-based clustering to optimize CTM, to select topic number, extract feature, estimate the reviews' utility. Moreover, our model analyzes the reviews from the user's preferences, review content and product properties in three dimensions. Based on the five indexes, such as informative attributes and sentiment attributes, the feature vector of the review data is constructed. We found that after adding user's preference feature in sentiment analysis and utility estimation, PURE achieves high accuracy and classification speed in the review-mixing Chinese and English processing, and the quality of selection is improved significantly by 21%. To the best of our knowledge, this is the first work to incorporate users' preference feature in optimized CTM to do the study of sentiment analysis, review selection and recommendation.

Keywords: Correlated topic model · Text mining · Content-based clustering · User preference · Sentiment analysis · Reviews' utility estimation

1 Introduction

With the rapid development of e-commerce sites, users would like to write reviews after they experienced some online services. The detail information in these reviews, on one hand, could show their using feelings and shopping experience, on the other hand, some reviews are valuable for customers to make informed purchase decisions, and for businesses to form an effective feedback to improve the quality of their services and products. Different users may choose to focus on different aspects of products. Even if users talk about the same aspects, their shopping cognition and, subsequently, experience of the product or service can differ dramatically. According to a survey, which is mentioned in [1], that half of users find reviews are useful when booking hotel or shopping online, and 88% of respondents agree that they always consult reviews before making a purchase.

J. Kim et al. (Eds.): PAKDD 2017, Part II, LNAI 10235, pp. 388–400, 2017.
DOI: 10.1007/978-3-319-57529-2_31

To predict the effect of online reviews, a number of studies have been conducted in this field. For example, [2, 3] use topic model to analyze implicit information in reviews, especially in model [4], it split the document into paragraphs and mine the implicit features from the sentence. [5–7] take sentiment analysis into consideration, and model in [5] constructs an unsupervised hierarchical Bayesian model which can classify sentiment in documents. [8–12]'s main idea is to mine and recommend the basic review of online datasets, and [14] proposed a novel unsupervised feature selection method which performed structure learning and feature selection simultaneously.

In this work, we provide an optimized correlated topic model. It is mainly based on content-based clustering, and focuses on the sentiment analysis and utility estimation, not only in the given product level and review level, but in the user level. It is objective to study the user preference feature and time trend about each facet from the reviews. In addition, our model provides a feature extraction method. It firstly offers the reasonable value of topic number, and generates the topic in each review's facet according to this number. Then it finds the representation in these facets, deduces the opinion orientation tuple of each facet, and then estimates the review utility score and recommend the most helpful reviews to users. We also use machine learning method to make the detection more accurate,

We summarize our contributions as follows:

(1) PURE focuses on the review's influence estimation in tripartite level, and considers the impact of time and content correlation in reviews.
(2) We mainly use correlated topic model while generating the model, and optimize it into content-based analysis. To improve the judgment of their sentiment orientation, we use "feature-sentiment" polynomial to target the review.
(3) Active learning strategies are being used to optimize the performance and reduce the error in review-mixing of English and Chinese processing.

The rest of the paper is organized as follows. We describe related work in Sect. 2, and define some descriptions and terminologies in Sect. 3. The detailed PURE model is proposed in Sect. 4. The experiments and analysis are shown in Sect. 5. Finally, we draw the conclusion in Sect. 6.

2 Related Work

Our work on review feature extraction and sentiment analysis has connections to text mining in information retrieval and text analysis in artificial intelligence. Great bulk of work has been focused on these fields at various levels.

Previous work typically focuses on extracting data features, and then formulating review's utility prediction as a regression [13] or classification problem, such as Sun and Zhou [7] based on TF-IDF word frequency recommendation, this model cannot mine the underlying feature of the document. To improve the performance of the model, Blei et al. [17] proposed the supervised topic model framework, in which a document is modeled as a latent set of topics, where each topic is modeled by a distribution over a set of words. There are many other topic models such as [3] based on topic model in text mining fields, and Lin et al. [4] consumed a probabilistic modeling framework and considered customer review as a correlation of "topic-phrase-sentence-paragraph-document". While considering sentiment analysis in review mining field, JST [3] overcomes these short-comings as it is based on topic model with a better statistical foundation by adding a sentiment layer. Further, models proposed by [6, 7, 10, 15] achieved fine results in review rating and recommendation, and make the review summarization more accurate.

3 Problem Formulation

In this section, we define some core terminologies and descriptions of the problems.

At Product Level

property: A proper p_i is used to describe product. $P = \{p_1, p_2, ..., p_n\}$, if a specified product itself has n properties. Noted that a product may have at least one property.

At User Level

user preference: U is a specific weighted by the user's facet-specific preference, like the user's expertise, gender, writing style, andshopping level. Assuming that a user with high rating start will be more likely to write fine-grained review.

At Review Level

facet: A facet f is a description of a property or component of an review entity. In our study, facet f often known as an implicit topic. Let a review R has a set of facets f_i, $f_i \in F$ with respect to which the review is to be evaluated.

facet-opinion pairs: Every facet expressed in the review is associated to one opinion, the pair $fo = \{fo_1, fo_2, \ldots, fo_n\} = \langle f, o \rangle$ indicates it. The words of a facet are always nouns, and the words of an opinion are often adjective.

sentiment: The sentiment score s is formulated in our study, which is defined to represent the sentiment words' distribution value. It is a threshold value to distinguish whether the review's sentiment s is positive, negative, or other.

time-related: We divided the year into four group according to the seasons and temperatures. T_1 = <Nov., Dec., Jan., Feb.> , T_2 = <Mar., Apr., May.> , T_3 = <Jun., Jul., Aug.> , T_4 = <Sep., Oct.>. For example, in terms of hotel accommodation, the season will have a certain impact on the users' reviews on the hotel conditions.

Figure 1 presents an example to illustrate the above key terms, and to show relations between them.

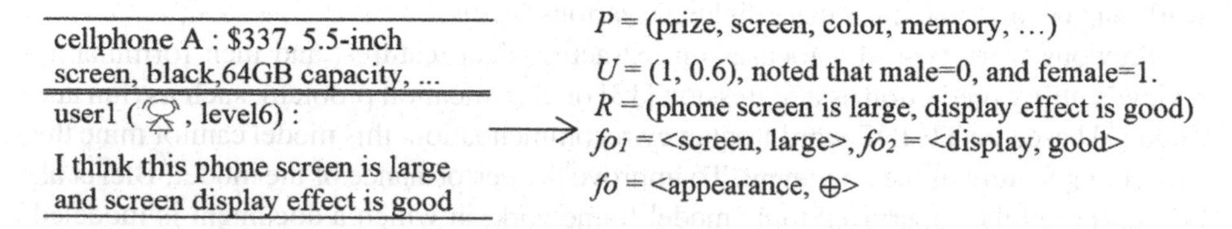

Fig. 1. An example of the key terms used in this problem and their relations.

4 Model and Inference

4.1 Proposed Model

We propose a graphical model, namely the product-user-review tripartite sentiment model(PURE), to address the problems defined above. The overview of PURE is shown in Fig. 2 and the description of each notation is in Table 1. From the overall perspective, PURE uses a parameter ξ as the background distribution to encode the domain white-noise in our experiment. This global distribution is drawn from a symmetric Dirichlet priors with concentration $\lambda_B = 0.2$.

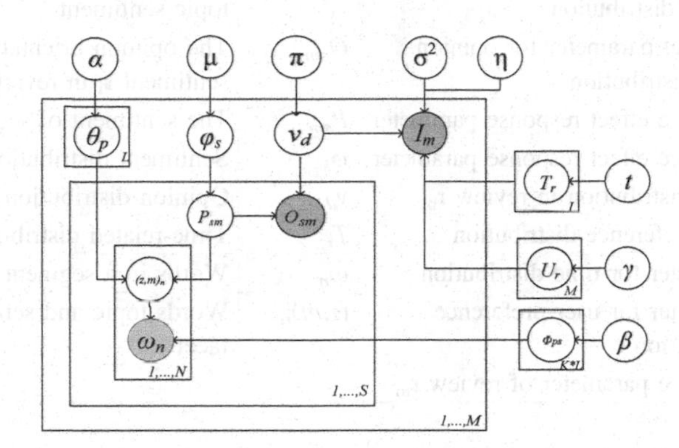

Fig. 2. Graphical model representation of PURE

The generative process of the graphical model is as follows:

1. For each facet topics $k \in [1, \ldots, K]$ and sentiments $l \in [1, \ldots, L]$ draw content's word correlated topic distribution $\phi_{k,l} \sim \text{Dir}(\beta)$.
2. For all review time $i \in [1, \ldots, I]$ draw time-related distribution $T_r \sim \text{Dir}(t)$.
3. For all user information $m \in [1, \ldots, M]$ draw user preference distribution $U_r \sim \text{Dir}(\gamma)$.
4. For all reviews r_m, $m \in [1, \ldots, M]$
   (1) draw facet sentiment distribution $\varphi_s \sim \text{Dir}(\mu)$.
   (2) draw facet opinion distribution $V_d \sim \text{Dir}(\eta)$.
   (3) for all segments $s \in [1, \ldots, S]$
      (1) draw facet distribution θ_{ms} in review r_m, $\theta_{ms} \sim \text{Dir}(\alpha)$.
      (2) for each word $n \in [1, \ldots, N]$ in segememt s
         (1) draw words' topic and sentiment distribution $(z,m)_s \sim \text{Multi}(\theta_{ms})$.
         (2) draw words distribution $\omega_n \sim \text{Multi}((z,m)_{s,n}), \phi_{(z,m)\ s})$.
         (3) draw opinion words O_{sm} distribution in s, $O_{sm} \sim \text{Multi}(P_{sm}, V_d)$.
   (4) draw influence factor $I_m \sim \text{N}(\overline{Z_m}, \sigma^2)$, where we define

Table 1. Notations of PURE

Symbols	Representation	Symbols	Representation
M	The number of reviews	N	Number of words in facet above a topic
S	The number of segments in reviews	K	Number of facets
L	The number of sentiment	I	Number of group of time-related partition
α	Parameter for topic correlation distribution	β	Dirichlet parameter for word's number distribution
μ	Dirichlet parameter for sentiment words' distribution	$\Phi_{p,s}$	Words distribution above correlated topic sentiment
π	Dirichlet parameter for opinion word distribution	O_{sm}	The opinion orientation of sentiment s_i in review r_m
σ^2	Influence effect response parameter	P_{sm}	The sentiment of segment s in r_m
η	Influence effect response parameter	φ_s	Sentiment distribution in reviewr_m
θ_p	Facet distribution in review r_m	v_d	Opinion distribution in reviewr_m
U_r	User preference distribution	T_r	Time-related distribution
t	Parameter for time distribution	ω_n	Words in a segment
γ	Parameter for user preference distribution	$(z,m)_n$	Words topic and sentiment in each facet
I_m	Influence parameter of review r_m		

$$\overline{Z_m} = \frac{1}{C}\sum_{s=1}^{S_m}(((C_1^T v_{rs}) * C_2^T T_{rs}) * C_3^T U_{rs}. \tag{1}$$

In above Eq. 1, $\overline{Z_m}$ represents the combined empirical frequencies of the implicit facets and sentiments of the review r_m. Moreover, C is a normalization constant, C_1, C_2 and C_3 are weight vector for sentiment orientation, time-related distribution and user preference, which can be obtained experimentally from data.

4.2 Inference and Prediction

In this section, we describe the inference and procedure for PURE. We introduce three main modules in our model, and describe how to apply this model in review utility estimation and review recommendation. The goal of PURE is to evaluate the posterior distribution P (T, U, Φ, ν, θ, φ, (z,m),P |ω, o, h), as shown below.

$$\mathrm{P}\ (\mathrm{T}, \mathrm{U}, \Phi, \nu, \theta, \varphi, (\mathrm{z}, \mathrm{m}), \mathrm{P}, \omega,\ \mathrm{O},\ \mathrm{I}) = \frac{P(T, U, \Phi, v, \theta, \varphi, (z, m), P, \omega, O, I)}{\sum_f \sum_s P(T, U, \Phi, v, \theta, \varphi, (z, m), P, \omega, O, I)}. \tag{2}$$

The exact inference of this distribution is intractable, due to the difficulty in the denominator of Eq. 2. According to Griffiths and Seyvers [16], a collapsed Gibbs sampling algorithm can be used to get the approximate inference of PURE. Then, Eq. 3 shows the joint probability of all latent variables and observable variables.

$$\begin{aligned} \mathrm{P} &= \mathrm{P}(\mathrm{T}_{1:I}, \mathrm{U}_{1:M}, \Phi_{(1,1):(k,l)}, \nu_{1:M}, \theta_{1:M}, \varphi_{1:M}, (\mathrm{z,m})_{1:M}, \omega_{1:M}, \mathrm{O}_{1:S}, \mathrm{I}_{1:M}) \\ &= \prod\nolimits_{1}^{I} P(T_p|t) * \prod\nolimits_{1}^{M} P(U_p|\gamma) * \prod\nolimits_{1,1}^{k,l} P(\phi_{(i,j)}|\beta) * \prod\nolimits_{d=1}^{M} P(v_d|\eta)P(\theta_d|\alpha)P(\varphi_d|\mu) \\ &\quad * (\prod\nolimits_{s=1}^{S} P(P_{s,m}|\varphi_M)P(O_{s,m}|v_M) \prod\nolimits_{n=1}^{N} P(\omega_{s,n}|\varphi_{(p,m)_{s,n}})P((z,m)_{s,n}|m_s, \theta_d)). \end{aligned} \tag{3}$$

PURE also draws the review facet distribution θ_{ps} to represent the topic correlation distribution, and it can solve the Φ_{ps} and φ_s as follow:

$$\widehat{\varphi}_s^{(i)} = \frac{n_s^{(i)} + \mu_s}{\sum_{s=1}^{L} n_s^{(i)} + \mu_s}, \widehat{\theta}_{p,s}^{(m)} = \frac{n_{p,s}^{(m)} + \alpha_s}{\sum_{s=1}^{K} n_{p,s}^{(m)} + \alpha_s}, \widehat{\Phi}_{p,s}^{(z)} = \frac{n_{p,s}^{(z)} + \beta_i}{\sum_{i=1}^{N} n_{p,s}^{(z)} + \beta_i}. \tag{4}$$

The parameter $\widehat{\theta}_{p,s}^{(m)}$ represents the probability estimation when the correlated topic is p and the sentiment score is s in the correlated topic distribution of review r_m. $\widehat{\varphi}_s^{(m)}$ represents the probability estimation of the distribution of review r_m's sentiment score s, and $\widehat{\phi}_{p,s}^{(m)}$ represents the probability estimation of words z distribution allocated above topic p and sentiment s. Additionally, we follow Blei and McAuliffe [19] to approximately evaluate the normal linear model parameters η and σ^2. Let X be the S × K matrix whose rows are the vectors $\bar{x}_m^T$. Then η and σ^2 is approximate in Eq. 5, where h indicates the influence factor response vector.

$$\widehat{\eta} \approx \left(X^T X\right)^{-1} X^T h, \widehat{\sigma}^2 \approx \frac{1}{S}\left[h^T h - h^T X \eta\right]. \tag{5}$$

We propose three closely related modules as described below:

Preprocessing Module

PURE firstly splits a long review into some segments by one or more sentences, and assumed that each segment has a single facet associated with it. Then, it models each segment and opinion words and detects the implicit facets and sentiments simultaneously. Let the review R consists of n segments s_i ($i = 1...n$), that is $R = \{s_1, s_2, ..., s_n\}$ where each segment has a facet and opinion. Then we can get joint probability of our model in Eq. 6.

$$
\begin{aligned}
\mathrm{P}(z_i, m_{s_i} | z_{N-i}, m_{R-s_i}, \omega_n) &= \frac{P(z, m, \omega)}{P(z_{N-i}, m_{R-s_i}, \omega_n)} \\
&= \frac{P(\omega | z, m) P(z, m)}{P(\omega_n | z_{N-i}, m_{R-s_i}) P(\omega_i) P(z_{N-i}, m_{R-s_i})} \\
&\propto \frac{B(n_{k,l} + \beta)}{B(n_{k,l,N-i} + \beta)} \frac{B(n_s + \mu)}{B(n_{s,R-s_i} + \mu)} \frac{B(n_{m,l} + \alpha)}{B(n_{m,l,N-i} + \alpha)}.
\end{aligned} \tag{6}
$$

In Eq. 6, $B(\alpha)$ is Beta function. We use N-i to represents the topic distribution number except i, and R-s_i means the remaining sentiment distribution except s_i.

Feature Extraction and Sentiment Analysis Module

Feature extraction by CTM in the general situation is to use the features as the input after removing the low-frequency words directly. It will have a huge computational cost, and the effect is not very friendly. So, our model not only considers low-frequency words, but takes sentiment property which is shown above as facet features, and uses list to record them. For positive sentiment, it can calculate each value in list <*PosSco, AvgPos, FluPos*> according to the Eq. 7. (For negative sentiment representation is similar with it, and to avoid repetition, it is not described in this part.)

$$
\begin{cases}
PosSco = \sum_{i-1}^{n} Pos_i \\
AvgPos = \frac{PosSco}{n} \\
FluPos = \sqrt{\frac{\sum_{i-1}^{n} Pos_i - AvgPos}{n}}
\end{cases} \tag{7}
$$

Through the calculation and processing, the final weight vector will get and the regression method is used to get the overall sentiment orientation score.

Review Utility Prediction Module

When the parameters in PURE are fitted, we can calculate the review impact and recommend the fine-grained reviews to users. Our idea for the overall review influence factor prediction is to infer these tripartite level—product level, user level, and review level—evaluated facets, and then approximately form the regression function on the posterior mean $\overline{x}_{m'}$. It is shown in Eq. 4. We have to note that the posterior mean $\overline{x}_{m'}$ is obtained by applying Gibbs sampling technique, and Eq. 8 is used to estimate the overall review utility scores of testing reviews.

$$
\widehat{I}_{m'} = a\eta^T \widehat{x}_{m'} + b\theta_k^{(m)} \frac{\widehat{\mathrm{I}}_m}{\sum_{\mathrm{m}} \widehat{\mathrm{I}}_m}. \tag{8}
$$

Note that parameter θ_k is described in the previous section in Eq. 4, and a and b is weight vector parameter according to their degree. The model can select the most influential one for each facet in terms of the estimated score based on correlated facet. To get the optimal results, a and b will be discussed in the next section.

5 Experiment and Evaluation

We evaluate PURE in following three parts: reviews' facets detection, sentiment recognition, and the useful review recommendation. We compare PURE against three well-established typical benchmark models: a supervised topic model called supervised latent Dirichlet allocation (sLDA) [17], unsupervised topic model called unsupervised text sentiment model (UTST) [7], and classic linear regression model (LR).

Data Sets
The online website review data from two product categories is used to test the performance of PURE. The datasets[1] are including 11,190 mobile phone reviews of ZOL and 10,530 book reviews of Douban, including the profiles of many users, and the contents of reviews. Note that almost reviews are in Chinese, and the rest of reviews are mixed in English and Chinese.

Preprocessing
We store data with excel, and firstly use jieba to do words' segmentation. And then, the related function is used to filter stop-words. After that, do frequency statistics of words in JAVA programming to statistic the number of different words, and different times of each word which appears in each review. Next, these datasets are divided into two parts, 75% data for training, and the rest for testing. After preprocessing the reviews' datasets, the statistics of these data are listed in Table 2.

Table 2. Statistics of data sets

Category	Mobile phones	Books
#review	10,070	8,500
#words	768,936	617,100
#average words	79	73
#reviewer	9,835	8,436

Noted that, the reason why the amount of data vanished a lot after preprocessing is that there are plenty of reviews are insignificant and useless. For example, some reviews only contain punctuation marks, or the words' length is less than two, although these reviews may contain the implicit sentiment, they are not qualified to the next process and analysis.

Topic Detection of Reviews
To get the topic number of our model, in this section, our study uses content-based CTM to cluster the training data, it can reduce the dimensionality and get the topic in each review. PURE can identify correlated facets and opinion words semantically, as a result, some adjective words which are close to the topic words can also be analyzed and separated by the same facet at the same time. According to the topic "system" in Table 3, the topic terms are discovered by our model. The column 1 are property words.

[1] http://data.bupt.edu.cn/xueyue/data/datasets/.

The opinion words, such as "amazing" and "consumption" are shown in column 2. The score in column 3 is the word-related-facet' value. The purpose of column 3 is to express the degree of correlation between the topic and review's facet more clearly. The situation in book reviews is shown in the right of Table 3, it is similar to the former.

Table 3. Topic and their related value on book reviews and mobile phone reviews.

"system" topic @ mobile phone			"plot" topic @ books		
Property	Opinion words	Related value	Property	Opinion words	Related value
Dual core system	Not crash	0.976	Climax	Charming	0.851
Speed	High	0.821	Storyline	Surprised	0.706
Picture	Good	0.738	Twists	Original	0.713
Power	Less charge	0.891	Puberty	Unfolding	0.479
Speaker	Consumption	0.901	Phenomenon	Bored	0.717
Wi-fi	Well	0.694	Beginning	Attractive	0.886
Voice	Better	0.832	Storyline	Unexpected	0.823
Sound	Amazing	0.723	Ending	Unfinished	0.769

Secondly, Pearson Correlation Co-efficient (PCC) is used to find correlation of the predicted facets. We evaluate PURE and other models via PCC versus the number of facets (*N*). We define that sentiment orientation count as two, which are positive and negative, for avoiding the ambiguity. Figure 3 plots the correlation curves against the facets' number of PURE versus the sLDA, UTST and LR models on mobile phones reviews and books reviews.

On the left of Fig. 3, we can see that LR has only one correlation value as it cannot mine the hidden topical structure of data. What's more, PURE performs better than other models. The average correlation of PURE over all the observations is 43.08%,

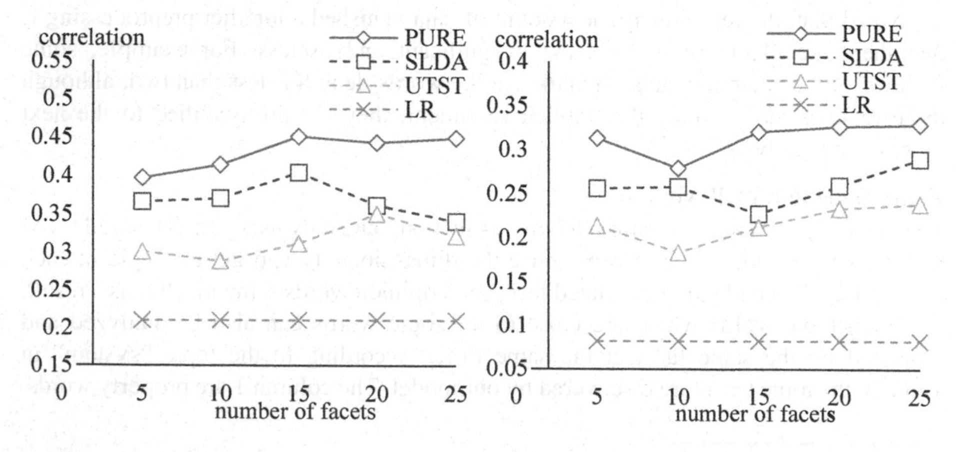

Fig. 3. Correlation versus number of facets on reviews of phones (right)/books(left).

which is better than 36.82% of sLDA, 31.42% of UTST and 21.03% of LR. The right curves of Fig. 3 plots the results on the books reviews. PURE again results in better performance compared to sLDA, UTST and LR. The average correlation score of PURE across all the five observations is 31.32%, better than sLDA, UTST and LR which are 25.78%, 21.61%, and 8.2%.

It can be concluded that the proposed PURE outperforms the state-of-the-art supervised topic model sLDA, and unsupervised model UTSU, as well as one classic linear regression model for the overall review sentiment classification. PURE benefits from the sentiment analysis module while sLDA could not mine the underlying sentiment in its structure. As to UTST, unsupervised topic model, the sentiment detection and analysis are not as accurate as our model.

In addition, PURE is also used to process the 3,298 mobile phones reviews data and 8,436 books reviews data to find the most favorite facet in each area. Figure 4 plots the facets specific preferences of top5 user in the corpus. Overall, the main facets preferences what we mining form large number of books reviews datasets have been found to be in the order *plot* > *author* > *type* > *ending*. At the same time, in mobile phones datasets, the order is *performance* > *brand* > *appearance* > *price*.

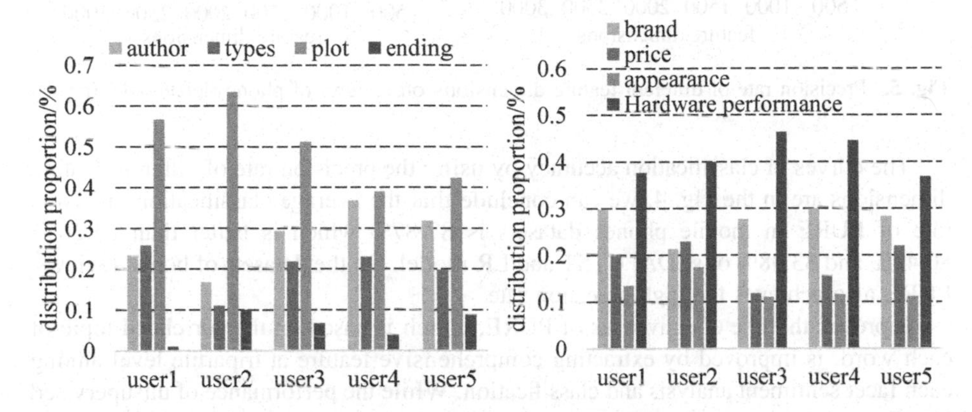

Fig. 4. Facet specific preference of different users on books (right)/phones (left).

Performance of Sentiment Classification

To classify the sentiment orientation, we select different feature dimensions, calculate the orientation score based on the Eq. 7, and detect the classification performance of these models. The precision rate, recall rate and F_1 in Eq. 9 can evaluate PURE and other models.

$$\text{Precision} = \frac{\sum_{r \in R} |R(r) \cap T(r)|}{\sum_{r \in R} |R(r)|}, \text{Recall} = \frac{\sum_{r \in R} |R(r) \cap T(r)|}{\sum_{r \in R} |T(r)|},$$
$$F_1 = \frac{2 \times \text{Precision} \times \text{Recall}}{\text{Precision} + \text{Recall}}. \tag{9}$$

Note that $R(r)$ is the training set of data, and $T(r)$ represents testing set of data. The experimental results are shown in Fig. 5. As the recall rate and precision rate are mutual influence, that is to say, the accuracy rate is high, the recall rate is low at the same time. Therefore, we propose precision rate as the final criterion to measure the performance. The 50% test data is used to do this experiment.

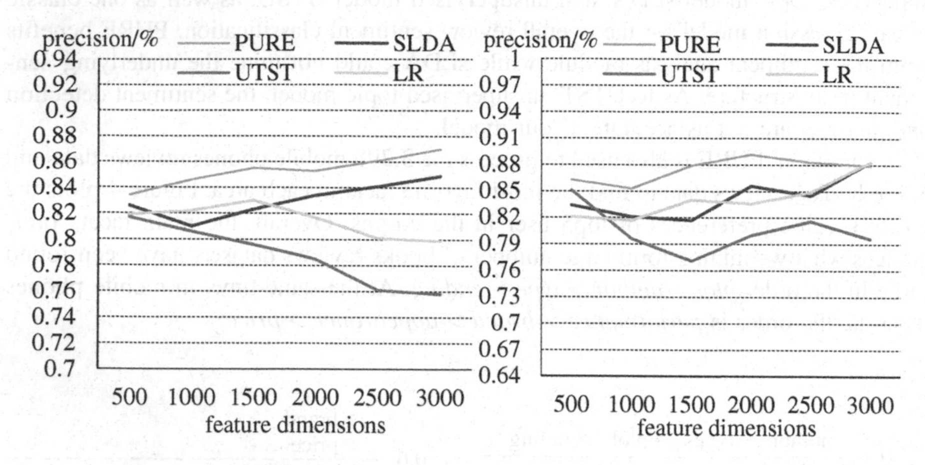

Fig. 5. Precision rate of different feature dimensions on reviews of phones(left)/books (right).

The curves of classification accuracy by using the precision rate of different feature dimensions are in the Fig. 4, we can conclude that the average classification precision rate of PURE in mobile phones datasets is 87.87% which is better than 84.26%, 80.47% and 83.98% of sLDA, UTST and LR model. On the datasets of books reviews, PURE also achieves the high precision rate.

It proves that the effectiveness of PURE, which is based on the correlated topic of each word, is improved by extracting comprehensive feature at tripartite level among each facet sentiment analysis and classification. While the performance of unsupervised UTST is slightly worse than that of supervised sentiment classification sLDA.

The Useful Reviews Recommendation

The final goal of our model is to recommend the most influential review for user. In order to make the review's utility score higher, the review must satisfy two conditions: the facet itself has to be relevant to the product properties, and organized specific; the sentiment that included in the review is clear, not ambiguous. The remaining 50% test data is used to evaluate our model, and for each facet, the influence score is estimated using Eq. 8. And we test PURE facet-based review impact estimation by using precision at top N (P @top - N) recommended reviews for each data set. The result is shown in Table 4.

From the facet "plot" on books reviews, all models achieved 100% precision, while at top-10 selected reviews, PURE achieved 98% precision, either supervised learning sLDA or unsupervised learning UTST have lower rate than our model. We omit the model LR preference as it cannot mine the hidden topical structure of data.

Table 4. Influential estimation of models.

Facet detected from model	Models	P@top-1	P@top-10
"plot" @Books	PURE	100%	98%
	sLDA	100%	79%
	UTST	100%	54%
"appearance" @mobile phone	PURE	100%	87%
	sLDA	100%	51%
	UTST	100%	43%

6 Conclusion

Our goal of this paper is to select the best reviews to the users by taking into account of tripartite levels, which are product property level, fine-grained review level and user preference level. Our key breakthrough is that the user's preference feature and time-related feature in the given facets do influence the review sentiment analysis and their order, which greatly help PURE achieve better performance. Moreover, we optimize the correlated topic model by using content-based clustering, the dimensionality of the datasets is reduced significantly. PURE is able to evaluate the review utility and recommend the most valuable review to customers at the same time.

As every bean has its black, our paper only applies PURE in review analysis. Further research may be able to promote the accuracy of our model in other areas.

References

1. Hai, Z., Cong, G., Chang, K., et al.: Coarse-to-fine review selection via supervised joint aspect and sentiment model. In: International Conference on Research on Development in Information Retrieval, pp. 617–626 (2014)
2. Ghose, A., Ipeirotis, P.G.: Estimating the helpfulness and economic impact of product reviews: mining text and reviewer characteristics. Soc. Sci. Electron. Publ. **23**(10), 1498–1512 (2011)
3. Sauper, C., Haghighi, A., Barzilay, R.: Incorporating content structure into text analysis applications. In: Conference on Empirical Methods in Natural Language Processing, pp. 377–387 (2010)
4. Lin, C., He, Y., et al.: Weakly supervised joint sentiment-topic detection from text. IEEE Trans. Knowl. Data Eng. **24**(6), 1134–1135 (2011)
5. Lin, C., He, Y.: Joint sentiment/topic model for sentiment analysis. In: ACM SIGKDD, pp. 375–384 (2009)
6. Mukherjee, S., Basu, G., Joshi, S.: Incorporating author preference in sentiment rating prediction of reviews. In: International Conference on World Wide Web Conference, pp. 47–48 (2013)
7. Sun, Y., Zhou, X.: Unsupervised topic and sentiment unification model for sentiment analysis. Acta Scientiarum Nat. Univ. Pekin. **19**(1), 102–108 (2013)
8. Nguyen, T.V., Karatzoglou, A., Baltrunas, L.: Gaussian process factorization machines for context-aware recommendations. In: ACM SIGIR, pp. 63–72 (2014)

9. Nguyen, T.S., Lauw, H.W., Tsaparas, P.: Review synthesis for micro-review summarization. In: ACM WSDM, pp. 169–178 (2015)
10. Wang, J., Srebro, N., Evans, J.: Active collaborative permutation learning. In: ACM SIGKDD, pp. 502–511 (2014)
11. Zoghi, M., Whiteson, S., De Rijke, M.: MergeRUCB: a method for large-scale online ranker evaluation. In: ACM WSDM, pp. 17–26 (2015)
12. Rayana, S., Akoglu, L.: Collective opinion spam detection: bridging review networks and metadata. In: ACM SIGKDD, pp. 985–994 (2015)
13. Chen, W., Chen, Y., Mao, Y., et al.: Density-based logistic regression. In: ACM SIGKDD, pp. 140–148 (2013)
14. Du, L., Shen, Y.D.: Unsupervised feature selection with adaptive structure learning. In: ACM SIGKDD, pp. 209–218 (2015)
15. Lu, Y., Tsaparas, P., Ntoulas, A., et al.: Exploiting social context for review quality prediction. In: International Conference on World Wide Web Conference, pp. 691–700 (2010)
16. Griffiths, T.L., Steyvers, M.: Finding scientific topics. Proc. Natl. Acad. Sci. **101**, 5228–5235 (2004)
17. Blei, D.M., Mcauliffe, J.D.: Supervised topic models. Adv. Neural. Inf. Process. Syst. **3**, 327–332 (2010)

Clustering and Matrix Factorization

Multi-View Matrix Completion for Clustering with Side Information

Peng Zhao, Yuan Jiang, and Zhi-Hua Zhou(✉)

National Key Laboratory for Novel Software Technology, Nanjing University,
Collaborative Innovation Center of Novel Software Technology and Industrialization,
Nanjing 210023, China
{zhaop,jiangy,zhouzh}@lamda.nju.edu.cn

Abstract. In many clustering applications, real world data are often collected from multiple sources or with features from multiple channels. Thus, multi-view clustering has attracted much attention during the past few years. It is noteworthy that in many situations, in addition to the data samples, there are some side information describing the relation between instances, such as must-links and cannot-links. Though side information has been well exploited in single-view clustering, they have rarely been studied in multi-view scenario. Considering that matrix completion has sound theoretical properties and demonstrates an excellent performance in single-view clustering, in this paper, we propose the first matrix completion based approach for multi-view clustering with side information. Instead of concatenating multiple views into a single one, we enforce the consistency of clustering results on different views as constraints for alternative optimization, and the global optimal solution is obtained since the objective function is jointly convex. The proposed Multi-View Matrix Completion (MVMC) approach exhibits impressive performance in experiments.

Keywords: Multi-view · Clustering · Matrix completion

1 Introduction

Data clustering is one of the most important tasks in machine learning and data mining. Aiming at grouping data instances into different clusters based on the similarity, clustering has plenty of real applications, such as data summarization [9], text mining [24], bioinformatics [8], etc.

In many applications, data are collected from multiple sources or with feature from different channels. For example, the content and hyperlink information can be thought of two views for webpage dataset [3]. Another example is that the representations in various languages can be regarded as different views for multilingual information retrieval [12]. Since feature information from different views are complementary to each other, multi-view clustering dedicates to leverage information from multiple views to improve the performance of clustering.

J. Kim et al. (Eds.): PAKDD 2017, Part II, LNAI 10235, pp. 403–415, 2017.
DOI: 10.1007/978-3-319-57529-2_32

It's noteworthy that while plenty of unsupervised clustering algorithms have been widely used, clustering with limited side (supervised) information has gradually obtained more attentions. In general, side information can be divided into two groups, instance-level and label-level. Usually, label-level one is difficult to gather. In contrast, it is often more convenient to collect instance-level information among which the pairwise constraint is one of the most common representations. Pairwise constraints are consisted of two parts: must-link($\mathcal{M}$) and cannot-link($\mathcal{C}$). A must-link (cannot-link) specifies that the pair of instances should (not) be assigned into the same cluster. Pairwise relationship occurs in a variety of applications and domains. For example, when clustering various movies, we may only know two of them should (not) be assigned into the same style which can be viewed as a must-link (cannot-link). Another example is our knowledge that two proteins always co-occur in the Database of Interacting Proteins (DIP) dataset, which can be regarded as a must-link when performing gene clustering [13]. Generally speaking, it is convenient to gather pairwise constraints along with collecting the unlabelled data. Thus, in this paper, we only consider pairwise constraints prototype side information.

Similarly, clustering with side information is also useful for data collected from multiple sources. Existing multi-view clustering approaches cannot directly handle side information properly. Admittedly, by concatenating all the features from multiple views into a single one, one can handle it with a semi-supervised clustering algorithm. However, a simple concatenation has several drawbacks. First, the dimension of concatenation feature matrices is usually high which may trigger the curse of dimensionality and result in a high computational cost. Secondly, the approach of concatenation, in fact, treats different views equally which is not appropriate since the difference between views is ignored. Thus, it's still difficult to efficiently utilize side information in multi-view clustering, due to the trade-off between diversity of feature in multiple views and consistency of side information constraints.

To address this issue, in this paper, we propose a novel clustering approach to utilize side information called **M**ulti-**V**iew **M**atrix **C**ompletion (**MVMC**). Firstly, MVMC constructs a pairwise similarity matrix S_v for the v-th view independently and cast clustering task into a matrix completion problem based on given pairwise constraints and feature information from multiple views. Then, the final pairwise similarity matrix S is learned by controlling S and S_v in different views to approach each other. The global optimal solution is obtained by projective alternative optimization since the objective function is jointly convex. Experimental results on benchmark datasets demonstrate that the proposed MVMC can efficiently utilize side information and outperform other state-of-the-art approaches. Our major contribution is the development of the first approach to tackle constrained multi-view clustering based on matrix completion.

In the following, we start with a brief review of some related work. Then, we propose our MVMC approach and examine the empirical performance of proposed method on several benchmark datasets. Finally, we conclude the paper.

2 Related Work

Multi-view learning has attracted much attention since many real world data are collected from multiple sources or intrinsically have multi-faceted feature representations. In general, various multi-view learning algorithms in different areas can be classified into three groups: (1) co-training, (2) multiple kernel learning, and (3) subspace learning [33]. Multi-view co-training constructs two learners each from one view, and then lets them to provide pseudo-labels for the other learner [3]. And some studies [28,29] show that the diversity of multiple views is the essence of co-training. Multiple kernel learning (MKL) is suitable for multi-view learning because kernels in MKL naturally correspond to different views to improve learning performance [1,15]. Subspace learning algorithms aim at obtaining a common subspace shared by multiple views and then learning models in that shared subspace [17,19].

Multi-view clustering aims at leveraging information from multiple views to improve clustering performance, various multi-view clustering algorithms have been proposed. Roughly, they can be categorized into spectral approaches, subspace approaches and late-fusion approaches. Spectral approaches extend spectral clustering [27] into multi-view data by constructing a measure of similarity between instances [18,23]. The subspace approaches assume that multiple views are generated from a common low-dimensional subspace where the representations of similar instances are close [6,30]. The late-fusion approaches learn a clustering solution from each single view, and then fuse all these intermediate outputs based on consensus [4,35]. The proposed approach in this paper belongs to the first stream.

Clustering with side information in single view scenario has been well developed. Inspired by the work proposed in [32], plenty of algorithms are proposed based on distance metric learning. For example, ITML proposed in [7] learns a metric matrix with side information based on information theory. MCCC proposed in [37] converts clustering to a matrix completion problem.

Matrix Completion (MC) problem was originally proposed for collaborative [10]. Assuming that the matrix to be recovered is low-rank, MC finds a matrix X that minimizes the difference with the given observation. However, it is still challenging because rank minimization problem is NP-hard. A major breakthrough in [5] states that minimizing $\mathsf{rank}(X)$, under broad conditions, can be achieved using the minimizer obtained with its convex envelope, the nuclear norm, $\|X\|_*$. In addition, [34] proposed an approach to speed up the process of MC by utilizing side information.

Due to a solid mathematical foundation of MC, it was recently exploited into clustering. For example, a graph-based clustering proposed by [11] identifies clusters from partially observed unweighted graphs via MC. In [36], a crowdsourced clustering is proposed to use the crowd information to recover a similarity metric, which can then be applied on large, growing collections. Besides, a related clustering approach proposed in [37] convert clustering into a MC problem based on side information, which performs well in single view scenario.

All these previous studies on clustering cannot efficiently handle the scene where some side information is provided for multiple views. To the best of our knowledge, this is the first study on multi-view clustering by matrix completion with side information.

3 Our Proposed MVMC Approach

In this section, the matrix completion multi-view clustering assisted with side information model is introduced. Let $\mathcal{D} = \{O_1, O_2, \cdots, O_n\}$ be n instances, and the feature of each instance is collected from m views (channels). Feature in the v-th view is denoted as $X_v = (\mathbf{x}_1^v; \mathbf{x}_2^v; \cdots; \mathbf{x}_n^v)$, where $\mathbf{x}_i^v \in \mathbb{R}^{1\times d_v}$ is the feature of O_i in the v-th view, and d_v is dimension of the v-th view. Let $\mathcal{M}$ ($\mathcal{C}$) denote the set of must-link (cannot-link) constraints, $(i,j) \in \mathcal{M}$ ($(i,j) \in \mathcal{C}$) implies O_i and O_j should (not) be assigned into the same cluster. We define $\Omega = \mathcal{M} \cup \mathcal{C}$ to represent all the pairwise constraints. Meanwhile, let r be the number of clusters.

3.1 Similarity Matrix Construction

For each view, let $\mathbf{u}_i^v \in \{0,1\}^n$ be the membership vector of the i-th cluster in the v-th view, where $\mathbf{u}_{i,j}^v = 1$ if O_j is assigned to the i-th cluster and zero, otherwise. Then the pairwise similarity matrix $S_v \in \{0,+1\}^{n\times n}$ is defined as

$$S_v = \sum\nolimits_{i=1}^{r} \mathbf{u}_i^v (\mathbf{u}_i^v)^{\mathrm{T}} \tag{1}$$

Evidently, $[S_v]_{i,j} = 1$ if O_i and O_j are assigned to the same cluster from the perspective of feature information provided in the v-th view, and zero, otherwise. Furthermore, it is easy to verify that $\mathsf{rank}(S_v) \leq r$, which implicates a low-rank property of similarity matrix.

3.2 Single-View Clustering by Matrix Completion

For a specific view (the subscribe v is omitted in this part for simplicity), finding the best data partition is equivalent to recovering the binary matrix S. Apparently, pairwise constraints are tightly associated with the similarity matrix. More specifically, $[S_v]_{i,j} = 1$ if $(i,j) \in \mathcal{M}$ and $[S_v]_{i,j} = 0$ if $(i,j) \in \mathcal{C}$ for $v = 1, \cdots, m$. Thus, clustering problem with pairwise constraints can be cast into a matrix completion problem, i.e., filling out the missing entries in binary similarity matrix S based on $\mathcal{M}$ and $\mathcal{C}$ (i.e., the partial observations, called S_{ob}) and the feature information from multiple views.

Formally, for a specific view, the binary similarity matrix S can be recovered from the following matrix completion problem,

$$\begin{aligned} \min_{S} \quad & \|S\|_* \\ \text{s.t.} \quad & \mathcal{R}_\Omega(S) = \mathcal{R}_\Omega(S_{\mathrm{ob}}) \end{aligned} \tag{2}$$

where $\|\cdot\|_*$ is nuclear norm, and $\mathcal{R}_\Omega(\cdot) : \mathbb{R}^{n\times n} \mapsto \mathbb{R}^{n\times n}$ is a linear operator which preserve the entry of S in Ω and 0 outside.

However, feature information is not utilized. To efficiently exploit feature information, let $Z = [\mathbf{z}_1, \cdots, \mathbf{z}_k]$ be the first k left singular vectors of X corresponding to the k largest singular values, where $k \geq r$. And we make an assumption to reveal the relationship between X and S:

Assumption: the cluster membership vectors $\{\mathbf{u}_i\}_{i=1}^r$ lie in the subspace of the first k left singular vectors of feature matrix $\{\mathbf{z}_i\}_{i=1}^k$.

A similar assumption is used by the spectral clustering algorithm [22], matrix completion [34] and some others. When assumption holds, i.e., $\mathbf{Span}(\mathbf{u}_1, \cdots, \mathbf{u}_r) \subseteq \mathbf{Span}(\mathbf{z}_1, \cdots, \mathbf{z}_k)$, we know that $\forall i = 1, \cdots, r, \mathbf{u}_i = Z\theta_i$, where $\theta_i \in \mathbb{R}^k$. Then the similarity matrix S can be derived as

$$S = \sum\nolimits_{i=1}^r \mathbf{u}_i\mathbf{u}_i^{\mathrm{T}} = \sum\nolimits_{i=1}^r Z\theta_i(Z\theta_i)^{\mathrm{T}} = ZMZ^{\mathrm{T}},$$

where $M = \sum_{i=1}^r \theta_i\theta_i^{\mathrm{T}} \in \mathbb{R}^{k\times k}$. Obviously, M is a symmetric positive semidefinite matrix, i.e., $M \in \mathcal{S}_+^k$, where $\mathcal{S}_+^k = \{X \in \mathbb{R}^{k\times k} | X = X^{\mathrm{T}} \text{ and } X \succeq 0\}$.

It's proved in [34] that $\|AXB\|_* = \|X\|_*$ holds when A and B are orthonormal matrices, i.e., $\mathbf{a_i}^{\mathrm{T}}\mathbf{a_j} = \delta_{i,j}$ and $\mathbf{b_i}^{\mathrm{T}}\mathbf{b_j} = \delta_{i,j}$ for any i and j, where $\delta_{i,j}$ is the Kronecker delta function that outputs 1 if $i = j$ and 0, otherwise. Hence, $\|S\|_* = \|ZMZ^{\mathrm{T}}\|_* = \|M\|_*$.

Besides, since pairwise constraints usually express a belief rather than certainty in many cases, soft constraints are introduced. Incorporating with feature information, Eq. 2 can be reformulated as follows:

$$\min_M \|M\|_* + C\|\mathcal{R}_\Omega(ZMZ^{\mathrm{T}}) - \mathcal{R}_\Omega(S_{\mathrm{ob}})\|_F^2 \tag{3}$$

where $C > 0$ is the regularization parameter introduced to trade off between low-rank property and the consistency of recovery and given side information.

In [37], the fast stochastic subgradient descent method is adopted to solve this optimization problem. And when S has been recovered, spectral clustering algorithm is applied to find the best data partition. This single-view clustering approach is referred as Matrix Completion Constrained Clustering (MCCC).

3.3 From Single-View to Multi-View

When managing to solve the multi-view clustering problem, a simple idea to come up with is to convert multi-view features to a single one. There are two types: the first one is concatenating all the features of multiple views, and then performing semi-supervised single-view algorithms directly on the concatenation; the second one is clustering on each view independently, and selecting the best one w.r.t. the preferred performance measurement index.

Besides, for MCCC, another approach based on the late-fusion arises naturally which performs clustering with pairwise constraints in each view independently, and then concatenates results above all views to obtain final clustering results. Concretely speaking, the pairwise similarity matrix S_v in each view can be recovered with side information and feature information independently. Then, $S_1, \cdots, S_m$ are fused into a final similarity matrix S as $S = \frac{1}{m}\sum_{v=1}^{m} S_v$. We refer to this approach as MCCC_fusion.

As the information from multiple views is usually complementary to each other, all above fail to combine feature information from multiple views efficiently nevertheless. To address this problem, we directly restrict pairwise similarity matrix S_v and learn the final S. Because the final clustering result should be consistent over all multiple views, the consistency of multiple similarity matrices S_v is enforced. To utilize multi-view feature information, we incorporate them via the assumption claimed previously. Then S_v is expanded as $Z_v M_v Z_v^{\mathrm{T}}$, where $M_v \in \mathbb{R}^{k\times k}$ and $Z_v = [\mathbf{z}_1^v, \cdots, \mathbf{z}_k^v]$, the first k left singular vectors of feature in the v-th view X_v. In fact, k is able to vary over different views. However, it does not make difference to the essence of the problem. Thus, we set k in various views the same in the following.

It's noteworthy to mention that the original nuclear norm term $\|S_v\|_*$ or $\|M_v\|_*$ is non-smooth, which implies that it is inevitable to adopt sub-gradient or proximal approach. Fortunately, since M_v is constrained as a positive semi-definite matrix, then $\|M_v\|_* = \sum_{i=1}^{k}|\sigma_i| = \sum_{i=1}^{k}\mathrm{eig}_i = \mathrm{tr}(M_v)$, where σ_i and eig_i are the i-th singular value and eigenvalue of M, respectively. Thus, the optimization problem can be formulated as follows:

$$\begin{aligned}
\min_{S,\{M_v\}_{v=1}^m} \quad & \sum_{v=1}^{m}\left(\mathrm{tr}(M_v) + C_1\|\mathcal{R}_\Omega(Z_v M_v Z_v^{\mathrm{T}} - S_{\mathrm{ob}})\|_F^2 + C_2\|Z_v M_v Z_v^{\mathrm{T}} - S\|_F^2\right) \\
\text{s.t.} \quad & 0 \le S_{i,j} \le 1,\ \forall i,j \in \{1,\cdots,N\}, \\
& M_v \in \mathcal{S}_+^k, \quad v = 1,\cdots,m.
\end{aligned} \tag{4}$$

where $C_1, C_2 > 0$ are two regularization parameters. The optimization object function is consisted of three terms, the first two terms are generated from single-view matrix completion, and the last term measures the difference among S_v from multiple views. If we split the Frobenius norm into the square sum of entries, in fact, it is the entry-variance of multiple similarity matrix.

After converting the non-smooth term $\|M_v\|_*$ to a smooth term $\mathrm{tr}(M_v)$, projected gradient descend is adopted which is pretty easy to implement.

3.4 Optimization

In Eq. 4, the constraint regions are convex sets and the objective function is jointly convex w.r.t S and $\{M_v\}_{v=1}^m$. Thus, we developed an iterative algorithm to find the global optimal solution. Firstly, $\{M_v\}_{v=1}^m$ and S are initialized by the given observation S_{ob}. Then the following two steps are repeated until convergence: minimizing $\{M_v\}_{v=1}^m$ over S; and then minimizing S over $\{M_v\}_{v=1}^m$.

(1) **Initialization** $\{M_v\}_{v=1}^m$ and S:
Since the observation S_{ob} is given, then $\{M_v\}_{v=1}^m$ and S can be initialized as follows:

$$M_v = Z_v^{\mathrm{T}} S_{\mathrm{ob}} Z_v, \quad S = S_{\mathrm{ob}}. \tag{5}$$

Because each pairwise constraint corresponds to a pair entries in S_{ob} and the value of each entry in S_{ob} is 0/1, this initialization meets the constraint condition in Eq. 4.

(2) **Minimizing** object over S with fixed $\{M_v\}_{v=1}^m$:

$$\hat{S} = \underset{0 \leq S_{i,j} \leq 1}{\arg\min}\, C_2 \sum\nolimits_{v=1}^{m} \|S - Z_v M_v Z_v^{\mathrm{T}}\|_F^2$$

Obviously, this sub-problem has a closed-form solution,

$$\hat{S} = \mathbf{Proj}_1\left(\frac{1}{m}\sum_{v=1}^{m} Z_v M_v Z_v^{\mathrm{T}}\right) \tag{6}$$

where $\mathbf{Proj}_1(\cdot)$ is defined as

$$[\mathbf{Proj}_1(X)]_{i,j} = \begin{cases} 0 & \text{if } X_{i,j} < 0; \\ 1 & \text{if } X_{i,j} > 1; \\ X_{i,j} & \text{otherwise.} \end{cases} \tag{7}$$

(3) **Minimizing** object over $M_v (v = 1, \cdots, m)$ with fixed S:
Obviously, when fixing S, each M_v can be solved independently. The objective function of sub-problem is

$$\mathcal{L}(M_v) = \mathrm{tr}(M_v) + C_1\|\mathcal{R}_\Omega(Z_v M_v Z_v^{\mathrm{T}} - S_{\mathrm{ob}})\|_F^2 + C_2\|Z_v M_v Z_v^{\mathrm{T}} - S\|_F^2 \tag{8}$$

And the optimal solution of sub-problem is

$$\hat{M}_v = \underset{M_v \in \mathcal{S}_+^k}{\arg\min}\, \mathcal{L}(M_v) \tag{9}$$

$\mathcal{L}(M_v)$ is differential and its gradient $\nabla\mathcal{L}(M_v)$ is

$$\nabla\mathcal{L}(M_v) = I + 2C_1 Z_v^{\mathrm{T}}(\mathcal{R}_\Omega(Z_v M_v Z_v^{\mathrm{T}} - S_{\mathrm{ob}}))Z_v + 2C_2 Z_v^{\mathrm{T}}(Z_v M_v Z_v^{\mathrm{T}} - S)Z_v$$

Besides, it's easy to verify that $\nabla\mathcal{L}(M_v)$ is Lipschitz continuous with constant $L = 2(C_1\|Z_v\|_F^4 + C_2)$. The projective gradient descend method is adopted, the update sequence is defined as:

$$M_v^{(\ell+1)} = \mathbf{Proj}_2\left(M_v^{(\ell)} - \eta\nabla\mathcal{L}(M_v^{(\ell)})\right). \tag{10}$$

where η is chosen as $1/L$ for a linear convergence referring to [21]. $\mathbf{Proj}_2$ is a operator projecting M_v back to semi-definite positive cone $\mathcal{S}_+^k$ defined as:

$$\mathbf{Proj}_2(X) = U \max(\sigma, 0) U^{\mathrm{T}} \tag{11}$$

where U and σ correspond to the eigenvectors and eigenvalues of X.

When obtaining final pairwise similarity matrix S, we apply spectral clustering algorithms [27] on S to find the best data partition. The proposed clustering approach above is referred as MVMC (Multi-View Matrix Completion), which is summarized in Algorithm 1.

Convergence Analysis: Because objective function in Eq. 4 is jointly convex with a convex constraints region, Algorithm 1 converges to a global optima.

Algorithm 1. MVMC (Multi-View Matrix Completion)

Input:

1) Multi-view feature: $\mathcal{X} = \{X_v\}_{v=1}^m$, where $X_v = (\boldsymbol{x}_1; \boldsymbol{x}_2; \cdots; \boldsymbol{x}_n)_v \in \mathbb{R}^{N \times d_v}$;
2) The set of pairwise constraints: $\Omega = \mathcal{M} \cup \mathcal{C}$;
3) Regularization parameters: C_1 and C_2;
4) The number of clusters: r.

Output:

Pairwise similarity matrix S and clustering results.

1: Initialize S and $\{M_v\}_{v=1}^m$ by Eq. 5;
2: **repeat**
3: Fixing $\{M_v\}_{v=1}^m$ to optimize the objective, update S by Eq. 6 ;
4: Fixing S to optimize the objective, update $\{M_v\}_{v=1}^m$ by Eq. 10;
5: **until** objective function in Eq. 4 converges.
6: Performing spectral clustering on S to obtain final clustering results.

4 Experiment

In this section, we compare the performance of proposed approach **MVMC** with several baseline methods over different real world datasets. The baseline methods are representations from two paradigms: multi-view clustering and semi-supervised clustering. The muti-view clustering algorithms are (a) **Co-Reg**, the co-regularized spectral clustering [14], (b) **MKKM**, the multi-view kernel k-means algorithm [26], (c) **RMSC**, robust multi-view spectral clustering based on Markov chain. [31]; The semi-supervised clustering algorithms are (d) **ITML**, the information theoretic metric learning algorithm [7], (e) **MCCC**, matrix completion based constraint clustering [37]. Since there are two ways, i.e., concatenation and best view selection, for semi-supervised algorithms to handle multiple views, ITML and MCCC are separated as ITML_best, ITML_concat and MCCC_best, MCCC_concat. Besides, as we mentioned before, MCCC_fusion is also added into comparison.

4.1 General Experiment Settings

Datasets: The *WebKB* dataset [3] has been widely used in multi-view learning, which contains webpages collected from four universities: Cornell, Texas, Washington and Wisconsin. The webpages are distributed over five clusters and described by two views: the content and citation view. BBCSport consists of 2 views from news articles [14]. The Reuters dataset [2] is built from the Reuters Multilingual test collection, multi-view information is created from different languages, i.e., English, French, German, Italian and Spanish [2]. Statistics of these datasets are summarized in Table 1.

Parameter Settings: There are two regularization parameters C_1 and C_2, cross-validation is applied because of the existence of side information [36,37]. To choose an appropriate k, a trade-off need to be balanced between computational efficiency and violation of assumption. It's noteworthy to mention that k in

Table 1. Statistics of six datasets, the first four datasets are subsets of *WebKB* and d_v denotes the dimension of the v-th view of datasets.

Data set	# size	# view	# cluster	dimension of each view $d_v(v = 1, \cdots, m)$
Cornell	195	2	5	1703, 195
Texas	187	2	5	1703, 187
Washington	230	2	5	1703, 230
Wisconsin	265	2	5	1703, 265
BBCSport	737	2	5	3183, 3208
Reuters	1600	5	6	2000 for each

each view, in fact, can be different. However, in our experiments, k is chosen as $\min(100, d_v)$ for convenience, where d_v is the dimension of the v-th view.

Side Information: In our experiments, we follow the typical routine of experiments with side information [38,39], where each pairwise constraint is generated by randomly selecting a pair of samples. A must-link constraint is formed if they belong to the same cluster, and cannot-link, otherwise. RATIO is used to measure quantity of side information, i.e. $|\Omega| = \text{RATIO} \cdot n^2$. We vary RATIO from $[0.01, 0.02, \cdots, 0.1]$.

Evaluation: In all the experiments, to evaluate the effectiveness of the proposed approach, we use six different and widely-used criteria to measure clustering performances: F-score, precision, recall, the normalized mutual information (NMI) [25], adjusted rand index(Adj-RI) [20] and average entropy. Note that all the other criteria except for average entropy lie in interval $[0, 1]$, and a higher value indicates a better performance. Meanwhile, a lower average entropy means a more competitive performance.

4.2 Results

Due to the space limitation, we only present Fig. 1 and Table 2 in experiments part to demonstrate MVMC approach. Figure 1 summarizes the results w.r.t NMI and Adj-RI on *WebKB* dataset. 10 test runs were conducted and the average performance as well as standard deviation are presented.

From Fig. 1 we can see that, for all four datasets, firstly, the performance of proposed approach MVMC is gradually better as RATIO increases, which means MVMC can handle side information efficiently. Secondly, comparing with the multi-view clustering, when the side information is extremely scarce, the behavior of MVMC is relatively poor. However, MVMC is able to demonstrate a much better performance with plenty of side information. The reason is that matrix completion cannot give a satisfying recovery with an exceedingly small amount of side information, and when a relatively large amount is given, MVMC can take advantage of side information while multi-view clustering approaches cannot. Thirdly, comparing with the semi-supervised clustering, MVMC almost

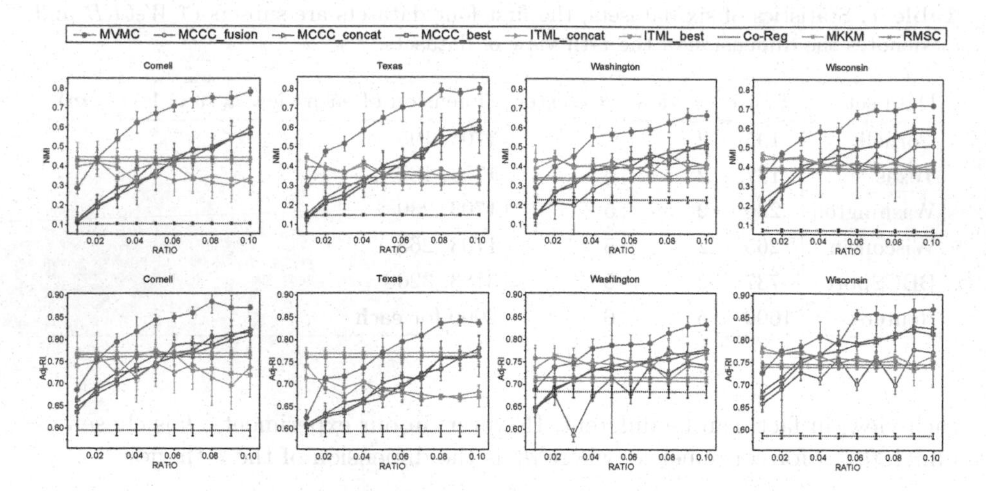

Fig. 1. Comparisons of clustering performance with other approaches on *WebKB* dataset (with 4 subsets) w.r.t. NMI and Adjust Rand-Index (the higher, the better). RATIO is used to measure amount of side information which varies from 0.01 to 0.1. On each dataset, 10 test runs were conducted and the average performance as well as standard deviation are presented.

Table 2. Comparisons of clustering performance on BBC (abbrv. for BBCSport), REU (abbrv. for Reuters) w.r.t six criteria (except that a lower entropy indicates a better performance, the others lie in [0,1] and the higher, the better). The number of pairwise constraints is chosen as 5,000. On each dataset, 10 test runs were conducted and the average performance as well as standard deviation are presented. Besides, • (∘) indicates that MVMC is significantly better (worse) than the compared method (paired t-tests at 95% significance level).

Dataset	Method	Fscore↑	Precision↑	Recall↑	NMI↑	Adj-RI↑	Avg Entropy↓
BBC	CoReg	.385 ± .002•	.285 ± .003•	.606 ± .011•	.173 ± .005•	.090 ± .005•	1.881 ± 0.010•
	MKKM	.745 ± .013•	.774 ± .020•	.719 ± .023•	.661 ± .016•	.669 ± .016•	0.724 ± 0.046•
	RMSC	.452 ± .017•	.472 ± .017•	.434 ± .021•	.297 ± .020•	.290 ± .020•	1.527 ± 0.043•
	ITML_concat	.681 ± .072•	.633 ± .097•	.742 ± .048•	.624 ± .056•	.568 ± .104•	0.882 ± 0.153•
	ITML_best	.560 ± .065•	.452 ± .072•	.740 ± .041•	.518 ± .054•	.373 ± .100•	1.198 ± 0.127•
	MCCC_concat	.823 ± .070•	.783 ± .085•	.869 ± .052•	.805 ± .053•	.772 ± .088•	0.476 ± 0.135•
	MCCC_best	.768 ± .057•	.721 ± .066•	.823 ± .047•	.750 ± .047•	.702 ± .072•	0.609 ± 0.112•
	MCCC_fusion	.861 ± .088•	.822 ± .109•	.906 ± .063•	.867 ± .053•	.822 ± .112•	0.336 ± 0.142•
	MVMC	**.990 ± .003**	**.989 ± .003**	**.991 ± .003**	**.982 ± .005**	**.987 ± .004**	**0.040 ± 0.011**
REU	CoReg	.346 ± .001•	.316 ± .004•	.384 ± .006•	.274 ± .002•	.200 ± .003•	1.902 ± 0.008•
	MKKM	.345 ± .002•	.319 ± .015•	.377 ± .020•	.274 ± .006•	.201 ± .009•	1.897 ± 0.028•
	RMSC	.369 ± .008	.342 ± .018•	.402 ± .020•	.303 ± .017•	.231 ± .014•	1.825 ± 0.054•
	ITML_concat	.360 ± .010•	.294 ± .017•	.466 ± .022•	.294 ± .021•	.197 ± .018•	1.895 ± 0.064•
	ITML_best	.362 ± .015•	.298 ± .015•	.464 ± .033	.305 ± .020•	.201 ± .020•	1.866 ± 0.051•
	MCCC_concat	.351 ± .033•	.359 ± .034•	.343 ± .033•	.246 ± .038•	.218 ± .041•	1.918 ± 0.097•
	MCCC_best	.334 ± .029•	.338 ± .029•	.331 ± .030•	.231 ± .033•	.200 ± .035•	1.976 ± 0.083•
	MCCC_fusion	.459 ± .051•	.489 ± .028•	.437 ± .071•	.377 ± .071•	.193 ± .036•	1.496 ± 0.081•
	MVMC	**.528 ± .030**	**.559 ± .024**	**.499 ± .037**	**.472 ± .027**	**.427 ± .038**	**1.294 ± 0.061**

outperforms all the time especially along with the growth of RATIO. This phenomenon implicates that simple concatenation or late-fusion does not leverage information from multiple views. By exploiting different views via minimizing variance of similarity matrix, MVMC is validated to be effective.

Table 2 summarizes the results w.r.t all the six criteria on BBCSport and Reuters. The number of pairwise constraints $|\Omega|$ is both chosen as 5,000. We can see that, MVMC demonstrates a surprisingly better performance than all the other approaches on almost all criteria. It's noteworthy to mention that the randomly sampled pairwise constraints, in fact, only accounts for about 0.9% and 0.2% for BBCSport and Reuters, respectively. It is encouraging that, with such a limited side information, MVMC can still yield a satisfying performance.

5 Conclusions

In this paper, we present MVMC, which is possibly the first attempt to efficiently handle multi-view clustering with side information based on matrix completion. By constructing similarity matrix for each view, we cast clustering into a matrix completion problem. Instead of concatenating multi-views into a single view, we enforce the consistency of clustering results on different views as constraints for alternative optimization, and the global optimal solution is obtained. The proposed MVMC approach exhibits impressive performance in experiments. Studying partial multi-view clustering [16] where each view suffers from some missing features assisted by side information will be an interesting future issue.

Acknowledgement. This research was supported by the National Science Foundation of China (61673201, 61333014).

References

1. Bach, F.R., Lanckriet, G.R., Jordan, M.I.: Multiple kernel learning, conic duality, and the SMO algorithm. In: ICML (2004)
2. Bisson, G., Grimal, C.: Co-clustering of multi-view datasets: a parallelizable approach. In: ICDM, pp. 828–833 (2012)
3. Blum, A., Mitchell, T.: Combining labeled and unlabeled data with co-training. In: COLT, pp. 92–100 (1998)
4. Bruno, E., Marchand-Maillet, S.: Multiview clustering: a late fusion approach using latent models. In: SIGIR, pp. 736–737 (2009)
5. Candès, E.J., Recht, B.: Exact matrix completion via convex optimization. Commun. ACM **55**(6), 111–119 (2012)
6. Chaudhuri, K., Kakade, S.M., Livescu, K., Sridharan, K.: Multi-view clustering via canonical correlation analysis. In: ICML, pp. 129–136 (2009)
7. Davis, J.V., Kulis, B., Jain, P., Sra, S., Dhillon, I.S.: Information-theoretic metric learning. In: ICML, pp. 209–216 (2007)
8. Dougherty, E.R., Barrera, J., Brun, M., Kim, S., Cesar, R.M., Chen, Y., Bittner, M., Trent, J.M.: Inference from clustering with application to gene-expression microarrays. Comput. Biol. **9**(1), 105–126 (2002)

9. Ganti, V., Gehrke, J., Ramakrishnan, R.: Cactus—clustering categorical data using summaries. In: KDD, pp. 73–83 (1999)
10. Goldberg, D., Nichols, D., Oki, B.M., Terry, D.: Using collaborative filtering to weave an information tapestry. Commun. ACM **35**(12), 61–70 (1992)
11. Jalali, A., Chen, Y., Sanghavi, S., Xu, H.: Clustering partially observed graphs via convex optimization. In: ICML, pp. 1001–1008 (2011)
12. Kim, Y.M., Amini, M.R., Goutte, C., Gallinari, P.: Multi-view clustering of multilingual documents. In: SIGIR, pp. 821–822 (2010)
13. Kulis, B., Basu, S., Dhillon, I.S., Mooney, R.J.: Semi-supervised graph clustering: a kernel approach. In: ICML, pp. 457–464 (2005)
14. Kumar, A., Rai, P., Daume, H.: Co-regularized multi-view spectral clustering. In: NIPS, vol. 24, pp. 1413–1421 (2011)
15. Lanckriet, G.R., Cristianini, N., Bartlett, P., Ghaoui, L.E., Jordan, M.I.: Learning the kernel matrix with semidefinite programming. JMLR **5**, 27–72 (2004)
16. Li, S.Y., Jiang, Y., Zhou, Z.H.: Partial multi-view clustering. In: AAAI (2014)
17. Li, S., Shao, M., Fu, Y.: Multi-view low-rank analysis for outlier detection. In: SDM, pp. 748–756 (2015)
18. Li, Y., Nie, F., Huang, H., Huang, J.: Large-scale multi-view spectral clustering via bipartite graph. In: AAAI, pp. 2750–2756 (2015)
19. Liu, M., Luo, Y., Tao, D., Xu, C., Wen, Y.: Low-rank multi-view learning in matrix completion for multi-label image classification. In: AAAI, pp. 2778–2784 (2015)
20. Manning, C.D., Raghavan, P., Schütze, H.: Introduction to Information Retrieval. Cambridge University Press, New York (2008)
21. Nesterov, Y.: Introductory Lectures on Convex Optimization: A Basic Course. Springer, Heidelberg (2013)
22. Ng, A.Y., Jordan, M.I., Weiss, Y., et al.: On spectral clustering: analysis and an algorithm. In: NIPS, vol. 15, pp. 849–856 (2002)
23. de Sa, V.R.: Spectral clustering with two views. In: ICML Workshop on Learning with Multiple Views, pp. 20–27 (2005)
24. Steinbach, M., Karypis, G., Kumar, V., et al.: A comparison of document clustering techniques. In: KDD Workshop on Text Mining, vol. 400, pp. 525–526 (2000)
25. Strehl, A., Ghosh, J.: Cluster ensembles - a knowledge reuse framework for combining multiple partitions. JMLR **3**, 583–617 (2002)
26. Tzortzis, G., Likas, A.: Kernel-based weighted multi-view clustering. In: ICDM, pp. 675–684 (2012)
27. Von Luxburg, U.: A tutorial on spectral clustering. Stat. Comput. **17**(4), 395–416 (2007)
28. Wang, W., Zhou, Z.-H.: Analyzing co-training style algorithms. In: Kok, J.N., Koronacki, J., Mantaras, R.L., Matwin, S., Mladenič, D., Skowron, A. (eds.) ECML 2007. LNCS (LNAI), vol. 4701, pp. 454–465. Springer, Heidelberg (2007). doi:10.1007/978-3-540-74958-5_42
29. Wang, W., Zhou, Z.H.: Multi-view active learning in the non-realizable case. In: NIPS 23, pp. 2388–2396 (2010)
30. Wang, Y., Zhang, W., Wu, L., Lin, X., Fang, M., Pan, S.: Iterative views agreement: an iterative low-rank based structured optimization method to multi-view spectral clustering. In: IJCAI, pp. 2153–2159 (2016)
31. Xia, R., Pan, Y., Du, L., Yin, J.: Robust multi-view spectral clustering via low-rank and sparse decomposition. In: AAAI, pp. 2149–2155 (2014)
32. Xing, E.P., Ng, A.Y., Jordan, M.I., Russell, S.: Distance metric learning with application to clustering with side information. In: NIPS, vol. 15, pp. 505–512 (2003)

33. Xu, C., Tao, D., Xu, C.: A survey on multi-view learning. arXiv preprint arXiv:1304.5634 (2013)
34. Xu, M., Jin, R., Zhou, Z.H.: Speedup matrix completion with side information: application to multi-label learning. In: NIPS, vol. 27, pp. 2301–2309 (2013)
35. Ye, H., Zhan, D., Miao, Y., Jiang, Y., Zhou, Z.H.: Rank consistency based multi-view learning: a privacy-preserving approach. In: CIKM, pp. 991–1000 (2015)
36. Yi, J., Jin, R., Jain, A.K., Jain, S., Yang, T.: Semi-crowdsourced clustering: generalizing crowd labeling by robust distance metric learning. In: NIPS, vol. 25, pp. 1772–1780 (2012)
37. Yi, J., Zhang, L., Jin, R., Qian, Q., Jain, A.K.: Semi-supervised clustering by input pattern assisted pairwise similarity matrix completion. In: ICML, pp. 1400–1408 (2013)
38. Zeng, H., Cheung, Y.: Semi-supervised maximum margin clustering with pairwise constraints. TKDE **24**(5), 926–939 (2012)
39. Zhang, X., Zong, L., Liu, X., Yu, H.: Constrained NMF-based multi-view clustering on unmapped data. In: AAAI, pp. 3174–3180 (2015)

Weighted NMF-Based Multiple Sparse Views Clustering for Web Items

Xiaolong Gong, Fuwei Wang, and Linpeng Huang(✉)

Department of Computer Science and Technology, Shanghai Jiao Tong University, Shanghai, China
{gxl121438,wfwzy2012,lphuang}@sjtu.edu.cn

Abstract. Many web items contain different types of information resources such as user profile, comments, users preference and so on. All these aspects can be seen as different views of real-world datasets and often admit same underlying clustering of the data. However, each view of dataset forming a huge sparse matrix results in the non-robust characteristic during matrix decomposition process, and further influences the accuracy of clustering results. In this paper, we attempt to use rating value given by the users as latent semantic information to handle those features that are unobserved in each data point so as to resolve the sparseness problem in all views matrices. To combine multiple views in our constructed corpus *Doucom*, we present WScoNMF (Weighted similarity co-regularized Non-negative Matrix Factorization), which provides an efficient weighted matrix factorization framework to further explore the sparseness problem in semantic space of data. The overall objective function is to minimize the loss function of weighted NMF under the $l_{2,1}$-norm and the co-regularized constraint under the F-norm. Experimental results on all datasets demonstrate the effectiveness of the proposed method.

Keywords: Multi-view clustering · WScoNMF · Co-clustering

1 Introduction

There are many types of media websites, like sport, music, movie, social network, etc. A main characteristic to all of these websites is containing a large number of resources such as comments, user profile, rating by users, which can be treated as different views of each web item in corresponding websites. However, a main problem is how to organize those large multi-view web resources accurately and automatically.

In previous years, NMF has been successfully used in unsupervised learning, such as text and document mining [20]. But a few clustering algorithms have been proposed to apply NMF on multi-view data before work [6] showed the connection between NMF and clustering methods. After that, the NMF methods have been widely used as one of the most important clustering methods because they can handle large numbers of unlabeled datasets, and have been applied

J. Kim et al. (Eds.): PAKDD 2017, Part II, LNAI 10235, pp. 416–428, 2017.
DOI: 10.1007/978-3-319-57529-2_33

to a lot of multi-view applications. [11,12] are among the first works proposed to solve the multi-view clustering problem via spectral projection. [1] enforces a shared coefficient matrix among different views. Recently, the use of nonnegative matrix factorization (NMF) with different regularization constraints for multi-views clustering has attracted many interests. [14] aims to find a unified low-dimensional space to fuse the multi-view representations in order to well explore the common latent structure shared by multi-views.

Other multi-view co-clustering algorithms [2,3,8,9,21] seek groupings that are consistent across different views which gives some pair-wise co-regularization constraints on the jointly factorizing matrices. However, such methods suffer from two challenges: (1) they do not care about the consistency on each item pair similarity. For instance, if item i is similar to item j, then the similarity of i and j should stay consistent after mapped to a new vector space in each single view. (2) they ignore the sparsity of the real-world datasets, indicating existence of an amount of unobserved features in latent semantic representation. For example, every item vector has a high-dimensional latent space in view matrix, but large numbers of features are unobserved for each item. i.e., document i has no word w means the value of i-th row and w-th column in item-word co-occurrence matrix equals to 0. Our intuition is since observed features in an item are too few to tell us what the item is about, unobserved features can tell us what the item is not about. We assume that the semantic spaces of both the observed and unobserved features make up the complete semantics profile of an item.

In this paper, we propose WScoNMF (Weighted similarity co-regularized Non-negative Matrix Factorization) to handle the sparseness of the views by integrating the joint weighted nonnegative matrix factorization and maintain the robustness by introducing the $l_{2,1}$-norm. However, most recently proposed multi-view co-clustering methods [16,19] presented algorithm to deal with the sparseness using the Frobenius norm based objectives, here we proposed a novel weighted NMF framework using $l_{2,1}$-norm that has been proved well performance in feature selection and also robust to residual in items [15]. We use similar pair-wise co-regularization constraint to solve the first challenge, and propose a new framework with restriction on weights to handle the second challenge so as to achieve a better effectiveness. The contributions of this paper can be summarized as following:

1. We first construct a large organized dataset with four views and corresponding user ratings, namely *Doucom*.
2. The proposed WScoNMF method uses similar pair-wise co-regularization constraint for multiple views and attempts to integrate user subjective preference (rating value) with associated view matrix, which we expect to enhance latent semantic information and weaken the sparseness of original view matrix.
3. An iterative optimization framework is used in WScoNMF, which is proved to converge and solve the proposed non-convex objective function.
4. Applying WScoNMF to four real-world datasets, and demonstrating the effectiveness of these results for multi-view clustering.

The rest of this paper is organized as follows. In the next section, problem formulation and some backgrounds are given. The details of the proposed WScoNMF framework are presented in Sect. 3. Extensive experimental results and analysis are shown in Sect. 4, and followed by the conclusion in Sect. 5.

2 Preliminaries

In this section, we first give the notations. Then the background knowledge on multi-view nonnegative matrix factorization will be introduced.

2.1 Notations

Before we describe the formulation of the problem, we summarize some notations used in this paper in Table 1. Assume we are given a set of data points, let $X = [x_1, x_2, \cdots, x_m]^T \in \mathbb{R}_+^{m \times d}$ be the original data points matrix of non-negative elements. Each row vector $x_i^T (1 \leq i \leq m)$ denotes a data point and each column represents one feature. The factorization is formulated as $X \approx UV^T$, where $U \in \mathbb{R}_+^{m \times K}$ represents the class indicators, indicating the final clustering result. $V \in \mathbb{R}_+^{d \times K}$ is termed the basis matrix. K denotes the desired reduced dimension. Further, let $\{X^{(1)}, X^{(2)}, \cdots X^{(n_v)}\}$ denote the data of n_v views, and each view $X^{(l)}$ is factorized as $X^{(l)} \approx U^{(l)}(V^{(l)})^T$. Here for different views, they have the same number of data points but allow for different number of features, which means $U^{(l)}$ are with the same dimension m-by-K for all views, while $V^{(l)}$ are of dimension K-by-d$^{(l)}$ for per view. The fundamental multi-view based on NMF function tries to minimize the joint problem over $U^{(l)}, V^{(l)}$:

$$\sum_{l=1}^{n^v} \|X^{(l)} - U^{(l)}(V^{(l)})^T\|_F^2, \quad s.t.\ U^{(l)}, V^{(l)} \geq 0 \tag{1}$$

Table 1. Summary of the notations

Notations	Description
m	Total number of data points
n_v	Total number of views
$X^{(l)}$	Data matrix for the l-th view
$d^{(l)}$	Dimension of features in the l-th view
$U^{(l)}$	Class indicator matrix for the l-th view
$V^{(l)}$	The basis matrix for the l-th view
λ_{ls}	Weight parameter for similar pair-wise co-regularization constraint

2.2 Multi-view NMF Model

Multi-view NMF [14] aims to search for a factorization that gives compatible clustering solutions across multiple views. The key idea is to formulate joint matrix factorization process and construct a softly regularized constraint between coefficient matrices of different views and common consensus matrix. To incorporate the regularized constraint with consensus matrix U^* for individual views, the final form of Multi-view NMF algorithm can be formulated as bellow:

$$\sum_{l=1}^{n^v} \|X^{(l)} - U^{(l)}(V^{(l)})^T\|_F^2 + \sum_{l=1}^{n^v} \lambda_l \|U^{(l)} - U^*\|_F^2 \tag{2}$$

$$s.t.\ \forall 1 \leq k \leq K, \|V_{\cdot,k}^{(l)}\| = 1 \ and\ U^{(l)}, V^{(l)}, U^* \geq 0$$

where λ_l is the only parameter tuning the relative weight among different views.

3 The WScoNMF Framework

In this section, we first present our similar pair-wise constraint, where we expect that the class indicator matrices learned from different views indicate the same class label for one item. Furthermore, we present our weighted strategy to model our sparse views of data points.

3.1 Inter-View Constraint on Similar Pair-Wise

Let $X^{(l)} = \{X_1^{(l)}, X_2^{(l)}, \cdots, X_m^{(l)}\}^T$ denotes the set of m items[1] in view l. We should note that $X_i^{(l)}$ and $X_i^{(s)}$ $(1 \leq i \leq m)$ represent the same item, which means the true class labels for $X_i^{(l)}$ and $X_i^{(s)}$ $(l \neq s)$ should be the same. The simple pair-wise co-regularization can capture the difference between two indicator matrices of two views and force the representations from different views to be similar, which can be formulated as

$$\sum_{l=1}^{n^v} \|X^{(l)} - U^{(l)}(V^{(l)})^T\|_F^2 + \sum_{l=1}^{n^v}\sum_{s=1}^{n^v} \lambda_{ls} \|U^{(l)} - U^{(s)}\|_F^2, \quad s.t. \quad U^{(l)}, V^{(l)} \geq 0 \tag{3}$$

where λ_{ls} is the regularization parameter controlling the importance of constraint among different indicator class matrices. But this constraint ignores the similarity of each data point in intra-view. So we propose similar pair-wise co-regularize constraint for further refinement, and the objective function (ScoNMF) is as follows,

$$\sum_{l=1}^{n^v} \|X^{(l)} - U^{(l)}(V^{(l)})^T\|_F^2 + \sum_{l=1}^{n^v}\sum_{s=1}^{n^v} \lambda_{ls} \|M^{(l)} - M^{(s)}\|_F^2, \quad s.t. \quad U^{(l)}, V^{(l)} \geq 0 \tag{4}$$

where $M^{(l)} = U^{(l)}(U^{(l)})^T$ denotes a similarity matrix between each two items in view l.

[1] In this paper, we use 'data point' and 'item' exchangeable.

3.2 Weighted ScoNMF Algorithm

ScoNMF framework presented above considers the residual by using the Frobenius norm, but this co-regularization approach ignores the sparsity of data in the semantic space, especially in our new dataset. To address the sparseness issue for our clustering task, we introduce a novel weighting strategy for the weight matrix and further present a completely NMF formulation with manifold regularization. The main idea is to use the $l_{2,1}$-norm to replace the Frobenius norm in NMF objective and solve:

$$\min_{U,V} \sum_{l=1}^{n^v} \|X^{(l)} - U^{(l)}(V^{(l)})^T\|_{2,1}, \quad s.t. \quad U^{(l)}, V^{(l)} \geq 0 \tag{5}$$

and Eq. 5 can be reformulated as following:

$$\min_{U,V} \sum_{l=1}^{n^v} \sum_{i=1}^{m} \|X_i^{(l)} - U_i^{(l)}(V^{(l)})^T\|_2 \tag{6}$$

We let $\mu^{(l)} = [\mu_1^{(l)}, \cdots, \mu_m^{(l)}]^T \in \mathbb{R}_+^{m \times d}$ be the weight matrix for view l and $\mu^{(l)}$ defines a weight for each cell in $X^{(l)}$, where $\mu_i^{(l)} = [\mu_{i,\cdot}^{(l)}]^T \in \mathbb{R}_+^{d \times 1}$ is the vector of data point i. And we define $w_i^{(l)} = \frac{1}{d}\sum_j \mu_{i,j}^{(l)}$ here. Also, we enforce the orthogonal constraint to $U^{(l)}$ to guarantee the uniqueness of the solution which was introduced in [14]. Finally, our ScoNMF (Eq. 4) can be rewritten to WScoNMF as,

$$\begin{aligned} &\min_{U,V} \sum_{l=1}^{n^v} \sum_{i=1}^{m} w_i^{(l)} \|X_i^{(l)} - U_i^{(l)}(V^{(l)})^T\|_2 + \sum_{l=1}^{n^v} \sum_{s=1}^{n^v} \lambda_{ls} \|M^{(l)} - M^{(s)}\|_F^2 \\ &s.t. \quad U^{(l)}, V^{(l)} \geq 0, \quad {U^{(l)}}^T U^{(l)} = I \end{aligned} \tag{7}$$

Weighting Scheme. Now we give the definition of $\mu_{i,j}^{(l)}$. If each item in dataset has the corresponding rating feature, we use f to represent their relevant weight, otherwise we use τ_i. Our weighting scheme is defined as:

$$\mu_{i,j}^{(l)} = \begin{cases} 1 & \text{if } x_{i,j}^{(l)} \neq 0 \text{ and } 'X^{(l)} \text{ is } a \text{ 0/1 matrix}' \\ f \text{ or } \tau_i & \text{if } x_{i,j}^{(l)} \neq 0 \\ \epsilon & \text{if } x_{i,j}^{(l)} = 0 \end{cases} \tag{8}$$

where $f = e^{-\frac{1}{r_i^{(l)}}}$ and $r_i^{(l)}$ means a rating score ($0 < r_i < 10$) for a data point $X_i^{(l)}$. $\tau_i = \max_j\{cos(U_i^{(l)}, U_j^{(l)})\}$ denotes a maximum value about the similarity of two data points and ϵ is a very small weight for penalty. The intuition of a small ϵ when $x_{i,j}^{(l)} = 0$ is to diminish the influence where dimension d_j is unobserved in the data point $X_i^{(l)}$. Our strategy can be applied to different types of web resources in many web applications, particularly those comment-based corpus.

3.3 Optimization Algorithm

For the sake of convenience in representing, we let $R = \sum_{l=1}^{n^v} \sum_{s=1}^{n^v} \lambda_{ls} \|M^{(l)} - M^{(s)}\|_F^2$ in Eq. 7. Now the objective function Eq. 7 is equivalent to the following:

$$J(U,V) = \min_{U,V} \sum_{l=1}^{n^v} \sum_{i=1}^{m} w_i^{(l)} \|X_i^{(l)} - U_i^{(l)} (V^{(l)})^T\|_2^2 + R \qquad (9)$$
$$s.t. \quad U^{(l)}, V^{(l)} \geq 0, \quad {U^{(l)}}^T U^{(l)} = I$$

So Eq. 9 can be rewritten as:

$$J(U,V) = \min_{U,V} \sum_{l=1}^{n^v} Tr((X^{(l)})^T D X^{(l)} - 2(U^{(l)})^T D X^{(l)} V^{(l)} + V^{(l)} (U^{(l)})^T D U^{(l)} (V^{(l)})^T) +$$
$$\sum_{l,s} \lambda_{ls} Tr(U^{(l)} (U^{(l)})^T U^{(l)} (U^{(l)})^T - 2U^{(l)} (U^{(l)})^T U^{(s)} (U^{(s)})^T + U^{(s)} (U^{(s)})^T U^{(s)} (U^{(s)})^T)$$
$$+ \sum_l Tr(\gamma^{(l)} ((U^{(l)})^T U^{(l)} - I))$$

where D is a diagonal matrix with the i-th diagonal element $D_{i,i} = w_i^{(l)}$ and $\gamma^{(l)}$ is the Lagrange symmetric matrix for the condition of constraint. Similar to the known solution for NMF, we can adopt alternation optimization to minimize the objective function when D is fixed.

Fixing $U^{(l)}$, Computing $V^{(l)}$. The third part in $J(U,V)$ is a constant, so the derivatives of $J(U,V)$ with respect to $V^{(l)}$ is:

$$\frac{\partial J(U,V))}{\partial V^{(l)}} = (-2(X^{(l)})^T D U^{(l)} + 2V^{(l)} (U^{(l)})^T D U^{(l)}) \qquad (10)$$

The Karush-Kuhn-Tucker (KKT) complementarity condition gives[2]

$$\frac{\partial J(U,V)}{\partial V^{(l)}} \odot V^{(l)} = 0$$

so the update solution of $V^{(l)}$ is:

$$V^{(l)} \leftarrow V^{(l)} \odot \frac{(X^{(l)})^T D U^{(l)}}{V^{(l)} (U^{(l)})^T D U^{(l)}} \qquad (11)$$

Fixing $V^{(l)}$, Computing $U^{(l)}$. Now, we analyze the stationary point $U^{(l)}$ in $J(U) = \Sigma_l \Sigma_i w_i^{(l)} \|X_i^{(l)} - U_i^{(l)} (V^{(l)})^T\|_2^2 + R + \Sigma_l Tr(\gamma^{(l)} ((U^{(l)})^T U^{(l)} - I))$ using the auxiliary function approach.

Lemma 1 [13]. *$A(U,U')$ is an auxiliary function of $J(U)$ if the conditions $A(U,U') \geq J(U)$ and $A(U,U) = J(U)$. If A is an auxiliary function for J then J is non-increasing under the update $U^{(t+1)} = \arg\min A(U,U')$.*

[2] $\odot$ in matrix denote element-wise multiplication.

Lemma 2 [5]. *For any matrices $A \in \mathbb{R}_+^{n\times n}, B \in \mathbb{R}_+^{r\times r}, S \in \mathbb{R}_+^{n\times r}, S' \in \mathbb{R}_+^{n\times r}$, with A and B symmetric, the following inequality holds:*

$$Tr(S^T ASB) \leq \sum_{i=1}^{n}\sum_{p=1}^{r} \frac{(AS'_{ip}B)S_{ip}^2}{S'_{ip}} \tag{12}$$

Lemma 3 [18]. *For any matrices $A \in \mathbb{R}_+^{n\times n}, B \in \mathbb{R}_+^{r\times r}, S \in \mathbb{R}_+^{n\times r}, S' \in \mathbb{R}_+^{n\times r}$, with A and B symmetric, the following inequality holds:*

$$Tr(SAS^T SBS^T) \leq \sum_{i=1}^{n}\sum_{p=1}^{r} (\frac{S' AS'^T S' B + S' BS'^T S' A}{2}) \frac{S_{ip}^4}{S'^3_{ip}} \tag{13}$$

For brevity, we let X, U, V represent $X^{(l)}, U^{(l)}, V^{(l)}$ and an appropriate auxiliary is defined as

$$A(U,U') = -2\sum_{ip}(DXV)_{ip}U'_{ip}(1 + log\frac{U_{ip}}{U'_{ip}}) + \sum_{ip}(DU' V^T V + U'\gamma)_{ip}\frac{U_{ip}^4 + U'^4_{ip}}{2U'^3_{ip}}$$
$$+\sum_{ls}\lambda_{ls}[\sum_{ip}(U' U'^T U')_{ip}\frac{U_{ip}^4}{U'^3_{ip}} - 2\sum_{ipq}(U^{(s)}(U^{(s)})^T)_{pq}U'_{ip}U'_{iq}(1 + log\frac{U_{ip}U_{iq}}{U'_{ip}U'_{iq}})]$$

where we ignore the irrelevant items $Tr(X^T DX)$ and $Tr(\gamma)$. Here we use the inequality $z \geq 1 + logz$ for all $z > 0$, and obtain $\frac{U_{ip}}{U'_{ip}} \geq 1 + log\frac{U_{ip}}{U'_{ip}}$. Because of Lemma 2 and the inequality $2ab \leq a^2 + b^2$, we also obtain the second term of $J(U)$, which is bounded by $Tr(U^T DUV^T V + U^T U\gamma) \leq \sum_{ip}(DU' V^T V + U'\gamma)_{ip}\frac{U_{ip}^2}{U'_{ip}} \leq \sum_{ip}(DU' V^T V + U'\gamma)_{ip}\frac{U_{ip}^4 + U'^4_{ip}}{2U'^3_{ip}}$.

To find stationary point of $A(U, U')$, we should take the derivative of $A(U, U')$ with respect to U and fix U' according to Lemma 1:

$$\frac{\partial A(U,U')}{\partial U_{ip}} = -2(DXV)_{ip}\frac{U'_{ip}}{U_{ip}} + 2(DU' V^T V + U'\gamma)_{ip}\frac{U_{ip}^3}{U'^3_{ip}}$$
$$+\sum_{s}\lambda_{ls}[4(U' U'^T U')_{ip}\frac{U_{ip}^3}{U'^3_{ip}} - 4(U^{(s)}(U^{(s)})^T U')_{ip}\frac{U'_{ip}}{U_{ip}}] \tag{14}$$

Following the same derivations as in previous work from others [4,17], we should check the Hessian matrix of $A(U, U')$, the Hessian matrix containing the second derivatives

$$\frac{\partial^2 A(U,U')}{\partial U_{ip}\partial U_{jq}} = \{2(DXV)_{ip}\frac{U'_{ip}}{U_{ip}^2} + 6(DU' V^T V + U'\gamma)_{ip}\frac{U_{ip}^2}{U'^3_{ip}} +$$
$$\sum_{s}\lambda_{ls}[12(U' U'^T U')_{ip}\frac{U_{ip}^2}{U'^3_{ip}} + 4(U^{(s)}(U^{(s)})^T U')_{ip}\frac{U'_{ip}}{U_{ip}^2}]\}\delta_{ij}\delta_{pq} \tag{15}$$

Algorithm 1. WScoNMF algorithm

Input: Multi-view datasets $X^{(l)}$; Weighting matrices $\mu^{(l)}$; Number of clusters K; Parameters λ_{ls};
Output: Class indicator matrices $U^{(l)}$; Basis matrices $V^{(l)}$;
1 Initialize $U^{(l)}$ and $V^{(l)}$ using the traditional k-means;
2 Compute $\mu^{(l)}$ by Eq. 8 ;
3 **repeat**
4 **for** *l to* n_v **do**
5 Update $V^{(l)}$ by Eq. 11;
6 Update $U^{(l)}$ by Eq. 16;
7 **end**
8 **return** $U^{(l)}$ and $V^{(l)}$
9 **until** *Eq. 9 is converged*;

Above Hessian matrix is a positive semidefinite diagonal matrix, thus $A(U, U')$ is a convex function of U. We can obtain the global minimum of $A(U, U')$ by setting the value of Eq. 14 equals 0, and our update rule for the stationary point is (we use U here for brevity),

$$U \leftarrow U \odot \left[\frac{DXV + 2\sum_s \lambda_{ls}(U^{(s)}(U^{(s)})^T U)}{DUV^T V + U\gamma + 2\sum_s \lambda_{ls}(UU^T U)} \right]^{\frac{1}{4}} \tag{16}$$

The whole procedure is summarized in Algorithm 1. It should be noted that we choose to use a traditional and efficient clustering method (k-means) as our initialization function in all views. In our WScoNMF framework, we use the constraint $U^T U = I$ to guarantee the uniqueness of our solution and the other advantage is to significantly reduce the computation cost for the optimization algorithm [10]. Actually, our orthogonal to U is a normalization process in which use the l_2 norm.

4 Experiments

In this section, we apply the proposed WScoNMF multi-view clustering algorithm to compare its performance with other multi-view clustering techniques. Extensive experiments are made on five real-world datasets.

4.1 Datasets and Settings

Table 2 summarizes the characteristics of those data sets used in experiments. We collect *Doucom*[3] consisting of four views which we call 'summary' (Sum), 'short-comment' (Short), 'long-review' (Long) and 'user' (User) respectively.

[3] https://developers.douban.com/wiki.

Similar to the Doucom, Last.fm[4], Yelp[5] and 3-Sources[6] have three views respectively.

Table 2. Description of the multi-view datasets

Dataset	# items size	# view	#cluster
Doucom	31297	4	39
Last.fm	9694	3	21
Yelp	2624	3	7
3-Sources	169	3	6

To evaluate the performance of the proposed method, we compare our method with the following algorithms.

1. **CoRe.** [12] proposed the objective functions to co-regularize the eigenvectors of all views' Laplacian matrices.
2. **MulitNMF.** [14] developed a solution on consensus-based regularization for NMF to group the multi-view data.
3. **PcoNMF.** This is a recent pair-wise co-regularization method [9] for clustering the whole mapped data.
4. **CMVNMF.** [21] proposed a novel small number of constraints on must-link sets and cannot-link sets based on the NMF framework.

In this work, λ_{ls} determines the weight of the similarity constraint in co-regularization, and we set $\lambda_{ls} = 1$ for each pair of view in all experiments. We also set our reduced dimension $K = 100$, empirical $\epsilon = 0.001$ and we calculate each weighting matrix μ for different views based on our weighting scheme before optimization. We run K-means 100 times and select the best clustering result to initialize all the NMF methods, and we iterate algorithm 20 rounds to achieve final average results. For all the used text datasets, we apply the TF-IDF transformation on all the item-word frequency matrices. It should be noted that our item-user matrix is a zero-one matrix and it have no rating feature. To evaluate the clustering performance, we use clustering accuracy (ACC) and normalized mutual information (NMI) [7] as our metrics.

4.2 Clustering Results

In Table 3, we present results of all methods measured by ACC and NMI for each dataset. Overall, it can be seen that our method WScoNMF is very competitive, always better than the other four methods. From the experimental comparisons,

[4] http://www.last.fm/api.

[5] http://www.yelp.com/dataset_challenge.

[6] http://mlg.ucd.ie/datasets.

Table 3. Mulit-view clustering performance on five real-world datasets (%) (Both mean value and standard deviation are reported, best results are formated in bold, while second best result are underlined)

	ACC(%)				NMI(%)			
Metric	Doucom	Last.fm	Yelp	3-Sources	Doucom	Last.fm	Yelp	3-Sources
CoRe	42.8 ($\pm$2.9)	51.7 ($\pm$2.3)	60.8 ($\pm$2.7)	47.9 ($\pm$0.3)	37.1 ($\pm$3.4)	48.6 ($\pm$3.1)	57.8 ($\pm$3.2)	41.6 ($\pm$0.2)
MultiNMF	40.1 ($\pm$4.7)	45.5 ($\pm$2.3)	30.2 ($\pm$2.6)	68.4 ($\pm$0.1)	37.6 ($\pm$4.2)	39.4 ($\pm$2.3)	34.7 ($\pm$1.9)	60.2 ($\pm$0.1)
PcoNMF	46.3 ($\pm$3.6)	51.8 ($\pm$2.5)	67.6 ($\pm$4.6)	73.3 ($\pm$1.8)	44.8 ($\pm$3.7)	47.6 ($\pm$2.1)	64.7 ($\pm$3.2)	72.8 ($\pm$3.6)
CMVNMF	54.4 ($\pm$7.1)	60.4 ($\pm$3.8)	71.2 ($\pm$3.8)	74.9 ($\pm$5.7)	53.7 ($\pm$6.1)	64.0 ($\pm$1.8)	74.4 ($\pm$2.7)	75.3 ($\pm$5.5)
ScoNMF	49.2 ($\pm$5.6)	54.6 ($\pm$1.8)	68.2 ($\pm$2.4)	73.6 ($\pm$2.1)	49.6 ($\pm$4.8)	53.3 ($\pm$2.4)	67.1 ($\pm$3.8)	73.4 ($\pm$1.6)
WScoNMF	**58.8** ($\pm$4.3)	**67.2** ($\pm$2.7)	**75.3** ($\pm$4.1)	**78.8** ($\pm$3.3)	**57.2** ($\pm$4.6)	**66.5** ($\pm$3.2)	**76.3** ($\pm$2.7)	**77.4** ($\pm$2.8)

we observe that: (1) The weighted NMF framework usually outperforms the standard NMF. This may indicate that, those standard NMF algorithms are usually used for non-sparseness learning and ignore the sparsity of data structure, especially in our large dataset Doucom. (2) WScoNMF outperforms the second best algorithm in terms of ACC/NMI as 4.4%/3.5% on Doucom dataset, 6.8%/2.5% on Last.fm, 4.1%/1.9% on Yelp and 3.9%/2.1% on 3-Sources. (3) Among the co-clustering method with different co-regularization constraint, the similar pair-wise constraint (ScoNMF) performs slightly better than the simple pair-wise constraint (PcoNMF), which validates that the algorithm based on our proposed co-regularization constraint framework might be a better way of capturing the difference in intrinsic connection between every two data points.

Table 4 demonstrates the clustering accuracy results for each single-view and Table 5 also shows the accuracy results of single-view when we use the different weighted strategy in our created corpus Doucom. In Table 4, we should note that the best accuracy results for each single-view of Doucom is long-reviews unlike the User in Last.fm. There are two possible explanations for above situation, first one is that the user view in Doucom is extremely sparse than in Last.fm; second reason is that the long-reviews in Doucom are too informative which helps make it less noisy than any other three views. Another important thing is that overall single-view clustering results are worse than their correspond multi-view clustering results, which means that incorporating views of data can improve clustering performance. Table 5 gives a direct comparison between two weighting schemes. "WSco-r" represents our algorithm combined with rating feature as the weighting scheme (Eq. 8) and "WSco-s" denotes that algorithm with similarity weighting method. Generally speaking, a data point has a high rate in one view means that the view is more valuable to this data point. Furthermore, we try to use utmost cosine similarity between latent semantic vectors U_i and U_j when datasets don't have rating characteristic. It should be understood that WSco-s is not fixed and changes with the iteration, which could lead to a little deviation in similarity between latent semantic vectors. Our experimental results indicate that WSco-r has a little better effectiveness than WSco-s and we should note that our results on user view are the same, because item-user is a zero-one matrix.

Parameter Study. In our WScoNMF framework, there is only one regularization parameter for each pair of views: λ_{ls}. Relative λ_{ls} determines the weight of the consistency on pair's similarity in co-regularization. Figure 1 shows the performance of WScoNMF when varing λ_{ls} for all views. As we can see, for two large datasets, WScoNMF performs best when λ_{ls} located in 1 or 2. We also studied the parameter on other small datasets like *Yelp* and 3-*Sources*, and all results indicate that performance is the best when λ_{ls} located around 1. This suggests that the parameter λ_{ls} can be set to 1.

Table 4. Single-view clustering results in terms of accuracy

Data	Doucom				Last.fm		
View	Sum	Short	Long	User	Des	Com	User
PcoNMF	31.3	38.7	46.4	41.1	33.2	42.4	51.9
CMVNMF	40.8	44.2	52.6	48.3	48.5	53.7	60.1
ScoNMF	33.4	39.8	47.8	40.5	35.4	46.8	52.7
WScoNMF	48.7	50.8	**57.3**	52.4	41.6	54.8	**62.9**

Table 5. Effect of two weighting schemes on the clustering accuracy of each single view

Data	Doucom			
View	Sum	Short	Long	User
WSco-r	48.7	50.8	57.3	52.4
WSco-s	42.4	46.8	53.8	52.4

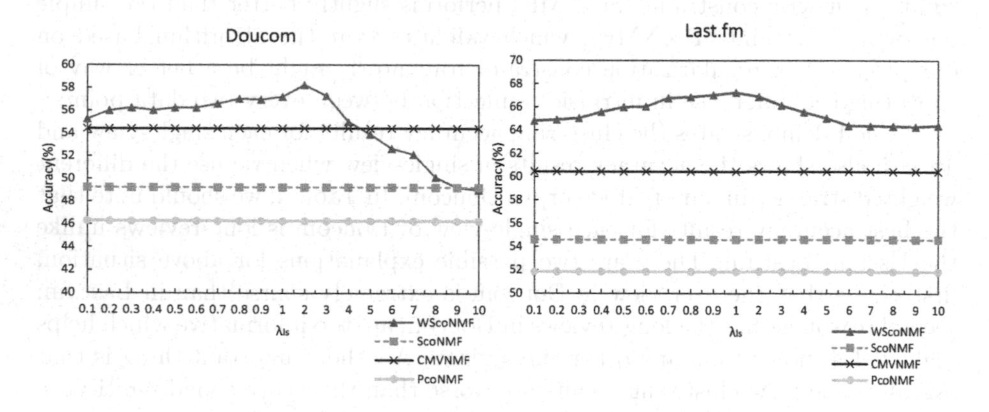

Fig. 1. Evalutaion on λ_{ls} for all views

5 Conclusion

In this paper, we have proposed a novel weighted multi-view clustering framework to cluster the multi-view data, which addressed the sparseness in real-world datasets. Also we have developed an iterative optimization algorithm. Extensive experiments have demonstrated that the proposed method is effective. In the future, we will study how to model any other features together generated by comment user such as list of user preference and investigate how to improve the algorithm efficiency when dealing with the huge real-world datasets.

Acknowledgements. This work is supported by the National Natural Science Foundation of China (No. 61472241) and the National High Technology Research and Development Program of China (No. 2015AA015303).

References

1. Akata, Z., Thurau, C., Bauckhage, C.: Nonnegative matrix factorization in multimodality data for segmentation and label prediction. In: Computer Vision Winter Workshop, pp. 1–8 (2011)
2. Cai, X., Nie, F., Huang, H.: Multi-view K-means clustering on big data. In: International Joint Conference on Artificial Intelligence, pp. 2598–2604 (2013)
3. Cheng, W., Zhang, X., Guo, Z., Wu, Y.: Flexible and robust co-regularized multidomain graph clustering. In: Proceedings of the 19th ACM SIGKDD International Conference on Knowledge Discovery and Data Mining, KDD 2013, vol. 1, pp. 320–328 (2013)
4. Ding, C., Li, T., Jordan, M.I.: Convex and semi-nonnegative matrix factorizations. IEEE Trans. Pattern Anal. Mach. Intell. **32**(1), 45–55 (2010)
5. Ding, C., Li, T., Peng, W., Park, H.: Orthogonal nonnegative matrix t-factorization for clustering. In Proceedings of the 12th ACM SIGKDD International Conference on Knowledge Discovery and Data Mining, pp. 126–135 (2006)
6. Ding, C.H., He, X., Simon, H.D.: On the equivalence of nonnegative matrix factorization and spectral clustering. In: SDM, vol. 5, pp. 606–610 (2005)
7. Du, L., Li, X., Shen, Y.D.: Robust nonnegative matrix factorization via half-quadratic minimization. In: Proceedings of the 12th International Conference on Data Mining, pp. 201–210 (2012)
8. Eaton, E., desJardins, M., Jacob, S.: Multi-view constrained clustering with an incomplete mapping between views. Knowl. Inf. Syst. **38**(1), 231–257 (2012)
9. He, X., Kan, M.Y., Xie, P., Chen, X.: Comment-based multi-view clustering of web 2.0 items. In: International Conference on World Wide Web, pp. 771–782 (2014)
10. Huang, J., Nie, F., Huang, H., Ding, C.: Robust manifold non-negative matrix factorization. ACM Trans. Knowl. Discov. Data (TKDD) **8**(3), 11:1–11:21 (2013)
11. Kumar, A., Iii, H.D.: A co-training approach for multi-view spectral clustering. In: ICML 2011, pp. 393–400 (2011)
12. Kumar, A., Rai, P., Iii, H.D.: Co-regularized multi-view spectral clustering. In: Proceedings of NIPS 2011, pp. 1413–1421 (2011)
13. Lee, L., Seung, D.: Algorithms for non-negative matrix factorization. In: Advances in Neural Information Processing Systems, vol. 13, pp. 556–562 (2001)
14. Liu, J., Wang, C., Gao, J., Han, J.: Multi-view clustering via joint nonnegative matrix factorization. In: Proceedings of SDM 2013, pp. 252–260 (2013)
15. Nie, F.P., Huang, H., Cai, X., Ding, C.: Efficient and robust feature selection via joint $l_{2,1}$-norms minimization. In: NIPS 2010 (2010)
16. Sun, J.W., Lu, J., Xu, T.Y., Bi, J.B.: Multi-view sparse co-clustering via proximal alternating. In: Proceedings of the 32th International Conference on Machine Learning, Lille, France, vol. 37 (2015)

17. Wang, H., Huang, H., Ding, C.: Cross-language web page classification via joint nonnegative matrix tri-factorization based dyadic knowledge transfer. In: Annual ACM SIGIR Conference, pp. 933–942 (2011)
18. Wang, H., Huang, H., Ding, C.: Simultaneous clustering of multi-type relational data via symmetric nonnegative matrix tri-factorization. In: CIKM 2011, pp. 279–284 (2011)
19. Wang, H., Nie, F.P., Huang, H.: Multi-view clustering and feature learning via structured sparsity. In: Proceedings of the 30th International Conference on Machine Learning, Atlanta, Georgia, USA, vol. 28 (2013)
20. Xu, W., Liu, X., Gong, Y.: Document clustering based on non-negative matrix factorization. In: SIGIR, pp. 267–273 (2003)
21. Zhang, X.C., Zong, L.L., Liu, X.Y., Yu, H.: Constrained nmf-based multi-view clustering on unmapped data. In: AAAI 2015, pp. 3174–3180 (2015)

Parallel Visual Assessment of Cluster Tendency on GPU

Tao Meng and Bo Yuan(✉)

Intelligent Computing Lab, Division of Informatics,
Graduate School at Shenzhen, Tsinghua University,
Shenzhen 518055, People's Republic of China
zdhmengtao@163.com, yuanb@sz.tsinghua.edu.cn

Abstract. Determining the number of clusters in a data set is a critical issue in cluster analysis. The Visual Assessment of (cluster) Tendency (VAT) algorithm is an effective tool for investigating cluster tendency, which produces an intuitive image of matrix as the representation of complex data sets. However, VAT can be computationally expensive for large data sets due to its $O(N^2)$ time complexity. In this paper, we propose an efficient parallel scheme to accelerate the original VAT using NVIDIA GPU and CUDA architecture. We show that, on a range of data sets, the GPU-based VAT features good scalability and can achieve significant speedups compared to the original algorithm.

Keywords: Cluster analysis · Cluster tendency · VAT · GPU

1 Introduction

Cluster analysis is an important task in pattern recognition and data mining. In general, it consists of three steps: (1) assessing the cluster tendency (e.g., how many groups to seek); (2) partitioning the data into groups; (3) validating the clusters discovered [1]. For data that can be directly projected onto a 2D or 3D Euclidean space (e.g., with a scatter plot), direct observation can provide good insight on the appropriate number of clusters. However, for high-dimensional data, or when only the pairwise relationship between objects is available, advanced techniques are necessary.

Visual Assessment of (cluster) Tendency (VAT) [2] is one of the popular methods widely used to assess the cluster tendency. Given the dissimilarity matrix D of a set of n objects, VAT represents D as an $n \times n$ image $I(D^*)$ where the objects are reordered to reveal the hidden cluster structure as dark blocks along the diagonal of the image.

VAT works well on relatively small data sets (e.g., 500 or fewer objects). However, for data sets of moderate sizes (e.g., 20,000 data points), the computing time of VAT, with time complexity $O(N^2)$, may become intolerable. In view of the high computing time of VAT, several extensions such as reVAT [3], bigVAT [4] and sVAT [5] have been proposed. reVAT performs quasi-ordering of the objects based on a threshold parameter and replaces the intensity image with a series of one-dimensional profile graphs. However, the profile graphs are not as interpretable as the images produced by VAT. To address this problem, bigVAT uses the profile graphs to select a sample of

J. Kim et al. (Eds.): PAKDD 2017, Part II, LNAI 10235, pp. 429–440, 2017.
DOI: 10.1007/978-3-319-57529-2_34

objects and displays the quasi-ordered dissimilarity data of the sampled objects as a VAT-like intensity image. However, the resulting image may not be as descriptive as the VAT-ordered image. sVAT selects a sample of (approximately) size n from the full set of objects $O = \{o_1, o_2, \ldots, o_N\}$, and performs VAT on the sample. The sample is chosen so that it contains similar cluster structure as the original data set. However, if the original data set contains many clusters, the value of n needs to increase accordingly, creating a computational issue again.

GPU (Graphics Processing Unit) is an inexpensive, energy efficient and highly efficient SIMT (Single Instruction, Multiple Thread) parallel computing device, which can be found in many mainstream desktop computers and workstations. In this paper, we propose to improve the computational efficiency of VAT using CUDA-enabled GPUs by exploiting their massively parallel computing capability and the potential of parallelism of VAT.

This paper is organized as follows. Section 2 gives a brief review of VAT and its variations as well as GPU computing. Section 3 presents the details of the proposed parallel VAT algorithm based on GPU. The main experimental studies are reported in Sect. 4, focusing on the comparison between CPU-based VAT and GPU-based VAT. This paper is concluded in Sect. 5 with some discussions on future work.

2 Related Work

2.1 A Brief Review of VAT

Let $O = \{o_1, o_2, \ldots, o_n\}$ denote n objects in the data set and D denote a matrix of pairwise dissimilarities between objects each element of which $d_{ij} = d(o_i, o_j)$ is the dissimilarity between objects o_i and o_j, with $0 \le d_{ij} \le 1; d_{ij} = d_{ji}; d_{ii} = 0$, for $1 \le i, j \le n$. Let K be the permutation of $\{1, 2, \ldots, n\}$ such that $K(i)$ is the index of the ith element in the list. The reordered list is represented as: $\{o_{K(1)}, o_{K(2)}, \ldots, o_{K(n)}\}$. Let P be the permutation matrix with $p_{ij} = 1$ if $j = K(i)$ and 0 otherwise. The matrix D^* for the reordered list is a similarity transform of D by P: $D^* = P^T DP$.

The key idea is to find P so that D^* is as close to a block diagonal form as possible. VAT reorders the row and columns of D using a modified version of Prim's minimal spanning tree (MST) algorithm [6], and displays D^* as a gray-scale image. The main difference is that VAT does not form a MST. Instead, it identifies the order in which vertices are added and the initial vertex is selected based on the maximum edge weight in the underlying complete graph [7]. The general procedure of the VAT algorithm is shown in Table 1.

An example of VAT is shown in Fig. 1. Figure 1(a) is the scatter plot of 2,000 data points in 2D. The 5 visually apparent clusters are reflected by the 5 distinct dark blocks along the main diagonal in Fig. 1(c), which is the VAT image of the data. Compared to the image of D in the original order as shown in Fig. 1(b), it is evident that reordering is necessary to reveal the underlying cluster structure of the data.

Table 1. Algorithm I: The VAT algorithm

Input: A $N \times N$ dissimilarity matrix D
1: Set $I = \emptyset, J = \{1, 2, \dots, N\}$ and $K = (0, 0, \dots, 0)$.
2: Select $(i, j) \in \arg_{p \in I, q \in J} \max\{d_{pq}\}$.
3: Set $K(1) = i, I \leftarrow \{i\}$ *and* $J \leftarrow J - \{i\}$.
4: **for** $t = 2 : N$ **do**
5: Select $(i, j) \in \arg_{p \in I, q \in J} \min\{d_{pq}\}$.
6: Set $K(t) = j$, update $I \leftarrow I \cup \{j\}$ *and* $J \leftarrow J - \{j\}$.
7: **end for**
8: Form the reordered matrix $D^* = [\, d^*_{ij} \,] = [\, d_{K(i)K(j)}]$, for $1 \leq i, j \leq n$.
Output: A gray-scale image $I(D^*)$ with $\max\{\, d^*_{ij} \,\}$: white and $\min\{\, d^*_{ij} \,\}$: black.

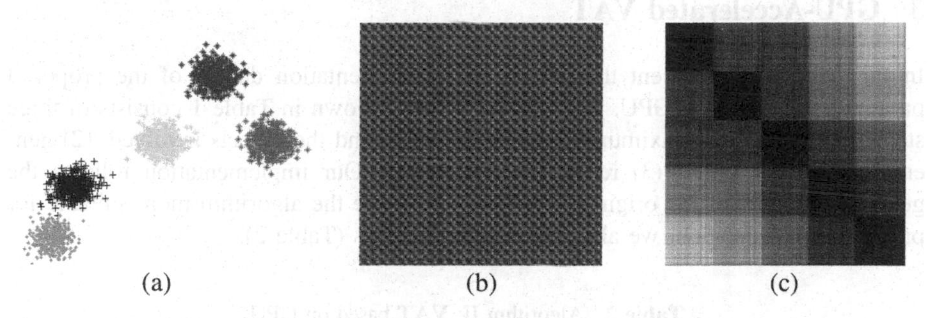

Fig. 1. An example of VAT: (a) the scatter plot of the 2D dataset; (b) the original dissimilarity image $I(D)$ and (c) the reordered VAT image $I(D^*)$

2.2 GPU High Performance Computing

In recent years, GPUs have evolved into highly parallel, multi-threaded, many-core processors and are widely used for general purpose computing [13]. Compared with CPU based distributed systems such as Hadoop, GPU based parallel computing systems are more lightweight, portable and energy-efficient. GPUs are well suited to problems that can be represented as data-parallel tasks where the same instruction is executed on massive data elements in parallel. It is also highly desirable that the arithmetic intensity is high, which is the ratio between the number of arithmetic operations and the number of memory operations.

CUDA (Compute Unified Device Architecture) is a general-purpose parallel computing platform and programming model that leverages the parallel computing engine in NVIDIA GPUs to solve challenging computational problems in a more efficient way than CPUs. It was introduced by NVIDIA in November 2006, which significantly reduces the difficulty faced by programmers for developing flexible parallel programs based on NVIDIA GPUs.

Threads and kernels are the core concepts in CUDA. Threads are lightweight processes executed on independent processors in GPU, and they are easy to be created and synchronized. Kernels are functions executed on the GPU in parallel by massive threads organized into blocks and grids [14].

There are different types of memory in GPUs, which can significantly affect the performance of GPU programs. Each thread has its private local memory called register, which is the fastest type of memory. Each thread block features shared memory accessible by all threads within the same block, which can be as fast as registers if accessed properly. All threads have access to the same global memory, which is the largest and slowest storage and the only memory visible to CPU. Constant memory and texture memory are two read-only memory spaces accessible by all threads [15]. The global, constant, and texture memory spaces are persistent across kernel launches by the same application.

In data science, examples of successful GPU applications include matrix multiplication [16], databases [17–19], data stream mining [20], FIMI mining [21], subsequence search [22] and GPU-based primitives for database applications [19, 23].

3 GPU-Accelerated VAT

In this section, we present the design and implementation details of the proposed parallel VAT based on GPU. The VAT algorithm shown in Table 1 consists of three steps: (1) finding the maximum dissimilarity value and the objects involved; (2) generating the new order; (3) reordering the matrix. Our implementation follows the general workflow of the original algorithm. To make the algorithm more suitable for parallel implementation, we also make some changes (Table 2).

Table 2. Algorithm II: VAT based on GPU

Input: An $N \times N$ dissimilarity matrix D.

1: $I = (0, 0, \dots, 0), J = (0, 0, \dots, 0), K = \{1, 2, \dots, N\}$ and $L = (0, 0, \dots, 0)$.

2: Select $(i, j) \in \arg_{p \in K, q \in K} \max\{d_{pq}\}$. // **parallel**

3: Set $L(k) = \infty$, for $1 \le k \le N$. // **parallel**

4: Set $J(i) = 1$ *and* $I(1) = \{ i \}$.

5: **for** $t = 2$: N **do**

6: $\quad \arg_{J(k)=0} L(k) = \min\{D_{[I(t-1)][k]}, L(k)\}$, for $1 \le k \le N$. // **parallel**

7: $\quad$ Select $i \in \arg_{J(k)=0} \min\{L(k)\}$. // **parallel**

8: $\quad$ Set $J(i) = 1$ *and* $I(t) = i$.

9: **end for**

10: Normalize D to D' as $d'_{pq} \in [0,\ 255]$, for $0 \le p, q \le N$. // **parallel**

11: Form the reordered matrix $D^* = [d^*_{ij}] = [\, d'_{I(i)I(j)}\,]$, for $1 \le i, j \le N$. // **parallel**

Output: A gray-scale image $I(D^*)$ with max$\{ d^*_{ij} \}$: white and min $\{ d^*_{ij} \}$: black.

3.1 Finding the Maximum Value

The reduction algorithm is a good choice for finding the maximum value of a matrix in GPU. Reduction refers to a class of parallel operations that pass over $O(N)$ input data and generate $O(1)$ result, computed by a binary associative operator $\oplus$. Examples of such operations include minimum, maximum, sum, sum of squares, AND, OR, and the dot product of two vectors [24]. Unless the operator $\oplus$ is extremely expensive to evaluate, reduction tends to be bandwidth-bound. Figure 2 shows an example of parallel reduction that computes the maximum of an 8-element array. There are four threads in use, which are marked in different colors.

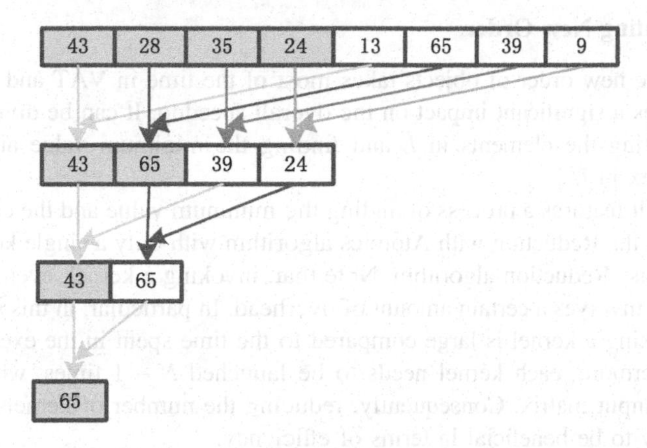

Fig. 2. An example of parallel reduction: finding the maximum value of a vector

Although Thrust, a popular library in CUDA, can find the maximum value efficiently, we employ a special reduction method as the index of the object with the maximum value is required. Furthermore, the input matrix itself is symmetric, which means that only half of the matrix needs to be processed. In this paper, we apply the Two-Pass Reduction algorithm [24] to find the maximum value and the maximum value's index in the matrix. The Two-Pass Reduction operates in two stages, as shown in Fig. 3. A kernel performs *NumBlocks* reductions in parallel, where *Numblocks* is the

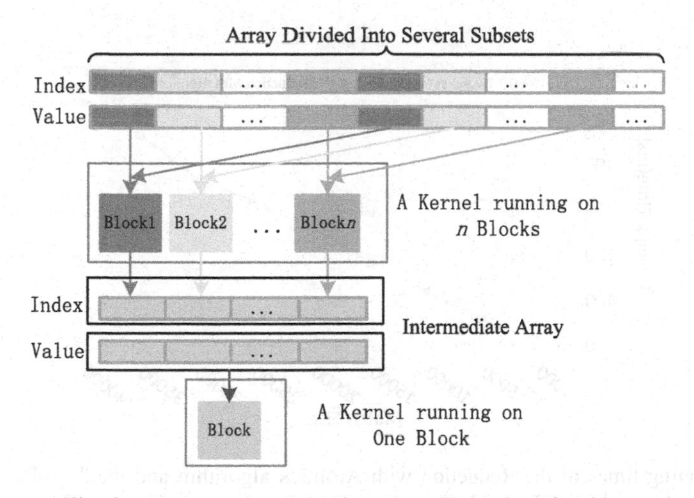

Fig. 3. An example of the Two-Pass Reduction algorithm

number of blocks used to invoke the kernel. Then, the results are stored into an intermediate array. The final result is generated by invoking the same kernel to perform a second pass on the intermediate array using a single block. Note that, this method imposes no requirement on the compute capability of GPUs, making it applicable to a wide range of GPU facilitates.

3.2 Generating New Order

Generating the new order of objects takes most of the time in VAT and its degree of parallelism has a significant impact on the overall speedup. It can be divided into two steps: computing the elements in L and finding the minimum value and the corresponding index in L.

Although it features a process of finding the minimum value and the corresponding index, we use the Reduction with Atomics algorithm with only a single kernel, instead of the Two-Pass Reduction algorithm. Note that, invoking a kernel, even a kernel that does nothing, involves a certain amount of overhead. In particular, in this step, the time spent in invoking a kernel is large compared to the time spent in the execution of the kernel. Furthermore, each kernel needs to be launched $N - 1$ times, where N is the width of the input matrix. Consequently, reducing the number of kernels in the algorithm is likely to be beneficial in terms of efficiency.

Similar to the Two-Pass Reduction algorithm, the Reduction with Atomics algorithm stores the result in an intermediate array. The difference is that the Reduction with Atomics algorithm uses a flag value for recording the number of exited blocks. As each block exits, it performs the `atomicAdd` function, a type of atomic operation in CUDA, to check whether it is the block that needs to perform the final reduction. Although the atomic operation does cost some extra time, the Reduction with Atomics algorithm is more efficient than Two-Pass Reduction when the size of data to be processed is small. Figure 4 shows the running times of the Reduction with Atomics algorithm and the Two-Pass Reduction algorithm.

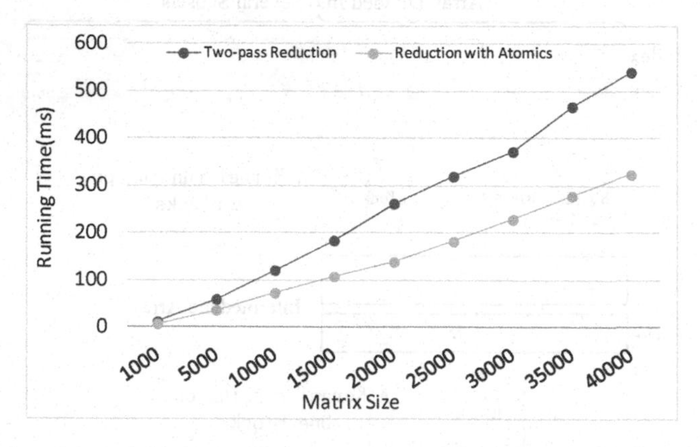

Fig. 4. Running times of the Reduction with Atomics algorithm and the Two-Pass Reduction algorithm on datasets of different sizes

3.3 Creating the Reordered Matrix

D needs to be transformed into $D^{'}$ where $d^{'}_{pq} \in [0, 255]$, to reflect the image density range [0, 255] in openCV [26]. Directly applying $N \times N$ threads may seem to be a straightforward way to create the reordered matrix. However, the memory in GPU is limited and in order to process more data, we need to transform the *double* type D to *unsigned char* type $D^{'}$ before applying $N \times N$ threads to execute $d^{*}_{ij} = d^{'}_{I(i)I(j)}$.

Since in GPU blocks are executed in an unordered manner [17], a memory space may be filled with new data before the original data has been read when multiple blocks are in use. To solve this issue, we perform the transformation of the first n elements using the same block where

$$n = \frac{N^2 \times size\ of\ (unsignedchar)}{size\ of\ (double)} \tag{1}$$

so that the rest elements can be safely processed in parallel with multiple blocks, as shown in Fig. 5.

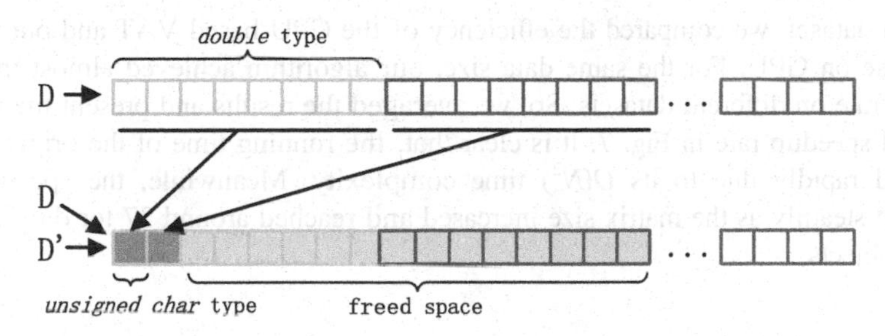

Fig. 5. Transformation of *double* type to *unsigned char* type

4 Experimental Results

We conducted the experiments on a workstation with two Intel Xeon E5-2640 v2 (2.00 GHz, 8 Cores) CPUs, 128 GB RAM and NVIDIA GeForce GTX TITAN X GPU. Powered by NVIDIA Maxwell architecture, the GeForce GTX TITAN X GPU features 3,072 CUDA cores and 12 GB GDDR5 memory. The programming environment was gcc-4.7 with CUDA 7.5 running on Ubuntu 15.04 (64 bit).

4.1 Test Datasets

We used a random dataset generator from *scikit-learn* [25]. Four different types of datasets (circles, moons, blobs and random) were generated (Fig. 6) and 10 instances (2D) were created for each type of dataset with 1,000 to 45,000 objects. We also used a dataset from UCI Machine Learning Repository [27] from which we sampled subsets

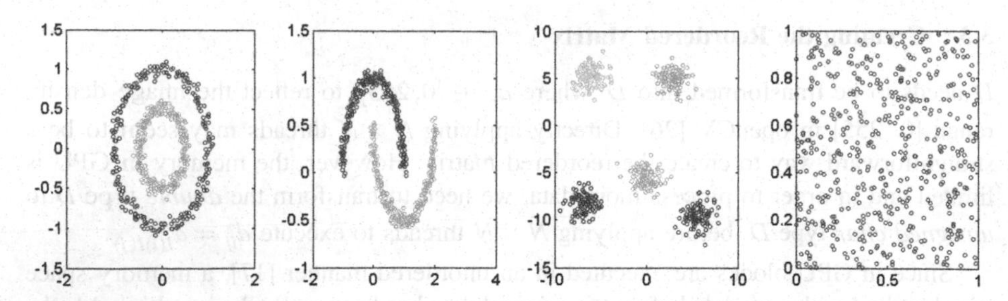

Fig. 6. Four different types of datasets used in the experiments. From left to right: circles, moons, blobs and random

with different sizes. Since the input of VAT is a dissimilarity matrix, once this matrix is given, the efficiency of VAT is fully determined, regardless of the dimension of the original dataset.

4.2 Results and Analysis

For each dataset, we compared the efficiency of the CPU-based VAT and our parallel VAT base on GPU. For the same data size, our algorithm achieved almost the same speedup rate on different datasets. So, we averaged the results and present the running time and speedup rate in Fig. 7. It is clear that, the running time of the original VAT increased rapidly due to its $O(N^2)$ time complexity. Meanwhile, the speedup rate increased steadily as the matrix size increased and reached around 37 for datasets with 40,000 objects.

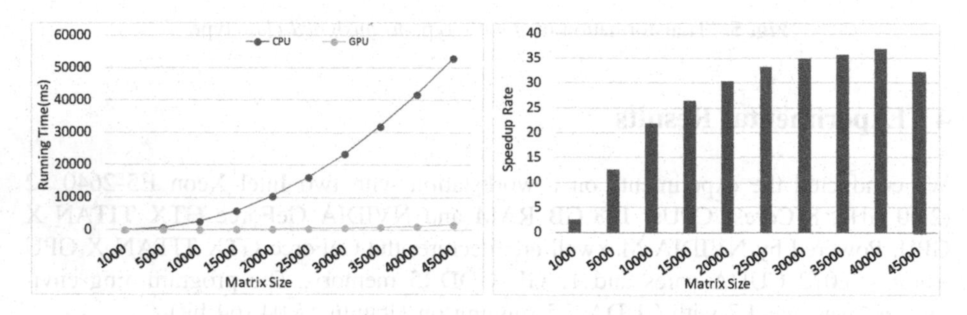

Fig. 7. Average running time and speedup rate on synthetic datasets of different sizes

Figures 8, 9 and 10 show the individual running time and speedup rate for each of the three major operations in VAT: finding the maximum value, generating the new order and creating the reordered matrix. Figure 11 shows the average time of data transmission between CPU and GPU. Figure 12 shows the average running time and speedup rate on real datasets of different sizes, showing an overall trend similar to Fig. 7. Note that, on datasets with 45,000 objects, we transformed the *double* type D to an *unsigned char* type D' to reduce the space requirement before applying $N \times N$ threads to execute $d^*_{ij} = d'_{I(i)I(j)}$. Due to the extra time cost, the overall speedup rates (e.g., Fig. 7) and the speedup rate for creating the reordered matrix (Fig. 10) both dropped slightly.

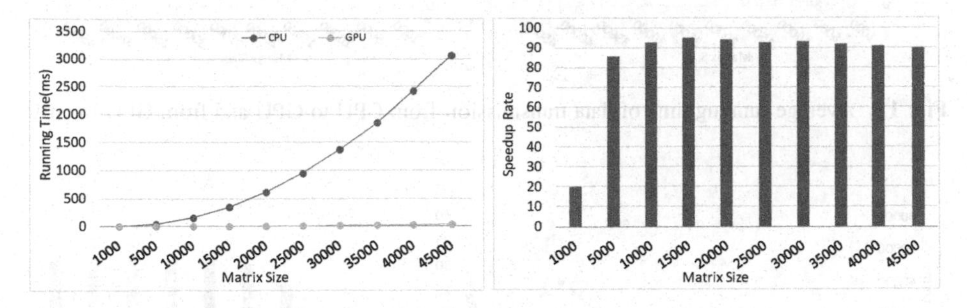

Fig. 8. Average running time and speedup rate of finding the maximum value on datasets of different sizes

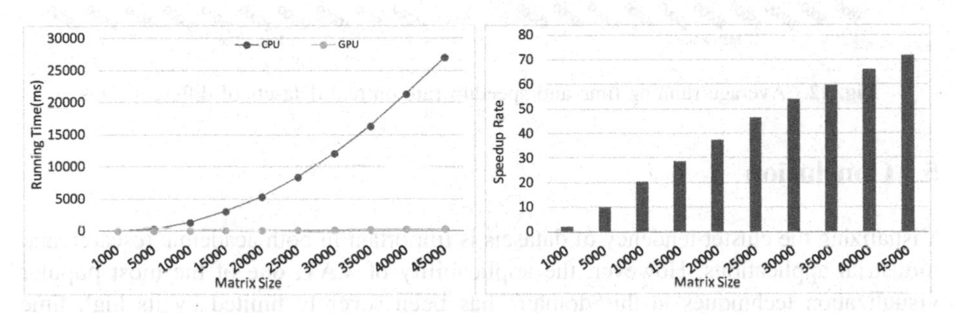

Fig. 9. Average running time and speedup rate of generating the new order on datasets of different sizes

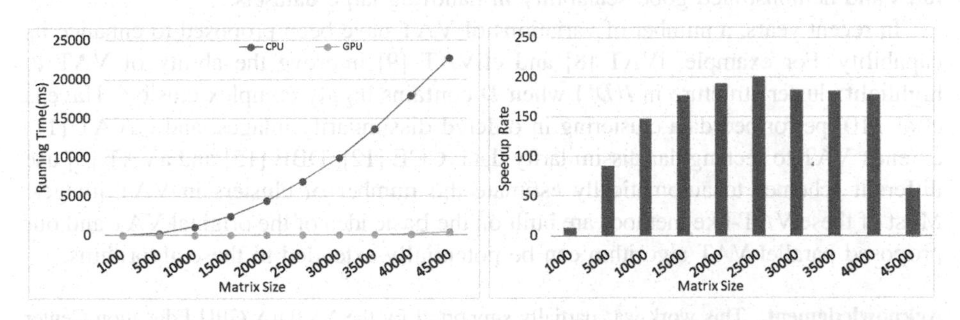

Fig. 10. Average running time and speedup rate of creating the reordered matrix on datasets of different sizes

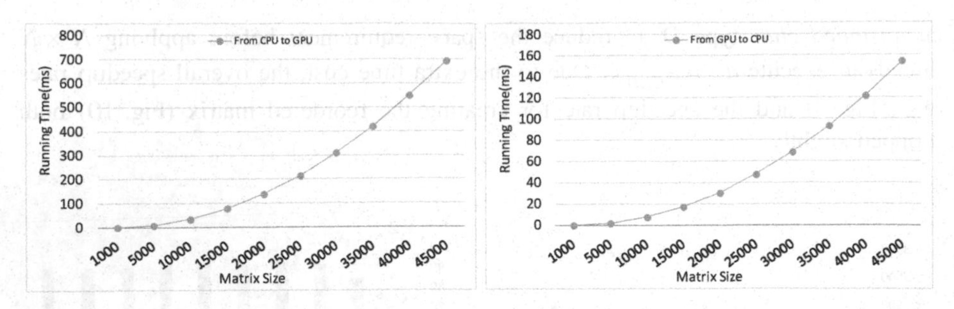

Fig. 11. Average running time of data transmission from CPU to GPU and from GPU to CPU

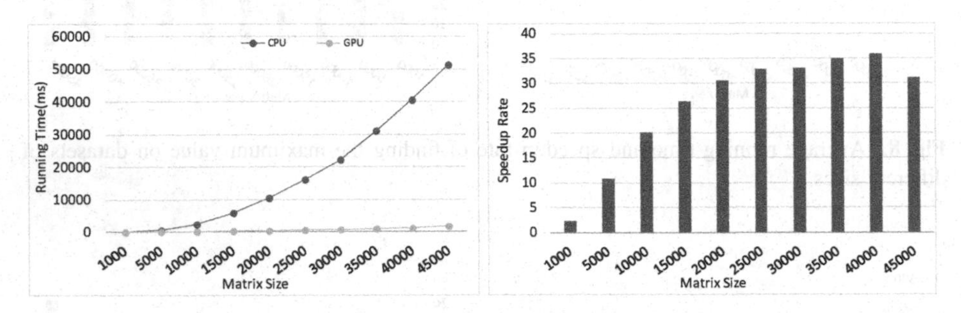

Fig. 12. Average running time and speedup rate on real datasets of different sizes

5 Conclusion

Visualizing the cluster tendency of datasets is important in both academic research and industrial applications. However, the applicability of VAT, one of the most popular visualization techniques in this domain, has been severely limited by its high time complexity. In this paper, we investigated the potential of parallelism of various components in VAT and proposed a GPU-based parallel VAT. Experiments on a variety of test datasets showed that the parallel VAT can achieve significant speedup rates and demonstrated good scalability in handling large datasets.

In recent years, a number of variations of VAT have been proposed to enhance its capability. For example, iVAT [8] and efiVAT [9] improve the ability of VAT to highlight cluster structure in $I(D^*)$ when D contains highly complex clusters. Havens et al. [10] performed data clustering in ordered dissimilarity images, and coVAT [11] extends VAT to rectangular dissimilarity data. CCE [12], DBE [15] and aVAT [8] use different schemes to automatically estimate the number of clusters in VAT images. Most of these VAT-like methods are built on the basic idea of the original VAT and our proposed parallel VAT algorithm can be potentially extended to these algorithms.

Acknowledgment. This work was partially supported by the NVIDIA GPU Education Center awarded to Tsinghua University.

References

1. Wang, L., Geng, X., Bezdek, J., Leckie, C., Kotagiri, R.: SpecVAT: enhanced visual cluster analysis. In: International Conference on Data Mining, pp. 638–647 (2008)
2. Bezdek, J.C., Hathaway, R.J.: VAT: a tool for visual assessment of (cluster) tendency. In: International Joint Conference on Neural Networks, vol. 3, pp. 2225–2230 (2002)
3. Huband, J.M., Bezdek, J.C., Hathaway, R.J.: Revised visual assessment of (cluster) tendency (reVAT). In: International Conference of the North American Fuzzy Information Processing Society, pp. 101–104 (2004)
4. Huband, J., Bezdek, J.C., Hathaway, R.: bigVAT: visual assessment of cluster tendency for large data sets. Pattern Recogn. **38**(11), 1875–1886 (2005)
5. Hathaway, R., Bezdek, J.C., Huband, J.: Scalable visual assessment of cluster tendency. Pattern Recogn. **39**(7), 1315–1324 (2006)
6. Prim, R.C.: Shortest connection networks and some generalizations. Bell Syst. Tech. J. **36** (6), 1389–1401 (1957)
7. Pakhira, M.K.: Finding number of clusters before finding clusters. Procedia Technol. **4**, 27–37 (2012)
8. Wang, L., Nguyen, U.T.V., Bezdek, J.C., Leckie, C.A., Ramamohanarao, K.: iVAT and aVAT: enhanced visual analysis for cluster tendency assessment. In: Zaki, M.J., Yu, J.X., Ravindran, B., Pudi, V. (eds.) PAKDD 2010. LNCS (LNAI), vol. 6118, pp. 16–27. Springer, Heidelberg (2010). doi:10.1007/978-3-642-13657-3_5
9. Havens, T.C., Bezdek, J.C.: An efficient formulation of the improved visual assessment of cluster tendency (iVAT) algorithm. IEEE Trans. Knowl. Data Eng. **24**(5), 813–822 (2012)
10. Havens, T.C., Bezdek, J.C., Keller, J.M., Popescu, M.: Clustering in ordered dissimilarity data. Int. J. Intell. Syst. **24**(5), 504–528 (2009)
11. Bezdek, J.C., Hathaway, R., Huband, J.: Visual assessment of clustering tendency for rectangular dissimilarity matrices. IEEE Trans. Fuzzy Syst. **15**(5), 890–903 (2007)
12. Sledge, I., Huband, J., Bezdek, J.C.: (Automatic) Cluster count extraction from unlabeled datasets. In: Joint International Conference on Natural Computation and International Conference on Fuzzy Systems and Knowledge Discovery, vol. 1, pp. 3–13 (2008)
13. CUDA Toolkit Documentation. http://docs.nvidia.com/cuda/index.html
14. Cook, S.: CUDA Programming: A Developer's Guide to Parallel Computing with GPUs. Newnes, Oxford (2012)
15. Farber, R.: CUDA Application Design and Development. Elsevier, Amsterdam (2012)
16. Larsen, E.S., McAllister, D.: Fast matrix multiplies using graphics hardware. In: Proceedings of the 2001 ACM/IEEE Conference on Supercomputing, no. 43 (2001)
17. Govindaraju, N.K., Lloyd, B., Wang, W., Lin, M., Manocha, D.: Fast computation of database operations using graphics processors. In: Proceedings of the 2004 ACM SIGMOD International Conference on Management of Data, pp. 215–226 (2004)
18. Govindaraju, N., Gray, J., Kumar, R., Manocha, D.: GPUTeraSort: high performance graphics co-processor sorting for large database management. In: Proceedings of the 2006 ACM SIGMOD International Conference on Management of Data, pp. 325–336 (2006)
19. He, B., Yang, K., Fang, R., Lu, M., Govindaraju, N., Luo, Q., Sander, P.: Relational joins on graphics processors. In: Proceedings of the 2008 ACM SIGMOD International Conference on Management of Data, pp. 511–524 (2008)
20. Govindaraju, N.K., Raghuvanshi, N., Manocha, D.: Fast and approximate stream mining of quantiles and frequencies using graphics processors. In: Proceedings of the 2005 ACM SIGMOD International Conference on Management of Data, pp. 611–622 (2005)

21. Fang, W., Lu, M., Xiao, X., He, B., Luo, Q.: Frequent itemset mining on graphics processors. In: Proceedings of the Fifth International Workshop on Data Management on New Hardware, pp. 34–42 (2009)
22. Sart, D., Mueen, A., Najjar, W., Keogh, E., Niennattrakul, V.: Accelerating dynamic time warping subsequence search with GPUs and FPGAs. In: 2010 IEEE International Conference on Data Mining, pp. 1001–1006 (2010)
23. He, B., Govindaraju, N.K., Luo, Q., Smith, B.: Efficient gather and scatter operations on graphics processors. In: Proceedings of the 2007 ACM/IEEE Conference on Supercomputing, no. 46 (2007)
24. Nicholas, W.: The CUDA Handbook: A Comprehensive Guide to GPU Programming. Addison-Wesley Professional, Boston (2013)
25. Pedregosa, et al.: Scikit-learn: machine learning in python. J. Mach. Learn. Res. **12**, 2825–2830 (2011)
26. OpenCV User Guide. http://docs.opencv.org/2.4.13/doc/user_guide/user_guide.html
27. Bache, K., Lichman, M.: UCI Machine Learning Repository (2013)

Clustering Complex Data Represented as Propositional Formulas

Abdelhamid Boudane(✉), Said Jabbour, Lakhdar Sais, and Yakoub Salhi

CRIL-CNRS, Université d'Artois, 62307 Lens Cedex, France
{boudane,jabbour,sais,salhi}@cril.fr

Abstract. Clustering has been extensively studied to deal with different kinds of data. Usually, datasets are represented as a n-dimensional vector of attributes described by numerical or nominal categorical values. Symbolic data is another concept where the objects are more complex such as intervals, multi-categorical or modal. However, new applications might give rise to even more complex data describing for example customer desires, constraints, and preferences. Such data can be expressed more compactly using logic-based representations. In this paper, we introduce a new clustering framework, where complex objects are described by propositional formulas. First, we extend the two well-known k-means and hierarchical agglomerative clustering techniques. Second, we introduce a new divisive algorithm for clustering objects represented explicitly by sets of models. Finally, we propose a propositional satisfiability based encoding of the problem of clustering propositional formulas without the need for an explicit representation of their models. Preliminary experimental results validating our proposed framework are provided.

1 Introduction

Clustering is a technique used to recover hidden structure in a dataset obtained by grouping data into clusters of similar objects. It is derived by several important applications ranging from scientific data exploration, to information retrieval, and computational biology (e.g. [1]). Such diversity in terms of application domains induces a variety of data types and clustering techniques (see. [2] for a survey). Indeed, data can be transactional, sequential, trees, graphs, texts, or even of a symbolic nature [6,7,10]. This last kind of data is particularly suitable for modeling complex and heterogeneous objects usually described by a set of multivalued variables of different types (e.g. intervals, multi-categorical or modal) (e.g. [3,4,8]). We can also mention conceptual clustering proposed more than thirty years ago by Michalski [14] and defined as a machine learning task. It accepts a set of object descriptions (events, facts, observations, ...) and produces a classification scheme over them. Conceptual clustering not only partitions the data, but generates clusters that can be summarized by a conceptual description. As a summary, conceptual and symbolic clustering are two paradigms proposed to deal with kinds of data other than those usually described by numerical values.

J. Kim et al. (Eds.): PAKDD 2017, Part II, LNAI 10235, pp. 441–452, 2017.
DOI: 10.1007/978-3-319-57529-2_35

In today's data-driven digital era, data might be even more complex and heterogenous. Such complex data might represent customers desires or preferences collected in different possible ways using surveys and quizzes. As an example, one can cite configuration systems usually designed to provide customized products satisfying the different requirements of the customer, usually modeled by constraints or logic-formulas (e.g. [11]). These customers requirements-data or the data-models provided by the configuration systems are some kind of complex data that we are interested in. These data can be represented by logic-formulas (requirements) or by models (the products satisfying the requirements). Data can also represent more complex entities such as transaction databases. Indeed, suppose that we collected several transaction databases from stores chain selling the same products, one can be interested in determining similar stores (clusters) or stores with the same behavior. This could help the manager of the stores chain to better define its trade policy. In the two previous examples, data can be better represented as a set of propositional formulas or as sets of models.

In this paper, we introduce a new clustering framework, where complex objects are described by propositional formulas. We first extend the two well known k-means and hierarchical agglomerative clustering techniques. Then, we introduce a new divisive algorithm for clustering objects represented explicitly by sets of models. Finally, we propose a propositional satisfiability based encoding of clustering propositional formulas without the need for an explicit representation of their models. Preliminary experimental results validating our proposed framework are provided before concluding.

1.1 Propositional Satisfiability

Let $\mathcal{P}$ be a countably infinite set of propositional variables. The set of *propositional formulas*, denoted $F_{\mathcal{P}}$, is defined inductively starting from $\mathcal{P}$, the constant $\bot$ denoting absurdity, the constant $\top$ denoting true, We use the greek letters ϕ, ψ to represent formulas. A *Boolean interpretation* $\mathcal{I}$ of a formula ϕ is defined as a function from $\mathcal{P}(\phi)$ to $\{0, 1\}$ (0 for $false$ and 1 for $true$). A *model* of a formula ϕ is a Boolean interpretation $\mathcal{I}$ that satisfies ϕ (written $\mathcal{I} \vDash \phi$), i.e. $\mathcal{I}(\phi) = 1$. We denote the set of models of ϕ by $\mathcal{M}(\phi)$. A formula ϕ is satisfiable (or consistent) if there exists a model of ϕ; otherwise it is called unsatisfiable (or inconsistent).

Let ϕ and ψ be two propositional formulas, we say that ψ is a logical consequence of ϕ, written $\phi \vDash \psi$, iff $\mathcal{M}(\phi) \subseteq \mathcal{M}(\psi)$. The two formulas ϕ and ψ are called equivalent iff $\phi \vDash \psi$ and $\psi \vDash \phi$, i.e. $\mathcal{M}(\phi) = \mathcal{M}(\psi)$.

A CNF formula is a conjunction ($\wedge$) of clauses, where a *clause* is a disjunction ($\vee$) of literals. A *literal* is a propositional variable (p), called positive literal, or ($\neg p$), called negative literal. The *SAT problem* consists in deciding whether a given CNF formula admits a model or not. Another problem related to SAT is the SAT model enumeration problem. Enumeration requires generating all models of a problem instance without duplicates. Models enumeration is related to #SAT, the problem of computing the number of models for a given propositional formula. Model counting is the canonical #P-complete problem. On the practical side, for model counting, *SampleCount* a sampling based approach proposed by Gomes et al. in [9], provides very good lower bounds with high confidence. Similarly, an efficient model enumeration algorithm has been proposed in [5,12].

2 Motivating Example

To motivate our proposed framework, let us consider a simple example of a car dealer selling different cars bands with several possible options. For each car brand, several colors and types of fuels are available. The car dealer collected the preferences of four customers through a survey questionnaire. The first customer does not want red cars. The second wants a car with a diesel fuel, while the third wants a red car with gasoline fuel. Finally, the fourth customer prefers brand Peugeot cars. In addition to these customer desires, we also consider mutual exclusion constraints (mutex), allowing to express that each car must have only one color, one type of fuel and one car brand.

To express the different customer desires in propositional logic, we consider the following propositional variables: r (resp. b) represents red (resp. black) colors, p (resp. c) represents the Peugeot (resp. Citroen) car brand and d (resp. g) represents cars with diesel (resp. gasoline) fuel.

The mutex constraints are expressed by the following formula: $\mu = [(r \wedge \neg b) \vee (b \wedge \neg r)] \wedge [(g \wedge \neg d) \vee (d \wedge \neg g)] \wedge [(p \wedge \neg c) \vee (c \wedge \neg p)]$.

In Fig. 1 (left hand side), for each customer c_i, we associate a propositional formula ϕ_{c_i} expressing its desires. We also provide the set of models satisfying both the desires of the customer and the mutex constraints ($\mathcal{M}(\phi_{c_i} \wedge \mu)$). The presentation of the models follows the variables ordering: $r \prec b \prec d \prec g \prec c \prec p$. In Fig. 1 (right hand side), we give a graphical representation of the preferences of the four customers. This illustrative example highlights the expressiveness of logic-based data representation while allowing the possibility to define both user and background constraints.

Customers	c_1	c_2	c_3	c_4
ϕ_{c_i}	$\neg r$	d	$g \wedge r$	p
$\mathcal{M}(\phi_{c_i} \wedge \mu)$	010101	011001		010101
	010110	011010		011001
	011001	101001	100101	100101
	011010	101010	100110	101001

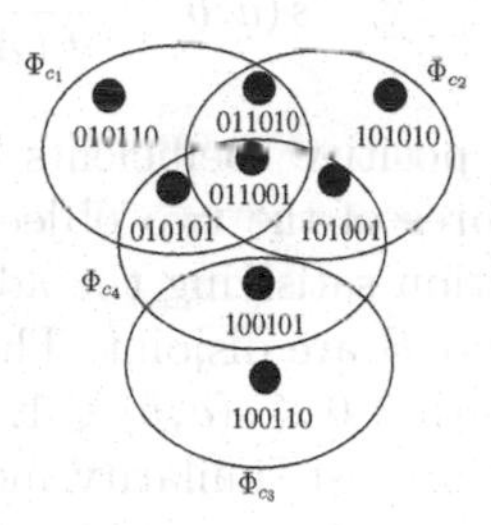

Fig. 1. Logical and graphical representation of customers preferences

3 Adapting Standard Clustering Algorithms

In this section, we present our extension of the well-known k-means and agglomerative hierarchical clustering algorithms to handle objects expressed as propositional formulas. Let us first fix some necessary notations and definitions.

We use $\mathcal{P}(k, \Phi)$ to denote the problem of clustering the set of propositional formulas $\Phi = \{\phi_1, \ldots, \phi_n\}$ into a set of k clusters with $k \leqslant n$. Let $\mathcal{C}$ be a family of sets over Φ. $\mathcal{C}$ is a solution of $\mathcal{P}(k, \Phi)$ if and only if $|\mathcal{C}| = k$, $\bigcup_{\mathcal{C}_i \in \mathcal{C}} \mathcal{C}_i = \Phi$ with $C_i \cap C_j = \emptyset$ for $1 \leqslant i < j \leqslant k$, and $\mathcal{M}(\bigwedge_{\phi \in \mathcal{C}_i} \phi) \neq \emptyset$ for every $\mathcal{C}_i \in \mathcal{C}$. We say that a clustering problem $\mathcal{P}(k, \Phi)$ is *consistent* if it admits a solution.

3.1 k-Means Algorithm for Propositional Formulas Clustering

Given a set of n data points in d-dimensional space $\mathbb{R}^d$ and a positive integer k, the k-means algorithm determines a set of k points in $\mathbb{R}^d$, called centers, so as to minimize an objective function such as the mean squared distance from each data point to its nearest center. To extend the k-means algorithm to clustering of objects described by propositional formulas, we need to define,

1. a distance between two formulas;
2. a centroid representing a given cluster;
3. an objective function to optimize.

Let us recall that a propositional formula ϕ can be equivalently expressed by its set of models $\mathcal{M}(\phi)$. With this representation in mind, one can consider that two formula ϕ_1 and ϕ_2 are similar if their set of common models $\mathcal{M}(\phi_1) \cap \mathcal{M}(\phi_2)$ is higher with respect to the remaining (distinctive) models $\mathcal{M}(\phi_1) \setminus \mathcal{M}(\phi_2) \cup \mathcal{M}(\phi_2) \setminus \mathcal{M}(\phi_1)$. This kind of similarity is related to the well-known contrast model of similarity proposed in a seminal paper by Tversky [15].

Definition 1 (Tversky [15]**).** *Let a and b be two objects described by two sets of features A and B respectively. Similarity between a and b, denoted s(a, b), is defined as:*

$$s(a,b) = \frac{f(A \cap B)}{f(A \cap B) + \alpha f(A - B) + \beta f(B - A)} \quad \alpha, \beta \geqslant 0$$

The positive coefficients α and β reflects the weights given to the distinctive features of the two objects a and b. We usually assume that f is a matching function satisfying the additivity property $f(A \cup B) = f(A) + f(B)$, whenever A and B are disjoint. The ratio model defines a normalized value of similarity such that $0 \leqslant s(a, b) \leqslant 1$.

Contrast similarity model is particularly suitable in our context. To extend Definition 1, we consider the relationship between set operations and logical connectives. Indeed, the set union (resp. intersection) corresponds to disjunction (resp. conjunction). The difference between sets can be expressed using both conjunction and negation connectives, while the symmetric difference between sets can be expressed using the xor ($\oplus$) logical connective. Indeed, we have $\mathcal{M}(\phi_1) \setminus \mathcal{M}(\phi_2) \cup \mathcal{M}(\phi_2) \setminus \mathcal{M}(\phi_1) = \mathcal{M}((\phi_1 \wedge \neg\phi_2) \vee (\phi_2 \wedge \neg\phi_1)) = \mathcal{M}(\phi_1 \oplus \phi_2)$.

Using these relationships, we derive the following extension of the ratio model [16].

Definition 2. *Let a and b be two objects described by two propositional formulas ϕ_1 and ϕ_2 respectively. Similarity between a and b is defined as:*

$$s(a,b) = \frac{f(\phi_1 \wedge \phi_2)}{f(\phi_1 \wedge \phi_2) + \alpha f(\phi_1 \wedge \neg\phi_2) + \beta f(\phi_2 \wedge \neg\phi_1)} \quad \alpha, \beta \geqslant 0$$

In our context, as no distinction is made between the measure of $\phi_1 \wedge \neg\phi_2$ and $\phi_2 \wedge \neg\phi_1$, we derive the following similarity measure.

Definition 3. *Let a and b be two objects described by two propositional formulas ϕ_1 and ϕ_2 respectively. Similarity between a and b is defined as:*

$$s(a,b) = \frac{f(\phi_1 \wedge \phi_2)}{f(\phi_1 \wedge \phi_2) + \gamma f(\phi_1 \oplus \phi_2)}, \gamma \geqslant 0$$

From Definition 2 (resp. Definition 3), instantiating $\alpha = \beta = 1$ (resp. $\gamma = 1$), we derive a logic-based variant of the well known Jaccard similarity coefficient (resp. distance) [13]:

Definition 4. *Let a and b be two objects described by two propositional formulas ϕ_1 and ϕ_2 respectively. Similarity and distance between a and b or between ϕ_1 and ϕ_2 are defined respectively as:*

$$s_J(a,b) = s_J(\phi_1,\phi_2) = \frac{f(\phi_1 \wedge \phi_2)}{f(\phi_1 \vee \phi_2)} \textit{ and } d_J(a,b) = 1 - s_J(a,b) = d_J(\phi_1,\phi_2)$$

As mentioned previously, considering the model based representation of propositional formulas, we define the function f as:

$$f : \left| \begin{array}{l} F_{\mathcal{P}} \longrightarrow \mathbf{N} \\ \phi \longmapsto |\mathcal{M}(\phi)| \end{array} \right.$$

Clearly, the function f satisfies the additive property. Indeed, we have $\mathcal{M}(\phi_1 \vee \phi_2) = \mathcal{M}(\phi_1) \cup \mathcal{M}(\phi_2)$. Computing f involves solving a #P-Complete model counting problem as discussed in Sect. 1.1.

Let us now define the representative of a cluster of propositional formulas.

Definition 5. *Let C_i be a cluster involving n_i formulas $\{\phi_{1_i}, \phi_{2_i}, \ldots, \phi_{n_i}\}$. We define the cluster representative (also called centroid) $\mathcal{O}_{C_i}$ of the cluster C_i as:*

$$\mathcal{O}_{C_i} = \phi_{1_i} \wedge \phi_{2_i} \wedge \ldots \wedge \phi_{n_i}$$

It is important to note that in our proposed extension, the goal is to group formulas into consistent clusters. Consequently, the *formula representing a given cluster must be consistent.*

We use the classical k-means objective function introduced in Definition 6.

Definition 6. *Let $\mathcal{P}(k, \Phi)$ be the problem of clustering a set of propositional formulas $\Phi = \{\phi_1, \ldots, \phi_n\}$ to $k (\leqslant n)$ clusters $\mathcal{C} = \{C_1, \ldots, C_k\}$. The objective function is defined using Absolute-Error Criterion (AEC):*

$$C^* = \arg\min_{\mathcal{C}} \sum_{i=1}^{k} \sum_{\phi \in C_i} d_J(\phi, \mathcal{O}_{C_i}) \tag{1}$$

Our clustering algorithm of a set of propositional formulas can now be derived from the classical k-means algorithm using the new components (distance, centroid and objective function) defined above.

3.2 Hierarchical Agglomerative Algorithm for Propositional Formulas Clustering

Hierarchical algorithms can behave better than the k-means. The base idea of hierarchical agglomerative algorithms is to build a dendrogram such that at each level the two closest clusters are merged. By applying a hierarchical algorithm, we will ensure that if there are two objects that are closest to each other, they will necessarily be in the same cluster. In this adaptation, the similarity between two clusters is identical to the similarity between their representatives. Similarly to Definition 5, the conjunction of all formulas in a cluster represents its centroid. To merge clusters, we combine the two clusters with the smallest centroid distance. Using this adaptation, we can applay a standard hierarchical agglomerative algorithm on data represented as boolean formulas as illustrated in Fig. 2. Note that this algorithm needs at least $\mathcal{O}(n^2)$ calls to a # SAT oracle.

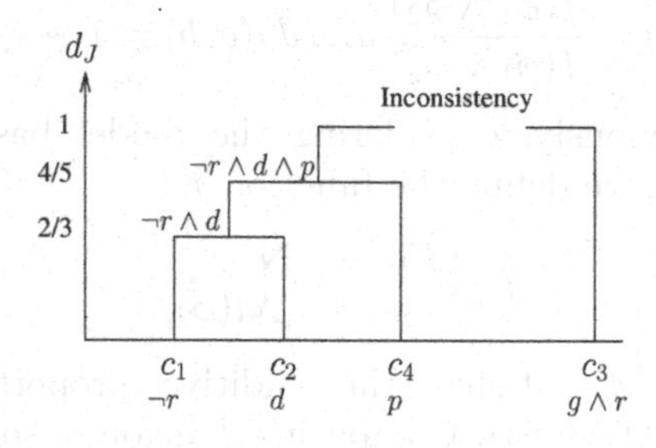

Fig. 2. Agglomerative clustering on the car dealer example

4 Divisive Algorithm for Model Based Representation

As mentioned previously, when we consider the problem of clustering a set of formulas $\Phi = \{\phi_1, \phi_1, \dots, \phi_n\}$ without common model, i.e., $\Phi \vdash \bot$, agglomerative algorithm and k-means can fail to find a clustering with the desired number of clusters. In the sequel, we propose a top-down hierarchical (or divisive) algorithm for clustering a set of propositional formulas. Our proposed adaptation makes use of the well-known minimum hitting sets problem, that we recall.

Definition 7. *H is a hitting set of a set of sets Ω if $\forall S \in \Omega$, $H \cap S \neq \emptyset$. A hitting set H is irreducible if there is no other hitting set H' s.t $H' \subset H$. H is called minimum hitting set if there is no hitting set H' such that $|H'| < |H|$.*

Example 1. Let $\Phi = \{\phi_1, \phi_2, \phi_3\}$ be a set of propositional formulas such that $\mathcal{M}(\phi_1) = \{m_1, m_2, m_3\}$, $\mathcal{M}(\phi_2) = \{m_1, m_4\}$ and $\mathcal{M}(\phi_3) = \{m_3, m_5\}$. The set $H = \{m_1, m_3\}$ is a minimum and irreducible hitting set of the models of Φ.

In our adaptation, we choose the worst cluster to divide according to the following quality measure.

Definition 8. *Let $\mathcal{C}_i = \{\phi_{1_i}, \ldots, \phi_{n_i}\}$ be a cluster of n_i propositional formulas. We define the quality of $\mathcal{C}_i$ as:*

$$\mathcal{Q}(\mathcal{C}_i) = \frac{|\mathcal{M}(\phi_{1_i} \wedge \cdots \wedge \phi_{n_i})|}{|\mathcal{M}(\phi_{1_i} \vee \cdots \vee \phi_{n_i})|}$$

The quality of a cluster is obtained by extending the similarity measure between two formulas to a set of formulas. Indeed, a cluster is qualified to be of poor quality, when its formulas admits a great number of models while sharing a small number of models. Consequently, the worst cluster is obtained as follows:

$$\mathcal{C}_i^* = \underset{\mathcal{C}_i \in \mathcal{C}}{\text{argmin}}\ \mathcal{Q}(\mathcal{C}_i)$$

Definition 9. *Let Φ be a set of propositional formulas and $\mathcal{I}$ a Boolean interpretation. We define the subset of formulas of Φ sharing the model $\mathcal{I}$ as $\mathcal{S}(\mathcal{I}, \Phi) = \{\phi \in \Phi | \mathcal{I} \models \phi\}$.*

To build consistent clusters, Algorithm 1 starts by computing a minimum hitting set H of the set of sets of models of the formulas in Φ (line 1). The main idea behind our algorithm is to use the models of the computed minimum hitting set to divide a cluster into several consistent clusters. Each cluster is obtained by selecting for each model m of the minimum hitting set, the set of formulas admitting m as a model. In this way, the formulas in the obtained clusters share at least one model. If the size of the minimum hitting set H is greater than k, then no clustering is possible, and the algorithm returns an empty set (line 3), otherwise a consistent clustering can be obtained. In this last case, the algorithm starts by a clustering $\mathcal{C}$ where all the formulas in Φ are grouped into a single cluster (line 6). We start an iterative top-down divisive process (lines 7–20), until generating k clusters. At each iteration, we choose a cluster to divide (line 8) which is one of those with the worst quality (see Definition 8). Then, we build Ω the set of sets of models of the formulas involved in the selected cluster, while removing the set of common models M (lines 9–10). A minimum hitting set H of Ω is then computed (line 11). It is important to note that by removing the common models M from the models of each formula of the selected cluster, we avoid the trivial minimum hitting sets of size 1. Now, we use the hitting set H to divide the chosen cluster $\mathcal{C}_i^*$ into $|H|$ clusters (line 12). Indeed, for each model m in H, we associate a cluster Ψ_m made of formulas of $\mathcal{C}_i^*$ sharing the model m. In this way, we maintain the consistency property on each new cluster Ψ_m. Now, we substitute in $\mathcal{C}$ the cluster of poor quality $\mathcal{C}_i^*$ with the new set of clusters (line 18). However, this is only done when the size of the new

Algorithm 1. Model-Based Divisive Algorithm for Clustering Boolean Formulas

Input: A set of formulas $\Phi = \{\phi_1, \ldots, \phi_n\}$ and an integer $k \geqslant 1$
Output: A set of clusters $\mathcal{C} = \{C_1, \ldots, C_k\}$

1 $H \leftarrow minHittingSet(\{\mathcal{M}(\phi_1), \ldots, \mathcal{M}(\phi_n)\})$;
2 **if** $(|H| > k)$ **then**
3   **return** $\emptyset$;
4 **end**
5 **else**
6   $\mathcal{C} \leftarrow \{\Phi\}$;
7   **while** $(|\mathcal{C}| \mathrel{!}= k)$ **do**
8     $C_i^* = \{\phi_{i_1} \ldots \phi_{i_{n_i}}\} \leftarrow \underset{C_i \in \mathcal{C}, |C_i| > 1}{\arg\min} \mathcal{Q}(C_i)$, $\triangleright\ n_i = |C_i^*|$;
9     $M = \mathcal{M}(\phi_{i_1}) \cap \cdots \cap \mathcal{M}(\phi_{i_{n_i}})$;
10     $\Omega = \{\mathcal{M}(\phi_{i_1}) \setminus M, \ldots, \mathcal{M}(\phi_{i_{n_i}}) \setminus M\}$;
11     $H \leftarrow minHittingSet(\Omega)$;
12     $\forall m \in H, \Psi_m \leftarrow \mathcal{S}(m, C_i^*)$;
13   **if** $(|\mathcal{C}| + |H| - 1 > k)$ **then**
14       $\Psi \leftarrow merge(\{\Psi_{m_1}, \ldots, \Psi_{m_{|\mathcal{C}|+|H|-1-k}}\})$;
15       $\mathcal{C} \leftarrow (\mathcal{C} \setminus C_i^*) \cup \{\Psi\} \cup \{\Psi_{m_{|\mathcal{C}|+|H|-k}}, \ldots, \Psi_{m_{|H|}}\}$
16   **end**
17   **else**
18       $\mathcal{C} \leftarrow (\mathcal{C} \setminus C_i^*) \cup \{\Psi_{m_1}, \ldots, \Psi_{m_{|H|}}\}$
19   **end**
20   **end**
21 **end**
22 $\mathcal{C} \leftarrow eliminateOverlap(\mathcal{C})$;
23 **return** $\mathcal{C}$

clustering does not exceed k (line 13); otherwise to obtain exactly k clusters, we merge (function `merge`) the first $|\mathcal{C}| + |H| - (k+1)$ of these new clusters (line 14) before applying substitution (line 15). Note that in the divisive step (line 12), a formula can belong to several new clusters. The reason comes from the fact that a given formula can share several models of the minimum hitting set. Consequently, a last step is then performed to produce non overlapping clusters (line 20 - function *eliminateOverlap*). To do this, for each formula occurring in several clusters, we keep it in the cluster with the best quality, while removing it in the remaining clusters. Obviously, depending on applications, overlapping clusters might be more suitable. In this case, one only need to skip the call to the overlap elimination function.

Algorithm 1, involves $\mathcal{O}(n)$ calls to model enumeration problem (line 1), $\mathcal{O}(k)$ calls to # SAT oracle (line 8) and $\mathcal{O}(k)$ calls to minimum hitting set problem (line 1 and 11).

Let us now gives some interesting properties of our propositional formulas based divisive algorithm. The first one states the correctness of our algorithm.

Proposition 1. *If* $\mathcal{P}(k, \Phi)$ *is consistent, then Algorithm* 1 *produces a clustering.*

The proof trivially follows from the previous detailed explanation on how the algorithm operates.

The second property allows us to establish that two equivalent formulas might be located in the same cluster when overlaps between clusters are allowed.

Proposition 2. *Let $\mathcal{P}(k, \Phi)$ be a clustering problem with overlaps, $\mathcal{C}$ a clustering of $\mathcal{P}(k, \Phi)$ and $\phi_1, \phi_2 \in \Phi$. If $\phi_1 \equiv \phi_2$ then $\forall \mathcal{C}_i \in \mathcal{C}$, $\phi_1 \in \mathcal{C}_i$ iff $\phi_2 \in \mathcal{C}_i$.*

The last property generalizes the previous property to the case of two formulas where one is a logical consequence of the other.

Proposition 3. *Let $\mathcal{P}(k, \Phi)$ be a clustering problem with overlaps, $\mathcal{C}$ a clustering of $\mathcal{P}(k, \Phi)$ and $\phi_1, \phi_2 \in \Phi$. If $\phi_1 \vdash \phi_2$ then $\forall \mathcal{C}_i \in \mathcal{C}$, if $\phi_1 \in \mathcal{C}_i$ then $\phi_2 \in \mathcal{C}_i$.*

5 SAT Encoding for a Bounded Consistent Clustering

As discussed in the previous section, when the propositional formulas are not represented by their models, our proposed model based divisive algorithm requires $\mathcal{O}(n)$ calls to model enumeration oracle, to compute the set of models of each formula. Such set of models might be of exponential size in the worst case. In addition to these limitations, one also need to compute a minimum hitting set of a set of sets of models ($\mathcal{O}(k)$ calls). In this section, we present an alternative approach that significantly reduces the overall complexity of our Algorithm. To this end, we introduce a SAT-based encoding that allows to find a bounded consistent clustering of a given set of propositional formulas.

Let $\Phi = \{\phi_1, \ldots, \phi_n\}$ be a set of propositional formulas and k a positive integer. To define our encoding, we associate to each propositional variable p appearing in Φ a set of k fresh propositional variables, denoted $p^1, \ldots, p^k$. Then, for every formula $\phi_i \in \Phi$ and $j \in \{1, \ldots, k\}$, we use ϕ_i^j to denote the formula obtained from ϕ_i by replacing each propositional variable p with the fresh variable p^j. The formula ϕ_i^j is used to model the fact that ϕ_i is in the jth cluster.

The following formula expresses that each formula in Φ has to be true in at least one consistent cluster:

$$\bigwedge_{i=1}^{n} \left(\bigvee_{j=1}^{k} \phi_i^j\right) \tag{2}$$

One can easily see that (2) is satisfiable if and only if Φ can be partitioned in k consistent clusters. It is worth noting that in a model of (2) a formula can belong to more than one cluster. To obtain a bounded consistent clustering from a model m, we only have to consider for each formula $\phi_i \in \Phi$ a single positive integer j in the set $\{1 \leqslant j \leqslant k \mid m(\phi_i^j) = 1\}$. This problem can be avoided by reformulation. To this end, we associate to each formula ϕ_i in Φ a set of k fresh propositional variables, denoted $q_{\phi_i}^1, \ldots, q_{\phi_i}^k$. The variable $q_{\phi_i}^j$ is used to represent the fact that ϕ_i is in the jth cluster by using the following formula:

$$\bigwedge_{i=1}^{n} \left(\bigwedge_{j=1}^{k} q_{\phi_i}^j \Leftrightarrow \phi_i^j\right) \tag{3}$$

Then, to express that each formula in Φ belongs to exactly one consistent cluster, we use the following formula:

$$\bigwedge_{i=1}^{n} \left(\sum_{j=1}^{k} q_{\phi_i}^j = 1\right) \tag{4}$$

Our second SAT encoding of the bounded consistent clustering problem $\mathcal{P}(k,\Phi)$ is defined by the formula $\mathcal{P}_{SAT}(k,\Phi) = (3) \wedge (4)$. From a model m of $\mathcal{P}_{SAT}(k,\Phi)$, a clustering can be easily extracted. Indeed, if $m(q^j_{\phi_i}) = true$ then $\phi_i \in \mathcal{C}_j$ otherwise $\phi_i \notin \mathcal{C}_j$.

Definition 10. *Let $\Phi = \{\phi_1, \dots, \phi_n\}$. C is called a minimum consistent clustering of Φ if there is no consistent clustering $\mathcal{C}'$ of Φ such that $|\mathcal{C}'| < |\mathcal{C}|$.*

As we can observe, clustering propositional formulas can be done using Algorithm 1 by replacing the computation of the minimum hitting set with the computation of the minimum consistent clustering (Definition 10) using $\mathcal{P}_{SAT}(k,\Phi)$. Similarly to Algorithm 1, Properties 1, 2 and 3 holds.

6 Experimentation

In this section, we carried out an experimental evaluation of the performance of our divisive and agglomerative algorithms for the clustering of a set of propositional formulas. Our goal is to assess the feasibility and effectiveness of our proposed framework.

We performed our experiments on a machine with Intel Core2 Quad CPU of 2.66 GHz and 8G of RAM. Our first aim is to compare the performance of our divisive and agglomerative algorithms. To this end, We consider two datasets `splice`, and `german-credit`[1]. We consider each data set as a set of transactions, where each transaction is a formula (a set of models). Consequently, an item is assimilated to a model.

Figure 3 shows the performances of agglomerative (Algorithm 1) and divisive (Algorithm 1) methods on the problem of clustering transaction databases. First, our divisive algorithm outperforms the agglomerative algorithm on `splice` and `german-credit`. Nevertheless, as illustrated in Sect. 3.2, the agglomerative algorithm is unable to find a clustering all the time. This is the case on `splice` data, where such approach can not provide clustering answer when the number of desired clusters is less than 84.

To further investigate the expressiveness and the ability of our approach to scale, we enlarge our experiments of the previous problem by studying the clustering of a set of formulas resulting from a random-generated poll with 100 to 1000 participants where each participant is invited to report its preferences. The questions of the poll are organized in four levels. At the first level, the participant is invited to select its 3 preferred options among 5. According to the preferences of the participant, she/he is invited to select other preferences from the second level and so on until the last level (level 4). For illustration, assume that in the first level we consider a set S of courses (e.g. Artificial Intelligence, Data Mining, Databases, Networks and Web Programming). A student selects three courses from S (level 1). Then, for each selected course, she/he chooses chapters (level 2), and so on. The preferences of each participant are encoded

[1] https://dtai.cs.kuleuven.be/CP4IM/.

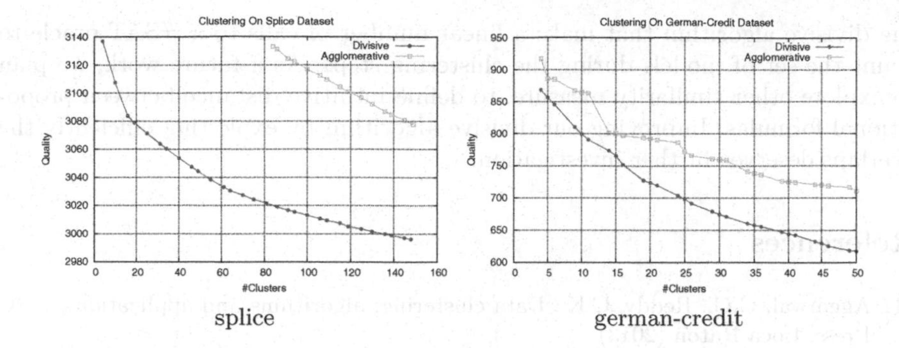

splice german-credit

Fig. 3. Model approach: Agglomerative vs Divisive

as a propositional formula (the resulting formulas have between 567 and 1813 models). Agglomerative approach is not considered since it can not guaranty to find a clustering solution if it exists.

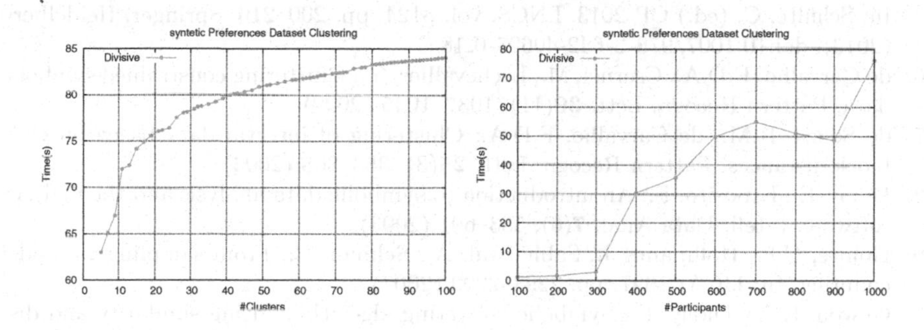

Fig. 4. Time vs #Clusters vs #Participants

The time needed to obtain a clustering, Fig. 4, does not exceed 100 s for all values of k. This shows that our approach scale well. Finally, we study the evolution of the time needed to find a clustering when the number of clusters is fixed to 20 and the number of participants is varied from 100 to 1000 (Fig. 4). Here again the time needed is reasonable, i.e., less than 100 s.

7 Conclusion et Perspectives

In this work we introduced the concept of consistent clustering propositional formulas. We show how well-known k-means, agglomerative and divisive algorithms can be adapted to this new framework. We then, propose two new solutions. The first one called model based, assume that the set of models of each formula are given. We then show how the hitting set notion is used to efficiently give a consistent clustering. In the second part, we propose an encoding into SAT of the divisive algorithm that make a linear number of calls to a #SAT oracle to count the set of models during the clustering steps. As a future work, we plan to explore other similarity measure, to define intuitive distance between propositional formulas. Improving our divisive algorithm by exploiting efficiently the overlaps deserves further investigation.

References

1. Aggarwal, C.C., Reddy, C.K.: Data clustering: algorithms and applications. CRC Press, Boca Raton (2013)
2. Berkhin, P.: A survey of clustering data mining techniques. In: Kogan, J., Nicholas, C.K., Teboulle, M. (eds.) Grouping Multidimensional Data - Recent Advances in Clustering, pp. 25–71. Springer, Heidelberg (2006)
3. Billard, L., Diday, E., Analysis, S.D.: Conceptual Statistics and Data Mining. Wiley, Hoboken (2012)
4. Bock, H.H.: Analysis of Symbolic Data: Exploratory Methods for Extracting Statistical Information from Complex Data. Springer, New York (2000)
5. Chakraborty, S., Meel, K.S., Vardi, M.Y.: A scalable approximate model counter. In: Schulte, C. (ed.) CP 2013. LNCS, vol. 8124, pp. 200–216. Springer, Heidelberg (2013). doi:10.1007/978-3-642-40627-0_18
6. de Carvalho, F.D.A., Csernel, M., Lechevallier, Y.: Clustering constrained symbolic data. Pattern Recogn. Lett. **30**(11), 1037–1045 (2009)
7. de Souza, R.M., de Carvalho, F.D.A.: Clustering of interval data based on city-block distances. Pattern Recogn. Lett. **25**(3), 353–365 (2004)
8. Diday, E., Esposito, F.: An introduction to symbolic data analysis and the SODAS software. Intell. Data Anal. **7**(6), 583–601 (2003)
9. Gomes, C.P., Hoffmann, J., Sabharwal, A., Selman, B.: From sampling to model counting. In: IJCAI 1997, pp. 2293–2299 (2007)
10. Gowda, K.C., Diday, E.: Symbolic clustering algorithms using similarity and dissimilarity measures. In: Diday, E., Lechevallier, Y., Schader, M., Bertrand, P., Burtschy, B. (eds.) New Approaches in Classification and Data Analysis, pp. 414–422. Springer, Heidelberg (1994)
11. Hotz, L., Felfernig, A., Stumptner, M., Ryabokon, A., Bagley, C., Wolter, K.: Configuration knowledge representation and reasoning. In: Knowledge-Based Configuration, chap. 6, pp. 41–72. Morgan Kaufmann (2014)
12. Jabbour, S., Lonlac, J., Sais, L., Salhi. Y.: Extending modern SAT solvers for models enumeration. In: IEEE-IRI 2014, pp. 803–810 (2014)
13. Jaccard, P.: The distribution of the flora of the alpine zone. New Phytol. **11**, 37–50 (1912)
14. Michalski, R.S.: Knowledge acquisition through conceptual clustering: a theoretical framework and an algorithm for partitioning data into conjunctive concepts. J. Policy Anal. Inf. Syst. **4**(3), 219–244 (1980)
15. Tversky, A.: Features of similarity. Psychol. Rev. **84**(4), 327–352 (1977)
16. Tversky, A.: Preference, Belief, and Similarity. The MIT Press, Cambridge (2003)

Deep Bayesian Matrix Factorization

Sotirios P. Chatzis(✉)

Department of Electrical Engineering, Computer Engineering and Informatics, Cyprus University of Technology, Limassol, Cyprus
sotirios.chatzis@cut.ac.cy

Abstract. Matrix factorization is a popular collaborative filtering technique, assuming that the matrix of ratings can be written as the inner product of two low-rank matrices, comprising latent features assigned to each user/item. Recently, several researchers have developed Bayesian treatments of matrix factorization, that infer posterior distributions over the postulated user and item latent features. As it has been shown, by allowing for taking uncertainty into account, such Bayesian inference approaches can better model sparse data, which are prevalent in real-world applications. In this paper, we consider replacing the inner product in the likelihood function of Bayesian matrix factorization with an arbitrary function that we learn from the data at the same time as we learn the latent feature posteriors; specifically, we parameterize the likelihood function using dense layer (DL) deep networks. In addition, to allow for addressing the cold-start problem, we also devise a model extension that takes into account item content, treated as side information. We provide extensive experimental evaluations on several real-world datasets; we show that our method completely outperforms state-of-the-art alternatives, without compromising computational efficiency.

1 Introduction

Latent variable models have been extensively used in collaborative filtering applications aimed at modeling user preferences. Their main rationale consists in the assumption that user preferences are determined by a small number of unobserved (latent) variables (factors) that characterize both users and items. Under this modeling framework, the rating function is typically considered to be a linear combination (inner product) of the latent factor vectors of each user and item pair. Such approaches are typically referred to as probabilistic matrix factorization (PMF) models [16].

The assumption of existing PMF-type models that the likelihood function mean is the *inner product* of the user and item latent variables classifies them into the family of linear latent variable (LLV) models. Unfortunately, LLV models cannot be considered realistic in most real-world data modeling scenarios. Traditionally, a solution towards the amelioration of these issues has been obtained by postulating mixtures of local LLV models. Recently though, a much more potent solution has been obtained in the context of deep learning techniques [7].

J. Kim et al. (Eds.): PAKDD 2017, Part II, LNAI 10235, pp. 453–464, 2017.
DOI: 10.1007/978-3-319-57529-2_36

Specifically, in the last couple of years, immense research interest has concentrated on the development of nonlinear latent variable models, where the inferred likelihood functions are parameterized via deep neural networks. This novel class of latent variable models is commonly referred to as deep generative models (DGMs) [4,5,13,14].

Inspired from these advances, in this work we propose a novel matrix factorization model, where the mean and the variance of the employed likelihood function are parameterized by means of dense layer (DL) networks. This way, our proposed model yields a nonlinear PMF scheme, which is expected to allow for a notable predictive performance improvement over existing linear models, in analogy to the existing work in DGMs. A difficulty with such a model formulation is that it naturally gives rise to a non-conjugate construction that prohibits application of conventional variational techniques to perform inference (e.g., [3]). We resolve this issue by resorting to a Monte-Carlo (MC) approximation of variational inference; to reduce the unacceptably high variance of MC estimators, we employ stochastic gradient variational Bayes (SGVB) [4].

In addition, to allow for addressing the cold-start problem, which is the problem of complete inadequacy of prior ratings to base recommendation generation upon, in this work we also consider a variant that allows for the utilization of side information. Indeed, many researchers have taken a similar path in the context of PMF-type algorithms. For instance, [12] fused side information into the PMF model by means of a simple linear regression term defined over the side information pertaining to the modeled users and items. More recently, [19] used a deep belief network (DBN) [15] extracting latent variable representations of music content to inform the postulated rating function of PMF. In the case of our proposed model, introduction of content-driven side information is effected by parameterizing the likelihood function via DL networks that are presented with both the latent feature vectors of each user and item, as well as *content-related latent variables* the posteriors over which are also parameterized via separate DL networks. We dub our approach deep Bayesian matrix factorization (DBMF).

The remainder of this paper is organized as follows: In Sect. 2 we provide a brief overview of the theoretical foundation of our work, i.e. Bayesian PMF (BPMF) models and the SGVB inference algorithm. In Sect. 3, we present the proposed DBMF model, as well as its content-driven extension. In Sect. 4, we perform the experimental evaluation of our approach, under a variety of scenarios that include ratings-only based inference, as well as inference that takes item content into account. Finally, in the concluding section of this paper, we summarize our contribution and discuss our results.

2 Theoretical Foundation

2.1 BPMF

Let us consider we are given a set of rankings $R = \{r_{ij}\}_{i,j}$ assigned by a set of users with indices $i \in \{1, \dots, N\}$ to a set of items with indices $j \in \{1, \dots, M\}$. PMF considers that the modeled users can be represented by means of the set of latent feature vectors $U = \{\boldsymbol{u}_i\}_{i=1}^N$, while the modeled items can be represented by means of the set of latent feature vectors $V = \{\boldsymbol{v}_j\}_{j=1}^M$. The observed ratings are considered to be conditionally independent given the latent feature vectors U and V, yielding [16]

$$p(R|U, V; \sigma^2) = \prod_{i=1}^{N} \prod_{j=1}^{M} [\mathcal{N}(r_{ij}|\boldsymbol{u}_i^T \boldsymbol{v}_j, \sigma^2)]^{I_{ij}} \tag{1}$$

where σ^2 is the variance of the model, and I_{ij} is an indicator variable equal to 1 if the ith user has rated the jth item, 0 otherwise.

To perform Bayesian inference for the PMF model (1), appropriate priors have to be imposed over the latent variable sets U and V. To this end, the BPMF method postulates [17]

$$p(\boldsymbol{u}_i|\boldsymbol{\mu}_U, \boldsymbol{\Lambda}_U) = \mathcal{N}(\boldsymbol{u}_i|\boldsymbol{\mu}_U, \boldsymbol{\Lambda}_U^{-1}) \tag{2}$$

$$p(\boldsymbol{v}_i|\boldsymbol{\mu}_V, \boldsymbol{\Lambda}_V) = \mathcal{N}(\boldsymbol{v}_i|\boldsymbol{\mu}_V, \boldsymbol{\Lambda}_V^{-1}) \tag{3}$$

where $\mathcal{N}(\cdot|\boldsymbol{\mu}, \boldsymbol{\Lambda}^{-1})$ is a multivariate Gaussian with mean $\boldsymbol{\mu}$ and covariance matrix $\boldsymbol{\Lambda}^{-1}$. In addition, to allow for a more comprehensive Bayesian inference treatment of the model, the prior hyperparameter sets $\{\boldsymbol{\mu}_U, \boldsymbol{\Lambda}_U\}$ and $\{\boldsymbol{\mu}_V, \boldsymbol{\Lambda}_V\}$ are also imposed Normal-Wishart hyper-priors. Inference for BPMF is performed by means of Gibbs sampling, as described in [17].

2.2 SGVB

Let us consider a dataset $X = \{\boldsymbol{x}_n\}_{n=1}^N$ consisting of N i.i.d. samples of some observed random variable $\boldsymbol{x}$. We assume that the observed random variable is generated by some random process, involving an unobserved continuous random variable $\boldsymbol{y}$. In this context, we introduce a conditional independence assumption for the observed variables $\boldsymbol{x}$ given the corresponding latent variables $\boldsymbol{y}$; we adopt the conditional likelihood function $p(\boldsymbol{x}|\boldsymbol{y}; \boldsymbol{\theta})$. To perform Bayesian inference for the postulated model, we impose some prior distribution $p(\boldsymbol{y}; \boldsymbol{\varphi})$. Under this formulation, the log-marginal likelihood of the model w.r.t. the dataset X yields

$$\log p(X) \geq \mathcal{L}(\boldsymbol{\theta}, \boldsymbol{\varphi}, \boldsymbol{\phi}|X) = \sum_{i=1}^{N} \Big\{ -\mathrm{KL}\big[q(\boldsymbol{y}_i; \boldsymbol{\phi})||p(\boldsymbol{y}_i; \boldsymbol{\varphi})\big] + \mathbb{E}_{q(\boldsymbol{y}_i;\boldsymbol{\phi})}[\log p(\boldsymbol{x}_i|\boldsymbol{y}_i; \boldsymbol{\theta})] \Big\} \tag{4}$$

where $\mathrm{KL}\big[q||p\big]$ is the KL divergence between the distribution $q(\cdot)$ and the distribution $p(\cdot)$, $q(\boldsymbol{y}; \boldsymbol{\phi})$ is the sought approximate (variational) posterior over the latent variable $\boldsymbol{y}$, while $\mathbb{E}_{q(\boldsymbol{y};\boldsymbol{\phi})}[\cdot]$ is the (posterior) expectation of a function w.r.t. the random variable $\boldsymbol{y}$, the distribution of which is taken to be the posterior $q(\boldsymbol{y}; \boldsymbol{\phi})$.

In case the form of the likelihood $p(\boldsymbol{x}|\boldsymbol{y}; \boldsymbol{\theta})$, prior $p(\boldsymbol{y}; \boldsymbol{\varphi})$, and posterior distribution, $q(\boldsymbol{y}; \boldsymbol{\phi})$, do not result in a conjugate overall model formulation, the analytical expression of $\mathbb{E}_{q(\boldsymbol{y}_i;\boldsymbol{\phi})}[\log p(\boldsymbol{x}_i|\boldsymbol{y}_i; \boldsymbol{\theta})]$ and its gradient become intractable. One could argue that this problem might be resolved by approximating this expectation using MC samples from the posterior $q(\boldsymbol{y};\boldsymbol{\phi})$. However, it is well-known that such an approximation would result in estimators with unacceptably high variance. SGVB is a recently proposed technique that resolves these issues by means of a smart reparameterization of the MC samples of $\boldsymbol{y} \sim q(\boldsymbol{y};\boldsymbol{\phi})$, obtained via an appropriate differentiable transformation $\boldsymbol{g}_{\boldsymbol{\phi}}(\boldsymbol{\epsilon})$ of an (auxiliary) random noise variable $\boldsymbol{\epsilon}$:

$$\boldsymbol{y} = \boldsymbol{g}_{\boldsymbol{\phi}}(\boldsymbol{\epsilon}) \quad \text{with} \quad \boldsymbol{\epsilon} \sim p(\boldsymbol{\epsilon}) \tag{5}$$

By adopting this reparameterization, SGVB allows for rewriting the variational lower bound (4) in the form

$$\mathcal{L}(\boldsymbol{\theta},\boldsymbol{\varphi},\boldsymbol{\phi}|X) = \sum_{i=1}^{N} \left\{ -\mathrm{KL}\big[q(\boldsymbol{y}_i;\boldsymbol{\phi})||p(\boldsymbol{y}_i;\boldsymbol{\varphi})\big] + \frac{1}{L}\sum_{l=1}^{L} \log p(\boldsymbol{x}_i|\boldsymbol{y}_i^{(l)};\boldsymbol{\theta}) \right\} \tag{6}$$

where L is the number of drawn samples. Hence, the key difference between SGVB and a naive MC estimator is that the drawn samples of $\boldsymbol{y}$, used to approximate the intractable posterior expectation $\mathbb{E}_{q(\boldsymbol{y}_i;\boldsymbol{\phi})}[\log p(\boldsymbol{x}_i|\boldsymbol{y}_i;\boldsymbol{\theta})]$, are now taken as *functions of the parameters* $\boldsymbol{\phi}$ *of the posterior* $q(\boldsymbol{y}_i;\boldsymbol{\phi})$ *that we seek to optimize*. As shown in [4], this formulation of the inference algorithm allows for yielding low variance estimators, under some mild conditions.

The derivatives $\nabla_{\boldsymbol{\theta},\boldsymbol{\varphi},\boldsymbol{\phi}}\mathcal{L}(\boldsymbol{\theta},\boldsymbol{\varphi},\boldsymbol{\phi}|X)$ can be used in conjunction with any stochastic optimization method to train the model. For instance, [4] suggest utilization of Adagrad, which constitutes a stochastic gradient descent algorithm with adaptive step-size [2]. As discussed in [4], Adagrad yields an efficient parameter optimization scheme for DGM-type models, with fast and proven convergence to a local optimum.

In the following, we will assume that the (conditional) likelihood function $p(\boldsymbol{x}|\boldsymbol{y};\boldsymbol{\theta})$ is a diagonal Gaussian distribution, the mean and (diagonal) covariance matrix of which are modeled as the outputs of DL networks with input $\boldsymbol{y}$ and parameters set $\boldsymbol{\theta}$. We also adopt the same assumptions for the sought variational posteriors; we consider an approximate Gaussian form, with mean $\boldsymbol{\mu}_i = \boldsymbol{\mu}(\boldsymbol{x}_i;\boldsymbol{\phi})$ and *diagonal* covariance matrix $\boldsymbol{\Sigma}_i = \boldsymbol{\Sigma}(\boldsymbol{x}_i;\boldsymbol{\phi})$. This way, the reparameterization trick (5) reduces to:

$$\boldsymbol{y}_i^{(l)} = \boldsymbol{g}_{\boldsymbol{\phi}}(\boldsymbol{\epsilon}_i^{(l)},\boldsymbol{x}_i) = \boldsymbol{\mu}_i + \boldsymbol{\Sigma}_i^{1/2}\boldsymbol{\epsilon}_i^{(l)} \tag{7}$$

where $\boldsymbol{\epsilon}_i^{(l)}$ is white random noise with unitary variance, i.e. $\boldsymbol{\epsilon}_i^{(l)} \sim \mathcal{N}(\boldsymbol{0},\boldsymbol{I})$.

3 Proposed Approach

3.1 The DBMF Model

Contrary to existing PMF-type approaches, DBMF postulates a conditional likelihood function for the ratings set R of the following form:

$$p(R|U,V;\sigma^2) = \prod_{i=1}^{N}\prod_{j=1}^{M} \left[\mathcal{N}\left(r_{ij}|\mu_{\boldsymbol{\theta}}(\boldsymbol{u}_i,\boldsymbol{v}_j),\sigma_{\boldsymbol{\theta}}^2(\boldsymbol{u}_i,\boldsymbol{v}_j)\right)\right]^{I_{ij}} \tag{8}$$

where I_{ij} is an indicator variable equal to 1 if the ith user has rated the jth item, 0 otherwise. In Eq. (8), both the mean $\mu_{\boldsymbol{\theta}}(\boldsymbol{u}_i, \boldsymbol{v}_j)$ and the variance $\sigma^2_{\boldsymbol{\theta}}(\boldsymbol{u}_i, \boldsymbol{v}_j)$ are taken as outputs of a DL network with parameters set $\boldsymbol{\theta}$, and inputs $(\boldsymbol{u}_i, \boldsymbol{v}_j)$. The network comprises one hidden layer, and is described by the following equations:

$$\mu_{\boldsymbol{\theta}}(\boldsymbol{u}_i, \boldsymbol{v}_j) = \boldsymbol{w}_\mu^T \boldsymbol{h}_{ij} + b_\mu \tag{9}$$

$$\sigma^2_{\boldsymbol{\theta}}(\boldsymbol{u}_i, \boldsymbol{v}_j) = \exp(\boldsymbol{w}_{\sigma^2}^T \boldsymbol{h}_{ij} + b_{\sigma^2}) \tag{10}$$

$$\boldsymbol{h}_{ij} \triangleq \boldsymbol{h}(\boldsymbol{u}_i, \boldsymbol{v}_j) = \gamma([\boldsymbol{W}_1 \boldsymbol{u}_i; \boldsymbol{W}_2 \boldsymbol{v}_j] + \boldsymbol{b}) \tag{11}$$

Here, $\gamma(\cdot)$ is a ReLU nonlinearity [9], $[\boldsymbol{\xi}_1; \boldsymbol{\xi}_2]$ is the concatenation of two vectors $\boldsymbol{\xi}_1$ and $\boldsymbol{\xi}_2$, $\boldsymbol{b} \in \mathbb{R}^{2K}$, $\boldsymbol{W}_1 \in \mathbb{R}^{K \times D}$, and $\boldsymbol{W}_2 \in \mathbb{R}^{K \times D}$ (hence, $\boldsymbol{\theta} = \{\boldsymbol{b}, \boldsymbol{W}_1, \boldsymbol{W}_2, \boldsymbol{w}_\mu, \boldsymbol{w}_{\sigma^2}, b_\mu, b_{\sigma^2}\}$).

To perform Bayesian inference for our model, we have to impose appropriate prior distributions over the user latent variables $\boldsymbol{u}_i$ and the item latent variables $\boldsymbol{v}_j$. Here, for simplicity, we consider Gaussian priors of the form:

$$p(\boldsymbol{u}_i | \boldsymbol{\mu}_U, \boldsymbol{\lambda}_U) = \mathcal{N}\left(\boldsymbol{u}_i | \boldsymbol{\mu}_U, \mathrm{diag}(\boldsymbol{\lambda}_U^{-1})\right) \tag{12}$$

$$p(\boldsymbol{v}_j | \boldsymbol{\mu}_V, \boldsymbol{\lambda}_V) = \mathcal{N}\left(\boldsymbol{v}_j | \boldsymbol{\mu}_V, \mathrm{diag}(\boldsymbol{\lambda}_V^{-1})\right) \tag{13}$$

where $\mathrm{diag}(\boldsymbol{\xi})$ denotes a square diagonal matrix with the elements of vector $\boldsymbol{\xi}$ on the main diagonal; the set $\boldsymbol{\varphi} = \{\boldsymbol{\mu}_U, \boldsymbol{\lambda}_U, \boldsymbol{\mu}_V, \boldsymbol{\lambda}_V\}$ of prior distribution (hyper-)parameters will be optimized as part of the model inference procedure. For convenience, we postulate corresponding posteriors of the same form, that read:

$$q(\boldsymbol{u}_i | \tilde{\boldsymbol{\mu}}_i, \tilde{\boldsymbol{\lambda}}_i) = \mathcal{N}\left(\boldsymbol{u}_i | \tilde{\boldsymbol{\mu}}_i, \mathrm{diag}(\tilde{\boldsymbol{\lambda}}_i^{-1})\right) \tag{14}$$

$$q(\boldsymbol{v}_j | \hat{\boldsymbol{\mu}}_j, \hat{\boldsymbol{\lambda}}_j) = \mathcal{N}\left(\boldsymbol{v}_j | \hat{\boldsymbol{\mu}}_j, \mathrm{diag}(\hat{\boldsymbol{\lambda}}_j^{-1})\right) \tag{15}$$

where the set $\boldsymbol{\phi} = \{\tilde{\boldsymbol{\mu}}_i, \tilde{\boldsymbol{\lambda}}_i, \hat{\boldsymbol{\mu}}_j, \hat{\boldsymbol{\lambda}}_j\}_{i,j}$ of posterior distribution (hyper-)parameters will be also optimized as part of the model inference procedure.

This concludes the formulation of our model. The variational lower bound expression of our model yields

$$\begin{aligned}\mathcal{L}(\boldsymbol{\theta}, \boldsymbol{\varphi}, \boldsymbol{\phi} | R) = &- \sum_{i=1}^{N} \mathrm{KL}\big[q(\boldsymbol{u}_i | \tilde{\boldsymbol{\mu}}_i, \tilde{\boldsymbol{\lambda}}_i) || p(\boldsymbol{u}_i | \boldsymbol{\mu}_U, \boldsymbol{\lambda}_U)\big] \\ &- \sum_{j=1}^{M} \mathrm{KL}\big[q(\boldsymbol{v}_j | \hat{\boldsymbol{\mu}}_i, \hat{\boldsymbol{\lambda}}_i) || p(\boldsymbol{v}_j | \boldsymbol{\mu}_V, \boldsymbol{\lambda}_V)\big] \\ + \sum_{l=1}^{L} \sum_{i=1}^{N} \sum_{j=1}^{M} \frac{I_{ij}}{L} &\log \mathcal{N}\left(r_{ij} | \mu_{\boldsymbol{\theta}}(\boldsymbol{u}_i^{(l)}, \boldsymbol{v}_j^{(l)}), \sigma^2_{\boldsymbol{\theta}}(\boldsymbol{u}_i^{(l)}, \boldsymbol{v}_j^{(l)})\right)\end{aligned} \tag{16}$$

where

$$\boldsymbol{u}_i^{(l)} = \tilde{\boldsymbol{\mu}}_i + \mathrm{diag}(\tilde{\boldsymbol{\lambda}}_i^{-1/2}) \epsilon_i^{(l)} \tag{17}$$

$$\boldsymbol{v}_j^{(l)} = \hat{\boldsymbol{\mu}}_j + \text{diag}(\hat{\boldsymbol{\lambda}}_j^{-1/2})\boldsymbol{\epsilon}_j^{(l)} \tag{18}$$

and

$$\boldsymbol{\epsilon}_i^{(l)}, \boldsymbol{\epsilon}_j^{(l)} \sim \mathcal{N}(\mathbf{0}, \boldsymbol{I}) \tag{19}$$

Optimization of the approximate variational lower bound (16) is performed via off-the-shelf stochastic optimization methods. Specifically, motivated by the discussions of Sect. 2.2, in this work we employ Adagrad.

3.2 Content-Driven DBMF

To allow for addressing the cold-start problem, we now consider a content-driven DBMF variant, capable of extracting useful high-level information from raw item content to inform recommendation generation. Let us consider that, for each available item $j \in \{1, \ldots, M\}$, our model is also presented with an additional, C-dimensional, raw content variable $\boldsymbol{x}_j \in \mathbb{R}^C$. To effectively integrate this information into the formulation of our model, we postulate the (conditional) likelihood function:

$$p(R|U, V, Z; \sigma^2) = \prod_{i,j=1}^{N,M} \mathcal{N}\left(r_{ij}|\mu_{\boldsymbol{\theta}}(\boldsymbol{u}_i, \boldsymbol{v}_j, \boldsymbol{z}_j), \sigma_{\boldsymbol{\theta}}^2(\boldsymbol{u}_i, \boldsymbol{v}_j, \boldsymbol{z}_j)\right)^{I_{ij}} \tag{20}$$

where the latent variable $\boldsymbol{z}_j \in \mathbb{R}^D$ is considered to encode a high-level, abstract representation of the raw content $\boldsymbol{x}_j$ of the item j, and will be inferred as part of the model inference procedure described next. In the conditional likelihood function (20), the mean $\mu_{\boldsymbol{\theta}}(\boldsymbol{u}_i, \boldsymbol{v}_j, \boldsymbol{z}_j)$ and the variance $\sigma_{\boldsymbol{\theta}}^2(\boldsymbol{u}_i, \boldsymbol{v}_j, \boldsymbol{z}_j)$ are taken as outputs of a DL network with parameters set $\boldsymbol{\theta}$, one hidden layer, and input $(\boldsymbol{u}_i, \boldsymbol{v}_j, \boldsymbol{z}_j)$:

$$\mu_{\boldsymbol{\theta}}(\boldsymbol{u}_i, \boldsymbol{v}_j, \boldsymbol{z}_j) = \boldsymbol{w}_\mu^T \boldsymbol{h}_{ij} + b_\mu \tag{21}$$

$$\sigma_{\boldsymbol{\theta}}^2(\boldsymbol{u}_i, \boldsymbol{v}_j, \boldsymbol{z}_j) = \exp(\boldsymbol{w}_{\sigma^2}^T \boldsymbol{h}_{ij} + b_{\sigma^2}) \tag{22}$$

$$\boldsymbol{h}_{ij} \triangleq \gamma([\boldsymbol{W}_1 \boldsymbol{u}_i; \boldsymbol{W}_2 \boldsymbol{v}_j; \boldsymbol{W}_3 \boldsymbol{z}_j] + \boldsymbol{b}) \tag{23}$$

whence $\boldsymbol{\theta} = \{\boldsymbol{b}, \boldsymbol{W}_1, \boldsymbol{W}_2, \boldsymbol{W}_3, \boldsymbol{w}_\mu, \boldsymbol{w}_{\sigma^2}, b_\mu, b_{\sigma^2}\}$.

Then, to perform Bayesian inference we have to define an appropriate prior distribution over the $\boldsymbol{z}_j$; we consider

$$p(\boldsymbol{z}_j|\boldsymbol{\mu}_V, \boldsymbol{\lambda}_V) = \mathcal{N}\left(\boldsymbol{z}_j|\boldsymbol{\mu}_Z, \text{diag}(\boldsymbol{\lambda}_Z^{-1})\right) \tag{24}$$

The resulting set $\boldsymbol{\varphi} = \{\boldsymbol{\mu}_U, \boldsymbol{\lambda}_U, \boldsymbol{\mu}_V, \boldsymbol{\lambda}_V, \boldsymbol{\mu}_Z, \boldsymbol{\lambda}_Z\}$ of prior distribution (hyper-)parameters will be optimized as part of the model inference procedure. Turning to the sought posteriors over the $\boldsymbol{z}_j$, we now assume:

$$q(\boldsymbol{z}_j|\bar{\boldsymbol{\mu}}(\boldsymbol{x}_j), \bar{\boldsymbol{\lambda}}(\boldsymbol{x}_j)) = \mathcal{N}\left(\boldsymbol{z}_j|\bar{\boldsymbol{\mu}}(\boldsymbol{x}_j), \text{diag}(\bar{\boldsymbol{\lambda}}^{-1}(\boldsymbol{x}_j))\right) \tag{25}$$

Here, the $\bar{\boldsymbol{\mu}}(\boldsymbol{x}_j)$ and $\bar{\boldsymbol{\lambda}}(\boldsymbol{x}_j)$ are parameterized via DL networks with one hidden layer, input $\boldsymbol{x}_j$, and described by the equations:

$$\bar{\boldsymbol{\mu}}(\boldsymbol{x}_j) = \boldsymbol{W}_{\bar{\mu}} \gamma(\boldsymbol{W}'_{\bar{\mu}} \boldsymbol{x}_j + \boldsymbol{b}'_{\bar{\mu}}) + \boldsymbol{b}_{\bar{\mu}} \tag{26}$$

$$\bar{\boldsymbol{\lambda}}(\boldsymbol{x}_j) = \exp(\boldsymbol{W}_{\bar{\boldsymbol{\lambda}}}\gamma(\boldsymbol{W}'_{\bar{\boldsymbol{\lambda}}}\boldsymbol{x}_j + \boldsymbol{b}'_{\bar{\boldsymbol{\lambda}}}) + \boldsymbol{b}_{\bar{\boldsymbol{\lambda}}}) \tag{27}$$

This assumption alleviates the need of extracting per-item parameters for the latent variable posteriors; instead, we compute a set of global variational parameters, valid for inference at both training and test time. Thus, we *amortize* the cost of inference by generalizing between the posterior estimates for the latent variables, $\boldsymbol{z}_j$, of all items through the parameters of the DL networks in Eqs. (26) and (27). Eventually, we yield $\boldsymbol{\phi} = \{\boldsymbol{W}_{\bar{\boldsymbol{\mu}}}, \boldsymbol{W}'_{\bar{\boldsymbol{\mu}}}, \boldsymbol{W}_{\bar{\boldsymbol{\lambda}}}, \boldsymbol{W}'_{\bar{\boldsymbol{\lambda}}}, \boldsymbol{b}_{\bar{\boldsymbol{\mu}}}, \boldsymbol{b}'_{\bar{\boldsymbol{\mu}}}, \boldsymbol{b}_{\bar{\boldsymbol{\lambda}}}, \boldsymbol{b}'_{\bar{\boldsymbol{\lambda}}}, \{\tilde{\boldsymbol{\mu}}_i, \tilde{\boldsymbol{\lambda}}_i, \hat{\boldsymbol{\mu}}_j, \hat{\boldsymbol{\lambda}}_j\}_{i,j=1}^{N,M}\}$; this set will be also optimized as part of the model inference procedure.

This concludes the formulation of the proposed *content-driven* variant of our DBMF model; its variational lower bound yields:

$$\begin{aligned}
\mathcal{L}(\boldsymbol{\theta}, \boldsymbol{\varphi}, \boldsymbol{\phi}|R) = &- \sum_{i=1}^{N} \mathrm{KL}\left[q(\boldsymbol{u}_i|\tilde{\boldsymbol{\mu}}_i, \tilde{\boldsymbol{\lambda}}_i)||p(\boldsymbol{u}_i|\boldsymbol{\mu}_U, \boldsymbol{\lambda}_U)\right] \\
&- \sum_{j=1}^{M} \mathrm{KL}\left[q(\boldsymbol{v}_j|\hat{\boldsymbol{\mu}}_i, \hat{\boldsymbol{\lambda}}_i)||p(\boldsymbol{v}_j|\boldsymbol{\mu}_V, \boldsymbol{\lambda}_V)\right] \\
&- \sum_{j=1}^{M} \mathrm{KL}\left[q(\boldsymbol{z}_j|\bar{\boldsymbol{\mu}}(\boldsymbol{x}_j), \bar{\boldsymbol{\lambda}}(\boldsymbol{x}_j))||p(\boldsymbol{v}_j|\boldsymbol{\mu}_Z, \boldsymbol{\lambda}_Z)\right] \\
&+ \sum_{l,i,j=1}^{L,N,M} \frac{I_{ij}}{L} \log \mathcal{N}\left(r_{ij}|\mu_{\boldsymbol{\theta}}(\boldsymbol{u}_i^{(l)}, \boldsymbol{v}_j^{(l)}, \boldsymbol{z}_j^{(l)}), \sigma_{\boldsymbol{\theta}}^2(\boldsymbol{u}_i^{(l)}, \boldsymbol{v}_j^{(l)}, \boldsymbol{z}_j^{(l)})\right)
\end{aligned} \tag{28}$$

where

$$\boldsymbol{z}_j^{(l)} = \bar{\boldsymbol{\mu}}(\boldsymbol{x}_j) + \mathrm{diag}(\bar{\boldsymbol{\lambda}}(\boldsymbol{x}_j)^{-1/2})\boldsymbol{\epsilon}_j^{(l)} \tag{29}$$

Optimization of the approximate variational lower bound (28) is again performed via an off-the-shelf stochastic optimization method, namely Adagrad.

Table 1. Obtained optimal DBMF model configurations in each experimental scenario.

Scenario	Latent space dimensionality	Hidden layer size
MovieLens 100K	50	100
MovieLens 1M	50	200
Content-based music recommendation	150	400

4 Experiments

Here, we evaluate the efficacy of our DBMF model and its content-driven variant considering a number of benchmark datasets. In all cases, we compare the performance of our approach to related PMF-type approaches with state-of-the-art performance. Our employed performance metric is the root mean square error (RMSE) of the generated predictions. Prediction generation for a user/item pair using our model is performed by drawing 100 MC posterior samples, and computing the corresponding average of the likelihood mean, $\mu_{\boldsymbol{\theta}}(\cdot)$. A similar MC sampling procedure is used by all PMF-type models.

Our source codes have been developed in Python, and made use of the Theano library[1] [1]. We run our experiments on an Intel Xeon 2.5 GHz Quad-Core server with 32 GB RAM and an NVIDIA Tesla K40 GPU accelerator. We use Adagrad to perform optimization of the variational lower bound. Following the suggestions in [4], we select the minibatch size in such a way that at least 1,000 mini-batches are available on each iteration. Similarly, following the suggestions of [2], the configuration of Adagrad, namely its global stepsize parameters, are chosen from the set $\{0.01, 0.02, 0.1\}$, based on the model performance on the training set in the first few iterations.

In all cases, the performance results reported below correspond to optimal configuration of the evaluated models. In the case of the DBMF model, this translates into *selection of the latent factor vector dimensionality, as well as of the hidden layer size of the DL networks* that parameterize the model posteriors and likelihood functions, such that we *maximize the obtained empirical performance.*

Table 2. MovieLens 100K: obtained RMSEs (mean and standard deviation) over the conducted 5 repetitions, for optimal model configuration.

Training data	99% of the whole	80% of the whole	50% of the whole
PMF	0.9164 ± 0.0261	0.9190 ± 0.0052	0.9506 ± 0.0024
Biased PMF	0.8923 ± 0.0150	0.9087 ± 0.0030	0.9337 ± 0.0020
SCMF	0.8891 ± 0.0146	0.9068 ± 0.0036	0.9331 ± 0.0021
BPMF	0.8807 ± 0.0139	0.8955 ± 0.0034	0.9203 ± 0.0021
DBMF	0.8720 ± 0.0134	0.8888 ± 0.0031	0.9142 ± 0.0020

Table 3. MovieLens 1M: obtained RMSEs (mean and standard deviation) over the conducted 5 repetitions, for optimal model configuration.

Training data	99% of the whole	80% of the whole	50% of the whole
PMF	0.8388 ± 0.0059	0.8512 ± 0.0017	0.8745 ± 0.0011
Biased PMF	0.8367 ± 0.0067	0.8493 ± 0.0020	0.8722 ± 0.0012
SCMF	0.8323 ± 0.0065	0.8465 ± 0.0018	0.8678 ± 0.0007
BPMF	0.8248 ± 0.0062	0.8339 ± 0.0015	0.8567 ± 0.0005
DBMF	0.8136 ± 0.0060	0.8209 ± 0.0012	0.8414 ± 0.0005

4.1 DBMF Evaluation

Here, we perform the experimental evaluation of the core DBMF model. We use two datasets commonly considered in the related literature, namely MovieLens

[1] http://deeplearning.net/software/theano/.

100K and MovieLens 1M. MovieLens 100K[2] contains 100,000 ratings of 1,682 movies provided by 943 users of the GroupLens website. MovieLens 1M[3] contains 1 million ratings of 6,040 users and 3,706 movies of the GroupLens website. In both cases, the ratings take in a set of 10 discrete values (1.0–5.0).

To obtain some comparative results, apart from our method we also evaluate in the same tasks some popular competitors, including PMF [16], Biased PMF [6], BPMF [17], and the recently proposed SCMF approach [18]. As is common in the literature, we evaluate our model and the considered competitors under scenarios where 99%, 80%, or 50% of the available ratings are used for training, and the rest for testing. The dataset is split in such a way that ensures the existence of ratings pertaining to all users and items in both the training set and the test set. To alleviate the effect of random selection of training and test data on the reported performance statistics, we run our experiments five times, with different data splits into a training and a test set each time.

We evaluate our model with the number of drawn samples set to $L = 20$; the used minibatches comprise 1,000 ratings each. These selections were obtained after trying multiple alternatives, with the goal of yielding the highest performance improvement with the least computational burden. The obtained results are provided in Tables 2 and 3; the DBMF model configuration that yields these outcomes is reported in Table 1. We observe that our approach yields a clear improvement over the competition. Application of the Student's-t test confirms the statistical significance of these performance differences.

4.2 Content-Driven DBMF Evaluation

To assess the efficacy and the performance of the proposed content-driven DBMF variant, we here consider the problem of content-based music recommendation. To this end, we use a publicly available dataset, namely the Echo Nest Taste Profile Subset [8], which is, to our knowledge, the largest publicly available music recommendation dataset. The original dataset comprises 1,019,318 users, 384,546 songs, and 48,373,586 listening histories. Using the Taste Profile Subset, we can determine the songs that a user has listened to. Thus, we assign a rating of 1 to each such user/item pair, and use them as the positive examples to train our model with. To generate an equal number of negative examples for each user, we employ the well-established User-Oriented Sampling method [10].

To obtain content-related side information, we follow the experimental setup of the recent related work presented in [19]. Specifically, we have been able to crawl preview audio clips with length of about half a minute from *7digital*[4], for 282,508 of the songs included in the used dataset. The so-obtained clips were first converted to WAV files with mono channel, 8 kHz sampling rate, and 16 bit depth. We then randomly sampled a 5-second continuous segment from each audio clip; directly using the half-minute clips becomes computationally

[2] http://grouplens.org/datasets/movielens/100k/.
[3] http://grouplens.org/datasets/movielens/1m/.
[4] http://7digital.com.

Table 4. Content-driven DBMF evaluation: statistics of the used datasets.

	# of users	# of songs	# of ratings
Total	100,000	282,508	28,258,926
Training set	100,000	262,508	18,382,954
Warm-start validation set	100,000	262,454	3,939,204
Warm-start test set	100,000	262,457	3,939,206
Cold-start validation set	99,963	10,000	1,025,654
Cold-start test set	99,933	10,000	971,908

inefficient, while segments shorter than 5 s may lose too much information. We next converted each 5-second segment into a 166×120 spectrogram (30 ms window, no overlap); this was further processed via principal component analysis (PCA), to transform the spectrograms into vectors. From these PCA-obtained dimensions, we retained only those that explain the (top) 95% of the variance of the data. These dimensions are finally normalized to have zero mean and unit variance, and are fed into our model as the raw content vectors $\boldsymbol{x}$.

We randomly split the so-obtained dataset into 5 disjoint sets: the training set, warm-start validation/test sets, and cold-start validation/test sets. All songs in the warm-start sets are also included in the training set. To simulate the new-song problem, songs in the cold-start validation/test sets do not exist in the training set. All users in both the cold-start and warm-start sets are also included in the training set. The overall statistics of the created datasets are provided in Table 4. To obtain some comparative results, apart from our method we also evaluate the side information-augmented BPMF approach presented in [12], hereafter referred to as BPMFSI, the state-of-the-art, DBN-driven, HLDBN approach of [19], and the recently proposed HBMFSI method [11]. We evaluate our model with the number of drawn samples set to just $L = 1$ sample; the used minibatches comprise 10,000 ratings each. We repeat our experiment five times, to account for the random selection of training and test data.

The obtained results are illustrated in Table 5; the DBMF model configuration that yields these outcomes is reported in Table 1. We observe that our approach performs better than the competition in all cases. Application of the Student's-t test confirms the statistical significance of these performance

Table 5. Content-based music recommendation: obtained RMSEs (mean and standard deviation) over the conducted 5 repetitions, for optimal model configuration.

Method	Warm-start valid.	Warm-start test	Cold-start valid.	Cold-start test
BPMFSI	0.31 (0.008)	0.31 (0.008)	0.50 (0.011)	0.50 (0.011)
HLDBN	0.32 (0.009)	0.33 (0.009)	0.48 (0.011)	0.48 (0.011)
HBMFSI	0.32 (0.009)	0.32 (0.009)	0.48 (0.010)	0.49 (0.010)
DBMF	0.26 (0.006)	0.26 (0.006)	-	-
Content-driven DBMF	0.26 (0.006)	0.26 (0.006)	0.39 (0.008)	0.39 (0.008)

differences. Interestingly enough, we also observe that the core formulation of our DBMF model performs comparably to the proposed content-driven DBMF variant in the case of the warm-start datasets. In our view, this is a strong indication that the capability of extracting high-level representations of item content is beneficial for the recommendation algorithm, especially in cold-start scenarios.

4.3 Computational Complexity

Let us now provide both some theoretical insights, as well as related empirical evidence, regarding the computational complexity of our method. We begin with prediction generation: DBMF computes predictions by merely feedforwarding some posterior samples through the DL network that parameterizes its likelihood mean, $\mu_{\boldsymbol{\theta}}(\cdot)$. This is similar to PMF-type alternatives, which also use posterior samples in the context of matrix multiplication computations. Hence, prediction generation using DBMF entails computations with complexity similar to state-of-the-art competitors. Our obtained empirical evidence has corroborated these theoretical intuitions; we have not observed statistically significant prediction time differences between our approach and the best performing PMF-type alternative, in any considered experimental scenario.

Turning to model training, DBMF entails more trainable parameters compared to PMF-type approaches, being a deep learning model. This is actually the only potential source of computational overhead for the training algorithm of our method compared to PMF-type alternatives. However, the devised stochastic gradient variational inference algorithm allows for DBMF training computations to be parallelized in a large-scale fashion. This way, one can compensate for the extra computational complexity (stemming from the training of a higher number of parameters), eventually ending up with similar computational times for model training. Indeed, the empirical evidence stemming from our experiments shows that, by proper utilization of a GPU accelerator, our model takes at most 1.34 times as long as the best performing PMF-type alternative to train. Since training is an offline procedure, this is a rather reasonable overhead given the obtained predictive performance improvement.

5 Conclusions

In this paper, we considered the problem of formulating matrix factorization models under a nonlinear perspective. Specifically, inspired from the literature of DGMs, we considered the case where the conventional inner product between the latent factor vectors is replaced with a nonlinear function, learned via DL deep networks. To allow for obtaining an efficient inference algorithm for our model, we resorted to variational Bayes. Specifically, to render variational Bayesian inference possible under the non-conjugate formulation of our model, we resorted to SGVB. This approach essentially consists in approximating the intractable posterior expectations of the variational lower bound of our model via MC samples, and reducing the resulting estimator variance by application of a reparameterization trick. To evaluate the efficacy of our approach, we performed extensive experimental evaluations using several benchmark datasets. As we observed, our approach yields a significant improvement over the competition in all the considered cases. As we also showed, both experimentally as well as through theoretical analysis, our approach manages to yield this improvement without any compromise of computational efficiency compared to the alternatives.

Acknowledgment. This work has been partially supported by the NVIDIA Corporation, as well as the EU H2020 NOTRE project (grant 692058).

References

1. Bastien, F., Lamblin, P., Pascanu, R., Bergstra, J., Goodfellow, I.J., Bergeron, A., Bouchard, N., Bengio, Y.: Theano: new features and speed improvements. In: Deep Learning and Unsupervised Feature Learning NIPS 2012 Workshop (2012)
2. Duchi, J., Hazan, E., Singer, Y.: Adaptive subgradient methods for online learning and stochastic optimization. JMLR **12**, 2121–2159 (2010)
3. Jaakkola, T., Jordan, M.: Bayesian parameter estimation via variational methods. Stat. Comput. **10**, 25–37 (2000)
4. Kingma, D., Welling, M.: Auto-encoding variational Bayes. In: Proceedings of the ICLR (2014)
5. Kingma, D.P., Rezende, D.J., Mohamed, S., Welling, M.: Semi-supervised learning with deep generative models. In: Proceedings of the NIPS (2014)
6. Koren, Y.: Factorization meets the neighborhood: a multifaceted collaborative filtering model. In: Proceedings of the ACM SIGKDD, pp. 426–434 (2008)
7. LeCun, Y., Bengio, Y., Hinton, G.: Deep learning. Nature **512**, 436–444 (2015)
8. McFee, B., Bertin-Mahieux, T., Ellis, D.P.W., Lanckriet, G.R.G.: The million song dataset challenge. In: Proceedings of the WWW, pp. 909–916 (2012)
9. Nair, V., Hinton, G.: Rectified linear units improve restricted Boltzmann machines. In: Proceedings of the ICML (2010)
10. Pan, R., Zhou, Y., Cao, B., Liu, N.N., Lukose, R., Scholz, M., Yang, Q.: One-class collaborative filtering. In: Proceedings of the ICDM, pp. 502–511 (2008)
11. Park, S., Kim, Y.D., Choi, S.: Hierarchical Bayesian matrix factorization with side information. In: Proceedings of the IJCAI (2013)
12. Porteous, I., Asuncion, A., Welling, M.: Bayesian matrix factorization with side information and Dirichlet process mixtures. In: Proceedings of the AAAI (2010)
13. Rezende, D.J., Mohamed, S., Wierstra, D.: Stochastic backpropagation and approximate inference in deep generative models. In: Proceedings of the ICML (2014)
14. Rezende, D.J., Mohamed, S.: Variational inference with normalizing flows. In: Proceedings of the ICML (2015)
15. Salakhutdinov, R., Murray, I.: On the quantitative analysis of deep belief networks. In: Proceedings of the ICML, pp. 872–879. ACM Press (2008)
16. Salakhutdinov, R., Mnih, A.: Probabilistic matrix factorization. In: Proceedings of the NIPS (2007)
17. Salakhutdinov, R., Mnih, A.: Bayesian probabilistic matrix factorization using Markov chain Monte Carlo. In: Proceedings of ICML (2008)
18. Shi, J., Wang, N., Xia, Y., Yeung, D.Y., King, I., Jia, J.: SCMF: sparse covariance matrix factorization for collaborative filtering. In: Proceedings of the IJCAI (2013)
19. Wang, X., Wang, Y.: Improving content-based and hybrid music recommendation using deep learning. In: Proceedings of the ACM MultiMedia (2014)

Dynamic, Stream Data Mining

Mining Competitive Pairs Hidden in Co-location Patterns from Dynamic Spatial Databases

Junli Lu[1,2], Lizhen Wang[1(✉)], Yuan Fang[1], and Momo Li[2]

[1] Department of Information Science and Engineering, Yunnan University, Kunming, China
ljl1982_3_6@126.com, lzhwang2005@126.com, fy1990825@163.com

[2] Department of Mathematics and Computer Science, Yunnan Minzu University, Kunming, China
limomo99@163.com

Abstract. Co-location pattern discovery is an important branch in spatial data mining. A spatial co-location pattern represents the subset of spatial features which are frequently located together in a geographic space. However, maybe some features in a co-location get benefit from the others, maybe they just accidentally share the similar environment, or maybe they competitively live in the same environment. In fact, many interesting knowledge have not been discovered. One of them is competitive pairs. Competitive relationship widely exists in nature and society and worthy to research. In this paper, competitive pairs hidden in co-locations are discovered from dynamic spatial databases. At first, competitive participation index which is the measure to show the competitive strength is calculated. After that, the concept of competitive pair is defined. For improving the course of mining competitive pairs, a series of pruning strategies are given. The methods make it possible to discover both competitive pairs and prevalent co-location patterns efficiently. The extensive experiments evaluate the proposed methods with "real + synthetic" data sets and the results show that competitive pairs are interesting and different from prevalent co-locations.

Keywords: Spatial data mining · Spatial co-location pattern · Competitive pair · Dynamic spatial database · Competitive co-location instance

1 Introduction

Advanced spatial database systems such as GPS have accumulated a growing number of large spatial databases. The spatial databases are considered to be full of valuable information bonanza. Spatial data mining can find previously unknown, interesting and potentially valuable patterns from spatial databases.

Spatial co-location pattern mining is one of the most important research directions in spatial data mining. A spatial co-location pattern represents the subset of spatial features which are frequently located together in a geographic space [1]. Yet there

J. Kim et al. (Eds.): PAKDD 2017, Part II, LNAI 10235, pp. 467–480, 2017.
DOI: 10.1007/978-3-319-57529-2_37

are many interesting knowledge, such as causal rules [2], symbiotic patterns, and competitive pairs, hidden in co-locations in spatial databases. They are even more powerful than co-locations. For example, many restaurants and ice cream shops both appear in the same community, university towns, or downtown streets. {Restaurants, ice cream shops} is a prevalent co-location, but ({restaurants}, {ice cream shops}) is not a competitive pair. Another example, invasive species and native species frequently live together, and the increase (decrease) of invasive/native species will induce the decrease (increase) of native/invasive species, so {invasive species, native species} is a prevalent co-location and ({invasive species}, {native species}) is a competitive pair.

In spatial database, features usually locate with their competitors together, which is exactly the reason competition is caused. Therefore, the competition is hard to recognize. Hoverer, competitive relationship widely exists in nature and society. E.g., the restaurants in one street, the similar services or products from different enterprises, invasive and native species, any adjacent two in biological chain, are competitive.

Most existing methods mined the competitors from the web data [3, 4, 14–17], and the representative method emphasized that "comparative entities often occur together more frequently" [3] and "the entity and its competitor usually have more co-occurrence than non-competitors" [4]. They used linguistic patterns to obtain competitive nature and co-occurrence between entity and its competitors to show competitive strength. This paper uses co-occurrence to measure competitive strength as well, and considers mutually exclusive change of competitors as competitive nature.

Scope: This paper focuses on mining competitive pairs hidden in co-locations giving a database with two time points. The following issues are outside the scope: (i) selecting two time points of database for getting more accurate result, (ii) determining thresholds for prevalent co-location and competitive pair mining, and (iii) indexing and query processing issues related to generate competitive co-location instances.

This paper mines competitive pairs hidden in co-locations from dynamic spatial databases. In comparison with previous works, three main contributions are made.

Firstly, competitive nature and competitive strength are expressed on dynamic spatial databases, competitive participation index which measures the competitive strength is calculated, and the formal definition of competitive pair is given.

Secondly, the algorithm with three pruning strategies for mining competitive pairs is proposed, analysis on the power of pruning strategies is conducted.

Finally, the experiments show that the proposed algorithms can efficiently discover competitive pairs and compare them with traditional prevalent co-locations.

The rest of paper is organized as follows. Related works are introduced in Sect. 2. In Sect. 3, we give some concepts and pruning lemmas of mining competitive pairs. Section 4 describes our algorithm. Section 5 is experimental evaluation. In Sect. 6 we conclude the paper and suggest future works.

2 Related Works

Co-location mining is a rising and promising field in spatial data mining. Huang and Shekhar et al. [1] firstly proposed a general framework for co-location mining and defined the minimum participation index to measure the prevalence of a co-location. It is a like Apriori [5] method. After that, many researchers did lots of works on improving the efficiency [6–8], fitting for different data types [9, 10] and for special cases [11–13]. Partial join [6] and joinless [7] avoid the expensive join operation [1]. A compact format, prefix-tree, is used in [8] to store star neighborhoods and help to prune candidates. Spatial co-locations for fuzzy and uncertain objects are studied in [9] and [10] respectively. The maximum participation index is proposed [11] for the co-locations with rare features, and it is improved by the weighted participation index proposed in [12]. A framework for mining regional co-locations is proposed in [13].

The above approaches mainly focused on discovery of the prevalent co-locations not competitive pairs. Most existing methods mining the competitors are focused on web data. Li and Bao et al. [3] mined competitors for a given entity from the web, and further mined the competitive fields and mined competitive evidence [4]. Both of research ranked the competitors by considering the co-occurrence between the entity and its competitor. Sun et al. [14] studied the comparative web search problem, in which the user inputs a set of entities and the system tries to find relevant and comparative information from the web. Yang et al. [15] used two data sources: text documents and social networks to mine competitive relationships by learning across heterogeneous networks for avoiding biased aspects in a single network. A formal definition of competitiveness between products sold on B2C Web sites was proposed until 2012 [16]. The competitiveness between two items in this paper is based on whether (competitive nature) they compete for the attention and business of the same group of users, and to what extent (competitive strength). However, it is still focused on web data. Ruan et al. [17] proposed a novel unsupervised approach to identify competitors from prospectuses. It considered the linguistic patterns [3, 4] cannot capture competitors that are expressed in different ways and used heuristic rules to identify competitors, and the competitive strength is still measured by the co-occurrence.

We have seen that the competitor mining from web has been extensively studied. However, the research on competitive pair mining from spatial dataset has not yet been investigated. With the explosive growth of spatial data and widespread use of spatial database, the competitive relationships from spatial data are more and more interesting. This paper mines the competitive pairs in dynamic spatial databases.

3 Concepts and Lemmas

3.1 Prevalent Co-location

Let $F = \{f_1, \ldots, f_m\}$ be a set of spatial features, and $S = \{o_1, \ldots, o_n\}$ be a set of their objects with geographic location. When the Euclidean metric is used for the neighbor relationship R, two objects o_i and o_j are neighbors of each other if the Euclidean distance between them is not greater than a neighbor distance threshold d. A **co-location** c is a set of features, $c \subseteq F$. The **size** of c is the number of features in c. **The co-location instance** I of c is defined as a set of objects, $I \subseteq S$, if (1) I contains instances of all the features in c and no proper subset of I does so, and (2) the objects in I form a clique relation under R. **The table instance** of c, $T(c)$, is the collection of all co-location instances of c.

The prevalence strength of a co-location $c = \{f_1, \ldots, f_k\}$ is often measured by **the participation index** [2] $(PI(c)), PI(c) = \min_{f_i \in c}\{PR(c, f_i)\}$, where $1 \leq i \leq k$. $PR(c, f_i)$ is **the participation ratio** of feature f_i in c, $PR(c, f_i) = |\pi_{f_i}(T(c))| / |T(\{f_i\})|$, where π is the relational projection operation with duplication elimination. If $PI(c)$ is greater than a given minimum prevalence threshold *min_prev*, we say c is a prevalent co-location. *min_prev* is 0.5 in all the examples of this paper.

Example 1. In Fig. 1, The feature set $F = \{$A, B, C$\}$, A, B and C have 5, 8 and 6 objects. A.1 is the first object of A. Two objects are connected with a line if they are neighbors. For co-location $c = \{$B, C$\}$, $T(c) = \{\{$B.1, C.4$\}$, $\{$B.3, C.6$\}$, $\{$B.4, C.5$\}$,. $\{$B.5, C.6$\}$, $\{$B.6, C.3$\}$, $\{$B.7, C.4$\}\}$. $PR(c$, B) = 6/8 since there are 6 different objects of B in $T(c)$. Similarly, $PR(c$, C) = 4/6. $PI(c) = min\{6/8, 4/6\} = 2/3 > min\_prev$, so c is prevalent.

3.2 Competitive Pair

In this subsection, we will give some definitions to introduce the formal definition of competitive pair.

Definition 1 (Subset pair). For a k-size co-location c_k, $c_k = X \cup Y, X \cap Y = \emptyset$, then we call ($X$, Y) is a **subset pair** of c_k.

Example 2. For a co-location $c = \{$A, B, C$\}$ in Fig. 1, ($\{$A$\}$, $\{$B, C$\}$), ($\{$A, C$\}$, B) and ($\{$A, B$\}$, $\{$C$\}$) are its subset pairs.

Definition 2 (Competitive co-location instance). For a k-size co-location c_k, its competitive co-location instance, $CCI(c_k)$ meets (1) $CCI(c_k)$ is a co-location instance of c_k, and (2) $CCI(c_k)$ is made of increased objects and decreased objects.

Example 3. Figure 2 is an updated dataset with data's increase and decrease compared to the original dataset in Fig. 1. The increased and decreased objects are marked by "*" and "+". For co-location $c = \{$A, B, C$\}$, the *CCI*s are $\{\{$A.5*, B.9$^+$, C.7$^+\}$, $\{$A.6$^+$, B.6*, C.3$^*\}$, $\{$A.8$^+$, B.3*, C.6$^*\}$, $\{$A.10$^+$, B.2*, C.8$^+\}\}$. However, $\{$A.10$^+$, B.2*, C.8$^+\}$ will not support the competition of $\{$A$\}$ and $\{$B, C$\}$. Therefore, for a subset pair (X, Y), we should use competitive pair co-location instances to support their competition.

Definition 3 (Competitive pair co-location instance). For a subset pair (X, Y) of k-size co-location c_k, the competitive pair co-location instance $CPCI(c_k, (X, Y))$ meets (1) $CPCI(c_k, (X, Y))$ is a co-location instance of c_k, and (2) the objects of X are increased (decreased), and the objects of Y are decreased (increased).

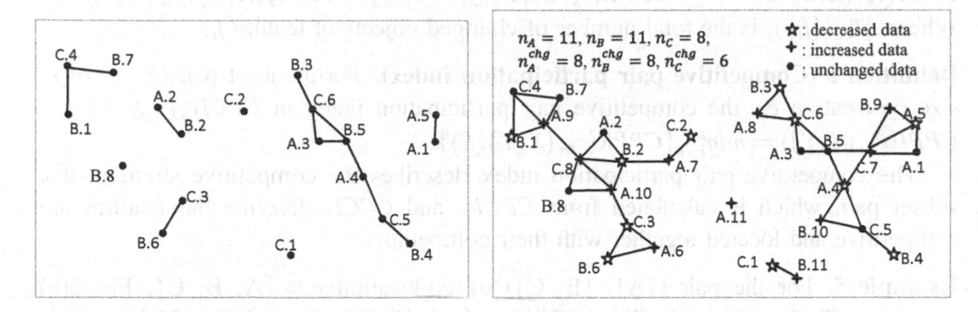

Fig. 1. An example of spatial dataset **Fig. 2.** An updated dataset compared to Fig. 1

For a co-location c, its *CCI*s can be got from its competitive star instances, *CSI*(c)s, by checking the cliqueness and deleting the ones made of all increased/decreased objects. The ***CSI*(*c*)s** are obtained by collecting the competitive star neighbors (*CSN*). ***CSN***($f_i.j$) is a set of changed objects which have neighbor relationships under R with center object $f_i.j$, and their features should be larger than $f_i.j$'s feature. The computing course of *CPCI*s of co-locations in Fig. 2 is shown in Fig. 3.

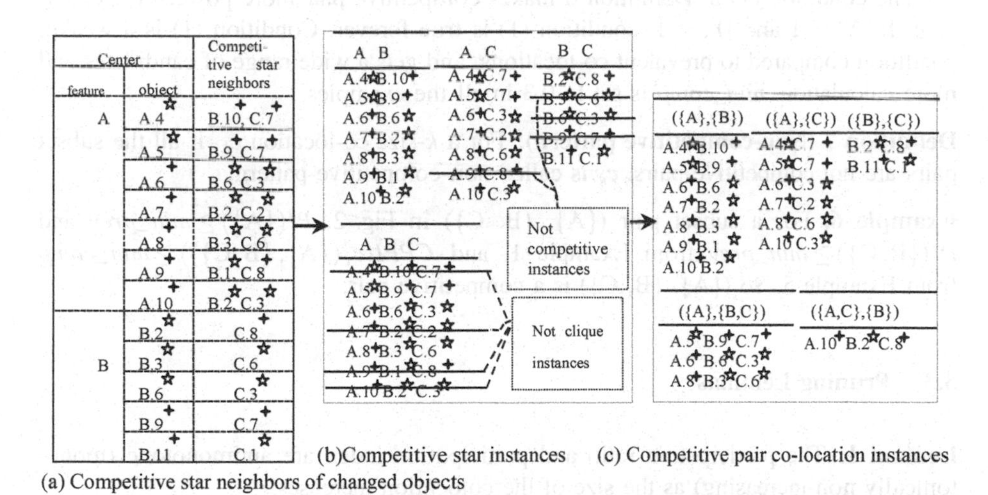

(a) Competitive star neighbors of changed objects (b)Competitive star instances (c) Competitive pair co-location instances

Fig. 3. Computing course of competitive pair co-location instances of co-locations in Fig. 2

Example 4. In Fig. 3(b), the *CSI* {A.9, C.8} is not a *CCI* of co-location {A, C}, since both A.9 and C.8 are increased objects. The *CSI* {A.7, B.2, C.2} is not a *CCI* of co-location {A, B, C} since they cannot form a clique.

Definition 4 (Competitive pair participation ratio). For a subset pair (*X*, *Y*) of *k*-size co-location c_k $(f_j \in c_k)(1 \leq j \leq k)$, the competitive pair participation ratio of f_i in $CPCI(c_k,(X,Y))$s is $CPPR(c_k,(X,Y),f_i) = |\pi_{f_i}(CPCI(c_k,(X,Y)))|/|T_{ch}(\{f_i\})|$. Where, $|T_{ch}(\{f_i\})|$ is the total number of changed objects of feature f_i.

Definition 5 (Competitive pair participation index). For a subset pair (*X*, *Y*) of *k*-size co-location c_k, the competitive pair participation index in $CPCI(c_k,(X,Y))$s is $CPPI(c_k,(X,Y)) = min_{i=1}^{k}\{CPPR(c_k,(X,Y),f_i)\}$.

The competitive pair participation index describes the competitive strength of a subset pair, which is calculated from *CPCI*s, and *CPCI*s describe the features are competitive and located together with their competitors.

Example 5. For the pair ({A}, {B, C}) of co-location $c = \{A, B, C\}$, Fig. 2(b) shows its $CPCI(c,(\{A\},\{B,C\}))$. $CPPR(c,(\{A\},\{B,C\}),A) = 3/8$. $CPPR(c,(\{A\},\{B,C\}),B) = 3/8$, $CPPR(c,(\{A\},\{B,C\}),C) = 3/6$, so $CPPI(c,(\{A\},\{B,C\})) = 3/8$.

Definition 6 (Competitive pair). Given a minimum prevalence threshold (*min_prev*), a minimum competition threshold (*min_comp*), a subset pair (*X*, *Y*) of *k*-size co-location c_k, is a competitive pair if it meets the following conditions.

(1) $PI(X) \geq min\_prev$ and $PI(Y) \geq min\_prev$ in original database,
(2) $CPPI(c_k,(X,Y)) \geq min\_comp$.

The condition (1) in Definition 6 makes competitive pair more powerful and reliable. If $|X| = 1$ and $|Y| = 1$, condition (1) is true forever. Condition (1) is a weaker condition compared to prevalent co-locations, and gets a wide range of candidates and more calculation. *min_comp* is set to 0.3 in all the examples.

Definition 7 (Non-competitive pattern). For a *k*-size co-location c_k, if all the subset pairs are not competitive pairs, c_k is called non-competitive pattern.

Example 6. For a subset pair ({A}, {B, C}) in Fig. 2, $PI(\{A\}) \geq min\_prev$ and $PI(\{B,C\}) \geq min\_prev$ from Example 1, and $CPPI(c,(\{A\},\{B,C\})) \geq min\_comp$ from Example 5. So ({A}, {B, C}) is a competitive pair.

3.3 Pruning Lemmas

Lemma 1. The participation ratio and participation index are antimonotone (monotonically non-increasing) as the size of the colocation increases.

Lemma 1 is proposed and proved in literature [1], which ensures that the participation index can be used to effectively prune the search space of co-location mining.

Lemma 2. A k-size co-location has $\frac{2^k-1}{2}$ subset pairs.

Proof. If k is an odd number, the number of the pairs is $C_k^1 + C_k^2 + \cdots + C_k^{\frac{k-1}{2}}$. Due to $C_k^1 + C_k^2 + \cdots + C_k^{\frac{k-1}{2}} + C_k^{\frac{k+1}{2}} + \cdots + C_k^{k-1} = 2^k - 1$, and $C_k^1 + C_k^2 + \cdots + C_k^{\frac{k-1}{2}} = C_k^{\frac{k+1}{2}} + \cdots + C_k^{k-1}$, so $C_k^1 + C_k^2 + \cdots + C_k^{\frac{k-1}{2}} = \frac{2^k-1}{2}$. If k is an even number, the number of the pairs is $C_k^1 + C_k^2 + \cdots + C_k^{\frac{k}{2}-1} + \frac{1}{2}C_k^{\frac{k}{2}}$. Due to $C_k^1 + C_k^2 + \cdots + C_k^{\frac{k}{2}-1} + C_k^{\frac{k}{2}} + C_k^{\frac{k}{2}+1} + \cdots + C_k^{k-1} = 2^k - 1$ and $C_k^1 + C_k^2 + \cdots + C_k^{\frac{k}{2}-1} = C_k^{\frac{k}{2}+1} + \cdots + C_k^{k-1}$, so $2\left(C_k^1 + C_k^2 + \cdots + C_k^{\frac{k}{2}-1}\right) + C_k^{\frac{k}{2}} = 2^k - 1$, $C_k^1 + C_k^2 + \cdots + C_k^{\frac{k}{2}-1} + \frac{1}{2}C_k^{\frac{k}{2}} = \frac{2^k-1}{2}$.

A k-size co-location contains $\frac{2^k-1}{2}$ subset pairs. It is time consuming to judge all the subset pairs, the lemmas below help to prune them.

Lemma 3. If a k-size co-location c_k is not prevalent, the pairs whose one part contains c_k are not a competitive pairs.

Proof. It is obvious according to Definition 6 and Lemma 1.

Example 7. In Fig. 1, co-location $c = \{\text{A, B}\}$ is not prevalent, since $T(c) = \{\{\text{A.2,B.2}\}, \{\text{A.3,B.5}\}, \{\text{A.4,B.5}\}\}$, and $PI(c) = min\{3/5, 2/8\} = 1/4$. So the pair ({A, B}, {C}) in Fig. 2 is not competitive and can be pruned.

Lemma 4. The competitive pair participation ratio and competitive pair participation index are anti-monotone as the size of the colocation increases.

Proof. Suppose (X, Y) is a subset pair of c_k, and $(X, Y \cup \{p\})$ and $(X \cup \{p\}, Y)$ are two subset pairs of c_{k+1}. For $f_i \in c_k$ and $f_i \in c_{k+1}$, suppose a spatial object e of f_i is included in $CPCI(c_{k+1}, (X, Y \cup \{p\}))$s or $CPCI(c_{k+1}, (X \cup \{p\}, Y))$s, then e must be included in $CPPI(c_k, (X, Y))$s. The opposite is not true. Therefore, the competitive pair participation ratio is monotonically non-increasing.

$$\begin{aligned} CPPI(c_{k+1}, (X, Y \cup \{p\})) &= min_{i=1}^{k+1}\{CPPR(c_{k+1}, (X, Y \cup \{p\}), f_i)\} \\ &\leq min_{i=1}^{k}\{CPPI(c_k \cup \{p\}, (X, Y \cup \{p\}), f_i)\} \\ &\leq min_{i=1}^{k}\{CPPR(c_k, (\{X, Y\}), f_i)\} = CPPI(c_k, (X, Y)) \end{aligned}$$

$CPPI(c_{k+1}, (X \cup \{p\}, Y))$ is similar. Therefore, $CPPI$ is anti-monotone.

Lemma 5. For a subset pair (X, Y) of k-size co-location c_k, if (X, Y) is not competitive, then for $P, Q, X \subset P$, $Y \subset Q$, (P, Y), (X, Q) and (P, Q) are not competitive pairs.

Proof. If (X, Y) does not meet condition (1) in Definition 6, then (P, Y), (X, Q) and (P, Q) do not meet condition (1) according to Lemma 1. If (X, Y) does not meet condition (2) in Definition 6, then (P, Y), (X, Q) and (P, Q) do not meet condition (2) either according to Lemma 4.

Example 8. In Fig. 2, ({B}, {C}) is not a competitive pair since it does not meet condition (2), so the pairs ({A, B}, {C}) and ({B}, {A, C}) must be not competitive.

Lemma 5 can efficiently prune many non-competitive pairs. However, Lifting up to co-location level to prune non-competitive pattern is much more helpful.

Lemma 6. For a k-size co-location $c_k = \{f_1,\ldots,f_k\}$, ① if only one $(k-1)$-size subset of c_k is a non-competitive pattern, then just one pair cannot be pruned in all the pairs of co-location c_k ② if at least two $(k-1)$-size subsets of c_k are non-competitive patterns, co-location c_k is a non-competitive pattern.

Proof. ① Suppose $c_{k-1} = \{f_1,\ldots,f_{k-1}\}$ is a non-competitive pattern, only the pair $(\{f_1,\cdots,f_{k-1}\},\{f_k\})$ in c_k cannot be pruned. ② Suppose any two subsets $c_{k-1} = \{f_1,\cdots,f_{i-1},f_{i+1},\cdots,f_k\}$ and $c'_{k-1} = \{f_1,\cdots,f_{j-1},f_{j+1},\cdots,f_k\}$ $(j \neq i)$ of c_k are non-competitive patterns. Only the pair $(\{f_1,\cdots,f_{i-1},f_{i+1},\cdots,f_k\},\{f_i\})$ of c_k cannot be pruned from c_{k-1} according to ①. However, it can be pruned by the non-competitive pair $(\{f_1,\cdots,f_{i-1},f_{i+1},\cdots,f_{j-1},f_{j+1},\cdots,f_k\},\{f_i\})$ of c'_{k-1}. So c_k is non-competitive.

Example 9. In Fig. 2, co-location $c = \{B,C\}$ is a non-competitive pattern since $CPPI(c,(\{B\},\{C\})) = min\{2/8,2/6\} = 1/4 < min\_comp$. For co-location {A, B, C}, only one subset {B, C} is non-competitive pattern, so only one pair, ({A}, {B, C}), of {A, B, C} cannot be pruned. It is truly a competitive pair from Example 6.

3.4 Pruning Power of the Pruning Lemmas

Suppose feature set $F = \{f_1,\ldots,f_m\}$, $|F| = m$. There are C_m^k k-size co-locations and $(2^k - 1)/2$ pairs for each co-location. Let $c_k = \{f_1,\ldots,f_k\}$ be a co-location candidate ① If c_k is not prevalent, it can prune $m-k$ pairs of $(k + 1)$-size co-locations according to Lemma 3. Suppose there are n_k k-size prevalent and $C_m^k - n_k$ not prevalent co-locations. $(C_m^k - n_k) * (m - k)$ pairs can be pruned totally. ② If a subset pair (X,Y) of c_k is not competitive, it can prune $2(m - k)$ pairs of $(k + 1)$-size co-locations according to Lemma 5. Suppose there are $\sum_{i=1}^{C_m^k} n_{ncom}^i$ non-competitive pairs in all k-size co-locations, n_{ncom}^i is the number of non-competitive pairs in the i-th co-location. There are totally $\sum_{i=1}^{C_m^k} 2 * (m - k) * n_{ncom}^i$ non-competitive pairs can be pruned. ③ Lemma 6 is in fact based on Lemmas 3 and 5, and cannot prune any pairs. The pairs or co-locations seem to be pruned by Lemma 6 have been pruned by Lemmas 3 or 5 in the last loop. Lemma 6 can help to terminate the algorithm in advance.

The non-competitive pairs pruned by Lemmas 3 and 5 will overlapped partly, the accurate number of overlapped pairs is hard to compute. E.g., the pair ({A, B}, {C}) is repeatedly pruned by ({A}, {B}) in Example 7 and ({B}, {C}) in Example 8.

4 Mining Competitive Pairs

In this section, we will discuss the algorithm of mining competitive pairs.

Algorithm: AMCP
Input: the original dataset S_{ori}, the updated dataset S_{upd}, distance threshold d, minimum prevalent threshold *min_prev*, minimum competitive threshold *min_comp*
Output: the set of competitive pairs (*CP*)
Variables: *k*: co-location size. C_k: a set of *k*-size candidate co-locations. CP_k: a set of competitive pairs w.r.t *k*-size co-locations. *CSN*: competitive star neighbors of changed objects. $CSI(c_k,(X,Y))$: competitive star instances of a pair (*X*, *Y*) in c_k.

```
(1)  CSN=gen_com_neib(S_ori, S_upd, d);
(2)  C_1=F; flag=0; PI(C_1) = 1 ≥ min_prev
(3)  for (k=2; flag<|C_{k-1}|; k++)
(4)      flag=0;
(5)      C_k=get_k-size_candi_co-location (C_{k-1});
(6)      For a k-size co-location c_k ∈ C_k,
(7)        If (k>2)
(8)          If (at least two subsets of c_k are non-competitive)
(9)              c_k is pruned; flag++; continue;
(10)           Else if (only one subset of c_k is non-competitive)
(11)               Only one pair cannot be pruned;
(12)       For a subset pair (X, Y) in the co-location c_k,
(13)           If (PI(X) ≥ min_prev and PI(Y) ≥ min_prev)
(14)               CSI(c_k,(X,Y))=gen_com_str_ins(c_k, CSN);
(15)               CPCI(c_k,(X,Y))=gen_spe_com_ins (c_k, CSI(c_k,(X,Y)));
(16)               CPPI(c_k,(X,Y))=compt_com_part_idx(CPCI(c_k,(X,Y)));
(17)               If (CPPI(c_k,(X,Y)) ≥ min_comp)
(18)                   CP_k= CP_k ∪ {(X, Y)};
(19)                   Computing PI(c_k);
(20)               Else
(21)                   Notated the pairs (X, Q), (P, Y) and (P, Q)
                        as visited and non-competitive, X ⊂ P, Y ⊂ Q;
(22)           Else if ((PI(X) < min_prev or PI(Y) < min_prev)
(23)               Notated the pair whose one part contains X or Y
                   as non-competitive;
(24)     If all the pairs w.r.t c_k are non-competitive, flag=flag+1;
```

Step (3)–Step (24) mines competitive pairs for *k*-size co-locations. Step (6)–Step (24) recognizes competitive pairs for a *k*-size co-location c_k. Step (7)–Step (11) prunes the candidate co-locations according to Lemma 6. Step (12)–Step (23) examines whether a subset pair (*X*, *Y*) is competitive. If it meets condition (1) in Definition 6 (Step (13)), computing $CPPI(c_k,(X,Y))$ (Step (14)–Step (16)), and judging whether it meets condition (2) in Definition 6 (Step (17)). If the pair does, putting it into the result (Step 18), and computing c_k's prevalence (Step 19). Or else, if (*X*, *Y*) does not meet condition (2), pruning some pairs according to Lemma 5 (Step (21)). If (*X*, *Y*) does not meet condition (1), pruning some pairs according to Lemma 3 (Step (23)).

5 Experimental Evaluation

Existing methods are mainly focused on mining competitors from web data, not competitive pairs from spatial data. It is hard to compare their efficiency, so we design a set of experiments to test the performance of pruning strategies. We get rid of Lemmas 3, 5 and 6 from AMCP to form basic algorithm (BA), get rid of Lemmas 5 and 6 to form BA with Lemma 3, and just get rid of Lemma 6 to form BA with Lemmas 3 and 5. The performance of 4 algorithms is compared on synthetic databases. Furthermore, competitive pairs and prevalent co-locations are compared on real databases.

5.1 Datasets

We use series of synthetic datasets and two real datasets in our experiments. All the algorithms are implemented using C# under Windows 8 and run on a normal PC with AMD A10 2.5 GHz CPU and 4 GB of memory.

We use two real datasets and series of synthetic datasets in our experiments. Real-1 contains 26 features and only 335 objects in a 10000 m × 30000 m area, which is from the rare plant data of the "Three Parallel Rivers of Yunnan Protected Areas". Real-2 is from the plant data of "Gong Shan", which owns 13349 objects and 14 features in a 1000 m × 1000 m area. Each object is depicted as <object-id, spatial feature, location>. Each plant type is a spatial feature, and each plant instance is an object.

5.2 On Synthetic Databases

This subsection examines the efficiency of proposed algorithms on series of synthetic datasets with several workloads, i.e., different thresholds *d*, *min_prev*, and *min_comp*, the size of original database $|S_{ori}|$, number of spatial features $|F|$, and increased size of changed objects $|S_{chg}|$. The results are shown in Figs. 4, 5, 6 and 7.

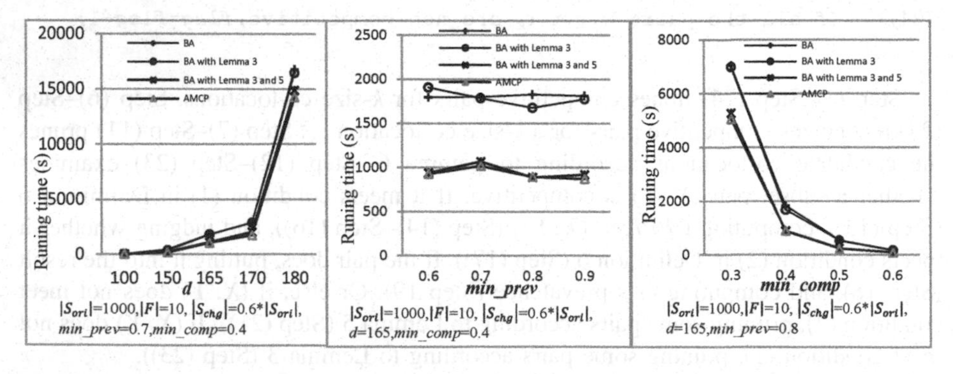

Fig. 4. Effect of *d*, *min_prev* and *min_comp*

5.2.1 The Performance of Four Algorithms on Varying Thresholds

Performance of four algorithms is shown in Fig. 4. We can observe (1) the four algorithms get worse with the increase of *d* and with the decrease of *min_comp*. The bigger *d* and smaller *min_comp* is, the more competitive pairs are. (2) the performance of the four algorithms does not change obviously with the decrease of *min_prev*. Because *min_prev* is not the direct threshold for a pair, it just confines its left and right part is prevalent. (3) AMCP and BA with Lemmas 3 and 5 are better than BA and BA with Lemma 3, which means Lemma 5 is the most efficient in three lemmas.

5.2.2 The Scalability of Four Algorithms on Varying Objects and Features

This subsection tests the scalability of the four algorithms with increase of the number of objects and features. From the results, shown in Fig. 5, we can get, (1) when $|F| = 15$, the four algorithms gets the best. When $|F| = 10$, the four algorithms get worse in that the number of objects belong to each feature get more when the feature set is small. When $|F| = 20$, the four algorithms get worse as well in that the more the features, the more candidate co-locations. (2) AMCP and BA with Lemmas 3 and 5 outperform BA and BA with Lemma 3. (3) The four algorithms get worse with the increase of object set, especially BA and BA with Lemma 3. BA and BA with Lemma 3 nearly cannot execute when the object set is 70000.

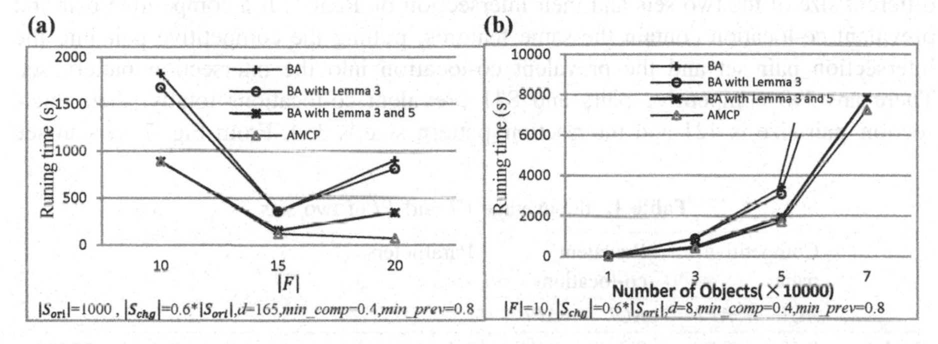

Fig. 5. (a) Effect of feature set. (b) Effect of $|S_{ori}|$

5.2.3 The Performance of 4 Algorithms on the Increased of Changed Objects

The performance of the proposed algorithms depends on the number of changed objects. We change the changed objects and the performance is shown in Fig. 6. We can see the algorithms get worse obviously with the increase of changed objects, and AMCP and BA with Lemmas 3 and 5 are better than BA and BA with Lemma 3.

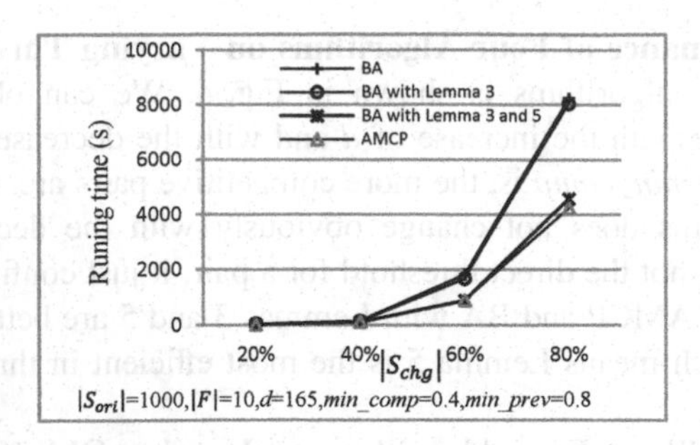

Fig. 6. Effect of changed objects

5.3 On Real Datasets

This section will statistically compare competitive pairs and prevalent co-locations mined by AMCP on real datasets.

Table 1 shows the average competitive index (*ACI*) and average participation index (*API*) of the competitive pairs and prevalent co-locations. The *API* of competitive pairs is the *API* of co-locations that the competitive pairs belonged to. The *ACI* of prevalent co-locations is the *ACI* of their subset pairs. The *ACI* of competitive pairs is bigger than *ACI* of prevalent co-locations, and the *API* of prevalent co-locations is bigger than *API* of competitive pairs. They are two different sets. Figure 7 compares the number of different size of the two sets and their intersection on Real-1. If a competitive pair and prevalent co-location contain the same features, putting the competitive pair into the intersection_pair set and the prevalent co-location into the intersection_pattern set. There are 590 competitive pairs and 841 prevalent co-locations totally, their intersection_pair size is 421 and intersection_pattern size is 354. From Fig. 7 we can see

Table 1. the average *CI* and *PI* of two sets

	Competitive pairs		Prevalent co-locations		Parameters
	ACI	*API*	*ACI*	*API*	
Real-1	0.58	0.70	0.31	0.84	min_prev = 0.6, min_comp = 0.4, d = 23000
Real-2	0.61	0.75	0.37	0.79	min_prev = 0.7, min_comp = 0.4, d = 50

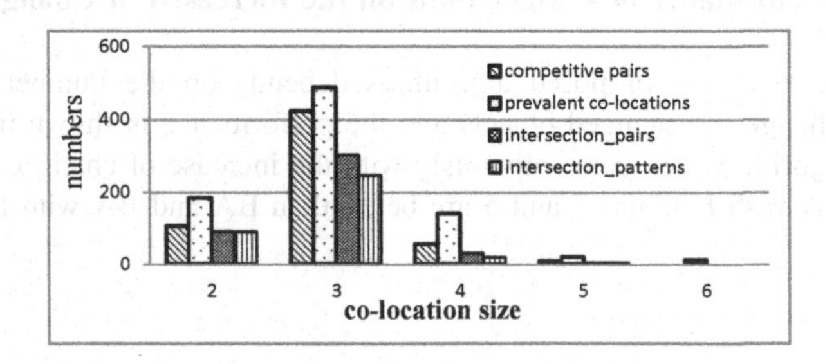

Fig. 7. The number of competitive pairs and prevalent co-locations over different sizes

that 71.4% (421/590) of competitive pairs are prevalent co-locations and 42.1% (354/841) of prevalent co-locations can form competitive pairs. So these competitive pairs are interesting. Importantly, they are accepted by domain experts.

6 Conclusion and Future Works

In this work, we gave the definition of competitive pair on spatial databases and presented some efficient pruning lemmas to improve the course. This work fills the gap of research of competition on the spatial database. Series of experiments on "synthetic + real" databases were conducted to verify our algorithms. The experiments showed the pruning lemmas are efficient and the discovered results are interesting.

Selecting databases with two times to get the convergence results is a future work. Mining other interesting patterns on dynamic spatial databases is another future work.

Acknowledgements. This work was supported partly by grants (No. 61472346, No. 61662086) from the National Natural Science Foundation of China and partly by grants (No. 2015FB149, No. 2016FA026) from the Science Foundation of Yunnan Province.

References

1. Huang, Y., Shekhar, S., Xiong, H.: Discovering co-location patterns from spatial data sets: a general approach. TKDE **16**(12), 1472–1485 (2004)
2. Lu, J., Wang, L., Fang, Y.: Mining causal rules hidden in spatial co-locations based on dynamic spatial databases. In: 2016 International Conference on Computer, Information and Telecommunication Systems, pp. 103–108 (2016)
3. Li, R., Bao, S., Wang, J., Yu, Y., Cao, Y.: CoMiner: an effective algorithm for mining competitors from the web. In: ICDM 2006, pp. 948–952 (2006)
4. Bao, S., Li, R., Yu, Y., Cao, Y.: Competitor mining with the web. IEEE Trans. Knowl. **20** (10), 1297–1310 (2008)
5. Agrawal, R., Imielinski, T., Swami, A.N.: Mining association rules between sets of items in large databases. In: SIGMOD Conference, pp. 207–216 (1993)
6. Yoo, J.S., Shekhar, S., Smith, J., Kumquat, J.P.: A partial join approach for mining co-location patterns. In: Proceedings of the 12th Annual ACM International Workshop on Geographic Information Systems (GIS 2004), pp. 241–249 (2004)
7. Yoo, J.S., Shekhar, S.: A joinless approach for mining spatial co-location patterns. TKDE **18** (10), 1323–1337 (2006)
8. Wang, L., Bao, Y., Lu, J., Yip, J.: A new join-less approach for co-location pattern mining. In: Proceedings of the 8th IEEE International Conference on Computer and Information Technology (CIT 2008), pp. 197–202 (2008)
9. Ouyang, Z., Wang, L., Chen, H.: Mining spatial co-Location patterns for fuzzy objects. Chin. J. Comput. **34**(10), 1947–1955 (2011)
10. Wang, L., Han, J., Chen, H., Lu, J.: Top-k probabilistic prevalent colocation mining in spatially uncertain data sets. Front. Comput. Sci. **10**(3), 488–503 (2016)
11. Huang, Y., Pei, J., Xiong, H.: Mining co-location patterns with rare events from spatial data sets. Geoinformatica **10**(3), 239–260 (2006)

12. Feng, L., Wang, L., Gao, S.: A new approach of mining co-location patterns in spatial datasets with rare feature. J. Nanjing Univ. (Nat. Sci.) **48**(1), 99–107 (2012)
13. Wang, S., Huang, Y., Wang, X.S.: Regional co-locations of arbitrary shapes. In: Nascimento, M.A., Sellis, T., Cheng, R., Sander, J., Zheng, Yu., Kriegel, H.-P., Renz, M., Sengstock, C. (eds.) SSTD 2013. LNCS, vol. 8098, pp. 19–37. Springer, Heidelberg (2013). doi:10.1007/978-3-642-40235-7_2
14. Sun, J.T., Wang, X., Shen, D., Zeng, H.J., Chen, Z.: Cws: a comparative web search system. In: Proceedings of the 15th International Conference on World Wide Web, pp. 467–476. ACM (2006)
15. Yang, Y., Tang, J., Keomany, J., Zhao, Y., Li, J., Ding, Y., Li, T., Wang, L.: Mining competitive relationships by learning across heterogeneous networks. In: ACM International Conference on Information & Knowledge Management, New York, 1432–1441 (2012)
16. Lappas, T., Valkanas, G., Gunopulos, D.: Efficient and domain-invariant competitor mining. In: Proceedings of 18th ACM SIGKDD International Conference on Knowledge Discovery and Data Mining, pp. 408–416 (2012)
17. Ruan, T., Lin, Y., Wang, H., Pan, J.Z.: A multi-strategy learning approach to competitor identification. In: Supnithi, T., Yamaguchi, T., Pan, J.Z., Wuwongse, V., Buranarach, M. (eds.) JIST 2014. LNCS, vol. 8943, pp. 197–212. Springer, Cham (2015). doi:10.1007/978-3-319-15615-6_15

Utility Aware Clustering for Publishing Transactional Data

Michael Bewong(✉), Jixue Liu, Lin Liu, and Jiuyong Li

ITMS, University of South Australia, Adelaide, SA 5095, Australia
michael.bewong@mymail.unisa.edu.au,
{jixue.liu,lin.liu,jiuyong.li}@unisa.edu.au

Abstract. This work aims to maximise the utility of published data for the partition-based anonymisation of transactional data. We make an observation that, by optimising the clustering *i.e.* horizontal partitioning, the utility of published data can significantly be improved without affecting the privacy guarantees. We present a new clustering method with a specially designed distance function that considers the effect of sensitive terms in the privacy goal as part of the clustering process. In this way, when the clustering minimises the total intra-cluster distances of the partition, the utility loss is also minimised. We present two algorithms *DocClust* and *DetK* for clustering transactions and determining the best number of clusters respectively.

1 Introduction

Transactional data is often used in data mining applications for its richness in embedded knowledge. Such data may contain sensitive terms and their direct publication may lead to privacy disclosures. Table 1a depicts an example of online shopping logs containing sensitive terms (*italicised*). The sensitive terms of individuals' transactions in the dataset can be learnt by an adversary even when the names are removed, by linking known individuals to non-sensitive terms and then the non sensitive terms to their sensitive terms [1,6]. For example, if Bob knows that Alice bought a drone, then Bob can identify Alice's record to be r_{10}, and further learn that Alice also bought *lice soap*, a sensitive term.

Several anonymisation methods have been proposed to deal with such privacy disclosures, including partition-based methods, which have the advantage of not adding, changing or deleting the words in the original transactions to preserve utility. Predominant amongst these methods is *PPD* (Privacy Preservation by Disassociation) [9], which publishes a set of transactions in two steps, horizontal (*HorPart*) and vertical (*VertPart*) partitioning. In *HorPart*, transactions are grouped into clusters by recursively using the most frequent terms until the cluster size is smaller than a user specified value. *VerPart* breaks the transactions of a cluster into multiple chunks (record chunks and a term chunk) so that m or less terms in a chunk identifies no less than K transactions. The produced publication is said to be K^m-anonymous. If sensitive terms are specified, these

J. Kim et al. (Eds.): PAKDD 2017, Part II, LNAI 10235, pp. 481–494, 2017.
DOI: 10.1007/978-3-319-57529-2_38

Table 1. Original dataset and its publications

(a) Dataset

ID	Transactions
r_1	hat, bag, *viagra*
r_2	hat, bag, *playboy*
r_3	hat, bag, *viagra*
r_4	bag, *playboy*
r_5	hat, mug
r_6	bag, mug
r_7	pants, mug *playboy*
r_8	pants, shirt, *playboy*
r_9	pants, shirt, *playboy*
r_{10}	drone, shirt, *lice soap*

(b) 2^2-anony. 2-diverse pub.

ID	Rec. Chunk	Term Chunk
Cluster 1		
r_1	hat, bag	*viagra*, *playboy*
r_2	hat, bag	
r_3	hat, bag	
Cluster 2		
r_4	bag	*playboy*, mug
r_6	bag	
Cluster 3		
r_7	pants, shirt	*playboy*, mug
r_8	pants, shirt	
r_9	pants	
Cluster 4		
r_5		*lice soap*, shirt, drone, mug, hat
r_{10}		

(c) Alt. 2^2-anony. 2-diverse pub.

ID	Rec. Chunk	Term Chunk
Cluster 1		
r_1	hat, bag	*viagra*, *playboy*
r_2	hat, bag	
Cluster 2		
r_3	bag	*playboy*, *viagra*, hat
r_4	bag	
Cluster 3		
r_5	mug	*playboy*, hat, pants
r_7	mug	
Cluster 4		
r_6		*playboy*, pants, bag, mug, shirt
r_8		
Cluster 5		
r_9	shirt	*lice soap*, pants, drone, *playboy*
r_{10}	shirt	

are placed in the term chunk (with set semantics) along with the rare non-sensitive terms to achieve l-diversity. Applying *PPD* with $K = m = 2$ to the data in Table 1a produces a publication in Table 1b. Any two or less terms in a cluster identifies no less than two records. Also, any sensitive term in the term chunk links to a record with a probability of no more than $1/2$.

Utility preservation is about maintaining the information contained in the original data in the published data. In partition-based methods like *PPD* it is often measured by comparing the preserved term associations in the published data to those in the original data [9] as illustrated in Example 1.

Example 1. In Table 1a the term association (*hat*, *viagra*) appears in two records, giving it a support of 2. In Table 2b the support must be calculated probabilistically. The chance for any of the *hat* records to be linked to the sensitive term *viagra* in Cluster 1 is $1/3$. Thus the support of (*hat*, *viagra*) in cluster 1 is $3 * 1/3 = 1$. Thus, the support to the term association in Table 1a is 2 but it is 1 in Table 1b and the loss is 50%. The utility loss due to the publication is the total loss of all term associations and this loss for Table 1b stands at 43% overall.

We realise that, for the same vertical partitioning but with different clustering, the same level of privacy can be achieved with better utility (Example 2).

Example 2. Table 1c presents an alternative publication of Table 1b. Both tables are 2^2-anonymous and 2-diverse, however following the same calculation, the published data in Table 1c has an overall utility loss of 13%, which is much lower than that of Table 1b.

From this example, we observe that clustering should be optimised to improve utility. This paper develops a framework for improved clustering for minimising utility loss without sacrificing the privacy guarantee. Our framework takes into consideration the privacy goal as part of the clustering process, in addition to the similarity of transactions within clusters. Specifically our contributions are:

1. We present two algorithms, *DocClust*, which employs a specially designed distance function that aims to put similar transactions with different sensitive

terms into a cluster; and *DetK*, which finds the optimal number of clusters k by using an effective pruning method.
2. We empirically demonstrate the effectiveness of our method on three real world benchmark datasets.

2 Preliminaries and the Problem

Let $\mathbf{D} = \{D_1, \cdots, D_n\}$ be a set of transactions called the corpus where each transaction $D_i \in \mathbf{D}$ is a set of terms. A subset $\mathbf{C} \subset \mathbf{D}$ is called a **cluster**. A partition $\mathbb{C}$ of the corpus is a set of clusters $\mathbb{C} = \{\mathbf{C}_1, \cdots, \mathbf{C}_k\}$ such that $\bigcup_{i=1}^{k} \mathbf{C}_i = \mathbf{D}$ and $\mathbf{C}_i \cap \mathbf{C}_j = \emptyset$ $(i \neq j)$. Also $D_{\cup} = D_1 \cup \cdots \cup D_n$ is the set of all terms in $\mathbf{D}$. S and $\bar{S} = D_{\cup} \setminus S$ is the set of all sensitive and non-sensitive terms in $D_{\cup}$ respectively. If a transaction contains a term t, it is called a t-transaction else it is a non-t-transaction.

2.1 Privacy Model

We adopt the same privacy model, K^m-anonymity, used in *PPD* [9]. Specifically, a cluster $\mathbf{C}$ ($|\mathbf{C}| \geq K$) is said to be K^m-anonymous, if an attacker whose background knowledge of up to m terms can be linked to at least K transactions of a cluster $\mathbf{C}$. Further, a K^m-anonymous cluster $\mathbf{C}$ satisfies l-diversity [8] if the probability of linking a sensitive term to a transaction $D \in \mathbf{C}$ is at most $1/l$.

Let $\mathcal{A}$ be an anonymisation transformation that takes as input $\mathbb{C}$, a partition of $\mathbf{D}$ and produces an anonymised counterpart $\mathcal{A}(\mathbb{C})$ consisting of anonymised clusters $\mathbf{C}^A \in \mathcal{A}(\mathbb{C})$. Also let $\mathcal{I}$ be the inverse transformation that takes as input the anonymised partition $\mathcal{A}(\mathbb{C})$ and outputs all possible $\mathbf{D}'$, reconstruction(s) of $\mathbf{D}$, *i.e.* $\mathcal{I}(\mathcal{A}(\mathbb{C})) = \{\mathbf{D}'|\mathcal{A}(\mathbb{C})\}$. We aim to enforce the following guarantee:

Definition 1 (Privacy Guarantee). *For an anonymised partition $\mathcal{A}(\mathbb{C})$ and a set $R \subseteq \bar{S}$ of up to m non-sensitive terms, (1) Applying $\mathcal{I}(\mathcal{A}(\mathbb{C}))$ always produces at least one dataset $\mathbf{D}'$ for which there are at least K records that contain all terms in R; (2) the probability of linking any sensitive term $s \in S$ to a transaction in any cluster of $\mathcal{A}(\mathbb{C})$ is no more than $1/l$.*

To satisfy the privacy guarantee, we adapt the *VerPart* method by considering only the non-sensitive terms of a cluster $\mathbf{C}$ in the vertical partitioning stage to enforce K^m-anonymity and introduce an operation called *sanitisation* to enforce diversity in the sensitive terms as follows.

As a first step, each cluster $\mathbf{C}$ is divided into two parts, a multiset of termsets called the non-sensitive segment $\mathbf{C}^{\bar{S}} = \{D \setminus S | D \in \mathbf{C}\}$ containing the non-sensitive part of the cluster; and a multiset of terms called the sensitive chunk $SC = \biguplus_{D \in \mathbf{C}}(D \cap S)$ containing all the sensitive terms of the cluster.

Given the non-sensitive segment $\mathbf{C}^{\bar{S}}$, *VerPart* produces multisets of term sets called record chunks $RC_1, \cdots, RC_u$ and a set of terms called the term chunk TC such that each record chunk is K^m-anonymous. The main points of the operation of the *VerPart* algorithm [9] on a cluster $\mathbf{C}^{\bar{S}}$ are as follows:

1. Move all terms with a support less than K in $\mathbf{C}^{\bar{S}}$ into TC; $\bar{S}_{remain} = \bar{S} \setminus TC$.

2. Create a record chunk RC_i with a projection on the most frequent term $t_f \in \bar{S}_{remain}$ in $\mathbf{C}^{\bar{S}}$; $\bar{S}_{remain} = \bar{S}_{remain} \setminus t_f$.
3. For every term $t \in \bar{S}_{remain}$ project t unto RC_i. If RC_i remains K^m-anonymous then $\bar{S}_{remain} = \bar{S}_{remain} \setminus t$; else remove projection.
4. Repeat step 2 and 3 until $\bar{S}_{remain} = \emptyset$.

Example 3. Table 2a is a cluster of transactions and Table 2b contains its non-s segment $\mathbf{C}^{\bar{S}}$ and sensitive chunk SC. Given $K = m = 2$, Table 2c is the results of *VerPart*. In Table 2c the terms *toy* and *spray* are placed in the term chunk TC since their support is less than 2. RC_1 and RC_2 are constructed by keeping the terms pen and lamp in RC_1 and moving the term book to RC_2 since the record chunk is not 2^2-anonymous otherwise.

Table 2. Anonymisation via VerPart

(a) Cluster $\mathbf{C}$

ID	Transactions
r_1	lamp, book *p.boy, viagra*
r_2	pen, book *p.boy, viagra*
r_3	lamp, toy *p.boy, lice soap*
r_4	pen, spray *p.boy, lice soap*

(b) $\mathbf{C}^{\bar{S}}$ and SC

$\mathbf{C}^{\bar{S}}$	SC
lamp, book	*p.boy, viagra,*
pen, book	*p.boy, lice soap,*
lamp, toy	*p.boy, viagra,*
pen, spray	*p.boy, lice soap*

(c) Results of VerPart

RC_1	RC_2	TC	SC
{lamp}			*p.boy, viagra,*
{pen}	{book}	toy,	*p.boy, lice soap,*
{lamp}	{book}	spray	*p.boy, viagra,*
{pen}			*p.boy, lice soap*

After *VerPart* the cluster is guaranteed to be K^m-anonymous and thus the first part of the privacy guarantee is assured. To satisfy the second part of the guarantee, all the sensitive terms can be moved to the term chunk [9], however, when $l > K$, the guarantee will not be satisfied. Also when $|\mathbf{C}| > l$, it may not be necessary to move all sensitive terms to TC. We therefore propose *sanitisation*, which firstly tests if it is possible for the cluster containing sensitive terms to satisfy l-diversity, if not(*i.e.* $|\mathbf{C}| < l$) then the cluster is merged with others. If $|\mathbf{C}| \geq l$, $h_s(\mathbf{C})$ copies of s in SC are moved into TC; $h_s(\mathbf{C}) = \lceil N(s, SC) - (|\mathbf{C}|/l) \rceil$. In the formular, $N(s, SC)$ is the number of s in SC and $(|\mathbf{C}|/l)$ is the number of s terms expected in SC to satisfy l-diversity for a given l. The difference between the two is the excess number of s terms that must be moved.

Definition 2 (Sanitisation). *Given a cluster* $\mathbf{C}$ *and its sensitive chunk* SC *(*$SC \neq \emptyset$*)* **sanitisation** *denoted by* λ *proceeds as follows: If* $|\mathbf{C}| < l$*,* λ *merges cluster with the next closest cluster; else if* $h_s(\mathbf{C}) > 0$*, move* $h_s(\mathbf{C})$ *copies of* s *from* SC *into the term chunk* TC*; Otherwise, do nothing.*

Closest cluster refers to the distance (defined later) between the cluster centres. Sanitisation is required to be applied to every cluster and for every sensitive term s until each cluster is l-diverse.

Table 3. Sanitisation

RC_1	RC_2	TC	SC
{lamp}		toy,	*viagra, lice*
{pen}	{book}	spray,	*soap,*
{lamp}	{book}	*p.boy*	*viagra, lice*
{pen}			*soap, p.boy*

Example 4. Given $l = 2$, Table 3 is the sanitisation of Table 2c. In the table, $h_{p.boy}(\mathbf{C}) = 2$, $h_{lice\ soap}(\mathbf{C}) = 0$, $h_{viagra}(\mathbf{C}) = 0$. Consequently 2 copies of the sensitive term *p.boy* (*playboy*) must be moved, however we move 3 copies to cater for the remainder in the term chunk TC. The probability of linking any record to a sensitive term is now no more than 1/2.

2.2 Utility

In the partition-based anonymisation techniques, utility is lost via the breaking of term associations. This is because the terms of a record are put into separate vertical partitions to reduce the chances of linking the information. A utility metric must therefore capture the amount of broken term associations.

We propose to use the well known *KL-divergence* metric for measuring the difference between two probability distributions to capture this loss.

Definition 3 (Term-Association). *Given the corpus* $\mathbf{D}$*, the terms* t_i *and* t_j $(t_i \neq t_j)$ *form an association* $\tau = (t_i, t_j)$*, if there exists a transaction* $D \in \mathbf{D}$ *such that* t_i *and* t_j *are in* D*. The* **support** *of* τ*,* $N(\tau, \mathbf{D})$*, is the number of all transactions containing* τ *in* $\mathbf{D}$*.* $\mathcal{T}$ *denotes all associations in* $\mathbf{D}$*.*

In our definition above, we consider only term pairs because we believe they are representative of the more complex term associations that may also be lost by anonymisation. This is analogous to the a priori rule on frequent itemsets, *i.e.* if a complex term association is lost, then its term pairs are also lost.

Definition 4 (Utility Loss). *Given a corpus* $\mathbf{D}$ *and its anonymisation* $\mathcal{A}(\mathbb{C})$*, the utility loss* UL*, is the KL divergence for all associations* $\tau \in \mathcal{T}$ *between* $\mathbf{D}$ *and* $\mathcal{A}(\mathbb{C})$*:*

$$UL(\mathbf{D}, \mathcal{A}(\mathbb{C})) = \sum_{\forall \tau \in \mathcal{T}} P(\tau) \cdot log \frac{P(\tau)}{P'(\tau)} \quad (1)$$

$P(\tau)$ *is the probability of* τ *in* $\mathbf{D}$*, and* $P'(\tau) \neq 0$ *is the probability of* τ *in* $\mathcal{A}(\mathbb{C})$*.*

For $\tau = (t_i, t_j) \in \mathcal{T}$, the probability $P(\tau)$ in the original corpus $\mathbf{D}$ is calculated from the support $N(\tau, \mathbf{D})$ and $P'(\tau)$ is calculated probabilistically as follows:

$$P(\tau) = \frac{N(\tau, \mathbf{D})}{|\mathbf{D}|}; \qquad P'(\tau) = \frac{1}{|\mathbf{D}|} \sum_{\mathbf{C}^A \in \mathcal{A}(\mathbb{C})} \left(N(t_i, \mathbf{C}^A) \times \frac{N(t_j, \mathbf{C}^A)}{|\mathbf{C}^A|} \right)$$

The calculation of $P'(\tau)$ in the anonymised corpus $\mathcal{A}(\mathbb{C})$ is the probability of linking t_i to t_j in a reconstruction $\mathbf{D}'$ of $\mathcal{A}(\mathbb{C})$. In this formula, $N(t_i, \mathbf{C}^A)$ is the number of t_i terms in the anonymised cluster $\mathbf{C}^A$, and $N(t_j, \mathbf{C}^A)/|\mathbf{C}^A|$ is the probability that any term will be associated with the term t_j in $\mathbf{C}^A$. The product returns the number of the term association (t_i, t_j) that can be recovered from $\mathbf{C}^A$. The sum of the number of recovered term associations over all the clusters will be the total (t_i, t_j) recovered from $\mathcal{A}(\mathbb{C})$; and the division by $|\mathbf{D}|$ gives us the probability of (t_i, t_j) in the whole anonymised corpus $\mathcal{A}(\mathbb{C})$. We see that $P'(\tau) \neq 0$ since our anonymisation does not suppress all copies of a term.

Definition 5 (Problem Definition). *Given a corpus* $\mathbf{D}$ *of transactions, a sensitive term set* S*, and an anonymisation transformation* $\mathcal{A}$ *that satisfies* K^m *anonymity and* l*-diversity for user specified privacy parameters of* K, m *and* l*, the problem is to find a partition* $\mathbb{C} = \{\mathbf{C_1} \cdots \mathbf{C_k}\}$ *of* $\mathbf{D}$ *such that* $\mathcal{A}(\mathbb{C})$ *minimises utility loss* UL *(Definition 4).*

3 Framework of the Solution

Our clustering framework, *Utility Aware Clustering* (UAC) contains three key components: a distance function for calculating the distance between transactions that relates to the utility; a clustering algorithm, *DocClust* for partitioning transactions towards a minimal total distance for a given number k of clusters; and a search algorithm, *DetK* for finding the best number of clusters k.

3.1 Distance Function

The first component of UAC, the distance function has two parts: (1) *homogeneity distance*, to reflect the number of transactions for each term $t \in D_\cup$ to be modified in $\mathbf{C}$ to make the transactions in $\mathbf{C}$ identical; (2) *sensitivity distance*, to reflect the number of non-s-transactions for each sensitive term $s \in S$ to be added to $\mathbf{C}$ to make the probability of s in $\mathbf{C}$ close to the l-diversity goal (Definition 1).

Definition 6 (Homogeneity distance). *Given a cluster* $\mathbf{C} \subseteq \mathbf{D}$*, and a transaction* $D \in \mathbf{D} \setminus \mathbf{C}$*, the* **homogeneity distance** *(Hdis) of adding* D *to* $\mathbf{C}$ *is the number of transactions in* $\mathbf{C} \cup \{D\}$ *to be modified to have the same terms:*

$$Hdis(D, \mathbf{C}) = \sum_{t \in C_\cup} [(|\mathbf{C}| + 1) - (N(t, \mathbf{C}) + x)] \tag{2}$$

$C_\cup$ *is the set of terms in* $\mathbf{C} \cup \{D\}$ *and* $N(t, \mathbf{C})$ *is the number of* t*-transactions in* $\mathbf{C}$*; Also,* $x = 1$ *if* $t \in D$ *else* $x = 0$*. The difference is the number of transactions that do not contain the term* t*.*

Definition 7 (Sensitivity distance). *Given a cluster* $\mathbf{C} \subseteq \mathbf{D}$*, the sensitive term set* S*, and a transaction* $D \in \mathbf{D} \setminus \mathbf{C}$ *the* **sensitivity distance** *(Sdis) of adding* D *to* $\mathbf{C}$ *is defined to be the sum of the number of non-s transactions required to make* $\mathbf{C}$ l*-diverse for every* $s \in S$*:*

$$Sdis(D, \mathbf{C}) = \sum_{s \in S} [(2x - 1)(M - \frac{|\mathbf{C}| + 1}{N(s, \mathbf{C}) + x})] \tag{3}$$

$x = 1$ *if* $s \in D$ *else* $x = 0$*;* $M = l - 1$ *for the privacy parameter* l *in* l*-diversity.*

In the formula, $(|\mathbf{C}| + 1)/(N(s, \mathbf{C}) + x)$ is the number of non-s-transactions available for each s-transaction in $\mathbf{C}$. M is the number of non-s-transactions required for each s-transaction to make $\mathbf{C}$ l-diverse. $(2x-1)$ ensures that $Sdis$ is positive if $s \in D$, else it is negative, thus if a cluster $\mathbf{C}$ requires more non-s transactions and the transaction D to be added is non-s, the distance is smaller, otherwise it is larger. When $N(s, \mathbf{C}) + x = 0$, $Sdis$ is set to M to avoid a division by 0 and ensure that clusters without any s terms are not encouraged.

The sum, $Hdis(D, \mathbf{C}) + w.Sdis(D, \mathbf{C})$ is the distance between D and $\mathbf{C}$, where the **balancing factor** w makes $Hdis$ and $Sdis$ contribute equally. Further details on w is given in Sect. 4. The total distance for the partition $\mathbb{C}$ that we seek to minimise in our clustering is therefore defined as follows.

Definition 8 (Total distance of a partition). *Given a corpus* $\mathbf{D}$ *and a partition* $\mathbb{C} = \{\mathbf{C_1}, \cdots, \mathbf{C_k}\}$ *of* $\mathbf{D}$, *the* ***total distance*** *of* $\mathbb{C}$, *denoted by* $\zeta(\mathbb{C})$, *is the sum of distances for all clusters:*

$$\zeta(\mathbb{C}) = \sum_{j=1}^{k} dist(\mathbf{C}_j) = \sum_{j=1}^{k} \Big(\sum_{D_i \in \mathbf{C}_j} Hdis(D_i, \mathbf{C}_j \setminus D_i) + w \cdot \sum_{D_i \in \mathbf{C}_j} Sdis(D_i, \mathbf{C}_j \setminus D_i) \Big) \tag{4}$$

The total distance ζ relates to the utility loss by the following observation. Let $n.RC(\mathbf{C}^A)$ be the number of record chunks in the anonymised cluster $\mathbf{C}^A$ and $n.TC(\mathbf{C}^A)$ the number of terms in its term chunk and $|\mathbf{C}^A|$ its cluster size.

Observation 1. *Given a corpus* $\mathbf{D}$, *if a partition* $\mathbb{C}$ *of* $\mathbf{D}$ *minimises the total distance* $\zeta(\mathbb{C})$ *then; (1)* $\sum_{\mathbf{C}^A \in \mathcal{A}(\mathbb{C})} n.RC(\mathbf{C}^A) \times n.TC(\mathbf{C}^A) \times |\mathbf{C}^A|$ *is minimised; and (2) utility loss is also minimised.*

Observation 1 is justified as follows: In Definition 4 utility loss is minimised when $P'(\tau) \xrightarrow{approaches} P(\tau)$. Three cases of the term association $\tau(t_i, t_j)$ arise $(1) t_i, t_j \in \bar{S}$ $(2) t_i, t_j \in S$ $(3) t_i \in \bar{S} \wedge t_j \in S$. In these cases, $P'(\tau) \xrightarrow{approaches} P(\tau)$ when $n.RC(\mathbf{C}^A)$, $n.TC(\mathbf{C}^A)$ and $|\mathbf{C}^A|$ are minimum which occurs when $Hdis$ and $Sdis$ are equally minimised. A smaller $n.TC()$ means less terms are moved to the term chunk and the associations between the terms in TC and the terms in other chunks are preserved. A smaller $n.RC()$ indicates less record chunks and thus less broken associations between non-sensitive terms. A smaller $|\mathbf{C}^A|$ means less false associations will be constructed when $P'(\tau)$ is calculated.

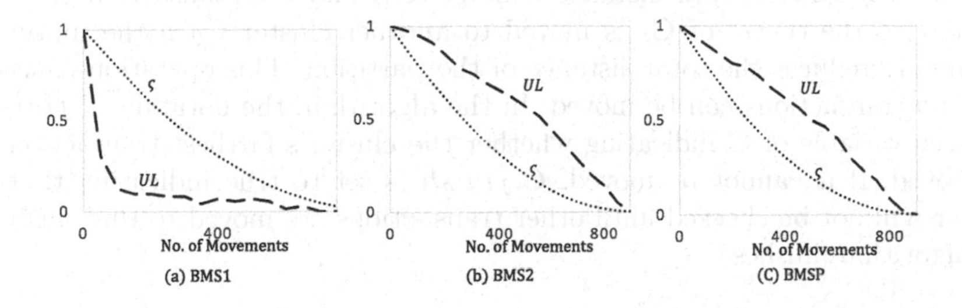

Fig. 1. Total distance and utility loss vs movements

We use experiments to demonstrate the relationship between our distance function and the utility loss as shown in Fig. 1. This was done by calculating the total distance ζ and total utility loss UL after each transaction movement in *DocClust* clustering (described shortly). We used a random sample of 1000 transactions from three different dataset (described in Sect. 4), with $K = m = l = 2$ and number of clusters $k = 50$. Figure 1 confirms our expectation that ζ and UL both decrease with each transaction move and eventually get to a minimum.

3.2 Clustering

Clustering, the second key component of our solution, UAC, aims to minimise the total distance (Definition 8). We note that existing well known algorithms like k-means do not apply to our scenario because of the special properties of our distance function. We use an example to illustrate this.

Function 1. initCluters (D, k)

1: Create k clusters: $\mathbb{C} = \{\mathbf{C}_1, ..., \mathbf{C}_k\}$
2: order the transactions in **D** by the number of terms in descending order.
3: select top k transactions from the sorted list as initial seeds for the clusters.
4: **for** each non-allocated $D \in \mathbf{D}$: **do**
5: find cluster $\mathbf{C}_j = \underset{\mathbf{C} \in \mathbb{C}}{argmin}\ dist(D, \mathbf{C})$; put D in $\mathbf{C}_j$; calculate the centroid $ctr(\mathbf{C}_j)$
6: **end for**; return $\mathbb{C}$

Example 5. For a transaction $D_1 = \{a, s_1\}$, and an empty cluster $\mathbf{C}\{\}$, if $\mathbf{C}$ is initially assigned D_1 as the cluster seed, conventional similarity based distance metrics, will always give a distance of 0 between D_1 and the seeded $\mathbf{C}$ since they are identical. By k-means D_1 will always be assigned to $\mathbf{C}$ in every round of iterations. In contrast, our metric, which considers the sensitive terms of D_1 and the seeded $\mathbf{C}$, will assign a large distance value. When used with k-means, D_1 is alternately added and removed from $\mathbf{C}$ in each iteration since D_1 and $\mathbf{C}$ will alternately have the same sensitive value s. This creates unstable assignments.

We develop a two-stage clustering algorithm, *DocClust* (Algorithm 1). The first stage, the *initialisation stage* (Function 1) uses transactions with the most number of terms as initial cluster seeds. We expect these to be in separate clusters in the final partition. The remaining transactions are then assigned to the clusters based on their closest distance to the centres.

In the second stage, the *update stage* (Lines 2-end of Algorithm 1), the cluster with the largest total distance denoted $\mathbf{C}_1$; and the transaction $D \in \mathbf{C}_1$, furthest to the centre of $\mathbf{C}_1$ is moved to another cluster $\mathbf{C}_2$ if the movement maximally reduces the total distance of the partition. This operation continues until no transactions can be moved. In the algorithm, the notation $\mathbf{C}.finish$ is a status variable of $\mathbf{C}$ indicating whether the cluster's farthest transaction can be moved. If it cannot be moved, $\mathbf{C}.finish$ is set to true indicating that the cluster will not be checked until other transactions are moved to this cluster or the algorithm finishes.

The algorithm converges. Each transaction D moved from $\mathbf{C}_1$ to $\mathbf{C}_2$ satisfies $f(\mathbf{C}) > 0$, hence every move reduces the total distance (ζ). The lower bound of ζ is 0 hence the move will stop when ζ reaches its lower bound or $f(\mathbf{C}) > 0$ is not true and the algorithm terminates. We remark that for K^m-anonymity, each cluster is compelled to have at least K transactions.

Algorithm 1. DocClust

Input: A corpus $\mathbf{D}$; number of clusters k.
Output: The partition of $\mathbf{D}$, $\mathbb{C} = \{\mathbf{C}_1, ..., \mathbf{C}_k\}$, with the least cost.
1: $\mathbb{C} = initClusters(\mathbf{D}, k)$
2: order clusters in $\mathbb{C}$ in descending order of by the total distance.
3: for every cluster $\mathbf{C} \in \mathbb{C}$, $\mathbf{C}.finish = false$.
4: **while** some clusters are not finished: **do**
5: let $\mathbf{C}_1$ be the 1st cluster of $\mathbb{C}$.
6: find the farthest $D \in \mathbf{C}_1$: $D = \underset{D \in \mathbf{C}_1}{argmax}\ dist(D, \mathbf{C}_1 - D)$
7: find cluster $\mathbf{C}_2$ to move D into s.t. $\zeta(\mathbb{C})$ is maximally reduced: $\mathbf{C}_2 = \underset{\mathbf{C} \in \mathbb{C}}{argmin}\ \{f(\mathbf{C}) : f(\mathbf{C}) > 0\}$ where $f(\mathbf{C}) = dist(\mathbf{C}_1) + dist(\mathbf{C}) - [dist(\mathbf{C}_1 - D) + dist(\mathbf{C} + D)]$
8: if $\mathbf{C}_2$ is null: $\mathbf{C}_1.finish = true$; continue.
9: move D to $\mathbf{C}_2$; reorder $\mathbb{C}$ by the total distance.
10: if $\mathbf{C}_2.finish = true$: $\mathbf{C}_2.finish = false$
11: **end while**

3.3 Search for the Best k

In the third component of UAC, finding the optimal k number of clusters, we propose a heuristic algorithm called *DetK*. *DetK* uses the total distance ζ of an already computed partition to prune some intervals of k. As more partitions are computed (with different k's), we prune more intervals and finally get to the best k that minimises ζ. We begin with some properties of our distance function.

Let $\kappa(\mathbf{D}, k) \Rightarrow (\mathcal{H}, \mathcal{S}, \zeta, \mathbb{C})$ be the clustering of $\mathbf{D}$ for k number of clusters resulting in: the total homogeneity distance $\mathcal{H}$, the total sensitivity distance $\mathcal{S}$, the total distance $\zeta = \mathcal{H} + \mathcal{S}$, and the partition $\mathbb{C}$ in order.

Observation 2. *The total homogeneity and sensitivity distances are monotonic w.r.t. k.* i.e. *if* $\kappa(\mathbf{D}, k) \Rightarrow (\mathcal{H}_k, \mathcal{S}_k, \zeta_k, \mathbb{C}_k)$ *and* $\kappa(\mathbf{D}, k+1) \Rightarrow (\mathcal{H}_{k+1}, \mathcal{S}_{k+1}, \zeta_{k+1}, \mathbb{C}_{k+1})$ *for any* $k \in [1, |\mathbf{D}| - 1]$ *then: (1)* $\mathcal{H}_k \geq \mathcal{H}_{k+1}$ *(2)* $\mathcal{S}_k \leq \mathcal{S}_{k+1}$.

We justify the observation as follows. As k increases the average number of transactions per cluster decreases. Smaller cluster sizes reduce the number of uncommon terms between transactions of a cluster making the total homogeneity distance $\mathcal{H}_k \geq \mathcal{H}_{k+1}$. Conversely, smaller cluster sizes cause an increase in the probability of the sensitive terms in the clusters making the number of transactions needed by each cluster for l-diversity larger, causing $\mathcal{S}_k \leq \mathcal{S}_{k+1}$.

These observations, also verified by experiments, are shown in Fig. 2. In the figure, each plot shows the total homogeneity distance $\mathcal{H}$, the total sensitivity distance $\mathcal{S}$ and the total distance ζ of clustering with different k's for three real world datasets. All three plots confirm the monotonicity of $\mathcal{H}$ and $\mathcal{S}$ with k.

In *DetK* (Algorithm 2) our approach is to: (1) find some k_m, (2) partition $\mathbf{D}$ at k_m to get the total distance ζ_m, (3) calculate the lower total distance bound *lbd* for the intervals $[k_l, k_m]$ and $[k_m, k_r]$, and (4) prune the intervals whose *lbd* is greater than ζ_m. For the remaining intervals, the division and search processes continue until the optimal k, k_o where $\zeta(\mathbb{C})$ gets a minimum, is found among all possible k's. This is illustrated by Fig. 3. In Fig. 3, the maximal $\mathcal{H}$ point is $(k_l = 1,\ h_l)$ and the minimal $\mathcal{H}$ point $(k_r = |D|,\ h_r)$. Similarly, the minimal $\mathcal{S}$ point is $(k_l = 1,\ s_l)$ and the maximal $\mathcal{S}$ point $(k_r = |D|,\ s_r)$. Then, k_m is

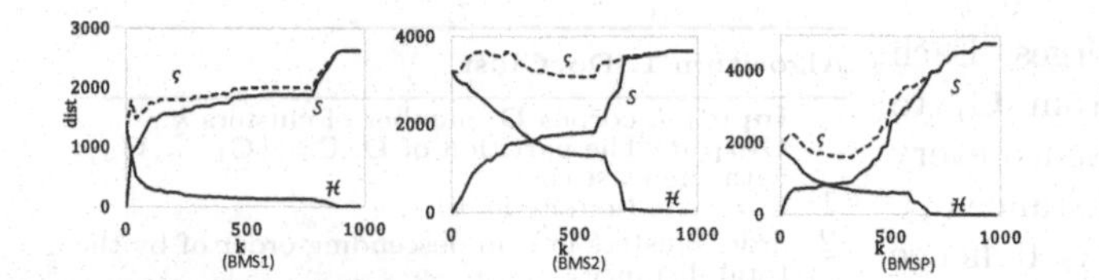

Fig. 2. Total distances vs k for three datasets

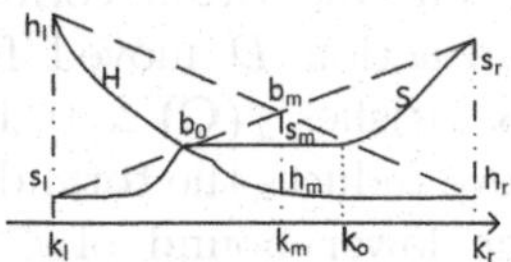

Fig. 3. Finding Optimal k_o

calculated geometrically from the cross point of the straight lines $(k_l, h_l), (k_r, h_r)$ and $(k_l, s_l), (k_r, s_r)$. The intuition is that, the minimal total distance point is often close to the cross point of $\mathcal{H}$ and $\mathcal{S}$ as depicted by Fig. 2. After the initial split of $[k_l, k_r]$, the resulting intervals are subsequently split at their mid-points.

The lower bound total distance for an interval is developed as follows:

Lemma 1. *Given an interval* $[k_l, k_r]$ *of* k *and the clusterings at* k_l *and* k_r *respectively:* $\kappa(\mathbf{D}, k_l) \Rightarrow (\mathcal{H}_l, \mathcal{S}_l, \zeta_l, \mathbb{C}_l)$ *and* $\kappa(\mathbf{D}, k_r) \Rightarrow (\mathcal{H}_r, \mathcal{S}_r, \zeta_r, \mathbb{C}_r)$*, the minimal total distance of the partitions for* $k \in [k_l, k_r]$ *is bounded by:* $lbd(k_l, k_r) = s_l + h_r$.

Lemma 2. *Given the total distance* ζ *at* k *and an interval* $[k_l, k_r]$*, the optimal* k *is not in* $[k_l, k_r]$ *if* $\zeta < lbd(k_l, k_r)$.

Lemma 1 is correct since $\mathcal{H}$ and $\mathcal{S}$ are monotonic to k and Lemma 2 is important to pruning the intervals of k to be searched.

Algorithm 2. DetK

Input: $\mathbf{D}$, stop thresholds θ_d, θ_k
Output: k_m (the optimal k), partition $\mathbb{C}_m$ of $\mathbf{D}$ for k_m

1: let $k_l = 1$, $k_r = |\mathbf{D}|$;
2: cluster: $\kappa(\mathbf{D}, k_l) \Rightarrow (h_l, s_l, \zeta_l, \mathbb{C}_l)$; $\kappa(\mathbf{D}, k_r) \Rightarrow (h_r, s_r, \zeta_r, \mathbb{C}_r)$;
3: calculate k_m;
4: cluster: $\kappa(\mathbf{D}, k_m) \Rightarrow (h_m, s_m, \zeta_m, \mathbb{C}_m)$;
5: let $lbd_l = s_l + h_m$; $lbd_r = s_m + h_r$; $\zeta_p = \zeta_m$ (pruner)
6: if $stop(k_l, k_m, \zeta_l, \zeta_m) = false \wedge lbd_l < \zeta_p$: insert $(k_l, k_m, \mathbb{C}_l, \mathbb{C}_m, h_l, s_l, h_m, s_m, lbd_l)$ to Iv
7: if $stop(k_m, k_r, \zeta_m, \zeta_r) = false \wedge lbd_r < \zeta_p$: insert $(k_m, k_r, \mathbb{C}_m, \mathbb{C}_r, h_m, s_m, h_r, s_r, lbd_r)$ to Iv
8: **while** $Iv.length > 0$ **do**
9:     let $e = Iv[0]$ and remove $Iv[0]$ from Iv
10:     let $(k_m, \zeta_m, \mathbb{C}_m) = bisearch(\mathbf{D}, e)$;
11:     if $\zeta_m < \zeta_p$: $\zeta_p = \zeta_m$
12:     for each $x \in Iv$: if $x.lbd \geq \zeta_p$: $Iv.remove(x)$// prun.
13: **end while**; return $(k_m, \mathbb{C}_m)$

For *DetK* (Algorithm 2) we define a data structure called the **interval sequence**, which is a sequence of elements Iv. An element $Iv[i] = (k_l, k_r, \mathbb{C}_l, \mathbb{C}_r, \mathcal{H}_l, \mathcal{S}_l, \mathcal{H}_r, \mathcal{S}_r, lbd)_i$ includes the interval $[k_l, k_r]$ of k, the partitions $\mathbb{C}_l$ and $\mathbb{C}_r$ from clustering at k_l and k_r respectively, the total $\mathcal{H}$ distances $\mathcal{H}_l$ and $\mathcal{H}_r$, and the total $\mathcal{S}$ distances $\mathcal{S}_l$ and $\mathcal{S}_r$ for the respective $\mathbb{C}_l$ and $\mathbb{C}_r$, and the lower total distance bound lbd of the interval. Elements in the sequence are ordered by the lbd.

In Algorithm 2, the input has two thresholds to terminate the search of the interval, if the interval is not very wide (θ_k, percentage of $|\mathbf{D}|$) and the relative difference of the total distances is small (θ_d, $\in [0, 0.1]$). The split point k_m of the input interval $[1, |\mathbf{D}|]$ and its clustering are determined (Lines 3–4). The pruner ζ_p and lower bounds of the intervals are calculated in Line 5. The interval sequence Iv is constructed in Lines 6–7 with stop condition (Function 3) checked, followed by pruning. The while loop takes the first element, which has the smallest lower bound among all un-searched intervals, in Iv to start the search (*bisearch* Function 2). After the search returns, the pruner ζ_p is updated as necessary based on the newly found minimal total distance. The new pruner is used to remove all intervals whose lower bound is more than the pruner (line 12). During *bisearch*, the new intervals that are not searched immediately are put into the interval sequence Iv for later pruning or searching. The search of an interval stops if there are no more k values to be searched (*i.e.*, $k_r - k_l \leq 1$) or the total distance difference at the two ends of the interval is very small while the interval itself is narrow (stop condition in Function 3).

DetK terminates when the intervals in Iv are all pruned or searched. The best k which is k_m and its clustering $\mathbb{C}$ are returned as the results. In the worst case *DetK* searches all k values in $[1, |\mathbf{D}|]$.

Function 2. biSearch $(\mathbf{D}, e)$

$(k_l, k_r, \mathbb{C}_l, \mathbb{C}_r, h_l, s_l, h_r, s_r, lbd) = e$, $\zeta_l = h_l + s_l$, $\zeta_r = h_r + s_r$
1: let $k_m = (k_l + k_r)/2$
2: cluster: $\kappa(\mathbf{D}, k_m) \Rightarrow (h_m, s_m, \zeta_m, \mathbb{C}_m)$
3: if $stop(k_l, k_m, \zeta_l, \zeta_m) \wedge stop(k_m, k_r, \zeta_m, \zeta_r)$: return $arg(\{(k_l, \zeta_l, \mathbb{C}_l), (k_m, \zeta_m, \mathbb{C}_m), (k_r, \zeta_r, \mathbb{C}_r)\}, min(\zeta_l, \zeta_m, \zeta_r))$ //returns $(k_x, \zeta_x, \mathbb{C}_x)$ with smallest ζ_x
4: else if $stop(k_l, k_m, \zeta_l, \zeta_m)$:
  insert $(k_m, k_r, \mathbb{C}_m, \mathbb{C}_r, h_m, s_m, h_r, s_r, s_m + h_r)$ in Iv
  return $arg(\{(k_l, \zeta_l, \mathbb{C}_l), (k_m, \zeta_m, \mathbb{C}_m)\}, min(\zeta_l, \zeta_m))$
5: else if $stop(k_m, k_r, \zeta_m, \zeta_r)$:
  insert $(k_l, k_m, \mathbb{C}_l, \mathbb{C}_m, h_l, s_l, h_m, s_m, s_l + h_m)$ in Iv
  return $arg(\{(k_m, \zeta_m, \mathbb{C}_m), (k_r, \zeta_r, \mathbb{C}_r)\}, min(\zeta_m, \zeta_r))$
  // now both intervals are open.
6: if $\zeta_l < \zeta_r$:
  insert $(k_m, k_r, \mathbb{C}_m, \mathbb{C}_r, h_m, s_m, h_r, s_r, s_m + h_r)$ in Iv
  let $k_r = k_m$; $\zeta_r = \zeta_m$; goto Line 1
7: else:
  insert $(k_l, k_m, \mathbb{C}_l, \mathbb{C}_m, h_l, s_l, h_m, s_m, s_l + h_m)$ in Iv
  let $k_l = k_m$; $\zeta_l = \zeta_m$; goto Line 1

Function 3. stop $(k_1, k_2, \zeta_1, \zeta_2)$

1: $\delta k = k_2 - k_1$; $\delta\zeta = abs(\zeta_2 - \zeta_1)/(\zeta_2 + \zeta_1)$
2: if $\delta k \leq 1 \vee (\delta\zeta \leq \theta_d \wedge \delta k/|\mathbf{D}| \leq \theta_k)$: return true
3: else return false

4 Empirical Study

We aim to demonstrate: (1) how the balancing factor w affects utility loss and the total distance and thus how to calculate an ideal w for Definition 8; (2) utility loss in actual queries; (3) and the scalability of the algorithm.

Three real transactional datasets, *BMS-Webview-1 (BMS1)*, *BMS-Webview-2 (BMS2)* and *BMS-POS (BMSP)* [11] (Table 4) were used. We randomly selected 10% of the terms for each dataset as sensitive terms for the experiments.

Table 4. Datasets

	BMS1	BMS2	BMSP
No. of trans.	59,602	77,512	515,597
No. of terms	497	3,340	1,657

(1) Ideal balancing factor w: We consider the effect of w in Definition 8 on the optimal k_{od} based on the total distance ζ in comparison the optimal k_{ou} based on the minimal utility loss UL. We randomly sampled 1000 transactions of BMSP. With representative privacy parameters $K = m = l = 2$ we clustered the dataset using multiple k values in $[1, |\mathbf{D}|/K]$. For each partition, ζ and UL were calculated. This experiment was repeated 3 times for 3 different w value. Figure 4 shows 3 plots for the 3 different w values. In each plot, the x-axis is k and the y-axis is the normalised total distance and utility loss. The dotted line represents the total distance and the dashed line the total utility loss. In (a) when w is a small value ($w = 0.5$), the contribution of the sensitivity distance to ζ is little, the optimal number of clusters that give the minimum UL

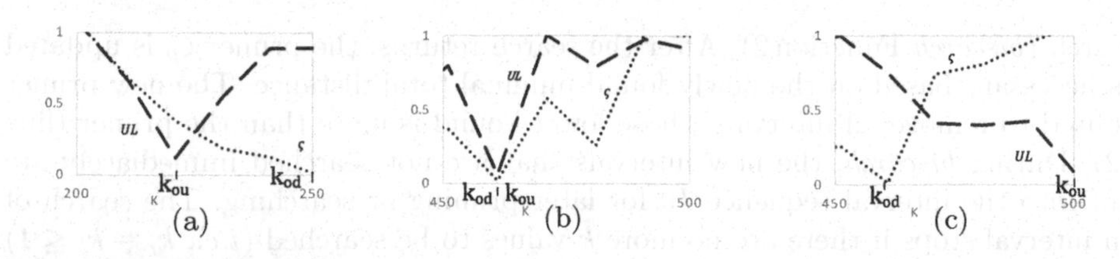

Fig. 4. Effect of balancing factor w on optimal k_o

k_{ou} is found to be at 220 which is to the left hand side of the optimal number of clusters that give a minimum total distance k_{od} which is 250 clusters. When w makes the sensitivity contribute twice as much as the homogeneity distance in plot (c) the positions of the two optimal k's are swapped. In plot (b) where w makes the homogeneity and sensitivity distances contribute equally to the total distance, the positions of the two optimal k's align well. This is the ideal w and it is decided in two rounds of initialisations for any given k. First, w is set to 1 and the transactions initially assigned to the seeded clusters. An actual w is then calculated by the total homogeneity $\mathcal{H}$ and sensitivity $\mathcal{S}$ distances ($w = \mathcal{H}/\mathcal{S}$). Second, the actual w is used to re-do the cluster initialisation, after which $w = \mathcal{H}/\mathcal{S}$ is updated and finally used in the update stage as the ideal w. The results in Fig. 4 were also replicated in BMS1 and BMS2.

(2) Utility loss in queries: We compare the utility loss of UAC to *HorPart* [9] using 10,000 transactions sampled from each dataset. The resulting partitions that were produced were then anonymised with *VerPart* and *sanitisation* (Sect. 2). In the experiments, three groups of term associations with relatively low (1–10), medium (10–20), and high (40–60) supports in the dataset were randomly selected as queries. For each anonymised data, 20 reconstructed datasets were generated by randomly linking entries in the record, sensitive and term chunks of a cluster. The same associations were searched in the reconstructed datasets to find their supports and the results averaged. The utility loss of an association (a, b) is computed by calculating the relative error [9] of the supports.

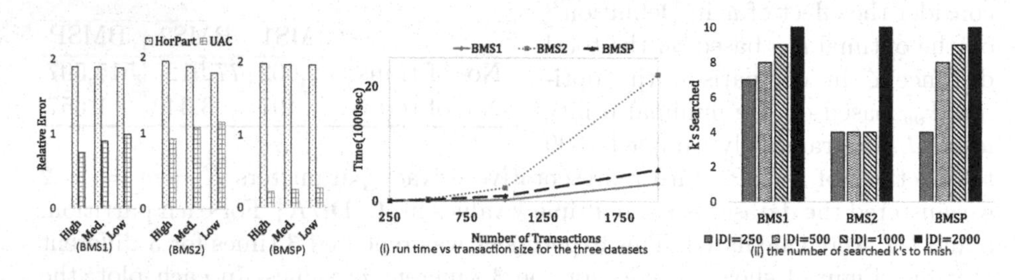

Fig. 5. Utility losses **Fig. 6.** Scalability and effectiveness of algorithm

Figure 5 shows that our method *UAC* has significant utility gains over *HorPart* on all three datasets with a typical improvement of over 50%.

(3) Scalability and Effectiveness: This experiment shows the scalability of our algorithm *DetK* which calls the clustering algorithm *DocClust*. There are two plots. Figure 6(i) indicates our method scales well for all three datasets. BMS2 which has much more terms (Table 4) required more time to compute than BMS1 and BMSP. Figure 6(ii), indicates the number of k values searched before *DetK* finds the optimal k. There are three groups of bars and each group is for a dataset. Within each group, different corpus sizes are used. The plot proves that the pruning technique in ***DetK*** is very effective. In the worst case (the black bars), the algorithm searched only 10 k values, less than 3% of all possible k values, before the optimal was found. We remark that the time cost is mostly from the clustering algorithm ***DocClust*** which is expected.

5 Related Work

Existing clustering methods for privacy preservation can be grouped as *similarity based* which consider only the similarity of transactions during the clustering or *constraint based* where the clustering is subject to privacy constraints. In *similarity based* methods, [10] uses Hamming distance while [2] considers the position of terms in a taxonomy to calculate the distance. [5] enforces K-anonymity by considering a semantic-similarity between transactions. Also [4] uses a sequential clustering while [12] uses density based clustering. In *constraint based* methods, [3] proposes NN-search to group transactions on their similarity restricting each cluster to $p-1$ other transactions for every s-transaction. [6,7] also require explicit utility constraints to be specified by the user.

These works differ from ours, particularly the privacy goal is not incorporated into the clustering objective function (total distance) to be minimised. In addition, our method UAC lends well to other stronger privacy models like t-closeness by suitably modifying M (Formular 3) to reflect the needed number of non-s transactions for each s in the corpus $\mathbf{D}$ *i.e.* $M = |\mathbf{D}|/N(s, \mathbf{D})$.

6 Conclusion

In this work, we presented a novel clustering framework *DocClust* based on a special distance function that considers both the similarity and disclosure risk of transactions. Hence, when the total distance of a clustering is minimised, anonymisation can be achieved with minimal utility loss. We also presented a search algorithm *DetK* with an effective pruning to solve the best k problem.

References

1. Barbaro, M., Zeller, T.: A face is exposed for AOL searcher no. 4417749. The New York Times (2006)
2. Byun, J., Kamra, A., Bertino, E., Li, N.: Efficient k-anonymization using clustering techniques. In: DASFAA, pp. 188–200 (2007)

3. Ghinita, G., Kalnis, P., Tao, Y.: Anonymous publication of sensitive transactional data. IEEE TKDE **23**(2), 161–174 (2011)
4. Goldberger, J., Tassa, T.: Efficient anonymizations with enhanced utility. TDP **3**(2), 149–175 (2010)
5. Liu, J., Wang, K.: Anonymizing bag-valued sparse data by semantic similarity-based clustering. KIS **35**(2), 435–461 (2013)
6. Loukides, G., Liagouris, J., Gkoulalas-Divanis, A., Terrovitis, M.: Disassociation for electronic health record privacy. JBI **50**, 46–61 (2014)
7. Loukides, G., Liagouris, J., Gkoulalas-Divanis, A., Terrovitis, M.: Utility-constrained electronic health record data publishing through generalization and disassociation. In: Medical Data Privacy Handbook, pp. 149–177 (2015)
8. Machanavajjhala, A., Gehrke, J., Kifer, D., Venkitasubramaniam, M.: l-Diversity: privacy beyond k-anonymity. In: ICDE, p. 24 (2006)
9. Terrovitis, M., Liagouris, J., Mamoulis, N., Skiadopoulos, S.: Privacy preservation by disassociation. PVLDB **5**(10), 944–955 (2012)
10. Terrovitis, M., Mamoulis, N., Kalnis, P.: Local and global recoding methods for anonymizing set-valued data. VLDB J. **20**(1), 83–106 (2011)
11. Zheng, Z., Kohavi, R., Mason, L.: Real world performance of association rule algorithms. In: ACM SIGKDD, pp. 401–406 (2001)
12. Zhu, H., Ye, X.: Achieving k-anonymity via a density-based clustering method. In: WAIM, pp. 745–752 (2007)

Self-tuning Filers — Overload Prediction and Preventive Tuning Using Pruned Random Forest

Kumar Dheenadayalan(✉), Gopalakrishnan Srinivasaraghavan, and V.N. Muralidhara

International Institute of Information Technology, Bangalore, India
d.kumar@iiitb.org

Abstract. The holy-grail of large complex storage systems in enterprises today is for these systems to be self-governing. We propose a self-tuning scheme for large storage *filers*, on which very little work has been done in the past. Our system uses the performance counters generated by a filer to assess its health in real-time and modify the workload and/or tune the system parameters for optimizing the operational metrics. We use a Pruned Random Forest based solution to predict overload in real-time — the model is run on every snapshot of counter values. Large number of trees in a random forest model has an immediate adverse effect on the time to take a decision. A large random forest is therefore not viable in a real-time scenario. Our solution uses a pruned random forest that performs as well as the original forest. A saliency analysis is carried out to identify components of the system that require tuning in case an overload situation is predicted. This allows us to initiate some 'action' on the bottleneck components. The 'action' we have explored in our experiments is 'throttling' the bottleneck component to prevent overload situations.

Keywords: Random forest · Pruning · Feature ordering · Storage · Filer · Self-tuning · Storage load

1 Introduction

Large-scale cluster-based storage systems [9,23,25] form an important part of any distributed infrastructure. This is true, because the data stored in these systems is often critical and these account for a large chunk of the overall infrastructure acquisition and maintenance cost. These systems typically take care of almost all the data storage needs of an enterprise and are expected to deliver consistent I/O performance across varied workload demands by a wide spectrum of data intensive applications. Traditionally these systems have been tuned based on workload characteristics that are assumed to be known in advance and predictable. Unfortunately, such assumptions do not hold in large organizations. We therefore focus on using the live system counters that are

J. Kim et al. (Eds.): PAKDD 2017, Part II, LNAI 10235, pp. 495–507, 2017.
DOI: 10.1007/978-3-319-57529-2_39

generated routinely in most storage systems and use them effectively to periodically evaluate and suggest reconfiguration. Counters in general can provide information on the performance of the system at various levels. These counters are critical for an accurate analysis of system behavior and tuning the system if required.

Research on self-organizing storage in the past [8] has largely focused on scaling storage systems dynamically in response to workloads. However, online optimization or self-tuning of storage systems based on system counter data has received little attention. A reasonably generic way to make a complex system self-sustaining in terms of performance and making it relatively immune to changing workload patterns is to enable dynamic reconfiguration/tuning based on the performance counters collected from within the system. Large scale filers are an assembly of a number of *filer components*/storage objects (physical, logical, protocol, software components). Some key components are buffer, cache, NVRAM, cpu, volume, aggregate, disk etc. Complex storage system architectures can include tens to hundreds of such interconnected components working together to provide high-levels of performance expected from such systems. Each component typically generates tens to hundreds of performance counters, each measuring different aspect of performance of the component. The total number of performance counters/features, across all components could be more than 100,000. When collected at intervals of 5–10 s, this counter data forms a rich source of information for system diagnostics and tuning.

Our contributions in this paper is to develop a self-tuning storage system. Towards this, we assume that the granularity at which the system takes actions during self-tuning is at a component-level. We also assume that components can be throttled, isolated, etc. towards improving the overall performance characteristics of the system as a whole. The primary objective in our experiments was to minimize the duration of overload without impacting the live workloads. Another key contribution is the use of component saliency analysis, enabling self-tuning system to identify the bottleneck components that lead to performance degradation.

The rest of this paper is organized as follows. The next section briefly presents the past literature followed by Sect. 3, which gives an overview of our system and introduces the key ideas in our self-tuning solution. Section 4 describes in detail the system we have implemented. We conclude in Sect. 6 with some possible extensions of our work and a discussion of some of broader challenges in building a self-tuning system of the kind we have envisaged in this paper.

2 Related Work

Use of learning algorithms for performance modeling [7,13] and failure predictions at a disk-level [11,16] has been researched for more than two decades. With the evolution of I/O intensive scientific applications and the prevalence of data redundancy and replication schemes, focus has gradually moved towards large scale storage systems. Research related to large scale storage systems [4,19] has gained importance in the recent years due the complexity of such systems increasing exponentially in the recent past and the associated difficulty in maintenance and management of such systems.

Self-* Storage systems was proposed in [8], where Self-* stands for self-configuring, self-organizing, self-tuning, self-healing and self-managing. Though, this is an ideal scenario, the work was carried out from a storage administrators' point of view. Architectural advancements in the past decade have ensured that cluster-based storage systems can provide a self-organizing system [23] to some extent but aspects such as self-tuning and self-healing continue to be rapidly evolving areas of research. Our current work is aimed at addressing the possibility of self-tuning using learning algorithms. Analysis of storage load in the past was restricted to latency analysis. More recently, there has been progress in measurement of the end-to-end load of the storage system at the client level [4]. The load is measured as a function of standard benchmark operation that makes it unique and user/administrator friendly for load analysis.

3 System Overview

Datasets used in our experiments are performance counters collected from storage system by periodically polling for and extracting counters related to all the active components. Every counter set (a snapshot of all the counter values collected together) is labeled with one of the three broad categories, viz., ZEROLOAD (**Z**), NORMAL (**N**), *OVERLOAD* (**O**). The labels are typically derived through thresholds implied by formal service-level-agreements (SLAs) on these systems. SLAs for storage systems include the total up-time of the system, time taken for read/write operations, number of retries and so on. Thresholds are defined to help maintain the agreed levels of latencies/up-time and a potential breach of the same is proactively avoided through actions taken by the proposed self-tuning system. The data collection in our experiments is similar to that collected by [4] but for real world application I/O patterns. The number of counters is typically too large for use in a model that will operate in real-time. We use a couple of simple strategies to bring the number of counters that will be considered for the final model down to a few hundreds. We use the terms counter and feature (of the current system state) interchangeably in the rest of this paper. A fraction (30%) of the dataset collected is held-out as a test set that we use for the final model selection based on a weighted class loss function.

Any model that predicts the state of the system based on performance counter data in real time (the predictions are useful only if they are actionable in real-time) needs to be fast, interpretable and accurate, in that order of priority. It needs to be fast because decisions using the model have to be taken in real time for them to be useful. Interpretability is important for being able to accommodate manual/automated interventions as required. Accuracy in this case may not be as important as the other two since we believe that some intervention is better than no intervention at all. The system performance cannot be significantly worse even if the model is not terribly accurate.

We propose a Pruned Random Forest (PRF) based solution augmented with a way to accommodate weighted classes. Our solution is built around a random forest classifier that will classify a performance counter vector into one of three classes — **Z**, **N**, **O**. Note that we assume the performance counter vector is a complete representation of the 'current' state of the system as a whole. The random forest constructed from the subset of counters is pruned based on a weighted *Matthews Correlation Coefficienct* (MCC) [20] ranking of the trees in the random forest. Optimal pruning is hard in general and pruning is probably not even worth it in most applications. However, several recent papers [5,6,15,21, 22,24] have shown the effectiveness of pruning using heuristics. It can drastically cut down the number of trees in the random forest without compromising on the accuracy/generalizability of the model. In many cases, the size of the random forest is down to nearly one tenth of its original size while improving the accuracy in most cases by explicitly preserving or enhancing the strength and diversity of the ensemble [5]. The most recent attempt in this area [5] makes pruning very attractive because it effectively eliminates the need to specify any limits on the size of the pruned ensemble. It makes pruning almost independent of the original (pre-pruning) size of the ensemble.

Pruning is important to ensure the first two criteria we laid out earlier — speed and interpretability. Unfortunately, random forests with hundreds of trees and several hundreds of features on which each random tree in the forest has been built can be very expensive in their decision making. Each decision requires the new input to be pushed through all the trees in forest before bagging them together. Hence, it is not quite suited for high throughput online scenarios like that of online tuning of large scale filers where prediction and real-time analysis are necessary for tuning the storage system. Cutting the size of the random forest by a factor of 10 directly cuts down on the decision making time by a factor of 10. Our algorithm generates several candidate pruned forests by varying a couple of hyperparameters. The candidate forests are evaluated based on a loss function that is sensitive to the fact that we are more concerned about misclassifications related to an **O**verload situation — either a normal load being classified as **O**verload or a genuine **O**verload situation not being recognized. We then annotate each node in every decision tree of the pruned random forest with exactly one system component label. A node is annotated with the component from which the feature (counter value) being tested at the node has originated. The final pruned, annotated random forest is the predictive model deployed in the system for self-tuning. The schematic in Fig. 1 illustrates the entire proposed system architecture.

While in deployment, the self-tuning system monitors a stream of sets of counter values. These counters are the same as those used during training and are collected at frequent (5–10 s) intervals. The predictive model is run on each snapshot of counter values in real-time. If the model predicts that it is potentially an overload situation, it carries out a *component saliency analysis* to determine, which component(s) is(are) the 'bottleneck(s)' leading to the predicted overload situation. The internal annotated structure of the pruned random forest is used

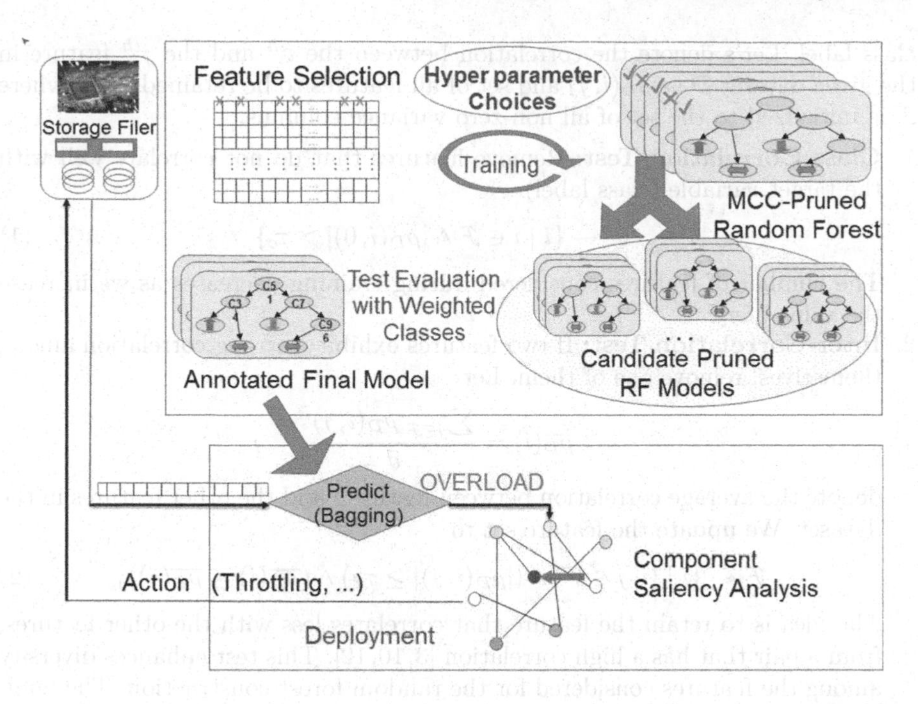

Fig. 1. System architecture

to arrive at a 'diagnosis' regarding the most likely components that need some action in case the system suffers from an overload situation. Experimental results reported in this paper are based on an implementation of self-tuning through throttling of I/O on specific components. Our experimental results adequately demonstrate the usefulness of our proposed scheme as a way to implement self-tuning on live storage systems. Short term reconfigurations such as throttling for short periods enables minimal impact on live workloads and reduce **O**verload scenarios.

4 Implementation Details

4.1 Feature Selection from Counter Data

Importance of good feature selection, both as a dimensionality reduction tool and as a way to improving accuracy (by removing features that are redundant and potentially misleading), in the design of machine learning algorithms is well known [1,18]. In our context, it was crucial to bring the dimensionality down significantly for filer dataset with more than 100,000 features. Even after the routine pre-processing task of removing columns with zero variance the time taken to generate a random forest model for our dataset was in the order of days on a standard high memory machine. We use two custom feature selection steps in our scheme. Suppose there are d features and the 0th 'feature' is the class label. Let's denote the correlation between the ith and the jth feature in the given dataset $\mathcal{D}$ as $\rho_{\mathcal{D}}(i,j)$ and set of all features to be retained as $\mathcal{F}$, where $\mathcal{F}$ is initialized to the set of all non-zero variance columns.

1. **Class Correlation Test:** Remove features that do not correlate well with the target variable (class label).

$$\mathcal{F} \leftarrow \{i \mid i \in \mathcal{F} \land |\rho_{\mathcal{D}}(i,0)| > \tau_c\} \quad (1)$$

   The number of features considered during training decreases as we increase the value of τ_c.

2. **Inter-Correlation Test:** If two features exhibit a strong correlation among themselves, remove one of them. Let

$$\overline{\rho_{\mathcal{D}}}(i) = \frac{\sum_{j \in \mathcal{F}} \rho_{\mathcal{D}}(i,j)}{d}$$

   denote the average correlation between feature i and the other features in the dataset. We update the feature set to

$$\mathcal{F} \leftarrow \{i \mid (i,j \in \mathcal{F}) \land (|\rho_{\mathcal{D}}(i,j)| \geq \tau_f) \land (\overline{\rho_{\mathcal{D}}}(i) \leq \overline{\rho_{\mathcal{D}}}(j))\} \quad (2)$$

   The idea is to retain the feature that correlates less with the other features, from a pair that has a high correlation [3,10,12]. This test enhances diversity among the features considered for the random forest construction. The number of features considered during training increases as we increase the value of τ_f.

4.2 Random Forest Pruning

We follow the MCC based pruning proposed in [5]. MCC, as a choice of performance measure is known to be an unbiased estimate of accuracy of a model [20]. One of the internal estimates known to act as a cross-validation step during forest construction is the Out Of Bag (OOB) error measure [2]. It helps in avoiding the need to maintain an exclusive hold-out set for testing or ever carry out explicit cross validation. MCC for each tree is defined on the OOB set for that tree. For a multiclass scenario like ours, the MCC is computed as the weighted sum of the class-MCCs computed for each class. Class-MCC is computed by taking the class against the rest, and computing the class-MCC as if it were a binary classification problem. Denoting the full random forest produced on the selected subset of features be $\mathcal{T}$, let class-MCC calculated in this manner for the kth tree $T_k \in \mathcal{T}$ and class c be denoted $m_c^{(k)}$. Also, let the fraction of the OOB set for T_k that belongs to class c be denoted $w_c^{(k)}$. The weighted MCC $m^{(k)}$ for T_k is then defined as

$$m^{(k)} = \sum_{c \in C} m_c^{(k)} \times w_c^{(k)}$$

It was shown empirically in [5] that removing trees whose weighted MCC is below the 80th percentile often works well. For our purposes we use a derived limit on the number of trees in the model to retain as many trees in the order of their weighted MCC values.

4.3 Hyperparameter Tuning for Weighted Classes

Our algorithm introduces two hyperparameters τ_c, τ_f in Eqs. 1 and 2. The effect of τ_c on the overall training time is the most pronounced. Notice that the inter-correlation test requires a pair-wise test on all features. Having a very low value of τ_c can therefore make the inter-correlation test unviable. Similarly, having a large value of τ_f can significantly increase the number of features on which the random forest needs to be built. We set a lower bound τ_c^* on $\tau_c \geq \tau_c^*$ and an upper bound τ_f^* on $\tau_f \leq \tau_f^*$ primarily from training time considerations. We set an upper limit ϵ^* on the OOB error that any of the models can commit. Another important performance criterion is the maximum time available to take a decision when the model is deployed. This time is directly dependent on the number of conditions that need to be checked on any test input for classifying it using a random forest. A rough estimate of this is clearly the product of the average depth of a tree $d_\mathcal{T}$ in the forest and the number of the trees $n_\mathcal{T}$ in the forest. We therefore require that $d_\mathcal{T} \times n_\mathcal{T} \leq \Delta$ for some constant Δ derived from the maximum time available for an online decision. We calculate the average depth of the random forest $\mathcal{T}$ consisting of trees $\{T_1, \ldots, T_{n_\mathcal{T}}\}$ as

$$d_i = \frac{\sum_{j=1}^{l_i} d_{ij}}{l_i}, \quad d_\mathcal{T} = \frac{\sum_{i=1}^{n_\mathcal{T}} d_i}{n_\mathcal{T}}$$

where d_i is the average depth of the tree T_i with l_i leaves, the depth of leaf j being d_{ij}. The pseudocode for the complete hyperparameter tuning scheme is shown in Algorithm 1. It generates a series of candidate models for evaluation.

Algorithm 1. Hyperparameter Tuning

```
Fix values for τc*, τf*, ε* and Δ.
for τc = τc* ... 1 step 0.1 do
    for τf = τf* ... 0 step −0.1 do
        # Remove features with low class-correlation
        F ← {i | i ∈ F ∧ ρD(i,0) < τc}
        # Remove features with high inter-correlation
        F ← {i | (i,j ∈ F) ∧ (|ρD(i,j)| ≥ τf) ∧ (ρ̄D(i) ≤ ρ̄D(j))}
        # Continue only if feature set is different from previous iteration
        # Build random forest
        T ←randomForest(D,F); εT ← OOB Error of T; dT ← Average Depth of T
        if εT > ε* then
            break                    ▷ εT is expected to increase for any smaller τf
        else
            T ← {Ti | r(Ti) ≤ Δ/dT}     ▷ r(Ti) = weighted MCC rank of Ti ∈ T
            Output T as a candidate model
        end if
    end for
    if τf = τf* then
        break  ▷ (τc, τf*) pair failed. So εT is expected to increase for any larger τc
        # No more candidate forests get generated
    end if
end for
```

Table 1. Confusion and loss matrices

Confusion Matrix

	Z	**N**	**O**
Z	c_{zz}	c_{zn}	c_{zo}
N	c_{nz}	c_{nn}	c_{no}
O	c_{oz}	c_{on}	c_{oo}

Loss Matrix

	Z	**N**	**O**
Z	0	1	λ
N	1	0	λ
O	λ	λ	0

The models generated by Algorithm 1 are evaluated on a separate (hold-out) test set using a weighted class loss function.

4.4 Weighted Class Loss Function Evaluation

We are concerned primarily about accurate predictions of **O**verload situations. It is important to control both false-positives and false-negatives in this case — false positives cause unnecessary 'tuning' actions that can degrade the system performance, false negatives make the self tuning system ineffective. This does not distort the overall performance of the random forests because the MCC based pruning carried out prior to the test evaluation ensures overfitting for the **O**verload class is avoided. Each candidate model is evaluated against the hold-out test set. The confusion matrix is computed for the model on the test set. Table 1 shows a representative confusion matrix and the loss matrix, where $\lambda > 1$ is a fixed constant. The loss matrix ensures that any mis-classification involving **O** is penalized much more than those not involving **O**. We can use an appropriately large value of λ to ensure that models performing better over **O** are preferred. The final model picked is the one that minimizes

$$c_{zn} + c_{nz} + \lambda(c_{oz} + c_{on} + c_{zo} + c_{no})$$

4.5 Component Saliency Analysis

The final model used for prediction is annotated by associating each node of the final random forest with the component to which the feature being tested at the node belongs. The idea is that an overload situation arising due to a component would invariably be indicated by counters belonging to that component. So one would cxpect that in case it is an overload situation, a number of counters from the component 'culprit' would have figured in the paths in a number of trees that the current test counter vector 'passed through' during the decision making. The key components responsible would often be represented in these decision paths. Notice that the number of components is of the same order as the number of trees in the forest — both are around 100. We therefore zero in on the salient component that could be 'responsible' for the overload situation as follows. Let the final model be the forest $\{T_1, \ldots, T_n\}$. Given a test counter vector $\mathbf{c}$, let v_i be a binary variable representing the verdict of $T_i(\mathbf{c})$. T_i on $\mathbf{c} \implies v_i$ is 1 if and only if T_i concludes that $\mathbf{c}$ indicates an **O**verload situation. Let $P_i(\mathbf{c})$ denote the path taken by $\mathbf{c}$ in T_i and S_i, a vector of length equal to the number of components in the system, for each $1 \leq i \leq n$. The pseudocode for the models' response for every new counter vector is shown in Algorithm 2.

Algorithm 2. Deployment Scenario — Component Saliency Analysis

```
New counter vector c, model {T_1, ..., T_n}.
for i = 1, ..., n do
    v_i = T_i(c)
    if v_i = 1 then
        for x ∈ P_i(c) do
            S_ij += 1 if x ∈ Component j
        end for
    end if
    if Σ_{i=1}^{n} v_i > n/2 then                          ▷ Bagging
        S* = Σ_{v_i=1} S_i
        # return the (index of) most salient component for tuning action
        return arg max_j S*_j              ▷ S*_j is the j^th component of S*
    end if
end for
```

5 Experiment Setup

We conducted our experiments on a NetApp Cluster storage system with workloads similar to those observed in the real world. These workloads were generated using the 'Standard Performance Evaluation Corporations' (SPEC) Solution File Server (SFS) tool, designed by a consortium of storage vendors to evaluate the performance of different storage systems for real-world workload patterns. The key characteristics of the four different datasets used in our experiments are shown in Table 2. During offline testing, each dataset was split with 70% used as training data and the rest used for testing. Random Forest pruning and the annotations required for our algorithm was implemented by patching the *randomForest* package in *R* [14]. The size of the random forests generated for each

Table 2. Training workload/dataset summary

Dataset	Size	Instances	Clients	Workload description
TXN	3.1 Gb	7,715	43	Online Txn Data (financial, telecom,...)
SVD	2.4 Gb	5,767	20	Streaming Video Data
BLD	3.0 Gb	7,489	39	Software Build — meta-data operations, file reads, source compilation and binary data generation
MIX	2.4 Gb	6,003	102	Mixture (of TXN, SVD and BLD) Workload

Table 3. Summary of implementation results for filer data. (*%Overload** – Percentage improvement in overload classification after pruning)

	$\lvert\mathcal{F}\rvert$	$min(t_c, t_f)$	$max(t_c, t_f)$	PRF	CRF	$d_{\mathcal{T}}$	$candidates$	k_{PRF}	$\%Overload^*$
TXN	177	(0.01, 0.65)	(0.21, 0.9)	25.42	25.52	15	7	88	0
SVD	101	(0.1, 0.70)	(0.5, 0.9)	3.29	3.35	9	9	153	0.65
BLD	221	(0.01, 0.7)	(0.23, 0.9)	16.55	16.93	12	5	107	0
MIX	12	(0.1, 0.75)	(0.5, 0.9)	2.27	2.47	8	6	185	0.9

(τ_c, τ_f) combination was 500, before they were pruned to the required size. Based on the filer configuration, $\Delta = 2,000$ ms was set to identify the size of the candidate pruned forest.

Discussion: The implementation results are summarized in Table 3. The table summarizes the hyperparameter ranges that were searched $\{min(t_c, t_f), max(t_c, t_f)\}$, the number of features that were retained ($\lvert\mathcal{F}\rvert$), number of forests produced as candidates (k_{PRF}) and the accuracy of the pruned (PRF) and unpruned (CRF) models. Improvements observed for **O**verload class due to pruning is also presented as $\%Overload^*$. $d_{\mathcal{T}}$ and *candidates* column in Table 3 indicate the average depth of trees and the number of alternated forests available for consideration. Hyperparameter tuning is illustrated in plots of the accuracy of the model against τ_f threshold for every fixed value of τ_c. The plots are shown in Fig. 2. Highlighted rectangular regions in the plot indicates the choices of τ_f and τ_c satisfying $\epsilon_{\mathcal{T}} \leq \epsilon^*$. The best candidate within the rectangular region is identified using the Loss matrix with $\lambda = 8$.

5.1 Online Self-tuning

Many possible configuration changes are available for administrators to optimize/tune/balance load on NetApp filer. Choosing the most appropriate configuration change depends on identifying the bottleneck components. Throttling the load on storage objects such as volumes is one such option [17]. Throttling helps in deterring load generated by users using a volume. This will lower the load affecting throughput of storage for a small set of users but a gain can be observed by rest of the users serviced by other volumes. Component saliency analysis helps in identifying if any single volume is accommodating huge I/O forming the bottleneck. Self-tuning of storage system was implemented on a live filer and tested on workloads for a period of 4 h. Throttling was initiated when **O**verload is predicted for 5 consecutive counter snapshots. It was enabled for a short interval (5 min). Also, throttling was carried out by restricting the I/O bandwidth of the volume to the average I/O load observed on the volume in the past 15 min. Figure 3 shows the Gantt chart of the various Overload durations encountered during the online testing phase. The figure clearly shows a significant reduction (40%) in the occurrence of **O**verload scenarios at the cost of loss

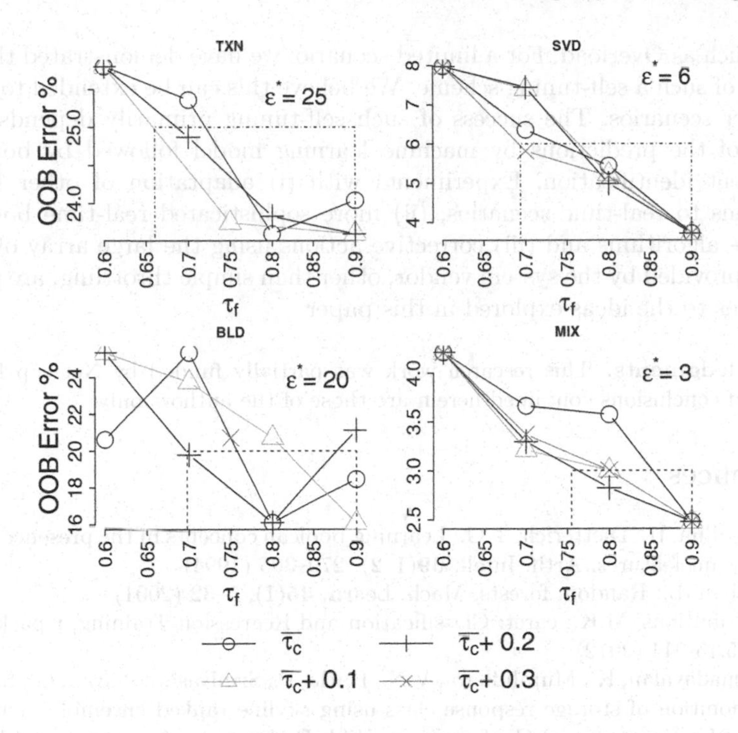

Fig. 2. Effect of (τ_c, τ_f) on accuracy

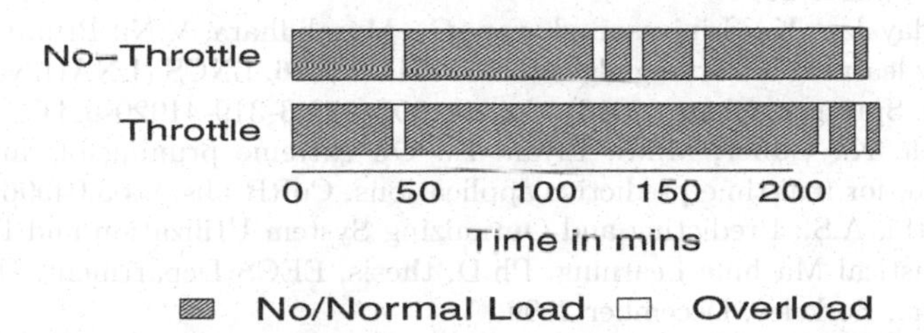

Fig. 3. Effect of self-tuning on load of a filer

of throughput of 1.33%. Self-tuning, when implemented using unpruned random forest will lead to 3 to 5 times increase in computation cost along with loss in prediction accuracy.

6 Conclusions and Future Work

Storage system counter data is very useful in identifying its health and the same has been demonstrated for different industry workload patterns. With appropriate modifications (like pruning) to standard machine learning algorithms such as random forests, it is possible to predict the impending health of the system in real-time and initiate corrective actions as a response to prevent unfavourable events such as Overload. For a limited scenario, we have demonstrated the effectiveness of such a self-tuning scheme. We believe this can be extended to numerous other scenarios. The success of such self-tuning primarily depends on the quality of the predictions by machine learning model followed by bottleneck component identification. Experiments with (i) adaptation of other learning algorithms to real-time scenarios, (ii) more sophisticated real-time bottleneck diagnosis algorithms and (iii) corrective actions using the large array of tuning options provided by the system vendor, other than simple throttling, are possible extensions to the ideas explored in this paper.

Acknowledgments. This research work was partially funded by NetApp Inc. The views and conclusions contained herein are those of the authors only.

References

1. Almuallim, H., Dietterich, T.G.: Learning boolean concepts in the presence of many irrelevant features. Artif. Intell. **69**(1–2), 279–305 (1994)
2. Breiman, L.: Random forests. Mach. Learn. **45**(1), 5–32 (2001)
3. Contributions, M.K.: caret: Classification and Regression Training, r package version 5.15-044 (2012)
4. Dheenadayalan, K., Muralidhara, V.N., Datla, P., Srinivasaraghavan, G., Shah, M.: Premonition of storage response class using skyline ranked ensemble method. In: 2014 21st International Conference on High Performance Computing (HiPC), pp. 1–10, December 2014
5. Dheenadayalan, K., Srinivasaraghavan, G., Muralidhara, V.N.: Pruning a random forest by learning a learning algorithm. MLDM 2016. LNCS (LNAI), vol. 9729, pp. 516–529. Springer, Cham (2016). doi:10.1007/978-3-319-41920-6_41
6. Fawagreh, K., Gaber, M.M., Elyan, E.: On extreme pruning of random forest ensembles for real-time predictive applications. CoRR abs/1503.04996 (2015)
7. Ganapathi, A.S.: Predicting and Optimizing System Utilization and Performance via Statistical Machine Learning. Ph.D. thesis, EECS Department, University of California, Berkeley, December 2009
8. Ganger, G.R., Strunk, J.D., Klosterman, A.J.: Self-*storage: Brick-based storage with automated administration. Technical report, Carnegie Mellon University, School of Computer Science, Technical report (2003)
9. Ghemawat, S., Gobioff, H., Leung, S.T.: The google file system. In: Proceedings of the Nineteenth ACM Symposium on Operating Systems Principles, pp. 29–43. ACM (2003)
10. Hall, M.A.: Correlation-based feature selection for discrete and numeric class machine learning. In: Proceedings of the Seventeenth International Conference on Machine Learning, pp. 359–366. Morgan Kaufmann Publishers Inc. (2000)
11. Hamerly, G., Elkan, C.: Bayesian approaches to failure prediction for disk drives, pp. 202–209. Morgan Kaufmann Publishers Inc. (2001)
12. Kohavi, R., John, G.H.: Wrappers for feature subset selection. Artif. Intell. **97**(1–2), 273–324 (1997)
13. Lee, E.K.: Performance Modeling and Analysis of Disk Arrays. Ph.D. thesis, EECS Department, University of California, Berkeley, August 1993
14. Liaw, A., Wiener, M.: Classification and regression by randomforest. R News **2**(3), 18–22 (2002)

15. Martinez-Munoz, G., Hernandez-Lobato, D., Suarez, A.: An analysis of ensemble pruning techniques based on ordered aggregation. IEEE Trans. Patt. Anal. Mach. Intell. **31**(2), 245–259 (2009)
16. Murray, J.F., Hughes, G.F., Kreutz-Delgado, K.: Machine learning methods for predicting failures in hard drives: a multiple-instance application. J. Mach. Learn. Res. **6**, 783–816 (2005)
17. NetApp Inc.: Managing workload performance by using storage qos. https://library.netapp.com/ecmdocs/ECMP1196798/html/GUID-660A6C00-6D7E-4EE5-B97E-9D33C0B706B5.html
18. Opitz, D.W.: Feature selection for ensembles. In: Proceedings of the Sixteenth National Conference on Artificial Intelligence, pp. 379–384. American Association for Artificial Intelligence (1999)
19. Pollack, K.T., Uttamchandani, S.M.: Genesis: a scalable self-evolving performance management framework for storage systems. In: 26th IEEE International Conference on Distributed Computing Systems, p. 33 (2006)
20. Powers, D.M.W.: Evaluation: from precision, recall and f-measure to roc., informedness, markedness & correlation. J. Mach. Learn. Technol. **2**(1), 37–63 (2011)
21. Schwing, A.G., Zach, C., Zheng, Y., Pollefeys, M.: Adaptive random forest - how many "experts" to ask before making a decision? In: Proceedings of the 2011 IEEE Conference on Computer Vision and Pattern Recognition, pp. 1377–1384. IEEE Computer Society (2011)
22. Tamon, C., Xiang, J.: On the boosting pruning problem. In: López de Mántaras, R., Plaza, E. (eds.) ECML 2000. LNCS (LNAI), vol. 1810, pp. 404–412. Springer, Heidelberg (2000). doi:10.1007/3-540-45164-1_41
23. Tang, H., Gulbeden, A., Zhou, J., Strathearn, W., Yang, T., Chu, L.: A self-organizing storage cluster for parallel data-intensive applications. In: Proceedings of the 2004 ACM/IEEE Conference on Supercomputing, p. 52. IEEE Computer Society (2004)
24. Tsoumakas, G., Partalas, I., Vlahavas, I.: An ensemble pruning primer. In: Okun, O., Valentini, G. (eds.) Applications of Supervised and Unsupervised Ensemble Methods. SCI, vol. 245, pp. 1–13. Springer, Heidelberg (2009). doi:10.1007/978-3-642-03999-7_1
25. Zhu, Y., Jiang, H., Wang, J., Xian, F.: Hba: distributed metadata management for large cluster based storage systems. IEEE Trans. Parallel Distrib. Syst. **19**(6), 750–763 (2008)

A Centrality-Based Local-First Approach for Analyzing Overlapping Communities in Dynamic Networks

Ximan Chen[1](✉), Heli Sun[1], Hongxia Du[1], Jianbin Huang[2], and Ke Liu[2]

[1] Department of Computer Science and Technology,
Xi'an Jiaotong University, Xi'an 710049, China
chenximan@gmail.com, hlsun@xjtu.edu.cn, duhx123@163.com

[2] School of Software, Xidian University, Xi'an 710049, China
jbhuang@xidian.edu.cn

Abstract. With the increasing demand of dynamic graph data analysis, mining communities in time-evolving data has been a research hotspot. However, traditional community detection methods have efficiency issue in the huge dynamic network data and rarely consider about overlapping communities. In this paper, we first propose a centrality-based local-first approach for overlapping community discovery in static network, called CBLF. Different with the traditional top-down approach, CBLF detects communities from central nodes and theirs neighbors which conforms to reality better. Then we present a novel evolutionary community detection approach called CBLFD based on this effective approach and sequence smoothing mechanism. Experimental results on real-world and synthetic datasets demonstrate that these algorithms achieve higher accuracy and efficiency compared with the state-of-art algorithms.

Keywords: Dynamic networks · Overlapping communities · Community detection · Evolutionary clustering

1 Introduction

Recently community detection has attracted extensive attention due to their ubiquity and generality of modeling. Furthermore, dynamic networks gains more and more attention due to the topology structures of the networks will change constantly over time in real world, such as Internet traffic data network, paper co-authorship network and dynamic social network. Therefore detecting communities on dynamic networks are of significant meanings for revealing important information hidden in networks. There are many studies about community detection. However, these methods detect communities on a global view, while it is difficult to obtain the total structure on a large scales. Besides, these methods rarely consider about detecting overlapping communities on dynamic networks and have poor performance in highly overlapping community.

J. Kim et al. (Eds.): PAKDD 2017, Part II, LNAI 10235, pp. 508–520, 2017.
DOI: 10.1007/978-3-319-57529-2_40

In this paper, we first propose a novel approach for analyzing overlapping communities in static networks called CBLF and then put forward its dynamic version—CBLFD. Our main contributions in this paper are as follows: (1) We design a novel metric called *ego centrality* to find out the central nodes, which are more likely to have multiple community membership and guide the evolution process of their neighbors. So we can use *ego centrality* to naturally detect overlapping communities in dynamic network. (2) To the best of our knowledge, our work is the first one that providing a novel local-first perspective from the ego network in dynamic network. (3) We design a new concept called *community relevancy* to locate the most relevancy community in previous timestep and reinforce the community results in current timestep. This concept also provides convenience to analyze the evolutionary clustering results on dynamic networks. (4) Accuracy improvement over other state-of-art methods for detecting highly overlapping communities in dynamic network.

The rest of this paper is organized as follows: we introduce the related work about community detection and evolutionary clustering in Sect. 2. Next, we introduce our algorithm in Sect. 3. The experimental results and analysis are provided in Sect. 4. Finally, we conclude the paper in Sect. 5.

2 Related Work

Since the importance of revealing functionality in social network, community detection has been one of the research hot spots in recent years. Besides classical static unoverlapping communities, numerous techniques have been developed for overlapping communities. In 2005, Palla [11] first put forward the significant meaning of overlapping community and propose a clique percolation algorithm (CPM) to detect overlapping partition by searching adjacent cliques. Ahn et al. proposed Link clustering [1] for hierarchical clustering. Many research addressed this problem by a wide spread algorithm—label propagation process [12] such as SLPA [17] and COPRA [5]. However, these methods above suffer from poor performance with highly overlapping density, and usually detect communities in a global perspective.

Dynamic networks have also attracted increasing interest for the great potential in capturing natural and social phenomena over time. And evolutionary clustering is an important method for dynamic community detection in the field of dynamic network research. Chi et al. extend the concept of similarity and propose an evolutionary spectral clustering method [3]. Tang et al. design an evolutionary clustering framework based on spectral clustering for detecting communities on multi-features networks, which can handle the evolution of edges between nodes [15]. Kim et al. propose an evolutionary clustering method based on grain and density [6].

The major differences between our method and the state-of-the-art approaches are that: (1) We emphasize local first which detect communities from the view of central nodes rather than the perspective of the whole network. (2) We can find the overlapping community in dynamic network at the same time.

3 Proposed Method

3.1 Preliminaries

The main problems that this paper need to solve are: (1) Detecting the overlapping communities in dynamic networks. (2) Tracing the evolving process of communities between adjacent timestamps and finding out their corresponding relationship.

In order to define community in an appropriate grain and describe the evolution process of communities, herein we introduce some basic concepts first. Given a dynamic network $G = <G_1, G_2, \ldots, G_k>$, which is constituted of a sequence of networks at k timestamps. And $G_t(V_t, E_t)$ is a snapshot of the dynamic network at time t, in which V_t represents the node set and E_t is the set of edges. We can detect the community structure $\mathcal{CR}_t$ of the network in the snapshot G_t by clustering algorithm. The *ego network* of node $v \in V_t$ is a subgraph $G'_t(V'_t, E'_t)$, denoted by $EN(v)$, in which $V'_t = \{x| \{v, x\} \in E_t, x \in V_t\} \bigcup \{v\}$, $E'_t = \{\{u, x\} | u, x \in V'_t, \{u, x\} \in E_t\}$. When we delete vertex v and all the edges attached to v from its ego network $EN(v)$, we got the *neighborhood network* $N(v) = \{V''_t, E''_t\}$. $V''_t = \{x| \{v, x\} \in E_t, x \in V_t\}$, $E''_t = \{\{u, x\} | u, x \in V''_t, \{u, x\} \in E_t\}$. Then the *neighborhood communities* $NC(v)$ of node $v \in V_t$ is the partitions of its neighborhood network $N(v)$. So the *ego communities* of node v is the set of its neighborhood communities including the ego node itself.

Similar ego communities can be overlap, and merged into a bigger global communities. So we need to describe the similarity of two communities. For two communities C_1 and C_2, the similarity $ComSim(C_1, C_2)$ is defined as follow

$$ComSim(C_1, C_2) = \frac{|C_1 \cap C_2|}{min(|C_1|, |C_2|)} \tag{1}$$

3.2 Ego Centrality

The global community could be obtained by merging similar ego communities with Eq. 1. However taking every node's ego communities into consideration is time consuming and may lead to redundant results. So we need a more effective approach. In 2014, Rodriguez found that cluster centers are characterized a higher density than their neighbors, and by a relatively large distance from points with higher densities [14]. Inspired by his idea, we propose a new metric called *ego centrality* to find community centers in network. We first introduce local density $\rho(v)$ for node v to pick up the most influent nodes.

$$\rho(v) = \frac{\sum_{u \in Neighbors(v)} w(u, v)}{N - 1} \tag{2}$$

where N is the number of nodes in G_t, $Neighbors(v) = \{u|\{u, v\} \in E_t\}$. When in unweighted network, it can consider as for $\forall u \in Neighbors(v), w(u, v) = 1$. So the local density degenerates to the normalized degree of node v. Then use local similarity $\delta(v)$to choose the representative central nodes:

$$\delta(v) = \begin{cases} \max(sim(u,v)) & \exists u \in Neighbors(v), \rho(u) > \rho(v) \\ \min(sim(u,v)) & \forall u \in Neighbors(v), \rho(u) < \rho(v) \end{cases} \tag{3}$$

where $sim(u,v)$ is the similarity between u and v. Here we use Jaccard similarity:

$$sim(u,v) = \frac{|\Gamma(u) \cap \Gamma(v)|}{|\Gamma(u) \cup \Gamma(v)|} \tag{4}$$

where $\Gamma(u) = \{x | x \in Neighbors(u)\} \cup \{u\}$. Based on this two metric, the ego centrality $\gamma(v)$ is defined as the ratio between $\rho(v)$ and $\delta(v)$.

$$\gamma(v) = \frac{\rho(v)}{\delta(v)} \tag{5}$$

3.3 The Detection Algorithm for Static Network

Now we can present our solution to the static overlapping community detection problem. Our method is based on this idea: the cluster centers have a large scope of influence on more nodes, and they may take part in more communities. So we filter these centers and make their neighbors vote for which communities the center have joined by label propagation algorithm [12]. Our Centrality Based Local First algorithm (CBLF) contains three main phases: calculating the ego centrality and sorting nodes; using improved label propagation algorithm to find local communities and merging these local communities to get global communities.

In the first phase, all the nodes' ego centrality are calculated. Firstly, the local density of every node and the Jaccard similarity between its neighbors are calculated. If node v is the most influent one which has the highest local density, we make its local similarity lowest to away from other centers; otherwise we make the maximum similarity among its neighbors as its local similarity. Then the ego centrality of every node is obtained by Eq. 5. Finally the nodes are sorted in a descending order according to this measure.

Based on this order, every unvisited central node is treated as an ego, then we extract the neighborhood networks of it. Next the label propagation algorithm is employed on its neighborhood network to find neighborhood communities. We only detect ego communities from every unvisited central nodes, which can prevent duplicate detection and reduce time complexity at the same time. Herein, we use the label propagation algorithm (LPA) [12] to detect local ego communities because of its simplicity and efficiency.

Finally in the third phase, this ego is added to the set of neighborhood communities to obtain its ego communities. Then the similarity between these ego communities and the already existing community sets are compared, and the similar local communities are merged to get a global community. We used Eq. 1 to compute the similarity.

3.4 The Detection Algorithm for Dynamic Network

From the discussion above, we can solve the overlapping community detection problem in static network by a local-first approach. It is not enough, because the real networks are dynamic over time. So in this section, we propose a novel dynamic algorithm based on CBLF algorithm.

In order to guarantee that the communities don't change dramatically over time, we also need to know the relationship of the two communities at previous timestamp in the sequence smoothing framework. So We first introduce a new concept called community relevancy to depict the relation of two ego communities between time $t-1$ and time t, then we put this relevancy into a sequence smoothing framework which trade off the history quality with snapshot quality.

The *community Relevancy* of two communities $C_{m,t}$ and $C_{n,t}$, denoted by $rel(C_{m,t}, C_{n,t})$, is defined as follows:

$$rel(C_{m,t}, C_{n,t}) = \frac{\max\limits_{C_{p,t-1} \in \mathcal{CR}_{t-1}} |C_{p,t-1} \cap (C_{m,t} \cup C_{n,t})|}{|C_{m,t} \cup C_{n,t}|} \quad (6)$$

where $C_{m,t}$ is a community m at time t. The community relevancy helps us locate the most relevant community $C_{p,t-1}$ with $C_{m,t} \cup C_{n,t}$ in previous timestamp.

Evolutionary clustering under the temporal smoothness framework usually uses a cost function traded of the history quality with snapshot quality. Here we use a parameter α to trade off the similarity between two communities in current snapshot t with the relevancy between them in previous snapshot $t-1$. And η is also a user-defined parameter for controlling the occurrence of community merging:

$$cq = \alpha \cdot sim(C_{m,t}, C_{n,t}) + (1-\alpha) \cdot rel(C_{m,t}, C_{n,t}) \geq \eta \quad (7)$$

From Eq. 7 we can know that when α is large, the community detection result will have a strong bias towards the community similarity and reflect the real topology structure of the network at the present timestamp to a great degree. While α is small, the result will put more weight on the community relevancy, namely the community result at present will be similar with the result at previous timestamp to a large extent. Specially, the equation only contains the community relevancy when $\alpha = 0$. When $\alpha = 1$, community similarity is the

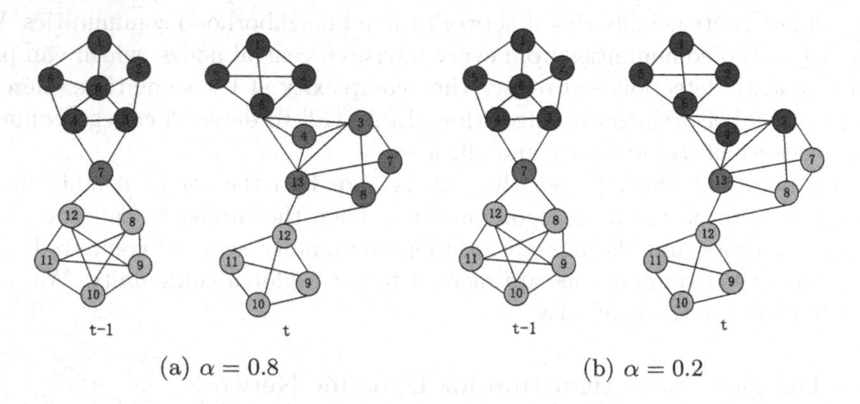

Fig. 1. The effect of parameter α for evolutionary clustering (Color figure online)

only item, causing the evolutionary clustering degraded into static network clustering because it just considers the topology structure of the present network. Figure 1 demonstrates the effect caused by different values of α. We can see from Fig. 1(a) that when $\alpha = 0.8$, clustering result reflects the real structure of the current network and a new community $\{3, 4, 7, 8, 13\}$ is found in green color. When α is small, like Fig. 1(b), the size of new local community becomes smaller or even vanishes when $\alpha = 0$. So various value of parameter α controls the level of preference to current similarity and previous relevancy. So we can firstly detect ego communities of the unvisited ego center, and then merge these communities by Eq. 7. Finally, the global communities can be obtained. The pseudo-code of the CBLFD are specified in Algorithms 1 and 2.

Algorithm 1. CBLFD

Input: Network $G_t = (V_t, E_t)$, $\mathcal{CR}_t = \emptyset$, $\mathcal{CR}_{t-1}$, Smoothing parameter $\alpha \in [0, 1]$, Merging parameter $\eta \in [0, 1]$

Output: Global community set $\mathcal{CR}_t$

1: $CaculateCentralitySort(V)$
2: **for** all $v \in V$ **do**
3: $visited.put(v, False)$
4: **end for**
5: **for** all $v \in V_t$ **do**
6: **if** $visited.get(v) == False$ **then**
7: $N(v) = EN(v) - v$ /*Delete the ego and all the edges attached to it*/
8: $NC(v) = LabelPropagation(N(v))$
9: **for** all $C \in NC(v)$ **do**
10: $C \leftarrow C \cup \{v\}$ /* Add the ego to local communities for overlapping communities*/
11: $\mathcal{CR}_t \leftarrow EvolutionMerge(\mathcal{CR}_t, \mathcal{CR}_{t-1}, C, \alpha, \eta)$ /*Use Sequence Smoothing to merge local communities into global communities*/
12: **end for**
13: **for** all $u \in EN(v)$ **do**
14: $visited.set(u, True)$ /*Set the visited flag*/
15: **end for**
16: **end if**
17: **end for**
18: **return** $\mathcal{CR}_t$

From Algorithm 1 we can know that each node in the network is just computed once. Taking each node as a center, its neighborhood network can be obtained through scanning, and label propagation algorithm is used to detect the local communities (Line 8). Then we put the central node into every community detected to form the ego communities (line 10). Next, we merge these ego communities into global communities according to Algorithm 2 (Line 11). In Algorithm 2, similarity between two communities in current snapshot (Line 7) and the relevancy in previous timestamp (Line 8) are calculated. Then a tradeoff between this two measures guides the merging phase. Eventually the final global communities $\mathcal{CR}_t$ are detected.

Algorithm 2. EvolutionMerge

Input: $\mathcal{CR}_t, \mathcal{CR}_{t-1}, C, \alpha, \eta \in [0,1]$
Output: $\mathcal{CR}_t$ after merging
1: MergeFlag = False
2: **for** every $C' \in \mathcal{CR}_t$ **do**
3: /*If two communities are the same, no need to merge*/
4: **if** $sim(C', C) == 1$ **then**
5: **return** $\mathcal{CR}_t$
6: **end if**
7: $sim(C', C) = \frac{|C' \cap C|}{min(|C'|,|C|)}$
8: $rel(C', C) = \frac{\max_{C_{t-1} \in \mathcal{CR}_{t-1}} |C_{t-1} \cap (C' \cup C)|}{|C' \cup C|}$
9: **if** $\alpha \cdot sim(C, C') + (1 - \alpha) \cdot rel(C, C') \geq \eta$ **then**
10: $U = C' \bigcup C$
11: $\mathcal{CR}_t = \mathcal{CR}_t - C'$
12: $\mathcal{CR}_t = \mathcal{CR}_t \bigcup U$
13: MergeFlag = True
14: **end if**
15: **end for**
16: /*Merge unsuccessfully*/
17: **if** MergeFlag == False **then**
18: $\mathcal{CR}_t = \mathcal{CR}_t \bigcup C$
19: **end if**
20: **return** $\mathcal{CR}_t$

For example, in Fig. 1, we compute the clustering result at time t on the condition that $\alpha = 0.8$. If we know that the ego community $\{1, 2, 5\}$ of node 1 has been in global communities, and the next node to handle is node 4. We can detect its ego community as $\{3, 4, 6, 13\}$. Now we compute the value of cq according to Eq. 7: $cq = 0.8 \times 1/4 + 0.2 \times 6/7 \approx 0.371$. So two ego communities can be merged into a new community $\{1, 2, 3, 4, 5, 6, 13\}$ as long as the parameter $\eta < 0.371$. Otherwise, the two communities exist in global communities simultaneously. When $\alpha = 0.2$, $cq = 0.2 \times 1/4 + 0.8 \times 6/7 \approx 0.736$. So only the parameter $\eta > 0.736$, the two communities will not be merged.

The merits of algorithm proposed in this paper are as follows: (1) Handling center nodes to avoid redundancy and unnecessary calculation. (2) Using the strategy of local first to avoid the constraint that the whole network structure should be clear before clustering and shows a kind of democracy that each node can vote for their center's community by label propagation algorithm. (3) Finding out ego communities for center nodes, so some nodes belonging to multiple communities can be divided into many local communities, resulting in overlapping communities by merging.

4 Experimental Results and Analysis

4.1 Datasets and Evaluation Metrics

Synthetic Datasets. For static network, we used the Lancichinetti-Fortunato-Radicchi (LFR) [8] benchmark. We set the number of nodes is 1000, the mixing parameter is $\mu = 0.1$ and the number of memberships of the overlapping nodes O_m is from 2 to 7. For dynamic networks, we adopt the tool in [4] to generate 5 groups of dynamic networks that contain 1000 nodes and 10 timestamps. The 5 groups of networks corresponding 5 evolution events: birth and death, expend and contract, merge and split, hide, switch.

Real-World Datasets. For static algorithms, we evaluate the performances on real-world network: karate, high school, books, dolphins, lesmis and netsci which are all available from Network Data Repository[1]. For dynamic networks, we used the dynamic annotated networks: DBLP[2] and Enron[3]. DBLP contains 2,723 nodes, 91,470 edges and 9 timesteps, and we used the number of publications as its *qualitative attribute* which mentioned in detail in evaluation metrics. Enron contains 2,356 nodes, 250,179 edges and 12 timestamps, and we used the TF-IDF of email key words as its *qualitative attribute*.

Evaluation Metrics. For synthetic networks whose communities are already know, we adopt expanded Normalized Mutual Information (NMI) for overlapping community [7]. For real world without ground truth, we use the overlap modularity Q_{ov} [10]. NMI and modularity are too strictly depend on graph structure and there is no consensus on the definition of what a community should look like in academic so far. It's unclear whether a particular mathematical definition is correct. So we used the atomic attributes attached with nodes which are regarded as quality attributes of nodes. Quality attributes don't belong to the network structure but they define the nodes in a better way. Here, we introduce the concept of *community quality* according to the quality attributes of nodes as:

$$CQ(P_t) = \frac{\sum_{(v_1,v_2)\in P_t} |QA(v_1) \cap QA(v_2)|}{\sum_{(v_1,v_2)\in E_t} |QA(v_1) \cap QA(v_2)|} \tag{8}$$

where P_t is a partition of G_t and $QA(v)$ denotes the quality attributes of node v. The greater the value of CQ, the bigger the possibility that similar nodes are in the same community. This equation aims at categorical attributes. When it comes to numeric attributes, we used the cosine similarity.

4.2 Experiments Settings

For static network, the compared algorithms are SLPA [17] and COPRA [5], which have better performance than the other state-of-the-art algorithms [16];

[1] https://networkdata.ics.uci.edu.
[2] http://www.informatik.uni-trier.de/~ley/db.
[3] https://www.cs.cmu.edu/~./enron/.

and Link, a representative hierarchical link clustering algorithm [1] and CPM [11]. For dynamic network, we select AFOCS [9] and iLCD [2]. AFOCS is a two-phase framework for detecting overlapping communities and also tracing the evolution of overlapping communities in dynamic mobile networks. iLCD is an efficient evolutionary clustering method via adding edges and merging similar edges for dynamic overlapping community detection. Our experiment is conducted on a Intel Core2 Quad CPU 64 bits @ 2.66 GHz, equipped with 4 GB of RAM.

4.3 Experimental Results on Static Network

In this section, we evaluate the performance of the static algorithm CBLF. We use the overlapping NMI, overlapping modularity and community quality to measure the algorithms on both synthetic networks and real world networks.

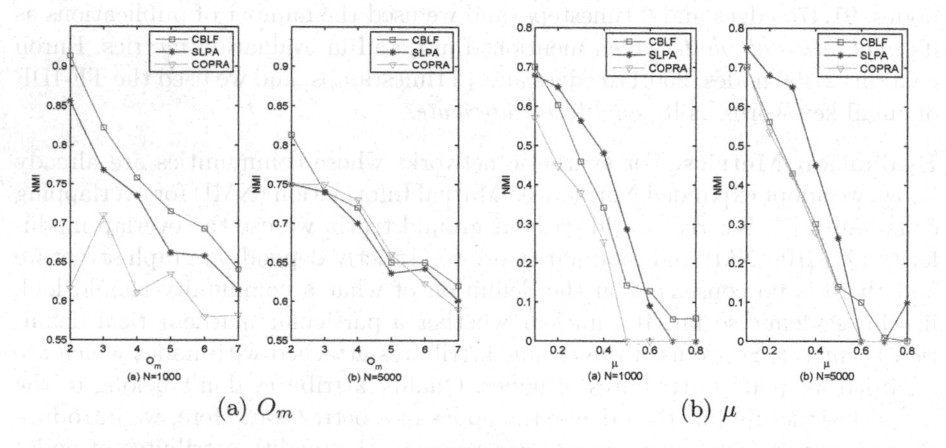

Fig. 2. The overlapping NMI values for LFR networks with different O_m and μ values

Firstly, we compare these algorithms in terms of NMI values on twelve synthetic networks with different O_m and μ values. Results are shown in Fig. 2(a) and (b) respectively. Figure 2(a) depicts the NMI values for three algorithms with $/mu = 0.1$ and O_m changes from 2 to 7. We can observe that CBLF achieves best performance among SLPA and COPRA when $n = 1000$ in Fig. 2(a). When $n = 5000$, CBLF achieves a comparable result with COPRA in Fig. 2(a), (b). In Fig. 2(b), when $O_m = 5$ and μ changes from 0 to 0.8, we can observe that CBLF achieves a comparable result in both $n = 1000$ and $n = 5000$.

Then, we evaluate the Q_{ov} scores of these algorithms in six real networks. From Fig. 3(a), we can observe that CBLF achieves comparable and relative stability performance compared with other algorithms. SLPA and COPRA have fluctuating Q_{ov} scores while CBLF is much more stable over different datasets.

Table 1. The community quality scores for annotated datasets and each static algorithms

Dataset	CBLF	SLPA	COPRA	Link
Congress	**5.49**	3.26	3.34	5.47
IMDB	**15.63**	9.85	2.43	1.69

To avoid the biased evolution metric as mentioned in Sect. 4.1, we also evaluate the *community quality* of different algorithms on two static annotated networks, as showed in Table 1. In this case, CBLF outperforms the other static overlapping algorithms.

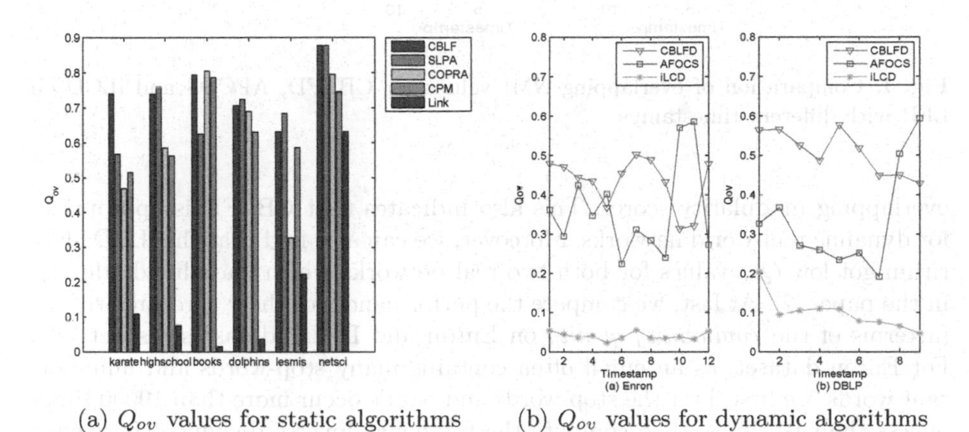

(a) Q_{ov} values for static algorithms (b) Q_{ov} values for dynamic algorithms

Fig. 3. Comparation of Q_{ov} values for static networks and dynamic networks on real-world datasets

4.4 Experimental Results on Dynamic Network

Next, we analysis the performance of CBLFD. In order to compare the accuracy of dynamic community detection algorithms, firstly, we apply CBLFD on 50 dynamic LFR networks. Figure 4 demonstrates the comparison of several algorithms on their overlapping NMI scores with different timestamps.

From Fig. 4, we can observe that our algorithm CBLFD can deal with overlapping communities in dynamic networks, comparable to AFOCS and better than iLCD as showed in Fig. 4. iLCD achieves a tolerable NMI scores for the birth and death stage, but fails for other stages. This phenomenon had already been mentioned in the paper [2] that it's not yet able to deal with the natural evolution of communities, such as merging and splitting. Then we compare the overlapping modularity scores for CBLFD, AFOCS and iLCD on two real-world networks, as showed in Fig. 3(b). From Fig. 3(b), we find that in both Enron and DBLP networks, CBLFD gains obvious advantage over AFOCS and iLCD in terms of

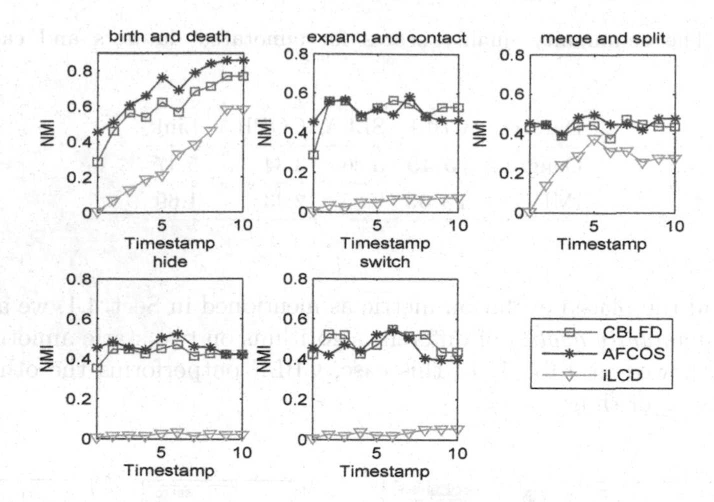

Fig. 4. Comparation of overlapping NMI values for CBLFD, AFCOS and iLCD on LFR with different timestamps

overlapping modularity scores. This also indicates that CBLFD is appropriate for dynamic real-world networks. Moreover, we can also find that the iLCD algorithm got low Q_{ov} values for both two real networks which had already discuss in the paper [2]. At last, we compare the performances of these three algorithms in terms of the *community quality* on Enron and DBLP datasets respectively. For Enron dataset, as an email often contains many stop-words and unimportant words, we first filter the stop-words and words occur more than 19000 times or less than 10 times, then compute the frequency-inverse document frequency (TF-IDF) [13] values for every remaining words. TF-IDF is a numerical statistic aims at reflecting how important a word to a document which is widely used in information retrieval and text mining. Then we use the TF-IDF values of the set of email words as the email sender's *qualitative attributes*. Finally, we use Eq. 8 to compute the partition's *community quality*. The results are shown in Table 2. From Table 2, we can see that CBLFD achieves a large margin over other algorithms on community quality. Because CBLFD detects communities in a local

Table 2. Community quality of three algorithms when using DBLP and Enron dataset

Dataset	Timestamp	1	2	3	4	5	6	7	8	9	10	11	12
DBLP	CBLFD	0.5	**10.3**	**41.5**	**11.2**	**12.4**	**13.5**	**15.7**	**16.7**	**16.8**	-	-	-
	AFOCS	**0.8**	1.5	0.8	0.8	0.9	1.7	1.2	7.0	11.6	-	-	-
	iLCD	0.5	0.7	0.7	0.9	0.7	0.9	1.0	9.3	1.2	-	-	-
Enron	CBLFD	**7.9**	**6.4**	**7.4**	**8.2**	**9.2**	**8.3**	**7.6**	**8.2**	**8.8**	**10.0**	**9.2**	**7.1**
	AFOCS	3.6	3.0	5.1	4.4	8.9	3.4	3.3	3.3	3.3	8.1	7.6	2.0
	iLCD	0.1	0.1	0.1	0.1	0.2	0.1	0.1	0.1	0.2	0.2	0.2	0.1

level and similar nodes are easier to be merged into one community, so the final partition results are overlapping with good community quality meanwhile.

In conclusion, our algorithm shows its superiority in real networks, proving it is more suitable to detect overlapping communities in dynamic networks and more suitable for real world datasets.

5 Conclusions

In this paper, we propose a novel evolutionary community detection algorithm. Different from the other traditional algorithms, this approach observes the network structure from a local-first approach via the central nodes and can find overlapping communities in dynamic network. The experimental results on synthetic and real-world networks indicate that our algorithm performs well with other state-of-art algorithms.

References

1. Ahn, Y.Y., Bagrow, J.P., Lehmann, S.: Link communities reveal multiscale complexity in networks. Nature **466**(7307), 761–764 (2010)
2. Cazabet, R., Amblard, F., Hanachi, C.: Detection of overlapping communities in dynamical social networks. In: Proceedings of the 2nd International Conference on Social Computing (SocialCom), pp. 309–314. IEEE (2010)
3. Chi, Y., Song, X., Zhou, D., Hino, K., Tseng, B.L.: Evolutionary spectral clustering by incorporating temporal smoothness. In: Proceedings of the 13th ACM SIGKDD International Conference on Knowledge Discovery and Data Mining (KDD), pp. 153–162. ACM (2007)
4. Greene, D., Doyle, D., Cunningham, P.: Tracking the evolution of communities in dynamic social networks. In: Proceedings of the 2010 IEEE/ACM International Conference on Advances in Social Networks Analysis and Mining (ASONAM), pp. 176–183. IEEE (2010)
5. Gregory, S.: Finding overlapping communities in networks by label propagation. New J. Phys. **12**(10), 103018 (2010)
6. Kim, M.S., Han, J.: A particle-and-density based evolutionary clustering method for dynamic networks. Proc. VLDB Endowment **2**(1), 622–633 (2009)
7. Lancichinetti, A., Fortunato, S., Kertész, J.: Detecting the overlapping and hierarchical community structure in complex networks. New J. Phys. **11**(3), 033015 (2009)
8. Lancichinetti, A., Fortunato, S., Radicchi, F.: Benchmark graphs for testing community detection algorithms. Phys. Rev. E **78**(4), 046110 (2008)
9. Nguyen, N.P., Dinh, T.N., Tokala, S., Thai, M.T.: Overlapping communities in dynamic networks: their detection and mobile applications. In: Proceedings of the 17th Annual International Conference on Mobile Computing and Networking (MobiCom), pp. 85–96. ACM (2011)
10. Nicosia, V., Mangioni, G., Malgeri, M., Carchiolo, V.: Extending modularity definition for directed graphs with overlapping communities. Technical report (2008)
11. Palla, G., Derényi, I., Farkas, I., Vicsek, T.: Uncovering the overlapping community structure of complex networks in nature and society. Nature **435**(7043), 814–818 (2005)

12. Raghavan, U.N., Albert, R., Kumara, S.: Near linear time algorithm to detect community structures in large-scale networks. Phys. Rev. E **76**(3), 036106 (2007)
13. Rajaraman, A., Ullman, J.D., Ullman, J.D., Ullman, J.D.: Mining of Massive Datasets, vol. 1. Cambridge University Press, Cambridge (2012)
14. Rodriguez, A., Laio, A.: Clustering by fast search and find of density peaks. Science **344**(6191), 1492–1496 (2014)
15. Tang, L., Liu, H., Zhang, J., Nazeri, Z.: Community evolution in dynamic multi-mode networks. In: Proceedings of the 14th ACM SIGKDD International Conference on Knowledge Discovery and Data Mining (KDD), pp. 677–685. ACM (2008)
16. Xie, J., Kelley, S., Szymanski, B.K.: Overlapping community detection in networks: the state-of-the-art and comparative study. ACM Comput. Surv. **45**(4), 43 (2013)
17. Xie, J., Szymanski, B.K., Liu, X.: SLPA: uncovering overlapping communities in social networks via a speaker-listener interaction dynamic process. In: Proceedings of the 11th International Conference on Data Mining Workshops (ICDMW), pp. 344–349. IEEE (2011)

Web-Scale Personalized Real-Time Recommender System on Suumo

Shiyingxue Li[1](✉), Shimpei Nomura[1], Yohei Kikuta[2,3], and Kazuma Arino[2]

[1] Recruit Sumai Company Ltd., Tokyo, Japan
{li_shiyingxue,nomunomu}@r.recruit.co.jp
[2] Freelance, Tokyo, Japan
diracdiego@gmail.com, kazuma.arino@gmail.com
[3] Cookpad Inc., Tokyo, Japan
yohei-kikuta@cookpad.com

Abstract. In this paper we investigate the performance of machine learning based recommender system with real-time log streaming on a large real-estate site, in the views of system robustness, business productivity and algorithm performance. Our proposed recommender system, providing personalized contents as opposed to item/query based recommendation, consists of a real-time log processor, auto-scaling recommender API and machine learning modules. System is carefully designed to let data scientists focus on improving core algorithms and features (instead of taking care of distributing systems) and achieves weekly release cycle in production environment. On Suumo, the largest real-estate portal site in Japan, the system returns more than 99.9% of the API calls successfully in real-time and shows finally a 250% improvement of conversion rate compared to the existing recommendation. With its flexible nature, we would also expect the system to be applied in various kinds of real-time recommendation in the near future.

Keywords: Recommender system · Real-time log processing · Real-estate

1 Introduction

It has become rather common to search the Internet for information to decide whether to purchase a product. In real-estate domain, web portal sites providing property description and reviews has come to play a rather significant role in a customer's purchasing process, too. Yet despite all the efforts on creating convenient search features, it could be pain-taking and confusing to collect all the necessary information as the site becomes enormous. Consequently, it is of great value for the site to automatically learn the user's preferences and recommend relevant products accordingly.

Suumo.jp (Suumo, see Fig. 1 for the interface) is the largest real-estate advertising site in Japan for buyers, sellers and renters. With approximately 10 million property listings, it delivers more than 270 million page views to 14 million unique visitors per month.

J. Kim et al. (Eds.): PAKDD 2017, Part II, LNAI 10235, pp. 521–538, 2017.
DOI: 10.1007/978-3-319-57529-2_41

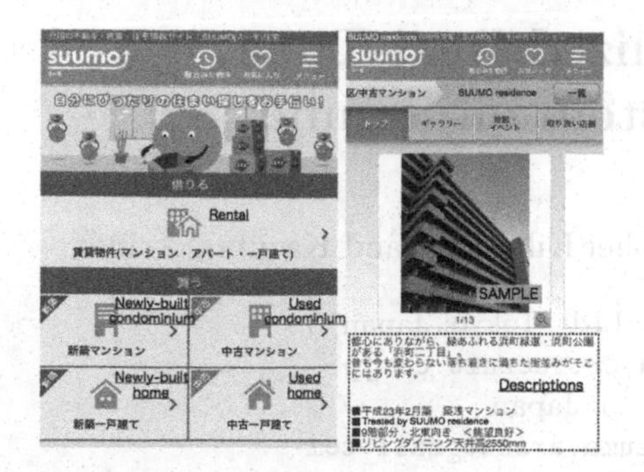

Fig. 1. Suumo interface on the smartphone site. The left figure is the top page and the right figure is an example of a detail page of property.

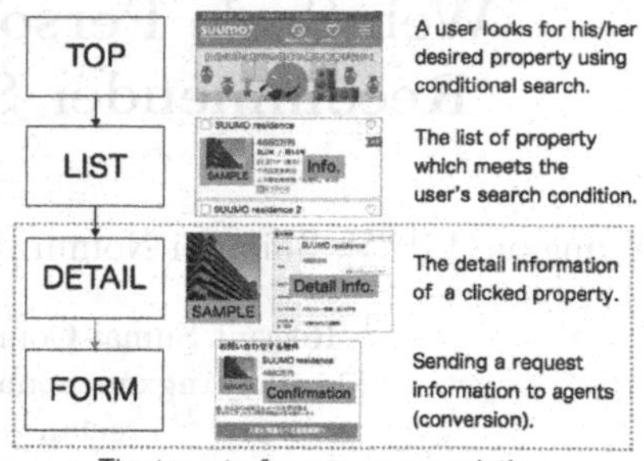

Fig. 2. Typical page transition on Suumo and our recommendation target.

A user usually queries multiple times and reads on many pages before they could find the best fit for themselves. It is such a hard process that a lot of users become frustrated and leave, and even some of those who do request on a purchase to agents are not sure whether they have found the most satisfactory home. Suumo tries to improve our users' experience by building a recommender system with data generated in the searching process. Such data, generally web access logs combined with property listing information, contains rich information about the user's preferences on residence and lifestyle, which could be used to improve matching precision.

Currently, more than 10% of conversions (CVs) at Suumo are directly resulted by said recommender system, making it an important part of the revenue structure; a typical page transition of users is shown in Fig. 2. Yet the existing system is mainly simple association analysis (Listing 1.2 is often viewed after Listing 1.1 etc.) or contents-based recommendation with similarity metrics (property listing A and B are in the same location and have similar price and size, so it is natural to recommend B to users viewing A). Sophisticated models were hard to deploy due to both system limitations as well as algorithm designing problems.

A huge reason of this comes from the lack of available transaction data. For example, a user of real-estate sites is less likely to repeat the purchase in the coming years due to the longevity of real-estate related-decision. Consequently, there are less CV data samples compared to those for e-commerce sites, and a large percentage of users exit the site before even reaching the detailed listing page. Such situation poses a lot of challenges for recommender systems.

In this paper, we address these challenges by building an recommendation API backed up with real-time log processing platform. The particular model adopted here is the gradient boosting method. We try to predict the property on which the user most likely requests information to agents (CV), and display said property in the recommendation widget. To evaluate the solution's performance, we also set up A/B tests which compare existing contents based recommender to our recommendation API. All above is developed as a site feature on Suumo.

Key contributions include:

- **Insights on Building Recommender System on Real-Estate Domain** Despite being a mainstream feature for e-commerce sites and contents streaming services, real-estate sites appears to have some trouble introducing machine learning based real-time recommender system. This could be caused by the lack of online purchase data, thus difficulty in learning the user preference. We provide insights on building machine learning based recommender system for such sites by generating recommendation based on the whole user history instead a single query. And these insights would also be applicable to other domains like car-sale or recruiting, which tend to share the same problem.
- **Collaboration of Engineers and Data Scientists for Short Release Cycles**
To improve the system leveraging user feedback, short release cycles are crucial. But building distributing algorithms for millions of users could be hard itself, and often slows down scientists' model designing. We adopt the architecture often used in micro services these days, using middle-wares to distribute API requests to single-thread programs containing core model description, on which scientists put great efforts improving.

The paper is organized as follows. In Sect. 2 we list related works. In Sect. 3 we present the overview of our approach, along with the system architecture explanation. The experiment description are provided in details in Sect. 4, and the results on both online and offline tests are demonstrated. We conclude our investigation and talk about the future work in Sect. 5.

2 Related Work

In this section we list the previous work done in real-time recommendation efforts, rapid release cycles, and the deployment of recommender systems in real-estate web sites.

2.1 Real-Time Recommendation

Modern web sites, such as e-commerce, news, video sharing, and music streaming service, come with recommender systems almost by default these days, to provide better usability in contents searching. User purchase logs (or the equivalent of it) are commonly collected to improve recommendation precision, as described in [1]. Since these logs provides rich information about user preferences, it is quite common to apply machine learning approaches (computation-expensive) to enhance recommendation quality, yet the nature of said sites expects low-latency for the recommender systems. Hopfgartner et al. [2] provides a tutorial for real-time recommender systems.

Freno et al. [3] discusses this problem for web scale, real-time ranking model. It is common practice to train a new model on fresh log at regular intervals, e.g. daily or weekly. Such requirements of training time of the model is critical in the industry scenario, meaning that the models to use are limited. Also, the real-time or quasi real-time response requirement (less than 2 s from business requirement in our case) poses limited access to databases on response time. This restricts information available at prediction. In this study, an one-pass ranking model is proposed which achieves enough prediction quality with frequent model updates. To tackle training time requirements for novel machine learning algorithms, Wang et al. [4] proposes online learning of a Boosting Tree. And Bottou et al. [5] provides stochastic optimization tools for ranking purpose.

2.2 Rapid Iteration

Carrying out new idea in quick iteration is considered important, especially for the real-world recommender systems where the site itself along with its users could be ever-changing.

Yet traditional approach of collaborating between data engineers and scientists doesn't work well for rapid iteration. O'Sullivan [6] approaches this problem by blending teams.

Schleier-Smith [7] applies Agile development process to data analysis too, and build an architecture which can deploy new idea within two weeks. Using existing frameworks is also a common practice, such as [8]. Our approach is similar to [6] in the sense we encourage deep collaboration between engineers and scientists, but we solve the scale problem independent of model used. Short release cycles mentioned in [7] is also achieved in our experiment.

2.3 Real-Estate Recommendation

Yuan et al. [9] develops an online homebuyer's search system based on case-base reasoning and ontology structure. Ho et al. [10] proposes fuzzy goal programming model with S-shaped utility function and use it to provide more powerful search system and conduct laboratory level testing. To our best knowledge, literature on real-estate sites with state-of-art machine learning recommender systems is scarce.

Correspondingly, deployment of state-of-art recommender systems is rarely seen on real-estate sites, and we consider the following reasons relevant. First, users don't repeat purchase frequently, leaving fewer data points for Customer Relation Management (RMF analysis for example), and algorithms based on it don't work properly. Secondly, data management on property itself appears less developed. In this work we put the goal of learning as information request, producing more data points for learning. In concern to data management issues, new data storage system is designed to integrate data for future learning.

3 Proposed Approach

3.1 Use Case: Recommendation Widget

On a typical smart-phone Suumo detail page, the property's images, description (price, size, distance from the nearest station etc.), and agent information are displayed. When a user finishes reading all of this, the recommendation widget (see Fig. 3) comes in with commonly 5–10 listings. This process is supposed to take 10–20 s on average, which leaves the recommender a few seconds to generate its contents.

We use following CONVERSION RATE (CVR) to measure recommendation performance. CV is an info request to the agent on a property listing (typically to show interest on purchasing) in Suumo's case.

$$\mathrm{CVR} = \frac{\text{info requests on recommendation}}{\text{recommendation clicks}} \tag{1}$$

For EC sites, this CV information (i.e. purchase logs) is utilized for modeling, yet as mentioned in previous section, real-estate sites have less data points and we have to use more explicit data, e.g. click-through logs, instead. Note that while less sparse than CV logs, click-through logs obviously provide less insights on predicting CV.

3.2 Early Attempts

In this section we introduce existing recommendation systems on Suumo.

Contents-Based Recommendation. This earlier recommendation, depicted in Fig. 3, is a hybrid model of collaborative filtering and contents-based filtering. On a daily batch, we first do association rule mining on weeks of web logs. The output contains a recommendation candidate lists (typically about 100 property

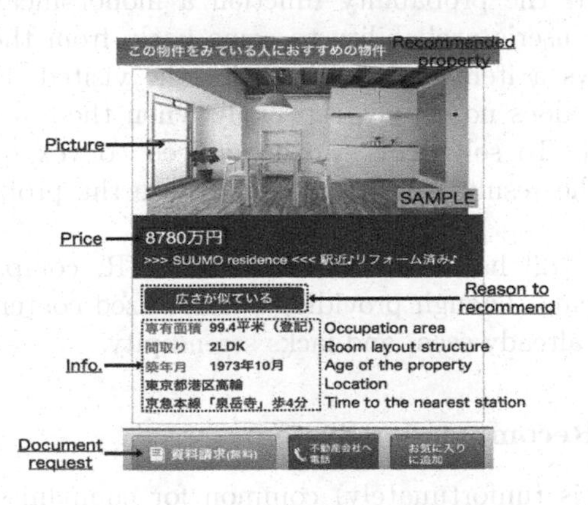

Fig. 3. Suumo interface for recommending relevant property.

listings or less) for each property listing based on confidence, support and lift. The recommender is a simple implementation of content-based filtering where we calculate all the numerical attributes' difference between property listings (Euclidean distance for geo-location) and simply sort the candidates by the average similarity. No specific weighting scheme is applied on attributes. We also could track which attribute is appealing and use it as the recommendation description (reason to recommend) shown in Fig. 3 (lower price, nearer to the station etc.). In the past A/B tests, we have found that recommendation description leads to more than 25% improvement on the click through rate.

But this model tends to circle around user's first queried location and has barely a chance to provide new insights to the user. It is also hard to tell how different this recommendation is from the user's List page. And for obvious reasons the recommendation is not personalized.

Incremental Collaborative Filtering. According to the methods described in [11], the similarity of users, items, and that of user-item pairs, could be described by "pheromones structures". By estimating what the authors call evaporation rate, we could build "pheromones structure" for each user with their viewed items, and generate recommendations to users with similar "pheromones structure".

We implemented this algorithm in comparison with contents-based recommendation described previously on another page of Suumo, where Incremental Collaborative Filtering produces 20% better of CVR. On the other hand, it requires certain amount of page views for proper estimation on each candidate item, thus new items have less chance to be recommended.

Frequency and Recency on Viewed Item. Another try on user log usage at Suumo is the calculation of user's Frequency and Recency on Viewed Item (FRVI). It is intuitive that the more page views a user has, and the more recently he has visited the site, it is more likely he would continue to view the items.

When we define the probability function a monotonically increasing one, we could infer the user's probability to come back, from the number of times that the user views a item and the last time he visited. For some segments such monotonicity does not apply, especially when there are too few samples to get valid result. To solve this, we introduced convex quadratic programming to smooth the results, and could finally use the probability to produce recommendations.

This model in [12] has a 50% increase in CVR, compared to displaying recently viewed items. Though providing personalized contents, it is limited to what the user has already seen, and lacks serendipity.

3.3 Proposed Recommender System

Background. It is (unfortunately) common for companies to have complex existing systems and code bases, sometimes completely outsourced, making it rather hard for a scientist to build recommendation models as site features, let alone scripts leveraging real-time logs. Suumo is in the same situation, and the scientists dealt with it mostly by creating static files and let developing teams take them in the existing systems on a regular basis. Deploying models described in the Subsect. 3.2 took more efforts than one would expect, resulting in less updating on built models and slow iterations for improving them. Needless to say this situation remarkably limited big data usage at Suumo.

To better make use of all the data we have, it is necessary to build a recommender system that is not tightly coupled with the existing ones. Since we have to generate contents dynamically, it should also be fast. With millions of users the system is expected to scale. And last but not least, to get rapid release cycles the system should be easy to use for scientists. To summarize, this loosely coupled recommender system should be fast, scaling and accessible to scientists.

Overview. The proposed system consists of a streaming processor, a large scale data warehouse for log storage, in-memory data storage for request time calculations, a machine learning module, and finally a recommendation API, as depicted in Fig. 4. With this design, we could hide all algorithmic parts from other components of Suumo site. And since there is no need to modify other components of the site, updates on the algorithms becomes more flexible and reliable.

Latency. When the recommendation API is called, the system asks Key Value Store (KVS) for not only user history, but also property information. At the beginning we simply used the property ID as key, and the necessary information as its value. Yet as information necessary to models increases, the processing time gets longer and footprint gets larger. When the system under development started to take more than ten seconds to process input and score them, we re-designed custom shared in memory data structure in consideration of the model design. Since recommendation algorithms typically have a spares matrix for input, custom columnar oriented data structure was created to keep property information.

This drastically reduced the response time, as described in Experiment section. Also, smaller footprint of input data allows us to put more data beforehand that model might require, making frequent modifications easier to achieve.

In this way, we can provide highly personalized recommendations while achieving quasi real-time response time as well as very short development iteration cycles.

Scaling. The recommendation cluster leverages user log and property information. Log, streaming down from the processor, is soon put in the in-memory KVS. (Fine-grained master log data is processed on a daily batch, consistency fixed later.) As described earlier, the property information DB, updated multiple times a day, is also maintained in said KVS. By keeping all the input data in memory, there is no need to access RDB to handle API calls.

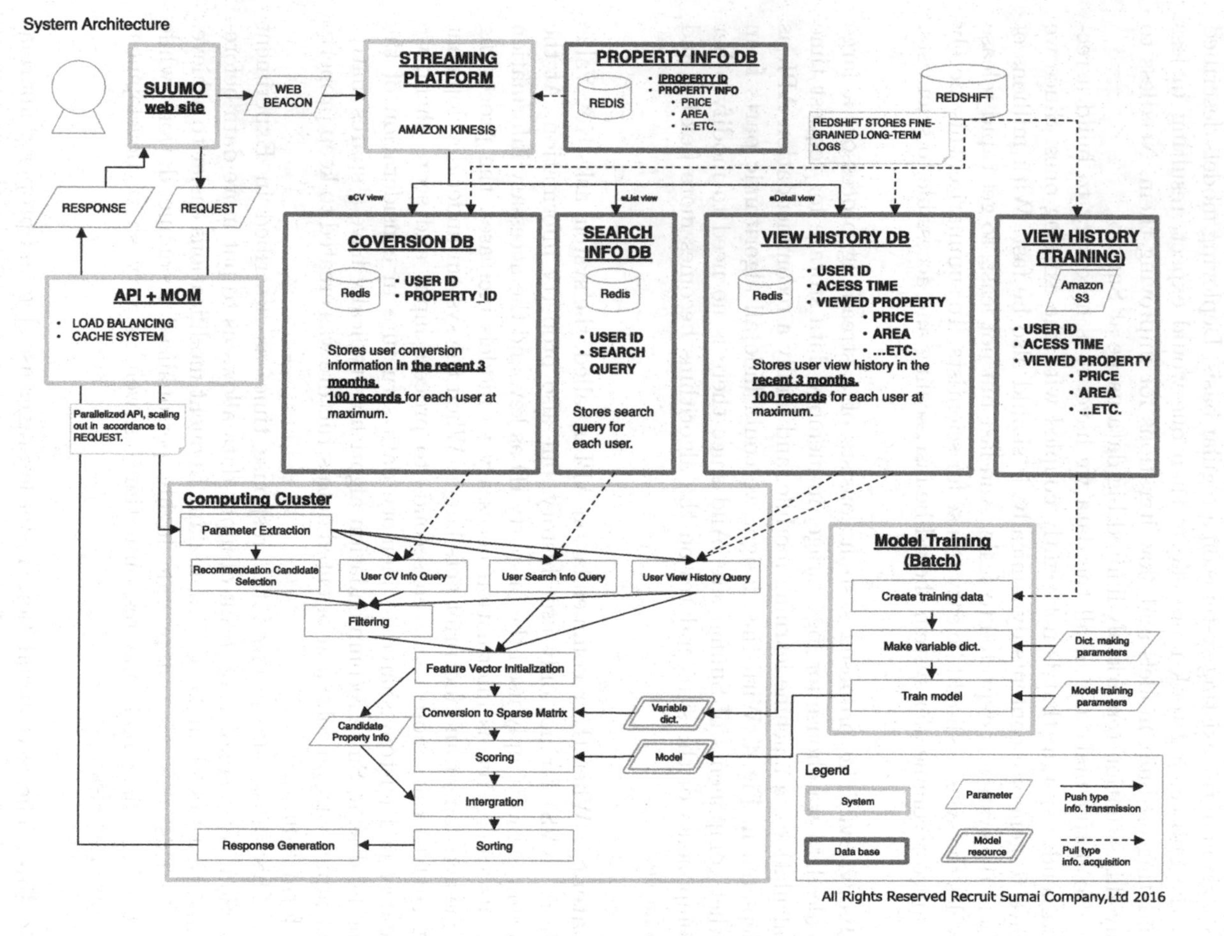

Fig. 4. Real-time recommender system on Suumo.

This enables the computing clusters to add nodes on demand, scaling-out in rush hours and shrinking in size for quiet hours. For this experiment we launched 9 instances with 81 processes for rush hours, and scale-out for other larger recommendation task in the future comes quite easy.

We choose Redis for in-memory KVS because of its speed. Master node, its read replica and a back-up node are provided. Recommender clusters only access to read replica, while master writable Redis node is only updated by streaming services and batch jobs, which are responsible for user log update. The data size for each node in our experiment is about 12 [GB].

In this way, KVS update does not affect front web experience and recommend response time while fresh log data is available for most of the time. Access and update to the data in KVS run parallel and asynchronously, make whole system robust to delay and achieve high throughput.

Accessible Environment for Scientists. We train models with large scale data warehouse using Hadoop technologies such as MapReduce and Spark, to take advantage of large cluster computational power. At first, we used MapReduce and Hive and later replaced part of system with Spark to construct higher abstraction and enhance speed of development iteration.

We have also introduced Jupyterhub as our developing environment, so scientists could manipulate months of user log data as they would on their local machine, using the familiar Notebook interface.

The main language used in our system by both scientists and engineers is Python, for its popularity in both fields.

There are many advantages of Python language like good libraries, active communities, simple language specifications that is easy to learn for scientists, general purpose nature that is powerful enough to construct complex software systems etc. But for our recommender system, Python's global interpreter lock poses some challenges for utilizing multi-core processors under our low latency high throughput response requirement.

Instead of banning the use of the language, to solve this problem and distribute the computation task, we introduce messaging oriented middle-ware (RabbitMQ in our case) to send the requests to multiple nodes with Python calculation processes. In this way, from front web side, the API has stable response time for large amount of inputs, whereas a scientist does not have to put much efforts on dealing with massive parallelism with strict real-time restriction, and could focus the improvement of the models.

Rapid Iteration. As one can infer from the (lack of) literature on real-estate recommendation engines, simply applying known algorithms doesn't guarantee performance. Delicate engineering on models is required often, and short iteration cycle is inevitable.

Fortunately, as the result of the efforts of making collaboration easier, we are now able to achieve such iteration cycles. Moreover, since our scientist team use Jupyter Notebook for their daily analysis, engineering team could integrate their infrastructure work to the same notebook file. Both members understand contents of this notebook. Not only the scientists but also the engineers reproduce scientists' result and understand what problem now WE are facing.

This style of deep collaboration enables engineering team to automate and optimizing analysis work, and reduce huge communication cost.

3.4 Model Design

In this paper we try to apply machine learning approach to the recommender system on Suumo (namely, the real-estate industry in Japan) with the aim to achieve the improvement of the CVR by means of personalizing the recommendation.

Using the system architecture described in the Subsect. 3.3, we can use the following three types of variables for model input.

- User's latest information, u_l.
  u_l is real-time search query information. When a user searches property under various search conditions, such as price range and room layout structure, this information is embedded in the URL. u_l is stored in SEARCH INFO DB in Fig. 4.
- User's recent log, u_r.
  u_r is recent log information of property that is browsed by a user on the detail pages. This information contains price, area, distance from the nearest station, and so on. VIEW HISTORY DB in Fig. 4 stores u_r for each user; u_r is available in recent 3 months or up to 100 records for each user.
- Property spec, p_s.
  p_s is spec information of property. p_s includes information of property itself like price and area; the latest and individual information is given to each property. All information of candidate property for recommendation is stored in PROPERTY INFO DB in Fig. 4.

We would like to construct a model whose input is a set of $\{u_l, u_r, p_s\}$ and output is the score that reflects the likelihood of CV.

$$\{u_l, u_r, p_s\} \xrightarrow{F} \mathbb{R} \tag{2}$$

where F expresses the model which assigns a score to the record composed of u_l, u_r and p_s; a higher score means higher chance of a CV. In the training phase, we perform model training using the stored data. The training is done in batch mode and described as

$$\arg\min_F \left[\sum_{i=1}^{D} loss(y_i, F(\{u_{l,i}, u_{r,i}, p_{s,i}\})) \right] \tag{3}$$

where D is the training data size, $loss$ is some loss function and y is the CV flag (0 or 1). In our case the data size D is $\mathcal{O}(10^6) \sim \mathcal{O}(10^7)$ and the typical loss function is the logarithmic loss function. With respect to the recommendation, variables u_l and u_r are given for each user when he/she reaches a detail page and the model scores all candidate property that exists in the same prefecture as that of the latest property browsed on the page. The recommendation provides a list of top N (typically N = 5) property.

Quite a few classification (or learning to rank) models are available for our purpose, for example recently-developed the gradient boosting model [13] and the Factorization Machines [14]. In order to include various user preferences, a boosting method is a powerful technique since each weak learner captures the local characteristic in the data. In addition, from the view point of the prediction stability and the system maintainability, a model that needs less pre-processing of variables is preferred; typically a tree-based model is suitable for this purpose. As a first attempt, therefore, we adopt a gradient boosting decision tree (GBDT) model which has high predictability and is robust to outliers and defects.

Gradient Boosting Method. Boosting is a well-developed method based on an ensemble of weak learners. Here we focus on the gradient boosting algorithm in which a weak learner is added in the functional gradient descent manner. The gradient boosting is a flexible non-parametric statistical learning method and shows a good performance in many applications [15]. In the predictive learning problem, one has a system consisting of an output variable y and a set of input variables $\mathbf{x} = \{x_1, x_2, \cdots, x_p\}$. The goal is to estimate a function that minimizes the expected value of some objective function $l(y, F(\mathbf{x}))$ over the joint distribution of all $(y, \mathbf{x})$

$$\hat{F} = \arg\min_{F} E_{y,\mathbf{x}} \left[l(y, F(\mathbf{x})) \right]. \tag{4}$$

In this paper we set the objective function as the binary log loss function, $l(y, F(\mathbf{x})) = -(1/N) \sum_j^D y_j \log(p_j)$, where $p_i = e^{score_i} / \sum_k e^{score_k}$. We also add L_1 and L_2 regularization terms to encourage generalization. The algorithm is summarized as below:

Listing 1.1. Gradient boosting

```
1  Initialize model with a constant:
2    F_0 = arg min_Δ Σ_i l(y_i, Δ))
3    For m = 1 to M:
4      Compute the steepest descent direction:
5        -g_m(x_i) = -[∂l(y_i, F(x_i))/∂F(x_i)]_{F(x)=F_{m-1}(x)}.
6      Find parameters by fitting a weak learner to pseudo residuals:
7        θ_m = arg min_{θ,β} Σ_{i=1}^N [-g_m(x_i) - βf(x_i; θ)]^2.
8      Find the weight via the line search:
9        β_m = arg min_β Σ_{i=1}^N l(y_i, F_{m-1}(x_i) + βf(x_i; θ_m)).
10     Update the model:
11       F_{m+1} = F_m + β_m f(;θ_m)).
12 Output F_M = F(x; {β_m, θ_m}_{m=1}^M).
```

We use the XGBoost package [16] for building the prediction model.

Feature Engineering. As we explained, we have three types of model inputs, u_l, u_r and p_s. We relate them to three mathematical variables, numerical, categorical and URL variables, in order to treat machine learning models. Adding the target variable to these, we handle the following four mathematical variables.

- Numerical variables.
  Numerical variables like price are included in u_r and p_s. Although the range of values for each variable is quite different, we use raw values from the view point of the maintainability and the simplicity at the real-time processing. A tree-based model can predict stable even in such a situation.
- Categorical variables.
  Categorical variables like room layout structure are included in u_r and p_s. Within categorical variables, the variables whose cardinality is too large (like user id) are excluded since they are useless for the prediction. All the other are converted into dummy variables in binary form; this is required by the implementation of XGBoost.
- URL variables.
  URL variables correspond to u_l. They are user's search conditions which are selected from various options, such as the price range and the room layout structure. These variables are parsed so that we can extract user's queries. We thereafter treat each query parameter as categorical variables.
- Target variables.
  Target variable is the CV flag which takes the binary value, $\{0, 1\}$. As we mentioned, the CV = 1 flag is assigned to the page-view where the request for property information is clicked.

We treat the data as dictionary-like object which is composed of a combination of key and value (key:value). One important thing is that the keys of the variables have to be matched between the training phase and deploy phase. Hence we create the dictionary that maps each variable at the recommendation to the correct key number of the training data, e.g. price → 4231. Following this mapping dictionary, we translate the real time data ([variable, value]) into the properly format (key:value), e.g. [price, 50000000] → 4231:50000000. The mapping process requires time which can be a bottle neck for the real-time recommendation. However, we optimize the processing, including the code and the previously mentioned columnar data structure, which enables the real-time recommendation. See Table 1 for examples of the data. The total number of variables is a few times 10^4.

Making CV = 0 Data. Machine learning algorithms for the supervised binary classification construct models which distinguish the difference between CV = 1 and CV = 0 data. In the early stage of the project, we naively prepare the data where CV = 1 is an actual request event for property information and CV = 0 is NOT. This is natural and no problem in many cases. However, this naive assignment does not work well in the real-estate industry because many users repeatedly watch similar property until CV, which makes it difficult to find the difference between CV = 1 and CV = 0 data.

After many trial and error steps, we find that it is valid to make CV = 0 data in the way user's information is identical to CV = 1 data but spec information

Table 1. Examples of the data

CV	Explanatory variables				
0	5:0	563:1	4231:43000000	10007:15	...
1	5:1	563:0	4231:45000000	13210:1	...
0	5:1	23:1	4231:0	13210:0	...
...	...	...	...	...	...

is that of random property; if we have $\{u_l, u_r, p_s : \text{CV} = 1\}$ data, we make corresponding CV = 0 data as $\{u_l, u_r, p'_s : \text{CV} = 0\}$ where p'_s is spec information of property which is randomly chosen from the same prefecture property as p_s. In the sense of the classification, CV = 0 data is unlikely to be a CV; our randomly sampling method for making CV = 0 effectively achieves the unlikeliness of CV. The ratio of CV = 0 to CV = 1 is good to be 1:1–10:1 in our case, namely we sample 1–10 p'_s for each p_s.

This idea may be generalized to the case where each user tends to show many similar behaviors until some sort of action.

Model Cascading. As we stated, we use numerical and categorical features; their treatments, however, are not equivalent. Categorical variables are tend to be sparse because of creating dummy variables using binarization, which leads the boosting trees to favor numerical variables as branches. Because the branching algorithm focuses on splitting numerical variables, the frequency of appearance of categorical variables which are equally important for users is relatively reduced.

In order to adequately include the effect of categorical variables, we prepare two functions; f_c is the function whose argument is only categorical and f_n only uses numerical variables as input. Hence we decompose the score function as

$$F(\{u_l, u_r, p_s\}) = \Phi(f_c(\{u_l, u_r, p_s\}_{cat}) + f_n(\{u_r, p_s\}_{num})) \quad (5)$$

where Φ is some linear sum function to compute the total score. Our concrete steps of the recommendation using the cascading model is expressed in the following listing:

Listing 1.2. Recommendation using cascading model

```
1  Given u_l and u_r for a user on a detail page.
2  Get {p_{s,i}} where i is the index of candidate property.
3  for each candidate i :
4     Compute scores f_c({u_l, u_r, p_{s,i}}_cat) = s_{c,i}.
5  Sort s_i in descending order.
6  Pick up~top 1000 property, denoting p_{s,i'}.
7  for each candidate i' :
8     Compute scores f_n({u_r, p_{s,i'}}_num) = s_{n,i'}.
9  Compute total scores as s_{c,i'} × (max{s_{n,i'}}/ max{s_{c,i'}}) + s_{n,i'}
10 Sort candidate property p_{s,i'} by the total score, descending.
11 Return top N (typically N = 5) candidates.
```

On line 6 we pick up top 1000 property since we would like to put emphasis on user's search conditions and area information that are encoded into the categorical variables, (because of business requirements). Because the scores of categorical model fluctuate largely among each set of candidate property, this is due to the sparseness of the categorical variables, they are re-scaled on line 9.

4 Experiment

4.1 Model Offline Test

Our goal in property recommendation is to recommend property in a way that maximizes the CVR defined in Eq. (1). Although the key performance indicator is the CVR, this is only evaluated on the A/B test and it is difficult to check the model performance contributing to the CVR on the offline test. Here we assume that the model showing a good performance for the offline data set is also useful for the recommendation. We divide the offline data set into the training data and the test data; the model is trained on the training data and evaluated on the test data using the ranking metrics.

In this paper we use the *nDCG* and *AUROC* on the offline test. For our purpose, these metrics can be used to evaluate the relevance of the page-view which results in CV, which means that we are able to evaluate the relevant combination of user information (u_l and u_r) and property information (p_s) for the CV.

Offline Performance. We evaluate the offline performance of the model with $\mathcal{O}(10^7)$ data. Although this result is obtained in the early stage of the project, it is not so changed through the project. According to the offline test, we find the required data size of the model and the desirable data refreshing frequency.

Figure 5 shows the dependency between the data size and the performance of the model. In this experiment, we train the model while increasing the data size by 1 million and produce predictions for 7 million test data records. Although the values itself are not so high due to the large number of the test data, we can see both *AUROC* and *nDCG* improve with data size. This result suggests the enough data size of the training is over about 15 million.

The effect of data freshness is also investigated. We first order the data in ascending order according to the date and time, then build the models using sliding date windows. The windows sizes are either one week or two weeks; the data size of a day is about 1 million. Predicting for the fixed data (we here used the last 4 million records of the data) using each model, we find that the model should be replaced with the new one within a week. The result is shown in Fig. 6.

We evaluate other machine learning models, especially the Support Vector Machine and the Random Forest. However it is difficult to build effective models which are compared to the random classification, even after varying the parameters. The GBDT is the prominent model for our purposes: the high predictability, stability and scalability.

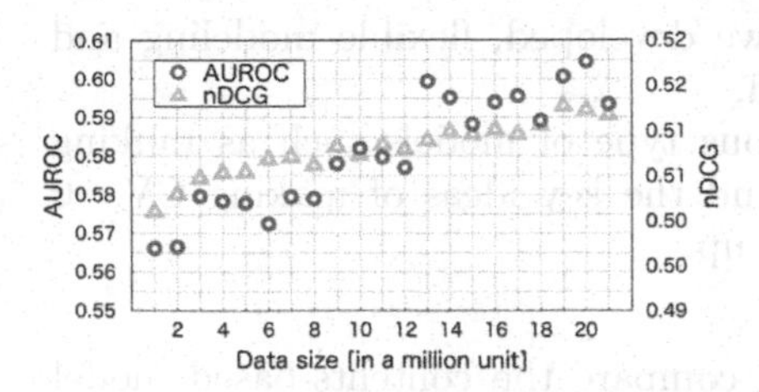

Fig. 5. Data size dependency of the model performances. The horizontal axis is the data size of the training and the test data is independent 7 million records.

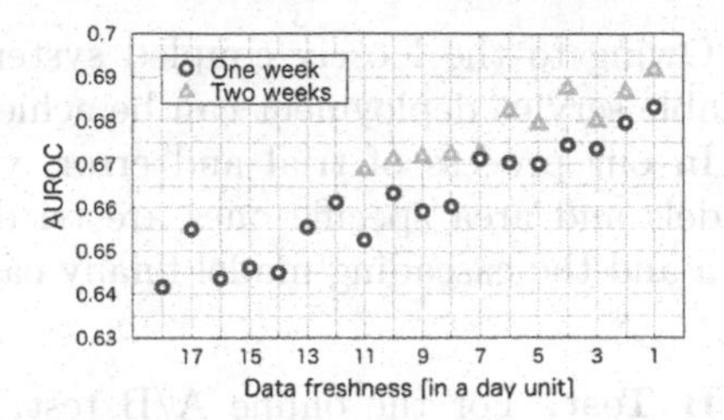

Fig. 6. Data freshness dependency of model predict. The horizontal axis is the freshness of the model; the left is old and the right is new. The plot of the two weeks model is disconnected due to the limitation of the total data size.

Model Parameter Tuning. The model hyperparameters are inferred using a sequential model-based optimization to minimize an objective function of the validation set. We use the hyperopt package [17], applying tree-structured Parzen estimator approach to the uniformly distributed XGBoost parameters, such as the learning rate, the regularization coefficients and so on.

One of examples of optimized model parameters is shown in Table 2. The parameters are for our final deployed model; we have two sets of parameters since using the cascading model.

Table 2. Examples of optimized model parameters

Parameter description	Category	Number
Learning rate used in update	0.3	0.575
Required minimum loss reduction	1.6	1.275
Maximum depth of a tree	9	7
Minimum sum of instance weight	8	8
Sample rate of training records	0.8	0.85
Sample rate of columns	1.0	0.95
L_1 regularization	0.775	0.675
L_2 regularization	0.625	0.725

4.2 Results

Deploy History. In our project we have had more than 10 deployments within three month, about once a week on average. The term "deployment" includes both new models and new systems.

Owing to the loosely coupled system we developed, flexible modeling and reliable service deployment can be achieved.

In our process of trial and error, various type of models such as ranking models and area specific ones are tried, and the key ideas of making CV = 0 data and the cascading model finally came up.

A/B Test. For the online A/B test, we compare the contents-based model (CBM) described in the Sect. 3.1 and the proposed real-time models (we simply call them GBDT).

In the early stage of the project, we did not achieve better result than the existing model, CBM.

However, as repeatedly emphasized in this paper, our rapid iteration cycle enables us to improve models, and so we can eventually construct the outperforming model adopting the cascading method and randomly sampled CV = 0 data.

We test the model for last 5 days of the project. The result is shown in Table 3, the value is the average CVR (in % unit) within the period. We here also put the results of the previous A/B tests where we compared the model of our early attempts in the Subsect. 3.2: random property in the same area (RPSA) recommendation and page-view ranking (PVR) recommendation. Our final proposed model managed to provide high quality recommendations to users, achieving 250% improvement when compared to CBM. Note that CVR in real-estate sites varies with seasons even for the same algorithm, which is reflected in the table.

Table 3. CVR [%] in the A/B test. B:A ratio means (CVR of the new model) ÷ (CVR of the previous one)

Model	Prev. 1	Prev. 2	This time	B:A Ratio
RPSA	0.105	-	-	-
PVR	0.121	0.146	-	115
CBM	-	0.265	0.310	181
GBDT	-	-	**0.776**	**250**

Personalized Site Experience. Following tables shows two recommendation users', picked up randomly, relocating processes using the recommender system.

We would expect the recommender system to help find the next neighborhood after learning the user's preferences, yet as shown in Table 4, user tends to stay similar area in CBM' case. The GBDT system depicted in Table 5, on the other hand, helps user to relocate after a while. When we a take a look at the average number of Cities that a user views, there is a slight increase for GBDT group of users.

Table 4. CBM

Time	Pref	City	Price	Via Rec.
06:20	Tokyo	Shibuya	36990000	0
22:12	Tokyo	Shinjyuku	29800000	0
15:04	Tokyo	Suginami	25800000	0
15:20	Tokyo	Suginami	25800000	0
15:23	Tokyo	Suginami	26800000	1
15:23	Tokyo	Suginami	28800000	1

Recommendation circles around the same place despite that the user would like to change location.

Table 5. GBDT

Time	Pref	City	Price	Via Rec.
20:42	Tokyo	Nakano	46800000	0
20:43	Tokyo	Mitaka	41800000	0
20:44	Tokyo	Meguro	57900000	0
20:44	Tokyo	Minato	53000000	0
20:46	Tokyo	Shinjyuku	49800000	1
20:47	Tokyo	Adachi	46800000	1

After a few clicks, the recommender learns the user's preference.

Response Time. During the A/B test, the system responses to 84% of the requests under 1 s, and 99.9% under 2.5 s. And process is distributed almost evenly across the computation clusters. As mentioned in 3.1, it usually takes more than 10 s before user scrolls down to the recommendation parts. Should the user read the page and get down to our recommendation widget, it should be perfectly ready to view.

5 Conclusion and Future Work

The proposed recommender system meets the response time requirement and shows a 250% performance improvement. Instead of providing query-based results, we generate unique recommendation contents to each user according to their click/viewing events. This result indicates that even the most subtle actions could reflect on a user's preferences and has a positive effect on predicting their needs.

The recommender system as a framework also provides better business productivity since it frees data scientists from dealing with distributing system, and allows them to focus on building better core algorithms. This is achieved by deliberate system design and deep collaboration with engineers.

For future work, we have found that in comparison to existing recommendations, ours appear to work effectively during the first session, but this effect decrease after the second session. It could be caused by some kind of over learning of user preference, leading the recommender to favor what the user have already seen after a few sessions. The future models should consider to lower the importance of viewed items in a algorithmic way. Extending this system to other pages and contents, such as real-estate news, is now under development.

Acknowledgments. We are grateful to Yoichi Maejima for useful discussions about the model specification. Special thanks to Iwao Watanabe and Nobuaki Oshiro for building the fast API, Kentaro Hashimoto and all the guys in the infrastructure team for indulging us with general help on the log processing platform.

References

1. Joachims, T.: Optimizing search engines using clickthrough data. In: Proceedings of the Eighth ACM SIGKDD International Conference on Knowledge Discovery and Data Mining, pp. 133–142. ACM, New York (2002)
2. Hopfgartner, F., Kille, B., Heintz, T., Turrin, R.: Real-time recommendation of streamed data. In: Proceedings of the 9th ACM Conference on Recommender Systems, pp. 361–362. ACM, New York (2015)
3. Freno, A., Saveski, M., Jenatton, R., Archambeau, C.: One-pass ranking models for low-latency product recommendations. In: Proceedings of the 21th ACM SIGKDD International Conference on Knowledge Discovery and Data Mining, pp. 1789–1798. ACM, New York (2015)
4. Wang, F., Yuan, C., Xu, X., van Beek, P.: Supervised and semi-supervised online boosting tree for industrial machine vision application. In: Proceedings of the Fifth International Workshop on Knowledge Discovery from Sensor Data, pp. 43–51. ACM, New York (2011)
5. Bottou, L., Le Cun, Y.: Large scale online learning. Adv. Neural Inf. Process. Syst. **16**, 217 (2004)
6. O'Sullivan, S.: Webinar: working together at the intersection of data science and data engineering (2015). https://datascience.berkeley.edu/blog/webinar-data-science-engineering/. Accessed 27 May 2016
7. Schleier-Smith, J.: An architecture for agile machine learning in real-time applications. In: Proceedings of the 21th ACM SIGKDD International Conference on Knowledge Discovery and Data Mining, pp. 2059–2068. ACM, New York (2015)
8. Huang, Y., Cui, B., Zhang, W., Jiang, J., Xu, Y.: Tencentrec: real-time stream recommendation in practice. In: Proceedings of the 2015 ACM SIGMOD International Conference on Management of Data, pp. 227–238. ACM, New York (2015)
9. Yuan, X., Lee, J.-H., Kim, S.-J., Kim, Y.-H.: Toward a user-oriented recommendation system for real estate websites. Inf. Syst. **38**(2), 231–243 (2013)
10. Ho, H.-P., Chang, C.-T., Cheng-Yuan, K.: House selection via the internet by considering homebuyers risk attitudes with s-shaped utility functions. Eur. J. Oper. Res. **241**(1), 188–201 (2015)
11. Wang, Y., Liao, X., Wu, H., Wu, J.: Incremental collaborative filtering considering temporal effects. arXiv preprint arXiv:1203.5415 (2012)
12. Iwanaga, J., Nabetani, K., Kajiwara, Y., Igarashi, K.: About the recommendation method based on the frequency and recency. J. Oper. Res. Soc. Jpn. **2013**, 194–195 (2013) (in Japanese)
13. Friedman, J.H.: Greedy function approximation: a gradient boosting machine. Ann. Stat. **29**, 1189–1232 (2001)
14. Rendle, S.: Factorization machines. In: 2010 IEEE 10th International Conference on Data Mining (ICDM), pp. 995–1000. IEEE (2010)
15. Distributed (Deep) Machine Learning Community. Xgboost (2016). https://github.com/dmlc/xgboost
16. Chen, T., He, T.: XGboost: extreme gradient boosting. R package version 0.4-2 (2015)
17. Bergstra, J., Yamins, D., Cox, D.D.: Hyperopt: a python library for optimizing the hyperparameters of machine learning algorithms. In: Proceedings of the 12th Python in Science Conference, pp. 13–20 (2013)

Modeling Contextual Changes in User Behaviour in Fashion e-Commerce

Ashay Tamhane(✉), Sagar Arora, and Deepak Warrier

Myntra, Bengaluru, India
{ashay.tamhane,sagar.arora,deepak.warrier}@myntra.com

Abstract. Impulse purchases are quite frequent in fashion e-commerce; browse patterns indicate fluid context changes across diverse product types probably due to the lack of a well-defined need at the consumer's end. Data from our fashion e-commerce portal indicate that the final product a person ends-up purchasing is often very different from the initial product he/she started the session with. We refer to this characteristic as a 'context change'. This feature of fashion e-commerce makes understanding and predicting user behaviour quite challenging. Our work attempts to model this characteristic so as to both detect and preempt context changes. Our approach employs a deep Gated Recurrent Unit (GRU) over clickstream data. We show that this model captures context changes better than other non-sequential baseline models.

1 Introduction

Understanding user behaviour is critical for any e-commerce platform in order to personalise products and induce the user to convert. This becomes easier if the user has a well-defined need and exhibits cohesive intent. Unlike other domains, purchases are often impulsive in fashion e-commerce. While some users do visit a fashion portal out of their need to purchase specific products, a huge number of folks are there just to explore and transact once they come across a product of their liking. This results in "fluid" browsing patterns that cut across different products categories. In other words, there are many context changes that happen in a typical user session. This makes modeling the user's behaviour, and hence personalisation complex [16, 26].

Consider real user sessions illustrated in Figs. 1 and 2. In the session shown in Fig. 1, the user browsed a Shirt, and on the very next click switched to Shorts. Again on the third click, we see a switch back to a Shirt (of a different style altogether) and then again a switch back to a few Shorts in the next clicks. After consistently viewing a few Shorts, the user switches to a T-shirt on his final click. Such changing contexts makes modeling the user behaviour quite difficult and simple heuristics will fall short. For example, if we predicted that the user would browse Shorts in the final click using majority voting, it would have failed. It is important to note that before browsing Shorts, the user also looked at a couple of Shirts and later changed context to Shorts. This is indicative of the fact that the user wasn't really looking for Shorts from the moment he started browsing

J. Kim et al. (Eds.): PAKDD 2017, Part II, LNAI 10235, pp. 539–550, 2017.
DOI: 10.1007/978-3-319-57529-2_42

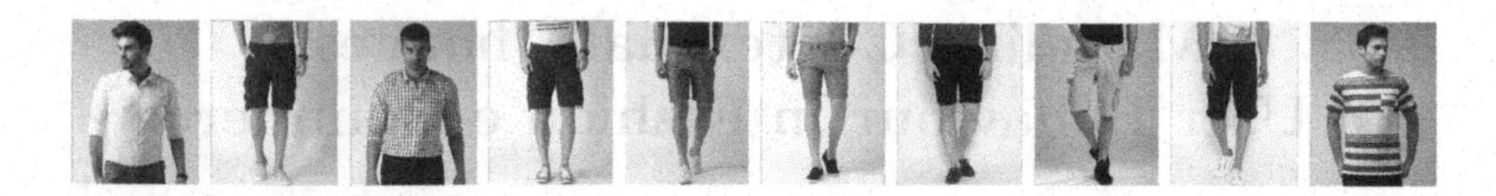

Fig. 1. Men categories session

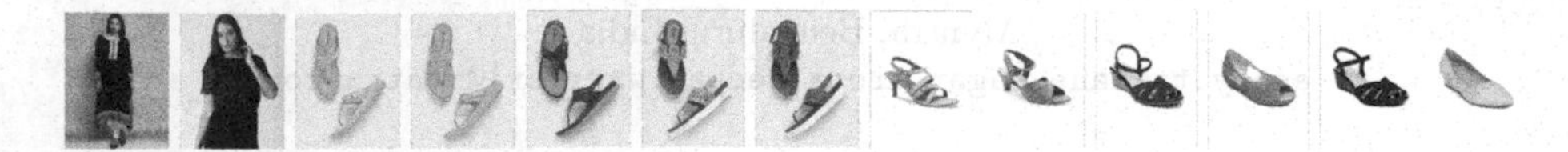

Fig. 2. Women categories session

but rather changed context during the session. Similarly, in Fig. 2, the user is seen switching between different article types from women categories.

On Myntra, a leading fashion e-commerce platform in India, daily about 1.3 million sessions have clicks on at least one product[1]. Only 35% of these sessions have a unique product category (considering sessions with more than one click). We also note that in 41% of the cases the final product a person ends up purchasing is from a different product category than what the user started the session with. This further establishes that context changes are common.

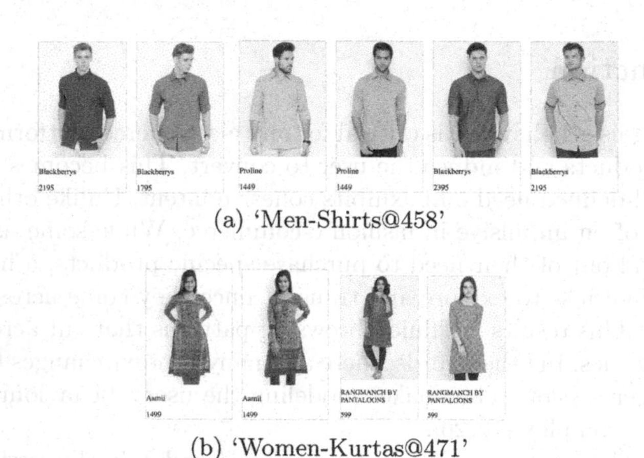

(a) 'Men-Shirts@458'

(b) 'Women-Kurtas@471'

Fig. 3. Samples from product groups, coded as <category@group_number>

Our work attempts to model such impulsive user behaviour so as to both detect and preempt change in context. Our approach employs a deep Gated Recurrent Unit (GRU) over the user sessions data (clickstream). Our dataset presents another challenging scenario of catalogue being highly dynamic due to the ephemeral nature of fashion products. In such a scenario, training any model

[1] Here sessions are all the products clicked by a user within a 30 min window.

directly at the product level is difficult due to data sparsity. In addition the diversity of a catalogue implies that context changes could happen at a more granular level even within a product category - people could change context from a graphic T-Shirt to a Polo. For both these reasons we need to work with a higher and stable abstraction of products. To tackle this problem, we leverage our earlier work [21] using word embeddings to create abstractions of products called product groups which represent homogeneous product clusters serving similar fashion contexts, as illustrated by Fig. 3. Each product in a session is replaced by its corresponding product group and the context change is modelled across products groups - both inter and intra category. Hereafter, we will use the terms product groups, product clusters and contexts interchangeably. We use anonymised user sessions data from our fashion portal for our experiments. We evaluate our trained model against two baseline algorithms - majority voting and "Product Group Graph" model. We show in Sect. 4 that GRU outperforms both these baseline methods.

2 Related Work

There is substantial research on understanding behaviour of online users by analysing their clickstream data. There are several studies that visualise clickstream data in an attempt to make the complex sequential data more intuitive to understand [2,14,26–28,30]. Others use cluster based approaches [23] to differentiate and understand different browsing patterns. There have also been qualitative studies to observe behavioural differences of buyer and non-buyer sessions [22]. While visualising and clustering approaches are directed towards descriptive analysis, there is also substantial amount of literature for predictive analysis. Studies have attempted to determine the intent of users on e-commerce websites in both offline [11,12,16,19,20,24] and real time settings [6]. There has been work [8,19] to distinguish user sessions according to the purpose of the visit as "buyer" or "window shopper". While much of this literature focusses on predicting if a user has intent to purchase, there seems to be relatively less focus on generating recommendations for users with multiple context changes within a session.

Modeling user behaviour and generating recommendations is common, especially in studies relevant to search engines. There has been work on recommending web pages in general [1,7,29] and also in specific cases like personalised news recommendation [15]. Several modeling approaches have been evaluated in past [10] for such web-personalisation. In this work, we propose modeling the user browsing behaviour with a GRU [3,18] for predicting the next user click. Our approach is similar to the recent work in [9] in the sense that we propose RNNs for predicting the next click. However, as mentioned earlier, our dataset presents a different challenge due to the dynamic catalogue leading to sparsity of styles. In the next section, we describe how we overcome this challenge by using product groups.

3 Modeling User Behaviour

As discussed earlier, we wish to model user behaviour that is often prone to context changing. Predicting user behaviour is especially useful for use cases like personalisation, where we try to identify products that will be of most interest to the user next. Therefore, the goal is to predict the user's next click, while making sure that the recent context of the browsing sequence is taken into account. Specifically, we want to predict the user's last clicked product group, given rest of the product groups he has seen in the session so far. We use three different models for this purpose that we will compare against each other in Sect. 4. In this section, we describe these models.

3.1 Majority Voting

In majority voting, we simply predict the next click based on the frequency of product groups in the session. As an example, consider the same session that we show in Fig. 1 - 'Men-Shirts@458', 'Men-Shorts@20', 'Men-Shirts@380', 'Men-Shorts@20', 'Men-Shorts@20', 'Men-Shorts@20', 'Men-Shorts@20', 'Men-Shorts@20', 'Men-Shorts@20', 'Men-Tshirts@788'. The predictions for next click for this session would be in the following order - Men-Shorts@20, Men-Shirts@458, Men-Shirts@380, Men-Tshirts@788. If we need more recommendations, we randomly pick from the top 50 product groups for that gender.

3.2 Product Group Graph (PG Graph)

We now describe our second baseline approach, that uses a product group graph for prediction. As earlier mentioned, we leverage our earlier work [21] to create a smaller number of time invariant product groups from the originally large number of products (there were 408,155 products in the catalogue for Men and 625,171 products for Women at the time of this work). Unlike a typical approach of using product as a document, we consider each session as a document; the words in the documents correspond to the attributes for all styles observed in that session. We then train a Word2Vec [17] model on this dataset to learn a vector representation of attributes. We aggregate these vector representations to arrive at vector representations of products, sessions (contexts) and users. These vectors are used for creating the final clusters, which we refer to as product groups. These product group embeddings are the centroid of products' embeddings. This results in 2813 homogenous and time invariant product groups for Men and 4763 product groups for Women.

From the training set, we build a directional graph where nodes are the product groups and edges capture the transition probabilities between product groups (normalised frequency counts) in all the sessions. We now use a bag of words approach to generate recommendations based on this graph. Specifically, for every click in a session, we get the transition probabilities to all the product groups from the graph. We then aggregate the transition probabilities incoming towards every product group and then normalise them. Continuing with the same example of Fig. 1, the predictions from this approach are in the following order - 'Men-Shorts@20', 'Men-Shirts@458', 'Men-Shirts@380', 'Men-Shorts@19', 'Men-Shirts@469'. Notice that 'Men-Shorts@19' and 'Men-Shirts@469' are now included in the recommendations even though the user did not browse them, as they have high transition frequencies from the browsed product groups as per the graph.

3.3 GRU

A major drawback of the above two models is that they do not handle sequential data. Taking into account the sequential information is critical to our data as it provides the context to the session. This sequential context becomes even more critical for sessions with longer duration, as we shall show in Sect. 4. GRUs are variants of RNNs, that are designed to model sequential data and have been shown to do this effectively [3,18].

Let x_t be the input in the sequence at time t. Let s_t be the hidden state at step t. This state acts as the "memory" of the network as it communicates the information the network has thus far to the next step. s_t is calculated based on s_{t-1} and input at time t using Eq. 1. o_t is the output at step t, and is calculated using Eq. 2. The network can be visualised as shown in Fig. 4 [13]. Note that the parameters U, V and W remain same throughout all the steps.

$$s_t = f(U * x_t + W * s_{t-1}) \tag{1}$$

$$o_t = softmax(V * s_t) \tag{2}$$

Instead of using Eq. 1 that is used by RNNs, GRUs calculate the hidden state s_t, using Eqs. 3, 4, 5 and 6 that can be visualised by logical gates as shown in Fig. 5 [4].

$$z = \sigma(x_t * U^z + s_{t-1} * W^z) \tag{3}$$

$$r = \sigma(x_t * U^r + s_{t-1} * W^r) \tag{4}$$

$$h = tanh(x_t * U^h + (s_{t-1} * r) * W^h) \tag{5}$$

$$s_t = (1 - z) * h + z * s_{t-1} \tag{6}$$

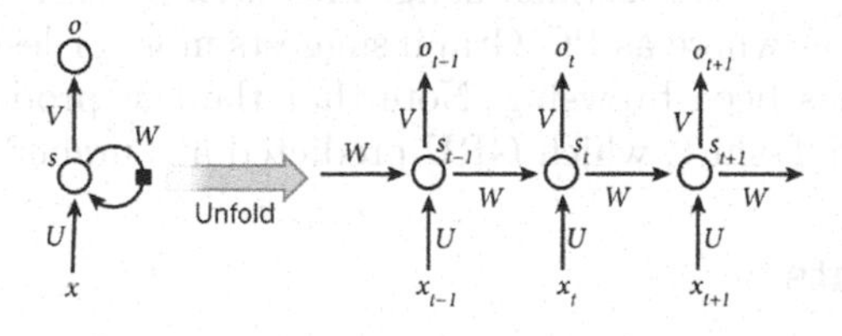

Fig. 4. Unfolding of RNN.

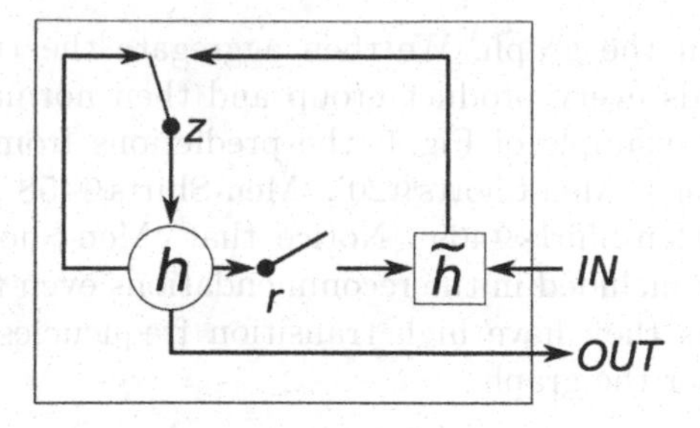

Fig. 5. GRU gating

We used a 2 layer deep GRU network for all our experiments. While training our model, we cross validated on various number of hidden units (50,100,200,500) and found 100 to be the best. We used rmsprop [5] for optimisation. Rmsprop used a learning rate of 0.001 and a decay rate of 0.9. We also tried SGD but found rmsprop to work better. In order to get a visual intuition of this deep learnt model, we generate sample sessions from the GRU in terms of the product groups and later randomly sample a product corresponding to each product group in the session. Figure 6 shows artificially generated samples as "imagined" by the GRU. As can be seen in Fig. 6, the men session starts with our popular value brands 'Roadster' and 'Mast & Harbour'. There is a context change to Levis in middle, which is a premium brand and then again the session switches back to the value brands. This is typical user behaviour on our platform where users make an aspirational click on a high value item, but later often purchase a lower value product. Another interesting trend which the model seems to capture well can be observed in the women session shown in Fig. 6. Here, the session changes context from western categories (Tshirts, Tops) to Indian ethnic categories (Kurta, Churidar Kurta). The intuition here is that if a user keeps browsing western categories, they are likely to eventually move towards Indian ethnic categories. Note that the artificial session for women actually had ten more products at the beginning, all belonging to western categories that we have truncated for the sake of brevity.

We show quantitative results in Sect. 4. Here, we show specific qualitative results for the sessions corresponding to Fig. 1 that were discussed earlier. As can be seen in Fig. 7, both GRU and PG Graph place shorts on number one ranking since the user has been clicking a lot of shorts. However, note that GRU seems to have detected the constant change in context of user and hence predicts different article types where as PG Graph suggests more or less similar products to what the user has been browsing. Note that the last product clicked by the user was a stripped T-shirt, which GRU predicted at number four in the list.

4 Experiments

We leverage the clickstream data store from our e-commerce platform for all our experiments in this paper. Specifically, for every session browsed on our

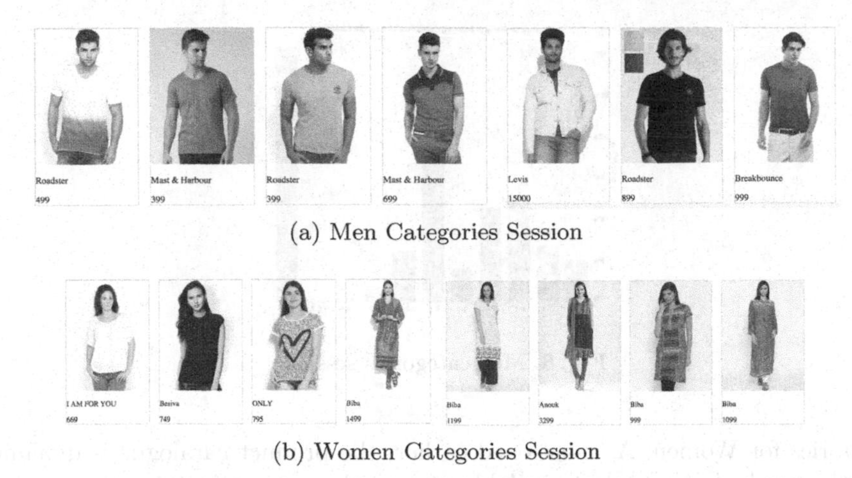

(a) Men Categories Session

(b) Women Categories Session

Fig. 6. Artificially generated sessions.

e-Commerce portal, we have the sequence of products clicked in the session by the user. If we compute the Jaccard Similarity between the sets of browsed products' categories and bought products' categories across these sessions, about 24% sessions have Jaccard similarity less than equal to 0.3 as shown in Fig. 8. The lower values of Jaccard similarity mean that difference between browsed categories and eventually bought categories is significant. This further establishes that context changing in our data is common, as described in Sect. 1.

We train separate models for Men and Women and hence firstly we filter out sessions that have both genders present. We then split the rest of the sessions into 'Men' and 'Women' sessions. This results in 7,010,289 Men sessions and 9,543,282 Women sessions that are spread across 45 categories for Men and 50

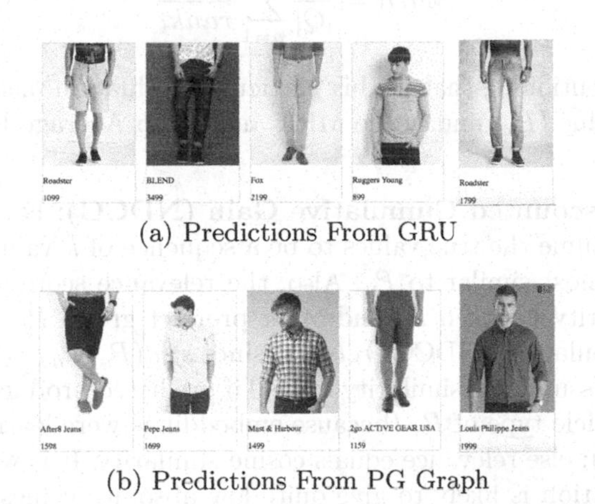

(a) Predictions From GRU

(b) Predictions From PG Graph

Fig. 7. Predictions corresponding to session in Fig. 1

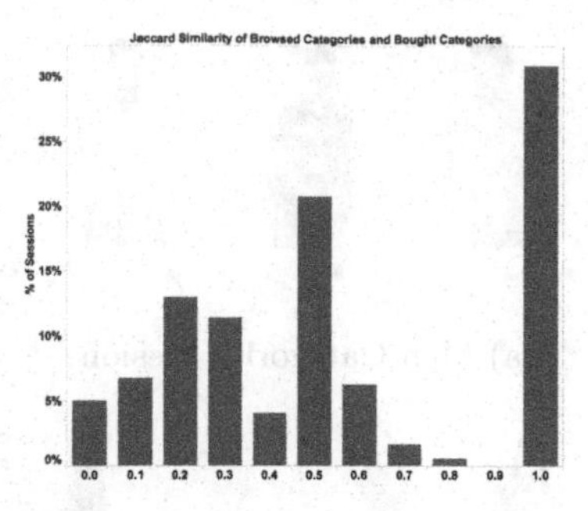

Fig. 8. Men categories session

categories for Women. As mentioned earlier, the product catalogue is dynamic and the products (or styles) available at any given point in time varies a lot. To tackle this problem, we replace each product by it's corresponding product group.

We use the following notations hereafter. Let Q be the set of all sessions/queries with a sequence of product groups $<P1, P2, ...P_n>$ (Pn being the true value to be predicted). All models described above output a sequence (sorted in descending order by the probabilities) R of length k $<R1, R2...R_k>$; k being the number of recommendations per query/session. To evaluate the models described in Sect. 3, we use the following metrics:

Mean Reciprocal Rank (MRR): The reciprocal rank of a query response is the multiplicative inverse of the rank of the first correct answer. The mean reciprocal rank is the average of the reciprocal ranks [25] of results for a sample of queries Q as shown in Eq. 7, where $rank_i$ refers to rank of P_n in R.

$$MRR = \frac{1}{|Q|} \sum_{n=1}^{|Q|} \frac{1}{rank_i} \tag{7}$$

It is worth mentioning that in this particular evaluation metric, we just consider one true value (P_n) and hence MRR and Mean Average Precision (MAP) are equivalent.

Normalised Discounted Cumulative Gain (NDCG): For the formulation of NDCG, we assume the true values to be a sequence of k values: P_n and $k-1$ product groups most similar to P_n. Also, the relevance score in the formula is the cosine similarity between P_n and other product group P_j. We use the true sequence for formulation of IDCG ($rel_1 = 1$ since $sim(P_n, P_n) = 1$). For the DCG formulation, we assume the similarity to be 0 if predicted product group does not belong to the article type of P_n (because embeddings were learnt separately for each article type); else relevance equals cosine similarity. It is worth mentioning that this formulation is likely to give quite low absolute values of NDCG since our sessions are quite noisy consisting of multiple article types; but the true ideal ranking just considers the similar product groups from the same article type (P_n 's article type) and hence high IDCG values. Equations 8 and 9 show the standard NDCG metric.

$$DCG_p = rel_i + \sum_{i=2}^{p} \frac{rel_i}{log_2 i} \quad (8)$$

$$NDCG_p = \frac{DCG_p}{IDCG_p} \quad (9)$$

Recall (R): We define recall@K as fraction of sessions wherein the true value P_n is retrieved in K recommendations.

Tables 1 and 2 show MRR and NDCG for different values of K (number of recommendations) for men and women respectively. The tuples in bracket show GRU improvement over PG-PG graph and Majority vote respectively. As can be clearly seen, GRU performs significantly better than Majority voting and PG-PG graph, specifically when K increases.

We also experimented by predicting just the category of last click (by ignoring the predicted product group). This showed a similar improvement in MRR. For instance, we report MRR for different values of K for Men and Women in Table 3 below. The subscript tuples show GRU improvement over PG-PG graph and Majority vote respectively.

We are especially interested in understanding how GRU performs for sessions with a high number of context changes. To evaluate this, we consider 20,853 men and 22,576 women sessions with more than two categories, and ones that differ in categories of the first and last click in the session. This was done to capture sessions where user started with a particular category, but ended on a totally different category (not just product group). As seen in Table 4, GRU gives a better recall than the other two baselines for these sessions.

Table 1. MRR and NDCG as a function of K for Men

	GRU		PG graph		Majority voting	
	MRR	NDCG	MRR	NDCG	MRR	NDCG
K = 3	$0.32_{(+15.16\%,+10.5\%)}$	$0.20_{(+26\%,+1.9\%)}$	0.28	0.16	0.29	0.20
K = 5	$0.34_{(+12.17\%,+8.7\%)}$	$0.18_{(+17.3\%,+1.2\%)}$	0.30	0.16	0.31	0.18
K = 10	$0.35_{(+10.34\%,+10.51\%)}$	$0.19_{(+13.87\%,+44.49\%)}$	0.31	0.17	0.31	0.13

Table 2. MRR and NDCG as a function of K for Women

	GRU		PG graph		Majority voting	
	MRR	NDCG	MRR	NDCG	MRR	NDCG
K = 3	$0.31_{(+17.94\%,+9.5\%)}$	$0.28_{(+0.99\%,+304\%)}$	0.26	0.28	0.28	0.07
K = 5	$0.32_{(+15\%,+7.6\%)}$	$0.28_{(+0.3\%,+123\%)}$	0.28	0.28	0.30	0.12
K = 10	$0.34_{(+12.72\%,+9.5\%)}$	$0.26_{(+13.64\%,+144\%)}$	0.30	0.23	0.31	0.11

Table 3. MRR for predicting on the category level for different values of K

	GRU		PG graph		Majority voting	
	Men	Women	Men	Women	Men	Women
K = 3	$0.65_{(+20.44\%,+9.9\%)}$	$0.62_{(+13.06\%,+3.8\%)}$	0.54	0.55	0.59	0.6
K = 5	$0.66_{(+19.11\%,+8.7\%)}$	$0.63_{(+12.45\%,+3.17\%)}$	0.55	0.56	0.6	0.61
K = 10	$0.67_{(+19.17\%,+9.6\%)}$	$0.64_{(+12.07\%,+3.5\%)}$	0.56	0.57	0.61	0.62

Table 4. Recall for different values of K

	GRU		PG graph		Majority voting	
	Men	Women	Men	Women	Men	Women
K = 3	$0.32_{(+20.03\%,+12.24\%)}$	$0.29_{(+29.79\%,+11.5\%)}$	0.26	0.22	0.28	0.26
K = 5	$0.37_{(+9.22\%,+3.01\%)}$	$0.34_{(+16.8\%,+0.8\%)}$	0.34	0.29	0.36	0.34
K = 10	$0.44_{(+0.6\%,+11.07\%)}$	$0.42_{(+5.12\%,+8.33\%)}$	0.44	0.4	0.4	0.39

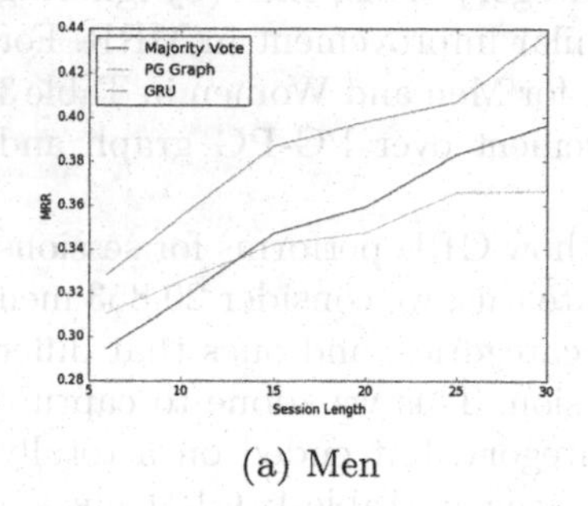

(a) Men

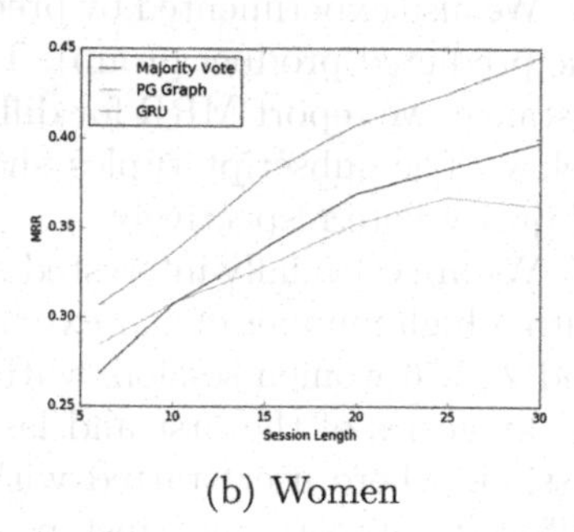

(b) Women

Fig. 9. MRR v/s session length

In Figs. 9a and b, we show how MRR varies as the session length increases (keeping K constant at 20). Clearly, GRUs handle the context change quite well in the sessions, and perform significantly better for longer sessions.

5 Conclusion

In this work, we attempted to model the behaviour of users in fashion e-commerce, that consists of multiple context changes while browsing. We built three models - majority vote, product group graph and GRU for predicting the next click of users. Through multiple metrics, we showed that GRUs outperform the other models. We also showed that as the session grows bigger in length, GRUs perform better. This suggests that with more available context, the performance of GRU improves.

References

1. Agichtein, E., Brill, E., Dumais, S., Ragno, R.: Learning user interaction models for predicting web search result preferences. In: Proceedings of the 29th Annual International ACM SIGIR Conference on Research and Development in Information Retrieval, pp. 3–10. ACM (2006)
2. Cadez, I., Heckerman, D., Meek, C., Smyth, P., White, S.: Model-based clustering and visualization of navigation patterns on a web site. Data Min. Knowl. Disc. **7**(4), 399–424 (2003)
3. Chung, J., Gulcehre, C., Cho, K., Bengio, Y.: Empirical evaluation of gated recurrent neural networks on sequence modeling. arXiv preprint arXiv:1412.3555 (2014)
4. Chung, J., Gülçehre, Ç., Cho, K., Bengio, Y.: Empirical evaluation of gated recurrent neural networks on sequence modeling. CoRR abs/1412.3555 (2014). http://arxiv.org/abs/1412.3555
5. Dauphin, Y.N., de Vries, H., Chung, J., Bengio, Y.: Rmsprop and equilibrated adaptive learning rates for non-convex optimization. CoRR abs/1502.04390 (2015). http://arxiv.org/abs/1502.04390
6. Ding, A.W., Li, S., Chatterjee, P.: Learning user real-time intent for optimal dynamic web page transformation. Inf. Syst. Res. **26**(2), 339–359 (2015). http://dx.doi.org/10.1287/isre.2015.0568
7. Gündüz, Ş., Özsu, M.T.: A web page prediction model based on click-stream tree representation of user behavior. In: Proceedings of the Ninth ACM SIGKDD International Conference on Knowledge Discovery and Data Mining, pp. 535–540. ACM (2003)
8. Guo, Q., Agichtein, E.: Ready to buy or just browsing?: detecting web searcher goals from interaction data. In: Proceedings of the 33rd International ACM SIGIR Conference on Research and Development in Information Retrieval, pp. 130–137. ACM (2010)
9. Hidasi, B., Karatzoglou, A., Baltrunas, L., Tikk, D.: Session-based recommendations with recurrent neural networks. CoRR abs/1511.06939 (2015). http://arxiv.org/abs/1511.06939
10. Kim, D.H., Atluri, V., Bieber, M., Adam, N., Yesha, Y.: A clickstream-based collaborative filtering personalization model: towards a better performance. In: Proceedings of the WIDM 2004 6th Annual ACM International Workshop on Web Information and Data Management, pp. 88–95. ACM, NY, USA (2004). http://doi.acm.org/10.1145/1031453.1031470
11. Kim, E., Kim, W., Lee, Y.: Combination of multiple classifiers for the customer's purchase behavior prediction. Decis. Support Syst. **34**(2), 167–175 (2003)
12. Lakshminarayan, C., Kosuru, R., Hsu, M.: Modeling complex clickstream data by stochastic models: theory and methods. In: Proceedings of the 25th International Conference Companion on World Wide Web. WWW 2016 Companion, International World Wide Web Conferences Steering Committee, Republic and Canton of Geneva, Switzerland, pp. 879–884 (2016). http://dx.doi.org/10.1145/2872518.2891070
13. LeCun, Y., Bengio, Y., Hinton, G.: Deep learning. Nature **521**(7553), 436–444 (2015). http://dx.doi.org/10.1038/nature14539
14. Lee, J., Podlaseck, M., Schonberg, E., Hoch, R.: Visualization and analysis of clickstream data of online stores for understanding web merchandising. In: Kohavi, R., Provost, F. (eds.) Applications of Data Mining to Electronic Commerce, pp. 59–84. Springer, New York (2001)

15. Li, Z., Shang, W.: Personalized news recommendation based on links of web. In: 2015 IEEE/ACIS 14th International Conference on Computer and Information Science (ICIS), pp. 581–584. IEEE (2015)
16. Lo, C., Frankowski, D., Leskovec, J.: Understanding behaviors that lead to purchasing: a case study of pinterest. In: Proceedings of the 22nd ACM SIGKDD International Conference on Knowledge Discovery and Data Mining, KDD 2016, NY, USA, pp. 531–540 (2016). http://doi.acm.org/10.1145/2939672.2939729
17. Mikolov, T., Chen, K., Corrado, G., Dean, J.: Efficient estimation of word representations in vector space. CoRR abs/1301.3781 (2013). http://arxiv.org/abs/1301.3781
18. Mikolov, T., Karafiát, M., Burget, L., Cernocký, J., Khudanpur, S.: Recurrent neural network based language model. In: 11th Annual Conference of the International Speech Communication Association. INTERSPEECH 2010, Makuhari, Chiba, Japan, 26–30 September 2010, pp. 1045–1048 (2010)
19. Moe, W.W.: Buying, searching, or browsing: differentiating between online shoppers using in-store navigational clickstream. J. Consum. Psychol. **13**(1), 29–39 (2003)
20. Montgomery, A.L., Li, S., Srinivasan, K., Liechty, J.C.: Modeling online browsing and path analysis using clickstream data. Mark. Sci. **23**, 579–595 (2004)
21. Sagar Arora, D.W.: Decoding fashion contexts using word embeddings. Machine Learning Meets Fashion, KDD 2016 Workshop (2016)
22. Senecal, S., Kalczynski, P.J., Nantel, J.: Consumers' decision-making process and their online shopping behavior: a clickstream analysis. J. Bus. Res. **58**(11), 1599–1608 (2005)
23. Su, Q., Chen, L.: A method for discovering clusters of e-commerce interest patterns using click-stream data. Electron. Commer. Res. Appl. **14**(1), 1–13 (2015)
24. Vieira, A.: Predicting online user behaviour using deep learning algorithms. arXiv preprint arXiv:1511.06247 (2015)
25. Voorhees, E.M.: The TREC question answering track. Nat. Lang. Eng. **7**(4), 361–378 (2001). http://dx.doi.org/10.1017/S1351324901002789
26. Wang, G., Zhang, X., Tang, S., Zheng, H., Zhao, B.Y.: Unsupervised clickstream clustering for user behavior analysis. In: Proceedings of the 2016 CHI Conference on Human Factors in Computing Systems, CHI 2016, NY, USA, pp. 225–236 (2016). http://doi.acm.org/10.1145/2858036.2858107
27. Waterson, S.J., Hong, J.I., Sohn, T., Landay, J.A., Heer, J., Matthews, T.: What did they do? understanding clickstreams with the webquilt visualization system. In: Proceedings of the Working Conference on Advanced Visual Interfaces, pp. 94–102. ACM (2002)
28. Wei, J., Shen, Z., Sundaresan, N., Ma, K.L.: Visual cluster exploration of web clickstream data. In: 2012 IEEE Conference on Visual Analytics Science and Technology (VAST), pp. 3–12. IEEE (2012)
29. Zhang, M., Chen, G., Wei, Q.: Discovering consumers' purchase intentions based on mobile search behaviors. Flexible Query Answering Systems 2015. AISC, vol. 400, pp. 15–28. Springer, Cham (2016). doi:10.1007/978-3-319-26154-6_2
30. Zhao, J., Liu, Z., Dontcheva, M., Hertzmann, A., Wilson, A.: MatrixWave: Visual comparison of event sequence data. In: Proceedings of the 33rd Annual ACM Conference on Human Factors in Computing Systems, pp. 259–268. ACM (2015)

Weighted Ensemble Classification of Multi-label Data Streams

Lulu Wang[1], Hong Shen[2,3](✉), and Hui Tian[4]

[1] School of Computer and Information Technology, Beijing Jiaotong University, Beijing, China
llwang_14@bjtu.edu.cn

[2] School of Data and Computer Science, Sun Yat-Sen University, Guangzhou, China
hongsh01@gmail.com

[3] School of Computer Science, University of Adelaide, Adelaide, Australia

[4] School of Electronics and Information Engineering, Beijing Jiaotong University, Beijing, China
htian@bjtu.edu.cn

Abstract. Many real world applications involve classification of multi-label data streams. However, most existing classification models mostly focused on classifying single-label data streams. Learning in multi-label data stream scenarios is more challenging, as the classification systems should be able to consider several properties, such as large data volumes, label correlations and concept drifts. In this paper, we propose an efficient and effective ensemble model for multi-label stream classification based on ML-KNN (Multi-Label KNN) [31] and propose a balance AdjustWeight function to combine the predictions which can efficiently process high-speed multi-label stream data with concept drifts. The empirical results indicate that our approach achieves a high accuracy and low storage cost, and outperforms the existing methods ML-KNN and SMART [14].

Keywords: Multi-label · Data stream · Classification

1 Introduction

Due to the recent advances in computer networks and data storage, many data are produced and accumulated at an ever increasing rate in the form of stream. Such as online shopping information, logistics information, online news, stock market data, emails, credit card transactions, *etc.* These data are real-time, continuous and orderly arrival, and need to be analyzed promptly and effectively. For example, in online mail systems, incoming emails need to be classified into different categories, like spams, business emails, personal emails, important emails, *etc.* This classification task, each stream example is associated with a single class label l from a set of labels L $(|L| > 1)$, is called single-label data stream classification, and has been extensively studied [3,4,11,20,33] in the literature.

J. Kim et al. (Eds.): PAKDD 2017, Part II, LNAI 10235, pp. 551–562, 2017.
DOI: 10.1007/978-3-319-57529-2_43

In many emerging applications, each stream record may carry multiple class labels. A good example is news reports in the online news systems, most news reports carry multiple news topics (e.g. entertainment, financial and politics), then this is called **M**ulti-**L**abel data **S**tream **Classification** (**MLSC**) [21]. Formally, the multi-label stream classification problem is to training a model to attach a label subset $Y \subseteq L$ to each instance in a high-speed data stream. Although, multi-label classification has been studied in traditional database mining scenarios, multi-label data stream classification is a relatively new concept and has not been fully addressed yet.

An intuitive approach to solving the multi-label classification problem is to transform it into one or more single-label classification problems. In this fashion, traditional single-label classifiers can be employed to make single-label classifications. Finally, the multi-label predictions can be produced by combining these multi-label predictions. On the other hand, an alternative category is to adopt the existing single-label classifiers directly to multi-label classification [18,23].

The multi-label data stream environment has different challenges from the traditional batch learning setting. As the instances in a multi-label data stream contain multi-labels (multi-concepts), dealing with the concept drift is the most important challenge to a classifier. Another challenge with regard to MLSC is that, it is possible that an arriving example will belong to a set of labels, some of which, will not have been previously observed because of the dynamic nature of the set of labels [21]. Besides, the learner must be able to handle the stream using limited memory in real time, because stream data flood in continuously at a high speed, which makes it impossible to be stored in the memory and processed offline [14].

To address the above issues, in this paper we propose an efficient and effective ML-KNN-based ensemble model for multi-label stream classification with a balance AdjustWeight function, called **S**treaming **W**eighted **ML-KNN**-based **E**nsemble **C**lassifier (SWMEC). More specifically, we first propose an ML-KNN based algorithm to build the basic classifier C_i. The ensemble classifiers $C =< C_1, C_2, \cdots, C_L >$ will be build at the beginning of the stream with randomly selected L test data chunks. This only needs to compute and save a small amount of information of the cluster center points. Thus the building process is highly efficient while consuming constant memory space. In addiction, each classifier C_i has its own weight w_i, and $C =< (C_1, w_1), (C_2, w_2), \cdots, (C_L, w_L) >$. As data flow in, the weights will be adjusted and the classifier C will be updated according to the weights. Thus the proposed SWMEC approach can work adaptively to evolving data and deal with concept drifts, and can efficiently classify the incoming data in real-time.

The main contributions of this paper are as follows. (1) an adaption of the existing multi-label methods to evolving data streams. (2) an effective weighting adjustment strategy for ensemble classifiers. (3) experimental results validating the performance of our method and benchmarks in predictive performance and space complexity.

The remainder of this paper is organized as follows. In Sect. 2, we discuss relevant work in multi-label classification and stream classification. Section 3 presents the n preliminaries about the problem. Section 4, describes the proposed framework in details, and the experimental results are presented in Sect. 5. Finally, Sect. 6 concludes the paper.

2 Related Work

Our work is related to multi-label classification and stream classification techniques. We will briefly review the existing work on both of them.

Multi-label classification is the problem to deal with such instances that may belong to multiple different classes simultaneously and focuses on offline settings [16,19]. Multi-label classification methods can be grouped into two categories, namely, *problem transformation* and *algorithm adaptation* [21]. Problem transformation methods transform the multi-label problem into multiple single-label problems. Problem transformation methods transform the multi-label problem into multiple single-label problems including Label Power-set (LP), Binary Relevance (BR) and Ranking by Pairwise Comparison (RPC [12]). Algorithm adaptation methods extend the traditional learning techniques to multi-label context, such as decision trees [5], neural networks [30], maximal margin methods [10], maximum entropy methods [25], and ensemble methods [25], etc. One well-known such approach is ML-KNN [31], which is derived from the popular lazy learning algorithm kNN. It's the most relevant approach to our model and will be introduced in the next section.

Many studies have also been done on single-label stream classification. There are two sets of solutions: single-model based and ensemble based. Single-model based approaches [1,2,6,9,26,27,32,33] use new data to incrementally update their model so that the model can scale to a large data volume. Ensemble based approaches [13,22,28], on the other hand, partition the data stream into equal sized chunks, and train multiple base models on different chunks of data. Then all the models are combined for prediction. The ensemble based approaches are easier to scale and parallelize, tend to achieve better accuracy and can also avoid over fitting than single classifier methods.

Recently, there are also some studies focusing on multi-label stream classification [14,15,17,18,21,29]. Kong et al. [14] builds an ensemble of classifiers on successive data chunks. It proposes a random-tree based algorithm to improve it's efficiency. Work also has been done on adopting the ensemble based strategy in handling multi-label streams [15,29].

We follow a similar strategy to design our classification model with ML-KNN-based ensemble methods. Our model builds streaming classifiers by extending MLkNN, which is just designed for multi-label static data classification.

3 Preliminaries

We first introduce the notations that will be used throughout this paper, and then briefly describe the techniques ML-KNN [31] to make this paper self-contained.

Consider a multi-label stream S with a label set $\mathcal{L} = \{1, 2, \cdots, q\}$, $S \subseteq \mathbb{R}^d$. Stream S consists of infinite data chunks, $\{D_1, \cdots D_n, \cdots\}$, where labeled chunk D_i is denoted by D_L and unlabeled chunk denoted by D_U. Each instance $x \in S$ has a label subset $Y = \{y_1, y_2, \cdots, y_q\} \in \{-1, 1\}^q$, where $Y[j] = \begin{cases} 1 & \text{if } y_j \in Y \\ -1 & \text{if } y_j \notin Y \end{cases}$, (x_i, Y_i) is an instance in the multi-label data stream. The task of multi-label stream classification based on ensemble solution is to train a classification model on the historical examples $F(\cdot)$, $F(\cdot) = g(f_1(\cdot), f_2(\cdot), \cdots, f_L(\cdot))$, where $f_i(\cdot)$ is the sub-classifier, $g(\cdot)$ is the combination function that combines the outputs of all $f(\cdot)$, L is the number of classifiers. Then it uses $F(\cdot)$ to predict a label set Y_i to the incoming data x_i.

Table 1. Summary of major mathematical notations

Notations	Mathematical meanings
S	Multi-label data stream with d-dimensional space $\mathbb{R}^d$
D	data chunk with size N
D_L	Labeled data chunk
D_U	Unlabeled data chunk
x	d-dimensional feature vector $(x_1, x_2, \cdots x_d)^{\top} (x \in S)$
$\mathcal{L}$	label space with q possible class labels $\{1, 2, \cdots, q\}$
Y	label subset associated with x $(Y \subseteq \mathcal{L})$

3.1 ML-KNN [31]

We briefly describe our model's basic approach ML-KNN, which is derived from the traditional k-Nearest Neighbor (kNN) algorithm and classify the traditional static multi-label data in a lazy learning way. There are three main steps in this approach. For convenience, several notations are summarized in Table 1. In addition, given a training set $T = \{(x_1, Y_1), (x_2, Y_2), \cdots, (x_n, Y_n)\}$ $(x_i \in \mathbb{R}^d, Y_i \in \mathcal{L})$, t is the test instance, s is the smoothing parameter with a default value 1.

Step 1: Computing the prior probabilities $P(H_b^l)$ according to Eqs. 1 and 2. Where $l \in Y$ is the l-th label, and $b \in \{0,1\}$, H_1^l represents the event the instance has label l. Conversely, H_0^l means the instance does not have label l.

$$P(H_1^l) = \left(s + \sum_{i=1}^{n} L_t(l)\right) / (s \times 2 + n) \tag{1}$$

$$P(H_0^l) = 1 - P(H_1^l) \tag{2}$$

Step 2: Computing the posterior probabilities $P(E_j^l \mid H_b^l)$ according to Eqs. 3 and 4. Where E_j^l $(j \in \{0, 1, \cdots, k\})$ denotes the event that, among the k nearest neighbors of t, there are exactly j instances which have label l. $c[j]$ counts the number of training instances with label l whose k nearest neighbors contain exactly j instances with label l. Correspondingly, $c^{'}[j]$ counts the number of training instances without label l whose k nearest neighbors contain exactly j instances with label l.

$$P\left(E_m^l \mid H_1^l\right) = \left(s + c\left[j\right]\right) / \left(s \times (k+1) + \sum_{p=0}^{k} c\left[p\right]\right) \tag{3}$$

$$P\left(E_m^l \mid H_0^l\right) = \left(s + c^{'}\left[j\right]\right) / \left(s \times (k+1) + \sum_{p=0}^{k} c^{'}\left[p\right]\right) \tag{4}$$

Step 3: Computing the output y_t (label subset) and r_t of t, where r_t is a real-valued vector calculated to rank labels in $\mathcal{L}$, according to Eqs. 5 and 6.

$$\vec{y_i}\left(l\right) = \arg\max_{b \in \{0,1\}} P\left(H_b^l\right) P\left(E_{C_t(t)}^l \mid H_b^l\right) \tag{5}$$

$$\begin{aligned} \vec{r_i}\left(l\right) &= P\left(H_1^l \mid E_{C_t(l)}^l\right) \\ &= \left(P\left(H_1^l\right) P\left(E_{C_t(l)}^l \mid H_1^l\right)\right) / P\left(E_{C_t(l)}^l\right) \end{aligned} \tag{6}$$

The detailed architecture of ML-KNN was given in [31].

4 Weighted Ensemble Classification

In this section we first give the main idea of our weighted ensemble classification approach, then we introduce the process of the ensemble classifiers' training and updating and the adjustment of their weights, finally we give the description of our algorithm.

4.1 Basic Idea

The data stream S is divided into a fixed number of chunks and each classification model in the ensemble is trained from a different chunk. Each classifier in the ensemble has it's own weight. The new arriving unlabeled data chunk is classified by the ensemble while the corresponding weight will be changed. According to the weights, the latest classified data chunk will be decided if to be trained to generate a new model and replace one of the existing models in the ensemble. In this way the ensemble can be maintained at a fixed size and kept up-to-date. The problems of data stream's infinite length and concept-drift can correspondingly be well addressed.

4.2 Classifier Training and Updating

The ML-KNN based multi-label ensemble classifier is built at the beginning of the data stream and timely updated over the data stream as follow: when a training data chunk in the data stream is arriving at time t, we build h clusters with the labeled data points by the application of XMeans technique. After the building of these clusters, we save each cluster's centroid $O = \{o_1, o_2, \cdots, o_h\}, o_i \in \mathbb{R}^d, i \in \{1, 2, \cdots, h\}$, and compute each centroid's label subset in the same way as ML-KNN. After that, all centroids and their label subset $C =< (o_1, y_1, r_1), (o_2, y_2, r_2), \cdots, (o_h, y_h, r_h) >, y_j \in Y$, y_i,r_j are respectively the label subset and the a real-valued vector calculated by ML-KNN, will be saved as a summary. At the same time, the current summary's arriving time t will also be recorded. Each summary's weight w then can be calculated according to o, $\vec{r}$ and t, and set to be 1 at beginning. After the process of L chunks, the ensemble classifier $C =< (C_1, w_1), (C_2, w_2), \cdots, (C_L, w_L) >$ (each model C_i is a collection of h summaries and the number of h is unfixed, w_i is the weight of model C_i) will be built. When an unlabeled data chunk D_U is arriving, for each instance $x_i \in D_U$, we find the $m_i = ((o_j, y_j), w_i) \in C_i, j \in [1, 2, \cdots h]$ whose centroid o_j is nearest from x_i. The corresponding weight w_i will then be determined by the distance between o_j and x_i, $\vec{r_i}$ and t_i. Then the x_i will be labeled according to the summaries $\{m_1, m_2, \cdots, m_L\}$. Each unlabeled data chunk will be classified as above. After a chunk D_U has been handled by the ensemble C, If the lowest $w_{lowest} \in w$ falls below a threshold value ϵ, the corresponding model C_j will be replaced by the new model that trained by D_U. This ensures that the ensemble will be updated with the passing of data chunk and the number of models in the ensemble remains constant.

4.3 Ensemble Weighting

In this paper, inspired by [8], we propose a combination function $g(\cdot)$ including three components ($\vec{\alpha_i}$, β_i and γ_i) to combine classifiers in the ensemble. They are: (1) label confidence, which is a vector measuring the confidence in the sub-classifier outputting each label; (2) time difference; (3) distance difference, both (2) and (3) describe how confident a sub-classifier is when making a classification decision.

We estimate $\vec{\alpha_i}$ from the ensemble model C by Eq. 7:

$$\vec{\alpha_i} = \frac{\vec{r_i}}{\sum_{i=1}^{L} \vec{r_i}} \tag{7}$$

where β_i, γ_i describe how confident a sub-classifier is when making a classification decision. β_i is estimated by the time difference between arriving new data chunk D_U and the sub-classifier C_i. $\Delta t_i = t_{D_U} - t_{C_i}$, where t_{D_U},t_{C_i} are the arriving times of data chunk D_U and D_i respectively. The longer the chunks are apart, the lower the β is.

$$\beta_i = \frac{e^{-\Delta t_i}}{\sum_{i=1}^{L} e^{-\Delta t_i}}; 0 < \beta_i < 1, \sum_{i=1}^{L} \beta_i = 1 \tag{8}$$

γ_i is estimated by the distance difference between x_i which is the instance from the arriving new data chunk D_U and it's nearest center o_j in sub-classifier C_i.

$$\gamma_i = \frac{e^{-d_i}}{\sum_{i=1}^{L} e^{-d_i}}; 0 < \gamma_i < 1; \sum_{i=1}^{L} \gamma_i = 1 \tag{9}$$

where $d_i = distance(x_i, o_j)$, the closer the distance between x_i and o_j the more confident the sub-classifier.

Finally, we combine $\vec{\alpha_i}$, β_i and γ_i to decide the weight $w_i = \vec{\alpha_i} \times \beta_i \times \gamma_i$. Consequently:

$$g(\cdot) = \sum_{i=1}^{L} \left(f_i(x_i) \times \vec{\alpha_i} \times \beta_i \times \gamma_i \right) \tag{10}$$

The output $Y = \{y_1, y_2, \cdots, y_q\} \in \{-1, 1\}^q$,

$$Y[j] = \begin{cases} 1 & \text{if } g(\cdot)[j] = \sum_i^L (f_i(\cdot)[j] \times \alpha_i \times \beta_i \times \gamma_i) > 0 \\ -1 & \text{if } g(\cdot)[j] = \sum_i^L (f_i(\cdot)[j] \times \alpha_i \times \beta_i \times \gamma_i) < 0 \end{cases}.$$

4.4 The Classification Algorithm

The pseudo code of our method for classifying multi-label data streams using the ML-KNN-based ensemble method is given in Algorithm 1.

Algorithm 1. KNN-based ensemble classification for multi-label data streams

Input: Data Stream $S = < D_1, D_2, \cdots, D_n, \cdots >$;
Initial ensemble classifiers:
$C = < (C_1, w_1), (C_2, w_2), \cdots, (C_L, w_L) >$;
Empty buf;
Latest chunk of unlabeled instances D_U;
Latest r labeled data chunks D_r;
Output: Classification Result.
(1) **while true do**
(2) **for all** $x_i \in D_U$ **do**
(3) **Classification**$(C, x_i) = D_L \to buf$
(4) **end for**
(5) **Evaluation&Adjust**(C, D_U)
(6) **if** the lowest $w_{lowest} \in (w_1, w_2, \cdots, w_L) < threshold$ ϵ **then**
(7) buf replaces the corresponding data chunk in D_r
(8) **Update**(C, D_r)
(9) **end if**
(10) new chunk of unlabeled data $\to D_U$
(11) **end while**

5 Experiments

In this section, we show the results of several experiments performed to evaluate the effectiveness of our proposed SWMEC for classification in real-world multi-label data streams TMC2007 [24] (see Table 2 in detail). All the experiments are conducted on a PC with Intel(R) Core(TM) i3-3220 3.30 GHz CPU and 4 GB RAM. We have implemented SWMEC in python2.7. Most importantly, in order to get more precise classification results, we firstly preprocess the data set. We remove the infrequent words that occur in less than 11% of the documents as [14] did.

Table 2. Summary of dataset TMC2007

Data set	Properties					
	N	d	q	Avg$\|Y\|$	IDens	Domain
TMC2007	28,596	204	22	2.158	0.098	Text

A. Classification quality comparison with ML-KNN

We compare the performance of our approach SWMEC against ML-KNN [31] as the base-line of our model. In order to apply this one of the state-of-the-art methods in offline context to the stream classification, we train a ML-KNN classifier on the latest chunk of data, and use it to classify the next chunk of data, which is similar to sliding window approaches. Since the data chunk size is the most important factors in the classification and training process, here we change the chunk sizes (sliding window sizes) to test the performance of basic ML-KNN. The Parameters are set as follows: SWMEC: L (the size of the ensemble classifier model) $= 4$, k (the number of nearest neighbors in ML-KNN) $= 4$, ϵ (the threshold about weight's adjustment) $= 0.001$. ML-KNN ($w = 100$): w (the size of the window) $= 100$. ML-KNN ($w = 200$): w (the size of the window) $= 200$. ML-KNN ($w = 400$): w (the size of the window) $= 400$.

Multi-label classification problems has many different metrics for evaluation. Such as Hamming-loss, F-measure, Log-Loss, Ranking-Loss [17]. Here we adopt two metrics to evaluate multi-label classification performance in a data stream. First, Micro F1: considers both micro average of Precision and Recall with equal importance, evaluates a classifier's label set prediction performance. The higher the value, the better the performance.

$$MicroF1\left(f_i, D_L\right) = \frac{2 \times \sum_{i=1}^{|D_L|} \left|f_i\left(x\right) \cap Y_i\right|}{\sum_{i=1}^{|D_L|} \left|f_i\left(x\right)\right| + \sum_{i=1}^{|D_L|} \left|Y_i\right|}$$

where $|D|$ is the number of instances in a multi-label data stream D, which contains (x_i, Y_i), where $Y_i \subseteq L(i = 1, \cdots, |D|)$, $f(x_i) \subseteq L$ denotes a multi-label classifier's predicted label set for x_i and $f(x_i, k)$ denotes the classifier's probability outputs for x_i on the k-th label (l_k).

Second, Ranking Loss [7]: compute the average number of label pairs that are incorrectly ordered given Y_i weighted by the size of the label set and the number of labels not in the label set. Evaluates the performance of classifier's probability outputs or real-value outputs $f(x_i, k)$. The best performance is achieved with a ranking loss of zero.

$$RankLoss(f, D) = \frac{1}{|D|} \sum_{i=1}^{|D|} \frac{1}{|Y_i||\bar{Y_i}|} loss\left(f, x_i, Y_i\right)$$

$$loss(f, x_i, Y_i) = \sum_{k \in Y_i} \sum_{k' \in \bar{Y_i}} I\left(f\left(x_i, k\right) \leq f\left(x_i, k'\right)\right)$$

where the $\bar{Y_i}$ denotes the complementary set of Y_i in $\mathcal{L}$.

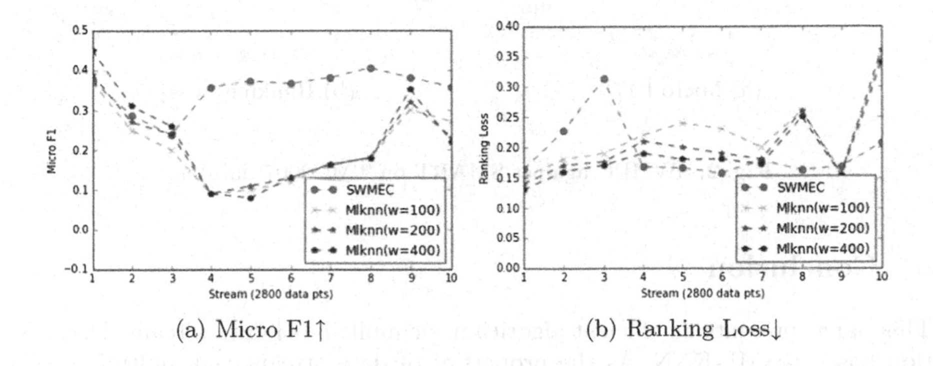

(a) Micro F1↑ (b) Ranking Loss↓

Fig. 1. SWMEC against ML-KNN on TMC2007 dataset.

Figure 1(a) shows the Micro F1 for the four algorithm throughout the stream in TMC2007 data-set. We report the average performance on every $|D|/10$ instances. For example, at X axis = 5, the Y values show the average Micro F1 of four classification models from the $|D| * 4/10$th instance of the stream to the $|D| * 5/10$th instance. At this point, the Micro F1 of SWMEC is 0.3729, the Y values in ML-KNN ($w = 100$), ML-KNN ($w = 200$) and ML-KNN ($w = 400$) are all below SWMEC.

We also calculate the Ranking loss of four classification models. Figure 1(b) show the Ranking loss of the four algorithms throughout the stream in TMC2007 data-set. For example, at X axis = 6, the ranking loss of SWMEC is 0.1571, the Y values in ML-KNN ($w = 100$), ML-KNN ($w = 200$) and ML-KNN ($w = 400$) are all higher than SWEMC.

B. Classification quality comparison with SMART [14]
We also compare the performance of our approach SWMEC against SMART [14], which also adopt the strategy of ensemble and gives us a great inspiration in this paper. As Fig. 2(a) and (b) shows, Our method SWMEC is slightly better than SMART in Micro F1 and Ranking Loss. Especially noteworthy is that SMART has a much bigger space overhead than SWMEC, because SMART

needs to maintain several tree structures while SWMEC only needs to store small amount of central points. In the future, we would like to combine both of these two methods' advantages to address the multi-label classification problem in data streams.

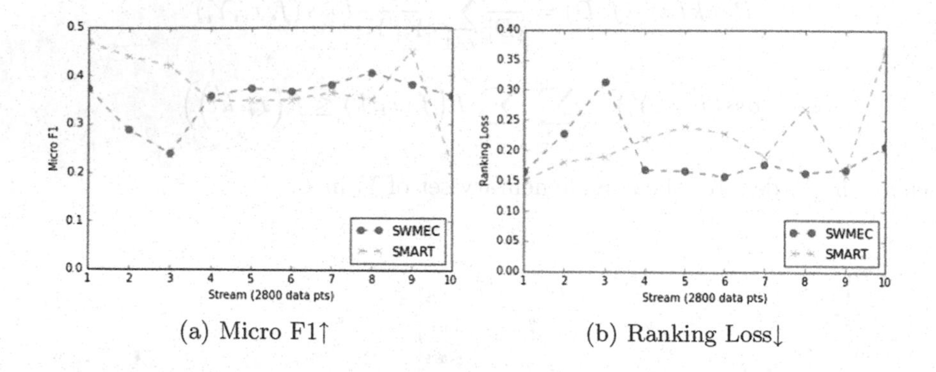

(a) Micro F1↑ (b) Ranking Loss↓

Fig. 2. SWMEC against SMART on TMC2007 dataset.

6 Conclusion

This paper presents an efficient algorithm for multi-label data stream classification based on ML-KNN. As the properties of data stream and multiple labels assigned to each instances. It becomes more challenging than the traditional static multi-label data classification problems and single-label data stream classification problems. To address these challenges, we propose an ensemble multilabel data stream classification approach, manly Streaming Weighting ML-KNN based Ensemble Classifier (SWMEC), to efficiently update the model with the multi-label data stream flows. Then our model can effectively and efficiently predict multiple labels for future data points. The experimental results on the real world validate that our multi-label data stream classification approach is very effective and efficient for multi-label stream classification.

Acknowledgment. This work is supported by Research Initiative Grant of Sun Yat-Sen University under Project 985 and Australian Research Council Discovery Projects funding DP150104871. The corresponding author is Hong Shen.

References

1. Aggarwal, C.C., Han, J., Wang, J., Yu, P.S.: On demand classification of data streams. In: Proceedings of the Tenth ACM SIGKDD International Conference on Knowledge Discovery and Data Mining, pp. 503–508. ACM (2004)
2. Aggarwal, C.C., Yu, P.S.: Locust: an online analytical processing framework for high dimensional classification of data streams. In: 2008 IEEE 24th International Conference on Data Engineering, pp. 426–435. IEEE (2008)

3. Boutell, M.R., Luo, J., Shen, X., Brown, C.M.: Learning multi-label scene classification. Patt. Recogn. **37**(9), 1757–1771 (2004)
4. Brinker, K., Fürnkranz, J., Hüllermeier, E.: A unified model for multilabel classification and ranking. In: Proceedings of the 2006 Conference on ECAI 2006: 17th European Conference on Artificial Intelligence, August 29–September 1, 2006, Riva del Garda, Italy, pp. 489–493. IOS Press (2006)
5. De Comité, F., Gilleron, R., Tommasi, M.: Learning multi-label alternating decision trees from texts and data. In: Perner, P., Rosenfeld, A. (eds.) MLDM 2003. LNCS, vol. 2734, pp. 35–49. Springer, Heidelberg (2003). doi:10.1007/3-540-45065-3_4
6. Domingos, P., Hulten, G.: Mining high-speed data streams. In: Proceedings of the Sixth ACM SIGKDD International Conference on Knowledge Discovery and Data Mining, pp. 71–80. ACM (2000)
7. Elisseeff, A., Weston, J.: A kernel method for multi-labelled classification. In: Advances in Neural Information Processing Systems, pp. 681–687 (2001)
8. Fung, G.P.C., Yu, J.X., Wang, H., Cheung, D.W., Liu, H.: A balanced ensemble approach to weighting classifiers for text classification. In: Sixth International Conference on Data Mining (ICDM 2006), pp. 869–873. IEEE (2006)
9. Gama, J., Rocha, R., Medas, P.: Accurate decision trees for mining high-speed data streams. In: Proceedings of the Ninth ACM SIGKDD International Conference on Knowledge Discovery and Data Mining, pp. 523–528. ACM (2003)
10. Godbole, S., Sarawagi, S.: Discriminative methods for multi-labeled classification. In: Dai, H., Srikant, R., Zhang, C. (eds.) PAKDD 2004. LNCS (LNAI), vol. 3056, pp. 22–30. Springer, Heidelberg (2004). doi:10.1007/978-3-540-24775-3_5
11. Gopal, S., Yang, Y.: Multilabel classification with meta-level features. In: Proceedings of the 33rd International ACM SIGIR Conference on Research and Development in Information Retrieval, pp. 315–322. ACM (2010)
12. Hüllermeier, E., Fürnkranz, J., Cheng, W., Brinker, K.: Label ranking by learning pairwise preferences. Artif. Intell. **172**(16), 1897–1916 (2008)
13. Kolter, J.Z., Maloof, M.A.: Using additive expert ensembles to cope with concept drift. In: Proceedings of the 22nd International Conference on Machine Learning, pp. 449–456. ACM (2005)
14. Kong, X., Yu, P.S.: An ensemble-based approach to fast classification of multi-label data streams. In: 2011 7th International Conference on Collaborative Computing: Networking, Applications and Worksharing (CollaborateCom), pp. 95–104. IEEE (2011)
15. Qu, W., Zhang, Y., Zhu, J., Qiu, Q.: Mining multi-label concept-drifting data streams using dynamic classifier ensemble. In: Zhou, Z.-H., Washio, T. (eds.) ACML 2009. LNCS (LNAI), vol. 5828, pp. 308–321. Springer, Heidelberg (2009). doi:10.1007/978-3-642-05224-8_24
16. Read, J.: Scalable multi-label classification (2010)
17. Read, J., Bifet, A., Holmes, G., Pfahringer, B.: Scalable and efficient multi-label classification for evolving data streams. Mach. Learn. **88**(1–2), 243–272 (2012)
18. Read, J., Bifet, A., Holmes, G., Pfahringer, B.: Efficient multi-label classification for evolving data streams (2010)
19. Read, J., Pfahringer, B., Holmes, G., Frank, E.: Classifier chains for multi-label classification. Mach. Learn. **85**(3), 333–359 (2011)
20. Sanden, C., Zhang, J.Z.: Enhancing multi-label music genre classification through ensemble techniques. In: Proceedings of the 34th International ACM SIGIR Conference on Research and Development in Information Retrieval, pp. 705–714. ACM (2011)

21. Spyromitros-Xioufis, E.: Dealing with concept drift and class imbalance in multi-label stream classification. Ph.D. thesis, Department of Computer Science, Aristotle University of Thessaloniki (2011)
22. Street, W.N., Kim, Y.: A streaming ensemble algorithm (sea) for large-scale classification. In: Proceedings of the Seventh ACM SIGKDD International Conference on Knowledge Discovery and Data Mining, pp. 377–382. ACM (2001)
23. Tsoumakas, G., Katakis, I.: Multi-label classification: An overview. Department of Informatics, Aristotle University of Thessaloniki, Greece (2006)
24. Tsoumakas, G., Spyromitros-Xioufis, E., Vilcek, J., Vlahavas, I.: Mulan: a java library for multi-label learning. J. Mach. Learn. Res. **12**, 2411–2414 (2011)
25. Tsoumakas, G., Vlahavas, I.: Random k-labelsets: an ensemble method for multilabel classification. In: Kok, J.N., Koronacki, J., Mantaras, R.L., Matwin, S., Mladenič, D., Skowron, A. (eds.) ECML 2007. LNCS (LNAI), vol. 4701, pp. 406–417. Springer, Heidelberg (2007). doi:10.1007/978-3-540-74958-5_38
26. Wang, H., Wu, J., Zhu, X., Zhang, C.: Time-variant graph classification (2016). arXiv preprint: arXiv:1609.04350
27. Wang, H., Zhang, P., Zhu, X., Tsang, I., Chen, L., Zhang, C., Wu, X.: Incremental subgraph feature selection for graph classification. IEEE Trans. Knowl. Data Eng. **29**, 128–142 (2016)
28. Wang, H., Fan, W., Yu, P.S., Han, J.: Mining concept-drifting data streams using ensemble classifiers. In: Proceedings of the Ninth ACM SIGKDD International Conference on Knowledge Discovery and Data Mining, pp. 226–235. ACM (2003)
29. Wang, P., Zhang, P., Guo, L.: Mining multi-label data streams using ensemble-based active learning. In: SDM, pp. 1131–1140. SIAM (2012)
30. Zhang, M.-L., Zhou, Z.-H.: Multilabel neural networks with applications to functional genomics and text categorization. IEEE Trans. Knowl. Data Eng. **18**(10), 1338–1351 (2006)
31. Zhang, M.-L., Zhou, Z.-H.: Ml-knn: a lazy learning approach to multi-label learning. Patt. Recogn. **40**(7), 2038–2048 (2007)
32. Zhang, P., Gao, B.J., Zhu, X., Guo, L.: Enabling fast lazy learning for data streams. In: 2011 IEEE 11th International Conference on Data Mining, pp. 932–941. IEEE (2011)
33. Zhu, X., Zhang, P., Lin, X., Shi, Y.: Active learning from data streams. In: Seventh IEEE International Conference on Data Mining (ICDM 2007), pp. 757–762. IEEE (2007)

Novel Models and Algorithms

Improving One-Class Collaborative Filtering with Manifold Regularization by Data-driven Feature Representation

Yen-Chieh Lien(✉) and Pu-Jen Cheng

Department of Computer Science and Information Engineering,
National Taiwan University, Taipei, Taiwan
r03922056@ntu.edu.tw, pjcheng@csie.ntu.edu.tw

Abstract. When considering additional features of users or items in a recommendation system, previous work focuses mainly on manually incorporating these features into original models. In this paper, manifold regularization is introduced to the well-known one-class collaborative filtering problem. To fully benefit from large unlabeled data, we design a data-driven framework, which learns a representation function by not only transferring raw features of users or items into latent ones but also directly linking the relation between the latent features and user behaviors. The framework is expected to bring cluster hypothesis from machine learning to recommendation, that is, more similar transferred features can bring more similar user behaviors. The experiments have been conducted on two real datasets. The results demonstrate that the learned representation through our framework can boost prediction performance significantly.

Keywords: Feature representation learning · Manifold regularization · One-class collaborative filtering

1 Introduction

One-class collaborative filtering (OCCF) [10] is an important research topic in the field of recommender systems. Unlike traditional collaborative filtering (CF), which pays attention to numerical or star ratings such as 1–5 scales rated by users for movies as an explicit feedback, OCCF handles data only containing implicit binary feedback such as users' purchasing (buy or not-buy) or web surfing (click or not-click) behaviors. It is especially useful for applications with only positive training data and fits the practical issue much better because not all of the users are willing to provide ratings after purchasing or browsing.

Recently, matrix factorization approaches [7] have shown their effectiveness in discovering relevant latent features for OCCF (and CF, of course). To boost the prediction performance, previous work [4,9] attempts to directly incorporate additional features of users or items into latent factor models and regards these features as an input of the hybrid models. Despite their success in better

J. Kim et al. (Eds.): PAKDD 2017, Part II, LNAI 10235, pp. 565–577, 2017.
DOI: 10.1007/978-3-319-57529-2_44

performance, such methods only improve optimization by additional features of training instances that have feedback. The additional features of the instances without feedback, i.e., unlabeled data, are not considered in the optimization process. For this issue, another line of work is manifold regularization, which this work focuses mainly on.

Manifold regularization (MR) [1] is originally designed for semi-supervised learning on classification problems. By assuming instances closer in a feature space are more likely to have the same label, it builds a regularization term from features of both labeled and unlabeled data to let the target function fit data distribution better. MR alleviates the problem of sparse labeled data by incorporating unlabeled data to constrain the learned function. In recommendation, it states the idea that more similar features *might* bring more similar user behaviors. For example, two movies with the common tag "love" would attract similar users who like this topic. Various features like user profile [2] and social relation [8] can be adopted in MR.

However, bringing MR into OCCF introduces two problems. The first one is feature mismatching when raw features of users or items are used to build regularization. For example, "love" and "romance" show similar topics, but movies tagged with "love" and movies tagged with "romance" cannot get closer via MR due to the independent dimensions of the two tags. The second problem is that features are not equally important to user behaviors. For example, tags of well-known actors are generally more influential than tags of unknown actors. Conventionally, features in MR are always determined according to domain knowledge or heuristic. The hybrid models cannot be fully benefited from additional features in the large unlabeled data. In fact, possible solutions are existing methods for representation learning, such as Principal Components Analysis (PCA) [6] and MF [7]. However, they intend to find semantics of features, but the goal of feature transformation should be coupling the features of users or items and implicit feedbacks.

To solve the problems, we propose a data-driven framework for learning an appropriate representation function that transforms raw features to new ones based on users' feedback. Through the representation, mismatching can solved and importance of features can be extracted according to the feedback. To link the representation to user behaviors, we model the preference by the similarity between transformed features and optimize the representation according to the observed feedback directly. We apply our framework to a state-of-the-art approach on the OCCF problem by using linear transformation as our representation function. The experiments have been conducted on two real datasets, MovieLens and Github. The results demonstrate that the transformed features (either user features or item features) achieve a significant improvement in the OCCF problem. We also show that the framework is able to reflect quality of selected feature in its own performance as a content-based model.

In the rest of the paper, we review related works to OCCF in Sect. 2, provide prerequisites in Sect. 3, present our framework in Sect. 4, discuss experimental results in Sect. 5 and finally conclude the paper in Sect. 6.

2 Related Work

2.1 One-Class Collaborative Flitering (OCCF)

In OCCF, training data only contain binary records to show users' implicit feedback instead of real-value rating. Many works have been proposed to improve MF in OCCF. Pan et al. [10] and Hu et al. [5] propose weight regularization matrix factorization (WRMF). It treats missing feedback as negative samples and gives lower weights to reduce the impact. Rendle et al. [11] propose Bayesian Personalized Ranking (BPR) to handle implicit feedback by optimizing pairwise errors between items with and without feedback. Shi et al. [12] propose collaborative less-is-more filtering (CLiMF) to specifically focus on top-k recommendation by directly optimizing reciprocal ranks. Here we select MF with BPR (BPRMF) (one of the state-of-the-art methods) as a base when applying our method.

2.2 Incorporating Feature into Recommendation

Model-based collaborative filtering methods focus on users' feedback and find latent preference. However, additional features like user profiles or item contents are not considered generally. Many content-based models show that such features are effective in recommendation. Zhang et al. [14] propose a feature-centric model to show the importance of item features like tags. Gu et al. [3] match users' and jobs' content for job recommendation.

There are several works extending CF models by incorporating additional features. In some works, model parameters are directly related to features. He and McAuley [4] append projected visual features of product images to latent item factors in e-commerce, and compute latent image preference for users. van den Oord et al. [9] train a deep neural network for transforming existing features to pre-trained latent factors. Wang and Wang [13] directly use the summation of CF-based and content-based models as an objective function in music recommendation.

Different from these methods, some works exploit additional features in manifold regularization. Du et al. [2] build regularization with user profiles. Ma et al. [8] consider social regularization of users in social recommendation. Zheng et al. [15] design a new dual similarity regularization for similar users and items. Different from ours, all these works simply make use of raw features.

3 Problem Definition and Prerequisites

3.1 Problem and Notations

We first give formal definition and notation of the OCCF problem. Suppose we have n users $\in U$, m items $\in I$ and observed implicit feedback (u, i), which means user u gives positive feedback to item i, in training data D. We also define U_i^+ as a set of users giving positive feedback to item i and I_u^+ as a set of items given positive feedback by user u according to D. The goal we attempt to achieve is to recommend a ranking list of item $i \in I \backslash I_u^+$ for each user u sorted by predicted preference score, and make the items with higher ranks are more possible to get positive feedback from users.

Besides, each item has corresponding features. We define $v_i \in \mathbb{R}^{d_1}$ as a feature vector of item i, where d_1 is the number of feature dimensions. For simplicity, we only discuss item features here. The proposed framework can actually be well applied to user features. Our experiments in Sect. 5 will examine both features carefully.

3.2 Bayesian Personalized Ranking

BPR [11] is an optimization criterion for personalized ranking. Its goal is to maximize the probability for latent preference of all users by optimizing model's parameters.

Here the probability that user u prefers item i to item j is defined as a sigmoid function of pairwise error in an chosen model. For maximizing the likelihood, the criterion of the optimization problem for personalized ranking can be written as follows:

$$\underset{\theta}{\text{argmin}}\ \text{BPR-OPT} = \underset{\theta}{\text{argmin}} \sum_{(u,i)\in D,(u,j)\notin D} - ln(\sigma(r_{uij})) + \frac{\gamma_\theta}{2}||\theta||^2 \quad (1)$$

where σ is a logistic function, r_{uij} is the pairwise error for the model, and γ_θ is the parameters of regularization, which can be optimized by stochastic gradient descent methods based on bootstrap sampling.

BPR can be applied for any models for ranking by giving an appropriate form of pairwise error to derive $\frac{\partial r_{uij}}{\partial \theta}$ according to the model. It is very appropriate for OCCF because implicit feedback for items can be regarded as a specific preference. A sample (u, i, j) can be created from preferred item i and unpreferred item j. In fact, MF with BPR (BPRMF) is one of the state-of-the-art methods in OCCF. In the paper, we choose BPRMF as our base model for MR. Besides, we also apply BPR to find a proper feature representation function because the training data are from implicit feedback.

3.3 Manifold Regularization

MR is a technique for classification problems in semi-supervised learning, where unlabeled data are exploited to help training with small amount of labeled data. In MR, it uses data-driven regularization to take advantage of geometry of distribution to constrain the learned function.

The objective of MR is to incorporate unlabeled data to consider the distribution of whole dataset as the constraint of optimization. It assumes that instances closer in feature space are more likely to share the same label because the learned function is smooth. Thus, the decision function is affected by the distribution of labeled and unlabeled data.

To make use of the assumption, a graph in which nodes are labeled and unlabeled data is built and the weight of an edge represents the closeness of data instances. The regularization term of this idea for the minimization problem is written as follows:

$$\sum_{i,j=1}^{l+u} W_{ij}(f(x_i) - f(x_j))^2 \tag{2}$$

where l is the number of labeled data, u is the number of unlabeled data, x_i means the ith instance, W_{ij} is the weight of edge between x_i and x_j, and $f(\cdot)$ is the learned function. For closer points, their difference in output of function becomes smaller because the corresponding factor is larger. In other words, closer points tend to have similar prediction result.

For the recommendation problem, the factor is about behavior of users and items. Although the idea of MR is generalized for any learning problem, we need to restate the objective to fit the specific problem. We, therefore, make an assumption as follows:

Assumption 1. *If users/items are similar in feature space, they might have similar behavior in a recommender system.*

Under this assumption, we can easily apply the concept of MR for those models that focus on user and item behaviors such as the latent factor model.

In the latent factor model, users' and items' latent behaviors are usually modeled as $P \in \mathbb{R}^{n\times k}$ and $Q \in \mathbb{R}^{m\times k}$. Referring the form shown in Eq. 3 and Assumption 1, we build a weighted graph for the items and design the regularization term as below:

$$\sum_{i,j=1}^{m} W_{ij}||q_i - q_j||^2 \tag{3}$$

where W_{ij} is the weight of an edge between item i and item j, and $||\cdot||_2$ is the Euclidean norm. We use the norm of the difference between latent factors to represent the difference of item behaviors, and make items having similar features tend to have closer latent factors. Note that it's also feasible to build a regularization term for users by using P here.

4 Manifold Regularized with Data-Driven Feature Representation

4.1 Manifold Regularized OCCF Model

First, we introduce the manifold regularized model BPRMF, which is performed as a base model we plan to improve.

Like all MF models, BPRMF predicts preference matrix R in $\mathbb{R}_{n\times m}$ by two matrices, where $P \in \mathbb{R}_{n\times k}$ represents users' latent factors while $Q \in \mathbb{R}_{m\times k}$ represents items' latent factors, as follows:

$$R = P \cdot Q^T \tag{4}$$

where k is the dimension of latent factors. Regarding the pairs of preferred item i and unpreferred item j for user u, we use $r_{uij} = r_{ui} - r_{uj}$ to model the pairwise error. Thus, by referring Eq. 1, the objective function BPR-OPT can be written below:

$$\underset{P,Q}{\operatorname{argmin}} \sum_{\substack{(u,i)\in D,\\ (u,j)\notin D}} -ln(\sigma(r_{uij})) + \frac{\gamma_\theta}{2}(||P||^2 + ||Q||^2) \tag{5}$$

where $r_{uij} = p_u q_i^T - p_u q_j^T$ and γ_θ is the regularization parameters for P and Q.

To exploit MR to incorporate item features, we add a regularization term stated in Sect. 3.3 to Eq. 5, and the optimization criterion is defined as:

$$\underset{P,Q}{\operatorname{argmin}} \text{ BPR-OPT} + \frac{\gamma_m}{2} \sum_{i,j=1}^{m} W_{ij}||q_i - q_j||^2 \tag{6}$$

where γ_m is the regularization parameter and W_{ij} comes from the similarity between feature vectors. There are many choices for the measurement of similarity. In this work, We just use dot product as our metric.

In implementation, we use stochastic gradient descent to optimize Eq. 6. In each iteration, we sample a tuple (u, i, j), and update its corresponding parameters.

4.2 Framework for Learning Data-Driven Feature Representation

Next, we describe the proposed framework for learning feature representation from feedback.

Our goal is to predict the user preference based on item similarity. The prediction of item i for user u comes from how similar the item i is with the item set user u has given feedback to. Formally, prediction r_{ui} can be defined as:

$$r_{ui} = \frac{1}{|I_u^+|} \sum_{j\in I_u^+} sim(v_i, v_j) \tag{7}$$

where $sim(\cdot)$ can be an arbitrary measurement for similarity. This is a naive method and no space is left for learning. Nevertheless, our intention is to find a better feature representation that tightly couples the original features to user preference. We, therefore, define our target function as f, and modify Eq. 7 by considering f as follows:

$$r_{ui} = \frac{1}{|I_u^+|} \sum_{j\in I_u^+} sim(f(v_i), f(v_j)) \tag{8}$$

Currently, the equation represents a content-based model for predicting preference, and the target is to find appropriate f. Because f is not limited to a certain type, we can choose a transformation, which can contain feature combination such as linear combination.

To optimize f, we need to learn from the implicit feedback data, which means that it is not suitable to use an evaluation metric for real values like RMSE in OCCF. So we apply BPR, which is effective for implicit feedback, on our objective. This model can still take $r_{uij} = r_{ui} - r_{uj}$ as the pairwise error. Combining Eq. 8 into Eq. 1, we finally define the optimization problem as follows:

$$\begin{aligned} &\underset{f}{\operatorname{argmin}} \sum_{\substack{(u,i)\in D,\\(u,j)\notin D}} - ln(\sigma(r_{uij})) + \frac{\gamma_\theta}{2}(||\theta_f||^2), \\ &r_{uij} = \frac{1}{|I_u^+|}\sum_{k\in I_u^+}(sim(f(v_i), f(v_k)) - sim(f(v_j), f(v_k))) \end{aligned} \tag{9}$$

where θ_f denotes the parameters of function f. The objective function can be optimized by the stochastic gradient descent method if we can derive $\frac{\partial sim(f(v_i), f(v_j))}{\partial \theta_f}$ by choosing proper forms for $sim(\cdot,\cdot)$ and f. Through optimizing it, we can find appropriate f without any human effort and link the features to implicit feedback.

For the case of user features, we can just exchange the position of users and items for the same process. However, because U_i^+ and U_j^+ have different size, we only use the most similar k users for user u from U_i^+ and U_j^+ in the objective function. In training process, we also randomly sample k users from the set when calculating gradient in each iteration. In our experiment, we set k as 10.

By combining our framework and manifold regularized model, the whole procedure can be regard as a two-staged algorithm. In the first stage, we find appropriate representation for original features, and get the similarity among items or users in the transformed feature space at the same time. In the second stage, we exploit the similarity we obtain previously, and perform the manifold regularized model based on it to solve the OCCF problem.

4.3 Example of Proposed Framework

In this paper, we implement our framework with some intuitive methods, dot product and linear combination, to validate its effectiveness. The definitions are mathematically given for vector $a, b \in \mathbb{R}^{1\times d_1}$ as follows:

$$f(a) = aK, \quad sim(a, b) = ab^T$$

where $K \in \mathbb{R}^{d_1\times d_2}$ and d_2 is the dimension we project to. By combining the above equations into Eq. 9, we can write our objective function as follows:

$$\begin{aligned} &\underset{f}{\operatorname{argmin}} \sum_{\substack{(u,i)\in D,\\(u,j)\notin D}} - ln(\sigma(r_{uij})) + \frac{\gamma_\theta}{2}||K||^2, \\ &r_{uij} = \frac{1}{|I_u^+|}\sum_{k\in I_u^+}((v_iK)(v_kK)^T - (v_jK)(v_kK)^T) \end{aligned}$$

Here stochastic gradient descent is used to update K. Here is the way to update the parameters:

$$K \leftarrow K - \alpha((1 - \sigma(x_{uij})) \sum_{k \in I_u^+} (v_i^T v_k - v_j^T v_k) \cdot K + \gamma_\theta K)$$

We run our experiment according to the above equation. Actually, it is possible to apply other more sophisticated methods as an alternative in our framework.

5 Experiments

5.1 Dataset and Setting

We run our experiments on the Movielens dataset collected from the Hetrec2011 workshop and the Github dataset. The first dataset is about movie recommendation. Besides the interaction between users and items, it also provides the information about movies such as genre, tag, actor and so on. We select tags as the raw features to incorporate. In the dataset, each movie is assigned some tags and corresponding weights, which contribute to our feature vectors. The second dataset is about repository recommendation of source coding. The data are collected from the Github archive in which there many kinds of actions between users and repositories. We regard some kinds of actions as positive feedback such as push and pull requests. Users' programming language preference of users are treated as the raw features. More specifically, a vector with binary values is adopted to indicate whether a user is able to write a certain language according to the training data. Different from the item features in previous dataset, languages belong to user features and only contain 0/1 values rather than real values.

To eliminate the effect of different length of vectors, we normalize all vectors to unit vectors. We perform data preprocessing to avoid biased result affected by other factors. First, we throw out inactive users and unpopular items (# feedback <10). Second, items with no features are ignored. Besides, we filter out the tags appearing in less than 10 items for the first dataset because the number of items is not much larger than the number of feature dimensions, and it is possible to overfit some scarce tags. For the second dataset, we only extract the subsample with 10,000 users and 10,000 items to save the time of evaluation. The final statistical information about the datasets is shown in Table 1.

We first take 20% of whole datasets as our testing sets, and then split the remaining into two parts with the same size for training sets and validation sets. For model parameters, we set the number of latent factor as 30 for all compared methods, and set the number of dimensions of a projected feature as 50 for PCA and our framework. For regularization, we set appropriate values according to our trials of experiments.

Table 1. The statistical information about two datasets

	Hetrec'11	Github
# users	2,103	10,000
# items	5,094	10,000
# feedback	646,728	616,633
# feature	956	129
Density	0.06037	0.00616

5.2 Baselines

We compare the proposed framework with some baselines, including the state-of-the-art MF method and its manifold regularized version. Besides, our framework can be regarded as a content-based method and we also give the result to show that the framework does learn better features representation for recommendation. The baselines include:

- **Most Popular (MP):** We recommend items according to popularity. Formally, we let r_{ui} equal to $|U_i^+|$ and predict the ranking according to it.
- **KNN:** We use how similar an item/user is with the items/users which users/items have had interaction with to make recommendation. In this baseline, we take dot product between normalized feature vectors, i.e., cosine similarity, as our measurement.
- **WRMF:** WRMF [10] applies implicit feedback with weighted regularization, and optimizes the square-loss. Following the user-oriented setting in [10], for a user, we set the weight for positive feedback as 1, and choose a value for negative feedback to let the sum equal the sum of positive feedback.
- **BPRMF:** BPRMF is regarded as the state-of-the-art method for OCCF, and the detail has been described in Sect. 4.1
- **MR:** This baseline is BPRMF with MR. We use raw features (denoted as MR(Raw)) and transformed features with PCA (denoted as MR(PCA)). PCA is one of famous representation learning and dimension reduction methods. Its goal is to find the semantics of raw features instead of the relation between raw features and user behaviors.
- **MR-R:** This is our proposed method, the manifold regularized model with learned representation depending on feedback data. We follow the example described in Sect. 4.3. We also use **KNN-R** to represent our representation learning framework as a content-based method.

By comparing these baselines, we validate the benefit of our framework with simple selected functions on the performance of item recommendation. To evaluate the quality of item recommendation, we report several ranking metrics, including AUC, NDCG@k, Precision@k and MRR.

Table 2. Performance of baselines. All improvements for NDCG, Prec, and MRR of our method are significantly different with the best baseline at least 95% level in a paired t-test

	Hetrec'11				Github			
	AUC	NDCG@10	Prec@10	MRR	AUC	NDCG@10	Prec@10	MRR
MP	0.8673	0.3698	0.3397	0.5904	0.7867	0.0530	0.0406	0.1502
KNN	0.7959	0.1983	0.1788	0.4185	0.8391	0.0586	0.0475	0.1605
WRMF	0.8797	0.4003	0.3685	0.6220	0.8708	0.0543	0.0476	0.1425
BPRMF	0.9035	0.4085	0.3843	0.6246	0.8741	0.0850	0.0694	0.2095
MR(Raw)	0.9068	0.4268	0.4005	0.6373	0.8822	0.0916	0.0746	0.2238
MR(PCA)	0.9058	0.4296	0.4055	0.6331	0.8763	0.0856	0.0699	0.2124
Proposed method								
KNN-R	0.8624	0.3776	0.3445	0.6347	0.8557	0.0656	0.0531	0.1743
MR-R	**0.9070**	**0.4408**	**0.4121**	**0.6553**	**0.8884**	**0.1037**	**0.0825**	**0.2516**

5.3 Experiment Results

Comparison with Baselines. Table 2 shows the results. From the table, some observations and corresponding explanations are given below:

- BPRMF outperforms simple baselines, MP and itemKNN, and WRMF. MR with all kind of applied features further enhance the performance of item recommendation. Although the improvement for AUC is limited, it can be inferred that MR benefits the quality of top ranking by observing the improvements of NDCG, Precision and MRR.
- Representation from the proposed method reaches the best performance among all of manifold regularized models. It shows that the data-driven representation depending on feedback exactly performs better than raw features and transformed features with PCA. Although PCA also slightly improves the quality for hetrec'11, its benefit is not as explicit as our method's. For Github, its performance even becomes worse.
- Comparing KNN and KNN-R, it is obvious that the framework can help the similarity-based method by transforming features to the space which is more related to user feedback.

Partial observation of the learned representation is shown in Table 3. For the chosen 6 tags, we show the tags which have similar transformation with them. In other words, they tend to be projected to the same latent dimensions. Overall, the experimental results reasonably explain the advantages of different models. Our transferred feature representation is most effective for MR in OCCF.

Judgment of Feature Quality. We also run our framework with randomly generated features (**KNN-ran**) on the two datasets to assure that our framework is capable of showing feature's quality in ranking performance as a content-based model. For the random features of item i, we use the same number of

Table 3. Partial observation of feature transformation

Musical	Romance	Cartoon
Music	Competition	Animation
Dance	Girlie movie	Sad
Wedding	France	Disney
Romance	Chick Flick	Redemption
Meryl Streep	Leonardo Dicaprio	Ben Stiller
Scifi	Action	Thriller
Virtual reality	Sweeping	Emotional
Cyberpunk	Mel Gibson	Assassination
Space	Samuel L Jackson	Disaster
Heroic mission	Thriller	Unrealistic
Alien	Rousing	Survival

Table 4. Performance of the framework with different features. All improvements are significant differences with the baseline at 99% level in a paired t-test.

	AUC	NDCG@10	Prec@10	MRR
Hetrec'11				
KNN-ran	0.8177	0.3305	0.2965	0.5653
KNN-R	0.8624	0.3776	0.3445	0.6347
Github				
KNN-ran	0.7799	0.0435	0.0342	0.1277
KNN-R	0.8557	0.0656	0.0531	0.1743

dimensions of original features, and randomly choose some dimensions to give a value generated from the uniform distribution ranged from 0 to 1 in hetrec'11 or assign 1 in Github. We also normalize all the feature vectors to unit ones. The results are shown in Table 4. It can be observed that the performance with the random features is weaker than one with the meaningful features. Its quality is even worse than MP, stating that we can use the performance of the content-based model to judge the quality of selected feature.

6 Conclusions

In this paper, we propose a data-driven framework of learning feature transformation function from user feedback for manifold regularized OCCF model. Through the optimization from feedback, we can find the representation linked with behavior of users or items, fitting the motivation of manifold regularization. In the experiment, we show that our framework with simple selected forms of functions enhances the quality of item recommendation. Compared with raw features or transformed features based on semantics, our method can help OCCF model to perform better. Besides, we also show that the framework is able to reflect quality of selected feature in its own performance as a content-based model.

In the future, some directions are still valuable to dig into. We can use other complex forms of representation, such as multi-layer neural network encoder, to validate our idea and further boost the performance. Besides, the current framework is limited by stochastic gradient descent method because it is based on BPR framework. The possible direction is to seek other optimization criterion for our preference score and break the constraint of gradient descent.

References

1. Belkin, M., Niyogi, P., Sindhwani, V.: Manifold regularization: a geometric framework for learning from labeled and unlabeled examples. J. Mach. Learn. Res. **7**, 2399–2434 (2006)
2. Du, L., Li, X., Shen, Y.-D.: User graph regularized pairwise matrix factorization for item recommendation. In: Tang, J., King, I., Chen, L., Wang, J. (eds.) ADMA 2011. LNCS (LNAI), vol. 7121, pp. 372–385. Springer, Heidelberg (2011). doi:10.1007/978-3-642-25856-5_28
3. Gu, Y., Zhao, B., Hardtke, D., Sun, Y.: Learning global term weights for content-based recommender systems. In: Proceedings of the 25th International Conference on World Wide Web, WWW 2016, pp. 391–400. International World Wide Web Conferences Steering Committee (2016)
4. He, R., McAuley, J.: VBPR: visual Bayesian personalized ranking from implicit feedback. In: AAAI Conference on Artificial Intelligence (2016)
5. Hu, Y., Koren, Y., Volinsky, C.: Collaborative filtering for implicit feedback datasets. In: Proceedings of the 2008 Eighth IEEE International Conference on Data Mining, ICDM 2008, pp. 263–272. IEEE Computer Society (2008)
6. Jolliffe, I.: Principal Component Analysis. Wiley Online Library, New York (2002)
7. Koren, Y., Bell, R., Volinsky, C.: Matrix factorization techniques for recommender systems. Computer **42**(8), 30–37 (2009)
8. Ma, H., Zhou, D., Liu, C., Lyu, M.R., King, I.: Recommender systems with social regularization. In: Proceedings of the Fourth ACM International Conference on Web Search and Data Mining, WSDM 2011, pp. 287–296. ACM (2011)
9. van den Oord, A., Dieleman, S., Schrauwen, B.: Deep content-based music recommendation. Adv. Neural Inf. Process. Syst. **26**, 2643–2651 (2013)
10. Pan, R., Zhou, Y., Cao, B., Liu, N.N., Lukose, R., Scholz, M., Yang, Q.: One-class collaborative filtering. In: 2008 Eighth IEEE International Conference on Data Mining, pp. 502–511 (2008)
11. Rendle, S., Freudenthaler, C., Gantner, Z., Schmidt-Thieme, L.: BPR: Bayesian personalized ranking from implicit feedback. In: Proceedings of the Twenty-Fifth Conference on Uncertainty in Artificial Intelligence, UAI 2009, pp. 452–461. AUAI Press (2009)
12. Shi, Y., Karatzoglou, A., Baltrunas, L., Larson, M., Oliver, N., Hanjalic, A.: CLiMF: learning to maximize reciprocal rank with collaborative less-is-more filtering. In: Proceedings of the Sixth ACM Conference on Recommender Systems, RecSys 2012, pp. 139–146. ACM (2012)

13. Wang, X., Wang, Y.: Improving content-based and hybrid music recommendation using deep learning. In: Proceedings of the 22nd ACM International Conference on Multimedia, MM 2014, pp. 627–636. ACM (2014)
14. Zhang, C., Wang, K., Lim, E.p., Xu, Q., Sun, J., Yu, H.: Are features equally representative? a feature-centric recommendation. In: Proceedings of the Twenty-Ninth AAAI Conference on Artificial Intelligence, AAAI 2015, pp. 389–395. AAAI Press (2015)
15. Zheng, J., Liu, J., Shi, C., Zhuang, F., Li, J., Wu, B.: Dual similarity regularization for recommendation. In: Bailey, J., Khan, L., Washio, T., Dobbie, G., Huang, J.Z., Wang, R. (eds.) PAKDD 2016. LNCS (LNAI), vol. 9652, pp. 542–554. Springer, Cham (2016). doi:10.1007/978-3-319-31750-2_43

Stable Bayesian Optimization

Thanh Dai Nguyen(✉), Sunil Gupta, Santu Rana, and Svetha Venkatesh

Center for Pattern Recognition and Data Analytics, Deakin University, Waurn Ponds 3216, Australia
{thanh,sunil.gupta,santu.rana,svetha.venkatesh}@deakin.edu.au

Abstract. Tuning hyperparameters of machine learning models is important for their performance. Bayesian optimization has recently emerged as a de-facto method for this task. The hyperparameter tuning is usually performed by looking at model performance on a validation set. Bayesian optimization is used to find the hyperparameter set corresponding to the best model performance. However, in many cases, where training or validation set has limited set of datapoints, the function representing the model performance on the validation set contains several spurious sharp peaks. The Bayesian optimization, in such cases, has a tendency to converge to sharp peaks instead of other more stable peaks. When a model trained using these hyperparameters is deployed in real world, its performance suffers dramatically. We address this problem through a novel stable Bayesian optimization framework. We construct a new acquisition function that helps Bayesian optimization to avoid the convergence to the sharp peaks. We conduct a theoretical analysis and guarantee that Bayesian optimization using the proposed acquisition function prefers stable peaks over unstable ones. Experiments with synthetic function optimization and hyperparameter tuning for Support Vector Machines show the effectiveness of our proposed framework.

1 Introduction

Bayesian optimization is a technique to sequentially optimize expensive black-box functions in a sample efficient manner. Recently, it has emerged as a *de-facto* method to tune complex machine learning algorithms [1]. In tuning, the goal is to train a classifier at the right complexity so that it neither overfits, nor underfits. Performance on a validation set is used as an indicator of the fitting, and it is expected to peak at the hyperparameters corresponding to the right complexity and exhibit lower values at other hyperparameters. Thus to tune a machine learning algorithm, Bayesian optimization is employed in the pursuit of the peak validation set performance. However, in some situations, especially when training or validation dataset is small, spurious peaks appear along the performance surface (eg. Fig. 1). These peaks tend to be distributed randomly over low performance region. They are characteristically different from the peak corresponding to the right complexity in two ways (a) they tend to be narrow and (b) they vanish when tested on a large test data, whereas the right peak remains stable. Due to the latter difference, a Bayesian optimization method that

J. Kim et al. (Eds.): PAKDD 2017, Part II, LNAI 10235, pp. 578–591, 2017.
DOI: 10.1007/978-3-319-57529-2_45

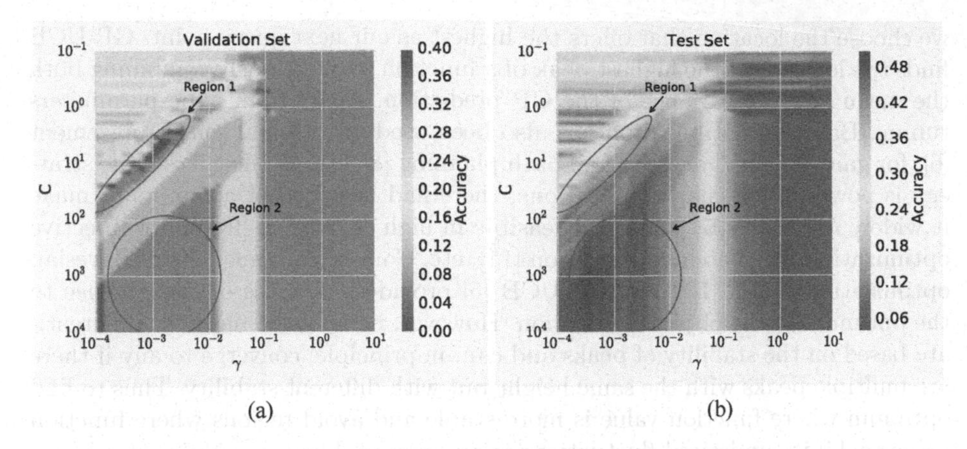

Fig. 1. Performance versus hyperparameters for a Support Vector Machine training as color coded images: (a) on a small validation set, and (b) on a large test set. Spurious peaks of region 1 seen for the validation set vanish for the test set while the stable peak of region 2 still remains. (Color figure online)

does not explicitly avoid these spurious peaks can converge to one of them and may result in a badly tuned system with inexplicably low performance during real world deployment. To the best of our knowledge, we are the first to identify and analyze this issue of spurious peaks and its serious downside.

Existence of multiple peaks with different widths along an optimization surface is prevalent in many real world systems. For some of them, the end result of optimization can get dramatically affected depending on whether the optimization has converged to a wide peak or a narrow peak. For example, in alloy design [2], one of the main goals is to find the mixing proportion of a set of elements with the highest physical property (eg. strength, ductility, etc.). However, alloy making is an imprecise process. Due to the impurities in the raw material, the elements can never be mixed at the desired proportion. Therefore, if the desired proportion is at a narrow peak then the performance of the alloy would not be stable when made repeatedly as even a small difference in impurities could result in dramatic loss in performance. Hence, being able to avoid narrow peaks in favor of more stable peaks is a critical factor of success for several different applications of Bayesian optimization. Unfortunately, till now the various downsides of reaching a narrow peak in the optimization of physical systems and processes have never been identified and attended to.

Bayesian optimization, in its simplest form, consists of a Gaussian process (GP) [3] to maintain a distribution on the objective function based on the observations so far, and a mechanism to select the next query point based on an optimistic exploration strategy. This strategy is more commonly known as *acquisition function* and can be of different types, such as Expected Improvement (EI) [4], or GP-UCB [5], etc. Based on the predictive distribution of the Gaussian process, EI computes the expected improvement over the current best observation and we choose the location that offers the highest as our next query point. GP-UCB finds the location of the highest peak of a function by judiciously combining both the mean and the variance of the GP prediction. Apart from hyperparameters tuning, Bayesian optimization has also been used for optimal sensor placement [6], for gait design [7] and optimal path planning [8], etc. While this simple strategy is powerful for many applications, there had been recent attempts to make it widely applicable by making it feasible in high dimension [9], multi-objective optimization [10], batch optimization [11], etc. Convergence analyses of Bayesian optimization for EI [12], and GP-UCB [5] provide guarantee of convergence to the optimum of the objective function. However, none of the methods differentiate based on the stability of peaks and can, in principle, converge to any if there are multiple peaks with the same height but with different stability. Thus to find optimum where function value is more stable and avoid regions where function values exhibits undesired fluctuations is an open problem.

To address the issues with spurious peaks, we propose a new acquisition function for Bayesian optimization that actively seeks stable peaks of the objective function. Based on our definition of stability, we show that it is possible to measure the stability of a peak by subjecting the underlying Gaussian process model with input perturbation. When faced with input perturbation, the predictive distribution of the Gaussian process changes. At any peak the mean of the distribution goes lower, and the variance goes higher. But more importantly, for two peaks of same height, the narrower peak will have lesser mean and more variance than the other peak. Further we show that the variance can effectively be decomposed as a sum of two parts: (a) *epistemic variance* due to the limited number of samples, and (b) *aleatoric variance* arising from the interaction between the curvature of the function with the noise. The narrower a peak is, the higher the aleatoric variance will be around that peak. Therefore, aleatoric variance can be used as a measure of the instability of a peak. An acquisition function is proposed in line with the GP-UCB that while exploiting the usual combination of mean and variance also penalize for instability. Theoretically, we prove that under mild assumptions, when two peaks are of same height, the proposed acquisition function would always favor the more stable peak. We compare our method with a standard Bayesian optimization implementation on both synthetic function optimization and real-world hyperparameter tuning. On synthetic function optimization, we create a function that has both stable region and spurious region. Experiments with synthetic function show that our proposed method converges to stable regions more often than the baseline. For real world application, we demonstrate tuning the hyperparameters of Support Vector Machine on two real datasets. Experimental results clearly demonstrate that our proposed method converges to a stable peak whereas the standard Bayesian optimization converges to an unstable peak, and hence the SVMs tuned by our method perform better on test sets.

2 Background

Bayesian optimization [13] is a well-known method to find the maximum value of a unknown blackbox function f. This optimization problem can be formally defined as

$$\mathbf{x}^* = \text{argmax}_{\mathbf{x}\in\mathcal{X}} f(\mathbf{x})$$

where $\mathcal{X}$ is domain of $\mathbf{x}$. It is assumed that although f is a blackbox function without a closed-form expression, it can be evaluated at any point $\mathbf{x}$ in the domain. The idea behind Bayesian optimization is to use all the information available from observations $\mathbf{x}$ and $f(\mathbf{x})$ for reasoning rather than rely on only local gradient.

Bayesian optimization consists of two main components. The first is a meta model that can be evaluated at any points with uncertainty. There are plenty of choices for this function such as Gaussian process, Bayesian neural networks, random forest, etc. The second component of a Bayesian optimization algorithm is the acquisition function that suggests where to evaluate the function next. This function demonstrates the trade-off between exploitation and exploration (high predicted value versus highly uncertain regions). In this work, we use Gaussian process [3] as the meta model and Upper Confidence Bound (UCB) as the acquisition function. GP-UCB [5] using Gaussian process and UCB has nice theoretical properties and is guaranteed to converge.

2.1 Gaussian Process

Gaussian process is a stochastic process such that every finite collection of its random variables is a multivariate Gaussian distribution. Intuitively, one can think of Gaussian process as a multivariate Gaussian distribution over infinite dimensional vectors. A function $f(\mathbf{x})$ drawn from a Gaussian process with mean $m(\mathbf{x})$ and covariance function $k(\mathbf{x}, \mathbf{x}')$ is denoted as follow $f(\mathbf{x}) \sim \mathcal{GP}(m(\mathbf{x}), k(\mathbf{x}, \mathbf{x}'))$. Assume that we have a dataset $\mathcal{D}_t = \{(\mathbf{x}_i, y_i)\}, i = 1, 2, \ldots, t$ where $y_i = f(\mathbf{x}_i)$ and $f(\mathbf{x})$ is drawn from a Gaussian process $\mathcal{GP}(m(\mathbf{x}), k(\mathbf{x}, \mathbf{x}'))$. We can make the Gaussian process depends only on the covariance function $k(\mathbf{x}, \mathbf{x}')$ by assuming mean function $m(\mathbf{x})$ to be zero. The covariance function $k(\mathbf{x}, \mathbf{x}')$ should be a valid kernel function in order to make the covariance matrix $\mathbf{K}$ valid, where $\mathbf{K}_{i,j} = k(\mathbf{x}_i, \mathbf{x}_j)$. To make a prediction using Gaussian process, we consider the joint distribution of old observations and the new observation $(\boldsymbol{x}_{t+1}, y_{t+1})$ as:

$$\begin{bmatrix} \mathbf{y}_{1:t} \\ y_{t+1} \end{bmatrix} \sim \mathcal{N}\left(0, \begin{bmatrix} \mathbf{K} & \mathbf{k} \\ \mathbf{k}^T & k(\mathbf{x}_{t+1}, \mathbf{x}_{t+1}) \end{bmatrix}\right)$$

where $\mathbf{k} = [k(\mathbf{x}_1, \mathbf{x}_{t+1}), k(\mathbf{x}_2, \mathbf{x}_{t+1}), \ldots, k(\mathbf{x}_t, \mathbf{x}_{t+1})]^T$. The predictive distribution of the function value at $\mathbf{x}_{t+1}$ can be written as:

$$p(y_{t+1}|\mathbf{y}_{1:t}, \mathbf{x}_{1:t+1}) = \mathcal{N}(\mu_t(\mathbf{x}_{t+1}), \sigma_t^2(\mathbf{x}_{t+1}))$$

where $\mu_t(\mathbf{x}_{t+1}) = \mathbf{k}^T\mathbf{K}^{-1}\mathbf{y}_{1:t}$ and $\sigma_t^2(\mathbf{x}_{t+1}) = k(\mathbf{x}_{t+1}, \mathbf{x}_{t+1}) - \mathbf{k}^T\mathbf{K}^{-1}\mathbf{k}$.

2.2 Upper Confidence Bound

Since the target function is unknown and expensive to evaluate, Bayesian optimization uses an acquisition function that can be optimized efficiently to determine the next sample to be evaluated. By doing so, instead of optimizing the target function, we find the maximum value of following function:

$$\mathbf{x}_{t+1}^{*} = \arg\max \alpha\left(\mathbf{x}; \mathcal{I}_t\right)$$

where $\mathcal{I}_t$ denotes the Gaussian process estimated using t observations. This optimization problem can be solved using standard optimization techniques such as local optimizers, sequential quadratic programming or DIRECT [14].

Normally, the acquisition function is defined such that it has high value at uncertain regions or high prediction or both. The trade-off between exploring highly uncertain region or exploiting promising area is also represented in acquisition function. Given the posterior Gaussian process, the Upper Confidence Bound acquisition function is defined as follow [5]:

$$\text{GP-UCB}(\mathbf{x}_{t+1}) = \mu_t(\mathbf{x}) + \kappa_t \sigma_t(\mathbf{x})$$

where κ_t is a positive parameter that balances exploitation and exploration. Maximizing GP-UCB acquisition function suggests the point where to next evaluate the function f. Srinivas et al. [5] proved that if $\kappa_t = 2\log\left(t^2 2\pi^2/3\delta\right) + 2d\log\left(t^2 dbr\sqrt{\log\left(4da/\delta\right)}\right)$, GP-UCB achieves an upper bound on the cumulative regret $\sum_{t=1}^{T}\left(f\left(\mathbf{x}^*\right) - f\left(\mathbf{x}_t\right)\right)$ that has the order $\mathcal{O}\left(\sqrt{T\gamma_T\kappa_T}\right) \forall T \geq 1$, with probability greater or equal $1-\delta$, where γ_T is the maximum information gain after T round, search space is a subset of $[0,r]^d$ with some $r > 0$ and $a, b > 0$ are constants.

3 The Proposed Framework

We present a new acquisition function for Bayesian optimization designed to maximize a blackbox function with behavior that the maxima from stable regions are preferred over the maxima from relatively unstable regions. We first discuss the notion of stability, then describe how a Gaussian process model gets modified in presence of any perturbation in the input variables. Next, we use the predictive distribution of the modified Gaussian process to formulate the novel acquisition function. We theoretically analyze the proposed acquisition function and prove that it is guaranteed to take higher values in more stable regions and thus Bayesian optimization using this acquisition function will have higher tendency to converge in more stable regions. Finally, we present an algorithm summarizing the proposed stable Bayesian optimization.

3.1 Stability of Gaussian Process Prediction

Given a set of observed data $\mathcal{D}_t = \{\mathbf{x}_i, y_i\}_{i=1}^t$ where $\mathbf{x}_i \in \mathbb{R}^D$ and $y_i = f(\mathbf{x}_i) + \epsilon$, we use a Gaussian process to model the function f. Using $\mathcal{D}_t$, for a new input $\mathbf{x}$, the predictive distribution of the corresponding output $y = f(\mathbf{x})$ is given as

$$P(y|\mathcal{D}_t, \mathbf{x}) = \mathcal{N}\left(\mu_t(\mathbf{x}), \sigma_t^2(\mathbf{x})\right),$$

where the predictive mean $\mu_t(\mathbf{x}) = \mathbf{k}^T\mathbf{K}^{-1}\mathbf{y}$ and the predictive variance $\sigma_t^2(\mathbf{x}) = k(\mathbf{x}, \mathbf{x}) - \mathbf{k}^T\mathbf{K}^{-1}\mathbf{k}$ with a notation $\mathbf{y} = y_{1:t}$. We define $\beta = \mathbf{K}^{-1}\mathbf{y}$ to be used later.

The above predictive mean and variance are instrumental to Bayesian optimization as they provide a way to estimate the function value at any point in the function support along with the model's uncertainty. The model uncertainty, also called "epistemic uncertainty", is used in the Bayesian optimization to express our belief in the estimation and guides efficient exploration of the function while keeping a balance on exploitation. This phenomenon is an instance of a general concept in reinforcement learning known as exploitation-exploration trade-off.

Since our goal is to develop a stable Bayesian optimization framework that prefers solutions insensitive to small perturbations in input data, we start from asking the question how does the predictive mean and variance of the function value change if the input is slightly perturbed. A large shift in the mean and/or a large increase in the variance indicates a fast varying function and can be used to detect the unstable regions. In an early work, Girard and Murray-Smith [15] analyze that if a test input is corrupted by a Gaussian noise, $\epsilon_{\mathbf{x}} \sim \mathcal{N}(\mathbf{0}, \boldsymbol{\Sigma}_{\mathbf{x}})$ such that $\mathbf{u} = \mathbf{x} + \epsilon_{\mathbf{x}}$, the predictive distribution is given as

$$p(y|\mathcal{D}_t, \mathbf{x}, \boldsymbol{\Sigma}_{\mathbf{x}}) = \int p(y|\mathcal{D}_t, \mathbf{u})p(\mathbf{u}|\mathbf{x}, \boldsymbol{\Sigma}_{\mathbf{x}})d\mathbf{u}.$$

This distribution, in general, is non-Gaussian. However, in [15], it is shown that Gaussian approximation is a fairly close approximation under the constraint of tractability. Let us use $\mu_t(\mathbf{x})$ and $\sigma_t^2(\mathbf{x})$ to denote the mean and variance of the Gaussian predictive distribution $p(y|\mathcal{D}_t, \mathbf{x})$ in the *perturbation-free* case. Also use $m_t(\mathbf{x}, \boldsymbol{\Sigma}_{\mathbf{x}})$ and $v_t(\mathbf{x}, \boldsymbol{\Sigma}_{\mathbf{x}})$ to denote the mean and variance of predictive distribution $p(y|\mathcal{D}_t, \mathbf{x}, \boldsymbol{\Sigma}_{\mathbf{x}})$ and use a Gaussian approximation as below:

$$p(y|\mathcal{D}_t, \mathbf{x}, \boldsymbol{\Sigma}_{\mathbf{x}}) \approx \mathcal{N}(m_t(\mathbf{x}, \boldsymbol{\Sigma}_{\mathbf{x}}), v_t(\mathbf{x}, \boldsymbol{\Sigma}_{\mathbf{x}})).$$

The predictive mean and variance can also become intractable in general for an arbitrary covariance function. Fortunately it is possible to express them in closed form for popular covariance functions such as linear and square-exponential. We demonstrate our framework using squared exponential covariance function. Nonetheless, our framework remains amenable to any valid covariance function and appropriate approximations arising due to an arbitrary covariance function can be easily incorporated. For the squared exponential covariance function, the predictive mean and variance are given as

$$m_t(\mathbf{x}, \mathbf{\Sigma_x}) = \sum_{i=1}^{t} \beta_i k(\mathbf{x}, \mathbf{x}_i) k_1(\mathbf{x}, \mathbf{x}_i) \tag{1}$$

$$v_t(\mathbf{x}, \mathbf{\Sigma_x}) = \sigma_t^2(\mathbf{x}) + \sigma_{t,a}^2(\mathbf{x}, \Sigma_x) \tag{2}$$

where $\sigma_t^2(\mathbf{x})$ is the variance as in the unperturbed case and the extra variance due to perturbation is given as

$$\sigma_{t,a}^2(\mathbf{x}, \Sigma_x) = \sum_{i,j=1}^{t} \mathbf{K}_{ij}^{-1} k(\mathbf{x}, \mathbf{x}_i) k(\mathbf{x}, \mathbf{x}_j)(1 - k_2(\mathbf{x}, \bar{\mathbf{x}}_{ij})) +$$
$$\sum_{i,j=1}^{t} \beta_i \beta_j k(\mathbf{x}, \mathbf{x}_i) k(\mathbf{x}, \mathbf{x}_j)(k_2(\mathbf{x}, \bar{\mathbf{x}}_{ij}) - k_1(\mathbf{x}, \mathbf{x}_i) k_1(\mathbf{x}, \mathbf{x}_j)). \tag{3}$$

In the above expressions, we have used the definitions:

$$k_1(\mathbf{x}, \mathbf{x}_i) = \left|\mathbf{I} + \mathbf{W}^{-1}\mathbf{\Sigma_x}\right|^{-1/2} \exp\left[\frac{1}{2}(\mathbf{x} - \mathbf{x}_i)^T \mathbf{S}(\mathbf{W}, \Sigma_{\mathbf{x}})(\mathbf{x} - \mathbf{x}_i)\right], \text{ and}$$

$$k_2(\mathbf{x}, \bar{\mathbf{x}}_{ij}) = \left|\mathbf{I} + \left(\frac{\mathbf{W}}{2}\right)^{-1} \mathbf{\Sigma_x}\right|^{-1/2} \exp\left[\frac{1}{2}(\mathbf{x} - \bar{\mathbf{x}}_{ij})^T (\mathbf{S}(\frac{\mathbf{W}}{2}, \Sigma_{\mathbf{x}})(\mathbf{x} - \bar{\mathbf{x}}_{ij})\right],$$

where $\bar{\mathbf{x}}_{ij} = \frac{\mathbf{x}_i + \mathbf{x}_j}{2}$ and $\mathbf{S}(\mathbf{W}, \Sigma_{\mathbf{x}}) = \mathbf{W}^{-1}(\mathbf{W}^{-1} + \mathbf{\Sigma}_{\mathbf{x}}^{-1})^{-1}\mathbf{W}^{-1}$.

In the following, we utilize the above analysis to define a novel acquisition function to propose a stable Bayesian optimization framework.

3.2 Stable Bayesian Optimization

Having a closed form expression for the predictive mean and variance as in (1) and (2) provides us the required tractability to formulate an acquisition function for "stable" Bayesian optimization. In the expression for predictive variance in (2), we note that the variance $v_t(\mathbf{x}, \mathbf{\Sigma_x})$ has two components: (1) the *epistemic* variance (uncertainty) term $\sigma_t^2(\mathbf{x})$, arising due to our lack of understanding about the function value, mainly due to finite set of observations and (2) the *aleatoric* variance term $\sigma_{t,a}^2(\mathbf{x}, \Sigma_x)$ (further detailed in (3)), arising due to the inherent variation in the function around $\mathbf{x}$. We associate the notion of the stability to this aleatoric variance which takes higher values in regions where the function has faster variations. In the remainder of this section, we use this property to define a new acquisition function that yields a stable Bayesian optimization which results in a solution where the function value is robust to small perturbations.

Denoting the epistemic and aleatoric variances at time t by $\sigma_t^2(\mathbf{x})$ and $\sigma_{t,a}^2(\mathbf{x}, \Sigma_x)$ respectively, our stable Bayesian optimization uses the following acquisition function

$$a_t(\mathbf{x}, \Sigma_x) = m_t(\mathbf{x}, \Sigma_x) + \kappa_t \sigma_t(\mathbf{x}, \Sigma_x) - \lambda \sigma_{t,a}(\mathbf{x}, \Sigma_x) \tag{4}$$

Algorithm 1. The proposed stable Bayesian optimization.

1:Input:
2: Initial observation set $\mathcal{D}_{t_0} = \{\mathbf{x}_{1:t_0}, y_{1:t_0}\}$.
3: Bounds for the search space $\mathcal{X}$.
4:Output: $\{\mathbf{x}_t, y_t\}_{t=1}^{T}$
5: **for** $t = t_{0+1}, \ldots, T$
6: Find optimizer of acquisition function (4): $\mathbf{x}_{t+1} = \arg\max_{\mathbf{x}\in\mathcal{X}} a_t(\mathbf{x}, \Sigma_x)$.
7: Evaluate the target function as $y_{t+1} = f(\mathbf{x}_{t+1}) + \epsilon$.
8: Augment the observation set: $\mathcal{D}_t = \mathcal{D}_t \cup \{\mathbf{x}_{t+1}, y_{t+1}\}$ and update the GP $\mathcal{I}_t$.
9: **end for**

where κ_t is a t-dependent weight that sets a balance between exploitation and exploration, and $\lambda > 0$ is a fixed weight that sets our penalty on the instability. In the above formulation, our intuition is to *penalize* the points where the function is varying fast with even small change in $\mathbf{x}$. We note that the above acquisition function extends the popular GP-UCB function [5] for which the convergence of Bayesian optimization is theoretically guaranteed. The addition of the aleatoric variance term in the acquisition function can be interpreted as a constrained optimization problem of blackbox function $f(\mathbf{x})$ under the constraint that aleatoric variance is smaller than a specified value. Stable Bayesian optimization maximizes the acquisition function $a_t(\mathbf{x}, \Sigma_x)$ to suggest the next function evaluation at each iteration. A step-by-step procedure is provided in Algorithm 1.

Theoretical Analysis

In this section, we analyze the proposed acquisition function to provide a theoretical guarantee that the acquisition function $a_t(\mathbf{x}, \Sigma_x)$ indeed prefers less sharper peaks of the function $f(\mathbf{x})$.

Definition 1 (Identical data topology): *Any two points* $\mathbf{x}, \mathbf{x}'$ *are said to have identical data topology if there exist a pair of observations* $\mathbf{x}_i$ *and* $\mathbf{x}_{i'}$ *such that* $||\mathbf{x} - \mathbf{x}_i|| = ||\mathbf{x}' - \mathbf{x}_{i'}||$.

A consequence of identical data topology is that for points $\mathbf{x}, \mathbf{x}'$, any distance based kernels induce Gram matrices that are equal up to a permutation. With increasing set of observations, it is not difficult to achieve identical data topology approximately.

Theorem 1: *If* $\mathbf{x}, \mathbf{x}'$ *are the two highest peaks in the support of function* f *such that* $|f(\mathbf{x}) - f(\mathbf{x}')| < \eta_0$ *for small* η_0, *and* f *locally varies faster around* $\mathbf{x}'$ *compared to* $\mathbf{x}$ *in a small* h_0*-neighborhood, i.e.* $|\frac{f(\mathbf{x}+\mathbf{h})-f(\mathbf{x})}{f(\mathbf{x}'+\mathbf{h})-f(\mathbf{x}')}| < 1$, $\forall h \in (-h_0, h_0)$, *the acquisition function in (4) satisfies the relation:* $a_t(\mathbf{x}, \Sigma_x) \geq a_t(\mathbf{x}', \Sigma_x)$ *under certain mild assumptions.*

Proof: To have no favor to any peak, let us assume that there are sufficiently many observations around both $\mathbf{x}, \mathbf{x}'$ so that the two points have identical data topology. Due to this mild assumption, we have a pair of observations $\mathbf{x}_i$ and $\mathbf{x}_{i'}$ such that $||\mathbf{x} - \mathbf{x}_i|| = ||\mathbf{x}' - \mathbf{x}_{i'}||$. Next consider the difference between the acquisition function values at $\mathbf{x}$, $\mathbf{x}'$ as

$$\Delta a_t = [m_t(\mathbf{x}, \Sigma_x) - m_t(\mathbf{x}', \Sigma_x)] + [\kappa_t (\sigma_t(\mathbf{x}, \Sigma_x) - \sigma_t(\mathbf{x}', \Sigma_x))] - [\lambda (\sigma_{t,a}(\mathbf{x}, \Sigma_x) - \sigma_{t,a}(\mathbf{x}', \Sigma_x))]$$

Our aim is to show that $\Delta a_t \geq 0$, *i.e.* $a_t(\mathbf{x}, \Sigma_x) \geq a_t(\mathbf{x}', \Sigma_x)$. We note that due to the identical data topology assumption around both peaks, we have equal epistemic uncertainties, *i.e.* $\sigma_t(\mathbf{x}, \Sigma_x) = \sigma_t(\mathbf{x}', \Sigma_x)$.

Next to show that $m_t(\mathbf{x}, \Sigma_x) \geq m_t(\mathbf{x}', \Sigma_x)$ consider the expression of (1). Once again using the identical data topology assumption, there exists a pair of observations $\mathbf{x}_i$ and $\mathbf{x}_{i'}$ such that $||\mathbf{x} - \mathbf{x}_i|| = ||\mathbf{x}' - \mathbf{x}_{i'}||$. This implies that the covariance values $k(\mathbf{x}, \mathbf{x}_i) = k(\mathbf{x}', \mathbf{x}_{i'})$ and $k_1(\mathbf{x}, \mathbf{x}_i) = k_1(\mathbf{x}', \mathbf{x}_{i'})$. By definition, $\beta = \mathbf{K}^{-1}\mathbf{y}$. Since the peak at $\mathbf{x}'$ is sharper than the peak at $\mathbf{x}$, meaning $y_{i'} \leq y_i$ and therefore $\beta_{i'} \leq \beta_i$. Hence, $\sum_{i=1}^{t} \beta_i k(\mathbf{x}, \mathbf{x}_i) k_1(\mathbf{x}, \mathbf{x}_i) \geq \sum_{i'=1}^{t} \beta_{i'} k(\mathbf{x}', \mathbf{x}_{i'}) k_1(\mathbf{x}', \mathbf{x}_{i'})$.

Finally, we show that $\sigma_{t,a}(\mathbf{x}, \Sigma_x) \leq \sigma_{t,a}(\mathbf{x}', \Sigma_x)$. For this, consider the aleatoric variance term in (3). As above, we have the following relations: $k(\mathbf{x}, \mathbf{x}_i) = k(\mathbf{x}', \mathbf{x}_{i'})$, $k_1(\mathbf{x}, \mathbf{x}_i) = k_1(\mathbf{x}', \mathbf{x}_{i'})$, $\beta_{i'} \leq \beta_i$ and additionally, $k_2(\mathbf{x}, \bar{\mathbf{x}}_{ij}) = k_2(\mathbf{x}', \bar{\mathbf{x}}_{i'j'})$. Using these relations, it is straightforward to show that $\sigma_{t,a}(\mathbf{x}, \Sigma_x) \leq \sigma_{t,a}(\mathbf{x}', \Sigma_x)$.

Combining the three separate inequalities, we can prove that $\Delta a_t \geq 0$, i.e. $a_t(\mathbf{x}, \Sigma_x) \geq a_t(\mathbf{x}', \Sigma_x)$. □

Remarks: The above theorem covers an important case that when the peaks in both stable and unstable regions are approximately equal in height, a Bayesian optimization algorithm using the acquisition function in (4) will prefer the peak from the stable region. In the case, when a peak of unstable region is higher than the peak of stable region, the two terms $(m_t(\mathbf{x}, \Sigma_x) - m_t(\mathbf{x}', \Sigma_x))$ and $(\sigma_{t,a}(\mathbf{x}, \Sigma_x) - \sigma_{t,a}(\mathbf{x}', \Sigma_x))$ would be acting against each other and their net difference will decide whether the algorithm suggest the point from the stable region or unstable region. Since the parameter λ is user specified, there exists a sufficiently large value of λ that always guarantees the suggestion of stable peak. In the case, when a peak of unstable region is lower than the peak of stable region, both algorithms will select the stable peak.

4 Experiments

In this section, we experiment on a set of synthetic and real datasets to demonstrate the efficacy of our stable Bayesian optimization algorithm (BO-STABLE). Experiments with synthetic dataset show the behavior of our proposed method with a known and complex function with multiple sharp peaks and one stable peak. We also conduct experiments with several hyperparameter tuning problems to show the utility of our method for real world applications.

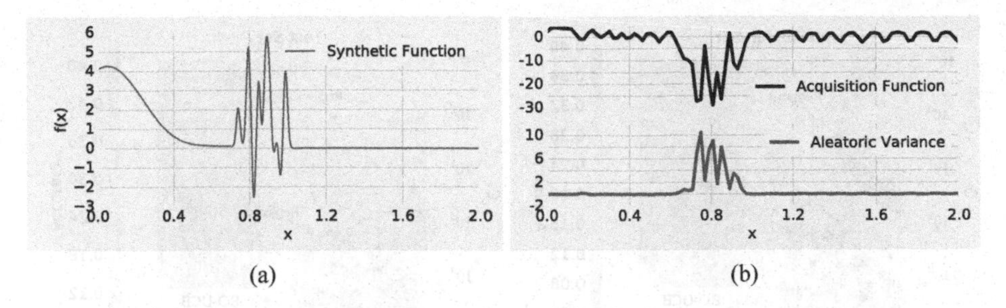

Fig. 2. (a) The synthetic function with one stable peak and multiple spurious peaks (b) The acquisition function and aleatoric variance after 30 iterations.

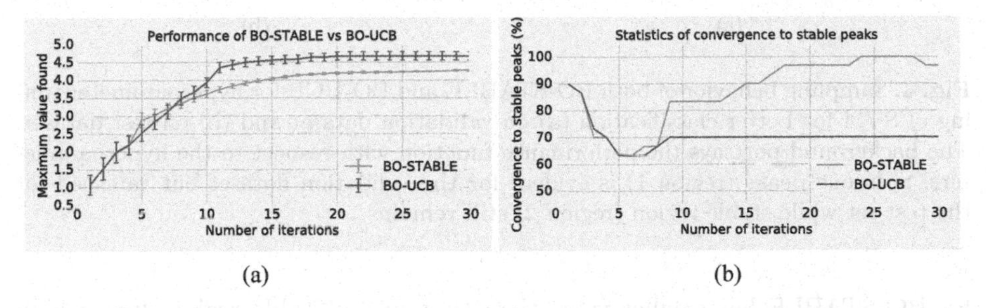

Fig. 3. Performance of BO-STABLE and BO-UCB with respect to number of iterations on Synthetic function. (a) Shows that BO-STABLE converges to 4.25 (stable peak) and BO-UCB converges to 4.7 (spurious peak). (b) Shows that BO-STABLE reaches stable peak more often than BO-UCB.

4.1 Baseline Method and Evaluation Measures

We compare the stable Bayesian optimization with standard Bayesian optimization using UCB acquisition function (BO-UCB). On synthetic data, we compare BO-STABLE with the baseline in two aspects: 'the maximum value found' and 'the number of times an algorithm visits around the stable peak' with respect to number of iterations. On real data, we show the performance of stable Bayesian optimization and standard Bayesian optimization on both validation and test sets.

4.2 Experiments with Synthetic Function

Data Generation: The synthetic function $f(x)$ is generated using a squared exponential kernel with two different parameters. To create a stable peak at the left side of the objective function, we use squared exponential kernel with length scale 0.2. The right side is generated using a squared exponential kernel with length scale 0.01 to simulate spurious peaks. The stable regions of $f(x)$ are $0 \leq x \leq 0.7$ and $1.1 \leq x \leq 2$, and the rest is the unstable region (See Fig. 2a). Figure 2b illustrates the value of acquisition function and aleatoric variance after 30 iterations. In the unstable region, the acquisition function used

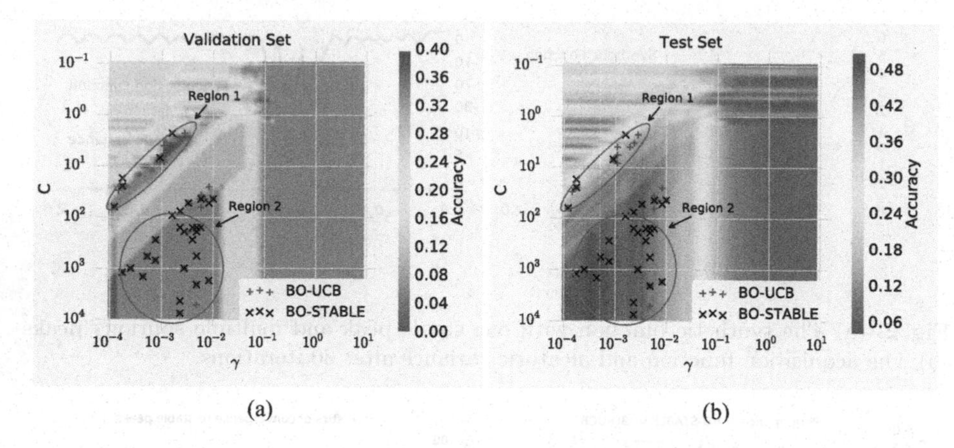

(a) (b)

Fig. 4. Sampling behavior of both BO-STABLE and BO-UCB for hyperparameter tuning of SVM for Letter classification (a) on validation dataset and (b) on test dataset. The background portrays the performance function with respect to the hyperparameters. Spurious peaks (region 1) is evident for the validation dataset but vanished for the test set while stable region (region 2) still remains.

for BO-STABLE has smaller value than that in the stable region due to high aleatoric variance capturing instability.

Experimental Results: We randomly initialize 2 observations for Bayesian optimization. Figure 3a shows the result of 'maximum value found' averaged over 30 different initializations. After 25 iterations, the proposed BO-STABLE converges to averaged maximum value at around 4.25 while BO-UCB converges to 4.7. This is because BO-STABLE often converges to stable region unlike BO-UCB which converges to unstable region. The number of peaks visited in the stable region by BO-STABLE and BO-UCB are compared in Fig. 3b. In 10 iterations, the percentage of peaks visited in the stable region by BO-STABLE and BO-UCB are 83% and 70%, respectively. In 20 iterations, more than 96% of peaks visited by the proposed BO-STABLE are stable whereas this number for BO-UCB is only at 70%, illustrating better stability behavior of BO-STABLE.

4.3 Experiments with Hyperparameter Tuning Problems

Dataset: We use Letter and Glass classification dataset from UCI repository (http://archive.ics.uci.edu/ml). Letter dataset contains 20,000 datapoints about the image characteristic of 26 capital letters in the English alphabet. Since spurious peaks occur mostly when (a) *the training set is inadequate* and (b) *the validation set is small*, we sample only 200 datapoints from the Letter dataset. Glass dataset consists of 214 datapoints represented using 10-features related to glass properties. Both datasets are divided into training set, validation set and test set.

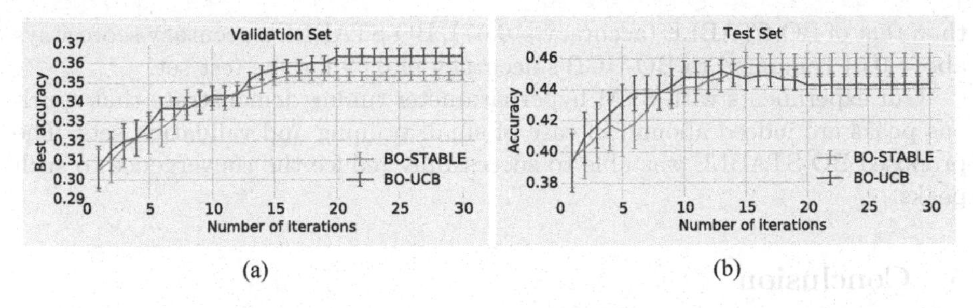

Fig. 5. Performance of BO-STABLE and BO-UCB using SVM with respect to number of iterations on Letter dataset. The performance of BO-UCB, due to convergence to spurious peaks on the validation set (a), degrades for the test set (b).

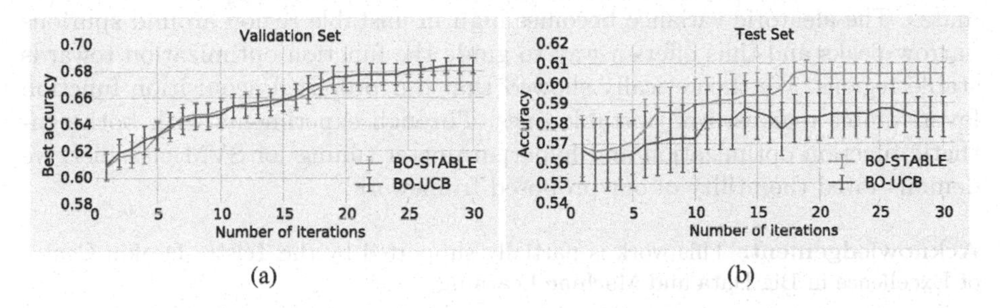

Fig. 6. Performance of BO-STABLE and BO-UCB using SVM with respect to number of iterations on Glass dataset. The performance of BO-UCB, due to convergence to spurious peaks on the validation set (a), degrades for the test set (b)

Experimental Results with Support Vector Machine (SVM): SVM is a popular machine learning algorithm for classification problem. Two main hyperparameters in SVM using RBF kernel are C and γ that represent the misclassification trade-off and the RBF kernel parameter respectively. We apply both BO-STABLE and BO-UCB for tuning C and γ. Figure 4 shows the converged peaks by our proposed BO-STABLE and BO-UCB over 30 different initializations. As seen from the figure, the number of times BO-UCB converges to spurious peaks is considerably higher than that of BO-STABLE. This behavior leads to the accuracy performance shown in Fig. 5. Figure 5a shows the performance of two methods on validation set. We note that this is a multi-class classification task, hence a random classifier would have a mean accuracy of only $1/26 = 0.0385$. After 20 iterations, BO-STABLE's best accuracy on the validation set is 0.35 whereas BO-UCB's best is 0.36. However, as we move to the test set and compare the performance of the two methods using the hyperparameters optimized using the validation set, we find that BO-STABLE performance is higher compared to BO-UCB (see Fig. 5b). After 20 iterations, BO-STABLE performance remains high at 0.46 whereas BO-UCB reaches only up to 0.44. We observed the similar behavior of BO-STABLE and BO-UCB for Glass dataset (Fig. 6). On the validation set, although BO-UCB converges to a higher accuracy (accuracy = 0.69) than that of BO-STABLE (accuracy = 0.67), BO-STABLE's accuracy score stays above 0.61 compared to BO-UCB's accuracy of 0.59 for the test set.

Our experiments with SVM hyperparameter tuning demonstrate that spurious peaks are indeed abound in case of small training and validation sets. The proposed BO-STABLE was able to successfully reduce the convergence to such peaks.

5 Conclusion

We proposed a stable Bayesian optimization framework to find stable solutions for hyperparameter tuning. We constructed a novel acquisition function combining the epistemic and aleatoric variances of the Gaussian process based estimates. The aleatoric variance becomes high in unstable region around spurious narrow peaks and thus offers a way to guide the function optimization towards stable regions. We theoretically showed that our proposed acquisition function favors stable regions over unstable ones. Through experiments with both synthetic function optimization and hyperparameter tuning for SVM classifier, we demonstrated the utility of our proposed framework.

Acknowledgement. This work is partially supported by the Telstra-Deakin Centre of Excellence in Big Data and Machine Learning.

References

1. Thornton, C., Hutter, F., Hoos, H.H., Leyton-Brown, K.: Auto-WEKA: combined selection and hyperparameter optimization of classification algorithms. In: ACM SIGKDD (2013)
2. Xue, D., et al.: Accelerated search for materials with targeted properties by adaptive design. Nat. Commun. **7**, 11241–11249 (2016)
3. Rasmussen, C.E.: Gaussian Processes for Machine Learning. Citeseer (2006)
4. Mockus, J., Tiesis, V., Zilinskas, A.: The application of Bayesian methods for seeking the extremum. Towards Glob. Optim. **2**(117–129), 2 (1978)
5. Srinivas, N., Krause, A., Seeger, M., Kakade, S.M.: Gaussian process optimization in the bandit setting: no regret and experimental design. In: ICML (2010)
6. Garnett, R., Osborne, M.A., Roberts, S.J.: Bayesian optimization for sensor set selection. In: IPSN (2010)
7. Lizotte, D.J., Wang, T., Bowling, M.H., Schuurmans, D.: Automatic gait optimization with Gaussian process regression. In: IJCAI, vol. 7, pp. 944–949 (2007)
8. Martinez-Cantin, R., et al.: A Bayesian exploration-exploitation approach for optimal online sensing and planning with a visually guided mobile robot. Auton. Robots **27**, 93–103 (2009)
9. Chen, B., Castro, R., Krause, A.: Joint optimization and variable selection of high-dimensional Gaussian processes. arXiv preprint arXiv:1206.6396 (2012)
10. Laumanns, M., Ocenasek, J.: Bayesian optimization algorithms for multi-objective optimization. In: Guervós, J.J.M., Adamidis, P., Beyer, H.-G., Schwefel, H.-P., Fernández-Villacañas, J.-L. (eds.) PPSN 2002. LNCS, vol. 2439, pp. 298–307. Springer, Heidelberg (2002). doi:10.1007/3-540-45712-7_29

11. Azimi, J., Fern, A., Fern, X.Z.: Batch Bayesian optimization via simulation matching. In: Advances in Neural Information Processing Systems, pp. 109–117 (2010)
12. Bull, A.D.: Convergence rates of efficient global optimization algorithms. J. Mach. Learn. Res. **12**(Oct.), 2879–2904 (2011)
13. Snoek, J., Larochelle, H., Adams, R.P.: Practical Bayesian optimization of machine learning algorithms. In: NIPS, pp. 2951–2959 (2012)
14. Jones, D.R., Perttunen, C.D., Stuckman, B.E.: Lipschitzian optimization without the Lipschitz constant. J. Optim. Theory Appl. **79**, 157–181 (1993)
15. Girard, A., Murray-Smith, R.: Gaussian processes: prediction at a noisy input and application to iterative multiple-step ahead forecasting of time-series. In: Murray-Smith, R., Shorten, R. (eds.) Switching and Learning in Feedback Systems. LNCS, vol. 3355, pp. 158–184. Springer, Heidelberg (2005). doi:10.1007/978-3-540-30560-6_7

An Exponential Time-Aware Recommendation Model for Mobile Notification Services

Chenglin Zeng[1(✉)], Laizhong Cui[2], and Zhi Wang[1]

[1] Shenzhen Key Laboratory of Information Science and Technology, Graduate School at Shenzhen, Tsinghua University, Shenzhen, China
cengcl14@mails.tsinghua.edu.cn, wangzhi@sz.tsinghua.edu.cn
[2] College of Computer Science and Software Engineering, Shenzhen University, Shenzhen, China
cuilz@szu.edu.cn

Abstract. Mobile notifications attract users' attention with minimum interruption. It is intriguing to study how to utilize such notifications for personal content recommendation. Recommendation for mobile notification services is nontrivial due to the following challenges: (1) A user may be bothered when receiving many irrelevant or uninterested notifications; (2) Notifications are newly produced without feedbacks before pushed out; (3) Notifications are time-sensitive, and are significantly affected by the time when users receive them. To address these challenges, we propose an exponential time-aware recommendation model. Firstly, based on traces covering 155,141 users receiving 1,464 notification messages provided by NextMedia (http://www.nextmedia.com/), we build an exponential-decaying model to reflect the timeliness of notifications. Secondly, we design a temporal preference model to capture users' willingness to open notifications over time. Finally, we use LDA to get users' content preferences and incorporate the two models to provide time-varying mobile notification services. Our experimental results show that our model achieves 15% improvement in precision against the vanilla LDA method.

Keywords: Exponential distribution · Time · Notification · Recommendation

1 Introduction

The rapid development of mobile networks and smart devices has enabled ubiquitous instant message delivery from content providers to users anytime and anywhere. A new type of message service—mobile notification—has emerged in recent years, which allows the content provider to broadcast to its users different messages. These messages will pop up on their devices' screens, so that they may be attracted to subscribe to more detailed information.

In conventional recommendation for mobile notification services, messages are usually delivered to users in a broadcasting manner, i.e., all users receive

J. Kim et al. (Eds.): PAKDD 2017, Part II, LNAI 10235, pp. 592–603, 2017.
DOI: 10.1007/978-3-319-57529-2_46

the same messages at the same time. However, such generic broadcast notifications are not appealing to users: only 3% of notifications are opened by users; even when users are divided into groups and each group receives different broadcasting messages, the open rate (fraction of users who actually open a notification message) can only be slightly improved to 7%, according to Localytics[1]. The notification messages then fail to satisfy users with their personal interests. Notification messages are eventually viewed by individuals, who have their own preferences, and such generic messages can not attract all users equally.

To this end, personal preferences should be taken into consideration in mobile notification services. However, conventional recommendation approaches end up with poor notification performance due to the following reasons: Firstly, broadcast notifications fail to push notifications to only users who are interested in messages. As a result, notification messages are likely to be discarded by users. This will result in high customer churn rate. Secondly, rating-based recommendations [10], such as Collaborative Filtering [11], are based on users' historical ratings of contents, but notifications do not provide any explicit feedbacks such as ratings. Besides, feedbacks are unavailable before notifications are pushed out. Therefore, these methods can not work well. Thirdly, content-based recommendations, including Latent Dirichlet Allocation (LDA) [4] fail to take users' temporal preferences into consideration, which is important for mobile recommendation [14]. Lastly, notifications are time-sensitive, and it has a very low chance to be opened after a period of time. So, for each newly produced notifications, we should immediately decide whether to push it or not.

To address these challenges, we propose an exponential time-aware recommendation model for mobile notification services. Our contributions can be summarized as follows:

- First of all, we design an exponential-decaying model, which reflects the timeliness of notifications to be pushed to users.
- Then, we build a temporal preference model to capture each user's willingness to open notifications varying over time.
- Next, we incorporate the aforementioned two models into LDA to compute each user's willingness to open each notification and only push notifications to users with willingness to open.
- Finally, we carry out comprehensive experiments on a real-world dataset to verify the effectiveness of our design. Our experimental results show that: (1) Our model has a significant improvement on the notification open rate, improving the precision by 15% against the baseline vanilla LDA method. (2) Our model also outperforms other conventional strategies.

The remainder of the paper is organized as follows. We view related works in Sect. 2. Then, we present our model for mobile notification services in Sect. 3. In Sect. 4, we conduct several experiments to evaluate the performance of our model. Finally, we conclude the paper in Sect. 5.

[1] http://info.localytics.com/blog/52-percent-of-users-enable-push-messaging.

2 Related Work

Mobile notification services are potential to improve the revenue if such notification messages can successfully attract users' attentions on their smart devices. Using recommendation techniques in notification message delivery is a promising way to improve the notification open rate.

As described in [18], content-based approaches have been widely used with the emergence of topic modeling techniques such as LDA. Generally, documents exhibit multiple topics. LDA extracts each document's distribution over topics and recommend similar documents by computing the similarity between documents over topic distributions. Lu et al. [9] unified Collaborative Filtering and Content-based Filtering by incorporating PLSA [6] model into Matrix Factorization [7] model. The Topics over Time (TOT) model was proposed by Fani [5] to discover topics and model users' temporal preferences towards these topics.

For mobile users, context can affect the recommendation performance [2], and context-aware recommendation can improve the recommendation accuracy [1]. Zheng et al. [15] pulled users' data together and applied Collaborative Filtering to find like-minded users and like-patterned activities at different locations. The study [16] mixed context similarity into Matrix Factorization to predict ratings of unrated items. Wang [12] proposed a joint social and content recommendation for User-Generated Videos in Online Social Network.

Temporal information is widely used among all the elements in context. It was demonstrated by Yuan et al. [13] that incorporating temporal information into Point-of-interest (POI) recommendations in Location-Based Social Networks could improve the performance. Zhong et al. [17] exploited LDA and time series prediction to recommend services for mashup creation. Liang et al. [8] made use of temporal information of microblogs to build users' time-drifting topic interests.

To the best of our knowledge, our study is the first to introduce temporal information into content recommendation for mobile notification services.

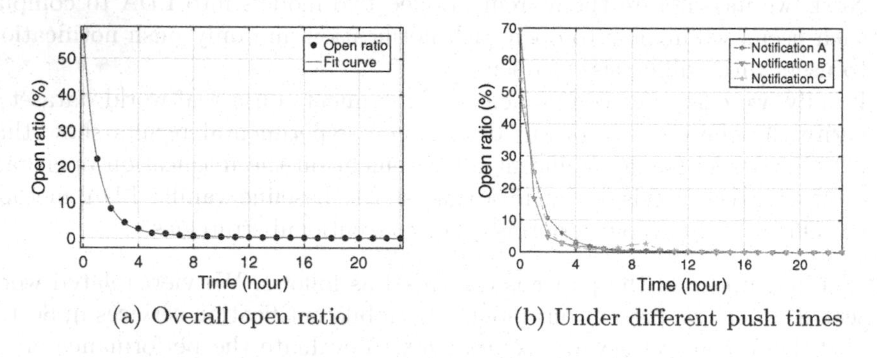

(a) Overall open ratio (b) Under different push times

Fig. 1. Open ratio vs. the elapsed time

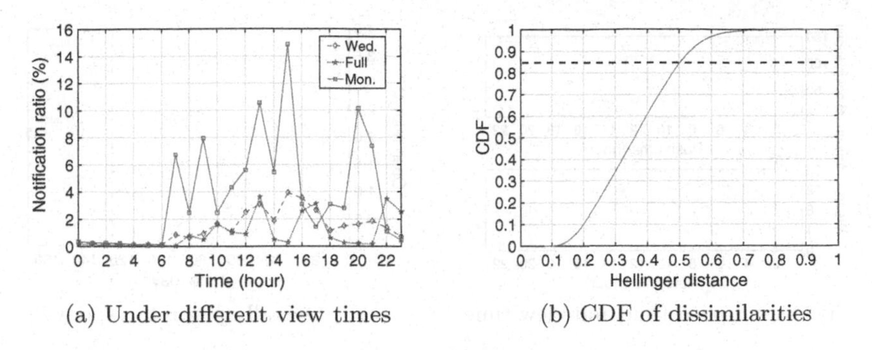

(a) Under different view times (b) CDF of dissimilarities

Fig. 2. Users' temporal preferences

3 Exponential Time-Aware Recommendation

3.1 Design Motivations

We use a real-world notification dataset (details illustrated in Sect. 4.1) to provide the motivations in our model as follows (DM is short for Design Motivation):

DM1. Notifications are time-sensitive. More than 99% of users open notifications in 24 hours after notifications are pushed out. We split the elapsed time after notifications are opened by users into hourly-based slots and draw Fig. 1a. The t-th slot denotes the period of hour $[t, t+1)$, similarly hereinafter. As shown in Fig. 1a, the open ratio obeys an exponential distribution. A simple 2-term exponential function can fit it well on history records. We randomly pick 3 representative notifications (Notification A–C) pushed at time 6:59, 11:43 and 21:52 respectively. They all obey an exponential distribution on the whole, as shown in Fig. 1b. Figure 3a shows views are about an hour behind the push time of notifications. It also demonstrates notification's timeliness.

DM2. Users are time-aware, and their open rates to notifications vary over time. We define notification ratio as the ratio of the views through notifications to the total views in each hourly-based slot of a day, to reflect the impact of notifications over time. As illustrated in Fig. 2a, we collect full records and records of just Wednesday or Monday for comparison. It shows that notification ratio obviously varies over time. The results indicate that we can infer users' temporal preferences by referring their previous records.

DM3. Users share similar temporal preferences. We choose users who have opened more than 20 records and split view time into hourly-based slots of a day to build each user's temporal preference distribution of open ratio versus time slot. We further compute the Hellinger distance [3] between the distribution of each user and the overall distribution and plot the CDF in Fig. 2b. We observe that about 85% of the distances are smaller than 0.5, indicating that many users share similar temporal preferences.

DM4. It will result in higher customer churn rate to push too many notifications to users without considering users' preferences. Figure 3b keeps track of each

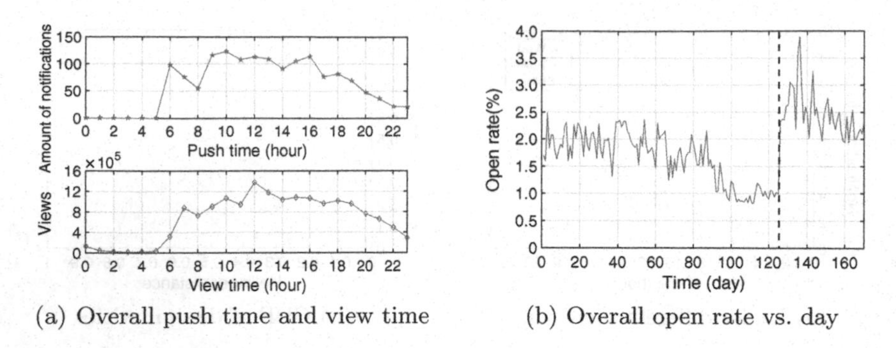

(a) Overall push time and view time (b) Overall open rate vs. day

Fig. 3. Overall statistics for mobile notifications

day's average notification open rate about half a year from Jan. 1, 2016. Before May 5, 2016 (shown as a dashed), the app pushed notifications to all users, and the open rate versus day was in decline. Since then, the app randomly chooses 30% of users for each notification to send. The simple strategy increases users' willingness to open notifications due to less disturbance to them, and achieves a good improvement.

3.2 Exponential Time-Aware (ETA) Model

Based on above motivations, we propose the Exponential Time-Aware (ETA) model to solve these problems in mobile notification services.

For DM1, we formulate the distribution of a user's open ratio versus time slot into a 2-term exponential function

$$d(u,t) = a_1 e^{-\beta_1 t} + a_2 e^{-\beta_2 t}, t \in \mathbb{N} \tag{1}$$

where u stands for a user, and t is the elapsed hourly-based time before user u opens a notification. $a_i, \beta_i, i \in \{1,2\}$ are parameters. When $t \geq 24$, the value of the function is very small. So we truncate the range of t into $[0,24)$ and discretize the function into a 24-dimension probability distribution $\hat{d}_u$.

For DM2, we extract the hourly-based time from each notification n user u opened and transform the hourly-based time into a 24-dimension one-hot vector $\omega_{un} = (0,\ldots,1,\ldots,0)$. Since user u's temporal preference may change over time, we assign the latest notifications with more importance. We list notifications in descending order by user u's opening time, and build his temporal preference distribution

$$\omega_u = norm(\sum_{n=1}^{R} \omega_{un} e^{-\beta_3 n}) \tag{2}$$

where $norm(\cdot)$ is a normalized function for probability distribution, R is the amount of notifications user u opened, β_3 is a decay factor. We use records of all users to build the overall distribution $\bar{\omega}$ the same way. The value of each dimension in ω_u or $\bar{\omega}$ is the open ratio of corresponding hourly-based time slot of a day.

The Hellinger distance can be used to compute the distance between two distributions and its value is normalized, so the dissimilarity can be calculated as follows:

$$h(d, f) = \frac{1}{\sqrt{2}}\sqrt{\sum_{i=1}^{D}(\sqrt{d_i} - \sqrt{f_i})^2} \tag{3}$$

where D is the dimension of the 2 discrete distributions d and f. d_i is the i-th dimension value of d. The value of $h(\cdot, \cdot) \in [0, 1]$ shows the dissimilarity between the 2 distributions. 0 indicates that they are identically distributed, while 1 shows totally different.

For DM3, we select out users that share similar temporal preferences. We regard user u as a similar user if $h(\omega_u, \bar{\omega}) \leq \tau$ (where $\tau \in [0, 1]$ is a threshold parameter), otherwise we consider him an outlier user. It is not easy to model outlier users in a simple generic method since they have different temporal preferences. So, for similar users, we consider their temporal information and content preferences, while for outlier users, we just consider their content preferences.

Combining $\hat{d}_u$ and ω_u, we compute the possibility $T_{u,n}$ for each similar user u (i.e. $h(\omega_u, \bar{\omega}) \leq \tau$) opens each notification n by

$$T_{u,n} = \sum_{t=t_p}^{t_p+23} \hat{d}_{u,t-t_p} * \omega_{u,t\%24} \tag{4}$$

where t_p denotes the push time of notification n. $\omega_{u,t}$ is the open ratio on the t-th hourly-based time of a day. We accumulate the open possibilities in the next 24 hours after the notification n is pushed out.

We use Eq. 4 to compute $T_{u,n}$ for each similar user. If $T_{u,n} \geq \eta$, we send notification n to user u. $\eta \in [0, 1]$ decreases the total amount of notifications sent to users. On one hand, it reduces disturbance to uninterested users; on the other hand, it improves open rate by sending notifications to only users who have willingness to open notifications.

3.3 Incorporating ETA into Content-Based Recommendation

LDA is a latent factor model, we use it to get each notification n's topic distribution $\theta_n = (\theta_{n,0}, \theta_{n,1}, \cdots, \theta_{n,K-1})$. K is the number of topics. We list notifications in descending order by user u's opening time, and assign the latest ones with more importance and build his topic preference distribution

$$\phi_u = norm(\sum_{n=1}^{R} \theta_n e^{-\beta_4 n}) \tag{5}$$

where R is the amount of notifications user u opened, β_4 is a decay factor.

For a notification to be recommended, we compute the similarity $C_{u,n}$ between ϕ_u and θ_n as follows:

$$C_{u,n} = 1 - h(\phi_u, \theta_n) \tag{6}$$

where $h(\cdot,\cdot)$ is the Hellinger distance.

$T_{u,n}$ considers temporal information, while $C_{u,n}$ captures users' content preferences. For a notification, not only content but also temporal information affects users' open rates. So, we combine ETA and LDA model to compute users' willingness to open notifications, denoted by $R_{u,n}$

$$R_{u,n} = (1-\alpha) * T_{u,n} + \alpha * C_{u,n} \tag{7}$$

where $\alpha \in [0,1]$ is a tuning parameter to decide the proportion of $C_{u,n}$. For outlier users, α equals 1. If $R_{u,n} > \delta$ (where $\delta \in [0,1]$ is a threshold parameter), we actually push notification n to user u.

4 Experiments

In this section, we evaluate the proposed model on a real-world dataset. We first describe the dataset we used in the following experiments, then introduce comparison methods we adopt and evaluation metrics. Finally, we compare the results of our proposed model with the comparison methods and analyze impacts of parameters and other factors.

4.1 Experimental Setup

Dataset. Since no open source datasets are available for evaluating the recommendation performance for mobile notification services, we use a notification dataset from a media company, NextMedia. One of its mobile applications, Apple Daily App, provides news video services. Each news video, well-edited, lasts no more than 3 min. The app sometimes pushes notifications of the latest news videos to users and tracks users' viewing records in log files.

We extract about 51 days' notification log files which are collected on Android devices from March 14, 2016 to May 4, 2016. Each log file trails the summary of the pushed news videos, including push time, view time, location, and device information when a user opens a notification to view a news video. Each log file contains about 60 fields, and we extract some main fields shown in Table 1.

Data Preprocessing. We remove records of users who opened less than 20 notifications. After preprocessing, the dataset contains 15,041,313 access records of 155,141 users on 1,464 notifications. In order to use LDA to predict the topic distribution of each notification, we use the online segmentation tool BosonNLP[2] to segment article titles into a bag of words, and remove stop words. Since some

[2] http://bosonnlp.com.

Table 1. Main fields in log files

Field	Description
push_time	The time when a notification is pushed out
view_time	The time when a user opens a notification
article_id	The ID of the news video
article_title	The title of the news video
category	The category of the news video
udid	Device ID, can be used as users' ID

words occur very few times in the dataset, we filter out these low-frequency words.

The preprocessed dataset contains 1,464 notifications. Due to retaining users' time-varying preferences, we divide the dataset into train set and test set by splitting notifications with a ratio of 4:1 in order of time.

Comparison Methods. Recommendation on mobile notification services is quite different from conventional recommendation problems. Many classic algorithms such as Collaborative Filtering, can not work well, since notifications are newly produced without feedbacks. So we adopt two simple but widely used methods in notification recommendations.

Besides, we use LDA as an instance of content-based recommendations. In order to study the impact of temporal information, we use ETA model alone as a contrast method. Our validation method is short for ETA-LDA. The parameters of all adopted methods are fine-tuned according to performances on train set by 5-fold cross validation. We compare the performances of these methods in test set.

RND: It is a recommendation model which recommends items to users at random. Since the preprocessed dataset just remains records of users that opened more than 20 notifications (if users do not open any notifications, no traces of them will be kept.), we directly use the notification open rate from the app's statistics with the mechanism that randomly chooses 30% of users for each notification to send.

POP: POP method recommends popular items to users. We can not tell if a notification will be popular before it is pushed out. Instead, we recommend notifications to active users who have opened more than 50 notifications.

LDA: It can calculate topic distributions of newly produced notifications. Gensim[3] provides a good Python implementation of LDA, so we use it as our LDA implementation.

ETA: It recommends notifications just considering temporal information.

ETA-LDA: ETA depicts users' willingness to open a notification in a certain time, while LDA captures users' content preferences to the notification.

[3] http://radimrehurek.com/gensim/.

ETA-LDA makes a trade-off between the two parts to improve the recommendation performance.

Evaluation Metrics. To evaluate the recommendation methods, we use metrics: P, and R to compare the performance of recommendation methods. P measures the ratio of the notifications pushed by recommendation methods that are really opened by users. Higher P will not only attain better user experience but also decrease customer churn rate. R measures the ratio of the notifications pushed by recommendation methods that are opened by users to the total opened notifications. Higher R means more users' involvements and may increase popularity. However, higher P will result in lower R. We use P as our key metric.

4.2 Recommendation Effectiveness

In this part, we first compare experimental results of the comparison methods and ETA-LDA on metrics of P and R. We then analyze impact of parameters and other factors.

Performance Analysis. Since RND uses open rate from the app's statistics of notifications, we do not compute R on it. Each algorithm has parameters fine-tuned. For POP, we recommend all items to each active user, and it achieves 58.8% on R. For latent factor models (LDA and ETA-LDA), we use the same hyper-parameters. The number of topics is tuned to be 12, and other hyper-parameters are set to default values. Besides, we set decay factors β_3 and β_4 to 1 equally. It is insufficient to train the parameters in function $d(\cdot, \cdot)$, due to many users just opened few notifications. Instead, we use the full records to train the parameters. As shown in Fig. 2b, it will filter out about 15% of users by setting parameter τ to be 0.5. For these outlier users, we use their content preferences to decide whether to push notifications or not. Then, we fix τ to be 0.5 to ensure temporal preferences of most (about 85% of) users will be taken into consideration in ETA and ETA-LDA methods. Besides, we tune η to be 0.054 in ETA. In ETA-LDA, we tune α, δ to be 0.94 and 0.21, respectively.

Table 2 lists the performance of the comparison methods and ETA-LDA on the test set. ETA just with temporal information achieves a good performance. Our proposed ETA-LDA method clearly outperforms the baseline methods. Without sufficient content information of news videos, LDA is not sufficiently trained. However, incorporating ETA into LDA achieves 15% improvement in P and significant improvement in R against the vanilla LDA method. Given more content information, ETA-LDA will improve much more.

Parameter Analysis. α determines the proportion of content preferences, τ decides the ratio of users who are similar or outlier users, and δ has the final say on whether to push notifications. We set α, δ and τ to be 0.94, 0.21 and 0.50, respectively. The three parameters impact the performance of ETA-LDA. By varying each parameter, we have the following result:

Table 2. Comparison results on test set.

Method	P (%)	R (%)
RND	2.2	-
POP	6.7	**58.8**
LDA	8.1	5.7
ETA	8.4	54.3
ETA-LDA	**9.3**	27.1

We analyze how α influences the performance of ETA-LDA. Figure 4a plots the performance of ETA-LDA with α varying from 0.80 to 0.99. We observe that P achieves the maximum (9.3%) by tuning α to be 0.94; R always decreases as α increases. These results demonstrate it is wise to set α to be 0.94.

We study how δ influences the performance of ETA-LDA. Figure 4b shows the performance of ETA-LDA by varying δ from 0.11 to 0.30. We see that P achieves the maximum (9.3%) when δ is 0.21; R simultaneously increases with δ. These observations figure out the optimized set up of δ (i.e., δ is set to be 0.21).

We further investigate the impact of τ in our model. As illustrated in Fig. 4c, the performance of ETA-LDA by varying τ in [0.35, 0.55] is not convergent. We can not get optimal recommendation performance through these metrics, due to the lack of appropriate metrics to balance P and R. Hence, we set τ to be 0.5, considering temporal information of 85% of users.

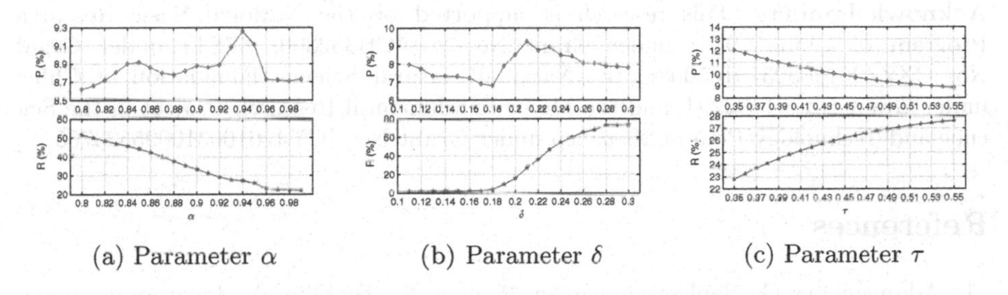

(a) Parameter α (b) Parameter δ (c) Parameter τ

Fig. 4. Impact of implementation parameters

Other Factor Analysis. First, we study how ETA-LDA performs on different categories. We extract notifications from some different categories. Figure 5a shows that ETA-LDA has relatively stable performance on different categories. Then we investigate how the push time of notifications performs in our model. We split test set into sub sets by the hour of push time. As Fig. 3a illustrates, the hours in prime time achieve better performance. We vary the minimum number of each user's records from 20 to 50. Figure 5c shows that denser datasets significantly improve P, but R is not sensitive.

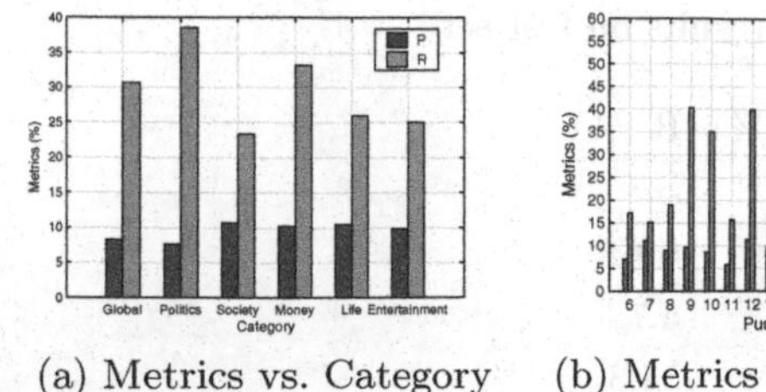

(a) Metrics vs. Category

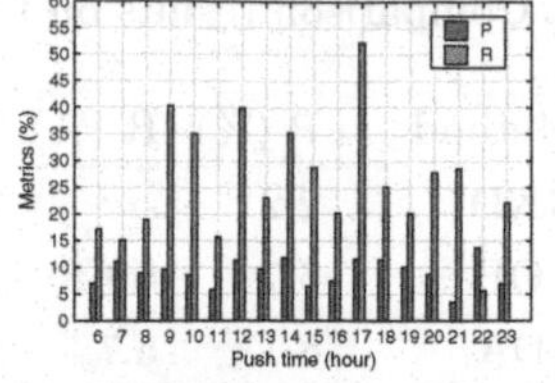

(b) Metrics vs. Push time

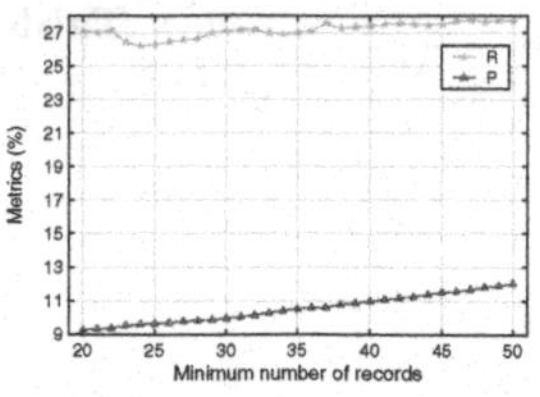

(c) Metrics vs. Minimum number of records

Fig. 5. Impact of other factors

5 Conclusion

In this paper, we study recommendation on mobile notification services. We first present the challenges in notification recommendation and give our design motivations. Based on these motivations, we propose an exponential time-aware recommendation model to tackle with these problems. We build an exponential-decaying model to reflect the timeliness of notifications, and design a temporal preference model to capture users' willingness to open notifications over time. Based on the two models, we incorporate LDA to provide time-varying mobile notification services. Our model integrates temporal information and users' content preferences. Our trace-driven experimental results have shown that our model is effective and outperforms several alternative methods.

Acknowledgments. This research is supported by the National Basic Research Program of China (973) under Grant No. 2015CB352300, SZSTI under Grant No. CXZZ20150323151850088, the National Natural Science Foundation of China under Grant No. 61402294, and the Major Fundamental Research Project in the Science and Technology Plan of Shenzhen under Grant No. JCYJ20160310095523765.

References

1. Adomavicius, G., Sankaranarayanan, R., Sen, S., Tuzhilin, A.: Incorporating contextual information in recommender systems using a multidimensional approach. ACM Trans. Inf. Syst. (TOIS) **23**(1), 103–145 (2005)
2. Adomavicius, G., Tuzhilin, A.: Context-aware recommender systems. In: Ricci, F., Rokach, L., Shapira, B. (eds.) Recommender Systems Handbook, pp. 191–226. Springer, New York (2015)
3. Birgé, L.: On estimating a density using hellinger distance and some other strange facts. Probab. Theory Relat. Fields **71**(2), 271–291 (1986)
4. Blei, D.M., Ng, A.Y., Jordan, M.I.: Latent dirichlet allocation. J. Mach. Learn. Res. **3**(Jan), 993–1022 (2003)
5. Fani, H., Zarrinkalam, F., Bagheri, E., Du, W.: Time-sensitive topic-based communities on Twitter. In: Khoury, R., Drummond, C. (eds.) AI 2016. LNCS (LNAI), vol. 9673, pp. 192–204. Springer, Cham (2016). doi:10.1007/978-3-319-34111-8_25

6. Hofmann, T.: Learning the similarity of documents: an information-geometric approach to (1999)
7. Koren, Y., Bell, R., Volinsky, C.: Matrix factorization techniques for recommender systems. Computer **42**(8) (2009)
8. Liang, H., Xu, Y., Tjondronegoro, D., Christen, P.: Time-aware topic recommendation based on micro-blogs. In: Proceedings of the 21st ACM International Conference on Information and Knowledge Management, pp. 1657–1661. ACM (2012)
9. Lu, Z., Dou, Z., Lian, J., Xie, X., Yang, Q.: Content-based collaborative filtering for news topic recommendation. In: AAAI, pp. 217–223. Citeseer (2015)
10. Pazzani, M.J., Billsus, D.: Content-based recommendation systems. In: Brusilovsky, P., Kobsa, A., Nejdl, W. (eds.) The Adaptive Web. LNCS, vol. 4321, pp. 325–341. Springer, Heidelberg (2007). doi:10.1007/978-3-540-72079-9_10
11. Schafer, J.B., Frankowski, D., Herlocker, J., Sen, S.: Collaborative filtering recommender systems. In: Brusilovsky, P., Kobsa, A., Nejdl, W. (eds.) The Adaptive Web. LNCS, vol. 4321, pp. 291–324. Springer, Heidelberg (2007). doi:10.1007/978-3-540-72079-9_9
12. Wang, Z., Sun, L., Zhu, W., Yang, S., Li, H., Wu, D.: Joint social and content recommendation for user-generated videos in online social network. IEEE Trans. Multimed. **15**(3), 698–709 (2013)
13. Yuan, Q., Cong, G., Ma, Z., Sun, A., Thalmann, N.M.: Time-aware point-of-interest recommendation. In: Proceedings of the 36th International ACM SIGIR Conference on Research and Development in Information Retrieval, pp. 363–372. ACM (2013)
14. Zhang, J.D., Chow, C.Y.: TICRec: a probabilistic framework to utilize temporal influence correlations for time-aware location recommendations. IEEE Trans. Serv. Comput. **9**(4), 633–646 (2016)
15. Zheng, V.W., Cao, B., Zheng, Y., Xie, X., Yang, Q.: Collaborative filtering meets mobile recommendation: a user-centered approach. AAAI **10**, 236–241 (2010)
16. Zheng, Y., Mobasher, B., Burke, R.: Similarity-based context-aware recommendation. In: Wang, J., Cellary, W., Wang, D., Wang, H., Chen, S.-C., Li, T., Zhang, Y. (eds.) WISE 2015. LNCS, vol. 9418, pp. 431–447. Springer, Cham (2015). doi:10.1007/978-3-319-26190-4_29
17. Zhong, Y., Fan, Y., Huang, K., Tan, W., Zhang, J.: Time-aware service recommendation for mashup creation in an evolving service ecosystem. In: 2014 IEEE International Conference on Web Services (ICWS), pp. 25–32. IEEE (2014)
18. Zhu, W., Cui, P., Wang, Z., Hua, G.: Multimedia big data computing. IEEE MultiMed. **22**(3), 96-c3 (2015)

Discovering Periodic Patterns in Non-uniform Temporal Databases

R. Uday Kiran[1(✉)], J.N. Venkatesh[2], Philippe Fournier-Viger[3], Masashi Toyoda[1], P. Krishna Reddy[2], and Masaru Kitsuregawa[1,4]

[1] Institute of Industrial Science, The University of Tokyo, Tokyo, Japan
{uday_rage,toyoda,kitsure}@tkl.iis.u-tokyo.ac.jp
[2] Kohli Center on Intelligent Systems (KCIS), International Institute of Information Technology Hyderabad, Hyderabad, India
jn.venkatesh@research.iiit.ac.in, pkreddy@iiit.ac.in
[3] Harbin Institute of Technology Shenzhen Graduate School, Shenzhen, China
philfv8@yahoo.com
[4] National Institute of Informatics, Tokyo, Japan

Abstract. A temporal database is a collection of transactions, ordered by their timestamps. Discovering periodic patterns in temporal databases has numerous applications. However, to the best of our knowledge, no work has considered mining periodic patterns in temporal databases where items have dissimilar *support* and *periodicity*, despite that this type of data is very common in real-life. Discovering periodic patterns in such non-uniform temporal databases is challenging. It requires defining (*i*) an appropriate measure to assess the periodic interestingness of patterns, and (*ii*) a method to efficiently find all periodic patterns. While a pattern-growth approach can be employed for the second sub-task, the first sub-task has to the best of our knowledge not been addressed. Moreover, how these two tasks are combined has significant implications. In this paper, we address this challenge. We introduce a model to assess the periodic interestingness of patterns in databases having a non-uniform item distribution, which considers that periodic patterns may have different *period* and minimum number of cyclic repetitions. Moreover, the paper introduces a pattern-growth algorithm to efficiently discover all periodic patterns. Experimental results demonstrate that the proposed algorithm is efficient and the proposed model may be utilized to find prior knowledge about event keywords and their associations in Twitter data.

Keywords: Data mining · Periodic pattern · Non-uniform temporal database

1 Introduction

Temporal databases are commonly used in many domains. A temporal database is a collection of transactions, ordered by their timestamps. A temporal database is said to be non-uniform if it contains items with dissimilar *support*

J. Kim et al. (Eds.): PAKDD 2017, Part II, LNAI 10235, pp. 604–617, 2017.
DOI: 10.1007/978-3-319-57529-2_47

Table 1. Some tweets produced during GEJE

Timestamp	Tweets
1301575750	#Discrimination n demagoguery 4 foreigners, mainly #Asian s by #Japanese in #Sendai. #earthquake #jishin http://htn.to/rhGtmd
1301575750	shhhhh dont tell @jimcramer El-Erian says recent Japanese earthquake is NOT like Kobe, won't have same V shaped recovery - Reuters #newsmkr
1301583131	Some people will never be able to go home. They lost their homes in the tsunami. #VOAAsiachat1
1301583579	Social media had a critical role in Japan during/after the quake/tsunami. Somewhat else can better assess overall than me. #VOAAsiachat1

and *periodicity*. Non-uniform temporal data is naturally produced in many real-world situations. For instance, disasters such as earthquakes and tsunami happen at irregular time intervals. Twitter data related to these disasters is thus non-uniform. For example, Table 1 shows a part of a temporal database generated from the tweets produced during the Great East Japan Earthquake (GEJE), which occurred on the 11th March 2011.

Discovering patterns in temporal databases is challenging because they not only allow time gaps between consecutive transactions, but also to have multiple transactions with the same timestamp. An important type of patterns that can be extracted from temporal databases is (Partial) periodic patterns. A periodic pattern is something persistent and predictable that appears in a database. Finding periodic patterns is thus useful to understand the data. For example, it was revealed in our present study on Twitter data related to the GEJE that over 80% of the event keywords found by a supervised event detection algorithm [1] can also be discovered as periodic patterns. The proposed study thus may be used as an unsupervised learning technique to generate some prior knowledge about event keywords and their associations in Twitter data.

The task of finding periodic patterns has two important sub-tasks: (*i*) determining the periodic interestingness of patterns, and (*ii*) finding all periodic patterns in a given database. While a variation of pattern-growth algorithms could be employed for the second sub-task, the first sub-task is non-trivial because of the following reasons:

1. Current periodic pattern models [2–4] do not take into account the information about the temporal occurrences of items in a dataset.
2. Since a temporal database allows transactions to share a common timestamp, the periodic interestingness of a pattern has to be determined by taking into account not only its *support*, but also its inter-arrival times in a database. Unfortunately, current measures assess the interestingness of a pattern by only taking its *support* into account. We need to investigate new measure(s) to assess the interestingness of patterns by taking into account both their *support* and *periodicity* in a database.

Moreover, how to combine the two aforementioned tasks has significant implications.

This paper addresses this challenge. It presents a model to discover periodic patterns in non-uniform temporal databases. The proposed model lets the user specify a different *maximum inter-arrival time* ($MIAT$) for each item. Thus, different patterns may satisfy different *period* depending on their items' $MIAT$ values. A new measure, *Relative Periodic-Support* (RPS), is proposed to determine the periodic interestingness of a pattern in a database. Unlike existing *support*-based measures, the proposed measure assess the interestingness of a pattern by taking into account its number of cyclic repetitions in the database. An inter-arrival time of a pattern is considered periodic (or cyclic) if it is no more than *period*. This measure satisfies the *null-invariant property* [5]. Thus, the usage of item specific $MIAT$ values and RPS allows the proposed model to capture the non-uniform distribution of items in a database. We also propose a pattern-growth algorithm that discovers the complete set of periodic patterns. Experimental results demonstrate that the proposed algorithm is efficient. We also demonstrate the usefulness of the proposed model by finding various event keywords and their associations in disaster related Twitter data.

The rest of paper is organized as follows. Section 2 describes the related work. Section 3 describes the proposed periodic pattern model. Section 4 introduces our algorithm to find all periodic patterns in a database. Section 5 reports on experimental results. Finally, Sect. 6 concludes the paper with future research directions.

2 Related Work

Frequent pattern mining is an important data mining task. Several *support* related measures have been discussed to determine the interestingness of a pattern in a transactional database. Each measure has a selection bias that justifies the significance of a knowledge pattern. As a result, there exists no universally acceptable best measure to judge the interestingness of a pattern in any given database. Researchers have proposed criteria to select an interestingness measure based on user and/or application requirements [5]. Recently, measures that satisfy the *null-invariant property* have became popular for finding frequent patterns. The reason is that this property guarantees finding genuine correlation patterns that are not influenced by object co-absence in a database. Unfortunately, current measures cannot be used to determine the periodic interestingness of a pattern in temporal databases. This is because these measures only take the *support* into account and completely ignore the temporal occurrence behavior of patterns in databases. We introduce a new null-invariant measure that assess the interestingness of a pattern by taking into account both the *support* and temporal occurrence information of patterns.

Periodic patterns are an important class of regularities that exist in a time series data. Since it was first introduced in [2], the problem of finding these patterns has received a great deal of attention [4]. A major limitation of these studies is that they consider time series as a symbolic sequence and ignore the temporal occurrence information about events in a series.

Similar to our problem, the mining of full periodic-frequent patterns in a transactional database has been studied in [6–8]. This problem of finding full periodic-frequent patterns greatly simplifies the design of the model because there is no need of any measure to determine the partial periodic interestingness of a pattern. More important, these studies also consider transactional database as a symbolic sequence of transactions (or itemsets) and ignore the temporal occurrence information of the transactions in a database. To the best of our knowledge, this is the first study that considers the problem of finding (partial) periodic patterns by taking into account the temporal occurrence information of the transactions in a database.

3 Proposed Model

Let $I = \{i_1, i_2, \cdots, i_n\}$ be the set of 'n' items appearing in a database. A set of items $X \subseteq I$ is called an itemset (or **a pattern**). A pattern containing k items is called a k-pattern. The length of this pattern is k. A transaction is a triplet $tr = (tid, ts, Y)$, where tid represents the transactional identifier, $ts \in \mathbb{R}$ represents the transaction time (or timestamp) and Y is an itemset. A temporal database TDB is an ordered set of transactions, i.e. $TDB = \{tr_1, tr_2, \cdots, tr_m\}$, where $m = |TDB|$ represents the database size (the number of transactions). Let ts_{min} and ts_{max} denote the minimum and maximum timestamps in TDB, respectively. For a transaction $tr = (tid, ts, Y)$, such that $X \subseteq Y$, it is said that X occurs in tr and such a timestamp is denoted as ts^X. Let $TS^X = (ts_a^X, ts_b^X, \cdots, ts_c^X)$, $a \leq b \leq c$, be the **ordered list of timestamps** of transactions in which X appears in TDB. The number of transactions containing X in TDB (i.e., the size of TS^X) is defined as the *support* of X and denoted as $sup(X)$. That is, $sup(X) = |TS^X|$.

Example 1. Table 2 shows a temporal database with $I = \{abcdefg\}$. The set of items 'a' and 'b,' i.e., 'ab' is a pattern. This pattern contains 2 items. Therefore, it is a 2-pattern. The length of this pattern is 2. In the first transaction, $tr_1 = (100, 1, ab)$, '100' represents the tid of the transaction, '1' represents the timestamp of this transaction and 'ab' represents the items occurring in this transaction. Other transactions in this database follow the same representation. The size of the database is $m = 12$. The minimum and maximum timestamps in this database are 1 and 14, respectively. Therefore, $ts_{min} = 1$ and $ts_{max} = 14$. The pattern 'ab' appears in the transactions whose timestamps are 1, 3, 6, 8, 10, 11 and 12. Therefore, $TS^{ab} = \{1, 3, 6, 8, 10, 11, 12\}$. The *support* of '$ab$,' i.e., $sup(ab) = |TS^{ab}| = 7$.

Definition 1 ***(Period of a pattern X).*** *Let $MIAT(i_j)$ be the user-defined maximum inter-arrival time $(MIAT)$ specified for an item $i_j \in I$. The period of a pattern X, denoted as $PER(X)$, represents the largest $MIAT$ value of all items in X. That is, $PER(X) = max(MIAT(i_j) | \forall i_j \in X)$. The items' $MIAT$ values can also be expressed in percentage of $(ts_{max} - ts_{min})$.*

Table 2. Running example: temporal database

tid	ts	items	tid	ts	items	tid	ts	items	tid	ts	items
100	1	*ab*	103	4	*cd*	106	8	*abcd*	109	11	*abf*
101	3	*acdg*	104	6	*abcd*	107	9	*ce*	110	12	*abcd*
102	3	*abef*	105	7	*efg*	108	10	*abef*	111	14	*acdeg*

Example 2. Let the $MIAT$ values for the items a, b, c, d, e, f and g be 2, 2, 2, 2, 3, 4 and 4, respectively. The *period* of the pattern '*ab*,' i.e., $PER(ab) = max(2, 2) = 2$.

The usage of items' $MIAT$ values enable us to achieve the goal of having lower *periods* for patterns that only involve frequent items, and having higher *periods* for patterns that involve rare items. The items' $MIAT$ values may be derived using the *period* determining functions, such as Fast Fourier Transformations (FFTs) and auto-correlation.

Definition 2 ***(Periodic occurrence of a pattern X).*** *Let* ts_j^X, $ts_k^X \in TS^X$, $1 \leq j < k \leq m$, *denote any two consecutive timestamps in* TS^X. *The time difference between* ts_k^X *and* ts_j^X *is referred as* ***an inter-arrival time*** *of* X, *and denoted as* iat^X. *That is,* $iat^X = ts_k^X - ts_j^X$. *Let* $IAT^X = \{iat_1^X, iat_2^X, \cdots, iat_k^X\}$, $k = sup(X) - 1$, *be the list of all inter-arrival times of* X *in* TDB. *An inter-arrival time of* X *is said to be* ***periodic*** *(or cyclic) if it is no more than* $PER(X)$. *That is, a* $iat_i^X \in IAT^X$ *is said to be* ***periodic*** *if* $iat_i^X \leq PER(X)$.

Example 3. The pattern '*ab*' has initially appeared at the timestamps of 1 and 3. The difference between these two timestamps gives an inter-arrival time of '*ab*.' That is, $iat_1^{ab} = 2\ (= 3 - 1)$. Similarly, other inter-arrival times of '*ab*' are $iat_2^{ab} = 3\ (= 6-3)$, $iat_3^{ab} = 2\ (= 8-6)$, $iat_4^{ab} = 2\ (= 10-8)$, $iat_5^{ab} = 1\ (= 11-10)$ and $iat_6^{ab} = 1\ (= 12 - 11)$. Therefore, $IAT^{ab} = \{2, 3, 2, 2, 1, 1\}$. If $PER(ab) = 2$, then iat_1^{ab}, iat_3^{ab}, iat_4^{ab}, iat_5^{ab} and iat_6^{ab} are considered as the periodic occurrences of '*ab*'. The iat_2^{ab} is considered as an aperiodic occurrence of '*ab*' because $iat_2^{ab} \nleq PER(ab)$.

Definition 3 ***(Relative periodic-support of a pattern X).*** *Let* $\widehat{IAT^X}$ *be the set of all inter-arrival times in* IAT^X *that have* $iat^X \leq PER(X)$. *That is,* $\widehat{IAT^X} \subseteq IAT^X$ *such that if* $\exists iat_k^X \in IAT^X : iat_k^X \leq PER(X)$, *then* $iat_k^X \in \widehat{IAT^X}$. *The relative periodic-support of* X, *denoted as* $RPS(X) = \frac{|\widehat{IAT^X}|}{|IAT^{i_j}|}$, *where* i_j *is an item that has the lowest support and maximum* $MIAT$ *value among all items in* X. *This measure satisfies the null-invariant property [5].*

Example 4. Continuing with the previous example, $\widehat{IAT^{ab}} = \{2, 2, 2, 1, 1\}$, the item '*b*' in the pattern '*ab*' has the lowest *support* and maximum $MIAT$ value. Therefore, the *relative periodic-support* of '*ab*,' i.e., $RPS(ab) = \frac{|\widehat{IAT^{ab}}|}{|IAT^b|} = \frac{5}{6} = 0.83$.

For brevity, we call $\widehat{IAT^X}$ as ***periodic-frequency***. The *periodic-frequency* determines the number of cyclic repetitions of a pattern in the data. The proposed measure enables us to achieve the goal of specifying a higher number of cyclic repetitions for patterns that only involve frequent items, and a lower number of cyclic repetitions for patterns that involve rare items. For a pattern X, $RPS(X) \in [0,1]$. If all inter-arrival times of X are more than $PER(X)$, then $RPS(X) = 0$. In other words, X is an irregular pattern. If all inter-arrival times of X are within $PER(X)$, then $RPS(X) = 1$. In other words, X is a full periodic pattern.

In the proposed model, we have considered an inter-arrival time of X as interesting if $iat^X \leq PER(X)$. However, our model is flexible and allows other ways to consider an inter-arrival time of a pattern as interesting. For instance, we can consider an inter-arrival time of a pattern as interesting if $iat^X \leq PER(X) \pm \Omega$, where $\Omega > 1$ is a constant that denotes time tolerance. We stick to the above definition for brevity.

Definition 4 ***(Periodic pattern X).*** *The pattern X is a periodic pattern if $RPS(X) \geq minRPS$, where $minRPS$ is the user-specified minimum relative periodic-support.*

Example 5. Continuing with the previous example, if the user-specified $min RPS = 0.6$, then 'ab' is a periodic pattern because $RPS(ab) \geq minRPS$.

Definition 5 ***(Problem definition).*** *Given a temporal database (TDB), set of items (I), user-defined minimum interval times of the items ($MIAT$) and minimum relative periodic-support ($minRPS$), the problem of finding periodic patterns involve discovering all patterns in TDB that have* relative periodic-support *no less than $minRPS$.*

The periodic patterns generated by the proposed model satisfy the **convertible anti-monotonic property** [9].

Property 1. Let $Z = \{i_1, i_2, \cdots, i_k\}$, $1 \leq k \leq |I|$, be a pattern with $MIAT(i_1) \geq MIAT(i_2) \geq MIAT(i_k)$. If $Y \subset Z$ and $i_1 \in Y$, then $RPS(Y) \geq RPS(Z)$ as $\frac{|\widehat{IAT^Y}|}{|IAT^{i_1}|} \geq \frac{|\widehat{IAT^Z}|}{|IAT^{i_1}|}$.

4 Periodic Pattern-Growth Algorithm

In this section, we describe the proposed PP-growth algorithm that discovers the complete set of periodic patterns. Our algorithm involves the following two steps: (i) compress the database into a periodic pattern tree (PP-tree) and (ii) recursively mine the PP-tree to find all periodic patterns. Before we discuss these two steps, we describe the PP-tree structure.

Structure of PP-Tree Structure. A PP-tree has two components: a PP-list and a prefix-tree. The PP-list consists of each distinct *item* (i) with *minimum interval time* ($MIAT$), *support* (S), *periodic-frequency* (PF) and a pointer pointing to the first node in the prefix-tree carrying the item. The prefix-tree in a PP-tree resembles that of the prefix-tree in a FP-tree [10]. However, to capture both *support* and inter-arrival times of the patterns, the nodes in the PP-tree explicitly maintain the occurrence information for each transaction by keeping an occurrence timestamp list, called a ***ts*-list**. To achieve memory efficiency, only the last node of every transaction maintains the *ts*-list. We now explain the construction and mining of PP-tree.

Construction of PP-tree. The procedure for constructing a PP-tree is shown in Algorithm 1. We illustrate the working of this algorithm using the database

---

Algorithm 1. Construction of PP-Tree(TDB: Time series database, I: Set of items, $MIAT$: minimum interval time, $minRPS$: minimum relative periodic-support)

1: Insert all items in TDB into the PP-list with their $MIAT$ values. Set the *support* and *periodic-frequency* values of all these items to 0. The *timestamps* of the last occurring transactions of all items in the PP-list are explicitly recorded for each item in a temporary array, called ts_l.
2: Let $t = \{ts_{cur}, X\}$ denote the current transaction with ts_{cur} and X representing the timestamp and pattern, respectively.
3: **for** each transaction $t \in TDB$ **do**
4:     **for** each item $i \in X$ **do**
5:         $S(i) + +;$
6:         **if** $((ts_l(i) \neq 0) \&\& (ts_{cur} - ts_l(i)) \leq MIAT(i))$ **then**
7:             $PF(i) + +;$
8:         $ts_l(i) = ts_{cur};$
9: All items in PP-list are sorted in ascending order of their $MIAT$ values. The items having a common $MIAT$ value are sorted in descending order of their *support*.
10: Measure the RPS value for the bottom most item in the PP-list. If the RPS value of this item is less than $minRPS$, then prune this item from the PP-list and repeat the same step for the next bottom most item in the PP-list. Stop this pruning process once the RPS value of the bottom most item in PP-list is no less than $minRPS$. Let CI denote this sorted list of items.
11: Create a root node in the prefix-tree, T, and label it as "*null.*"
12: **for** each transaction $t \in TDB$ **do**
13:     Sort the items in X according to the order of CI. Let the sorted candidate item list in t be $[p|P]$, where p is the first item and P is the remaining list. Call *insert_tree*($[p|P], ts_{cur}, T$), which is performed as follows. If T has a child N such that *N.item-name* $\neq$ *p.item-name*, then create a new node N, Let its parent link be linked to T. Let its node-link be linked to nodes with the same *item-name* via the node-link structure. Remove p from P. If P is nonempty, call *insert_tree*(P, ts_{cur}, N) recursively; else add ts_{cur} to the leaf node.

Algorithm 2. PP-growth($Tree$, α)

1: **for** each a_i in the header of $Tree$ **do**
2: **if** $\frac{PF(a_i)}{S(a_i)-1} \geq minRPS$ **then**
3: Generate pattern $\beta = a_i \cup \alpha$. Traverse $Tree$ using the node-links of β, and construct an array, TS^{β}, which represents the timestamps at which β has appeared in TDB. Construct β's conditional pattern base and β's conditional PP-tree $Tree_{\beta}$ by calling calculateRPS(β, TS^{β}, $MIAT(a_i)$). The calculateRPS function calculates the *periodic-frequency* of β from TS^{β}, and returns RPS value by dividing the *periodic-frequency* with $S(a_i) - 1$.
4: **if** $Tree_{\beta} \neq \emptyset$ **then**
5: call PP-growth($Tree_{\beta}$, β);
6: Remove a_i from the $Tree$ and push the a_i's ts-list to its parent nodes.

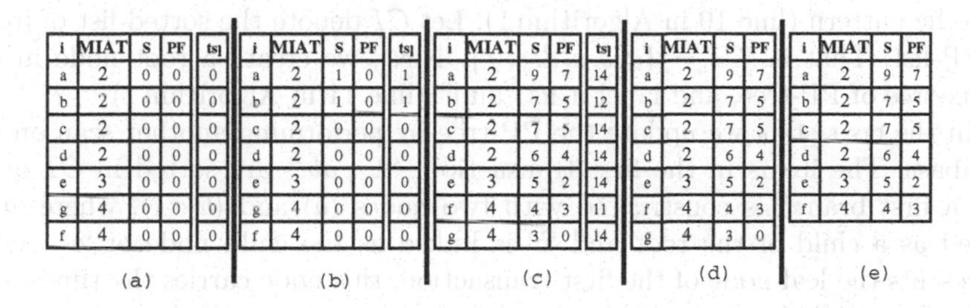

(a)

i	MIAT	S	PF	tsl
a	2	0	0	0
b	2	0	0	0
c	2	0	0	0
d	2	0	0	0
e	3	0	0	0
g	4	0	0	0
f	4	0	0	0

(b)

i	MIAT	S	PF	tsl
a	2	1	0	1
b	2	1	0	1
c	2	0	0	0
d	2	0	0	0
e	3	0	0	0
g	4	0	0	0
f	4	0	0	0

(c)

i	MIAT	S	PF	tsl
a	2	9	7	14
b	2	7	5	12
c	2	7	5	14
d	2	6	4	14
e	3	5	2	14
f	4	4	3	11
g	4	3	0	14

(d)

i	MIAT	S	PF
a	2	9	7
b	2	7	5
c	2	7	5
d	2	6	4
e	3	5	2
f	4	4	3
g	4	3	0

(e)

i	MIAT	S	PF
a	2	9	7
b	2	7	5
c	2	7	5
d	2	6	4
e	3	5	2
f	4	4	3

Fig. 1. Construction of PP-List. (a) Before scanning the database. (b) After scanning the first transaction. (c) After scanning the entire database. (d) Updated PP-list. (e) Final PP-list with sorted list of items

shown in Table 2. (**Please note that we ignore the *tid* information of transactions for brevity**).

For the construction of PP-list, we insert all items into the PP-list with their $MIAT$ values. The *support* and *periodic-frequency* of all these items are simultaneously set to 0. Figure 1(a) shows the PP-list generated before scanning the database (line 1 in Algorithm 1). The scan on the first transaction, "1:ab," updates the *support* and ts_l values of a and b to 1 and 1, respectively. Figure 1(b) shows the PP-list generated after scanning the first transaction. This process is repeated for other transactions in the database and PP-list is updated accordingly. Figure 1(c) shows the PP-list generated after scanning the entire database (lines 2 to 8 in Algorithm 1). The items in PP-list are sorted in ascending order of their $MIAT$ values. Items having a common $MIAT$ value are sorted in descending order of their *support* (to achieve memory efficiency). Figure 1(d) shows the sorted PP-list (line 9 in Algorithm 1). We calculate RPS for the item 'g,' which is the bottom-most item in the PP-list. As $RPS(g) \not\geq minRPS$, the item 'g' is pruned from the PP-list. Next, we calculate the RPS value for the item f, which is the current bottom-most item in the PP-list. As $RPS(f) \geq minRPS$, we consider f as a periodic 1-pattern and stop the process of pruning other aperiodic 1-patterns from the PP-list. Figure 1(e) shows the final PP-list after pruning some of the aperiodic 1-patterns whose supersets can never produce any

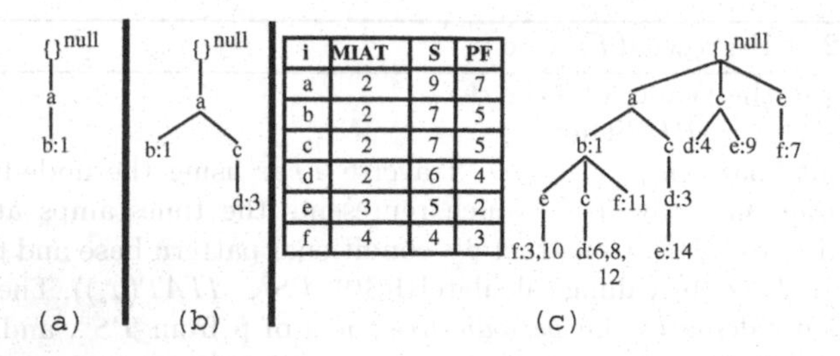

i	MIAT	S	PF
a	2	9	7
b	2	7	5
c	2	7	5
d	2	6	4
e	3	5	2
f	4	4	3

Fig. 2. Construction of PP-Tree. (a) After scanning the first transaction. (b) After scanning the second transaction. (c) After scanning the entire database

periodic pattern (line 10 in Algorithm 1). Let CI denote the sorted list of items in PP-list. That is, $CI = \{a, b, c, d, e, f\}$. Next, we create a root node in the prefix-tree of PP-tree, and label it as "null" (line 11 in Algorithm 1).

In the next step, we update the PP-tree by performing another scan on the database. The items in the first transaction, "1 : *ab*," are sorted in CI order and a first branch is constructed with two nodes $\langle a \rangle$ and $\langle b : 1 \rangle$, where '*a*' is linked as a child of the root and '*b*' is linked as the child node of '*a*'. As '*b*' represents the leaf node of the first transaction, this node carries the timestamp of 1. Figure 2(a) shows the PP-tree updated after scanning the first transaction. This process is repeated for the remaining transactions in the database and the PP-tree is updated accordingly. Figure 2(b) shows the PP-tree generated after scanning the second transaction. Figure 2(c). shows the PP-tree generated after scanning the entire database (lines 12 and 13 in Algorithm 1).

Recursive Mining of PP-Tree. The PP-tree is mined as follows. Start from length-1 pattern (as an initial suffix pattern). If the RPS value of this pattern satisfies the $minRPS$, then consider this pattern as a periodic item (or 1-pattern), construct its conditional pattern base (a sub-database, which consists of the set of prefix paths in the PP-tree with the suffix pattern), then construct its conditional PP-tree, and recursively mine that tree. Pattern-growth is achieved by concatening the suffix pattern with the periodic patterns generated from a conditional PP-tree. Next, the initial suffix pattern is pruned from the original PP-tree by moving its *ts*-lists to the corresponding parent nodes.

Algorithm 2 describes the procedure for finding periodic patterns in a PP-tree. We do not discuss this algorithm in detail as it is straightforward to understand. Mining the PP-tree is summarized in Table 3. It can be observed that conditional pattern bases have not been constructed for the item '*e*,' because it is an aperiodic 1-pattern with $RPS(e) \not\geq minRPS$. The above bottom-up mining technique is efficient, because it shrinks the search space dramatically as the mining process progresses. Some of the improvements discussed for FP-growth [10] can be straight forward extended to PP-growth. We are unable to discuss these improvements due to the page limitation.

Table 3. Mining the PP-tree by creating conditional (sub-)pattern bases

item	*support*	*MIAT*	Conditional pattern base	Conditional PP-tree	Periodic patterns
f	4	4	$\{abe : 3, 10\}$, $\{ab : 11\}$, $\{e : 7\}$	$\langle e : 3, 7, 10\rangle$	$\{ef : 0.66\}$
e	5	3	–	–	–
d	6	2	$\{abc : 6, 8, 12\}$, $\{ac : 3, 14\}$, $\{c : 4\}$	$\langle c : 3, 4, 6, 8, 12, 14\rangle$	$\{cd : 0.8\}$
c	7	2	$\{ab : 6, 8, 12\}$, $\{a : 3, 14\}$	–	–
b	7	2	$\{a : 1, 3, 6, 8, 10, 11, 12\}$	$\langle a : 1, 3, 6, 8, 10, 11, 12\rangle$	$\{ab : 0.83\}$

5 Experimental Results

Since there exists no algorithm to find periodic patterns in temporal databases, we only evaluate the proposed algorithm and show that our algorithm is memory and runtime efficient. We also show that PP-tree consumes less memory than the FP-tree for many databases. Finally, we discuss the usefulness of the proposed model by demonstrating that over 80% of the event keywords found by a supervised event detection system [1] in Twitter data can also be discovered as periodic patterns. (Similar to FP-growth [9,10], PP-growth also scales linearly with the increase of database size. Unfortunately, we are unable to present these results due to page limitation.)

The algorithms PP-growth and FP-growth are written in GNU C++ and run on a 2.66 GHz machine having 16 GB of memory. Ubuntu 14.04 is the operating system of our machine. The event detection system is written in python and java, and available for download at https://github.com/aritter/twitter_nlp. The experiments have been conducted using both synthetic (**T10I4D100K**) and real-world (**FAA-accidents** and **Twitter**) databases. The synthetic database, **T10I4D100K**, is generated by using the IBM data generator [11]. This data generator is widely used for evaluating association rule mining algorithms. The **T10I4D100K** database contains 870 items with 100,000 transactions. The **FAA-accidents** database is constructed from the accidents data recorded by the Federal Aviation Authority (FAA) from 1-January-1970 to 31-December-2014. Only categorical attributes have been taken into account while constructing the database. This database contains 9,290 items and 98,864 transactions. The Twitter database constitutes of 2,680,896 tweets collected from 10-march-2011 to 31-march-2011. These tweets are related to GEJE. We have created temporal database by considering top 4000 frequent english words.

Figure 3(a)–(c) present scatter plots about the inter-arrival times of items in the T10I4D100K, FAA-accidents and Twitter databases, respectively. The X-axis represents the items ranked in descending order of their *support* and

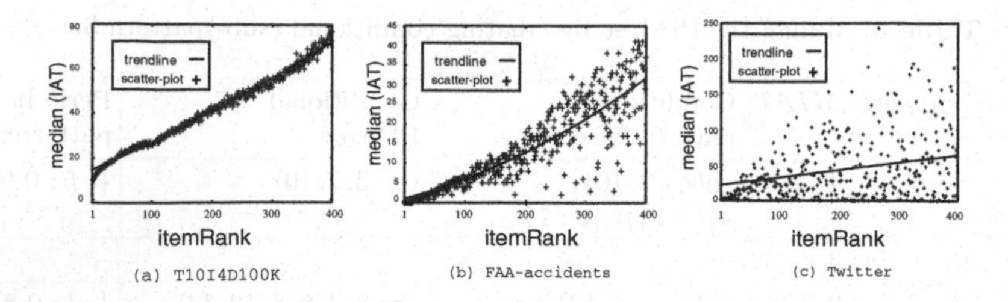

Fig. 3. The median of inter-arrival times of items in a database

Y-axis represents the median of inter-arrival times of an item in a database. The thick line in these figures denote the trend line. The equations of these trend lines and R^2 values are shown in Table 4. It can be observed from the trend lines that rare items not only have low *support*, but also have high inter-arrival times as compared against the frequent items. This experiment clearly demonstrates the importance of enabling every pattern to satisfy a different *period* and minimum number of cyclic repetitions to be a periodic pattern.

The performance of PP-growth has to be evaluated by varying the items' $MIAT$ values. Unfortunately, popular *period* identification functions (e.g. FFTs and auto-correlation) do not help us vary items' $MIAT$ values. In this context, we employ the following methodology to specify the items' $MIAT$ values. For each database, we use the equation of trend line as a reference, and specify the items' $MIAT$ values by multiplying the equation of the trend line with a constant β. That is, $MIAT(i_j) = \beta \times f(x)$, where $\beta \geq 1$ is a user-specified constant and $f(x)$ is the equation of trend line in which x denotes the rank of an item in support descending order.

Table 4. Trend line equations for various databases

Database	Equation of trend line ($f(x)$)	R^2
T10I4D100K	$y = 7.06E{-}05x^2 + 0.12x + 14.70$	0.9892
FAA-Accidents	$y = -9.67E{-}05x^2 + 0.04x + 1$	0.9122
Twitter	$y = 3.32E{-}06x^2 + 0.11x + 21.39$	0.0777

Figure 4(a)–(c) shows the number of periodic patterns generated for different $minRPS$ and β values in T10I4D100K, FAA-accidents and Twitter databases, respectively. The following two observations can be drawn from these figures: (i) Increase in β value may increase the number of periodic patterns. The reason is that higher β values tend to increase $MIAT$ values of items. (ii) Increase in $minRPS$ may decrease the number of periodic patterns. The reason is that increasing $minRPS$ increases the minimum number of cyclic repetitions necessary for a pattern to be a periodic pattern.

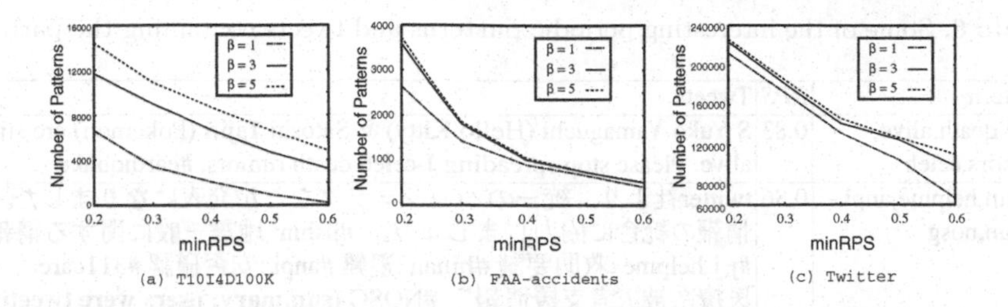

Fig. 4. The periodic patterns generated at different *minRPS* and β values

Figure 5(a)–(c) show the runtime requirements of PP-growth at different *minRPS* and β values in T10I4D100K, FAA-accidents and Twitter databases, respectively. It can be observed that varying the β and *minRPS* values has similar influence on runtime than on the generation of periodic patterns.

Table 5 lists the maximum memory usage of PP-tree and FP-tree on T10I4100K, FAA-accidents and Twitter databases, respectively. Both trees are constructed with every item in the database. It can be observed from the results that PP-tree consumes less memory than FP-tree if number of nodes in a *tree* exceed the database size, otherwise, PP-tree consumes more memory than FP-tree.

5.1 A Case Study: Evaluation of Periodic Patterns Discovered from Twitter Data

While investigating the usefulness of periodic patterns discovered from Twitter data, we have observed that many generated periodic 1-patterns (and their asso-

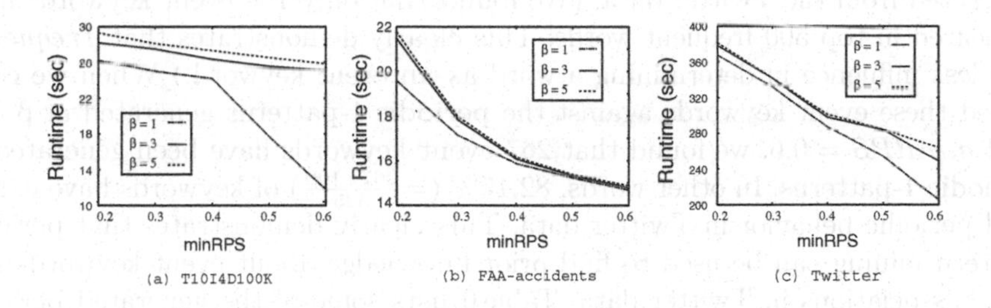

Fig. 5. Runtime requirements of PP-growth at different *minRPS* and β values

Table 5. Memory comparison of FP-tree and PP-tree

Data set	FP-tree (in MB)	PP-tree (in MB)	No. of nodes
T10I4D100K	10.906	8.561	714,739
FAA-accidents	5.898	4.801	316,935
Twitter	7.172	15.606	470,040

Table 6. Some of the interesting periodic patterns and tweets containing the patterns

Pattern	*RPS*	Tweets
still,death,alive,-rumors,celeb	0.83	S Yuko Yamaguchi (Hello Kitty) & Satoshi Tajiri (Pokemon) are still alive. Please stop spreading J-celeb death rumors. #earthquake
jishin,helpme,anpi,-hinan,nosg	0.86	twitter社より。統一のハッシュタグなどが発表になりました。情報の統合に協力しましょう。 #jishin: 地震一般に関する情報 #j_j_helpme :救助要請 #hinan :避難 #anpi :安否確認 #311care: 医療系被災者支援情報" #NOSG (**summary: users were tweeting the list of hashtags provided by Twitter for GEJE)**
tsunami,earthquake,-warning,massive,-widened	0.83	RT @bbcbreaking: #Tsunami warning is widened to incl rest of Pacific coast, incl #Australia and #South America massive #earthquake in #Japan

ciations) were interesting as they were referring to the event GEJE. This motivated us to study the following: (*i*) Do event keywords in Twitter exhibit periodic behavior? and (*ii*) If event keywords exhibit periodic behavior, then what would be their percentage? The significance of this study is that if we find many event keywords exhibiting periodic behavior, then one can use the proposed model as an unsupervised learning technique to derive some prior knowledge about event keywords and their associations in Twitter data.

Ritter et al. [1] discussed a supervised learning model to discover event keywords from tweets. We use this model for our experiment. This model annotates tweets using natural language processing techniques, generates a model from the training set of tweets and uses the model to extract event keywords from the test set of tweets. As the authors have already trained their model to identify event keywords in tweets, we have simply provided our Twitter data as the test set and extracted event keywords. A total of 325 event keywords have been extracted from the Twitter data. (We found that only 106 event keywords have appeared in top 500 frequent words. This clearly demonstrates that *frequency* has less influence in determining a word as an event keyword.) When we compared these event keywords against the periodic 1-patterns generated at $\beta = 3$ and $minRPS = 0.6$, we found that 267 event keywords have been generated as periodic 1-patterns. In other words, 82.15% ($=\frac{267 \times 100}{325}$) of keywords have exhibited periodic behavior in Twitter data. This clearly demonstrates that periodic pattern mining can be used to find prior knowledge about event keywords and their associations in Twitter data. Table 6 lists some of the generated periodic patterns and their associated tweets.

6 Conclusions and Future Work

We have proposed a model to find periodic patterns in temporal databases. It enables every pattern to satisfy different *period* and minimum number of cyclic repetitions depending on its items. A null-invariant measure, *relative periodic-support*, was discussed to determine the periodic interestingness of a pattern in a database. A pattern-growth algorithm has also been presented to find periodic patterns. Experimental results show that the proposed model can find useful information and that the algorithm is efficient.

Our study has been confined to mining periodic patterns in a static temporal database. The method developed here can be extended to incremental mining of temporal databases.

References

1. Ritter, A., Mausam, Etzioni, O., Clark, S.: Open domain event extraction from twitter. In: KDD, pp. 1104–1112 (2012)
2. Han, J., Gong, W., Yin, Y.: Mining segment-wise periodic patterns in time-related databases. In: KDD, pp. 214–218 (1998)
3. Aref, W.G., Elfeky, M.G., Elmagarmid, A.K.: Incremental, online, and merge mining of partial periodic patterns in time-series databases. IEEE TKDE **16**(3), 332–342 (2004)
4. Chen, S.-S., Huang, T.C.-K., Lin, Z.-M.: New and efficient knowledge discovery of partial periodic patterns with multiple minimum supports. J. Syst. Softw. **84**(10), 1638–1651 (2011)
5. Tan, P.-N., Kumar, V., Srivastava, J.: Selecting the right interestingness measure for association patterns. In: Knowledge Discovery and Data Mining, pp. 32–41 (2002)
6. Tanbeer, S.K., Ahmed, C.F., Jeong, B.-S., Lee, Y.-K.: Discovering periodic-frequent patterns in transactional databases. In: Theeramunkong, T., Kijsirikul, B., Cercone, N., Ho, T.-B. (eds.) PAKDD 2009. LNCS (LNAI), vol. 5476, pp. 242–253. Springer, Heidelberg (2009). doi:10.1007/978-3-642-01307-2_24
7. Uday Kiran, R., Krishna Reddy, P.: Towards efficient mining of periodic-frequent patterns in transactional databases. In: Bringas, P.G., Hameurlain, A., Quirchmayr, G. (eds.) DEXA 2010. LNCS, vol. 6262, pp. 194–208. Springer, Heidelberg (2010). doi:10.1007/978-3-642-15251-1_16
8. Venkatesh, J.N., Uday Kiran, R., Krishna Reddy, P., Kitsuregawa, M.: Discovering periodic-frequent patterns in transactional databases using all-confidence and periodic-all-confidence. In: Hartmann, S., Ma, H. (eds.) DEXA 2016. LNCS, vol. 9827, pp. 55–70. Springer, Cham (2016). doi:10.1007/978-3-319-44403-1_4
9. Pei, J., Han, J., Lakshmanan, L.V.: Pushing convertible constraints in frequent itemset mining. DMKD **8**, 227–252 (2004)
10. Han, J., Cheng, H., Xin, D., Yan, X.: Frequent pattern mining: current status and future directions. DMKD **14**(1), 55–86 (2007)
11. Agrawal, R., Imieliński, T., Swami, A.: Mining association rules between sets of items in large databases. In: SIGMOD, pp. 207–216 (1993)

Discovering Both Explicit and Implicit Similarities for Cross-Domain Recommendation

Quan Do[1(✉)], Wei Liu[1], and Fang Chen[2]

[1] Advanced Analytics Institute, University of Technology Sydney, Sydney, Australia
{Quan.Do,Wei.Liu}@uts.edu.au

[2] Analytics Research Group, Data61 Commonwealth Scientific and Industrial Research Organisation (CSIRO), Sydney, Australia
Fang.Chen@data61.csiro.au

Abstract. Recommender System has become one of the most important techniques for businesses today. Improving its performance requires a thorough understanding of latent similarities among users and items. This issue is addressable given recent abundance of datasets across domains. However, the question of how to utilize this cross-domain rich information to improve recommendation performance is still an open problem. In this paper, we propose a cross-domain recommender as the first algorithm utilizing both explicit and implicit similarities between datasets across sources for performance improvement. Validated on real-world datasets, our proposed idea outperforms the current cross-domain recommendation methods by more than 2 times. Yet, the more interesting observation is that both explicit and implicit similarities between datasets help to better suggest unknown information from cross-domain sources.

Keywords: Cross-domain learning · Recommendation system · Matrix Factorization

1 Introduction

Recommender systems have been received increasing attention and popularity from many products and services providers. Two approaches have been widely used for building recommender systems: content based [10] and collaborative filtering (CF) based [7]. A content based approach focuses on users' profile or items' information for making prediction whereas a CF method bases on latent similarities among users and items for recommending items for particular users. This paper focuses on improving CF based approaches.

As CF based methods rely on latent similarities among users and items for making recommendation, they require to have sufficient ratings to achieve a reliable result. There are two scenarios that may occur. Firstly, newly established services may take time to acquire enough ratings. Secondly, even they have enough ratings for making reliable recommendations, how can recommendation performance be improved using external data sources? Solving these two problems is our main focus.

J. Kim et al. (Eds.): PAKDD 2017, Part II, LNAI 10235, pp. 618–630, 2017.
DOI: 10.1007/978-3-319-57529-2_48

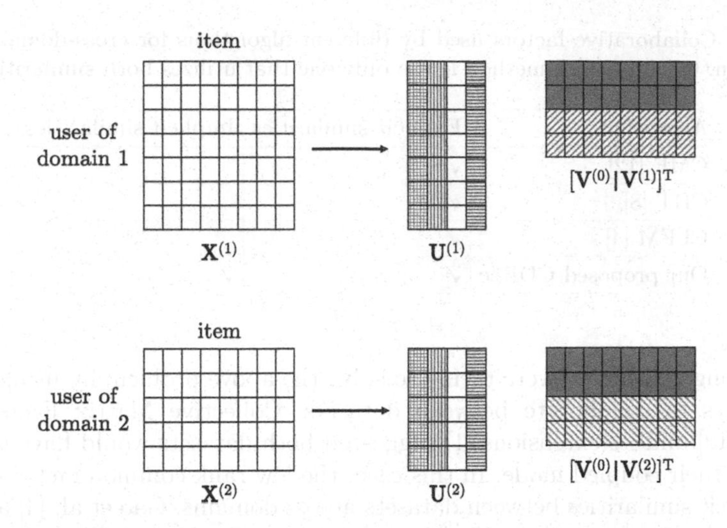

Fig. 1. Our Cross-Domain Recommender model. $\mathbf{X}^{(1)}$ matches $\mathbf{X}^{(2)}$ in their item mode. Our proposed cross-domain factorization decomposes $\mathbf{X}^{(1)}$ into $\mathbf{U}^{(1)}$, $[\mathbf{V}^{(0)}|\mathbf{V}^{(1)}]$ and $\mathbf{X}^{(2)}$ into $\mathbf{U}^{(2)}$, $[\mathbf{V}^{(0)}|\mathbf{V}^{(2)}]$ where $\mathbf{V}^{(0)}$ is a common part in coupled mode of both $\mathbf{X}^{(1)}$ and $\mathbf{X}^{(2)}$ while $\mathbf{V}^{(1)}$ and $\mathbf{V}^{(2)}$ are domain specific parts. Note that we also propose to utilize the columns of non-coupled $\mathbf{U}^{(1)}$ and $\mathbf{U}^{(2)}$ such that clusters of similar users (denoted by the same color patterns in their columns) in them are as close as possible.

The above issues are addressable given recent innovations on Internet and social media that have made many datasets publicly available [2,5,9,11]. It is therefore easy to find a correlated dataset from another domain. For example, New South Wales (NSW) state's crime statistics report can find NSW's demography closely related. These correlations across domains possess explicit similarities, e.g., NSW crime report and NSW demography contain information of the same local areas, which are conventionally used to coupled analyze them [1,12]. Nevertheless, datasets across domains also have implicit similarities, for example demography and crime behavior may have similar hidden patterns. Both similarities, if utilized properly, can provide rich insights to improve recommendation performance on the above mentioned two scenarios.

Formally, suppose we have rating matrices $\mathbf{X}^{(1)}$ and $\mathbf{X}^{(2)}$ from two different domains with only a few entries observed. Suppose $\mathbf{X}^{(1)}$ and $\mathbf{X}^{(2)}$ are explicitly coupled in one dimension, i.e., one mode of $\mathbf{X}^{(1)}$ and one mode of $\mathbf{X}^{(2)}$ are coupled. This is a reasonable assumption as many datasets across domains possess this characteristic. For instance, $\mathbf{X}^{(1)}$ contains population profiles of a country's cities and $\mathbf{X}^{(2)}$ includes those cities' crime reports; $\mathbf{X}^{(1)}$ and $\mathbf{X}^{(2)}$ are coupled in their first mode (city mode). By joint analyzing cross-domain $\mathbf{X}^{(1)}$ and $\mathbf{X}^{(2)}$, we want to learn ratings from observed $\mathbf{X}^{(1)}$ and $\mathbf{X}^{(2)}$ to predict their missing entries with high accuracy. In other words, our question is how to utilize the ratings in a domain to help predicting unknowns in another one and vice versa.

Table 1. Collaborative factors used by different algorithms for cross-domain recommendation. Our proposed method is the only one that utilizes both similarities.

Algorithm	Explicit similarities	Implicit similarities
CMF [12]	✓	
CBT [8]	✓	
CLFM [4]	✓	
Our proposed CDRec	✓	✓

Existing algorithms were trying to solve the above problem by using explicit similarities to collaborate between datasets. Collective Matrix Factorization (CMF) [12] and its extensions [1] suggested both datasets would have the same factor in their coupled mode. In this case, the low rank common factor captures the explicit similarities between datasets across domains. Gao et al. [4] and Li et al. [8] assumed cross-domain datasets would share explicit latent rating patterns. The ratings' similarities were then used to collaborate between them. Nevertheless, just only explicit is not enough to really improve the accuracy.

We propose a Cross-Domain Recommender (**CDRec**) as the first method that analyzes both explicit and implicit similarities (Table 1). One of our key hypothesis, extended from CMF where both datasets have the same factor in their coupled mode, is that two datasets across domains also possess their own specific patterns. Our idea is to find a way to combine these unique patterns into the common factor. One plausible solution is to allow the coupled factors to have both common and domain-specific parts. Figure 1 illustrates an example of common $\mathbf{V}^{(0)}$ and unique $\mathbf{V}^{(1)}$, $\mathbf{V}^{(2)}$ in coupled factors.

In addition, our another key hypothesis for implicit similarities, extended from the concept of factorization as a clustering method [3], is that non-coupled factors share common clusters. For example, in Fig. 1, even though users in $\mathbf{X}^{(1)}$ and $\mathbf{X}^{(2)}$ are different, their behaviors may be grouped by latent similarities. Because of that, we want to align common clusters in $\mathbf{U}^{(1)}$ and $\mathbf{U}^{(2)}$ to be closer. One reasonable solution is to regularize the centroids of common clusters in $\mathbf{U}^{(1)}$ and $\mathbf{U}^{(2)}$. This solution matches the fundamental concept of CF in which similar users rate similarly.

In short, our main contributions are:

(1) **Preserving common and specific parts in the coupled factor** (Subsect. 4.1): We extend existing common coupled factor by introducing specific parts (e.g., factor **V** in Fig. 1.) These common and specific parts better capture the true explicit characteristics of datasets across domains.

(2) **Aligning non-coupled factors' similarities** (Subsect. 4.2): We present a method to utilize implicit similarities. CDRec is the first factorization method utilizing both explicit and implicit similarities for cross domain recommender systems' performance improvement (Fig. 1).

(3) **Proposing an algorithm that solves optimization problem** (Subsect. 4.3): We optimize the process of utilizing both similarities following

Alternating Least Squared (ALS) (Algorithm 1). Our empirical results on real-world datasets suggest our proposed algorithm the best choice for cross-domain recommendation (Sect. 5).

2 Notations

We denote matrices by boldface capitals, e.g., $\mathbf{X}$; $\mathbf{I}$ is the identity matrix. Boldface lowercases are for vectors, i.e., $\mathbf{u}$ is a column vector and $\mathbf{u}^T$ is a row vector. A boldface capital and lowercase with indices in its subscript are used for an entry of a matrix and a vector, respectively. Transpose of $\mathbf{X}$ is denoted by $\mathbf{X}^T$.

3 Related Work

Joint analyzing cross-domain datasets has attracted huge research effort to extract more meaningful insights. Many methods were proposed for making accurate recommendations. Some popular algorithms are being discussed below.

Collective Matrix Factorization (CMF). To deal with two datasets coupled in one of their modes, Singh et al. [12] and later Acar et al. [1] assumed two datasets had a common low rank subspace in their coupled dimension. Suppose $\mathbf{X}_1$ and $\mathbf{X}_2$ are joint in their first mode, the authors modeled this CMF with a coupled loss function: $\mathcal{L} = \|\mathbf{X}_1 - \mathbf{U}\mathbf{V}_1^T\|^2 + \|\mathbf{X}_2 - \mathbf{U}\mathbf{V}_2^T\|^2$ where common $\mathbf{U}$ represents explicit similarities between two datasets.

CodeBook Transer (CBT). Targeting on improving recommendation on one domain by utilizing latent rating patterns from another domain, Li et al. [8] suggested one as a source domain $\mathbf{X}_{src}$ and the other one as a target domain $\mathbf{X}_{tgt}$. Then the $\mathbf{X}_{src}$ was decomposed into tri-factor: $\mathbf{X}_{src} \approx \mathbf{U}_{src}\mathbf{S}_{src}\mathbf{V}_{src}^T$. Rating patterns ($\mathbf{S}_{src}$) were used as the CodeBook to be transfered from $\mathbf{X}_{src}$ to $\mathbf{X}_{tgt}$. Thus, $\mathbf{X}_{tgt}$ became $\mathbf{X}_{tgt} \approx \mathbf{U}_{tgt}\mathbf{S}_{src}\mathbf{V}_{tgt}^T$. This explicit knowledge transfered from the source improved the accuracy of recommendation in the target domain.

Cluster-Level Latent Factor Model (CLFM). The assumption that two datasets from different domains have the same rating patterns is unrealistic in practice. They may share some common patterns while possess their own characteristics. This motivated Gao et al. [4] to propose CLFM for cross-domain recommendation. In specific, the authors partitioned the rating patterns across domains into common and domain-specific parts:

$$\mathbf{X}_1 \approx \mathbf{U}_1[\mathbf{S}_0|\mathbf{S}_1]\mathbf{V}_1^T$$
$$\mathbf{X}_2 \approx \mathbf{U}_2[\mathbf{S}_0|\mathbf{S}_2]\mathbf{V}_2^T$$

where $\mathbf{S}_0 \in \mathbb{R}^{R_1 * C}$ is the common patterns and $\mathbf{S}_1,\ \mathbf{S}_2 \in \mathbb{R}^{R_1*(R_2-C)}$ are domain-specific parts; C is the number of common columns.

This model allows CLFM to learn the only shared latent space $\mathbf{S}_0$, having two advantages. Firstly, as $\mathbf{S}_0$ captures the similar rating patterns across domains, it helps to overcome the sparsity of each datasets. Secondly, domain-specific $\mathbf{S}_1$ and $\mathbf{S}_2$ contain domains' discriminant characteristics. As a result, diversity of ratings in each domain is still preserved, improving recommendation performance.

4 Our Proposed Cross-Domain Recommender (CDRec)

We propose a model that utilizes both explicit and implicit similarities between datasets across domains. Without loss of generality, assume $\mathbf{X}^{(1)}$ and $\mathbf{X}^{(2)}$ are coupled in their second mode, i.e. $\mathbf{X}^{(1)}$ is a rating matrix from I users for J items and $\mathbf{X}^{(2)}$ is another rating matrix from K users for the same J items. We follow the CMF model [12] to extract low rank user factors and item ones: $\mathbf{X}^{(1)} \approx \mathbf{U}^{(1)}\mathbf{V}^{\mathrm{T}}$ and $\mathbf{X}^{(2)} \approx \mathbf{U}^{(2)}\mathbf{V}^{\mathrm{T}}$. Nevertheless, we make two key extensions:

4.1 Preserving Common and Domain-Specific Parts in V

As $\mathbf{X}^{(1)}$ and $\mathbf{X}^{(2)}$ come from different domains, it is implausible to suggest them to have the same $\mathbf{V}$. They are highly correlated in a sense that they have something in common, yet also possess their own domain-specific parts. We, therefore, propose to include both common and domain-specific parts in $\mathbf{V}$ factors in the coupled loss function:

$$\min \mathcal{L} = \left\| \mathbf{X}^{(1)} - \mathbf{U}^{(1)} \left[\mathbf{V}^{(0)} | \mathbf{V}^{(1)}\right]^{\mathrm{T}} \right\|^2 + \left\| \mathbf{X}^{(2)} - \mathbf{U}^{(2)} \left[\mathbf{V}^{(0)} | \mathbf{V}^{(2)}\right]^{\mathrm{T}} \right\|^2$$

where $\mathbf{V}^{(0)} \in \mathbb{R}^{J*C}$ is the common part and $\mathbf{V}^{(1)}, \mathbf{V}^{(2)} \in \mathbb{R}^{J*(R-C)}$ are domain-specific parts; C is the number of common columns.

Figure 1 illustrates an example of common $\mathbf{V}^{(0)}$ and unique $\mathbf{V}^{(1)}$, $\mathbf{V}^{(2)}$ in coupled factors. Common $\mathbf{V}^{(0)}$ and domain-specific $\mathbf{V}^{(1)}$ and $\mathbf{V}^{(2)}$ better capture the characteristics of datasets across domains.

4.2 Utilizing Implicit Similarities in $\mathbf{U}^{(1)}$ and $\mathbf{U}^{(2)}$

Besides explicit similarities in coupled mode as in Subsect. 4.1, cross-domain datasets also correlate in non-coupled mode. Following the concept of factorization as a clustering method [3], user groups in $\mathbf{X}^{(1)}$ and $\mathbf{X}^{(2)}$ are captured in $\mathbf{U}^{(1)}$ and $\mathbf{U}^{(2)}$. Although users in $\mathbf{X}^{(1)}$ and those in $\mathbf{X}^{(2)}$ are different, their behaviors or preferences can be grouped together. This idea is the fundamental concept of CF where users rated similarly in observed items will also rate similarly for unobserved items. As a result, we suggest common clusters among columns of $\mathbf{U}^{(1)}$ and $\mathbf{U}^{(2)}$ to be closer. One reasonable solution is to regularize the centroids of clusters across $\mathbf{U}^{(1)}$ and $\mathbf{U}^{(2)}$.

$$\min \mathcal{L} = \left\| \mathbf{X}^{(1)} - \mathbf{U}^{(1)} \left[\mathbf{V}^{(0)} | \mathbf{V}^{(1)}\right]^{\mathrm{T}} \right\|^2 + \left\| \mathbf{X}^{(2)} - \mathbf{U}^{(2)} \left[\mathbf{V}^{(0)} | \mathbf{V}^{(2)}\right]^{\mathrm{T}} \right\|^2 + \left\| \mathbf{u}_c^{(1)\mathrm{T}} - \mathbf{u}_c^{(2)\mathrm{T}} \right\|^2$$

where $\mathbf{u}_c^{(1)\mathrm{T}}$ and $\mathbf{u}_c^{(2)\mathrm{T}}$ denotes row vectors of columns' centroids in $\mathbf{U}^{(1)}$ and $\mathbf{U}^{(2)}$, respectively.

Moreover, we also employ weighted λ-regularization [13] to our model below to prevent overfitting.

$$\min \mathcal{L} = \left\| \mathbf{X}^{(1)} - \mathbf{U}^{(1)} [\mathbf{V}^{(0)} | \mathbf{V}^{(1)}]^{\mathrm{T}} \right\|^2 + \left\| \mathbf{X}^{(2)} - \mathbf{U}^{(2)} [\mathbf{V}^{(0)} | \mathbf{V}^{(2)}]^{\mathrm{T}} \right\|^2 + \left\| \mathbf{u}_c^{(1)\mathrm{T}} - \mathbf{u}_c^{(2)\mathrm{T}} \right\|^2 + \lambda\theta \tag{1}$$

where θ is the L2 regularization term such that

$$\theta = \|\mathbf{U}^{(1)}\|^2 + \|\mathbf{U}^{(2)}\|^2 + \|\mathbf{V}^{(0)}\|^2 + \|\mathbf{V}^{(1)}\|^2 + \|\mathbf{V}^{(2)}\|^2$$

4.3 Optimization

Even though (1) is a non-convex function with respect to all parameters, it is convex with respect to any of them when the others are fixed. Thus, we apply Alternating Least Square (ALS) algorithm [6] to alternately optimize the function with respect to one factor while fixing the others as in Algorithm 1. Moreover, to achieve efficiency, we perform our model optimization on each row of $\mathbf{U}$ and $\mathbf{V}$ factors instead of full matrix computation. So we can rewrite (1) as

$$\min \mathcal{L} = \sum_{\mathrm{i,j}}^{\mathrm{I,J}} \left(\mathbf{X}_{\mathrm{i,j}}^{(1)} - \mathbf{u}_{\mathrm{i}}^{(1)\mathrm{T}} \begin{bmatrix} \mathbf{v}_{\mathrm{j}}^{(0)} \\ \mathbf{v}_{\mathrm{j}}^{(1)} \end{bmatrix} \right)^2 + \sum_{\mathrm{k,j}}^{\mathrm{K,J}} \left(\mathbf{X}_{\mathrm{k,j}}^{(2)} - \mathbf{u}_{\mathrm{k}}^{(2)\mathrm{T}} \begin{bmatrix} \mathbf{v}_{\mathrm{j}}^{(0)} \\ \mathbf{v}_{\mathrm{j}}^{(2)} \end{bmatrix} \right)^2 + \|\mathbf{u}_c^{(1)\mathrm{T}} - \mathbf{u}_c^{(2)\mathrm{T}}\|^2 + \lambda\theta \tag{2}$$

Solving $\mathbf{U}^{(1)}$ and $\mathbf{U}^{(2)}$

Let $\mathbf{v}_{\mathrm{j}}^{(01)} = \begin{bmatrix} \mathbf{v}_{\mathrm{j}}^{(0)} \\ \mathbf{v}_{\mathrm{j}}^{(1)} \end{bmatrix}$ and $\mathbf{v}_{\mathrm{j}}^{(02)} = \begin{bmatrix} \mathbf{v}_{\mathrm{j}}^{(0)} \\ \mathbf{v}_{\mathrm{j}}^{(2)} \end{bmatrix}$, then (2) becomes

$$\mathcal{L} = \sum_{\mathrm{i,j}}^{\mathrm{I,J}} \left(\mathbf{X}_{\mathrm{i,j}}^{(1)} - \mathbf{u}_{\mathrm{i}}^{(1)\mathrm{T}} \mathbf{v}_{\mathrm{j}}^{(01)} \right)^2 + \sum_{\mathrm{k,j}}^{\mathrm{K,J}} \left(\mathbf{X}_{\mathrm{k,j}}^{(2)} - \mathbf{u}_{\mathrm{k}}^{(2)\mathrm{T}} \mathbf{v}_{\mathrm{j}}^{(02)} \right)^2 + \|\mathbf{u}_c^{(1)} - \mathbf{u}_c^{(2)}\|^2 + \lambda\theta$$

Optimal $\mathbf{u}_{\mathrm{i}}^{(1)\mathrm{T}}$ can be achieved by setting the derivative of $\mathcal{L}$ with respect to $\mathbf{u}_{\mathrm{i}}^{(1)\mathrm{T}}$ to zero.

$$\begin{aligned} \frac{\delta \mathcal{L}}{\delta \mathbf{u}_{\mathrm{i}}^{(1)\mathrm{T}}} &= -2 \sum_{\mathrm{j}}^{\mathrm{J}} (\mathbf{X}_{\mathrm{i,j}}^{(1)} - \mathbf{u}_{\mathrm{i}}^{(1)\mathrm{T}} \mathbf{v}_{\mathrm{j}}^{(01)}) \mathbf{v}_{\mathrm{j}}^{(01)\mathrm{T}} + 2\lambda \mathbf{u}_{\mathrm{i}}^{(1)\mathrm{T}} + 2(\mathbf{u}_{\mathrm{i}}^{(1)\mathrm{T}} - \mathbf{b}^{\mathrm{T}}) \\ &= -2\mathbf{x}_{\mathrm{i},*}^{(1)\mathrm{T}} \mathbf{V}^{(01)} + 2\mathbf{u}_{\mathrm{i}}^{(1)\mathrm{T}} \mathbf{V}^{(01)\mathrm{T}} \mathbf{V}^{(01)} + 2\lambda \mathbf{u}_{\mathrm{i}}^{(1)\mathrm{T}} + 2\mathbf{u}_{\mathrm{i}}^{(1)\mathrm{T}} - 2\mathbf{b}^{\mathrm{T}} \end{aligned}$$

where $\mathbf{b}^{\mathrm{T}} = \mathbf{u}_c^{(1)\mathrm{T}} + \mathbf{u}_c^{(2)\mathrm{T}} - \mathbf{u}_{\mathrm{i}}^{(1)\mathrm{T}}$ and $\mathbf{x}_{\mathrm{i},*}^{(1)\mathrm{T}}$ is a row vector of all observed $\mathbf{x}_{\mathrm{i,j}}^{(1)}$, $\forall \mathrm{j} \in \mathrm{J}$.

Let $\frac{\delta\mathcal{L}}{\delta\mathbf{u}_i^{(1)^T}} = 0$, we can achieve the update rule for $\mathbf{u}_i^{(1)^T}$:

$$\mathbf{u}_i^{(1)^T} = \left(\mathbf{V}^{(01)^T}\mathbf{V}^{(01)} + (\lambda+1)\mathbf{I}\right)^{-1}\left(\mathbf{x}_{i,*}^{(1)^T}\mathbf{V}^{(01)} + \mathbf{b}^T\right) \tag{3}$$

Similarly, optimal $\mathbf{u}_k^{(2)^T}$ can be derived by:

$$\mathbf{u}_k^{(2)^T} = \left(\mathbf{V}^{(02)^T}\mathbf{V}^{(02)} + (\lambda+1)\mathbf{I}\right)^{-1}\left(\mathbf{x}_{k,*}^{(2)^T}\mathbf{V}^{(02)} + \mathbf{b}^T\right) \tag{4}$$

where $\mathbf{b}^T = \mathbf{u}_c^{(1)^T} + \mathbf{u}_c^{(2)^T} - \mathbf{u}_k^{(2)^T}$ and $\mathbf{x}_{k,*}^{(2)^T}$ is a row vector of all observed $\mathbf{x}_{k,j}^{(2)}$, $\forall j \in J$. $\mathbf{I}$ is the identity matrix.

Solving common $\mathbf{V}^{(0)}$

Let $\mathbf{u}_i^{(1)^T} = [\mathbf{u}_i^{(10)}|\mathbf{u}_i^{(11)}]^T$ and $\mathbf{u}_k^{(2)^T} = [\mathbf{u}_k^{(20)}|\mathbf{u}_k^{(22)}]^T$ where $\mathbf{u}_i^{(10)^T}, \mathbf{u}_k^{(20)^T} \in \mathbb{R}^{1*C}$ and $\mathbf{u}_i^{(11)^T}, \mathbf{u}_k^{(22)^T} \in \mathbb{R}^{1*R-C}$, then (1) can be rewritten as:

$$\mathcal{L} = \sum_{i,j}^{I,J}\left(\mathbf{X}_{i,j}^{(1)} - \mathbf{u}_i^{(10)^T}\mathbf{v}_j^{(0)} - \mathbf{u}_i^{(11)^T}\mathbf{v}_j^{(1)}\right)^2 + \sum_{k,j}^{K,J}\left(\mathbf{X}_{k,j}^{(2)} - \mathbf{u}_k^{(02)^T}\mathbf{v}_j^{(0)} - \mathbf{u}_k^{(22)^T}\mathbf{v}_j^{(2)}\right)^2$$
$$+ \|\mathbf{u}_c^{(1)} - \mathbf{u}_c^{(2)}\|^2 + \lambda\theta$$

Analogy to solving $\mathbf{U}^{(1)}$ and $\mathbf{U}^{(2)}$, optimal $\mathbf{v}_j^{(0)}$ can be achieved by setting the derivative of $\mathcal{L}$ with respect to $\mathbf{v}_j^{(0)}$ to zero.

$$\frac{\delta\mathcal{L}}{\delta\mathbf{v}_j^{(0)}} = -2\sum_i^I(\mathbf{Y}_{i,j}^{(1)} - \mathbf{u}_i^{(10)^T}\mathbf{v}_j^{(0)})\mathbf{u}_i^{(10)} - 2\sum_k^K(\mathbf{Y}_{k,j}^{(2)} - \mathbf{u}_k^{(20)^T}\mathbf{v}_j^{(0)})\mathbf{u}_k^{(20)} + 2\lambda\mathbf{v}_j^{(0)}$$
$$= -2\mathbf{U}^{(1)^T}\mathbf{y}_{*,j}^{(1)} + 2\mathbf{U}^{(1)^T}\mathbf{U}^{(1)}\mathbf{v}_j^{(0)} - 2\mathbf{U}^{(2)^T}\mathbf{y}_{*,j}^{(2)} + 2\mathbf{U}^{(2)^T}\mathbf{U}^{(2)}\mathbf{v}_j^{(0)} + 2\lambda\mathbf{v}_j^{(0)}$$

The update rule for $\mathbf{v}_j^{(0)}$ can be derived as:

$$\mathbf{v}_j^{(0)} = \left(\mathbf{U}^{(1)^T}\mathbf{U}^{(1)} + \mathbf{U}^{(2)^T}\mathbf{U}^{(2)} + \lambda\mathbf{I}\right)^{-1}\left(\mathbf{U}^{(1)^T}\mathbf{y}_{*,j}^{(1)} + \mathbf{U}^{(2)^T}\mathbf{y}_{*,j}^{(2)}\right) \tag{5}$$

Solving domain-specific $\mathbf{V}^{(1)}$ and $\mathbf{V}^{(2)}$

$$\frac{\delta\mathcal{L}}{\delta\mathbf{v}_j^{(0)}} = -2\mathbf{U}^{(1)^T}\mathbf{y}_{*,j}^{(1)} + 2\mathbf{U}^{(1)^T}\mathbf{U}^{(1)}\mathbf{v}_j^{(1)} + 2\lambda\mathbf{v}_j^{(1)} = 0$$

Then the update rule for $\mathbf{v}_j^{(1)}$ can be derived as:

$$\mathbf{v}_j^{(1)} = \left(\mathbf{U}^{(1)^T}\mathbf{U}^{(1)} + \lambda\mathbf{I}\right)^{-1}\mathbf{U}^{(1)^T}\mathbf{y}_{*,j}^{(1)} \tag{6}$$

Algorithm 1. CDRec

Input : $\mathbf{X}^{(1)}, \mathbf{X}^{(2)}, \mathcal{E}$
Output: $\mathbf{U}^{(1)}, \mathbf{V}^{(0)}, \mathbf{V}^{(1)}, \mathbf{U}^{(2)}, \mathbf{V}^{(2)}$

1 Randomly initialize all factors
2 Initialize $\mathcal{L}$ by a small number
3 **repeat**
4 Pre$\mathcal{L} = \mathcal{L}$
5 Solve $\mathbf{U}^{(1)}$ while fixing all other factor by minimizing (3)
6 Solve $\mathbf{U}^{(2)}$ while fixing all other factor by minimizing (4)
7 Solve common $\mathbf{V}^{(0)}$ while fixing all other factor by minimizing (5)
8 Solve domain-specific $\mathbf{V}^{(1)}$ while fixing all other factor by minimizing (6)
9 Solve domain-specific $\mathbf{V}^{(2)}$ while fixing all other factor by minimizing (7)
10 Compute $\mathcal{L}$ following (1)
11 **until** $(\frac{\text{Pre}\mathcal{L}-\mathcal{L}}{\text{Pre}\mathcal{L}} < \mathcal{E})$

In a similar way, the update rule for $\mathbf{v}_{\mathrm{j}}^{(2)}$ can be achieved by:

$$\mathbf{v}_{\mathrm{j}}^{(2)} = \left({\mathbf{U}^{(2)}}^{\mathrm{T}}\mathbf{U}^{(2)} + \lambda\mathbf{I}\right)^{-1}{\mathbf{U}^{(2)}}^{\mathrm{T}}\mathbf{y}_{*,\mathrm{j}}^{(2)} \tag{7}$$

5 Performance Evaluation

To evaluate our proposed idea, we compare CDRec[1] with existing algorithms on how well they utilize observed ratings to recommend unknown information. This section summaries our experiments' settings and their results.

5.1 Data for the Experiments

Two publicly available datasets: census data from Australian Bureau of Statistics (ABS) on different states[2] and crime statistics from Bureau of Crime Statistics and Research (BOCSAR)[3] are used for our evaluation. Table 2 summarizes these datasets' distribution.

Australian Bureau of Statistics (ABS) publishes a comprehensive data about people and families for all Australia geographic areas. This ABS dataset has population and family profile within 154 New South Wales (NSW) state' areas, so-called "local government areas" (LGA), and within 81 Victoria (VIC) state' LGAs. We form these into a matrix $\mathbf{X}^{(1)}$ of (LGA, population and family profile)

[1] CDRec's source code is available at https://github.com/quanie/CDRec.
[2] ABS: http://www.abs.gov.au/websitedbs/censushome.nsf/home/datapacks.
[3] BOCSAR: http://www.bocsar.nsw.gov.au/Pages/bocsar_crime_stats/bocsar_crime_stats.aspx.

Table 2. Dimension and number of known entries for training, validation and testing of census data on New South Wales (NSW)($\mathbf{X}^{(1)}$) and Victoria (VIC) ($\mathbf{X}^{(2)}$) states as well as crime data of NSW ($\mathbf{X}^{(3)}$).

Characteristics	$\mathbf{X}^{(1)}$	$\mathbf{X}^{(2)}$	$\mathbf{X}^{(3)}$
Dimension	$154 \times 7{,}889$	$81 \times 7{,}889$	154×62
Training	91,069	47,900	661
Validation	4,793	2,521	34
Testing	23,965	12,605	173

of 154 by 7889 for NSW and a matrix $\mathbf{X}^{(2)}$ of 81 by 7889 for VIC. We randomly select 10% of the data in our experiment and use about its 80% for training and 20% for testing.

BOCSAR's crime data reports criminal incidents. There are 62 specific offences within 154 LGAs of New South Wales. The counting unit is the rate of criminal incidents per 100 population. We randomly select 10% of the data and include them in a matrix $\mathbf{X}^{(3)}$ of (LGA, offences) of 154 by 62. Among this 10%, we use 80% for training and the rest for testing.

5.2 Baselines and Metric for Evaluation

We compare our proposed CDRec with existing cross-domain factorization algorithms: CMF [12], CBT [8] and CLFM [4] that leverage explicit similarities. Our goal is to assess how well these algorithms suggest unknown information based on the observed cross-domain ratings. For this purpose, we compare them with a popular Root Mean Squared Error (RMSE) metric.

5.3 Experimental Setting

Two scenarios are thoroughly tested with the following settings:

Case #1. States' demographic similarities in latent sense can help to collaboratively suggest unknown information in these states

We use $\mathbf{X}^{(1)}$ and $\mathbf{X}^{(2)}$ which are from different LGAs in NSW and VIC states. Nevertheless, their demography may share some common characteristics. We would like to assess how well both explicit similarities in demography dimension and implicit ones in LGA dimension can help to collaboratively suggest unknown information in both NSW and VIC.

Case #2. LGAs' demographic similarities in latent sense can help to collaboratively suggest unknown crime information

In this case, we cross factorize NSW's demography $\mathbf{X}^{(1)}$ matrix and NSW's crime $\mathbf{X}^{(3)}$ matrix. Both $\mathbf{X}^{(1)}$ and $\mathbf{X}^{(3)}$ explicitly have the same LGAs. Nevertheless, their hidden similarities in demography and crime rate can also help to improve

Table 3. Tested RMSE on ABS NSW and ABS VIC data with different algorithms. Best results for each rank are in bold.

Dataset	Rank	CMF	CBT	CLFM	CDRec
ABS NSW	5	0.0226 ± 0.0026	0.0839 ± 0.0002	0.0838 ± 0.0002	**0.0132** ± 0.0002
	7	0.0222 ± 0.0009	0.0836 ± 0.0004	0.0842 ± 0.0006	**0.0131** ± 0.0003
	9	0.0241 ± 0.0011	0.0841 ± 0.0002	0.0848 ± 0.0009	**0.0143** ± 0.0004
	11	0.0265 ± 0.0026	0.0846 ± 0.0007	0.0841 ± 0.0007	**0.0143** ± 0.0003
	13	0.0237 ± 0.0024	0.0851 ± 0.0002	0.0850 ± 0.0005	**0.0151** ± 0.0004
	15	0.0229 ± 0.0029	0.0853 ± 0.0005	0.0847 ± 0.0005	**0.0150** ± 0.0000
ABS VIC	5	0.0364 ± 0.0031	0.0844 ± 0.0003	0.0845 ± 0.0004	**0.0266** ± 0.0030
	7	0.0428 ± 0.0020	0.0845 ± 0.0004	0.0849 ± 0.0004	**0.0239** ± 0.0025
	9	0.0476 ± 0.0040	0.0852 ± 0.0003	0.0848 ± 0.0003	**0.0221** ± 0.0019
	11	0.0501 ± 0.0029	0.0858 ± 0.0005	0.0851 ± 0.0007	**0.0242** ± 0.0015
	13	0.0489 ± 0.0032	0.0860 ± 0.0003	0.0852 ± 0.0003	**0.0227** ± 0.0015
	15	0.0514 ± 0.0041	0.0862 ± 0.0002	0.0854 ± 0.0006	**0.0215** ± 0.0005

recommendation performance. We want to assess how these explicit as well as implicit similarities can help to collaboratively suggest unknown crime rate.

Other than that, all algorithms stop when changes are less than 10^{-5} which indicates convergence. In all algorithms, rank of the decomposition is set from 5 to 19. Each algorithm was run 5 times and we report their results' mean and standard deviation in the next subsection.

5.4 Empirical Results

Table 3 shows RMSE performance of all models on ABS data for New South Wales and Victoria states. Both CBT and CLFM that assume two states' demography similarities in latent sense clearly perform the worst. The results demonstrate that explicit similarities in latent sense does not help both CBT and CLFM to improve performance. CMF applies another approach to take advantages of explicit correlations between NSW state's population and family profile and those of VIC state. As a result, CMF's assumption on the same population and family profile factor between NSW and VIC helps improve CMF's performance over CBT's and CLFM's almost 4 times in NSW data and 2 times in VIC data. Nevertheless, the prediction accuracy can be improved even more as illustrated with our proposed idea of explicit and implicit similarities discovery. Utilizing them help our proposed CDRec to achieve about 2 times higher accuracy compared with CMF.

The advantages of both explicit and implicit similarities are further confirmed in Table 4. In this case, they are applied to other cross domains: ABS NSW demography and NSW Crime. These datasets have explicit similarities in their LGA latent factors. At the same time, implicit similarities in demography profile and crime behaviors are also utilized to collaborate between datasets.

Table 4. Tested RMSE on ABS NSW demography and BOCSAR NSW crime data with different algorithms. Best results for each rank are in bold.

Dataset	Rank	CMF	CBT	CLFM	CDRec
Demography	5	0.0209 ± 0.0016	0.0840 ± 0.0001	0.0840 ± 0.0001	$\mathbf{0.0174} \pm 0.0015$
	7	0.0223 ± 0.0024	0.0840 ± 0.0002	0.0855 ± 0.0006	$\mathbf{0.0143} \pm 0.0004$
	9	0.0199 ± 0.0027	0.0838 ± 0.0002	0.0850 ± 0.0008	$\mathbf{0.0143} \pm 0.0003$
	11	0.0212 ± 0.0049	0.0839 ± 0.0001	0.0843 ± 0.0004	$\mathbf{0.0146} \pm 0.0003$
	13	0.0194 ± 0.0022	0.0837 ± 0.0001	0.0837 ± 0.0003	$\mathbf{0.0149} \pm 0.0003$
	15	0.0173 ± 0.0014	0.0835 ± 0.0001	0.0834 ± 0.0002	$\mathbf{0.0149} \pm 0.0002$
Crime	5	0.2796 ± 0.0204	0.3411 ± 0.0035	0.3422 ± 0.0071	$\mathbf{0.2697} \pm 0.0073$
	7	0.2907 ± 0.0265	0.3432 ± 0.0021	0.3912 ± 0.0188	$\mathbf{0.2716} \pm 0.0029$
	9	0.2813 ± 0.0261	0.3562 ± 0.0134	0.3722 ± 0.0249	$\mathbf{0.2648} \pm 0.0058$
	11	0.2689 ± 0.0143	0.3539 ± 0.0061	0.3712 ± 0.0199	$\mathbf{0.2618} \pm 0.0012$
	13	0.2700 ± 0.0150	0.3481 ± 0.0070	0.3500 ± 0.0135	$\mathbf{0.2623} \pm 0.0024$
	15	0.2647 ± 0.0031	0.3485 ± 0.0038	0.3580 ± 0.0099	$\mathbf{0.2625} \pm 0.0015$

Our proposed CDRec leveraging both of the similarities outperforms existing algorithms by achieving the lowest RMSEs.

We also show how CDRec works with different number of common column C parameter in Figs. 2 and 3. When there is no explicit similarities (C = 0), the accuracy of VIC (Fig. 2b) produced by our proposed method is almost the same as CMF's performance whereas that of Crime (Fig. 3a) is much worse than CMF's one. Nevertheless, as C is larger, both explicit and implicit similarities help to further improve recommendation performance of them. Specifically, CDRec achieves the best result with C = 6 in Fig. 2b and C = 9 in Fig. 3a. Moreover, it is interesting to observe in both figures that RMSE of NSW demography reduces a bit to significantly improve that of VIC and NSW crime. This confirms both explicit and implicit similarities between cross-domain datasets can be collaborative used to improve both of their recommendation performance.

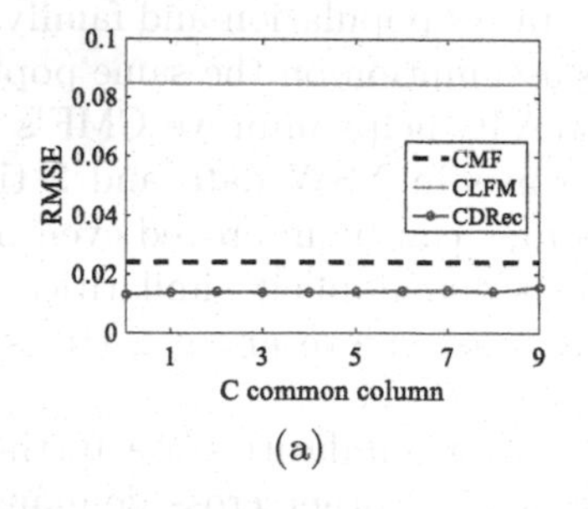

(a)

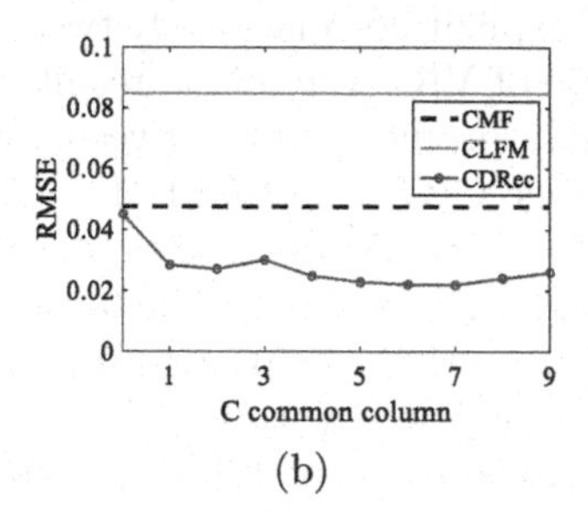

(b)

Fig. 2. Tested RMSEs under different number of common column C in CDRec. (a) Results on ABS NSW dataset; (b) Results on ABS VIC dataset.

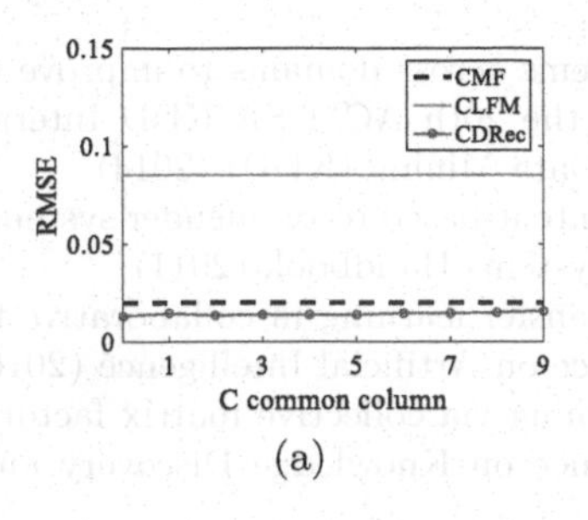

(a)

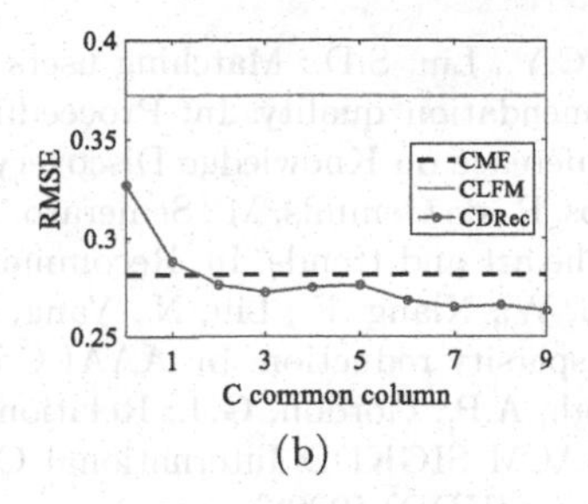

(b)

Fig. 3. Tested RMSEs under different C in CDRec. (a) Results on ABS NSW demography; (b) Results on BOCSAR NSW Crime. At C = 9, RMSE of NSW demography reduces a little to further improve the accuracy of predicting crime information.

6 Conclusion

We have discovered both explicit and implicit similarities between datasets across domains. In this paper we propose a method to preserving common and specific parts in coupled factor as well as aligning non-coupled factors' similarities. Moreover, an algorithm that solves the optimization problem of this method is also introduced. The advantages of our ideas, validated with real-world datasets, suggest combining both explicit and implicit similarities is the best way to improve cross-domain recommendation.

References

1. Acar, E., Kolda, T.G., Dunlavy, D.M.: All-at-once optimization for coupled matrix and tensor factorizations. arXiv preprint arXiv:1105.3422 (2011)
2. Chen, W., Hsu, W., Lee, M.L.: Making recommendations from multiple domains. In: Proceedings of the 19th ACM SIGKDD International Conference on Knowledge Discovery and Data Mining (KDD) (2013)
3. Ding, C., Li, T., Peng, W., Park, H.: Orthogonal nonnegative matrix tri-factorizations for clustering. In: Proceedings of the 12th ACM SIGKDD International Conference on Knowledge Discovery and Data Mining (KDD) (2006)
4. Gao, S., Luo, H., Chen, D., Li, S., Gallinari, P., Guo, J.: Cross-domain recommendation via cluster-level latent factor model. In: Blockeel, H., Kersting, K., Nijssen, S., Železný, F. (eds.) ECML PKDD 2013. LNCS (LNAI), vol. 8189, pp. 161–176. Springer, Heidelberg (2013). doi:10.1007/978-3-642-40991-2_11
5. Jiang, M., Cui, P., Chen, X., Wang, F., Zhu, W., Yang, S.: Social recommendation with cross-domain transferable knowledge. IEEE TKDE **27**, 3084–3097 (2015)
6. Kolda, T.G., Bader, B.W.: Tensor decompositions and applications. SIAM Rev. **51**, 455–500 (2009)
7. Koren, Y., Bell, R.: Advances in collaborative filtering. In: Recommender Systems Handbook (2011)
8. Li, B., Yang, Q., Xue, X.: Can movies and books collaborate? Cross-domain collaborative filtering for sparsity reduction. In: Proceedings of the 21st International Joint Conference on Artificial Intelligence (IJCAI) (2009)

9. Li, C.Y., Lin, S.D.: Matching users and items across domains to improve the recommendation quality. In: Proceedings of the 20th ACM SIGKDD International Conference on Knowledge Discovery and Data Mining (KDD) (2014)
10. Lops, P., de Gemmis, M., Semeraro, G.: Content-based recommender systems: state of the art and trends. In: Recommender Systems Handbook (2011)
11. Pan, W., Xiang, E., Liu, N., Yang, Q.: Transfer learning in collaborative filtering for sparsity reduction. In: AAAI Conference on Artificial Intelligence (2010)
12. Singh, A.P., Gordon, G.J.: Relational learning via collective matrix factorization. In: ACM SIGKDD International Conference on Knowledge Discovery and Data Mining (KDD) (2008)
13. Zhou, Y., Wilkinson, D., Schreiber, R., Pan, R.: Large-scale parallel collaborative filtering for the netflix prize. In: Fleischer, R., Xu, J. (eds.) AAIM 2008. LNCS, vol. 5034, pp. 337–348. Springer, Heidelberg (2008). doi:10.1007/978-3-540-68880-8_32

Mining Recurrent Patterns in a Dynamic Attributed Graph

Zhi Cheng, Frédéric Flouvat, and Nazha Selmaoui-Folcher(✉)

PPME, University of New Caledonia, BP R4, 98851 Nouméa, New Caledonia
{zhi.cheng,frederic.flouvat,nazha.selmaoui}@univ-nc.nc

Abstract. A great number of applications require to analyze a single attributed graph that changes over time. This task is particularly complex because both graph structure and attributes associated with each node can change. In the present work, we focus on the discovery of recurrent patterns in such a graph. These patterns are sequences of subgraphs which represent recurring evolutions of nodes w.r.t. their attributes. Various constraints have been defined and an original algorithm has been developed. Experiments performed on synthetic and real-world datasets have demonstrated the interest of our approach and its scalability.

Keywords: Dynamic attributed graph · Patterns · Recurrent evolutions

1 Introduction

Graphs are more and more playing a prominent role in modeling complex structures. A large number of graph mining algorithms have been developed [1,8]. They have been used in various application domains such as remote sensing, social networks, epidemiology and bioinformatics [4,15,17]. Recently, mining evolutions of graphs over time has received much attention [2,3,5,6,9,12,14,16]. For example, [2] mined frequent coevolving relational motifs in a dynamic labeled network, i.e. set of vertices whose relations evolve in a similar way. [3] adopted an incremental tensor analysis approach to discover transient and periodic communities in a large network (unlabeled graph). [5] introduced novel absolute-time subgraph patterns and extracted rules in time-evolving graphs (where labels did not change w.r.t. times). [6] mined subgraphs in dynamic labeled graphs by performing edge insertions and deletions over time. [12] proposed a method to mine frequent and relevant subgraph patterns from a set of labeled graph sequences. [14] developed an algorithm to discover correlated sequential subgraphs from a sequence of labeled graphs. [16] designed a method to uncover evolving patterns that are pseudo-cliques which appear slightly modified in consecutive timestamps. Most of these works focus on labeled graphs, i.e. graphs with a single attribute per node. Few methods have been proposed to mine a dynamic attributed graph, i.e. a single graph where edges, vertices and several attributes can change over time. Mining such graphs is a complex task because every vertex is associated to a set of attributes (instead of a single label).

J. Kim et al. (Eds.): PAKDD 2017, Part II, LNAI 10235, pp. 631–643, 2017.
DOI: 10.1007/978-3-319-57529-2_49

[9] extracted cohesive co-evolutions in a dynamic attributed graph. These patterns represented a set of vertices with same values for a subset of attributes and a similar neighborhood over a set of timestamps (vertices and attributes were fixed). The authors extended their work in [10] to integrate constraints on the graph topology and on the attribute values.

In this paper we introduce a more general pattern domain, called recurrent patterns, which describes frequent evolutions in a dynamic attributed graph (Sect. 2). It enables to capture vertices having similar values for periods of time (such as in [9,10]), but also to capture evolutions of values and vertices over time. These patterns represent connected subgraph sequences satisfying topological, frequency and non-redundancy constraints in the input data. We develop a novel algorithm, called *RPminer*, based on graph intersections and a progressive extension of patterns over time (Sect. 3). Experiments performed on artificial and real-world data demonstrate the scalability of the algorithm and the interest of extracted patterns (Sect. 4).

2 Notations and Definitions

2.1 Dynamic Attributed Graph

The input database is a single dynamic attributed graph $\mathcal{G} = \langle G_{t_1}, G_{t_2}, \ldots, G_{t_{max}} \rangle$ which represents the evolution of a graph over a set of time $\mathcal{T} = \{t_1, \ldots, t_{max}\}$. The set of vertices of $\mathcal{G}$ is denoted $\mathcal{V}$. Each vertex is labelled by a set of attributes A (numerical or categorical). Each attribute $a \in A$ is associated with a domain value $\mathbb{D}_a$. For each time $t \in \mathcal{T}, G_t = (V_t, E_t, \lambda_t)$ is an attributed undirected graph where $V_t \subseteq \mathcal{V}$ is the set of vertices at time t, $E_t \subseteq V_t \times V_t$ is the set of edges at time t and $\lambda_t : V_t \longrightarrow 2^{A\mathbb{D}}$ is a function that associates each vertex of V_t with a set of values $A\mathbb{D} = \bigcup_{a \in A}(a \times \mathbb{D}_a)$. In the following, we consider $\mathbb{D}_a = \{+, -, 0\}$ in order to simplify examples ($\mathcal{G}$ then represents a graph of trends). Figure 1 presents an example of dynamic attributed graph. As shown in this example, such a graph is not necessarily connected at a given time.

$G = (V, E, \lambda)$ is an attributed subgraph of a graph $G' = (V', E', \lambda')$, denoted $G \sqsubseteq G'$, iff (1) $V \subseteq V'$, (2) $E \subseteq E'$, and (3) $\forall v \in V : \lambda(v) \subseteq \lambda'(v)$. G is a connected attributed graph of G', denoted $G \sqsubseteq_{conn} G'$, iff $G \sqsubseteq G'$ and for all $u, v \in V$, there exists a path between u and v in G.

2.2 A New Pattern Domain and Its Constraints

Recurrent Evolutions of Vertices. Let (V, λ) be a subset of attributed vertices of $\mathcal{G}$ with $V \subseteq \mathcal{V}$ and $\lambda : V \rightarrow 2^{A\mathbb{D}}$. (V, λ) can be considered as an attributed graph without edges. The definition of attributed subgraph presented in the previous section can be easily extended to a set of attributed vertices. We then have $(V', \lambda') \sqsubseteq (V, \lambda)$, iff $V' \subseteq V$ and $\forall v' \in V' : \lambda'(v') \subseteq \lambda(v')$. To facilitate the reading of examples, (V, λ) can also be denoted $(v_1 : \lambda(v_1) \mid v_2 : \lambda(v_2) \mid \ldots), \forall v_1, v_2 \cdots \in V$. As shown in Fig. 1, $(1 : a_1+a_2- \mid 2 : a_1+a_2- \mid 3 : a_1-a_2- \mid 4 : a_1-a_2+ \mid 5 : a_1-a_2-)$ is a set of attributed vertices at time t_1.

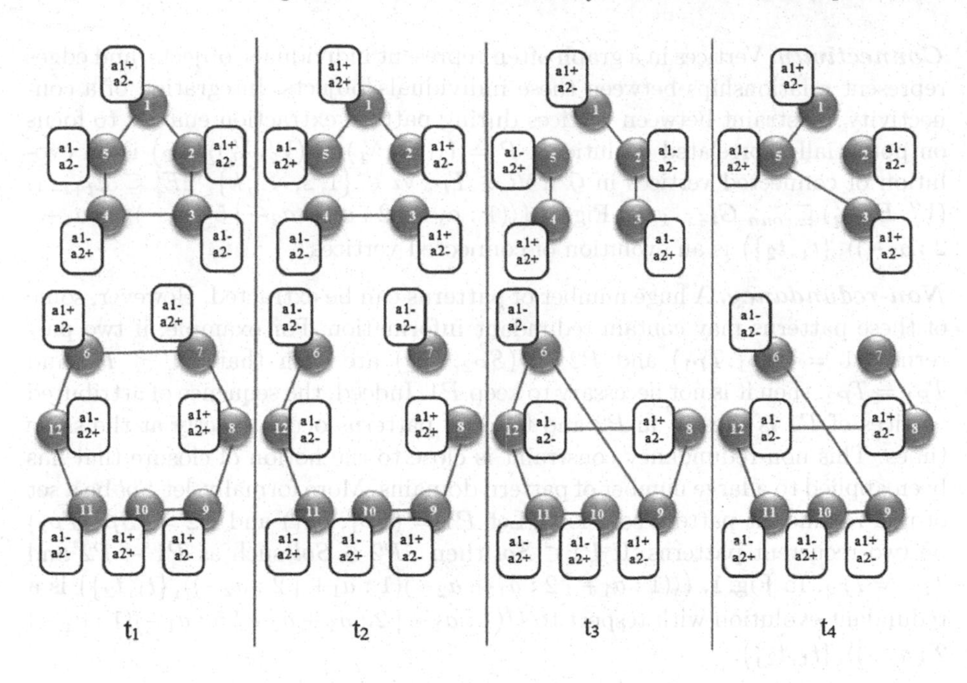

Fig. 1. An example of dynamic attributed graph $\mathcal{G}$

An evolution of a subset of vertices of $\mathcal{G}$ starting at time $t \in \mathcal{T}$ is a sequence $S = \langle (V_1', \lambda_1') \dots (V_k', \lambda_k') \rangle$, such as $\forall i \in \{1, 2, \dots, k\}$, $\exists E_i' \subseteq E_{t+i-1}$, $(V_i', E_i', \lambda_i') \sqsubseteq G_{t+i-1}$. For example, in Fig. 1, $\langle (1 : a_1 + a_2- \mid 2 : a_1 + a_2- \mid 3 : a_1 - a_2- \mid 4 : a_1 - a_2+ \mid 5 : a_1 - a_2-)\ (1 : a_1 + a_2+ \mid 2 : a_1 + a_2- \mid 5 : a_1 - a_2+) \rangle$ is an evolution starting at time t_1.

Let $T_P = \{t_{i_1}, \dots, t_{i_m}\}$ be a set of times associated with the evolution $S_P = \langle (V_1', \lambda_1') \dots (V_k', \lambda_k') \rangle$. A recurrent evolution of a subset of vertices of $\mathcal{G}$ starting at times T_P, according to the sequence S_P, is denoted $P = (S_P, T_P)$. In this case, the size of P is k. In Fig. 1, $(\langle (1 : a_1+ \mid 2 : a_1 + a_2- \mid 5 : a_1-)(1 : a_1+ \mid 2 : a_2-) \rangle, \{t_1, t_2\})$ is an example of recurrent pattern starting at times t_1 and t_2.

A relation of specialization/generalization can be defined on this pattern domain. Let $P1 = (\langle (V_1', \lambda_1') \dots (V_k', \lambda_k') \rangle, T_{P1})$ and $P2 = (\langle (V_1'', \lambda_1'') \dots (V_l'', \lambda_l'') \rangle, T_{P2})$ be two patterns representing two recurrent evolutions of $\mathcal{G}$. $P1$ is a recurrent evolution more general (resp. more specific) than $P2$, denoted $P1 \preceq P2$ (resp. $P1 \succeq P2$), if there exists $j \in \{0, \dots, l-k\}$, such as $\forall i \in \{1, \dots, k\}$, $(V_i', \lambda_i') \sqsubseteq (V_{i+j}'', \lambda_{i+j}'')$. In Fig. 1, $(\langle 1 : a_1+ \mid 2 : a_1 + a_2- \mid 5 : a_1-)(1 : a_1+ \mid 2 : a_2-) \rangle, \{t_1, t_2\})$ is a recurrent pattern more specific than $(\langle (1 : a_1+ \mid 2 : a_1 + a_2-)(1 : a_1+) \rangle, \{t_1, t_2\})$.

Interesting Measures and Constraints. We defined several measures and constraints which allow users to filter interesting patterns.

Connectivity. Vertices in a graph often represent individuals/objects, and edges represent relationships between these individuals/objects. Integration of a connectivity constraint between vertices during pattern extraction enables to focus on potentially correlated evolutions. $P = (\langle(V'_1, \lambda'_1)\dots(V'_k, \lambda'_k)\rangle, T_P)$ is an evolution of connected vertices in $\mathcal{G}$ if $\forall t \in T_P$, $\forall i \in \{1, 2, \dots, k\}$, $\exists E'_i \subseteq E_{t+i-1}$, $(V'_i, E'_i, \lambda'_i) \sqsubseteq_{conn} G_{t+i-1}$. In Fig. 1, $(\langle(1 : a_1+ \mid 2 : a_1 + a_2- \mid 5 : a_1-)(1 : a_1+ \mid 2 : a_2-)\rangle, \{t_1, t_2\})$ is an evolution of connected vertices.

Non-redundancy. A huge number of patterns can be extracted. However, some of these patterns may contain redundant information. For example, if two patterns $P1 = (S_{P1}, T_{P1})$ and $P2 = (S_{P2}, T_{P2})$ are such that $P1 \preceq P2$ and $T_{P1} = T_{P2}$, then it is not necessary to keep $P1$. Indeed, the sequence of attributed vertices of $P1$ is present in $P2$ and the two patterns occur exactly at the same times. This non-redundancy constraint is close to the notion of closure that has been applied to a large number of pattern domains. More formally, let *Sol* be a set of non-redundant pattern solutions. Let $P1 = (S_{P1}, T_{P1})$ and $P2 = (S_{P2}, T_{P2})$ be two recurrent patterns. If $P1 \in Sol$ then $\nexists P2 \in Sol$ such as $P1 \prec P2$ and $T_{P1} = T_{P2}$. In Fig. 1, $(\langle(1 : a_1+ \mid 2 : a_1 + a_2-)(1 : a_1+ \mid 2 : a_2-)\rangle, \{t_1, t_2\})$ is a redundant evolution with respect to $(\langle(1 : a_1+ \mid 2 : a_1 + a_2- \mid 5 : a_1-)(1 : a_1+ \mid 2 : a_2-)\rangle, \{t_1, t_2\})$.

Frequency. Minimum frequency is one of the most widely used constraints. It aims to filter patterns which occur more than a minimum number of times. It is commonly applied when a database is a collection of transactions. However, defining a frequency constraint is generally more challenging in a single graph context [7, 11], mainly because of the presence of embedded overlappings. Nevertheless, frequency is easy to calculate in our case because of the nature of the extracted patterns. Indeed, the frequency of a pattern is simply the number of times at which the evolution begins. It represents the number of recurrences of the evolution. Let $P = (S_P, T_P)$ be a pattern. The frequency of P is $sup(P) = |T_P|$. Consequently, P is a frequent evolution iff $sup(P) \geq minsup$, where $minsup$ is a user-defined threshold. For example, in Fig. 1, the frequency of $(\langle(6 : a_2- \mid 11 : a_1- \mid 12 : a_1-)(11 : a_1 - a_2+ \mid 12 : a_1 - a_2-)\rangle, \{t_1, t_2, t_3\})$ is 3 since it begins at t_1, t_2 and t_3.

Volume. Volume is another measure commonly applied in the context of graph mining. It is defined as the number of vertices of a graph. It can represent, for instance, the size of a community in a social network - assuming that vertices are individuals and edges are friendship relations. Let $vol(P) = min_{\forall i \in \{1\dots k\}}(|V'_i|)$ be the volume of the pattern $P = (\langle(V'_1, \lambda'_1)\dots(V'_k, \lambda'_k)\rangle, T_P)$. P is a sufficiently voluminous pattern iff $vol(P) \geq minvol$, where $minvol$ is a user-defined threshold. For example, the pattern $(\langle(1 : a_1+ \mid 2 : a_1 + a_2- \mid 5 : a_1-)(1 : a_1+ \mid 2 : a_2-)\rangle, \{t_1, t_2\})$ has a volume of 2.

Temporal Continuity. By default, an evolution may include totally different vertices at each step. In other words, if $P = (\langle(V'_1, \lambda'_1)\dots(V'_k, \lambda'_k)\rangle, T_P)$, then it is possible to have $\bigcap_{\forall i \in 1\dots k} V'_i = \emptyset$. Interpreting such evolutions can be difficult for end users because there is actually no direct relation between individuals/objects (represented by vertices). We propose a new constraint to target patterns which describe evolutions around a common core of individuals. Such a constraint allows to follow evolutions of a number of vertices over time while taking into account neighboring vertices (directly or indirectly). Let $P = (\langle (V'_1, \lambda'_1) \dots (V'_k, \lambda'_k) \rangle, T_P)$ be a pattern. Let $com(P) = |\bigcap_{\forall i \in 1 \dots k} V'_i|$ be the number of vertices occurring at all times in T_P. P is a continuous pattern over time iff $com(P) \geq mincom$, where $mincom$ is a user-defined threshold. For example, the pattern $(\langle (1 : a_1+ \mid 2 : a_1 + a_2- \mid 5 : a_1-)(1 : a_1+ \mid 2 : a_2-) \rangle, \{t_1, t_2\})$ has two common vertices at t_1 and t_2, i.e. $comp(P) = 2$.

Problem Setting. Given a dynamic attributed graph $\mathcal{G}$, the problem is to enumerate the complete set of recurrent evolutions in $\mathcal{G}$, denoted Sol, such that $\forall P \in Sol$: (1) vertices of P are connected at each time; (2) P is not redundant in Sol; (3) P is frequent (i.e. $sup(P) \geq minsup$); (4) P is sufficiently voluminous (i.e. $vol(P) \geq minvol$); and (5) P is centered around a core of vertices sufficiently large (i.e. $com(P) \geq mincom$), where $minsup$, $minvol$ and $mincom$ are user-defined thresholds.

3 Mining Recurrent Patterns Under Constraints

In this section, we present our enumeration strategy to extract recurrent patterns satisfying the constraints defined in the previous section. Unlike a number of pattern mining algorithms, our approach is not based on a generate-test strategy (where candidate patterns are generated, tested and then combined). It performs neither a breadth-first nor a depth-first search. It is not based on a projection strategy either (such as *PrefixSpan*). Instead, our method is an incremental approach based on successive intersections and extensions of connected components occurring over time. We thus get a set of solutions of different sizes at each iteration (at each time). The main advantage of this approach it to avoid generating a large number of patterns which do not satisfy the constraints. In the following subsection, we introduce the notion of intersection between attributed graphs and explain its interest w.r.t. our pattern mining problem.

3.1 Intersection of Attributed Graphs

Intersection and Frequency. Let us consider two times $i, j \in \mathcal{T}$. The intersection of two attributed graphs $G_i = (V_i, E_i, \lambda_i)$ and $G_j = (V_j, E_j, \lambda_j)$, denoted by $G_i \sqcap G_j$, is an attributed graph $G = (V, E, \lambda)$ such as $V = V_i \cap V_j, E = E_i \cap E_j$, $\forall v \in V$, $\lambda(v) = \lambda_i(v) \cap \lambda_j(v)$. The result is a subgraph composed of vertices, edges and attribute values common to the two initial graphs. We can notice that every subgraph of G occurs at least two times in $\mathcal{G}$. In Fig. 2, the subgraph $c \sqsubset G_1 \sqcap G_3$ occurs at least 2 times (at t_1 and t_3).

This definition can be generalized to the intersection of k graphs, with $k \in \{2, 3, \dots |\mathcal{T}|\}$. Let $T^k \subseteq \mathcal{T}$ be a subset of times in $\mathcal{G}$ such that $|T^k| = k$. The

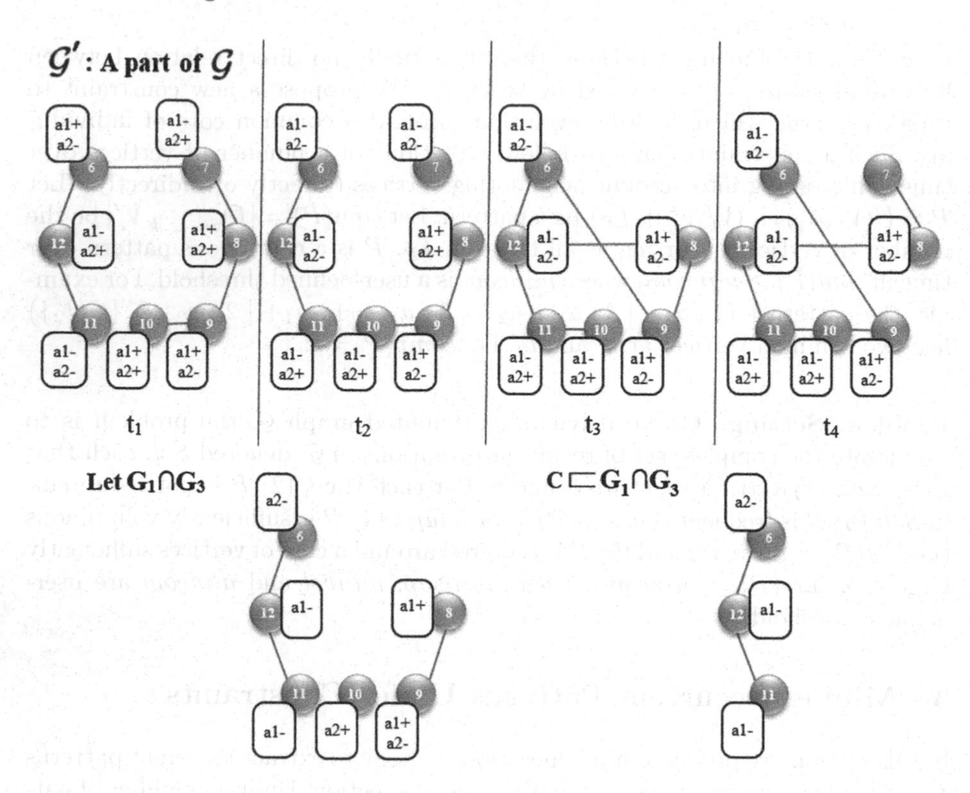

Fig. 2. Example of graph intersection

intersection of graphs in $\mathcal{G}$ at the k times of T^k, denoted by $\bigsqcap_{i \in T^k} G_i$, is a graph $G = (V, E, \lambda)$, with $V = \bigcap_{i \in T^k} V_i$, $E = \bigcap_{i \in T^k} E_i$, $\forall v \in V$, $\lambda(v) = \bigcap_{i \in T^k} \lambda_i(v)$. The minimum frequency in $\mathcal{G}$ of all subgraphs of $\bigsqcap_{i \in T^k} G_i$ is k. Consequently, all patterns constructed from the intersection of *minsup* graphs of $\mathcal{G}$ will satisfy the minimum frequency constraint.

Intersection and Non-redundancy. Intersections also have other properties. Let us study in particular connected components (i.e. maximal connected subgraphs) resulting from intersection of several graphs. We denote $\mathbb{C}_{i \sqcap j}$ the set of connected components obtained after intersection of graphs in $\mathcal{G}$ at times i and j, i.e. $G_i \sqcap G_j$. More formally, $\mathbb{C}_{i \sqcap j} = \{(V, E, \lambda) \mid (V, E, \lambda) \sqsubseteq_{conn} G_i \sqcap G_j$ and $\nexists (V', E', \lambda'), (V, E, \lambda) \sqsubset (V', E', \lambda')$ s.t. $(V', E', \lambda') \sqsubseteq_{conn} G_i \sqcap G_j\}$.

Let us consider two connected components c and c' obtained after intersection of graphs in $\mathcal{G}$ at times $\{i, j\}$ and $\{k, l\}$ respectively, i.e. $c \in \mathbb{C}_{i \sqcap j}$ and $c' \in \mathbb{C}_{k \sqcap l}, \forall i, j, k, l \in \mathcal{T}$. Let $T_c = \{t \in \mathcal{T} \mid c \sqsubseteq G_t\}$ (resp. $T_{c'}$) be the subset of times in $\mathcal{T}$ when the connected component c (resp. c') occurs. It is not possible to have $c \sqsubset c'$ and $\boldsymbol{T_c} = \boldsymbol{T_{c'}}$. Indeed it would imply that c' occurs at times $\{k, l\}$ but also at $\{i, j\}$. So, we would have $c' \sqsubseteq G_i \sqcap G_j$, which is impossible as c is a connected component of $G_i \sqcap G_j$ (thus it is maximal). In Fig. 2, the connected component $c_1 = (6 : a_2- \mid 11 : a_1- \mid 12 : a_1-)$ is in G_1, G_2 and G_3. Consequently, it appears in $G_1 \sqcap G_2$ and $G_1 \sqcap G_3$. In addition, there is no superset of vertices occurring at the same times. On the other hand, the subset $c_2 = (11 : a_1- \mid 12 : a_1-)$ can be obtained by performing $G_1 \sqcap G_4$, but it is not redundant to c_1 because it occurs at four times (t_1, t_2, t_3 and t_4). To conclude, if $c = (V, E, \lambda)$, then the pattern $(\langle(V, \lambda)\rangle, T_c)$ satisfies the connectivity constraint (since it is a connected component), but also the non-redundancy constraint (w.r.t. size-1 patterns). In other words, this pattern will be either a solution or a fragment of solution.

This property can be generalized to any set $T, T' \subseteq \mathcal{T}$. We note $\mathbb{C}_{\sqcap T}$ (resp. $\mathbb{C}_{\sqcap T'}$), the set of connected components obtained after intersection of graphs at the times T (resp. T'), i.e. $\bigsqcap_{i \in T} G_i$. If $c \in \mathbb{C}_{\sqcap T}$, then $\nexists c' \in \mathbb{C}_{\sqcap T'}$ such as $c \sqsubset c'$ and $T_c = T_{c'}$. Size-1 patterns associated with those connected components satisfy both connectivity and non-redundancy constraints. The inverse of this proposition is also true. All size-1 solutions and all size-1 pattern fragments can be derived from connected components obtained after intersecting graphs in $\mathcal{G}$. These intersections provide the 'building blocks' to construct solutions.

The interest of these intersections is to avoid performing a large number of inclusion tests during pattern enumeration (to verify the frequency and non-redundancy constraints). The number of intersections is $2^{|\mathcal{T}|}$. Thus, it depends only on the number of times in $\mathcal{G}$, whereas the number of inclusion tests depends on the number of patterns generated, which is much more important.

3.2 Generation of Size-1 Patterns

As shown in the previous section, size-1 patterns resulting from graph intersections directly satisfy frequency, connectivity and non-redundancy constraints. To extract final solutions, it is sufficient to verify volume and temporal continuity constraints. These constraints are simple and not costly to calculate as they are based on the studied pattern structure. The set of size-1 solutions and size-1 fragments of solutions can be defined as follows: $\{(\langle(V, \lambda)\rangle, T) \mid T \subseteq \mathcal{T}, |T| \geq minsup, |V| \geq minvol, \text{ and } \exists c = (V, E, \lambda) \text{ such as } c \in \mathbb{C}_{\sqcap T}\}$.

Preprocessing of graphs before intersections can reduce connectivity tests. Intersections are then not performed on initial graphs but on their connected components. In the end, firstly connected components are identified in graphs G_i, and secondly pattern extraction is performed on the intersection of these connected components.

3.3 Pattern Extension

Next, size-1 patterns extracted in the intersections can be combined, according to the times when they occur, to built the solutions. This extension can be done by processing times incrementally. Figure 3 illustrates this incremental generation starting from times t_1 and t_2. It displays the parallel extensions of a pattern which occurs at t_1 and t_2. As the frequency constraint is directly related to the number of "intersected" times, we can conclude that the minimum frequency in this example is 2. Let $\mathbb{C}_i$ and $\mathbb{C}_j$ be the sets of connected components of G_i and G_j. Let c, c' and c^* be connected components extracted in these intersections. Let us consider that there exists a solution $P = (\langle(V'_1, \lambda'_1)\dots(V'_n, \lambda'_n)\rangle, \{t_1, t_2, t_3\})$. Intersection between $\mathbb{C}_1$ and $\mathbb{C}_2$ results in a graph composed of several connected components, s.t. $c = (V'_1, E'_1, \lambda'_1)$, occurring at times t_1 and t_2. Pattern $P = (\langle(V'_1, \lambda'_1)\rangle, \{t_1, t_2\})$ can be generated based on this intersection. The first occurrence of this pattern is at time t_1, and the second one at time t_2. Candidate extensions for these occurrences can only be at t_2 and t_3 respectively (since gaps are not allowed). Now let us consider times $\{t_2, t_3\}$. Let us suppose that $c' = (V'_2, E'_2, \lambda'_2)$ is a connected component of $\mathbb{C}_2 \sqcap \mathbb{C}_3$. If c and c' share a sufficient number of vertices (temporal continuity constraint), then we can extend the pattern P to obtain $(\langle(V'_1, \lambda'_1)(V'_2, \lambda'_2)\rangle, \{t_1, t_2\})$. This process continues until no more extension can be performed. At each iteration, connected components can be used to extend patterns from the previous iteration, but they can also be "starting points" for new patterns. Thus, these successive extensions generate all solutions starting at time t_1, then all solutions starting at time t_2, etc.

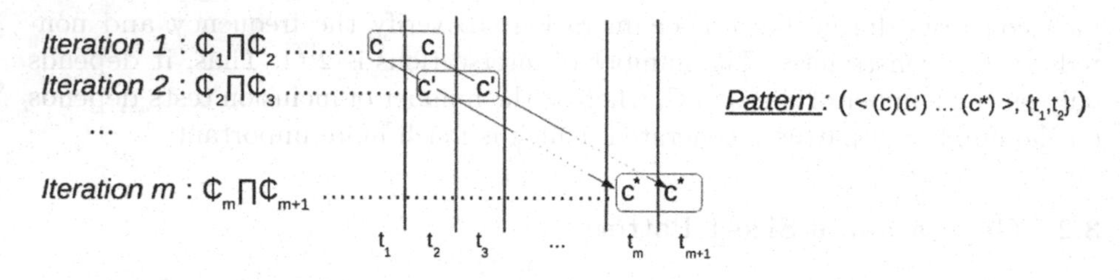

Fig. 3. Intersections and extensions in parallel of patterns from $\{t_1, t_2\}$

With this approach, the pattern P will be generated and extended four times (from $\{t_1, t_2\}$, from $\{t_1, t_3\}$, from $\{t_2, t_3\}$, and from $\{t_1, t_2, t_3\}$). For each generation, the pattern starting times are updated. Notice that even if the processing of $\{t_2, t_3\}$ and $\{t_1, t_2, t_3\}$ do not provide any new information w.r.t. P, it can lead to the generation of other patterns. All those combinations of intersections are thus necessary. That highlights the importance of our preprocessing to guarantee the scalability of this approach.

3.4 Algorithm RPMiner

Our method is detailed in Algorithm 1. Line 1 corresponds to the extraction of connected components for each graph. Lines 3–7 construct size-1 patterns starting at time t_1, and whose frequency is higher than the minimum threshold. For this purpose, the algorithm firstly calculates all time combinations containing t_1 (T_1^k, line 4), then generates size-1 patterns by performing intersections of connected components occurring at these times (line 5, method *ExtractIntersect*). After that, the other times are processed incrementally. For each time t_i,

RPMiner constructs all time combinations containing t_i (T_i^k, line 11), and extracts size-1 patterns P_i from intersections of connected components (line 12). Next, it tries to extend each pattern P generated in the previous iteration with these size-1 patterns (lines 13–14). If pattern P' resulting from the extension of P with P_i satisfies the temporal continuity constraint, it is added to the set of patterns generated at time t_i (lines 15–16). Otherwise, P is added to the set of solutions, and P_i is saved for future extensions. In the end (line 26), all solutions generated at each time are put together and associated times are updated.

Algorithm 1. *RPMiner*: mining recurrent evolutions

Require: $\mathcal{G}$: a dynamic attributed graph , $minsup, minvol, mincom$
Ensure: Sol: set of evolutions satisfying the constraints
1: $\mathbb{C} = \{\mathbb{C}_i$ set of connected components of $G_i \mid \forall c \in \mathbb{C}_i, c = (V, E, \lambda), |V| \geq minvol\}$
2: $Cand_i = \emptyset, \forall i \in \{1, 2, \ldots, |\mathcal{T}|\}$
3: **for** $k = minsup$ to $|\mathcal{T}|$ **do**
4:   for each $T_1^k \subseteq \mathcal{T}$ such as $\|T_1^k\| = k$ and $t_1 \in T_1^k$ **do**
5:     $Cand_1 = Cand_1 \cup \{P_1 \in ExtractIntersect(\mathbb{C}, T_1^k) \mid vol(P) \geq minvol\}$
6:   **end for**
7: **end for**
8: $Sol_i = \emptyset, \forall i \in \{1, 2, \ldots, |\mathcal{T}|\}$
9: **for** $i = 2$ to? $|\mathcal{T}|$ **do**
10:   **for** $k = minsup$ to $|\mathcal{T}|$ **do**
11:     for each $T_i^k \subseteq \mathcal{T}$ such as $\|T_i^k\| = k$ and $t_i \in T_i^k$ **do**
12:       for each $P_i \in ExtractIntersect(\mathbb{C}, T_i^k)$ such as $vol(P) \geq minvol$ **do**
13:         **for** each $P = (S, T_P)$ such as $P \in Cand_{i-1}$ and $T_P = T_i^k$ **do**
14:           $P' = ExtendWith(P, P_i)$
15:           **if** $com(P') \geq mincom$ **then**
16:             $Cand_i = Cand_i \cup \{P'\}$
17:           **else**
18:             $Sol_{i-1} = Sol_{i-1} \cup \{P\}$
19:             $Cand_i = Cand_i \cup \{P_i\}$
20:           **end if**
21:         **end for**
22:       **end for**
23:     **end for**
24:   **end for**
25: **end for**
26: $Sol = MergeUpdate(\bigcup_{\forall i \in \mathcal{T}} Sol_i)$

4 Experimental Results

The algorithm was implemented in C++. Experiments were performed on a PC (CPU: Intel(R) Core(T:) 3.5 GHz) with 8 GB of main memory. We used two real-world datasets and twenty eight synthetic datasets for our tests.

Synthetic Datasets. Graph sequences were randomly generated by varying different parameters such as number of vertices per timestamp, number of attributes, number of edges and sequence size.

DBLP Dataset. This dataset used in [9] represents DBLP authors (with more than 10 publications) and their co-publications between 1990 and 2010. It is composed of 2,723 vertices per timestamp (authors), 10,737 edges in average (co-publications), 43 attributes (a set of selected conferences/journals) and 9 timestamps ([1990–1994][1992–1996]...[2006–2010]).

Domestic US Flight Dataset. This dataset used in [13] represents airport traffic in the US during the Katrina hurricane period (from 01/08/2005 to 25/09/2005). It is composed of 280 vertices per timestamp (airports), 1206 edges in average (flight connections), 8 attributes (e.g. number of departures/arrivals, number of canceled flights), and 8 timestamps (data are aggregated by weeks).

Quantitative Results. Figure 4(i) presents execution times and number of solutions for twelve synthetic datasets with a growing number of vertices and edges (number of attributes is set to 50 and number of timestamps to 8). As shown by these results, our approach remains relatively scalable for large graphs (20,000 vertices and 320,000 edges at each timestamp) and low thresholds. Figure 4(ii) shows the impact of the number of timestamps on our algorithm. This impact is important but performances of our incremental approach remains comparable with the ones proposed in [9,10], while our approach extracts more general and more complex patterns. Note that larger graphs were already mined in the literature but they were only labelled. Adding several attributes per vertex and studying their joint evolution is far more complex. Figure 4(iii) reports experiment results with different number of attributes. The execution time remains almost the same when the number of attributes increase. Figure 4(iv) and (v) shows performances for the DBLP dataset w.r.t. different frequency and volume thresholds. **RPMiner** is still efficient on this real-world dataset even for low thresholds. The impact of the volume threshold is less important than the impact of the support threshold, because volumes of connected components are large (numerous co-publications). Execution times for US Flight data are not reported here due to space limitation but they are very low (120 sec. in the worst case).

Qualitative Interpretation. We have also curried out a qualitative analysis of patterns extracted in the two real-world datasets ($minvol = 2$, $minsup = 2$ and $mincom = 1$). An example of pattern extracted in the DBLP dataset is $(\langle(MasahikoTsukamoto : TKDE- \mid ShojiroNishio : TKDE-)(ShojiroNishio : DASFAA+)(MasahikoTsukamoto : SAC- \mid ShojiroNishio : DASFAA+)\rangle, \{[96-00], [98-02]\})$. It highlights a sequence of publications of Shojiro Nishio in the TKDE journal and the DASFAA conference. This is a sequence of size 3 which represents an evolution during a period over a 8 year period (i.e. 3 timestamps). This sequence is repeated twice from 1996 to 2004 (i.e. timesptamps [96–00], [98–02] and [00–04]), and from 1998 to 2006 (i.e. timesptamps [98–02], [00–04] and [02–06]). It shows that publications of this author in TKDE decreased, while they increased later in DASFAA. This pattern also shows that during the same periods one of its co-authors, Masahiko Tsukamoto, also had a decreasing number of publications in TKDE and SAC conferences. Another example of pattern is $(\langle(NingZhong : TKDE+, PAKDD-, PKDD- \mid SetsuoOhsuga : PAKDD-, PKDD-)(NingZhong :$

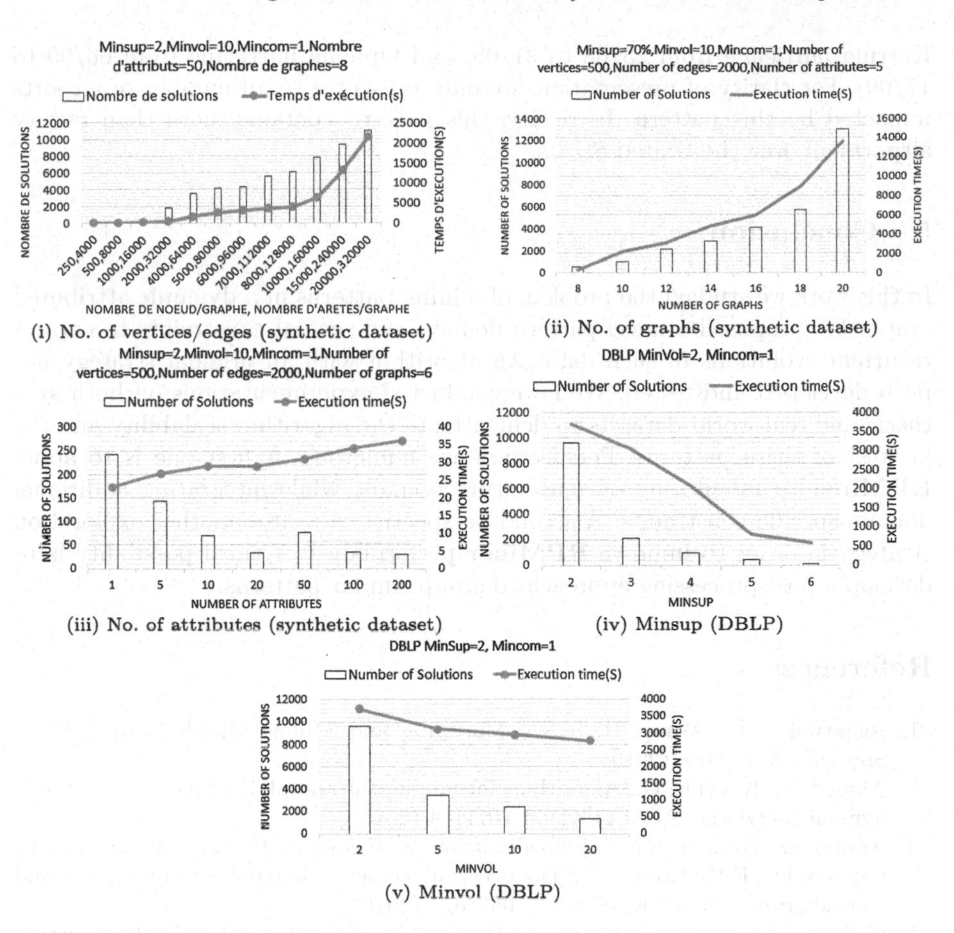

(i) No. of vertices/edges (synthetic dataset)
(ii) No. of graphs (synthetic dataset)
(iii) No. of attributes (synthetic dataset)
(iv) Minsup (DBLP)
(v) Minvol (DBLP)

Fig. 4. Impacts of parameters on execution times and number of solutions

$DMKD{+}, ICDM{-}, PAKDD{-}, JIntellInfSys{-})(NingZhong : IEEEInt Sys{+}, PAKDD{-}, KDD{+})\rangle, \{[98{-}02], [00, 04]\})$. It shows an evolution of Ning Zhong and Setsuo Ohsuga's publications, occurring at two times (from 1998 to 2006 and from 2000 to 2008).

For the US Flight dataset, extracted patterns highlight the impact of hurricanes on the US airport traffic. For example, the pattern $(\langle(Bangor : Delay Departure{+} \mid Boston : DelayArrival{+} \mid NewportNews : DelayDeparture{+}) (Augusta : Cancelled{-} \mid Bangor : Cancelled{-} \mid Boston : Cancelled - Diverted{-})\rangle, \{01/08, 08/08, 29/08, 05/09\})$ shows the impact of hurricanes on delays, cancellation and diverted flights. First, delays increased at destination and arrival airports. Then, cancellations and diverted flights decreased the following week when the hurricane became weaker. This pattern occurred at the beginning of August and then again at the end of the month, because there was three hurricanes during this period (Irene hurricane from 04/08 to 18/08,

Katrina hurricane from 23/08 to 31/08, and Ophelia hurricane from 06/09 to 17/09). For clarity of presentation, we only present a small number of airports impacted by this pattern. In reality, this pattern contains more than twenty airports all over the United States.

5 Conclusion

In this work we studied the problem of mining patterns in a dynamic attributed graph. We proposed a novel pattern domain and several constraints to extract recurrent evolutions in such data. An algorithm with an original strategy has been developed and tested. We have conducted experimentations on both synthetic and real-world datasets to demonstrate the algorithm scalability and the interest of these patterns. Perspectives are numerous. A first one is to apply **RPMiner** to monitoring of aquaculture pounds, while integrating additional domain-specific constraints. A second perspective is to use another exploration strategy in order to improve **RPMiner** performance. A third possibility is to develop a post-processing approach to group similar patterns.

References

1. Aggarwal, C.C., Wang, H. (eds.): Managing and Mining Graph Data, vol. 40. Springer, New York (2010)
2. Ahmed, R., Karypis, G.: Algorithms for mining the coevolving relational motifs in dynamic networks. ACM (TKDD) **10**(1), 4 (2015)
3. Araujo, M., Günnemann, S., Papadimitriou, S., Faloutsos, C., Basu, P., Swami, A., Papalexakis, E.E., Koutra, D.: Discovery of "comet" communities in temporal and labeled graphs Com^2. KaIS **46**(3), 657–677 (2016)
4. Berlingerio, M., Coscia, M., Giannotti, F., Monreale, A., Pedreschi, D.: Foundations of multidimensional network analysis. In: ASONAM 2011, pp. 485–489 (2011)
5. Berlingerio, M., Bonchi, F., Bringmann, B., Gionis, A.: Mining graph evolution rules. In: Buntine, W., Grobelnik, M., Mladenić, D., Shawe-Taylor, J. (eds.) ECML PKDD 2009. LNCS (LNAI), vol. 5781, pp. 115–130. Springer, Heidelberg (2009). doi:10.1007/978-3-642-04180-8_25
6. Borgwardt, K.M., Kriegel, H., Wackersreuther, P.: Pattern mining in frequent dynamic subgraphs. In: ICDM 2006, pp. 818–822 (2006)
7. Bringmann, B., Nijssen, S.: What is frequent in a single graph? In: Washio, T., Suzuki, E., Ting, K.M., Inokuchi, A. (eds.) PAKDD 2008. LNCS (LNAI), vol. 5012, pp. 858–863. Springer, Heidelberg (2008). doi:10.1007/978-3-540-68125-0_84
8. Cook, D.J., Holder, L.B.: Mining Graph Data. Wiley, Hoboken (2006)
9. Desmier, E., Plantevit, M., Robardet, C., Boulicaut, J.-F.: Cohesive co-evolution patterns in dynamic attributed graphs. In: Ganascia, J.-G., Lenca, P., Petit, J.-M. (eds.) DS 2012. LNCS (LNAI), vol. 7569, pp. 110–124. Springer, Heidelberg (2012). doi:10.1007/978-3-642-33492-4_11
10. Desmier, E., Plantevit, M., Robardet, C., Boulicaut, J.-F.: Trend mining in dynamic attributed graphs. In: Blockeel, H., Kersting, K., Nijssen, S., Železný, F. (eds.) ECML PKDD 2013. LNCS (LNAI), vol. 8188, pp. 654–669. Springer, Heidelberg (2013). doi:10.1007/978-3-642-40988-2_42

11. Fiedler, M., Borgelt, C.: Subgraph support in a single large graph. In: Workshops Proceedings of the 7th IEEE (ICDM 2007), pp. 399–404 (2007)
12. Inokuchi, A., Washio, T.: FRISSMiner: mining frequent graph sequence patterns induced by vertices. IEICE Trans. **95–D**(6), 1590–1602 (2012)
13. Kaytoue, M., Pitarch, Y., Plantevit, M., Robardet, C.: Triggering patterns of topology changes in dynamic graphs. In: ASONAM 2014, pp. 158–165 (2014)
14. Ozaki, T., Ohkawa, T.: Discovery of correlated sequential subgraphs from a sequence of graphs. In: Huang, R., Yang, Q., Pei, J., Gama, J., Meng, X., Li, X. (eds.) ADMA 2009. LNCS (LNAI), vol. 5678, pp. 265–276. Springer, Heidelberg (2009). doi:10.1007/978-3-642-03348-3_27
15. Prakash, B.A., Vreeken, J., Faloutsos, C.: Efficiently spotting the starting points of an epidemic in a large graph. KaIS **38**(1), 35–59 (2014)
16. Robardet, C.: Constraint-based pattern mining in dynamic graphs. In: IEEE ICDM, pp. 950–955 (2009)
17. Sanhes, J., Flouvat, F., Pasquier, C., Selmaoui-Folcher, N., Boulicaut, J.: Weighted path as a condensed pattern in a single attributed DAG. In: IJCAI 2013, pp. 1642–1648 (2013)

SS-FIM: Single Scan for Frequent Itemsets Mining in Transactional Databases

Youcef Djenouri[1], Marco Comuzzi[1](✉), and Djamel Djenouri[2]

[1] Ulsan National Institute of Science and Technology, Ulsan, Republic of Korea
{ydjenouri,mcomuzzi}@unist.ac.kr
[2] DTISI, CERIST Center Research, Algiers, Algeria
ddjenouri@acm.org

Abstract. The quest for frequent itemsets in a transactional database is explored in this paper, for the purpose of extracting hidden patterns from the database. Two major limitations of the Apriori algorithm are tackled, (i) the scan of the entire database at each pass to calculate the support of all generated itemsets, and (ii) its high sensitivity to variations of the minimum support threshold defined by the user. To deal with these limitations, a novel approach is proposed in this paper. The proposed approach, called Single Scan Frequent Itemsets Mining (SS-FIM), requires a single scan of the transactional database to extract the frequent itemsets. It has a unique feature to allow the generation of a fixed number of candidate itemsets, independently from the minimum support threshold, which intuitively allows to reduce the cost in terms of runtime for large databases. SS-FIM is compared with Apriori using several standard databases. The results confirm the scalability of SS-FIM and clearly show its superiority compared to Apriori for medium and large databases.

Keywords: Frequent itemsets mining · Apriori heuristic · Support computing

1 Introduction

Frequent Itemsets Mining (FIM) aims to extract highly correlated items from a large transactional database. It is defined as follows: Let T be a set of m transactions, $\{T_1, T_2, \ldots, T_m\}$ a transactional database, and I a set of n different items or attributes $\{I_1, I_2, \ldots, I_n\}$. An itemset X is a subset of the set of items $(X \subseteq I)$. The support of X is the number of transactions that contains X divided by the number of all transactions in T. The itemset X is called frequent if its support is no less than a user's predefined minimum support threshold [1].

Two categories of approaches have been proposed for solving the FIM problem. Approaches in the first category are based on the Apriori heuristic [1]. They first generate the k-sized candidate itemsets from the $(k-1)$-sized frequent itemsets and then test the frequency of the generated candidate itemsets.

J. Kim et al. (Eds.): PAKDD 2017, Part II, LNAI 10235, pp. 644–654, 2017.
DOI: 10.1007/978-3-319-57529-2_50

Approaches in the second category are based on the FPgrowth heuristic [2]. They compress the transactional database in the main memory using an efficient tree structure, then they apply recursively the mining process to find the frequent itemsets. Although this second heuristic reduces the number of database scanning as compared to Apriori, these approaches consume a high amount of memory, particularly when dealing with large database instances.

We propose in this paper a different approach called SS-FIM (Single Scan Frequent Itemsets Mining), which solves the FIM problem with only one scan of the database T. In SS-FIM, candidates itemsets are first generated from each transaction and stored in a hash table to maintain information about their support. When generating from a new transaction an itemset that already exists in the hash table, then its entry counter is simply incremented. Otherwise, if the itemset does not exist, then a new entry is created with the counter initiated to one. In the end, the frequencies of itemsets' occurrences in the hash table are compared to the minimum support to determine which itemsets to retain (considered as frequent). The proposed approach has been tested on several well known FIM instances. The results show that SS-FIM outperforms the Apriori heuristic for medium size and large size databases. They also show the scalability of SS-FIM compared to the Apriori heuristic when varying the minimum support.

The remainder of the paper is organized as follows. Section 2 reviews existing FIM algorithms. In Sect. 3, the Apriori heuristic is presented in detail, followed by the proposed SS-FIM approach in Sect. 4. The performance evaluation is presented in Sect. 5, while Sect. 6 draws the conclusions.

2 Related Work

Deterministic optimal strategies for solving the FIM problem can be divided into two categories. The first one is the *generate and test* strategy, where the itemsets are first generated and then their frequency is tested. The second one is the *divide and conquer* strategy. Solutions based on this strategy compress the database in an efficient tree structure and then apply recursively the mining process to extract the frequent itemsets. In the following, we discuss more in detail the existing FIM approaches of both categories.

The first algorithm we cite within the *generate and test* category is Apriori, by Agrawal et al. [1]. In this reference algorithm, candidate itemsets are generated incrementally and recursively. To generate candidates of k-sized itemsets, the algorithm calculates and combine the frequent $(k-1)$-sized itemsets. This process is repeated until an empty candidate itemsets is obtained in an iteration. Many FIM algorithms are based on Apriori. The Dynamic Itemsets Counting (DIC) algorithm has been proposed by Brin et al. [3] as a generalization of Apriori where the database is split into P equally sized partitions such that each of them fits in memory. DIC then gathers support of single items for the first partition. Locally found frequent items are used to generate candidate 2-sized itemsets. Then, the second partition is read to find support of all current candidates. This process is repeated for the remaining partitions. DIC terminates if no new candidates are generated from the current partition and all previous candidate have been counted. Mueller [4] has proposed a sequential FIM algorithm that is similar to Apriori, except that it stores candidates in a prefix tree instead of a hash tree. This structure enables fast testing of whether subsets of prospective candidates are frequent or not. However, both candidates and frequent itemsets are stored in the same structure, which degrades the performance of the algorithm in terms of memory footprint. Zaki et al. [5] have proposed the Eclat algorithm, which uses vertical tidlists of itemsets. Frequent k-sized itemsets are organized into disjoint equivalence classes by common $(k-1)$-sized prefixes, so that candidate $(k+1)$-sized itemsets can be generated by joining pairs of frequent k-sized itemsets from the same classes. The support of a candidate itemsets can then be computed simply by intersecting the tid-lists of the two component subsets. In [6], a data structure is proposed to store and compress the transactions in an efficient tidlist. With this structure, the number of scans of the transactional database is reduced. However, only regular frequent itemsets can be extracted.

For the divide and conquer strategy, we start with the FPgrowth algorithm [2], which uses a compressed FP-tree structure for mining a complete set of frequent itemsets without candidate itemsets generation. The algorithm is divided into two phases: (i) construct a FP-tree that encodes the dataset by reading the database and mapping each transaction onto a path in the FP-tree, while simultaneously counting the support of each item, and (ii) extract frequent itemsets directly from the FP-tree using a bottom-up strategy to find all possible frequent itemsets that end with a particular item. Cerf et al. [8] have proposed the NFP-growth algorithm. It improves the original FP-growth by constructing an independent head table, which allows creating a frequent pattern tree only once. This dramatically increases the processing speed. In [7], the authors proposed a new FPGrowth algorithm for mining uncertain data. They develop a tree structure to store uncertain data, in which the occurrence count of a node is at least the sum of occurrence counts of all its children nodes. This allows to count rapidly the support of each candidate itemset. In [9], an FP-array technique that reduces the need to traverse FP-trees is proposed. This structure is adopted to mine several types of frequent itemsets, such as maximal, closed and categorical frequent itemsets. A more detailed survey of most existing FIM algorithms can be found in [10].

The generate and test strategy requires multiple scanning of the database to generate all frequent itemsets, whereas the divide and conquer requires only two scans of the database. Divide and conquer approaches, however, are highly memory consuming because of the need to compress the database into a tree structure. Nowadays, transactional databases are very large and possibly extends to several million transactions [11]. Storing these transactions into an efficient tree structure is a very challenging problem. This makes the divide and conquer approaches inefficient for large transactional databases. Recently, some bio-inspired approaches have been proposed to reduce the number of scans of the transactional database. Among these, we cite BSO-ARM [12], PeARM [13] and

PGARM [14], to quote just a few. These approaches deal with FIM in reasonable time. However, the quality of their mining is limited, i.e., they discover only a part of frequent itemsets, and miss many.

3 Apriori Heuristic

The goal of the Apriori heuristic is to reduce the search space of frequent itemsets by exploring recursively the candidate itemsets. In the Apriori heuristic, an itemset of size k is frequent iff all its subsets are frequent. Thus, at each iteration k, the candidates itemsets of size k are generated by joining two frequent itemsets of size $(k-1)$. This process is repeated until the set representing the candidate itemsets of size k is empty. To determine the frequent itemsets at each iteration from the candidates, the support of every candidate itemset is computed. If it is greater than the minimum support threshold, then it is added to the set of frequent itemsets.

The support of each itemset, t, is calculated as the ratio between the number of transactions that contain t, and the total number of transactions in T, i.e., the frequency of a transaction t in the database T. To compute the support of t, the entire transactional database T is scanned, such that t is verified against each transaction T_i. If t belongs to T_i, then the numerator of the frequency ratio is incremented by one.

Let us consider the example of a transactional database with 5 transactions $\{T_1, T_2, T_3, T_4, T_5\}$ and 5 items $\{a, b, c, d, e\}$, as illustrated in Table 1.

Table 1. Illustrative example of a transactional database

TID	T_1	T_2	T_3	T_4	T_5
Items	a, b	b, c, d	a, b, c	e	c, d, e

Figure 1 illustrates the results of the Apriori algorithm when applied with minimum support σ_{sup} set to 0.4. The transactional database is first scanned to calculate the support of each candidate itemset of size 1 (candidate itemsets containing only one item). The frequent itemsets of size 1 are then extracted. In this example, all candidates itemsets are frequent because their supports exceeds 0.4. In the second iteration, the candidate itemsets of size 2 are extracted by joining the frequent itemsets of size 1. The support of each candidate itemsets of size 2 is computed and then the frequent itemsets of size 2 are extracted, i.e. $\{ab, bc, cd\}$. The itemsets $\{abc, abd, bcd\}$ are candidates for the size 3, but as their support is less than 0.4, they are not considered, and the process terminates. The set of frequent itemsets with minimum support greater than 40% is the union of the frequent itemsets of size 1 and 2, that is, $\{a, b, c, d, e, ab, bc, cd\}$.

The Apriori algorithm has two limitations:

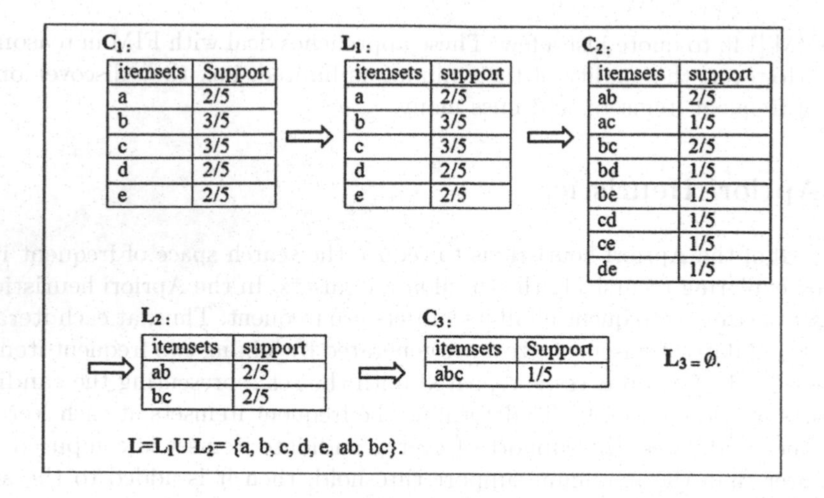

C1:

itemsets	Support
a	2/5
b	3/5
c	3/5
d	2/5
e	2/5

L1:

itemsets	support
a	2/5
b	3/5
c	3/5
d	2/5
e	2/5

C2:

itemsets	support
ab	2/5
ac	1/5
bc	2/5
bd	1/5
be	1/5
cd	1/5
ce	1/5
de	1/5

L2:

itemsets	support
ab	2/5
bc	2/5

C3:

itemsets	Support
abc	1/5

$L_3 = \emptyset$.

$L = L_1 \cup L_2 = \{a, b, c, d, e, ab, bc\}$.

Fig. 1. Apriori heuristic illustration

1. Multiple scanning of the transactional database is required: To compute the support of candidate itemsets, all existing approaches based on the Apriori heuristic scan the entire transactional database. Thus, the number of database scans is proportional to the number of generated candidate itemsets, which tends to be high for large databases.
2. Setting the minimum support user's threshold is challenging: Apriori heuristic is very sensitive when varying the minimum support. When low minimum support is chosen, a high number of candidate itemsets is obtained, which worsens the runtime of the algorithm, as each candidate itemset requires a scan of the entire database.

4 Single Scan Frequent Itemset Mining (SS-FIM)

This section presents our proposed algorithm, i.e., Single Scan Frequent Itemset Mining (SS-FIM). The algorithm description is followed by a theoretical analysis of SS-FIM in comparison to the Apriori heuristic.

4.1 SS-FIM Algorithm Description

The aim of SS-FIM is to minimize the number of database scans and the number of generated candidates while discovering frequent itemsets. This to overcome the limitations of the Apriori heuristic. The main idea of SS-FIM is to generate all possible itemsets for each transaction. If a generated itemset t has already been created when processing a previous transaction, then its support is incremented by one. Otherwise, its support is created and initialized to one. The process is repeated until all the transactions in the database have been processed.

SS-FIM allows to find all frequent itemsets by performing a single scan of the transactional database. SS-FIM is also *complete*, because the frequent itemsets are extracted directly from the transactional database and, a given itemset is frequent iff it is found ($\sigma_{sup} \times m$) times in the transactional database. Consequently, no information is lost in the itemset generation process. Algorithm 1 describes SS-FIM in detail.

Algorithm 1. SS-FIM Algorithm

```
1: Input: T: Transactional database. σ_sup: user's minimum support threshold.
2: Output :F: The set of frequent Itemsets.
3: for each Transaction T_i do
4:    S ← GenerateAllItemsets(T_i).
5:    for each itemset t ∈ S do
6:       if t ∈ h then
7:          h(t) ← h(t)+1.
8:       else
9:          h(t) ← 1.
10:      end if
11:   end for
12: end for
13: F ← ∅.
14: for itemset t ∈ h do
15:   if h(t) ≥ σ_sup then
16:      F ← F ∪ t.
17:   end if
18: end for
19: return F
```

SS-FIM has as input the transactional database, T, and the minimum support value, σ_{sup}. It also uses an internal data structure represented by a hash table h to store all generated itemsets with their partial number of occurrences. The algorithm returns the set of all frequent itemsets, F.

First, the set of itemsets, S, is computed from each transaction in T. For instance, if the transaction T_i contains the items a, b, and c, then S contains the itemsets a, b, c, ab, ac, bc, and abc. Afterwards, each itemset, $t \in S$, is stored in the hash table h. If t already exists as a key in h, then the entry with key t in h, i.e., $h(t)$ is increased by one. Otherwise, a new entry with key, t, is created in h and initialized to one. Finally, each entry, $t \in h$ with support exceeding the minimum support σ_{sup} is added to the set of the frequent itemsets F.

4.2 Illustration

Figure 2 shows the SS-FIM algorithm execution using the example of Table 1 with σ_{sup} set to 0.4. SS-FIM starts by scanning the first transaction $\{a, b\}$ and extracting from it all possible candidates itemsets, i.e., $\{a, b, ab\}$. The hash table, h, is empty at this stage, so for each candidate itemset, an entry in h is created and initialized to one. For the second transaction $\{b, c, d\}$, SS-FIM determines all possible candidate itemsets, i.e., $\{b, c, d, bc, bd, cd, bcd\}$. The itemset $\{b\}$ already exists in h, hence its entry is increased by one. As the remaining candidate itemsets are not in h, their entries are created and initialized to one. The same process is repeated for all remaining transactions $\{T_3, T_4, T_5\}$. In the end, the itemsets in h with supports no less 0.4 are selected. The returned set of frequent itemsets in this example is $\{a, b, c, d, e, ab, bc, cd\}$, the same result as of the Apriori heuristic.

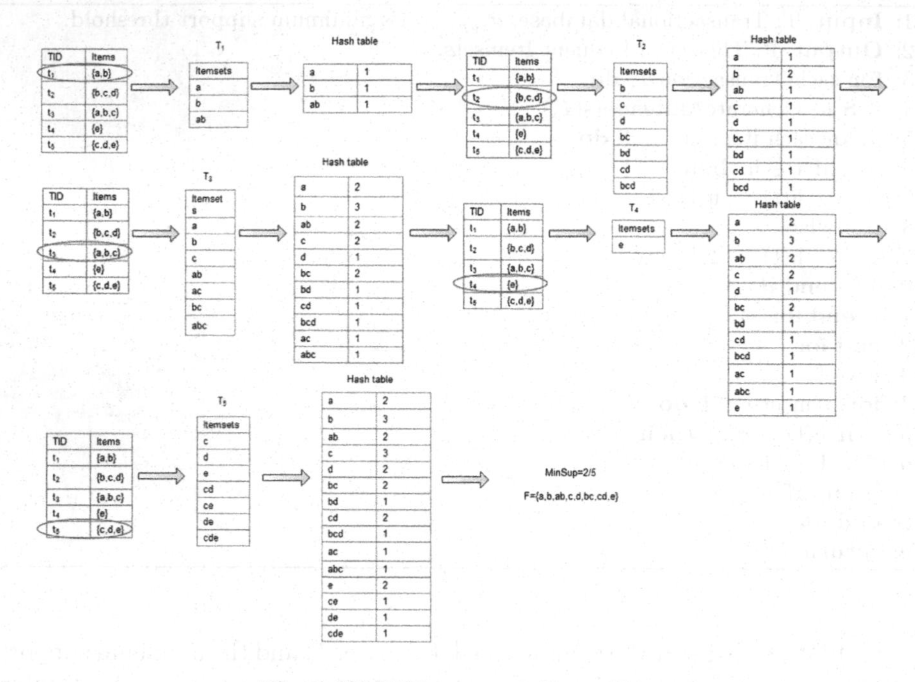

Fig. 2. SS-FIM approach illustration

4.3 Theoretical Analysis

The runtime cost of SS-FIM is the sum of (i) the cost of generating itemsets, and (ii) the cost of determining the frequent itemsets. Regarding the former, the number of all candidates generated from a transaction T_i is $2^{|T_i|} - 1$, where $|T_i|$ represents the number of items of T_i. The total number of generated candidate itemsets is thus $\sum_{i=1}^{m} 2^{|T_i|} - 1$, where m is the number of transactions in the database T. If p is the maximum number of items generated per transaction, then the number of candidates itemsets is at most $m(2^p - 1)$. The complexity of the operations needed for the generation of itemsets is then $O(m(2^p - 1))$.

For determining the frequent itemsets, the hash table has to be scanned for each candidate itemset, to evaluate its frequency against σ_{sup}. This operation is $O(m(2^p - 1))$.

Consequently, the runtime cost of SS-FIM is:

$$O(2m \times (2^p - 1)) = O(m2^p). \tag{1}$$

According to the theoretical study of Hegland [15], the complexity of Apriori algorithm is:

$$O(m \times n^2), \quad (2)$$

where n is the number of items in the database.

Although Eq. 1 has exponential form, while Eq. 2 has polynomial form, Eq. 1 generally yields lower values compared to Eq. 2 for most existing transactional databases. In fact, Eq. 1 is exponential with respect to the parameter p, that is, the maximum number of itemsets generated from a transactions, not the problem size, i.e., the number of transactions in the database. In practice, the value of p is usually much lower than the number of items in the database n. For instance, for the well known case of supermarket basket analysis, the number of products sold by a supermarket can be several thousands whereas the average number of products bought by each client hardly exceeds a few dozens.

Table 2. Theoretical runtime complexity comparison of SS-FIM and Apriori using standard database.

Data set type	Data set name	m	n	p	SS-FIM cost/m	Apriori cost/m
Small	Bolts	40	8	8	510	64
Small	Sleep	56	8	8	510	64
Small	Pollution	60	16	16	131070	256
Small	Basket ball	96	5	5	62	25
Small	Quake	2178	4	4	30	16
Average	BMS-WebView-1	59602	497	2.5	14	247009
Average	BMS-WebView-2	77512	3340	5	62	11155600
Average	retail	88162	16469	10	2046	271227961
Average	Connect	100000	999	10	2046	998001
Large	BMP POS	515597	1657	2.5	14	2745649

Table 2 presents a comparison between SS-FIM and Apriori using the standard FIM datasets described in [16]. The columns "SS-FIM cost" and "Apriori-cost", in particular, show an estimate of the number of CPU operations required based on the theoretical study of the two algorithms presented in this section. The table reveals that for small instances, the Apriori algorithm gives better results compared to SS-FIM in terms of number of CPU operations. However, for medium and large instances, SS-FIM clearly outperforms Apriori. These results are confirmed by the experimental study presented in the next section. To conclude, SS-FIM is more scalable than the Apriori and has lower computation cost for databases with medium and large number of items n.

5 Experimental Results

To evaluate the SS-FIM algorithm, several experiments have been carried out using three types of well known database instances [16]. The first one is a collection of 5 small instances with number of transactions ranging between 40 and 2178, number of items ranging between 8 and 16 items, and the average size of transactions between 8 and 16.

The second instance is a collection of 4 medium-sized database instances, with number of transactions ranging between 59000 and 100000 transactions, the number of items between 500 and 16000 items, and the average size of transactions between 2 and 10 items.

The third type of instance is a large-sized database instance, named *BMP-POS*, which contains more than 500000 transactions and more than 1600 items, with average size of transaction equal to 2.5.

All algorithms in the experiments have been implemented in C++ and experiments run on a desktop machine equipped with Intel *I3* processor and 4 GB memory.

Table 3. Runtime (Sec) of SS-FIM and Apriori using standard database.

Data set Name	SS-FIM	Apriori
Bolts	140	4
Sleep	110	6
Pollution	821	20
Basket ball	18	15
Quake	50	29
BMS-WebView-1	45	1002
BMS-WebView-2	80	3985
retail	525	4895
Connect	1285	2600
BMP POS	500	9825

Table 3 presents the runtime performance of the Apriori heuristic and SS-FIM using the standard FIM datasets described above. This table shows that, for small instances, the Apriori algorithm outperforms SS-FIM. However, for medium and large instances, SS-FIM clearly outperforms Apriori. This result confirms that our approach is better than Apriori when dealing with non dense and large transactional database. Apriori, however, outperforms our approach when dealing with dense but small transactional database.

The second experiment focuses on the sensitivity of both approaches to variations of the minimum support. Figure 3 shows the runtime performance of the

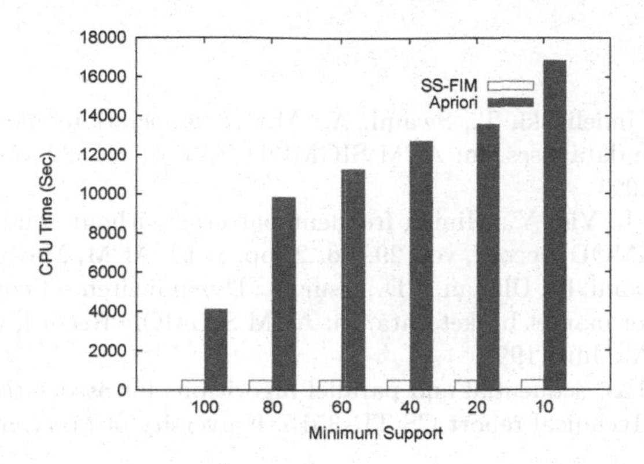

Fig. 3. Runtime (Sec) of SS-FIM and Apriori approaches for different minimum support (%) using the BMP-POS instance.

Apriori and SS-FIM approaches using the BMP-POS instance with variable minimum support. By varying the minimum support (from 100% to 10%), the execution time of the Apriori algorithm highly increases, while the one of SS-FIM remains stable.

These results confirm that SS-FIM is not sensitive to variations of the minimum support threshold. This can be explained by considering that SS-FIM is a transaction-based approach, in which the number of generated candidates itemsets is fixed no matter the support used in the input. Conversely, the Apriori heuristic is an item-based approach, in which the number of generated candidates increases when the minimum support is reduced.

6 Conclusions

This paper has proposed SS-FIM, a new intelligent frequent itemsets mining algorithm. SS-FIM extracts frequent itemsets with only one scanning of the database. Candidate itemsets are first generated from each transaction and a hash table is used to keep track of the partial frequency of occurrence of candidate itemsets while processing transactions.

Both the theoretical and the experimental evaluation reveal that SS-FIM outperforms the Apriori heuristic for large and non dense database instances. The scalability of SS-FIM also has been proven when varying the minimum support constraint.

Motivated by the promising results shown in this paper, we plan to extend SS-FIM for solving domain specific big data related problems, such as in the fields of business intelligence, e.g., process mining based on process event logs, or Internet of things, e.g., mining of real-time sensor data.

References

1. Agrawal, R., Imielinski, T., Swami, A.: Mining association rules between sets of items in large databases. In: ACM SIGMOD Record, vol. 22, no. 2, pp. 207–216. ACM, June 1993
2. Han, J., Pei, J., Yin, Y.: Mining frequent patterns without candidate generation. In: ACM SIGMOD Record, vol. 29, no. 2, pp. 1–12. ACM, May 2000
3. Brin, S., Motwani, R., Ullman, J.D., Tsur, S.: Dynamic itemset counting and implication rules for market basket data. In: ACM SIGMOD Record, vol. 26, no. 2, pp. 255–264. ACM, June 1997
4. Mueller, A.: Fast sequential and parallel algorithms for association rule mining: a comparison. Technical report CS-TR-3515, University of Maryland, College Park, August 1995
5. Zaki, M.J., Parthasarathy, S., Ogihara, M., Li, W.: New algorithms for fast discovery of association rules. In: Third International Conference Knowledge Discovery and Data Mining (1997)
6. Amphawan, K., Lenca, P., Surarerks, A.: Efficient mining top-k regular-frequent itemset using compressed tidsets. In: Cao, L., Huang, J.Z., Bailey, J., Koh, Y.S., Luo, J. (eds.) PAKDD 2011. LNCS (LNAI), vol. 7104, pp. 124–135. Springer, Heidelberg (2012). doi:10.1007/978-3-642-28320-8_11
7. Leung, C.K.-S., Mateo, M.A.F., Brajczuk, D.A.: A tree-based approach for frequent pattern mining from uncertain data. In: Washio, T., Suzuki, E., Ting, K.M., Inokuchi, A. (eds.) PAKDD 2008. LNCS (LNAI), vol. 5012, pp. 653–661. Springer, Heidelberg (2008). doi:10.1007/978-3-540-68125-0_61
8. Cerf, L., Besson, J., Robardet, C., Boulicaut, J.F.: Closed patterns meet n-ary relations. ACM Trans. Knowl. Discov. Data (TKDD) **3**(1), 3 (2009)
9. Grahne, G., Zhu, J.: Fast algorithms for frequent itemset mining using FP-trees. IEEE Trans. Knowl. Data Eng. **17**(10), 1347–1362 (2005)
10. Borgelt, C.: Frequent itemset mining. Wiley Interdisc. Rev.: Data Min. Knowl. Discov. **2**(6), 437–456 (2012)
11. Djenouri, Y., Bendjoudi, A., Mehdi, M., Nouali-Taboudjemat, N., Habbas, Z.: GPU-based bees swarm optimization for association rules mining. J. Supercomput. **71**(4), 1318–1344 (2015)
12. Djenouri, Y., Drias, H., Habbas, Z.: Bees swarm optimisation using multiple strategies for association rule mining. Int. J. Bio-Inspired Comput. **6**(4), 239–249 (2014)
13. Gheraibia, Y., Moussaoui, A., Djenouri, Y., Kabir, S., Yin, P.Y.: Penguins search optimisation algorithm for association rules mining. CIT J. Comput. Inf. Technol. **24**(2), 165–179 (2016)
14. Luna, J.M., Pechenizkiy, M., Ventura, S.: Mining exceptional relationships with grammar-guided genetic programming. Knowl. Inf. Syst. **47**(3), 571–594 (2016)
15. Hegland, M.: The apriori algorithm tutorial. Math. Comput. imaging Sci. Inf. Process. **11**, 209–262 (2005)
16. Guvenir, H.A., Uysal, I.: Bilkent university function approximation repository (2000). http://funapp.CS.bilkent.edu.tr/DataSets. Accessed 12 Mar 2012

Multi-view Regularized Gaussian Processes

Qiuyang Liu and Shiliang Sun(✉)

Department of Computer Science and Technology, East China Normal University, 3663 North Zhongshan Road, Shanghai 200062, People's Republic of China
qiuyangliu2014@gmail.com, slsun@cs.ecnu.edu.cn

Abstract. Gaussian processes (GPs) have been proven to be powerful tools in various areas of machine learning. However, there are very few applications of GPs in the scenario of multi-view learning. In this paper, we present a new GP model for multi-view learning. Unlike the existing methods, it combines multiple views by regularizing marginal likelihood with the consistency among the posterior distributions of latent functions from different views. Moreover, we give a general point selection scheme for multi-view learning and improve the proposed model by this criterion. Experimental results on multiple real world data sets have verified the effectiveness of the proposed model and witnessed the performance improvement through employing this novel point selection scheme.

Keywords: Gaussian process · Multi-view learning · Posterior consistency · Co-regularization · Supervised learning

1 Introduction

Gaussian processes (GPs) [10] are flexible and popular Bayesian nonparametric tools for probabilistic modeling. Without giving concrete functional forms, they can be employed to define distributions over functions. As effective probabilistic models, they provide estimations of the uncertainty of predictions. With many convenient properties, GPs are widely used in machine learning and statistics. For instance, GPs have made great progress in semi-supervised learning [9,14,23], active learning [8,22], multi-task learning [3,19], reinforcement learning [5,11], and time series modeling [4,21].

Standard GPs only deal with single view data. However, in practice, many data involve multi-view information, which may come from different feature extractors or different domains. For example, in web-page classification, a web-page can be described by its content and hyperlink structure. In image classification, an image can be represented by its color, texture, shape, and so on. Therefore, recently, multi-view learning has drawn great attention in machine learning field. There are increasing number of algorithms proposed for multi-view learning, which can be mainly divided into two major categories [15]: co-training style algorithm [2,16] and co-regularization style algorithms [6,20]. However, GPs, as efficient and

J. Kim et al. (Eds.): PAKDD 2017, Part II, LNAI 10235, pp. 655–667, 2017.
DOI: 10.1007/978-3-319-57529-2_51

elegant methods in machine learning, have very few applications in multi-view learning [13,20]. Our work intends to apply the GPs in multi-view learning.

Existing multi-view learning methods involving the GPs can be classified into two categories: Bayesian co-training [20] and subspace learning [13,17]. The Bayesian co-training approach [20] is a Bayesian undirected graphical model, which pays attention to semi-supervised multi-view learning. The conditional independence between the output y of each data and latent functions f_j for each view is ensured by involving a latent function f_c [18]. The subspace learning methods [13,17] use the GPs as tools to construct a latent variable model which could tackle the task of non-linear dimensional reduction. Compared to the existing methods, our work, inspired by the thought of co-regularization, directly extends the GPs to the context of the multi-view learning via the posterior consistency regularization, leading to elegant inference and optimization.

Our method models the classifier of each separated view as a Gaussian process. We optimize hyperparameters of the GPs by maximizing weighted average of marginal likelihood on each view and minimizing the discrepancy among the posterior distributions of the latent function on each view. As the Kullback-Leibler (KL) divergence [7] is frequently used in describing the difference between two probability distributions, we employ it to characterize the discrepancy among the posterior distributions. Moreover, as data sets in real world are complex and may be seriously contaminated by noises, the sufficiency assumption, i.e., each view is sufficient for classification on its own, and the compatibility assumption, i.e., the target functions of all the views predict the same labels with a high probability [18], may fail in some cases. Considering these situations, we improve our model by using a selective regularization idea, which is different from the previous multi-view methods. In the experiments, we have compared the improved method with the original model to verify the effectiveness of the idea of the selective regularization on real world data sets.

The highlights of our work can be summarized as follows. First, we present a new GP model for multi-view learning, which extends the GPs to the scenario of multi-view learning by simple and elegant posterior consistency regularization. Secondly, our models can automatically learn which views of the data should be trusted more when predicting class labels. Finally, we give a general point selection scheme for multi-view learning to deal with the situations where the sufficiency and compatibility assumptions fail, and propose the multi-view GPs with selective posterior consistency inspired by this criterion.

The remainder of this paper is organized as follows. Section 2 reviews the Gaussian processes. In Sect. 3, we present the multi-view GPs with posterior consistency (MvGP1), our first algorithm, covering the principles and detailed inference and learning in the proposed model. Moreover, we improve the MvGP1 to the multi-view GPs with selective posterior consistency (MvGP2) based on a general idea of the consistent set in Sect. 4. Experimental results are provided in Sect. 5. Finally, we conclude this paper and discuss the future work in Sect. 6.

2 Gaussian Processes

This section briefly reviews the Gaussian process (GP) model.

GPs are powerful tools for classification and regression. A Gaussian process is a collection of random variables, any finite number of which have a joint Gaussian distribution [10]. The GP is widely used to describe a distribution over functions, and can be completely specified by its mean function and covariance function. Formally, suppose that the training set has N examples $\{(\boldsymbol{x}_i, y_i)\}_{i=1}^N$, where $\boldsymbol{x}_i \in R^M$ is the ith input, and $y_i \in R$ is the corresponding label. Denote $\boldsymbol{X} = [\boldsymbol{x}_1, \boldsymbol{x}_2, \ldots, \boldsymbol{x}_N]^{\mathbf{T}}$, and $\boldsymbol{y} = [y_1, y_2, \ldots, y_N]^{\mathbf{T}}$. Following standard notations for GPs, the prior distribution for the latent functions $\boldsymbol{f}$ is assumed to be Gaussian, $\boldsymbol{f}|\boldsymbol{X} \sim \mathcal{N}(\mathbf{0}, \boldsymbol{K})$, with a zero mean and a covariance matrix $\boldsymbol{K}$, whose element K_{ij} is determined by the covariance function $k(\boldsymbol{x}_i, \boldsymbol{x}_j)$. Diverse covariance functions can be employed in GPs. In this paper, we select a commonly used covariance function, the squared exponential kernel,

$$k(\boldsymbol{x}_i, \boldsymbol{x}_j) = s_f^2 \exp(-\frac{1}{2l^2} \sum_{d=1}^{N} (x_{id} - x_{jd})^2), \tag{1}$$

where s_f^2 is the signal variance, and l is the length-scale of the covariance.

The Gaussian likelihood for regression can be written as $\boldsymbol{y}|\boldsymbol{f} \sim \mathcal{N}(\boldsymbol{f}, \sigma^2 \boldsymbol{I})$, and after integrating out the hidden variables, the marginal likelihood is $\boldsymbol{y}|\boldsymbol{X} \sim \mathcal{N}(0, \boldsymbol{K} + \sigma^2 \boldsymbol{I})$.

Under these settings, the posterior of the latent functions should be

$$\boldsymbol{f}|\boldsymbol{y} \sim \mathcal{N}(\boldsymbol{\mu}, \boldsymbol{\Sigma}), \tag{2}$$

where $\boldsymbol{\mu} = \boldsymbol{K}(\boldsymbol{K} + \sigma^2 \boldsymbol{I})^{-1} \boldsymbol{y}$ and $\boldsymbol{\Sigma} = \boldsymbol{K} - \boldsymbol{K}(\boldsymbol{K} + \sigma^2 \boldsymbol{I})^{-1} \boldsymbol{K}$ are the mean and the covariance of the posterior distribution, respectively.

We use Θ to denote the hyperparameters in the Gaussian process regression model, that is, $\Theta = \{s_f^2, l, \sigma^2\}$. These hyperparameters can be obtained by generalized maximum likelihood. In generalized maximum likelihood, we calculate the negative logarithmic marginal likelihood of the samples, $L(\Theta) = -\log p(\boldsymbol{y}|\boldsymbol{X}, \Theta)$, and then minimize $L(\Theta)$ with respect to Θ.

For a new point $\boldsymbol{x}^*$, the prediction is also Gaussian,

$$f_*|\boldsymbol{X}, \boldsymbol{y}, \boldsymbol{x}^* \sim \mathcal{N}(\bar{f}_*, \mathrm{cov}(\mathrm{f}_*)), \tag{3}$$

where $\bar{f}_* = \boldsymbol{k}_*{}^{\mathbf{T}}[\boldsymbol{K} + \sigma^2 \boldsymbol{I}]^{-1} \boldsymbol{y}$, $\mathrm{cov}(f_*) = k(\boldsymbol{x}_*, \boldsymbol{x}_*) - \boldsymbol{k}_*{}^{\mathbf{T}}[\boldsymbol{K} + \sigma^2 \boldsymbol{I}]^{-1} \boldsymbol{k}_*$. Here, k is the covariance function, and $\boldsymbol{k}_*$ is the vector of covariance function values between $\boldsymbol{x}_*$ and the training data $\boldsymbol{X}$.

Standard GPs only handle single view data. However, collected data sets in the real world can often be represented by multiple views which may come from different feature extractors or various measurement modalities. As GPs are popular tools in machine learning, we propose to develop the GPs to multi-view learning.

3 Multi-view GPs with Posterior Consistency

In this section, we present the formulation of the multi-view GPs with posterior consistency (MvGP1) and show the corresponding inference and optimization in the proposed model. We pay attention to two views learning tasks, and then give illustrations about the extensions to the scenario concerning more than two views.

3.1 Model Representation

Assume that the two views training set $\boldsymbol{D}$ has N examples $\{(\boldsymbol{x}_i, \boldsymbol{z}_i, y_i)\}_{i=1}^{N}$, where $\boldsymbol{x}_i \in R^{M_1}$ is the ith input on the first view, $\boldsymbol{z}_i \in R^{M_2}$ is the ith input on the second view, and $y_i \in \{+1, -1\}$ is the corresponding label. Denote $\boldsymbol{X} = [\boldsymbol{x}_1, \ldots, \boldsymbol{x}_N]^{\mathbf{T}}$, $\boldsymbol{Z} = [\boldsymbol{z}_1, \ldots, \boldsymbol{z}_N]^{\mathbf{T}}$, and $\boldsymbol{y} = [y_1, \ldots, y_N]^{\mathbf{T}}$.

First, on account of leveraging the information in the separated single view, we simply assume that each view of data is modeled by a GP. That is, the prior distribution for the latent functions $\boldsymbol{f}_1$ on the first view and $\boldsymbol{f}_2$ on the second view are supposed to be Gaussian, i.e. $p(\boldsymbol{f}_1|\boldsymbol{X}) = \mathcal{N}(\boldsymbol{0}, \boldsymbol{K}_1)$, and $p(\boldsymbol{f}_2|\boldsymbol{Z}) = \mathcal{N}(\boldsymbol{0}, \boldsymbol{K}_2)$, where $\boldsymbol{K}_1$ and $\boldsymbol{K}_2$ are covariance matrixes determined by the corresponding covariance functions of two views, respectively. In our model, the covariance function is the squared exponential kernel as mentioned in (1). Although the Gaussian noise model is originally developed for regression, it has also been proved effective for classification, and its performance is typically comparable to the more complex probit and logit likelihood models used in classification problems [1]. Therefore, we use the Gaussian regression likelihood in our classification task to enjoy the elegant exact inference. The Gaussian likelihood for regression on the first view is $p(\boldsymbol{y}|\boldsymbol{f}_1) = \mathcal{N}(\boldsymbol{f}_1, \sigma_1^2\boldsymbol{I})$, and the likelihood on the second view is $p(\boldsymbol{y}|\boldsymbol{f}_2) = \mathcal{N}(\boldsymbol{f}_2, \sigma_2^2\boldsymbol{I})$. Secondly, we also need leverage the consistence between two views. The KL divergence [7] can measure the discrepancy between two distributions. Thus, we use the KL divergence between the posterior distributions on two views to regularize the objective function in MvGP1, enforcing the consistence between two views.

Suppose the posterior distribution of the latent function $\boldsymbol{f}_1$ on the first view is

$$p_1 = p(\boldsymbol{f}_1|\boldsymbol{X}, \boldsymbol{y}) = \mathcal{N}(\boldsymbol{\mu}_1, \boldsymbol{\Sigma}_1), \tag{4}$$

and the posterior distribution of the latent function $\boldsymbol{f}_2$ on the second view is

$$p_2 = p(\boldsymbol{f}_2|\boldsymbol{Z}, \boldsymbol{y}) = \mathcal{N}(\boldsymbol{\mu}_2, \boldsymbol{\Sigma}_2). \tag{5}$$

Based on the above setting, our objective function of MvGP1 is

$$\min \mathrm{L}_1 = \min\{-[a \log p(\boldsymbol{y}|\boldsymbol{X}) + (1-a)\log p(\boldsymbol{y}|\boldsymbol{Z})] + \frac{b}{2}[KL(p_1||p_2) + KL(p_2||p_1)]\}, \tag{6}$$

where

$$\log p(\boldsymbol{y}|\boldsymbol{X}) = -\tfrac{1}{2}\boldsymbol{y}^{\mathbf{T}}(\boldsymbol{K}_1 + \sigma_1^2\boldsymbol{I})^{-1}\boldsymbol{y} - \tfrac{1}{2}\log|\boldsymbol{K}_1 + \sigma_1^2\boldsymbol{I}| - \tfrac{N}{2}\log 2\pi \tag{7}$$

is the marginal likelihood on the first view,

$$\log p(\boldsymbol{y}|\boldsymbol{Z}) = -\tfrac{1}{2}\boldsymbol{y}^{\mathrm{T}}(\boldsymbol{K}_2 + \sigma_2^2\boldsymbol{I})^{-1}\boldsymbol{y} - \tfrac{1}{2}\log|\boldsymbol{K}_2 + \sigma_2^2\boldsymbol{I}| - \tfrac{N}{2}\log 2\pi \quad (8)$$

is the marginal likelihood on the second view, and the KL divergences between the posterior distributions $\boldsymbol{f}_1$ and $\boldsymbol{f}_2$ are

$$\begin{aligned} KL(p_1||p_2) = \tfrac{1}{2}[&\log|\boldsymbol{\Sigma}_2| - \log|\boldsymbol{\Sigma}_1| + tr(\boldsymbol{\Sigma}_2^{-1}\boldsymbol{\Sigma}_1) \\ &+(\boldsymbol{\mu}_2 - \boldsymbol{\mu}_1)^{\mathrm{T}}\boldsymbol{\Sigma}_2^{-1}(\boldsymbol{\mu}_2 - \boldsymbol{\mu}_1) - N], \end{aligned} \quad (9)$$

and

$$\begin{aligned} KL(p_2||p_1) = \tfrac{1}{2}[&\log|\boldsymbol{\Sigma}_1| - \log|\boldsymbol{\Sigma}_2| + tr(\boldsymbol{\Sigma}_1^{-1}\boldsymbol{\Sigma}_2) \\ &+(\boldsymbol{\mu}_1 - \boldsymbol{\mu}_2)^{\mathrm{T}}\boldsymbol{\Sigma}_1^{-1}(\boldsymbol{\mu}_1 - \boldsymbol{\mu}_2) - N]. \end{aligned} \quad (10)$$

Since the KL divergence is not a symmetrical quantity, that is, $KL(p_1||p_2) \neq KL(p_2||p_1)$, and we have no general method to determine which one is better for measuring the discrepancy between the two posterior distribution p_1 and p_2, we construct a symmetrical quantity based on the above KL divergences, i.e., $\frac{1}{2}[KL(p_1||p_2) + KL(p_2||p_1)]$.

Parameters $\boldsymbol{\mu}_1$, $\boldsymbol{\mu}_2$, $\boldsymbol{\Sigma}_1$, and $\boldsymbol{\Sigma}_2$ in (9) and (10) are provided as follows. The mean and covariance of the posterior distribution for the latent function $\boldsymbol{f}_1$ are $\boldsymbol{\mu}_1 = \boldsymbol{K}_1(\boldsymbol{K}_1 + \sigma_1^2\boldsymbol{I})^{-1}\boldsymbol{y}$ and $\boldsymbol{\Sigma}_1 = \boldsymbol{K}_1 - \boldsymbol{K}_1(\boldsymbol{K}_1 + \sigma_1^2\boldsymbol{I})^{-1}\boldsymbol{K}_1$. The mean and covariance of the posterior distribution for the latent function $\boldsymbol{f}_2$ are $\boldsymbol{\mu}_2 = \boldsymbol{K}_2(\boldsymbol{K}_2 + \sigma_2^2\boldsymbol{I})^{-1}\boldsymbol{y}$ and $\boldsymbol{\Sigma}_2 = \boldsymbol{K}_2 - \boldsymbol{K}_2(\boldsymbol{K}_2 + \sigma_2^2\boldsymbol{I})^{-1}\boldsymbol{K}_2$.

3.2 Inference and Optimization

In our model, we consider the hybrid prediction function $sign(a\boldsymbol{f}_1 + (1-a)\boldsymbol{f}_2)$, $a \in [0,1]$. For a new point $\{\boldsymbol{x}^*, \boldsymbol{z}^*\}$, the prediction distribution $\boldsymbol{f}_{1*}$ of the first view and the prediction distribution $\boldsymbol{f}_{2*}$ of the second view are also Gaussian. The mean of $\boldsymbol{f}_{1*}$ is ${\boldsymbol{k}_{1*}}^{\mathrm{T}}(\boldsymbol{K}_1 + \sigma_1^2\boldsymbol{I})^{-1}\boldsymbol{y}$, and the covariance is $k_1(\boldsymbol{x}_*, \boldsymbol{x}_*) - {\boldsymbol{k}_{1*}}^{\mathrm{T}}(\boldsymbol{K}_1 + \sigma_1^2\boldsymbol{I})^{-1}\boldsymbol{k}_{1*}$. Here, k_1 is the covariance function, and $\boldsymbol{k}_{1*}$ is the vector of covariance function values between $\boldsymbol{x}_*$ and the training data $\boldsymbol{X}$. The mean of $\boldsymbol{f}_{2*}$ is ${\boldsymbol{k}_{2*}}^{\mathrm{T}}(\boldsymbol{K}_2 + \sigma_2^2\boldsymbol{I})^{-1}\boldsymbol{y}$, and the covariance is $k_2(\boldsymbol{z}_*, \boldsymbol{z}_*) - {\boldsymbol{k}_{2*}}^{\mathrm{T}}(\boldsymbol{K}_2 + \sigma_2^2\boldsymbol{I})^{-1}\boldsymbol{k}_{2*}$. Here, k_2 is the covariance function, and $\boldsymbol{k}_{2*}$ is the vector of covariance function values between $\boldsymbol{z}_*$ and the training data $\boldsymbol{Z}$. According to the hybrid prediction function, we give our prediction on the new data point $\{\boldsymbol{x}^*, \boldsymbol{z}^*\}$.

As GPs are nonparametric models, the related hyperparameters need to be determined. In the context of a single view, we often obtain the hyperparameters via generalized maximum likelihood. In the case of two views, we optimize the hyperparameters collaboratively by two views. Co-regularization approaches often expect the predictions for the same observation of different views to be consistent to optimize the parameters. Similarly, in the context of Bayesian learning, we want the posterior distributions of the latent functions of the same observation across different views to be consistent to obtain the hyperparameters, which can be realized via minimizing the above objective function shown in (6).

The hyperparameters in our model can be divided into two classes: the trade-off hyperparameters a, b and the GP related hyperparameters, which include the hyperparameters s_f^2, l in the covariance functions and the noise hyperparameters σ^2 in the likelihood. We use $\Theta = \{s_{f1}^2, l_1, \sigma_1^2, s_{f2}^2, l_2, \sigma_2^2\}$ to denote hyperparameters in the second group, where $s_{f1}^2, l_1, \sigma_1^2$ are hyperparameters related to the first view, and others are hyperparameters related to the second view. The GP related hyperparameters Θ are optimized by the gradient descent method. As our model has an elegant formulation, the gradients with respect to Θ also have graceful forms, and the code can be easily implemented by the existing toolbox [12]. Following the above parameter notations, the gradient w.r.t the s_{f1} is

$$
\begin{aligned}
\frac{\partial \mathrm{L}_1}{\partial s_{f1}} &= \frac{a}{2}\left\{-\boldsymbol{y}^{\mathbf{T}}(\boldsymbol{K}_1+\sigma_1^2\boldsymbol{I})^{-1}\frac{2\boldsymbol{K}_1}{s_{f1}}(\boldsymbol{K}_1+\sigma_1^2\boldsymbol{I})^{-1}\boldsymbol{y} + tr\left[(\boldsymbol{K}_1+\sigma_1^2\boldsymbol{I})^{-1}\frac{2\boldsymbol{K}_1}{s_{f1}}\right]\right\} \\
&+\frac{b}{2}tr\left\{\boldsymbol{\Sigma}_2^{-1}\left[\frac{2\boldsymbol{K}_1}{s_{f1}} - \frac{2\boldsymbol{K}_1}{s_{f1}}(\boldsymbol{K}_1+\sigma_1^2\boldsymbol{I})^{-1}\boldsymbol{K}_1 - \boldsymbol{K}_1(\boldsymbol{K}_1+\sigma_1^2\boldsymbol{I})^{-1}\frac{2\boldsymbol{K}_1}{s_{f1}} + \right.\right. \\
&\left.\boldsymbol{K}_1(\boldsymbol{K}_1+\sigma_1^2\boldsymbol{I})^{-1}\frac{2\boldsymbol{K}_1}{s_{f1}}(\boldsymbol{K}_1+\sigma_1^2\boldsymbol{I})^{-1}\boldsymbol{K}_1\right] - \boldsymbol{\Sigma}_1^{-1}\left[\frac{2\boldsymbol{K}_1}{s_{f1}} - \frac{2\boldsymbol{K}_1}{s_{f1}}(\boldsymbol{K}_1+\sigma_1^2\boldsymbol{I})^{-1}\boldsymbol{K}_1\right. \\
&\left.\left.+\boldsymbol{K}_1(\boldsymbol{K}_1+\sigma_1^2\boldsymbol{I})^{-1}\frac{2\boldsymbol{K}_1}{s_{f1}}(\boldsymbol{K}_1+\sigma_1^2\boldsymbol{I})^{-1}\boldsymbol{K}_1 - \boldsymbol{K}_1(\boldsymbol{K}_1+\sigma_1^2\boldsymbol{I})^{-1}\frac{2\boldsymbol{K}_1}{s_{f1}}\right]\boldsymbol{\Sigma}_1^{-1}\boldsymbol{\Sigma}_2\right\} \\
&+\frac{b}{2}\left\{\left[\frac{2\boldsymbol{K}_1}{s_{f1}}(\boldsymbol{K}_1+\sigma_1^2\boldsymbol{I})^{-1}\boldsymbol{y} - \boldsymbol{K}_1(\boldsymbol{K}_1+\sigma_1^2\boldsymbol{I})^{-1}\frac{2\boldsymbol{K}_1}{s_{f1}}(\boldsymbol{K}_1+\sigma_1^2\boldsymbol{I})^{-1}\boldsymbol{y}\right]^{\mathbf{T}}\right. \\
&(\boldsymbol{\Sigma}_1^{-1}+\boldsymbol{\Sigma}_2^{-1})(\boldsymbol{\mu}_1-\boldsymbol{\mu}_2) - (\boldsymbol{\mu}_1-\boldsymbol{\mu}_2)^{\mathbf{T}}\boldsymbol{\Sigma}_1^{-1}\left[\frac{2\boldsymbol{K}_1}{s_{f1}} - \frac{2\boldsymbol{K}_1}{s_{f1}}(\boldsymbol{K}_1+\sigma_1^2\boldsymbol{I})^{-1}\boldsymbol{K}_1\right. \\
&\left.+\boldsymbol{K}_1(\boldsymbol{K}_1+\sigma_1^2\boldsymbol{I})^{-1}\frac{2\boldsymbol{K}_1}{s_{f1}}(\boldsymbol{K}_1+\sigma_1^2\boldsymbol{I})^{-1}\boldsymbol{K}_1 - \boldsymbol{K}_1(\boldsymbol{K}_1+\sigma_1^2\boldsymbol{I})^{-1}\frac{2\boldsymbol{K}_1}{s_{f1}}\right]\boldsymbol{\Sigma}_1{}^{-1} \\
&(\boldsymbol{\mu}_1-\boldsymbol{\mu}_2) + (\boldsymbol{\mu}_1-\boldsymbol{\mu}_2)^{\mathbf{T}}(\boldsymbol{\Sigma}_1^{-1}+\boldsymbol{\Sigma}_2^{-1})\left[\frac{2\boldsymbol{K}_1}{s_{f1}}(\boldsymbol{K}_1+\sigma_1^2\boldsymbol{I})^{-1}\boldsymbol{y}\right. \\
&\left.\left.-\boldsymbol{K}_1(\boldsymbol{K}_1+\sigma_1^2\boldsymbol{I})^{-1}\frac{2\boldsymbol{K}_1}{s_{f1}}(\boldsymbol{K}_1+\sigma_1^2\boldsymbol{I})^{-1}\boldsymbol{y}\right]\right\}
\end{aligned} \tag{11}
$$

The gradients with respect to other hyperparameters in Θ are similar with (11), and hence we omit them here. The trade-off hyperparameters a and b are obtained through grid search.

We summarize MvGP1 in Algorithm 1.

3.3 Extension to Multiple Views

In the above sections, we take two views as an example to illustrate our model MvGP1. This model can be easily extended to multiple views because of the elegant formulation. The posterior distribution of the latent function on each view is the Gaussian distribution. Moreover, the KL divergence [7] between two Gaussian distributions can be calculated analytically, i.e., for two Gaussian distributions $N_0 = \mathcal{N}(\boldsymbol{\mu}_0, \boldsymbol{\Sigma}_0)$ and $N_1 = \mathcal{N}(\boldsymbol{\mu}_1, \boldsymbol{\Sigma}_1)$, we have $KL(N_0||N_1) = \frac{1}{2}\{\log(\frac{|\boldsymbol{\Sigma}_1|}{|\boldsymbol{\Sigma}_0|}) + tr(\boldsymbol{\Sigma}_1^{-1}\boldsymbol{\Sigma}_0) + (\boldsymbol{\mu}_1-\boldsymbol{\mu}_0)^{\mathbf{T}}\boldsymbol{\Sigma}_1^{-1}(\boldsymbol{\mu}_1-\boldsymbol{\mu}_0) - N\}$. Therefore, we can

Algorithm 1. Multi-view GPs with Posterior Consistency

Input: training data $\{\boldsymbol{x}_i, \boldsymbol{z}_i, y_i\}_{i=1}^{N_1}$, test samples $\{\boldsymbol{x}_i^*, \boldsymbol{z}_i^*, y_i^*\}_{i=1}^{N_2}$.
Output: accuracy *acc*, trade-off parameters a, b, and GP related hyperparameters Θ.

1: initialize Θ randomly.
2: **for** $k = 1$ to 10 **do**
3: Divide the training data $\{\boldsymbol{x}_i, \boldsymbol{z}_i, y_i\}_{i=1}^{N_1}$ into the training set $\{\boldsymbol{x}_i^t, \boldsymbol{z}_i^t, y_i^t\}_{i=1}^{N_1^t}$ and the validation set $\{\boldsymbol{x}_i^v, \boldsymbol{z}_i^v, y_i^v\}_{i=1}^{N_1^v}$.
4: **for** a, b in the search grids **do**
5: **while** termination conditions are not satisfied **do**
6: Update Θ by gradient descent to minimize L_1 in (6).
7: **end while**
8: Calculate the predictions by $sign(a\boldsymbol{f}_1 + (1-a)\boldsymbol{f}_2)$ on the validation set.
9: Calculate the accuracy acc_v on the validation set $\{\boldsymbol{x}_i^v, \boldsymbol{z}_i^v, y_i^v\}_{i=1}^{N_1^v}$.
10: **if** acc_v is larger than the accuracy on the last iteration **then**
11: record the trade-off parameters a, b.
12: **end if**
13: **end for**
14: **end for**
15: **while** termination conditions are not satisfied **do**
16: Update Θ by gradient descent to minimize L_1 in (6).
17: **end while**
18: Calculate the predictions by $sign(a\boldsymbol{f}_1 + (1-a)\boldsymbol{f}_2)$ on the test samples.
19: Calculate the accuracy *acc* on the test samples $\{\boldsymbol{x}_i^*, \boldsymbol{z}_i^*, y_i^*\}_{i=1}^{N_2}$.

extend MvGP1 to multiple views by regularizing the weighted logarithm with the KL divergences between every pair of the distinct posterior distributions. Following the notations in MvGP1, given a data set which involves K views, $p(\boldsymbol{y}|\boldsymbol{X}^k)$ denotes the marginal distribution of the kth view, and p_i represents the corresponding posterior distribution of the latent function $\boldsymbol{f}_i$. The objective function in multiple views is

$$\min -\sum_{k=1}^{K} a_k \log p(\boldsymbol{y}|\boldsymbol{X}^k) + \sum_{i=1}^{K}\sum_{j>i}^{K}\{b_{ij}[KL(p_i||p_j) + KL(p_j||p_i)]\}. \quad (12)$$

4 Multi-view GPs with Selective Posterior Consistency

When the predictions from different views are not consistent on some data points, the multi-view sufficiency assumption and compatibility assumption cannot be satisfied on the whole data set. In this context, it is not appropriate to enforce the posteriors on the whole data set across different views as similar as possible. For instance, in some cases, the input data of some views may be largely affected by the noises, which may make the predictions of these points in these views totally different from those in other views. In these cases, the multi-view assumptions fail, and enforcing the predictions of all views on these data points consistent seems to be improper.

Considering the above problem, we improve MvGP1 and present multi-view GPs with selective posterior consistency (MvGP2). In this model, we modify the regularization term in the objective function to make the posterior distributions across different views on a subset of the data set other than the whole one as similar as possible. In order to find the proper subset, namely the consistent set, we first optimize the hyperparameters through MvGP1 on the training set and give the predictions for the training data on each view. Next, we select data points whose predictions on each view are all consistent and are also consistent with the true label to construct the consistent set. Finally, we optimize the hyperparameters with the analogous procedure to MvGP1 except that we regularize the posteriors only on the chosen consistent set.

In fact, the key idea of MvGP2 is to construct the consistent set and constrain the multi-view assumptions on the consistent set. We construct the consistent set by selecting the data points whose label predictions on each view are all equal and the same as true labels. Formally, given two views of the input data $\boldsymbol{X} = [\boldsymbol{x}_1, \ldots, \boldsymbol{x}_N]^{\mathbf{T}}$, $\boldsymbol{Z} = [\boldsymbol{z}_1, \ldots, \boldsymbol{z}_N]^{\mathbf{T}}$ and the corresponding label data $\boldsymbol{y} = [y_1, \ldots, y_N]^{\mathbf{T}}$, we find a index set $T = [t_1, t_2 \ldots, t_k]$, $(k <= N)$ such that for each $t \in T$, the predictions for $\boldsymbol{X}_t$, $\boldsymbol{Z}_t$ and the corresponding $\boldsymbol{y}_t$ are all agreed. Then the consistent set is $\{\boldsymbol{x}_i, \boldsymbol{z}_i, y_i\}_{i \in T}$. Let $\boldsymbol{X}_T, \boldsymbol{Z}_T$, and $\boldsymbol{y}_T$ denote the corresponding data matrix of the consistent set. After constructing the consistent set, we modify the objective function to only restrict the KL divergence of the posterior distributions of the latent functions on the consistent set to be minimized,

$$\min \mathrm{L}_1' = \{-[a \log p(\boldsymbol{y}|\boldsymbol{X}) + (1-a) \log p(\boldsymbol{y}|\boldsymbol{Z})] + \frac{b}{2}[KL({p_1}'||{p_2}') + KL({p_2}'||{p_1}')]\}, \tag{13}$$

where ${p_1}' = p(\boldsymbol{f}_1'|\boldsymbol{X}_T, \boldsymbol{y}_T)$ is the posterior distribution of the latent function $\boldsymbol{f}_1{}'$ on the set $\{\boldsymbol{X}_T, \boldsymbol{y}_T\}$, and ${p_2}' = p(\boldsymbol{f}_2{}'|\boldsymbol{Z}_T, \boldsymbol{y}_T)$ is the posterior distribution of the latent function $\boldsymbol{f}_2'$ on the set $\{\boldsymbol{Z}_T, \boldsymbol{y}_T\}$.

We summarize MvGP2 in Algorithm 2. The idea of constraining the multi-view assumption on the consistent set other than the whole data set is a novel view in multi-view learning. In the real world, data are complex and noisy. It is likely that not all the data satisfy the requirement that the predictions on different views should be equal. Moreover, we have verified this selective idea through the experiments on real world data sets in the following section.

5 Experiments

In this section, we performed experiments with our proposed MvGP1 and MvGP2 on multiple real world data sets. For comparison, we use three single-view methods corresponding to GPs named GP1, GP2, GP3, which use the first view, the second view, and the combination of two views, i.e., concatenating two

Algorithm 2. Multi-view GPs with Selective Posterior Consistency

Input: training data $\{\boldsymbol{x}_i, \boldsymbol{z}_i, y_i\}_{i=1}^{N_1}$, test samples $\{\boldsymbol{x}_i^*, \boldsymbol{z}_i^*, y_i^*\}_{i=1}^{N_2}$.
Output: accuracy *acc*, trade-off parameters a, b, GP related hyperparameters Θ and the consistent index set T.

1: $T = \{\}$, initialize Θ randomly.
2: run MvGP1.
3: **for** $i = 1$ to N_1 **do**
4: Calculate the predictions by f_{i1}, f_{i2} on the training data $\boldsymbol{x}_i$ and $\boldsymbol{z}_i$, respectively.
5: **if** $f_{i1} = f_{i2}$ && $f_{i2} = y_i$ **then**
6: add i into the set T
7: **end if**
8: **end for**
9: **for** $k = 1$ to 10 **do**
10: Divide the training data $\{\boldsymbol{x}_i, \boldsymbol{z}_i, y_i\}_{i=1}^{N_1}$ into the training set $\{\boldsymbol{x}_i^t, \boldsymbol{z}_i^t, y_i^t\}_{i=1}^{N_1^t}$ and the validation set $\{\boldsymbol{x}_i^v, \boldsymbol{z}_i^v, y_i^v\}_{i=1}^{N_1^v}$.
11: **for** a, b in the search grids **do**
12: **while** termination conditions are not satisfied **do**
13: Update Θ by gradient descent to minimize L_1' in (13).
14: **end while**
15: Calculate the predictions by $sign(af_1 + (1-a)f_2)$ on the validation set.
16: Calculate the accuracy acc_v on the validation set $\{\boldsymbol{x}_i^v, \boldsymbol{z}_i^v, y_i^v\}_{i=1}^{N_1^v}$.
17: **if** acc_v is larger than the accuracy on the last iteration **then**
18: record the trade-off parameters a, b.
19: **end if**
20: **end for**
21: **end for**
22: **while** termination conditions are not satisfied **do**
23: Update Θ by gradient descent to minimize L_1' in (13).
24: **end while**
25: Calculate the predictions by $sign(af_1 + (1-a)f_2)$ on the test samples.
26: Calculate the accuracy *acc* on the test samples $\{\boldsymbol{x}_i^*, \boldsymbol{z}_i^*, y_i^*\}_{i=1}^{N_2}$.

views to construct new high-dimensional data, respectively. In addition, we also compare our algorithms with a multi-view method SVM-2K [6], which is a two-view version of SVMs and is also inspired by the thought of co-regularization.

5.1 Data Sets

Web-Page. The web-page data sets have been extensively used in multi-view learning, which consist of two-view web pages collected from computer science department web sits at four universities: Cornell university, university of Washington, university of Wisconsin, and university of Texas. The two views are words occurring in a web page and words appearing in the links pointing to that page. The documents are described by 1703 words in the content view, and by 569 links between them in cites views. We list the statistical information about the four data sets in Table 1. The web pages are classified into five classes: student, project, course, staff and faculty. We set the category with the greatest size to be the positive class (denoted as class 1), and all the other categories as the negative class (denoted as class 2) in each data set.

Table 1. Statistical information of four web-page data sets.

Data set	Size	View size	Content dimension	Cite dimension	Class 1 size	Class 2 size
Cornell	195	2	1703	195	83	112
Washington	230	2	1703	230	107	123
Wisconsin	265	2	1703	265	122	143
Texas	187	2	1703	187	103	84

Ionosphere. Downloaded from UCI,[1] the ionosphere data set is collected by a system in Goose Bay, Labrador. This system involves a phased array of 16 high-frequency antennas with a total transmitted power on the order of 6.4 kW. The targets are free electrons in the ionosphere. Those showing evidence of some type of structure in the ionosphere are good radar returns, while those which do not show the above phenomenon and whose signals pass through the ionosphere are bad returns. The data set consists of 351 examples in which 225 are "good" instances and 126 are "bad" instances. There is only one view in this data set, and we generate the other view via principal component analysis, resulting in two views, which have 35 and 24 dimensions, respectively.

5.2 Experimental Setting

In the experiments, we select 60% data in each data set as the training set, and the rest as the test set. Multiple values of the hyperparameters a and b in MvGP1 and MvGP2 are explored in all the experiments. Given a division of the training and test set, we use cross validation with 10 folds and 20% training set as the validation set for the selection of the hyperparameters a and b in MvGP1 and MvGP2. The considered grid ranges are $a \in \{0, 0.1, 0.2, 0.3, 0.4, 0.5, 0.6, 0.7, 0.8, 0.9, 1\}$ and $b \in \{2^{-18}, 2^{-12}, 2^{-8}, 2, 2^3, 2^8\}$. Other hyperparameters in MvGP1, MvGP2 and hyperparameters in GP1, GP2, and GP3 are initialized randomly. As for the kernel function in GP1, GP2, GP3, MvGP1, and MvGP2, we all use the squared exponential kernel as mentioned in (1). We repeat the experiments for all the data sets five times and record the average accuracies and the corresponding standard deviations.

We compared our proposed MvGP1, MvGP2 with multi-view method SVM-2K, and three single view methods GP1, GP2, and GP3. For SVM-2K, besides the prediction functions $sign(f_1)$ and $sign(f_2)$ from the separated views, we also consider the hybrid prediction function $sign((f_1 + f_2)/2)$.

[1] Data sets are available at http://archive.ics.uci.edu/ml/.

5.3 Results

We present the average accuracies and standard deviations of all the methods on the webpage data sets and ionosphere data set in Table 2.

It is clearly shown in Table 2 that our proposed methods MvGP1 and MvGP2 are superior to GP1, GP2, GP3 and SVM-2K. We can also observe that MvGP2 further improves the performance over MvGP1, which benefits from the idea of using the selective posterior regularization other than the posterior regularization on the whole data sets to ensure the consistency.

Table 2. The average accuracies and standard deviations (%) of six methods on real world data sets.

Data set	Cornell	Washington	Wisconsin	Texas	Ionosphere
GP1	80.26 ± 14.52	67.61 ± 14.71	72.64 ± 16.02	56.01 ± 6.32	84.75 ± 2.15
GP2	62.56 ± 7.87	74.78 ± 1.42	61.51 ± 6.27	73.52 ± 12.66	98.72 ± 1.37
GP3	77.95 ± 14.66	73.04 ± 14.89	75.85 ± 17.15	62.35 ± 16.98	97.87 ± 4.76
SVM-2K	73.68 ± 5.04	74.78 ± 4.38	75.28 ± 5.75	75.15 ± 9.44	99.72 ± 0.39
MVGP1	85.64 ± 6.87	86.30 ± 6.08	$\mathbf{91.32} \pm 1.55$	78.33 ± 15.14	99.29 ± 1.24
MVGP2	$\mathbf{87.18} \pm 5.94$	$\mathbf{87.18} \pm 5.94$	$\mathbf{91.32} \pm 2.04$	$\mathbf{81.29} \pm 10.81$	$\mathbf{100} \pm 0$

6 Conclusion

In this paper, we have proposed MvGP1 which extends GPs to the scenario of learning with multiple views via the methods of posterior consistency regularization. This approach is very intuitive, resulting in an elegant objective function formulation. Experimental results on real-world web-page classification validate the effectiveness of the proposed MvGP1. Moreover, considering that the multi-view assumptions may not be met on all data points, we have proposed MvGP2, which constructs a consistent set and constrains the posterior consistency regularization on the consistent set other than the whole data set, leading to further improvements of the performance. In fact, the idea of constraining the multi-view assumptions on a selective consistent set other than the whole data set is general. It not only can be applied to GPs, but also can inspire other multi-view learning methods.

In the future, we will attempt to apply the proposed models to big data, which may use Nyström methods or other approximate approaches.

Acknowledgments. The corresponding author Shiliang Sun would like to thank supports from the National Natural Science Foundation of China under Projects 61673179 and 61370175.

References

1. Ashish, K., Grauman, K., Urtasun, R., Darrell, T.: Gaussian processes for object categorization. Int. J. Comput. Vis. **88**, 169–188 (2010)
2. Blum, A., Mitchell, T.: Combining labeled and unlabeled data with co-training. In: Proceedings of the 11th Annual Conference on Computational Learning Theory, pp. 92–100 (1998)
3. Bonilla, E.V., Chai, K.M.A., Williams, C.K.I.: Multi-task Gaussian process prediction. In: Proceedings of the 21th Annual Conference on Neural Information Processing Systems, pp. 153–160 (2007)
4. Damianou, A.C., Titsias, M.K., Lawrence, N.D.: Variational Gaussian process dynamical systems. In: Proceedings of the 25th Annual Conference on Neural Information Processing Systems, pp. 2510–2518 (2011)
5. Engel, Y., Mannor, S., Meir, R.: Reinforcement learning with Gaussian processes. In: Proceedings of the 22th International Conference on Machine Learning, pp. 201–208 (2005)
6. Farquhar, J.D.R., Hardoon, D.R., Meng, H., Shawe-Taylor, J., Szedmák, S.: Two view learning: SVM-2K, theory and practice. In: Proceedings of the 19th Annual Conference on Neural Information Processing Systems, pp. 355–362 (2005)
7. Joyce, J.M.: Kullback-Leibler Divergence. Springer, New York (2011)
8. Krause, A., Guestrin, C.: Nonmyopic active learning of Gaussian processes: an exploration-exploitation approach. In: Proceedings of the 24th International Conference on Machine Learning, pp. 449–456 (2007)
9. Lawrence, N.D., Jordan, M.I.: Semi-supervised learning via Gaussian processes. In: Proceedings of the 18th Annual Conference on Neural Information Processing Systems, pp. 753–760 (2004)
10. Rasmussen, C.E., Williams, C.K.I.: Gaussian Processes for Machine Learning. MIT Press, Cambridge (2006)
11. Rasmussen, C.E., Kuss, M.: Gaussian processes in reinforcement learning. In: Proceedings of the 17th Annual Conference on Neural Information Processing Systems, pp. 751–758(2003)
12. Rasmussen, C.E., Nickisch, H.: Gaussian processes for machine learning (GPML) toolbox. J. Mach. Learn. Res. **11**, 3011–3015 (2010)
13. Shon, A.P., Grochow, K., Hertzmann, A., Rao, R.P.N.: Learning shared latent structure for image synthesis and robotic imitation. In: Proceedings of the 19th Annual Conference on Neural Information Processing Systems, pp. 1233–1240 (2005)
14. Sindhwani, V., Chu, W., Keerthi, S.S.: Semi-supervised Gaussian process classifiers. In: Proceedings of the 20th International Joint Conference on Artificial Intelligence, pp. 1059–1064 (2007)
15. Sun, S.: A survey of multi-view machine learning. Neural Comput. Appl. **23**, 2031–2038 (2013)
16. Sun, S., Jin, F.: Robust co-training. Int. J. Pattern Recognit Artif Intell. **25**, 1113–1126 (2011)
17. Xu, C., Tao, D., Li, Y., Xu, C.: Large-margin multi-view Gaussian process for image classification. In: Proceedings of the 5th International Conference on Internet Multimedia Computing and Service, pp. 7–12. ACM (2013)
18. Xu, C., Tao, D., Xu, C.: A survey on multi-view learning. arXiv preprint arXiv:1304.5634 (2013)

19. Yu, K., Tresp, V., Schwaighofer, A.: Learning Gaussian processes from multiple tasks. In: Proceedings of the 22th International Conference on Machine Learning, pp. 1012–1019 (2005)
20. Yu, S., Krishnapuram, B., Rosales, R., Rao, R.B.: Bayesian co-training. J. Mach. Learn. Res. **12**, 2649–2680 (2011)
21. Zhao, J., Sun, S.: Variational dependent multi-output Gaussian process dynamical systems. J. Mach. Learn. Res. **17**, 1–36 (2016)
22. Zhao, J., Sun, S.: Gaussian process versus margin sampling active learning. Neurocomputing **167**, 122–131 (2016)
23. Zhu, X., Ghahramani, Z., Lafferty, J.D.: Semi-supervised learning using Gaussian fields and harmonic functions. In: Proceedings of the 20th International Conference on Machine Learning, pp. 912–919 (2003)

A Neural Network Model for Semi-supervised Review Aspect Identification

Ying Ding(✉), Changlong Yu, and Jing Jiang

School of Information Systems, Singapore Management University, Singapore, Singapore
{ying.ding.2011,jingjiang}@smu.edu.sg, changlong.ycl@gmail.com

Abstract. Aspect identification is an important problem in opinion mining. It is usually solved in an unsupervised manner, and topic models have been widely used for the task. In this work, we propose a neural network model to identify aspects from reviews by learning their distributional vectors. A key difference of our neural network model from topic models is that we do not use multinomial word distributions but instead embedding vectors to generate words. Furthermore, to leverage review sentences labeled with aspect words, a sequence labeler based on Recurrent Neural Networks (RNNs) is incorporated into our neural network. The resulting model can therefore learn better aspect representations. Experimental results on two datasets from different domains show that our proposed model can outperform a few baselines in terms of aspect quality, perplexity and sentence clustering results.

1 Introduction

Sentiment analysis of online customer reviews has been well studied for over a decade. One of the key tasks in mining customer reviews is aspect identification [15]. Here aspects refer to features, components and other criteria on which a product or service may be evaluated by online users. Since the seminal work in [10], aspect identification has been recognized as a central problem in mining and summarizing customer reviews. Given a collection of reviews from the same domain (e.g., reviews of restaurants), aspect identification aims to discover a set of aspects, each associated with a set of aspect terms (or a distribution over such terms). For example, from restaurant reviews, we may expect to discover an aspect on service, with aspect terms such as "waiter" and "serve," and another aspect on food, with aspect terms such as "pizza" and "burger." The aspect identification task is useful for downstream tasks such as aspect-based review summarization [32] and product comparison [17].

Aspect identification is generally treated as an unsupervised task and a commonly adopted solution is based on topic models such as LDA (Latent Dirichlet Allocation) [1]. Here each aspect is modeled as a topic, which is essentially a multinomial distribution over words, and reviews are modeled as mixtures of these topics. A number of special topic models have been proposed for aspect identification [9,21,31].

J. Kim et al. (Eds.): PAKDD 2017, Part II, LNAI 10235, pp. 668–680, 2017.
DOI: 10.1007/978-3-319-57529-2_52

With recent advances in neural networks and representation learning for natural language processing, embedding words in a low-dimensional hidden space to capture their distributional behaviors has shown to be effective for a number of data mining tasks [7,28,30]. In this paper, we explore how neural network models can be used to address the review aspect identification problem and whether they can outperform standard topic models. Our work is motivated by two observations: (1) Compared with the traditional multinomial word distribution based language models, neural language models constructed in a continuous space may better handle low-frequency words in reviews and address the data sparsity problem. (2) Sometimes review sentences with aspect terms annotated are available. For example, the Aspect Based Sentiment Analysis task in SemEval-2014 provides such annotated data. It has been shown that neural network models can achieve strong results on the supervised aspect term extraction task [16,29]. We would like to explore how these trained neural network models can be used to help the aspect identification task.

In this work, we propose a neural network model for review aspect identification. Different from existing topic model based approaches to aspect identification, our model is based on continuous space language models, and it uses a small amount of labeled review sentences to train an RNN model for semi-supervised learning. Using reviews from two different domains, we show that our model improves the quality of the identified aspects compared with some baseline models, and both components of our proposed model contribute to the improved performance.

2 Related Work

Unsupervised topic models are one of the most popular techniques used for aspect identification. They have the advantages of requiring no supervision and being easy to extend. A model that jointly considers aspect words and sentiment words was proposed in [14]. Simple prior information based on sentiment lexicons is used in this work. Zhao et al. [31] developed a more advanced model by using a Maximum Entropy classifier to separate words belonging to different types. To further improve the performance of unsupervised topic model, some distant supervision based on domain knowledge or prior information has been incorporated [4–6]. With both users' ratings and reviews available from online review websites, aspect identification based on topic models is jointly studied with many other tasks such as rating prediction [24] and item recommendation [19,27]. While these studies have advanced aspect identification effectively, they do not take advantage of new emerging techniques like neural networks and word embeddings.

Neural networks and word embeddings have been proven to be effective in various data mining tasks, especially supervised learning problems. They have been applied to information retrieval [23], opinion mining [8], recommender systems [11], online advertising [8] and many other various tasks. In recent years, neural networks for unsupervised learning have also been invented. Autoencoder is one representative model among them [12,25]. However, these models lack interpretability. So neural network based topic models are proposed to overcome this shortcoming [2,13,22]. However, no one has combined supervised neural networks and unsupervised neural networks for aspect identification, which is what we study in this paper.

3 Method

In this section, we present our neural network model for aspect identification.

3.1 Problem Formulation

The setup of our aspect identification task is as follows. We assume that we have a set of unlabeled reviews $\mathcal{R}$ from the same domain, e.g., a set of restaurant reviews. In addition, we have a set of review sentences $\mathcal{S}$ from the same domain annotated with aspect terms, as shown in Table 1. Our goal is to discover K aspects from $\mathcal{R}$ and $\mathcal{S}$, where each aspect is associated with some parameter $\mathbf{v}_k$ and from $\mathbf{v}_k$ we can understand the meaning of the kth aspect. In traditional topic model-based approaches to aspect discovery, each $\mathbf{v}_k$ would be a distribution over the words in the vocabulary, and the words with the highest probabilities in $\mathbf{v}_k$ would well represent the aspect. In our work, we do not constrain $\mathbf{v}_k$ to be a probability distribution, as we will explain below.

Table 1. Examples of annotated sentences. Aspect words are highlighted and enclosed with brackets.

From the [**appetizers**] we ate, the [**dim sum**] and other variety of [**food**], it was impossible to criticize.
The [**design**] and [**atmosphere**] are just so good.

3.2 Model Overview

The general idea behind our model is as follows. We aim to re-construct the reviews in $\mathcal{R}$ from a set of parameters capturing various properties of the reviews. To re-construct a review, we treat the review as a bag of sentences and generate the sentences one by one in a probabilistic way. Each sentence will probabilistically be assigned an aspect, and then be treated as a bag of words sharing the same aspect.

Different from standard topic models, however, we also model the context of each word using a recurrent neural network (RNN) and the context will be used to influence the probability of generating the word. Specifically, the probability of generating a word comes from a combination of a number of vectors representing different aspect models and a background model. This kind of a mixture model is inspired by [31]. However, our model has notably the following differences from [31]: (1) Unlike [31], which is an extension of LDA, we do not use multinomial distributions to model topics (i.e., aspects in this case). Instead, we use a neural networks with continuous vectors to derive the probabilities of generating different words. This treatment is similar to a number of recent work on neural topic models [2,22]. (2) Unlike [31], which uses a Maximum Entropy model to incorporate the context of word into its probabilistic modeling, we use an RNN to incorporate the context, which presumably is more effective given the recent success of using RNN models for sequence modeling problems.

3.3 Review Generation Process

Modeling Aspects. We assume that there are K underlying aspects. Similar to [31], which assumes that each aspect has two word distributions, namely an aspect word distribution and an opinion word distribution, we assume that each aspect k has two embedding vectors associated with it: $\mathbf{v}_k \in \mathbb{R}^d$ and $\mathbf{c}_k \in \mathbb{R}^d$. Here $\mathbf{v}_k$ is meant to capture words that directly describe the aspect, such as "pizza" and "cake" for the aspect on food or "waiter" and "waitress" for the aspect on service. $\mathbf{c}_k$ is meant to capture other words closely associated with the aspect but are not considered opinion target terms (as those highlighted terms in Table 1). These may include "delicious" and "tasty" for the aspect on food or "friendly" for the aspect on service. Note however that neither $\mathbf{v}_k$ nor $\mathbf{c}_k$ is a distribution over the words in the vocabulary, and we will explain later how they are used to generate words.

Modeling Background Words. We assume that there is a background distribution over words, which we denote with $\boldsymbol{\theta}^{\mathrm{b}}$. This distribution represents how reviews may contain words not related to any aspect.

Modeling Documents. Similar to [31], we assume that each review has a multinomial distribution over the K aspects. Let us use $\boldsymbol{\beta}_r$ to represent this distribution for the rth review. We also assume that there is a document-independent probability λ that controls how likely a word is associated with an aspect or with the background model $\boldsymbol{\theta}^{\mathrm{b}}$.

Modeling Word Context. We use $w_{r,s,n}$ to represent the nth word in the sth sentence in the rth review. Here $1 \leq w_{r,s,n} \leq V$ is an index in the vocabulary and V is the vocabulary size. We assume that this word has a vector $\mathbf{h}_{r,s,n}$ that encodes its context using an RNN model we will describe later. With this vector $\mathbf{h}_{r,s,n}$ and the RNN model, there is a probability $\pi_{r,s.n}$ associated with word $w_{r,s,n}$ to indicate how likely this word is an opinion target term rather than an opinion term, i.e., how likely $w_{r,s,n}$ is going to be generated from some $\mathbf{v}_k$ or from some $\mathbf{c}_k$.

Review Generation. With the various embedding vectors and probabilities defined above, we now describe the re-construction loss function which we try to minimize in order to learn the parameters. We use the negative log likelihood of generating the words inside all the reviews in $\mathcal{R}$ as our objective function. The overall objective function is as follows:

$$-\log p(\mathcal{R}) = -\sum_{r=1}^{|\mathcal{R}|} \log p(\mathbf{w}_r) = -\sum_{r=1}^{|\mathcal{R}|}\sum_{s=1}^{M_r} \log \sum_{k=1}^{K} \beta_{r,k} p(\mathbf{w}_{r,s}|k),$$

$$p(\mathbf{w}_{r,s}|k) = \prod_{n=1}^{N_{r,s}} p(w_{r,s,n}|k)$$

$$= \prod_{n=1}^{N_{r,s}} \Big[(1-\lambda)\theta^{\mathrm{b}}_{w_{r,s,n}} + \lambda\Big(\pi_{r,s,n}\phi_{k,w_{r,s,n}} + (1-\pi_{r,s,n})\psi_{k,w_{r,s,n}}\Big)\Big],$$

where M_r is the number of sentences in the rth review, $N_{r,s}$ is the number of words in the sth sentence in the rth review, $\mathbf{w}_r$ represents all the words in the rth review, $\mathbf{w}_{r.s}$ represents all the words in the sth sentence in the rth review, and $\boldsymbol{\phi}_k$ and $\boldsymbol{\psi}_k$ are two distributions corresponding to aspect terms and opinion terms, which we will explain below.

Basically the loss function above shows that to generate a review r, for each sentence in the review we pick an aspect k according to the distribution $\boldsymbol{\beta}_r$. Then for each word in this sentence, we generate it either from the background model $\boldsymbol{\theta}_{\mathrm{b}}$ or one of the two models $\boldsymbol{\phi}_k$ and $\boldsymbol{\psi}_k$.

So far the model above is very similar to [31]. However, $\boldsymbol{\phi}_k$ and $\boldsymbol{\psi}_k$ are modeled differently from [31]. Instead of treating these as multinomial distributions and directly learning the probabilities, we assume that they are derived from the embedding vectors $\mathbf{v}_k$ and $\mathbf{c}_k$ as follows:

$$\boldsymbol{\phi}_k = \mathrm{softmax}(\mathbf{v}_k \cdot \mathbf{W}_A),$$
$$\boldsymbol{\psi}_k = \mathrm{softmax}(\mathbf{c}_k \cdot \mathbf{W}_C).$$

$\mathbf{W}_A \in \mathbb{R}^{d\times V}$ and $\mathbf{W}_C \in \mathbb{R}^{d\times V}$ are two matrices to model the semantic representations of words, which are initialized with pre-trained Google word2vec.[1] Each column in them is used to encode one word type.

3.4 RNN to Incorporate Context

We now explain how we obtain $\pi_{r,s,n}$ for each word $w_{r,s,n}$ by making use of the annotated review sentences. Our method is again inspired by the MaxEnt-LDA model [31], in which a Maximum Entropy model was trained on some labeled data to help separate aspect words, opinion words and background words.

[1] https://code.google.com/archive/p/word2vec/.

Algorithm 1. Gibbs-EM algorithm for learning

1: **for** $i \leftarrow 1, maxEpoch$ **do** ▷ $maxEpoch$ is the maximum number of epochs.
2: **E-step:**
3: **for** $r \leftarrow 1, |\mathcal{R}|$ **do**
4: **for** $s \leftarrow 1, M_r$ **do**
5: Sample an aspect $t^i_{r,s}$ according to Formula 1.
6: **end for**
7: **end for**
8: **M-step:**
9: Keep $\mathbf{T}_i$ fixed. Compute the gradient $\frac{\partial \mathcal{L}_i}{\partial \mathbf{\Theta}}$ by back-propagation.
10: Use the gradient to update all parameters $\mathbf{\Theta}$.
11: **end for**

The same idea applies to our problem, but here we use a Recurrent Neural Network (RNN) model, which represents the state of the art for aspect term extraction [16].

The motivation of making use of the labeled review sentences is that there are some patterns we can learn to locate aspect terms. For example, nouns following adjectives which are sentiment words, such as the word "service" in the phrase "excellent service," are more likely to be aspect terms. We can try to learn such patterns from the labeled review sentences, even though the labels only indicate which words are aspect terms but do not group them into aspects.

Because usually there is only a small amount of such labeled review sentences, to address the data sparsity problem, here we again make use of dense vector representations to train a classifier. Specifically, we use Recurrent Neural Network (RNN) models. Let us assume that $(\mathbf{l}_1, \mathbf{l}_2, \ldots, \mathbf{l}_n)$ is the sequence of words in a labeled sentence, where each $\mathbf{l}_i \in \mathbb{R}^d$ is a dense word embedding vector. Let $(y_1, y_2, \ldots, y_n)$ represent the corresponding labels marking the positions of the aspect terms. We can build an RNN model from the sequence $(\mathbf{l}_1, \mathbf{l}_2, \ldots, \mathbf{l}_n)$ as follows:

$$\mathbf{h}_i = f(\mathbf{U}\mathbf{h}_{i-1} + \mathbf{V}\mathbf{l}_i + \mathbf{e}),$$

where $f(\cdot)$ is a non-linear activation function, $\mathbf{U} \in \mathbb{R}^{d_o \times d_o}$, $\mathbf{V} \in \mathbb{R}^{d_o \times d}$ and $\mathbf{e} \in \mathbb{R}^{d_o}$ are parameters to be learned, d_o is the output dimension and $\mathbf{h}_i$ is the hidden state at position i. We can then use $\mathbf{h}_i$ to predict the label y_i through a softmax layer. While there exist some other RNN structures like LSTM(Long Short Term Memory), Bidirectional-RNN, Bidirectional-LSTM and so on, RNN has simpler structure and competitive performance [16]. So we only use RNN to predict $\pi_{r,s,n}$ in this work.

To train this model, we maximize the probabilities of the observed labels in the training dataset $\mathcal{S}$. Given a new sentence, we can use the trained RNN model to obtain the hidden states $\mathbf{h}$, and for each word in the sentence, we can use its corresponding hidden state to obtain a probability $\pi_{r,s,n}$ for the word to be an aspect term.

3.5 Connections with Topic Models

With certain configurations, our model is closely connected with traditional topic models. However, our model learns aspect vectors and uses a linear transformation followed by the softmax function to model topic-word dependencies. Compared with multinomial distributions, which are typically used in topic models, our model can incorporate more information, like semantic meanings of words and topics. In recent years, neural network based topic models have been invented to incorporate pre-trained word embeddings [2,13,22]. Compared with these models, our model is a more general framework. Each component of it can be replaced with other suitable options. So it is easier to extend and adapt to different tasks. Besides this, we uses RNN to separate aspect words from context words, which can potentially help us learn better topics. This has not been used in existing neural topic models.

3.6 Learning

To learn our model, we need to find the optimal values of $\mathbf{v}_k$, $\mathbf{c}_k$, $\boldsymbol{\theta}^b$, $\boldsymbol{\beta}_r$, $\mathbf{W}_A$, $\mathbf{W}_c$ and λ that can minimize the objective function $-\log p(\mathcal{R})$.

Back-propagations cannot be directly used to learn our neural network as there are some constraints placed on $\mathbf{h}_d$. To deal with this, one alternative is variational-EM algorithm. However, it is not an exact estimation algorithm as it tries to optimize the lower bound of the objective function. Instead of using variational inference to approximate posterior distributions at the E-step, we adopt Gibbs sampling to sample an aspect for the sth sentence in the rth review according to

$$p(t_{r,s} = k) = \frac{\beta_{r,k} p(\mathbf{w}_{r,s}|k)}{\sum_{k'} \beta_{r,k'} p(\mathbf{w}_{r,s}|k')}. \tag{1}$$

Then, in the M-step, we apply back-propagation to update all parameters in our neural network with the sampled aspect for sentence fixed. The objective function for the M-step in the ith epoch is

$$\mathcal{L}_i = -\log p(\mathcal{R}|\mathbf{T}_i) = -\sum_{r=1}^{|\mathcal{R}|} \sum_{s=1}^{M_r} \log p(\mathbf{w}_{r,s}|t^i_{r,s}), \tag{2}$$

where $\mathbf{T}_i$ is the sampled aspects of all sentences in epoch i and $t^i_{r,s}$ is the sampled aspect in epoch i for the sth sentence in the rth review. An overview of the learning process can be found in Algorithm 1, where $\boldsymbol{\Theta}$ represents all parameters to be learned: $\boldsymbol{\Theta} = \{\mathbf{v}_k, \mathbf{c}_k, \boldsymbol{\theta}^b, \boldsymbol{\beta}_r, \mathbf{W}_A, \mathbf{W}_C, \lambda\}$, $k \in \{1, 2, \cdots, K\}, r \in \{1, 2, \cdots, |\mathcal{R}|\}$.

4 Experiments

In this section, we evaluate our proposed model from different angles. Through the evaluation we mainly want to test if our neural network model using aspect and context vectors to generate words work better than traditional topic models based on multinomial unigram word distributions for aspect identification. In addition, we also look at the generative ability and the effectiveness of clustering sentences using our model.

We consider the following different models for comparison.

- **LDA:** Latent Dirichlet Allocation. This is a classical topic modeling technique proposed in [1].
- **JST:** Joint Sentiment/Topic Model. It is an extension of LDA that models both sentiments and topics [14].
- **ME-LDA:** LDA with Maximum Entropy classifier [31]. This models uses both traditional topic models based on multinomial unigram word distributions and Maximum Entropy models for supervision.
- **RNN-LDA:** LDA with RNN.
  We replace the maximum entropy classifier in ME-LDA with the trained RNN model to estimate the probability of each word being an aspect word or not. By comparing with this model, we can evaluate the effect of using aspect and context vectors together with softmax to generate words.
- **ME-NA:** Neural network for aspect identification with Maximum Entropy. This is a variation of our model. We replace LDA in ME-LDA with our neural network model.
  By comparing with this model, we can evaluate the usefulness of using RNN instead of standard linear classifiers for the supervision.
- **RNN-NA:** Neural network for aspect identification with RNN. This is our complete model as presented in Sect. 3, where we use both unlabeled and labeled data for aspect identification. We do not fine tune $\mathbf{W}_A$ and $\mathbf{W}_C$, i.e., the word embeddings are not updated during training.
- **RNN-NA-t:** This is also our complete model RNN-NA. However, we initialize $\mathbf{W}_A$ and $\mathbf{W}_C$ with word embeddings and fine-tune them during training.

To compare the models above, we first conduct three experiments to evaluate the quality of identified aspects. Then we do a quantitative evaluation based on perplexity to check the model's ability to predict words in unseen reviews. We also do another quantitative evaluation using sentence clustering to evaluate each model's effectiveness in grouping review sentences into different aspects.

4.1 Data

We use two datasets for our experiments. The first one contains restaurant reviews from the Yelp academic dataset.[2] As the original dataset contains millions of reviews from different businesses, we only keep the restaurant reviews and randomly sample 20,000 from them. The other dataset is a laptop dataset crawled from Amazon, used by [26].[3] For the set of labeled training sentences, we use the sentences tagged with aspect terms from SemEval competitions.

[2] https://www.yelp.com.sg/dataset_challenge.

[3] http://www.cs.virginia.edu/~hw5x/dataset.html.

For the restaurant domain, the training sentences are from SemEval 2014 and 2015, and for the laptop domain, the training sentences are from SemEval 2015.

To pre-process the review data, we remove stop words and words with no pre-trained embeddings. Sentences with less than 3 words are also removed. After preprocessing, the Yelp dataset contains 17948 reviews, with each document containing 9.1 sentences on average and each sentence containing 5.8 words on average. In the Laptop dataset, there are 31,363 documents, where each document has 8.8 sentences on average and each sentence has 7.6 words on average.

4.2 Aspect Quality

Word Intrusion. To evaluate the quality of aspects identified by our models, we conduct the word intrusion experiment [3]. For each discovered aspect, we extract 5 most probable words. We also extract another intrusion word that has a high probability in some other aspect but low probability in the current aspect. There words are then mixed and presented to the annotators to pick out the intrusion word. We ask four graduate students for the annotation. Fleiss' Kappa, which is a standard way to measure agreement among more than two annotators, shows that the inter-annotator agreement is 0.353 for the Yelp dataset and 0.487 for the Laptop dataset. These two scores indicate fair agreement and moderate agreement respectively. Model Precision (MP) is used as the evaluation metric, which is defined as

$$MP = \frac{1}{N}\sum_{a=1}^{N}\frac{M_a}{T}.$$

Here, N is the number of annotators, T is the number of aspects, M_a is the number of intrusion words that are correctly identified by annotator a.

The performances of all models with aspect number set to be 10 and 20 are shown in Table 2. We can see that RNN-NA-t performs the best most of the time, which demonstrates that our model is effective in mining aspects with high quality. RNN-NA can only outperform RNN-NA-t in one case. It proves that fine-tuning word embeddings in our model is important.

Coherence. Besides human evaluation, we also evaluated our models with topic coherence, which is a metric measuring aspect quality based on co-occurrence of words [20]. It is defined as

Table 2. Model precision (MP) of word intrusion by various models.

Dataset	#Aspect	JST	LDA	ME-LDA	RNN-LDA	ME-NA	RNN-NA	RNN-NA-t
Yelp	10	0.63	0.50	0.65	0.45	0.50	0.53	**0.65**
	20	0.44	0.45	0.51	0.40	0.50	**0.63**	0.55
Laptop	10	0.40	0.33	0.50	0.70	0.70	0.58	**0.73**
	20	0.64	0.44	0.59	0.74	0.65	0.55	**0.75**

Table 3. Topic coherence.

Dataset	#Aspect	JST	LDA	ME-LDA	RNN-LDA	RNN-NA	ME-NA	RNN-NA-t
Yelp	10	−3.589	−2.854	−4.421	−4.110	−0.757	−0.639	**−0.363**
	20	−3.579	−2.833	−4.319	−4.129	−0.698	−0.628	**−0.443**
Laptop	10	−3.218	−3.424	−5.476	−5.591	−1.090	−1.077	**−0.866**
	20	−3.236	−3.459	−5.514	−5.698	−1.186	−1.111	**−0.787**

$$C(t, V^{(t)}) = \frac{2}{M(M+1)} \sum_{m=2}^{M} \sum_{l=1}^{m-1} \log \frac{D(v_m^{(t)}, v_l^{(t)}) + 1}{D(v_l^{(t)})},$$

where $V^{(t)}$ contains the M most probable words in topic t. $v_m^{(t)}$ and $v_l^{(t)}$ are the mth and lth words in $V^{(t)}$. $D(v_l^{(t)})$ is the number of documents containing word $v_l^{(t)}$ and $D(v_m^{(t)}, v_l^{(t)})$ is the number of documents containing both $v_m^{(t)}$ and $v_l^{(t)}$.

Table 3 displays the averaged topic coherence of different models. All models based on our proposed neural network can get better performance than others. Meanwhile, RNN-NA-t consistently gets the best performance. It proves that aspects discovered by our models are more coherent than those discovered by the competitors.

Qualitative Evaluation. To qualitatively study the quality of aspects identified by our proposed model, we show 4 sample aspects of the laptop dataset identified by RNN-NA-t and ME-LDA in Table 4. The top 10 most probable words of each aspect are displayed. Words that are closely related to the aspect are emphasized in bold font. From the tables we can see that aspects learned by RNN-NA-t look more coherent and more words are closely related to the topic. The qualitative evaluation shows the advantage of our neural network for aspect identification in discovering meaningful and coherent aspects.

Table 4. Sampled learned aspects from the Laptop dataset.

RNN-NA-t				ME-LDA			
Network	Display	OS	Support	Network	Display	OS	Support
Wifi	**Screen**	**Windows**	**Support**	Windows	**Screen**	**Windows**	**Warranty**
Wireless	**Display**	**OS**	**Service**	Screen	Keyboard	**System**	**Service**
Connection	**Resolution**	**System**	**Customer**	Support	Windows	**OS**	**Customer**
Internet	Keyboard	**Operating**	**Warranty**	**Wireless**	Battery	Screen	**Support**
Windows	**Color**	Software	**Tech**	**Wifi**	Quality	**Operating**	Drive
Driver	**Size**	**XP**	**Shipping**	**Connection**	**Display**	Software	Screen
Card	Quality	**Vista**	Samsung	System	Sound	Use	Hard
Network	**Colors**	Use	Screen	**Internet**	Price	Keyboard	Windows
Drivers	**Brightness**	Works	Battery	Battery	Touch	Drive	Battery
Support	**Retina**	Hardware	System	Keyboard	Drive	Battery	**Shipping**

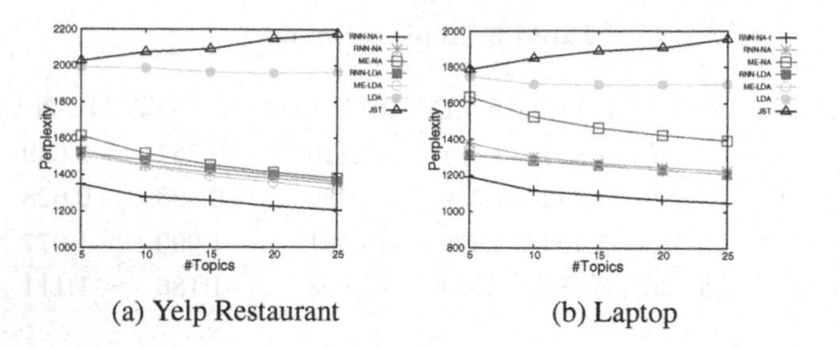

(a) Yelp Restaurant (b) Laptop

Fig. 1. Perplexities over different numbers of aspects for different models.

4.3 Perplexity

We evaluate all models' generative abilities using perplexity, which is a commonly used metric to evaluate the quality of language models and topic models. The definition of perplexity is as follows:

$$\text{perplexity} = \exp(-\frac{1}{N}\sum_{s \in T} P(s)), \tag{3}$$

where T is our held-out test dataset, N is the total number of sentences in it and $P(s)$ is the probability of generating sentence s. In our experiment, we leave 20% of our dataset for testing and train the models based on the remaining 80% dataset. Perplexities over different numbers of aspects are shown in Fig. 1.

We can see that our complete model with fine tuning of word embeddings is performing the best over various numbers of aspects on both datasets. Meanwhile, using RNN models to help separate aspect words from the rest performs better than using Maximum Entropy based models most of the time. Both findings verify that using neural networks in our model can improve generalization capabilities.

Sentence Clustering

To show how topical embeddings learned by different models benefit downstream tasks, we compare the different models in terms of sentence clustering. We manually labeled 100 sentences from the Yelp dataset and 100 sentences from the Laptop dataset. Normalized mutual information [18], which is a popular metric in text clustering, is used to measure performances in our experiment. As topics discovered by JST are sentiment oriented, we do not include it in this evaluation.

The results are shown in Fig. 2. We can see that our proposed neural network models outperform all other competitors. As all sentences are from the same domain, it is uneasy to effectively discover clear aspects and cluster sentences by using co-occurrence statistics. So traditional topic models perform poorly. By learning topic embeddings, our models can improve a lot. Figure 2 also shows that using RNN to help separate out aspect words is much more effective than Maximum Entropy classifier.

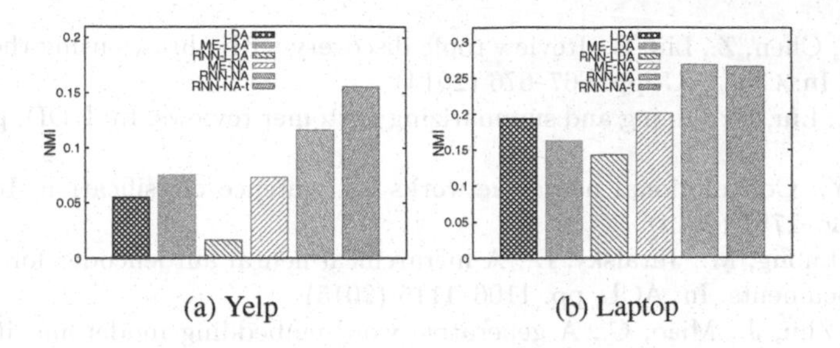

Fig. 2. Normalized mutual information.

5 Conclusions

We explored aspect identification from reviews by proposing a novel neural network model. Our model is able to associate aspects and words using distributional vectors. An RNN model trained on labeled sentences is embedded into our model, which helped the model learn cleaner and more discriminative topics. Experiments on two datasets from different domains show that our model is effective in discovering meaningful aspects, predicting words and benefiting downstream applications such as sentence clustering. In the future, we will explore more complex neural network layers to model aspects and documents, and to jointly train the RNN with the neural network model for aspect identification.

Acknowledgement. This research is supported by the National Research Foundation, Prime Ministers Office, Singapore under its International Research Centres in Singapore Funding Initiative.

References

1. Blei, D.M., Ng, A.Y., Jordan, M.I.: Latent Dirichlet allocation. JMLR **3**, 993–1022 (2003)
2. Cao, Z., Li, S., Liu, Y., Li, W., Ji, H.: A novel neural topic model and its supervised extension. In: AAAI, pp. 2210–2216 (2015)
3. Chang, J., Boyd-Graber, J.L., Gerrish, S., Wang, C., Blei, D.M.: Reading tea leaves: how humans interpret topic models. In: NIPS, pp. 288–296 (2009)
4. Chen, Z., Mukherjee, A., Liu, B.: Aspect extraction with automated prior knowledge learning. In: ACL, pp. 347–358 (2014)
5. Chen, Z., Mukherjee, A., Liu, B., Hsu, M., Castellanos, M., Ghosh, R.: Discovering coherent topics using general knowledge. In: CIKM, pp. 209–218 (2013)
6. Chen, Z., Mukherjee, A., Liu, B., Hsu, M., Castellanos, M., Ghosh, R.: Exploiting domain knowledge in aspect extraction. In: EMNLP, pp. 1655–1667 (2013)
7. Collobert, R., Weston, J., Bottou, L., Karlen, M., Kavukcuoglu, K., Kuksa, P.P.: Natural language processing (almost) from scratch. JMLR **12**, 2493–2537 (2011)
8. Du, H., Xu, X., Cheng, X., Wu, D., Liu, Y., Yu, Z.: Aspect-specific sentimental word embedding for sentiment analysis of online reviews. In: WWW, pp. 29–30 (2016)

9. Fei, G., Chen, Z., Liu, B.: Review topic discovery with phrases using the Pólya urn model. In: COLING, pp. 667–676 (2014)
10. Hu, M., Liu, B.: Mining and summarizing customer reviews. In: KDD, pp. 168–177 (2004)
11. Kim, Y.: Convolutional neural networks for sentence classification. In: EMNLP, pp. 1746–1751 (2014)
12. Li, J., Luong, M., Jurafsky, D.: A hierarchical neural autoencoder for paragraphs and documents. In: ACL, pp. 1106–1115 (2015)
13. Li, S., Zhu, J., Miao, C.: A generative word embedding model and its low rank positive semidefinite solution. In: ACL (2016)
14. Lin, C., He, Y.: Joint sentiment/topic model for sentiment analysis. In: CIKM, pp. 375–384 (2009)
15. Liu, B.: Sentiment analysis and opinion mining. Synth. Lect. Hum. Lang. Technol. **5**(1), 1–167 (2012)
16. Liu, P., Joty, S., Meng, H.: Fine-grained opinion mining with recurrent neural networks and word embeddings. In: EMNLP, pp. 1433–1443 (2015)
17. Lu, Y., Zhai, C., Sundaresan, N.: Rated aspect summarization of short comments. In: WWW, pp. 131–140 (2009)
18. Manning, C.D., Raghavan, P., Schütze, H.: Introduction to Information Retrieval. Cambridge University Press, New York (2008)
19. McAuley, J., Leskovec, J.: Hidden factors and hidden topics: understanding rating dimensions with review text. In: RecSys, pp. 165–172 (2013)
20. Mimno, D., Wallach, H.M., Talley, E., Leenders, M., McCallum, A.: Optimizing semantic coherence in topic models. In: EMNLP, pp. 262–272 (2011)
21. Mukherjee, A., Liu, B.: Aspect extraction through semi-supervised modeling. In: ACL, pp. 339–348 (2012)
22. Nguyen, D.Q., Billingsley, R., Du, L., Johnson, M.: Improving topic models with latent feature word representations. TACL **3**, 299–313 (2015)
23. Severyn, A., Moschitti, A.: Learning to rank short text pairs with convolutional deep neural networks. In: SIGIR, pp. 373–382 (2015)
24. Wang, H., Ester, M.: A sentiment-aligned topic model for product aspect rating prediction. In: EMNLP, pp. 1192–1202 (2014)
25. Wang, H., Wang, N., Yeung, D.Y.: Collaborative deep learning for recommender systems. In: KDD, pp. 1235–1244 (2015)
26. Wang, H., Lu, Y., Zhai, C.: Latent aspect rating analysis without aspect keyword supervision. In: KDD, pp. 618–626 (2011)
27. Wu, Y., Ester, M.: FLAME: a probabilistic model combining aspect based opinion mining and collaborative filtering. In: WSDM, pp. 199–208 (2015)
28. Zhai, S., Chang, K., Zhang, R., Zhang, Z.M.: Deepintent: learning attentions for online advertising with recurrent neural networks. In: KDD, pp. 1295–1304 (2016)
29. Zhang, M., Zhang, Y., Vo, D.: Neural networks for open domain targeted sentiment. In: EMNLP, pp. 612–621 (2015)
30. Zhang, Q., Gong, Y., Wu, J., Huang, H., Huang, X.: Retweet prediction with attention-based deep neural network. In: CIKM, pp. 75–84 (2016)
31. Zhao, W.X., Jiang, J., Yan, H., Li, X.: Jointly modeling aspects and opinions with a MaxEnt-LDA hybrid. In: EMNLP, pp. 56–65 (2010)
32. Zhu, J., Wang, H., Zhu, M., Tsou, B.K., Ma, M.: Aspect-based opinion polling from customer reviews. IEEE Trans. Affect. Comput. **2**(1), 37–49 (2011)

Behavioral Data Mining

Unsupervised Embedding for Latent Similarity by Modeling Heterogeneous MOOC Data

Zhuoxuan Jiang[1(✉)], Shanshan Feng[2], Weizheng Chen[1], Guangtao Wang[3], and Xiaoming Li[1]

[1] School of EECS, Peking University, Beijing, China
{jzhx,cwz,lxm}@pku.edu.cn
[2] School of CSE, Nanyang Technological University, Singapore, Singapore
sfeng003@e.ntu.edu.sg
[3] University of Michigan, Ann Arbor, USA
gtwang@umich.edu

Abstract. Recent years have witnessed the prosperity of Massive Open Online Courses (MOOCs). One important characteristic of MOOCs is that video clips and discussion forum are integrated into a one-stop learning setting. However, discussion forums have been in disorder and chaos due to 'Massive' and lack of efficient management. A technical solution is to associate MOOC forum threads to corresponding video clips, which can be regarded as a problem of representation learning. Traditional textual representation, e.g. Bag-of-words (BOW), do not consider the latent semantics, while recent semantic word embeddings, e.g. Word2vec, do not capture the similarity between documents, i.e. latent similarity. So learning distinguishable textual representation is the key to resolve the problem. In this paper, we propose an effective approach called No-label Sequence Embedding (NOSE) which can capture not only the latent semantics within words and documents, but also the latent similarity. We model multiform MOOC data in a heterogeneous textual network. And we learn the low-dimensional embeddings without labels. Our proposed NOSE owns some advantages, e.g. course-agnostic, and few parameters to tune. Experimental results suggest the learned textual representation can outperform the state-of-the-art unsupervised counterparts in the task of associating forum threads to video clips.

Keywords: Unsupervised embedding · Latent similarity · Heterogeneous · MOOC

1 Introduction

As a new paradigm of online learning environments, Massive Open Online Courses (MOOCs) are rapidly developed in recent years, e.g. Coursera, edX and Udacity. Millions of learners have been benefited from the free and open courses. Compared with traditional online education, an important characteristic of MOOCs is the one-stop learning setting which integrates video clips and

J. Kim et al. (Eds.): PAKDD 2017, Part II, LNAI 10235, pp. 683–695, 2017.
DOI: 10.1007/978-3-319-57529-2_53

a discussion forum. However, due to 'Massive' and lack of efficient management, there are overload and chaos in majority of MOOC forums [21]. Based on our statistic of a MOOC forum as shown in Fig. 1, the various categories include question of concept understanding, enquiry and advice of course arrangement, feedback, and etc. In order to archive the threads, there are several methods which have limited effect [3], e.g. defining sub-forums in advance empirically according to weeks and asking learners to tag threads. Some machine learning methods have been studied recently, e.g. content-related thread identification [21], confusion classification [1], sentiment classification [15,20], and so on. But they are developed for specific research problems and cannot be applied for other tasks.

Another idea is to associate threads to corresponding knowledge points, i.e. video clips. The feasibility relies on some MOOCs' natures. Firstly, the pace of a MOOC is consistent to its off-line counterpart, i.e. learning contents are regularly opened to learners by each week [2]. The temporal information can be leveraged as a constraint. Secondly, learning contents are broken down to small video clips, where each one usually lasts about 10 min and just contains one piece of knowledge point in most situations. This makes the association result educationally meaningful. As the snapshot showed in Fig. 2, the syllabus (or knowledge points) of a MOOC is usually organized in a two-level structure. The first level corresponds to weeks and the second are in the form of several video clips. As to corresponding forums, by mining rich information, e.g. behavioral log, textual content and social relationship [8], it is probable to fulfil the task.

The association between threads and video clips is profound. Both the learners and teachers can readily seek what they want, such that the learning efficiency and teaching quality can be improved. After completing the association, the result can be used as the prerequisite of subsequential applications, such as knowledge management, learning guidance, and question answering system.

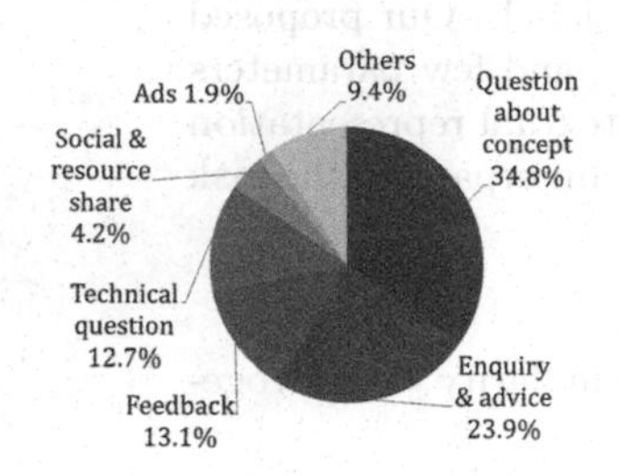

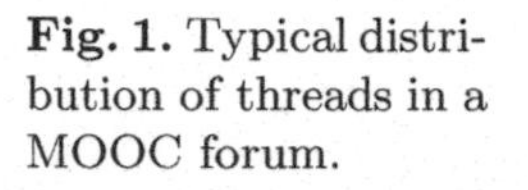
Fig. 1. Typical distribution of threads in a MOOC forum.

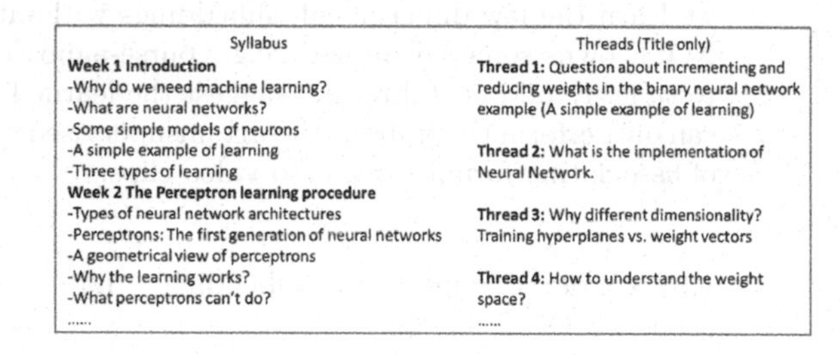

Fig. 2. A snapshot of video clips and threads.

The task of association between threads and course video clips, actually, can be regarded as a document ranking problem in the view of information retrieval. An intuitive idea is to use the weighted Bag-of-words (BOW) [16] as features to calculate the similarities between threads and video clips, and then rank them by the computed similarities. However, BOW cannot effectively capture semantic knowledge. Moreover, recent studied semantic word embeddings, e.g. Word2vec [13], cannot capture the similarity of documents, and we call it latent similarity. The latent similarity is crucial to distinguish whether a document can be associated to the right target. In our task, the textual representation is expected to preserve both the latent semantics and latent similarity.

On the other hand, as a kind of real-world data, it is very expensive to obtain labeled training dataset, since associating MOOC threads to video clips requires expertise. So we have to leverage more information to compensate the lack of labels during learning word embeddings.

In this paper, we focus on the technical problem that how to discover latent similarity knowledge from heterogeneous data. There are several challenges to solve this problem: (1) how to leverage characteristics of data to compensate the lack of labels, (2) how to build a model which can integrate heterogeneous information, and (3) how to learn the latent similarity from the new data model.

By our observation, we find four kinds of information helpful: (1) word sequence in threads and posts, (2) word sequence in subtitles of video clips, (3) learners' behavioral sequence of clickstream about videos and threads, and (4) timestamp sequence of video clips (publishing time) and threads (posting time). The reason that they are chosen will be discussed in Sect. 3.1.

Our basic idea is to first model all the sequential information into a heterogeneous textual network, and then embed the nodes of the network, i.e. words and documents, to low-dimensional vectors which can preserve both latent semantics and latent similarity. At last the learned vectors are used to rank video clips by calculating similarities between threads and them under a constraint of time. Inspired by methods of distributional representation which have been proven successful in a mass of natural language processing tasks, e.g. word analogy [13], text classification [10], sentiment analysis [11] and POS tagging [5], we propose an approach called No-label Sequence Embedding (NOSE).

To summarize, our contributions in this paper include:

- We design a novel embedding model to integrate heterogeneous sequential information, e.g. word sequence and clickstream sequence, in order to learn latent semantics and latent similarity simultaneously.
- We develop a new computing framework to learn the textual representation based on the proposed model.
- We collect two real-world MOOC data and conduct thorough experiments. Results confirm that our proposed approach can effectively learn the latent similarity without labels by modeling heterogeneous sequential information. And our novel textual embeddings can outperform the state-of-the-art counterparts in our task.

2 Related Work

It is a fundamental research question that how to better represent text. Existing approaches can be generally classified into two aspects: unsupervised and supervised. Supervised embeddings, such as deep neural network, need to be fed with labels [9,12]. However, in the association task, we do not have the label information. Thus, we review related work merely within this field of unsupervised learning textual representation.

Unsupervised word embeddings are usually universal and can be applied to various tasks. The process of learning them is often efficient to scale up to millions of documents. Other advantages of unsupervised embeddings include no label required and few parameters to tune. However, unsupervised embeddings are lightly under-performed compared with supervised ones in most specific tasks [18]. Among most learning methods of unsupervised embeddings, the information of textual co-occurrence in local context at different levels are leveraged. For example, CBOW and Skip-gram [13] learn word embeddings based on word co-occurrence. Para2vec [10] utilizes word and document co-occurrence to learn their embeddings. Hierarchical Document Vector (HDV) [6] leverages both document co-occurrence and contents to achieve better representation. Latest Predictive Text Embedding (PTE) [18] models text to a uniform heterogeneous network and obtains the state-of-the-art performance on textual classification. However, it leverages labels to guarantee the performance of classification. It cannot be adopted directly to solve the proposed problem in this paper.

Another series of methods for representing text without labels are discussed in the task of Dataless Classification [4,17]. These methods commonly require large-scale world knowledge, e.g. Wiki data, to extract textual features. But world knowledge is hard to obtain and process in many cases. Also similar to BOW and Word2vec, the latent similarity still cannot be embodied in the embeddings. By the way, this is also a classification problem, which is different from ours.

3 No-Label Sequence Embedding and Ranking

In this section, we introduce our approach for learning word embeddings and document ranking. Firstly, we introduce the motivation to extract the information of latent similarity from four kinds of sequential data. Then we state the data model to integrate heterogeneous information. Then the method of embedding the heterogeneous network to low-dimensional vectors is described. At last we introduce the algorithm to rank documents within a constraint of timestamps sequence. We can place the problem definition after we have introduced the data. Otherwise, people may not understand the definition.

3.1 Data Model

In order to compensate the missing of labels, the data model should integrate as much as information. Besides, it also should capture the information of latent similarity. Based on our observation of multiform MOOC data, subtitles of video clips and contents of threads are essential for associating them. The subtitles of video clips are well-organized and formal, since they are generated by instructors. While the contents of threads are written by various learners, thus they are colloquial and informal. We should separately learn their embeddings because this way can preserve their linguistic peculiarity, i.e. latent similarity. This intuition is also confirmed by our experimental analysis.

In addition, we find learners' clickstream logs, that is records of watching videos, reading threads and posting threads in chronological order, indicate that the adjacent entities, especially a video clip and a thread, are semantically relevant. An intuitive explanation is that a learner may jump between videos and threads either he wants to seek some further discussion when he is watching a video, or he wants to review the relevant video contents when he is reading a thread. So behavioral logs contain the latent similarity.

In order to integrate all these different sources of text, we design a heterogeneous network which can facilitate learning embeddings both separately and cooperatively. We leverage the co-occurrence of different textual levels and construct four sub-networks which comprise a comprehensive heterogeneous network as Fig. 3 shows.

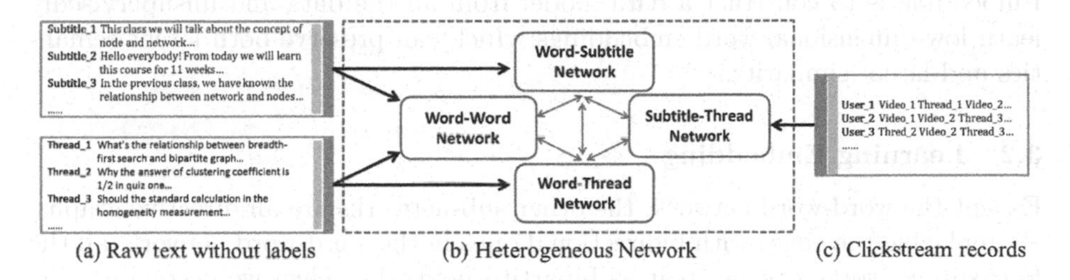

Fig. 3. Data model constructed from raw sequential MOOC data.

Word-Word Network. A word-word network is based on word co-occurrence in word sequence and denoted as $\mathcal{G}_{ww} = (\mathcal{V}, \mathcal{E}_{ww})$. The $\mathcal{V}$ is the word vocabulary and $\mathcal{E}_{ww}$ is the edge between words. The edge weight w_{ij} is defined as the co-occurrence times of words v_i and v_j within a fixed-size window. Word co-occurrence is basic information to capture semantics between words.

Word-Subtitle Network. A word-subtitle network is based on word sequence in document level contexts and denoted as $\mathcal{G}_{ws} = (\mathcal{V} \cup \mathcal{S}, \mathcal{E}_{ws})$. The $\mathcal{V}$ is the word vocabulary and $\mathcal{S}$ is the set of subtitles. $\mathcal{E}_{ws}$ is the edge between words and subtitles. The edge weight w_{ij} is defined as the occurrence times of word v_i in subtitle s_j. Word-document co-occurrence is to capture semantics between words and documents.

Word-Thread Network. Similar to word-subtitle network, a word-thread network is also based on word sequence in document level contexts and denoted as $\mathcal{G}_{wt} = (\mathcal{V} \cup \mathcal{T}, \mathcal{E}_{wt})$. The only difference is that $\mathcal{T}$ is the set of threads and $\mathcal{E}_{wt}$ is the edge between words and threads. We treat them as two different networks due to their different latent linguistic styles as mentioned before.

Subtitle-Thread Network. A subtitle-thread network is based on document sequence in local context of learners' clickstream records and denoted as $\mathcal{G}_{st} =$

$(\mathcal{S} \cup \mathcal{T}, \mathcal{E}_{st})$. The $\mathcal{S}$ is the set of subtitles and $\mathcal{T}$ is the set of threads. $\mathcal{E}_{st}$ is the edge between subtitles and threads. The edge weight w_{ij} is defined as the co-occurrence times of subtitle s_i and thread t_j in learners' clickstream logs within a window size. This network may capture knowledge of both latent semantics and latent similarity between subtitles and threads.

Heterogeneous Network. Combining all the four sub-networks via common nodes, we can get a heterogeneous network. Note that we only study the heterogeneous network with four sub-network by considering the latent semantics information. The heterogeneous network can be extended by adding more sub-networks.

Based on the heterogeneous networks, we define the No-label Sequence Embedding problem as follows.

No-label Sequence Embedding. Given multiform sequential data, e.g. raw behavioral sequence and textual sequence, the problem of No-label Sequence Embedding is to construct a data model from all the data and unsupervisedly learn low-dimensional word embeddings which can preserve both latent semantics and latent similarity.

3.2 Learning Embedding

Except the word-word network, the other sub-networks are all directive graphs. By replacing the edges with bidirectional ones in the word-word network, all the four sub-networks can be treat as bipartite networks. Then we decompose the task of heterogeneous network embedding to four sub-tasks of bipartite network embedding.

There are several network embedding, such as DeepWalk [14], LINE [19] and node2vec [7]. However, all of them are proposed for homogeneous network, and cannot be simply used for the heterogeneous network. Inspired by LINE, we develop a new algorithm to learn the embedding of heterogeneous networks.

Bipartite Network Embedding. Given an arbitrary bipartite network $\mathcal{G} = (\mathcal{V}_A \cup \mathcal{V}_B, \mathcal{E})$, where $\mathcal{V}_A$ and $\mathcal{V}_B$ represent two separate sets of nodes and $\mathcal{E}$ is the set of edges between nodes. The conditional probability of observing node v_i in set $\mathcal{V}_A$ given the node v_j in set $\mathcal{V}_B$ is defined as:

$$p(v_i|v_j) = \frac{\exp(\boldsymbol{u}_i^T \cdot \boldsymbol{u}_j)}{\sum_{i' \in A} \exp(\boldsymbol{u}_{i'}^T \cdot \boldsymbol{u}_j)} \tag{1}$$

where $\boldsymbol{u}_i$ is the embedding vector of node v_i in $\mathcal{V}_A$, and $\boldsymbol{u}_j$ is the embedding vector of node v_j in $\mathcal{V}_B$. Equation (1) actually computes the conditional probability of any v_j in $\mathcal{V}_B$ over all nodes in $\mathcal{V}_A$. Thus the second-order proximity can be preserved if let $p = (\cdot|v_j)$ close to its empirical distribution $\hat{p} = (\cdot|v_j)$. So the objective function to be minimize is:

$$O = \sum_{j \in B} \lambda_j d(\hat{p}(\cdot|v_j), p(\cdot|v_j)) \tag{2}$$

where $d(\cdot,\cdot)$ is the KL-divergence between two probability distribution, λ_j is the importance of vertex v_j in the network and can be defined as the degree $degree_j = \sum_i w_{ij}$, and the empirical distribution can be defined as $\hat{p}(v_i|v_j) = \frac{w_{ij}}{degree_j}$. Equation (2) can be simplified by removing some constants as:

$$O = -\sum_{(i,j)\in\mathcal{E}} w_{ij} \log p(v_j|v_i) \tag{3}$$

To optimize (3), we utilize stochastic gradient descent method of edge sampling [19] and negative sampling [13] which are fast and efficient, instead of computing the summation of the entire set of nodes in $\mathcal{V}_A$. Firstly a positive sample is randomly selected based on probability proportional to its weight w_{ij}, and then a fixed size of negative samples are selected based on a noise distribution $p_n(j)$. By using negative sampling, Equation (3) is in detail replaced by:

$$O = \sum_{(i,j)\in\mathcal{E}} \left\{ \log \sigma(\boldsymbol{u}_i^T \cdot \boldsymbol{u}_j) + \sum_{j=1}^{K} E_{v_n \sim P_n(v)} [\log \sigma(-\boldsymbol{u}_n^T \cdot \boldsymbol{u}_j)] \right\} \tag{4}$$

where $\sigma(x) = 1/(1+\exp(-x))$ is the sigmoid function; the first term means positive samples and the second term means negative samples of which K is the fixed size. Noise distribution $P_n(v) \propto d_v^{3/4}$ is the same as in [13] where d_v is the out degree of node v.

The reason for adopting the edge sampling is to reduce the large variance of edge weights, because very-large-weight edges may dominate the optimization process. An alternative idea is to duplicate the edges proportional to their weights. However, this will deteriorate the efficiency greatly due to the great increase of edges.

Heterogeneous Network Embedding. In order to learn the word embeddings of four sub-networks separately and cooperatively, a simple strategy is that each time one bipartite network is selected to learn eambeddings and the intermediate results are transited to other bipartite networks. Although subtitle ids and thread ids are two different kinds of nodes, for simplicity, we regard them as the same one since there is no overlapping between them. Thus, in total we set two kinds of nodes, document id (d) and words (v). They are shared between bipartite networks. The whole objective function is:

$$O_{NOSE} = O_{ww} + O_{ws} + O_{wt} + O_{st} \tag{5}$$

where

$$O_{ww} = -\sum_{(i,j)\in E_{ww}} w_{ij} \log p(v_i|v_j) \tag{6}$$

$$O_{ws} = -\sum_{(i,j)\in E_{ws}} w_{ij} \log p(v_i|d_j) \tag{7}$$

$$O_{wt} = -\sum_{(i,j)\in E_{wt}} w_{ij} \log p(v_i|d_j) \tag{8}$$

$$O_{st} = -\sum_{(i,j)\in E_{st}} w_{ij} \log p(d_i|d_j) \tag{9}$$

Although we can assign weights to each bipartite network for proportionally selecting edges to learn each once, we choose to equally and sequentially learn embeddings for simplicity. As shown in Algorithm 1, the algorithm is very efficient because the time complexity is just $O(KN)$ and K is a small number.

3.3 Document Ranking

After learning word embeddings, we simply use them to calculate the similarity between threads and subtitles with metric of Cosine distance. Considering the trait of MOOC data that the time of threads posted are usually behind their corresponding video clips. So we have a constraint to make the ranking more reasonable. The algorithm detail is shown in Algorithm 2. The reason for why we do not integrate the subtitle sequence and thread sequence based on timestamps into the heterogeneous network is that we believe they may contain little semantics between any two entities.

Algorithm 1. Learning embeddings of heterogeneous network

INPUT: $\mathcal{G}_{ww}$, $\mathcal{G}_{ws}$, $\mathcal{G}_{wt}$, $\mathcal{G}_{st}$, number of samples N, number of negative samples K
OUTPUT: word embeddings $\boldsymbol{w}$
1:while *iteration* $\leq N$:
2: sample an edge e_{ij} and K negative edges from E_{ww}, update $\boldsymbol{u}_{v_i}$ and $\boldsymbol{u}_{v_j}$
3: sample an edge e_{ij} and K negative edges from E_{ws}, update $\boldsymbol{u}_{d_i}$ and $\boldsymbol{u}_{v_j}$
4: sample an edge e_{ij} and K negative edges from E_{wt}, update $\boldsymbol{u}_{d_i}$ and $\boldsymbol{u}_{v_j}$
5: sample an edge e_{ij} and K negative edges from E_{st}, update $\boldsymbol{u}_{d_i}$ and $\boldsymbol{u}_{d_j}$
6:return word embeddings

Algorithm 2. Document Ranking

INPUT: thread embeddings T, subtitle embeddings S, timestamps $Time$
OUTPUT: rankings of subtitles for each thread L
1:for t_i in T:
2: for s_j in S
3: if $Time_{t_i} > Time_{s_j}$:
4: $L_{t_i}.add(s_j, \cos(t_i, s_j))$
5: $L_{t_i}.rank()$
6:return L

4 Experiment

4.1 Data Sets

We collect the sequential data of two MOOCs from Coursera[1] and China University MOOC[2] respectively. The former is an interdiscipline called *People and Network* which involves computer science, social science and economics, while the other is a conceptual course called *Introduction of MOOC* which introduces a new concept. From both course, we collect subtitles of video clips, forum contents, learners' sequential clickstream log and timestamps sequence. Since our textual data is mostly in Chinese, we preprocess the raw data with a word segment tool consistently before evaluating our method.

As to evaluation, we invite the course TAs to tag some threads in advance. Due to the chaos of MOOC forum threads as mentioned before, we discard the threads about technical operation, social contact, advice to the course, thoughts and others that are irrelevant to course contents. At a result we have 103 valid threads of *People and Network* and 254 valid threads of *Introduction to MOOC* in total. They are our whole available test samples. In some situations, there may be more than one topics being asked in an initial post, we allow to tag the test sample with the two most possible labels. The numbers of double-labeled test samples in each course are 8/103 and 119/254 respectively. Table 1 shows the course information.

Table 1. Statistics of two MOOC datasets.

Course name	#users	#video clips	#threads	#posts
People and Network	10,807	60	219	1,206
Introduction to MOOC	3,949	19	557	7,177

4.2 Baselines and Parameters

We only compare our embeddings with unsupervised rivals. Consistently, the representation of subtitles is set as the average of all the word vectors which belong to that document. While the representation of threads only use the word vectors which belong to the title and the initial post of a thread. However, all posts are kicked in for learning word embeddings.

- BOW: the classical text representation which firstly builds a vocabulary V with the whole textual contents and then each document is represent as $|V|$-dimensional vector like a bag of words. We remove stop words from the vocabulary and TFIDF weights are set to each dimension.

[1] https://www.coursera.org, which is an educational technology company that offers MOOCs worldwide.

[2] http://www.icourse163.org, which is a leading MOOCs platform in China. Supported by Ministry of Education of the People's Republic of China and NetEase, Inc.

- CBOW: the state-of-the-art word embeddings proposed by [13] which uses context to predict the target word embedding.
- Skip-gram: another version of the state-of-the-art word embeddings proposed by [13] which uses the target word to predict its context.
- Pare2vec: the state-of-the-art word embeddings which considers the document-level context information [10].
- LINE: the large-scale information network embeddings which can be used to textual network, but only homogeneous network. Here we simply combine different kinds of bipartite networks to one and see how LINE performs without separately treating them.
- NOSE: our proposed no-label sequence embeddings which leverages all information. We also try to remove one sub-network ordinally to see their contribution degree.

The dimension of word vectors is empirically set as 100. In CBOW, Skip-gram, Para2vec, the window sizes are all set as 5. In LINE and NOSE, the number of negative samples are also set as 5, while the window size used in constructing $\mathcal{G}_{st}$ is set as 3 because this can get best performance shown in later experiments. Especially, in LINE and NOSE, the total number of edge samples, N, is set 50 million since we find larger number may cause over-fitting.

4.3 Results and Analysis

As a ranking result, the measure metric we use is precision averaged by 10 times of runs. From Table 2, we can find traditional sematic word embeddings, CBOW, Skip-gram and Para2vec, are comparable to BOW in terms of precision @1. Separate learning the four sub-networks plays a crucial role since NOSE performs better than LINE. This suggests latent similarity is captured by our learning approach. Among the four sub-networks, $\mathcal{G}_{st}$ contributes the most while $\mathcal{G}_{ww}$ degrades the performance both in the series of LINE and NOSE, confirming that $\mathcal{G}_{st}$ contains the latent similarity while $\mathcal{G}_{ww}$ only contains latent semantics. Building a data model in such a form of textual network may also strengthen the distinctive relationship between documents, comparing LINE and NOSE with CBOW, Skip-gram and Para2vec, indicating that the network can capture both information of word-word semantics and word-document semantics simultaneously. Although BOW is competitive on results of @3 and @5, we concern more about precision @1 because our task is to associate threads to videos rather than a real search problem. We find NOSE ($\mathcal{G}_{ws}+\mathcal{G}_{wt}+\mathcal{G}_{st}$) algorithm can achieve the best result @1. Table 3 shows that the constraint of time plays another crucial role during document ranking.

Parameter Sensitivity. In this part, LINE and NOSE utilize information of $\mathcal{G}_{ws}$, $\mathcal{G}_{wt}$ and $\mathcal{G}_{st}$ consistently. Figure 4 shows the results of different vector dimensions used during learning word embeddings. We find NOSE is better than the others with various number of dimensions. Considering the trade-off between learning efficiency and precision, number of dimensions is set as 100.

Table 2. Ranking precision result of two MOOC datasets.

Algorithm	*People and network*			*Introduction to MOOC*		
	@1	@3	@5	@1	@3	@5
BOW	0.583	0.845	**0.903**	0.449	0.811	**0.906**
CBOW	0.563	0.718	0.786	0.512	0.697	0.795
Skip-gram	0.592	0.738	0.786	0.551	0.744	0.866
Para2vec	0.583	0.670	0.777	0.524	0.713	0.827
LINE($\mathcal{G}_{ws} + \mathcal{G}_{wt}$)	0.621	0.845	0.883	0.535	0.807	0.898
LINE($\mathcal{G}_{ws} + \mathcal{G}_{wt} + \mathcal{G}_{ww}$)	0.592	0.786	0.854	0.394	0.728	0.846
LINE($\mathcal{G}_{ws} + \mathcal{G}_{wt} + \mathcal{G}_{st}$)	0.680	0.825	0.883	0.646	0.799	0.902
LINE(all)	0.650	0.767	0.845	0.406	0.728	0.862
NOSE($\mathcal{G}_{ws} + \mathcal{G}_{wt}$)	0.738	0.805	0.874	0.654	0.803	0.890
NOSE($\mathcal{G}_{ws} + \mathcal{G}_{wt} + \mathcal{G}_{ww}$)	0.699	0.835	0.874	0.657	**0.827**	0.886
NOSE($\mathcal{G}_{ws} + \mathcal{G}_{wt} + \mathcal{G}_{st}$)	**0.776**	**0.845**	0.883	**0.693**	0.803	0.870
NOSE(all)	0.767	0.825	0.874	0.685	0.803	0.874

Table 3. Ranking precision @1 result without time constrain by NOSE.

Algorithm	*People and network*	*Introduction to MOOC*
NOSE($\mathcal{G}_{ws} + \mathcal{G}_{wt}$)	0.631	0.520
NOSE($\mathcal{G}_{ws} + \mathcal{G}_{wt} + \mathcal{G}_{ww}$)	0.621	0.512
NOSE($\mathcal{G}_{ws} + \mathcal{G}_{wt} + \mathcal{G}_{st}$)	0.670	0.567
NOSE(all)	0.641	0.547

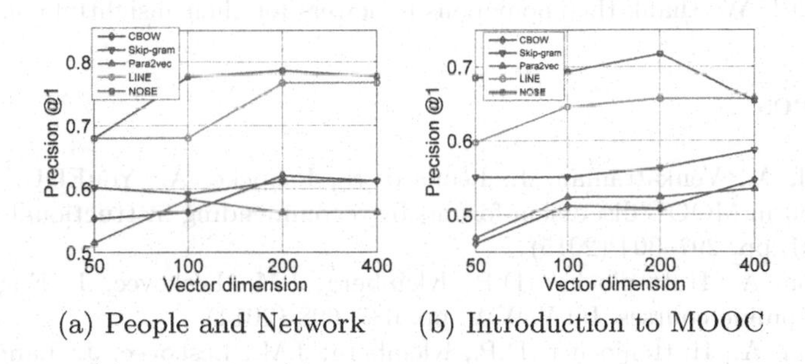

Fig. 4. Precision @1 of different vector dimensions. $\mathcal{G}_{ws}$, $\mathcal{G}_{wt}$ and $\mathcal{G}_{st}$ are utilized.

Figure 5 shows the results of different window sizes used when constructing $\mathcal{G}_{st}$. We set window sizes as 3, 5, 7 and 9, and find NOSE performs best with size of 3 while LINE is best with 5. This result may stem from the guess that small local context already can capture the latent similarity during separately learning embeddings, while LINE needs larger contexts.

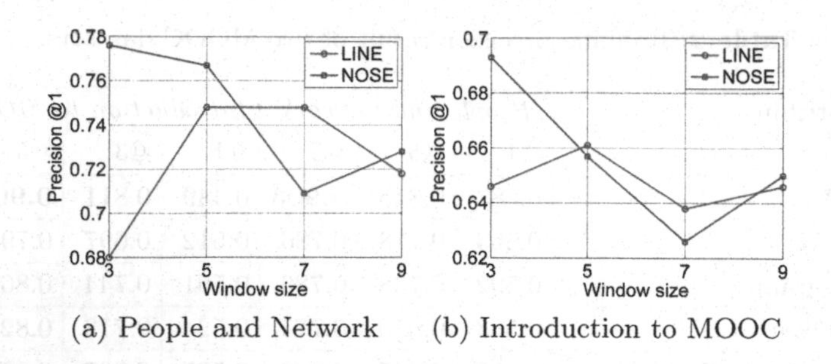

(a) People and Network (b) Introduction to MOOC

Fig. 5. Precision @1 of different window sizes for constructing $\mathcal{G}_{st}$.

5 Conclusion

Archiving forum threads to video clips is meaningful both for instructors and learners. This task is technically regarded as a problem of document ranking. In order to solve the problem, we propose an approach to learn latent semantics and latent similarity simultaneously. Our approach can also overcome the constraints of no label and heterogeneous data, which often happen in real-world datasets. The experimental results perform well and confirm the effectiveness of our approach for learning latent similarity. Last but not the least, there is still room to improve the precision by leveraging more effective information, e.g. in the view of instructor's behavior.

Acknowledgments. This research is supported by the National Research Foundation, Prime Ministers Office, Singapore under its IDM Futures Funding Initiative, China NSFC with Grant No.61532001 and No.61472013, and China MOE-RCOE with Grant No.2016ZD201. We thank the anonymous reviewers for their insightful comments.

References

1. Agrawal, A., Venkatraman, J., Leonard, S., Paepcke, A.: YouEDU: Addressing confusion in MOOC discussion forums by recommending instructional video clips. In: EDM, pp. 297–304 (2015)
2. Anderson, A., Huttenlocher, D.P., Kleinberg, J.M., Leskovec, J.: Engaging with massive online courses. In: WWW, pp. 687–698 (2014)
3. Anderson, A., Huttenlocher, D.P., Kleinberg, J.M., Leskovec, J.: Language independent analysis and classification of discussion threads in coursera MOOC forums. In: IRI, pp. 654–661 (2014)
4. Chang, M.W., Ratinov, L.A., Roth, D., Srikumar, V.: Importance of semantic representation: dataless classification. In: AAAI, pp. 830–835 (2008)
5. Collobert, R., Weston, J., Bottou, L., Karlen, M., Kavukcuoglu, K., Kuksa, P.: Natural language processing (almost) from scratch. J. Mach. Learn. Res. **12**, 2493–2537 (2011)
6. Djuric, N., Wu, H., Radosavljevic, V., Grbovic, M., Bhamidipati, N.: Hierarchical neural language models for joint representation of streaming documents and their content. In: WWW, pp. 248–255 (2015)

7. Grover, A., Leskovec, J.: node2vec: Scalable feature learning for networks. In: KDD, pp. 855–864 (2016)
8. Huang, J., Dasgupta, A., Ghosh, A., Manning, J., Sanders, M.: Superposter behavior in MOOC forums. In: L@S, pp. 117–126 (2014)
9. Kim, Y.: Convolutional neural networks for sentence classification. In: EMNLP, pp. 1746–1751 (2014)
10. Le, Q.V., Mikolov, T.: Distributed representations of sentences and documents. In: ICML, pp. 1188–1196 (2014)
11. Mesnil, G., Mikolov, T., Ranzato, M., Bengio, Y.: Ensemble of generative and discriminative techniques for sentiment analysis of movie reviews (2014), arXiv preprint arXiv:1412.5335
12. Mikolov, T., Karafit, M., Burget, L., Cernocký, J., Khudanpur, S.: Recurrent neural network based language model. In: INTERSPEECH, pp. 1045–1048 (2010)
13. Mikolov, T., Sutskever, I., Chen, K., Corrado, G.S., Dean, J.: Distributed representations of words and phrases and their compositionality. In: NIPS, pp. 3111–3119 (2013)
14. Perozzi, B., Al-Rfou', R., Skiena, S.: Deepwalk: Online learning of social representations. In: KDD, pp. 701–710 (2014)
15. Ramesh, A., Kumar, S.H., Foulds, J.R., Getoor, L.: Weakly supervised models of aspect-sentiment for online course discussion forums. In: ACL, pp. 74–83 (2015)
16. Salton, G., Buckley, C.: Term-weighting approaches in automatic text retrieval. Inf. Process. Manage. **24**(5), 513–523 (1988)
17. Song, Y., Roth, D.: On dataless hierarchical text classification. In: AAAI, pp. 1579–1585 (2014)
18. Tang, J., Qu, M., Mei, Q.: Hierarchical neural language models for joint representation of streaming documents and their content. In: KDD, pp. 1165–1174 (2015)
19. Tang, J., Qu, M., Wang, M., Zhang, M., Yan, J., Mei, Q.: Line: large-scale information network embedding. In: WWW, pp. 1067–1077 (2015)
20. Wen, M., Yang, D., Rosé, C.P.: Sentiment analysis in MOOC discussion forums: what does it tell us?. In: EDM, pp. 130–137 (2014)
21. Wise, A.F., Cui, Y., Vytasek, J.: Bringing order to chaos in MOOC discussion forums with content-related thread identification. In: LAK, pp. 188 197 (2016)

Matrix-Based Method for Inferring Variable Labels Using Outlines of Data in Data Jackets

Teruaki Hayashi(✉) and Yukio Ohsawa

Department of Systems Innovation, School of Engineering, The University of Tokyo, Hongo 7-3-1, Bunkyo-ku, Tokyo, Japan
teru-h.884@nifty.com, ohsawa@sys.t.u-tokyo.ac.jp

Abstract. Data Jacket (DJ) is a technique for sharing information about data and for considering the potential value of datasets, with the data itself hidden, by describing the summary of data in natural language. In DJs, variables are described by variable labels (VLs), which are the names/meanings of variables, and the utility of data is estimated through the discussion about combinations of VLs. However, DJs do not always contain VLs, because the description rule of DJs cannot force data owners to enter all the information about their data. Due to the lack of VLs in some DJs, even if DJs are related to each other, the connection cannot be made through string matching of VLs. In this paper, we propose a method for inferring VLs in DJs whose VLs are unknown, using the texts in outlines of DJs. We specifically focus on the similarity of the outlines of DJs and created two models for inferring VLs, i.e., the similarity of the outlines and the co-occurrence of VLs. The results of experiments show that our method works significantly better than the method using only the string matching of VLs.

Keywords: Data Jacket · Variable label · Meta-data · Co-occurrence

1 Introduction

The potential benefits of reusing and analyzing massive quantities of data have been discussed among various stakeholders from diverse domains. The discussion involves privacy and security of data. Acquisti and Gross raised awareness that the combination of public databases may cause a serious violation of privacy [1]. Xu et al. reviewed the privacy issues related to data mining, by differentiating the responsibilities of different users [2]. From the overviews of the current situations of data utilization and exchange, the cost of data management and security issues discourage private companies and individuals to open or share their datasets. In order to overcome these problems, Data Jacket (DJ) has been developed as a technique for sharing information of data and for considering the potential value of data, with the data itself hidden, by describing the summary of data in natural language [3]. The idea of DJ is to share "a summary of data" as meta-data without sharing data itself, which reduces the risk of data management cost and privacy and enables stakeholders to discuss the combination of data.

J. Kim et al. (Eds.): PAKDD 2017, Part II, LNAI 10235, pp. 696–707, 2017.
DOI: 10.1007/978-3-319-57529-2_54

In the communication about data utilization and exchange using DJs, stakeholders start from discussing variable labels (VLs). VL is the name/meaning of variables in data. Variables and values in data are summarized as VLs in DJs. For example, the dataset "daily weather data in March 2016 in Tokyo" (Fig. 1) includes variables "yy," "mm," "dd," "highest temperature," and "lowest temperature," and each variable contains values. The VLs are the summary of variables and values in the dataset. Even if the data itself is not open, we can learn and evaluate which data is useful for your decision making from the summary of data described in DJs. Some data include private information, i.e., values and variables, such as "name," "address," or "ID." The values cannot be shared, but the VLs may be shared. Introducing DJs with VLs, stakeholders can learn the meaning of variables in data, by leading the hypotheses about possible combinations of VLs, reducing the risks of data management and privacy.

The workshop-styled methods introducing DJs have been proposed for discussions and generation of the feasible plans of data analyses. Once different stakeholders recognize the utility of data, they can negotiate conditions for exchanging their data. In the gamified workshops Innovators Marketplace on Data Jackets (IMDJ) [4,5] and Action Planning (AP) [6], data owners provide DJs representing their data, data analysts create solutions for solving data users' problems stated as requirements. In the process of IMDJ and AP, participants negotiate for data exchange or buying/selling to create new businesses. As a result of this discussion and evaluation among participants, data owners are expected to learn how to use their data from a possible combination of DJs proposed by data analysts. Users are expected to learn how their requirements can be satisfied with proposed plans. However, DJs do not always contain VLs, because the description rule of DJs does not force data owners to enter all the information about their data. In other words, only the information written by data owners is registered as DJs, therefore, due to the lack of descriptions about VLs, DJs essentially related to each other may not have linkage via VLs, which makes it difficult to think of plans for data analyses and combinations. In this paper, we propose a method for inferring variable labels not explicitly included in the outline of data. Focusing on the similarity of outlines of data and the co-occurrence of VLs, we construct models according to the following two features.

1. When a pair of datasets whose similarity of outlines is high, the pair of datasets is considered to be similar and should have similar VLs.
2. When a pair of VLs (vl_i and vl_j) frequently appears in datasets, and if vl_i appears, vl_j is considered to appear.

By modeling the features of VLs and using stored DJs as training data, even if a new DJ misses the VLs, it is possible to infer the VLs from the outlines. In the previous study of DJs, the co-occurrence of words in the outline of data [3,4] has been used for discussing the combination of DJs, e.g. using the visualization tool such as KeyGraph [7]. Our method suggests a possible connection between DJs whose VLs are missing, via inferred VLs. The significance of our approach and the contributions of our paper can be summarized as follows. It is the first

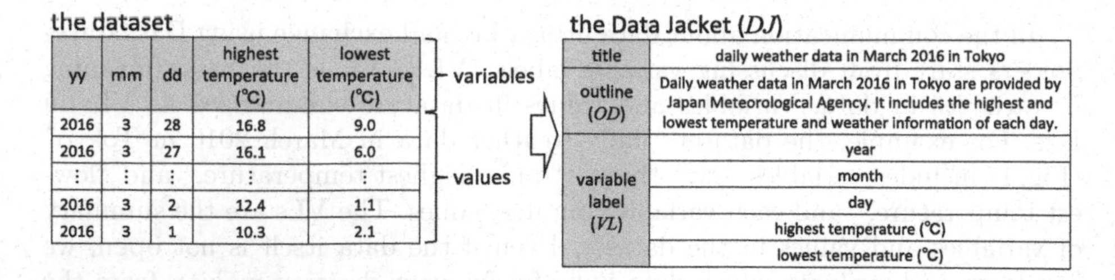

Fig. 1. Example of the dataset and the Data Jacket

approach for inferring VLs focusing on the similarity of outlines of datasets and the co-occurrence of VLs using DJs. The method for showing the related VLs from the outlines of data in DJs may be useful for encouraging data utilization. In particular, it is important not only to be useful for the purpose of knowledge discovery from their data, but also for decision makers who want to acquire new data. Our method can show them what kinds of set of variables to obtain are useful. Furthermore, proposed models have extensibility for various calculation methods. In this paper, in addition to the similarity of the data, we show the performance of the model considering the co-occurrence of VLs.

2 Inference of Variable Labels

2.1 Our Approach

The purpose of this study is to infer VLs of DJs whose variable labels are unknown. Because data itself is not open, it is impossible to know about the VLs by observing the data itself. Therefore, we consider tackling the problem, using the information about the data described in DJs. We assume that (1) datasets are similar when the information for explaining data is similar, and (2) datasets should have similar VLs when the similarity of datasets is high.

In this study, we introduce the outline of data (OD) as an indicator of the similarity of DJs. OD represents a description written in natural language for explaining data. For example, the OD of data in Fig. 1 is "Daily weather data in March 2016 in Tokyo are provided by Japan Meteorological Agency. It includes the highest and lowest temperature and weather information of each day." Although there are 12 items in the description of DJs (the title, the outline, variable labels, the sharing policy of data, the format, and so forth), we empirically consider ODs to be appropriate as the characteristic of datasets, because ODs are provided with textual data. Data portal sites such as DATA.GOV.UK[1] or DATA.GOV[2] provide datasets with the outlines of data in natural language, and users can search datasets from queries in free texts. Data Jacket Store[3],

[1] https://data.gov.uk/.
[2] https://www.data.gov/.
[3] http://www.panda.sys.t.u-tokyo.ac.jp/hayashi/djs/djs4ddi/.

a recommender system for DJs, also allows free text queries for searching the information about data using outlines in DJs. By the above discussion, in order to infer VLs of data whose VLs are unknown, we propose the method to obtain a set of likely VLs from the outlines of data.

2.2 Models

The expected function is to obtain sets of likely VLs ($\{vl \in V | f_n(vl, OD_x)\}$) stored in training data of DJs and VLs (V) by inputting ODs (OD_x) as queries. $f_n(vl, OD_x)$ represents a condition that a set of the top n variable labels (vl) are associated with a query (OD_x). In order to achieve above function, our models are conducted as follows.

The Similarity of DJs from the Outlines (Model 1): This model is based on the assumption that when a pair of datasets whose ODs are similar, the pair of datasets has similar VLs. By this model, a scored set of VLs are obtained considering the similarity between DJs with VLs and OD whose VLs are unknown.

The Co-occurrence of Variable Labels (Model 2): This model takes into account the co-occurrence of VLs. The co-occurrence of VLs is a feature that there may be a high frequent pair of VLs appearing at the same time, e.g., "year" and "day," or "name" and "gender." By introducing this model with Model 1, a scored set of VLs are obtained from the similarity between DJs with VLs and OD whose VLs are unknown.

2.3 Inference Process for Obtaining VLs

We show the inference process of VLs from ODs. In this study, we introduce bag-of-words and the vector space model [9,11]. In the pre-processing steps, we conduct morphological analysis of the text of ODs, (1) extracting words, (2) removing stop words, and (3) restoring words to their original forms.

Term-VL Matrix *E* (Model 1). Based on Model 1, we consider an algorithm to calculate the similarity among training data of ODs. After conducting the pre-processing steps to ODs, the ODs are converted into a matrix representation. Using the outlines of data as a corpus, a Term-OD matrix M ($|W| \times |D|$) are obtained, consisting of D-dimensional term vectors as rows, and W-dimensional OD vectors as columns, with each element v_{ij} in an OD vector ($\boldsymbol{od_j}$) corresponding to the frequency with which a term (a row i) occurs in an OD (a column j) as shown in (1) and (2). Note that the subscript T on the upper-right corner of vectors represents the transposition, and the vectors are highlighted in bold.

$$M = (\boldsymbol{od_1}, \cdots, \boldsymbol{od_j}, \cdots, \boldsymbol{od_D}) \tag{1}$$

$$\boldsymbol{od_j} = (v_{1j} \cdots v_{ij} \cdots v_{Wj})^{\mathrm{T}} \tag{2}$$

In the second step, a set of VLs included in DJs is converted into a VL-OD matrix. In the training data of DJs, ODs and VLs are linked when they appear in the same DJs. A VL-OD matrix R ($|V| \times |D|$) consists of V-dimensional VL vectors as rows, and D-dimensional OD vectors as columns, with each element r_{ij} in the jth OD vector ($\boldsymbol{od'_j}$) corresponding to the frequency (0 or 1) with which the ith VL occurs in the jth OD as shown in (3) and (4).

$$R = (\boldsymbol{od'_1}, \cdots, \boldsymbol{od'_j}, \cdots, \boldsymbol{od'_D}) \tag{3}$$

$$\boldsymbol{od'_j} = (r_{1j} \cdots r_{ij} \cdots r_{Vj})^{\mathrm{T}} \tag{4}$$

In the third step, we create a Term-VL matrix E ($= MR^T$) ($|W| \times |V|$) from a Term-OD matrix M ($|W| \times |D|$) and a VL-OD matrix R ($|V| \times |D|$) obtained in the second step. This process is equivalent to mapping the ith ($1 \leq i \leq |V|$) D-dimensional VL vector in the OD space into W-dimensional term space, by the Term-OD matrix M. The Term-VL matrix E is represented as follows:

$$MR^{\mathrm{T}} = (\boldsymbol{vl_1}, \cdots, \boldsymbol{vl_j}, \cdots, \boldsymbol{vl_V}) \tag{5}$$

$$\boldsymbol{vl_j} = (e_{1j} \cdots e_{ij} \cdots e_{Wj})^{\mathrm{T}} \tag{6}$$

$$e_{ij} = \sum_{k=1}^{|D|} v_{ik} r_{kj} \tag{7}$$

which means the sum of the product of the frequency (v_{ik}) with which the ith term (t_i) occurs in the kth OD (od_k) and the frequency (r_{kj}) with which the jth VL (vl_j) links with the kth OD (od_k). In other words, e_{ij} represents the number of OD related to both a term (t_i) and a VL (vl_j). Moreover, the Term-VL matrix E is equivalent to the adjacency matrix of the 3-partite graph, which consists of 3-disjoint sets of nodes, i.e., terms, ODs, VLs (Fig. 2). The element e_{ij} of the Term-VL matrix E also represents the number of passes from the ith term (t_i) to the jth VL (vl_j) by way of OD nodes.

Through the above process, Model 1 was implemented as the Term-VL matrix E. With this matrix, a scored set of VLs are obtained considering the similarity between ODs in the matrix E and OD_x whose VLs are unknown. When OD_x is given, a W-dimensional feature vector of OD_x ($\boldsymbol{od_x}$) is obtained after the pre-processing of morphological analysis. By comparing the similarity of $\boldsymbol{od_x}$ and each W-dimensional feature vector of VL ($\boldsymbol{vl_j}$ ($1 \leq j \leq |V|$)) in the matrix E, a scored set of VLs are obtained.

Term-VL Matrix *EC* (Model 1 and 2). We combine Model 2 to Model 1, considering the co-occurrence of VLs. First, we assume that any pair of VLs in the same DJ occurs once. In order to combine with the Term-VL matrix E created in Model 1, we conduct the VL co-occurrence matrix C ($= RR^T$ ($|V| \times |V|$)) whose element c_{ij} represents the number of DJs which include a pair of VLs vl_i and vl_j (8). In other words, when we define the frequency of co-occurrences of a pair of VLs (vl_i, vl_j) as $co(vl_i, vl_j)$, an element c_{ij} in the VL

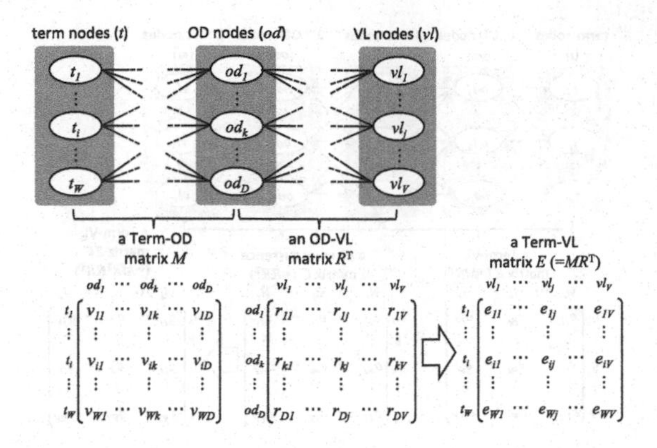

Fig. 2. Term-VL matrix E in the 3-partite graph

co-occurrence matrix C is represented as (9), where $|vl_i|_{od_s}$ means the frequency of vl_i in od_s $(1 \leq s \leq |D|)$.

$$c_{ij} = \sum_{k=1}^{|D|} r_{ik} r_{kj} = co(vl_i, vl_j) \quad (8)$$

$$= \sum_{s=1}^{|D|} |vl_i|_{od_s} |vl_j|_{od_s} \quad (9)$$

Finally, a Term-VL matrix EC is generated by a product of the Term-VL matrix E (5) and the VL co-occurrence matrix C, considering the co-occurrence of VLs. The Term-VL matrix EC consists of V-dimensional term vectors as rows, and W-dimensional VL vectors as columns, which has the same structure as the Term-VL matrix E. The difference between E and EC is whether the co-occurrences of VLs (Model 2), i.e., the elements of the matrices, are considered.

The element e_{ij} of the matrix E is given as (7), which represents the number of ODs related to both a term (t_i) and a VL (vl_j). On the other hand, the element g_{ij} of matrix EC is given as follows:

$$g_{ij} = \sum_{m=1}^{|V|} \left(\sum_{k=1}^{|D|} v_{ik} r_{km} \right) \left(\sum_{l=1}^{|D|} r_{ml} r_{lj} \right) \quad (10)$$

which represents the value considered the similarities of ODs and queries (the function of the matrix E), and the co-occurrence of VLs (the function of the matrix C). In other words, the Term-VL matrix EC is equivalent to the adjacency matrix of the 5-partite graph, which consists of 5-disjoint sets of nodes, i.e., terms, ODs, VLs, ODs, VLs (Fig. 3). The element g_{ij} represents the number of passes from the ith term (t_i) to the jth VL (vl_j) in the second VL nodes, by way of the first OD nodes, the first VL nodes, and the second OD nodes.

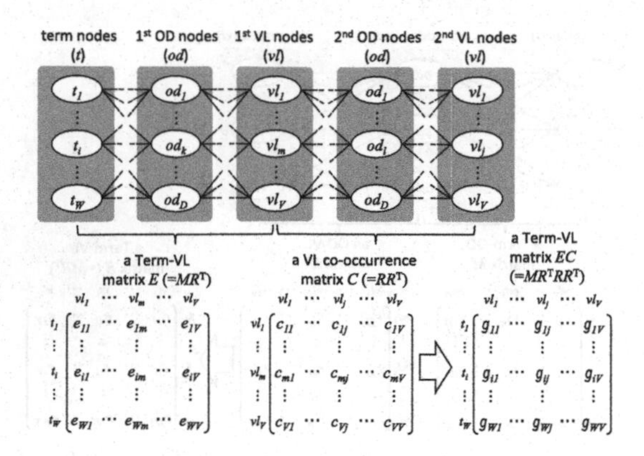

Fig. 3. Term-VL matrix EC in the 5-partite graph

When OD_x whose VLs are unknown is given, a W-dimensional feature vector of OD_x ($\boldsymbol{od_x}$) is obtained. By comparing the similarity of $\boldsymbol{od_x}$ and each W-dimensional feature vector of VL ($\boldsymbol{vl_j}$ $(1 \leq j \leq |V|)$) in the matrix EC, a scored set of VLs are obtained.

2.4 Example

Table 1 shows the list of top 10 inferred VLs for an OD "Data on consumption and amount of beer consumed by foreign tourists visiting Japan at a restaurant." Whose VLs are unknown (the experimental conditions for obtaining the inferred result will be explained in detail in the following section). Moreover, the OD does not exist in training data of DJs. The inference with the matrix E, using only the similarity of ODs, and with the matrix EC, considering both the co-occurrence of VLs and the similarity of ODs, seems to be highly related VLs in the OD. Looking at the example of the result in Table 1, it may be possible to infer the related VLs which may be included in ODs whose VLs are unknown, by introducing the models based on the similarity of ODs and the co-occurrence of VLs.

3 Experimental Details

In this paper, we used 799 DJs including both ODs and VLs, which were collected from business persons, researchers, and data holders who are interested in data utilization in various domains. Each DJ is constructed from an OD and several VLs. There are 3,215 unique VLs in total. The corpus and the dictionary were constructed from all the words in OD texts. We removed punctuation marks and symbols in the texts as stop words, restored words to their original forms, and extracted nouns, verbs, adverbs, and adjectives which appear more than one. The OD corpus consists of approximately 2,000 unique words. We used MeCab[4] for

[4] http://taku910.github.io/mecab/.

Table 1. The example of inferred result

The term-VL matrix E		The term-VL matrix EC	
Inferred VL	Similarity	Inferred VL	Similarity
Languages which can be offered	0.381758	Languages which can be offered	0.381758
Languages understood by foreigners	0.381758	Languages understood by foreigners	0.381758
National origin	0.317921	Satisfaction level through visiting	0.277441
Attractions in Tokyo	0.277441	Attractions in Tokyo	0.277441
The number of visits by foreigners	0.277441	The number of visits by foreigners	0.277441
The number of visitors	0.277441	The number of visitors	0.272505
Experience with or without activity	0.272505	Experience with or without activity	0.272505
Attribution of visitors (age)	0.272505	Attribution of visitors (age)	0.272505
Consumption amount	0.272505	Consumption amount	0.272505
Purchase	0.272505	Purchase	0.272505

Table 2. Training data (corpus) statistics

Number of Data Jackets	799
Average number of terms in each OD	39.5
Average number of VLs in each Data Jacket	5.34
Total number of terms in ODs	30,767
Unique terms in ODs	1,935
Total number of VLs	4,160
Unique variable labels	3,216

the morphological analysis [8], which is one of the common tools for analyzing morphemes of Japanese texts. The detail information of the training data is shown in Table 2. For weighting the discriminative terms in DJs, we introduced tf-idf in weighting scheme [10], which is reliable in identifying distinctive terms in each DJ. The term frequency (tf) is the number of times a term appears in a document, and the inverse document frequency (idf) diminishes the weight of frequent terms in all the documents and increases the weight of terms which appear rarely. As the test data, we collect 50 DJs from Open Data of Shizuoka prefecture in Japan[5], which publishes governmental records on the web. We collected DJs with ODs and VLs of them. The detail information of the test data is shown in Table 3.

The purpose of this experiment is to evaluate the inference ability of VLs from ODs whose VLs are unknown, using the similarity of ODs and the co-occurrence of VLs. We introduce the string matching as a comparative method with the Term-VL matrix E and EC. It is because when someone retrieves data from a description about data, a method using the string matching with VLs using the outlines of data as a query can be considered. The function of the string matching is to obtain sets of VLs which match the terms included in ODs,

[5] http://open-data.pref.shizuoka.jp/.

Table 3. Test data statistics

Number of Data Jackets	50
Average number of terms in each OD	36.7
Average number of VLs in each Data Jacket	4.70

from the training data of VLs. Inputted ODs are converted into bag-of-words in the same manner as our proposed method. The obtained VLs are scored in descending order of the number of times of acquisition.

We prepare the 50 DJs as test data, and extract ODs from them. Using these ODs as queries, we compare each feature vector of ODs with feature vectors of VLs in the Term-VL matrix E and EC, and obtain the sets of VLs in descending order. The similarity scores of OD_x and vl_j are calculated as cosine similarities shown as $sim(\boldsymbol{od_x}, \boldsymbol{vl_j}) = \boldsymbol{od_x} \cdot \boldsymbol{vl_j}/|\boldsymbol{od_x}||\boldsymbol{vl_j}|$. For the evaluation of this experiment, we use Precision, Recall, and F measure. We define Precision as $P = TP/(TP + FP)$ and Recall as $R = TP/(TP + FN)$, using the top 15 VLs returned as the inferred results scored by similarities, where TP =true positives, FP =false positives, FN =false negatives. F measure is defined as $F = 2PR/(P + R)$. Finally, by calculating the average of F measure of each query, we compare the performance of the matrix E, the matrix EC, and the string matching.

For the second criterion, we define Average Similarity (AS) shown as (11), considering the relationships of ODs and VLs by similarities, to compare the performance of the matrix E and the matrix EC. Here, V_{od_q} means the set of correct VLs included in od_q, and $rel(od_q, vl_p)$ is an indicator function equivalent to 1 if vl_p is the correct VL, i.e., $vl_p \in V_{od_q}$, 0 otherwise. As well as F measure, by calculating AS of each query, we compare the performance of the matrix E and EC using a paired t-test. Although there is MAP (Mean Average Precision) for evaluating the ranked inferred results [12], which is the method for evaluating the order of the results, our method AS focuses on the similarity of the results. In a DJ, each VL is equally linked with an OD, i.e., there is no order among VLs in DJs. For example, the VLs "day," "month," and "weather" equally exist in the weather data. Therefore, in this experiment we do not evaluate the inferred results with MAP, but with AS.

$$AS_{od_q} = \frac{1}{|V_{od_q}|} \sum_{p=1}^{|V|} (sim(\boldsymbol{od_q}, \boldsymbol{vl_p}) \cdot rel(od_q, vl_p)) \tag{11}$$

4 Result and Discussion

We got the top 15 VLs returned as the inferred results scored by similarities from each query using the matrix E and EC, and obtained the top 15 VLs from the string matching of ODs and VLs. Comparing the F measures calculated from Precision and Recall of each method, the inferred results using the matrix E and EC

Table 4. The evaluation of results (average scores ± standard deviation)

	F measure	Precision	Recall
String matching	0.075 ± 0.104	0.104 ± 0.148	0.059 ± 0.083
Matrix E	0.207 ± 0.124	0.137 ± 0.083	0.462 ± 0.306
Matrix EC	0.190 ± 0.136	0.127 ± 0.091	0.424 ± 0.320

Table 5. Average similarity (average scores ± standard deviation)

	Mean AS
Matrix E	0.329 ± 0.113
Matrix EC	0.399 ± 0.095
p-value	**

**: $p < 0.01$

got higher performance than using only the string matching. Especially, The performance of the matrix E was 2.76 times better in the score of F measure than that of the string matching. This result shows that although the outline of data is one of the important attributes for characterizing the data, they do not always include the information about VLs. In other words, the string matching of ODs and VLs is not enough to infer VLs in data. In this experiment, it is suggested that the model that "when a pair of datasets whose similarity of outlines is high, the pair of datasets is considered to be similar and should have similar VLs" works well for inferring VLs. In other words, the information of other datasets (the relationship between ODs and VLs) may compensate the missing terms for explaining data and work well to discover VLs from the outlines of data whose VLs are unknown. In addition, this result shows that the evaluations of inferred sets of VLs are almost the same in the Term-VL matrix E and EC (Table 4).

On the other hand, comparing the average of AS for evaluating the similarities of the inferred VLs, we found the significant differences in the Term-VL matrix E and EC ($t(98) = 9.52, p < 0.01$). Although the evaluation of inferred results is almost the same in a comparison of F measure, the evaluation values of AS, which is the criterion of similarities of VLs with ODs, got higher marks in the results introducing the Term-VL matrix EC than the matrix E (Table 5). In terms of the number, the similarity of correct sets of VLs increase in 48 of 50 test data when introducing the matrix EC. This result shows that the model considering the co-occurrence of VLs (when a pair of VLs (vl_i and vl_j) frequently appears in datasets, and if vl_i appears, vl_j is considered to appear) may work well to improve the similarity of VLs to ODs.

However, there is a possibility that the similarity of incorrect VL sets with queries also increases by the matrix EC. Therefore, we define the average similarity of incorrect VL sets to $\overline{AS}$ (12). $|V \cap \overline{V_{od_q}}|$ represents the number of VLs which are not included in od_q, and $unrel(od_q, vl_p)$ is an indicator function equivalent to 1 if vl_p is the incorrect VL, i.e., $vl_p \notin V_{od_q}$, 0 otherwise.

$$\overline{AS_{od_q}} = \frac{1}{|V \cap \overline{V_{od_q}}|} \sum_{p=1}^{|V|} (sim(\boldsymbol{od_q}, \boldsymbol{vl_p}) \cdot unrel(od_q, vl_p)) \tag{12}$$

Applying (12) to 50 test data, we compare $\overline{AS}$ values of the Term-VL matrix E and EC. We found the evaluation values of $\overline{AS}$ of the Term-VL matrix EC are significantly higher than those of the matrix E ($t(98) = 30.7, p < 0.01$). It shows that the similarity of incorrect VL sets with queries also increases by the matrix EC. However, comparing the points of increase of AS (the correct VL sets) and $\overline{AS}$ (the incorrect VL sets) in each query with a paired t-test, the points of increase of AS are significantly higher than the points of increase of $\overline{AS}$ ($AS : 0.0707, \overline{AS} : 0.0422, t(98) = 4.47, p < 0.01$). This result shows that the similarities of the correct VL sets with ODs significantly increase by introducing the Term-VL matrix EC.

5 Conclusion

In this paper, we proposed a method for inferring variable labels from the outline of data whose variable labels are missing or unknown. Focusing on the similarity of the outlines of data in DJs and the co-occurrence of variable labels, we constructed two models according to the features of DJs. By modeling the features of variable labels and the outlines of data, we found that even if a query DJ misses the variable labels, it is possible to infer the variable labels from the outline of the DJ. The result of the experiment suggests that the model that "when a pair of datasets whose similarity of outlines is high, the pair of datasets is considered to be similar and should have similar variable labels" works well for inferring variable labels. In addition, when we consider of not only the similarity of outlines, but also the co-occurrence of variable labels may improve the similarity of variable labels to the outlines of data. When someone retrieves variable labels from a description about data, it seems that a method using the string matching with variable labels can be considered. However, outlines of data do not always include the terms corresponding to variable labels. There is the problem that decision makers who want to acquire new data cannot discover the information about what kinds of data (set of variables) should be obtained. Our proposed method using the outlines of data in DJs may be helpful for encouraging data acquisition and utilization for the purpose of knowledge discovery.

In this study, because the outlines of data are small but include a certain amount of terms, it was possible to discuss and compare the similarities in the vector space model by creating the term-document matrix. However, a variable label is a very small element composed of one or several words. Because the description of DJs allows variable labels written in natural language, even if the variable labels have the same meaning, they are sometimes presented in different descriptions, e.g., "location" and "address," "the number of births" and "fertilities," or "the number of death" and "fatalities." In our future work, we aim at constructing a model considering the meaning of variable labels and synonyms, even if they have small descriptions. In addition, this study has been developed as a technique for supporting decision making in data utilization and exchange. It is important to validate the performance of the application using our proposed method in the workshops of IMDJ or AP.

Acknowledgments. This study was partially supported by JST-CREST, and JSPS KAKENHI Grant Number JP16J06450. Also we would like to thank all the staff members of Kozo Keikaku Engineering Inc. for supporting our research.

References

1. Acquisti, A., Gross, R.: Predicting social security numbers from public data. Proc. Nat. Acad. Sci. **106**(27), 10975–10980 (2009)
2. Xu, L., Jiang, C., Wang, J., Yuan, J., Ren, Y.: Information security in big data: privacy and data mining. IEEE Access **2**, 1149–1176 (2014)
3. Ohsawa, Y., Kido, H., Hayashi, T., Liu, C.: Data Jackets for synthesizing values in the market of data. In: 17th International Conference in Knowledge Based and Intelligent Information and Engineering Systems, vol. 22, pp. 709–716 (2013)
4. Ohsawa, Y, Liu, C., Suda, Y., Kido, H.: Innovators marketplace on Data Jackets for externalizing the value of data via stakeholders' requirement communication. In: Proceedings of AAAI 2014 Spring Symposium on Big Data Becomes Personal: Knowledge into Meaning, AAAI Technical report, pp. 45–50 (2014)
5. Ohsawa, Y., Kido, H., Hayashi, T., Liu, C., Komoda, K.: Innovators marketplace on Data Jackets, for valuating, sharing, and synthesizing data. In: Tweedale, J.W., Jain, L.C., Watada, J., Howlett, R.J. (eds.) Knowledge-Based Information Systems in Practice. SIST, vol. 30, pp. 83–97. Springer, Cham (2015). doi:10.1007/978-3-319-13545-8_6
6. Hayashi, T., Ohsawa, Y.: Processing combinatorial thinking: innovators marketplace as role-based game plus action planning. Int. J. Knowl. Syst. Sci. **4**(3), 14–38 (2013)
7. Ohsawa, Y., Benson, N.E., Yachida, M.: KeyGraph: automatic indexing by co-occurrence graph based on building construction metaphor. In: Proceedings of Advanced Digital Library Conference, pp. 12–18 (1998)
8. Kudo, T., Matsumoto, Y.: Japanese dependency structure analysis based on support vector machines. In: Proceedings of EMNLP, pp. 18–25 (2000)
9. Salton, G., Wong, A., Yang, C.S.: A vector space model for automatic indexing. Commun. ACM **18**(11), 613–620 (1975)
10. Salton, G., Buckley, C.: Term-weighting approaches in automatic text retrieval. Inf. Process. Manage. **24**(5), 513–523 (1988)
11. Turney, P.D., Pantel, P.: From frequency to meaning: vector space models of semantics. J. Artif. Intell. Res. **37**, 141–188 (2010)
12. Buckley, C., Voorhees, E.M.: Evaluating evaluation measure stability. In: Proceedings of SIGIR, pp. 33–40 (2000)

Integrating Reviews into Personalized Ranking for Cold Start Recommendation

Guang-Neng Hu[1] and Xin-Yu Dai[2](✉)

[1] Department of Computer Science and Engineering,
Hong Kong University of Science and Technology, Kowloon, Hong Kong
njuhgn@gmail.com

[2] National Key Laboratory for Novel Software Technology,
Nanjing University, Nanjing 210023, China
daixinyu@nju.edu.cn

Abstract. Item recommendation task predicts a personalized ranking over a set of items for individual user. One paradigm is the rating-based methods that concentrate on explicit feedbacks and hence face the difficulties in collecting them. Meanwhile, the ranking-based methods are presented with rated items and then rank the rated above the unrated. This paradigm uses widely available implicit feedback but it usually ignores some important information: item reviews. Item reviews not only justify the preferences of users, but also help alleviate the cold-start problem that fails the collaborative filtering. In this paper, we propose two novel and simple models to integrate item reviews into matrix factorization based Bayesian personalized ranking (BPR-MF). In each model, we make use of text features extracted from item reviews via word embeddings. On top of text features we uncover the review dimensions that explain the variation in users' feedback and these review factors represent a prior preference of a user. Experiments on real-world data sets show the benefits of leveraging item reviews on ranking prediction. We also conduct analyses to understand the proposed models.

1 Introduction

Users confront with the "information overload" dilemma and it is increasingly difficult for them to choose the preferred items over others because of the growing large item set, e.g., hundreds of millions products at Amazon.com and tens of thousands videos at Netflix.com [1,6]. Recommender systems (RSs) assist users in tackling this problem and help them make choices by ranking the items based on their past behavior history. Item recommendation predicts a personalized ranking over a set of items for individual user and hence alleviates the dilemma.

The rating-based (or point-wise) methods predict ratings that a user will give to items and then rank the items according to their predicted ratings. Many methods are proposed and matrix factorization based models are most popular due to their scalability, simplicity, and flexibility [2,4,5,10]. This paradigm concentrates on explicit feedback and it faces the difficulties in collecting them.

J. Kim et al. (Eds.): PAKDD 2017, Part II, LNAI 10235, pp. 708–720, 2017.
DOI: 10.1007/978-3-319-57529-2_55

Meanwhile, the ranking-based (pair-wise) methods are presented with seen items and then rank the seen above the unseen. Bayesian personalized ranking (BPR-MF) and collaborative item selection are typical representatives [11,14]. This paradigm takes advantage of widely available implicit feedback but it usually ignores a kind of important information: item reviews.

Related Works. Item reviews justify the preferences of users and help alleviate the cold-start problem; they are a diverse and complementary data source for recommendation beyond the user-item co-rating information. The CMF method [15] can be adapted to factorize the user/item-word matrix constructed from the item reviews. The CTR [16] and HFT [7] models integrate explicit ratings with item content/reviews to build better rating predictors; they employ topic modeling to learn hidden topic factors which explain the variations of users' preferences. The CTRank model [17] also adopts topic modeling to exploit item meta-data like article title and abstract via bag-of-words representation for one-class collaborative filtering [12], while the CDR [18] and CKE [19] models adopt stacked denoising autoencoders. Nevertheless, integrating item reviews into the ranking-based methods presents both opportunities and challenges for traditional Bayesian personalized ranking. There are few works on leveraging item reviews to improve personalized ranking.

In this paper we propose two novel and simple models to incorporate item reviews into BPR-MF. Like HFT, they integrate item reviews and unlike HFT they generate a ranked list of items for individual ranking. Like CTRank, they focus on personalized ranking and unlike CTRank they are based on matrix factorization and using word embeddings to extract features. Like BPR-MF, they rank preferred items over others and unlike BPR-MF they leverage the information from item reviews. In each of the two models, we make use of text features extracted from item reviews using word embeddings. And on top of text features we uncover the review dimensions that explain the variation in users' feedback. These review factors represent a prior preference of a user. One model treats the review factor space independent of the latent factor space; another connects implicit feedback and item reviews through the shared item space.

The contributions of this work are summarized as follows.

1. We propose two novel models to integrate item reviews into matrix factorization based Bayesian personalized ranking (Sects. 3.2 and 3.3). They generate a ranked list of items for individual user by leveraging the information from item reviews.
2. For exploiting item reviews, we build the proposed models on the top of text features extracted from them. We demonstrate a simple and effective way of extracting features from item reviews by averagely composing word embeddings (Sect. 4).
3. We empirically evaluate the proposed models on multiple real-world datasets which contains over millions of feedback in total. The experimental results show the benefit of leveraging item reviews on personalized ranking prediction. We also conduct analyses to understand the proposed models including the training efficiency and the impact of the number of latent factors.

2 Notation and Problem Statement

Before proposing our models, we briefly review the personalized ranking task and then describe the problem statement. To this end, we first introduce the notations used throughout the paper. Suppose there are M users $\mathcal{U} = \{u_1, ..., u_M\}$ and N items $\mathcal{I} = \{i_1, ..., i_N\}$. We reserve u, v for indexing users and i, j for indexing items. Let $X \in \mathbb{R}^{M \times N}$ denote the user-item binary implicit feedback matrix, where $x_{u,i}$ is the preference of user u on item i, and we mark a zero if it is unknown. Define N_u as the set of items on which user u has an action: $N_u \equiv \{i | i \in \mathcal{I} \wedge x_{u,i} > 0\}$. Rating-based methods [5,10] and ranking-based methods [3,14] are mainly to learn the latent user factors $P = [P_1, ..., P_M] \in \mathbb{R}^{F \times M}$ and latent item factors $Q = [Q_1, ..., Q_N] \in \mathbb{R}^{F \times N}$ from partially observed feedback X.

Item i may have text information, e.g., review d_{ui} commented by user u. We aggregate all reviews of a particular item as a 'doc' $d_i = \cup_{u \in \mathcal{U}} d_{ui}$. Approaches like CTR and HFT [7,16] integrate item content/reviews with explicit ratings for rating prediction using topic modeling. Another approach is to learn word embeddings and then compose them into document level as the item text features; we adopt this way of extracting text features $f_i \in \mathbb{R}^D$ from d_i (see Sect. 4).

2.1 Problem Statement

Our work focuses on the item recommendation or personalized ranking task where a ranked list of items is generated for each individual user. The goal is to accurately rank the unobserved items which contain both truly negative items (e.g., the user dislikes the Netflix movies or is not interesting in buying Amazon products) and missing ones (e.g., the user wants to see a movie or buy a product in the future when she knows it).

Instead of accurately predicting unseen ratings by learning a model from training samples $(u, i, x_{u,i})$ where $x_{u,i} > 0$, personalized ranking optimizes for correctly ranking item pairs by learning a model from training tuples $D_S \equiv \{(u, i, j) | u \in \mathcal{U} \wedge i \in N_u \wedge j \in \mathcal{I} \backslash N_u\}$. The meaning of item pairs of a user (u, i, j) is that she prefers the former than the latter, i.e., the model tries to reconstruct parts of a total order $>_u$ for each user u. From the history feedback X we can infer that the observed items i are ranked higher than the unobserved ones j; and for both observed items i_1, i_2 or both unobserved items j_1, j_2 we can infer nothing. Random (negative) sampling is adopted since the number of such pairs is huge. See the original BPR paper [14] for more details.

Problem 1. Personalized Ranking with Item Reviews.

Input: (1) A binary implicit feedback matrix X, (2) an item reviews corpus C, and (3) a user u in the user set $\mathcal{U}$.

Output: A ranked list $>_u$ over the unobserved items $\mathcal{I} \backslash N_u$.

In Problem 1, to generate the ranked list, we have item reviews to exploit besides implicit feedback.

3 The Proposed Models

In this section, we propose two models as a solution to Problem 1 which leverage item reviews into Bayesian personalized ranking. One model treats the review factor space independent of the latent factor space (Sect. 3.2). Another model connects implicit feedback and item reviews through the shared item space (Sect. 3.3). In each of the two proposed models, we make use of text features extracted from item reviews via word embeddings (Sect. 4). On top of text features we uncover the review dimensions that explain the variation in users' feedback and these review factors represent a prior preference of a user. Both models are based on basic matrix factorization (Sect. 3.1) and learned under the Bayesian personalized ranking framework (Sect. 3.4).

3.1 Basic Matrix Factorization

The basic matrix factorization (Basic MF) is mainly to find the latent user-specific feature matrix $[P_u]_1^M$ and item-specific feature matrix $[Q_i]_1^N$ to approximate the partially observed feedback matrix X in the regularized least-squares (or ridge regression) sense by solving the following problem.

$$\min_{P,Q} \sum\nolimits_{x_{u,i}\neq 0} (x_{u,i} - \hat{x}_{u,i})^2 + \lambda(\|P\|_F^2 + \|Q\|_F^2), \tag{1}$$

where λ is the regularization parameter to avoid over-fitting. The predicted scores $\hat{x}_{u,i}$ can be modeled by various forms which embody the flexibility of matrix factorization. A basic form is $\hat{x}_{u,i}^{Basic} = \alpha + \beta_u + \beta_i + P_u^{\mathsf{T}} Q_i$, where α, β_u and β_i are biases [5].

3.2 Integrating Item Reviews into Basic MF: Different Space Case

In this section, we propose our first model *TBPR-Diff* to integrate item reviews with implicit feedback. Analogical to the Basic MF which factorizes the ratings into user- and item- *latent* factors, we can factorize the reviews into user- and item- *text* factors (see the illustration in Fig. 1—Up). The TBPR-Diff model sharpens this idea and teases apart the rating dimensions into latent factors and text factors:

$$\hat{x}_{u,i}^{Diff} = \alpha + \beta_u + \beta_i + P_u^{\mathsf{T}} Q_i + \theta_u^{\mathsf{T}}(Hf_i) + \beta'^{\mathsf{T}} f_i, \tag{2}$$

where the term $\theta_u^{\mathsf{T}}(Hf_i)$ is newly introduced to capture the text interaction between user u and item i. To exploit item reviews, text features $f_i \in \mathbb{R}^D$ are firstly extracted from item reviews via word embeddings (hence they are known and fixed). The shared embedding kernel $H \in \mathbb{R}^{K\times D}$ linearly transforms original text features f_i from high-dimensional space (e.g., 200) into a lower text rating space (e.g., 15), and then it (Hf_i) interacts with text factors of user $\theta_u \in \mathbb{R}^K$. A text bias vector β' is also introduced to model users' overall preferences towards the item reviews. The details of text features extracted from item reviews using word embeddings are described later (see Sect. 4).

Since the text factors of user θ_u and of item (Hf_i) are *independent* of latent factors P_u and Q_i, there is no deep interactions between the information sources of observed feedback and item reviews, and hence they cannot benefit from each other. Also additional parameters increase the model complexity. Based on these observations, we propose another model to alleviate the above challenges.

3.3 Integrating Item Reviews into Basic MF: Shared Space Case

In this section, we propose our second model *TBPR-Shared* to integrate item reviews with implicit feedback more compactly. For an item i, its latent factors Q_i learned from feedback can be considered as characteristics that it processes; meanwhile, these characteristics are probably discussed in its reviews and hence exhibit in its text factors Hf_i (see the illustration in Fig. 1—Down). For user u, if we let Q_i and $\{Hf_k|k \in N_u\}$ be in the same space then it leads to deep interactions between text factors of user u and the latent factor of item i. The TBPR-Shared model sharpens this idea and enables the deep interactions between text factors and latent factors as well as reduces complexity of the model:

$$\hat{x}_{u,i}^{Shared} = Q_i^{\mathsf{T}}(P_u + |N_u|^{-1/2} \sum\nolimits_{k \in N_u} Hf_k) + \alpha + \beta_u + \beta_i + {\beta'}^{\mathsf{T}} f_i. \quad (3)$$

On the right hand, the last four terms are the same with the TBPR-Diff model. Different from the TBPR-Diff model, the shared item factors Q_i now have two-fold meanings: one is item latent factors that represent items' characteristics; another is to interact with item text factors that capture items' semantics from item reviews. Also different from the TBPR-Diff model, the preferences of a user now have a prior term which shows the 'text influence of her rated items' captured by the text factors of corresponding items. In summary, on top of text features the TBPR-Shared model uncovers the review dimensions that explain the variation in users' feedback and these factors represent a prior preference of user.

Remarks I. The VBPR model [3] proposed an analogical formulation with Eq. (2). It exploits visual features extracted from item images and we leverage item features extracted from item reviews. The SVD++ and NSVD [5,13] models proposed similar formulas with Eq. (3). They learn an implicit feature matrix to capture implicit feedback and we learn a text correlation matrix to capture text factors; note that they didn't exploit item reviews and hence they had no the text bias term. **II.** There can be an adjustable weight on the term of text (i.e., $\theta_u^{\mathsf{T}}(Hf_i)$ in Eq. (2) and $Q_i^{\mathsf{T}}|N_u|^{-1/2}\sum_{k \in N_u} Hf_k$ in Eq. (3)) to balance the influence from feedback and from reviews, but here we just let feedback and reviews be equally important.

Before we delve into the learning algorithm, the preference predictors of TBPR-Diff and of TBPR-Shared models are shown in Fig. 1.

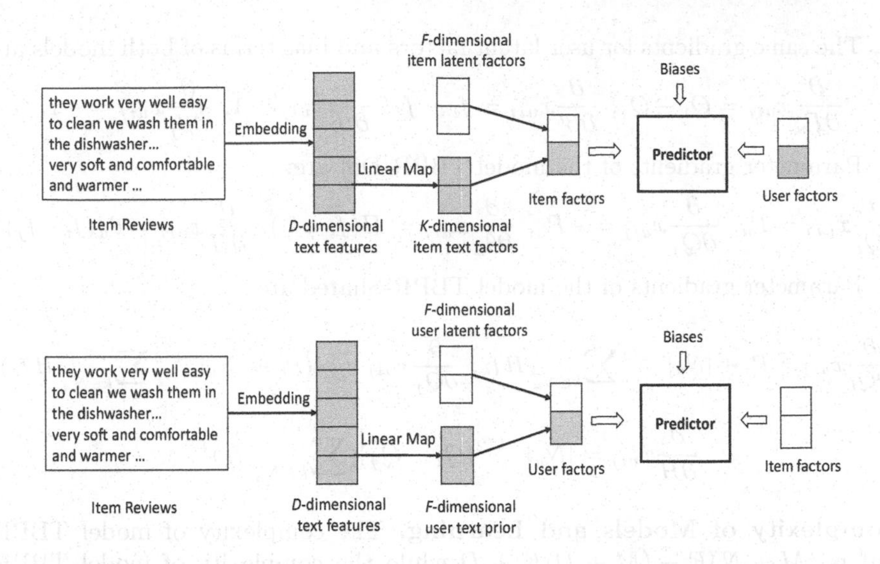

Fig. 1. Illustrating the preference predictors of our proposed two models. Up—TBPR-Diff model: The rating dimensions are to tease apart into text factors and latent factors for both user and item. Down—TBPR-Shared model: The rating dimensions are to tease apart into text factors and latent factors for only users where the text factors transformed from text features act as prior preference and show the 'text influence of her rated items'.

3.4 Model Learning with BPR

Revisit Problem 1, we need to generate a ranked list of items for individual user. Bayesian personalized ranking [14] is a generic pair-wise optimization framework that learns from the training item pairs using gradient descent. Denote the model parameters as Θ and let $\hat{x}_{uij}(\Theta)$ (for simplicity we omit model parameters, and the notation x_{ui} is the same with $x_{u,i}$) represent an arbitrary real-valued mapping under the model parameters. Then the optimization criterion for personalized ranking BPR-OPT is

$$\mathcal{L}(\Theta) \equiv \sum\nolimits_{(u,i,j)\in D_S} \ln \sigma(\hat{x}_{uij}) - \lambda \|\Theta\|^2, \quad (4)$$

where $\hat{x}_{uij} \equiv \hat{x}_{ui} - \hat{x}_{uj}$, and the sigmoid function is defined as $\sigma(x) = 1/(1+\exp(-x))$. The meaning behind BPR-OPT requires ranking items accurately as well as using a simple model.

Under the generic BPR-OPT framework, we derive the learning process for our proposed models TBPR-Diff and TBPR-Shared by embodying $\hat{x}_{ui}$ with $\hat{x}_{ui}^{Diff}$ and $\hat{x}_{ui}^{Shared}$, respectively. The BPR-OPT defined in Eq. (4) is differentiable and hence gradient ascent methods can be used to maximize it. For stochastic gradient ascent, a triple (u, i, j) is randomly sampled from training sets D_S and then update the model parameters by:

$$\Theta \leftarrow \Theta + \eta(\sigma(-\hat{x}_{uij})\frac{\partial \hat{x}_{uij}}{\partial \Theta} - \lambda\Theta). \quad (5)$$

The same gradients for user latent factors and bias terms of both models are:

$$\frac{\partial}{\partial P_u}\hat{x}_{uij} = Q_i - Q_j,\ \frac{\partial}{\partial \beta'}\hat{x}_{uij} = f_i - f_j,\ \frac{\partial}{\partial \beta_i}\hat{x}_{uij} = 1,\ \frac{\partial}{\partial \beta_j}\hat{x}_{uij} = -1.$$

Parameter gradients of the model TBPR-Diff are:

$$\frac{\partial}{\partial Q_i}\hat{x}_{uij} = P_u,\ \frac{\partial}{\partial Q_j}\hat{x}_{uij} = -P_u,\ \frac{\partial}{\partial \theta_u}\hat{x}_{uij} = H(f_i - f_j),\ \frac{\partial}{\partial H}\hat{x}_{uij} = \theta_u (f_i - f_j)^{\mathsf{T}}.$$

Parameter gradients of the model TBPR-Shared are:

$$\frac{\partial}{\partial Q_i}\hat{x}_{uij} = P_u + |N_u|^{-1/2}\sum\nolimits_{k\in N_u} Hf_k,\ \frac{\partial}{\partial Q_j}\hat{x}_{uij} = -(P_u + |N_u|^{-1/2}\sum\nolimits_{k\in N_u} Hf_k),$$

$$\frac{\partial}{\partial H}\hat{x}_{uij} = |N_u|^{-1/2}(Q_i - Q_j)(\sum\nolimits_{k\in N_u} f_k)^{\mathsf{T}}.$$

Complexity of Models and Learning. The complexity of model TBPR-Diff is $(M+N)F + (M+D)K + D$ while the complexity of model TBPR-Shared is $(M+N)F + (D+1)K$. We can see that the latter model reduces the complexity by $\mathcal{O}(MK)$, i.e., the parameters $[\theta_u]_1^M$. For updating each training sample $(u, i, j) \in D_S$, the complexity of learning TBPR-Diff is linear in the number of dimensions (F, K, D) while the complexity of learning TBPR-Shared is also linear provided that the scale of rated items of users is amortizing constant, i.e., $\sum_{u\in\mathcal{U}} |N_u|/|\mathcal{U}| \approx const \ll |\mathcal{I}|$, which holds in real-world datasets because of sparsity (see Table 1).

4 Feature Representations of Item Reviews

Recall that when generating the ranked list of items for individual user, we have item reviews to exploit besides implicit feedback. To exploit item reviews, we extract text features from them, i.e., there is a feature vector for each item. Our proposed two models are both built on the top of text features ($[f_i]_{i=1}^N$) and hence they are important for improving personalized ranking. In this section, we give one simple way to extract text features from reviews of item—word embedding.

The SGNS model [9] is an architecture for learning continuous representations of words from large corpus; these representations, or word embeddings, can capture the syntactic and semantic relationships of words. We first run the Google word2vec code on Amazon reviews corpus (see Table 1) using the default setting (particularly, dimensionality $D = 200$) to learn a vector $\mathbf{e}_w$ for each word w. And then we directly sum up all of the embeddings in an item's reviews (excluding stop words) and get a composition vector as the text feature for this item:

$$f_i \equiv \frac{1}{|d_i|}\sum\nolimits_{w\in d_i} \mathbf{e}_w. \tag{6}$$

To get f_i, we can also use complex methods (e.g., tensor networks to compose word embeddings or learning the doc representation directly); they are left for future work.

Table 1. Statistics of datasets

Datasets	#Users	#Items	#Feedback	#Words	#ColdUser	#ColdItem	Density (%)
Girls	778	3,963	5,474	302M	572	3,946	0.177
Boys	981	4,114	6,388	302M	787	4,080	0.158
Baby	1,238	4,592	8,401	302M	959	4,482	0.147
Men	21,793	55,647	157,329	302M	15,821	52,031	0.013
Women	62,928	157,656	504,847	302M	41,409	143,444	0.005
Phones	58,741	77,979	420,847	210M	43,429	67,706	0.009

5 Experiments

We evaluate our two models on multiple Amazon.com datasets in terms of ranking performance (Sect. 5.1). They integrate item reviews into Bayesian personalized ranking optimization criterion and we want to know the benefit from them. So we compare with BPR-MF [14] which ignores them and also with the most popular (POP) baseline that doesn't show personalized ranking (Sect. 5.2). We report the results in different settings (Sect. 5.3) and analyse the proposed methods (Sect. 5.4).

5.1 Datasets and Evaluation Protocol

Datasets. We evaluate our models on six Amazon datasets [8] http://jmcauley.ucsd.edu/data/amazon/. They consist of five from clothing and shoes category, and one from cell phones and accessories. We use the review history as implicit feedback and aggregate all users' reviews to an item as a doc for this item. We draw the samples from original datasets such that every user has rated at least five items (i.e., $\forall u \in \mathcal{U} : |N_u| \geq 5$) and the statistics of final evaluation datasets are show in Table 1. From the table we can see that: (1) the observed feedback is very sparse, typically less than 0.01%; (2) the average feedback events for users are typical about ten, i.e., $\sum_{u\in\mathcal{U}} |N_u|/|\mathcal{U}| \approx 10 \ll |\mathcal{I}|$ holds; (3) more than half of the users and of the items are cold and have feedback less than seven. Note that the cold-users/-items are those that have less than seven feedback events, and the feedback Density $= \#Feedback/(\#Users * \#Items)$.

We split each of the whole datasets into three parts: training, validation, and test. In detail, for each user $u \in \mathcal{U}$, we randomly sample two items from her history feedback for test set $Test_u$, two for validation set $Valid_u$, and the rest for training set $Train_u$; and hence $N_u = Train_u \cup Valid_u \cup Test_u$. This is the reason that we discard users who rated items less than five to ensure that there is at least one training sample for her.

Evaluation Protocol. For item recommendation or personalized ranking, we need to generate a ranked list over the unobserved items. Therefore for the hold-out test item $i \in Test_u$ of individual user u, the evaluation calculates how accurately the model rank i over other unobserved items $j \in \mathcal{I} \backslash N_u$. The widely used measure Area Under the ROC Curve (AUC) sharpens the ranking correctness intuition:

$$AUC = \frac{1}{|\mathcal{U}|} \sum_{u \in \mathcal{U}} \frac{1}{|E(u)|} \sum_{(i,j) \in E(u)} \delta(\hat{x}_{u,i} > \hat{x}_{u,j}), \tag{7}$$

where $E(u) = \{(i,j)|i \in Test_u \wedge j \in \mathcal{I} \wedge j \notin N_u\}$ and the $\delta(\cdot)$ is an indicator function. A higher AUC score indicates a better recommendation performance.

The validation set $\mathcal{V} = \cup_{u \in \mathcal{U}} Valid_u$ is used to tune hyperparameters and we report the corresponding results on the test set $\mathcal{T} = \cup_{u \in \mathcal{U}} Test_u$.

5.2 Comparing Methods

We compare our proposed models TBPR-Diff (see Eq. (2)) and TBPR-Shared (see Eq. (3)) with the Most Popular **(POP)** and **BPR-MF** [14] baselines. The difference of models lies in their preference predictors.

Reproducibility. We use the released code in [3] to implement the comparing methods and our proposed models. The hyperparameters are tuned on the validation set. Referring to the default setting, for the BRP-MF model, the norm-penalty $\lambda = 11$, and learning rate $\eta = 0.005$. As with our proposed models TBPR-Diff and TBPR-Shared, the norm-penalty $\lambda_{latent} = 11$ for latent factors and $\lambda_{text} = 5$ for text factors, and learning rate $\eta = 0.001$. For simplicity, the number of latent factors equals to the number of text factors; the default values for them are both fifteen (i.e., $F = K = 15$). Since the raw datasets, comparing code, and parameter setting are given publicly, we confidently believe our experiments are easily reproduced.

Table 2. AUC performance results (#factors = 15, **best** result is boldfaced).

Datasets	Setting	POP	BPR-MF	TBPR-Diff	TBPR-Shared	Improv1	Improv2
Girls	All	0.1699	0.5658	0.5919	**0.5939**	4.966	7.097
Boys	All	0.2499	0.5493	0.5808	**0.5852**	6.535	11.99
Baby	All	0.3451	0.5663	0.5932	**0.6021**	6.321	16.18
Men	All	0.5486	0.6536	0.6639	**0.6731**	2.983	18.57
	Cold	0.4725	0.5983	0.6114	**0.6225**	4.044	19.23
Women	All	0.5894	0.6735	0.6797	**0.6842**	1.588	12.72
	Cold	0.4904	0.6026	0.6110	**0.6152**	2.090	11.22
Phones	All	0.7310	0.7779	0.7799	**0.7809**	0.386	6.396
	Cold	0.5539	0.6415	0.6464	**0.6467**	0.811	5.936

5.3 Performance Results

The AUC performance results on eight Amazon.com datasets are shown in Table 2 where the last but one column is $(AUC_{\text{TBPR-Shared}} - AUC_{\text{BPR-MF}}) / AUC_{\text{BPR-MF}} \times 100\%$, and the last column is $(AUC_{\text{TBPR-Shared}} - AUC_{\text{BPR-MF}}) / (AUC_{\text{BPR-MF}} - AUC_{\text{POP}}) \times 100\%$. For each dataset there are three evaluation settings: The *All Items or All* setting evaluates the models on the full test set $\mathcal{T}$; the *Cold Start or Cold* setting evaluates the models on a subset $\mathcal{T}_{cold} \subseteq \mathcal{T}$ such that the number of training samples for each item within $\mathcal{T}_{cold}$ is no greater than three (i.e., $|Train_u| \leq 3$ or $|N_u| \leq 7$); the *Warm* setting evaluates the models on the difference set of All and Cold. Revisit the Table 1 we can see that: (1) almost all of the items are cold-item for datasets Girls, Boys, and Baby; and hence the results of Cold setting are almost the same with All and the results of Warm setting is not available to get a statistical reliable results; and (2) for other three datasets, the percent of cold-items is also more than 86% which requires the model to address the inherent cold start nature of the recommendation problem.

There are several observations from the evaluation results.

1. *Under the All setting*, TBPR-Shared is the top performer, TBPR-Diff is the second, with BPR-MF coming in third and POP the weakest. These results firstly show that leveraging item reviews besides the feedback can improve the personalized ranking; and also show that the personalization methods are distinctly better than the user-independent POP method. For example, TBPR-Shared averagely obtains relative 4.83% performance improvement compared with BPR-MF on the first three smaller datasets in terms of AUC metric, and 2.74% in total six datasets. This two figures show, to some extent, that transferring the knowledge from auxiliary data source (here item reviews) helps most when the target data source (here rating feedback) is not so rich.
2. *Under the Cold setting*, TBPR-Shared is the top performer, TBPR-Diff is the second, with BPR-MF coming in third and POP is also the weakest. These results firstly show that leveraging item reviews besides the feedback can improve the personalized ranking even in the cold start setting; and also show that the personalization methods are distinctly better than the user-independent POP method since the cold items are not popular. In detail, TBPR-Shared averagely obtains relative 2.31% performance improvement compared with BPR-MF in terms of AUC metric. Furthermore, TBPR-Shared compared with BPR-MF, the relative improvement in the *cold start setting* is about 1.6 times than that in the *All setting* which implies that integrating item reviews more benefits when observed feedback is sparser. As with the results on the Phones dataset, revisiting Table 1 we can see that the ratio of cold items over all item is 86.8% which is far less than those on other two datasets (~92.2%). And in this case adding auxiliary information doesn't help much.

We also evaluate on *the Warm setting* (not shown in Table 2), and all of the personalized, complex methods are worse than the user-independent, simple

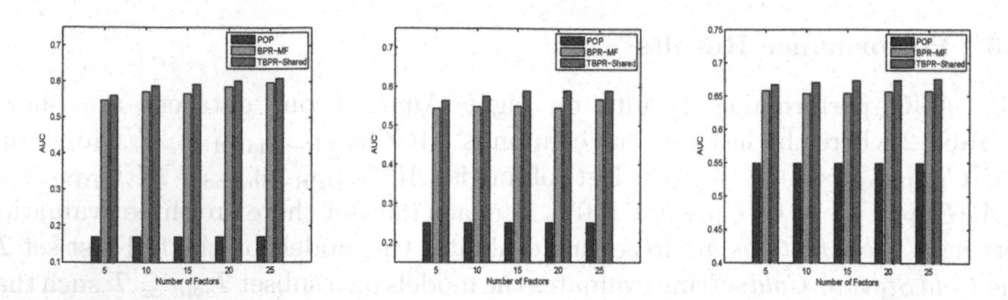

Fig. 2. Performance results of AUC on the test set with varying the number of latent factors. From top to bottom and left to right, the datasets are Girls, Boys, and Men (due to limited space we omit the results on Baby, Women and Phones). For clarity, we omit the results of TBPR-Diff where they are slightly better than BPR-MF and slightly worse than TBPR-Shared.

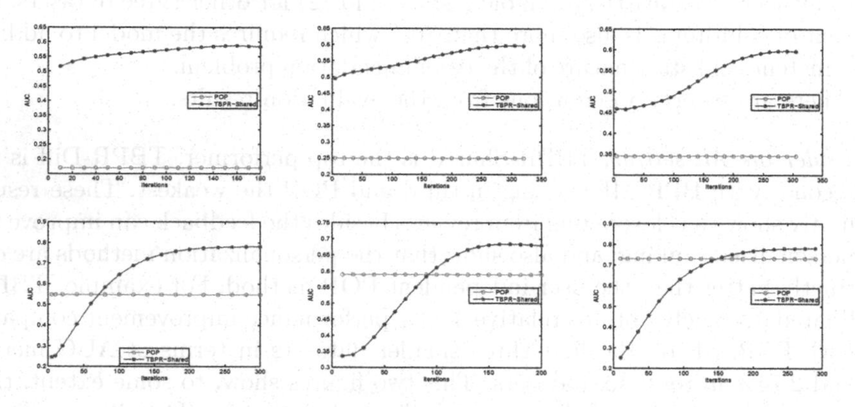

Fig. 3. Performance results of AUC on the validation set with training iterations (#factors = 15). From top to bottom and left to right, the datasets are Girls, Boys, Baby, Men, Women, and Phones. For clarity, we omit the results of TBPR-Diff where they are slightly worse than TBPR-Shared. As a reference, BPR-MF model usually converges in 50 iterations. Due to limited space, we only give the validation results for #factors = 15 and omit 5, 10, 20, 25.

method POP. Warm items are more likely to be popular and show less personalized characteristics. It reminds us the commonplace that recommendation plays an important role in long-tailed items.

5.4 Analysis of the Proposed Models

After demonstrating the benefits of leveraging item reviews, we analyse the proposed models from two points; one is the impact of number of latent factors, and one is the training efficiency and convergence analysis. More depth investigation like the impact of embedding dimensionality and of corpus source to train the embeddings, is left to future work.

Impact of the Number of Latent Factors. The two proposed models TBPR-Shared and TBPR-Diff have two important hyperparameters; one is the number of latent factors F and one is the number of text factors K. For simplicity, we let the two values equal. We vary the number of latent factors $\#factors = \{5, 10, 15, 20, 25\}$ to observe the performance results of different methods. The test AUC scores are shown in Fig. 2. On the Girls and Boys datasets, both of the personalized models are to perform better as the number of factors increases; on the other datasets, the performance improves as the number of factors increases to around fifteen; then it doesn't go up and may even downgrade. We set the default value as 15.

Also the plots visually show the benefits of integrating item reviews (TBPR-Shared vs. BPR-MF) and of generating a personalized ranking item list for individual user (TBPR-Shared and BPR-MF vs. POP).

Training Efficiency and Convergence Analysis. The complexity of learning is approximately linear in the number of parameters of our proposed models. Figure 3 shows the AUC scores of the TBPR-Shared model on validation sets with increasing training iterations. In summary, our models take 3–4 times more iterations to converge than BPR-MF. On three smaller datasets (Girls, Boys, and Baby), the first five iterations are enough to get a better score than POP; and on the other larger datasets (Men, Women, and Phones), it takes longer.

As a reference, the BPR-MF model usually converges in 50 iterations. As another reference, all of our experiments are completed in about one week using one server that has 65 GiB memory and 12 cores with frequency 3599 MHz.

6 Conclusion and Future Work

We proposed two models to integrate item reviews into Bayesian personalized ranking based on matrix factorization for cold start recommendation. In each of the two models, we make use of text features extracted from item reviews via word embeddings. On top of text features we uncover the review dimensions that explain the variation in users' feedback. These review factors represent a prior preference of a user and show the 'text influence of her rated items'. Empirical results on multiple real-world datasets demonstrated the improved ranking performance under the All and Cold start setting. And the shared space model is slightly better than the different space one which shows the benefits of considering the interactions between latent factors and text factors. Training efficiency is analyzed.

Since we investigate the benefits of leveraging item reviews, we only compare our models with BPR-MF (and POP); and to know the effectiveness, comparing with more baselines is needed. The construction strategy of positive/negative samples is also worth further investigating because it deeply affects the modeling design, the learning results, and the evaluation performance.

Acknowledgments. The work is supported by HKPFS PF15-16701, NSFC (61472183), and 863 Program (2015AA015406).

References

1. Bennett, J., Lanning, S.: The netflix prize. In: Proceedings of KDD Cup and Workshop (2007)
2. Cremonesi, P., Koren, Y., Turrin, R.: Performance of recommender algorithms on top-n recommendation tasks. In: Proceedings of ACM RecSys (2010)
3. He, R., McAuley, J.: VBPR: visual Bayesian personalized ranking from implicit feedback. In: Proceedings of AAAI (2016)
4. Hu, G., Dai, X., Song, Y., Huang, S., Chen, J.: A synthetic approach for recommendation: combining ratings, social relations, and reviews. In: Proceedings of IJCAI (2015)
5. Koren, Y.: Factorization meets the neighborhood: a multifaceted collaborative filtering model. In: Proceedings of SIGKDD (2008)
6. Linden, G., Smith, B., York, J.: Amazon.com recommendations: item-to-item collaborative filtering. IEEE Internet Comput. **7**(1), 76–80 (2003)
7. McAuley, J., Leskovec, J.: Hidden factors and hidden topics: understanding rating dimensions with review text. In: Proceedings of ACM RecSys (2013)
8. McAuley, J., Targett, C., Shi, Q., Van Den Hengel, A.: Image-based recommendations on styles and substitutes. In: Proceedings of SIGIR (2015)
9. Mikolov, T., Sutskever, I., Chen, K., Corrado, G.S., Dean, J.: Distributed representations of words and phrases and their compositionality. In: Advances in NIPS (2013)
10. Mnih, A., Salakhutdinov, R.: Probabilistic matrix factorization. In: Advances in NIPS (2007)
11. Mnih, A., Teh, Y.W.: Learning label trees for probabilistic modelling of implicit feedback. In: Advances in NIPS (2012)
12. Pan, R., Zhou, Y., Cao, B., Liu, N.N., Lukose, R., Scholz, M., Yang, Q.: One-class collaborative filtering. In: Proceedings of IEEE ICDM (2008)
13. Paterek, A.: Improving regularized singular value decomposition for collaborative filtering. In: Proceedings of KDD Cup and Workshop (2007)
14. Rendle, S., Freudenthaler, C., Gantner, Z., Schmidt-Thieme, L.: BPR: Bayesian personalized ranking from implicit feedback. In: Proceedings of UAI (2009)
15. Singh, A.P., Gordon, G.J.: Relational learning via collective matrix factorization. In: Proceedings of SIGKDD (2008)
16. Wang, C., Blei, D.M.: Collaborative topic modeling for recommending scientific articles. In: Proceedings of SIGKDD (2011)
17. Yao, W., He, J., Wang, H., Zhang, Y., Cao, J.: Collaborative topic ranking: leveraging item meta-data for sparsity reduction. In: Proceedings of AAAI (2015)
18. Ying, H., Chen, L., Xiong, Y., Wu, J.: Collaborative deep ranking: a hybrid pairwise recommendation algorithm with implicit feedback. In: Proceedings of PAKDD (2016)
19. Zhang, F., Yuan, N., Lian, D., Xie, X., Ma, W.: Collaborative knowledge base embedding for recommender systems. In: Proceedings of SIGKDD (2016)

Taste or Addiction?: Using Play Logs to Infer Song Selection Motivation

Kosetsu Tsukuda(✉) and Masataka Goto

National Institute of Advanced Industrial Science and Technology (AIST), Tsukuba, Japan
{k.tsukuda,m.goto}@aist.go.jp

Abstract. Online music services are increasing in popularity. They enable us to analyze people's music listening behavior based on play logs. Although it is known that people listen to music based on topic (*e.g.*, rock or jazz), we assume that when a user is addicted to an artist, s/he chooses the artist's songs regardless of topic. Based on this assumption, in this paper, we propose a probabilistic model to analyze people's music listening behavior. Our main contributions are three-fold. First, to the best of our knowledge, this is the first study modeling music listening behavior by taking into account the influence of addiction to artists. Second, by using real-world datasets of play logs, we showed the effectiveness of our proposed model. Third, we carried out qualitative experiments and showed that taking addiction into account enables us to analyze music listening behavior from a new viewpoint in terms of how people listen to music according to the time of day, how an artist's songs are listened to by people, etc. We also discuss the possibility of applying the analysis results to applications such as artist similarity computation and song recommendation.

1 Introduction

Among various leisure activities such as watching movies, reading books, and eating delicious food, listening to music is one of the most important for people [14]. In terms of the amount of accessible music, the advent of online music services (*e.g.*, Last.fm[1], Pandora[2], and Spotify[3]) has made it possible for people to access millions of songs on the Internet, and it has become popular to play music using such services rather than physical media like CDs [8]. When users play music online, such services record personal musical play logs that show when users listen to music and what they listen to.

Since personal music play logs have become available, it has become popular to use *session* information to analyze and model people's music listening behavior [2,4,13,18]. Here, a session is a sequence of logs within a given time frame. Zheleva *et al.* [18] were the first to model listening behavior using a topic model

[1] http://www.last.fm.
[2] http://www.pandora.com.
[3] http://www.spotify.com.

J. Kim et al. (Eds.): PAKDD 2017, Part II, LNAI 10235, pp. 721–733, 2017.
DOI: 10.1007/978-3-319-57529-2_56

based on session information. They revealed that a user tends to choose songs in a session according to the session's specific topic such as rock or jazz. However, it is not always correct to assume that a user chooses songs according to the session's topic. For example, after a user buys an artist's album or temporarily falls in love with an artist, s/he will be *addicted* to the artist and repeatedly listen to the artist's songs regardless of topic.

In light of the above, this paper proposes a model that can deal with both a session topic and addiction to artists. Our proposed model uses the model proposed by Zheleva *et al.* [18] as the starting point. We present each song-listening instance in terms of the corresponding song artist. In our model, each user has a distribution over topics that reflects the user's usual taste in music and a distribution over artists that reflects the user's addiction to artists. In addition, each user has a different ratio between usual taste and addiction, and probabilistically chooses a song in a session based on this ratio. That is, if a user has a high addiction ratio, s/he will probably choose a song of an artist from his/her artist distribution for addiction. Modeling people's music listening behavior by considering addiction is worth studying from various viewpoints:

- Our model can show topic characteristics (*e.g.*, the rock topic has a high ratio of addiction) and artist characteristics (*e.g.*, most users choose an artist's songs when addicted to that artist). It is important to understand such characteristics from the social scientific viewpoint.
- Our model can also show user characteristics (*e.g.*, a user chooses songs based on addiction in a session). There are many applications that could use this data such as advertisements and recommendation systems. For example, if a user chooses songs of an artist based on addiction in a session, it would be useful to recommend songs of that artist; if s/he chooses songs based on a topic, it would be better to recommend other artists' songs in the same topic.

Our main contributions in this paper are as follows.

- To the best of our knowledge, this is the first study modeling music listening behavior by considering both the usual taste in music and the addiction to artists.
- We quantitatively evaluated our model by using real-world music play logs of two music online services. Our experimental results show that the model adopting both factors achieves the best results in terms of the perplexity computed by using test data.
- We carried out qualitative experiments in terms of user characteristics, artist characteristics, and topic characteristics and show that our model can be used to analyze people's music listening behavior from a new viewpoint.

The remainder of this paper is organized as follows. Section 2 presents related work on analyzing music play logs and on modeling music listening behavior. Section 3 describes the model that extends the model by Zheleva *et al.* [18] by considering the addiction phenomenon. Section 4 presents a procedure to infer the parameters. Sections 5 and 6 report on our quantitative and qualitative experiments, respectively. Finally, Sect. 7 concludes this paper.

2 Related Work

2.1 Analysis of Music Listening Behavior

Analyzing people's music listening behavior has attracted a lot of attention because (1) understanding how people listen to music is important from the social scientific viewpoint and (2) the analysis results can give useful insight into various applications such as music player interfaces and recommender systems.

People's music listening behavior has been analyzed from various viewpoints. Rentfrow and Gosling [14] carried out a questionnaire-based survey and revealed the correlations between music preferences and personality, self-views (*e.g.*, wealthy and politically liberal), and cognitive ability (*e.g.*, verbal skills and analytical skills). Renyolds *et al.* [15] made an online survey and reported that environmental metadata such as the user's activity, weather, and location affect the user's music selection. Analysis by Berkers [3] using Last.fm play logs showed the significant differences between male and female in terms of their music genre preferences. More recently, Lee *et al.* [10] collected responses from users of commercial cloud music services and reported the criteria for generating playlists: personal preference, mood, genre/style, artists, etc. Among various factors, time information has received a lot of attention. Herrera *et al.* [6] analyzed play counts from Last.fm and discovered that a non-negligible number of listeners listen to certain artists and genres at specific moments of the day and/or on certain days of the week. Park and Kahng [12] used log data of a commercial online music service in Korea and showed that there existed seasonal and time-of-day effects on users' music preference. Baur *et al.* [2] also showed the importance of seasonal aspects, which influence music listening, using play logs from Last.fm.

In spite of the variety of listening behavior analyses, to the best of our knowledge, no work has focused on users' addiction to, for example, songs and artists. In this work, we deal with this factor and analyze people's music listening behavior from a new perspective.

2.2 Application Based on Music Listening Logs

Listening logs have been used for various applications, including the detection of similar artists. Schedl and Hauger [17] crawled Twitter[4] for the hash tag #nowplaying and computed artist similarity using co-occurrence-based methods. Their experimental results showed that listening logs can be used to derive similarity measures for artists. Another application is playlist generation. Liu *et al.* [11] proposed a playlist generation system informed by time stamps of a user's listening logs in addition to the user's music rating history and audio features such as wave forms. The most popular application is music recommendation. Since personal music play logs have become available, it has become popular to use session information to recommend songs. Park *et al.* [13] proposed Session-based Collaborative Filtering (SSCF), which extends traditional collaborative filtering

[4] http://twitter.com/.

techniques by using preferred songs in the similar session. Dias and Fonseca [4] proposed temporal SSCF, where for each session, a feature vector is created consisting of five properties including time of day and song diversity. The work closest to ours is that of Zheleva *et al.* [18], who proposed a statistical model to describe patterns of song listening. They showed that a user tends to choose songs in a session according to the session's specific topic. We will describe the details of their model in Sect. 3.2.

Although none of these applications used addiction information, we believe that this information could improve the usefulness of these applications. We discuss the possibility of using our analysis results to improve these applications in Sect. 6.

3 Model

As was mentioned earlier, our model builds on the one proposed by Zheleva *et al.* [18]. After summarizing the notations used in our model in Sect. 3.1, we first describe the model by Zheleva *et al.* [18] in Sect. 3.2 and then propose our model in Sect. 3.3.

3.1 Notations

Given a music play log dataset, let U be a set of users in the dataset. Let $l_{un} = (u, a, t_{un})$ denote the nth play log of $u \in U$. More specifically, user u plays a song of artist $a \in A$ at time t_{un}. Here, A is the set of artists in the dataset. Without loss of generality, we assume that play logs are sorted in ascending order of their timestamps: $t_{un} < t_{un'}$ for $n < n'$.

To capture user's listening preferences over time, we divide user's play logs into sessions. Following Zheleva *et al.* [18] and Baur *et al.* [2], we use the time gap approach to generate sessions. If the gap between t_{un} and t_{un+1} is less than 30 min, l_{un} and l_{un+1} belong to the same session; otherwise, they belong to different sessions. Let S_{ur} be the rth session of u where S_{ur} consists of one or more of u's logs. Let R_u be the total number of u's sessions; then the set of u's sessions is given by $D_u = \{S_{ur}\}_{r=1}^{R_u}$. Hence, the set of sessions of all users is given by $D = \{D_u\}_{u \in U}$.

3.2 Session Model

The model proposed by Zheleva *et al.* [18], which is called the session model, is a probabilistic graphical model based on the Latent Dirichlet Allocation (LDA) [1]. The session model assumes that for each session, there is a latent topic (*e.g.*, rock or love song) that guides the choice of songs in the session. Figure 1(a) shows the graphical model of the session model, where shaded and unshaded circles represent observed and unobserved variables, respectively. In the figure, K is the number of topics, V_{ur} is the number of logs in the rth session of u, θ is the user-topic distribution, and ϕ is the topic-artist distribution. We assume that θ

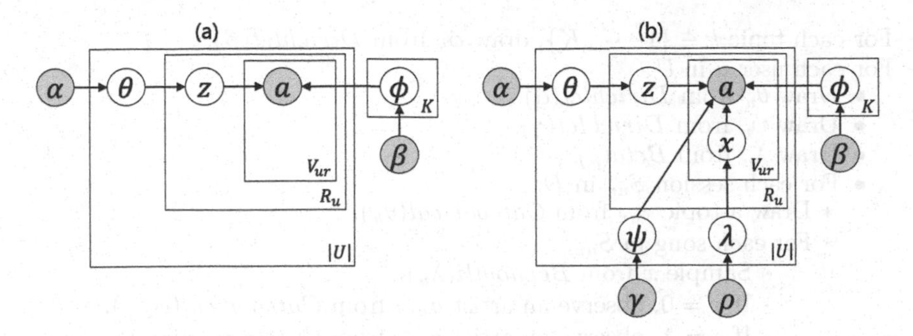

Fig. 1. Graphical models of (a) session model and (b) session with addiction model.

and ϕ have Dirichlet priors of α and β, respectively. The generative process of the session model is as follows:

- For each topic $k \in \{1, \cdots, K\}$, draw ϕ_k from $Dirichlet(\beta)$.
- For each user u in U,
  - Draw θ_u from $Dirichlet(\alpha)$.
  - For each session S_{ur} in D_u,
    - * Draw a topic z_{ur} from $Categorical(\theta_u)$.
    - * For each song in S_{ur}, observe an artist a_{urj} from $Categorical(\phi_{z_{ur}})$.

In the generative process, a_{urj} represents the jth song's artist in the rth session of u.

3.3 Session with Addiction (SWA) Model

Although Zheleva *et al.* [18] reported the usefulness of generating played songs based on a session's topic, we hypothesize that users can choose a song independently of topic. For example, after a user buys an artist's album or temporarily falls in love with an artist, s/he will repeatedly listen to the artist's songs regardless of the topic. In other words, the user can be *addicted* to some artists. In such an addiction mode, we assume that the user directly chooses a song without going through the topic.

In light of the above, our model takes both session-topic-based and addiction-based choices of songs. Figure 1(b) shows the graphical model of our proposed model. Each user has a Bernoulli distribution λ that controls the weights of influence for a session topic and addiction. To be more specific, when user u chooses a song in a session, we assume that the choice is influenced by the session topic with probability λ_{u0} $(x = 0)$ and by u's addiction to the artist with probability λ_{u1} $(x = 1)$, where $\lambda_{u0} + \lambda_{u1} = 1$. When $x = 0$, a song is generated through the same process of the session model, while when $x = 1$, a song is directly generated from a user-artist distribution ψ. The generative process of the SWA model is as follows:

- For each topic $k \in \{1, \cdots, K\}$, draw ϕ_k from *Dirichlet*(β).
- For each user u in U,
  - Draw θ_u from *Dirichlet*(α).
  - Draw ψ_u from *Dirichlet*(γ).
  - Draw λ_u from *Beta*(ρ).
  - For each session S_{ur} in D_u,
    - Draw a topic z_{ur} from *Categorical*(θ_u).
    - For each song in S_{ur},
      - Sample x from *Bernoulli*(λ_u).
      - If $x = 0$, observe an artist a_{urj} from *Categorical*($\phi_{z_{ur}}$).
      - If $x = 1$, observe an artist a_{urj} from *Categorical*(ψ_u).

4 Inference

To learn the parameters of our proposed model, we use collapsed Gibbs sampling [5] to obtain samples of hidden variable assignment. Since we use a Dirichlet prior for θ, ϕ, and ψ and a Beta prior for λ, we can analytically calculate the marginalization over the parameters. The marginalized joint distribution of D, latent variables $Z = \{\{z_{ur}\}_{r=1}^{R_u}\}_{u \in U}$, and latent variables $X = \{\{\{x_{urj}\}_{j=1}^{V_{ur}}\}_{r=1}^{R_u}\}_{u \in U}$ is computed as follows:

$$P(D, Z, X|\alpha, \beta, \gamma, \rho) = \iiiint P(D, Z, X|\boldsymbol{\Theta}, \boldsymbol{\Phi}, \boldsymbol{\Psi}, \boldsymbol{\Lambda}) P(\boldsymbol{\Theta}|\alpha) P(\boldsymbol{\Phi}|\beta) P(\boldsymbol{\Psi}|\gamma) P(\boldsymbol{\Lambda}|\rho) d\boldsymbol{\Theta} d\boldsymbol{\Phi} d\boldsymbol{\Psi} d\boldsymbol{\Lambda}, \tag{1}$$

where $\boldsymbol{\Theta} = \{\theta_u\}_{u \in U}$, $\boldsymbol{\Phi} = \{\phi_k\}_{k=1}^{K}$, $\boldsymbol{\Psi} = \{\psi_u\}_{u \in U}$, and $\boldsymbol{\Lambda} = \{\lambda_u\}_{u \in U}$. By integrating out those parameters, we can compute Eq. (1) as follows:

$$\begin{aligned} &P(D, Z, X|\alpha, \beta, \gamma, \rho) \\ &= \left(\frac{\Gamma(2\rho)}{\Gamma(\rho)^2}\right)^{|U|} \prod_{u \in U} \frac{\Gamma(\rho + N_{u0})\Gamma(\rho + N_{u1})}{\Gamma(2\rho + N_u)} \left(\frac{\Gamma(\gamma|A|)}{\Gamma(\gamma)^{|A|}}\right)^{|U|} \prod_{u \in U} \frac{\prod_{a \in A} \Gamma(N_{u1a} + \gamma)}{\Gamma(N_{u1} + \gamma|A|)} \\ &\times \left(\frac{\Gamma(\beta|A|)}{\Gamma(\beta)^{|A|}}\right)^{K} \prod_{k=1}^{K} \frac{\prod_{a \in A} \Gamma(N_{ka} + \beta)}{\Gamma(N_k + \beta|A|)} \left(\frac{\Gamma(\alpha K)}{\Gamma(\alpha)^K}\right)^{|U|} \prod_{u \in U} \frac{\prod_{k=1}^{K} \Gamma(R_{uk} + \alpha)}{\Gamma(R_u + \alpha K)}. \end{aligned} \tag{2}$$

Here, N_{u0} and N_{u1} are the number of u's logs such that $x = 0$ and $x = 1$, respectively, and $N_u = N_{u0} + N_{u1}$. The term N_{u1a} represents the number of times that user u chooses artist a's song under the condition of $x = 1$, and $N_{u1} = \sum_{a \in A} N_{u1a}$. Furthermore, $N_k = \sum_{a \in A} N_{ka}$ where N_{ka} is the number of times artist a is assigned to topic k under the condition of $x = 0$. Finally, R_{uk} is the number of times u's session is assigned to topic k, and $R_u = \sum_{k=1}^{K} R_{uk}$.

For the Gibbs sampler, given the current state of all but one variable z_{ur}, the new latent assignment of z_{ur} is sampled from the following probability:

$$\begin{aligned} &P(z_{ur} = k|D, X, Z_{\backslash ur}, \alpha, \beta, \gamma, \rho) \\ &\propto \frac{R_{uk\backslash ur} + \alpha}{R_u - 1 + \alpha K} \frac{\Gamma(N_{k\backslash ur} + \beta|A|)}{\Gamma(N_{k\backslash ur} + N_{ur} + \beta|A|)} \prod_{a \in A} \frac{\Gamma(N_{ka\backslash ur} + N_{ura} + \beta)}{\Gamma(N_{ka\backslash ur} + \beta)}, \end{aligned} \tag{3}$$

where $\backslash ur$ represents the procedure excluding the rth session of u. Moreover, N_{ur} and N_{ura} represent the number of logs in rth session of u and the number of a's logs in rth session of u, respectively.

In addition, given the current state of all but one variable x_{urj}, the probability at which $x_{urj} = 0$ is computed as follows:

$$P(x_{urj} = 0|D, X_{\backslash urj}, Z, \alpha, \beta, \gamma, \rho) \propto \frac{\rho + N_{u0\backslash urj}}{2\rho + N_u - 1} \frac{N_{z_{ur}a_{urj}\backslash urj} + \beta}{N_{z_{ur}\backslash urj} + \beta|A|}, \quad (4)$$

where $\backslash urj$ represents the procedure excluding the jth song in the rth session of u. Similarly, the probability at which $x_{urj} = 1$ is computed as follows:

$$P(x_{urj} = 1|D, X_{\backslash urj}, Z, \alpha, \beta, \gamma, \rho) \propto \frac{\rho + N_{u1\backslash urj}}{2\rho + N_u - 1} \frac{N_{u1a_{urj}\backslash urj} + \gamma}{N_{u1\backslash urj} + \gamma|A|}. \quad (5)$$

Finally, we can make the point estimates of the integrated out parameters as follows:

$$\theta_{uk} = \frac{C_{uk} + \alpha}{C_u + \alpha K}, \quad \phi_{ka} = \frac{N_{ka} + \beta}{N_k + \beta|A|}, \quad \psi_{ua} = \frac{N_{u1a} + \gamma}{N_{u1} + \gamma|A|}. \quad (6)$$

$$\lambda_{u0} = \frac{N_{u0} + \rho}{N_u + 2\rho}, \quad \lambda_{u1} = \frac{N_{u1} + \rho}{N_u + 2\rho}, \quad (7)$$

where remind that λ_{u0} and λ_{u1} represent the ratio of usual taste in music and addiction when u chooses songs, respectively.

5 Quantitative Experiments

In this section, we answer the following research question based on our quantitative experimental results: is adopting two factors, which are users' daily taste in music and addiction to artists, effective to model music listening behavior?

5.1 Dataset

To examine the effectiveness of the proposed model, we constructed two datasets. The first one is created from music play logs on a music download service in Japan. On the service, users can buy a single song and an album and listen to them. For this evaluation, we obtained 10 weeks of log data between 1/1/2016 and 10/3/2016. We call this dataset JPD. The second one consists of logs on Last.fm. To guarantee the repeatability, we used a publicly available music play log data on Last.fm provided by Schedl [16]. Similar with JPD, we extracted 10 weeks of log data between 1/1/2013 and 11/3/2013; we call the dataset LFMD.

From the 10 weeks of data of JPD, we created two pairs of training and test datasets as follows. In the first/second dataset, the training dataset consists of logs of the first four/eight weeks and the test dataset consists of the next two

Table 1. Statistics of our datasets

	4WJPD	8WJPD	4WLFMD	8WLFMD
Number of users	7,230	13,986	2,501	2,850
Number of artists	3,441	6,431	7,899	12,360
Number of logs in training data	141,381	331,437	400,410	872,614
Number of sessions in training data	35,780	82,427	50,106	106,840
Number of logs in test data	48,837	57,126	179,983	201,966
Number of sessions in test data	11,767	13,516	23,167	24,958

weeks. For each dataset, we excluded artists whose songs were played by ≤3 users and created session data as described in Sect. 3.1. Let the first and second dataset be 4WJPD (4W means four weeks) and 8WJPD, respectively. As for LFMD, we also created two pairs of training and test datasets 4WLFMD and 8WLFMD in the same manner as we created the 4WJPD and 8WJPD datasets. Table 1 shows the statistics of the four datasets.

5.2 Settings

In terms of hyperparameters, in line with other topic modeling work, we set $\alpha = \frac{1}{K}$ and $\beta = \frac{50}{|A|}$ in the session model and the session with addiction (SWA) model. In addition, in the SWA model, we set $\gamma = \frac{50}{|A|}$ and $\rho = 0.5$.

To compare the performance of the session model and the SWA model, we use the perplexities of the two models. Perplexity is a widely used measure to compare the performance of statistical models [1] and the lower value represents the better performance. The perplexity of each model on the test data is given by:

$$perplexity(D_{test}) = \exp\left(-\frac{\sum_{u\in U}\sum_{r=1}^{R_u^{test}}\sum_{j=1}^{V_{ur}^{test}} p(a_{urj})}{\sum_{u\in U}\sum_{r=1}^{R_u^{test}} |V_{ur}^{test}|}\right), \tag{8}$$

where R_u^{test} and V_{ur}^{test} represent the number of u's sessions and the number of logs in rth session of u in the test data, respectively. The $p(a_{urj})$ is computed based on the estimated parameters obtained by Eqs. (6) and (7) as follows:

$$p(a_{urj}) = \lambda_{u0}\sum_{k=1}^{K}\theta_{uk}\phi_{ka_{urj}} + \lambda_{u1}\psi_{ua_{urj}}. \tag{9}$$

In terms of the number of topics, we compute the perplexity for $K = 5$, 10, 20, 30, 40, 50, 100, 200, and 300.

5.3 Results

Figure 2 shows the perplexity for each dataset. In any dataset, regardless of the amount of training data and the number of topics, the SWA model outperformed

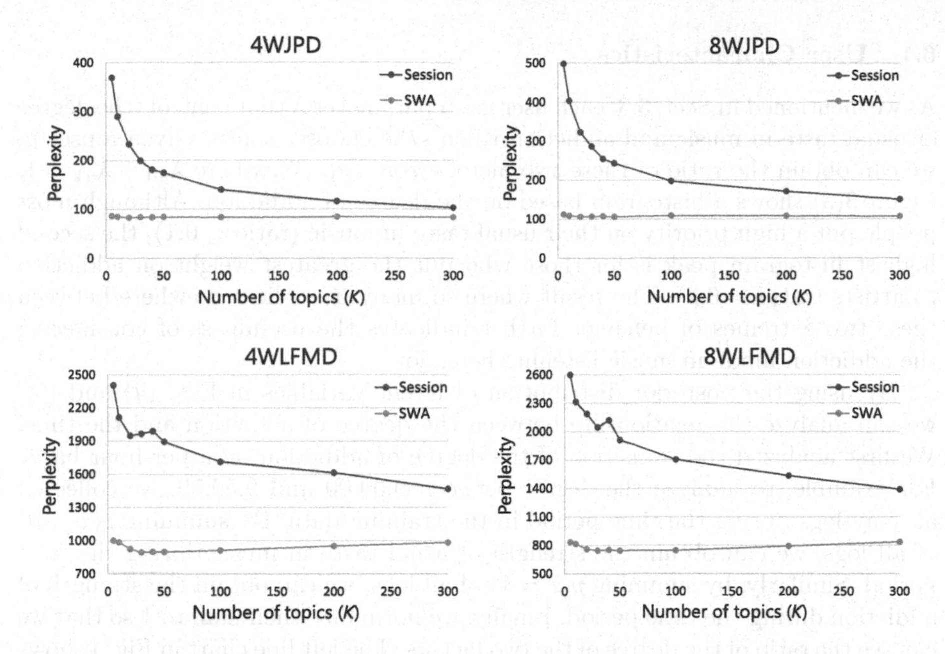

Fig. 2. Perplexity for 4WJPD, 8WJPD, 4WLFMD, and 8WLFMD.

the session model. If we set the number of topics to be larger than 300, the session model might outperform the SWA model; but we set the maximum value of K to 300 for the following two reasons. The first reason is due to the expended hours for the learning process. For example, when the session model learns parameters for $K = 300$ using 8WJPD, it takes 9.8 times longer than the SWA model does for $K = 30$ using 8WJPD (1,713 min for the session model and 175 min for the SWA model). In data analysis, the expended hours is an important factor; if it takes a long time to learn the parameters for a model, the model is inappropriate for data analysis. The second reason is due to the understandability of topics. When the number of topics becomes too large, it is difficult to understand the difference between topics because there are many similar topics. As we will show in Sect. 6.3, analyzing the characteristics of each topic is useful to understand people's music listening behavior. Hence, it is undesirable to set K to a large value. For these reasons, we conclude that the SWA model is a better model than the session model.

6 Qualitative Experiments

In this section, we report on the qualitative analysis results in terms of user characteristics, artist characteristics, and topic characteristics. Due to the space limitation, we only show the results for the training data of 8WJPD with $K = 30$. We not only analyze people's music listening behavior but discuss how we can apply the analysis results.

6.1 User Characteristics

As we mentioned in Sect. 3.3, each user has a parameter λ that controls the degree of usual taste in music and addiction when s/he chooses songs. Given a user u, we can obtain the ratio of these two factors from Eq. (7), where $\lambda_{u0} + \lambda_{u1} = 1$. Figure 3(a) shows a histogram based on the degree of addiction. Although most people put a high priority on their usual taste in music (ratio ≤ 0.1), the second highest histogram peak is for those who put the greatest weight on addiction to artists (ratio > 0.9). The result where so many users lie somewhere between these two extremes of behavior further indicates the usefulness of considering the addiction mode in music listening behavior.

By using the posterior distribution of latent variables in Eqs. (4) and (5), we can analyze the relationship between the degree of addiction and the time. We first analyzed the transition of the degree of addiction on a per-hour basis. For example, to analyze the degree between 9:00:00 and 9:59:59, we collected all play logs during the time period in the training data. By summing $p(x = 0)$ of all logs, we can obtain the strength of usual taste in music during the time period. Similarly, by summing $p(x = 1)$ of all logs, we can obtain the strength of addiction during the time period. Finally, we normalize their sum to 1 so that we can see the ratio of the degree of the two factors. The left line chart in Fig. 4 shows the results. It can be observed that the degree of addiction is high in the early morning (*i.e.*, at 5, 6, and 7 am), while it is low at night (*i.e.*, at 9, 10, and 11 pm).

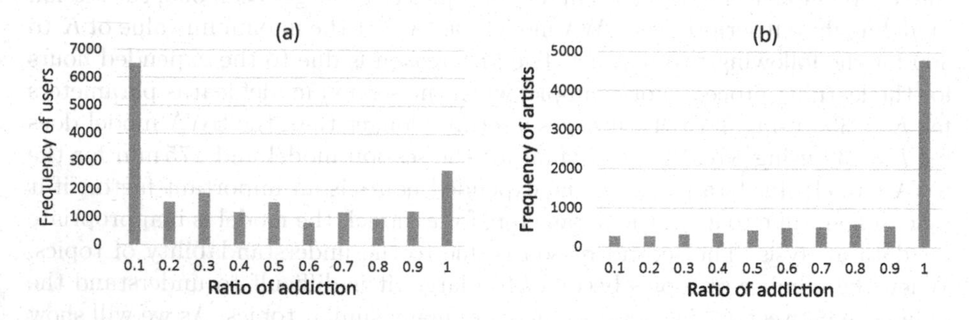

Fig. 3. Histogram based on ratio of addiction among (a) users and (b) artists.

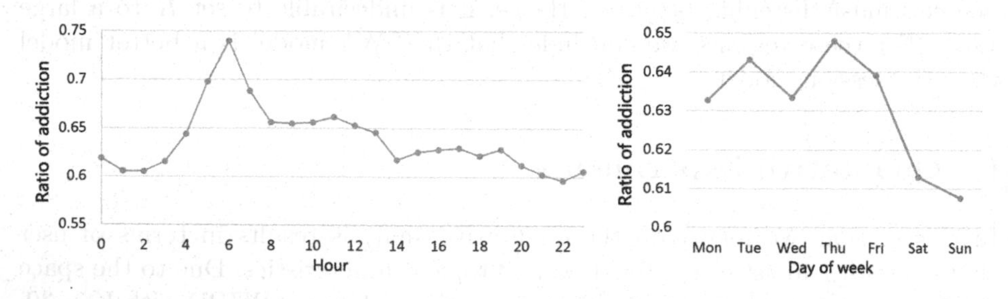

Fig. 4. Time-dependent ratio of addiction: per-hour analysis result and per-weekday analysis result.

We can estimate that people tend to be short on time in the morning, and as a result, they listen to a specific artist's songs rather than choosing various songs according to a topic. On the other hand, at night, people have time to spare and tend to listen to various artists' songs by choosing from a topic. These results indicate that the transition of the degree of addiction on a per-hour basis enables us to analyze people's music listening behavior from a new viewpoint. In addition, we propose applying the knowledge to music recommendation. For example, it would be more appropriate to recommend unknown songs to the user at night rather than in the morning because s/he would have time to try listening to new songs.

In the same manner as the above analysis, we also analyzed the transition of the degree of addiction on a day of the week basis. The right line chart in Fig. 4 shows the result. It can be observed that the degree of addiction is high on weekdays, while it is low on weekends. We can also estimate that the degree of addiction is high on weekdays because people are busy working on weekdays, while the degree is low on weekends because people have more time. These results would also be useful to recommend music.

6.2 Artist Characteristics

In the same way as Sect. 6.1, given an artist, by summing $p(x = 0)$ and $p(x = 1)$ of all the artist's logs, we can obtain the strength of usual taste and addiction during the time period, respectively. Then their sum is normalized to 1 to compute the ratio of each factor of the artist. Figure 3(b) shows a histogram based on the degree of addiction. It can be observed that most artists have a high degree of addiction. From these results, we can estimate whether the artist's songs are repeatedly played by users who are enthusiastic admirers of the artist or by various users who listen to the artist's songs with other artists' songs. In addition, we believe that the results could be used as one of the features to compute the similarity between artists by assuming that similar artists have similar degrees of addiction.

6.3 Topic Characteristics

Finally, we show that our model can also be used for topic analysis. Given a topic k, we collected representative artists in the category. To be more specific, the top 20 artists in terms of ϕ_k were extracted. For each of the 20 artists, we collected all logs in the training data and computed the ratio of the degree of taste in music and addiction as described in Sect. 6.1. We then computed the average values of each degree over 20 artists and normalized their sum to 1. Figure 5 shows the ratio of 30 topics, where topics are sorted in ascending order of addiction ratio. As can be seen, the ratio between two factors is largely different from one topic to another: the addiction ratio ranged from 0.297 (10th topic) to 0.620 (17th topic). As for the low addiction topics, the 10th topic has the lowest value of 0.297. This topic is related to songs created by using VOCALOID [9], which is popular singing synthesizer software in Japan. The 8th topic has the

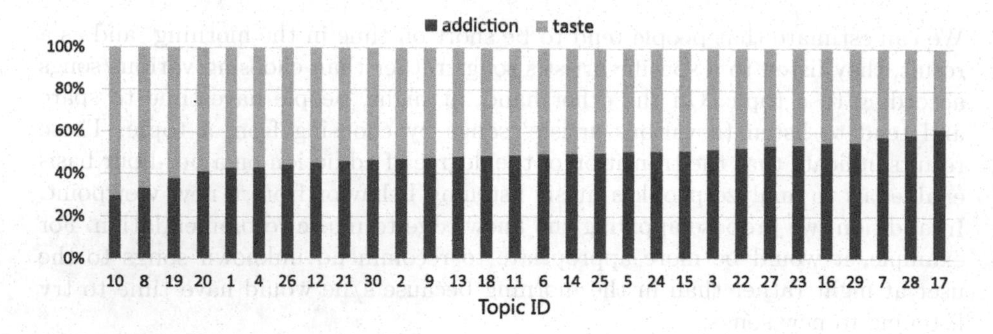

Fig. 5. Ratio of taste in music and addiction for each topic.

second lowest value of 0.334 and its topic is related to anime songs. From these results, we can estimate that when people listen to music related to popular culture, they tend to listen to various artists' songs in the topic. As for the high addiction topic, the 17th topic, which is related to Western artists, and the 28th topic, which is related to old Japanese artists, have the highest values of 0.620 and 0.592, respectively. These results indicate the possibility of applying the knowledge to playlist generation. In topics with a high addiction degree, it would be useful to generate a playlist that consists of songs of a specific artist; while in topics with a low addiction degree, it would be useful to generate a playlist that consists of various artists' songs.

7 Conclusion

In this paper we proposed a probabilistic model for analyzing people's music listening behavior. The model incorporates the user's usual taste in music and addiction to artists. Our experimental results using real-world music play logs showed that our model outperformed an existing model that considers only the user's taste in terms of perplexity. In our qualitative experiments, we showed the usefulness of our model in various aspects: time-dependent play log analysis (*e.g.*, the degree of addiction is high in the early morning and on weekdays), topic-dependent play log analysis (*e.g.*, the degree of addiction is low in an anime song topic), etc.

For future work, we are interested in applying the knowledge obtained from log analysis to applications such as artist similarity computation and song recommendation as discussed in Sect. 6. We are also interested in extending our model by considering the time transition of addiction. For example, a user who is addicted to some artists in summer may be addicted to largely different artists in autumn. Considering such time dependency by using the topic tracking model [7] is one possible direction to take to extend our model.

Acknowledgements. This work was supported in part by ACCEL, JST.

References

1. Airoldi, E.M., Blei, D.M., Fienberg, S.E., Xing, E.P.: Mixed membership stochastic blockmodels. J. Mach. Learn. Res. **9**, 1981–2014 (2008)
2. Baur, D., Büttgen, J., Butz, A.: Listening factors: a large-scale principal components analysis of long-term music listening histories. In: CHI, pp. 1273–1276 (2012)
3. Berkers, P.: Gendered scrobbling: listening behavior of young adults on last.fm. Interact. Stud. Commun. Culture **2**(3), 279–296 (2012)
4. Dias, R., Fonseca, M.J.: Improving music recommendation in session-based collaborative filtering by using temporal context. In: ICTAI, pp. 783–788 (2013)
5. Griffiths, T.L., Steyvers, M.: Finding scientific topics. PNAS **101**(Suppl. 1), 5228–5235 (2004)
6. Herrera, P., Resa, Z., Sordo, M.: Rocking around the clock eight days a week: an exploration of temporal patterns of music listening. In: WOMRAD, pp. 7–10 (2010)
7. Iwata, T., Watanabe, S., Yamada, T., Ueda, N.: Topic tracking model for analyzing consumer purchase behavior. In: IJCAI, pp. 1427–1432 (2009)
8. Kamalzadeh, M., Baur, D., Möller, T.: A survey on music listening and management behaviours. In: ISMIR, pp. 299–305 (2012)
9. Kenmochi, H., Ohshita, H.: Vocaloid - commercial singing synthesizer based on sample concatenation. In: INTERSPEECH, pp. 4009–4010 (2007)
10. Lee, J.H., Kim, Y.-S., Hubbles, C.: A look at the cloud from both sides now: an analysis of cloud music service usage. In: ISMIR, pp. 299–305 (2016)
11. Liu, N.-H., Hsieh, S.-J., Tsai, C.-F.: An intelligent music playlist generator based on the time parameter with artificial neural networks. Expert Syst. Appl. **37**(4), 2815–2825 (2010)
12. Park, C.H., Kahng, M.: Temporal dynamics in music listening behavior: a case study of online music service. In: ACIS-ICIS, pp. 573–578 (2010)
13. Park, S.E., Lee, S., Lee, S.G.: Session-based collaborative filtering for predicting the next song. In: CNSI, pp. 353–358 (2011)
14. Rentfrow, P., Gosling, S.: The do re mi's of everyday life: the structure and personality correlates of music preferences. J. Pers. Soc. Psychol. **84**(6), 1236–1256 (2003)
15. Reynolds, G., Barry, D., Burke, T., Coyle, E.: Interacting with large music collections: towards the use of environmental metadata. In: ICME, pp. 989–992 (2008)
16. Schedl, M.: The lfm-1b dataset for music retrieval and recommendation. In: ICMR, pp. 103–110 (2016)
17. Schedl, M., Hauger, D.: Mining microblogs to infer music artist similarity and cultural listening patterns. In: WWW, pp. 877–886 (2012)
18. Zheleva, E., Guiver, J., Mendes Rodrigues, E., Milić-Frayling, N.: Statistical models of music-listening sessions in social media. In: WWW, pp. 1019–1028 (2010)

Understanding Drivers' Safety by Fusing Large Scale Vehicle Recorder Dataset and Heterogeneous Circumstantial Data

Daisaku Yokoyama[1(✉)], Masashi Toyoda[1], and Masaru Kitsuregawa[2]

[1] Institute of Industrial Science, The University of Tokyo, Meguro, Japan
{yokoyama,toyoda}@tkl.iis.u-tokyo.ac.jp

[2] Data Integration and Analysis System (DIAS), The University of Tokyo, Meguro, Japan
kitsure@tkl.iis.u-tokyo.ac.jp

Abstract. We present a method of analyzing the relationships between driver characteristics and driving behaviors on the basis of fusing heterogeneous datasources with large-scale vehicle recorder data. It can be used, for example, by fleet managers to classify drivers by their skill level, safety, physical/mental fatigue, aggressiveness, and so on. Previous studies relied on precise data obtained in only critical driving situations and did not consider their circumstances, such as road width and weather. In contrast, our approach takes into account not only a large-scale (over 100 fleet drivers) and long-term (one year's worth) records of driving operations, but also their circumstances. In this study, we focused on classifying drivers by their accident history and examined the correlation between having an accident and driving behavior. Our method was able to reliably predict whether a driver had recently experienced an accident (f-measure = 72%) by taking into account both circumstantial information and velocity at the same time. This level of performance cannot be achieved using only the drivers' demographic information or kinematic variables of operation records.

Keywords: Vehicle recorder · Fusing data from heterogeneous datasources · Driving safety · Accident history · Individual driving behavior

1 Introduction

Driver management has been an important issue for the transportation industry. Keeping drives safe and at the same time efficient is still a hard problem; transport companies typically manage their drivers by using demographic information to estimate their safety; however, such information overlooks the current condition and improvements in skill of the driver.

We have developed a method for analyzing the relationships between driver characteristics and driving behaviors on the basis of vehicle recorder data combined with other datasources such as weather reports and road maps. It can be

J. Kim et al. (Eds.): PAKDD 2017, Part II, LNAI 10235, pp. 734–746, 2017.
DOI: 10.1007/978-3-319-57529-2_57

used, for example, by fleet managers to classify drivers by their skill level, safety, physical/mental fatigue, aggressiveness, and so on. Our method manages drivers by not *who they are*, but rather *how do they drive*.

Several studies [1,3,5] have analyzed driving behaviors. They relied, however, on detailed and precise data on a small number of drivers, so it is difficult to extrapolate their results to the general driver population. Many transportation companies have introduced dashboard cameras (dashcams) and/or vehicle data recorders (which collect GPS, velocity, and acceleration data) into their fleets. Although the amount of data collected tends to be sparse due to storage limitations, data can be collected on a large number of drivers. Many kinds of transportation related information, such as weather, road structure, degree of traffic congestion, are also available nowadays. Utilizing such heterogeneous datasources would improve the preciseness of the management's understanding of each driver's characteristic.

Our method classifies drivers on the basis of long-term records of kinematic variables (maximum velocity, acceleration, etc.) related to their driving operations (braking, steering, etc.). It is based on the assumption that the distributions of these variables differs from driver to driver. Our method takes into account the factors of driving circumstances by fusing various heterogeneous datasources. We focused on classifying drivers who had recently been involved in accidents and examined the correlation between having an accident and driving behavior. Our findings are useful both for educating drivers and preventing accidents.

Many studies [4,13] have analyzed driving behaviors as a means of estimating driver risks. However, they only used driving operation information and tended to focus on extreme case of driving operation. Driver characteristics such as driving skill are reflected in all situations, not only in critical ones; for example, a skillful driver will brake smoothly on slippery roads during heavy rainfall. The previous studies thus overlooked the information to be obtained from operations performed in non-critical situations. By contrast, in this study, we used *all* driving information derived from many heterogeneous datasources to better estimate a driver's characteristics.

Our main contributions are:

- An intensive examination of large-scale vehicle recorder data covering all driving operations demonstrated the effectiveness of our method for analyzing the relationships between driver characteristics and driving behaviors. It was able to reliably predict whether a driver had recently experienced an accident (f-measure = 72%). This level of performance cannot be achieved by using only drivers' demographic information or kinematic variables of operation records.
- It showed that fusing heterogeneous data is essential to depicting driver behavior precisely. When we only used kinematic variables of driving records as the features of drivers, classification performance was poor (f-measure < 66%).
- We found an appropriate way to combine circumstantial information. When we fused operation records and other non-kinematic information and took into account these information separately, the classification performance was almost same as using kinematic features. Performance improved after adding

features that took into account both velocity and circumstantial information at the same time.

In Sect. 2, we overview related work. In Sect. 3.1, we explain our analysis. We explain the driving operation dataset we used in Sect. 3.2 and describe other dataset to take into account driving circumstances in Sect. 3.3. In Sect. 3.4, we present our method for analyzing the relationships between driver characteristics and driving behaviors and evaluate its effectiveness. This article ends in Sect. 4 with a summary and a look at future work.

2 Related Work

There has been research on using vehicle recorded data, such as velocity and location, for various purposes [6,9,11]. The studies can be grouped into two categories: those that utilize large-scale vehicle location data [2] and those that investigate a small amount of driving operation data. We believe that ours is the first study to investigate both driving operation and its circumstances on a large-scale (more than 1000 drivers).

The 100-Car Naturalistic Driving Study [7] is one of the largest studies on the use of vehicle recorded data. It used many types of precise driving information and driver demographic data (age, gender, personality, etc.) and thoroughly analyzed the driver information statistically. Several studies have used the driving information in this archive to assess driver risk. For example, Guo et al. [4] reported an effective model for identifying high-risk drivers by using driver demographic information and the occurrence of critical-incident events. Their model mainly uses demographic information. Zheng et al. [13] collected data on naturalistic driving and analyzed the relationship between the kinematic information and driver risk-taking behavior. Their analysis focused on kinematic information for critical driving operations involving large accelerations. Yokoyama et al. [12] investigated the relationship between kinematic information and drivers' accident histories; however they did not utilize driving circumstances.

Some studies have tried to classify drivers on the basis of the aggressiveness of their driving behavior, with the aim of improving driving safety. Higgs et al. [5] analyzed the car-following behaviors of three drivers and identified the differences among them. Dang et al. [3] focused on the lane-changing behaviors of 12 drivers driving on a highway and found differences among them. Miyajima et al. [10] used data on 276 drivers and tried to identify drivers on the basis of their car-following behaviors and pedal operations. However, their data collection required the use of pedals with specially designed sensors. Their study and the other previous research relied on precise information on driving behavior, which is not always available.

3 Classification of Drivers' Accident Histories

3.1 Approach

Our research purpose is to identify the characteristics of drivers through their driving behaviors. In this study, we focused on classifying drivers as either safe

Table 1. Summary of vehicle recorder dataset

	All data	Driving days ≥ 20, driving hours ≥ 20
Number of drivers	1469	320
Driving duration in total	77,450 h	60,190 h

or unsafe on the basis of their driving records. Instead of using only critical operation records, we used a large amount of vehicle recorder data that included all driving operations and investigated how effective such data is for classifying drivers.

A driver performs various driving operations (braking, steering, etc.), each associated with several variables (maximum velocity, acceleration, etc.). A driver can be characterized by the distributions of these variables. We investigated ways to derive features from these variable distributions for use in classifying drivers as either safe or unsafe by using Support Vector Machine (SVM).

Each driving operation is affected by factors of the moment, such as the weather condition, road condition, degree of congestion, and time of day. We need to take into account the effects of these factors in order to derive good features from the operation records. These factors cannot be observed from the vehicle recorded operation records alone. Therefore, we should combine other datasources such as weather data to reconstruct other factors. Here, we focus on two circumstances: rainfall information and road width. We derived several features from the distributions of operation variables, taking into account the factors of the moment, and evaluated the effectiveness of our method.

3.2 Dataset

Vehicle Recorder Dataset. In our experiments, we used a large number of actual driving records[1] collected by a parcel delivery service company (transport company). The data were for about 1450 drivers working in the Tokyo area and covered one year (from 21 July 2014). A multifunctional data recorder in each delivery vehicle recorded longitudinal accelerometer, lateral accelerometer, gyro compass, and GPS data.

Since we focused on long-term driving behavior, we eliminated the data of drivers who had driven on fewer than 20 days or for less than 20 h in total. A summary of the data is shown in Table 1. The driving duration does not include the time during which the engine was turned off.

The vehicle data recorder automatically detected four basic driving operations: braking, steering, turning, and stopping. Several variables, including maximum velocity and acceleration, during each operation were recorded. The operation variables are listed in Table 2. The numbers of recorded operations per driver are summarized in Table 3. As mentioned, our dataset contained data on all driving operations, while those used in previous studies contained data only on critical operations involving high acceleration.

[1] The vehicle recorder data was provided by Datatec Co., Ltd.

Table 2. Operation record variables

Operation	Variables
Braking	Velocity (V), longitudinal acceleration (Gx), and jerk (derivative of acceleration with respect to time, Jx)
Steering	V, yaw velocity (Yr), yaw acceleration, and lateral acceleration (Gy)
Turning	{Gx, V} before turn, {V, centrifugal force (CG), yaw acceleration} during turn, and {V, CG} after turn
Stopping	V, Gx, and stopping duration

Table 3. Operation record statistics

Operation	No. of records per driver		No. of records
	(min)	(max)	(total)
Braking	114	45,861	1,993,341
Steering	239	46,452	2,783,723
Turning	121	21,027	1,218,957
Stopping	418	40,625	2,221,166

Driver Histories. With the cooperation of the transport company, we accessed their drivers' histories, including the traffic violations they had received and the accidents in which they had been involved. We used their histories to define their *accident experience* and *driving experience.*

Accident experience. Drivers who had at least one accident during a certain time period were defined as an *accident* driver. Even though some accidents were only small ones without any responsibility being assigned, we treated all accidents the same.

Driving experience. To estimate how long a driver had been driving, we used the oldest record in the driver's history to estimate the minimum number of driving years.

Using these definitions and the estimates, we investigated the differences in driving operation between the accident and no-accident drivers. The no-accident drivers, however, are not necessarily safe drivers. For example, a reckless driver may simply have been lucky enough to avoid an accident over the course of a year. We therefore focused on drivers who had at least five years' worth of driving experience. We defined a driver who had at least five years' worth of driving experience without any accidents in the previous five years as safe and otherwise as unsafe. There were 82 safe drivers and 43 unsafe drivers.

3.3 Fusing the Driving Circumstances with Operation Records

To understand each driver's driving behavior, we focused on the distributions of variables for driving operations. Each driving operation is affected by factors of the moment, such as the weather condition, road structure, and degree of traffic congestion. Therefore, we combined other datasources with the operation records to reflect the effect of these factors. Each operation record contains GPS data and time information; thus we could perform spatial- and temporal- matching with the other datasources.

To take into account driving circumstances, we created two different variable distributions: (a) splitting up operation records by circumstance and (b) splitting up operation records by the combination of two circumstances. We selected several factors that represent driving circumstances; they are as follows.

Velocity. Operation variables are correlated due to kinematic restrictions for both safe and unsafe drivers. For example, steering at a high velocity tends to cause a low yaw rate. We therefore treated velocity as the basic variable for each operation and split up the operation records according to their velocity values. For example, we divided the braking operation records into six bins on the basis of velocity and estimated the longitudinal jerk densities for each bin. We found that the shapes of the distributions differ among the velocity bins.

Time of Day. The degree of traffic congestion heavily affects driving behavior. To reflect this factor, we used the occurrence time of each operation record. We separated the operation records into several time ranges, and compared the variable distributions. We found the operation distributions in the morning and evening differ from at other time, which seems to be the result of traffic congestion. Time is not kinematic information; however, it surely affects kinematic variables of operations.

Road Properties. Driving operations are also affected by the road width. For example, turning onto a narrower road tends to require more deceleration than turning onto a wider road. We could match each operation location with a point on a digital road map[2]. We simply searched for the road segment nearest the operation location. If the nearest segment was more than 30 m away (due, for example, to being on a private site such as a factory or university), we considered that the location could not be matched to a point on the map and ignored that record. The road map contains information about the road width, represented in several ranks, and whether the road is bi-directional or not. If the road was bi-directional, we assumed that the width of the segment was one rank narrower. We used four road width ranges: >13 m, $13 > w > 5.5$, <5.5 m, and unknown.

Rainfall. Weather heavily affects road conditions and driving operations. When it is raining, for example, the accident rate is eight times higher when the weather is dry[3]. We used X-band Multi Parameter Radar information collected by the

[2] We used the "Advanced Digital Road Map Database" developed by Sumitomo Electric System Solutions Co., Ltd. The database was provided by the Center for Spatial Information Science at the University of Tokyo.

[3] From discussions with an Expressway company.

Ministry of Land, Infrastructure, Transport and Tourism[4]. It detects rainfall in a 250 m mesh every minute. This fine-grained weather radar can detect sudden rain showers that happen frequently in Japan. Since every operation record contained GPS data and time information, we could match each operation location with the rainfall information at that time.

3.4 Features

Derivation. We used all 17 dataset variables listed in Table 2 to derive the driver features. We also used driver demographic information known to be related to driving safety.

First, we created basic features that represent demographic characteristics or distributions of kinematic variables:

- Demographic features: We used the driver's age, gender, and time since obtaining a driver's license as three demographic features. This information is commonly used by insurance companies to set auto insurance rates.
- License feature: In Japan, a driver who has not had any accidents and has not been cited for a driving violation during the preceding five years is categorized as a "gold license" driver and is generally considered to be a safe driver. We thus defined a binary feature for whether a driver had a gold license or not. The license category is updated when one's license is renewed, and the renewal interval is three to five years. Therefore, a gold license does not always mean an accident-free driver; many drivers have had accidents in recent years and still hold a gold license. When we classified drivers as safe or unsafe by using their license category information alone, we achieved only a 35% precision, which is virtually the same performance as with a random classifier.
- Operation frequency features: We counted the number of instances for each of the four driving operations for each driver and normalized it by the driving duration.
- Variable distribution features: We defined the shapes of the variable distributions as features. Each variable value was binned into one of ten intervals; the maximum and minimum bin breakpoints were chosen by hand, and the other bins were defined to have the same width. Therefore, each variable distribution was represented by ten values. There were thus 170 variable distribution features (17 variables $\times$ 10 values).

Second, we consider the relationship between circumstances and basic kinematic variables (as described in Sect. 3.3, approach (a)):

- Variable distribution by velocity features: Driving operations are strongly affected by the vehicle's velocity. We therefore selected six velocity-related variables for use in separating the operation records, and combined them with other variables, as shown in Table 4. The operation records were separated by the corresponding velocity-related variable, and the distributions of the other

[4] XRAIN: http://www.river.go.jp/kawabou/ipXAreaMap.do.

Table 4. Combination patterns of operation variables

Operation	Velocity-related variable (number of bins)	Other variables combined with velocity-related variable
Braking	Velocity (6)	Gx, Jx
Steering	Velocity (5)	Yr, yaw acceleration, Gy
Turning	Velocity before turn (4)	Gx before turn
Turning	Velocity during turn (4)	CG, yaw acceleration during turn
Turning	Velocity after turn (5)	CG after turn
Stopping	Velocity (5)	Gx

variables were calculated separately. The velocity-related variables were digitized into b values by intervals with a constant width (5 km/h). The other variable distributions were digitized with ten intervals, so the feature of a variable is represented by $b \times 10$ values.
- Variable distribution by road width features: We defined each of the four road width ranges as an indicator of a circumstance, and use it to split operation records.
- Variable distribution by rainfall features: We decided the raining condition to be when rainfall is larger than 5.0 mm/h. Thus we split up the operation records into three rainfall ranges: >5.0, ≤5.0, unknown (that is caused by the lack of observation).
- Variable distribution by time of day features: We defined five time ranges to capture the different traffic conditions of the operation records: [6:00–9:00], [9:00–12:00], [12:00–18:00], [18:00–21:00], [21:00–6:00].

Finally, we considered two of the above circumstances at the same time (as described in Sect. 3.3, approach (b)). In this study, we limited the number of sets of combination to three. Among the circumstance features, velocity has the largest possibility to restrict vehicle's motion. Thus we selected velocity as the fixed feature, and combined it with the other three circumstance features as follows:

- Variable distribution by velocity and road width features
- Variable distribution by velocity and rainfall features
- Variable distribution by velocity and time of day features

Increasing the number of combinations improved the accuracy of the depicted variable distribution for each driver. Although this helped to describe the difference between driving behaviors precisely, it may cause data sparsity because it reduces the number of operation occurrence in each bin, which means the features will be more strongly affected by noise.

Feature Expression. We tested two methods of expressing the variable distributions as features.

Table 5. Feature settings

Feature category (no.)	a	b	c	d	e	f
Demographic (3)	✓	✓		✓	✓	✓
License (1)		✓		✓	✓	✓
Operation frequency (4)			✓	✓	✓	✓
Variable distribution (170)					✓	✓
Variable distribution by velocity (540)						✓
Number of available features	3	4	4	8	178	718
Number of frequent features	2	3	4	7	172	601

Feature category (no.)	g	h	i	j	k	l	m
Features of setting f (718)	✓	✓	✓	✓	✓	✓	✓
Variable distribution by road width (680)	✓			✓	✓		✓
Variable distribution by rainfall (510)		✓		✓		✓	✓
Variable distribution by time of day (850)			✓		✓	✓	✓
Number of available features	1398	1228	1568	1908	2248	2078	2758
Number of frequent features	1160	1032	1337	1591	1896	1768	2327

Feature category (no.)	n	o	p	q
Features of setting m (2758)	✓	✓	✓	✓
Variable dist. by velocity and road width (2160)	✓			✓
Variable dist. by velocity and rainfall (1620)		✓		✓
Variable dist. by velocity and time of day (2700)			✓	✓
Number of available features	4918	4108	5008	8158
Number of frequent features	3550	3295	3986	6177

Probability method. We denoted each driver's frequency for each bin as N_i and computed each driver's occurrence probability P_i, which is N_i normalized by the number of operation instances for the driver. We used P_i itself as a feature.

KL divergence method. We described the difference between two distributions, P and Q. The KL divergence [8] is a representative definition of the distance between two distributions: $KL(P||Q) = \sum_i P_i \log \frac{P_i}{Q_i}$. We used $P_i \log \frac{P_i}{Q_i}$ of each bin as the feature.

Performance Evaluation. We tested 17 combinations of features, as shown in Table 5. The feature settings are categorized into four groups; (a) to (d) use only demographic and statistical information on the driver; (e) and (f) introduce the variable distributions of the driving operations; (g) to (m) introduce driving circumstance information from other datasources or non-kinematic information in the operation records; (n) to (q) take into account the effects of the combination of the velocity and other circumstantial information.

We evaluated the performance by 10-fold cross validation. Features that appeared in the driving records of less than 30 drivers were eliminated. The number of remaining features of each combination is shown in Table 5, as the "Number of frequent features". All remaining features were normalized beforehand. Three types of kernel functions (linear, polynomial, Gaussian) with hyperparameters (Table 6) were evaluated in a grid-search manner to achieve the best AUC (area under the ROC curve) value. We also used feature selection based on the χ^2 value. The best number of features was determined from the grid search.

Table 6. Parameters for grid search

Kernel	Hyperparameter
Linear	$C : [2^{-5}, ..., 2^{10}]$, $w_{accident} : \{1, 2, 3, 5, 10\}$
Polynomial	$C : [2^{-5}, ..., 2^{10}]$, $\gamma : [2^{-10}, ..., 2^{3}]$, $degree : \{2, 3\}$, $w_{accident} : \{1, 2, 3, 5, 10\}$
Gaussian	$C : [2^{-5}, ..., 2^{10}]$, $\gamma : [2^{-10}, ..., 2^{3}]$, $w_{accident} : \{1, 2, 3, 5, 10\}$

Table 7. Classification performance

Setting	Method	No. of selected features	Precision	Recall	F-measure	AUC
a	-	2	0.36	1.00	0.52	0.57
b	-	3	0.43	0.93	0.58	0.64
c	-	4	0.36	1.00	0.53	0.45
d	-	5	0.38	0.88	0.53	0.59
e	p	50	0.47	0.88	0.62	0.71
f	p	20	0.57	0.79	0.66	0.79
g	KL	20	0.67	0.67	0.67	0.80
h	KL	40	0.53	0.77	0.63	0.76
i	KL	30	0.70	0.74	0.72	0.81
j	KL	50	0.55	0.84	0.66	0.77
k	KL	30	0.70	0.70	0.70	0.80
l	KL	50	0.59	0.77	0.67	0.81
m	KL	60	0.56	0.81	0.67	0.81
n	KL	80	0.59	0.79	0.67	0.83
o	KL	80	0.57	0.93	0.71	0.81
p	KL	90	0.72	0.77	0.74	0.85
q	KL	40	0.60	0.88	0.72	0.85
Random classifier			0.37			0.50

Figures 1 and 2 show the best f-measure and AUC for each setting, respectively. Representative results are shown in Table 7. The random classifier was used as a baseline; it had a precision of 37% (= 43/125).

The demographic information was not so helpful in classifying drivers, although it was slightly better than the random classifier: the AUC values for settings (a) and (b) were greater than 0.5. Since all the drivers were well-trained professionals, the demographic information may not have reflected their driving skills so well.

The use of the kinematic information obtained from vehicle recorders was helpful in classifying the drivers, as we can see from the results for setting (e). When we took into account the velocity at which the operation was performed, performance improved slightly (see results for (e) and (f)). Adding circumstantial information (road width, rainfall, and time of day) to the kinematic information resulted in almost same performance ((g) to (i)). This circumstantial information was of much help when it was combined with the velocity ((n) to (q)).

Figure 3 shows the ROC curves of representative results. Taking into account the velocity of driving operations improved performance ((e) and (f)). Adding circumstantial information improved performance; it was not so helpful when we combined it with simple variable distributions (m); however, it greatly improved performance when it was combined with both variable distributions and velocity (q).

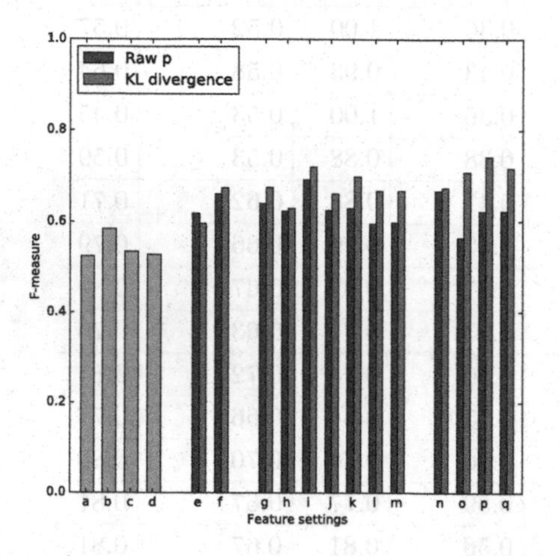

Fig. 1. F-measure for different feature settings

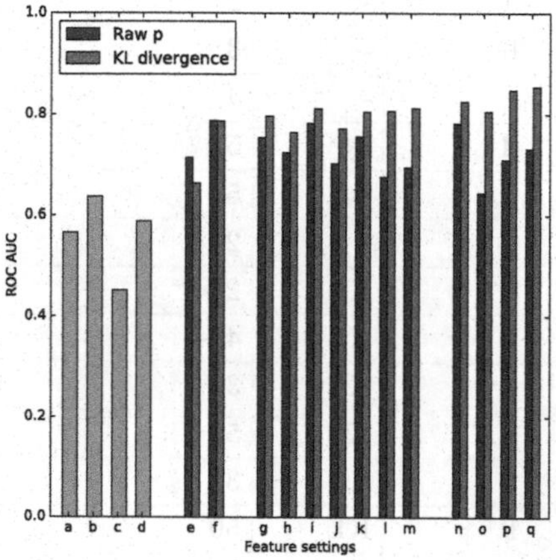

Fig. 2. AUC under the ROC curve for different feature settings

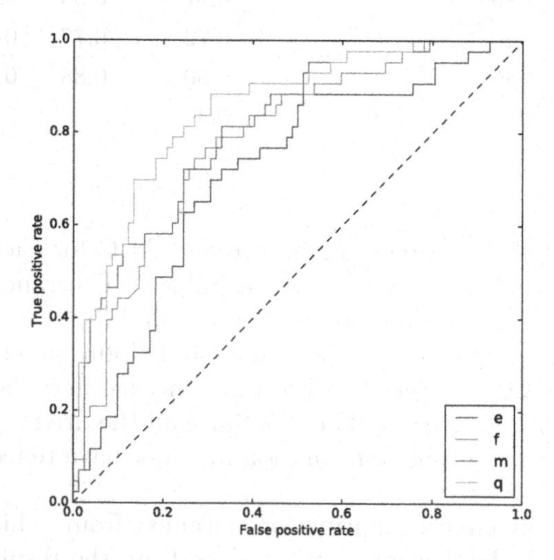

Fig. 3. ROC curves of representative results

4 Conclusion

We thoroughly examined a large-scale archive of recorded vehicle data in order to clarify the relationship between safety and driver behavior. We used multiple datasources to compensate for driving circumstances in operation records and successfully classified drivers as either safe or unsafe (f-measure = 72%). Methods that use only driver demographic information or kinematic variables of operation records have not achieved this level of performance.

This is the first step toward a better understanding of the relationship between safe driving and driver behavior. Although this study considered only past accidents, the knowledge acquired will be helpful in investigating driver safety and preventing future accidents. We thus plan to apply our method to predicting accidents. Our findings on the characteristics of drivers through their driving behaviors will be helpful in educating drivers.

References

1. Castignani, G., Frank, R., Engel, T.: Driver behavior profiling using smartphones. In: 16th International IEEE Conference on Intelligent Transportation Systems (ITSC 2013), pp. 552–557, October 2013
2. Castro, P.S., Zhang, D., Chen, C., Li, S., Pan, G.: From taxi GPS traces to social and community dynamics: a survey. ACM Comput. Surv. **46**(2), 17:1–17:34 (2013)
3. Dang, R., Zhang, F., Wang, J., Yi, S., Li, K.: Analysis of Chinese driver's lane change characteristic based on real vehicle tests in highway. In: ITSC 2013, pp. 1917–1922, October 2013
4. Guo, F., Fang, Y.: Individual driver risk assessment using naturalistic driving data. Accid. Anal. Prev. **61**, 3–9 (2013). http://www.sciencedirect.com/science/article/pii/S0001457512002382
5. Higgs, B., Abbas, M.: A two-step segmentation algorithm for behavioral clustering of naturalistic driving styles. In: ITSC 2013, pp. 857–862, October 2013
6. Johnson, D., Trivedi, M.: Driving style recognition using a smartphone as a sensor platform. In: ITSC 2011, pp. 1609–1615 (2011)
7. Klauer, S.G., Dingus, T.A., Neale, V.L., Sudweeks, J.D., Ramsey, D.J.: The impact of driver inattention on near-crash/crash risk: an analysis using the 100-car naturalistic driving study data. Technical report DOT HS 810 594, National Highway Traffic Safety Administration (2006)
8. Kullback, S., Leibler, R.A.: On information and sufficiency. Ann. Math. Statist. **22**(1), 79–86 (1951). doi:10.1214/aoms/1177729694
9. Liu, W., Zheng, Y., Chawla, S., Yuan, J., Xie, X.: Discovering spatio-temporal causal interactions in traffic data streams. In: SIGKDD 2011, August 2011
10. Miyajima, C., Nishiwaki, Y., Ozawa, K., Wakita, T., Itou, K., Takeda, K., Itakura, F.: Driver modeling based on driving behavior and its evaluation in driver identification. Proc. IEEE **95**(2), 427–437 (2007)
11. Wu, W., Ng, W.S., Krishnaswamy, S., Sinha, A.: To taxi or not to taxi? - enabling personalised and real-time transportation decisions for mobile users. In: IEEE 13th International Conference on Mobile Data Management, pp. 320–323, July 2012

12. Yokoyama, D., Toyoda, M.: Do drivers' behaviors reflect their past driving histories? - large scale examination of vehicle recorder data. In: 2016 IEEE International Congress on Big Data, pp. 361–368. IEEE (2016). doi:10.1109/BigDataCongress.2016.58
13. Zheng, Y., Wang, J., Li, X., Yu, C., Kodaka, K., Li, K.: Driving risk assessment using cluster analysis based on naturalistic driving data. In: ITSC 2014, pp. 2584–2589. IEEE (2014)

Graph Clustering and Community Detection

Query-oriented Graph Clustering

Li-Yen Kuo(✉), Chung-Kuang Chou, and Ming-Syan Chen

Department of Electrical Engineering, National Taiwan University, Taipei, Taiwan
{lykuo,ckchou}@arbor.ee.ntu.edu.tw, mschen@cc.ee.ntu.edu.tw

Abstract. There are many tasks including diversified ranking and social circle discovery focusing on the relationship between data as well as the relevance to the query. These applications are actually related to query-oriented clustering. In this paper, we firstly formulate the problem, query-oriented clustering, in a general form and propose the two measures, *query-oriented normalized cut* (QNCut) and *cluster balance* to evaluate the results for query-oriented clustering. We develop a model, *query-oriented graph clustering* (QGC), that combines QNCut and the balance constraint based on cluster balance in a quadratic form. In the experiments, we show that QGC achieves promising results on improvement in query-oriented clustering and social circle discovery.

Keywords: Query · Graph clustering · Laplacian eigenmaps

1 Introduction

Although most of the applications [1–5] are related to ranking (i.e., given a query, vertices in a graph are ranked according to their relevance to the query), the information requirement often goes beyond it. For instance, algorithms for diversified ranking capture both the relevance to the query and the diversity of the ranking result. As for social circle discovery [4], one may consider the graph structure among the top ranked vertices to evaluate not only their relevance to the query vertex (i.e., ego user) but also the similarity between them. Though both ranking and clustering have been well studied respectively, it still needs to be explored to consider the two objectives simultaneously. This issue is called *query-oriented clustering* (QC) [16]. Different from the bicriteria objective function in [16], we merge relevance and clustering into a graph-based objective function. Given an undirected graph and a query vertex, the graph can be partitioned into clusters, each of which has strong intra-cluster edges and weak inter-cluster edges, where these edges are weighted by the relevance of their endpoints to the query. Unlike *classical clustering* (CC) algorithms which treat all vertices on a graph with equal weights, such as k-means, query-oriented clustering pays more attention to the subgraph around the query.

Consider a toy graph illustrated in Fig. 1(a). We obtain a clustering result depicted in Fig. 1(b) by a CC method. Two different results for the QC problem are shown in Fig. 1(c) and (d), where vertices with high relevance are painted

J. Kim et al. (Eds.): PAKDD 2017, Part II, LNAI 10235, pp. 749–761, 2017.
DOI: 10.1007/978-3-319-57529-2_58

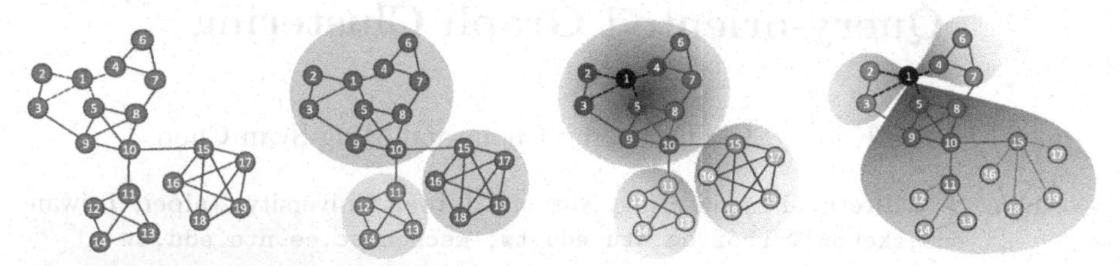

(a) An illustrative graph. (b) An ideal CC result. (c) Clustered by CC methods. (d) An ideal QC result.

Fig. 1. An illustration of difference between classical clustering (CC) and query-oriented clustering (QC). Inter-cluster edges are painted red. (Color figure online)

dark colors and the inter-cluster edges are painted red. Before partitioning the graph, the edges attached on the query vertex will be removed, which are shown in dotted lines. If we use a CC method to tackle query-oriented clustering, given query vertex v_1, the clustering result is shown in Fig. 1(c). We can see that top ranked vertices (i.e., v_2, v_3, v_4 and v_5) belong to the same cluster and the subgraph around v_1 can not be partitioned well. An ideal query-oriented clustering result given query vertex v_1 is shown in Fig. 1(d) since all three clusters are close to v_1. The top ranked vertices are partitioned into 3 clusters at the cost of 2 edges: (v_3, v_9) and (v_7, v_8). Notice that the result in Fig. 1(d) is desirable though its cost (i.e., sum of the weights of the inter-cluster edges) may be higher than the cost of the result in Fig. 1(c). Hence, the CC methods are not adaptive to the QC problem since there are two clusters.

The sum of the relevance of all the vertices in a cluster, called *cluster relevance* in this paper, is the other evaluation criterion. In Fig. 1(c), the yellow cluster and the green cluster are composed of vertices with low relevance while all the high-relevance vertices belong to the blue cluster. Ideally, a good result should contain clusters with similar relevance and we say that the result has high *cluster balance* which will be formulated more clearly in Sect. 3. Figure 1(d) shows an ideal result where the three clusters evenly cover the high ranked vertices. Thus, an ideal QC result should appear "centripetally" around the query.

In this paper, we firstly formulate the problem by two evaluation criteria: *query-oriented normalized cut* (QNCut) and *cluster balance*. Next, we propose a model, query-oriented graph clustering (QGC), that combines QNCut and cluster balance into a quadratic form by using Lagrange multiplier. For summarization, our contributions are as follows:

- We formulate the query-oriented clustering problem by the two evaluation criteria: QNCut and cluster balance.
- Considering QNCut minimization and entropy maximization simultaneously, we propose a novel model, QGC, that is based on the objective function which combines QNCut and the balance constraint into a quadratic form.

- In the experiments, QGC outperforms other QC algorithms for subtopic discovery. Moreover, QGC also outperforms the state-of-the-art methods for the discovery of social circles in ego networks.

The remainder of the paper is organized as follows. In Sect. 2, we discuss the related works including weighted clustering and graph clustering methods related to this problem. In Sect. 3, the problem of QC is formulated and then we define a novel measure, QNCut. In Sect. 4, our approach, QGC, containing the objective function and the algorithm for cluster assignment, are formally introduced. In Sect. 5, a comprehensive empirical analysis is described, followed by our conclusion in Sect. 6.

2 Related Works

Our work has connections to existing works in diversified ranking and clustering. We also discuss similar works that consider query and clustering simultaneously in the section.

2.1 Weighted Clustering

The idea of weighted clustering, in which some data points should have a greater effect on the utility of clustering than others, has been proposed [18–20]. In [22], they observe that some algorithms are point proportional admissible (i.e., the output of an algorithm should be consistent if any point is duplicated). By extending the concept, Ackerman et al. [21] conduct the analysis on the influence of weighted data on classical clustering algorithms.

2.2 Clustering on Networks

To simplify the hierarchical structure, prior works discuss clustering on the subgraph constructed by 1-hop neighbors of the query, called the ego network. Park et al. [7] re-rank retrieval results by HAC for content-based image retrieval. The rank of the results is adjusted according to the distance of a cluster from a query. Crabtree et al. [8] define the distance between webs according to the web-tag bipartite relationship and also apply HAC to group webs. Schwander et al. [9] propose a reranking framework based on contextual dissimilarity measures. On social circles discovery, Mcauley et al. [4] propose a learning model not only considering the 1-hop local graph structure and node features, but also uncovering "mixed-memberships" of vertices. Huang et al. [15] propose an algorithm based on the k-truss concept. Hence, compared with prior works, the formulated problem considers the whole graph and its structure.

3 Preliminary

We firstly define the notation in this paper followed by the formulation of query-oriented clustering. Then, we give two measures, QNCut and entropy, for evaluation.

3.1 Notation and Problem Formulation

Graph. Given a homogeneous undirected graph $\mathcal{G} = (\mathcal{V}, \mathcal{E})$, where $\mathcal{V} = \{v_1, v_2, ..., v_N\}$ denotes the vertex set and $\mathcal{E} = \{(v_i, v_j)|v_i, v_j \in \mathcal{V}\}$ denotes the edge set. Let $\mathbf{W}$ be the weighted adjacency matrix of $\mathcal{G}$. Given a query vertex v_q, we can obtain the relevance vector $\widehat{\mathbf{r}}$ by a ranking algorithm.
Vertex Similarity. Similarity between two adjacent vertices is used to evaluate their relationship so that similarity usually is customized and varies with datasets and needs. We can construct an adjacent matrix $\mathbf{W}$ where the value of W_{ij} depends on the definition of similarity between vertex v_i and vertex v_j.

Query. Given a homogeneous network, we submit a vertex $v_q \in \mathcal{V}$ as the query. A ranking algorithm should return vertices relevant to v_q.

Vertex Relevance. The *relevance* of vertex v_i, denoted by r_i, is defined as the relevance between v_i and query vertex v_q. We denote $\mathbf{r} = \{r_1, r_2, ..., r_N\}$ as the *relevance vector*. In this paper, we use well-known personalized PageRank [14] to calculate $\mathbf{r}$ toward the given query vertex v_q. We denote $\mathbf{p} = \{p_1, p_2, ..., p_N\}$ as the preference vector where $p_i = 1$ if $i = q$ or $p_i = 0$ otherwise. Thus, $\mathbf{r}$ can be obtained by the recursion defined as $\mathbf{r} = (1-\alpha)\mathbf{W}\widehat{\mathbf{D}}^{-1}\mathbf{r} + \alpha\mathbf{p}$, where α denotes the damping factor and $\widehat{\mathbf{D}}$ denotes a diagonal matrix with $[\widehat{d}_1, \widehat{d}_2, ..., \widehat{d}_N]$ on the diagonal, where $\widehat{d}_i = \sum_{j=1}^{N} W_{ij}$.
Cluster Relevance. The relevance of cluster $\mathcal{A}_k$, denoted by $r(\mathcal{A}_k)$, can be defined as $r(\mathcal{A}_k) = \sum_{v_i \in \mathcal{A}_k} r_i$. Notice that $r(\mathcal{A}_k)$ represents the size of $\mathcal{A}_k$ when $r_i = 1$ for all $i \in [1, N]$.

Problem 1. *Given query vertex v_q, let f be a query-oriented clustering function $f(\mathcal{G}, v_q) = \{\mathcal{A}_1, \mathcal{A}_2, ..., \mathcal{A}_K\}$, where $\mathcal{A}_k$ denotes the k-th cluster for all $k \in [1, K]$. Each cluster should consist of inter-cluster edges as less as possible and include as many vertices close to v_q as possible. In other words, the balance of cluster relevance should be kept.*

The task of f is to group vertices in $\mathcal{V} - \{v_q\}$ into clusters, where the similarity between adjacent vertices is determined by the weight of the edge linked between them as well as their relevance. In Fig. 1(d), considering vertices v_3, v_4 and v_5 which are assigned to different clusters, any two of them have no common neighbors except for v_1 and they have similar relevance to v_1 since they are the 1-hop neighbors of v_1. In such a way, each cluster should include vertices with high relevance to v_q.

3.2 Measure: Query-oriented Normalized Cut

According to Problem 1, an ideal query-oriented clustering result should achieve the two objectives stated as follows. First, some edges between high-relevance endpoints must be cut to make each cluster include vertices with high relevance; otherwise, there is at least one cluster composed entirely of irrelevant vertices.

Second, like the minimum cut problem, edges with large weight should be cut as few as possible. Now, we will introduce the measure, QGC, which extends the concept of NCut [10] on the basis of the two objectives.

Given query vertex v_q and relevance vector $\mathbf{r}$, the cut between cluster $\mathcal{A}_l$ and its complement $\overline{\mathcal{A}}_l$ is defined as $\text{cut}(\mathcal{A}_l, \overline{\mathcal{A}}_l|v_q) = \sum_{v_i \in \mathcal{A}_l, v_j \in \overline{\mathcal{A}}_l} r_i W_{ij} r_j$. The sum of the weights of the edges in A_l is defined as $\text{vol}(\mathcal{A}_l|v_q) = \sum_{v_i \in \mathcal{A}_l, v_j \in \mathcal{V}} r_i W_{ij} r_j$. We propose a novel measure, query-oriented normalized cut (QNCut), which is defined as

$$\text{QNCut}(\mathcal{A}_1, \mathcal{A}_2, ..., \mathcal{A}_K|v_q) = \sum_{l=1}^{K} \frac{\text{cut}(\mathcal{A}_l, \overline{\mathcal{A}}_l|v_q)}{\text{vol}(\mathcal{A}_l|v_q)}. \tag{1}$$

A smaller value of QNCut indicates a better query-oriented clustering quality.

3.3 Measure: Cluster Balance

According to Problem 1, the relevance of the produced clusters should be similar to make sure that each cluster contains the vertices with high relevance to v_q. *Cluster balance* is the measure to evaluate how close the relevance scores of the produced clusters are. We use *entropy* as the measure which can be defined as

$$H(\mathcal{A}) = -\sum_{k=1}^{K} p(\mathcal{A}_k)\text{log}p(\mathcal{A}_k), \tag{2}$$

where $p(\mathcal{A}_k) = r(\mathcal{A}_k)/\sum_{l=1}^{K} r(\mathcal{A}_l)$ denotes the normalized relevance of cluster $\mathcal{A}_k$. When all the relevance scores of clusters are the same, i.e., $p(\mathcal{A}_1) = p(\mathcal{A}_2) = ... = p(\mathcal{A}_K) = 1/K$, we have $H(\mathcal{A}) = \text{log}K$ which is maximized. On the contrary, the minimum, i.e., $H(\mathcal{A}) \approx 0$, is obtained when only one cluster, denoted by $\mathcal{A}_i$ without loss of generality, relevant to v_q, i.e., $p(\mathcal{A}_i) \approx 1$.

Considering the case when $K = 2$, we denote $\mathbf{a}_1$ and $\mathbf{a}_2$ as a N dimensional vector where each element can only be 1 or 0. When the i-th element in $\mathbf{a}_1$ is 1, we have two facts: $v_i \in \mathcal{A}_1$; i-th element in $\mathbf{a}_2$ must be 0 i.e., $v_i \notin \mathcal{A}_2$. When $H(\mathcal{A})$ is maximized, we have $\mathbf{r}^\top \mathbf{a}_1 = r(\mathcal{A}_1) = r(\mathcal{A}_2) = \mathbf{r}^\top \mathbf{a}_2$. Let $\widehat{\mathbf{y}} \approx \mathbf{a}_1 - \mathbf{a}_2$ and then we have the equation $\mathbf{r}^\top \widehat{\mathbf{y}} \approx 0$ which is the concept of the *balance constraint* in (6) in Sect. 4.

4 Query-oriented Graph Clustering

We will propose the objective function of QGC and derive a quadratic form for dimensionality reduction.

Our idea is stated as follows. We will find the H partitions each of which should minimize the sum of the weights of the edges crossing the cut. As such, K clusters can be produced from the H partitions. Let $\mathbf{Y} = (\widehat{\mathbf{y}}_1, ..., \widehat{\mathbf{y}}_H) \in \mathbb{R}^{N \times H}$ be the *partition matrix* where $H \geq K$ and each column vector $\widehat{\mathbf{y}}_h$, called the *partition vector*, partitions $\mathcal{G}$ into two clusters according to the sign of each element in $\widehat{\mathbf{y}}_h$. The positive elements and the negative ones in $\widehat{\mathbf{y}}_h$ belong to different clusters. The i-th row vector of $\mathbf{Y}$, denoted by $\mathbf{y}_i$, represents which partitions v_i belongs to. $\mathbf{y}_i$ is called the *feature vector* of v_i. We define the objective function as

$$\mathcal{F} = \sum_{i,j=1}^{N} W_{ij} r_i r_j \| \frac{\mathbf{y}_i}{\sqrt{d_i}} - \frac{\mathbf{y}_j}{\sqrt{d_j}} \|^2, \tag{3}$$

subject to the constraints

$$\widehat{\mathbf{y}}_h^\top \widehat{\mathbf{y}}_h = 1, \quad \forall h \tag{4}$$

$$\widehat{\mathbf{y}}_h^\top \widehat{\mathbf{y}}_l = 0, \quad \forall h \neq l \tag{5}$$

$$\mathbf{r}^\top \widehat{\mathbf{y}}_h = 0, \quad \forall h, \tag{6}$$

where $\|\|$ denotes the L2-norm and d_i is defined as $d_i = r_i \sum_{j=1}^{N} W_{ij} r_j$. (3) is also called the *smoothness constraint*, which means that nearby vertices will belong to the same cluster with high probability by a good clustering method [12]. In other words, a good clustering method is prone to cut weak edges to minimize the objective function. The idea is inspired by normalized cut minimization.

We have three constraints with respect to the partition vectors. First, the partition vectors should be normalized in (4). Second, (5) is the *discriminative constraint*, which makes a partition as different as possible to one another by minimizing the cosine similarity between their partition vectors. Third, (6) is the *balance constraint*. Let $\mathcal{C}_1$ and $\mathcal{C}_2$ be the clusters divided by $\widehat{\mathbf{y}}$. According to the explanation in Sect. 3.3, the cluster balance between $\mathcal{C}_1$ and $\mathcal{C}_2$ is maximized when $\mathbf{r}^\top \widehat{\mathbf{y}} = 0$.

By expanding (3), we have

$$f(\mathbf{Y}) = \sum_{h=1}^{H} \widehat{\mathbf{y}}_h^\top (\mathbf{I} - \mathbf{D}^{-\frac{1}{2}} \mathbf{RWRD}^{-\frac{1}{2}}) \widehat{\mathbf{y}}_h, \tag{7}$$

where $\mathbf{D}$ is a diagonal matrix with $[d_1, d_2, ..., d_N]$ on the diagonal and $\mathbf{R}$ is also a diagonal matrix with $\mathbf{r}$ on the diagonal. When the relevance of all vertices are equal, (7) can be reduced to the CC problem. By using the Lagrange duality, (6) and (7) can be integrated as a dual function

$$g(\boldsymbol{\tau}, \boldsymbol{\mu}) = \min_{\mathbf{Y}} \Lambda(\mathbf{Y}, \boldsymbol{\tau}, \boldsymbol{\mu}) = \min_{\mathbf{Y}} \sum_{h=1}^{H} \widehat{\mathbf{y}}_h^\top \mathbf{L} \widehat{\mathbf{y}}_h - \tau_h (\widehat{\mathbf{y}}_h^\top \widehat{\mathbf{y}}_h - 1) + 2\mu_h \mathbf{r}^\top \widehat{\mathbf{y}}_h,$$

where τ_h and μ_h are Lagrange multipliers and $\mathbf{L} = (\mathbf{I} - \mathbf{D}^{-\frac{1}{2}}(\mathbf{RWR})\mathbf{D}^{-\frac{1}{2}})$ denotes a weighted Laplacian. To simplify the problem, without loss of generality, we will demonstrate how to minimize $\mathbf{y}_h$ by following the work in [13]. The Lagrangian is defined as

$$\Lambda(\widehat{\mathbf{y}}_h, \tau, \mu) = \widehat{\mathbf{y}}_h^\top \mathbf{L} \widehat{\mathbf{y}}_h - \tau \widehat{\mathbf{y}}_h^\top \widehat{\mathbf{y}}_h + \tau + 2\mu \mathbf{r}^\top \widehat{\mathbf{y}}_h. \tag{8}$$

Lemma 1. *The minimum of $\Lambda(\widehat{\mathbf{y}}_h, \tau, \mu)$ equals to the smallest eigenvalue of* $\mathbf{BL}$, *where* $\mathbf{B} = \mathbf{I} - \mathbf{r}\mathbf{r}^\top$.

Proof. Setting the first derivative of (8) to zero, we obtain $\mathbf{L}\widehat{\mathbf{y}}_h - \tau\widehat{\mathbf{y}}_h + \mu\mathbf{r} = \mathbf{0}$. Since $\mathbf{r}^\top\mathbf{r} = 1$, by multiplying on the left by $\mathbf{r}^\top$, we have $\mu = -\mathbf{r}^\top\mathbf{L}\widehat{\mathbf{y}}_h$. By replacing μ, we have

$$\mathbf{BL}\widehat{\mathbf{y}}_h = \tau\widehat{\mathbf{y}}_h, \tag{9}$$

where $\mathbf{B} = (\mathbf{I} - \mathbf{r}\mathbf{r}^\top)$ is the projection matrix (i.e., $\mathbf{B} = \mathbf{B}^2$) which can project a vector to the orthogonal complement of $\mathbf{r}$. From (9), $\widehat{\mathbf{y}}_h$ must be an eigenvector of $\mathbf{BL}$ associated with eigenvalue τ. Since $\widehat{\mathbf{y}}_h$ satisfies the constraint in (6), the Lagrangian in (8) can be written as $\Lambda(\widehat{\mathbf{y}}_h, \tau_h, \mu) = \widehat{\mathbf{y}}_h^\top\mathbf{L}\widehat{\mathbf{y}}_h - \widehat{\mathbf{y}}_h^\top\mathbf{LB}\widehat{\mathbf{y}}_h + \tau = \widehat{\mathbf{y}}_h^\top\mathbf{r}\mathbf{r}^\top\widehat{\mathbf{y}}_h + \tau = \tau$. The value of the Lagrangian $\Lambda(\widehat{\mathbf{y}}_h, \tau, \mu)$ must be one of the eigenvalue of $\mathbf{BL}$. Hence, the minimum of $\Lambda(\widehat{\mathbf{y}}_h, \tau, \mu)$ is the smallest eigenvalue of $\mathbf{BL}$, which completes the proof.

From Lemma 1, the Lagrangian can be written as

$$\Lambda(\widehat{\mathbf{y}}_h, \tau_h, \mu) = \frac{\widehat{\mathbf{y}}_h^\top\mathbf{BL}\widehat{\mathbf{y}}_h}{\widehat{\mathbf{y}}_h^\top\widehat{\mathbf{y}}_h}. \tag{10}$$

We already know that $\widehat{\mathbf{y}}_h$ is an eigenvector of $\mathbf{BL}$. Since $\mathbf{BL}$ is not symmetric, the following task is to rewrite (10) as a quadratic form.

Lemma 2. *Let λ and $\mathbf{v}$ be an eigenvalue and the associated eigenvector of* $\mathbf{BLB}$ *respectively.* $\mathbf{v}$ *must be* $\mathbf{r}$ *or orthogonal to* $\mathbf{r}$.

Proof. First, we have $\mathbf{BLBr} = \mathbf{BL}(\mathbf{r} - \mathbf{r}) = \mathbf{0}$. Second, $\mathbf{BLB}$ is symmetric so that any two of the eigenvectors are orthogonal to each other. Since $\mathbf{r}$ is an eigenvector of $\mathbf{BLB}$, the other eigenvectors must be orthogonal to $\mathbf{r}$, which completes the proof.

Lemma 3. *Let λ and $\mathbf{v}$ be an eigenvalue and the associated eigenvector of* $\mathbf{BLB}$ *respectively. When $\widehat{\mathbf{y}}_h = \mathbf{Bv}$, where $\mathbf{v} \neq \mathbf{r}$, $\Lambda(\widehat{\mathbf{y}}_h, \lambda, \mu) = \lambda$.*

Proof. First, we have $\mathbf{BLBv} = \lambda\mathbf{v}$. By multiplying $\mathbf{B}$ on the left, we obtain

$$\lambda\mathbf{Bv} = \mathbf{B}(\mathbf{BLBv}) = (\mathbf{BB})\mathbf{LBv} = \mathbf{BLBv} = \lambda\mathbf{v}.$$

Since $\mathbf{BL}(\mathbf{Bv}) = \lambda(\mathbf{Bv})$, (9) is satisfied when $\widehat{\mathbf{y}}_h = \mathbf{Bv}$. In other words, $\mathbf{Bv}$ is an eigenvector of $\mathbf{BL}$. Since $\mathbf{v} \neq \mathbf{r}$, by using Lemma 2, (10) can be

$$\Lambda(\mathbf{Bv}, \lambda, \mu) = \frac{\mathbf{v}^\top\mathbf{BBLBv}}{\mathbf{v}^\top\mathbf{BBv}} = \frac{\mathbf{v}^\top\mathbf{BLBv}}{\mathbf{v}^\top\mathbf{v}}, \tag{11}$$

which completes the proof.

Theorem 1. $g(\tau, \mu) = \min_{\mathbf{V} \subseteq \mathbf{r}^\perp} \sum_{h=1}^{H} \frac{\mathbf{v}_h^\top\mathbf{BLBv}_h}{\mathbf{v}_h^\top\mathbf{v}_h}$, *where $\mathbf{r}^\perp$ denotes the orthogonal complement of* $\mathbf{r}$ *and* $\mathbf{V} = \{\mathbf{v}_1, ..., \mathbf{v}_H\}$.

Proof. Let $\widehat{\mathbf{y}}_h = \mathbf{B}\mathbf{v}_h$ for $k \in [1, K]$. By using Lemma 3, we have the quadratic objective function

$$g(\boldsymbol{\tau}, \boldsymbol{\mu}) = \min_{\mathbf{V} \subseteq \mathbf{r}^\perp} \sum_{h=1}^{H} \Lambda(\mathbf{B}\mathbf{v}_h, \lambda_h, \mu_h) = \min_{\mathbf{V} \subseteq \mathbf{r}^\perp} \sum_{h=1}^{H} \frac{\mathbf{v}_h^\top \mathbf{BLB}\mathbf{v}_h}{\mathbf{v}_h^\top \mathbf{v}_h}, \tag{12}$$

which completes the proof.

From Theorem 1, we can use (12) to solve the objective function defined in (3). Let $\lambda_{min} = \lambda_1 \leq \lambda_2 \leq ... \leq \lambda_N = \lambda_{max}$ be the eigenvalues of $\mathbf{BLB}$ in increasing order. We select the eigenvectors $[\boldsymbol{\phi}_1, \boldsymbol{\phi}_2, ..., \boldsymbol{\phi}_H]$ associated with the H smallest eigenvalues $[\lambda_1, \lambda_2, ..., \lambda_H]$. Then we obtain the partition matrix $\mathbf{Y} = [\widehat{\mathbf{y}}_1, \widehat{\mathbf{y}}_2, ..., \widehat{\mathbf{y}}_H] = [\mathbf{B}\boldsymbol{\phi}_1, \mathbf{B}\boldsymbol{\phi}_2, ..., \mathbf{B}\boldsymbol{\phi}_H]$. Notice that the orthogonal complement of $\mathbf{r}$ is the column space of $\mathbf{B}$, so that we have $\mathbf{Br} = \mathbf{0}$. As such, $\mathbf{r}$ will not be in $\mathbf{Y}$. To reduce the computational cost, the Lanczos algorithm [11] is used to find the eigenvectors associated with the H smallest eigenvalues approximately. The time complexity is linear to the number of non-zero elements in $\mathbf{W}$. Next, we will show that $\mathbf{Y}$ satisfies the constraint in (5).

Theorem 2. *For each two partition vectors $\widehat{\mathbf{y}}_l$ and $\widehat{\mathbf{y}}_m$, we have $\widehat{\mathbf{y}}_l^\top \widehat{\mathbf{y}}_m = 0$.*

Proof. Let $\widehat{\mathbf{y}}_l = \mathbf{B}\boldsymbol{\phi}_l$ and $\widehat{\mathbf{y}}_m = \mathbf{B}\boldsymbol{\phi}_m$ be two partition vectors, where $\boldsymbol{\phi}_l$ and $\boldsymbol{\phi}_m$ are eigenvectors of $\mathbf{BLB}$. Then we have

$$\widehat{\mathbf{y}}_l^\top \widehat{\mathbf{y}}_m = \boldsymbol{\phi}_l^\top \boldsymbol{\phi}_m - (\mathbf{r}^\top \boldsymbol{\phi}_l)(\mathbf{r}^\top \boldsymbol{\phi}_m). \tag{13}$$

Since $\mathbf{BLB}$ is positive semi-definite, we have $\boldsymbol{\phi}_l^\top \boldsymbol{\phi}_m = 0$. In addition, $\boldsymbol{\phi}_l$ and $\boldsymbol{\phi}_m$ are both in the orthogonal complement of $\mathbf{r}$ so that the second term on the right side of (13) is zero. Thus, the proof is completed.

So far, we have described the proposed objective function, the method for eigenmaps and cluster assignment. The whole approach is shown in Algorithm 1 where PPR denotes the personalized PageRank algorithm.

5 Experimental Results

In this section, we evaluate the effectiveness of QGC empirically. Firstly, we demonstrate the performances on QC problem. Then we conduct experiments on social circle discovery. The experiment is conducted on Macbook Pro with 2.6 GHz Core i5 CPU and 8 GB main memory.

5.1 Query-oriented Clustering

We experiment with the weighted spectral clustering (`WSC`) [21], query-oriented clustering (`QC`) [16] and RankComplete (`RC`) [23]. `QC` is a multi-objective approach based on genetic algorithms considering two evaluation criteria: relevance and within-cluster sum of similarity. `RC` considers clustering and ranking simultaneously.

Algorithm 1. Query-oriented Graph Clustering

Require: Weighted adjacent matrix $\mathbf{W}$, query vertex v_q, damping factor α and number of clusters K
Ensure: Cluster assignment $\mathbf{l}$
1: Compute relevance $\widehat{\mathbf{r}} = \text{PPR}(\mathbf{W}, v_q, \alpha)$; $\mathbf{r} = \widehat{\mathbf{r}}$;
2: Set $r_q = 0$ and normalize $\mathbf{r} = \mathbf{r}/\|\mathbf{r}\|_2$;
3: Construct $\mathbf{R}$ with $\mathbf{r}$ on the diagonal;
4: Construct $\mathbf{D}$ with $d_1, .., d_N$ on the diagonal;
5: Obtain the eigenvectors $\mathbf{z}_1, ..., \mathbf{z}_H$ associated with the H smallest eigenvalues $\lambda_1, ..., \lambda_H$ of $\mathbf{BLB}$, where $H \geq K$, $\mathbf{B} = \mathbf{I} - \mathbf{r}\mathbf{r}^\top$ and $\mathbf{L} = \mathbf{I} - \mathbf{D}^{-\frac{1}{2}}\mathbf{RWRD}^{-\frac{1}{2}}$
6: Compute $\mathbf{Y} = [\lambda_1 \mathbf{B}\mathbf{v}_1, ..., \lambda_H \mathbf{B}\mathbf{v}_H]$.
7: Compute clustering algorithm (i.e., K-means, hierarchical agglomerative) with $\mathbf{Y}$ as the input.

Dataset. We experiment on two datasets which are described as follows.

ODP239. The dataset consists of 239 topics, each of which contains 100 subtopics and about 100 documents associated with single subtopics. The topics, subtopics, and their associated documents were selected from the Open Directory Project[1]. By extracting words from the title and the document snippets and pruning words with frequency 1, we construct a document-word matrix $\mathbf{X}$ with dimension $25,580 \times 43,831$. Then we use TF-IDF to re-weight entries in $X_{i,w} = f_{i,w} \log \frac{|D|}{|\{X_{i,w}|X_{i,w}>0\}|}$, where $f_{i,w}$ denotes the frequency of word w in document i and $|D|$ denotes the number of documents. To run the graph-based algorithms, we construct the undirected graph according to the cosine similarity $W_{i,j} = \frac{\mathbf{x}_i^\top \mathbf{x}_j}{|\mathbf{x}_i||\mathbf{x}_i|}$. The entries are set to 0 if their values are smaller than μ, where is set to 0.03 heuristically. As such, we build a document graph containing 4,587,568 edges and 25,880 vertices.

TR30. The dataset selected from the Open Directory Project consists of 30 topics, each of which contains 10 subtopics and about 100 documents associated with single subtopics. Following the matrix constructing procedure aforementioned in ODP239, we build a document-word matrix $\mathbf{X}$ with dimension $2,957 \times 9,973$ and graph $\mathbf{W}$ containing 343,616 edges and 2,957 vertices.

In the two datasets, each topic is viewed as a query and its subtopics are viewed as the clustering result. Given topic t, we introduce an augmented vertex v_t as the query vertex into $\mathcal{G}$ and link v_t to its associated documents. As such, QGC is run on the graph $\mathcal{G} = (\mathcal{V}', \mathcal{E}')$, where $\mathcal{V}' = \mathcal{V} \cup \{v_t\}$ and $\mathcal{E}' = \mathcal{E} \cup \{(v_t, v)|v$ is associated with $v_t\}$. After clustering $\mathcal{G}'$, we retain the documents associated with v_t and drop the others. When evaluating the effectiveness of the algorithms, we view the set of documents associated with the same subtopic as a cluster. `QC` and `RC` run on matrix $\mathbf{X}$ while `QGC` and `WSC` run on $\mathbf{W}$ (i.e., graph $\mathcal{G}$).

[1] http://www.dmoz.org.

Evaluation Criteria. Let $\mathcal{C} = \{C_1, ..., C_K\}$ be the predicted clusters and $\overline{\mathcal{C}} = \{\overline{C}_1, ..., \overline{C}_{\overline{K}}\}$ be the ground-truth clusters. C should align closely to $\overline{C}$. We use F_1 score as the measure. Since the correspondence between predicted clusters and ground-truth clusters is unknown, we follow the works in [4,17] to evaluate C by finding the optimal match: $\max_{f:\mathcal{C}\to\overline{\mathcal{C}}} \frac{1}{|f|} \sum_{c\in \mathrm{dom}(f)} F_1(C, f(C))$, where f is an *injective function* from $\mathcal{C}$ to $\overline{\mathcal{C}}$. In other words, each ground-truth cluster $\overline{C} \in \overline{\mathcal{C}}$ at most has one match $C \in \mathcal{C}$. When $|\mathcal{C}| < |\overline{\mathcal{C}}|$, each predicted cluster will be assigned to a ground-truth cluster respectively by f.

Table 1. The performance of query-oriented clustering on ODP239 and TR30

		QNCut	Entropy	Accuracy	Pecision	Recall	F1score
ODP239	QGC-B	**0.361**	0.162	0.677	**0.456**	0.427	0.376
	QGC	0.373	0.184	**0.684**	0.443	**0.432**	**0.38**
	WSC	0.441	0.214	0.681	0.399	0.431	0.371
	QC	n/a	0.035	0.663	0.417	0.397	0.36
	RC	n/a	**0.218**	0.657	0.353	0.391	0.329
TR30	QGC-B	**0.445**	0.206	0.732	0.505	0.530	0.471
	QGC	0.493	0.211	**0.738**	**0.51**	**0.543**	**0.475**
	WSC	0.501	**0.229**	0.718	0.451	0.503	0.435
	QC	n/a	0.15	0.672	0.427	0.419	0.367
	RC	n/a	0.215	0.69	0.411	0.459	0.383

Performance. The performances of the five methods are shown in Table 1. QGC outperforms the other methods in terms of both accuracy and F_1 score in the two datasets. QGC achieves slightly higher accuracy and F_1 scores than WSC since QGC is based on normalized cut (ncut) minimization so that it is prone to find clumps in the graph while WSC, based on ratio cut (rcut) minimization, tends to find splits according to the observation in [10]. Since documents associated with the same subtopic have stronger intra-cluster relationship (i.e., more like a clump), QGC can output more precise clustering results. For the same reason, WSC which finds splits reaches higher entropy than QGC which finds clumps.

Considering the ncut-based algorithms, QGC-B reaches lower QNCut and entropy scores than QGC, since QGC-B does not consider the balance constraint. Considering a graph is perturbed by adding/deleting edges, QGC-B may output clusters with tiny cluster relevance, while QGC is more insensitive to noise, since QGC can avoid from outputting a result that consists of clusters with excessively variant relevance and can preserve the property of ncut-based algorithm (i.e., finding clumps).

5.2 Social Circle Discovery

We evaluate the efficiency and effectiveness of QGC applied to social circle discovery on two real-world networks.

Dataset. We conduct experiments on two data sets described as follows.

Facebook. The first dataset is Facebook which is downloaded from the Stanford Large Network Dataset Collection[2]. We only use the friend-to-friend feature to construct the ego network containing 4,039 vertices and 88,234 edges.

Twitter. The Twitter dataset is also downloaded from the Stanford Large Network Dataset Collection. We only use the friend-to-friend feature to construct the ego network containing 81,306 vertices and 1,768,149 edges.

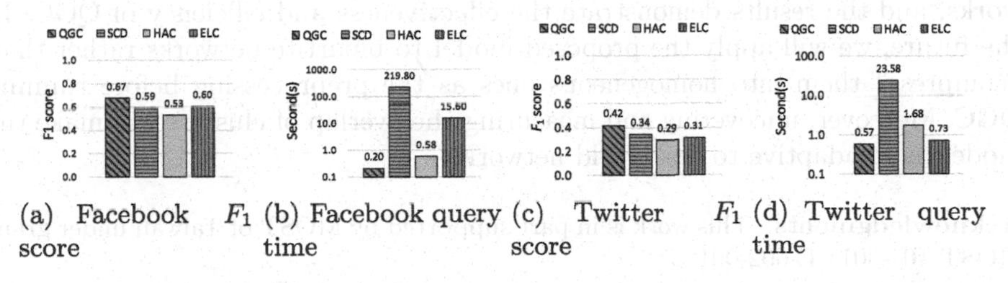

(a) Facebook F_1 score (b) Facebook query time (c) Twitter F_1 score (d) Twitter query time

Fig. 2. Performance of social circles detection on Facebook and Twitter, in terms of the F_1 score (higher is better) and the query time.

Evaluation Criteria. The evaluation criteria are the same with Sect. 5.1.

Performance. We experiment with the social circle discovery model (SCD) [4] and Enhanced link clustering (ELC) [17]. We also consider hierarchical agglomerative clustering algorithm (HAC). Since only the information of ego networks are completed in the two datasets, we run QGC on ego networks. OGC is implemented in Matlab and the other algorithms are implemented in C++.

The performance is shown in Fig. 2. QGC outperforms the other models and achieves F_1 scores of 0.67 on Facebook and 0.42 on Twitter. Compared to SCD, QGC improves on the F_1 scores by 13.6% on Facebook and 23.5% on Twitter. SCD takes more time than the other methods while OGC costs less query time which are approximately proportional to the number of edges. There are two explanations that QGC run on the graph constructed merely from the friend-to-friend features can reach the highest F_1 score. First, introducing user profiles some of which are uninformative may trap an algorithm into a local optimum. Second, according to [17], the similarity between users is transitive (i.e., two users in one circle may share few common features, if both of them are similar to a common node in the same circle). A method based on user similarity graph will perform better since graph-based methods can learn the transitive similarities [6,12].

[2] https://snap.stanford.edu/data/.

According to the explanation in [4], Facebook data is more complete than Twitter data where only publicly-visible circles can be observed. The tweet-based profiles are not so informative as the profile categories from Facebook. Furthermore, Twitter data only encodes the follower relationships rather than the friendships in Facebook.

6 Conclusion

In this paper, we precisely define the problem of query-oriented clustering and the corresponding evaluation criteria, QNCut and cluster balance. We also define the objective function and introduce the constraint of cluster balance to propose a novel model, QGC, which can find more representative clusters by considering the balance constraint. We conduct extensive experiments on real-world networks, and the results demonstrate the effectiveness and efficiency of QGC. In the future, we will apply the proposed model to bipartite networks rather than "compress" them into homogeneous ones as the preprocessing before running QGC. Moreover, uncovering and measuring the overlap of clusters will make the model more adaptive to real-world networks.

Acknowledgments. This work is in part supported by MOST of Taiwan under grant MOST 105-3011-E-002-001.

References

1. Carbonell, J., Goldstein, J.: The use of MMR, diversity-based reranking for reordering documents and producing summaries. In: SIGIR, pp. 335–336 (1988)
2. Küçüktunç, O., Saule, E., Kaya, K., Çatalyürek, Ü.V.: Diversified recommendation on graphs: pitfalls, measures, and algorithms. In: WWW (2013)
3. Mei, Q., Guo, J., Radev, D.: DivRank: the interplay of prestige and diversity in information networks. In: KDD (2010)
4. Mcauley, J.J., Leskovec, J.: Learning to discover social circles in ego networks. In: NIPS (2012)
5. Kuo, L.Y., Chen, M.S.: Diversified ranking on graphs from the influence maximization viewpoint. In: DSAA (2014)
6. Belkin, M., Niyogi, P., Sindhwani, V.: Manifold regularization: a geometric framework for learning from labeled and unlabeled examples. JMLR **7**, 2399–2434 (2006)
7. Park, G., Baek, Y., Lee, H.K.: Re-ranking algorithm using post-retrieval clustering for content-based image retrieval. Inf. Process. Manage. **41**(2), 177–194 (2005)
8. Crabtree, D., Andreae, P., Gao, X.: Query directed web page clustering. In: Proceedings of the 2006 IEEE/WIC/ACM International Conference on Web Intelligence (2006)
9. Schwander, O., Nielsen, F.: Reranking with contextual dissimilarity measures from representational Bregman k-Means. In: VISAPP (1) (2010)
10. Shi, J., Malik, J.: Normalized cuts and image segmentation. TPAMI **22**(8), 888–905 (2000)
11. Lanczos, C.: An iteration method for the solution of the eigenvalue problem of linear differential and integral operators. United States Government Press Office, Washington, DC (1950)

12. Zhou, D., Bousquet, O., Lal, T.N., Weston, J., Schölkopf, B.: Learning with local and global consistency. In: NIPS (2004)
13. Golub, G.H.: Some modified matrix eigenvalue problems. SIAM Rev. **15**(2), 318–334 (1973)
14. Haveliwala, T.H.: Topic-sensitive PageRank. In: WWW (2002)
15. Huang, X., Cheng, H., Qin, L., Tian, W., Yu, J.X.: Querying k-truss community in large and dynamic graphs. In: SIGMOD (2014)
16. Lamprier, S., Amghar, T., Saubion, F., Levrat, B.: Query-oriented clustering: a multi-objective approach. In: Proceedings of the 2010 ACM Symposium on Applied Computing (2010)
17. Hu, Y., Yang, B.: Enhanced link clustering with observations on ground truth to discover social circles. Knowl.-Based Syst. **73**, 227–235 (2015)
18. Chatterjee, M., Sajal, D.K., Damla, T.: WCA: a weighted clustering algorithm for mobile ad hoc networks. Cluster Comput. **5**(2), 193–204 (2002)
19. Long, B., Zhang, Z.M., Wu, X., Yu, P.S.: Spectral clustering for multi-type relational data. In: ICML (2006)
20. Tseng, G.C.: Penalized and weighted k-means for clustering with scattered objects and prior information in high-throughput biological data. Bioinformatics **23**, 2247–2255 (2007)
21. Ackerman, M., David, S.B., Branzei, S., Loker, D.: Weighted clustering. In: AAAI (2012)
22. Fisher, L., Van Ness, J.W.: Admissible clustering procedures. Biometrika **58**, 91–104 (1971)
23. Cao, L., Jin, X., Yin, Z., Del Pozo, A., Luo, J., Han, J., Huang, T.S.: RankCompete: simultaneous ranking and clustering of information networks. Neurocomputing **95**, 98–104 (2012)

CCCG: Clique Conversion Ratio Driven Clustering of Graphs

Prathyush Sambaturu(✉) and Kamalakar Karlapalem

Data Science and Analytics Centre IIIT Hyderabad, Hyderabad 500 032, India
prathyush.sambaturu@research.iiit.ac.in, kamal@iiit.ac.in

Abstract. Networks have become ubiquitous in many real world applications and to cluster similar networks is an important problem. There are various properties of graphs such as clustering coefficient (CC), density, arboricity, etc. We introduce a measure, Clique Conversion Coefficient (CCC), which captures the clique forming tendency of nodes in an undirected graph. CCC could either be used as a weighted average of the values in a vector or as the vector itself. Our experiments show that CCC provides additional information about a graph in comparison to related measures like CC and density. We cluster the real world graphs using a combination of the features CCC, CC, and density and show that without CCC as one of the features, graphs with similar clique forming tendencies are not clustered together. The clustering with the use of CCC would have applications in the areas of Social Network Analysis, Protein-Protein Interaction Analysis, etc., where cliques have an important role. We perform the clustering of ego networks of the YOUTUBE network using values in CCC vector as features. The quality of the clustering is analyzed by contrasting the frequent subgraphs in each cluster. The results highlight the utility of CCC in clustering subgraphs of a large graph.

1 Introduction

In many real world applications, data is naturally organized in the form of networks such as social networks [5], road networks [7], collaboration networks [3], communication networks, protein-protein interaction networks [4] and web graphs. In graph theory various measures exist such as diameter, clustering coefficient, density, arboricity [6], k-core number [2] and betweenness centrality [13] which describe certain properties of graphs. An interesting problem corresponding to data mining is to find similarity [14,15] between graphs. One of the main challenges here is to list the properties of graphs that can be used as their features in finding similarity.

Problem Statement: To determine the "goodness" of features for graph similarity.

To list all the graph properties is near to impossible. Our focus in this paper is on–Clique Conversion Ratio–the ratio at which the cliques of a particular order expand into the cliques of a higher order. In this paper, we define Clique Conversion Coefficient (CCC) to be a measure of the aforementioned tendency

J. Kim et al. (Eds.): PAKDD 2017, Part II, LNAI 10235, pp. 762–773, 2017.
DOI: 10.1007/978-3-319-57529-2_59

in graphs. Also, we show that the existing measures such as *clustering coefficient* [1] and *density* do not capture the idea of the Clique Conversion Ratio as well as CCC. To motivate the need for a measure like *CCC* we provide the following example. Let A, B, and C be three people in a large Social Network G who are mutually connected to each other. Let D be another person who forms at least one triangle in G. The expansion of the clique $\{A, B, C\}$ into the clique $\{A, B, C, D\}$ would require the presence of the cliques $\{A, B, D\}$, $\{A, C, D\}$ and $\{B, C, D\}$ in G. Although D forms more than one triangle it need not be connected to any of A, B and C. The clique $\{A, B, C\}$ could expand into multiple such cliques of size four. CCC, in fact, measures the ratio of actual conversions of all cliques of a particular number of people into cliques having one more person to the maximum number of such conversions possible. We could construct dense graphs with many cliques of size three with only a few of those actually expanding into cliques of size four. The Clustering Coefficient and Density would not be able to differentiate such graphs from the ones that have high clique conversion ratio. CCC is also useful in applications where the existence and expansion of cliques plays an important role. For example, an advertiser might want to advertise his product in a Social Network which is not only dense and well-connected but also has higher clique forming tendency. This would help his advertisement to percolate to more people in comparison to as in those networks having lesser clique forming tendency. In this paper, we define CCC, explore its properties, further provide experimental evidence that CCC provides new information in comparison to the existing measures. The remainder of this paper is organized as follows: definition of CCC in Sect. 2, heuristics to compute CCC are given in Sect. 3, experimental results showing the utility of CCC are presented in Sects. 4 and 5 and the last section provides some conclusions.

2 Clique Conversion Ratios and CCC

2.1 Notations

Let $G = (V, E)$ be an undirected graph where V is the set of nodes and E is the set of edges such that $|V| = n$ and $|E| = m$.

- C_p denotes the number of cliques of size p in G such that $2 \leq p \leq n$. C_2 is m.
- n_p denotes the number of nodes in G that participate in the formation of at least one clique of size p. We need at least p nodes to form at least one clique of size p. Hence $n_p \geq p$, if $C_p \geq 1$.
- r_{p+1} denotes the conversion ratio of cliques of size p to the cliques of size $p+1$ in G.

2.2 Conversion Ratios

We define r_{p+1} for $2 \leq p \leq n-1$ as

$$r_{p+1} = \begin{cases} \frac{C_{p+1}\,(p+1)}{C_p\,(n_p-p)} & \text{if } C_p \geq 1 \\ 0 & \text{if } C_p = 0 \end{cases} \tag{1}$$

It could be seen that a complete graph with k nodes has $r_{p+1} = 1, \forall p \in [2, k-1]$.

Combinatorial Justification of Ratios: In this section, we present a justification that the conversion ratio r_{p+1} correctly computes the ratio of conversion of cliques of size p to cliques of size $p+1$. As per our notation the number of cliques of size p are C_p and n_p nodes participate in at least formation of one clique of size p. For a clique $\{v_1, ..., v_p\}$ of size p there are at most $(n_p - p)$ nodes as choices to expand into a clique of size $p+1$. But for such an expansion to actually happen the node picked should have all p cliques of size p with each subset of $\{v_1, ..., v_p\}$ with size $(p-1)$. Therefore, C_p cliques each having $(n_p - p)$ nodes to expand into a maximum of $\frac{C_p(n_p-p)}{(p+1)}$ cliques of size $p+1$. The denominator comes from the observation that the node picked to expand a clique of size p could actually be forming all possible cliques of size p with subset of nodes in this clique. Therefore the resultant clique of size $(p+1)$ could at most be over counted $(p+1)$ times. We know that C_{p+1} is the number of cliques of size $p+1$. Therefore, the ratio of the number of cliques of size $p+1$ to the maximum number of possible cliques of size $p+1$ given C_p and n_p is our notion of conversion ratio, i.e.,

$$\frac{C_{p+1}}{\frac{C_p\,(n_p-p)}{p+1}} = \frac{C_{p+1}\,(p+1)}{C_p\,(n_p-p)} = r_{p+1} \tag{2}$$

Range of the Ratios: The definition of the conversion ratio can be viewed as the conditional probability of the event, where the formation of certain cliques of size $p+1$ is the event which is conditioned on the occurrence of an event in which certain cliques of size p are already formed by a set of n_p vertices. Therefore, the conversion ratios naturally have a range $[0, 1]$.

2.3 Global, Local and Average CCC

The Global CCC denoted by CCC_g is defined as the weighted average of the conversion ratios in a given graph defined for the following parameters:

1. A graph $G = (V, E)$ where $|V| = n$ and $|E| = m$.
2. A vector $\alpha = <\alpha_3, ..., \alpha_n>$ where α_i corresponds to the weight of the respective r_i and $\forall i \in [3, n]$, $\alpha_i \in [0, 1]$, $\sum_{i=3}^{n} \alpha_i = 1$.

$$CCC_g(G, \alpha) = \sum_{i=3}^{n} \alpha_i r_i$$

We now present global CCC values for two special graphs. Let $\alpha = \{\frac{1}{3}, \frac{1}{3}, \frac{1}{3}, 0, ..., 0\}$.

1. In a *Petersen Graph* [9], we have 10 vertices and 15 edges connected as shown in Fig. 1(a). Let us denote the Petersen graph with H. In spite of its good connectivity it has no Cliques, i.e., all conversion ratios equal to 0. Hence, $CCC_g(H, \alpha)$ value is 0.

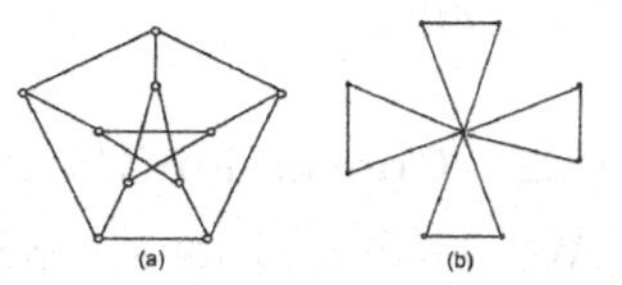

Fig. 1. (a) Petersen graph (b) F_4

2. A *Friendship Graph* [12], F_n has n triangles joined at a common vertex. Therefore, it has $2n+1$ vertices and $3n$ edges. This graph has no cliques of size greater than 3. Therefore, $CCC_g(F_n, \alpha) = \frac{1}{3(2n-1)}$ for $n \geq 2$. F_4 shown in Fig. 1(b), $CCC_g(F_4, \alpha) = \frac{1}{21}$.

Let $u \in V$ and $N(u)$ denote the neighborhood of u such that every node in it is connected by an edge to u. The Local CCC denoted by CCC_l for u is calculated with respect to the subgraph H induced by $N(u)$. Assume H has p nodes. CCC_l is defined for the following parameters:

1. A graph $G = (V, E)$ where $|V| = n$ and $|E| = m$.
2. A node $u \in V$. Let us assume the neighborhood $N(u)$ induced subgraph H has p nodes.
3. A vector $\alpha = <\alpha_2, ..., \alpha_p>$ where α_i corresponds to the weight of the respective r_i and $\forall i \in [2, p]$, $\alpha_i \in [0, 1]$, $\sum_{i=2}^{p} \alpha_i = 1$.

In local CCC, r_2 is also considered, because every edge in H corresponds to a clique of size three in G (as all nodes in H have an edge to u).

$$CCC_l(G, u, \alpha) = \sum_{i=2}^{p} \alpha_i r_i \tag{3}$$

The average CCC denoted by CCC_a is defined as an average on the local CCC of all nodes of G and a given vector $\alpha = <\alpha_2, ..., \alpha_n>$ where α_i corresponds to the weight of the respective r_i and $\forall i \in [2, n]$, $\alpha_i \in [0, 1]$, $\sum_{i=2}^{n} \alpha_i = 1$.

$$CCC_a(G, \alpha) = \frac{1}{n} \sum_{u \in V} CCC_l(G, u, \alpha) \tag{4}$$

The measure CCC in all its three variants has range $[0, 1]$. This follows from the fact that all the conversion ratios are defined to be in $[0, 1]$ and the choice that the sum of weights is to be 1.

2.4 Selection of Vector α

As per our experiments, different choices of vector α could strongly impact the value of CCC. One simple choice of α is to give each r_i equal weight. An example where this choice may not seem ideal for all requirements is as follows: Let us assume that a graph G consists of many triangles, only five cliques of size 4 and one clique of size 5. In this case, r_5 of G would be 1. This is because of all the Cliques of size 4 join to form a clique of size 5. In other words, the conversion of Cliques of size 4 if they exist to Cliques of size 5 is complete in G. If G has very low r_3 and r_4, r_5 would still skew the CCC value to be moderate to high. If such phenomenon is not wanted the weights in α could be set in decreasing order with increasing i values.

3 Computation of CCC

Computing CCC is a computationally intensive task. In all our experiments the heuristics presented in this section are used to compute conversion ratios and consequently CCC of any graph. The two heuristics to make the computation of CCC effective:

1. All vertices with degree less than $p-1$ can never form a clique of size p. Hence, these vertices can be pruned iteratively, so that the remaining vertices are selected as candidates that could possibly form cliques of size p.
2. Any vertex that can not form a clique of size $p-1$ also would not be able to form a clique of size i, when $i > p-1$. Therefore, only those vertices that form at least a clique of size $p-1$ are considered as candidates to check if they could form cliques of size p.

Let $G=(V,E), \alpha$ be the graph and the weight vector respectively which are the parameters required for computation of global CCC, $CCC_g(G,\alpha)$.

Complexity Analysis of the Algorithm Based on Heuristics: Let $\delta = \Delta(G)$ be the maximum degree in G. Let $\gamma = \delta+1$ be the size of maximum clique possible in G. The upper bound on the running time of the algorithm is given by $\sum_{i=3}^{\gamma}\left(n_{i-1}\binom{\delta}{i-1}\right) = O(n\ \delta^{\delta})$, where n_{i-1} is the number of nodes that form at least one clique of size $i-1$. Nevertheless, in the worst case, where G is a complete graph, $\forall i,\ n_i = n$ and $\delta = n-1$. But on sparse graphs the heuristics presented however help to reduce the running time, when $\forall i,\ n_i << n$, and $\delta << n$.

4 Clustering of Graphs Using CCC as a Feature

Experimental Setup: In our experiments, we use the large networks of the SNAP Dataset [8]. These graphs are selected from four different categories: road networks, networks with ground truth communities, collaboration networks, and peer-to-peer networks. We find the conversion ratios and global CCC values of each network with the parameter α set as

$$\alpha_i = \begin{cases} \frac{1}{3} & \text{if } i \in \{3,4,5\} \\ 0 & \text{otherwise} \end{cases} \tag{5}$$

Table 1 presents the $CCC_g(.,\alpha)$ for each network where n, m represent the number of nodes and edges in the network respectively. The Road Networks(9, 10, 11) are almost planar and hence have very low conversion ratios, consequently low global CCC values. This is also the case with Peer-to-peer Gnutella networks(5, 6, 8) except the fact that these are relatively very small networks and that they have a high conversion of edges into triangles in comparison to the road networks. Interestingly, in the Amazon network(7) the conversion ratios all have

Table 1. Table presenting values of conversion ratios and global CCC values of each network with CCC values in decreasing order

ID	Dataset	n	m	r_3	r_4	r_5	$CCC_g(.,\alpha)$
1	CA-GrQc	5242	14496	0.19×10^{-2}	0.71×10^{-2}	0.14×10^{-1}	0.77×10^{-2}
2	CA-HepTh	9877	25998	0.33×10^{-3}	0.12×10^{-2}	0.48×10^{-2}	0.21×10^{-2}
3	CA-CondMat	23133	93497	0.24×10^{-3}	0.33×10^{-3}	0.53×10^{-3}	0.37×10^{-3}
4	com-dblp	317080	1049866	0.20×10^{-4}	0.11×10^{-3}	0.40×10^{-3}	0.18×10^{-3}
5	P2p-Gnuetella05	8846	31839	0.12×10^{-4}	0.15×10^{-3}	0	0.53×10^{-4}
6	P2p-Gnutella06	8717	31525	0.12×10^{-4}	0.13×10^{-3}	0	0.46×10^{-4}
7	com-amazon	334863	925872	0.65×10^{-5}	0.62×10^{-5}	0.69×10^{-5}	0.65×10^{-5}
8	p2p-Gnuetella04	10876	39994	0.64×10^{-5}	0.74×10^{-5}	0	0.46×10^{-5}
9	RoadNet-PA	1088092	1541898	0.12×10^{-6}	0.70×10^{-8}	0	0.42×10^{-7}
10	RoadNet-TX	1379917	1921660	0.94×10^{-7}	0.69×10^{-8}	0	0.34×10^{-7}
11	RoadNet-CA	1965206	2766607	0.67×10^{-7}	0.44×10^{-8}	0	0.24×10^{-7}

similar values. Also, this phenomenon happens in CA-CondMat(3) adding to the observation that this network is dense and has higher clique conversion in comparison to the Amazon network. In both these cases, any weight vector α would give more or less a similar CCC_g value for each network respectively. The collaboration networks CA-GrQc(1) and CA-HepTh(2) along with the ground truth community com-dblp(4) have increasing conversion ratios r_p as the p increases. Hence, CCC_g for these networks is sensitive to the changes in the weight vector α. All these three networks are dense and have very high clique conversion ratios.

Comparison Between CCC, Clustering, and Density for Graphs in our Dataset: Table 2 presents the normalized scores of $CCC_g(.,\alpha)$, clustering

Table 2. Z-normalization scores [10] of values of CCC, CC, and Density for each graph

ID	Dataset	$CCC_g(.,\alpha)$	CC	Density
1	CA-GrQc	2.9106	1.0074	1.3241
2	CA-HepTh	0.4851	0.7868	1.3241
3	CA-CondMat	−0.2511	1.3750	0.4712
4	com-dblp	−0.3319	1.3750	−0.9765
5	p2p-Gnuetella05	−0.3860	−0.9147	0.7058
6	p2p-Gnutella06	−0.3889	−0.9165	0.7484
7	com-amazon	−0.4057	0.5294	−0.9851
8	p2p-Gnuetella04	−0.4066	−0.9184	0.4286
9	RoadNet-PA	−0.4085	−0.7684	−1.0158
10	RoadNet-TX	−0.4085	−0.7684	−1.0171
11	RoadNet-CA	−0.4085	−0.7721	−1.0183

coefficient(CC) and density. Among the collaboration networks, CA-GrQc(1) and CA-HepTh(2) both have high density and a very different clique conversion nature. Also, CA-GrQc(1), CA-CondMat(3) and com-dblp(4) all have high CC values, but only CA-GrQc among them has high CCC. All the peer-to-peer networks(5, 6, 8) have moderate density values but very low CCC values. This table shows that networks could still be dissimilar with respect to their clique forming tendency in spite of having similar CC and density values.

Clustering Results: We perform three hierarchical clusterings with single linkage of the eleven networks. All values are taken from Table 2. The three clusterings are performed with: a) $CCC_g(.,\alpha)$ and CC as features, b) CC and $Density$ as features and c) $CCC_g(.,\alpha)$ and $Density$ as features. Euclidean distance is used to compute the similarity between any two networks in all clusterings. Figure 2 shows the clustering of the networks. The green dotted ovals represent the clusters, while the red dotted lines represent the level at which we break the dendrogram to obtain clusters. Table 3 gives an overview of the clustering results. Figure 2(a) shows the clustering obtained when $CCC_g(.,\alpha)$ and CC are used as features. The first cluster in the result is just CA-GrQc(1) which has both high CCC and CC. The second cluster consists of the networks CA-HepTh(2), CA-CondMat(3), com-dblp(4) and com-amazon(7) which have moderate to high values of CC but moderate CCC values. Only com-amazon(7) has a low CCC

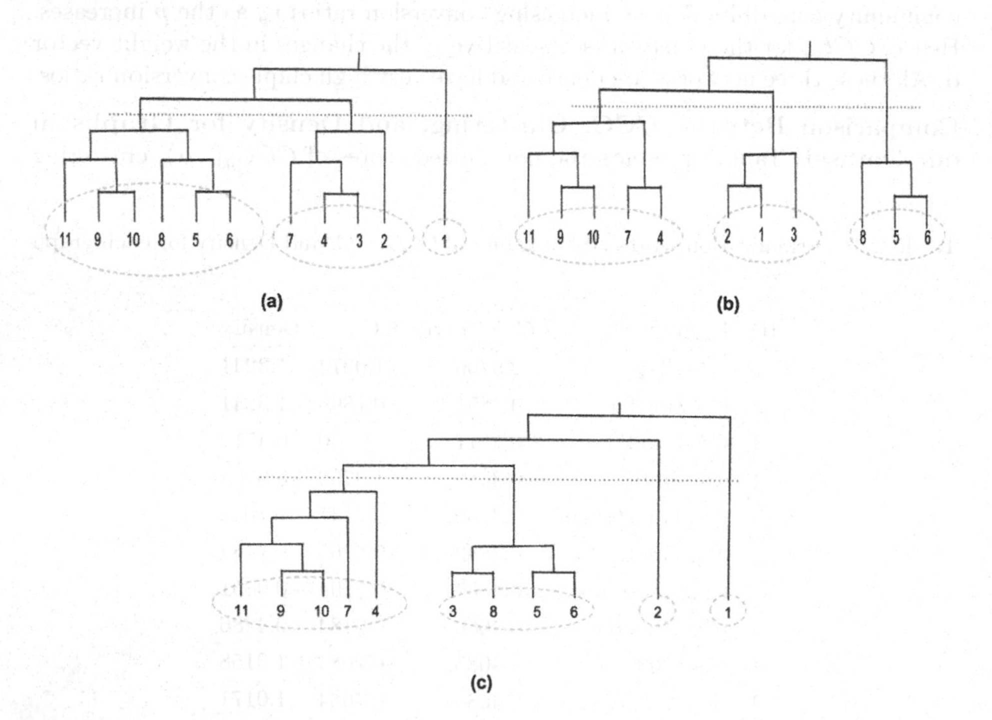

Fig. 2. The dendrogram for the three hierarchical clusterings of the networks. (Color figure online)

in this cluster. The third cluster is the peer-to-peer networks and the road networks which have very low CCC and CC values. Figure 2(b) shows the clustering obtained when CC and Density are used as features. The first cluster consists of the three collaboration networks(1, 2, 3) which have high CC and Density values. The second cluster consists of all peer-to-peer networks(5, 6, 8) which have high density and low CC. The third cluster consists of the rest of networks(4, 7, 9, 10, 11) which have low density and low CC. The only exception to this is com-dblp(4) which has the highest CC and very low Density. Interestingly, the CCC value of com-dblp(4) is in the range of moderate to low which is because its conversion ratios have almost same values. Figure 2(c) shows the clustering obtained when $CCC_g(., \alpha)$ and Density are used as features. The first cluster consists of only CA-GrQC(1) which has very high density and CCC. The second cluster consists of just CA-HepTh(2) which has high density and positive CCC. The third cluster consists of networks(3, 5, 6, 8) which have positive density and moderate CCC. The fourth cluster consists of networks(4, 7, 9, 10, 11) which have low values for both CCC and density. The important observation is that when CCC is used as a feature, the clusters are formed taking into account the similarity based on clique conversion ratio of graphs. But, in the absence of CCC, the clusters do not take into account such similarity. For example, the highly clique forming tendency in CA-GrQc is not considered in the clustering obtained when CCC is not used as a feature. As CCC gives additional insights into the clique forming tendency of graphs, we propose the use of CCC as one of the features in clustering to get better clustering results.

Table 3. Clustering results

Features used	Clusters obtained
CCC, CC	{{1}, {2, 3, 4, 7}, {5, 6, 8, 9, 10, 11}}
CC, Density	{{1, 2, 3}, {5, 6, 8}, {4, 7, 9, 10, 11}}
CCC, Density	{{1}, {2}, {3, 5, 6, 8}, {4, 7, 9, 10, 11}}

5 Clustering of Subgraphs of YOUTUBE Network

Experimental Setup: The main idea in this section is to cluster the subgraphs of a large YOUTUBE graph using the Conversion Ratios as features. We also cluster the same subgraphs using CC and Density as features. We find the frequent subgraphs in each cluster of the clustering obtained when Conversion Ratios (resp. CC and Density) are used as features. The two clusterings are compared based on the frequent subgraphs in each cluster. The YOUTUBE graph [17] has 1157827 users (nodes) and 4945382 user-to-user links (edges). In our experiments, we generate 200 subgraphs such that each subgraph is induced by the neighbors of a particular node, say, the ego node, of the corresponding subgraph. All ego nodes are selected to have exactly 20 neighbors.

We find the local CC and Density of the 200 subgraphs. The clustering of the subgraphs is performed by using DBSCAN [19] algorithm in Weka [18].

The attributes of the clustering are CC and Density. The parameters of DBSCAN are set as follows: $\epsilon = 0.04$ and $minPoints = 5$. The result is two clusters of subgraphs as presented in Table 4. Around 10% of the subgraphs–the *Outliers*–are not assigned to any cluster. Cluster 1 has subgraphs with low CC and high density whereas the Cluster 2 has subgraphs with high CC and density. Cluster 2 is a collection of subgraphs with nodes having a high tendency to group with each other. From the cluster centers in Table 4 we can note that density does not play a major role in Clustering. Majority of subgraphs belong to the Cluster 1 and have nodes with less tendency to group with eachother inspite of some subgraphs having high density.

Table 4. Clustering of YOUTUBE subgraphs using CC and Density

Cluster ID (Size)	Average $\{CC, Density\}$ in cluster
1 (167)	$\{0.0814, 0.3125\}$
2 (23)	$\{0.2453, 0.3416\}$

Frequent Subgraphs in the Clusters: To analyze the quality of the clustering lets look at the frequent subgraphs in each cluster ignoring the outliers. Each cluster is viewed as a Graph Database and each subgraph in the cluster as a transaction in it. Finding the frequent subgraphs in a cluster is then similar to finding frequent itemsets in a set of transactions. All vertices in each subgraph are labeled same. The Support S of each frequent subgraph I in a cluster is defined to be the number of subgraphs in the cluster that consist of an isomorphic substructure to I. We use GASTON (a frequent subGrAph, Sequences, and Tree ExtractiON)[16] to find the frequent subgraphs in each cluster. The minimum support is set as 40%. We find frequent subgraphs with at most 6 vertices. GASTON outputs the frequent subgraph patterns of three types: (i) frequent cyclic graphs, (ii) real trees (iii) paths. The frequent cyclic graphs are the patterns which consist of at least one cycle. The real trees are the patterns which are tree kind of structures that are not just simple paths from a node to another via intermediate nodes. We are mostly interested in analyzing the frequent cyclic graphs as they correspond to clique forming tendency. The frequent subgraphs in cluster 1 are 21 cyclic graphs and 8 real trees. The interesting frequent cyclic graphs in cluster 1 are presented in Fig. 3 along with their support. We can note

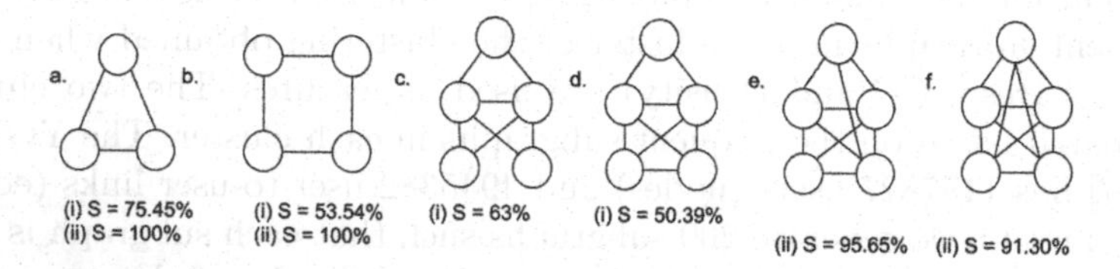

Fig. 3. Support for the presented frequent subgraphs in Cluster (i) and Cluster (ii) when CC and density are features.

that in spite of low CC values the support for cliques of size 3 (a. i) and four (b.i) in this cluster is over 75% and 50% respectively. More than 50% of the subgraphs in Cluster 1 have good clique forming tendencies in spite of low CC. The frequent subgraphs in cluster 2 are 23 cyclic graph and 8 real trees. The cliques of size 3 (a.ii) and 4 (b.ii) both have a support of 100% in Cluster 2. Also, a clique of size 5 (f.ii) has support over 90% in cluster 2. As cluster 2 has highly dense subgraphs with high CC this result is not surprising, but only 23 subgraphs are in this cluster out of the 200 subgraphs. The support for frequent subgraphs in the clusters shows us that both the clusters have many subgraphs that have high clique forming tendencies. To contrast, let us now use conversion ratios to find the clustering and analyze the frequent subgraphs in each cluster obtained. The conversion ratios r_2, r_3, r_4, r_5 of the 200 subgraphs are calculated and DBSCAN algorithm is used for the clustering with conversion ratios as the attributes of clustering. The parameters of DBSCAN are set as follows: $\epsilon = 0.07$ and $minPoints = 3$. DBSCAN outputs three Clusters with about 13% of the subgraphs in the 200 subgraphs being outliers. The details of the three Clusters obtained are presented in Table 5. Cluster 1 consists of 79 subgraphs with high clique forming tendency, while Cluster 2 consists of 91 subgraphs with low clique forming tendency. Cluster 3 consists of just four subgraphs with moderate values of r_2, but very high r_3, r_4 and r_5. Cluster 3 appeared because each of these subgraphs has fewer edges among the neighbors of the ego nodes in comparison to the subgraphs in Cluster 1. Most of these edges join to form at least a triangle. Also, most of those triangles join to form at least one clique of size 4. In other words, these subgraphs have nodes which once form at least an edge have high tendency to form Cliques of higher orders as well. This could be noticed from the Average Conversion Ratios values of the Cluster in Table 5.

Table 5. Clustering results of YOUTUBE subgraphs using the sequence of Clique Conversion Ratios

Cluster ID (Size)	Average $\{r_2, r_3, r_4, r_5\}$
1 (79)	{0.1744, 0.1545, 0.1422, 0.0875}
2 (91)	{0.0462, 0.0499, 0.0, 0.0}
3 (4)	{0.1158, 0.2679, 0.3929, 0.3611}

Frequent Subgraphs in the Clusters: Again, using GASTON we find the frequent subgraphs in each cluster obtained. We ignore the Cluster 3 (since it has only 4 subgraphs) and the Outliers from this discussion. The frequent subgraphs in Cluster 1 are 22 cyclic graphs and 8 real trees. Some interesting frequent cyclic graphs in Cluster 1 and Cluster 2 are presented in Fig. 4. In Cluster 1 the cliques of size 3, 4 and 5 (a.i, c.i and e.i) have support 100%, 94.94% and 41.77% respectively. This shows that the subgraphs in the cluster have higher clique forming tendency. The frequent subgraphs in Cluster 2 are 13 cyclic graphs and 8 real trees. In Cluster 2 the clique of size 3 (a.ii) has support over 54%. Cliques

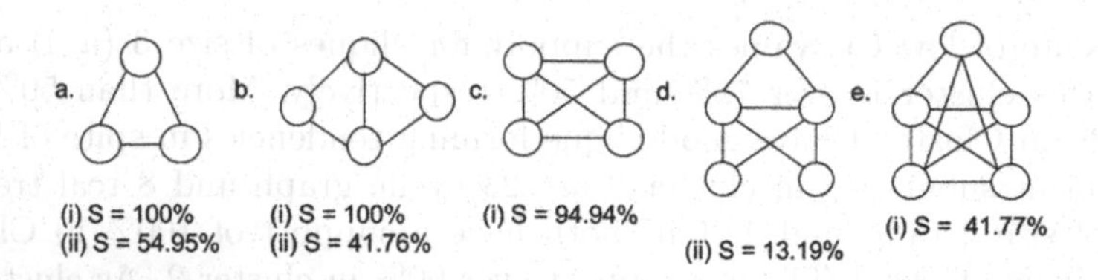

Fig. 4. Support for the presented frequent subgraphs in Cluster (i) and Cluster (ii) when conversion ratios are features

of size 4 and 5 are infrequent and have support less than 20%. Cluster 2 has subgraphs which do not have good clique conversions whereas Cluster 1 has subgraphs which have very good clique forming tendency.

6 Summary

The main contribution of this work is the measure CCC, which unlike existing measures, does not focus on the tendency of nodes to cluster alone, but also focuses on the tendency of nodes to form cliques. This measure gives new insights into graph properties, say, graphs with similar clustering coefficient or density, might have a lot of variation in their clique conversion ratios. We compare the clustering results of some real world graphs in the presence and absence of CCC. Our results show that clustering with CCC as a feature helps to Cluster the graphs with similar clique conversion ratios, while without CCC as a feature this is not always possible. This highlights the need for a measure like CCC. Also, we show the utility of CCC in clustering subgraphs of a large graph. The quality of the clusters obtained is verified using the frequent subgraph patterns in the clusters. This work could be further explored to find faster algorithms to compute Clique Conversion Ratios either deterministically or approximately if possible, to use the conversion ratios in the generation of synthetic graphs with desired clique forming tendencies. Also, the exploration of the uses of CCC in many applications areas has a good scope.

References

1. Luce, R.D., Perry, A.D.: A method of matrix analysis of group structure. Pschometrica **14**(1), 95–116 (1949)
2. Altaf-Ul-Amin, M., Nishikata, K., Koma, T., Miyasato, T., Shinbo, Y., Arifuzzaman, M., Wada, C., Maeda, M., Oshima, T.: Prediction of protein functions based on k-cores of protein-protein interaction networks and amino acid sequences. Genome Inf. **14**, 498–499 (2003)
3. Leskovec, J., Kleinberg, J.M., Faloutsos, C.: Graph evolution: densification and shrinking diameters. TKDD **1**(1) (2007)
4. Schwikowski, B., Uetz, P., Fields, S.: A network of protein-protein interactions in yeast. Nat. Biotechnol. **18**(12), 1257–1261 (2000)

5. McAuley, J.J., Leskovec, J.: Discovering social circles in ego networks. TKDD **8**, 4:1–4:28 (2014)
6. Chen, B., Matsumoto, M., Wang, J., Zhang, Z., Zhang, J.: A short proof of Nash-Williams' theorem for the arboricity of a graph. Graphs Comb. **10**(1), 27–28 (1994)
7. Leskovec, J., Lang, K.J., Dasgupta, A., Mahoney, M.W.: Community structure in large networks: natural cluster sizes and the absence of large well-defined clusters. Internet Math. **6**(1), 29–123 (2009)
8. Leskovec, J., Krevl, A.: SNAP Datasets: Stanford Large Network Dataset Collection, June 2014. http://snap.stanford.edu/data
9. Holton, D.A., Sheehan, J.: The Petersen Graph. Cambridge University Press, Cambridge (1993). doi:10.2277/0521435943. ISBN 0-521-43594-3
10. Kreyszig, E.: Advanced Engineering Mathematics, 4th edn. Wiley, New York (1979)
11. Rosen, K.H.: Discrete Mathematics and Its Applications, 7th edn. McGraw-Hill (2011). p. 655
12. Erdos, P., Renyi, A., Sos, V.: On a problem of graph theory. Stud. Sci. Math. **1**, 215–235 (1966)
13. Freeman, L.: A set of measures of centrality based on betweenness. Sociometry **40**, 35–41 (1977)
14. Awodey, S.: Isomorphisms. Oxford University Press, Category theory (2006)
15. Gao, X., Xiao, B., Tao, D., Li, X.: A survey of graph edit distance. Pattern Anal. Appl. **13**(1), 113–129 (2010)
16. Nijssen, S., Kok, J.: A quickstart in frequent structure mining can make a difference. In: Proceedings of the SIGKDD (2004). http://www.liacs.nl/home/snijssen/gaston
17. Mislove, A., Marcon, M., Gummadi, K.P., Druschel, P., Bhattacharjee, B.: Measurement and analysis of online social networks. In: Proceedings of the 5th ACM/Usenix Internet Measurement Conference (IMC 2007), San Diego, CA, October 2007
18. Hall, M., Frank, E., Holmes, G., Pfahringer, B., Reutemann, P., Witten, I.H.: The WEKA data mining software: an update. SIGKDD Explor. **11**(1), 10–18 (2009)
19. Ester, M., Kriegel, H.-P., Sander, J., Xu, X., Simoudis, E., Han, J., Fayyad, U.M. (eds.): A density-based algorithm for discovering clusters in large spatial databases with noise. In: Proceedings of the Second International Conference on Knowledge Discovery and Data Mining (KDD-96), pp. 220–231. AAAI Press (1996)

Mining Cohesive Clusters with Interpretations in Labeled Graphs

Hongxia Du[1](✉), Heli Sun[1], Jianbin Huang[2], Zhongbin Sun[1], Liang He[1], and Hong Cheng[3]

[1] Department of Computer Science and Technology, Xi'an Jiaotong University, Xi'an, China
duhx123@outlook.com, {hlsun,zhongbin725,lhe}@xjtu.edu.cn
[2] School of Software, Xidian University, Xi'an, China
jbhuang@xidian.edu.cn
[3] Department of Systems Engineering and Engineering Management, The Chinese University of Hong Kong, Hong Kong, China
hcheng@se.cuhk.edu.hk

Abstract. In recent years, community detection on plain graphs has been widely studied. With the proliferation of available data, each user in the network is usually associated with additional attributes for elaborate description. However, many existing methods only focus on the topological structure and fail to deal with node-attributed networks. These approaches cannot extract clear semantic meanings for communities detected. In this paper, we combine the topological structure and attribute information into a unified process and propose a novel algorithm to detect overlapping semantic communities. The proposed algorithm is divided into three phases. Firstly, we detect local semantic subcommunities from each node's perspective using a greedy strategy. Then, a supergraph which consists of all these subcommunities is created. Finally, we find global semantic communities on the supergraph. The experimental results on real-world datasets show the efficiency and effectiveness of our approach against other state-of-the-art methods.

Keywords: Semantic community · Community detection · Node-attributed graph

1 Introduction

In recent years, community detection on plain graphs has been widely studied [5,14,17]. In literature, a community or a cluster is a subgraph containing nodes which are more densely linked to each other than to the rest of the graph [11]. With the proliferation of available data, each user in the network is usually associated with additional attributes. However, most of existing community detection methods only focus on plain graphs and fail to handle situations with attributes. The underlying knowledge behind communities is hidden so that we cannot interpret the communities we find, which limits our insight into the graph

J. Kim et al. (Eds.): PAKDD 2017, Part II, LNAI 10235, pp. 774–785, 2017.
DOI: 10.1007/978-3-319-57529-2_60

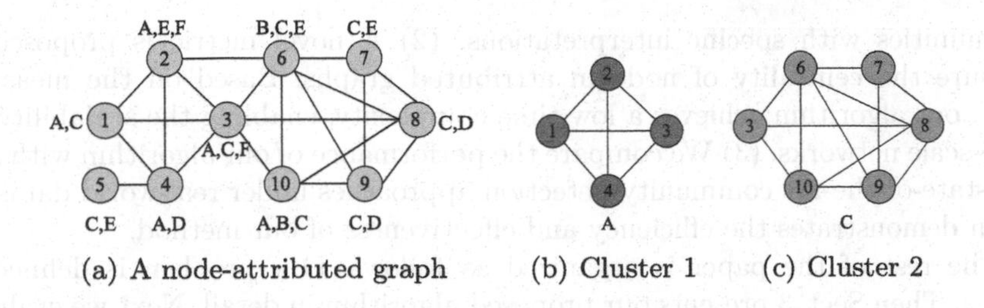

Fig. 1. The original graph and its semantic communities returned

structure. So detecting communities on node-attributed graphs becomes a critical problem. The communities we want to obtain should have cohesive structure and specific common interpretations, which are called semantic communities.

Semantic communities have wide application scenarios in real world. For example, each cluster has well-matched descriptions, so we can provide more effective and precise recommendations for users on social platforms or make proper market positioning strategy for target people. Furthermore, each object can be grouped into various communities for its different attributes [12,16]. Thus, finding overlapping semantic communities in node-attributed networks is meaningful and realistic. Figure 1 illustrates a toy node-attributed graph. Each node is associated with a set of attributes (interchangeably, features). And our goal is to find out underlying overlapping semantic communities shown in Fig. 1(b) and (c), which are two communities possessing attributes $\{A\}$ and $\{C\}$, respectively.

Although some approaches have been proposed to handle community detection problem on node-attributed networks, there are still some problems unsolved. First, some of these methods [10,18,21] can only deal with numerical attributes while some [20] can handle categorical attributes but suffer a high time complexity. Next, the clusters detected by some algorithms [1,13] do not consider attribute homogeneity and cohesive structure at the same time. Moreover, others [6,7] do not combine structural and attribute information into a unified process for detecting communities. All of them above cannot give precise interpretations to the results at the semantic level [2,15,19]. Furthermore, few of the existing methods pay attention to the overlapping situations [8,21]. So we aim to combine the available attributes and the topological structure to detect meaningful overlapping semantic communities.

In this paper, we design a novel metric to integrate the topological structure and attribute information to reveal communities leveraging a local-first strategy. The local perspective of each individual node [3] instead of the global view of the network is used to uncover semantic communities. Our method is capable of finding out high-quality semantic communities that is compact in structure and homogeneous in attribute. Also, this algorithm averts the trouble of tuning user-defined parameters.

The main contributions of our work are as follows. (1) We study the problem of semantic community detection on node-attributed networks and propose a parameter-free method based on a local-first strategy to find out overlapping communities with specific interpretations. (2) A novel metric is proposed to measure the centrality of node in attributed graphs. Based on the measurement, our algorithm achieves a low time complexity, enabling the scalability for large-scale networks. (3) We compare the performance of our algorithm with several state-of-the-art community detection approaches under real-world datasets, which demonstrates the efficiency and effectiveness of our method.

The rest of the paper is organized as follows. Our problem is defined in Sect. 2. Then Sect. 3 presents our proposed algorithm in detail. Next we evaluate our method against several approaches and present the experimental results in Sect. 4. Finally, we make a conclusion in Sect. 5.

2 Problem Statement

Formally, we define a node-attributed graph as $G = (V, E, \mathcal{A}, f)$, where V is the set of vertices, $E \subseteq V \times V$ is the set of edges, $\mathcal{A} = \{a_1, a_2, \cdots, a_k\}$ is the set of k attributes and $f : V \to 2^{\mathcal{A}}$ is the mapping function from the vertices to the attribute subsets of $\mathcal{A}$. Figure 1(a) presents an example of node-attributed graph. For one node $v_i \in V$, it is annotated with a set of attributes $f(v_i) = \{a_{i1}, a_{i2}, \cdots, a_{im}\}$, where m denotes the number of attributes node v_i has and $f(v_i) \subseteq \mathcal{A}$.

Given a graph $G = (V, E, \mathcal{A}, f)$, the problem studied in this paper is to find out all overlapping communities $\mathcal{C} = \{C_i\}_{i=1}^{k}$, in which one cluster C_i may overlap another cluster C_j with $i \neq j$, such that: (1) nodes within clusters are densely connected, and (2) all nodes in each cluster have some attributes in common. Those clusters can be called semantic communities. If a cluster $C(U, S)$ is a semantic community, where $U \subseteq V$, then $S = \bigcap_{v \in U} f(v)$ and $S \neq \emptyset$. For example, Fig. 1(b) and (c) are two semantic communities uncovered from Fig. 1(a).

3 The Algorithm

In this section, we present our Overlapping Semantic Community detection algorithm OSCom in detail, which includes three phases.

3.1 Detecting Local Semantic Subcommunities

The first phase is to find local semantic subcommunities from the perspective of each node. These subcommunities are obtained from centerless ego networks.

Definition 1 (Centerless Ego Network). *Let $G = (V, E, \mathcal{A}, f)$ be a node-attributed graph, the centerless ego network of a node $v \in V$ is a subgraph which deletes the node v and its adjacent edges from the ego network of node v, i.e.,*

$$CEN(v, G) = (\Gamma(v), E', \mathcal{A}, f), \tag{1}$$

where $E' = \{(u, w) | u, w \in \Gamma(v), (u, w) \in E\}$.

Definition 2 (Structural Neighborhood). *Let $G(V, E, \mathcal{A}, f)$ be a attributed graph, the structural neighborhood of node $v \in V$ is the node set containing the adjacent nodes to node v, i.e.,*

$$\Gamma(v) = \{u | (u, v) \in E\}. \tag{2}$$

Definition 3 (Attributed Neighborhood). *In a attributed graph $G(V, E, \mathcal{A}, f)$, the attributed neighborhood of a semantic community $C(U, S)$ is the node set containing nodes which satisfy: (1) locating outside of U but linking to any one of nodes in U; (2) having common attributes with cluster attributes S, i.e.,*

$$\Gamma_A(C) = \{u | \Gamma(u) \cap C \neq \emptyset, f(u) \cap S \neq \emptyset, u \notin U\}. \tag{3}$$

Based on Definitions 2 and 3, we define the hybrid density.

Definition 4 (Hybrid Density). *Given a node-attributed graph $G(V, E, \mathcal{A}, f)$, the hybrid density of a semantic community $C(U, S)$ is the ratio of inner degrees to all degrees that are related to attributes S in C, i.e.,*

$$\sigma(C) = \frac{k_{in}}{k_{in} + k_{out}^A}, \tag{4}$$

where k_{in} is the total internal degrees of nodes in cluster C, and k_{out}^A is the number of links between node set U and its attributed neighborhood $\Gamma_A(C)$.

On centerless ego networks, a greedy strategy is used to detect local semantic subcommunities based on the hybrid density. We first introduce the concept of density gain $\Delta\sigma$. When a node P is added into a semantic community C, the gain in hybrid density is the difference between before and after, i.e.,

$$\Delta\sigma = \sigma(C + \{P\}) - \sigma(C). \tag{5}$$

Initially, we randomly select a node O as an original community C, which expands from node O: (1) scan all the attributed neighborhood $\Gamma_A(C)$ of community C; (2) add the neighbor which has the largest $\Delta\sigma$ into C to form a larger cluster; (3) repeat from step (1) until $\Gamma_A(C)$ is empty or all attributed neighbors have negative $\Delta\sigma$. After these steps, the semantic subcommunity of node O is obtained. Then we choose randomly another node which does not yet belong to any group and do the same operation until there is no isolated node outside any cluster. For example, we can obtain the local clusters $\{1, 2, 3, 4\}$ and $\{3, 6, 8, 10\}$ on $CEN(3, G)$ by adding central node into the results, which are shown in Fig. 2.

Instead of scanning all nodes and doing the same operation, we propose a heuristic solution to decrease the iteration times. A new metric to value the centrality of nodes in node-attributed graphs is introduced.

Definition 5 (Centrality of Node). *Let $G = (V, E, \mathcal{A}, f)$ be a node-attributed graph, the centrality of a node $v \in V$ is defined as*

$$\theta(v) = \frac{\sum_{u \in \Gamma(v)} |f(v) \cap f(u)|}{|\Gamma(v)|}, \tag{6}$$

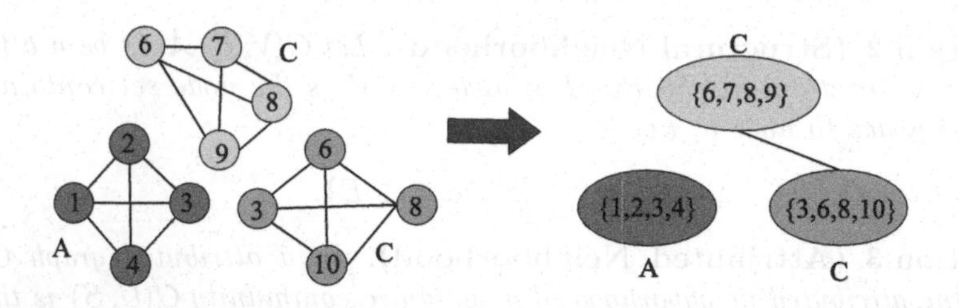

Fig. 2. Local semantic subcommunities and the subcommunity graph

According to Eq. (6), we can sort all nodes by the centrality. In Fig. 1, the final sequence of nodes after sorting is $\{3, 10, 1, 6, 7, 2, 9, 8, 4, 5\}$. We can begin from the node with the maximal centrality. If one is selected as the central node, the structural neighborhood and itself are labeled as visited so that we can skip them in the ordered sequence. Thus all the nodes we need to consider are $\{3, 7, 5\}$.

3.2 Creating Subcommunity Graph

After the first phase, we obtain a set of local subcommunities, which are limited by the central nodes' perspectives. Hence, we create a subcommunity graph H. We firstly introduce the concept of average degree for a cluster.

Definition 6 (Average Degree). *Given a cluster C in a graph $G = (V, E, \mathcal{A}, f)$, the average degree of the cluster C is defined as:*

$$d(C) = \frac{|\{(u,v)|u, v \in C, (u,v) \in E\}|}{|C|}, \tag{7}$$

where the numerator is the number of edges in cluster C and $|C|$ denotes the number of nodes in cluster C.

Definition 7 (Subcommunity Graph). *A subcommunity graph is a supergraph $H = (V(H), E(H))$, where*

(1) each node $v \in V(H)$ represents a local semantic subcommunity $C = (U, S)$, which can be called a supernode;
(2) an edge $(u, v) \in E(H)$ with $u : C = (U, S)$ and $v : C' = (U', S')$ iff the two supernodes meet the following conditions: (a) $U \cap U' \neq \emptyset$; (b) $S \cap S' \neq \emptyset$; (c) $d(U \cup U') > d(U)$ and $d(U \cup U') > d(U')$, where $d(U)$ represents the average degree of U.

According to Definition 7, we utilize the average degree to decide whether two subcommunities have connection or not. Take Fig. 2 for example, we have obtained three local semantic subcommunities after the first phase: $\{\{A\} : \{1, 2, 3, 4\}, \{C\} : \{\{3, 6, 8, 10\}, \{6, 7, 8, 9\}\}\}$. Then two subcommunities $\{3, 6, 8, 10\}$ and $\{6, 7, 8, 9\}$ can build a link according to conditions. The structure of the subcommunity graph is illustrated in right of Fig. 2.

3.3 Identifying Communities in Subcommunity Graph

After the subcommunity graph is created, we need to merge supernodes to obtain final semantic communities of the global view. The subcommunity graph is also a node-attributed network because the cluster attributes can be seen as the attributes of corresponding supernode. So we can detect semantic communities on the subcommunity graph using the same greedy method which is utilized in the first phase.

For example, in Fig. 2, communities $\{3, 6, 8, 10\}$ and $\{6, 7, 8, 9\}$ can be merged into a larger semantic community $\{3, 6, 7, 8, 9, 10\}$ because their merging can increase the hybrid density. Therefore, we can obtain the final semantic communities: $\{\{A\} : \{1, 2, 3, 4\}, \{C\} : \{3, 6, 7, 8, 9, 10\}\}$ that are presented in Fig. 1(b) and (c). Note that node 3 appears twice in different semantic communities, which demonstrates that our algorithm can detect overlapping communities.

3.4 Complexity Analysis

The pseudo code of OSCom is described in Algorithm 1. Lines 1–13 are the first phase of finding local semantic subcommunities. The second phase is specified in Lines 14–21. Lines 22–25 present the final phase of detecting the global semantic communities. We assume that there is a network with n nodes and m edges and analyze the time complexity of each phase. In the first phase, the time complexity of the speed-up strategy is $\mathcal{O}(n \log n)$. Then we leverage the greedy method to uncover local semantic subcommunities on selected central nodes, and the worst time complexity is $\mathcal{O}(n_c \cdot \bar{k}^2)$, where n_c is the number of central nodes selected and $\bar{k}$ is the average degree. In the second phase, we suppose there are l_c subcommunities from the first phase. The time complexity is $\mathcal{O}(l_c^2)$ because we need to scan each pair of supernodes to build the subcommunity graph. For the third phase, the worst time complexity is $\mathcal{O}(n_s \cdot \bar{k'}^2)$, where n_s is the number of selected central supernodes and $\bar{k'}$ is the average degree of these supernodes. Thus, the worst-case time complexity of OSCom algorithm is $\mathcal{O}(n \log n + n_c \cdot \bar{k}^2 + l_c^2 + n_s \cdot \bar{k'}^2)$, where the number of selected nodes n_c, selected supernodes n_s and subcommunities l_c are much smaller than the number of nodes n.

4 Experiments

In this section, we evaluate the performance of our algorithm on three real-world datasets. All experiments are implemented in Python on a computer with a 3.2 GHz CPU and 12 GB memory.

4.1 Datasets, Compared Algorithms and Evaluation Metrics

We use three real node-attributed graphs obtained from different sources in our experiments: LastFM, Delicious and DBLP datasets, in which LastFM and

Algorithm 1. The overlapping semantic community detection algorithm OSCom

Input: Node-attributed network $G = (V, E, \mathcal{A}, f)$, result set $\mathcal{C} = \emptyset$
Output: set of overlapping semantic communities $\mathcal{C}$

(01) set of local communities $LC = \emptyset$;
(02) $CalculateNodeCentrality(V)$;
(03) $SV = Sort(V)$; //sort all nodes in descending order of the centrality
(04) **for all** $v \in SV$ **do**
(05)   **if** $v.visited == False$ **then**
(06) $G(v) = RefineAttribute(CEN(v, G))$;
(07) $\mathcal{C}(v) = HybridDensityGreedy(G(v))$;
(08) $LC.add(\mathcal{C}(v))$;
(09)   **for all** $u \in \Gamma(v)$ **do**
(10) $u.visited = True$;
(11)   **end for**
(12)   **end if**
(13) **end for**
(14) Create a new network H, where $V(H) \rightarrow LC$;
(15) **for all** $v_i \in V(H)$ **do**
(16)   **for all** $v_j \in V(H)$ **do**
(17)   **if** $AveDegree(v_i \cup v_j) > AveDegree(v_i)$ and $AveDegree(v_i \cup v_j) > AveDegree(v_j)$ **then**
(18)   Add (v_i, v_j) to $E(H)$;
(19)   **end if**
(20)   **end for**
(21) **end for**
(22) $tmpC = HybridDensityGreedy(H)$;
(23) **for all** $C(U, S) \in tmpC$ **do**
(24) $\mathcal{C}(S).add(C)$;
(25) **end for**
(26) **return** $\mathcal{C}$

Delicious come from the HetRec 2011 workshop[1], DBLP dataset extracts from DBLP[2] co-authorship network. Some basic statistics are listed in Table 1, in which $|\mathcal{A}|$ is the total number of attributes that the network has and Attr.density is the average number of attributes per node has.

To evaluate the performance of our approach, we compare it with three existing algorithms. DEMON [4] can detect overlapping communities on plain graphs. DBP [9] uncovers overlapping communities using matrix factorization on node-attributed networks while only considering vertex attributes. LDense [6] reveals overlapping communities in labeled graphs based on a generic greedy scheme, which considers both structure and attribute information.

[1] http://ir.ii.uam.es/hetrec2011/.
[2] http://www.informatik.uni-trier.de/~ley/db.

Table 1. Real-world datasets and their statistics

Dataset	$\|V\|$	$\|E\|$	$\|\mathcal{A}\|$	Attr.density
LastFM	1,892	12,717	11,946	17.9
Delicious	1,861	7,664	1,350	52.5
DBLP	108,030	276,658	23,285	14.3

Suppose that the final community result is $\mathcal{C} = \{C_i\}_{i=1}^k$, where $C_i = (U_i, S_i)$ and k is the total number of semantic communities, we adopt several metrics as follows to quantitatively evaluate the quality of clustering results.

- *Density.* It measures the compactness of clusters in structure, formally,

$$density(\mathcal{C}) = \frac{1}{k}\sum_{i=1}^{k}\rho(C_i) = \frac{1}{k}\sum_{i=1}^{k}\frac{2|\{(u,v)|u,v \in U,(u,v) \in E\}|}{|U| \times (|U|-1)}. \quad (8)$$

- *Entropy.* It measures the randomness of attributes on semantic clusters.

$$entropy(\mathcal{C}) = \frac{1}{k}\sum_{i=1}^{k} entropy(f(U_i), U_i), \quad (9)$$

$$entropy(f(U_i), U_i) = \sum_{a_j \in f(U_i)} p_j log_2 p_j, p_j = \frac{|\{v|a_j \in f(v), v \in U_i\}|}{|U_i|}. \quad (10)$$

- *Community quality.* This metric evaluates the quality of clusters by integrating the structure and attribute, which also be use to select top-k communities for comparison. Given a semantic cluster $C(U, S)$, its quality is defined as:

$$Quality(C) = |U| \times |S| \times \rho(C). \quad (11)$$

The quality of one semantic community depends on the scale of the community $|U|$, the number of cluster attributes $|S|$ and the density $\rho(C)$. Generally, a good semantic community should have high density and low entropy.

4.2 Experimental Results and Discussion

In the following, the clustering performance on real-world networks using metrics above is reported. Note that Demon can automatically determine the number of communities, while LDense and DBP need input parameter k to control the number of output communities. Thus we analyze the clustering performance from two aspects: one aspect is to evaluate performance on whole community results, another is to select top-k communities for comparison.

Quality Evaluation on Overall Results. Since LDense and DBP need user-defined parameters to determine the number of communities they return, we set

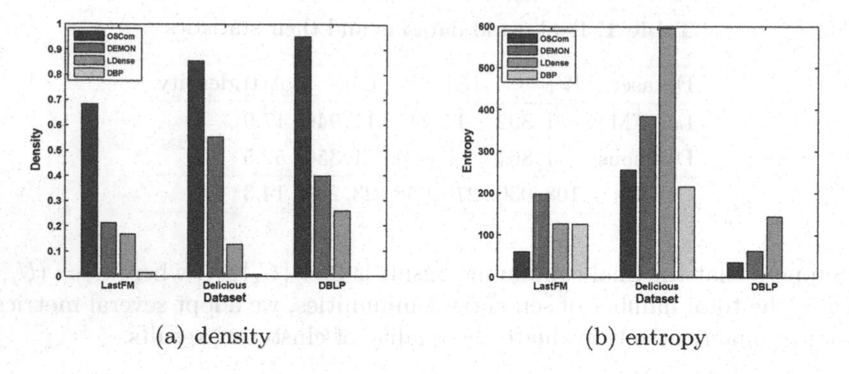

(a) density (b) entropy

Fig. 3. Overall performance comparison about density and entropy on three datasets

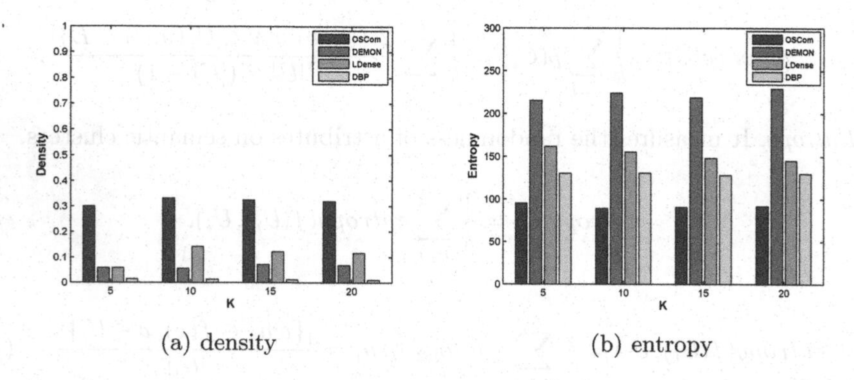

(a) density (b) entropy

Fig. 4. The performance of top-k communities of algorithms on LastFM dataset

the parameter k in LDense and DBP as the number of semantic communities our algorithm output for fair comparison.

Figure 3 presents the density and entropy performance of these algorithms, respectively. Specially, Fig. 3(a) shows that our algorithm outperforms all other competitors in density on all datasets. Furthermore, OSCom achieves the lowest values in entropy except for Delicious dataset according to Fig. 3(b). On the other hand, DBP achieves better performance than DEMON and LDense in the aspect of entropy on LastFM and Delicious datasets. Because DBP concentrates on the attribute uniformity and ignores structural information, the communities returned are extremely sparse in structure but competitive in attribute homogeneity. As for LDense, although it can also detect semantic communities, the performance of LDense is not remarkable. This is because LDense does not take the compactness of community into account.

Quality Evaluation on Top-k Results. We leverage the community quality to select top-k communities for comparison. We respectively set the cluster number $k = 5, 10, 15, 20$ as the inputs of LDense and DBP. For DEMON, we ignore the influence of attributes to select top-k results.

Figure 4(a) presents the density comparison among four algorithms on LastFM dataset. From the figure, we can find that our method gets the highest

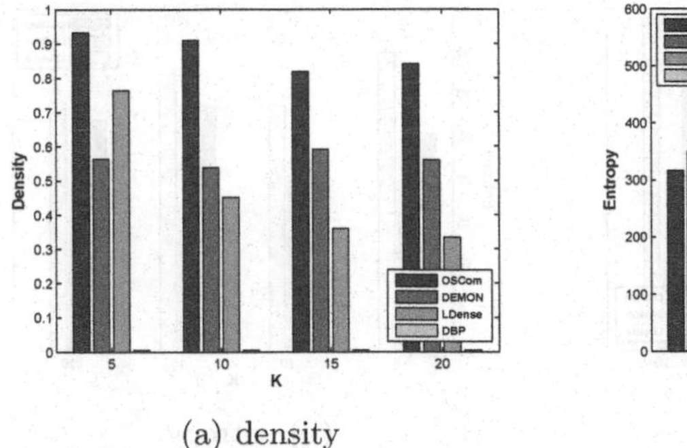

(a) density

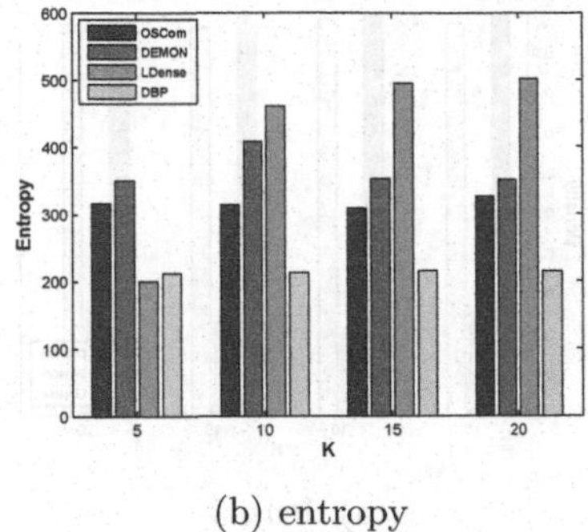

(b) entropy

Fig. 5. The performance of top-k communities of algorithms on Delicious dataset

density no matter what the value of k is. This is because OSCom pays more attention to the compactness of structure than DBP and LDense, which results in clusters of moderate size and cohesive structure. Figure 4(b) shows the entropy results of algorithms on LastFM dataset. OSCom outperforms other methods apparently. Moreover, we find that OSCom retains stable performance in term of entropy with the change of k. DEMON achieves the highest entropy, which is comprehensible since it finds clusters without considering attributes of nodes. Because DBP only focuses on attribute homogeneity, DBP keeps its advantages compared with DEMON and LDense, but still obtains higher entropy than OSCom.

Figure 5 shows the comparisons among algorithms on Delicious dataset. Figure 5(a) presents the density performance with different values of k. Our method OSCom achieves the best result while the performance of DBP is extremely poor. Because DBP has no concern for graph structure. Similar with the occasion in Fig. 4, DBP almost achieves the lowest entropy in Fig. 5(b) because it merely focuses on the attribute information, resulting in high homogeneity of attributes. OSCom also outperforms DEMON in entropy. Furthermore, except for $k = 5$, OSCom almost achieves better performance than LDense in term of entropy.

Since DBP is incapable of handling DBLP, Fig. 6 only shows three algorithms in terms of density and entropy on DBLP dataset. Obviously, in Fig. 6(a), OSCom and LDense achieve much better performance than DEMON in density. Besides, OSCom obtains a slight advantage compared with LDense. As for entropy, OSCom performs better than LDense, but worse than DEMON. Because local-first strategy exploited by DEMON detects communities on ego networks, which in fact returns communities with vague interpretations, causing that the entropy of DEMON is the lowest beyond expectation as shown in Fig. 6(b).

Running Time. Table 2 shows the running time of these algorithms on three datasets. We can observe that OSCom is the most efficient method compared with other approaches because OSCom adopts the speed-up strategy to decrease the number of iterations. DEMON needs to scan each node to detect communities on corresponding ego networks. LDense should operate on the entire network

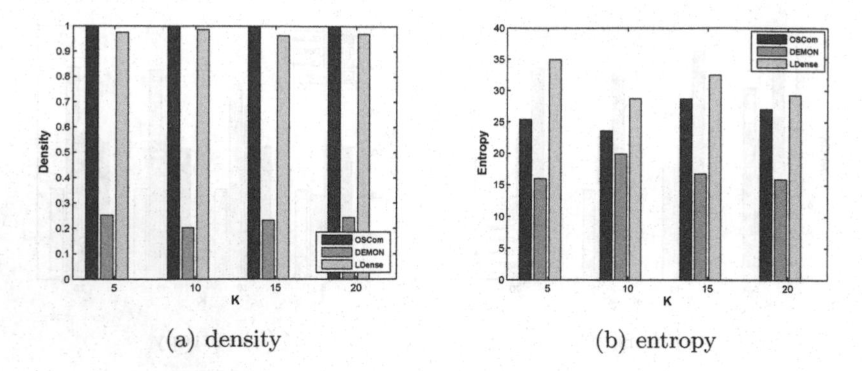

(a) density (b) entropy

Fig. 6. The performance of top-k communities of algorithms on DBLP dataset

Table 2. Running time of four algorithms on the three datasets (sec.)

Algorithms	LastFM	Delicious	DBLP
OSCom	**8.5**	**10.2**	**346.5**
DEMON	20.4	14.5	964.6
LDense	80.5	166.8	58555.0
DBP	541.6	38.4	–

over and over again to reveal each community. DBP adopts matrix factorization method, the time of which is determined by the number of attributes. So DBP is the most computationally expensive in LastFM.

5 Conclusion

In this paper, we study the problem of overlapping semantic community detection in node-attributed networks. A novel method OSCom which combines structural and attribute information is designed. The OSCom algorithm is divided into three phases, whose results are cohesive in structure and homogenous in attribute. Experimental results on real-world datasets demonstrate that our approach outperforms state-of-the-art methods in efficiency and effectiveness.

Acknowledgement. The work was supported in part by the National Science Foundation of China grants 61602354, 61672417 and 61472299, the Fundamental Research Funds for the Central Universities of China. Any opinions, findings and conclusions expressed here are those of the authors and do not necessarily reflect the views of the funding agencies.

References

1. Akoglu, L., Tong, H., Meeder, B., Faloutsos, C.: Pics: parameter-free identification of cohesive subgroups in large attributed graphs. In: Proceedings of the SIAM International Conference on Data Mining, pp. 439–450. SIAM (2012)

2. Atzmueller, M., Doerfel, S., Mitzlaff, F.: Description-oriented community detection using exhaustive subgroup discovery. Inf. Sci. **329**, 965–984 (2016)
3. Bagrow, J.P., Bollt, E.M.: Local method for detecting communities. Phys. Rev. E **72**(4), 046108 (2005)
4. Coscia, M., Rossetti, G., Giannotti, F., Pedreschi, D.: Demon: a local-first discovery method for overlapping communities. In: International Conference on Knowledge Discovery and Data Mining, pp. 615–623. ACM (2012)
5. Fortunato, S.: Community detection in graphs. Phys. Rep. **486**(3), 75–174 (2010)
6. Galbrun, E., Gionis, A., Tatti, N.: Overlapping community detection in labeled graphs. Data Min. Knowl. Discov. **28**(5–6), 1586–1610 (2014)
7. Gunnemann, S., Farber, I., Boden, B., Seidl, T.: Subspace clustering meets dense subgraph mining: a synthesis of two paradigms. In: 2010 IEEE International Conference on Data Mining, pp. 845–850. IEEE (2010)
8. Huang, X., Cheng, H., Yu, J.X.: Dense community detection in multi-valued attributed networks. Inf. Sci. **314**, 77–99 (2015). Elsevier
9. Miettinen, P., Mielikäinen, T., Gionis, A., Das, G., Mannila, H.: The discrete basis problem. IEEE Trans. Knowl. Data Eng. **20**(10), 1348–1362 (2008)
10. Moser, F., Colak, R., Rafiey, A., Ester, M.: Mining cohesive patterns from graphs with feature vectors. In: Proceedings of the SIAM International Conference on Data Mining, SDM, pp. 593–604 (2009)
11. Newman, M.E., Girvan, M.: Finding and evaluating community structure in networks. Phys. Rev. E **69**(2), 026113 (2004)
12. Palla, G., Derényi, I., Farkas, I., Vicsek, T.: Uncovering the overlapping community structure of complex networks in nature and society. Nature **435**(7043), 814–818 (2005)
13. Pool, S., Bonchi, F., Leeuwen, M.V.: Description-driven community detection. ACM Trans. Intell. Syst. Technol. **5**(2), 28 (2014)
14. Raghavan, U.N., Albert, R., Kumara, S.: Near linear time algorithm to detect community structures in large-scale networks. Phys. Rev. E **76**(3), 036106 (2007)
15. Silva, A., Meira Jr., W., Zaki, M.J.: Mining attribute-structure correlated patterns in large attributed graphs. Proc. VLDB Endowment **5**(5), 466–477 (2012)
16. Whang, J.J., Gleich, D.F., Dhillon, I.S.: Overlapping community detection using neighborhood-inflated seed expansion. IEEE Trans. Knowl. Data Eng. **28**(5), 1272–1284 (2016)
17. Xie, J., Kelley, S., Szymanski, B.K.: Overlapping community detection in networks: the state-of-the-art and comparative study. ACM Comput. Surv. (CSUR) **45**(4), 43 (2013)
18. Xu, Z., Ke, Y., Wang, Y., Cheng, H., Cheng, J.: A model-based approach to attributed graph clustering. In: Proceedings of the 2012 ACM SIGMOD International Conference on Management of Data, pp. 505–516. ACM (2012)
19. Yang, J., McAuley, J., Leskovec, J.: Community detection in networks with node attributes. In: 2013 IEEE 13th International Conference on Data Mining, pp. 1151–1156. IEEE (2013)
20. Zhou, Y., Cheng, H., Yu, J.X.: Graph clustering based on structural/attribute similarities. Proc. VLDB Endowment **2**(1), 718–729 (2009)
21. Zhou, Y., Cheng, H., Yu, J.X.: Clustering large attributed graphs: an efficient incremental approach. In: 2010 IEEE International Conference on Data Mining, pp. 689–698. IEEE (2010)

A SAT-Based Framework for Overlapping Community Detection in Networks

Said Jabbour[1], Nizar Mhadhbi[1], Badran Raddaoui[2(✉)], and Lakhdar Sais[1]

[1] CRIL - CNRS UMR 8188, University of Artois, Lens, France
[2] SAMOVAR, Télécom SudParis, CNRS, University of Paris-Saclay, Evry, France
badran.raddaoui@telecom-sudparis.eu

Abstract. In this paper, we propose a new approach to detect overlapping communities in large complex networks. We first introduce a parametrized notion of a community, called *k-linked community*, allowing us to characterize node/edge centered k-linked community with bounded diameter. Such community admits a node or an edge with a distance at most $\frac{k}{2}$ from any other node of that community. Next, we show how the problem of detecting node/edge centered k-linked overlapping communities can be expressed as a Partial Max-SAT optimization problem. Then, we propose a post-processing strategy to limit the overlaps between communities. An extensive experimental evaluation on real-world networks shows that our approach outperforms several popular algorithms in detecting relevant communities.

1 Introduction

Many complex interactions can be represented by networks, which are set of nodes connected by edges. Such connections might represent different type of relations between individuals or entities. Nodes in networks can be organized into *communities*, which often correspond to groups of nodes that share common properties, roles or functionnalities, such as functionally related proteins, social communities, or topically related webpages.

One of the most important task when studying networks is that of identifying communities. Indeed, detecting and analyzing communities is of great interest in several application domains, including clustering web clients who have similar interests, identifying clusters of customers in the network of customers-products purchase relationships of online retailers (e.g. Amazon), etc. Several efficient algorithm for discovering communities in complex networks have been proposed. Let us mention for example, the most popular algorithm based on non-negative matrix factorisation [12], the spectral clustering methods [17], the edge betweenness based approach [8], and the seed set expansion algorithm [21]. Some of them recuire several parameters such as the number of expected communities [12,17], while others involve for example the computation of the shortest paths between pairs of nodes [8].

In this paper, we introduce a parametrized notion of communities, called k-linked community, allowing us to characterize node/edge centered k-linked

J. Kim et al. (Eds.): PAKDD 2017, Part II, LNAI 10235, pp. 786–798, 2017.
DOI: 10.1007/978-3-319-57529-2_61

community admitting a node or an edge with a distance at most $\frac{k}{2}$ from any other node of the community. This can be seen as a way to look for communities of bounded diameter. Our approach is only dependent on this single parameter k, and does not require any other knowledge about the network or about the number of expected communities.

Our proposed overlapping communities detection framework is based on an appropriate encoding of the centered k-linked community detection task as a partial maximum satisfiability (Partial Max-SAT) optimisation problem. It allows us to benefit from the recent advances in propositional satisfiability and its optimisation variants. Finally, we propose a post-processing strategy to limit the overlaps between communities. Our proposed framework follows the recent data mining research trend exploiting two powerful declarative models, namely constraint programming and propositional satisfiability. Indeed, several data mining tasks including pattern mining [10] and clustering [7] have been modeled and solved using these two well-known declarative and flexible models.

2 Formal Preliminaries

2.1 Propositional Logic and SAT Problem

Let $\mathcal{L}$ be a propositional language defined inductively from a finite set $\mathcal{PS}$ of propositional symbols, the boolean constants $\top$ (*true* or 1) and $\bot$ (*false* or 0) and the standard logical connectives $\{\neg, \wedge, \vee, \rightarrow, \leftrightarrow\}$ in the usual way. We use the letters x, y, z, etc. to range over the elements of $\mathcal{PS}$. Formulas of $\mathcal{L}$ are denoted by A, B, C, etc. A *literal* is a propositional variable (x) of $\mathcal{PS}$ or the negation of a variable ($\neg x$). The two literals x and $\neg x$ are called complementary. A *clause* is a (finite) disjunction of literals, i.e., $a_1 \vee \ldots \vee a_n$. For every propositional formula $\mathcal{A}$ from $\mathcal{L}$, $\mathcal{P}(\mathcal{A})$ denotes the symbols of $\mathcal{PS}$ occurring in $\mathcal{A}$. A *Boolean interpretation* $\mathcal{I}$ of a formula $\mathcal{A}$ is a truth assignment of $\mathcal{PS}$, that is, a total function from $\mathcal{P}(\mathcal{A})$ to $\{0, 1\}$. A *model* of a formula $\mathcal{A}$ is a Boolean interpretation $\mathcal{I}$ that satisfies $\mathcal{A}$, i.e. $\mathcal{I}(\mathcal{A}) = 1$. A formula $\mathcal{A}$ is satisfiable if there exists a model of $\mathcal{A}$. We denote by $\mathcal{M}(\mathcal{A})$ is the set of all models of $\mathcal{A}$.

As usual, every finite set of formulas is considered as the conjunctive formula whose conjuncts are the elements of the set. A formula in *conjunctive normal form* (CNF) is a (finite) conjunction of clauses. The SAT problem consists in deciding whether a given CNF formula admits a model or not. This well-known NP-Complete problem has seen spectacular progress these recent years.

SAT has seen many successful applications in various fields such as electronic design automation, debugging of hardware designs, artificial intelligence, and data mining. Several SAT extensions have been proposed to deal with optimisation problems. For example, the Max-SAT Problem seeks the maximum number of clauses that can be satisfied. In this paper, we consider one of these optimisation variants referred to as Partial Max-SAT problem. Partial Max-SAT sits between SAT and Max-SAT problems. While SAT requires all clauses to be satisfied, Partial Max-SAT relaxes this requirement by considering two kind of clauses, hard and soft. Partial MaxSAT is the problem of finding an optimal assignment to the variables that satisfies all the hard clauses, while satisfying the maximum number of soft clauses.

2.2 Overlapping Community Detection

In this subsection, we discuss the classic problem of detecting overlapping community structure in networks.

A network is an undirected graph $\mathcal{N} = (V, E)$ where V is a set of nodes and $E \subseteq V \times V$ is a set of edges. We denote by n (respectively m) the number of nodes (respectively edges) in $\mathcal{N}$. The *degree* of a node $u \in V$, denoted d_u, is the number of edges connected to it. The length of the shortest path between two nodes $u, v \in V$ is called the *distance* between the nodes, noted $dist(u, v)$. Given an edge $e = (u, v) \in E$ and a node $w \in V$, the distance between e and w is defined as $dist(e, w) = \min\{dist(u, w), dist(v, w)\}$. In graph theory, a *community* is described as a set of nodes densely connected internally. In real-world networks, nodes are organized into densely linked sets of nodes that are commonly referred to as *network communities*, clusters or modules. Notice that communities in networks often overlap as nodes can belong to multiple communities at once. Network *overlapping community detection* problem consists in dividing a network of interest into (overlapping) communities for intelligent analysis. It has recently attracted significant attention in diverse application domains. Identifying the community structure is crucial for understanding structural properties of the real-world networks. Various methods have been proposed to identify the community structure of complex networks (see [6,15] for an overview).

Quality Metrics: Several measures have been proposed for quantifying the quality of communities in networks (see [13] for a comparative study of quality measures). In this paper, we adopt two popular metrics to assess the performance of our method:

Modularity. The most widely used metric for measuring the quality of network's partition into communities (without a ground-truth) is Newman's *modularity* function [18]. Modularity quantifies the community strength by comparing the fraction of edges within the community with such fraction when random connections between the nodes are made. Networks with high modularity have dense connections between the nodes within communities but sparse connections between nodes in different communities. We use the following equation of modularity, an extension of Newman's modularity function designed to support overlapping communities proposed in [19]. For the given community partition of a network $\mathcal{N} = (V, E)$ with m edges, an extended modularity EQ is given by:

$$EQ = \frac{1}{2m} \sum_{C \in C_{\mathcal{N}}} \sum_{u,v \in C} \frac{1}{O_u O_v} \left[A_{uv} - \frac{d_u d_v}{2m} \right] \quad (1)$$

with $C_{\mathcal{N}}$ the set of communities in $\mathcal{N}$; O_u the number of communities to which the node u belongs and A_{uv} is the element of the adjacency matrix representing the network.

F1 Score. Let $\mathcal{N} = (V, E)$ be a network, and $\hat{C}$ (respectively C^*) the set of (respectively ground truth) communities associated to $\mathcal{N}$. The average F1 score measure aims to quantify the level of correspondence between C^* and $\hat{C}$. More precisely, we need to determine which $C_i \in C^*$ corresponds to which $\hat{C}_i \in \hat{C}$. The F1 score is defined as the average of F1 score of the best matching ground-truth community to each detected community, and the F1 score of the best matching detected community to each ground-truth community [24]. More formally:

$$\frac{1}{2}\left(\frac{1}{|C^*|}\sum_{C_i \in C^*} F_1(C_i, \hat{C}_{g(i)}) + \frac{1}{|\hat{C}|}\sum_{\hat{C}_i \in \hat{C}} F_1(C_{g(i)}, \hat{C}_i)\right) \tag{2}$$

where the best matching g and g' is defined as follows: $g(i) = \arg\max_j F_1(C_i, \hat{C}_j)$, $g'(i) = \arg\max_j F_1(C_j, \hat{C}_i)$, and $F_1(C_i, \hat{C}_j)$ is the harmonic mean of Precision and Recall.

3 A SAT-Based Framework for Community Detection

Fundamentally, communities allow us to discover groups of interacting objects and the relations between them. A community (also referred to as a cluster) is a set of cohesive nodes that have more connections inside the set than outside. In this section, we propose an appropriate encoding of the community detection task as a SAT optimization problem. Proximity between nodes have been expressed as direct edges expressing formally a direct relation. Individuals can be grouped into the same cluster even if they are not linked directly. Relationships between individuals can be expressed via some proximity conditions. For instance, individuals having much common friends could be considered as very closed to each other. Consequently, the definition of individuals proximity is clearly a fundamental issue, as it have a great impact on the outcome. Next, we establish the main definitions which will be used to formulate our problem.

Definition 1 (k-linked community). *A community is k-linked if the nodes are pairwise k-linked, i.e., the distance between each two nodes is less or equal than k.*

According to Definition 1, a k-linked community has a diameter less or equal than k. Now, to simplify the encoding of the problem of discovering overlapping communities, we focus on the following kinds of k-linked communities called k-linked *centered* communities: those having a centroid node or centroid edge that possesses a distance at most $\frac{k}{2}$ from each other node of the community.

Definition 2 (Node/Edge Centered k-linked Community). *Let $\mathcal{N} = (V, E)$ be a network and $k > 1$ a positive integer. A community $C \subseteq V$ is node (resp. edge) centered k-linked community of $\mathcal{N}$ iff there exists $c \in C$ (resp. $e = (u, v) \in E$ with $u, v \in C$) s.t. $\forall\, w \in C$, $dist(c, w) \leqslant \frac{k}{2}$ (resp. $dist(e, w) \leqslant \frac{k}{2}$).*

Obviously, a node centered k-linked community is an edge centered k-linked community, while the converse is not true. Note also that a k-linked community is not necessarily a centered k-linked community. A counter-example consists of the network $\mathcal{N} = (V, E)$ where $V = \{1, 2, \ldots, 8\}$ and $E = \{(1,2), (2,3), (3,4), (5,6), (6,7), (7,8), (1,5), (4,8)\}$. Then, $C = V$ is a 4-linked community, while there is neither a node $v \in V$ ($v \in \{1, \ldots, 8\}$) nor edge $e \in E$ with distance at most 2 from all the remaining nodes of C.

Lemma 1. *Let $\mathcal{N} = (V, E)$ be a network, $C \subseteq V$ a community and an integer $k > 1$. If C is a centered k-linked community, then C is also a k-linked community.*

Now, based on the notion of centered k-linked community, community detection is defined as an optimization problem, solving Partial Max-SAT. To do so, our starting point is to find a set of centroids S in the given network. The next step is to formed the communities around the centroids based on a predefined parameter k which represents the diameter of the communities. Clearly, we distinguish the following two cases: k-linked node (resp. edge) centered communities corresponding to an even (resp. odd) value of k.

Next, we propose two appropriate reformulations as an optimization problem for the community detection problem corresponding to node and edge centered k-linked communities, respectively. To achieve this, propositional variables are used for representing the network. Indeed, we associate each node u (resp. edge e) with a propositional variable denoted x_u (resp. y_e) where $x_u, y_e \in \{0, 1\}$. The key idea is that the variables assigned to 1 represent the centroids nodes (resp. edges), i.e., $S_v = \{u \in V \mid \mathcal{I}(x_u) = 1\}$ (resp. $S_e = \{e \in E \mid \mathcal{I}(y_e) = 1\}$). We now describe our SAT-based encodings using such propositional variables.

Node Centered k-linked Community: Our encoding consists of a set of constraints. The first propositional formula expresses the fact that if a node u is a centroid ($\mathcal{I}(x_u) = 1$), then the nodes with a distance at most $\frac{k}{2}$ from u are placed to the same community that possesses u as a centroid.

$$\bigwedge_{u \in V} (x_u \rightarrow \bigwedge_{v \in V \mid dist(u,v) \leqslant \frac{k}{2}} \neg x_v) \tag{3}$$

Let us remark that constraint (3) can be expressed by a set of binary clauses:

$$\bigwedge_{u \in V} \bigwedge_{v \in V \mid dist(u,v) \leqslant \frac{k}{2}} (\neg x_u \vee \neg x_v)$$

After finding the centroids, we still have to determine whether a node u belongs to community C or not depending on the value of k. To achieve this, we use the following formula that affects nodes of the network to communities where they belong to, i.e., nodes that have a distance at most of $\frac{k}{2}$ from the centroid.

$$\bigwedge_{u \in V} \bigvee_{v \in V \mid dist(u,v) \leqslant \frac{k}{2}} x_v \tag{4}$$

Proposition 1. *If the constraints (3) $\wedge$ (4) are satisfied, then for all $u \notin S_v$ there exists $v \in S_v$ s.t. $dist(u,v) \leqslant \frac{k}{2}$.*

Proposition 1 ensures that if (3) $\wedge$ (4) admits a model $\mathcal{I}$, then the nodes corresponding to the variables assigned to 1 ($\{u \in V \mid \mathcal{I}(x_u) = 1\}$) are the centroids and the network can be partitioned into $|S|$ communities. The communities can then be constructed by finding the nodes with a distance at most $\frac{k}{2}$ from each centroid. Obviously, the formula (3) $\wedge$ (4) may admits many candidate solutions (i.e. models). However, choosing an arbitrary model do not always guarantee a best partition of the network into communities. To alleviate this problem, we will consider an objective function to optimize over the space of solutions. Then, the node centered k-linked community detection problem can be formulated as the following optimisation problem:

$$\min/\max \sum_{u \in V} x_u \qquad \text{subject to (3)} \wedge \text{(4)} \tag{5}$$

Edge Centered k-linked Community: Now, to derive the formulation of edge centered k-linked community detection problem, we use similar reasoning as for node centered k-linked community, except that we consider centroid edges instead of centroid nodes. To do so, a community is built around an edge $e = (u,v)$ by considering nodes with a distance at most $\frac{k}{2}$ from the edge e. This is equivalent to partition the set of edges into modules and from that modules we can deduce the set of communities of nodes.

In the same way as for centroid nodes, the following formula expresses the fact that if an edge $e = (u,v)$ is a centroid edge ($\mathcal{I}(y_e) = 1$), then the nodes with a distance at most $\frac{k}{2}$ from u or v are assigned to 0.

$$\bigwedge_{e=(u,v)\in E} (y_e \rightarrow \bigwedge_{e' \in E \mid dist(e',u) \leqslant \frac{k}{2} \mid\mid dist(e',v) \leqslant \frac{k}{2}} \neg y_{e'}) \tag{6}$$

Let us now introduce the following formula that affects nodes of the network to their associated communities, i.e. nodes that have a distance of $\frac{k}{2}$ from the centroid edge e.

$$\bigwedge_{e=(u,v)\in E} \quad \bigvee_{e' \in E \mid dist(e',u) \leqslant \frac{k}{2} \mid\mid dist(e',v) \leqslant \frac{k}{2}} y_{e'} \tag{7}$$

After fixing the centroids edges, the constraint 7 allows to identify whether a node u belongs to a community C or not from the value of k.

Similarly, to improve the quality of the detected communities, our edge centered k-linked community detection problem is formulated as the following optimisation problem:

$$\min/\max \sum_{e \in E} y_e \qquad \text{subject to (6)} \wedge \text{(7)} \tag{8}$$

We will use the notation $\text{CDSAT}^k_{\min/\max}$ to denote the optimization problems (5) and (8).

Example 1. Let us consider the undirected network $\mathcal{N} = (V, E)$ depicted in Fig. 1. Setting $k = 4$ can lead to the following solution of $\text{CDSAT}^4_{\max}$: $\mathcal{I} = \{\neg x_1, \neg x_2, \neg x_3, \neg x_4, \neg x_5, x_6, \neg x_7, x_8, \neg x_9, \neg x_{10}, \neg x_{11}\}$. So for that solution, $\mathcal{N}$ can be partitioned into the two communities $C_1 = \{1, \ldots, 6, 7, 11\}$ and $C_2 = \{1, 2, 5, 6, 7, \ldots, 11\}$. In contrast, $\text{CDSAT}^4_{\min}$ leads to one community with centroid x_1 and containing all the nodes of $\mathcal{N}$.

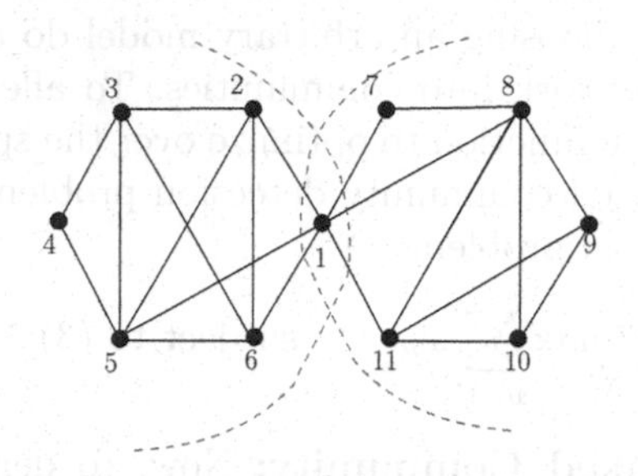

Fig. 1. A simple undirected network

Overlapping Enhancement: As said before, once the node/edge centroids are found, the communities are formed arround them based on a predefined parameter k. As a result, some nodes can belong to multiple communities as illustrated in Example 1. However, such overlapping can be huge and not significant enough w.r.t. real communities. To overcome this drawback and to allow for an accurate partition of the network, we propose a simple but effective overlaps reduction technique in order to correctly identify dense community overlaps. Starting from a set of communities, each overlapping node will be assigned to its closest communities according to its distance from the centroids of these communities.

Example 2. Let us consider again the network $\mathcal{N} = (V, E)$ of Fig. 1. By enhancing the overlapping, the two communities are reduced to $C_1 = \{1, \ldots, 6\}$ and $C_2 = \{1, 7, \ldots, 11\}$.

Algorithm 1 describes the general feature of our SAT-based node centered k-linked community detection procedure[1]. The algorithm takes as input the network and even integer k and returns a set of overlapping communities. It proceeds as follows: First, we generate the corresponding optimization problem that can be represented as a Partial MaxSAT problem (line 1). Then, a state-of-the-art Weighted Partial MaxSAT solver `WPM3` is used to get an optimal solution (i.e. model) $\mathcal{I}$. Next, the centroids are determined from the obtained model (lines 4–7). Using such centroids, the next step is to build communities by finding the nodes with a distance at most $\frac{k}{2}$ from each centroid. Finally, the cleaning step is called to improve the quality of detected communities (lines 11–13).

[1] Algorithm 1 can be slightly modified to deal with edge centered k-linked community detection problem.

Algorithm 1. CDSAT$^k_{\min/\max}$

```
Input: A network N = (V, E) and an integer k > 1
Output: A set of overlapping communities S
1  Φ = encodeToOpt(k, G);
2  I = solve(Φ) ;
3  S ← ∅;
4  for u_x ∈ I do
5    if I(u_x) == 1 then
6      C_u ← {u};
7      S ← S ∪ C_u
8    end
9  end
10 for v_x ∈ I do
11   for C_u ∈ S do
12     if dist(u, v) ⩽ k/2 then C_u ← C_u ∪ {v}
13   end
14 end
15 for C_u, C_v ∈ S × S do
16   for w ∈ V do
17     if dist(w, u) < dist(w, v) then C_v ← C_v \ {w}
18   end
19 end
20 return S
```

4 Performance Evaluation

4.1 Experiment Settings

In this section, we present an experimental evaluation of our proposed approach. It was conducted on fourteen networks that cover a variety of application areas and are briefly described in Table 1 (columns 1 and 2). Some of these networks have ground-truth communities as presented in column 2 of Table 2. We have also chosen three large networks (`Facebook`, `DBLP`, and `Amazon` taken from SNAP [14]) to show the scalability of our model.

We evaluate the performance of our approaches by comparing them with the following most prominent state-of-the-art overlapping community detection algorithms: (i) *Community-Affiliation Graph Model* (`AGM`) [23], (ii) *Clique Percolation Method* (`CPM`) [1], (iii) *Cluster Affiliation Model for Big Networks* (`BIGCLAM`) [24], and (iv) *Communities from Edge Structure and Node Attributes* (`CESNA`) [25]. For the `CPM` algorithm, we use the cliques of size equal to 3. For `BIGCLAM` method, user can specify the number of communities to detect, or let the program determine the number of communities from the topology of the network. We opt for the case where the number of communities is not fixed in advance.

The proposed system, referred to as CDSAT$^k_{\min/\max}$, was written in Python. Given an input network as a set of edges, our algorithm starts by generating the corresponding optimization problem represented as a Partial MaxSAT problem. To solve this problem, we consider the state-of-the-art Weighted Partial MaxSAT solver `WPM3` (best solver at the last MaxSAT competition[2]) [2]. As finding the

[2] http://maxsat.ia.udl.cat/introduction/.

optimal solution is NP-hard, in our experiment, we consider the first solution (not necessarily optimal) returned by the solver WPM3. For our experimental study, all algorithms have been run on a PC with an Intel Core 2 Duo (2 GHz) processor and 2 GB memory. We imposed 1 h time limit for all the methods. Last, we use the symbol (–) in Tables 1 and 2 to indicate that the method is not able to scale on the considered network under the time limit.

4.2 Choosing the Best Value of the Diameter

Our $\text{CDSAT}^k_{\text{min/max}}$ algorithms take as input a network and a positive integer k and return a set of overlapping communities. In order to determine the best diameter k, we run $\text{CDSAT}^k_{\text{min/max}}$ on the fourteen considered networks, while varying k from 3 to 6. The Fig. 2 summarises the relationship between the average modularity and k. As Fig. 2 reveals, the best average modularity is obtained by $\text{CDSAT}^4_{\text{min}}$ and $\text{CDSAT}^4_{\text{max}}$ with a value of 0.421 and 0.432, respectively. We also observe that the average modularity obtained by both algorithms decreases beyond $k = 4$. Overall, for both algorithms the best average modularity is obtained for $k = 4$. These performances are relatively close. This can be explained by the fact that real-world social networks possess small (average or effective) diameters (e.g. [5]). This can be related to the property of the small-world phenomenon observed by several authors on real graphs (e.g. [20]).

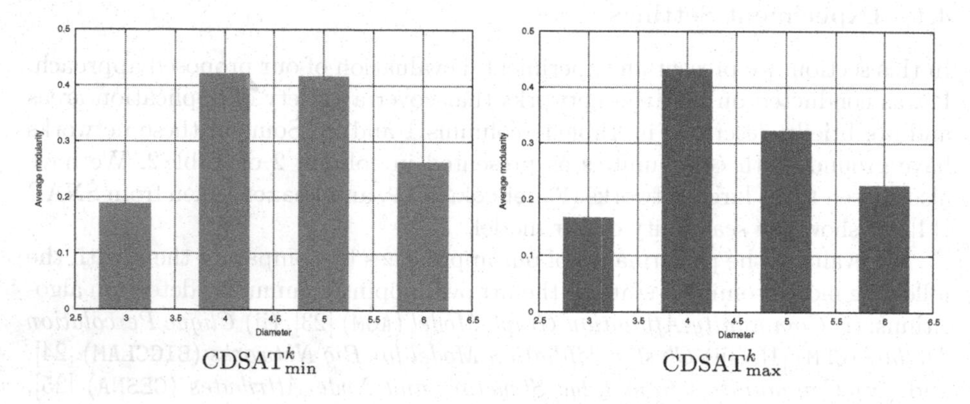

Fig. 2. Average modularity for $\text{CDSAT}^k_{\text{min/max}}$

4.3 Comparison with Baseline Algorithms

Results on Modularity Metric. Table 1 reports the performance comparison between our $\text{CDSAT}^4_{\text{min/max}}$ approaches and the considered methods. Experiments show that our methods outperform every baseline, in most cases, by an interesting margin as shown by the average modularity reported in the last line of Table 1. We observe that across all datasets and modularity metric, $\text{CDSAT}^4_{\text{min}}$ yields the best performance in 8 out of 14 networks. We also note that $CDSAT^4_{min}$ shows a high margin in performance gain against the baselines in two large networks DBLP and Amazon, and in a collaborations network such as Coauthorship. In terms of average performance, $CDSAT^4_{min}$ outperforms CPM by 111.55%, BIGCLAM by 26.42%, and CESNA by 40.80%. Similarly, we note that $CDSAT^4_{max}$ outperforms all the other methods in 7 out of 14 datasets. In terms of average performance, $CDSAT^4_{max}$ outperforms CPM by 117.08%, BIGCLAM by 29.72%, and CESNA by 44.48%. We also observe that $CDSAT^4_{max}$ gives an important improvement against the baselines in two large networks Facebook, DBLP, and also in a collaborations network like Coauthorship. On the Lemis, Power grid, Pilgrim, and Jazz datasets, our methods remain relatively competitive with the best baseline. A possible explanation for this phenomenon is that the WPM3 solver don't return the optimal solution for these datasets. Overall, our methods outperform BIGCLAM, which is the most competing algorithm, on all large real datasets.

Table 1. Modularity based performance on fourteen datasets

Networks	Nodes/Edges	AGM	CPM	BIGCLAM	CESNA	$CDSAT^4_{min}$	$CDSAT^4_{max}$
Dolphin [16]	62/159	−0.040	0.304	0.053	0.095	**0.438**	0.297
Karate [22]	34/78	0.200	0.230	0.195	0.180	**0.310**	**0.311**
Risk map [4]	42/83	0.415	0.488	0.194	0.504	**0.571**	**0.528**
Lemis [11]	77/254	0.162	0.205	**0.444**	0.311	0.064	0.419
Word-adj [17]	112/425	0.139	0.031	0.154	0.111	**0.175**	0.098
Football [8]	115/615	0.222	0.199	0.343	0.390	0.286	**0.404**
Facebook [14]	4039/88234	–	–	0.391	0.539	0.449	**0.701**
DBLP [14]	317080/1049866	–	0.293	0.216	0.202	**0.520**	**0.436**
Amazon [14]	334863/925872	–	0.195	0.341	0.430	**0.616**	**0.502**
Books [8]	105/441	0.366	0.265	0.308	0.255	**0.439**	0.345
Power grid [20]	4941/6594	–	0.007	**0.840**	0.586	0.679	0.547
Coauthership [17]	1462/2742	0.619	0.456	0.679	0.031	**0.923**	**0.852**
Pilgrim [3]	34/128	0.368	0.096	**0.415**	0.321	0.312	0.407
Jazz [9]	196/2742	**0.310**	0.022	0.099	0.231	0.112	0.208
Average	N/A	N/A	0.199	0.333	0.299	**0.421**	**0.432**

Results on Ground-Truth Communities. After finding communities in a given network, we can gauge the performance of each community that an algorithm has discovered and whether a ground-truth community has been successfully identified. Table 2 summarizes the evaluation results, with F1 scores of all algorithms on each network. Interestingly, it can be seen that $CDSAT^4_{min}$ and $CDSAT^4_{max}$ produce more accurate average w.r.t. the ground-truth setting than all the other baseline algorithms. In terms of average performance, $CDSAT^4_{min}$ outperforms CPM by 16%, BIGCLAM by 22%, and CESNA by 35.05%. Moreover, notice that $CDSAT^4_{max}$ outperforms CPM by 6.49%, BIGCLAM by 12%, and CESNA by 23.98%. In the cases of Karate, Risk map and DBLP data instances, $CDSAT^4_{min}$ and $CDSAT^4_{max}$ achieves a closely gain in the F1 score compared to the best baseline (CPM in this case).

Table 2. F1 Score using ground truth

Networks	Communities	AGM	CPM	BIGCLAM	CESNA	$CDSAT^4_{min}$	$CDSAT^4_{max}$
Dolphin	2	0.120	0.579	0.628	0.100	**0.749**	**0.659**
Karate	2	0.864	**0.857**	0.629	0.663	0.847	0.851
Risk map	6	0.641	**0.884**	0.694	0.842	0.779	0.769
DBLP	13477	–	**0.596**	0.370	0.310	0.470	0.483
Amazon	75149	–	0.519	0.498	0.642	**0.695**	0.399
Books	3	0.684	0.557	0.549	0.591	**0.804**	0.652
Pilgrim	4	0.773	0.427	0.835	0.652	0.785	**0.892**
Average	N/A	N/A	0.631	0.600	0.542	**0.732**	**0.672**

As a summary, experimental results confirm that $CDSAT^4_{min/max}$ methods achieve the overall best performance in terms of the accuracy of the detected overlapping communities.

Evaluating Scalability. Finally, we evaluate the scalability of the different community detection methods by measuring the CPU time (see Table 3). From the results, it can be seen that our algorithms make few seconds to generate all communities for small networks. However, the CPM, BIGCLAM and CESNA baselines are faster than our methods for small networks (up to 200 nodes). We can observe that $CDSAT^4_{min}$ and $CDSAT^4_{max}$ are third-fastest method overall, when the network becomes larger. Interestingly, we also notice that our algorithms are the second-fastest methods, next BIGCLAM, for DBLP and Amazon.

Table 3. Comparison in terms of running Time (s)

Networks	AGM	CPM	BiGCLAM	CESNA	$CDSAT^4_{max}$	$CDSAT^4_{min}$
Dolphin	6.77	0.09	0.24	0.07	14	8
Karate	35	0.07	0.29	0.07	11	7.15
Risk map	62	0.09	2.84	0.59	38	17
Lemis	200	0.10	0.55	0.09	16	12
Word-adj	60.35	0.09	0.97	0.13	60.60	11
Football	47.71	0.08	1.78	0.13	120.20	420
Facebook	>1 h	>1 h	240.38	4.81	360.30	480.7
DBLP	>1 h	3240	60.56	900.34	720.50	780.40
Amazon	>1 h	>1 h	60.09	1200.49	780.20	900
Books	19.83	0.12	2.71	0.10	14.35	60.20
Power grid	>1 h	0.66	0.81	4.48	420.15	480.25
Coauthership	360.17	0.07	14.08	0.05	360.58	360.20
Pilgrim	0.61	0.09	0.35	0.07	8.2	7
Jazz	60.02	0.09	2.84	0.59	360.12	120

5 Conclusion

In this paper, we developed a new framework for detecting overlapping community structure of real-world networks. Our method is based on a partition of the network into modules with bounded diameters. We have shown that the problem of centered k-linked community detection can be expressed as a Partial Max-SAT optimization problem. Experimental results showed that our approach outperforms the state-of-the-art methods in accurately discovering network communities. These performances are obtained while looking for the first non necessarily optimal solution of the underlying optimisation problem. As a future work, we intend to develop a parallel version to even improve the performance of our optimisation based approach. We also plan to extend our proposed framework to deal with dynamic community detection.

References

1. Adamcsek, B., Palla, G., Farkas, I.J., Derényi, I., Vicsek, T.: Cfinder: locating cliques and overlapping modules in biological networks. Bioinformatics **22**(8), 1021–1023 (2006)
2. Ansótegui, C., Didier, F., Gabàs, J.: Exploiting the structure of unsatisfiable cores in MaxSAT. In: IJCAI, pp. 283–289 (2015)
3. Dickinson, B., Valyou, B., Hu, W.: A genetic algorithm for identifying overlapping communities in social networks using an optimized search space. Soc. Networking **2**(4), 1–9 (2013)
4. Cheng, J., Leng, M., Li, L., Zhou, H., Chen, X.: Active semi-supervised community detection based on must-link and cannot-link constraints. PLoS **9**(10), 1–18 (2014)
5. Comellas, F., Ozón, J., Peters, J.G.: Deterministic small-world communication networks. Inf. Process. Lett. **76**(1), 83–90 (2000)
6. Fortunato, S.: Community detection in graphs. CoRR, abs/0906.0612 (2009)
7. Gilpin, S., Davidson, I.N.: Incorporating SAT solvers into hierarchical clustering algorithms: an efficient and flexible approach. In: KDD, pp. 1136–1144 (2011)
8. Girvan, M., Newman, M.E.J.: Community structure in social and biological networks. Proc. Natl. Acad. Sci. **99**, 7821 (2002)
9. Gleiser, P., Danon, L.: Community structure in jazz. Adv. Complex Syst. **6**, 565 (2003)
10. Guns, T., Nijssen, S., Raedt, L.D.: Itemset mining: a constraint programming perspective. Artif. Intell. **175**(12–13), 1951–1983 (2011)
11. Knuth, D.E.: The Stanford GraphBase - A Platform for Combinatorial Computing. ACM, New York (1993)
12. Lee, D.D., Seung, H.S.: Algorithms for non-negative matrix factorization. In: Advances in Neural Information Processing Systems, vol. 13, pp. 556–562 (2001)
13. Leskovec, J., Huttenlocher, D.P., Kleinberg, J.M.: Predicting positive and negative links in online social networks. In: WWW, pp. 641–650 (2010)
14. Leskovec, J., Krevl, A., Datasets, S.: Stanford large network dataset collection, June 2014. http://snap.stanford.edu/data
15. Leskovec, J., Lang, K.J., Mahoney, M.W.: Empirical comparison of algorithms for network community detection. In: WWW, pp. 631–640 (2010)

16. Lusseau, D., Schneider, K., Boisseau, O., Haase, P., Slooten, E., Dawson, S.: The bottlenose dolphin community of Doubtful Sound features a large proportion of long-lasting associations. Behav. Ecol. Sociobiol. **54**(4), 396–405 (2003)
17. Newman, M.E.J.: Finding community structure in networks using the eigenvectors of matrices. Phys. Rev. E **74**, 036104 (2006)
18. Newman, M.E.J., Girvan, M.: Finding and evaluating community structure in networks. Phys. Rev. E **69**(2), 026113 (2004)
19. Shen, H., Cheng, X., Cai, K., Hu, M.: Detect overlapping and hierarchical community structure in networks. Phys. A **388**(8), 1706–1712 (2009)
20. Watts, D.J., Strogatz, S.H.: Collective dynamics of small-world networks. Nature **393**(6684), 440–442 (1998)
21. Whang, J.J., Gleich, D.F., Dhillon, I.S.: Overlapping community detection using neighborhood-inflated seed expansion. TKDE **28**(5), 1272–1284 (2016)
22. Zachary, W.W.: An information flow model for conflict and fission in small groups. J. Anthropol. Res. **33**, 452–473 (1977)
23. Yang, J., Leskovec, J.: Community-affiliation graph model for overlapping network community detection. In: ICDM, pp. 1170–1175 (2012)
24. Yang, J., Leskovec, J.: Overlapping community detection at scale: a nonnegative matrix factorization approach. In: WSDM, pp. 587–596 (2013)
25. Yang, J., McAuley, J.J., Leskovec, J.: Community detection in networks with node attributes. In: ICDM, pp. 1151–1156 (2013)

Dimensionality Reduction

Denoising Autoencoder as an Effective Dimensionality Reduction and Clustering of Text Data

Milad Leyli-Abadi(✉), Lazhar Labiod, and Mohamed Nadif

LIPADE, Paris Descartes University, 75006 Paris, France
mleyliabadi@gmail.com, {lazhar.labiod,mohamed.nadif}@parisdescartes.fr

Abstract. Deep learning methods are widely used in vision and face recognition, however there is a real lack of application of such methods in the field of text data. In this context, the data is often represented by a sparse high dimensional document-term matrix. Dealing with such data matrices, we present, in this paper, a new denoising auto-encoder for dimensionality reduction, where each document is not only affected by its own information, but also affected by the information from its neighbors according to the cosine similarity measure. It turns out that the proposed auto-encoder can discover the low dimensional embeddings, and as a result reveal the underlying effective manifold structure. The visual representation of these embeddings suggests the suitability of performing the clustering on the set of documents relying on the Expectation-Maximization algorithm for Gaussian mixture models. On real-world datasets, the relevance of the presented auto-encoder in the visualisation and document clustering field is shown by a comparison with five widely used unsupervised dimensionality reduction methods including the classic auto-encoder.

Keywords: Auto-encoder · Deep learning · Cosine similarity · Neighborhood · Document clustering · Unsupervised learning · Dimensionality reduction

1 Introduction

Analyzing sparse high-dimensional point clouds is a classical challenge in visualization. Principal component analysis (PCA), one of the traditional techniques, is certainly the best known. More efficient in nonlinear cases, a number of techniques have been proposed, including Isometric Feature Mapping (Isomap), Locally Linear Embedding (LLE), and Stochastic Neighbor Embedding (SNE). Nevertheless these nonlinear techniques tend to be extremely sensitive to noise, sample size, choice of neighborhood and other parameters (for details see for instance [1]). On the other hand, t-SNE [2] and its parametric version [3] is better than existing techniques at creating a single map that reveals structure at many different scales. Parametric t-SNE learns the parametric mapping in such

J. Kim et al. (Eds.): PAKDD 2017, Part II, LNAI 10235, pp. 801–813, 2017.
DOI: 10.1007/978-3-319-57529-2_62

a way that the local structure of the data is preserved as well as possible in the latent space. Generally, it works better in the case of image datasets but it is very dependent on the adjustments of the hyper parameters, e.g. learning rate noted η. Laplacian Eigenmap (LE) [4] is another interesting method where the laplacian graph is used and has relatively the same objective as t-SNE, i.e. preserving the local structure of data.

The auto-encoders, a special method of deep learning architecture, have received more attention recently for dimensionality reduction tasks; their abilities to adapt to different domains are promising. They make it possible to embed the high dimensional data in a latent space of lower dimensionality while preserving the original structure of the data. In its traditional version, each data point is used to reconstruct itself from the code layer. If we have the same number of neurons in the code layer as in the input layer, the method learns the identity function. In order to avoid this trivial solution, there are many different approaches. The two most used consist in (1) using fewer number of neurons in the code layer so as to force the auto-encoder to compress the features in a lower space, (2) introducing some noise to data, for instance with a Gaussian noise applied to the whole data or randomly replacing with zeros a percentage of data entries. It is proved that some denoising auto-encoders (DAEs) correspond to a Gaussian RBM (Restricted Boltzmann Machine) in which minimizing the denoising reconstruction error estimates the energy function [5,6]. They generally give better results in comparison to classic auto-encoders without any denoising step. We make use of the former type of auto-encoders in the following. In this paper, we concentrate on the case of sparse high dimensional data and in particular on document-term matrices. The cells of such matrices contain the frequency counts of the terms belonging to the corresponding documents. We known that the auto-encoders aim to find the low dimensional embeddings in data by preserving the structure of the data as well as possible. Herein, with the proposed auto-encoder, we aim to capture the relations among documents while preserving the original structure. Therefore, the proposed method focuses on the dimensionality reduction and the main contributions of the paper, presented schematically in Fig. 1, are as follows:

- we propose a suitable normalization of document-term matrices;
- we introduce a weighted criterion where the weights rely on cosine similarity, and derive an appropriate autoencoder able to effectively reduce the dimension;
- finally in order to cluster the set of documents, we perform the Expectation-Maximization [7] algorithm for Gaussian mixture models on the reduced space instead of the k-means algorithm which is commonly used, and assess the number of clusters relying on the *Bayes Information Criterion* [8].

The rest of the paper is organised as follows. In Sect. 2, we first introduce different types of preprocessing that are mandatory in order to get the auto-encoders work and then the role of the denoising procedure is described. Section 3, is devoted to the introduction of the mathematical formulation of the proposed auto-encoder. The experimental results on different text datasets and

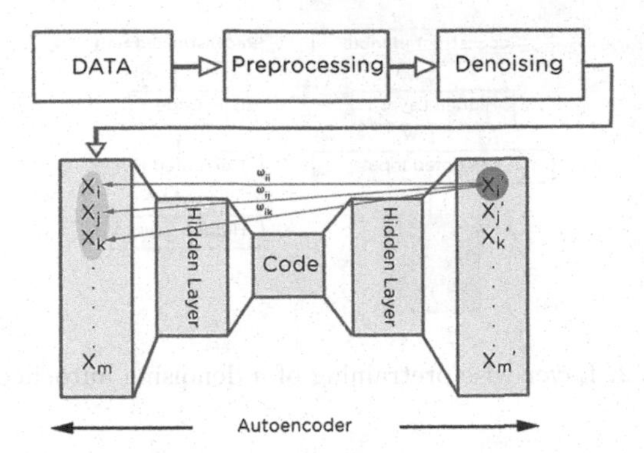

Fig. 1. Proposed method scheme (Color figure online)

clustering are presented in Sect. 4. Section 5 concludes the paper and presents directions for future research.

2 Data Pre-processing

Let $\mathbf{x}_1, \ldots, \mathbf{x}_m$ be a set of m objects where $\mathbf{x}_i$ is a n-dimensional vector on $\mathbb{R}^n$. In practice, $\mathbf{x}_i$ contains the variables corresponding to p measurements made on the ith recording of some features on the phenomenon under study. Then data will be denoted by an m by n data matrix $\mathbf{x} = (x_{ij})$. Before applying of any clustering or dimensionality reduction algorithm, a preprocessing step is necessary in order to reduce the effect of the outliers and prepare the data for a better and more faithful analysis. In the context of dimensionality reduction of document-term data, different normalization methods are available which increase the performance of such methods e.g. TF-IDF, mutual information or χ^2 normalization. While this type of normalizations is not widely employed for deep neural networks, the use of TF-IDF normalization followed by centering, yields good results. On the other hand it is shown in [9] that the other types of normalizations could contribute to the deep architectures to work better and that consists in (1) centering (2) applying KL-Expansion or PCA (3) using covariance equalization. The two steps (2) and (3) can be combined by applying a PCA with *whitening*; see for instance [10]. As the networks learn the fastest from the most unexpected sample, it is recommended to choose a sample at each iteration which is the most unfamiliar to the system [9]. In order to rely on this hypothesis, after the normalization step above, training data are shuffled to ensure that the successive examples are not drawn from the same class. Finally we train the deep auto-encoder as a pretraining step demonstrated in [11], by considering each layer separately as a simple auto-encoder. The activations of a layer below become the input for the next layer as in Fig. 2. In each layer we add some noise to the data by replacing randomly 30% of the data by zeros, a procedure called corruption or denoising. With this approach, we approximate the performance of RBMs with a lower cost in terms of complexity.

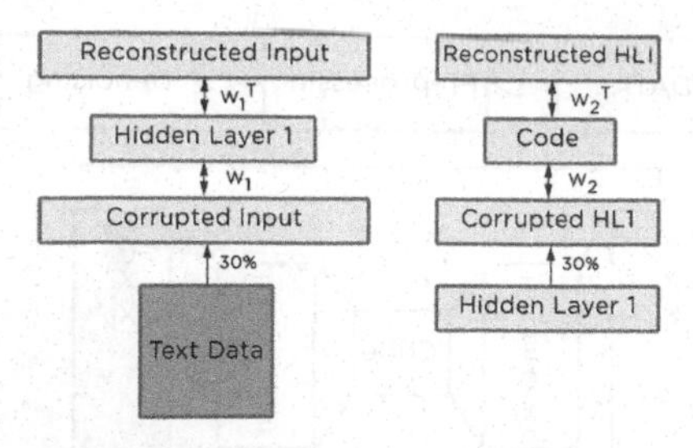

Fig. 2. Layer wise pretraining of a denoising autoencoder

3 Auto-encoder for Text Analysis

3.1 Classic Auto-encoders

Auto-encoders are traditionally composed of two parts, an encoder where the high dimensional feature space of data is encoded and compressed to a lower dimensional feature space by a function h as $\mathbf{y}_i = h(W\tilde{\mathbf{x}}_i + b)$ where $\tilde{\mathbf{x}}$ is a corrupted input obtained by following the denoising procedure explained in the last section, W is the weight matrix between the input and hidden layer as in Fig. 2 where $W \in \mathbb{R}^{d \times p}$ (d number of neurons in hidden layer, p number of input features), b is the bias term where $b \in \mathbb{R}^{d \times 1}$ and $\mathbf{y}_i$ contains hidden layer values in middle layers and code values in the last layer. The parameters W and b are estimated using mini batch gradient descent algorithm in each iteration of optimization and h is an activation function that can be a linear or non linear such as *sigmoid* or *hyperbolic tangent*. In this paper we opt for the *hyperbolic tangent* which generally provides better results. The second part of the auto-encoder consists of a decoder which tries to reconstruct the original data from the code layer $\mathbf{y}_i$ by $\mathbf{x}'_i = g(W'\mathbf{y}_i + b')$. The layer of reconstructions has the same dimensionality as the input layer. As the encoder part, the decoder layer has also the same type of parameters ($W' \in \mathbb{R}^{p \times d}$ and $b' \in \mathbb{R}^{p \times 1}$) that must be adjusted, and an activation function g; like h it can be a linear or non linear function. In classic auto-encoders, each example $\mathbf{x}'_i$ tries to reconstruct the original input $\mathbf{x}_i$ from the code or hidden layer $\mathbf{y}_i$. Therefore, the cost function takes the following form

$$C(\theta) = \arg\min \sum_{i=1}^{m} ||g(W'h(W\tilde{\mathbf{x}}_i + b) + b') - \mathbf{x}_i||^2 \quad (1)$$

where C is a cost function with $\theta = (W, b, W', b')$ the unknown parameters to estimate by minimizing this function. The symbol $||.||$ denotes an euclidean distance between the reconstructed examples and original input.

3.2 The Proposed Unsupervised Auto-encoder

The classic auto-encoder, presented above and referred to as `C-autoencoder` in the sequel, is not able to capture the original structure in the data, and to reveal the latent structure in the case of complex data. In order to achieve this, one can modify the cost function (1) in a way where each example $\mathbf{x}'_i$, in addition to reconstructing the correspondent original input $\mathbf{x}_i$, also reconstructs the data points that are in the neighborhood of $\mathbf{x}_i$, using *cosine similarity* metric. Moreover, each reconstruction term has a weight; this leads to the construction of a weighted graph with edges connecting nearby documents to each other. At first glance the idea is relatively similar to that in [12], but the prior normalization step following a novel auto-encoder configuration and regularization make this approach more relevant to cluster sparse text data, where the sparsity is regularized in order to avoid overfitting. This procedure is depicted in Fig. 1, where the example $\mathbf{x}'_i$ designated by a red circle reconstructs its correspondent in input layer i.e. $\mathbf{x}_i$ and its k nearest neighbors in input layer i.e. $\{\mathbf{x}_j, \mathbf{x}_k\}$, that are marked by a blue ellipse. The weights between these reconstruction terms are denoted by ω. So the cost function for the proposed auto-encoder becomes,

$$C(\theta) = \arg\min \sum_{i=1}^{m} \sum_{\ell \in \Psi_i} \omega_{i\ell} ||\mathbf{x}'_i - \mathbf{x}_\ell||^2. \tag{2}$$

where Ψ_i denotes the set of the k nearest neighbors of the document $\mathbf{x}_i$ and ω (not to be confused with W) is the weight associated to document $\mathbf{x}_i$ and document $\mathbf{x}_\ell$ belonging to Ψ_i. The set of parameters θ of the network in Eq. (1) holds also for the new loss function. The weight draws on *Laplacian Eigenmaps* where heat kernels are used to choose the weight decay function (parameter $t \in \mathbb{R}$) and the cosine between two documents is used as a similarity measure between them. It takes the following form

$$\omega_{i\ell} = \exp -(\frac{\cos(\mathbf{x}_i, \mathbf{x}_\ell)}{t}) \tag{3}$$

where t is a hyper parameter to adjust. The details on the choice of t is discussed in [4]. Note that with $t = 1$, two very similar documents lead to $\omega_{i\ell} \approx 1/e$, and so similar documents in embeddings are less penalized; while two distinct documents lead to $\omega_{i\ell} \approx 1$ and so they are more penalized. Furthermore, we have considered the sparsity regularization term in the cost function as follows,

$$C_{sparse}(\theta) = C(\theta) + \beta \sum_{j=1}^{s} KL(\rho||\hat{\rho}_j) \tag{4}$$

where β controls the importance of the regularization term and s is the number of neurons in hidden layer. $KL(\rho||\hat{\rho_j})$ is the Kullback-Leibler divergence between ρ a sparsity parameter and $\hat{\rho}_j$ its approximation by $\hat{\rho}_j = \frac{1}{m}\sum_{i=1}^{m} \mathbf{y}_i^{(j)}$ the average activation of hidden unit j; $KL(\rho||\hat{\rho_j})$ can be thought of as a measure of the information lost when $\hat{\rho}_j$ is used to approximate ρ. The details of this regularization are available in [13] and given by

$$KL(\rho||\hat{\rho}_j) = \rho \log \frac{\rho}{\hat{\rho}_j} + (1-\rho) \log \frac{1-\rho}{1-\hat{\rho}_j}. \tag{5}$$

Hereafter, we describe in Algorithm 1, referred as T-autoencoder, the main steps of the method, optimizing (4); we assume that $W' = W^\top$ as is often the case in the literature.

Algorithm 1. Unsupervised auto-encoder for Text data (T-autoencoder)

Input training set $\{\mathbf{x}_i\}_1^m$
Hypothesis: Tied weights i.e. $W' = W^\top$; fixed sparsity level $\rho = 0.05$
Parameters: θ=(W, b, b')
Notation: Ψ_i reconstruction set for $\mathbf{x}_i$
ω weights between $\mathbf{x}_i$ and Ψ_i

1. Compute the cosine similarity between documents and determine reconstruction set Ψ by k-nearest neighbor algorithm for each example.
2. Compute the weights ω between each example $\mathbf{x}_i$ and its reconstruction set Ψ_i, as in (3).
3. Update θ minimizing the Cost function C_{sparse} in (4).
4. Update reconstruction set Ψ and weights ω with respect to each hidden layer $\{\mathbf{y}_i\}_1^m$ separately.
5. Repeat 3 and 4 until convergence.

The time complexity of T-autoencoder is $\mathcal{O}(n^2)$ for cosine similarity computation, $\mathcal{O}(n \log n)$ for finding k-nearest neighbors and $\mathcal{O}(batch\_size \times k)$ to calculate each weighted reconstruction term in (2) in addition to the time complexity of neural networks; where k is the number of neighbors. As $n \to \infty$ the added time complexity approaches $\mathcal{O}(n^2)$.

4 Experiments

In order to evaluate the performance of T-autoencoder, we performed experiments on different document-term datasets. Our implementations are based on *python* and **R** languages, and the *theano* library in order to use the performance of GPU for accelerating computations.

4.1 Experimental Setup

The characteristics of datasets[1] used in experimentation are presented in Table 1. Each dataset presented has its own complexity e.g. excessive number of variables (Curse of dimensionality), high number of clusters or the complexity pertaining to the data structure. We compared the T-autoencoder with a linear method (PCA), and three non-linear methods (Isomap, LE and t-SNE). We run the methods which require the hyper parameters as number of neighbors or learning rate

[1] http://dataexpertise.org/research.

to be adjusted e.g. `Isomap`, `t-SNE` and `T-autoencoder`, with diverse configurations, and finally we picked the values corresponding to a minimum reconstruction error. As an example, for `t-SNE` we have opted for $\eta = 100$, as it provides better results than other configurations. In addition to the state of the art methods, `C-autoencoder` is also considered in experiments, in order to point out the improvement attained using `T-autoencoder`. We evaluated the performance of these methods by means of three-dimensional plots of embeddings and also by measuring different metrics such as Normalized Mutual Information (NMI) [14], Adjusted Rand Index (ARI) [15] and Purity after applying the Expectation-Maximization algorithm [7] for Gaussian Mixtures Models (GMM) instead of k-means which is based on a restricted Gaussian mixture.

Table 1. Datasets used for experimentation, # denotes the cardinality.

Dataset	#Documents	#Terms	Number of clusters
Classic3	3891	4303	3
CSTR	475	1000	4
20news	3970	8014	4
NG5	500	2000	5
Reviews	4069	18483	5
TR45	690	8261	10
TDT2_10	653	36771	10

For the auto-encoder part a $n - \frac{n}{2} - \frac{n}{2^2} - \ldots - d$ architecture is used, where n represents the dimensionality of the data and d represents the dimensionality of the latent space that should be attained in code layer. In this experimentation we opt for three dimensional latent space, so $d = 3$. After extensive numerical experiment trials, $t = 0.5$ in (3) appears appropriate, so we did not get involved as much with tuning of such hyper parameters. The auto-encoders were trained using the layer-wise pretraining procedure explained before, and are fine-tuned by performing back-propagation such as minimizing the weighted sum of squared errors between each example and its reconstruction set. We used a decreasing learning rate, starting from a large value in higher layers and reducing it gradually in lower layers. Weight decay was set to 0.0001 for all the layers. In our experiments, we opted for $\beta = 0.01$ and $\rho = 0.05$ for regularization term in (4); where ρ controls the sparseness of representation, and has a fixed value obtained via experimentation for all the datasets. Furthermore mini batch gradient descent method was used to adjust weights and biases; the batch size was fixed at 100.

4.2 Results

In this section, the results of the above mentioned methods are presented by means of the visualization of the embeddings (Figs. 3, 4 and 5). Furthermore

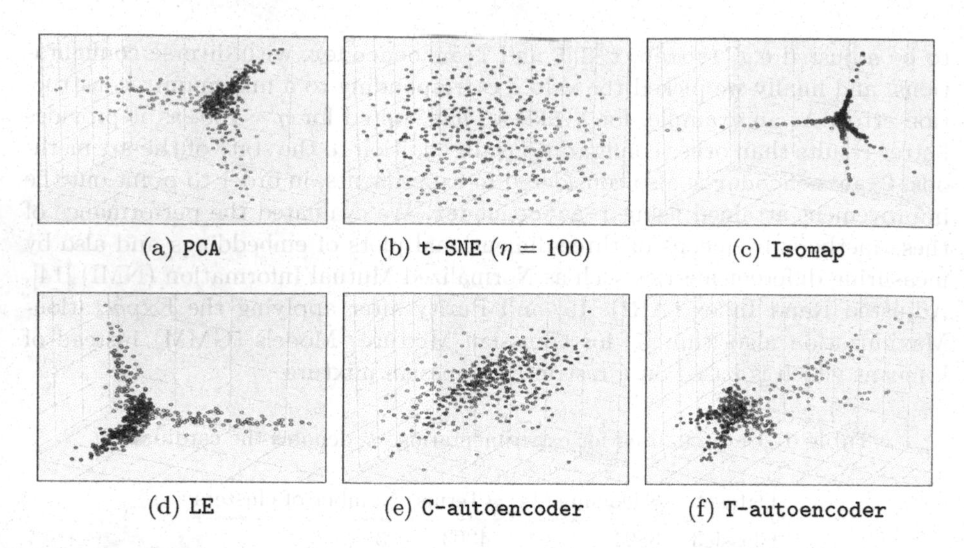

Fig. 3. Visualization of 500 documents from NG5 dataset by different unsupervised dimensionality reduction methods

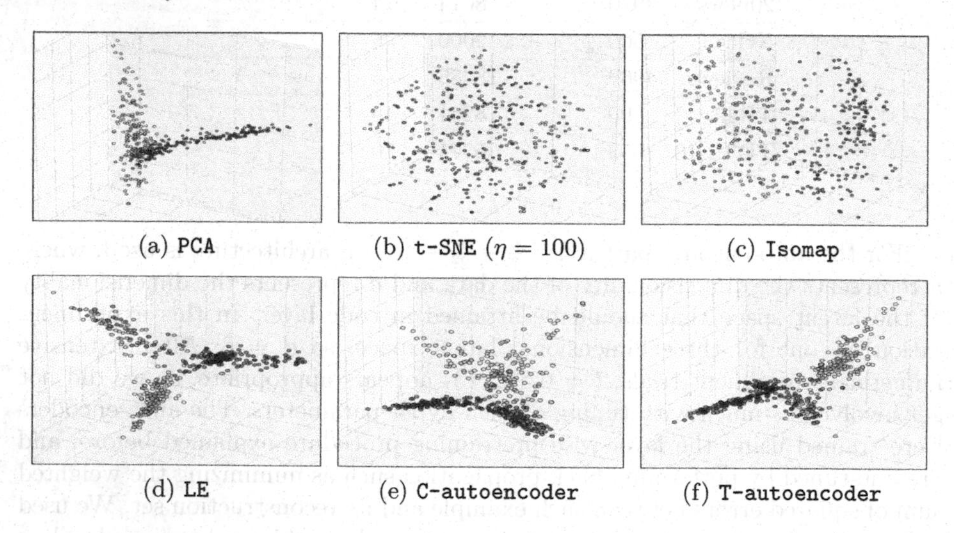

Fig. 4. Visualization of 475 documents from CSTR dataset by different unsupervised dimensionality reduction methods

in order to have more precise comparisons, the embeddings are clustered in homogeneous groups and are compared with true labels (Table 2).

Visualisation and Clustering. To illustrate the interest of T-autoencoder in terms of visualization and clustering, because of the lack of space, we chose to only present the visualizations of NG5, CSTR and TDT2_10 datasets obtained by all the presented methods in Figs. 3, 4 and 5. In Table 2 comparisons are

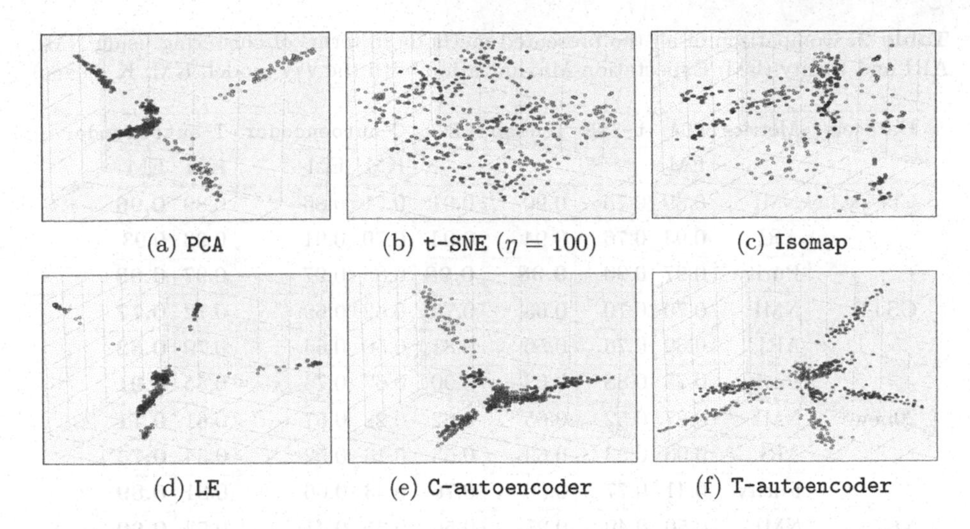

Fig. 5. Visualization of 653 documents from TDT2_10 dataset by different unsupervised dimensionality reduction methods

reported for all the datasets in Table 1 using NMI, ARI and purity metrics after applying the EM algorithm [7]. For each experiment the best performance is highlighted in bold type. Note that EM is conducted considering the General Gaussian Mixture (GMM) Model noted VVV in the sequel [16]. In the first step we consider that the number of clusters is known. The different Gaussian models are based on the cluster proportions and three characteristics of clusters (volume, shape, orientation) that can be equal (E) or variable (V) (for details see, [17]). For instance EVV corresponds to the model where the clusters have the same volume but the shapes and orientations are different. In Fig. 3 we observe that the latent structure of NG5 dataset is relatively complex and all the presented methods have difficulties in distinguishing the five existing clusters. For example PCA and LE can distinguish only three clusters and two remaining clusters are mixed together while Isomap can only recognize some documents from the four clusters while fifth group and the rest of documents are mixed in the center. Furthermore it gives the worst performance (see Table 2). Using t-SNE, we can see that in almost all the visualizations, the data are more dispersed than with other methods, but the examples of the same cluster remain close to each other; t-SNE gives the second best result. C-autoencoder cannot identify the frontier between clusters whereas the result from Fig. 3(f) and Table 2 reveals a good performance of T-autoencoder compared to the others. Using this method, we can observe a good separability between the five groups of documents and the best results in terms of purity, NMI and ARI.

In Fig. 4 we observe that two clusters are mixed together using most of the methods, proving a complex latent structure of them. Considering PCA and C-autoencoder, this complexity can be clearly observed. On the other hand t-SNE is not able to capture the existing relations with other clusters too while

Table 2. Comparison of all the presented methods in terms of clustering using NMI, ARI and Purity (EM: Expectation-Maximization with the VVV model, KM: K-means)

Datasets	Metrics	PCA	t-SNE	Isomap	LE	C-autoencoder		T-autoencoder	
		EM				KM	EM	KM	EM
Classic3	NMI	0.89	0.75	0.90	0.91	0.71	0.86	0.89	**0.96**
	ARI	0.93	0.76	0.94	0.94	0.70	0.91	0.93	0.93
	Purity	0.97	0.90	**0.98**	**0.98**	0.88	0.97	0.97	**0.98**
CSTR	NMI	0.70	0.70	0.63	0.76	0.62	0.69	0.72	**0.77**
	ARI	0.69	0.76	0.56	0.81	0.50	0.64	0.70	**0.82**
	Purity	0.75	0.88	0.63	0.90	0.67	0.74	0.85	**0.91**
20news	NMI	0.27	0.52	0.65	0.62	0.22	0.57	0.61	**0.71**
	ARI	0.06	0.53	0.69	0.53	0.16	0.52	0.54	**0.74**
	Purity	0.41	0.77	0.86	0.76	0.43	0.66	0.81	**0.89**
NG5	NMI	0.50	0.49	0.25	0.56	0.31	0.41	0.54	**0.60**
	ARI	0.38	**0.51**	0.07	0.49	0.20	0.33	0.38	**0.51**
	Purity	0.56	0.76	0.38	0.66	0.53	0.55	0.72	**0.79**
Reviews	NMI	0.31	0.31	0.50	**0.59**	0.31	0.46	0.45	0.54
	ARI	0.13	0.24	0.43	**0.54**	0.16	0.44	0.28	0.48
	Purity	0.44	0.49	0.61	0.65	0.47	**0.66**	0.60	**0.66**
TR45	NMI	0.49	0.63	0.54	0.56	0.46	0.56	0.64	**0.66**
	ARI	0.36	0.51	0.48	0.43	0.30	0.37	0.52	**0.57**
	Purity	0.48	0.59	0.61	0.46	0.49	0.58	0.65	**0.67**
TDT2_10	NMI	0.70	0.96	0.81	0.80	0.71	0.87	0.80	**0.98**
	ARI	0.30	0.95	0.70	0.57	0.43	0.81	0.66	**0.98**
	Purity	0.48	0.97	0.80	0.70	0.52	0.85	0.75	**0.99**

Isomap shows good performance on this dataset in terms of separability of clusters and clustering. Although the documents are dispersed in different directions there is not a clear separation between four existing groups of documents. The best visualizations are obtained using LE and T-autoencoder, where we can clearly see that each projected cluster has its own direction. In Table 2 we can also observe their higher performances in terms of all the metrics used.

In Fig. 5, the number of clusters is higher than in the two previous examples. We note that PCA, Isomap and LE are not able to recognize all the ten obtained clusters while LE has shown a good performance and t-SNE provides a good result in terms of clustering but not in terms of visualisation (Table 2). Finally, the good performance of T-autoencoder in terms of visualisation and clustering is easy to observe; T-autoencoder clearly outperforms C-autoencoder.

Assessing the Number of Clusters. Another reason why we used the GMM for clustering instead of a simple k-means algorithm is that, this approach offers the flexibility to fit the data, using an appropriate model. As we know, estimating the number of clusters for the input of a clustering algorithm is essential and hard to achieve. So to estimate it, we have used the *Bayesian Information Criterion* (BIC) [8] given by $BIC_{M,k} = -2L_{M,k} + \upsilon_{m,K} \log m$, where M is the model and k is the number of components. $L_{M,k}$ is the maximum likelihood for M and k and υ is the number of free parameters in the model M with k components. This criterion penalizes the number of parameters in model and maximizes the likelihood of data simultaneously; it is efficient on a practical ground. To choose the best model for mixture model with an appropriate number of clusters, we have considered the BIC criterion with different numbers of components (clusters) K on the latent space obtained. Due to the lack of space, we propose to illustrate the contribution of BIC on two data sets TDT2-10 and NG5. In Fig. 6(a), BIC takes the maximum value when the number of components is 10 with the `VVV` model (see Fig. 6(c)). In Fig. 7(a), we observe that the highest value of BIC is attained for the model `VVV` with 6 clusters (marked by red vertical dotted line) instead of 5 clusters (marked by green vertical dotted line) with the `EVV` model; it is also an ellipsoidal model which considers the same volume, but different shape and orientation for clusters. Notice that both suggestions of BIC are interesting; Fig. 7 reinforces them. In Fig. 7(c), we simulated the scheme of NG5 visualization depicted in Fig. 7(b), we have relatively five clusters with different shapes and orientations. The orientations are shown by arrows, and shapes by dotted ellipses.

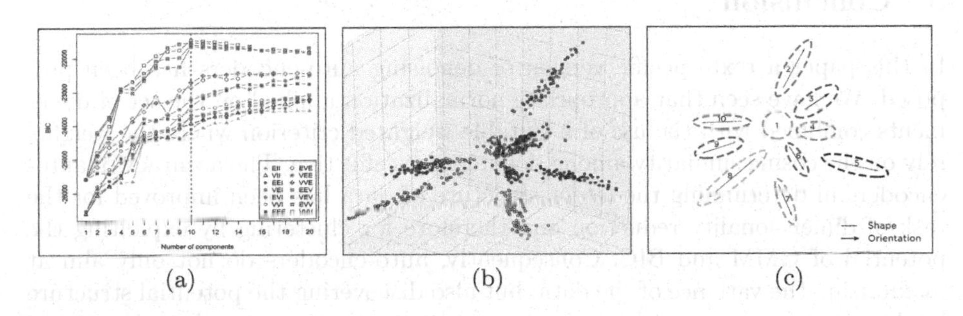

Fig. 6. TDT2_10 dataset: Bic plot related to the latent space obtained by `T-autoencoder` in (a), Latent space obtained by `T-autoencoder` with ground truth clusters in (b), Estimated clusters scheme by mixture model using BIC criterion in (c).

In short, we observe that clustering on embeddings obtained via `T-auto encoder` often outperforms all the presented methods including `C-auto encoder`. This is due to the main difference in the architecture of the proposed method in comparison with `C-autoencoder`. As mentioned earlier, `T-autoencoder` is trained to reconstruct data from the corrupted input. This procedure increases its ability to be less dependent on training data while

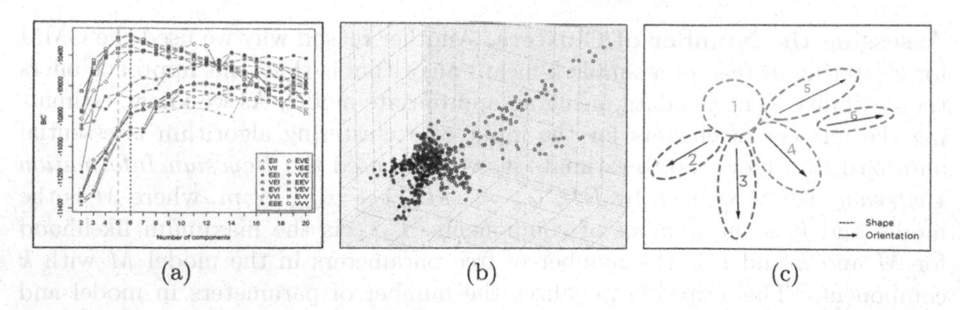

Fig. 7. NG5 Dataset: Bic plot related to the latent space obtained by `T-autoencoder` in (a), Latent space obtained via `T-autoencoder` with ground truth clusters in (b), Estimated clusters scheme by mixture model using BIC criterion in (c). (Color figure online)

promoting close documents; the GMM via EM confirms this performance by providing a better clustering of documents. On the other hand, the autoencoders generally do not construct low-dimensional data representations in which the natural clusters are widely separated. This could also be due to shortcoming of auto-encoders where latent relations in data cannot be discovered. Unlike `C-autoencoder`, `T-autoencoder` learns these relations by reconstructing an example from its k-nearest neighbors according to the more suitable cosine similarity for document-term matrices.

5 Conclusion

In this paper a text specific version of denoising auto-encoders has been proposed. We have seen that appropriate normalization applied on the set of documents combined with the use of a suitable weighted criterion where the weights rely on the cosine similarity among documents is effective. The accuracy of auto-encoders in determining the latent structure of data has been improved for the task of dimensionality reduction and therefore for clustering by exploiting the potential of GMM and BIC. Consequently, auto-encoders do not only aim at maximizing the variance of the data, but also discovering the potential structure in clusters.

The interest of our approach is to demonstrate the accuracy of the proposed method in the buoyant field of visualization and document clustering [18,19]. The efficiency in terms of time complexity is, however, another issue that could be considered in future works. Although we have used the GPU performance using the *theano* library, the efficiency should be improved by using more recent optimization methods such as BFGSs which converge faster than gradient descent.

References

1. Gittins, R.: Canonical Analysis - A Review with Applications in Ecology. Springer, Heidelberg (1985)
2. van der Maaten, L., Hinton, G.: Visualizing data using t-sne. J. Mach. Learn. Res. **9**(Nov), 2579–2605 (2008)
3. van der Maaten, L.: Learning a parametric embedding by preserving local structure. RBM, 500:500 (2009)
4. Belkin, M., Niyogi, P.: Laplacian eigenmaps and spectral techniques for embedding and clustering. NIPS **14**, 585–591 (2001)
5. Bengio, Y.: Learning deep architectures for ai. Found. Trends Mach. Learn. **2**(1), 1–127 (2009)
6. Vincent, P.: A connection between score matching and denoising autoencoders. Neural Comput. **23**(7), 1661–1674 (2011)
7. Dempster, A.P., Nan Laird, M., Rubin, D.B.: Maximum likelihood from incomplete data via the em algorithm. J. Roy. Stat. Soc. Ser. B (methodological) **39**, 1–38 (1977)
8. Schwarz, G., et al.: Estimating the dimension of a model. Ann. Stat. **6**(2), 461–464 (1978)
9. LeCun, Y.A., Bottou, L., Orr, G.B., Müller, K.-R.: Efficient BackProp. In: Montavon, G., Orr, G.B., Müller, K.-R. (eds.) Neural Networks: Tricks of the Trade. LNCS, vol. 7700, pp. 9–48. Springer, Heidelberg (2012). doi:10.1007/978-3-642-35289-8_3
10. Jégou, H., Chum, O.: Negative evidences and co-occurences in image retrieval: the benefit of PCA and whitening. In: Fitzgibbon, A., Lazebnik, S., Perona, P., Sato, Y., Schmid, C. (eds.) ECCV 2012. LNCS, pp. 774–787. Springer, Heidelberg (2012). doi:10.1007/978-3-642-33709-3_55
11. Hinton, G.E., Salakhutdinov, R.R.: Reducing the dimensionality of data with neural networks. Science **313**(5786), 504–507 (2006)
12. Wang, W., Huang, Y., Wang, Y., Wang, L.: Generalized autoencoder: a neural network framework for dimensionality reduction. In: IEEE Conference on Computer Vision and Pattern Recognition Workshops, pp. 490–497 (2014)
13. Ng, A.: Sparse autoencoder. CS294A Lecture Notes, vol. 72, pp. 1–19 (2011)
14. Strehl, A., Ghosh, J.: Cluster ensembles–a knowledge reuse framework for combining multiple partitions. J. Mach. Learn. Res. **3**, 583–617 (2003)
15. Hubert, L., Arabie, P.: Comparing partitions. J. Classif. **2**, 193–218 (1985)
16. Banfield, J.D., Raftery, A.E.: Model-based gaussian and non-gaussian clustering. Biometrics **49**, 803–821 (1993)
17. Fraley, C., Raftery, A.E.: Mclust version 3: an R package for normal mixture modeling and model-based clustering. Technical report (2006)
18. Priam, R., Nadif, M.: Data visualization via latent variables and mixture models: a brief survey. Pattern Anal. Appl. **19**(3), 807–819 (2016)
19. Allab, K., Labiod, L., Nadif, M.: A semi-NMF-PCA unified framework for data clustering. IEEE Trans. Knowl. Data Eng. **29**(1), 2–16 (2017)

Gradable Adjective Embedding for Commonsense Knowledge

Kyungjae Lee[1], Hyunsouk Cho[2], and Seung-won Hwang[1(✉)]

[1] Yonsei University, Seoul, South Korea
{lkj0509,seungwonh}@yonsei.ac.kr
[2] POSTECH, Pohang, South Korea
prory@postech.ac.kr

Abstract. Adjective understanding is crucial for answering qualitative or subjective questions, such as "is New York a big city", yet not as sufficiently studied as answering factoid questions. Our goal is to project adjectives in the continuous distributional space, which enables to answer not only the qualitative question example above, but also comparative ones, such as "is New York bigger than San Francisco?". As a basis, we build on the probability P(New York—big city) and P(Boston—big city) observed in Hearst patterns from a large Web corpus (as captured in a probabilistic knowledge base such as Probase). From this base model, we observe that this probability well predicts the graded score of adjective, but only for "head entities" with sufficient observations. However, the observation of a city is scattered to many adjectives – Cities are described with 194 adjectives in Probase, and, on average, only 2% of cities are sufficiently observed in adjective-modified concepts. Our goal is to train a distributional model such that any entity can be associated to any adjective by its distance from the vector of 'big city' concept. To overcome sparsity, we learn highly synonymous adjectives, such as big and huge cities, to improve prediction accuracy. We validate our finding with real-word knowledge bases.

Keywords: Adjective understanding · Commonsense knowledge · Word embedding

1 Introduction

In recent years, database and search engines have shown the effectiveness in answering quantitative questions on entities, such as "what is the population of New York". However, they are still limited in answering qualitative or subjective questions, often represented in adjective, such as "is New York a big city?" or "is New York bigger than San Francisco". This gets even harder for more subjective adjectives such as "is New York beautiful?". Adjectives, by modifying or elaborating the meaning of other words, are studied in linguistics [6] to play important roles in determining the semantic orientation of attributes, but existing computational approaches have the following limitations.

J. Kim et al. (Eds.): PAKDD 2017, Part II, LNAI 10235, pp. 814–827, 2017.
DOI: 10.1007/978-3-319-57529-2_63

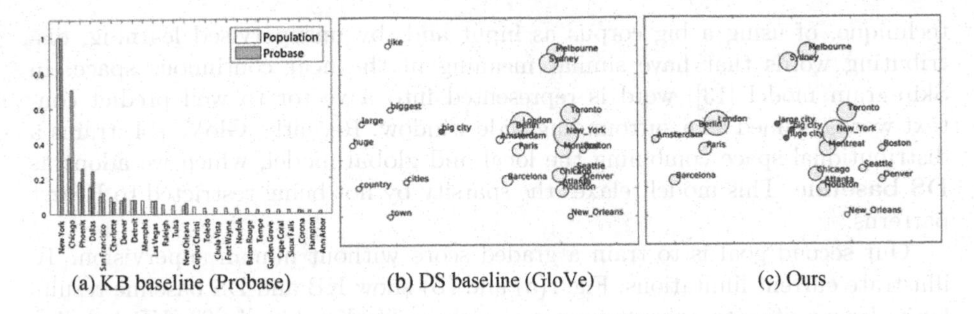

(a) KB baseline (Probase) (b) DS baseline (GloVe) (c) Ours

Fig. 1. The relation between population and the score of "big city". In (b) and (c), the size of circle is proportional to the population of the city

Existing work focuses on mining textual patterns to identify if 'New York' is frequently observed with 'big city' in Hearst patterns, like 'big city such as New York' and 'New York is a big city', defining an is. A relationship between New York and big city. Specifically, Probase [19] knowledge base (KB) captures P(New York|big city) from a large web corpus, which we adopt as **KB baseline**. However, in this KB, concept *city* is modified by 194 adjectives, such that textual observations of New York are scattered over these adjective-modified concepts. Such scattering makes lesser known, or tail entities, to be scarcely observed especially in adjective-modified concepts, which we call a **observation sparsity** problem – if Urbana is not observed in the 'big city' pattern, does this mean it is not big or simply unobserved?

Trummer et al. [17] alleviate this problem by extending observations to include not only positive isA patterns, but also negative isA patterns such as 'Urbana is not a big city'. They use a provided threshold to map the given entity and adjective pair, into positive state or negative state (binary condition). However, this still cannot handle the sparsity of entities observed in neither polarity. Iwanari et al. [9] later generalize the binary classification into an ordering, using textual patterns as evidences.

We summarize the limitations as follow:

- **Observation sparsity:** As New York can be associated with virtually infinite adjectives, only few head entities are sufficiently observed in adjective-modified concepts. (For example, New York is observed as a big city, but may not be observed as a large city).
- **Human intervention:** Existing work requires human intervention to decide a score threshold or provide human-generated ordering as training data. Our goal is to give a graded score without human intervention, to not only classify whether it is big, but also to compare how big it is with respect to another entity.

Our first goal is thus to overcome **observation sparsity**, by answering "Is New York a big city?" even when no Hearst co-occurrence pattern is observed. A naive solution is adopting an existing distributed word representation technique, of using a big corpus as input and, by unsupervised learning, distributing words that have similar meaning in the near continuous space. In Skip-gram model [13], word is represented into a vector to well predict context words defined as a surrounding slide window. Recently, GloVe [14] trains a distributional space combining the local and global model, which we adopt as **DS baseline**. This model relaxes the sparsity by not being restricted to Hearst patterns.

Our second goal is to train a graded score without human supervision. To illustrate current limitations, Fig. 1(a) and (b) show KB and DS baseline results for ordering cities to answer questions such as "Is X a big city?". KB baseline successfully grades the degree of "big city" (the bigger circle suggests higher population), but includes only a few head entities actually observed in Fig. 1(a). Meanwhile, though DS baseline overcomes the observation sparsity in Fig. 1(b) by placing all cities in the space, close vectors to 'big city' are not necessarily cities with higher population. Quantifying this failure requires a linear ordering of all cities, which requires costly human-generated labels.

We combine the strength of the two models. First, we build on sparse but highly precise Hearst patterns for training distributed word space. As a result, we obtain Fig. 1(c) where the distance from big city preserves the correlation with popularity. Meanwhile, DS baseline has higher recall but lower precision by treating all co-occurrences as equal: Highly frequent co-occurrence of 'big city' may include noisy words such as 'where', 'like', 'small'. Second, we capture the distributed similarity between adjective vectors. For example, as shown in Fig. 1(c), big and huge cities are nearly synonymous, such that scattered observations from two concepts can be combined to enhance the correlation. In other words, we can consider the distance to either vector (or the combination of the two) to predict adjective grade more robustly.

We quantify the improved performance by comparing with a total order generated by attributes (for objective adjectives), or by a total ordering generated by textual patterns (for subjective adjectives), as [17] confirms the quality of such ordering. This enables to include up to 250 concepts and 500K entities in evaluation.

2 Related Work

We categorize existing work for adjective understanding into implicit and explicit modeling. Lastly, we describe how our work complements both approaches, and describes other related attribute-related tasks.

2.1 Explicit Model

This approach considers textual patterns as explicit representation, to train a graded adjective score. Probase [19] considers Hearst patterns to extract isA relationship between concept and the given entity, observed from billions of web documents. For our purpose of adjective understanding, we can consider

Probase score for adjective-modified concepts, which we adopt as **KB baseline**. Alternatively, Trummer et al. [17] consider both positive and negative isA patterns, such as 'New York is a big city' and 'Urbana is not a big city', to train a binary classifier given the ratio of positive and negative statements. Iwanari et al. [9] use four textual patterns for finding various evidence between adjective and concept, aggregated into an ordering trained from supervised methods. This ordering is evaluated against human-generated ordering, which limits the scalability of evaluation. Our contribution is establishing Probase probability as an ordering proxy, evaluating against data attributes (for objective adjective) and missing probability (for subjective). WebChild Knowledge-Base [16] associates entity with adjectives for fine-grained relations like hasShape, hasTaste, evokesEmotion, etc.

The strength of explicit model in general is its high precision, but its weakness is missing observation. However, as there are virtually infinite combinations of adjective with concept, observations for adjective-modified concepts are typically scarce, especially for lesser known entities, for which we cannot predict the score.

2.2 Implicit Model

Meanwhile, implicit approaches leverage a neural network model and large corpus data to model latent semantic similarity between entities. For example, the continuous bag-of-words model (CBOW) and the skip-gram model [12,13] approaches predict semantic similarity between New York and Chicago based on the similarity of surrounding words, such as mayor, city, population, etc. In this space, the distance or similarity between every word can be calculated (or, achieves high recall) even if the two words did not co-occur in sentence, and the similarity will be high for two words with similar meaning. This helps infer Chicago as a big city, even when it is not explicitly observed in the Hearst pattern of "big city such as Chicago", unlike New York being frequently observed.

Similarly, LSA [5] predicts two entities being similar, based on word co-occurrence matrix. This model transforms a large co-occurrence matrix to low dimensional vectors using a dimensional reduction technique. More recently, GloVe [14] combines the strength of LSA and Skip-gram to train words into the distributed space (DS), which we adopt as **DS baseline**. Huang et al. [8] similarly predict similarity between query and document through deep learning, but this line of approach shares a common weakness of compromising precision for the increased recall.

Our goal is to increase recall without compromising precision. We thus use high precision signals from explicit model to train a distributed entity space, then infer missing scores based on its similarity with (possibly multiple synonymous) adjectives.

2.3 Joint Model and Other Attribute Work

Existing joint approach of combining implicit and explicit models can be categorized into two directions: First, we can use explicit model as a supervision to train word embedding, such as syntactic or lexical knowledge [2,15] to improve the quality of word embedding. Second, explicit knowledge can be projected onto an embedding space [3,11,18], to enable the inference between relations. We take the advantage of both approaches, by using explicit probability as supervision for high-quality embedding, while projecting concepts in the space to enable the inference of concept-concept or concept-entity similarity.

Our work is also related to attribute understanding, as adjective is often viewed as a qualitative and subjective attributes describing the concept. First, to understand a likely set of attributes describing the concept, [10] mines "the [attribute] of [concept]" patterns. Proposed method derives attributes for millions of concepts and predicts the score of the attributes with regard to the corresponding concepts. Second, to understand similar attributes, [7] discusses how to automatically discover attribute synonyms to integrate hundreds of web tables describing the same concept.

More recently, instead of textual data, several images of objects are used for inferring the size or to predict whether the object is relatively big or small [1]. This work can capture graded property of size and complement our work, for finding 'big animal' that can be captured in the photo, but not 'big city' which cannot be photographed.

3 Proposed Model

This section first overviews existing approaches for quantifying the graded score of the given entity for adjective-modified concepts. We then propose our approach combining the strength of the two existing models.

3.1 Preliminary

Explicit Model. Probase [19] used a pattern-based method to estimate the probability between the entity and its concept from billions of Web pages. We selected only the adjective-modified concepts among various concepts in Probase and used the probability as our score. The probability between concept and entity was calculated by counting how frequently the pair of two word are found in corpus, and can be defined as:

$$P(e|c) = \frac{n(e,c)}{\sum_{e' \in E(c)} n(e',c)} \tag{1}$$

where e, c are respectively the entity and adjective-modified concept, $E(c)$ is the set of sub-entity of the adjective-modified concept c and $n(e,c)$ is the number of times (e,c) discovered by Hearst pattern. In Probase data, when an adjective-modified concept "big city" is given, the probability renders a correct size ordering, such as Chicago > London > Dublin > Washington DC, with probability 3.82%, 3.58%, 1.42%, and 0.04% respectively. Though this signal is highly precise, their coverage is limited – only 304 cities in USA (40.1%) are observed in the Heart patterns with 'big city', though their probability score does meaningfully correlate with actual population with correlation score 0.75. However, this does not cover the rest 60% of big cities of comparable population. It is thus difficult to decide whether the unseen city is not big or simply unobserved.

Implicit Model. GloVe constructs a word embedding by using word co-occurrence data. This model trained co-occurred word vector as following equation.

$$F(w_i, \tilde{w_k}) = exp({w_i}^T \tilde{w_k}) = P(j|i) = \frac{n(i,k)}{n(i)} \tag{2}$$

where w_i is vector of word i, $\tilde{w_k}$ is separate vector of context word j, and $P(j|i)$ is the conditional probability that word j appear in the context of word i. F denotes a function that encode two vectors to real value and is used as exponential function in this model.

A naive adoption of implicit model is to train a Glove embedding and use the distance of words from the adjective-modified concept, such as 'big city. Such a naive adoption has two limitations. First, co-occurrence is more prominent with non-entity words, such as "like", "where" and "small", compared to which co-occurrence with city entities forms a long tail. This would work as a noise in generating a robust ordering among the city entities. Second, eliminating non-entity words in the space modeling cannot solve the problem either, as entity co-occurrence may bear different meanings as well. As [17] pointed out, co-occurrence of 'New York is a big city' and 'Urbana is not a big city' reflects the opposite meaning.

We discuss the joint modeling overcoming the limitations of the two models.

3.2 Embeddings for Adjectives

In semantic space represented as vectors, the distance or similarity between every word can be calculated even when two words do not co-occur in corpus. However, as training with simple co-occurrence is too noisy, we use Probase probability into the vector cosine distance in the range of −1 to 1. Therefore, we propose a model that uses word embedding and cosine distance to overcome sparsity and binary classification problem.

Loss Function for Concept. The proposed model trains adjective-modified concept into semantic word space by applying the scores to cosine similarity from the entity vector, instead of Glove model using co-occurrence. Our objective is thus to find the vector of adjective-modified concept satisfying the following condition.

$$F(v_e, v_c) = P(e|c) \tag{3}$$

where v_c, $v_e \in \mathbf{R^d}$ are the vectors of adjective-modified concept and its entity respectively. A simple way to obtain F is by inner product:

$$F(v_e, v_c) = v_c \cdot v_e + b_e = P(e|c) \tag{4}$$

where b_e is bias of entity. As we normalize entity vectors to have size 1, this corresponds to the cosine similarity of v_c and v_e, being proportional to the probability $P(e|c)$. Suppose P(New York|big city) is higher than P(Boston|big city). Then we want to train the vector "big city" to be located closer to the vector, "New York" than "Boston". F by inner product is the same as *linear regression model.* $P(e_{1:n}|c)$ are dependent variables of (n by 1), $v_{e_{1:n}}$ are independent variables of (n by d) and v_c is intercept of (d by 1), where n is the number of data, d is the dimension of vector.

However, as motivated in Fig. 2, the frequency is showing a power-law distribution, such that F cannot fit the frequency very well. We show the errors in Fig. 2(left), contrasting with how we can improve in the right figure.

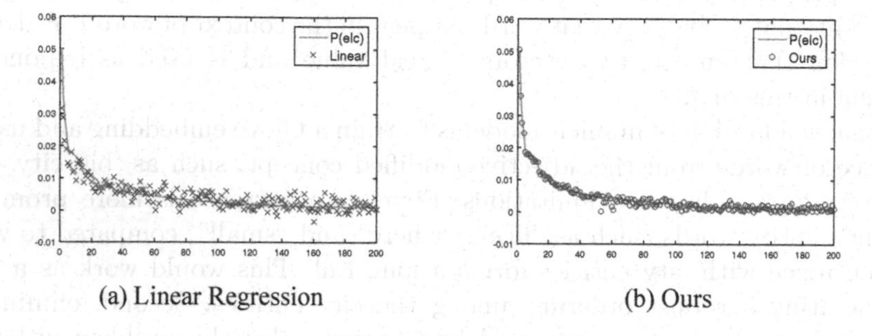

(a) Linear Regression (b) Ours

Fig. 2. Error comparison of the two models

More specifically, we modify the equation such that the inner product of two words in Eq. 2 equals to the "logarithm" of co-occurrence frequency showing Fig. 2. In other words, we can train "exponential" of the inner product to refer to the co-occurrence frequency as we reformulate as below:

$$F'(v_{e_k}, v_c) = \frac{exp(v_e \cdot v_c)}{\sum_{e_k \in E(c)} exp(v_{e_k} \cdot v_c)} = P(e|c) \tag{5}$$

where $E(c)$ is the entity set in concept c.

In Eq. 5, as the denominator is constant, $P(v_e|v_c)$ is proportional to $exp(v_e \cdot v_e)$. As a result, the entity vector placed closer to 'big city' can be bigger. For satisfying Eq. 5, the loss function of proposed method is:

$$\mathcal{L}(c) = \sum_{e_k \in E(c)} \left(P(e_k|c) - F'(v_{e_k}, v_c) \right)^2 \tag{6}$$

Global Loss Function. To optimize loss function for all adjective-modified concepts, a simple approach is minimizing $\sum_{t=i}^{T} \mathcal{L}(c_i)$, where T is the size of whole adjective-modified concept. This function considers only positively labeled data, as entities with high $P(e|c)$ to adjective-modified concept c. However, due to the limitation of explicit model, it is unclear whether unlabeled data $e' \notin E(c)$ is missing because it is a negative evidence or simply unobserved.

To apply negative evidence, a naive method randomly samples some unlabeled data as negative data. However, it may lead to false positive of selecting unobserved big city as a negative label or insignificant effects by extracting irrelevant data. To alleviate these problems, we firstly select entities which are included in noun concept out of the adjective, but excluded in the adjective-modified concept. For example, "Urbana" is included in "city" but excluded in "big city". And secondly, we weighted entities which likely to be more negative. Our hypothesis is that, those entities that are frequently observed with city, but not particularly with big city, are more likely not to be mentioned because it is a negative evidence. Based on this observation, we define our global loss function to consider the distance with negatively unlabeled data. To avoid false positive, we use a weighted function. Our global loss function is:

$$Loss = \sum_{t=i}^{T} \left(\mathcal{L}(c_i) + \sum_{e'_k \in N(c_i)} \frac{\log(n(e'_k))}{\log(\max n(e'))} F'(v_{e'_k}, v_{c_i})^2 \right) \tag{7}$$

where $N(c_i)$ is the sampled set of unlabeled entities which are excluded in adjective-modified concept c_i, but included in noun concept out of the adjective. And $n(e')$ is the sum of the frequency of entity e'. Through this approach, we can enhances the accuracy, as our empirical results confirm in Table 4 (precision improves by 6.9%).

3.3 Finding Adjective Synonym

This section reports how distributed space can be used to detect semantic relationship between adjectives. In Fig. 1(c), "big city" is placed near "large city", "huge city". We can observe that closest adjectives are all highly semantically related. This suggests that using highly related adjectives as a cluster can aggregate scattered observations of "big company" to "large company" or "huge company".

We can aggregate the closest pair at each iteration, until they converge to synonym clusters, by adopting a bottom-up agglomerative hierarchical clustering method [4]. Specifically, we compute a pairwise distance matrix using cosine similarity in word embedding and use it for clustering: We can continue iterative merges until the number of adjectives in one group is 4 or less.

Then, we combine the adjective set in cluster, *i.e.* (big, large, huge city), by using the average score of synonyms, instead of the score of one adjective. As shown in Table 4, we can show whether combining the statistical evidences from similar adjectives can enhance the quality of the graded score prediction. This average score indeed enhances the accuracy, as our empirical results confirm in Table 4 (precision improves by 19.2%).

4 Experiments

This section is organized to answer the following research questions respectively.

- RQ1: Some adjective can be used when people want to express objective properties such as the size of the country. Therefore, we select some obvious qualitative adjective and check the correlation with objective statistics to show how our model captures such correlation.
- RQ2: Meanwhile, there exist non-measurable or subjective adjectives such as great, valuable, or beautiful. We evaluate our model for these properties by using human-made gold standard ordering.
- RQ3: We validate whether our model overcomes the limitation of explicit models by extending prediction of $P(e|c)$ to unseen objects.

4.1 RQ1: Interpreting Qualitative Adjective with Statistics

Some adjectives naturally correlate with objective statistics, such as big city with statistics of population or area. We first demonstrate whether such correlation confirms commonsense understanding of humans. We generalize our observation to qualitative adjectives in Table 1, by using 8 field statistics. In these fields, we show Spearman correlation between statistics and the graded scores calculated by cosine similarity in word embedding.

In Table 1, we observe $P(e|c)$ in KB baseline reflects human-perceived correlation, but covers only a limited number of data. For example, KB baseline grades "big city" only for 186 cities, but our model scores them for 278 cities. Our model obtains high coverage, calculating the similarity of the entities that cannot be extracted by specific pattern. The last column shows that our model expands the coverage while preserving correlation.

Table 1. Spearman's rho between the graded score and the statistics

Adjective-modified concept	Statistics type	KB baseline		Ours	
		Correlation	# of data	Correlation	# of data
Big city	Population	0.705	186	0.706	278 (149%)
	Land area	0.460	186	0.446	278 (149%)
Expensive city	Big mac Index	0.630	54	0.648	70 (130%)
	Cost of living	0.571	282	0.508	444 (157%)
Large country	Population	0.651	119	0.741	191 (161%)
	Land area	0.799	119	0.803	191 (161%)
Rich country	GDP	0.690	120	0.690	169 (141%)
	PPP GDP	0.717	120	0.708	169 (141%)

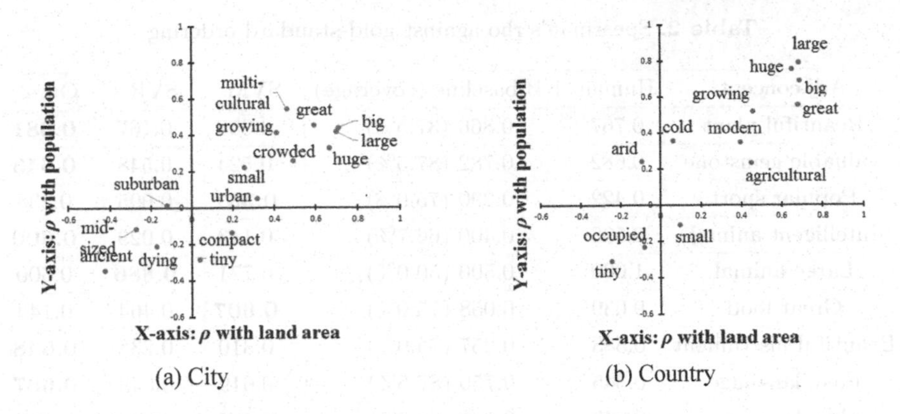

(a) City (b) Country

Fig. 3. Correlations between adjectives and population/land area

We observed that the correlations between adjective and statistics are different depending on the combined concepts. As shown in Fig. 3, “large country” is more highly correlated with population than land area, but “large city” is more highly correlated with population than land area. The opposite meaning of “large” is “ancient” or “dying” in city, but “tiny” in country. “dying” and “ancient” are rarely used in country, and human tend to represent “small town” for negative correlation word for population and land area not “tiny city”. The use of “large city” correlates more with population than land area, while “large country” correlates more with land area. This also confirms the human perception of considering countries such as Russia, China, or the US with large area as big countries, while considering metropolis with high population as big cities.

4.2 RQ2: Comparing Correlation with Human-Made Gold-Standard

To evaluate our model in terms of correlation, we adopt the gold standard orderings made by human. Iwanari et al. [9] release evaluation dataset including 35 adjective-modified concepts and average 7 entities per each concept. They asked multiple volunteers to order some entities set on attribute intensity expressed by adjective. Then, they pick the ordering that achieved the best average Spearman's correlation and use the ordering as gold standard. However, unlike our English dataset, the model was built on Japanese corpus and evaluation set. Because the domain of data we use is English, we excluded 16 specific concepts related to Japanese, such as cartoon, alcohol, temple, corner store, and town. Finally, we use ordering between 19 concepts and 134 entities for comparing our model. Additionally, Iwanari et al. [9] translated the concept and adjective words into English. However, the translated words are in less general form, we changed the words to synonyms that are more frequently used. For example, we chose the word “intelligent animal”, instead of “clever mammal” in dataset.

The experimental results are listed in Table 2. SVM and SVR refer to the methods proposed by Iwanari et al. [9]. KB baseline refers to the pattern-based

Table 2. Spearman's rho against gold-standard ordering

Adj.concept	Human	KB baseline (coverage)	SVM	SVR	Ours
Beautiful plant	0.767	0.866 (37.5%)	0.357	0.167	**0.381**
Valuable gemstone	0.682	0.782 (87.5%)	0.524	0.548	**0.643**
Popular sport	0.422	0.290 (75.0%)	**0.381**	−0.095	0.238
Intelligent animal	0.598	0.400 (66.7%)	0.143	0.029	**0.600**
Large animal	1.000	0.500 (50.0%)	0.771	**0.886**	0.600
Great food	0.639	0.058 (75.0%)	**0.607**	0.464	0.143
Beautiful instrument	0.583	0.257 (75.0%)	0.310	0.238	**0.548**
Easy language	0.845	0.750 (87.5%)	0.619	0.643	**0.667**
Slow language	0.840	0.100 (62.5%)	**0.381**	0.238	−0.167
Lovely animal	0.806	1.000 (37.5%)	0.548	0.595	**0.738**
Great vegetable	0.462	0.696 (75.0%)	**0.524**	0.476	0.429
Sweet fruit	0.729	0.783 (71.4%)	0.607	0.607	**0.821**
Great tool	0.772	0.300 (71.4%)	0.393	**0.500**	0.464
Good protein	0.662	0.900 (71.4%)	0.143	−0.286	**0.964**
Safe country	0.804	0.300 (100%)	−0.200	0.000	**0.500**
Warm country	0.961	0.866 (60.0%)	0.700	0.700	**1.000**
Well-known brand	0.659	0.743 (87.5%)	0.619	0.286	**0.900**
Nice browser	0.856	0.600 (80.0%)	−0.600	−0.600	**0.429**
Safe city	0.655	0.378 (100%)	0.357	0.250	**0.762**
Average	0.723	0.556 (72.2%)	0.378	0.297	**0.561**

method by the probability in Probase. While KB baseline has a coverage of only 72.2%, our model has not only 100% coverage, but also preserves precision.

We see that the correlation between our score and the gold standard ordering is less than 0.4 for "popular sport", "beautiful plant", "popular sport", "great food", and "slow language". The reasons for this result are that the coverage at extracted positive evidence is low or human's agreement is inconsistent due to its subjective property.

4.3 RQ3: Generalizing Beyond Implicit and Explicit Models

In this section, we evaluate further on how we predict $P(e|c)$ for unseen pairs during the training. Table 3 shows how we expand the observation made for four adjective-modified concepts into 250 concepts.

More specifically, to validate our model for unobserved entities, we set some $P(e|c)$ to test set and estimate that probability. By Eq. 5, the probability $P(e|c)$ and cosine similarity between e and c are monotonically increasing. Therefore, we evaluate the Spearman correlation of the cosine similarity and $P(e|c)$ that were not used in the training.

Table 3. Datasets

	Concept	Adjective-modified concept	Entity–Adj.concept pair
# of data	37	250	498,007
Example	City, Country, Company, Sport, Movie	Big city, Rich country, Great sport, Funny movie, Big company	Big city-New York, Great sport-Tennis, Big company-Apple

Table 4. Experimental results

Model	ρ
KB (5)	0.434
KB (10)	0.477
DS	0.484
KB+DS (5)	0.454
KB+DS (10)	0.461
Ours (Eq4)	0.469
Ours (Eq6)	0.535
Ours (Eq6+7)	0.572
Ours (Eq6+8)	0.641
Ours (Eq6+7+8)	**0.682**

For experiment, we split the entities which have the probability $P(e|c)$ to 9/10 training set and 1/10 testing set. Because the distribution of $P(e|c)$ is skewed, random sampling selects mostly tail entities with low probability similar to each other. Such sampling is inappropriate for comparing the correlation with the actual and predicted ranking, as it contains mostly tail entities with tied ranks. We thus sample more on head entities using stratified sampling, dividing sample size n into 5 section by rank and select $\frac{n}{2^i}$ data from the highest rank ($i = 1$) to lowest ($i = 5$).

We compare with KB and DS baselines, and consider its combination as well. It is to show how each component technique we proposed contribute to overall performance.

Baselines:

- **KB baseline:** KB baseline itself cannot be used for estimating missing $P(e|c)$, but we can extend by averaging the probability of the nearest 5 or 10 concepts, which we denote as $KB(5)$ and $KB(10)$ respectively.
- **DS baseline:** In DS baseline, we estimate $P(e|c)$ by averaging the vector of adjective and noun and computing the distance to this vector.
- **KB + DS baseline:** In KB+DS baseline, we estimate $P(c|c)$ by averaging the probability of 5 and 10 closest entities e' in the word embedding, denoted as $KB + DS(5)$ and $KB + DS(10)$.

Our model outperforms all these baselines. To show how we each equation contributes to the overall performance, we denote the complete model as Ours (Eq. 6+7+8) in the last line, which we compare with our model applying only some of such equations.

5 Conclusions

This paper studied the problem of understanding adjective by predicting a graded score for the given entity and adjective pair. Specifically, we train a distributed space to reflect Probase probability as distance. Semantic similarity with unseen objects is then used to predict missing Probase probability. Semantic similarity between adjectives contributes to enhance recall by collapsing the scattered observations of an entity with synonymous adjectives. Our extensive analysis using real-life data validates that we can predict adjective for unseen entity with comparable quality to seen ones, and thus improves the coverage to all adjective entity pairs.

Acknowledgment. This work was supported by Institute for Information & communications Technology Promotion (IITP) grant funded by the Korea government (MSIP) (No. B0101-16-0307, Basic Software Research in Human-level Lifelong Machine Learning (Machine Learning Center)).

References

1. Bagherinezhad, H., Hajishirzi, H., Choi, Y., Farhadi, A.: Are elephants bigger than butterflies? Reasoning about sizes of objects. arXiv preprint arXiv:1602.00753 (2016)
2. Bian, J., Gao, B., Liu, T.-Y.: Knowledge-powered deep learning for word embedding. In: Calders, T., Esposito, F., Hüllermeier, E., Meo, R. (eds.) ECML PKDD 2014. LNCS (LNAI), vol. 8724, pp. 132–148. Springer, Heidelberg (2014). doi:10.1007/978-3-662-44848-9_9
3. Bordes, A., Weston, J., Collobert, R., Bengio, Y.: Learning structured embeddings of knowledge bases. In: Conference on Artificial Intelligence (2011)
4. Day, W.H., Edelsbrunner, H.: Efficient algorithms for agglomerative hierarchical clustering methods. J. Classif. **1**(1), 7–24 (1984)
5. Deerwester, S., Dumais, S.T., Furnas, G.W., Landauer, T.K., Harshman, R.: Indexing by latent semantic analysis. J. Am. Soc. Inf. Sci. **41**(6), 391 (1990)
6. Hatzivassiloglou, V., McKeown, K.R.: Predicting the semantic orientation of adjectives. In: Proceedings of the Eighth Conference on European Chapter of the Association for Computational Linguistics. Association for Computational Linguistics (1997)
7. He, Y., Chakrabarti, K., Cheng, T., Tylenda, T.: Automatic discovery of attribute synonyms using query logs and table corpora. In: International World Wide Web Conferences Steering Committee, WWW (2016)
8. Huang, P.-S., He, X., Gao, J., Deng, L., Acero, A., Heck, L.: Learning deep structured semantic models for web search using clickthrough data. In: CIKML. ACM (2013)
9. Iwanari, T., Yoshinaga, N., Kaji, N., Nishina, T., Toyoda, M., Kitsuregawa, M.: Ordering concepts based on common attribute intensity. In: IJCAI (2016)
10. Lee, T., Wang, Z., Wang, H., Hwang, S.-W.: Attribute extraction and scoring: a probabilistic approach. In: ICDE. IEEE (2013)
11. Lin, Y., Liu, Z., Sun, M., Liu, Y., Zhu, X.: Learning entity and relation embeddings for knowledge graph completion. In: AAAI (2015)
12. Mikolov, T., Chen, K., Corrado, G., Dean, J.: Efficient estimation of word representations in vector space. arXiv preprint arXiv:1301.3781 (2013)
13. Mikolov, T., Dean, J.: Distributed representations of words and phrases and their compositionality. In: Advances in Neural Information Processing Systems (2013)
14. Pennington, J., Socher, R., Manning, C.D.: Glove: global vectors for word representation. In: EMNLP (2014)
15. Rothe, S., Schütze, H.: Autoextend: extending word embeddings to embeddings for synsets and lexemes. arXiv preprint arXiv:1507.01127 (2015)

16. Tandon, N., de Melo, G., Suchanek, F., Weikum, G.: Webchild: harvesting and organizing commonsense knowledge from the web. In: WSDM. ACM (2014)
17. Trummer, I., Halevy, A., Lee, H., Sarawagi, S., Gupta, R.: Mining subjective properties on the web. In: SIGMOD. ACM (2015)
18. Wang, Z., Zhang, J., Feng, J., Chen, Z.: Knowledge graph embedding by translating on hyperplanes. In: AAAI. Citeseer (2014)
19. Wu, W., Li, H., Wang, H., Zhu, K.Q.: Probase: a probabilistic taxonomy for text understanding. In: SIGMOD. ACM (2012)

Combining Dimensionality Reduction with Random Forests for Multi-label Classification Under Interactivity Constraints

Noureddine-Yassine Nair-Benrekia[1](✉), Pascale Kuntz[2], and Frank Meyer[3]

[1] Technicolor, 975 Avenue des Champs Blancs, 35576 Cesson-sévigné, Rennes, France
NoureddineYassine.NairBenrekia@technicolor.com

[2] Laboratoire d'Informatique de Nantes Atlantique, Site Polytech'Nantes, La Chantrerie, BP 50609, 44360 Nantes Cedex, France
pascale.kuntz@univ-nantes.fr

[3] Orange Labs, Avenue Pierre Marzin, 22307 Lannion Cedex, France
franck.meyer@orange.com

Abstract. Learning from multi-label data in an interactive framework is a challenging problem as algorithms must withstand some additional constraints: in particular, learning from few training examples in a limited time. A recent study of multi-label classifier behaviors in this context has identified the potential of the ensemble method "Random Forest of Predictive Clustering Trees" (RF-PCT). However, RF-PCT has shown a degraded performance in terms of computation time for large feature spaces. To overcome this limit, this paper proposes a new hybrid multi-label learning approach IDSR-RF (Independent Dual Space Reduction with RF-PCT) which first reduces the data dimension and then learns a predictive regression model in the reduced spaces with RF-PCT. The feature and the label spaces are independently reduced using the fast matrix factorization algorithm Gravity. The experimental results on nine high-dimensional datasets show that IDSR-RF significantly reduces the computation time without deteriorating the learning performances. To the best of our knowledge, it is currently the most promising learning approach for an interactive multi-label learning system.

1 Introduction

Interactive machine learning, based on a close interaction loop between a human and a learner, knows an increasing development today [1]. In particular, several interactive classification systems have been developed for various real-world applications: e.g. image classification [6], document classification [5], profile classification in social networks [2] and so on. In addition to the Human-Computer Interaction (HCI) aspects, which are not considered in this paper, the further developments of these systems face two major issues: (i) their extension to multi-label classification, and (ii) the choice of an appropriate learning algorithm.

The vast majority of the current approaches are based on a single-label classification that constrains items to span one label at a time. This simplifying

J. Kim et al. (Eds.): PAKDD 2017, Part II, LNAI 10235, pp. 828–839, 2017.
DOI: 10.1007/978-3-319-57529-2_64

framework significantly limits the user's expressiveness while he/she interacts with data that are inherently multi-label. Multi-label classification has received significant attention over the past few years and a large number of algorithms have been proposed in the literature [9,17]. However, the integration of these approaches into a human-centered interactive system is hampered by a double constraint: learning from few training examples in a limited time. A user can only annotate a very restricted set of examples and interactive systems are often required to provide a response in a time shorter than 100 ms [4]. Motivated by the importance of selecting a suitable classifier for an interactive multi-label classification system, Nair Benrekia et al. [10] have recently proposed the first extensive comparative study of the behavior of multi-label learning algorithms in an interactive framework. Their experiments showed that the ensemble classifier Random Forest of Predictive Clustering Trees (RF-PCT) [8] is the most efficient classifier. But, RF-PCT has shown a degraded performance in terms of time computation for high-dimensional datasets.

In this paper, we propose to improve the computational efficiency of RF-PCT by a preliminary dimensionality reduction of the data. From the pioneering work of Yu et al. [15] in the mid-2000's, several dimensionality reduction approaches have been proposed for multi-label classification. The vast majority of them arise from well-known approaches in data analysis (Principal Component Analysis, Latent Semantic Indexing, Linear Discriminant Analysis, Partial Least Squares, etc.) and few of them are inspired by compressive techniques in image processing [7]. Roughly speaking, the approaches can be classified in two families: the single space reduction approaches reduce the dimensionality of either the feature or the label space while the dual space ones reduce both of them. A recent comparative study [11] tends to show that, by considering both the curse of dimensionality in the feature space and the sparseness in the label space, the dual approaches are the most efficient. However, in most of the experiments considered in the literature, the learning models that are used are classical Binary Relevance (with Linear Regression or Support Vector Machine as base learners) and ML-kNN (Multi-label k nearest neighbors). And they are not the best ones in an interactive framework [17].

Consequently, in this paper, we combine an efficient dual space reduction with the most promising classifier for interactive multi-label learning. Our algorithm called IDSR-RF (Independent Dual Space Reduction combined with the RF-PCT classifier) is composed of two stages: an independent dual space reduction computed with the fast matrix factorization algorithm Gravity proposed by Takacs et al. [14] for the Netflix challenge and a predictive regression model learned with RF-PCT in the latent spaces. We have evaluated IDSR-RF on nine high-dimensional multi-label datasets to measure the learning time reduction and check that the dimensionality reduction does not damage the prediction quality of RF-PCT. The number of training example was set to 100 examples which is a maximal bound for a manually-operated labelling. Our experimental results show that IDSR-RF achieves a major reduction in the dimensionality of the data in a short time (less than a minute) and that it is able to both build a model and predict labels of a new example in less than half a second. Moreover, this significant reduction is obtained without learning performance degradation: IDSR-RF is as accurate as the original RF-PCT even with an advantage for a dataset.

The rest of the paper is organized as follows. Section 2 briefly reviews the dimension reduction algorithm. Section 3 defines the first phase of the interactive multi-label learning on which this paper is focused and describes the IDSR-RF algorithm. Section 4 presents the evaluation datasets, the evaluation criteria and the experimental protocol. Comparative results with the original RF-PCT are given in Sect. 5.

2 A Brief Review of Gravity for Dimensionality Reduction

We here recall the main steps of Gravity [14] which was the co-winner of the Netflix competition organized to develop a collaborative filtering algorithm to predict the ratings of users for unseen films [3]. Let us denote by $\mathbf{X} = [x_{ij}]$ the $n \times m$ matrix describing a set of n examples in a m-dimensional feature space. The objective is to redefine the original examples $x_i \in \mathcal{R}^m$ in a latent space of dimension $k << m$.

Unlike the traditional learning algorithms which learn from data vector-by-vector, Gravity learns from one cell x_{ij} at a time. The matrix $\mathbf{X}_{n\times m}$ can be approximated by the product of two sub rectangular matrices $\mathbf{P}_{n\times k}$ and $\mathbf{Q}_{m\times k}$:

$$\mathbf{X} \approx \mathbf{P} \times \mathbf{Q}^T$$

The matrices $\mathbf{P}$ and $\mathbf{Q}$ are learned by minimizing the Root Mean Squared Error (RMSE) which is the square root of the average square error e_{ij}^2 of the model approximation for each cell x_{ij}:

$$RMSE_{\mathbf{X}} = \sqrt{\frac{\sum_{\forall x_{ij} \in \mathbf{X}} (x_{ij} - \hat{x}_{ij})^2}{|\mathbf{X}|}} \quad \text{where } \hat{x}_{ij} = \sum_{w=1}^{k} p_{iw} \times q_{wj}.$$

The RMSE yields an analytically easy-to-derive quadratic error function which can be efficiently optimized with a descend gradient algorithm. Gravity computes the quadratic difference e_{ij}^2 between the model prediction $\hat{x_{ij}}$ and the ground-truth value x_{ij} and back-propagates its gradient in the factor vectors $\mathbf{P}_i$ and $\mathbf{Q}_j^T$. The gradient of e_{ij}^2 according to each row factor and each column factor is respectively computed in the following equations:

$$\frac{\sigma}{\sigma p_{iw}} e_{ij}^2 = -2 \times e_{ij} \times q_{wj} \quad \text{and} \quad \frac{\sigma}{\sigma q_{wj}} e_{ij}^2 = -2 \times e_{ij} \times p_{iw}.$$

To decrease the model mispredictions and better approximate the values x_{ij}, the factor vectors are updated in the opposite direction to the gradient:

$$p'_{iw} = p_{iw} + \eta \times 2 \times e_{ij} \times q_{wj} \quad \text{and} \quad q'_{wj} = q_{wj} + \eta \times 2 \times e_{ij} \times p_{iw}$$

where η is the learning rate which usually takes values smaller than 0.1. And to prevent overfitting and divergence of the factor values, a regularization rate λ is introduced during the learning process:

$$p'_{iw} = p_{iw} + \eta \times 2 \times e_{ij} \times q_{wj} - \lambda \times p_{iw} \text{ and } q'_{wj} = q_{wj} + \eta \times 2 \times e_{ij} \times p_{iw} - \lambda \times q_{wj}.$$

At the end of each learning iteration, the RMSE is evaluated on a small validation set: if it does not decrease during a fixed number of iterations, the learning process stops and the latest matrices **P** and **Q** are considered as optimal.

The complexity of Gravity is linear with respect to the size of the matrix **X** and is function of the number of passes $NbPasses$ required to insure its convergence: it is equal to $O(n \times m \times NbPasses)$. However, experiments have proven that $NbPasses$ is generally limited.

3 Multi-label Learning from Latent Data

After a brief presentation of our interactive framework, we describe the two phases of the IDSR-RF algorithm: (i) the dual dimensionality reduction with a double matrix factorization, and (ii) the learning and predicting process in the obtained latent spaces with RF-PCT.

3.1 Interactive Multi-label Learning

In the interactive process, the user initially labels a very small set of examples from which a learning algorithm computes a predictive model. For instance, to query a Video on Demand catalog for a film, a user defines his/her target concepts such as "Funny", "Masterpiece", and "Fairytale" and with an adapted interface he/she labels a small set of familiar films. The algorithm provides the user with relevant labels for a selected film or with relevant films from the catalog for one or a selection of labels. If the predictions do not align well with his/her preferences, he/she can boost the performance by adding few more examples and the learning process is run again. From the user's requests, the model keeps on refining its understanding of the desired concepts until his/her intervention is no longer required.

We here restrict ourselves to the evaluation of the classifier predictive and computation-time performances at the beginning of the learning task where few examples are available. In practice, the efficiency of this phase is crucial for catching the user's interest and confidence in the system. More precisely, at the beginning of the process, a user defines a set $\mathcal{L}$ of q desired labels and he/she labels a small set $\mathcal{X}_L$ of n_L examples either positively or negatively. Let y_i be the binary vector of size q which describes the labels given to an example $x_i \in \mathcal{X}_L$: $y_{ij} = 1$ (resp. 0) if the jth label is positively (resp. negatively) associated to x_i. From the multi-label training set $\mathcal{X}_L$, a learned model h predicts the most likely label set $\hat{y}_i = h(x_i)$ for each selected example x_i in a test set $\mathcal{S}$ much larger than $\mathcal{X}_L$.

3.2 Dual Dimensionality Reduction

Instead of considering the **X** factorization only like in the Gravity basic schema, we apply two matrix factorizations in order to reduce both the feature and the label spaces. Let us denote by **D** the data matrix of size $n \times (m+q)$ which contains the small number n_L of the labelled examples $x_i \in \mathcal{X}_L$ described by their m features and the q labels given by the user, and the $n - n_L$ unlabelled examples. The matrix **D** contains two sub-matrices: an example-feature matrix **F** of size $n \times m$ and an example-label matrix **L** of size $n_L \times q$. More precisely, **F** includes the m features of both labelled and unlabelled examples and **L** includes the q labels of the labelled examples only. The remaining part of size $(n - n_L) \times q$ is the part to predict. Two matrix factorizations are independently applied to the matrices **F** and **L**:

1. $\mathbf{F} \approx \mathbf{F'} \times \mathbf{C}^T$ where **F'** and **C** are respectively of size $n \times k$ and $m \times k$
2. $\mathbf{L} \approx \mathbf{L'} \times \mathbf{G}^T$ where **L'** and **G** are respectively of size $n_L \times k'$ and $q \times k'$

Let us note that in practice the unlabelled data are available before the beginning of the classification task. Consequently, we factorize the labelled and unlabelled examples at the same time to take advantage of the useful information contained in the unlabelled examples. This allows to learn more accurate feature factors. By combining the two example-factor matrices **F'** and **L'**, a new training set is obtained where the original training examples are re-described in a smaller k-dimensional latent feature space and labelled in a smaller k'-dimensional latent label space.

3.3 Learning and Predicting with RF-PCT in the Latent Spaces

The predictive model from the new training set is learned with the classifier RF-PCT (Random Forest of Predictive Clustering Trees) [8]. Let us recall that RF-PCT is an ensemble classifier based on a set of predictive clustering trees where each tree is considered as a hierarchy of clusters of increasingly small sizes. The trees are built with a standard top-down induction of decision trees algorithm. For diversity, RF-PCT trains a fixed number of decision trees with a bagging strategy and by selecting a random feature subset at each tree node. The size s of the feature subsets is function of the initial number k of latent features: $s = 0.1 \times k + 1$.

Here, RF-PCT is applied on the latent spaces: the criterion for selecting the best features in each tree is the minimization of the sum of the label variances and the leaves are associated with real-valued vectors containing the mean vote for each label. Each regression tree provides a prediction of the latent labels of the examples $x_i \in \mathcal{S}$ from their description in the latent feature space. The predictions of all trees are then averaged for each latent label. To bring the predictions back to the original label space $\mathcal{L}$, the vector of predictions $\hat{y}_i \in \mathcal{R}^{k'}$ is multiplied by the latent matrix of labels $\mathbf{G}^T$. As the obtained predictions are real-valued, we transform the real values into binary ones with the fast threshold

Pcut (Proportional Cut Method) [12]. It chooses the threshold which minimizes the label cardinality difference between the training data set $\mathcal{X}_L$ and the classified test data set $\mathcal{S}$.

The complexity of RF-PCT in the latent spaces is $O(N \times n_L \times s \times q)$ where N is the number of trees (originally set to 100).

4 Experimental Study

In this section, we first present the selected datasets and the evaluation criteria, and then we describe the experimental protocol.

4.1 Evaluation Datasets

We use nine textual multi-label datasets already selected in previous multi-label learning studies (Table 1). The feature dimensionality is of the same order as data in real-life applications: it varies from 21925 to 49060. The label number (from 22 to 101) is slightly greater than that of the current interactive uses where users would only define a limited number of labels to express their individual preferences. But as far as we know, there are still no published specific data sets in our context where the number of features is very high while the number of labels is small and the order-of-magnitude remains compatible with our applicative objectives.

Four datasets represent web pages collected through the hyperlinks from Yahoo!'s top directory. They are associated with the four Yahoo!'s top categories ("Arts & Humanities", "Business & Economy", "Computers & Internet", "Health"), and each page is labelled with one or more second level subcategories. The five other datasets come from the original Reuters dataset: they describe newswire stories which can be labelled by various categories (e.g. agriculture, fishing).

4.2 Evaluation Criteria

The prediction quality is evaluated with five complementary measures selected according to the requirements considered in multi-label learning:

1. **Ranking labels by relevance** (evaluated with the Ranking Loss (RL)): as users are mostly interested in a label ranking for a selected example, the classifier only presents its most likely labels at the top of the prediction list;
2. **Ranking examples by relevance** (evaluated with the macro-Ranking Loss (m-RL)): as users can also be interested in an example ranking for one or a set of labels, the classifier only presents the most likely examples at the top of the prediction list;
3. **Label classification** (evaluated with the Accuracy, F_1-score, multi-label Balanced Error Rate (BER)): if a label ranking is essential in practice, a label classification may be also desired. When an example is selected, the classifier only presents its most likely labels.

Table 1. Basic statistics of the selected multi-label datasets (m: number of features, n: number of examples, q: number of labels, $LCard$: label cardinality, $PUniq$: proportion of unique label combinations, $LDens$: label density).

Dataset	m	q	n	$LCard$	$PUniq$	$Dens$
Arts	23146	24	1000	1.66	0.18	0.07
Business	21925	28	1000	1.55	0.07	0.05
Health	30605	25	1000	1.63	0.11	0.06
Computers	34097	30	1000	1.44	0.10	0.05
Reuters S_4	47229	101	1000	2.48	0.14	0.03
Reuters S_5	47235	101	1000	2.64	0.16	0.03
Reuters S_1	47236	101	1000	2.88	0.17	0.03
Reuters S_2	47236	101	1000	2.63	0.16	0.03
Reuters S_3	47236	101	1000	2.61	0.16	0.03

The computation time is measured by the number of seconds required for learning the predictive model and for predicting labels of the unlabelled examples.

4.3 Experimental Protocol

Each dataset is partitioned into ten folds and each fold (10%) in turn is used for training while the nine remaining ones (90%) are used for the evaluation. Consequently, ten different data subsets are created. Each data subset is associated with two matrices: an example-feature matrix (with both labelled and unlabelled data) and an example-label matrix (with labelled data only). Each matrix is approximated with Gravity by the product of two latent matrices. From the training set, RF-PCT learns a predictive regression model in the latent spaces. Its predictive performance mainly depends on the choice of the Gravity parameters: the factor numbers k and k' of the two latent spaces, the learning rate η and the regularization rate λ. Our practical experience with Gravity has shown us that η and λ have less impact on the quality of the latent representation than k and k'. Thus, we have selected the same η and λ for the two factorizations. Precisely, four values $\{16, 32, 64, 128\}$ (resp. $\{2, 4, 8, 16\}$) are selected for k and k' and four values are randomly drawn in the interval $[0.01; 0.1]$ (resp. $[0.001; 0.01]$) for η (resp. λ). Let us note that additional experiments have shown that larger values for k and k' do not significantly improve the classifier predictive performance. The parameter selection is made by an exhaustive search in a limited space: RF-PCT is evaluated with each parameter combination on a validation set (10% of the training set) for the accuracy criterion. The predictive model with the highest accuracy is then used to predict labels of the test examples. The performance of IDSR-RF is an average over the ten data subsets.

To measure the contribution of the information contained in the unlabelled examples to the predictive performance of IDSR-RF, we consider a variant

IDSR*-RF where the feature space reduction is based on the training set only. A third matrix factorization is then performed to transfer the unlabelled examples to the latent feature space with the parameters used for the reduction of the training set feature space (i.e., k, η and λ). The performances of IDSR-RF, IDSR*-RF and RF-PCT are obviously compared on the same ten data subsets.

5 Experimental Results

We first compare the original size of the feature and label spaces of each dataset with the size of their respective latent spaces. Table 2 shows that IDSR-RF is able to perform a strong reduction (more than 99%) and the computation time required to provide the best latent representation is less than a minute which is reasonable for the beginning of the classification task. Let us remark that, for large training sets, we plan to re-run our approach without the time-consuming parameter search by keeping the parameters selected for the previous reduction. For each dataset, the factor numbers (k and k') change across the ten data subsets and we here report the most frequent factor combination.

Table 2. The reduction rate and computation time (in seconds) of IDSR-RF.

Dataset	q	m	k′	k	Reduction (%)	Time (s)
Arts	24	23146	**4**	**16**		**17.32**
Business	28	21924	**8**	**32**		**16.77**
Health	25	30605	**2**	**64**		**21.34**
Computers	30	34096	**16**	**16**		**23.86**
Reuters S_1	101	47236	**2**	**16**	**> 99%**	**29.96**
Reuters S_2	101	47236	**2**	**16**		**30.68**
Reuters S_3	101	47236	**4**	**32**		**30.39**
Reuters S_4	101	47229	**8**	**16**		**28.97**
Reuters S_5	101	47235	**4**	**16**		**34.02**

5.1 Learning and Prediction Speed in the Reduced Spaces

The computation time comparisons (Table 3) for IDSR-RF and RF-PCT show that the learning time of IDSR-RF is very significantly shorter than that of RF-PCT: the reduction is above 99% for all datasets and especially for the largest datasets such as Reuters where IDSR-RF is 9 min ahead of RF-PCT. While both approaches involve the same number of decision trees, the measured computation times indicate that IDSR-RF not only reduces the learning time but also the prediction time by over 98% for each dataset and especially for Reuters datasets where its prediction time is ahead by more than 2.20 s on RF-PCT (more than 99% reduction). The decision trees built from the latent data are less complex than those learned from the original data: one latent feature replaces a set of correlated features.

Table 3. The average learning and prediction times (in seconds) of IDSR-RF and RF-PCT.

Dataset		Learning time (s) ↓	Prediction time (s) ↓
Arts	IDSR-RF	0.22	0.02
	RF-PCT	70.76	1.17
Business	IDSR-RF	0.24	0.02
	RF-PCT	61.50	1.23
Health	IDSR-RF	0.25	0.02
	RF-PCT	89.26	1.38
Computers	IDSR-RF	0.25	0.02
	RF-PCT	124.09	1.65
Reuters S_1	IDSR-RF	0.23	0.02
	RF-PCT	516.57	2.49
Reuters S_2	IDSR-RF	0.27	0.02
	RF-PCT	572.88	2.30
Reuters S_3	IDSR-RF	0.23	0.02
	RF-PCT	514.61	2.25
Reuters S_4	IDSR-RF	0.22	0.02
	RF-PCT	581.02	2.52
Reuters S_5	IDSR-RF	0.22	0.02
	RF-PCT	568.25	2.34

In real-life applications, the classification system must provide personalized predictions at any time: when the user adds new examples, the predictive model must compute predictions more adapted to his/her preferences. By compressing the data, IDSR-RF significantly reduces the learning and prediction times of RF-PCT. This allows to update the model with a low cost, and most importantly to provide users with reliable predictions in a time compatible with the HCI recommendations.

5.2 Predictive Performances

The comparison of the average predictive performances (Table 4) shows that IDSR-RF and RF-PCT have a similar behavior. A Wilcoxon statistical test confirms that there are no significant differences between the two approaches except for the macro-ranking loss where RF-PCT remains on average slightly more efficient. For the dataset Computers, IDSR-FR outperforms RF-PCT for the majority of the criteria. This case shows that IDSR-RF not only maintains the information processing quality but can also improve it by taking advantage of the information contained in the unlabelled data and by exploiting the underlying correlations between both the features and the labels. The predictive

Table 4. The average predictive performances of IDSR-RF, IDSR*-RF and RF-PCT for the major quality criteria.

Dataset	Approach	RL (-)	m-RL (-)	Accuracy (+)	F_1-score (+)	BER (-)
Arts	IDSR-RF	0.23	0.48	0.17	0.22	0.40
	IDSR*-RF	0.22	0.50	0.17	0.22	0.40
	RF-PCT	0.19	0.40	0.25	0.30	0.35
Business	IDSR-RF	0.06	0.41	0.61	0.70	0.13
	IDSR*-RF	0.06	0.48	0.60	0.69	0.13
	RF-PCT	0.06	0.36	0.64	0.72	0.12
Health	IDSR-RF	0.12	0.45	0.38	0.44	0.29
	IDSR*-RF	0.13	0.47	0.35	0.41	0.29
	RF-PCT	0.10	0.38	0.44	0.51	0.24
Computers	IDSR-RF	0.13	0.39	0.48	0.53	0.24
	IDSR*-RF	0.14	0.49	0.38	0.45	0.27
	RF-PCT	0.12	0.41	0.42	0.48	0.25
Reuters S_1	IDSR-RF	0.22	0.46	0.11	0.18	0.42
	IDSR*-RF	0.24	0.47	0.10	0.16	0.43
	RF-PCT	0.13	0.28	0.23	0.32	0.32
Reuters S_2	IDSR-RF	0.24	0.45	0.14	0.21	0.39
	IDSR*-RF	0.25	0.48	0.14	0.21	0.39
	RF-PCT	0.15	0.32	0.21	0.29	0.34
Reuters S_3	IDSR-RF	0.26	0.47	0.12	0.18	0.40
	IDSR*-RF	0.24	0.49	0.12	0.18	0.41
	RF-PCT	0.15	0.33	0.20	0.28	0.34
Reuters S_4	IDSR-RF	0.23	0.44	0.13	0.20	0.39
	IDSR*-RF	0.22	0.46	0.13	0.20	0.39
	RF-PCT	0.13	0.31	0.21	0.28	0.33
Reuters S_5	IDSR-RF	0.23	0.45	0.15	0.23	0.38
	IDSR*-RF	0.25	0.47	0.14	0.21	0.39
	RF-PCT	0.15	0.30	0.20	0.29	0.34

performances of ISDR*-RF on all datasets and especially Computers clearly validate this interpretation. Further experiments for five additional classical quality criteria (Exact match, Coverage, Average precision, Hamming Loss and One error) confirm the conclusions drawn from Table 4.

6 Conclusion

The efficiency of an interactive multi-label learning system closely depends on the performances of the selected classifier both in prediction quality and in computation time. We have here proposed a new hybrid approach which improves the running time of the classifier RF-PCT which recently obtained the best prediction performances in an interactive framework. Our approach, based on a preliminary dual space reduction, takes advantage of the sparsity of the feature space and learns the latent features by only exploiting the informative values in the example-feature matrix. The experimental results on nine datasets with several tens of thousands of features have shown that IDSR-RF very significantly reduces the learning time of RF-PCT while maintaining its prediction quality. The experiments have also shown that IDSR-RF is significantly better than its version which does not take into account the available unlabelled data. Moreover, IDSR-RF reduces the dimensionality of any of our datasets in a very short time, which allows an interactive system to dynamically update its predictive model while the user is producing new preferences.

The numerical results obtained by IDSR-RF are very promising but we believe that there is still room for improvement in its performances. In the near future, we plan to follow two complementary research directions: (i) improving the parameter selection of IDSR-RF, and (ii) extending the numerical comparisons for the dimensionality reduction. It is well known that parameter tuning for a learning algorithm is a hard task. We here have simplified the problem by considering a limited set of parameter combinations selected from our strong practical experience with Gravity. However, parameter tuning has been investigated for many years in discrete optimization and we are currently discussing with optimization researchers how to build a new protocol which combines automatic search and expert knowledge.

Moreover, we have here restricted ourselves to the choice of Gravity for the matrix factorization because of its low complexity compared to that of the approaches previously developed in the literature for the multi-label learning. Nevertheless, new results [13] have recently allowed avoiding the eigendecomposition for the multi-label dimensionality reduction via dependence maximization (MDDM) initially proposed by [18] in an attempt to capture a shared information among different labels. The approach has been tested with the multi-label k-nearest neighbor classifier [16] whose performances are degraded in an interactive framework [10]. Nevertheless, we plan to investigate the properties of the new version of MDDM very soon to assess its potential use for an interactive learning and to compare it with Gravity in our IDSR-RF approach.

References

1. Amershi, S., Cakmak, M., Knox, W.B., Kulesza, T.: Power to the people: the role of humans in interactive machine learning. AI Mag. **35**(4), 105–120 (2014)
2. Amershi, S., Fogarty, J., Weld, D.: Regroup: interactive machine learning for on-demand group creation in social networks. In: Proceedings of the SIGCHI Conference on Human Factors in Computing Systems, pp. 21–30. ACM, New York (2012)
3. Bennett, J., Lanning, S.: The netflix prize. In: Proceedings of KDD Cup and Workshop, vol. 2007, p. 35 (2007)
4. Dabrowski, J.R., Munson, E.V.: Is 100 milliseconds too fast? In: CHI 2001 Extended Abstracts on Human Factors in Computing Systems, pp. 317–318. ACM, New York (2001)

5. Drucker, S.M., Fisher, D., Basu, S.: Helping users sort faster with adaptive machine learning recommendations. In: Campos, P., Graham, N., Jorge, J., Nunes, N., Palanque, P., Winckler, M. (eds.) INTERACT 2011. LNCS, vol. 6948, pp. 187–203. Springer, Heidelberg (2011). doi:10.1007/978-3-642-23765-2_13
6. Fogarty, J., Tan, D., Kapoor, A., Winder, S.: Cueflik: interactive concept learning in image search. In: Proceedings of the SIGCHI Conference on Human Factors in Computing Systems, pp. 29–38. ACM, New York (2008)
7. Hsu, D., Kakade, S., Langford, J., Zhang, T.: Multi-label prediction via compressed sensing. In: Advances in Neural Information Processing Systems 22: 23rd Annual Conference on Neural Information Processing Systems 2009, Proceedings of a Meeting Held 7–10 December, pp. 772–780. Vancouver, British Columbia (2009)
8. Kocev, D., Vens, C., Struyf, J., Džeroski, S.: Ensembles of multi-objective decision trees. In: Kok, J.N., Koronacki, J., Mantaras, R.L., Matwin, S., Mladenič, D., Skowron, A. (eds.) ECML 2007. LNCS (LNAI), vol. 4701, pp. 624–631. Springer, Heidelberg (2007). doi:10.1007/978-3-540-74958-5_61
9. Madjarov, G., Kocev, D., Gjorgjevikj, D., Dzeroski, S.: An extensive experimental comparison of methods for multi-label learning. Pattern Recogn. **45**(9), 3084–3104 (2012)
10. Nair-Benrekia, N.Y., Kuntz, P., Meyer, F.: Learning from multi-label data with interactivity constraints: an extensive experimental study. Expert Syst. Appl. **42**(13), 5723–5736 (2015)
11. Pacharawongsakda, E., Theeramunkong, T.: A comparative study on single and dual space reduction in multi-label classification. In: Skulimowski, A.M.J., Kacprzyk, J. (eds.) Knowledge, Information and Creativity Support Systems: Recent Trends, Advances and Solutions. AISC, vol. 364, pp. 389–400. Springer, Cham (2016). doi:10.1007/978-3-319-19090-7_29
12. Read, J.: Scalable multi-label classification. Ph.D. thesis. University of Waikato (2010)
13. Shu, X., Lai, D., Xu, H., Tao, L.: Learning shared subspace for multi-label dimensionality reduction via dependence maximization. Neurocomputing **168**, 356–364 (2015)
14. Takacs, G., Pilaszy, I., Nemeth, B., Tikk, D.: On the gravity recommendation system. In: Proceedings of KDD Cup and Workshop, vol. 2007 (2007)
15. Yu, K., Yu, S., Tresp, V.: Multi-label informed latent semantic indexing. In: Proceedings of the 28th Annual International ACM SIGIR Conference on Research and Development in Information Retrieval, pp. 258–265. ACM, New York (2005)
16. Zhang, M.L., Zhou, Z.H.: Ml-KNN: a lazy learning approach to multi-label learning. Pattern Recogn. **40**(7), 2038–2048 (2007)
17. Zhang, M.L., Zhou, Z.H.: A review on multi-label learning algorithms. IEEE Trans. Knowl. Data Eng. **26**(8), 1819–1837 (2014)
18. Zhang, Y., Zhou, Z.H.: Multilabel dimensionality reduction via dependence maximization. ACM Trans. Knowl. Discov. Data (TKDD) **4**(3), 1–21 (2010)

A Generalized Model for Multidimensional Intransitivity

Jiuding Duan([✉]), Jiyi Li, Yukino Baba, and Hisashi Kashima

Department of Intelligence Science and Technology,
Kyoto University, Kyoto 606-8501, Japan
dj@ml.ist.i.kyoto-u.ac.jp, {jyli,baba,kashima}@i.kyoto-u.ac.jp

Abstract. Intransitivity is a critical issue in pairwise preference modeling. It refers to the intransitive pairwise preferences between a group of players or objects that potentially form a cyclic preference chain, and has been long discussed in social choice theory in the context of the dominance relationship. However, such multifaceted intransitivity between players and the corresponding player representations in high dimension are difficult to capture. In this paper, we propose a probabilistic model that joint learns the d-dimensional representation ($d > 1$) for each player and a dataset-specific metric space that systematically captures the distance metric in $\mathbb{R}^d$ over the embedding space. Interestingly, by imposing additional constraints in the metric space, our proposed model degenerates to former models used in intransitive representation learning. Moreover, we present an extensive quantitative investigation of the wide existence of intransitive relationships between objects in various real-world benchmark datasets. To the best of our knowledge, this investigation is the first of this type. The predictive performance of our proposed method on various real-world datasets, including social choice, election, and online game datasets, shows that our proposed method outperforms several competing methods in terms of prediction accuracy.

Keywords: Representation learning · Preference · Matchup · Intransitivity

1 Introduction

The *transitivity* of pairwise comparison and matchup between individual objects is a fundamental principle in both social choice theory [25,26] and preference data modeling [24].

In pairwise comparison, two participants in a single round are evaluated by a third-party judge or an objective rule that judges the discriminative *win/lose* result for each player. Examples of applications of such a comparison include recommender systems [17], social choice systems [14,19,26], and so on. In pairwise matchup, two participants are each other's competitive opponents, and therefore the discriminative win/lose result is a reflection of their strength in the game. Examples of such matchup applications are sports tournaments [4] and online

J. Kim et al. (Eds.): PAKDD 2017, Part II, LNAI 10235, pp. 840–852, 2017.
DOI: 10.1007/978-3-319-57529-2_65

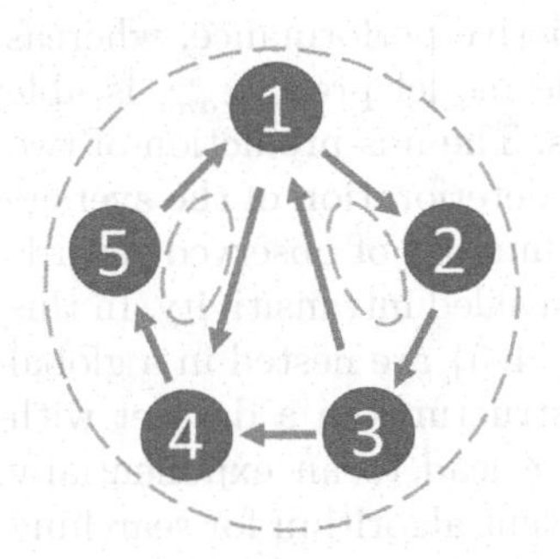

Fig. 1. Directed asymmetric graph illustration of the observed game in Table 1

Table 1. Toy model demonstrating the subtle deterioration in terms of test accuracy

Winner ID	Loser ID	#wins	#loses	GT	$pred_{trans}$	$pred_{intrans}$
1	2	10	5	✓	✓	✓
1	3	1	2	✓	x	✓
1	4	10	5	✓	✓	✓
1	5	1	2	✓	x	✓
2	3	10	5	✓	✓	✓
3	4	10	5	✓	✓	✓
3	5	10	5	✓	✓	✓
4	5	10	5	✓	✓	✓
Test Accuracy					0.6458	0.6667

games [6]. In both cases, the hidden winning ability of each individual object can be quantitatively profiled by parametric probabilistic models [2,25].

However, in addition to the thorough theoretical justifications of these parametric probabilistic models that assume certain levels of transitivity, the existence of *intransitivity*, which overrides the transitivity of preference in the real world, has been argued in ecometrics, behavior economics, and social choice theory for decades [21,26]. Intransitivity refers to the property of binary relations (i.e., win/loss or like/dislike) that are not transitive. For instance, in a rock-paper-scissors game, the pairwise matchup result is judged by three rules: $\{o_{paper} \succ o_{rock}, o_{rock} \succ o_{scissors}$, and $o_{scissors} \succ o_{paper}\}$. A transitive model results in a transitive dominance $o_{paper} \succ o_{scissors}$, that violates the third rule $o_{scissors} \succ o_{paper}$. In other words, the binary relations in the rock-paper-scissors game are not transitive. Such intransitivity in the real world exists in the form of cyclic dominance that implies the non-existence of a local dominant winner in the local preference loop. In many applications, the presence of a nested local intransitive preference loop results in systematically intransitive comparisons and matchups, and therefore predictive modeling is challenging. Intuitively, this situation occurs when objects have multiple features or views of judgment and each of these views dominates a corresponding pairwise comparison. The underestimation of such cyclic dominance is subtle in the numerical testing scores in terms of prediction accuracy, but critical for the cost-sensitive decision making based on the prediction results, as illustrated in the toy model in Fig. 1 and Table 1.

Figure 1 shows a directed asymmetric graph (DAG) to illustrate the toy game records in Table 1; the numbered nodes represent the corresponding player, the arrows demonstrate the dominant relationship between players, and the three dotted circles demonstrate the existing cyclic intransitive dominance relationships in the observed game records. In Table 1, the last two columns are exemplar predictions derived from transitive and intransitive models. The prediction of a transitive model $pred_{trans}$ cannot fully capture the intrinsic intransitivity in the dataset, leading to a deterioration in terms of predictive performance, whereas the prediction made by an intransitivity-compatible model $\text{pred}_{intrans}$ is able to accurately capture all the deterministic matchups. The mis-prediction of two out of the eight relationships results in only a subtle deterioration of the average test accuracy by 0.0208. Moreover, a growth in the number of observed records leads to a further difficulty in the evaluation of the unveiled intransitivity. In this toy model, the local intransitive sets $\{1, 2, 3\}$ and $\{1, 4, 5\}$ are nested in a global intransitive set $\{1, 2, 3, 4, 5\}$. Such c locally nested structures in a dataset with a large number of players n and active dominance e lead to an exponentially growing number of intransitive cycles. The most efficient algorithm for searching all such cycles yields a time complexity bounded by $O((n+e)(c+1))$ [18], which is intractable for stochastically observed dense matchups with large numbers of participants. Thus, the approach of modeling the multidimensional intransitive embedding by ensemble learning of all the possible views is blocked. A detailed quantitative exploration of the cyclic intransitivity in a variety of real-world datasets is presented in later sections.

The challenge presented by intransitivity motivated the alternative approach of learning the intransitivity-compatible multidimensional embedding from the parametric probabilistic models for pairwise comparisons. Without loss of generality, we attribute both pairwise comparison and matchup to the single notion of *matchup* and denote the individuals in the matchup as *players* in the following context, and discuss only the non-tie case for simplicity.

Existing work in this line of research includes studies on the seminal Bradley-Terry (BT) pairwise comparison model [2] and its extensions and applications in various real-world data science applications, e.g., matchup prediction [4], social choices [14,19,26], and so on. In the BT model, the strength of the players is parameterized as a single scalar value, by which the matchups between players always remain transitive. Other attempts to meet the challenge include extending the scalar into a 2-dimensional vector representation through a non-linear logistic model [5], and the more recently proposed Blade-Chest (BC) model with a multidimensional embedding scheme that imitates the offense and defense ability of a player in two independent multidimensional spaces [6,7]. However, the BC model, which was extended directly from the seminal BT model, is limited in its expressiveness of intransitivity by the arbitrary separation of the two representation metric spaces and an unexpected numerical conjugation drawback.

In this paper, we address the problem of predictive modeling of the intransitive relationships in real-world datasets by learning the multidimensional intransitivity representation for each player, i.e., items in a recommender system, tennis players in a tennis tournament, game players in online game platforms, or candidates in a political election. We focus on joint learning of the d-dimensional representation ($d > 1$) for each player and a dataset-specific metric space that systematically captures the distance metric in $\mathbb{R}^d$ over the embedding space. The joint modeling of the multidimensional embedding representation and the metric space is achieved by involving two types of covariate matrices, one to capture the interactive battling result between two players on the metric space, and a second to capture the intrinsic strength of each player. Through an analysis of the symmetry and expressiveness of our proposed embedding formulation, we further argue that the constrained optimization problem that is induced by our proposed multidimensional embedding formulation can be indentically transformed into an unconstrained form, thus allowing a generic numerical solution of the proposed model by using a stochastic gradient descent method [1]. Finally, we evaluate the effectiveness of our proposed method on a variety of real-world datasets, and demonstrate its superiority over other competitive methods in terms of predictive performance.

Our contributions are as follows:

- An extensive investigation of the wide existence of intransitive relationships between objects in many prevalent real-world benchmark datasets. This investigation required that special attention be paid to intransitive relationships. To the best of our knowledge, this is the first quantitative exploration of the existing intransitive relationships in these prevalent benchmark datasets, and even the first in the data mining research community.
- The proposal of a generalized embedding formulation for learning the intransitivity-compatible representation from pairwise matchup data, and an efficient solution to the induced optimization problem, together with a systematic characterization of the model, bridging the proposed generalized model and the former multidimensional representation learning methods.
- An empirical evaluation of the proposed method on various real-world datasets, which demonstrates the superior performance of the proposed method in terms of prediction accuracy.

The rest of the paper is organized as follows. Section 2 presents the related work on modeling intransitive relationships from pairwise comparison data. Section 3 defines the representation learning problem and presents our generalized formulation of the multidimensional embedding. In Sect. 4 we describe our investigation of the existence of intransitivity in the real-world datasets and present the experimental results for both synthetic and real-world datasets. Section 5 concludes our paper.

2 Related Work

Existing work on parametric models for pairwise matchups data, which originate from seminal work performed decades ago and include the Thurstone model [25] and the Bradley-Terry-Luce model [2], were surveyed extensively. The BT model [2] is based on maximum likelihood estimation and was further generalized to multiparty matchups [16] and adapted to comparisons involving a tie [10]. The first BT model generalized to multi-dimensional representation was limited to the 2-dimensional case with a non-linear logistic function, inspired by classical multidimensional scaling [5]. In real-world matchups, the ranking of the players' ability is an issue that is closely related to our parametric modeling for pairwise matchup data. Especially in sports tournaments [3,15,20] and online games [13], the Elo ratings system [11] and the TrueSkill ratings system [9,13] are noteworthy. In addition, instead of modeling the matchups between individual players, some methods concentrate on group matchup [15,20], rating individual players from the group matchup records [22], or alternatively model the belief of each collected record [8]. These methods are different from ours in that they were all developed according to the principles of transitivity.

In the context of modeling intransitivity, by extending the BT model, a 2-dimensional vector can be employed as the ability of players in matchups [5], with no verification of the modeling of the intransitive relationships on large datasets. The state-of-the-art model for intransitive modeling is the BC model [6], which imitates the offense and defense characteristics of a player and learns the corresponding multidimensional representations from matchup records. The BC model was then further extended to contexture-aware settings [7] with an improvement in the performance.

3 Proposed Model

Assume a given set of candidate players $\mathbf{P}$ with $|\mathbf{P}| = M$. The dataset $\mathbf{D}$ contains N pairwise matchup records $x_i(a_i, b_i) \in \{0, 1\}$, $i = [1{:}N]$, where the players a_i and $b_i \in \mathbf{P}$. An ordinal matchup record $o_a \succ o_b$ is the matchup record between player a and player b, meaning a beats b, and $o_a \prec o_b$, vice versa. The observed record $x(a, b)$ can be represented in a 4-tuple: either $x(a, b) = (a, b, 1, 0)$ meaning $o_a \succ o_b$ or $x(a, b) = (a, b, 0, 1)$ meaning $o_a \prec o_b$. The identical deterministic events can be aggregated, resulting in a collapsed dataset $\mathbf{D}^{collapse}$. The data entry $x_{aggregate}(a, b) \in \mathbf{D}^{collapse}$ is given by 4-tuples in $x_{aggregate}(a, b) = (a, b, n_a, n_b)$, where n_a is the total count of observed event $o_a \succ o_b$, and n_b of $o_a \prec o_b$, accordingly.

The goal is to predict the result of matchups by learning the interpretable multidimensional representation of the players, that reflects their ability in multiple views.

3.1 Bradley-Terry Model and Blade-Chest Model

In the BT model, each player $p \in \mathbf{P}$ is parameterized by a scalar $\gamma_p \in \mathbb{R}$ as the indicator of his/her ability to win. Following the probability axiom, the probability of the event is modeled as

$$Pr(o_a \succ o_b) = \frac{\exp(\gamma_a)}{\exp(\gamma_a) + \exp(\gamma_b)} \quad (1)$$

$$= \frac{1}{1 + \exp(-M_{ab})} \quad (2)$$

where $M_{ab} = \gamma_a - \gamma_b$ is the symmetric *matchup function* for player a and player b, with property

$$M_{ab} = -M_{ba} \quad (3)$$

and

$$Pr(o_a \prec o_b) = 1 - Pr(o_a \succ o_b)$$

The scalar-valued ability indicator of players γ_p is not intransitivity-aware and this has been shown in various datasets [6,21]. The parameter estimation of the BT model can be conducted by applying an EM algorithm for maximum likelihood or more generalized techniques [16]. Note that the matchup function $M_{ab}, a, b \in \mathbf{P}$ is the learning oracle that accesses the latent metric of players' ability, and therefore, it can be further extended to a multidimensional setting, named the BC model [6]. Intransitivity is then embraced by the BC model, where blade and chest vectors imitate the offense and defense, respectively.

Formally, in the BC model, the ability of player $p \in \mathbf{P}$ is parameterized by $\mathbf{a}_{blade}$ and $\mathbf{a}_{chest} \in \mathbb{R}^d$ and the corresponding matchup function is formulated by

- the Blade-Chest-Inner (BCI) embedding $M^{BCI}(a, b)$

$$M^{BCI}(a, b) = \mathbf{a}_{blade}^T \cdot \mathbf{b}_{chest} - \mathbf{b}_{blade}^T \cdot \mathbf{a}_{chest} \tag{4}$$

- the Blade-Chest-Distance (BCD) embedding $M^{BCD}(a, b)$

$$M^{BCD}(a, b) = \|\mathbf{b}_{blade} - \mathbf{a}_{chest}\|_2^2 - \|\mathbf{a}_{blade} - \mathbf{b}_{chest}\|_2^2$$

These formulations of the matchup function naturally ensure the symmetry property denoted in Condition (3), and therefore are compatible with the scalar-valued representation of the players' strength in the BT model. The connection between these two formulations can also be evidenced under a mild condition [6]. Assembled by this multidimensional formulation, the BC model is state-of-the-art in both predictive modeling and representation learning for the players' intransitivity.

3.2 Generalized Intransitivity Model

We propose a generic formulation of the matchup function that jointly captures a d-dimensional representation ($d > 1$) for each player and a dataset-specific distance metric for the learned representation in $\mathbb{R}^d$ over the embedded dimensions. Let us assume we have a d-dimensional representation $\mathbf{a} \in \mathbb{R}^d$ for player $a \in \mathbf{P}$; then, we formulate the generalized intransitivity embedding $M^G(a, b)$ as,

$$M^G(a, b) = \mathbf{a}^T \Sigma \mathbf{b} + \mathbf{a}^T \Gamma \mathbf{a} - \mathbf{b}^T \Gamma \mathbf{b} \tag{5}$$

where $\mathbf{a}$ and $\mathbf{b}$ are the d-dimensional representation for player a and player b, respectively, and $\Sigma, \Gamma \in \mathbb{R}^{d \times d}$ are the *transitive matrices*. The model parameters we attempt to learn are $\theta^G := \{\mathbf{a}, \mathbf{b}, \Sigma, \Gamma\}$. In the proposed formulation, the first term $\mathbf{a}^T \Sigma \mathbf{b}$ reflects the interaction between players, and the latter term $\mathbf{a}^T \Gamma \mathbf{a} - \mathbf{b}^T \Gamma \mathbf{b}$ reflects the intrinsic strength of each individual. The embedding is proposed to model the pairwise preference, in which two properties should be preserved, i.e., preference symmetry and expressiveness.

3.3 Properties

We characterize the detailed properties of the proposed formulation in terms of symmetry and expressiveness in comparison with the BC model, and show that the BC model is a specialized formulation in a family of our generalized formulation.

Symmetry. Since we discuss the matchup result between two players, the symmetry must be preserved [25]. This is different from other problems, such as link prediction in social networks, where the directed preference between items is naturally asymmetric [23].

Obviously, the two numerical computations of the first term $\mathbf{a}^T \Sigma \mathbf{b}$ and the latter term $\mathbf{a}^T \Gamma \mathbf{a} - \mathbf{b}^T \Gamma \mathbf{b}$ are independent given randomized d-dimensional embeddings $\mathbf{a}$ and $\mathbf{b}$. Without intuition of the specific design of $\mathbf{a}$ and $\mathbf{b}$, a.k.a. random initialization, the sufficient condition to preserve the symmetry of the first term is

$$\Sigma = -\Sigma^T \tag{6}$$

which is difficult to regularize given the gradient $\nabla_\Sigma M^G(a, b)$:

$$\nabla_\Sigma M^G(a, b) = \mathbf{a}\mathbf{b}^T$$

However, if we introduce it as a constraint in the optimization, the induced constrained optimization problem is difficult to solve. Alternatively, we devise an efficient solution which transforms the constrained optimization problem into an unconstrained optimization by reparameterizing Σ with Σ' by

$$\Sigma = \Sigma' - \Sigma'^T \tag{7}$$

where Σ' is a free matrix having the same shape as Σ. To this end, it is trivial to show that the symmetry of $\mathbf{a}^T \Sigma \mathbf{b}$ is preserved. Together with the fact that the symmetry of the self-regulation term $\mathbf{a}^T \Gamma \mathbf{a} - \mathbf{b}^T \Gamma \mathbf{b}$ in M^G holds constantly, we conclude that the symmetry of the proposed matchup function formulation is guaranteed.

Expressiveness. We further characterize the superior expressiveness of our proposed intransitive representation learning technique. Interestingly, we show that the BC model is a specialized formulation within a family of our proposed formulation.

Suppose that we have blade and chest vectors for player a, $\mathbf{a}_{blade}$ and $\mathbf{a}_{chest}$ $\in \mathbb{R}^{d'}$, where $d' = 3$; then, we integrate them into a generalized vector $\mathbf{a}_{general}$ $\in \mathbb{R}^{2d'}$ defined by

$$\mathbf{a}_{general} = \begin{bmatrix} \mathbf{a}_{blade} \\ \mathbf{a}_{chest} \end{bmatrix} = \begin{bmatrix} blade_1 \\ blade_2 \\ blade_3 \\ chest_1 \\ chest_2 \\ chest_3 \end{bmatrix} \tag{8}$$

This metaphorical definition is derived from the BC model, and therefore the $2d'$-dimensional generalized $\mathbf{a}_{general}$ has two distinct subspaces $\mathbf{a}_{blade}$ and $\mathbf{a}_{chest}$, which explicitly indicate the physical strength and weakness of player a, respectively.

Theorem 1 (Expressiveness). *Given the proposed matchup formulation in $2d'$-dimensional space, the proposed model degenerates to a BCI model in d'-dimensional space, under mild condition*

$$\|\mathbf{a}\|_2^2 = \|\mathbf{b}\|_2^2 \tag{9}$$

$$\|\Gamma\|_F \to 0$$

and,

$$\Sigma = \begin{bmatrix} 0 & I_{d'\times d'} \\ -I_{d'\times d'} & 0 \end{bmatrix}$$

Proof 1. *On the one hand, by the identified sufficient Condition (6) for the symmetry of $\mathbf{a}^T\Sigma\mathbf{b}$, given $I_{d'\times d'}$ as a d'-dimensional identity matrix, a fixed transitive matrix Σ with*

$$\Sigma = \begin{bmatrix} 0 & I_{d'\times d'} \\ -I_{d'\times d'} & 0 \end{bmatrix}$$

is a sufficient condition to preserve the symmetry of $\mathbf{a}^T\Sigma\mathbf{b}$, and results in

$$\mathbf{a}^T\Sigma\mathbf{b} = \begin{bmatrix} \mathbf{a}_{blade} \\ \mathbf{a}_{chest} \end{bmatrix}^T \begin{bmatrix} 0 & I_{d'\times d'} \\ -I_{d'\times d'} & 0 \end{bmatrix} \begin{bmatrix} \mathbf{b}_{blade} \\ \mathbf{b}_{chest} \end{bmatrix} \tag{10}$$

$$= \mathbf{a}_{blade}^T \cdot \mathbf{b}_{chest} - \mathbf{b}_{blade}^T \cdot \mathbf{a}_{chest} \tag{11}$$

On the other hand, given $\|\mathbf{a}\|_2^2 = \|\mathbf{b}\|_2^2 = c$, the inequality $\left\|\mathbf{a}^T\Gamma\mathbf{a} - \mathbf{b}^T\Gamma\mathbf{b}\right\| \leq 2c\,\|\Gamma\|$ holds. Thus, $\mathbf{a}^T\Gamma\mathbf{a} - \mathbf{b}^T\Gamma\mathbf{b} \to 0$ holds by $\|\mathbf{a}\|_2^2 = \|\mathbf{b}\|_2^2 = c$ and $\|\Gamma\|_F \to 0$. Therefore, the BCI model can be recovered by our proposed model. □

Base on the fact that BCI formulation M^{BCI} achieves better predictive performance than its variant M^{BCD} in practice, and our proposed formulation M^G degenerates into M^{BCI} by imposing additional conditions, we argue that the proposed method is superior in terms of expressiveness over the BC model and the former models [2,6].

3.4 Training

Without loss of generality, given a set of players $\mathbf{P}$ and a collapsed training dataset $\mathbf{D}^{collapse}$ with pairwise matchup between players in 4-tuple (a, b, n_a, n_b), as exemplified previously, our goal is to estimate the intransitivity parameters $\theta^G := \{\mathbf{a}, \mathbf{b}, \Sigma, \Gamma\}$ so that the predictive model can better predict unseen matchups. Following Eq. (7), we reparameterize the transitive matrix Σ as Σ'

and optimize $\theta^{G'} := \{\mathbf{a}, \mathbf{b}, \Sigma', \Gamma\}$ instead. In line with the BT model, we train the model by maximum likelihood. The overall likelihood is given by

$$L(D|\theta^{G'}) = \prod_{(a,b,n_a,n_b)\in\mathbf{D}^{\mathbf{collapse}}} Pr(o_a \succ o_b)^{n_a} \cdot Pr(o_a \prec o_b)^{n_b}$$

where $Pr(o_a \succ o_b)$ is the probability of the event $o_a \succ o_b$.

We take the log-likelihood and optimize it with a stochastic gradient descent method [1], and randomly sample one 4-tuple from $\mathbf{D}^{collapse}$ in each epoch, and then update the model parameters $\theta^{G'}$ w.r.t. the corresponding sample, until convergence.

Regularization. We choose the regularization terms as follows:

$$R_1(D|\theta^{G'}) = \sum_{a\in\mathbf{P}} \frac{1}{2} \|\mathbf{a}\|_2^2$$

$$R_2(D|\theta^{G'}) = \|\Sigma'\|_F$$

$$R_3(D|\theta^{G'}) = \|\Gamma\|_F$$

where $\|\cdot\|_2$ is L_2 norm and $\|\cdot\|_F$ is Frobenius norm. R_1 regularizes the scale of our embedding by intuition, as well as the scale of the blade and chest jointly, since they are integrated into our embedding. R_2 regularizes the scale of the free matrix Σ' as well as the scale of the symplectic matrix Σ, because $\|\Sigma\|_F = \|\Sigma' - \Sigma'^T\|_F$ is upper bounded by $2\|\Sigma\|_F$. R_3 regularizes the scale of the free matrix Γ, in line with Condition (9) given in Theorem 1.

Therefore, the regularized training objective for a given training dataset is

$$Q(D, \theta^{G'}) = L(D|\theta^{G'}) - \sum_i \lambda_i R_i(\theta^{G'}) \tag{12}$$

where $\theta^{G'} := \{\mathbf{a}, \mathbf{b}, \Sigma', \Gamma\}$ denotes the model parameters and λ controls the regularization.

4 Experiments

In this section, we first summarize the datasets with a quantitative investigation of the existence of intransitivity. Then, we report the experimental results of our proposed method on several challenging real-world benchmark datasets that consist of pairwise comparisons in social choice and matchups between individual players.

We used cross validation for parameter tuning in the experiments. Given the dataset in 4-tuple format, we first split the dataset randomly into three folds for cross validation and then identified the unique pairwise interactions and aggregated them. The hyperparameters were the dimensionality of the embedding

d and the regularization coefficient λ. The performance was measured by the average test accuracy $A(\mathbf{D}_{test}|\theta)$, defined by

$$A(\mathbf{D}_{test}|\theta) = \frac{1}{|\mathbf{D}_{test}|} \sum_{(a,b,n_a,n_b)\in\mathbf{D}_{test}} n_a \cdot \mathbb{1}(\hat{o}_a \succ \hat{o}_b) + n_b \cdot \mathbb{1}(\hat{o}_a \prec \hat{o}_b)$$

where $\mathbb{1}(\cdot)$ is the indicator function of an event.

We compared our proposed method with three competitive methods, namely the naïve method, BT model, and BC model. The **naïve method** estimates the winning probability of each player based on the empirical observations, with $Pr(o_a \succ o_b) = \frac{n_a+1}{(n_a+1)+(n_b+1)}$. If $n_a = n_b$, one player is randomly assigned as the winner. The **BT model** estimates player ability with a scalar representation. The **BC model** estimates player ability with two multidimensional vectors that are independent of each other.

4.1 Datasets

We investigated several challenging benchmark datasets from diversified areas. The datasets are commonly grounded on pairwise comparisons or matchups between objects or players. SushiA and SushiB [19] are food preference datasets. Jester [12] and MovieLens100K [14] are collective preference datasets in an online recommender system. ElectionA$_5$ [26] is an election dataset for collective decision making. Within the area of online games, SF4$_{5000}$ [6] is a dataset collected from professional players and is used to profile the characters in the virtual world. Dota [6] is a dataset of game records produced by a large number of players on an online RPG game platform.

Intransitivity in Datasets. Quantitative statistics of intransitive relationships in these datasets are presented in Table 2. *isIntrans* indicates the existence of the intransitivity relationships. *Intrans@3* indicates the percentage of intransitive loops that are analogous to the rock-paper-scissors game, where the number of involved players equals 3. In *Intrans@3*, the denominator is the total number of directed length-3 loops given by $2\binom{N}{3}$ for a fully observed pairwise dataset. *PlayerIntrans@3* is the number of players who are involved in a rock-paper-scissors-like relationship. Both *Intrans@3* and *PlayerIntrans@3* characterize the intensity of intransitivity, and a higher score indicates more intensive intransitivity in the dataset. In the majority of the seven datasets we investigated, an intransitive relationship exists. Moreover, in five out of the seven datasets, more than half of the players are involved in local intransitive relationships. To this end, we highlight the necessity of modeling the intransitivity, and to the best of our knowledge, this is the first quantitative exploration of the existing intransitive relationships in these prevalent benchmark datasets.

4.2 Experiments on Real Datasets

Table 3 shows the experimental results of our proposed method. For all of the four transitivity-rich datasets, SushiB, Jester, ElectionA$_5$, and SF4$_{5000}$, we observe

Table 2. Summary of real-world datasets

Dataset	No. of Players	No. of Records	isIntrans	Intrans@3	No. PlayerIntrans@3
SushiA	10	100000	x	0.00%	0/10
SushiB	100	25000	✓	26.87%	92/100
Jester	100	891404	✓	1.77%	97/100
MovieLens100K	1682	139982	✓	0.19%	1130/1682
ElectionA_5	16	44298	✓	0.44 %	6/16
SF4_{5000}	35	5000	✓	23.86%	34/35
Dota	757	10442	✓	97.58%	550/757

improvement in terms of the average test accuracy. In addition to the predictive performance, two practical facts are noteworthy. (a) The observed pairwise interactions in all these datasets are rich, and a K-fold cross validation procedure with no data augmentation results in a set of data bins, each of which contains identical players. Therefore, it is guaranteed that the representation of each player in the validation and test bin will be learned by a training set with a size of $K-2$ bins. However, as the number of players grows, the number of records required to accommodate such a cross validation procedure grows quickly. For instance, in the case of the SushiB dataset with 100 players, 25000 pairwise records, Intrans@3 $=$ 26.87%, and PlayerIntrans@3 $= 92/100$, the empirical down sampling for 3-fold cross validation is sufficient to perform a fully-evidenced prediction of the dominance for all possible player pairs, instead of a random guess caused by the existence of non-observed players in the validation and test bins. (b) Given a sampling scheme that is sufficiently stable to allow the model to give a fully evidenced prediction, a K-fold cross validation results in sparser interactions in the bins, which can be indicated by the connectivity of the matchup network, i.e., Borda count or Copeland count for directed graphs. However, in the challenging MovieLens100K and Dota datasets, the resultant heterogeneous interactions between players prevent us from providing evidenced dominance prediction from the observed sparse networks. A trivial solution for

Table 3. Test accuracy on real-world datasets

Dataset	Naïve	Bradley-Terry	Blade-Chest	Proposed Model
SushiA	0.6549 ± 0.0044	0.6549 ± 0.0021	0.6551 ± 0.0038	$\mathbf{0.6551 \pm 0.0027}$
SushiB	0.6466 ± 0.0042	0.6582 ± 0.0077	0.6591 ± 0.0051	$\mathbf{0.6593 \pm 0.0058}$
Jester	0.6216 ± 0.0006	0.6236 ± 0.0028	0.6242 ± 0.0035	$\mathbf{0.6243 \pm 0.0019}$
ElectionA_5	0.6507 ± 0.0031	0.6531 ± 0.0038	0.6533 ± 0.0043	$\mathbf{0.6535 \pm 0.0055}$
SF4_{5000}	0.5297 ± 0.0102	0.5329 ± 0.0044	0.5329 ± 0.0062	$\mathbf{0.5355 \pm 0.0080}$

such a case is a random guess, which is meaningless for intransitivity recovery. The above two facts hold for all the competitive methods.

5 Conclusion

In this paper, we focused on the issue of modeling intransitivity and representation learning for players involved in pairwise interactions. We proposed a generalized embedding formulation for learning the intransitivity-compatible representation from pairwise matchup data, and provided a theoretical characterization of the properties of the proposed formulation in terms of symmetry and expressiveness. We also tailored an efficient solution to the constraint optimization problem and verified the expressiveness of the proposed model by bridging it to former models. A thorough quantitative statistics analysis of the existing intransitivity in various real-world datasets was presented. To the best of our knowledge, it is the first of this kind in the data mining community. The results of the experiments based on real-world datasets show that our method achieves a better performance than the competitive models, including the state-of-the-art BC model.

References

1. Bottou, L.: Large-scale machine learning with stochastic gradient descent. In: Lechevallier, Y., Saporta, G. (eds.) Proceedings of COMPSTAT 2010, pp. 177–186. Springer, Heidelberg (2010)
2. Bradley, R.A., Terry, M.E.: Rank analysis of incomplete block designs: I. The method of paired comparisons. Biometrika **39**(3/4), 324–345 (1952)
3. Cao, Z., Qin, T., Liu, T.Y., Tsai, M.F., Li, H.: Learning to rank: from pairwise approach to listwise approach. In: Proceedings of the 24th International Conference on Machine Learning, pp. 129–136. ACM (2007)
4. Cattelan, M., Varin, C., Firth, D.: Dynamic Bradley-Terry modelling of sports tournaments. J. Roy. Stat. Soc.: Ser. C (Appl. Stat.) **62**(1), 135–150 (2013)
5. Causeur, D., Husson, F.: A 2-dimensional extension of the Bradley-Terry model for paired comparisons. J. Stat. Plan. Infer. **135**(2), 245–259 (2005)
6. Chen, S., Joachims, T.: Modeling intransitivity in matchup and comparison data. In: Proceedings of the 9th ACM International Conference on Web Search and Data Mining, pp. 227–236. ACM (2016)
7. Chen, S., Joachims, T.: Predicting matchups and preferences in context. In: Proceedings of the 22nd ACM SIGKDD International Conference on Knowledge Discovery and Data Mining, KDD 2016, NY, USA, pp. 775–784. ACM, New York (2016)
8. Chen, X., Bennett, P.N., Collins-Thompson, K., Horvitz, E.: Pairwise ranking aggregation in a crowdsourced setting. In: Proceedings of the Sixth ACM International Conference on Web Search and Data Mining, WSDM 2013, NY, USA, pp. 193–202. ACM, New York (2013)
9. Dangauthier, P., Herbrich, R., Minka, T., Graepel, T.: Trueskill through time: revisiting the history of chess. In: Advances in Neural Information Processing Systems, pp. 337–344 (2007)

10. Davidson, R.R.: On extending the Bradley-Terry model to accommodate ties in paired comparison experiments. J. Am. Stat. Assoc. **65**(329), 317–328 (1970)
11. Elo, A.E.: The Rating of Chess Players, Past and Present. Arco Pub., New York (1978)
12. Globerson, A., Chechik, G., Pereira, F., Tishby, N.: Euclidean embedding of co-occurrence data. J. Mach. Learn. Res. **8**, 2265–2295 (2007)
13. Herbrich, R., Minka, T., Graepel, T.: Trueskill: a Bayesian skill rating system. In: Advances in Neural Information Processing Systems, pp. 569–576 (2006)
14. Herlocker, J.L., Konstan, J.A., Borchers, A., Riedl, J.: An algorithmic framework for performing collaborative filtering. In: Proceedings of the 22nd Annual International ACM SIGIR Conference on Research and Development in Information Retrieval, pp. 230–237. ACM (1999)
15. Huang, T.K., Lin, C.J., Weng, R.C.: Ranking individuals by group comparisons. J. Mach. Learn. Res. **9**, 2187–2216 (2008)
16. Hunter, D.R.: MM algorithms for generalized Bradley-Terry models. Ann. Stat. **32**, 384–406 (2004)
17. Jamali, M., Ester, M.: A transitivity aware matrix factorization model for recommendation in social networks. In: International Joint Conference on Artificial Intelligence (IJCAI) (2011)
18. Johnson, D.B.: Finding all the elementary circuits of a directed graph. SIAM J. Comput. **4**(1), 77–84 (1975)
19. Kamishima, T.: Nantonac collaborative filtering: recommendation based on order responses. In: Proceedings of the 9th ACM SIGKDD International Conference on Knowledge Discovery and Data Mining, pp. 583–588. ACM (2003)
20. Keener, J.P.: The Perron-Frobenius theorem and the ranking of football teams. SIAM Rev. **35**(1), 80–93 (1993)
21. May, K.O.: Intransitivity, utility, and the aggregation of preference patterns. Econometrica: J. Econ. Soc. **22**, 1–13 (1954)
22. Menke, J.E., Martinez, T.R.: A Bradley-Terry artificial neural network model for individual ratings in group competitions. Neural Comput. Appl. **17**(2), 175–186 (2008)
23. Ou, M., Cui, P., Pei, J., Zhang, Z., Zhu, W.: Asymmetric transitivity preserving graph embedding. In: Proceedings of the 22nd ACM SIGKDD International Conference on Knowledge Discovery and Data Mining, KDD 2016, NY, USA, pp. 1105–1114. ACM, New York (2016)
24. Regenwetter, M., Dana, J., Davis-Stober, C.P.: Transitivity of preferences. Psychol. Rev. **118**(1), 42 (2011)
25. Thurstone, L.L.: A law of comparative judgement. Psychol. Rev. **34**(4), 273–286 (1927)
26. Tideman, N.: Collective Decisions and Voting: The Potential for Public Choice. Ashgate Publishing Ltd., Farnham (2006)

Author Index

Printed in the United States
By Bookmasters